SCIENCE WITH A NEXT-GENERATION VERY LARGE ARRAY

ngvla

ASTRONOMICAL SOCIETY OF THE PACIFIC
CONFERENCE SERIES

A SERIES OF BOOKS ON RECENT DEVELOPMENTS IN ASTRONOMY AND ASTROPHYSICS

Monograph 7

ASPCS volumes may be found online with color images at http://www.aspbooks.org.
ASP Monographs may be found online at http://www.aspmonographs.org.

For a complete list of ASPCS Volumes, ASP Monographs, and
other ASP publications see http://www.astrosociety.org/pubs.html.

All book order and subscription inquiries should be directed to the ASP at
800-335-2626 (toll-free within the USA) or 415-337-2126,
or email service@astrosociety.org

ASTRONOMICAL SOCIETY OF THE PACIFIC
CONFERENCE SERIES

Monograph 7

SCIENCE WITH A NEXT-GENERATION VERY LARGE ARRAY

A collection of peer-reviewed research papers supporting the development of a next-generation Very Large Array prepared by members of the astronomical community and compiled by the National Radio Astronomy Observatory in preparation for the U.S. 2020 Astronomy and Astrophysics Decadal Survey conducted by the National Research Council of the National Academy of Sciences.

Edited by

Eric J. Murphy
National Radio Astronomy Observatory, Charlottesville, Virginia, USA

and

the Next-Generation Very Large Array Science Advisory Council

SAN FRANCISCO

ASTRONOMICAL SOCIETY OF THE PACIFIC

390 Ashton Avenue
San Francisco, California, 94112-1722, USA

Phone: 415-337-1100
Fax: 415-337-5205
E-mail: service@astrosociety.org
Web site: www.astrosociety.org
E-books: www.aspbooks.org

First Edition

ASP Conference Series

ISBN: 978-1-58381-919-7
e-book ISBN: 978-1-58381-920-3

Library of Congress (LOC) Cataloging in Publication (CIP) Data:
Main entry under title
Library of Congress Control Number (LCCN): 2018912587

Printed in the United States of America by Sheridan Books, Ann Arbor, Michigan.
This book is printed on acid-free paper.

Contents

Preface xiii
E. J. Murphy and the ngVLA Science Advisory Council

Part I. A Current Summary of the Project

The ngVLA Science Case and Associated Science Requirements 3
E. J. Murphy, A. Bolatto, S. Chatterjee, C. M. Casey, L. Chomiuk, D. Dale, I. de Pater, M. Dickinson, J. Di Francesco, G. Hallinan, A. Isella, K. Kohno, S. R. Kulkarni, C. Lang, T. J. W. Lazio, A. K. Leroy, L. Loinard, T. J. Maccarone, B. C. Matthews, R. A. Osten, M. J. Reid, D. Riechers, N. Sakai, F. Walter, and D. Wilner

The ngVLA Reference Design 15
R. J. Selina, E. J. Murphy, M. McKinnon, A. Beasley, B. Butler, C. Carilli, B. Clark, S. Durand, A. Erickson, W. Grammer, R. Hiriart, J. Jackson, B. Kent, B. Mason, M. Morgan, O. Ojeda, V. Rosero, W. Shillue, S. Sturgis, and D. Urbain

Reaching Communities and Creating New Opportunities with the ngVLA 37
L. von Schill and S. Gurton

Part II. The Solar System

Potential for Solar System Science with the ngVLA 49
I. de Pater, B. Butler, R. J. Sault, A. Moullet, C. Moeckel, J. Tollefson, K. de Kleer, M. Gurwell, and S. Milam

High-Resolution Imaging of Comets 73
M. A. Cordiner and C. Qi

Spacecraft Telecommunications 81
T. J. W. Lazio and S. Asmar

ngVLA Observations of the Solar Wind 87
T. S. Bastian

Radio Observations of Solar Flares 99
D. E. Gary, T. S. Bastian, B. Chen, G. D. Fleishman, and L. Glesener

Planetary Bistatic Radar 113
M. Brozović, B. J. Butler, J.-L. Margot, S. P. Naidu, and T. .J W. Lazio

ngVLA Observations of Coronal Magnetic Fields 125
G. D. Fleishman, G. M. Nita, S. M. White, D. E. Gary, and T. S. Bastian

Part III. Planet Formation

Resolved Substructures in Protoplanetary Disks with the ngVLA 137
S. M. Andrews, D. J. Wilner, E. Macías, C. Carrasco-González, and A. Isella

Imaging Planetary Systems in the Act of Forming with the ngVLA 147
L. Ricci, A. Isella, S.-F. Liu, and H. Li

Disk Winds and the Evolution of Planet-Forming Disks 155
I. Pascucci, S. Andrews, C. Chandler, and A. Isella

Debris Disks: Structure and Composition 161
B. C. Matthews, J. S. Greaves, G. Kennedy, L. Matrà, D. Wilner, and M. C. Wyatt

Stellar Emission as a Source of Flux Bias in Debris Disks 171
J. A. White, J. Aufdenberg, A. C. Boley, P. Hauschildt, A. M. Hughes, B. Matthews, A. Moór, and D. J. Wilner

Resolving the Radio Complexity of EXor and FUor-type Systems with the ngVLA 177
J. A. White, M. Audard, P. Ábrahám, L. Cieza, F. Cruz-Sáenz de Miera, M. M. Dunham, J. D. Green, M. Güdel, N. Grosso, A. Hales, L. Hartmann, K. Kadam, J. H. Kastner, Á. Kóspál, S. Perez, A. Postel, D. Ruiz-Rodriguez, C. Rab, E. I. Vorobyov, and Z. Zhu

Indirect Detection of Extrasolar Planets via Astrometry 185
B. Butler and B. C. Matthews

Detection of Circumplanetary Disks 193
Z. Zhu, S. M. Andrews, and A. Isella

Dust Growth and Dust Trapping in Protoplanetary Disks 199
N. van der Marel, B. Matthews, R. Dong, T. Birnstiel, and A. Isella

Tracing the Water Snowline in Protoplanetary Disks with the ngVLA 209
K. Zhang, E. A. Bergin, J. P. Williams, P. Pinilla, and S. M. Andrews

Part IV. Planetary Systems and Life

Observing the Effects of Chemistry on Exoplanets and Planet Formation 217
B. A. McGuire, E. Bergin, G. A. Blake, A. M. Burkhardt, L. I. Cleeves, R. A. Loomis, A. J. Remijan, C. N. Shingledecker, and E. R. Willis

Organics in Disk Midplanes with the ngVLA 223
K. I. Öberg, L. I. Cleeves, and R. Loomis

Exo-space Weather in the Era of the ngVLA 229
R. A. Osten, M. K. Crosley, and G. Hallinan

Prebiotic Molecules 245
B. A. McGuire, P. B. Carroll, and R. T. Garrod

Complex Organic Molecules in Hot Molecular Cores/Corinos: Physics and Chemistry 249
M. T. Beltrán and V. M. Rivilla

SETI Searches for Evidence of Intelligent Life in the Galaxy 257
S. Croft, A. P. V. Siemion, J. M. Cordes, I. Morrison, Z. Paragi, and J. Tarter

Part V. Stellar Systems and Formation

Stellar Activity on Red Giant and Supergiant Stars: Mass Loss and the Evolution of the Stellar Dynamo 265
G. M. Harper

ngVLA Studies of Classical Novae 271
J. D. Linford, L. Chomiuk, and M. P. Rupen

Evolved Stars 281
L. D. Matthews and M. J. Claussen

Detecting Infall in High-Mass Protostellar Objects 293
M. T. Beltrán, Q. Zhang, and V. M. Rivilla

Star-Forming Filaments and Cores on a Galactic Scale 299
J. Di Francesco, J. Keown, R. Friesen, T. Bourke, and P. Caselli

Resolving the Structure and Kinematics of the Youngest HII Regions and Radio Jets from Young Stellar Objects 309
R. Galván-Madrid, M. Beltrán, A. Ginsburg, C. Carrasco-González, H. B. Liu, L. F. Rodríguez, and S. Kurtz

Understanding Massive Star Formation through Maser Imaging 321
T. R. Hunter, C. L. Brogan, A. Bartkiewicz, J. O. Chibueze, C. J. Cyganowski, T. Hirota, G. C. MacLeod, A. Sanna, and J.-M. Torrelles

New Frontiers in Protostellar Multiplicity with the ngVLA 333
J. J. Tobin, P. Sheehan, and D. Johnstone

Exploring Protostellar Disk Formation with the ngVLA 345
J. J. Tobin, P. Sheehan, D. Johnstone, and R. Sharma

Magnetic Fields in Forming Stars with the ngVLA 357
C. L. H. Hull, C. Carrasco-González, P. K. G. Williams, J. M. Girart, T. Robishaw, R. Galván-Madrid, and T. Bourke

Imaging Stellar Radio Photospheres with the Next Generation Very Large Array . 369
C. L. Carilli, B. Butler, K. Golap, M. T. Carilli, and S. M. White

Deuteration in Starless and Protostellar Cores 381
R. K. Friesen, M. T. Beltrán, P. Caselli, and R. T. Garrod

Massive and Evolved Stars with the ngVLA 389
T. J. Maccarone, S. Caballero-Nieves, N. Smith, and N. Lützgendorf

Planetary Nebulae 395
J. H. Kastner, A. Zijlstra, B. Balick, and R. Sahai

Probing Strong Binary Interactions and Accretion in AGB stars with the ngVLA . 403
R. Sahai

A Six-Dimensional Tomographic View of Galactic Star-Formation 411
L. Loinard and M. J. Reid

Part VI. Galaxy Ecosystems

Radio Continuum Emission from Galaxies: An Accounting of Energetic Processes 421
E. J. Murphy, J. J. Condon, A. Alberdi, L. Barcos-Muñozarcos, R. J. Beswick, E. Brinks, D. Dong, A. S. Evans, K. E. Johnson, R. C. Kennicutt Jr., S. T. Linden, T. W. B. Muxlow, M. Pérez-Torres, E. Schinnerer, M. T. Sargent, F. S. Tabatabaei, and J. L. Turner

Radio Recombination Lines from HII Regions 431
D. S. Balser, L. D. Anderson, T. M. Bania, J. M. Dickey, D. A. Roshi, T. V. Wenger, and T. L. Wilson

How Do Cold Gas Outflows Shape Galaxies? 441
A. D. Bolatto, L. Armus, A. K. Leroy, S. Veilleux, F. Walter, and R. Mushotzky

Accreting Supermassive Black Holes in Nearby Low-Mass Galaxies 451
K. Nyland and K. Alatalo

Radio Jet-ISM Feedback on Sub-Galactic Scales 457
K. Nyland, J. .J Harwood, D. Mukherjee, I. Prandoni, K. Alatalo, and A. D. Kapińska

Unwinding Star Formation with a Next-Generation Galactic Plane Survey 465
E. Rosolowsky, A. Ginsburg, and A. Leroy

Neutral Atomic Hydrogen in the Local Universe 471
D. J. Pisano, F. Walter, and S. Stanimirović

Gas Density Across the Universe 483
A. K. Leroy, A. Bolatto, A. Kepley, D. Meier, and T. Wong

Extreme Scattering Events and Symmetric Achromatic Variations 491
H. K. Vedantham

Molecular Clouds in Galaxies 499
A. K. Leroy, A. Bolatto, E. Rosolowsky, and E. Schinnerer

Precision Gas-Dynamical Mass Measurement of Supermassive Black Holes with the ngVLA 509
B. D. Boizelle, K. Nyland, and T. A. Davis

Accurate Probes of Density and Temperature in Star-Forming Gas 517
T. Wong, D. S. Meier, and E. Rosolowsky

Astrometry and Long Baseline Science 523
M. Reid, L. Loinard, and T. Maccarone

Part VII. Galaxy Formation and Evolution

Imaging Molecular Gas at High Redshift 535
C. L. Carilli and Y. Shao

The Star Formation History of the Universe 551
A. J. Barger, K. Kohno, E. J. Murphy, M. T. Sargent, and J. J. Condon

[CII] 158μm Emission from $z \geq 10$ Galaxies 557
C. L. Carilli, E.J. Murphy, A. Ferrara, and P. Dayal

Cold Gas in High-z Galaxies: The Molecular Gas Budget 565
R. Decarli, C. Carilli, C. Casey, B. Emonts, J.A. Hodge, K. Kohno, D. Narayanan, D. Riechers, M.T. Sargent, and F. Walter

Cold Gas in High-z Galaxies: The Dense ISM 575
R. Decarli, C. Carilli, C. Casey, B. Emonts, J.A. Hodge, K. Kohno, D. Narayanan, D. Riechers, M.T. Sargent, and F. Walter

Cold Gas in High-z Galaxies: CO as Redshift Beacon 581
R. Decarli, C. Carilli, C. Casey, B. Emonts, J.A. Hodge, K. Kohno, D. Narayanan, D. Riechers, M.T. Sargent, and F. Walter

The Molecular High-z Universe on Large Scales: Low-Surface-Brightness CO and the Strength of the ngVLA Core 587
B. Emonts, C. Carilli, D. Narayanan, M. Lehnert, and K. Nyland

Young Radio AGNs in the ngVLA Era 595
P. Patil, K. Nyland, J. J. Harwood, D. Mukherjee, and A. D. Kapińska

Probing Obscured Massive Black Hole Accretion and Growth since Cosmic Dawn 603
W. Rujopakarn, K. Nyland, A. E. Kimball, and I. Prandoni

From Megaparsecs To Milliparsecs: Galaxy Evolution and Supermassive Black Holes with NANOGrav and the ngVLA 611
S. R. Taylor and J. Simon

High-resolution Imaging of Radio Jets Launched by Active Galactic Nuclei: New Insights on Formation, Structure, and Evolution Enabled by the ngVLA . . 619
M. L. Lister, K. I. Kellermann, and P. Kharb

Imaging Cold Gas to 1 kpc Scales in High-Redshift Galaxies with the ngVLA . . 629
C. M. Casey, D. Narayanan, C. Carilli, J. Champagne, C.-L. Hung, R. Davé, R. Decarli, E. J. Murphy, G. Popping, D. Riechers, R. S. Somerville, and F. Walter

An ngVLA Wide Area AGN Survey 639
A. Kirkpatrick, K. Hall, K. Nyland, M. Lacy, and I. Prandoni

The Sunyaev-Zeldovich Effect from Quasar and Starburst Winds 649
M. Lacy, S. Chatterjee, A. Chakraborty, B. Mason, C. Sarazin, A. Kimball, K. Nyland, G. Rocha, and B. Rowe

Characterizing Feedback Through Molecular Outflows Across Cosmic Time . . . 657
J. Spilker and K. Nyland

Probing Dark Matter and Galaxy Evolution with Proper Motions of Galaxies . . . 663
T. J. Maccarone and A. H. Gonzalez

A Formaldehyde Deep Field 669
J. Darling

Part VIII. The Formation and Evolution of Black Holes in the Era of Multi-Messenger Astronomy

Supermassive Black Hole Pairs and Binaries 677
S. Burke-Spolaor, L. Blecha, T. Bogdanović, J. M. Comerford, J. Lazio, X. Liu, T. J. Maccarone, D. Pesce, Y. Shen, and G. Taylor

Compact Binary Mergers as Traced by Gravitational Waves 689
A. Corsi, D. A. Frail, D. Lazzati, D. Carbone, E. J. Murphy, B. J. Owen, D. J. Sand, and R. O'Shaughnessy

Radio Emission from Short Gamma-ray Bursts in the Multi-messenger Era 701
N. Lloyd-Ronning

Revealing the Galactic Population of Black Holes 711
T. J. Maccarone, L. Chomiuk, J. Miller-Jones, G. Sivakoff, and J. Strader

Local Constraints on Supermassive Black Hole Seeds from a Next Generation Very Large Array 719
R. M. Plotkin and A. E. Reines

Accretion and Jets in Local Compact Objects 727
D. L. Coppejans, J. C. A. Miller-Jones, E. G. Körding, G. R. Sivakoff, and M. P. Rupen

Tidal Disruption Events 737
S. van Velzen, G. C. Bower, and B. D. Metzger

Intermediate-Mass Black Holes in Globular Cluster Systems 743
J. M. Wrobel, J. C. A. Miller-Jones, K. E. Nyland, and T. J. Maccarone

Science with Pulsar Timing Arrays and the ngVLA 751
The North American Nanohertz Observatory for Gravitational Waves Collaboration

Offset Active Galactic Nuclei 763
L. Blecha, W. Brisken, S. Burke-Spolaor, F. Civano, J. Comerford, J. Darling, T. J. W. Lazio, and T. J. Maccarone

Serendipitous Fast Transient Science with the ngVLA 773
C. J. Law, G. C. Bower, S. Burke-Spolaor, B. J. Butler, P. Demorest, T. J. W. Lazio, and J. D. Linford

Flares from Coalescing Black Holes in the Centimeter-Wavelength Transient Sky 781
V. Ravi

Part IX. Cosmology and Fundamental Physics

Galactic Center Pulsars with the ngVLA 793
G. C. Bower, S. Chatterjee, J. Cordes, P. Demorest, J. S. Deneva, J. Dexter, M. Kramer, J. Lazio, S. Ransom, L. Shao, N. Wex, and R. Wharton

Weak Gravitational Lensing with CO Galaxies 803
P. Bull, I. Harrison, and E. Huff

Extragalactic Proper Motions: Gravitational Waves and Cosmology 813
J. Darling, A. Truebenbach, and J. Paine

H_2O Megamaser Cosmology with the ngVLA 821
J. Braatz, D. Pesce, J. Condon, and M. Reid

Author Index 827

Preface

Modern radio astronomy is playing a crucial role in compelling discoveries that span nearly all fields of astrophysics. The focus of the astronomy community on the priority objectives presented in *New Worlds, New Horizons* has fostered a number of these discoveries over the past decade. It is also true that many transformational findings are not fully anticipated, and their uncovering occurs because astronomers are equipped with the tools necessary to observe them. Two prime examples of this phenomenon include the discovery of pulsars and the Cosmic Microwave Background. More recent examples include: the exquisite structure of protoplanetary disks, which is challenging current theories of planet formation; the detection of radio-wavelength emission from gravitational wave sources; the discovery and localization of powerful but short-lived radio-wave transients, known as Fast Radio Bursts; and the observation of large masses of dust and molecular gas in galaxies well into the epoch of reionization.

Inspired by these dramatic discoveries – many enabled by the Karl G. Jansky Very Large Array (VLA), the Atacama Large Millimeter/submillimeter Array (ALMA), and the Very Long Baseline Array (VLBA) – the National Radio Astronomy Observatory (NRAO) and the international astronomy community are proposing to open a path to a vast unexplored discovery space with a new large collecting area radio telescope. Building on the superb centimeter-wave observing conditions and VLA site infrastructure in the southwest U.S., the current vision for this next generation VLA (ngVLA) is an interferometric array with more than ten times the sensitivity and spatial resolution of the existing VLA and ALMA, operating at frequencies spanning $\sim$ 1.2 – 116 GHz (25 – 0.26 cm) with baselines extending across North America and beyond. Such an advanced facility would build upon the remarkable scientific legacy of the VLA, which is nearing the half-century mark, and ensure the U.S. remains a world-leader in radio astronomy into the coming decades.

An integral part of science is to plan for and realize the advanced tools required for fundamental discovery. Astronomers are reaping the benefit of investments in premier facilities and new technologies, with ALMA in full science operations and the Square Kilometre Array Phase-1 (SKA-1) pathfinders nearing completion. Filling the wavelength gap between these instruments, partially served by the capable, but aging, VLA and VLBA, is a natural next step. The ngVLA focuses on precisely this domain with a dramatic improvement in both sensitivity and angular resolution over current capabilities, making it not only the premier facility in its wavelength range and discovery space, but a key complement to ALMA and the SKA-1, as well as other existing and planned facilities across the full electromagnetic spectrum.

The ngVLA science case and reference design are under development in a close collaboration between the NRAO and the international astronomy community. This effort is led by the ngVLA Science Advisory Council (SAC),[1] whose goal is the clear elucidation of the most compelling science case and the appropriate steering of the planning efforts. Input from the community has indicated a strong interest in an ngVLA capable of addressing a broad range of topics in planetary science, star formation, Galactic and extragalactic astronomy, fundamental physics, and much more. These form the basis of the developing ngVLA Key Science Goals and drive the corresponding reference design. The Key Science Goals were chosen to satisfy three criteria: (1) each addresses an important, unanswered question in astrophysics that has broad scientific and societal implications; (2) progress in each area is uniquely addressed by the capabilities of the ngVLA; (3) each exhibits key synergies and complementarity with science goals being pursued by existing or planned facilities in the 2025 and beyond time frame.

With nearly 90 chapters written by more than 285 unique authors, this volume is a snapshot of the community vision for an ngVLA and its enormous scientific potential, built around its five Key Science Goals: (1) unveiling the formation of Solar System analogs on terrestrial scales; (2) probing the initial conditions for planetary systems and life with astrochemistry; (2) characterizing the assembly, structure, and evolution of galaxies from the first billion years to the present; (4) using Galactic Center pulsars to make a fundamental test of gravity; and (5) understanding the formation and evolution of stellar and supermassive blackholes in the era of multi-messenger astronomy. This book convincingly shows that the capabilities brought forth by the ngVLA ultra-sensitive imaging of thermal continuum and line emission down to milli-arcsecond resolution, and unprecedented broad band continuum polarimetric imaging of non-thermal processes – will open a new and revealing window on the cosmos.

While the ngVLA roots are more than a decade old, the past three years have seen the rapid construction of an ambitious science case and a well-grounded reference design, all supported by more than a dozen scientific and technical meetings of the community. The most recent was held in June 2018 in Portland, Oregon, and brought together over 200 members of the multi-wavelength astronomy community to discuss how to effectively address the highest priority astrophysical questions with a new generation of facilities, including the ngVLA. Throughout this entire process, the community has demonstrated strong interest in having a flexible facility that is capable of supporting a broad scientific program over the course of its expected long lifetime. Consequently, the ngVLA is a different style of instrument than many others on the horizon, which are either very focused on specific experiments or will service large surveys.

In the near future, the ngVLA will be brought forth for consideration by the 2020 Astronomy and Astrophysics Decadal Survey (Astro2020), led by the National Academies of Science, as the next U.S. flagship radio observatory. Our hope is that the ngVLA scientific vision, well-understood technology, and credible costing will be recognized and promoted during the Astro2020 process. Beyond Astro2020, the NRAO looks forward to continuing its close partnership with the astronomy community and industry to refine the ngVLA science case and mature the reference design and costing in an effort to make it a high priority candidate for National Science Foundation Major Research Equipment and Facility Construction funding. In parallel with this effort, the

[1]The ngVLA SAC is a group of 24 leading scientists of the community outside the NRAO, with a wide range of interests and expertise, who are mostly based in U.S. institutions but also have international representation; `http://ngvla.nrao.edu/page/sciencecouncil`.

NRAO will seek additional international participation in the project, along with U.S. multi-agency support.

Finally, it is only with the substantial support of our scientific and technical advisory councils, science working groups, a large fraction of the astronomy community, and the National Science Foundation that the publication of this volume, describing the transformational science and discovery achievable with the ngVLA in the next decade and beyond, is possible. The NRAO and the ngVLA Project greatly appreciate working hand-in-hand with such a strong and vibrant community, especially those volunteering their service and ideas. We look forward to continuing this close and fruitful partnership through the realization of the ngVLA and its unparalleled science potential.

Eric J. Murphy and the ngVLA Science Advisory Council

The ngVLA Science Advisory Council

Name	Affiliation	Email Address	Role
Alberto Bolatto†	University of Maryland	*bolatto@astro.umd.edu*	SAC co-Chair
Shami Chatterjee	Cornell University	*shami@astro.cornell.edu*	
Caitlin Casey	University of Texas at Austin	*cmcasey@utexas.edu*	
Laura Chomiuk	Michigan State University	*chomiuk@pa.msu.edu*	
Daniel Dale†	University of Wyoming	*ddale@uwyo.edu*	SWG 2 Chair‡
Imke de Pater	University of California, Berkelely	*imke@berkeley.edu*	
Mark Dickinson	NOAO	*med@noao.edu*	
James Di Francesco†	NRC-Victoria	*James.DiFrancesco@nrc-cnrc.gc.ca*	
Gregg Hallinan	Caltech	*gh@astro.caltech.edu*	
Andrea Isella†	Rice University	*isella@rice.edu*	SAC co-Chair
Kotaro Kohno	University of Tokyo	*kkohno@ioa.s.u-tokyo.ac.jp*	
Shri Kulkarni†	Caltech	*srk@astro.caltech.edu*	
Cornelia Lang	University of Iowa	*cornelia-lang@uiowa.edu*	
Joseph Lazio†	JPL/Caltech	*Joseph.Lazio@jpl.nasa.gov*	SWG 4 Chair‡
Adam Leroy	The Ohio State University	*leroy.42@osu.edu*	
Laurent Loinard	UNAM	*l.loinard@irya.unam.mx*	
Thomas Maccarone	Texas Tech University	*Thomas.Maccarone@ttu.edu*	
Brenda Matthews†	NRC-Victoria	*bcmatthews.herzberg@gmail.com*	SWG 1 Chair‡
Rachel Osten	STScI	*osten@stsci.edu*	
Mark Reid	Harvard-Smithsonian CfA	*reid@cfa.harvard.edu*	
Dominik Riechers†	Cornell University	*riechers@astro.cornell.edu*	SWG 3 Chair‡
Nami Sakai	RIKEN	*nami.sakai@riken.jp*	
Fabian Walter	Max-Planck-Institut für Astronomie	*walter@mpia.d e*	
David Wilner	Harvard-Smithsonian CfA	*dwilner@cfa.harvard.edu*	

† Executive Committee Member

‡ SWG 1 – The Cradle of Life; SWG 2 – Galaxy Ecosystems/Baryon Cycling; SWG 3 – Galaxy Assembly Through Cosmic Time; SWG 4 – Time Domain, Cosmology, and Physics

Part I

A Current Summary of the Project

Science with a Next-Generation Very Large Array
ASP Conference Series, Monograph 7
Eric J. Murphy, ed.

The ngVLA Science Case and Associated Science Requirements

Eric J. Murphy,[1] Alberto Bolatto,[2] Shami Chatterjee,[3] Caitlin M. Casey,[4] Laura Chomiuk,[5] Daniel Dale,[6] Imke de Pater,[7] Mark Dickinson,[8] James Di Francesco,[9,10] Gregg Hallinan,[11] Andrea Isella,[12] Kotaro Kohno,[13] S. R. Kulkarni,[11] Cornelia Lang,[14] T. Joseph W. Lazio,[15] Adam K. Leroy,[16] Laurent Loinard,[17,18] Thomas J. Maccarone,[19] Brenda C. Matthews,[9,10] Rachel A. Osten,[20] Mark J. Reid,[21] Dominik Riechers,[3] Nami Sakai,[22] Fabian Walter,[23] and David Wilner[21]

Abstract. The science case and associated science requirements for a next-generation Very Large Array (ngVLA) are described, highlighting the five key science goals developed out of a community-driven vision of the highest scientific priorities in the next decade. Building on the superb cm observing conditions and existing infrastructure of the VLA site in the U.S. Southwest, the ngVLA is envisaged to be an interferometric array with more than 10 times the sensitivity and spatial resolution of the current VLA and ALMA, operating at frequencies spanning $\sim 1.2 - 116$ GHz with extended baselines reaching across North America. The ngVLA will be optimized for observations at wavelengths between the exquisite performance of ALMA at submm wavelengths, and the future SKA-1 at decimeter to meter wavelengths, thus lending itself to be highly complementary with these facilities. The ngVLA will be the only facility in the world that can tackle a broad range of outstanding scientific questions in modern astronomy by simultaneously delivering the capability to: (1) unveil the formation of Solar System analogues; (2) probe the initial conditions for planetary systems and life with astrochemistry; (3) characterize the assembly, structure, and evolution of galaxies from the first billion years to the present; (4) use pulsars in the Galactic center as fundamental tests of gravity; and (5) understand the formation and evolution of stellar and supermassive blackholes in the era of multi-messenger astronomy.

1. Introduction

The Very Large Array (VLA) has had a major impact on nearly every branch of astronomy, and the results of its research are abundant in the pages of scientific journals and textbooks. Five years after the completion of the Expanded Very Large Array (EVLA) Project, and more than 40 years since the first VLA antenna was commissioned, the VLA has strengthened its position as the most versatile, widely-used radio telescope in the world. Rededicated as the Karl G. Jansky VLA in March 2012, the array continues to make cutting-edge discoveries across a broad range of disciplines including exoplanet formation, galaxy formation in the nearby and distant Universe, and the rapidly growing field of time-domain astronomy. More than 3,000 researchers from around the world have used the VLA to carry out more than 11,000 observing projects.

Inspired by the VLA's ability to perennially deliver high-impact scientific results, and to prepare for the changing landscape in science priorities and capabilities of facilities at other wavelengths, the National Radio Astronomy Observatory (NRAO) recently started to consider ways to continue the legacy of the VLA as one of the most powerful radio telescopes to be included in the next generation of the world's suite of cutting edge astronomical observatories. By teaming with the greater astronomical community in an exercise to develop a cogent science case requiring observations at cm–mm wavelengths, NRAO is now currently pursuing a large collecting area radio interferometer that will open completely new discovery space by delivering an angular resolution and sensitivity that is each an order of magnitude larger than that of the VLA and the Atacama Large Millimeter Array (ALMA), allowing it to address fundamental questions in all major areas of astrophysics.

This chapter describes the community response following a solicitation from the NRAO to develop key science cases for a future U.S.-led radio telescope, the next generation Very Large Array (ngVLA). The resulting list of more than 80 science use cases received by NRAO represent some of the fundamental astrophysical problems that require observing capabilities at cm–mm wavelengths well beyond those of existing, or already planned, telescopes. The summary of this exercise has resulted in a transformative radio facility having roughly 10 times the sensitivity of the VLA and ALMA, frequency coverage from $\sim 1.2-116$ GHz with up to 20 GHz of instantaneous sampled bandwidth, a compact core for good sensitivity to low surface-brightness emission, and extended baselines reaching across North America for extremely high-resolution imaging. The ngVLA is being built on the scientific and technical legacies of the VLA and ALMA, and is being designed to provide the next major leap forward in our understanding of planets, galaxies, black holes, and the dynamic sky. As such, the ngVLA will open a new window on the universe through ultra-sensitive imaging of thermal line and continuum emission down to milliarcecond resolution, as well as deliver unprecedented broad band continuum polarimetric imaging of non-thermal processes.

2. Developing the ngVLA Key Science Goals

The ngVLA Science Advisory Committee (SAC)[1], a group of 24 leading scientists with a wide range of interests and expertise appointed by NRAO, in collaboration with the broader international astronomical community, recently developed a series of than 80 compelling science cases requiring $\approx$200 unique observations between $\sim 1.2-116$ GHz with sensitivity, angular resolution, and mapping capabilities far beyond those provided by the VLA, ALMA, and the Square Kilometre Array Phase 1 (SKA-1). The science cases submitted spanned a broad range of topics in the fields of planetary science, Galactic and extragalactic astronomy, as well as fundamental physics, and formed the basis for developing the ngVLA Key Science Goals (KSGs).

Given the overwhelmingly large spread of compelling science cases generated by the community, it is clear that the primary science requirement for the ngVLA is to be flexible enough to support the wide breadth of scientific investigations that will be proposed by its highly creative user base over the full lifetime of the instrument. This mandate is also made obvious given the breadth of scientific endeavors included in this volume, ranging from studies of planet formation and understanding the conditions for habitability in other star systems to rigorous testing of the theory of gravity using pulsars immersed in the space-time potential of the Galaxy's supermassive black hole.

[1] `http://ngvla.nrao.edu/page/sciencecouncil`

This in turn makes the ngVLA a different style of instrument than many other facilities on the horizon (e.g., SKA-1, LSST, etc.), which are heavily focused on carrying out large surveys.

In the next stage of the process, each of the individual science cases were objectively reviewed and thoroughly discussed by the different Science Working Groups within the ngVLA-SAC. The ultimate goal of this exercise was to distill the top scientific goals for a future radio/mm telescope. KSGs were identified as projects which satisfied three criteria: (1) the science case addresses an important and currently unanswered question in astrophysics that has broad implications to communities outside of radio astronomy; (2) progress in this area is uniquely addressed by the capabilities of the ngVLA; and (3) the science case exhibits strong synergies/complementarity with science being pursued by other existing/planned facilities in the $\gtrsim 2025$ time frame.

The resulting initial KSGs, along with the results from the entire list of over 80 science use cases (Selina, Murphy & Erickson 2017) were then presented and discussed with the broader community at the ngVLA Science and Technology Workshop June 26 – 29, 2017 in Socorro NM in an attempt to build consensus around a single vision for the key science missions of the ngVLA. Here we describe the five KSGs to come out of this community-driven science use case capture process (see Table 1) along with their corresponding requirements that in turn drive the Reference Design described by Selina et al. (this volume, p. 15).

3. The ngVLA Key Science Goals and Associated Requirements

In this section we briefly describe each of the five highest-priority ngVLA KSGs that are expected to be carried out during the lifetime of the ngVLA (see Table 1). We also provide a brief description of their basic requirements that in turn form the foundation used to construct the ngVLA Reference Design described in Selina et al. (this volume, p. 15). A more detailed description of the ngVLA KSGs can be found in (Bolatto et al. 2017), and the corresponding full description of the ngVLA Level 0 Science Requirements can be found in Murphy et al. (2017).

Table 1. The ngVLA Key Science Goals

KSG	Title
1	Unveiling the Formation of Solar System Analogs on Terrestrial Scales
2	Probing the Initial Conditions for Planetary Systems and Life with Astrochemistry
3	Charting the Assembly, Structure, and Evolution of Galaxies from the First Billion Years to the Present
4	Using Pulsars in the Galactic Center to Make a Fundamental Test of Gravity
5	Understanding the Formation and Evolution of Stellar and Supermassive Black Holes in the Era of Multi-Messenger Astronomy

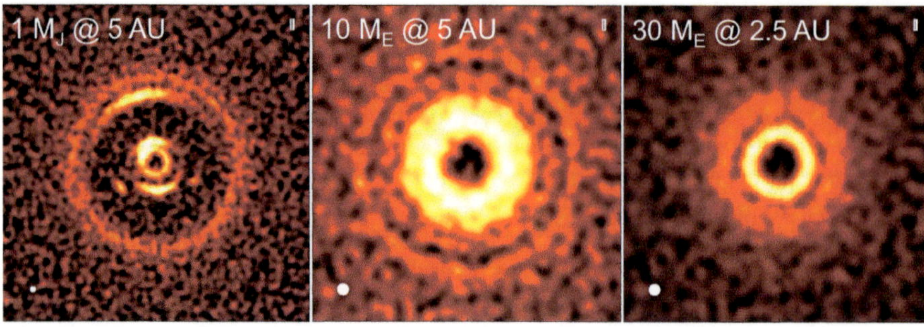

Figure 1. Simulated ngVLA observations of protoplanetary disk continuum emission perturbed by a Jupiter mass planet at 5 au (left), a 10 Earth mass planet at 5 au (center), and a 30 Earth mass planet at 2.5 au (right). The ngVLA observations at 100 GHz were simulated with 5 mas angular resolution and 0.5 μJy/bm rms (Ricci et al. 2018).

3.1. KSG 1: Unveiling the Formation of Solar System Analogs on Terrestrial Scales

Planets are thought to be assembled in disks around pre-main sequence stars, but the physical processes responsible for their formation are poorly understood. Only recently, optical, infrared, and (sub-) millimeter telescopes have achieved the angular resolution required to spatially resolve the innermost regions of nearby protoplanetary disks, unveiling morphological features with characteristic sizes of >20 au suggestive of gravitational perturbations of yet unseen giant planets. This in turn provides a powerful tool to measure planet masses, orbital radii, study the circumplanetary environment, and investigate how forming planets interact with the circumstellar material. The angular resolution, frequency coverage, and sensitivity of current disk imagery is limited to probing for the presence of planets more massive than Neptune at orbital radii larger than 20 – 30 au. The next step forward in the study of planet formation is the ability to image the formation of super-Earths and giant planets across the entire disk, particularly within 10 au from the central star.

To achieve this science goal requires that the ngVLA have the frequency coverage, sensitivity, and angular resolution to be able to measure the planet initial mass function down to a mass of 5 – 10 Earth masses. This capability will unveil the formation of planetary systems similar to our own Solar System by probing the presence of planets on orbital radii as small as 0.5 au at the distance of $\approx$140 pc. The ngVLA shall also be able to reveal circumplanetary disks and sub-structures in the distribution of mm-size dust particles created by close-in planets and measure the orbital motion of these features on monthly timescales.

This in turn requires continuum observations for center frequencies between 20 – 110 GHz with angular resolution better than 5 mas. This requirement will enable studies on the formation of planets in the innermost 10 au of nearby ($\lesssim$140 pc) proto-planetary disks. Extensive simulations of the disks perturbed by planets (see Figure 1; Ricci et al. 2018), suggest that a sensitivity of 0.2 μJy/bm in the continuum at 100 GHz is required to routinely map structures in the dust distribution created by planets of mass down to 10 Earth-masses and orbital radius of 2.5 au. Matching resolution (i.e., 5 mas) and achieving a continuum sensitivity of order 0.02 μJy/bm at 30 GHz will map the planet-disk interactions where the disk emission is expected to be optically thin.

3.2. KSG 2: Probing the Initial Conditions for Planetary Systems and Life with Astrochemistry

One of the most challenging aspects in understanding the origin and evolution of planets and planetary systems is tracing the influence of chemistry on the physical evolution of a system from a molecular cloud to a solar system, while also trying to determine the potential for habitability. To make significant progress in this area requires that the ngVLA has the frequency coverage and sensitivity to be able to detect predicted, but as yet unobserved, complex prebiotic species that are the basis of our understanding of chemical evolution toward amino acids and other biogenic molecules. In doing so, the ngVLA will also allow us to detect and study chiral molecules, testing ideas on the origins of homochirality in biological systems. The detection of such complex organic molecules will provide the chemical initial conditions of forming solar systems and individual planets.

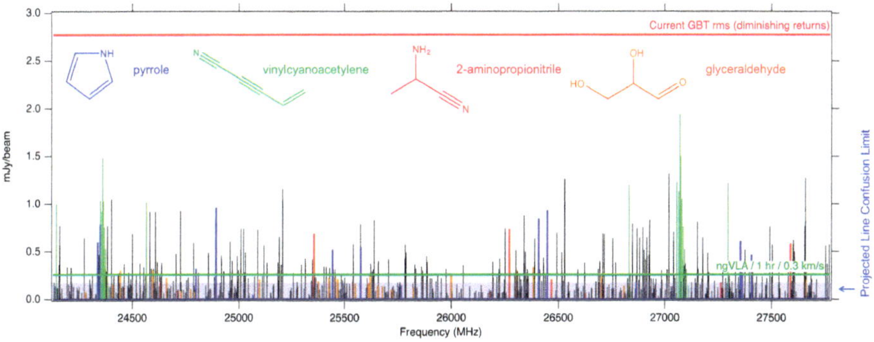

Figure 2. A conservative simulation of 30 as-yet-undetected complex interstellar molecules (black) likely to be observed by the ngVLA above the confusion limit around hot cores with typical sizes of $\sim 1 - 4''$. Key molecules are highlighted in color. (Credit: B. McGuire)

Presently, existing observations of complex organic (prebiotic) molecules using ALMA and the Great Bank Telescope (GBT) are hitting the limit of what can be accomplished due to a combination of achievable sensitivity and line confusion at higher frequencies. Both problems can be solved by sensitive observations in the cm-wave regime with the ngVLA. State-of-the-art models predict these molecules will display emission lines with intensities that are easily detectable with the ngVLA, but well below the current detectability thresholds of existing telescopes including ALMA, GBT, and IRAM. Figure 2 shows simulations of a representative set of the types of molecules whose discovery will be enabled by the ngVLA: N, O, and S-bearing small aromatic molecules, direct amino acid precursors, biogenic species such as sugars, chiral molecules, and, possibly amino acids themselves. The simulation assumes column densities of $10^{12} - 10^{14}\,\mathrm{cm}^{-2}$ (with more complex molecules being assigned lower column densities), a temperature of 200 K, and 3 km/s linewidth.

To achieve this science goal requires an angular resolution on the order of 50 mas at 50 GHz along with an rms sensitivity of 30 μJy/bm/km/s for frequencies between

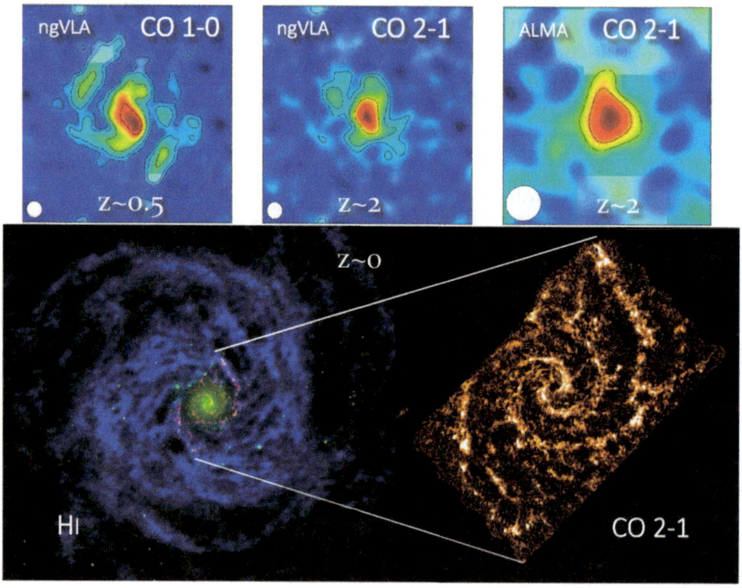

Figure 3. *Top Panels:* Simulations based on M 51 with molecular mass scaled by 1.4× ($z = 0.5$) and 3.5× ($z = 2$) to match the lowest molecular mass galaxies observable by ALMA and NOEMA (Carilli & Shao 2017). The synthesized beam shown in the bottom left corner is (left to right) θ_s = 0.″19, 0.″20, and 0.″43 corresponding to linear scales $L = 1.2, 1.7$, and 3.7 kpc, respectively. Integration times are 30 hr. *Bottom Panels:* The spiral galaxy M 74 illustrating the CO molecular disk imaged by ALMA (red; Schinnerer in prep.), the stellar disk at 4.5 μm imaged by *Spitzer* (green; Kennicutt et al. 2003), and the atomic disk imaged in HI by the VLA (blue; Walter et al. 2008), showing the gas phases to which the ngVLA will be sensitive. *Bottom Right Panel:* A zoom in showing the CO $J = 2 \rightarrow 1$ map at 1″ resolution.

16 – 50 GHz. Further, spectral resolution of 0.1 km/s is required, preferably concurrent with broadband (4+ GHz) observations.

3.3. KSG 3: Charting the Assembly, Structure, and Evolution of Galaxies from the First Billion Years to the Present

To make substantial progress in the field of galaxy formation and evolution requires that the ngVLA have the sensitivity to survey cold gas in thousands of galaxies back to early cosmic epochs, while simultaneously enabling routine sub-kiloparsec scale resolution imaging of their gas reservoirs. In doing so, the ngVLA will afford a unique view into how galaxies accrete, process, and expel their gas through detailed imaging of their extended atomic/molecular reservoirs and circumgalactic regions. To reveal the detailed physical conditions for galaxy assembly and evolution throughout the history of the universe requires that the ngVLA also have enough sensitivity to map the physical and chemical properties of molecular gas over the entire local galaxy population.

To carry out detailed studies of CO kinematics of high-z galaxies and blind CO searches of > 1000 galaxies requires a line sensitivity of ~46 μJy/bm/km/s at 0.″1 and 1″angular resolution between 10 – 50 GHz with a spectral resolution of 5 km/s. This is illustrated by a simulation of M 51 (the Whirlpool galaxy) shown in the top three panels of Figure 3 (see Carilli & Shao 2017). The spatial and kinematic information recovered by the ngVLA allows for the measurement of a precise rotation curve, which would only be possible to obtain from ALMA with an extraordinarily large ($\sim$ 1000 hr) time investment. Furthermore, a large instantaneous bandwidth (i.e., a minimum 1.6:1 BW ratio, up to 20 GHz instantaneous bandwidth) is required to conduct wide band observations at 5 km/s resolution to efficiently perform blind surveys of large cosmic volumes in a single observation and provide routine access molecular species in addition to CO (e.g., HCN, HCO^+, or N_2H^+).

Thermal imaging of 0.1 – 0.2 K sensitivity of CO (115 GHz) at $0\farcs1$ angular resolution and 1 km/s spectral resolution is required for detailed studies of molecular gas in the nearby universe (see bottom panels of Figure 3). Thermal imaging of 1 – 5 mK sensitivity between 70 and 116 GHz at $1-5''$ angular resolution and $1-5$ km/s spectral resolution is required to support studies of gas density across the local universe.

Full 1.2–116 GHz frequency coverage is required to obtain accurate, simultaneous measurements of star formation rates from free-free continuum and radio recombination line (RRL) emission. Angular resolutions of $0\farcs1$ – $1''$ for continuum imaging at all available frequencies are required. A continuum sensitivity of 0.15 μJy/bm at 33 GHz for a $1''$ synthesized beam is required to robustly study star formation within large samples of nearby galaxies. For studies of galaxies in the local universe, accurate recovery of flux density for extended objects on arcminute scales at all frequencies is required, along with the ability to make large mosaics or conduct on-the-fly line and/or continuum mappings of galaxies that extend beyond the area of a single primary beam.

Finally, a brightness dynamic range of $\approx$50 and 40 db is required at 10 GHz for deep field continuum studies of MW-like galaxies at Cosmic Noon to not be dynamic range limited in total and polarized intensity, respectively. Such deep field observations will be sensitive to $\gtrsim$90% of all stars formed since $z \lesssim 3$.

3.4. KSG 4: Using Pulsars in the Galactic Center to Make a Fundamental Test of Gravity

Testing theories of gravity requires probing as close as possible to the strong field regime, for which pulsars near the Galactic Center offer a powerful path forward. However, only a handful of pulsars in the central half-degree of the Galaxy are currently known (see Figure 4), which may be the result of enhanced radio-wave scattering toward the inner Galaxy that further decreases the effective sensitivity of searches, by increasing both the dispersion measure smearing and pulse broadening. Observing at higher frequencies than currently planned for the SKA can mitigate radio-wave scattering, but by itself the benefits are limited because of the generally steep radio spectra of pulsars. Thus, to achieve this goal, the ngVLA must deliver a combination of sensitivity and frequency range, enabling it to probe much deeper into the likely Galactic Center pulsar population to address fundamental questions in relativity and stellar evolution. The ability to address these questions is afforded by the fact that pulsars in the Galactic Center represent clocks moving in the space-time potential of a super-massive black hole and allow for qualitatively new tests of theories of gravity. More generally, they offer the opportunity to constrain the history of star formation, stellar dynamics, stellar evolution, and the magneto-ionic medium in the Galactic Center.

To carryout this science requires that the ngVLA be able to support pulsar search and timing observations from $\sim$ 1 – 30 GHz for Galactic Center pulsars. Pulsar search-

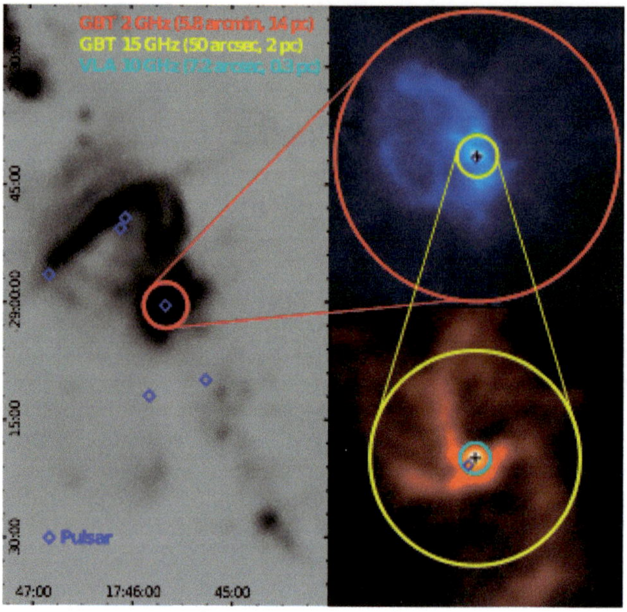

Figure 4. The pulsar distribution near the Galactic Center. Despite being the highest density in the Galaxy and multiple searches at sensitivities comparable to the VLA, only a few pulsars are known though ~ 1000 are predicted. (Credit: R. Wharton)

ing requires the ability to search on $100\,\mu$s scales ($20\,\mu$s scales desired), while timing requires $20\,\mu$s resolution. A continuum sensitivity equivalent of order 50 nJy/bm is desired at 20 GHz, which is a significant improvement compared to existing 100 m-class radio telescopes that have found few pulsars, indicating that substantial additional sensitivity is necessary. The system timing accuracy also must be better than 10 ns (1 ns desired) over periods correctable to a known standard from 30 min to 10 yr. To efficiently time time multiple pulsars, the array must have the ability to make multiple (minimum 10) beams (i.e., phase centers within the primary beam) within a single subarray, or distributed amongst multiple subarrays. Timing multiple pulsars within a single primary beam is desirable. Support for 5 or more independent de-dispersion and folding threads is desired.

3.5. KSG 5: Understanding the Formation and Evolution of Stellar and Supermassive Black Holes in the Era of Multi-Messenger Astronomy

While we now know that black holes exist on practically all mass scales, the astrophysics of how these objects form and grow remains a mystery. The Laser Interferometer Gravitational-wave Observatory (LIGO) is now detecting black holes that are substantially more massive than previously known stellar mass black holes, and observing black hole-black hole mergers, although we do not know how black hole binaries form. While supermassive black holes (SMBHs) are thought to be widespread in galaxy centers, we do not understand how their growth was seeded or how (and how often) these extreme objects merge. To address these questions requires that the ngVLA have the

Figure 5. Two tiny, but very dense neutron stars merge and explode as a kilonova. Such a very rare event produces gravitational waves and electromagnetic radiation, as observed on 17 August 2017. The ngVLA will play a pivotal role in characterizing the physics of such events in the era of multi-messenger astronomy. (Artist's impression, Credit: ESO/L. CalÇada/M. Kornmesser).

combination of sensitivity and angular resolution to be able to survey everything from the remnants of massive stars to the supermassive black holes that lurk in the centers of galaxies, making it the ultimate black hole hunting machine. High-resolution imaging abilities are required to separate low-luminosity black hole systems in our local Universe from background sources, thereby providing critical constraints on the formation and growth of black holes of all sizes and mergers of black hole-black hole binaries.

To become the ultimate black-hole survey instrument requires that the ngVLA have high angular resolution (mas – μas) imaging with relative astrometric accuracy that is $< 1\%$ of the synthesized beam FWHM or equal to the positional uncertainty in the reference frame, for a bright ($SNR \gtrsim 100$) point source. Such high-resolution imaging will enable proper motion separation of local black holes (both Galactic and in nearby galaxies, out to 15 Mpc) from background sources. Long baselines are required to enable the ngVLA to image the SMBH binaries that will be detected in gravitational waves by LISA and pulsar timing arrays. These astrometric science goals benefit from the implementation of very long baselines ($\gtrsim$ 1000 km for mas – μas accuracy). Associated VLBI recording capabilities shall be available for 3 or more beams (for 2 calibrators and the science target).

The field of multi-messenger astronomy continues to mature as we continue to open new astronomical windows through gravitational waves and neutrino observations. However, to progress further in our understanding of the physics associated with these phenomena requires the ability to localize and characterize the sources. Only the detection of the electromagnetic radiation associated with these energetic, and often cataclysmic events, can provide precise localization, establish energetics and allow us to understand how such events interact with their surrounding environments. Thus, to have a transformational impact in the growing era of multi-messenger astronomy, the ngVLA must also be able to identify the radio counterparts to transient sources discovered by gravitational wave, neutrino, and optical observatories (see Figure 5). This requires high-resolution, fast-mapping capabilities to make it the preferred instrument to pinpoint transients associated with violent phenomena such as supermassive black hole mergers and blast waves.

Specifically, mapping a ~10 square degree region (i.e., the localization uncertainty expected by gravitational wave detectors when ngVLA is operational) to a depth of $\sim 1\,\mu$Jy/bm at 2.5 GHz for detection of NS-NS and NS-BH mergers is required. Completing the on-the-fly mapping of each epoch within $\sim$ 10 hr is desirable. Similarly, mapping a $\sim$ 10 square degree region at 28 GHz to a depth of $\sim 10\,\mu$Jy/bm with on-the-fly mapping is required for localization of LISA-detected SMBH mergers. Again, completing the on-the-fly mapping of each epoch within $\sim$ 10 hr is desirable. Furthermore, an rms sensitivity of $\approx 0.23\,\mu$Jy/bm at 10 GHz for a 0.6 mas beam (i.e., on continental-scale baselines) is required to detect a source like GW170817 with a $SNR \approx 10$ at the Adv. LIGO horizon distance of 200 Mpc, which will in turn allow for measurements (movies) of its expansion at the 5σ level.

The ability to receive and respond to external triggers rapidly is also an essential requirement to enable multi-messenger science. Triggered response times not to exceed 10 minutes is required, while response times of better than 3 minutes is desired. The ability to perform time-domain transient searchers (e.g., for Fast Radio Bursts) requires a search capability on $100\,\mu$s scales, with $20\,\mu$s scales desired.

4. Summary

The ngVLA is being designed to tap into the astronomical community's intellectual curiosity by providing them with a world-class instrument that will enable a broad range of scientific discoveries (e.g., planet formation, signatures of pre-biotic molecules, cosmic cycling of cool gas in galaxies, tests of gravity, characterizing the energetics of gravitational wave counterparts, etc.). Based on community input to date, the ngVLA is the obvious next step to build on the VLA's legacy and continue the U.S.'s place as a world leader in radio astronomy. The ultimate goal of the ngVLA is to give the U.S. and international communities a highly capable and flexible instrument to pursue their science in critical, yet complementary ways, with the large range of multi-wavelength facilities that are on a similar horizon. Presently, there have been no major technological risks identified. However, the project is continually looking to take advantage of major engineering innovations, seeking to optimize the performance and operational efficiency of the facility. As the project continues to move forward and mature, the project will continue to work with the community to refine the ngVLA science mission and instrument specifications/performance. This science book acts as a major milestones in this effort.

Acknowledgments. The authors thank the 100's of community members that have contributed to the definition of the ngVLA science mission and corresponding science requirements. The National Radio Astronomy Observatory is a facility of the National Science Foundation operated under cooperative agreement by Associated Universities, Inc. Part of this work was carried out at the Jet Propulsion Laboratory, California Institute of Technology, under contract with the National Aeronautics and Space Administration.

References

Bolatto, A. D., Chatterjee, S., Casey, C. M., et al. 2017, arXiv:1711.09960
Carilli C. L. & Shao, Y. 2017, "Image Capabilities: High redshift CO," Next Generation VLA Memo 13
Kennicutt, R. C., Jr., Armus, L., Bendo, G., et al. 2003, PASP, 115, 928
Murphy, E. J. et. al. 2017, "ngVLA Science Requirements" Next Generation VLA Document No. 020.10.15.00-0001-REQ
Ricci, L., Liu, S.-F., Isella, A., & Li, H. 2018, ApJ, 853, 110
Selina, R. et al. 2018, ASP Monograph 7, Science with a Next-Generation VLA, ed. E. J. Murphy (San Francisco, CA: ASP), 15
Selina, R. & Murphy E. J. 2017, "ngVLA Reference Design Development & Performance Estimates," Next Generation VLA Memo 17
Selina, R., Murphy, E., Erickson, A. 2017, "Summary of the Science Use Case Analysis," Next Generation VLA Memo 18
Walter, F., Brinks, E., de Blok, W. J. G., et al. 2008, AJ, 136, 2563

[1]*NRAO, Charlottesville, VA, 22903, USA;* emurphy@nrao.edu

[2]*Univ. of Maryland, Dept. of Astronomy, College Park, MD 20742, USA*

[3]*Dept. of Astronomy, Cornell University, Ithaca, NY 14853, USA*

[4]*Astronomy, UT Austin, 2515 Speedway Blvd Stop C1400, Austin, TX 78712*

[5]*Dept. of Physics & Astronomy, MSU, East Lansing, MI 48824, USA*

[6]*Dept. of Physics & Astronomy, University of Wyoming, Laramie WY, USA*

[7]*Dept. of Astronomy and Earth & Planetary Sciences, UCB, Berkeley, CA, USA*

[8]*NOAO, 950 N Cherry Avenue, Tucson, AZ 85719, USA*

[9]*NRC Herzberg, 5071 W. Saanich Rd., Victoria, BC, V9E 2E7, Canada*

[10]*Univ. of Victoria, 3800 Finnerty Road, Victoria, BC, V8P 5C2, Canada*

[11]*Cahill Center for Astronomy & Astrophysics, Caltech, Pasadena CA, USA*

[12]*Physics & Astronomy, Rice 6100 Main Street, MS-108, Houston, Texas 77005*

[13]*Inst. of Astronomy, Univ. of Tokyo, Osawa, Mitaka, Tokyo 181-0015, Japan*

[14]*Physics & Astr., 703 Van Allen Hall, U. Iowa, Iowa City, IA 52242, USA*

[15]*JPL, Caltech, Pasadena, CA 91109, USA*

[16]*Dept. of Astronomy, OSU, 140 West 18th Avenue, Columbus, OH 43210*

[17]*Instituto de Radioastronomía y Astrofísica, Universidad Nacional Autónoma de México, Morelia 58089, Mexico*

[18]*Univ. Nacional Autónoma de México, Apartado Postal 70-264, 04510 Ciudad de México, México*

[19]*Dept. of Physics, Box 41051, Sci. Building, TTU, Lubbock TX, 79409, USA*

[20]*STScI 3700 San Martin Drive, Baltimore, MD 21218, USA*

[21] *Harvard-Smithsonian CfA 60 Garden Street, Cambridge MA 02138, USA*

[22]*RIKEN, 2-1 Hirosawa, Wako-shi, Saitama 351-0198, Japan*

[23]*MPIfA, Königstuhl 17, Heidelberg D-69117, Germany*

Science with a Next-Generation Very Large Array
ASP Conference Series, Monograph 7
Eric J. Murphy, ed.

The ngVLA Reference Design

Robert J. Selina,[1] Eric J. Murphy,[2] Mark McKinnon,[1] Anthony Beasley,[2] Bryan Butler,[1] Chris Carilli,[1] Barry Clark,[1] Steven Durand,[1] Alan Erickson,[1] Wes Grammer,[1] Rafael Hiriart,[1] James Jackson,[1] Brian Kent,[2] Brian Mason,[2] Matthew Morgan,[2] Omar Yeste Ojeda,[2] Viviana Rosero,[1] William Shillue,[2] Silver Sturgis,[1] and Denis Urbain[1]

[1]*National Radio Astronomy Observatory, Socorro, NM, USA*

[2]*National Radio Astronomy Observatory, Charlottesville, VA, Country*

Abstract. The next-generation Very Large Array (ngVLA) is an astronomical observatory planned to operate at centimeter wavelengths (25 to 0.26 centimeters, corresponding to a frequency range extending from 1.2 to 116 GHz). The observatory will be a synthesis radio telescope constituted of approximately 244 reflector antennas each of 18 meters diameter, and 19 reflector antennas each of 6 meters diameter, operating in a phased or interferometric mode. We provide a technical overview of the Reference Design of the ngVLA. This Reference Design forms a baseline for a technical readiness assessment and the construction and operations cost estimate of the ngVLA. The concepts for major system elements such as the antenna, receiving electronics, and central signal processing are presented.

1. Introduction

As part of its mandate as a national observatory, the National Science Foundation's (NSF) National Radio Astronomy Observatory (NRAO) is looking toward the long-range future of radio astronomy and fostering the long-term growth of the U.S. and global astronomical community. With NSF support, NRAO has sponsored a series of science and technical community meetings to define the science mission and concept for a next-generation Very Large Array (ngVLA; McKinnon 2016) that builds on the legacies of the Atacama Large Millimeter/submillimeter Array (ALMA) and the Jansky Very Large Array (VLA).

Based on input solicited from the astronomical community, the ngVLA is planned as an astronomical observatory that will operate at centimeter wavelengths (25 to 0.26 centimeters, corresponding to a frequency range extending from 1.2 to 116 GHz). The observatory will be a synthesis radio telescope consisting of:

- A main array of 214 reflector antennas each of 18 meters diameter, operating in a phased or interferometric mode. The main array is distributed to sample a wide range of scales from 10s of meters to 1000 km. A dense core and spiral arms provide high surface brightness sensitivity, with mid-baseline stations enhancing angular resolution.
- A short baseline array (SBA) of 19 reflector antennas of 6 m aperture will be sensitive to a portion of the larger angular scales undetected by the main array.

The SBA may be combined with 4 18 m (main-array) antennas used in a total power mode to completely fill in the central hole in the (u, v)-plane left by the 6 m dishes.

- A long baseline array (LBA) will add an additional 30 reflector antennas each of 18 m diameter in 10 clusters providing continental scale baselines ($B_{MAX} \sim$ 8860 km). The LBA is designed to sample a broad range of scales for stand-alone sub-array use, as well as for integrated operation with the main array.

It total, the ngVLA will have approximately ten times the sensitivity of the VLA and ALMA, continental-scale baselines providing sub-milliarcsecond-resolution, and a dense core on km-scales for high surface brightness sensitivity. Such an array bridges the gap between ALMA, a superb sub-mm array, and the future SKA1, optimized for longer wavelengths.

The dense core and the signal processing center of the array will be located at the Very Large Array site, on the plains of San Agustin, New Mexico. The high desert plains of the Southwest U.S., at over 2000 m elevation, provide excellent observing conditions for the frequencies under consideration, including reasonable phase stability and opacity at 3 mm wavelength over a substantial fraction of the year.

The array will also include stations in other locations throughout the state of New Mexico, west Texas, eastern Arizona, and northern Mexico. Long baseline stations are located in Hawaii, Washington, California, Iowa, Massachusetts, New Hampshire, Puerto Rico, the US. Virgin Islands, and Canada.

Operations will be conducted from both the VLA Site and the Array Operations and Repair Centers in Socorro, NM. A Science Operations Center and Data Center will be collocated in a large metropolitan area and will be the base for science operations and support staff, software operations and related administration. Research and development activities will be split amongst these centers as appropriate.

The facility will be operated as a proposal-driven instrument. The fundamental data products delivered to ngVLA users will be science-ready data products (i.e., images and cubes) generated using calibration and imaging pipelines created and maintained by the project. Both the pipeline products and the "raw" visibilities and calibration tables will be archived, retaining the option of future re-processing and archival science projects.

The ngVLA project is developing a Reference Design for the array as a baseline for construction and operation costing, and future design trade-off decisions. This Reference Design is intended to be low technical risk in order to provide a degree of conservativism in the estimates. However, leading-edge concepts and techniques that may improve the performance and/or reduce cost are being developed in parallel, and will be evaluated in the conceptual design phase of the facility. This paper provides an overview of the Reference Design as the project approaches the Astro2020 Decadal Survey.

2. Key Science & Technical Requirements

The Key Science Goals and all other science use cases were parameterized and analyzed (Selina, Murphy & Erickson 2017) to determine the science requirements for the ngVLA (Murphy 2017). While this aspect of the requirements definition is top-down and mission-driven, some judicious adjustment of the requirements is still appropriate.

A primary science requirement for the ngVLA is to be flexible enough to support the breadth of scientific investigations that will be proposed by its creative scientist-users over the decades-long lifetime of the instrument. The requirements have therefore been adjusted to provide a balanced, flexible, and coherent complement of capabilities. The primary requirements that drive the design are described below:

- **Frequency Coverage:** The ngVLA should be able to observe in all atmospheric windows between 1.2 and 116 GHz. These frequency limits are bracketed by spectral line emission from HI and CO respectively.
- **Continuum Sensitivity:** A continuum sensitivity of better than 0.02 μJy/bm at 30 GHz and 0.2 μJy/bm 100 GHz is required for studying protoplanetary disks. This requires a combination of large collecting area and wide system bandwidth.
- **Line Sensitivity:** A line sensitivity of 30 μJy/bm/km/s for frequencies between 10 and 50 GHz is simultaneously required to support both astrochemistry studies and deep/blind spectral line surveys. A line sensitivity of $1 - 100$ mK at $0\farcs1 - 5''$ angular resolution and $1 - 5$ km/s spectral resolution between 70 and 116 GHz is required to simultaneously support detailed studies of CO and variations in gas density across the local universe. The spectral line cases push the system design towards quantum-limited noise performance at the expense of bandwidth above 10 GHz.
- **Angular Resolution:** A synthesized beam having a FWHM better than 5 mas with uniform weights is required at both 30 and 100 GHz, while meeting the continuum sensitivity targets.
- **Largest Recoverable Scale:** Angular scales of $> 20'' \times (100\,\text{GHz}/\nu)$ must be recovered at frequencies $\nu < 100$ GHz. A more stringent desire is accurate flux recovery on arcminute scales at all frequencies. These scales approach the size of the primary beam of an 18 m dish, so both shorter baselines and a total power capability are necessary to completely fill in the central hole in the (u, v)-plane.
- **Surface Brightness Sensitivity:** The array must provide high-surface brightness sensitivity over the full range of angular scales recoverable with the instrument. This leads to a centrally condensed distribution of antennas.
- **Brightness Dynamic Range:** The system brightness dynamic range shall be better than 50 dB for deep field studies. This requirement pushes a number of systematic requirements including pointing, gain, and phase stability.
- **Survey Speed:** The array shall be able to map a $\sim$10 square degree region to a depth of $\sim$ 1 μJy/bm at 2.5 GHz and a depth of $\sim$ 10 μJy/bm at 28 GHz within a 10 hr epoch for localization of transient phenomena identified with other instruments. Holding collecting area and receiver noise constant, this favors smaller apertures.
- **Beamforming for Pulsar Search, Pulsar Timing and VLBI:** The array shall support no less than 10 beams spread over 1 to 10 subarrays that are transmitted, over the full available bandwidth, to a VLBI recorder/correlator, pulsar search engine or pulsar timing engine. The pulsar search and timing engine must be integral to the baseline design.

- **Science Ready Data Products:** The primary data product delivered to users shall be calibrated images and cubes. Uncalibrated, "raw" visibilities shall be archived to permit reprocessing. Producing these higher-level data products requires some standardization of the initial modes/configurations that the system is used in (e.g., limited tuning options), and repeatability/predictability from the analog system to reduce the calibration overheads.

3. Site Selection & Performance

The VLA site on the plains of San Agustin was originally chosen as the location for the array because of its desirable properties: large, relatively flat, undeveloped (to minimize RFI) yet not too remote (for accessibility), at low latitude (for sky coverage), and at high elevation (to minimize atmospheric effects) (Thompson et al. 1980). These properties still hold true, and motivate examination of the VLA site as the center of the ngVLA. Furthermore, with extensive existing infrastructure, the VLA site leverages an already-existing system of power, fiber, and buildings, which will reduce cost. The three main environmental or atmospheric quantities that may affect data, and what is known about them at the VLA site, are discussed in the following sections.

3.1. RFI

The VLA site is remote enough that Radio Frequency Interference (RFI) is not a debilitating problem, so it will be possible to observe at the lower frequencies of ngVLA (Stewart 2005). Furthermore, the ngVLA will benefit by advanced studies of RFI detection and excision that are currently ongoing (Burnett et al. 2018). The degree of RFI characterization of the site reduces the risk in site selection, and leveraging existing infrastructure could create significant cost savings for both the construction and operation of the array. Given the large extent of ngVLA ($B_{MAX} \sim$ 8860 km), it is clear that the antennas which are outside the plains will experience different RFI environments than that at the site. However, there are locations which are relatively free of locally generated RFI (downward RFI from orbiting satellites is ubiqitous and nearly site-independent), and the U.S. southwest has many such locations (Li et al. 2004).

3.2. Atmospheric Phase Stability

Analysis of data from the VLA site atmospheric phase monitor shows that fast switching phase calibration at 3 mm should be viable for most of the year with a 30 s total calibration cycle time (Carilli 2015). This analysis was based on one year of atmospheric phase monitoring at the VLA site (Butler & Desai 1999). A much longer time base of these values is now available. Figure 1 shows median values of the rms phase on the 300 m E-W baseline of the atmospheric phase monitor from 1995 through 2017, plotted as a function of UTC hour, and month. It is easy to see that these fluctuations are small for much of the time, and only become greater than 10° (rms @ 11.7 GHz, over 10 minutes) in the summer during daytime. Little information is available on phase fluctuations at locations outside the plains; this is a topic to be studied to determine the ability to use the remote sites at the highest frequencies of ngVLA. Note that there should also be a 25 mJy calibrator source within 2° in 98% of observed fields, ensuring short slews. Such a calibrator is adequate to ensure that the residual rms phase noise due to the signal-to-noise ratio on the phase calibrator is much less than that due to the

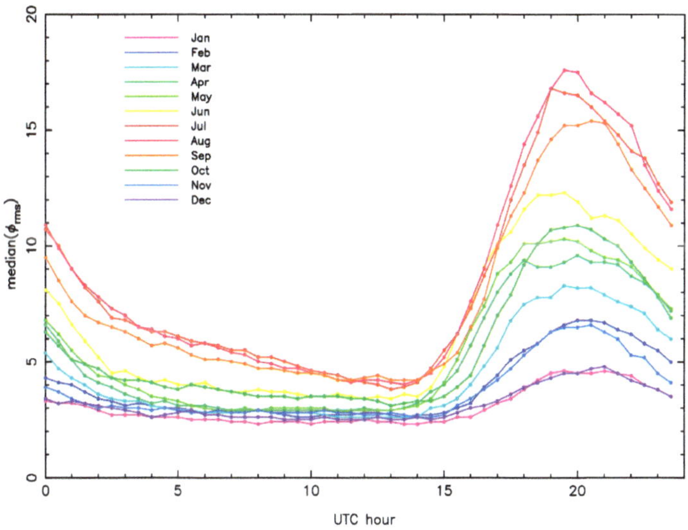

Figure 1. The median rms phase measured with the atmospheric phase monitor at the VLA (300 m E-W baseline, 11.7 GHz beacon), from 1995 to 2017. Measurements are calculated over a 10 minute period after subtracting any linear trend. Different months are plotted as different colors, as shown in the legend.

troposphere, even for a 30 s cycle time with only 3 s on the calibrator each visit (Carilli 2015; Clark 2015). The project is also investigating radiometric phase correction techniques as part of the ngVLA project to increase the total phase calibration cycle time.

3.3. Atmospheric Opacity

While at centimeter wavelengths atmospheric opacity is a relatively minor issue compared to phase stability, it becomes a much bigger issue at millimeter wavelengths. Similar to the atmospheric phase stability data, there is a long-time baseline of surface weather data at the VLA site. This can be used to estimate the atmospheric Precipitable Water Vapor (PWV), which is the main contributor to the fluctuating part of atmospheric opacity (Butler 1998). Figure 2 shows this value for the years 2010 through 2017. In winter months, the median over all hours is around 3 mm, and over the entire year the median over all hours is 5.4 mm. Vertical opacity for 5.4 mm PWV at 90 GHz is less than 7%, so opacity should not be a major problem for ngVLA. As with RFI and phase stability, there is little information on atmospheric opacity at other locations, though it is almost always clear that higher sites have less opacity. The project does have access to surface weather data, and to radiosonde launch data (twice per day) from NOAA for some tens of sites across the southwest US, which will be the subject of a future study to determine opacity properties across the extent of the ngVLA.

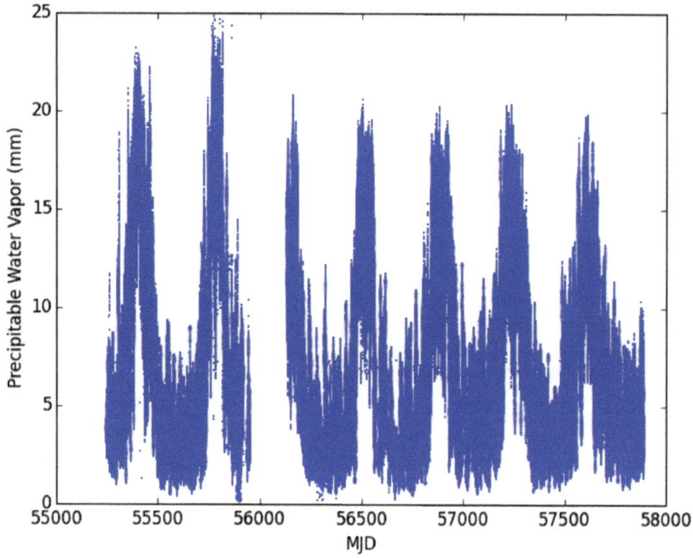

Figure 2. PWV at the VLA site, estimated using surface weather measurements, from 2010 to 2017. Note that a PWV of 6 mm produces an opacity of less than 7% at 90 GHz.

3.4. Final Site Selection

Because of the quality of the site for both low- and high-frequency observing, and the existing infrastructure, the ngVLA is centered near the current VLA. The southwest U.S. and northern Mexico are sparsely populated and the antennas within 1000 km of the VLA are sited to select remote, radio quiet, and dry sites, while still considering the logistics of site access, electrical infrastructure and fiber optic network topology. The long baseline array sites were selected to minimize site impact and leverage shared infrastructure of other existing observatories, so sites operated by the VLBA or other observatories are preferred. Note that the VLA site was used for acceptance testing of the original ALMA antennas, including observations up to 230 GHz, and the experience was that the VLA site, at 2124 m elevation is a high-quality 90 GHz site - comparable to the Plateau de Bure site in overall performance (Thompson, Moran & Swenson 2004).

4. Array Configuration

The ngVLA array design includes three fundamental subarrays providing a wide range of angular scales: a main interferometric array, a short baseline array, and a long-baseline array.

The main array configuration will consist of 214 18 m antennas at the approximate locations shown in Figure 3. The array collecting area (see Table 1) is distributed to provide high surface brightness sensitivity over a range of angular scales spanning from approximately 1000 to 100 mas while providing high point source sensitivity on scales

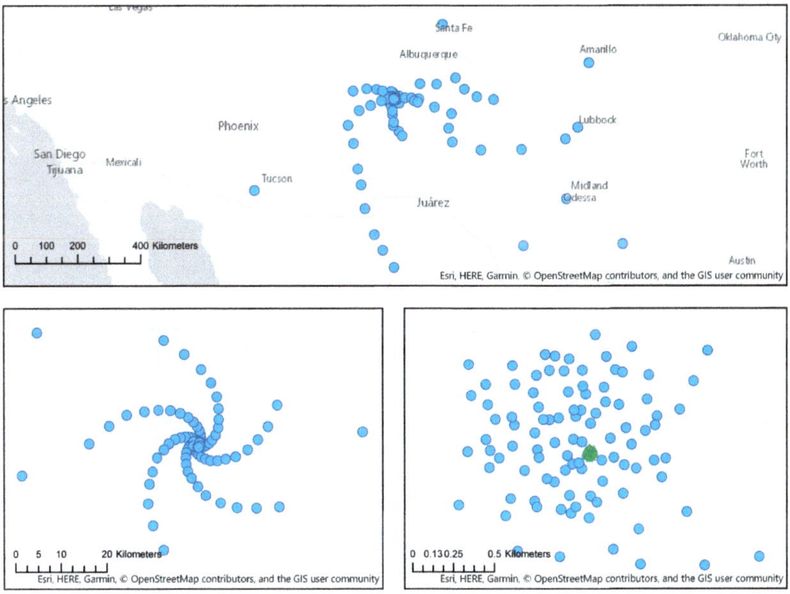

Figure 3. (*Top:*) ngVLA Main Array Configuration Rev. B (Spiral-214). The antenna positions are still notional, but are representative for performance quantification and cost estimation. (*Bottom Left*): Zoom view of the plains of San Agustin. (*Bottom Right:*) Zoom view of the compact core. SBA antennas are shown in green.

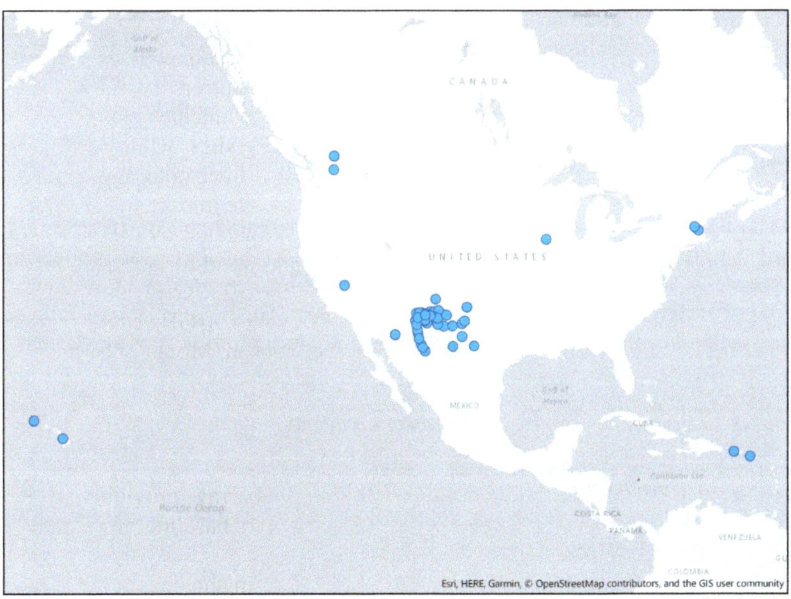

Figure 4. View of the Main Array and Long Baseline Array stations. Multiple antennas are located at each LBA site.

Radius	Collecting Area Fraction	Quantity
0 km< R < 1.3 km	$\approx 44\%$	94
1.3 km< R < 36 km	$\approx 35\%$	74
36 km< R <1000 km	$\approx 21\%$	46

Table 1. Radial distribution of collecting area for the main array (214 antennas).

up to 10 mas. A large fraction of the collecting area is in a randomly distributed core to provide high snapshot imaging fidelity and there are arms extending asymmetrically out to ~1000 km baselines to fill the (u, v)-plane via Earth rotation and frequency synthesis.

The design has been extended from the main interferometric array to include both a short spacing array and total power dishes (Mason 2018b). This was necessary after a review of the key science cases, as these are dependent on the recovery of large scale structure that approaches the size of the antenna primary beam.

The auxiliary short baseline array (SBA) of 19 reflector antennas of 6 m aperture will be sensitive to a portion of the larger angular scales undetected by the main array. The SBA will provide spacings from ~ 11 m to 56 m, providing comparable surface brightness sensitivity to the main array, in equal observing time, when the main array is (u, v)-tapered to the natural resolution of the SBA. This allows for commensal observing, and more importantly, full cross-correlation and cross-calibration of the SBA and main array. The array distribution is semi-randomized to improve the point spread function (Mason et al. 2018a).

The SBA will be combined with four 18 m (main array) antennas used in total power (TP) mode to completely fill in the central hole in the (u, v)-plane left by the 6 m dishes. It is a design goal to share the mount design of the 18 m interferometric array antennas and the TP antennas, but this will require further study.

In response to community feedback, a long baseline array (LBA) has also been added to the configuration (see Figure 4). The long baseline array adds 30 antennas of 18 m diameter at 10 additional sites. The LBA provides continental scale ($B_{MAX} \sim$ 8860 km) baselines while also providing scales from ~ 30 m to 1000 km within the subarray. This will enable the LBA to function effectively as a stand-alone array or as an integrated part of the main array.

The ngVLA array configuration elements are summarized in Table 2. The design of the array configuration is practical, accounting for logistical limitations such as topography, utility access, local RFI sources and land management/availability. An analysis of different weighting schemes (i.e., Briggs, (u, v)-taper) for specific science applications (Carilli 2017) found that the current configuration provides a reasonable

Array Element	Aperture Diameter [m]	Quantity	B_{MIN} [m]	B_{MAX} [km]
Long Baseline Array	18	30	32.6	8856
Main Interferometric Array	18	214	30.6	1005
Short Baseline Array	6	19	11.0	0.06
Total Power / Single Dish	18	$4^{\dagger}$	-	-

†These 4 dishes are included as part of the 214 main array.

Table 2. Summary of elements within the ngVLA array configuration.

compromise and baseline for further iteration. The configuration will be a primary area of study in the coming years, e.g., investigations are underway to improve the imaging fidelity and quality of the synthesized beam.

5. Array Calibration

The calibration strategy for ngVLA is being developed early in the design so that it may guide the design of the hardware elements. The size and complexity of the calibration and imaging pipeline requires that the system design be responsive to its needs, and it should inform the design where possible.

A secondary concern is the efficiency of the calibration process. Algorithms used must be suitable for parallel processing, antennas must not require much individual attention, and minimal human intervention should be generally required for routine operation. The calibration overheads applied will vary with the science requirements of a given observation, and less computationally or time intensive calibration approaches will be applied when possible.

The operations plan calls for guaranteed time on source to each observer, with calibration overheads being the responsibility of the facility (Ford et al. 2018). This enables the reuse of calibration observations for adjacent observations when their requirements are sufficiently similar, further improving observation efficiency.

The general calibration strategies under consideration for the reference design are summarized below.

- **Fast Atmospheric Phase Calibration:** Rapid atmospheric phase fluctuations will be mitigated by a combination of relative water vapor radiometry (WVR) and antenna switching cycles to astronomical phase calibrators. The switching cycle time will depend on empirical validation of the strategy, but is expected to be necessary on one to ten minute scales. The antenna will be designed to both house the WVR and move 4° on sky and settle to within the pointing specification with 10 seconds for elevation angles $< 70°$ (Selina 2017).

- **Slow Atmospheric & Electronic Phase Calibration:** Slow atmospheric and electronic phase calibration will be achieved by traditional approaches, with astronomical phase calibrator observations bracketing all observations. Several astronomical calibrators may be used to map the slow varying terms, including ionospheric fluctuations.

- **Amplitude Calibration:** A list of known astronomical amplitude calibrators will be used to correct for system gain fluctuations within an observation and between observations taken over an extended period of time. The calibration pipeline will maintain a history of recent solutions to enable look-up of prior values.

- **Bandpass Calibration:** At a minimum, the system will correct for digital effects, given the predictable bandpass ripple from finite impulse response filters. The number of setups in the analog portions of the system will be limited, so typical calibration can also correct for analog bandpass effects based on historical look-up tables that are updated as the configuration of the system changes (i.e., when an antenna is serviced).

- **Polarization Calibration:** The use of linear feeds will require polarization calibration for most observations. Feeds may be placed at different (but known) position angles in the various antennas, so a single observation of a point source can solve simultaneously for the polarization leakage terms and the source polarization. Calibration for polarization as a function of position within the antenna beam will be assumed to be time invariant and corrected based on look-up tables for most observations.

- **Relative Flux Density Calibration:** This calibration is used to tie together observations of a source taken over an extended period. The system will model atmospheric opacity based on barometric pressure and temperature monitored at the array core and each outlying station. A temperature stabilized noise diode will provide a flux reference, and when combined with corrections for modeled atmospheric opacity, we can assume a constant ratio in power from the switched noise calibrator and the source.

- **Absolute Flux Density Calibration:** Absolute flux scale calibration will employ similar methods to relative calibration, with two notable changes. First, atmospheric tipping scans will be used to empirically determine atmospheric opacity, with improved fidelity. Second, observations of astronomical flux density calibrators will be used, along with the switched power system, to determine the absolute flux density of the source.

The ngVLA will need to maintain multiple lists of calibrators by calibration intent. The flux density calibrator list can be relatively small and based on the one built and maintained by the VLA. An extensive grid of sources will be required for phase and amplitude calibration. The large range of baselines present on the ngVLA means that it cannot be assumed that the source is unresolved at all scales, and the calibrators themselves will have to be imaged before use in the calibration process.

6. Antenna

The antenna concept strikes a balance between competing science requirements and the programmatic targets for life cycle cost. Sensitivity goals will be met, in part, by the total effective collecting area of the array. The reference design includes 244 antennas of 18 m aperture (main array and long baseline array) and 19 antennas of 6 m aperture (short baseline array) using an offset Gregorian optical design.

The inclusion of frequencies down to 1.2 GHz, when combined with the operational cost targets, significantly constrain the optical configuration. The use of feeds with wide illumination angles decreases their size such that they can be mounted within shared cryostats. This choice constrains the secondary angle of illumination to a degree that only Gregorian optical designs are practical. However, with a science priority of high imaging dynamic range in the 10 – 50 GHz frequency range, an offset Gregorian is near optimal. The unblocked aperture will minimize scattering, spillover and sidelobe pickup. Maintenance requirements favor antenna optical configurations where the feed support arm is on the “low side” of the reflector.

The optimization for operations and construction cost suggests that a smaller number of larger apertures is preferable to larger numbers of small apertures. Survey speed requirements push the opposite direction, and a compromise value of 18 m diameter is

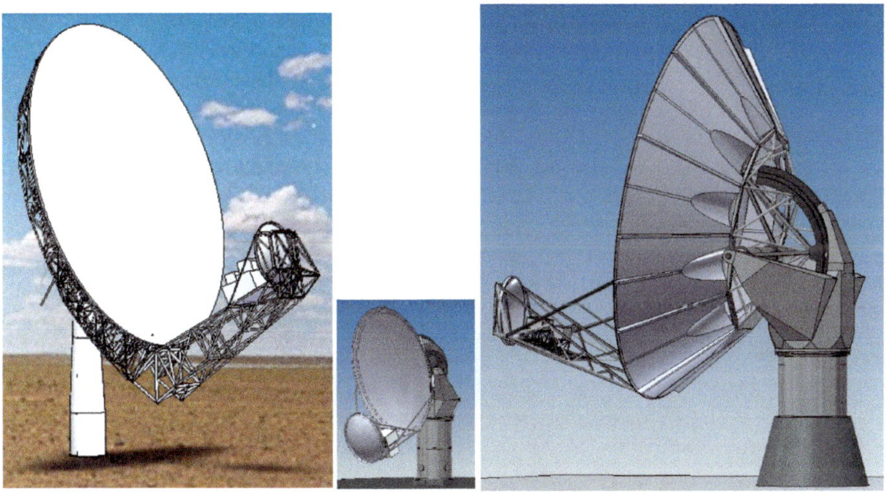

Figure 5. (*Left:*) ngVLA 18 m antenna reference design concept prepared by GDMS. Center: 6 m short spacing array antenna concept prepared by NRCC. (*Right:*) ngVLA 18 m antenna composite design concept prepared by NRCC.

adopted for the reference design. The design aims for Ruze performance to 116 GHz, with a surface accuracy of 160 μm rms ($\lambda/16$ @ 116 GHz) for the primary and subreflector combined under precision environmental conditions. The antenna optics are optimized for performance above 5 GHz with some degradation in performance accepted at the lowest frequencies due to diffraction, in exchange for more stiffness in the feed arm to improve pointing performance.

Since the ngVLA is envisioned as a general purpose, proposal-driven, pointed instrument (rather than a dedicated survey telescope), the optics will be shaped to optimize the illumination pattern of single pixel feeds, increasing antenna gain while minimizing spillover. High pointing accuracy will also be necessary to provide the required system imaging dynamic range. With an unblocked aperture, variations in the antenna gain pattern are expected to be dominated by pointing errors. Preliminary requirements are for absolute pointing accuracy of 18 arc-seconds rms, with referenced pointing of 3 arc-seconds rms, during the most favorable environmental conditions (Mason 2018b).

The mechanical and servo design is a typical altitude-azimuth design, Figure 5. Initial studies suggest pedestal designs are expected to have lower life cycle cost while meeting pointing specifications. The antenna mechanical and servo design will be optimized for rapid acceleration and a fast settling time, in order to manage the switching overhead associated with short slews.

The project is pursuing a reference design to specifications for the 18 m antenna with General Dynamics Mission Systems (GDMS). A parallel study into a composite design concept with the National Research Council of Canada (NRCC) is also underway, and NRCC are also preparing a reference design for the 6 m short baseline array antenna.

The 6 m aperture for the SBA antenna was chosen to provide overlapping (u, v)-plane coverage with the 18 m dishes when the later are used in both interferometric and total power mode. The optical design is inherently more offset than the 18m design in order to maintain a suitable minimum subreflector aperture (2.7 m) for 1.2 GHz operation, but shares the same feed illumination angle. Maintaining common interfaces ensures that the 6 m design can share the majority of the antenna electronics, including feeds and receivers.

The 6 m design employs a composite singe-piece reflector, and composite segmented backup structure on a steel pedestal mount. The mount includes space to house the digital electronics, power supplies and servo system, Figure 5.

7. Receiver Configuration

The ngVLA will provide continuous frequency coverage from 1.2 - 50.5 GHz and 70 - 116 GHz in multiple bands. Receivers will be cryogenically-cooled, with the receiver cryostats designed to integrate multiple receiver bands to the extent possible. Limiting the number of cryostats will reduce both maintenance and electrical power costs. The total number of bands required strongly depends on their fractional bandwidths: maximizing bandwidths will reduce the number of cryostats, with a possible penalty in sensitivity. Feeds for all receiver bands are cooled, and fully contained within the cryostat(s).

Band #	f_L (GHz)	f_M (GHz)	f_H (GHz)	BW (GHz)	Aptr. Eff., ηA @ f_L	@ f_M	@ f_H
1	1.2	2.0	3.5	2.3	0.80	0.79	0.74
2	3.5	6.6	12.3	8.8	0.80	0.78	0.76
3	12.3	15.9	20.5	8.2	0.84	0.87	0.86
4	20.5	26.4	34	13.5	0.83	0.86	0.83
5	30.5	39.2	50.5	20	0.81	0.82	0.78
6	70	90.1	116	46	0.68	0.61	0.48

Table 3. Band definitions and aperture efficiency of the baseline receiver concept.

The baseline ngVLA receiver configuration consists of the low-frequency receiver (1.2 - 3.5 GHz) in one cryostat, and five receivers spanning from 3.5 to 116 GHz in a second cryostat. Bands 1 and 2 employ wideband feed horns and LNAs, each covering L+S bands, and C+X bands. Quad-ridged feed horns (QRFHs) are used, having dual coaxial outputs. Due to improved optical performance (improving illumination efficiency and reducing T_{spill}), cooled feeds, and the simplified RF design sensing linear polarization, the T_{sys} is lower than current VLA L, S bands and comparable for C and X bands. Overall aperture efficiency and T_{sys} are slightly degraded from optimal due to the wider bandwidths spanned, but this permits a compact package that can be affordably constructed and operated.

The four high-frequency bands (12.3 - 116 GHz) employ waveguide-bandwidth (~1.67:1) feeds & LNAs, for optimum aperture efficiency and noise performance. Axially corrugated feed horns with circular waveguide output ensure uniform illumination over frequency, with minimum spillover and resistive loss.

Band #	T_{spill} (K) @f_L	T_{spill} (K) @f_M	T_{spill} (K) @f_H	T_{RX} (K) @f_L	T_{RX} (K) @f_M	T_{RX} (K) @f_H	T_{sys} (K) @f_L	T_{sys} (K) @f_M	T_{sys} (K) @f_H
1	12.8	10.1	4.0	9.9	10.3	13.8	27.1	24.9	22.4
2	12.8	7.0	3.9	13.4	15.4	14.4	30.8	27.1	23.6
3	4.1	4.1	4.1	13.9	16.9	18.6	23.3	27.3	36.3
4	4.1	4.1	4.1	15.4	16.2	18.6	33.1	32.4	36.0
5	4.1	4.1	4.1	19.1	20.4	26.5	34.0	41.0	101
6	4.1	4.1	4.1	50.6	49.0	72.6	123	68	189

Table 4. Noise performance of the baseline receiver concept. Assumes 1 mm PWV for band 6, and 6 mm PWV for others; 45° elev. on sky for all.

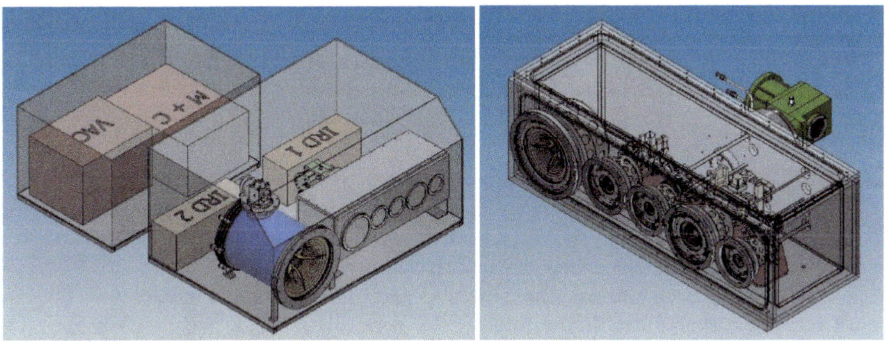

Figure 6. Front end component packaging at the secondary focus of the antenna. Band selection and focus are achieved with a dual-axis translation stage.The integrated receiver packages (labeled IRD 1 and IRD 2) are located in close proximity to the cryostats. Bands 2 – 6 are housed within in single cryostat.

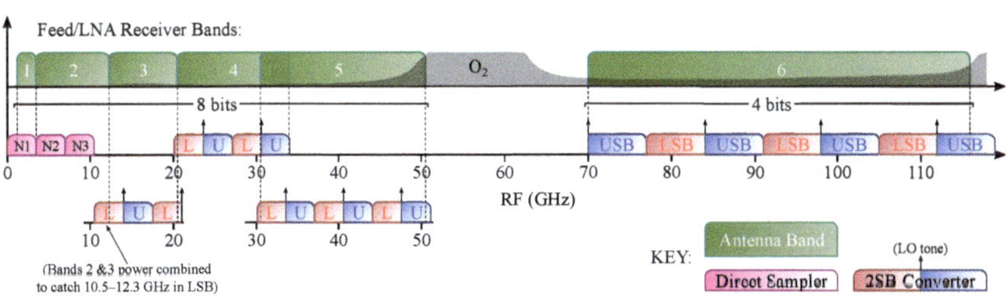

Figure 7. Sampling concept employing integrated receiver technology for both direct and dual sideband converter/samplers. Direct single side-band 8-bit sampling is used for the first three Nyquist zones. Dual sideband 8-bit samplers are then used up to 50 GHz. The 70-116GHz band is spanned by 4-bit samplers, due to the reduced risk of persistent RFI at these frequencies.

The electronics concept relies on integrated receiver packages (Morgan & Wunduke 2017) to further amplify the signals provided by the cryogenic stage, down convert them if necessary, digitize them, and deliver the resultant data streams by optical fiber to a moderately remote collection point (typically the antenna pedestal) where they can be launched onto a conventional network for transmission back to the array central processing facility. Interfaces are provided for synchronization of local oscillators (LO's) and sampler clocks, power leveling, command and control, health and performance monitoring, and diagnostics for troubleshooting in the event of component failure.

The integrated receiver concept is central to the antenna electronics concept for the ngVLA. Compact, fully-integrated, field-replaceable, warm electronic modules support single-stage, direct-to-baseband downconversion (when needed), followed by a very low-power, low-overhead digitization scheme and an industry-standard fiber optic interface carrying unformatted serial data. The frequency plan is shown in Figure 7.

8. Reference Distribution & Data Transmission

Given the large extent of the array, multiple time and frequency reference distribution concepts will be required to optimize for cost and performance. The array will be built as a combination of two different design methodologies.

A large number of antennas are located on the Plains of San Agustin, and each of these will be connected directly to the central processing facility by dedicated, buried fiber optics. Roughly 70% of the ngVLA antennas will be within this region. Clocks and local oscillator signals will be generated locally at the antenna and locked to a central reference with round trip phase correction.

The remainder of the antennas, the mid and long baseline antennas, will fall into a VLBI[1] station model with a number of local oscillator (LO) and data transmission stations located beyond the central core. These stations will be linked to the central timing system, correlator, and monitor and control system via long haul fiber optics. The most remote of these sites will have independent precision timing and frequency references, such as GPS-disciplined active hydrogen masers. Intermediate sites will used a mixed model dependent on the site logistics.

9. Central Signal Processor

The Central Signal Processor (CSP) ingests the voltage streams recorded and packetized by the antennas and transmitted via the data transmission system, and produces a number of low-level data products to be ingested by the archive. Among its many functionalities, the CSP is responsible for compensating for the large transmission delays from the remote stations, correcting the I/Q channel imbalance of the receiver and improve the separation of the upper and lower sidebands, tracking the delay and phase differences between antennas, flagging the spectral channels corrupted with RFI at a pre-correlation stage, selecting the spectral window of interest within the digitized bandwidth, offsetting the different frequency standards used by the remote stations, and achieving the desired spectral resolution.

In addition to synthesis imaging, the CSP will support other capabilities required of modern telescopes to enable VLBI and time-domain science. The functional capabilities of the CSP include full-polarization auto- and cross-correlation computation, as

[1]Very Long Baseline Interferometry.

well as beamforming capabilities for pulsar timing, pulsar/transient search, and VLBI recording. The CSP data products will vary by operation mode. The most common will be raw/uncalibrated visibilities, recorded in a common data model. The CSP will include all necessary "back end" infrastructure to average visibilities and package them for the archive, where they will be recorded to disk in a standard format. Calibration of these data products will be carried out through asynchronous data post-processing pipelines.

The CSP will support multiple sub-arrays operating simultaneously and fully independent from each other. Two key requirements for the system are the degree of commensality supported within a sub-array and the desired capabilities for sub-arrays operating simultaneously. At a minimum, the CSP will be able to compute auto- and cross-correlation products within a sub-array, as well as simultaneous cross-correlation and either pulsar timing, pulsar search or VLBI capabilities for different sub-arrays. Enabling correlation and beamforming products simultaneously within a sub-array is also under evaluation. Such a mode would reduce calibration overheads of the beamformer, and provide for localization/imaging concurrent with time-domain observations. The degree of commensality is expected to be a cost/complexity driver in the system and will be optimized on a best value basis.

The ngVLA correlator will employ an FX architecture, and will process an instantaneous bandwidth of up to 20 GHz per polarization. The correlator-beamformer Frequency Slice Architecture (Rupen 2017) developed by NRC Canada for the SKA Phase 1 mid-frequency telescope in South Africa (Dewdney et al. 2015) is well suited to ngVLA demands and is adopted for the reference design. This architecture will scale to the additional ngVLA apertures, bandwidth, and commensal mode requirements. Adopting this architecture could significantly reduce the non-recurring engineering costs during the design phase, while additional improvements in electrical efficiency can be expected from one additional FPGA manufacturing process improvement cycle due to ngVLA's later construction start date as compared to SKA Phase 1. Key performance requirements for the correlator are summarized in Table 5.

Requirement Description	**Specification**
Number of Connected Antennas	263 total
Maximum Baseline Length	10,000 km
Maximum Instantaneous Bandwidth	20 GHz per polarization
Maximum Number of Channels	$\geq 750,000$ channels
Highest Frequency Resolution	400 Hz, corresponding to 0.1 km/s resolution at 1.2 GHz.
Pulsar Search Beamforming	≥ 10 beams, $\geq 1''$ coverage, 60 km diameter sub-array
Pulsar Timing Beamforming	≥ 5 independent sub-arrays ≥ 1 beam per sub-array

Table 5. Central signal processor key specifications.

10. Post Processing System

The software architecture for ngVLA will leverage NRAO's existing algorithm development in reducing VLA and ALMA data and the CASA software infrastructure. The array will have a progressive series of data products suitable to different users groups. The data products may also change based on how well supported a mode is - common modes will have higher level data products that add value to the user, while clearly not all permutations can benefit from such a degree of automation. As with the VLA and ALMA, the fundamental data product that will be archived are uncalibrated visibilities, enabling future reprocessing. The online software system will also produce flags to be applied to the visibilities that would identify known system problems such as antennas being late on source, or the presence of RFI.

Automated post-processing pipelines will calibrate the raw data and create higher-level data products (typically image cubes) that will be delivered to users via the central archive. Calibration tables that compensate for large-scale instrumental and atmospheric effects in phase, gain, bandpass shapes, polarization and flux scale will be provided. Data analysis tools will allow users to analyze the data directly from the archive, reducing the need for data transmission and reprocessing at the user's institution.

The VLA and ALMA "Science Ready Data Products" project will be an ngVLA pathfinder to identify common high-level data products that will be delivered to the Principal Investigator and to the data archive to facilitate data reuse. This model will also enable the facility to support a broader user base, possibly catering to astronomers who are not intimately aware of the nuances of radio interferometry, thereby facilitating multi-wavelength science.

11. Overall System Performance

The predicted performance of the array is summarized in Table 6. This is an update to the performance estimates originally documented in Selina & Murphy (2017)[2].

The continuum and line rms values in Table 6 are for point source sensitivity with a naturally weighted beam. Imaging sensitivity is estimated based on a similar procedure as shown in Carilli (2017) and provided as a function of angular resolution in Table 7. The table is by necessity a simplification and the imaging sensitivity will vary from these reported values depending on the quality of the (sculpted) synthesized beam required to support the science use case. Herein, quality is defined as the ratio of the power in the main lobe of the sculpted beam attenuation pattern to the power in the entire beam attenuation pattern as a function of the FWHM of the synthesized beam (Murphy 2017).

The brightness sensitivity of an array is critically dependent on the array configuration. The ngVLA has the competing aims of both good point source sensitivity at full resolution and good surface brightness sensitivity on a range of larger scales. Different array configurations that might provide a reasonable compromise through judicious weighting of the visibilities for a given application have been explored (Clark & Brisken 2015) – see Lal (2011) for similar studies for the SKA. It is important to recognize that for any given observation, from full resolution imaging of small fields,

[2]`http://ngvla.nrao.edu/page/refdesign`

to imaging structure on scales approaching that of the primary beam, some compromise will have to be accepted to enable a practical and flexible general purpose facility.

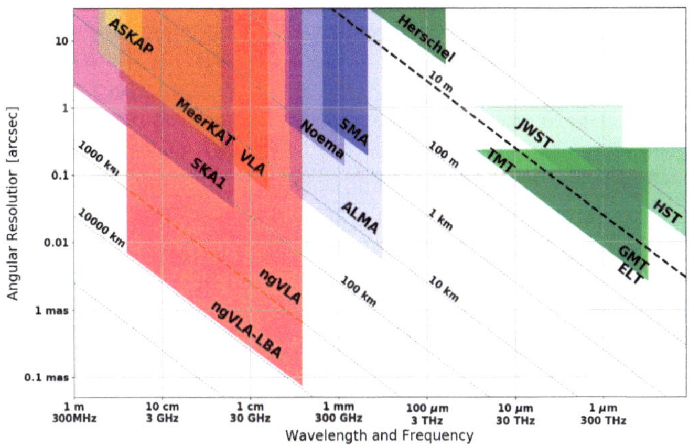

Figure 8. Spatial resolution versus frequency set by the maximum baselines of the ngVLA as compared to that of other existing and planned facilities.

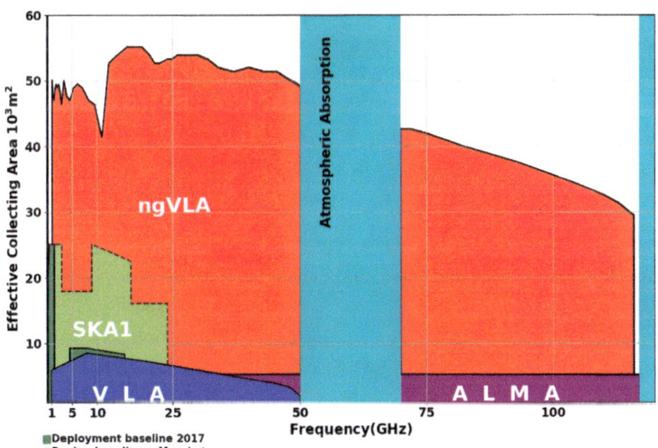

Figure 9. Effective collecting area versus frequency for the ngVLA as compared to that for other existing or planned facilities. Both the SKA1 'deployment baseline' (dark green) and 'design baseline' (light green) are shown, inclusive of the MeerKAT array (Dewdney 2013).

Figure 8 shows a slice through the parameter space, resolution versus frequency, covered by the ngVLA along with other existing and planned facilities that are expected in the 2030s at radio to optical wavelengths. The maximum baselines of the ngVLA support a resolution of better than 0.5 mas at 1 cm. Coupled with the high sensitivity of the array, this resolution provides a unique window into the formation of planets in disks on scales of our own Solar system at the distance of the nearest active star forming regions.

Figure 9 shows a second slice through parameter space: effective collecting area versus frequency. A linear-linear plot highlights the parameter space opened by the ngVLA. Note that the SKA-1 will extend to below 100 MHz while ALMA extends up to almost a THz. We note that there are other aspects of telescope phase space that are relevant, including field of view, mapping speed, surface brightness sensitivity, bandwidth, system temperature, dynamic range, etc. We have presented the two principle and simplest design goals, namely, maximum spatial resolution and total effective collecting area (as a gross measure of system sensitivity).

Imaging sensitivity will be dependent on the required resolution and imaging fidelity. Figures 10 and 11 show the effects of adjusting imaging weights to vary the resolution and PSF quality. These figures are based on a 4 hour simulation at 30 GHz using the 244 antenna array configuration, for a source at +24° Declination observed during transit. The reported beam size is the geometric mean of the major and minor axes full width at half maximum (FWHM) of the synthesized beam as parameterized by Gaussian fitting in the CASA `tclean` task. The highly centrally condensed antenna distribution leads to a naturally weighted beam that is not well characterized by a Gaussian function. Specific science applications may need to adjust the (u, v)-weighting and image parameters to 'sculpt' a synthesized beam that is adequate for the particular science goal being considered (Carilli 2017). The results in Figures 10 and 11 should be considered representative of the possibilities, and optimizing sensitivity vs. resolution will be a major area of investigation during telescope development.

In order to account for the change in sensitivity due to use of imaging weights (relative to the naturally weighted rms σ_{NA}), we have adopted an efficiency factor η_{weight} such that the expected image rms after weighting is $\eta_{weight}\,\sigma_{NA}$. The sensitivity calculations in Table 7 include η_{weight}, estimated using the blue and red data series in Figure 10 and by scaling $\theta_{1/2}$ with frequency ($\theta_{1/2} \times \nu/30\,\mathrm{GHz}$).

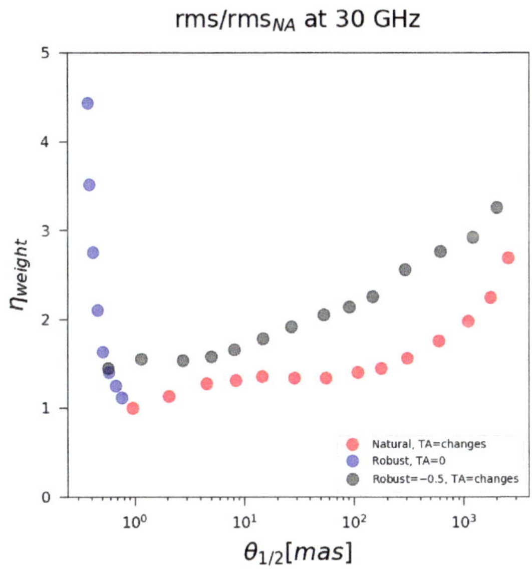

Figure 10. Image noise (rms) at different angular resolutions (FWHM) achieved by varying the imaging weights, simulated at 30 GHz. The noise has been scaled relative to that of the naturally weighted image (rms$_{NA}$). The red symbols correspond to use of a (u, v)-taper and natural weights, and the blue symbols to Briggs robust weighting without a taper. The gray symbols are for Briggs robust = -0.5 and a varying (u, v)-taper, which has a large effect on beam quality (see Figure 11).

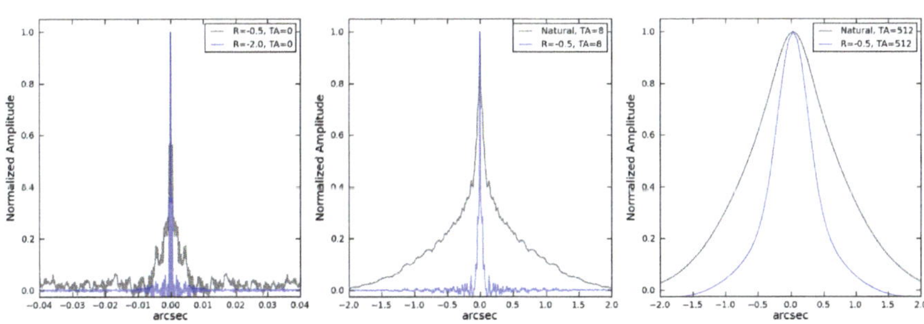

Figure 11. Simulated 30 GHz PSFs over a range of resolutions, showing the effect of different imaging weights (TA: (u, v)-taper in mas, R: Briggs robust parameter). The PSFs are a selection of the data presented in Figure 10: left panel (blue circles), central and right panels (gray and red circles). These examples illustrate how combinations of robustness and tapering allow for a beam of much higher quality at the expense of sensitivity.

Center Frequency [GHz]	2.4	8	16	27	41	93	Notes
Band Lower Frequency [GHz]	1.2	3.5	12.3	20.5	30.5	70.0	a
Band Upper Frequency [GHz]	3.5	12.3	20.5	34.0	50.5	116.0	a
Field of View FWHM [arcmin]	24.3	7.3	3.6	2.2	1.4	0.6	b
Aperture Efficiency	0.77	0.76	0.87	0.85	0.81	0.58	b
Effective Area, A_{eff}, x 10^3 [m^2]	47.8	47.1	53.8	52.6	50.4	36.0	b
System Temp, T_{sys} [K]	25	27	28	35	56	103	a, e
Max Inst. Bandwidth [GHz]	2.3	8.8	8.2	13.5	20.0	20.0	a
Sampler Resolution [Bits]	8	8	8	8	8	4	
Antenna SEFD [Jy]	372.3	419.1	372.1	485.1	809.0	2080.5	a, b
Resolution of Max. Baseline [mas]	2.91	0.87	0.44	0.26	0.17	0.07	c
Continuum rms, 1 hr [μJy/beam]	0.38	0.22	0.20	0.21	0.28	0.73	d
Line Width, 10 km/s [kHz]	80.1	266.9	533.7	900.6	1367.6	3102.1	
Line rms, 1 hr, 10 km/s [μJy/beam]	65.0	40.1	25.2	25.2	34.2	58.3	d

Table 6. ngVLA Key Performance Metrics. Notes: (a) 6-band 'baseline' receiver configuration. (b) Reference design concept of 244 18 m aperture antennas. Unblocked aperture with 160 μm surface. (c) Current reference design configuration, including LBA. Resolution in E-W axis. (d) Point source sensitivity using natural imaging weights, dual polarization and all baselines (main array + LBA). (e) Averaged over the band. Assumes 1 mm PWV at 93 GHz, 6 mm PWV for other bands, 45° elevation on sky.

Center Frequency [GHz]	2.4	8	16	27	41	93
Resolution [mas] : 1000						
Continuum rms, 1 hr, Robust [μJy/beam]	0.52	0.34	0.35	0.39	0.59	2.24
Line rms 1 hr, 10 km/s Robust [μJy/beam]	88.9	61.1	43.3	47.9	70.9	179.6
Brightness Temp. (T_B) rms continuum, 1 hr, Robust [K]	0.110	6.4E-3	1.7E-3	0.7E-3	0.4E-3	0.3E-3
T_B rms line, 1 hr, 10 km/s, Robust [K]	18.76	1.16	0.21	0.08	0.05	0.03
Resolution [mas] : 100						
Continuum rms, 1 hr, Robust [μJy/beam]	0.50	0.30	0.27	0.28	0.40	1.14
Line rms 1 hr, 10 km/s Robust [μJy/beam]	85.0	53.6	33.6	34.8	48.4	91.3
Brightness Temp. (T_B) rms continuum, 1 hr, Robust [K]	10.58	0.56	0.13	0.05	0.03	0.02
T_B rms line, 1 hr, 10 km/s, Robust [K]	1794.1	101.9	15.9	5.8	3.5	1.3
Resolution [mas] : 10						
Continuum rms, 1 hr, Robust [μJy/beam]	0.41	0.27	0.26	0.27	0.38	0.97
Line rms 1 hr, 10 km/s Robust [μJy/beam]	69.9	48.3	32.4	33.2	46.3	77.7
Brightness Temp. (T_B) rms continuum, 1 hr, Robust [K]	870.58	50.51	12.42	4.53	2.77	1.36
T_B rms line, 1 hr, 10 km/s, Robust [K]	1.5E5	9173	1540	555	335	109
Resolution [mas] : 1						
Continuum rms, 1 hr, Robust [μJy/beam]	-	20.87	0.31	0.21	0.29	0.90
Line rms 1 hr, 10 km/s Robust [μJy/beam]	-	3789.8	38.2	25.7	34.7	72.0
Brightness Temp. (T_B) rms continuum, 1 hr, Robust [K]	-	4.0E5	1466	350	207	126
T_B rms line, 1 hr, 10 km/s, Robust [K]	-	7.2E7	1.8E5	4.3E4	2.5E4	1.0E4
Resolution [mas] : 0.1						
Continuum rms, 1 hr, Robust [μJy/beam]	-	-	-	-	-	20.96
Line rms 1 hr, 10 km/s Robust [μJy/beam]	-	-	-	-	-	1683.2
Brightness Temp. (T_B) rms continuum, 1 hr, Robust [K]	-	-	-	-	-	2.9E5
T_B rms line, 1 hr, 10 km/s, Robust [K]	-	-	-	-	-	2E7

Table 7. Projected image sensitivity as a function of angular resolution. These calculations include η_{weight} and are scaled by frequency, as described in the text.

Acknowledgments. The ngVLA is a project of the National Science Foundation's National Radio Astronomy Observatory which is operated under cooperative agreement by Associated Universities, Inc.

References

Akiyama, K. et al. 2015, "230 GHz VLBI Observations of M87: Event-Horizon-Scale Structure During an Enhanced Very-High-Energy-Ray State in 2012" ApJ, 807, 150

Bolatto, et. al. 2017, "Key Science Goals for the Next Generation Very Large Array (ngVLA): Report from the ngVLA Science Advisory Council", Next Generation VLA Memo 19

Brisken, W. 2017, "Major Option: Continent-scale baselines" U.S. RMS Futures III Conference

Burnett, M.C. 2018, "Subarray Processing for Projection-based RFI Mitigation in Radio Astronomical Interferometers", AJ, 155, 146

Butler, B. 1998, "Precipitable Water at the VLA – 1990-1998," VLA Scientific Memo 176

Butler, B. & Desai, K. 1999, "Phase Fluctuations at the VLA Derived from One Year of Site Testing," VLA Test Memo 222

Carilli, C. 2015, "Fast Switching Phase Calibration at 3 mm at the VLA site," ngVLA Memo Series 1

Carilli, C. 2017, "More on Synthesized Beams and Sensitivity" ngVLA Memo 16

Clark, B. & Brisken, W. 2015, "Possible Configurations for the ngVLA," Next Generation VLA Memo 3

Clark, B. 2015, "Calibration Strategies for the Next Generation VLA," ngVLA Memo Series 2

Dewdney, P. 2013, "SKA1 System Baseline Design," SKA Doc No. SKA-TEL-SKO-DD-001

Dewdney P. et al. 2015, "SKA1 System Baseline V2 Description," Tech. Rep. SKA-TEL-SKO-0000308

Ford, E. et al. 2018, "ngVLA Operations Concept," Tech. Rep. 020.10.05.00.00-0001-PLA

Lal, D., Lobanov, A., Jimenez-Monferrer, S. 2011, SKA Design Studies Technical Memo 107

Li, L. et al. 2004, "A Preliminary Survey of Radio-Frequency Interference Over the U.S. in Aqua AMSR-E Data," IEEE TGRS, 42, 380

Mason, B. S., Cotton, W., Condon, J., et al. 2018a, BAAS 231, #342.11

Mason, B. et al. 2018b, "The ngVLA Short Baseline Array," ngVLA Memo 43

McKinnon, M. 2016, "The Next Generation Very Large Array," Proc. SPIE 9906, 10.1117/12.2234941

McLaughlin, M. A. 2013, "The North American Nanohertz Observatory for Gravitational Waves," Classical and Quantum Gravity, 30(22):224008

Morgan, M. & Wunduke, S. 2017, "An Integrated Receiver Concept for the ngVLA," Next Generation VLA Memo 29

Murphy, E. et. al. 2017, "ngVLA Science Requirements," Next Generation VLA Document 020.10.15.00-0001-REQ

Rupen, M. 2017, "TMs View of the Mid.CBF Frequency Slice Approach"

Selina, R. 2017, "ngVLA Antenna: Preliminary Technical Specifications," Next Generation VLA Document 020.25.00.00.00-0001-SPE, Version B

Selina, R. and Demorest P. 2018, "A Dedicated Pulsar Timing Array Telescope," ngVLA Memo Series (in prep.)

Selina, R. & Murphy E. 2017, "ngVLA Reference Design Development & Performance Estimates," Next Generation VLA Memo 17

Selina, R., Murphy, E., Erickson, A. 2017, "Summary of the Science Use Case Analysis," Next Generation VLA Memo 18

Stewart, K. P. et al. 2005, "An RFI Survey at the Site of the Long Wavelength Demonstration Array (LWDA)," BAAS 37, 1389

Taylor, G. et. al. 2017, "A Next Generation Low Band Observatory: A Community Study Exploring Low Frequency Options for ngVLA," arXiv:1708.00090

Thompson, A. R. et al. 1980, "The Very Large Array," ApJSS, 44, 151

Thompson A., Moran, J. & Swenson, G. 2004, "Interferometry and Synthesis in Radio Astronomy," Wiley-VCH

Science with a Next-Generation Very Large Array
ASP Conference Series, Monograph 7
Eric J. Murphy, ed.

Reaching Communities and Creating New Opportunities with the ngVLA

Lyndele von Schill and Suzanne Gurton

National Radio Astronomy Observatory, Charlottesville, VA USA;
lvonschi@nrao.edu , sgurton@nrao.edu

Abstract. The Office of Diversity and Inclusion (ODI) and the Education and Public Outreach (EPO) Department serve the strategic goal of the National Radio Astronomy Observatory (NRAO) to broaden public awareness of, support for, and participation in Science, Technology, Engineering, and Mathematics (STEM). ODI operates a suite of programs designed to support underrepresented minority undergraduate, graduate students in pursuit of careers in STEM. EPO highlights the discoveries, technologies, and careers pioneered and exemplified by the NRAO via multipurpose engagement strategies that include face-to-face and standalone learning programs, products, and public services for the general public and K-12 students, with attention to reaching diverse audiences. These established and diverse programs are described, along with proposals for new, unique opportunities enabled by the development and realization of a next-generation Very Large Array (ngVLA).

1. Broadening Participation

As a facility under the management of the National Radio Astronomy Observatory (NRAO), the next-generation Very Large Array (ngVLA) has full access to NRAO's suite of widely well-regarded Broadening Participation programs. NRAO's programs address the key priorities identified by the National Science Foundation (NSF) that include:

- Preparing a diverse, globally engaged science, technology, engineering, and mathematics (STEM) workforce;
- Integrating research with education, and building capacity;
- Expanding efforts to broaden participation from underrepresented groups and diverse institutions across all geographical regions in all NSF activities; and
- Improving processes to recruit and select highly qualified reviewers and panelists.

The NRAO has adopted a comprehensive, observatory-wide, "pipeline" approach to the development of STEM capacity, with particular emphasis on broadening participation of underrepresented groups and diverse institutions. This approach begins with engaging K-12 students in a seamless flow of STEM activities that relate to the full spectrum of fields associated with radio astronomy, and continues through providing undergraduate, graduate, and post-doctoral research, training, education, and mentoring programs (described below). The ngVLA will vastly increase the opportunities

Figure 1. The NRAO's Broadening Participation "Pipeline".

that the NRAO can offer to the U.S. and international public, student, and scientific communities.

In addition to its "outward-facing" broadening participation efforts, the NRAO has developed a set of policies and practices that are designed to recruit and select highly qualified, and diverse, reviewers, panelists, and search committees, with the goal of ensuring that selections are fair and equitable across all populations.

Examples of the NRAO's existing, significant investment in broadening participation efforts appear in this section (see Figure 1). Discussions about new opportunities generated by the ngVLA appear in Section 4 (Emerging Opportunities for Broader Impact and Broader Participation through the VLA).

K-12 Education Programs: The NRAO's education programs are designed to introduce the hidden universe revealed by radio astronomy by engaging program participants in age appropriate inquiry based activities and research. Examples of K-12 education programs aimed at broadening participation include:

- **Radio Astronomy and Physics in New Mexico (RAP-NM):** A one-week residential summer camp experience on the NM Tech campus for rising 9th graders from around the state of New Mexico. Student mentors are first recruited from the National Astronomy Consortium (NAC) program (see below).

- **Sister Cities and Observatories:** A 10-day international exchange recruiting from high school youth in areas near the Atacama Large Millimeter/submillimeter Array (ALMA) and VLA introducing them to the world class observatories and the diverse range of careers possible at each. This program will continue under ngVLA.

National Astronomy Consortium (NAC) Program: Opportunities for undergraduate research will be coordinated through the NRAO's NAC program, which is designed to provide research opportunities to underrepresented minority (URM) students. Students are recruited for participation in this program from an established network of Historically Black Colleges and Universities (HBCUs) and Hispanic-serving Institutions (HSIs). The ngVLA will open more opportunities for traditionally underrepresented students to participate in cutting-edge research which, in turn, increases opportunities for the students to attend graduate school and/or begin careers in radio astronomy.

National and International Non-Traditional Exchange (NINE) Program: Collaborative opportunities also exist with the NRAO NINE, which trains the next generation of scientists and engineers from countries in which radio astronomy expertise is limited, but needed. The NINE program focuses on training traditionally underrepresented populations in skills that result in meaningful contributions to the astronomical science body of knowledge. The NINE program includes a network of ever-growing national and international "Hubs" where regional populations learn state-of-the art science and technology relevant to astronomy. The ngVLA will provide cutting edge research opportunities to the NINE program, while benefiting from access to skilled scientists and technicians from the broad network of NINE Hub communities.

2. Public Engagement

Public Website: In addition to programs designed with face-to-face interactions, the ngVLA will have a strong presence on the NRAO website that will serve as a portal for the public to explore the engineering advances and astronomical discoveries made possible by ngVLA. Clear information for visitors wishing to visit the site will be on the website, as well as opportunities to explore the site virtually. An ngVLA Explorer[1], combining video and augmented reality, will be created to give those who cannot visit the facility a virtual tour. The Role Model[2] series will be expanded to include ngVLA staff. It currently reflects the diversity of jobs and experiences that are needed to run a national observatory.

Visitor Center: The planned upgrade of the VLA Visitor Center (VC) reflects modern interpretive methods to explore the intersection of three realms: The stories arising from the resource, visitors' intrinsic interests, and the mission and goals of NRAO. The goals for the ngVLA VC will echo those for the VLA VC, with a substantial update to be inclusive of the international collaborations that will help make the ngVLA possible. These interpretive goals will be expressed through an overarching theme that the ngVLA serves humanity's deep curiosity and drive to explore the universe and our relationship to it:

- Interpretation will make information understandable and relevant for scientists and non-scientists alike.

[1]e.g., `https://public.nrao.edu/special-features/vla-explorer/`

[2]`https://public.nrao.edu/special-features/role-models/`

- Non-scientist visitors will understand in general terms the basics of radio astronomy and the ngVLA - and feel happily surprised that they can.
- Scientists will appreciate the message clarity, without over-simplification to the point of introducing inaccuracies.

- People of diverse backgrounds will feel welcome and actively included as the VC will actively:
  - Provide a venue for diverse cultural voices.
  - Include women, persons of color.
  - Encourage Navajo and Puebloans to share their stories.
- Educators will value and use the ngVLA as a resource.
- Visitors and residents will understand how natural conditions here support the VLA, and will feel inspired to help maintain those conditions.
- Visitors will understand why the ngVLA exists, and feel the program is useful, relevant and worth supporting.
  - People will understand how the ngVLA fits into the bigger picture of astronomy and human discovery.
  - Users will feel they are an integral part of the universe, not separate from it.
  - Local residents and visitors will feel pride and ownership in the ngVLA's role in unlocking the secretes of the universe.
- Visitors will share their positive experiences with others back home.

Media Relations NRAO's EPO department includes a full media relations team with public information officers, artists and graphic designers to best represent the discoveries made possible by ngVLA observations in the popular science media. News and information about ngVLA will be featured in such national outreach venues as the U.S.A. Science and Engineering Festival and National Astronomy Night on the Mall.

3. Broader Impacts: Training the Next Generation

In addition to the NAC and NINE programs that focus on providing research, engineering, and other "full spectrum" astronomy fields, to underrepresented students, the NRAO has provided substantial internship opportunities for undergraduate and graduate students through its REU, graduate research, and NM Ops undergraduate internship programs. The NM Ops internship program, for example, has provided engineering students experience by working in the Observatory's electronics lab. The ngVLA will provide an opportunity to significantly increase the number of students with access to state-of-the-art research and engineering projects.

For the last 2 years EPO has funded a media intern at NRAO headquarters each summer, with the inauguration of ngVLA a media intern will also be stationed at the ngVLA to supplement the efforts of the Public Information Officer and assist with major media efforts.

4. Emerging Opportunities for Broader Impact and Broader Participation through the ngVLA

The NRAO recognizes that the ngVLA project will span more than 10 years; this longevity offers a unique opportunity for the Observatory to build a pipeline for future employees from the regions in which the ngVLA will be present. Our BI and BP plans, then, include a proactive, intentional plan to include workforce development as an integral part of the entire ngVLA project, from beginning construction through and beyond first science. We expect to continue to ask the community for input into this important component of the ngVLA project; we have, however, identified the following opportunities for new, substantial, outcome-oriented Broader Impact (BI) and Broadening Participation (BP) activities (see Table 1).

These ideas represent our commitment to building an ngVLA that is responsible to all members of the scientific and broader communities. We recognize and acknowledge that the ngVLA is indebted to and responsible to taxpayers, local and regional communities, members of underrepresented communities in STEM, K-12 and high education students, postdocs, faculty, scientists, engineers, technical experts, Education and Public Outreach professionals, and all other professions that support the ability to deliver cutting-edge science and technology to the scientific and broader communities.

5. Summary

The ngVLA will build on NRAO's existing, well-designed, and implemented, suite of EPO programs, including the creation of new opportunities to educate the general public about the exciting science resulting from ngVLA observations. The ngVLA will make use of, and significantly enhance, the Observatory's flagship National Astronomy Consortium (NAC) and National and International Non-traditional Exchange (NINE) programs, by providing cutting-edge research and engineering opportunities for a new generation of astronomers, engineers, and technicians. In addition to these existing programs, ngVLA will create innovative new BI and BP opportunities that enhance the value of the ngVLA well beyond the scientific community. Importantly, the ngVLA will embrace the NRAO's commitment to providing these opportunities to populations that are unrepresented in STEM fields, and in the field of astronomy in particular.

Table 1. Potential Broader Impact/Participation Opportunities Enabled by the ngVLA

Emerging Opportunities through ngVLA	**BI/BP**	**Communities Impacted**
Enhancement of broadband Internet access infrastructure, including fiber optics, as an intentional by-product of ngVLA internet needs.	BI	small, rural communities along antenna paths
Enhancement of emergency services (e.g., fire response units) through collaboration with local communities.	BI	small, rural communities along antenna paths
Educational opportunities to engage and inform local stakeholders about radio astronomy and the importance of 'their backyard' to important national/international science. • Educational sessions will be presented by a diverse group of students and post-doc to raise visibility of both radio astronomy and diversity in STEM. • Role model videos for grade school girls and URM students in all fields that support an Observatory.	BI & BP	small, rural communities along antenna paths
Opportunity for partnership with URM college or university to develop educational and historical films: • Filming of the ngVLA build for EPO and historical reference. • Filming at each site to document community stories and engagement with the ngVLA. • K-12 lesson plans developed to accompany films and film segments.	BI & BP	K-12, Undergraduate and graduate students; faculty; and members of underrepresented STEM groups (e.g., African-American, Hispanic, and Native American students and faculty); the general public

Table 1. Continued

Opportunities for collaborative, cross-discipline research during archaeological site excavations during the pre-construction period: • Research opportunities with multiple departments (e.g., archeology/environmental sciences, water quality) at community colleges and universities (e.g., Dry Lake Bed). • Efforts will be made to identify URM students for participation in the research opportunities. • Development of lesson plans centered around the archeology and environment of ngVLA sites. • Educational material produced for local communities and the general public, with a focus on connection to the land on which ngVLA antennas are located.	BI & BP	K-12, Undergraduate and graduate students; faculty; and members of underrepresented STEM groups (e.g., African-American, Hispanic, and Native American students and faculty); the general public; the scientific community
Opportunities to incorporate non-Western cosmologies into the education about the ngVLA: • Invite indigenous community members (expert and local) to organized talks about indigenous cosmology. • STEAM projects (e.g., murals, stories, videos) to capture non-Western cosmologies in local schools and community centers. • Collaborate with the Society for Advancement of Chicanos/Hispanics and Native Americans in Science (SACNAS) to engage experts and students in these projects, with a plan to publish comparisons of western and indigenous cosmologies in the context of the ngVLA locations. • Development and delivery of Star Parties that incorporate different cosmologies.	BI & BP	Local community members; the general public; K-12, undergrad and graduate students; post-docs and faculty; the scientific community
Investigate additional benefits/applications of radio frequency interference (RFI) mitigation: • Providing research opportunities to URM students. • Publications. • Inventions.	BI & BP	General public; industry; local communities; K-12, undergrad and grad students; faculty; scientific community

Table 1. Continued

Develop "road shows" to provide information – along the project timeline – about the ngVLA, its progress, scientific outcomes, careers, and its incorporation into the landscape: • Use of URM students as guides and speakers to provide opportunities for the students to gain experience speaking about their work, raise their visibility, and allow them to serve as role models for members of the audiences. • Material developed for the road show can also be distributed more broadly through social media, and traditional EPO routes. • Lesson plans for delivery to local/state schools and after school groups that incorporate radio astronomy and the ngVLA into state standards.	BI & BP	General public; local communities, major towns along the array; K-12, undergrad and grad students; faculty
Opportunities for connections to international colleges and universities to offer post-doctoral exchanges.	BI & BP	Graduate students
Opportunities for developing materials and resources that are tri-lingual (partners: Mexico and Canada): • Engage URM students and faculty in the creation of materials.	BI & BP	General public; local communities, K-12, undergrad and grad students; faculty; scientific community
Exploration of opportunities to explore uses and sharable benefits of renewable energy (e.g., solar): • Partnerships with energy collection and storage industries that may include co-op experiences for URM students and community members.	BI & BP	General public; local communities, K-12, undergrad and grad students; faculty; scientific community
Infrastructure improvement in Magdalena, NM: • Opportunities for employment and training of local community members.	BI & BP	Local community; workforce
Vocational training for regional students and community members: • Beta team to conduct remote diagnostics (quick telescope fixes and data back to A-team at HQ). • Partnerships with regional community colleges.	BI & BP	Undergrad and graduate students at Minority Serving Institutions (MSIs) and other URM-serving institutions

Table 1. Continued

Expanded National Astronomy Consortium (NAC) program: • Opportunity to develop new cohorts of colleges and universities in ngVLA regions.	BP	Undergrad and graduate students at MSIs and other URM-serving institutions
Identification of new outlets for dissemination of technical, engineering, and scientific information: • National Astronomy Day. • Engineers Week. • DragonCon and other similar conferences.	BI	General public, scientific community
Greater presence at SACNAS meetings: • Student and professional recruitment. • Opportunity to share information about radio astronomy and the ngVLA to a large population of URM students and professionals.	BP	undergrad and grad students; faculty; scientific community
Co-op opportunities for students from URM schools to work with the world-class NRAO Central Development and Electronics labs.	BP	undergrad and grad students; faculty; scientific community

Part II

The Solar System

Science with a Next-Generation Very Large Array
ASP Conference Series, Monograph 7
Eric J. Murphy, ed.

Potential for Solar System Science with the ngVLA

Imke de Pater,[1] Bryan Butler,[2] R. J. Sault,[3] Arielle Moullet,[4] Chris Moeckel,[5] Joshua Tollefson,[6] Katherine de Kleer,[7] Mark Gurwell,[8] and Stefanie Milam[9]

[1]*Departments of Astronomy and of Earth and Planetary Science, University of California, Berkeley, CA 94720, USA;* `imke@berkeley.edu`

[2]*National Radio Astronomy Observatory, Socorro, NM 87801, USA;* `bbutler@nrao.edu`

[3]*School of Physics, University of Melbourne, Victoria, 3010, Australia;* `rsault@nrao.edu`

[4]*SOFIA / USRA, NASA Ames Building N232, Moffett Field, CA 94035, USA;* `amoullet@usra.edu`

[5]*Department of Earth and Planetary Science, University of California, Berkeley, CA 94720, USA;* `chris.moeckel@berkeley.edu`

[6]*Department of Earth and Planetary Science, University of California, Berkeley, CA 94720, USA;* `jtollefs@berkeley.edu`

[7]*Division of Geological and Planetary Sciences, California Institute of Technology, Pasadena, CA 91125, USA;* `dekleer@caltech.edu`

[8]*Harvard-Smithsonian Center for Astrophysics, Cambridge, MA 02478 USA;* `mgurwell@cfa.harvard.edu`

[9]*NASA Goddard Space Flight Center, Astrochemistry Laboratory, Greenbelt, MD 20771 USA;* `stefanie.n.milam@nasa.gov`

Abstract. Radio wavelength observations of solar system bodies are a powerful method of probing many characteristics of those bodies. From surface and subsurface, to atmospheres (including deep atmospheres of the giant planets), to rings, to the magnetosphere of Jupiter, these observations provide unique information on the current state, and sometimes history, of the bodies. The ngVLA will enable the highest sensitivity and resolution observations of this kind, with the potential to revolutionize our understanding of some of these bodies. In this article, we present a review of state-of-the-art radio wavelength observations of a variety of bodies in our solar system, varying in size from ring particles and small near-Earth asteroids to the giant planets. Throughout the review we mention improvements for each body (or class of bodies) to be expected with the ngVLA. A simulation of a Neptune-sized object is presented in Section 6. Section 7 provides a brief summary for each type of object, together with the type of measurements needed for all objects throughout the Solar System.

1. The Solar System as a Link between Protoplanetary Disks and Exoplanets

It is widely accepted that planets form in protoplanetary disks surrounding stars. However, the details of such processes remain vague. Several thousand exoplanets have been detected, varying in scale from larger than Jupiter-sized down to almost moon-sized, and in stellar distance from less than 0.01 AU to many AU's from the star. It is not clear, though, where these planets formed, since they may well have formed at a very different location from that where they have been detected today. In particular, it has been difficult to explain the presence of giant planets at stellar distance < 1 AU, which triggered development of theories about planetary migration. Large planets may have formed further out, and subsequently migrated inwards as a consequence of angular momentum exchange with the disk. Planets could also migrate at a later stage if/when clearing the area near them of planetesimals. Planet migration is a difficult topic in planet formation theories, and hence no consensus has yet been reached on the effect of such migrations, not even for our own Solar System. Formation models for the latter may be hardest, since such models have to explain not only the present orbits of both large and small bodies, but also their composition and possible presence of satellites and rings. Composition of bodies in our Solar System provides one key element in formation models that would link protoplanetary disks to exoplanets. For example, the composition of comets and bodies in the outer Solar System is dominated by water ice, while bodies closer to the Sun have larger fractions of rock and metal. However, such bodies are not devoid of water, so how did the water get there; i.e., where did the planets and the planetesimals that make up the planets form originally?

As mentioned above, composition is a key parameter for planetary formation models, and forms a common thread throughout this chapter. In order to determine bulk composition of a body and/or its atmosphere, one needs to understand the entire body, i.e., the physics, dynamics and chemistry in a body's interior and atmosphere. Observations at radio wavelengths provide unique information as they probe regions inaccessible by nearly all other remote sensing techniques and wavelengths. For example, at radio wavelengths one can probe depths up to many tens of bars in atmospheres of the giant planets, and centimeters to meters below the surface of a solid body like a terrestrial planet, a tiny asteroid, or a planetary ring. Magnetic fields can be probed through emissions by electrons in such fields. In order to extract the most information, high spatial and high spectral resolution is essential, as well as sensitivity to large scale structure. Since objects in our Solar System are highly time variable, due e.g., to plume activity (comets, satellites), volcanic eruptions, atmospheric winds, and simply the rotation of a body, excellent sensitivity and imaging characteristics on short timescales (minutes or less) is highly desirable. These are attributes being planned for the ngVLA. To strengthen the case to build such an array in the near-future, in this chapter we highlight several cases where the ngVLA can make a difference.

Good prior reviews on radio observations of our Solar System are provided by Butler et al. (2004) and de Pater & Kurth (2014). We will build upon these reviews. In this chapter we will omit comets, as the ngVLA case for comets is provided in a separate chapter in this volume (Cordiner & Qi 2018, p. 73). Radar measurements are discussed in this volume by Lazio et al. (2018, p. 81) and Brozović et al. (2018, p. 113). We will also not consider observations of the Sun, as they are different enough to deserve their own treatment. Lastly, we will not investigate the newly-approved long baselines ($\gtrsim 1000$ km) of ngVLA, as while they have some uses for solar system investigations (the aforementioned radar, and spacecraft tracking, for instance), in general there is not enough brightness temperature sensitivity to be useful for observations of solar system bodies.

2. Giant Planets

Observations of the giant planets in the frequency range of the ngVLA (1.2–116 GHz) are sensitive to both thermal and non-thermal emissions. These emissions are received simultaneously, and can be distinguished from each other by examination of their different spatial, polarization, time (e.g., for lightning), and spectral characteristics. Given the sensitivity and resolution of the ngVLA, detailed maps of both of these types of emissions will be possible. In the following subsections we highlight some recent results obtained with the VLA, CARMA, and ALMA, and discuss how these can be improved with the ngVLA.

2.1. Deep Atmospheres

Understanding the coupling of gas abundances, temperature, and dynamics of the deep atmospheres of the giant planets is vital to our understanding of these planets as a whole, and to our understanding of extrasolar giant planets.

2.1.1. Radio Spectra

The atmospheres of the giant planets all emit thermal (blackbody) radiation. At radio wavelengths sources of opacity are collision induced absorption (CIA) by hydrogen, and absorption by NH_3, H_2S, PH_3, and H_2O gases (opacity by clouds can probably be ignored; de Pater et al. 2018). For near-solar composition atmospheres[1], most of the atmospheric opacity has been attributed to ammonia gas, which has a broad absorption band near 22 GHz. Indeed, for Jupiter and Saturn NH_3 gas indeed dominates, but not for Uranus and Neptune. The signal-to-noise ratio (SNR) on these planets appears to be considerably enhanced above the solar value, and spectra can be modeled well by including opacity from H_2S gas, and selectively increase the H_2S abundance considerably (factors of $\sim$ 10 and 30 for Uranus and Neptune, resp.) (e.g., de Pater & Mitchell 1993, and references therein). Example spectra for Jupiter and Neptune are shown in Figure 1.

2.1.2. Radio Maps

The thermal emission from all four giant planets has been imaged with the VLA. To construct high signal-to-noise images, the observations need to be integrated over several hours, resulting in maps that are smeared in longitude and only reveal brightness variations in latitude. In order to discern longitudinal structures on radio maps, such as the Great Red Spot (GRS) on Jupiter, Sault et al. (2004) developed an algorithm to essentially take out a planet's rotation. This algorithm has recently been applied to data of all planets obtained with the VLA after its upgrade. Below we highlight a few results.

---

[1]For a solar-composition atmosphere we adopt the proto-solar values of Asplund et al. (2009): C/H_2 = 5.90×10^{-4}; $N/H_2 = 1.48 \times 10^{-4}$; $O/H_2 = 1.07 \times 10^{-3}$; $S/H_2 = 2.89 \times 10^{-5}$; $Ar/H_2 = 5.51 \times 10^{-6}$.

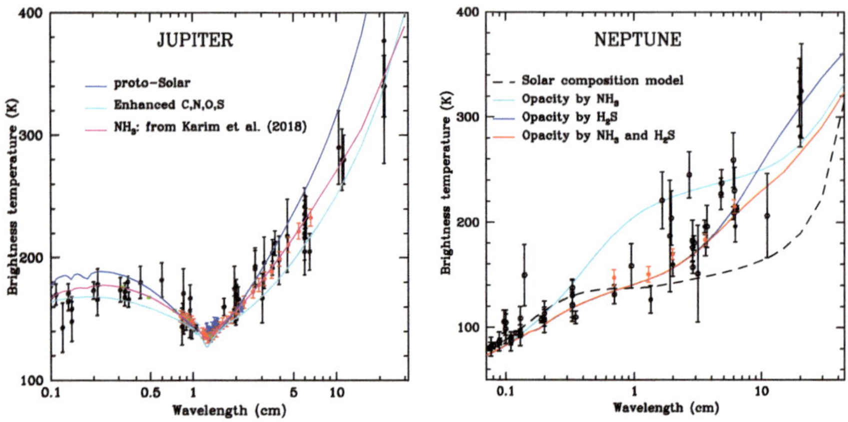

Figure 1. Disk-averaged spectra for Jupiter and Neptune, with various models superposed. *Left*–Jupiter: The red data points are recent VLA values. Various model calculations are superposed: blue: solar abundances for all gases; cyan: NH_3 and H_2S are enhanced by a factor of 3.2 compared to the solar values, and H_2O and CH_4 by a factor of 4; magenta: Karim et al. (2018) decreased NH_3 decreased compared to the cyan curve to match the data at 0.5–2 cm. This panel is adapted from de Pater et al. (2018); for more clarifications, the reader is referred to that paper. *Right*–Neptune: The red data points are VLA values from 2003 (de Pater et al. 2014). The curves are model calculations, as indicated. Opacity by all gases was included in the red curve. In the cyan curve, opacity due to H_2S was ignored, and in the black curve absorption by NH_3 was ignored.

Jupiter. A longitude-smeared image of Jupiter is shown in Figure 2a. This composite multi-wavelength radio image of Jupiter reveals numerous bright bands across the disk. These bands are roughly co-located with the brown belts seen at visible wavelengths. A longitude-resolved map is shown in panel b. The image shows an incredible amount of detailed structure, including the GRS and Oval BA, but also numerous small vortices, hot spots and plumes. The observed variations have been attributed to spatial variations in NH_3 gas, caused by a combination of atmospheric dynamics and condensation at higher altitudes (de Pater et al. 2016).

Spectra of zones and belts from the longitude-smeared images, and spectra of individual features on the longitude-resolved maps have been used together with radiative transfer calculations to determine the altitude distribution of ammonia gas at these locations. These calculations show that ammonia gas is brought up from the deep atmosphere to the cloud condensation levels in the plumes (and equatorial zone), and that the dry air is descending in the hot spots and North Equatorial Belt down to 20 bar or deeper. The hot spots coincide with hot spots seen at a wavelength of 5 μm; the plumes have been hypothesized to form the counterpart of the equatorial Rossby wave (de Pater et al. 2016) theorized to produce the 5 μm hot spots (e.g., Showman & Dowling 2004).

Saturn. Images of Saturn's microwave emission reveal, in addition to the planet itself, its rings. Using data from the upgraded VLA, Figure 3 shows the rings and planet in exquisite detail. The planet itself is visible through its thermal emission, and displays zones and belts as on Jupiter. Surrounding the north pole one can discern a hexagonal pattern, which has been seen at many other wavelengths, and both by the

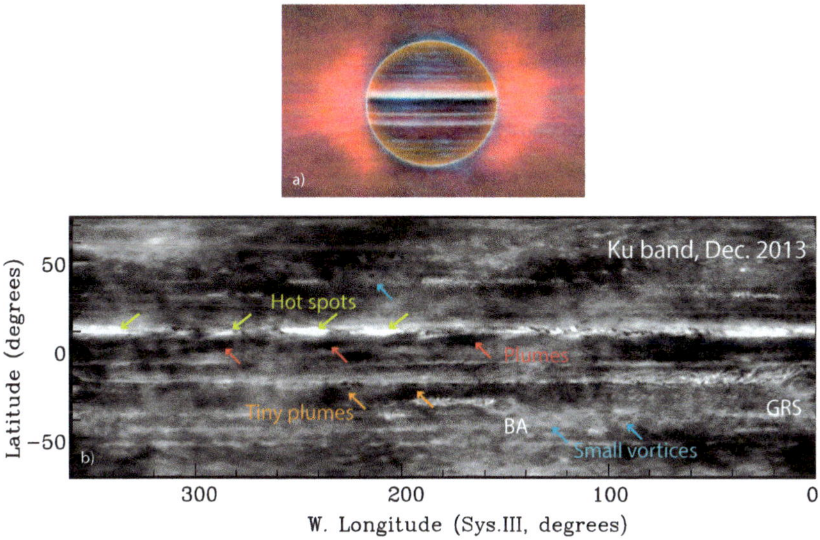

Figure 2. a) Radio image of Jupiter constructed from VLA data taken between December 2013 and May 2014 at three wavelengths: 2 cm in blue, 3.5 cm in gold, and 6 cm in red. A uniform disk had been subtracted to better show the fine banded structure on the planet. The pink glow surrounding the planet is synchrotron radiation produced by spiraling electrons trapped in Jupiter's magnetic field. This image is averaged from several 10-hr observing sessions, so any longitudinal structure is smeared by the planet's rotation (de Pater et al. 2016). b) Longitude-resolved map of Jupiter at a wavelength of 2 cm (Ku band, 12-18 GHz), from panel a). As in panel a), bright features indicate a high brightness temperature, or low NH_3 abundance, so deeper warmer layers are probed. Much finescale structure can be discerned. The Great Red Spot (GRS) and Oval BA are indicated, as well as hot spots (yellow arrows), ammonia plumes (red), small vortices (cyan), and tiny ammonia plumes (orange). [Adapted from de Pater et al. (2018).]

Voyager (1981) and Cassini (2004–2017) spacecraft. This pattern is the meandering path of an atmospheric jet stream. The hexagon itself remains nearly stationary in the rotating frame of Saturn, and has been interpreted as a westward (retrograde) propagating Rossby wave (Allison et al. 1990).

The emission from the planet's rings is dominated by Saturn's thermal radiation reflected off the ring particles. Only a small fraction of the radiation at centimeter wavelengths is thermal emission from the rings themselves. Water ice comprises the bulk of Saturn's rings, yet it is the small fraction of non-icy material that is key in revealing clues about the system's origin and age. Using the Monte Carlo Simrings package (Dunn et al. 2002) to fit multi-wavelength (0.7–13 cm) VLA and 2-cm Cassini/RADAR data of the rings, Zhang et al. (2017, 2018) show that the non-icy fraction of the rings varies from 0.1–0.5% in the B ring, to 1–2% in the C ring, and that the particles overall are quite porous (75%-90%, depending on location in the rings). They further showed that there is a band in the middle C ring where the intrinsic thermal emission is almost constant with wavelength, and which has an anomalously high non-icy material fraction (6–11%), This has been interpreted by the presence of large particles, composed of rocky cores covered by porous, icy mantles. Assuming that the non-icy fraction is due to continuous impacts by micrometeorites, the rings have been estimated to be no older than 200 Myr, while the middle C ring might have been hit by a rocky Centaur 10–20 Myr ago.

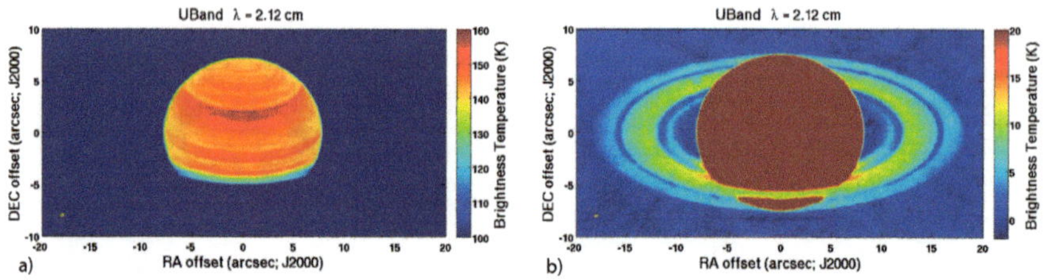

Figure 3. a) Radio image of Saturn at a wavelength of 2 cm (Ku band, 12–18 GHz). The color range is set from 100K to 160K to highlight Saturn's radio bands. b) The same image as in panel a), but here the color is set from -3 K to 20 K to show the structure of the rings. The yellow ellipses on the lower left of each panel indicate the synthesized beam sizes and corresponding position angles. (Zhang et al. 2018)

A longitude-resolved image of Saturn is shown in Figure 4b. Since the observations were obtained over time slots of only 2 hours, we did not cover a full rotation of the planet at each wavelength. As shown, by combining the X and Ku band data we could cover most of a rotation. A comparison with similarly obtained Cassini maps between 2009 and 2011 (Janssen et al. 2013) shows that the bright turbulent band at 30–40°N is caused by the giant storm system at that latitude in 2010-2011 (Figure 4a), which at the time caused that band to be anomalously bright at a wavelength of 2 cm. Even in May 2015 that band is still anomalously bright at radio wavelengths.

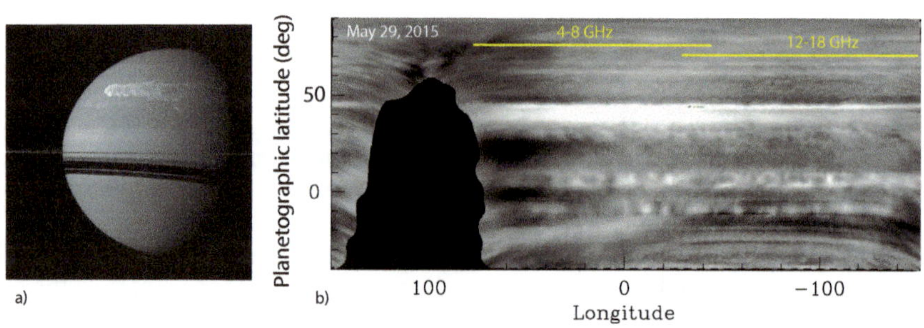

Figure 4. a) Image of Saturn taken by the Cassini spacecraft on 25 February 2011 about 12 weeks after a powerful storm was first detected in Saturn's northern hemisphere. This storm is seen overtaking itself as it encircles the entire planet. (NASA/JPL/Space Science Institute, PIA12826). b) VLA longitude-resolved map from May 2015, after subtraction of a uniform disk. As in Figure 2b, bight features indicate a high brightness temperature, or low NH_3 abundance. Observations at X (8-12 GHz) and Ku (12-18 GHz) were combined; together they cover almost a full rotation of the planet. (Courtesy of Sault & de Pater.)

Uranus. Radio observations of Uranus obtained between the early 60's and 80's revealed a steady increase in its disk-averaged brightness temperature, which was attributed to the poles potentially being warmer than its equator, a hypothesis confirmed using the imaging capabilities of the VLA (e.g., de Pater & Gulkis 1988); several images are shown in Figure 5, which clearly show that both poles are much brighter than the equator. Both the 2005 and 2015 images show several broad bands on Uranus. Uranus pole has an intriguing structure as well, though not dissimilar to that seen at other poles: a bright polar dot, surrounded by a slightly darker (cooler) region, encircled by a bright ring, like the polar collar seen at infrared wavelengths. Longitude-resolved images using the technique from Sault et al. (2004) revealed no longitudinal structure.

ALMA images at 3 mm show a similar banded structure, and these data reveal Uranus's ϵ ring as well.

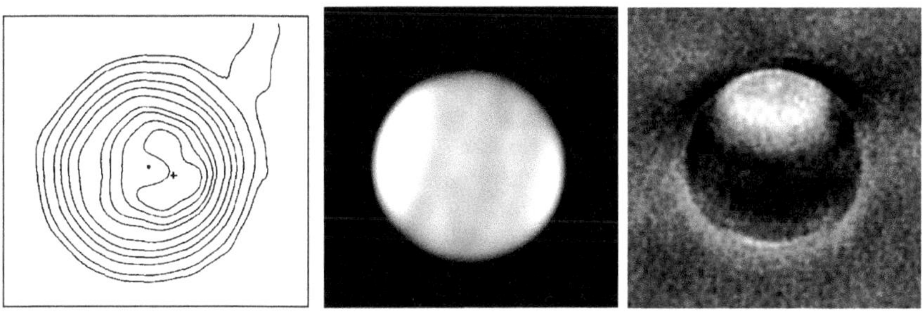

Figure 5. Radio images of Uranus taken at in different years and at different wavelengths: 1982 at 6 cm, when the south pole (indicated by a +) was near the subsolar point (the dot) (de Pater & Gulkis 1988). b) 2005 at a wavelength of 1.3 cm, (Hofstadter & Butler). c) 2015, a preliminary image at 2 cm (Courtesy of de Pater & Sault). In all images the planet is roughly 3.5–4" across. All maps were constructed from many hours of data, so longitudinal features are smeared by the planet's rotation. The bright regions near the poles imply a relative lack of absorbing gases (H_2S, NH_3) above both poles.

Neptune. Multi-wavelength observations obtained in 2003 showed that Neptune's south pole is considerably enhanced in brightness at both mid-infrared and radio wavelengths, i.e., from $\sim$ 0.1 mbar levels in the stratosphere down to tens of bars in the troposphere (de Pater et al. 2014). The enhanced brightness observed at mid-infrared wavelengths is interpreted to be due to adiabatic heating by compression in the stratosphere, and the enhanced brightness temperature at radio wavelengths reveals that the subsiding air over the pole is very dry; at altitudes above the NH_4SH cloud at $\sim$ 40 bar it is only $\sim$ 5% compared to that at deeper layers or other latitudes. This low humidity region extends from the south pole down to latitudes of 66°S, the latitude of the planet's south polar prograde jet. Based upon the combination of near– and mid-IR and radio data, the authors suggested a global circulation pattern where air is rising above southern and northern midlatitudes, from the troposphere up well into the stratosphere, and subsidence of dry air over the pole and equator. Although this model is corroborated by more recent VLA data (Figure 6a), it does not explain all observations (see, e.g., Tollefson et al. 2018).

Figure 6b shows a longitude-resolved map in Ku band. No obvious longitudinal structure can be discerned, despite the fact that there was a large storm system in the south that was connected to a dark vortex, features expected to be rooted in the deep atmosphere. This feature should be located near 340° W. longitude and 30–40°S latitude. Alhough we do see a dark, i.e., cold, area potentially caused by rising gas, this region does not stand out over other dark regions at that latitude.

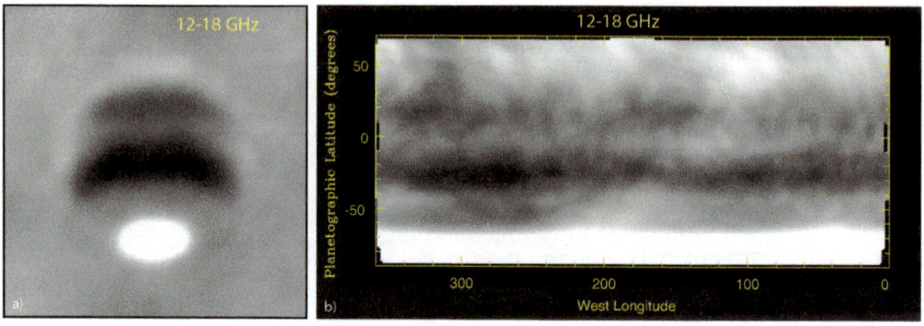

Figure 6. a) Longitude-smeared image of Neptune in Ku (2 cm; 12–18 GHz) band, taken in 2015. A uniform limb-darkened disk has been subtracted from the data. Brighter areas indicate higher brightness temperatures, likely caused by a low opacity (H_2S gas), whereas darker areas indicate cold areas. b) Longitude-resolved map from the same data. The undulating banding pattern may be caused by artifacts. (Courtesy of Sault & de Pater.)

2.2. Stratosphere

Several molecules are prevalent in the stratospheres of the giant planets, such as hydrocarbons C_mH_n (formed e.g., via photochemistry from methane gas), H_2O, CO and HCN (resulting from infalling materials, and/or brought up from the deep atmosphere). The distribution and line profiles of these species will help distinguish between species (or fractions thereof) brought up from the deep atmospheres versus those brought in from the outside. The data can also be used to derive the wind profiles as a function of latitude and depth.

Prominent emission lines of CO and HCN were detected on Neptune in the early 1990s (Marten et al. 1993), with abundances ~ 1000 times higher than predicted from thermochemical models. While HCN cannot have been brought up as such from Neptune's deep interior (it would condense), it is still unclear whether a fraction of the CO in Neptune's stratosphere is brought up from below, or if all of it has an external origin. Since the production of CO in Neptune's interior depends on the water abundance ($CH_4 + H_2O \rightarrow CO + H_2$) and the vertical mixing rate, an accurate measurement of CO constrains the water abundance in the planet's interior. The CO 1-0 line is the most sensitive to a potential internal source through accurate measurements of the wings of the line, which are seen in absorption (originate in the troposphere), in contrast to the emission line at its center (originating in the stratosphere). One of the challenges is to observe the entire line, which is several GHz wide, with a narrow emission peak $\lesssim 10$ MHz.

2.3. Synchrotron Radiation

Synchrotron radiation has only been detected from Jupiter. It is emitted by high energy (~ 1–100 MeV) electrons trapped in the planet's radiation belts. The synchrotron emission morphology and intensity is changing over time due to the planet's rotation, as well as in response to impacts (e.g., the impact of comet Shoemaker-Levy 9; de Pater et al. 1995), changes in the solar wind ram pressure, and any other phenomena (internal or external to the magnetosphere) that induce changes in the energetic electron distribution in Jupiter's magnetosphere.

Jupiter's synchrotron radiation has been imaged at frequencies between 74 MHz and 22 GHz (e.g., de Pater & Kurth 2014, and references therein), usually in all 4 Stokes parameters, which helps to better constrain the models (e.g., magnetic field field geometry). A recent VLA image of the planet's radio emission at 20 cm is shown in Figure 7. Because the radio emission is optically thin, and Jupiter rotates in 10 h, one can use tomographic techniques to map the 3D radio emissivity, assuming the emissions are stable over 10 h. The shape of Jupiter's radio spectrum is determined by the intrinsic spectrum of the synchrotron radiating electrons, the spatial distribution of the electrons and Jupiter's magnetic field. There are substantial variations over time in the spectrum. Changes in the radio spectrum most likely reflect a change in either the spatial or intrinsic energy distribution of the electrons. With the ngVLA we may begin investigating the cause of such variability through its imaging capabilities at high angular resolution at different wavelengths (quasi)-simultaneously, while at the same time being sensitive to short spacings.

Synchrotron radiation has not been detected from any of the other giant planets; a sensitive search might reveal such radiation, or put stringent constraints on such emissions.

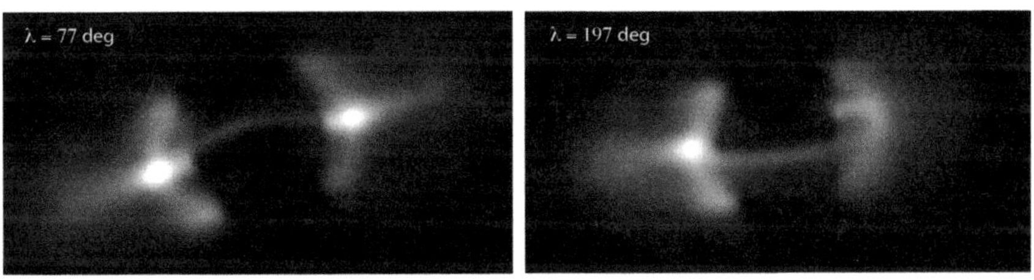

Figure 7. Jupiter's synchrotron radiation at two different viewing aspects (indicated by the longitude λ in Jupiter's System III coordinate system) as observed with the VLA in Jan. 2014. The thermal emission (limbdarkend disk) was subtracted from the data so only synchrotron radiation is visible. (Courtesy of de Pater & Sault.)

2.4. Outlook for the ngVLA

With an instrument that is 10 × more sensitive and has a 30 × higher spatial resolution, maps of the giant planets should be superb. We show a simulation and comparison with the VLA and ALMA later in Section 6.

3. Terrestrial Planets

Radio wavelength observations of the terrestrial planets (Mercury, Venus, the Moon and Mars) are important tools for determining atmospheric, surface and subsurface properties. For atmospheric studies, such observations can help determine temperature, composition and dynamics, similar to observations of the giant planets discussed in Sections 2.1 and 2.2. For surface and subsurface studies, such observations can help determine temperature, layering, thermal and electrical properties, and texture.

Thermal emission from a body's surface can be described as greybody Planck radiation, departing from the blackbody radiation spectrum by a wavelength-dependent factor representing the ability of a surface to re-emit radiation at a given wavelength (spectral emissivity). Since surfaces are not opaque at those wavelengths, 'surface' thermal emission originates in part below the surface, and needs to be determined by performing full radiative transfer modeling including vertical variations of soil properties. The strongest contribution comes from the electrical skin depth which corresponds to the level at which opacity reaches ~1. A commonly used rule of thumb is that observations mostly probe 10-20 wavelengths deep into the subsurface, for typical soil properties.

Although thermal emission is intrinsically unpolarized, it becomes partially linearly polarized as it passes through the surface because of the disconinuity in the dielectric constant at the surface-space (or atmosphere) interface. As observed from Earth, the polarization is zero at the center, and increases towards a maximum value at the limb, which depends on the dielectric constant and the surface roughness. The 3-D temperature distribution of a body is itself determined by its orbital and rotational properties and topography (which determines local insolation), and surface thermal and radiative properties such as thermal conductivity, thermal inertia, dielectric constant, bolometric Bond albedo - which are indicative of composition and physical nature (such as porosity and roughness).

Mercury. A prime example of the thermal (surface) emission displaying the importance of the history of solar insolation and topography is shown by the 3.6-cm radio image of Mercury displayed in Figure 8a, probing a typical depth of ~ 70 cm. Two "hot" regions are visible, one of them almost opposite to the direction of the Sun. This hot-cold pattern results from Mercury's 3/2 spin-orbit resonance combined with its large orbital eccentricity (Mitchell & de Pater 1994). While the surface temperature responds almost instantaneously to changes in illumination, the subsurface layers do not, and this variation in solar insolation remains imprinted at depths well below the surface, as shown in the figure.

Models of the emission revealed that the surface was largely basalt-free, as later confirmed by observations with the *MESSENGER* spacecraft. The map in Figure 8b shows the difference between the map in panel a) and the thermal model. The negative (blue) temperatures near the poles and along the terminator are indicative of areas colder than predicted in the model, likely caused by surface topography, which causes permanent shadowing in craters at high latitudes and transient effects in the equatorial regions, where crater floors and hillsides are alternately in shadow and sunlight as the day progresses.

Venus. Venus presents a different thermal forcing scenario than any of the other terrestrial planets. With a thick atmosphere (~100 bars at the surface), solar insolation does not penetrate directly to the surface, except in very restricted wavelength ranges.

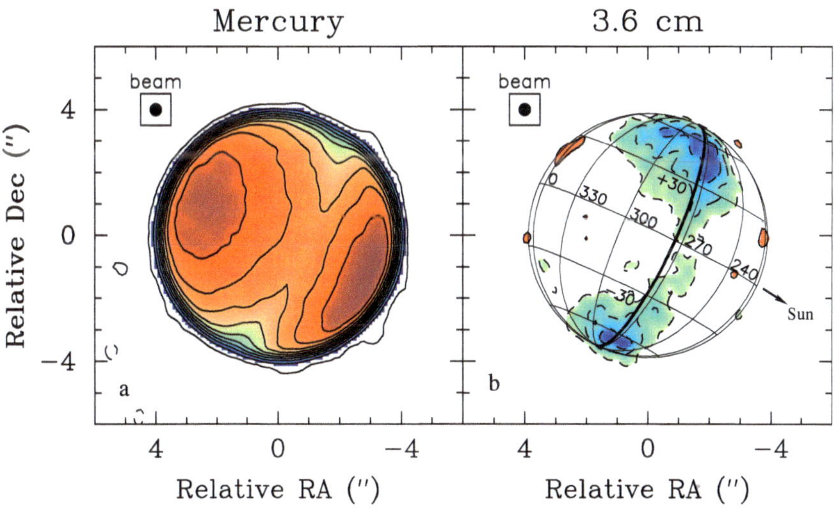

Figure 8. a) The 3.6-cm thermal emission from Mercury observed with the VLA. The beam size is 0.″4 or 1/10 of a mercurian radius. Note the two "hot" regions, discussed in the text. b) A residual map after subtracting a model image from the map in panel a). Mercury's geometry is indicated, including the direction to the Sun and the morning terminator (heavy line). The hot regions have been modeled well, since they do not show up in this residual map. Instead, large negative (blue) temperatures near the poles and along the morning side of the terminator are visible. (Mitchell & de Pater 1994)

The thick atmosphere acts like a thermal blanket, keeping the surface at nearly constant temperature as a function of latitude and longitude. This is moderated by surface topography, of course, where temperatures are much cooler at the tops of the mountains. Previous observations, including those from the *Magellan* spacecraft, have allowed for a detailed understanding of the composition of the atmosphere and top layers of the surface/subsurface - a model can be made which is a good match to data from mm to short cm wavelengths (Butler et al. 2001). However, it is now established that the model breaks down at longer wavelengths (Butler & Sault 2003; Mohan et al. 2017). A good physical explanation for that breakdown is still lacking, however, and high resolution, high sensitivity images, as could be made by ngVLA, could help address this lack of understanding.

The atmosphere of Venus is also a rich area of study. With complex dynamics and chemistry, there is a long history of studying its properties at radio wavelengths (see Encrenaz et al. 2015; Moullet et al. 2012, for recent examples). Since the ngVLA will have access to the CO(1-0) transition, doppler wind observations will yield much information about the distribution of that molecule as a function of latitude and longitude (local time of day). In addition, continuum absorption by sulfur-bearing molecules allows lower frequency observations to be used to determine constraints on their abundances (Butler et al. 2001; Jenkins et al. 2002). It is known from *Venus Express* that the abundance of these molecules is variable both spatially and temporally (Vandaele et al. 2017), and to understand this over long timescales (when *Venus Express* will have completed its mission), we need Earth-based observations like those that can be provided by the ngVLA.

Mars. By contrast, the atmosphere of Mars is four orders of magnitude less abundant than that of Venus, but no less interesting. In fact, understanding the distribution of water on the surface and subsurface of that planet (liquid or frozen) necessarily requires an understanding of the water in the atmosphere as well, and understanding water on Mars is one of the important scientific questions in the solar system. While there have been extensive spacecraft observations of the water (and other molecules) in the atmosphere of Mars (McConnochie et al. 2018; Wolkenberg et al. 2018; Montmessin et al. 2017; Clancy et al. 2017), there is no guarantee that spacecraft will continue to be sent to the planet to make such measurements. Ground-based observations of atmospheric water are needed to provide a continuous record of the abundance of water vapor in the martian atmosphere. The ground-state transition of water vapor near 22 GHz is observable in this way, though sufficient resolution and sensitivity are required. The VLA has been used for such observations (Clancy et al. 1992; Butler et al. 2005). While the line cannot be measured across the entire disk, there is enough path length at the limb to yield a detectable signal. Figure 9 shows observations of the Mars atmospheric water vapor with the VLA in 2003.

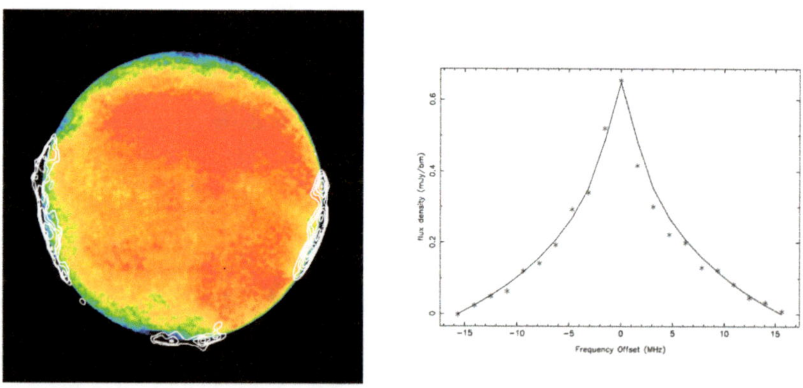

Figure 9. *Left:* Water vapor distribution around the limb of Mars observed in 2003 with the VLA. The colored background is continuum (thermal) emission from the surface and subsurface of the planet, with hottest regions red, through yellow, green, blue, to black for the coldest. Note the colder south polar regions to the lower left (the image has not been rotated so that North is exactly up). Integrated water vapor line emission is shown in white contours. Note the lack of water vapor at northern latitudes, and the concentration at the south pole. *Right:* Water vapor line shape, averaged around the limb (stars) along with a best-fit model (solid line). Total precipitable water vapor ranges from 0 to 20 μm. (Butler et al. 2005)

4. Satellites

The outer solar system's major satellites all host rocky and/or icy surfaces. While Titan is unique in our Solar System as the only moon with a dense (~ 1.5 bar) atmosphere, the other major satellites have, at best, extremely tenuous atmospheres. Continuum thermal emission arising from the satellites' 35-150 K surfaces and subsurfaces dominates their signal at radio frequencies, while emission lines atop this continuum provide a fingerprint of molecular species in their atmospheres.

4.1. Titan

Titan's atmosphere is dominated by N_2, with CH_4 (at a few % level) the only other major species. Like Earth, it has a well-defined troposphere and stratosphere (e.g. Fulchignoni et al. 2005). By mass, ~90% of the atmosphere lies below the tropopause. While *Cassini*/CIRS and its INMS system have made extensive observations of Titan's stratosphere and above, the troposphere and surface are difficult to sense due to hazes and molecular absorption (N_2 CIA and CH_4). These regions can only be probed at a few specific infrared wavelengths (away from strong CH_4 absorption bands) and at radio wavelengths. Yet the troposphere and surface are regions of high interest, and since the arrival of *Cassini* with its cloud penetrating Ku-band RADAR, one of consistent wonder. Surface conditions (pressure, temperature) are near the triple point of methane, like water on Earth, allowing liquid methane on the surface and gaseous methane in the atmosphere. *Cassini*/RADAR data revealed numerous lakes and seas (Stofan et al. 2007) confirming persistent liquid on its surface. The surface is shaped by aeolian and fluvial processes, and methane rain is strongly suspected from rapid equatorial changes (Turtle et al. 2011).

The tropopause is an ineffective cold trap for CH_4; it diffuses upward (Niemann et al. 2010) where it contributes to photochemistry and haze production. In fact, the troposphere contains the bulk of Titan's methane, not the surface lakes, and the stability of the methane cycle requires replacement over geologic timescales (Hörst 2017). The methane cycle is thus tightly tied to the temperature of the lower atmosphere, which has only been measured via *Voyager 1* (Lindal et al. 1983) and *Cassini* radio occultations (Schinder et al. 2012), and during the *Huygens probe* descent (Fulchignoni et al. 2005). Thus the lower atmosphere is poorly measured, despite holding most of the atmospheric mass. Broad latitudinal temperature seasonal variability, at the level of several K, is seen at the tropopause (Anderson et al. 2014) and the surface (Jennings et al. 2016; Le Gall et al. 2016), but the situation within the troposphere is not well sampled, while this information is crucial for climate models (Tokano 2014).

Cassini's 13.8 GHz (Ku-band) RADAR and radiometry data show that Titan's surface displays large scale variations in brightness, due in part to variations in surface roughness (enhancing scattering) and in absorptivity (dielectric constant) of the surface materials. The reflectivity of Xanadu, an extended and peculiar low latitude region, is anomalously high, indicative of coherent scattering on scales of a few wavelengths. In contrast, the dunes and plains which cover most of Titan are radar-dark, with an effective (low) dielectric constant that is indicative of a cover of organic material precipitated from the atmosphere. How these differences vary with seasonal or longer timescales is unknown. While *Cassini* leaves a rich scientific legacy, it only covered half of a Titan year; continued investigation of longterm climate and seasonal variability must rely on new facilities and methods.

Broadband observations of Titan at cm, mm, and submm wavelengths are affected primarily by CIA from N_2-N_2 and N_2-CH_4, with the strength scaling roughly as ν^2. Thermal emission from Titan at frequencies below 40-50 GHz mostly originates from the (sub)surface, while higher frequencies are mostly sensitive to the atmosphere; with increasing frequency, peak sensitivity rises in altitude. Thus sensitive broadband observations spanning the cm-mm bands allow for sounding the troposphere from the (sub)surface to near the tropopause, providing a full characterization of the troposphere and surface brightness temperature. The ngVLA, combining sensitivity and linear resolution gains of order 10 over the current VLA, will allow high precision surface brightness temperature maps for exploration of surface limb darkening, localized emissivity, and latitudinal temperature gradients, with sensitivity in a few hours to under 0.3K in a 35 mas beam at Ka band (225 km on Titan, the size of moderate geologic regions). At Ku band the sensitivity is superb, allowing comparison of the brightness temperatures of the leading and trailing hemispheres, and a direct high resolution tie to the *Cassini* 13.8 GHz RADAR and radiometry data. Adding higher frequencies, the ngVLA will allow detection of latitudinal and/or regional variations in tropospheric temperature, or potentially methane abundance. In the 3mm band the stratosphere can be probed via observations of several nitriles (e.g., HCN, HC_5N, HC_7N), providing information on the stratospheric chemistry and dynamics.

4.2. (Sub)-surface Structure of Other Satellites

As for the terrestrial planets and Titan, in the 1-100 GHz range observations of satellites are sensitive to regions from the upper few cm down to meters within the subsurface and provide quantitative information on temperature and material properties at these depths.

The galilean satellites' disk-integrated brightness temperatures have been measured with single-dish telescopes and arrays including the VLA from 5-230 GHz (Ulich et al. 1984; Muhleman & Berge 1991; de Pater et al. 1984; Butler 2012). The observations of Ganymede and Europa require a drop in emissivity from near 1 at 100 GHz to closer to 0.5 at 15 GHz, with the latter emissivity consistent with that of cold ice. However, Callisto's high brightness temperature across this frequency range deviates from this pattern and remains a mystery. More recently, spatially-resolved observations with the IRAM interferometer and ALMA have provided brightness temperature maps of the galilean satellite subsurfaces at millimeter wavelengths, some of which are shown in Figure 10 (Moullet et al. 2008; de Kleer et al. 2018). The maps demonstrate the expected correlations with surface albedo, but also reveal localized regions with temperatures and thermal properties distinct from the surrounding area.

While ALMA has provided unprecedented spatial resolution at ~cm depths in the galilean satellite subsurfaces, lower frequencies are needed to access the deeper (tens of cm to meters) subsurface regions. Multi-frequency studies utilizing both ALMA and the ngVLA could provide 3D thermal maps of the subsurfaces of the galilean satellites, yielding the first detailed information on the subsurface material properties over a large depth range, and associate specific thermal properties with surface compositional units. The improved sensitivity of the ngVLA will also enable brightness temperature measurements at these frequencies of satellites farther out in the Solar System, while the improved spatial resolution will aid in detecting satellites in close angular proximity to their primaries. Observations of multiple satellite systems over a range of frequencies would enable comparisons between the surfaces of icy satellites at different solar distances.

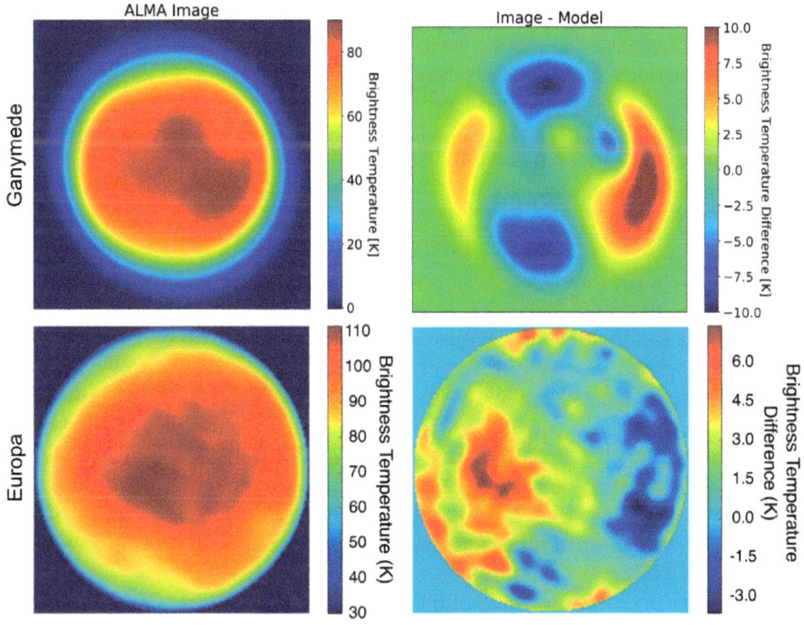

Figure 10. ALMA continuum images of Ganymede and Europa at 230 GHz before and after subtraction of a model. The Ganymede model is a simple disk model and the difference image demonstrates surface temperature variations on the order of 10 K, which largely correlate with surface albedo (de Kleer et al. 2018). The Europa observations were made at a higher spatial resolution, and the residual image is based on a thermal model, demonstrating that such models can account for much but not all of the surface temperature variations (Trumbo et al. 2017). While ALMA frequencies are sensitive to the ~cm subsurface, ngVLA observations of these satellites would probe deeper regions.

4.3. Atmospheres & Plumes of Satellites other than Titan

The tenuous atmospheres of the outer Solar System's satellites (excluding Titan) are sourced by the sublimation and sputtering of surface ices, with a contribution from geological activity in the case of volcanic or cryovolcanic moons. These processes generate atmospheres that vary spatially, dictated by the distribution of surface frost, pattern of particle bombardment, or locations of gas plumes. Orbital and seasonal changes in particle environment and solar illumination drive variability on a range of timescales. Investigations into the nature and variability of thin satellite atmospheres inform our understanding of the creation and evolution of surface conditions on these worlds, and shed light on their potential habitability.

Radio frequencies trace the composition and dynamics of thin satellite atmospheres through rotational transitions of molecular species. Data that resolve these narrow lines have the potential to constrain the atmospheric temperature and its diurnal and seasonal variations, which is a largely unexplored arena. High-sensitivity data may detect new species, or improve our understanding of the gas sources through the simultaneous measurement of multiple gas species. Long-baseline arrays with high sensitivity can even map the distribution of gas species and constrain their sources through spatial correlations with properties such as the surface frost distribution, bombardment pattern, or geological formations. Doppler shift measurements allow for wind speed mapping and characterization of atmospheric dynamics. Io's sulfur-based atmosphere is the densest (~nbar) and best-studied of the jovian satellites. A variety of gas species have been detected and mapped in Io's bulk atmosphere at millimeter wavelengths, including SO_2, $^{34}SO_2$, and SO (Moullet et al. 2010, 2013), and the SO_2 emissions have been used to map wind speeds (Nowling & Moullet 2015). The ngVLA will provide a tool that extends the high sensitivity and spatial resolution of ALMA down to lower frequencies, enabling searches for new species as well as mapping of a wider range of lines for known species, improving gas temperature estimates.

On satellites with volcanic or cryovolcanic activity, radio observations are directly sensitive to molecular gases within the plumes themselves. Radio-frequency studies of these active worlds have the potential to detect plumes and determine their composition, providing direct information on the geophysical processes driving the activity as well as the properties of the source reservoir (water or magma). Using ALMA, Moullet et al. (2015) have mapped alkali gases in Io's volcanic plumes (see Figure 11); the data resolve the distribution of these gases into distinct localized areas. With the recent end of the *Cassini* mission, in-system observations of the Enceladus plume have now ended, hindering efforts to determine the dependence of plume strength on Enceladus' long-period (multi-year) orbital evolution (Ingersoll & Ewald 2017). Enceladus' plume is dominated by water, but molecules such as ammonia, methanol, formaldehyde, and even more complex organics have been detected (Postberg et al. 2018). Radio frequencies are uniquely sensitive to these molecules, and spectral line observations present a potential avenue for continuing to track the plume's variability, as well as for detecting and determining the abundance of species in the plumes of Enceladus and of other satellites that may have ongoing plume activity such as Triton and Europa.

5. Minor Bodies

Observations of small Solar System bodies are an essential component of planetary studies at large. On one hand, characterization of individual bodies (including composition of surfaces, interiors and -in some cases - atmospheres) provides insight into processes affecting surfaces, as demonstrated by the extensive observations of dwarf planets Ceres and Pluto by the *Dawn* and *New Horizons* spacecraft, respectively. On the other hand, global population studies on the distribution of orbital, compositional and physical properties can bring clues pertaining to the thermal, dynamical and collisional history of the early Solar System (see e.g., DeMeo & Carry 2014), for example constraining planetary migration and the origin of some of the giant planets' moons. These investigations are especially relevant for objects orbiting in the outer Solar System, such as Trans-Neptunian Objects (TNOs), which are considered to be some of the most pristine remnants of the protoplanetary disk, and akin to debris disks observed in other planetary systems.

As mentioned in Section 3, the thermal emission from a body's (sub)surface depends on a large number of variables. Several thermal models have been developed to

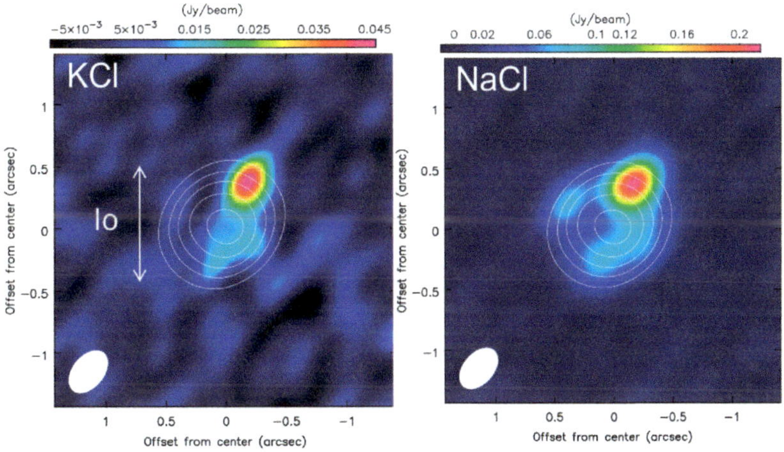

Figure 11. Spatial inhomogeneities in gas species are easily seen in observations that resolve the disk, as shown in this example of spatially-resolved ALMA maps of alkalis in Io's volcanic plumes. The data were taken in June 2015, at a spatial resolution of $0\farcs25 \times 0\farcs4$. (Moullet et al. 2015)

interpret thermal emission from asteroids and TNOs, and allow to classify bodies' thermal behavior based on the value of a single parameter, the beaming factor η, representing the effective thermal inertia (e.g., Lagerros 1996). In general, a low thermal inertia is indicative of a powdery nature while high thermal inertia can be linked to rock-like surfaces. Large-scale studies have shown that asteroids tend to display a much smaller effective thermal inertia than TNOs (Delbo et al. 2015), which suggests that regolith depth (and overall surface processing) decreases with distance to the Sun. However relatively few measurements are available for TNOs (observations are challenging due to large distances and low temperatures), and more sources must be observed to lower observational biases and firmly establish correlations between properties and differences across populations.

Thermal emission mapping, especially combined with monitoring on relevant timescales (diurnal/ seasonal), provides the most informative thermal data as it allows one to disentangle the effects of local hour and latitude from geographically-tied features. Since only a few dozens of objects are larger than $0\farcs1$ in apparent size, this technique has rarely been performed. In the absence of maps, disk-averaged thermal observations combined with optical photometry allows one to retrieve the bolometric Bond albedo and effective diameter through the radiometric method (Lebofsky 1989). While the results from this method depend on the assumed thermal model, it is the most effective technique to estimate sizes of a large number of otherwise spatially unresolved bodies, which can be used to deduce the collisional history for different populations of small bodies (Schlichting & Sari 2011). In addition, combination of thermal photometry at different wavelengths is essential to adjust thermal models for a given body, further improving the robustness of models (Brown & Butler 2018).

ngVLA's sensitivity is necessary to obtain cm-wave measurements with a SNR better to what can be obtained with th VLA or ALMA, giving access to objects with smaller apparent sizes. The specificity of cm-wavelengths lies in the ability to access deeper depths - down to a meter below the surface and beyond the diurnal skin depth. The combination of ngVLA measurements across the full spectral range with other thermal measurements is essential to capture the complete thermal structure in the upper meter of surface material, and determine variations of properties with depth which can be related to ongoing processing, differentiation or formation history. An excellent spatial resolution would surpass the ability of ALMA to detect thermal spatial features (including at depths not probed by ALMA), on smaller targets, and to separate the thermal emission from contact binary systems, which are relatively frequent amongst TNOs.

In addition to measuring continuum thermal emission, ngVLA's spectral coverage is also adequate to target the OH 18-cm maser emission, which has been regularly observed in comets (Crovisier et al. 2002). OH is of particular interest as it has a relatively long lifetime and is the photodissociation product of water, and hence can be used to derive a comet's water outgassing rate. It should be noted that this OH line corresponds to a solar-pumped maser transition, for which excitation strongly depends on its heliocentric velocity. Water atmospheres are not expected in the outer Solar System, but water outgassing has been suggested for Ceres, and for some objects of the Centaur class. OH detection would provide a strong proof of the reality of transient outgassing events, which would carry significant implications for the composition and surface processes on those bodies.

6. Simulations

To highlight the capabilities of the ngVLA, we have constructed a toy model of a planet with typical features: banded structure indicating zonal features, and gaussian features such as upwelling plumes and downwelling hotspots. We used Neptune as a sample case and superimposed structure on the planet. The features are constructed to be smallest at the limb with increasing size and spacing towards the center of the planet. The simulation was done for various ngVLA configurations: using antennas just in the core, using the core and the spiral arms extending outwards to 10 km (plains) and using the full 1000 km array (revB). Furthermore, we simulate our toy Neptune models at 100 GHz and compare the resulting sensitivities and resolutions to the extended VLA-A and ALMA configurations. All simulations were created in CASA with the task SIMOBSERVE using an integration time of 60 s and a total time of 1200 s (similar results are obtained with an integration time of only a few minutes). This creates a visibility file from which we create an image with the CASA task TCLEAN using multiscale and natural weighting. The results for the different configurations are shown in Figure 12. Our toy model in panel a) is shown after subtraction of a limb-darkened disk (disk-averaged brightness temperature $\sim$ 120 K); it is populated with both bright and dark bands and spots, that have a contrast of $\sim$ 1% of the background, or $1-2$K. Panel b) shows the sky as seen by the compact inner array (plains). The lower resolution leads to a smearing of many of these features, however, the large hotspots can still be seen protruding through the bands due to the very high sensitivity. Panel c) shows the structure imaged with the full array, where features that are 1K in magnitude and about $0\farcs01$ in size can be distinguished.

Panels d-f) shows the toymodel from panel a) as it would be seen using the extended configurations for ALMA d), the VLA e), and the ngVLA f), which is equal to c). This comparison highlights the scale of features we could detect, where Table 1 outlines the resulting beam size and rms for each configuration. The banded structure can be seen across all simulations, where the VLA substantially smears the bands and cannot pick up the fine-scale structure at the limb. ALMA can pick up the banded structure relatively well, however, fails at resolving the point sources in the upper right quadrant of the planet. Only the ngVLA is able to pick up a substantial number of the small point sources, including features as small as $0\farcs01$. This can be generalized to resolving structures down to a size of order $< 10\,\mathrm{km/AU}$, and with that an order of magnitude better than the VLA and Juno in the case of Jupiter. In terms of sensitivity, the large collecting area of the ngVLA results in the lowest rms in the beam, as seen by the lack of structure off the disk.

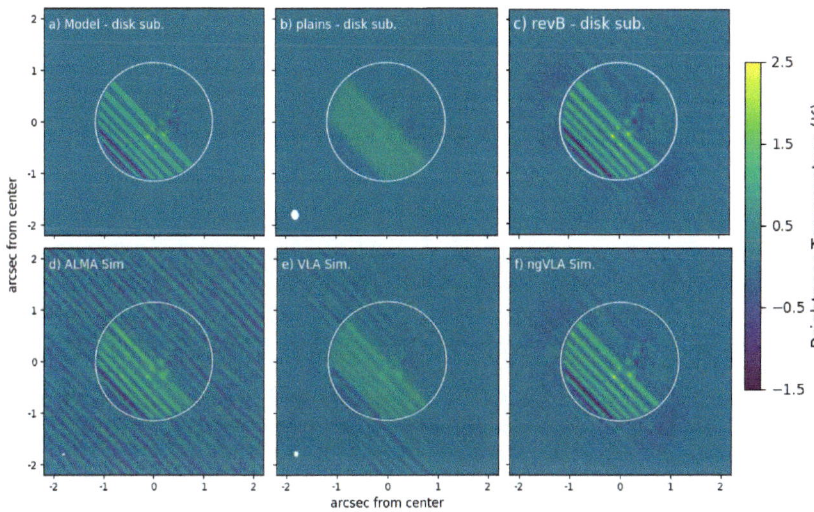

Figure 12. Model ngVLA simulations for various configurations of a toy Neptune outlined by the white disk. The brightness temperature of the various features is of order 1-2 K. The beam size is shown as a white oval in the bottom left of each simulation. a) The original sky model, after subtraction of a limb-darkened disk. Note the positive (bright) and negative (dark) stripes in the southern hemisphere and spots near the center, and extending up into the northern hemisphere up to ~ 1/4 of a radius from the northern limb. b) Model as seen by the compact inner array (plains). c) Model as seen using the fully extended array (revB). d) Toymodel as seen by ALMA in its most extended configuration. e) Toymodel as seen with the VLA in its most extended (A) configuration. f) Same as in panel c. (Courtesy of Moeckel & Tollefson.)

7. Conclusion and Recommendations

• *Giant Planet Atmospheres:* To determine the composition and unravel the dynamics in planetary atmospheres we need to map the 3D distribution of gases in the troposphere (e.g., NH_3, H_2S) through mapping at continuum wavelengths across the

Table 1. Information about the model simulations

Image	Configuration file	Beam Size	rms (K)	TCLEAN niter
ALMA sim	alma.cycle5.10.cfg	$0''.05 \times 0''.05$	0.20	1000
VLA sim	vla.a.cfg	$0''.11 \times 0''.08$	0.03	5000
ngVLA sim	ngvla-revB.cfg	$0''.02 \times 0''.02$	0.02	15000

entire ngVLA band at high spatial resolution ($\sim 0''.01$), and by mapping the stratosphere of the planets at particular molecular line transitions (e.g., CO, HCN, H_2O) at high spatial and spectral resolution ($\sim 0''.01$ and $\sim$ 100 kHz) to spectrally resolve the emission core of the lines and spatially the limb of the planets.

• *Giant Planet Synchrotron Radiation:* We distinguish two goals: 1) Map the 3D distribution of Jupiter's synchrotron radiation from 0.3 – 30 GHz. 2) Search for synchrotron radiation from the other giant planets. Measurements of all four Stokes parameters are key to interpret the data in terms of magnetic field geometry and electron (spatial + energy) distributions.

• *Planetary Rings*: High precision maps of rings at high spatial resolution and at different frequencies (2 – 100 GHz) are needed to determine the thermal and scattered (planet) light emission from the rings, which provides information on e.g., the particle size distribution and composition of rings. The mass fraction of non-icy material in Saturn's rings provides an estimate of the age of the ring system. Polarization will help interpret the data.

• *Terrestrial Planets*: Atmospheres and the (sub)surface structure can be studied in detail (composition, structure, dynamics) with high spatial resolution maps across the entire ngVLA band, in total intensity and, for surfaces, in linear polarization. Both low and high spectral resolution is required to study molecular line transitions throughout a planet's troposphere and stratosphere.

• *Satellites*: To determine the composition and dynamics in their atmospheres and (sub)surface properties (e.g., dielectric constant) we need to map the satellites across the entire ngVLA band both in total intensity and linear polarization, and map the atmospheres and plumes at high spectral and spatial resolution (similar to giant planet stratospheres).

• *Small Bodies:* To understand the dynamical and physical evolution of the small body populations, ngVLA observations across the entire ngVLA band can be carried out for a large number of small bodies (asteroids, Centaurs, Kuiper-Belt objects), which provide us with the necessary fundamental surface properties such as thermal inertia, conductivity and albedo, as well as equivalent size.

7.1. Measurements Required

Frequency coverage: A wide coverage in frequency ($\lesssim$ 1–116 GHz at minimum) is needed for atmospheric and (sub)surface mapping. At low frequencies, we typically probe deeper layers in atmospheres and surfaces, and shallower levels at higher (continuum) frequencies.

Field of View and Mosaicking: For Jupiter and Saturn a field-of-view of order 2 – 3 arcmin is needed to map the planet, rings and Jupiter's synchrotron radiation.

Angular Resolution: Any improvement in spatial resolution with respect to the VLA would directly translate into the ability to distinguish smaller features. For most solar system studies, a high angular resolution is required ($0\farcs01$), but simultaneous coverage of large scale structures. An angular resolution of order $\sim 0\farcs1$ would match *JWST* at $\sim 5\mu$m; $0\farcs01$ would match near-IR observations with future ELTs equipped with adaptive optics.

Imaging: Observations on short and long baselines are essential to capture the large range of spatial scales at play, and must be obtained simultaneously due to the relatively short timescales of atmospheric dynamics (compare to weather on Earth), volcanic eruptions (Io), plumes, and cometary outgassing.

Sensitivity: Since all planets rotate (Jupiter & Saturn: $\sim$ 10 hrs; Uranus & Neptune: 16-17 hrs), and winds with typical velocities of $\sim$ 100 m/s will move features with respect to each other, one ideally would like to get instantaneous high SNR. With the VLA we observe over a full rotation of the planet, and "derotate" the planet in the (u, v)-plane so as to create a longitude-resolved map of a planet (assuming features do not move with respect to each other over 10-20 hrs. For ngVLA we should ideally be able to image an object within a min of time. For Jupiter, a 5-min integration leads to a smearing of $1''$. Typical desired sensitivities are of order 0.1 K.

Frequency Switching: Due to the fast timescales involved in studies of time-variable phenomena, it is essential to obtain observations at different frequencies essentially simultaneously, i.e., by being able to switch frequencies often and fast.

Flexible Scheduling Capabilities, Cross Coordination, & ToO: Solar System observations often require simultaneous observations at different telescopes. For example, studying a giant planet's atmosphere, simultaneous data at mid-IR wavelengths can help determine the temperature structure in an atmosphere, a key parameter to model radio spectra. Data at near-IR, optical, and UV wavelengths provide information on aerosols and hazes, while the 5-μm wavelength range probes deeper layers in the atmospheres of Jupiter and Saturn, providing information complementary to that at radio wavelengths. It is therefore important to schedule observations simultaneously with other telescopes (e.g., *HST*, *JWST*, 8-10 m telescopes, and ELTs in the future). ToO observations are extremely important as well, such as a quick response to impacts on a planet, a comet or near-Earth asteroid apparition, or a volcanic eruption.

Tracking of Objects: All objects in the solar system require non-sideral tracking. In order to track objects such as near-Earth asteroids and satellites orbiting a (itself moving) planet, a minimum rate of $1''$/s is required.

Polarization: Measurements of all 4 Stokes parameters are highly desired, as these contribute much to our understanding of Jupiter's synchrotron radiation, scattering off planetary rings, and thermal (sub)surface emissions.

7.2. Synergies with HST, JWST, ALMA, SKA, LOFAR

Complementary information can be obtained by simultaneously observing at the ngVLA with, e.g., *HST/JWST*, ALMA, SKA, and/or LOFAR. For example, the deep atmospheres (below the optical cloud deck) of giant planets can only be observed at radio wavelengths, although for some planets (Jupiter, Saturn) one can probe down to 7 bar in cloud-free regions at 5 μm wavelength. Complementary information is obtained at 5 μm and radio wavelengths (e.g., Sault et al. 2004), which would be useful to better understand the atmospheres' composition and dynamics. Ideally high spectral resolution measurements in the 5-μm range are obtained, so that the abundance profiles of molecules such at PH_3, NH_3, H_2O, and CH_3D can be determined (Bjoraker et al. 2015). *JWST* will cover visible, near- and mid-IR wavelengths at superb sensitivity, though not necessarily at a high spatial nor spectral resolution. Visible and near-IR wavelength observations provide information on clouds and storm regions in planetary atmospheres; with ngVLA one can relate these features to the deeper atmospheric layers, and probe, perhaps, the cause of such systems (caveat: the observations need to be conducted simultaneously!). At mid-IR wavelengths the temperature structure of the atmosphere in the upper troposphere and stratosphere can be determined. Combining observations at all different wavelengths has the potential to unravel the 3D temperature and composition throughout the atmosphere.

SKA will be limited to lower frequencies and shorter baselines; even if the highest frequency is 15 GHz for that telescope, the resolution will not be near to what ngVLA can achieve. LOFAR covers still lower frequencies, down to ~ 10 MHz. These low frequencies are interesting for a number of cases (e.g., Jupiter's non-thermal radiation).

With ALMA observations at higher frequencies can be obtained; simultaneous ngVLA and ALMA coverage of several spectral lines from planets and satellites would be useful to fully characterize the atmospheres (and at continuum wavelengths to probe just below the surface of solid bodies).

7.3. Synergies with Spacecraft

A number of other synergies can be envisioned with direct planetary missions including past/current/future spacecraft, landers, and sample return missions. For the giant planets, continuous long-term studies to follow-up the extensive knowledge obtained from the *Cassini* mission towards Saturn and Titan should be acquired at all wavelengths. Should the New Frontiers 4 mission concept *Dragonfly* to Titan be selected, complimentary studies will be necessary for global measurements for the interpretation of data acquired by the spectrometers aboard the spacecraft. Other missions such as *Juno*, *Juice*, and *Europa Clipper* in the Jupiter system will also benefit from long term studies of the system's magnetosphere from ngVLA. Accessibility to thermal measurements and lines such as OH maser emission will provide essential complementary data for missions to small bodies such as the New Frontiers 4 mission concept for a comet sample return from Comet 67P, the *Lucy* mission to asteroids, and long term studies to follow-up *Dawn*'s results.

References

Allison, M., Godfrey, D. A., Beebe, R. F. 1990, Science, 247, 1061.

Anderson, C.M., et al. 2014. Icarus, 243, 129

Asplund, M., Grevesse, N., Sauval, A.J., Scott, P., 2009. Ann. Rev. Astron. Astrophys. 47, 481-522.

Bjoraker, G.L., Wong, M.H., de Pater, I., Ádámkovics, M., 2015. ApJ, 810, # 122, 10 pp.

Brozović, M., et al. 2018, Science with a Next-Generation Very Large Array, ASP Monograph 7, ed. E. J. Murphy (San Francisco, CA: ASP), 113

Butler, B.J., Steffes, P.G., Suleiman, S.H., Kolodner, M.A., Jenkins, J.M., 2001. Icarus 154, 226

Butler, B.J., Sault, R.J., 2003. IAUSS IE, 17B

Butler, B.J., Campbell, D.B., de Pater, I., Gary, D.E., 2004. *New Astronomy Reviews*, 48, 1511-1535.

Butler, B.J., Johnston, J.G., Clancy, R.T., Gurwell, M.A., 2005. BAAS, 37, 27.08

Butler, B., 2012. ALMA Memo 594.

Brown, M. E., & Butler, B. J. 2018, arXiv:1801.07221

Clancy, R.T., et al. 2017, Icarus, 293, 132.
Clancy, R.T., Grossman, A.W., Muhleman, D.O., 1992. Icarus 100, 48
Cordiner, M. A. & Qi, C., 2018, Science with a Next-Generation Very Large Array, ASP Monograph 7, ed. E. J. Murphy (San Francisco, CA: ASP), 73
Courtin, R., 1988, Icarus 75, 245
de Kleer, K., Butler, B., de Pater, I., Gurwell, M., Moreno, R., Moullet, A., 2018. 49th LPSC Meeting, No. 2083, ID 2567.
Crovisier, J., Colom, P., Gérard, E., Bockelée-Morvan, D., Bourgois, G. 2002, A&A, 393, 1053
de Pater, I., Brown, R.A., Dickel, J.R., 1984. Icarus 57, 93-101.
de Pater, I., et al., 1995. Science, 268, 1879-1883.
de Pater, I., Gulkis, S., 1988. Icarus, 75, 306-323.
de Pater, I., Kurth, W.S., 2014. The Solar System at radio wavelengths. *Encyclopedia of the solar system, 3ed.*, eds. T. Spohn, D. Breuer, and T.V. Johnson. pp. 1107-1132.
de Pater, I., Mitchell, D. L., 1993. *J. Geophys. Res. Planets*, 98, 5471-5490
de Pater, I., et al., 2014. Icarus, 237, 211-238.
de Pater, I., Sault, R. J., Butler, B., DeBoer, D., Wong, M. H., 2016. Science, 352, 1198-1201.
de Pater, I., Sault, R. J., Wong, M. H., Fletcher, L. N., DeBoer, D., Butler, B., 2018. Icarus, submitted.
Delbo, M., Mueller, M., Emery, J. P., Rozitis, B., Capria, M. T. 2015, Asteroids IV, 107
DeMeo, F. E., Carry, B. 2014, Nat, 505, 629
Dunn, D. E., Molnar, L. A., Fix, J. D., 2002. Icarus, 160, 132-160.
Encrenaz, T., Moreno, R., Moullet, A., Lellouch, E., Fouchet, T., 2015. P&SS, 113, 275.
Fulchignoni, M., et al. 2005, Nature, 438, 785.
Hörst, S. M. 2017, J. Geophys. Res. (Planets), 122, 432.
Ingersoll, A.P., Ewald, S.P., 2017. Icarus 282, 260-275.
Janssen, M. A., et al., 2013. Icarus, 226, 522-535.
Jenkins, J.M., Kolodner, M.A., Butler, B.J., Suleiman, S.H., Steffes, P.G., 2002, Icarus, 158, 312.
Jennings, D. E., et al. 2016, ApJ, 816, L17.
Karim, R. L., deBoer, D., de Pater, I., Keating, G. K., 2018. Astron. J., 155, #129, 8 pp
Lagerros, J. S. V. 1996, A&A, 310, 1011
Lazio, J., et al., 2018, Science with a Next-Generation Very Large Array, ASP Monograph 7, ed. E. J. Murphy (San Francisco, CA: ASP), 81
Le Gall, A., et al. 2016, J. Geophys. Res. (Planets), 121, 233.
Lebofsky, L. A. 1989, Icarus, 78
Lellouch, E., Bézard, B., Flasar, F. M., et al. 2014, Icarus, 231, 323.
Lindal, G. F., et al. 1983, Icarus, 53, 348.
Marten, A. et al., 1993, Astrophys. J. 406, 285.
Mitchell, D. L., de Pater, I. 1994. Icarus, 110, 2-32.
Mohan, N., et al., 2017. Icarus 297, 119
Moullet, A., Lellouch, E., Moreno, R., Gurwell, M.A., Moore, C., 2008. A&A 482, 279-292.
Moullet, A., Gurwell, M.A., Lellouch, E., Moreno, R., 2010. Icarus 208, 353-365.
Moullet, A., Lellouch, E., Moreno, R., Gurwell, M., Sagawa, H., 2012. A&A 546, A102.
Moullet, A., Lellouch, E., Moreno, R., Gurwell, M., Black, J.H., Butler, B., 2013. ApJ 776:32 (9pp).
Moullet, A., Lellouch, E., Gurwell, M., Moreno, R., Black, J., Butler, B., 2015. AAS DPS Meeting #47, Abstract 311.31
McConnochie, T.H., et al., 2018. Icarus, 307, 294.
Montmessin, F., et al., 2017. Icarus, 297, 195.
Muhleman, D.O., Berge, G.L., 1991. Icarus 92, 263-272.
Niemann, H. B., et al. 2010, J. Geophys. Res. (Planets), 115, E12006.
Nowling, M., Moullet, A., 2015. AAS Meeting #225, Abstract 137.11.
Postber, F., et al., 2018. Nature 558, 564-568.
Radebaugh, J., et al. 2010, Geomorphology, 121, 122.
Sault, R.J., Engel, C., de Pater, I., 2004. Icarus, 168, 336-343.

Schinder, P. J., et al. 2012, Icarus, 221, 1020.
Schlichting, H. E., Sari, R. 2011, ApJ, 728, 68
Showman, A.P., Dowling, T.E., 2000. Science 289, 1737-1740.
Stofan, E. R., et al. 2007, Nature, 445, 61.
Tokano, T. 2014, Icarus 231, 1
Tollefson, J., et al., 2018. Icarus, 311, 317-339.
Trumbo, S.K., Brown, S.E., Butler, B.J. AJ 154:148.
Turtle, E. P., et al. 2011, Science, 331, 1414.
Ulich, B.L., Dickel, J.R., de Pater, I., 1984. Icarus 60, 590-598.
Vandaele, A.C., et al. 2017, Icarus 295, 16.
Vinatier, S., et al. 2015, Icarus 250, 95
Vuitton, V., Yelle, R. V., McEwan, M. J. 2007, Icarus 191, 722
Wolkenberg, P., Smith, M.D., Formisano, V., Sindoni, G. 2018, Icarus 215, 628
Zhang, Z., et al., 2017. Icarus, 281, 297-321.
Zhang, Z., et al., 2018. Icarus, in press.

Science with a Next-Generation Very Large Array
ASP Conference Series, Monograph 7
Eric J. Murphy, ed.

High-Resolution Imaging of Comets

M. A. Cordiner[1,2] and C. Qi[3]

[1]*NASA Goddard Space Flight Center, 8800 Greenbelt Road, MD 20771, USA.*
martin.cordiner@nasa.gov

[2]*Institute for Astrophysics and Computational Sciences, The Catholic University of America, Washington, DC 20064, USA.*

[3]*Harvard-Smithsonian Center for Astrophysics, 60 Garden Street, MS 42, Cambridge, MA 02138, USA* cqi@cfa.harvard.edu

Abstract. Detailed mapping of the distributions and kinematics of gases in cometary comae at radio wavelengths can provide fundamental advances in our understanding of cometary activity and outgassing mechanisms. Furthermore, the measurement of cometary molecular abundances provides new insights into the chemical composition of some of the Solar System's most primitive materials. Here we investigate the opportunities for significant progress in these fields of study using a very large radio interferometer. The ngVLA concept will enable the detection and mapping a range of key coma species in the 1.2 – 116 GHz range, and will allow mapping of fundamental cometary molecules including OH and NH_3 for the first time at high ($\sim$1″) angular resolution. The extremely high angular resolution and continuum sensitivity of the proposed ngVLA will allow the possibility of imaging the thermal emission from the nucleus itself, as well as large dust/ice grains in the comae of comets passing within $\sim$1 au of Earth.

1. Comets: Messengers from the Birth of the Solar System

Comets are thought to have accreted in the Solar System at around the same time as the planets. Depending on the degree of subsequent thermal and radiative processing, cometary nuclei may contain pristine material from the proto-Solar disk or prior interstellar cloud. Studies of cometary ices therefore provide unique information on the physical and chemical properties of the early Solar System and pre-solar Nebula (Mumma & Charnley 2011). By virtue of their organic-rich composition, cometary impacts could have been important for initiating prebiotic chemistry on the early Earth, and examining their molecular content provides crucial details on the relationship between interstellar and planetary material (Ehrenfreund & Charnley 2000; Alexander et al. 2018).

To a first-order approximation, the structure of cometary comae can be interpreted using a uniform, spherically-symmetric outflow model, which defines volatile molecules as either 'primary' (parent) species, sublimated directly from the nucleus or photo-chemical 'product' (daughter) species produced in the coma (Haser 1957). Molecules commonly detected in the coma using ground-based telescopes include H_2O, CO, CH_4, CH_3OH, NH_3, HCN, C_2H_2, C_2H_6, H_2S and CS but detections of more complex molecules such as HC_3N, as well as above-average abundances of HNC, H_2CO, CN, C_2 and C_3 in some comets imply that other organic parents are likely present (Crovisier et al. 2004; Cottin & Fray 2008; Lis et al. 2008), probably in the form of organic polymers or macro-molecules. Distributed coma sources are known or suspected for several commonly-observed species (including CN, HNC and H_2CO), originating from presently unknown precursor material(s) at distances up to $\sim 10^5$ km from the nucleus (Biver et al. 1999; Cordiner et al. 2014, 2017). The clearest measure of release from the nucleus (*vs.* production in the coma) requires measurement of the detailed spatial distribution of a species, especially its variation with nucleocentric distance in the innermost coma, within a few hundred to a few thousand km of the nucleus.

2. Elucidating the Origins of Cometary Gases Through Radio Interferometry

A fundamental goal of cometary investigations is to understand the origin and nature of cometary nuclei, in order to explore the chemical link between mature planetary systems and the material present in the accretion disk from which they formed. Of particular interest are the primary volatiles, as measurement of their production can reveal relative abundances of species originally incorporated into a particular cometary nucleus, and therefore information about the conditions under which the comet formed.

The Rosetta mission to comet 67P demonstrated unequivocally the complex nature of cometary outgassing and activity. Strong chemical differentiation was observed for outgassing from different parts of the nucleus (Bockelée-Morvan et al. 2016), and narrow, highly directional jets were observed during outburst events as the comet approached perihelion (see Figure 1). Such features typically span kilometer (milliarcsecond) scales, which can only be resolved from the ground using the high-sensitivity, long-baseline interferometry.

Figure 1. Optical images of comet 67P/Churyumov-Gerasimenko from Rosetta's OSIRIS camera, obtained near perihelion on 12 August 2015, showing strong, narrow jets of gas and dust. Credit: ESA/Rosetta/MPS for OSIRIS Team MPS/UPD/LAM/IAA/SSO/INTA/UPM/DASP/IDA

Radio interferometers can map the spatial distributions of multiple primordial molecules released directly from the interior of the nucleus. Heterodyne receiver technology enables the derivation of high-resolution line-of-sight velocity information, from which 3-D structures can be obtained for the observed species, assisting in investigations of the chemical heterogeneity and true composition of the nucleus (Qi et al. 2015). The first reported interferometric observation of a comet was of the 18 cm OH line in

1P/Halley, detected and mapped using the VLA by de Pater et al. (1986). Formaldehyde (H_2CO) was then detected with the VLA by Snyder et al. (1990) in comet Machholz, and subsequently mapped in comet Hale-Bopp by Milam et al. (2006) using the BIMA interferometer. Figure 2 shows OVRO observations of molecular emission from the inner coma of comet Hale-Bopp (Blake et al. 1999), in which the HNC, DCN, and HDO are seen to lie in arcs or partial shells, offset from the nucleus of the comet. This was likely caused by asymmetric outgassing/jet activity, which lifted a population of icy grains from the nucleus that sublimated to release the observed gases. Interferometric mapping is thus required to properly elucidate the production mechanisms of cometary species, to enable the derivation of accurate mixing ratios.

The importance of radio interferometry to distinguish between molecules arising from different parts of the nucleus and coma was confirmed by Cordiner et al. (2014, 2017). Observations of two comets [C/2012 S1 (ISON) and C/2012 F6 (Lemmon)] were made using the Atacama Large Millimeter/submillimeter Array (ALMA) at frequencies 339 – 365 GHz (0.82 – 0.89 mm) with baselines 15-2700 m, resulting in an angular resolution of approximately $0''.5$. Figure 1 shows spectrally-integrated maps for three different molecules (HCN, HNC and H_2CO) in these two comets, and dramatic differences between their spatial distributions are evident.

Modeling of the interferometric visibility amplitudes (*e.g.*, Boissier et al. 2007) enables the coma origins of the observed species to be derived. In both ISON and Lemmon, HCN was found to originate from (or very close to) the nucleus, with a spatial distribution largely consistent with spherically-symmetric, uniform outflow. The HNC and H_2CO distributions, on the other hand, were found to be consistent with release as products of coma chemistry — *i.e.* they do not originate directly from the nucleus, but most likely arise from the breakdown of large organic molecules/polymers in the coma. The observed molecular distributions were found to be variable on a timescale of days to minutes (Cordiner et al. 2017). Thus, the importance of 'snapshot' interferometry of cometary comae is demonstrated.

The proposed ngVLA concept will provide important improvements for cometary studies compared with currently available facilities. In particular, the unique centimeter-wave capabilities will open up the possibility of high-resolution OH and NH_3 mapping for the first time. The exceptional *uv* coverage and surface brightness sensitivity of ngVLA will provide an order of magnitude improvement in signal-to-noise for mapping these species compared with existing state-of-the-art interferometers such as the JVLA.

3. Measuring the Chemical Composition of the Nucleus

While the chemistry of interstellar clouds and collapsing protostellar envelopes is known from radio and IR spectroscopy (Herbst & van Dishoeck 2009), our knowledge of protoplanetary disk compositions is comparatively sparse (Walsh et al. 2016). Theoretical models show that disk chemistry is very sensitive to the thermal, radiative and dynamical history (Drozdovskaya et al. 2016), which are not well constrained for the protosolar nebula. Understanding the chemical evolution of matter as it passes from interstellar to planetary phases is crucial for theories concerning the accretion and composition of planetary bodies and their atmospheres, but this work is hindered by the extreme difficulty of disk mid-plane ice (and gas) observations. Measurements of cometary compositions provide a unique method for probing the chemistry of the protosolar disk

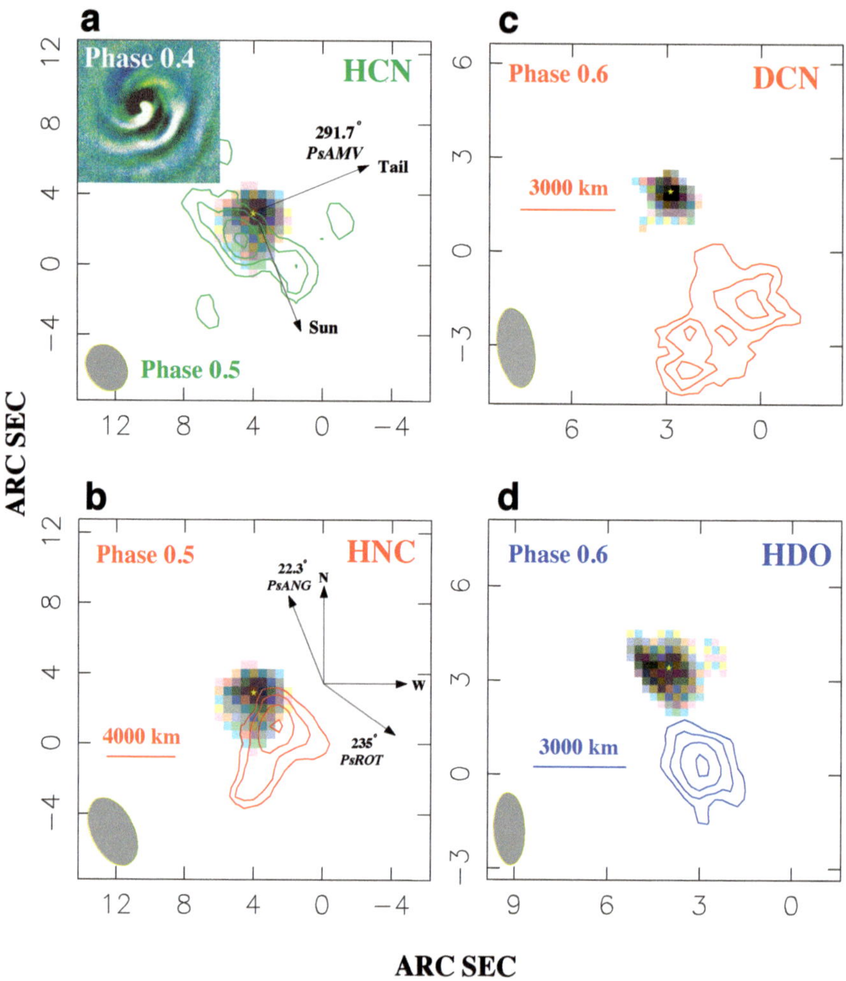

Figure 2. OVRO 3 mm observations of molecular emission from the inner coma of comet Hale-Bopp (Blake et al. 1999). The scattered light optical image (upper left) is dominated by a precessing nuclear dust jet. The star marks the position of the nucleus, and the grey scale presents the continuum emission. Contours in the various panels depict molecular emission from the species listed in the upper right corner. Arrows show the direction of the sun, the cometary rotation axis, and the motion of the comet on the sky. Note that the HCN and DCN emissions were observed on different days but are compared at the same rotational phase.

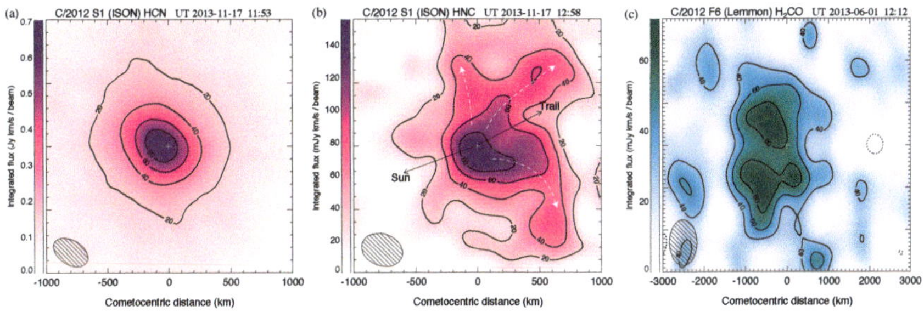

Figure 3. Contour maps of spectrally-integrated molecular line flux observed in comets S1/ISON and F6/Lemmon using ALMA. Contour intervals in each map are 20% of the peak flux (the 20% contour has been omitted from panel (c) for clarity). On panel (b), white dashed arrows indicate HNC streams/jets. The peak position of the (simultaneously observed) 0.9 mm continuum is indicated with a white '+'. See Cordiner et al. (2014) for further details.

mid-plane, and may be used to test theories regarding the chemical evolution of the early solar system.

To determine capabilities of the proposed ngVLA concept with regard to cometary composition studies, we generated spectral line flux simulations for a range of commonly-observed coma molecules. These simulations are based on a spherically-symmetric (Haser-type) outflow model in LTE at 75 K, for a typical (moderately bright) comet at 1 au from Earth, with a water production rate $Q(H_2O) = 2 \times 10^{29}$ s^{-1} and outflow velocity 0.8 km s^{-1}. Assumed abundances (with respect to the dominant volatile H_2O) are based on typically-observed values (Mumma & Charnley 2011). The coma morphology is dominated by a $1/\rho$ brightness distribution (where ρ is the sky-projected distance), and our calculated fluxes are for a nucleus-centered ngVLA beam, tapered to a resolution of $1''$.

Results of these calculations are given for the strongest line of each species (in the nominal ngVLA frequency range 1.2 – 116 GHz) in Table 1; for many species, additional lines will be observable. We have only included molecules for which detections will be readily achievable based on current estimates for the ngVLA sensitivity (with line fluxes $\gtrsim$ 1 mJy per $1''$ beam per 1.6 km s^{-1} unit bandwidth), assuming 1 hour on-source. Accurate, instantaneous OH and NH_3 abundance measurements will thus be possible at spectral line sensitivities 50 – 100 μJy, which are expected to be achievable using ngVLA, but would take weeks of integration with the JVLA. In addition to these fundamental cometary species, other species such as sulphur-bearing molecules and more complex organics may also be detectable, permitting detailed studies of cometary compositions beyond what is possible using currently-available instruments.

4. Revealing the Coma Structure Through High-Resolution Molecular Imaging

Cometary activity occurs as a result of solar radiation that heats the volatile ices in the nucleus as the comet approaches the inner solar system. Despite decades of remote and in-situ studies, the physical outgassing mechanisms and primary drivers of cometary activity remain to be fully understood (Gundlach et al. 2015).

Table 1. Calculated fluxes for representative cometary lines using ngVLA concept (1″ beam)

Species	Transition	Frequency (MHz)	Abund. (%)	Flux (mJy)
OH	$J = 3/2, F = 2 - 2$	1667	86	1.7
NH_3	$J_K = 3_3 - 3_{-3}$	23870	1.0	1.0
H_2CO	$J_{K_aK_c} = 1_{01} - 0_{00}$	72838	0.5	1.7
HCN	$J = 1 - 0$	88632	0.2	13.9
HNC	$J = 1 - 0$	90664	0.02	2.7
HC_3N	$J = 10 - 9$	90979	0.02	3.0
CS	$J = 2 - 1$	97981	0.1	7.2
CH_3OH	$J_K = 3_1 - 4_0\,A^+$	107014	2.0	4.9
CH_3CN	$J_K = 6_0 - 5_0$	110383	0.02	1.5
CO	$J = 1 - 0$	115271	5.0	2.3

Flux calculations are based on a Haser parent model for a typical comet at 1 au (see text). For OH, the predicted flux was scaled from the VLA observation of comet Wilson by Palmer et al. (1989), with a representative inversion parameter of 0.4.

The unique capability of ngVLA to map OH and NH_3 at high spatial resolution (see Section 2), will allow a leap forward in our ability to probe the detailed outgassing behaviour of comets. Hydroxyl (OH) and ammonia (NH_3) are among the most abundant cometary molecules, but their centimeter-wave transitions have never before been mapped in a cometary coma at high ($\sim 1''$) resolution. The spatial distribution of NH_3 in the coma is largely unknown, so detailed mapping is expected to reveal important new insights into the nature of the main reservoir of cometary nitrogen. Ammonia will be detectable in the proposed ngVLA receiver band 4 through its inversion transitions around 23.7 – 23.9 GHz, whereas OH has strong fine structure transitions around 1.7 GHz. Release of NH_3 from the nucleus may be mapped in active comets within a few astronomical units of Earth, allowing the detailed structure of the inner coma to be measured on scales $\sim 1''$ ($\sim$ 1000 km). This will reveal jets and other outgassing features, similar to those detectable by ALMA in other molecules (Figure 1). By contrast, OH is produced further from the nucleus as a result of H_2O photolysis, and can be used to map the larger-scale coma structure (*e.g.* de Pater et al. 1986). To fully capitalize on the unique sensitivity and mapping capability of ngVLA for comets, the need for long ($\sim$ 100 km) as well as short ($\sim$ 50 m) baselines is emphasized, to allow measurement of the coma structure on (milli-)arcsecond to arcminute scales. Cometary comae tend to be centrally peaked (with a $\sim 1/\rho$ brightness distribution), but can extend to over $100''$, so our science case would also benefit from high-sensitivity, simultaneous single-dish observations to fill in lost flux from the largest coma scales.

Resolved spectral line profiles for the observed gases (at ~ 0.1 km s^{-1} resolution) will reveal the detailed coma kinematics, and the gas temperatures may be obtained by measuring multiple NH_3 lines from the different fine structure levels of this molecule. This will enable new tests for our understanding of coma physics.

5. Resolving the Nucleus Using Long-Baseline Interferometry

Cometary nuclei are among the most difficult objects of the solar system to detect and characterize (Lamy et al. 2004). Rendezvous missions such as Giotto, Deep Impact and Rosetta have provided our only close-up views of a handful cometary nuclei, revealing diverse morphologies and surface properties that provide information on their origins as well as thermal and collisional histories. Direct imaging of cometary nuclei using traditional ground-based observing strategies is extremely difficult due to their faintness and small sizes (typically $\lesssim$ 10 km, or 14 mas at 1 au). In addition, dust emission (and reflected sunlight) tends to swamp the weak signal from the nucleus in the optical and infrared. Although thermal emission from large dust grains in the coma contributes significantly to the cometary continuum signal in the centimeter and millimeter bands, the much broader extent of the coma dust emission allows it to be identified and separated from the compact nucleus contribution.

Detections of nucleus thermal emission have previously been possible only in the case of rare, extremely favourable cometary apparitions (*e.g.* Boissier et al. 2014). The ngVLA will provide, for the first time, the ability to routinely observe the nuclei of comets and other near-Earth objects. Assuming an albedo of 0.05, the (black body) continuum flux from a 5 km-diameter comet within 0.5 au of Earth will be easily detectable at around 0.4 mJy. Observed with ngVLA at an angular resolution of 2 mas at 115 GHz (assuming 300 km baselines), the nucleus of such a comet would be well-resolved (with 7 resolution elements across), allowing accurate size and shape measurements.

6. Conclusion

Due to their relatively low surface brightness, cometary comae are extremely challenging to image at radio wavelengths. Only since the advent of ALMA has spectral mapping of moderately bright comets become routine, resulting in a paradigm shift in our understanding of the chemistry and physics of the molecular coma. However, the limited frequency coverage of ALMA prohibits the study of fundamental cometary molecules such as OH and NH_3, which will be uniquely accessible for high resolution mapping by ngVLA. Furthermore, in this article we have demonstrated the possible power of ngVLA for detecting and mapping a range of key coma species in the 1.2-116 GHz range, as well as the possibility of detecting dust, and thermal emission from the nucleus itself. Interferometry of comets with sub-millijansky sensitivity for narrow lines in the centimeter-millimeter spectral region will enable new insights into the composition of the nucleus, and consequently, the early history of our Solar System.

Acknowledgments. This work was supported by the National Science Foundation under Grant No. AST-1614471.

References

Alexander, C. M., McKeegan, K. D., Altwegg, K. 2018, Space Sci. Rev. 214, 36

Biver, N., Bockelée-Morvan, D., Crovisier, J. et al. 1999, AJ, 118, 1850

Blake, G. A., Qi, C., Hogerheijde, M. R., Gurwell, M. A., Muhleman, D. O. 1999, Nature, 398, 213

Bockelée-Morvan, D., Crovisier, J., Erard, S. et al. 2016, MNRAS, 426, 170

Boissier, J., Bockelée-Morvan, D., Biver, N. et al. 2007, A&A, 475, 1131

Boissier, J., Bockelée-Morvan, D., Biver, N. et al. 2014, Icarus, 228, 197
Cordiner, M. A., Remijan, A. J., Boissier, J. et al. 2014, ApJ, 792, L2
Cordiner, M. A.; Boissier, J.; Charnley, S. B. et al. 2017, ApJ, 838, 147
Cottin, H., Fray, N. 2008, Space Sci. Rev. 138, 179
Crovisier, J., Bockelée-Morvan, D., Colom, P. et al. 2004, A&A, 418, 1141
de Pater, I., Palmer, P., Snyder, L. E. 1986, ApJ, 304, 33
Drozdovskaya, M. N., Walsh, C., van Dishoeck, E. F. et al. 2016, MNRAS, 462, 977
Ehrenfreund, P., Charnley, S. B. 2000, ARA&A, 38, 427
Gundlach, B., Blum, J., Keller, H. U., Skorov, Y. V. 2015, A&A, 583, 12
Haser, L. 1957, Bulletin de la Societe Royale des Sciences de Liege, 43, 740
Herbst, E. & van Dishoeck, E. 2009, ARA&A, 47, 427
Lamy, P. L., Toth, I., Fernandez, Y. R., Weaver, H. A. 2004, Comets II, Eds. M. C. Festou, H. U. Keller, and H. A. Weaver, University of Arizona Press, Tucson, 223
Lis, D. C., Bockelée-Morvan, D., Boissier, J. et al. 2008, ApJ, 675, 931
Milam, S. N., Remijan, A. J., Womack, M. et al. 2006, ApJ, 649, 1169
Mumma, M. J., Charnley, S. B. 2011, ARA&A, 49, 471
Palmer, P., de Pater, I., Snyder, L. E. 1989, AJ, 97, 1791
Qi, C., Hogerheijde, M. R., Jewitt, D., Gurwell, M. A., Wilner, D. J., ApJ, 799, 110
Snyder, L., Palmer, P., de Pater, I. 1990, Icarus, 86, 289
Walsh, C., Loomis, R. A., Öberg, K. I. et al. 2016, ApJ, 823, L10

Science with a Next-Generation Very Large Array
ASP Conference Series, Monograph 7
Eric J. Murphy, ed.

Spacecraft Telecommunications

T. Joseph W. Lazio and Sami Asmar

Jet Propulsion Laboratory, California Institute of Technology, Pasadena, CA, USA; `Joseph.Lazio@jpl.nasa.gov`

Abstract. There is a long history of radio telescopes being used to augment the radio antennas regularly used to conduct telemetry, tracking, and command of deep space spacecraft. Radio telescopes are particularly valuable during short-duration mission critical events, such as planetary landings, or when a mission lifetime itself is short, such as a probe into a giant planet's atmosphere. By virtue of its high sensitivity and frequency coverage, the next-generation Very Large Array would be a powerful addition to regular spacecraft ground systems. Further, the science focus of many of these deep-space missions provides a "ground truth" in the solar system that complements other aspects of the ngVLA's science case, such as the formation of planets in proto-planetary disks.

1. Introduction

In addition to the multitude of astronomical applications described elsewhere in this volume, the next generation Very Large Array (ngVLA) could provide valuable benefits in other fields such as the telecommunications with robotic spacecraft on scientific missions throughout the Solar System. There have been many cases of radio astronomical telescopes participating in and enabling planetary missions by augmenting the receiving capabilities of traditional spacecraft tracking facilities, with notable examples including the VLA's participation in Voyager 2's flyby of the planet Neptune (Figure 1, Layland & Brown 1985) and the use of the Green Bank Telescope to track the descent of the Huygens probe to the surface of Titan and enabling the Doppler Wind Experiment (Folkner et al. 2006). The sensitivity and frequency coverage of the ngVLA has the potential to exceed considerably that of any existing or planned dedicated spacecraft radio tracking facility. This contribution is focused on the relevance of the ngVLA for potential NASA missions, but, with the increasing number of space agencies conducting deep space missions,[1] the ngVLA could be of benefit to many space agencies for increasing the science return from their space missions.

[1] In addition to NASA, the European Space Agency (ESA), the Japanese Aerospace Exploration Agency (JAXA), the Indian Space Research Organization (ISRO), and the Roscosmos State Corporation for Space Activities (Roscosmos) send robotic spacecraft into deep space, and potentially Korean and United Arab Emirates will do so in the future.

Figure 1. Neptune as viewed by Voyager 2 during its encounter in 1989. This image is a combination of images acquired with the green and orange filters from the narrow angle camera, acquired on 1989 August 20, just under five days from closest approach. Clearly visible is the Great Dark Spot. These images were likely downloaded via a combination of telemetry downlinks from the Very Large Array (VLA) and NASA's Deep Space Network. In a similar manner, the ngVLA could enable future mission critical events by robotic spacecraft throughout the solar system.

2. Scientific Context

Historically, NASA's most demanding application for highly sensitive ground receiving stations has been planetary science missions, in which spacecraft have been sent to other bodies in the solar system, often at distances of multiple astronomical units from the Earth. Increasingly, however, heliophysics and astrophysics missions have begun to demand similarly sensitive ground receiving stations. For instance, the Solar TErrestrial RElations Observatory (STEREO) mission has two spacecraft, one in an Earth-leading orbit and one in an Earth-trailing orbit, that are approximately 1 au distant,[2] and the Parker Solar Probe[3] will dip to within a few solar radii of the Sun (i.e., travel to approximately 1 au from the Earth). In order to reduce interference from the Earth and operate in a stable environment, many astrophysics missions (e.g., *Planck*, *Spitzer* Space Telescope, *Kepler*, *James Webb Space Telescope*) are being launched into orbits such as Earth-trailing orbits or Earth-Sun Lagrange 2 point halo orbits.

In this landscape, the ngVLA would complement NASA and ESA's network of deep space antennas by providing additional sensitivity for short-duration, high priority events. NASA's Deep Space Network (DSN) is a network of 34 m and 70 m diameter antennas at three longitudes around the world that provides routine collection of telemetry data (a.k.a. "downlink") from spacecraft beyond geosynchronous orbits; ESA has three 35 m antennas at similar longitudes with similar functionality. However, it is standard practice to consider augmenting these antennas during mission critical events, such as the entry, descent, and landing of a spacecraft on another planet. Such augmentation provides resiliency for the mission; for instance, for the Doppler wind experiment on the Huygens probe to Titan, the receiver on-board the *Cassini* spacecraft was not configured correctly and only the data from Earth-based radio telescopes assured the success of that experiment (Folkner et al. 2006). In addition, with its higher sensitivity, the ngVLA may enable the collection of additional data.

Table 1 compares the relative performance of the current DSN antennas (DSN 810-005 Modules 101, 104) with that potentially achievable with a fraction of the ngVLA, using the antenna gain at two standard deep space communications bands, X band (receiving at 8.24 GHz) and Ka band (receiving at 32 GHz); a 34 m antenna operating at X band is taken to be the reference. For the ngVLA, only the Core Array (Carilli 2018), consisting of 94 antennas of 18 m diameter, is considered; the antennas are assumed to have an efficiency of 65%. While the actual performance of the ngVLA may differ from that shown in Table 1, it is clear that even a fraction of the full ngVLA offers substantial improvements. All other things being equal, the relative performance of Table 1 is equivalent to the signal-to-noise ratio achieved on the downlink signal, as the DSN antennas have system temperatures of approximately 20 K and we assume similar values for the ngVLA antennas. In turn, signal-to-noise ratio translates approximately linearly to downlink data rate; alternately, the higher sensitivity can be considered additional margin against unexpected events. We now discuss a few examples of mission concepts that could benefit from having extremely high sensitivity ground receiving stations.

The *Visions and Voyages for Planetary Science in the Decade 2013–2022* Decadal Survey describes two New Frontier[4] mission concepts that would necessarily be short-lived. The Saturn Probe concept would deploy a probe into Saturn's atmosphere to measure its structure and composition. While the materials used to construct such a probe may have improved since the construction of the Galileo probe (Young et al. 1996, and references within), the Saturn probe would suffer the same eventual fate, crushed under the rising pressures as it descends. The Venus In Situ Explorer concept would land on the surface of Venus in order to sample the composition of the crust. Again, while material properties and thermal control systems have improved since the

[2]At the time of writing, the STEREO-B spacecraft is not operational.

[3]At the time of writing, scheduled for launch in 2018 August.

[4]NASA's medium-class missions, with a cost cap of approximately $1 billion.

Table 1. DSN and ngVLA Receiving Antenna Relative Performance

System	Relative Performance	
	X band	Ka band
DSN 34 m	1.0	11.7
DSN 70 m	4.3	...
DSN 4 × 34 m	3.7	46.8
ngVLA Core Array (94 antennas, 18 m diameter)	17.1	248

construction Soviet Venera landers, the surface of Venus is such a harsh environment that a lander will have a necessarily finite lifetime.

The *Ice Giants Pre-Decadal Survey Mission Study Report* considers flagship-class mission concepts to Uranus, Neptune, or both. Four mission architectures were studied in detail, with three of the four including an atmospheric probe. (Two architectures involved an atmospheric probe for Uranus, one for Neptune.) The expected lifetime of such a probe in the atmosphere of either planet is approximately 1 hr.

During the last perihelion passage of Comet 1/P Halley, a figurative armada of spacecraft were sent to conduct close fly-bys of its nucleus. Because of the high relative velocities between the Earth and a long-period comet such as Halley, the durations of the fly-bys were short. Inspired by both the success of the spacecraft that flew by Halley and the recent discovery of the first interstellar object (1I/2017 U1 'Oumuamua, Meech et al. 2017), there is active consideration of what kind of mission concept or concepts might be possible to enable a similar encounter. While the relative velocities could be even higher than in the case of the Halley encounters, the potential science return from a close encounter with an interstellar object would likely be significant.

These mission concepts are intended to be illustrative only, and specific instruments have not been selected. Nonetheless, the higher data rates enabled by the use of the ngVLA could enable more capable instruments (e.g., higher resolution mass spectrometers). Further, these example mission concepts are certainly not an exhaustive list of all possible short-lived mission concepts, but they illustrate how high-priority missions might nonetheless be short-lived and benefit greatly from additional sensitivity such as could be provided by the ngVLA.

An additional benefit of the ngVLA may be for use during spacecraft emergencies. It is standard practice to include low-gain antennas on deep space spacecraft for the purposes of enabling communications even if the spacecraft's attitude is unknown or uncontrolled. Low-gain antennas have the obvious benefit of a wide field of view at the cost of sensitivity. The received signal from a spacecraft in "safe mode" would most likely be (much) weaker than planned from its high-gain antenna, for which the ngVLA's high sensitity could be of benefit in capturing engineering data and understanding the problem with the spacecraft.

3. Design Considerations

The primary requirement for the ngVLA to receive spacecraft telemetry is for it to have frequency coverage of the allocated space-to-Earth downlink bands (Table 2). The relevance of the ngVLA is immediately obvious, as its planned frequency coverage encompasses all of the frequency allocations.

Table 2. Frequency Bands for Deep-Space Spacecraft Telemetry

Name	Telemetry (Downlink) (space-Earth) (GHz)
S band	2.29 – 2.30
X band	8.40 – 8.45
Ka band	31.8 – 32.3

Two additional bands warrant comment. There is a space-to-Earth allocation in the K band (25.5 GHz–27.0 GHz). This band is allocated for spacecraft with orbits that take them no farther than 2×10^6 km from Earth. This distance includes the Earth-Sun L2 point, which is used increasingly by astrophysics missions such as the *James Webb Space Telescope.* While such missions would also benefit from high sensitivity, they have regular telemetry requirements, which are more likely to be met by NASA's DSN (or ESA's Deep Space Antennas). However, should such a spacecraft enter "safe mode," the ngVLA could assist with the recovery of the spacecraft.

There is also an allocation in the ultra-high frequency (UHF) portion of the radio spectrum used for spacecraft-to-spacecraft communications ("proximity links"). This band is not within the ngVLA's frequency coverage, and the SKA1-Mid or a similar facility would be more appropriate.

Like pulsars, spacecraft are point sources, and full field interferometric synthesis is not required. Beamforming suffices. In general, the requirements for pulsar beam-formed observations would be sufficient for spacecraft telemetry reception, with the added benefit that spacecraft signals are typically quite narrow band relative to astronomical sources. A specific concern with the legacy VLA for both pulsar observations and spacecraft telemetry was the "waveguide switch cycle," which resulted in a 1.5 ms data gap every 50 ms (Deutsch 1982; Hankins 1999); modern digital electronics and fiber optic transmission should obviate this particular concern.

4. Conclusion

In conclusion, the ngVLA would be well suited for occasional use of high priority space missions, particularly those with intrinsically short lifetimes. Not only can these missions be considerable investments by a space agencies (with costs in excess of $1 billion), the science motivations for them often complement the larger ngVLA science case. Most notably, the planets in our solar system form the end states of the proto-planetary disks that the ngVLA will image and only by studying both will we understand planet formation and evolution.

Acknowledgments. We thank D. Jones, L. Deutsch, and R. Preston for helpful discussions and guidance on this topic. Some of the information presented is pre-decisional and is intended for informational purposes only. Part of this work was carried out at the Jet Propulsion Laboratory, California Institute of Technology, under contract with the National Aeronautics and Space Administration.

References

Carilli, C. L. 2018, "Next Generation Very Large Array Memo No. 47: Resolution and Sensitivity of ngvla-revB," http://library.nrao.edu/public/memos/ngvla/NGVLA_47.pdf

Deutsch, L. J. 1982, "The Performance of VLA as a Telemetry Receiver for Voyager Planetary Encounters," Telecommunications and Data Acquisition Prog. Report, TDA PR 42-71, 1982 July–September (Jet Propulsion Laboratory, California Institute of Technology: Pasadena, CA) pp. 27–39

Deep Space Network Telecommunications Link Design Handbook, DSN No. 810-005, Module 104, Rev. J, 2018 January 12, JPL D-19379 (Jet Propulsion Laboratory, California Institute of Technology: Pasadena, CA) https://deepspace.jpl.nasa.gov/dsndocs/810-005/

Deep Space Network Telecommunications Link Design Handbook, DSN No. 810-005, Module 101, Rev. F, 2015 August 5, JPL D-19379 (Jet Propulsion Laboratory, California Institute of Technology: Pasadena, CA) https://deepspace.jpl.nasa.gov/dsndocs/810-005/

Folkner, W. M., Asmar, S. W. Border, J. S., Franklin, G. W., Finley, S. G., Gorelik, J., Johnston, D. V., Kerzhanovich, V. V., Lowe, S. T., Preston, R. A., Bird, M. K., Dutta-Roy, R., Allison, M., Atkinson, D. H., Edenhofer, P., Plettemeier, D., & Tyler, G. L. 2006, "Winds on Titan from ground-based tracking of the Huygens probe," J. Geophys. Res., 111, E07S02; doi: 10.1029/2005JE002649

Hankins, T. H. 1999, "Pulsar Observing at the VLA," in *Synthesis Imaging in Radio Astronomy II, A Collection of Lectures from the Sixth NRAO/NMIMT Synthesis Imaging Summer School*, eds. G. B. Taylor, C. L. Carilli, & R. A. Perley, Astronomical Society of the Pacific Conference Series, Vol. 180 (Astronomical Society of the Pacific: San Francisco, CA) p. 613

Ice Giants Pre-Decadal Study Final Report, Solar System Exploration Directorate, Jet Propulsion Laboratory for Planetary Science Division, Science Mission Directorate, National Aeronautices and Space Administration, June 2017, JPL D-100520

Layland, J. W., & Brown, D. W. 1985, "Planning for VLA/DSN Arrayed Support to the Voyager at Neptune," Telecommunications and Data Acquisition Report 42-82, Jet Propulsion Laboratory, California Institute of Technology

Meech, K. J., Weryk, R., Micheli, M., et al. 2017, "A brief visit from a red and extremely elongated interstellar asteroid," Nature, 552, 378; doi: 10.1038/nature25020

Vision and Voyages for Planetary Science in the Decade 2013–2022, Committee on the Planetary Science Decadal Survey, National Research Council (National Academies Press: Washington, DC) ISBN: 0-309-20955-2

Young, R. E., Smith, M. A., & Sobeck, C. K. 1996, "Galileo Probe: In situ Observations of Jupiter's Atmosphere," Science, 272, 837; doi: 10.1126/science.272.5263.837

Science with a Next-Generation Very Large Array
ASP Conference Series, Monograph 7
Eric J. Murphy, ed.

ngVLA Observations of the Solar Wind

T. S. Bastian
National Radio Astronomy Observatory, Charlottesville, VA, USA
tbastian@nrao.edu

Abstract. The ngVLA has the potential to play a significant role in characterizing properties of the outer corona and the heating and acceleration of the solar wind into the inner heliosphere. In particular, using distant background sources to transilluminate the foreground corona and solar wind, a variety of radio propagation phenomena can be used to map plasma properties as a function of solar elongation and position angle throughout the solar cycle. These include angular broadening, interplanetary scintillations, and differential Faraday rotation, which can be used to map the solar wind velocity, determine properties of solar wind turbulence, and constrain the solar wind magnetic field. These observations will provide a global characterization of the solar wind that will be highly complementary to *in situ* observations made by various spacecraft. In addition, such observations can be used to probe disturbances in the solar wind – coronal mass ejections, for example – that may impact the near-Earth environment.

1. Introduction

Although the idea of mass loss from the Sun dates back more than a century, it was Parker (1958) who predicted the existence of a supersonic "solar wind" outflow from the Sun, subsequently confirmed in the early 1960s by spacecraft measurements. To date, most of what we know about the solar wind is based on *in situ* measurements of particles and fields from various vantage points in the solar system: from Earth orbit (e.g., the IMP satellites), missions near the L1 Langrangian point (e.g., ISEE-3, WIND, and ACE), those sent into the inner heliosphere (Helios 1 and 2), those sent out of the ecliptic (*Ulysses*), and those sent to the outer reaches of the heliosphere and into the local interstellar medium (Voyager 1 and 2) . A new chapter of the exploration of the solar wind opens in the coming decade with the launch of the NASA *Parker Solar Probe* (PSP), which will measure solar wind parameters at radial distances as close as 9.8 $R_\odot$ from the Sun, or 0.045 AU. The PSP is complemented by the ESA *Solar Orbiter* (SO), which will orbit the Sun at 60 $R_\odot$ (0.28 AU) at an inclination of up to 34°.

Remote sensing observations are an important complement to *in situ* measurements in the solar wind. Particularly important are white light coronagraphs (SOHO, STEREO/SECCHI) to trace solar wind outflows and disturbances therein (e.g., coronal mass ejections - CMEs), and UV coronagraph spectrometers (e.g., SOHO/UVCS) which have placed key constraints on electron, proton, and ion temperatures in the solar corona. Another powerful remote sensing technique is observations of radio propagation phenomena from the ground; that is, to observe a distant background radio source, spacecraft beacon, or other coherent radio source to measure the effects of the foreground solar wind plasma on the signal. These phenomena can be used to provide global context to *in situ* measurements and can, moreover, measure solar wind properties in regions that are otherwise inaccessible to *in situ* and other remote sensing techniques. While radio propagation phenomena have been exploited since the earliest days of coronal (Erickson 1964) and solar wind exploration (Hollweg & Harrington 1968) they have generally been limited to experiments performed when a spacecraft was at an opportune location (superior conjunction), or when a bright background source (e.g., Tau A) was at a favorable solar elongation. Even Venus, illuminated by a radar signal, has been used as a background source (Coles & Harmon 1989). These experiments have demonstrated the power and utility of radio propagation phenomena as probes of the solar wind and have placed valuable constraints on its properties. However, these techniques have not been systematically exploited. Satellite beacons and radar signals are rarely available as probes and, until recently, radio instruments on the ground have not been sensitive enough to observe many background sources. With the advent of the ngVLA, the sensitivity barrier to the use of background sources is overcome. The sensitivity of the ngVLA will allow a large number of naturally occurring background sources to be used to systematically explore the heliosphere as discussed further in §4.

In this chapter, the potential of the ngVLA to exploit several scattering phenomena as probes of the outer solar corona and the solar wind is outlined. The subject is vast and so the focus will be on just three propagation phenomena: angular broadening, interplanetary scintillations, and Faraday rotation.

2. Motivation

Remarkable progress has been made in understanding the nature of the solar wind over the past fifty years (see, e.g., the monograph by Bruno & Carbone 2016) but mysteries remain. It is widely believed that the solar wind undergoes spatially extended heating and that it plays a significant role in solar wind acceleration. Several heating mechanisms have been considered in detail including magnetic reconnection, wave dissipation, and turbulence. Turbulence has attracted significant attention in recent years as a means of not only heating the solar wind, but also mediating energy transport within the wind and modulating the transport of energetic particles. Briefly, the Sun is the source of outward propagating long-period Alfvén waves (AW). Sunward propagating AW can result from non-WKB wave reflection, velocity shear instabilities, and/or parametric instabilities which, in turn, interact with outward propagating AW to produce an (anisotropic) energy cascade to smaller spatial scales; i.e., a turbulent spectrum. When the cascade reaches scales of order the proton gyroradius the waves become compressive (kinetic Alfvén waves - KAW) and the wave energy is dissipated, thereby heating the ambient plasma (see, for example, Hollweg 1999; Dmitruk et al. 2002; Chandran et al. 2009; Salem et al. 2012). A detailed understanding of these processes nevertheless remains elusive and the fundamental questions remain germane:

- how are the corona and solar wind heated and how is the solar wind accelerated?
- how does strong MHD turbulence develop and evolve in the solar wind?
- what are the turbulence dissipation mechanisms?
- what role does solar turbulence play in mediating energy deposition and transport?

- how does solar wind turbulence modulate the transport of energetic particles (both solar energetic particles and cosmic rays)?

These questions are relevant to other stars, of course. More broadly, the loss of mass and angular momentum due to outflows affect the evolution of a given star, and the extension of the wind from a star can profoundly influence the interstellar medium in its neighborhood. In order to address questions posed by the solar wind and their implications for other stars, our understanding of the solar wind must be placed on a comprehensive observational footing. With the launch of PSP and SO, the inner heliosphere is a new frontier. For the first time, measurements will be made of the solar wind and solar wind turbulence close to the source. Nevertheless, there will still be many locations - the high latitude solar wind, for example - that will not be accessible to either spacecraft. Placing PSP/SO measurements into a broader context will be critically important. Particularly interesting is the region out to the Alfvén radius ($\sim 10-20$ $R_\odot$) where the wind becomes super-Alfvénic which, until the PSP mission, has been inaccessible to *in situ* measurements. The nature of the solar wind, solar wind turbulence, and solar wind heating on either side of the Alfvén surface remains somewhat speculative at this time and awaits detailed characterization by both *in situ* and remote sensing observations.

The ngVLA will be positioned to make significant and complementary contributions to this new frontier. In particular, the ngVLA can leverage a number of radio propagation phenomena to provide comprehensive and complementary diagnostic observations of the solar wind in the inner heliosphere. Electron density fluctuations act as a "passive scalar" that traces solar wind turbulence through the spatial spectrum of electron density fluctuation, $\Phi_n(\kappa)$. The spectrum of density fluctuation cause variations in the plasma refractive index over a broad range of spatial scales. For radio waves propagating through a turbulent plasma medium, the fluctuations in the refractive index result in a variety of phenomena collectively referred to as "scattering" phenomena. We discuss three such phenomena here that the ngVLA will be able to effectively exploit.

3. Angular Broadening

Interferometric observations of angular broadening are particularly straightforward to perform and to interpret. The relevant observable is the mutual coherence function, $\Gamma(s) = \langle E(\mathbf{r})E^*(\mathbf{r}+\mathbf{s})\rangle / \langle |E|^2\rangle$, where $E(\mathbf{r})$ is the electric field measured by an antenna at location $\mathbf{r}$, and $\mathbf{s}$ is the projected baseline to another antenna located at $\mathbf{r}+\mathbf{s}$. For a point-like radio source, $\Gamma(s)$ is related to the visibility function $V(s)$ and the wave structure function, $D_\circ(s) = \langle[\phi(\mathbf{r}) - \phi(\mathbf{r}+\mathbf{s})]^2\rangle$, as $\Gamma(s) = V(s)/V(0) = \exp[-D_\circ(s)/2]$, which is related to the angular spectrum of the source through a Fourier transform (Figure 1, left panel). A loss of phase coherence with increasing interferometer baseline, s, due to plasma turbulence leads to greater values of $D(s)$ which manifests itself as angular broadening of the source in the image plane. A convenient form for the spatial spectrum of turbulence (Coles et al. 1987) is a power law with an inner scale at $s = l_\circ$, given by $\Phi_n(\kappa) = C_n^2 \kappa^{-\alpha-2} \exp[-(\kappa l_\circ/2)^2]$, where C_n^2 is the turbulence level (Rickett 1977). In this case, $D_\circ(s)$ has particularly simple asymptotic limits: when $s > l_\circ$, $D_\circ(s) \propto C_n^2 \lambda^2 s^\alpha \Delta z$, and for $s < l_\circ$ we have $D_\circ(s) \propto C_n^2 \lambda^2 l_\circ^{\alpha-2} s^2 \Delta z$ where λ is the wavelength of the radio radiation and Δz is the thickness of the turbulent slab. The angular size of the scattered source is given by $\theta_s \sim \lambda / r_{diff}$, where $s = r_{diff}$ is the coherence scale (or diffractive scale) defined by $D_\circ(r_{diff}) = 1$.

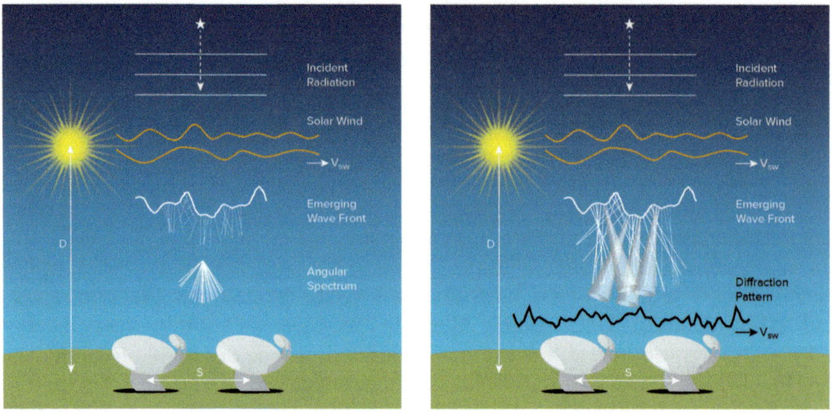

Figure 1. Left: A cartoon illustrating the interaction of radio waves from a distant point source interacting with foreground solar wind turbulence. The incident plane waves emerge as a corrugated wave front and propagate $D = 1$ AU to the observer. The wave normals represent different angles of arrival of the radiation observed on the ground as described by the angular spectrum. Right: the solar wind effectively scatters radiation into a cone of angular width λ/r_{diff} that illuminates a patch on the ground of size $\sim D\lambda/r_{diff}$. These patches constructively and destructively interfere to create a diffraction pattern that sweeps over the observer with the solar wind velocity V_{sw}.

Angular broadening measurements of a background source or, equivalently, measurements of the structure function $D_\circ(s)$ can therefore be used to deduce properties of solar wind turbulence through $\Phi_n(\kappa)$ for the range of transverse scales s measured by an interferometer. When many background sources are observed over a range of solar elongations and position angles, a map of the solar wind properties in the inner heliosphere can be constructed. An advantage of angular broadening measurements is that they can be made under conditions of strong or weak scattering (see below).

Such work was undertaken by the VLA in the 1990s (e.g., Armstrong et al. 1990; Anantharamaiah et al. 1994; Bastian 1999) using small numbers of VLA calibrator sources to transilluminate the solar wind. These observations show that the solar wind turbulence in the inner heliosphere is highly anisotropic, with turbulent eddies drawn out along the magnetic field. The degree of anisotropy may decrease at elongations $> 6\ R_\odot$ and may also decrease on spatial scales > 100 km (Grall et al. 1997), but this has not been systematically studied. Intriguingly, there are indications that the degree of anisotropy may be smaller on short spatial scales than on longer spatial scales, perhaps indicative of an isotropic dissipation mechanism (Armstrong et al. 1990). Since the density inhomogeneities are anisotropic along the magnetic field, the scattering is enhanced perpendicular to the direction of the magnetic field. Angular broadening observations can therefore be used to infer the orientation of the coronal and solar wind magnetic field in the plane of the sky. Furthermore, while well described by a power-law on spatial scales corresponding to VLA baselines in the A configuration (kilometers to 10s of kilometers), VLA observations showed that the power-law index α on these scales is significantly smaller than the Kolmogorov value of 5/3 observed on much larger spatial scales using *in situ* spacecraft measurements and VLBI phase scintillation measurements (Coles & Harmon 1989; Coles et al 1991). The scale on which the spectrum flattens is of order 100 – 300 km in the inner solar wind. The flat part of the spectrum represents a significant excess of turbulent power relative to that expected for an extrapolation of the Kolmogorov spectrum observed on large spatial scales (1000s of km to 10^5 km), consistent with the onset of KAW turbulence on small spatial scales (Chandran et al. 2009).

VLA observations were strongly constrained by its limited sensitivity to a handful of strong sources. The JVLA has much greater sensitivity, allowing a significant increase in the number of sources that may be exploited for "solar wind tomography" (Figure 2). However, the JVLA remains limited in its baseline coverage to spatial scales of just a few 10s of km. The ngVLA will provide both enormous sensitivity and baseline coverage out to ~ 1000 km which will allow not only angular broadening measurements to be fully exploited (see §6), but also enables additional propagation phenomena to be more fully exploited of which two will now be briefly discussed.

4. Interplanetary Scintillations

A second scattering phenomenon is interplanetary scintillation (IPS) intensity variations as a function of time. It is useful to distinguish between two regimes: weak scattering and strong scattering, demarcated by comparing r_{diff} to the Fresnel scale $r_F = \sqrt{\lambda D}$, where $D = 1$ AU. In weak scattering, $r_{diff} \gg r_F$ and intensity fluctuations are caused by weak focusing and defocusing of density fluctuations on the Fresnel scale and $\sigma_I/\langle I\rangle \ll 1$. The time scale of IPS in the weak scattering regime is therefore of order $r_F/V_{sw} \sim$ few $\times 0.1$s where V_{sw} is the solar wind speed in the plane of the sky. In strong scattering $r_{diff} \ll r_F$. Inhomogeneities with a scale r_{diff} scatter radiation into a diffractive cone of angular width λ/r_{diff} which, at 1 AU illuminates a patch on the ground of size $L = \lambda D/r_{diff}$. A given point on the ground is therefore illuminated by many phase-independent patches which interfere with one another, producing a diffraction pattern on the ground (Figure 1, right panel). The solar wind motion carries the diffraction pattern over an observer at the solar wind speed, causing intensity variations on a time scale $r_{diff}/V_{sw} \sim$ few $\times$ 0.01s.

Two observers separated by a distance s, oriented parallel to the outflow, can independently measure the time series of scintillations and cross correlate them to deduce the solar wind speed at a given solar elongation and position angle. A range of baselines and orientations can be used to deduce the solar wind velocity, including non-radial velocities, projected onto the plane of the sky. Observations of many background sources can map out the solar wind velocity and acceleration in the inner heliosphere. The solar wind is in fact comprised of fast solar wind streams (~ 700 km/s) that emerge from magnetically open regions in the solar corona (coronal holes), and slow solar wind (~ 400 km/s). Hence, two different velocities may be present along a given line of sight. Klinglesmith (1997), Breen et al. (2006), and others have shown that IPS measurements made on sufficiently well-separated baselines (~ 500 – 1000 km) can be used to resolve and measure the speeds of both types of solar wind. In addition, since the solar wind carries Alfvén wave fluctuations these, too, can be constrained (Klinglesmith 1997). IPS is most useful as a diagnostic in the weak scattering regime, where the IPS amplitude $\sigma_I/\langle I\rangle \propto n^2\lambda^2 \propto \lambda^2/r^4$; in strong scattering, it saturates with $\sigma_I/\langle I\rangle \approx 1$. Observations in the weak scattering regime can be assured by selecting appropriate values for λ and/or r.

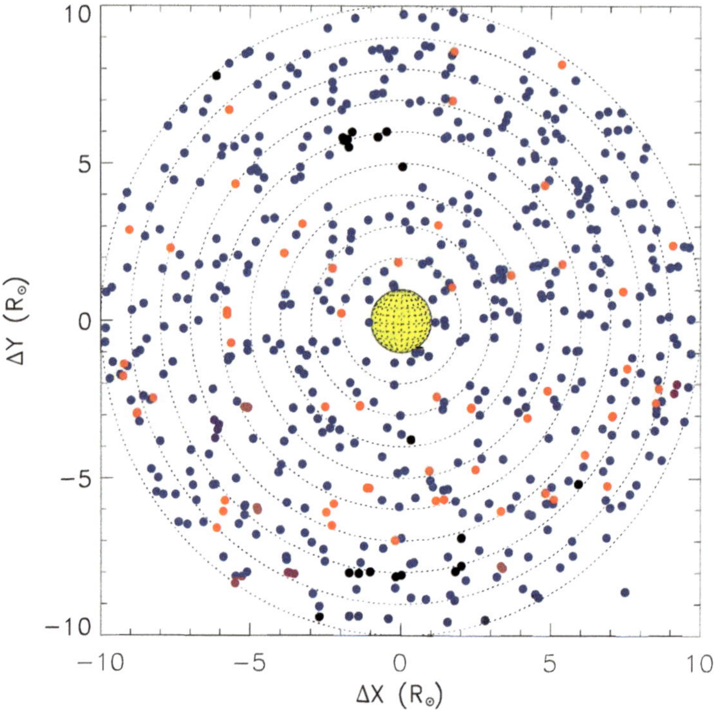

Figure 2. An illustrative plot of background sources that could be used as probes of the Sun's corona and solar wind. The blue points show daily positions (local noon) of *Jodrell Bank - VLA Astrometric Survey* 8.4 GHz sources (Patnaik et al. 1992; Browne et al. 1998; Wilkinson et al. 1998; Winn et al. 2003) that pass within 10 $R_\odot$ of the Sun during the course of one year. These are compact, flat-spectrum sources dominated by milliarcsec cores that have flux densities > 50 mJy. The red points show all known millisecond pulsars (MSPs) that likewise pass within 10 $R_\odot$ of the Sun (Manchester et al. 2005). Note that the bulk of the MSPs are near the galactic plane and are therefore accessible during a relatively short window of time each year.

A second type of scintillation, referred to as "refractive" scintillation, occurs in the strong scattering regime. It results from weak focusing and defocusing effects on the scale of the scatter-broadened source at 1 AU: $r_{ref} = D\theta_s = \lambda D/r_{diff} = r_F^2/r_{diff}$. Refractive scintillation occurs on a time scale $r_{ref}/V_{sw} \sim 10$ s. In the strong scattering regime we have $r_{diff} < r_F < r_{ref}$. Hence, refractive scintillation provides a second means of leveraging a scattering phenomenon to probe solar wind properties – in particular, those of $\Phi_n(\kappa)$ – on scales of ~ 1000s km. Specifically, Goodman & Narayan (1989) show that refractive scintillation yields a power law tail in mean square visibility $\langle|V(s)|^2\rangle$ on baselines $s > r_{diff}$ in the so-called "average" visibility regime for shallow spectra ($\alpha < 2$). This can be measured as part of any angular broadening observing program, allowing simultaneous constraints to be placed on $\Phi_n(\kappa)$ on scales of order r_{diff} and r_{ref} as long as the source is observed in the strong scattering regime (Bastian 1999).

5. Faraday Rotation

Faraday rotation (FR) is the phenomenon where the plane of polarization of a linearly polarized source – a spacecraft beacon or a naturally occurring background source – rotates as it traverses a magnetized plasma. The rotation $\Delta\chi$ is in the amount

$$\Delta\chi = \lambda^2 \left[\left(\frac{e^3}{2\pi m_e^2 c^4} \right) \int n(z)\mathbf{B}(z) \cdot dz \right] = \lambda^2 \text{ RM} \tag{1}$$

where the term in brackets is referred to as the *rotation measure* (RM). FR can be used to constrain both the coronal and solar wind magnetic field and density. The most straightforward experiment is to measure the gross properties of the magnetic field by making a differential measurement: the intrinsic position angle of the polarization vector compared with that after passage through the coronal plasma or the solar wind. Pätzold et al. (1987) observed the linearly polarized carrier signal of the Helios spacecraft during occultations for nearly a decade, allowing them to constrain the magnetic field and its time variation between 3 – 10 $R_\odot$. Ingleby et al. (2007) used the VLA to observe a number extragalactic radio sources with known polarization properties through the solar wind plasma at elongations ranging from approximately 5 – 10 $R_\odot$. In order to extract magnetic field constraints, however, independent observations of the solar wind density and/or careful modeling of the solar wind density are required. A second, but similar use of differential FR measurements is to observe the passage of a CME across a polarized background source, again using spacecraft beacons (Bird et al. 1985) or ground-based radio telescopes (Jensen & Russell 2008, Kooi et al. 2017). Ancillary observations or models are needed here too, to disentangle the number density from the magnetic field.

Another intriguing use of FR measurements is to constrain the spectrum of Alfvén waves in the solar wind turbulence, believed to play a role in heating and momentum deposition. Two approaches have been attempted: i) observations of temporal FR fluctuations (FRF); and, ii) FR depolarization as a result of turbulence introducing spatial FRF across the beam of a radio telescope. The first of these approaches was exploited by Efimov et al. (1993) and Andreev et al. (1997) who used FRF measurements of the Helios-2 beacon to measure FRF spectra, which they interpreted in terms of long-period Alfvén waves. Mancuso & Spangler (1999) used the VLA to search for FRF as evidence for long-period Alfvén wave, finding a marginal result of lower than expected magnitude. Spangler & Mancuso (2000) also used this approach to search for short-period Alfvén waves. However, no evidence for depolarization due to FRF was found although the sensitivity of the experiment was marginal.

Finally, the scientific potential of millisecond pulsars (MSPs) as a probe of the solar wind is strong. The advantage of using pulsars is that they provide both a polarized signal with which to measure the RM and also a precision pulse from which the group delay or, equivalently, the *dispersion measure* (DM) can be deduced:

$$\tau_G = \lambda^2 \left[\left(\frac{e^2}{2\pi m_e c^3} \right) \int n(z) dz \right] = \lambda^2 \text{ DM} \tag{2}$$

Together, the RM and DM can be used to impose strong constraints on the solar wind density and magnetic field. You et al. (2012) used the Parkes Pulsar Timing Array to observe several MSPs and were able to make simultaneous RM and DM measurements at elongations ranging from 6 – 10 $R_\odot$.

5.1. ngVLA Observations

The key capability offered by the ngVLA is global mapping of critical solar wind properties in two dimensions over the course of the solar activity cycle. This will provide critical benchmarks against which models of solar wind heating, acceleration, and turbulence must be measured. Here, necessary ngVLA capabilities are briefly summarized and the key science goals are reiterated under the assumption that ngVLA performance will be closely aligned with the reference instrument described by Selina et al. (2018).

The key capabilities that will allow the ngVLA to make unique contributions to solar wind studies are its wide and continuous frequency coverage, its sensitivity, and its fixed baseline coverage. Past experience has shown that it is unlikely that frequencies much larger than 20 GHz will be needed for solar wind studies. Hence, the emphasis will be on ngVLA reference Bands 1 – 3. The sensitivity of the ngVLA will be critical for mapping a variety of solar wind parameters in two dimensions. Such mapping depends on having adequate numbers of point-like background sources (or sources for which there are excellent models) to probe the foreground corona and solar wind. Figure 2 displays suitable sources that are currently accessible by the JVLA based on the JVAS survey of compact, flat spectrum sources >50 mJy (blue points) at 8.4 GHz during the course of one year. The sensitivity of the JVLA is such that a 10 min observation (Goodman & Narayan's "ensemble average" regime) yields a typical rms of 10-20 μJy in bands up to 20 GHz, more than sufficient to observe JVAS sources with high SNR. The ngVLA will be 5-10 times more sensitive than the JVLA at a given frequency and similar bandwidth and will therefore be able to use background sources of 10 mJy or lower as effective probes although these will need to be surveyed prior to use.

Also important is the greatly expanded baseline coverage provided by the ngVLA. The reference ngVLA provides 214 antenna with spacings up to $\sim$ 1000 km, nearly a factor of 30 longer than is currently possible with the JVLA. These will enable measurements of the spatial spectrum $\Phi_n(\kappa)$ over a much larger range of scales, including the inflection point in the spectrum at spatial scales of $\sim$ 100 km. The ngVLA will also provide unparalleled opportunity for IPS measurements. The reference instrument places 46 ngVLA antennas on baselines greater than the current footprint of the JVLA providing more than 500 baselines sampling scales >200 km. Hence, a long baselines can be found that are parallel to the solar wind vector for a source at essentially any elongation and position angle. It is important to note that the fixed antenna configuration allows observations at any time of the year. They will no longer be constrained by the VLA configuration cycle, allowing a continuous record of solar wind properties to be measured. While the reference antenna configuration includes baselines out to 1000 km the distribution is weighted more heavily toward the EW quadrants – the number of baselines exceeding 500 km in the NS quadrants is relatively small and further optimization of the array configuration therefore may be needed.

To address solar wind science objectives, the following ngVLA measurements are key:

- *Solar wind turbulence*: The angular size of scatter-broadened background sources depends on wavelength and elongation as $\sim \lambda^2/R^2$ where R is measured in units of $R_\odot$, and the angular resolution of the array varies as $\theta \propto \lambda$. Scaling from results reported by Anantharamaiah et al (1994) at $\lambda = 6$ cm for R=1.7 – 16 $R_\odot$, one finds that the ngVLA will be able to probe angular broadening out to elongations R$\sim$ 60 $R_\odot$. Hence, unlike the VLA/JVLA, the ngVLA will allow observations over a greater range of elongations and over a much greater range of s (and, hence, κ), providing measurements of the turbulence level, the evolution of $\Phi_n(\kappa)$ from a Kolmogorov-like spectrum to a flatter-than-Kolmogorov spectrum, its degree of anisotropy, the inner scale l_{in} - all as functions of solar elongation and position angle. Particularly important to wave-heating theories will be the use of the ngVLA as probe of solar wind properties as it transits the Alfvén radius, at $\sim 10 - 20\ R_\odot$, where the solar wind speed becomes super-Alfvénic . Models based on Alfvénic turbulence (Cranmer & van Ballegooijen 2005) suggest that both velocity and density fluctuations reach a maximum near the Alfvén surface. IPS observations through the transsonic region (e.g., Lotova 1988) also suggest enhanced scattering there.

- *Solar wind velocity and acceleration*: IPS observations offer the means of mapping solar wind velocities throughout the inner heliosphere. Unlike previous measurements for which point measurements were occasionally made of solar wind speed, the ngVLA will enable two dimension mapping observations of the solar wind speed and acceleration in the inner heliosphere for a given phase in the solar cycle, critical for understanding the relationship between coronal heating and solar wind heating and acceleration. For measurements of IPS, total power measurements will be needed at each antenna at a sampling rate of 100 Hz. It will be desirable to phase up proximate antennas and to cross-correlate antenna groups rather than individual antennas as a means of improving the sensitivity of the IPS measurements.

- *Solar wind magnetic field*: Since solar wind inhomogeneities are drawn out along the local magnetic field they are highly anisotropic. This results in anisotropic scattering which, in turn, enables allows observers to determine the mean line-of-sight orientation of the magnetic field in the plane of the sky toward the source in question. By observing many sources, a map of the the magnetic field orientation in the inner heliosphere can be constructed. When MSPs are available for observation, angular broadening measurement plus RM and DM measurements will be available to constrain both the magnetic field vector and magnitude in the inner heliosphere. For FR/FRF measurements, polarimetric measurements are needed over wide bandwidths, with the lower two bands favored (1.2 – 3.5 GHz and 3.5 – 12.3 GHz). The excellent sensitivity and the anticipated fidelity of polarimetric imaging will allow far deeper measurements to be made of FRF and Faraday depolarization.

- *Space weather*: All of the above phenomena may be used to probe space weather drivers such as co-rotating interaction regions (CIRs) and CMEs. CIRs are due to long-lived fast solar wind stream that are experienced as recurrent structures at Earth as the Sun rotates with its Carrington period of 27.3 days. CMEs involve the eruption of a magnetic flux rope, driving up to $\sim 10^{16}$ gm of material at speeds exceeding 2000 km/s into the interplanetary medium. Both CIRs and CMEs can interact with the Earth's magnetosphere, producing geomagnetic storms. Large CMEs drive shocks that accelerate solar energetic particles to high energies that are detrimental to both manned and unmanned spacecraft. Studies of CIRs and CMEs in the inner heliosphere will provide new insights into so-called "research to operations" aspects of space weather forecasting and "now-casting".

With few exceptions, the reference ngVLA concept is already an excellent match to requirements for solar wind and space weather studies. Support of IPS observations require care in avoiding self-generated RFI signals and system time constants that could affect cross-correlations between antennas or subarrays defined by groups of antennas. Requirements for supporting these studies otherwise require careful planning, pilot studies, and surveys to supplement and characterize JVAS and other background sources, and to identify larger numbers of polarized sources for FR/FRF observations. While a given observation may not require significant telescope time, many observations of background sources will be needed to support comprehensive mapping of solar wind properties in the inner heliosphere. The most fruitful work will likely require systematic studies through large proposals.

Acknowledgments. I thank Dustin Madison for providing a curated list of MSPs drawn from the *ATNF Pulsar Catalogue* and both Dustin Madison and Scott Ransom for discussion about the use of pulsars for RM and DM measurements.

References

Anantharamaiah, K. R., Gothoskar, P., & Cornwell, T. J. 1994, Journal of Astrophysics and Astronomy, 15, 387

Andreev, V. E., Bird, M. K., Efimov, A. I., Chashei, I. V., & Samoznaev, L. N. 1997, Astronomy Reports, 41, 227

Armstrong, J. W., Coles, W. A., Kojima, M., & Rickett, B. J. 1990, ApJ, 358, 685

Bastian, T. S. 1999, Magnetic Fields and Solar Processes, 448, 1131

Bird, M. K., Volland, H., Howard, R. A., et al. 1985, Solar Phys., 98, 341

Breen, A. R., Fallows, R. A., Bisi, M. M., et al. 2006, Journal of Geophysical Research (Space Physics), 111, A08104

Browne, I. W. A., Wilkinson, P. N., Patnaik, A. R., & Wrobel, J. M. 1998, MNRAS, 293, 257

Bruno, R., & Carbone, V. 2016, Turbulence in the Solar Wind, LNP 928, Springer, Switzerland, 267 pp.

Chandran, B. D. G., Quataert, E., Howes, G. G., Xia, Q., & Pongkitiwanichakul, P. 2009, ApJ, 707, 1668

Coles, W. A., Frehlich, R. G., Rickett, B. J., & Codona, J. L. 1987, ApJ, 315, 666

Coles, W. A., & Harmon, J. K. 1989, ApJ, 337, 1023

Coles, W. A., Harmon, J. K., & Martin, C. L. 1991, J. Geophys. Res., 96, 1745

Cranmer, S. R., & van Ballegooijen, A. A. 2005, ApJS, 156, 265

Cranmer, S. R., van Ballegooijen, A. A., & Woolsey, L. N. 2013, ApJ, 767, 125

Dmitruk, P., Matthaeus, W. H., Milano, L. J., et al. 2002, ApJ, 575, 571

Efimov, A. I., Chashej, I. V., Shishov, V. I., & Bird, M. K. 1993, Astronomy Letters, 19, 57

Erickson, W. C. 1964, ApJ, 139, 1290

Goodman, J., & Narayan, R. 1989, MNRAS, 238, 995

Grall, R. R., Coles, W. A., Spangler, S. R., Sakurai, T., & Harmon, J. K. 1997, J. Geophys. Res., 102, 263

Hollweg, J. V., & Harrington, J. V. 1968, J. Geophys. Res., 73, 7221

Hollweg, J. V. 1999, J. Geophys. Res., 104, 14811

Ingleby, L. D., Spangler, S. R., & Whiting, C. A. 2007, ApJ, 668, 520

Jensen, E. A., & Russell, C. T. 2008, Geophys. Res. Lett., 35, L02103

Klinglesmith, M. 1997, Ph.D. Thesis, 7586

Kooi, J. E., Fischer, P. D., Buffo, J. J., & Spangler, S. R. 2017, Solar Phys., 292, 56

Lotova, N. A. 1988, Solar Phys., 117, 399

Manchester, R. N., Hobbs, G. B., Teoh, A., & Hobbs, M. 2005, AJ, 129, 1993

Mancuso, S., & Spangler, S. R. 1999, ApJ, 525, 195

Parker, E. N. 1958, ApJ, 128, 664

Patnaik, A. R., Browne, I. W. A., Wilkinson, P. N., & Wrobel, J. M. 1992, MNRAS, 254, 655

Pätzold, M., Bird, M. K., Volland, H., et al. 1987, Solar Phys., 109, 91

Rickett, B. J. 1977, ARA&A, 15, 479

Salem, C. S., Howes, G. G., Sundkvist, D., et al. 2012, ApJ, 745, L9
Selina, R., J., Murphy, E. J., McKinnon, M., Beasley, A., Butler, B., et al. 2018, Science with a Next-Generation Very Large Array, ASP Monograph 7, ed. E. J. Murphy (San Francisco, CA: ASP), p. 15
Spangler, S. R., & Mancuso, S. 2000, ApJ, 530, 491
Wilkinson, P. N., Browne, I. W. A., Patnaik, A. R., Wrobel, J. M., & Sorathia, B. 1998, MNRAS, 300, 790
Winn, J. N., Patnaik, A. R., & Wrobel, J. M. 2003, ApJS, 145, 83

Science with a Next-Generation Very Large Array
ASP Conference Series, Monograph 7
Eric J. Murphy, ed.

Radio Observations of Solar Flares

D. E. Gary,[1] T. S. Bastian,[2] B. Chen,[1] G. D. Fleishman,[1] and L. Glesener[3]

[1]*New Jersey Institute of Technology, Newark, NJ, USA;*
dgary@njit.edu, bin.chen@njit.edu, gfleishm@njit.edu

[2]*National Radio Astronomy Observatory, Charlottesville, VA, USA;*
tbastian@nrao.edu

[3]*University of Minnesota, Minneapolis, MN, USA;* glesener@umn.edu

Abstract. Solar flares are due to the catastrophic release of magnetic energy in the Sun's corona, resulting in plasma heating, mass motions, particle acceleration, and radiation emitted from radio to γ-ray wavelengths. They are associated with global coronal eruptions of plasma into the interplanetary medium—coronal mass ejections—that can result in a variety of "space weather" phenomena. Flares release energy over a vast range of energies, from $\sim 10^{23}$ ergs (nanoflares) to more than 10^{32} ergs. Solar flares are a phenomenon of general astrophysical interest, allowing detailed study of magnetic energy release, eruptive processes, shock formation and propagation, particle acceleration and transport, and radiative processes. Observations at radio wavelengths offer unique diagnostics of the physics of flares. To fully exploit these diagnostics requires the means of performing time-resolved imaging spectropolarimetry. Recent observations with the Jansky Very Large Array (JVLA) and the Expanded Owens Valley Solar Array (EOVSA), supported by extensive development in forward modeling, have demonstrated the power of the approach. The ngVLA has the potential to bring our understanding of flare processes to a new level through its combination of high spatial resolution, broad frequency range, and imaging dynamic range—especially when used in concert with multi-wavelength observations and data at hard X-ray energies.

1. Introduction

Solar flares are produced by the catastrophic release of magnetic energy in the Sun's corona. The sporadic explosive release of excess (i.e., non-potential) magnetic energy is a ubiquitous and fundamental property of magnetized astrophysical plasmas, which is generally accompanied by shocks, turbulence, and acceleration of charged particles. Such energy releases may be driven by a supernova explosion or other gamma-ray burst progenitor. Acceleration of charged particles in solar flares and their transport share common mechanisms with the acceleration and propagation of the cosmic rays produced in such powerful explosions. Other cases may be powered by gas accretion onto a massive central object such as a black hole. In particular, this includes Galactic microquasars/X-ray binaries produced by accretion from a normal star onto a compact stellar mass object (black hole or neutron star) or Active Galactic Nuclei driven by a similar yet even more powerful phenomenon. The accretion forms a disk whose interaction with the central object produces, via unknown mechanisms, a single or pair of oppositely directed highly collimated relativistic jets. These jets produce most of the observed radio emission and contribute essentially to emission at other wavelength ranges, implying that the jets are the main or substantial source of the accelerated particles. Potentially similar jet phenomena are seen and can be studied in great detail in the solar case (e.g. Glesener & Fleishman 2018).

Solar flares result in plasma heating, the acceleration of electrons and ions to high energies, mass motions, shocks, and electromagnetic radiation across the entire spectrum, from radio to γ-ray wavelengths. Flares are often associated with global eruptions, coronal mass ejections (CMEs), that can eject $\gtrsim 10^{15}$ gm of material at speeds up to 2000 km s^{-1}, as well as smaller-scale collimated jets. Jets, flares, and CMEs can drive interplanetary disturbances and produce electromagnetic and particle radiation that can have a direct and adverse effects on Earth and near-Earth environment ("space weather"). Hence, while solar flares are of considerable astrophysical interest as an energetic phenomenon in their own right, flares and CMEs are of practical interest because of their potential for affecting modern technological infrastructure and astronaut safety. In this chapter, we touch on just two outstanding problems—magnetic energy release in flares and particle acceleration—and discuss the potential role of the ngVLA for addressing them. The range of solar and space physics science topics that can be addressed by the ngVLA is much broader, however (cf., Bastian et al. 2018; Fleishman et al. 2018, this volume p. 87 and 125).

2. Background and Motivation

Solar flares are efficient particle accelerators, converting several tenths of the available free energy stored in stressed magnetic fields to high-energy particles. Electrons can be promptly accelerated to energies of 10s of MeV and ions to energies ~1 GeV/nuc (Miller et al. 1997). Energetic particles on the Sun emit radiation at radio, X-ray, γ-ray wavelengths. Extreme ultraviolet (EUV) and soft X-ray (SXR) emission mainly arise from thermal plasma heated to ~ 20 MK. Hard X-ray (HXR) emission typically arises from nonthermal bremsstrahlung from electrons with energies of ~10 keV to a few × 100 keV interacting with dense chromospheric material or coronal plasma. Continuum γ-ray emission ($E >$ few × 100 keV) arises from electron-proton and electron-electron bremsstrahlung and pion decay, and gamma-ray lines are caused by electron-positron annihilation, neutron capture, and nuclear de-excitation. Thermal plasma produces radio waves via bremsstrahlung and gyroresonance emission. Suprathermal electrons (few keV to a few 10s of keV) may produce intense coherent radio emission and energetic electrons (10s of keV to many MeV) emit nonthermal gyrosynchrotron (GS) radio radiation. Together, HXR, γ-ray, and radio observations provide a powerful and complementary suite of perspectives from which to study magnetic energy release and particle acceleration, addressing questions such as:

- Where does magnetic energy release occur?
- Where are electrons accelerated? By what mechanisms?
- What is the electron energy distribution function and what is its spatiotemporal evolution?
- What are the dominant transport effects for the energetic electrons?

With their sensitivity to non-equilibrium particle distributions and magnetic fields, radio observations offer a number of unique diagnostics of flare physics. A revolutionary advance in radio instrumentation is the ability to map the Sun at radio wavelengths, with high spectral and temporal resolution over a broad range of frequencies. The combination of spatially resolved imaging and high spectral resolution is called imaging spectroscopy or, when circular polarization is added, imaging spectropolarimetry. When coupled with high time resolution, *dynamic* imaging spectropolarimetry becomes possible, bringing with it the ability to image every frequency-time bin in the broadband dynamic spectrum.

Broadband imaging spectroscopy with high temporal resolution (0.01–1 s) is key for resolving the dynamic behavior of solar radio bursts. To fully cover the range of phenomena of interest to solar physicists, a very broad bandwidth is needed. For example, the main microwave component of solar flares is the inherently broadband mechanism of gyrosynchrotron (GS) radiation, and the spectrum is further broadened by non-uniformities in magnetic field strength, density, and particle energy in the flaring region. The typical GS spectrum peaks in the range 5–10 GHz (Guidice & Castelli 1975), but the peak can move to much higher frequencies (> 35 GHz in some cases; see, e.g., Kaufmann et al. 1986; Raulin et al. 2004; Lysenko et al. 2018), or to lower frequencies (< 1 GHz; e.g., Lee et al. 1994; Fleishman et al. 2016), reflecting the wide ranges of magnetic field strength (few G to few kG) and particle energy (few keV to few MeV) that are possible. In addition, coherent bursts occur in a large fraction of solar flares, and although a discrete burst may be intrinsically narrow band, multitudes of features can appear simultaneously across a wide range of frequency, or rapidly drift over many octaves of frequency.

Two existing radio instruments are pioneering the exploitation of these new capabilities: the *Jansky Very Large Array* (JVLA) and the *Expanded Owens Valley Solar Array* (EOVSA). We draw on examples of observations from these instruments here. The next generation Very Large Array (ngVLA), if designed to take solar observing requirements into account, can expand greatly on these capabilities by providing improved frequency coverage and dynamic range. A cursory discussion of solar science with the Square Kilometre Array (SKA) was done by Butler et al. (2004), but in its current design state its importance for solar observations has not been established.

3. Probing Magnetic Energy Release with Coherent Radio Bursts

Coherent radio emission occurs at resonant frequencies of the coronal medium via a coherent radiation process: the electron plasma frequency ν_{pe}, the electron gyrofrequency Ω_{De}, or combinations and harmonics thereof—e.g., the upper hybrid frequency. Coherent radio bursts are frequently observed in solar flares of all sizes. They can dominate the radio spectrum at decimeter and meter wavelengths (Bastian et al. 1998; Nita et al. 2004), and are sometimes seen at MW frequencies as high as ~ 9 GHz.

There are two main radiation mechanisms for solar coherent radio emission at decimetric wavelengths (dm-λ hereafter). One of them is plasma radiation, which is due to the nonlinear conversion of plasma Langmuir waves to transverse radio waves. It occurs at the fundamental and/or second harmonic of the electron plasma frequency $\nu_{pe} = \sqrt{n_e e^2/\pi m_e} \sim n_{10}^{1/2}$ GHz, providing a direct measurement of the local electron density in the radio source n_{10} (units 10^{10} cm^{-3}). The other is electron cyclotron maser emission (ECME), which arises around the lowest harmonics ($s = 1, 2, 3$) of the elec-

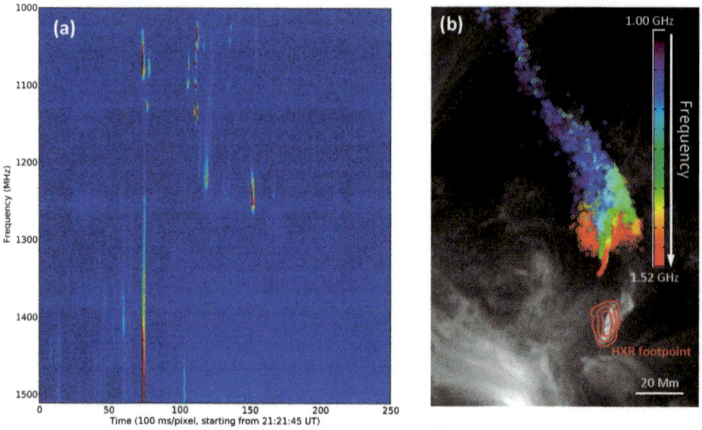

Figure 1. (a) Type III radio bursts observed by the JVLA associated with a solar jet. The bursts are visible in the radio dynamic spectrum (radio intensity as a function of time and frequency, colored from blue to red in increasing intensity) as bright, nearly vertical short-lived features. (b) Emission centroids of dm-λ bursts shown in (a), colored from blue to red in increasing frequency. The ensemble of the centroids show electron beam trajectories in projection. Red contours are HXR emission due to downward-moving nonthermal electrons colliding with the dense solar chromosphere. The background EUV image from SDO/AIA shows coronal loops and the jet event. (Adapted from Chen et al. 2013.)

tron gyrofrequency yet the source has to be reasonably tenuous to let the radiation to escape. Bursts due to ECME can provide valuable information about the magnetic field in the corona. Dm-λ coherent radio bursts correspond to plasma density and magnetic field characteristic of the low solar corona, where the primary energy release and particle acceleration in solar flares are presumed to occur (Bastian et al. 1998; White et al. 2011). Therefore, observations of dm-λ coherent bursts provide the means of tracing the presence and motion of electrons accelerated in and around the energy release site. Most previous studies have depended on single dish radio dynamic spectral observations of dm-λ bursts (Melrose 2017, and references therein). In the absence of spatial information, a definitive association of these coherent radio bursts to the larger magnetic environment, needed to establish their role in particle acceleration, is lacking. Dynamic imaging spectroscopy surmounts this limitation.

An excellent example is dm-λ type III radio bursts (Figure 1). These bursts are the signature of flare-accelerated, fast electron beams (~0.1–0.5c) propagating along newly-reconnected magnetic field lines. An electron beam propagating upward in the corona encounters plasma with a decreasing density, thus generating radio waves at progressively lower frequencies. In a dynamic spectrum, a type III dm burst is seen as a fast-drift feature with a negative frequency slope ($d\nu/dt < 0$). Similarly, a beam propagating downward will result in a fast-drift burst in the dynamic spectrum with a positive slope ($d\nu/dt > 0$). As an electron beam propagates along newly reconnected magnetic field lines, its trajectory traces the instantaneous magnetic topology of the on-going reconnection, which is usually invisible at other wavelengths unless the loops have been pre-heated or populated with dense plasma. Interestingly, sometimes both

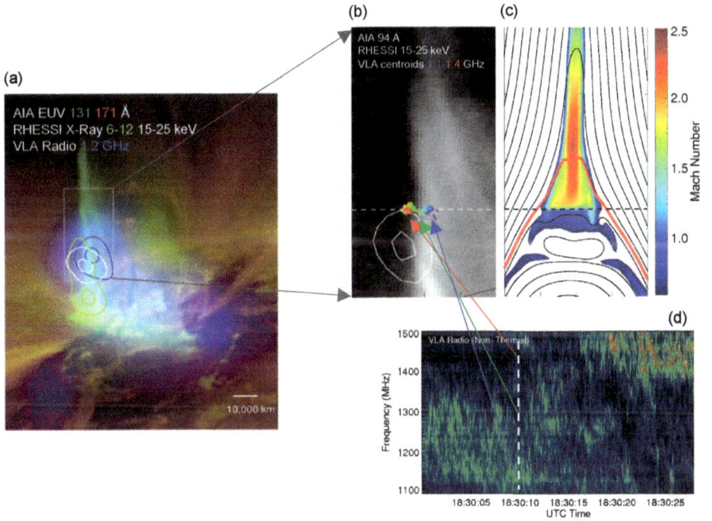

Figure 2. Radio observation interpreted as a termination shock in an eruptive solar flare. (a) Composite image of an eruptive solar flare in radio (JVLA; blue), EUV (SDO/AIA, green and red), and HXR [*Reuven Ramaty High Energy Solar Spectroscopic Imager* (RHESSI), white contours]. (b) A dynamic surface-like feature revealed by the spatial distribution of radio source centroids in different frequencies. It is interpreted as the front of a fast-mode termination shock formed at the top of flaring arcades seen in EUV and HXR. Each colored dot corresponds to the centroid location of a spike burst in the radio dynamic spectrum at a given time and frequency (panel d). This shock is well-reproduced by magnetohydrodynamics simulations (panel c) as a sharp transition layer through which super-magnetosonic reconnection outflows become sub-magnetosonic when they encounter the underlying, obscuring dense loops. (Adapted from Chen et al. 2015.)

the positive- and negative-drift type III bursts are seen simultaneously. These burst pairs are typically interpreted as oppositely-directed fast electron beams that emerge from a common region, presumably the energy release site. These radio bursts have the potential to pinpoint the particle acceleration site, diagnose its physical properties (Aschwanden & Benz 1997; Reid & Ratcliffe 2014), and allow the wider magnetic field connectivity to be mapped.

Dynamic structures such as shocks, magnetic islands, and turbulence are also believed to exist in or around the magnetic energy release region. Some, if not all, are thought to play an important role in the the particle acceleration process (Forbes 1986; Miller et al. 1996; Drake et al. 2006). Certain coherent radio burst types, such as dm-λ spike bursts (Figure 2d) and drifting pulsating structures, have been suggested to arise from interactions between accelerated electrons and dynamic structures in or around the energy release region (Benz 1985; Fleishman & Mel'nikov 1998; Kliem et al. 2000; Rozhansky et al. 2008). The location and dynamics of such radio bursts, obtained by radio dynamic imaging spectroscopy, provide unique access to these dynamic structures. This is demonstrated in a recent study by Chen et al. (2015) based on JVLA observations of an eruptive solar flare, in which dynamic spectral imaging of dm-λ spike bursts was utilized to map the radio source as a function of frequency and time

(Figure 2b). The spatial distribution of the burst centroids outlines a dynamic surface at the top of the flare arcades, interpreted by Chen et al. (2015) as the front of a fast-mode termination shock (Figure 2c), which likely contributed to accelerating electrons to nonthermal energies. Given that individual spikes are short-lived (ms to tens of ms) and narrowband (at the order of 1%), the spatiotemporal distribution of some important physical parameters in the source region, e.g., plasma density or magnetic field as well as their level of fluctuations (possibly due to turbulence), may be derived or inferred from the data with high accuracy, provided that the responsible emission mechanism is identified.

The ngVLA will be capable of imaging these coherent radio bursts over the 1.2-3.5 GHz band with high spectral resolution (0.1–1%) and time resolution (10 ms), with excellent dynamic range. One of the limitations of the current VLA is its relatively low dynamic range, which in the presence of bright coherent emissions renders the weaker incoherent emission undetectable. Combined with simultaneous observations in the higher frequency bands (cf., Fleishman et al. 2003), the ngVLA will be able to link coherent dm-λ emissions to incoherent GS emission from much higher energy electrons, for which quantitative diagnostics are available for parameters in the flaring region, as described in the following section.

4. Particle Acceleration and Transport

Radio emission processes (e.g., Dulk 1985; Bastian et al. 1998; Gary & Hurford 2004; Fleishman & Toptygin 2013) are known to depend strongly on the magnetic field strength in the source region due to two effects: (1) the Lorentz force causes the emitting electrons to spiral in the magnetic field to produce gyroemission, which takes the form of gyroresonance (GR) emission for thermal electrons in the hot corona of active regions, and nonthermal GS emission for mildly-relativistic electrons in flaring magnetic loops; (2) the plasma itself is birefringent, with differing index of refraction (e.g., Fleishman & Toptygin 2013) for the ordinary (o) and extraordinary (x) magnetoionic modes, which leads to a magnetic-field-dependent polarization of each type of emission, including free-free (FF) emission (Gelfreikh 2004). All three emission mechanisms, GR, GS, and FF, have been exploited in countless studies of radio emission from the Sun to constrain properties of the thermal plasma, the distribution function of energetic electrons, and the magnetic field strength in the source. However, such estimates based on either spatial or spectral information alone generally suffer from a large number of ambiguities that require varying degrees of assumptions, or else require special conditions where the assumptions and ambiguities can be minimized. These ambiguities can be largely eliminated when dynamic imaging spectropolarimetry is available.

Nonthermal GS emission is of particular interest. The sensitivity of its emissivity to the magnetic field means that it is visible wherever nonthermal electrons are present and have access to coronal magnetic loops. Figure 3 shows an illustrative example. This model representation of a snapshot (1-s) data cube has both spatial and spectral information in both senses of circular polarization, which provides spatially-resolved, polarized brightness-temperature spectra at each point in the field of view.

The model is obtained by placing a realistic distribution of nonthermal electrons and background thermal plasma in a magnetic loop (derived from an extrapolation of photospheric magnetic field measurements) and calculating the resulting emission including both GS and FF contributions. The calculation of the microwave emission is

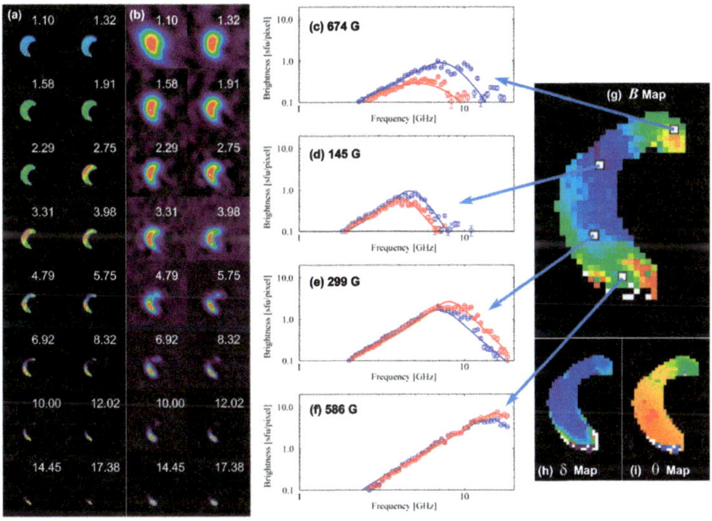

Figure 3. (a) Model images of a flaring loop at 16 frequencies in the range 1-18 GHz. (b) Reconstructed images after sampling with the EOVSA snapshot array coverage. Note the transformation from a uniform source at 1.2 GHz to a loop-top dominated source at 2.75 GHz to a footpoint dominated source above 12 GHz. (c-f) Polarized brightness (intensity) spectra at 64 frequencies from the reconstructed images (points with error bars), overlaid with red (RCP) and blue (LCP) 6-parameter fits. (g) Pixel-by-pixel magnetic field map derived from spectral fits, with the four specific locations shown that correspond to the plots in c-f. (h-i) Maps of power-law index (constant $\delta = 5$ was used), and line-of-sight angle to B ($64° < \theta < 123°$ along loop), derived from spectral fits. (Adapted from Gary et al. 2013.)

performed using the best currently available Fast Gyrosynchrotron Codes (Fleishman & Kuznetsov 2010), which include both accurate treatment of the GS and FF processes in each voxel along the given line of sight and the solution of the radiation transfer along the line of sight, taking into account optical-depth and Razin suppression effects as well as frequency-dependent mode coupling. These model images are then sampled with EOVSA's array configuration, with suitable thermal noise added, to obtain the images in Figure 3b. By stacking the multi-frequency images into a data cube, spatially-resolved spectra at every point in the flaring region are obtained in right and left circular polarization (RCP and LCP), shown with red and blue symbols, respectively, in Figure 3c-f. Multi-parameter fits to the spectra (solid red and blue curves) are performed as described by Fleishman et al. (2009). One of the parameters, magnetic field strength B, is indicated in each panel.

When such fits are done at every pixel, parameter maps such as those shown in panels Figure 3g-i can be constructed. Figure 3g shows a distribution of the magnetic field strength in the loop and Figure 3i shows the distribution of the magnetic field angle to the line of sight. Figure 3h shows the spectral index δ of the power-law electron distribution used for the simulation. In the more general case, the *spatiotemporal* evolution of the electron distribution function will be inferred from the brightness temperature spectra. These measurements will place *new* and powerful constraints on the

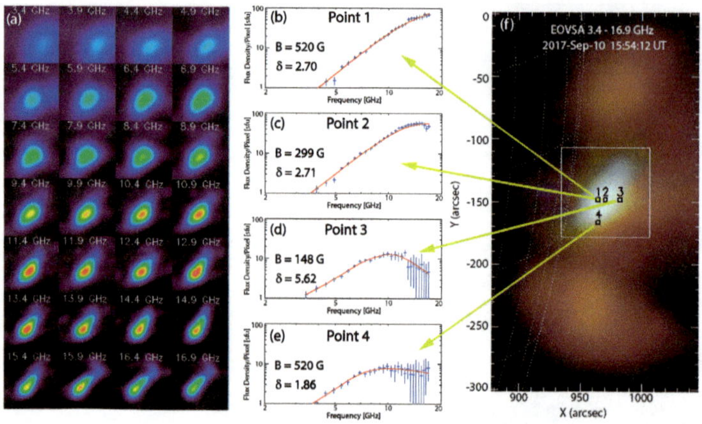

Figure 4. A figure analogous to Figure 3, but this time with real data, from the event of 2017-Sep-10. (a) Individual images at 28 frequencies, from the location of the white box in the overview image in panel f. (b-e) Measured flux-density spectra (points with error bars) in single pixels of the images in panel a, corresponding to locations 1-4 marked in panel f, and corresponding multi-parameter fits (red lines). (f) A "true-color" representation of the EOVSA data cube, combining images at the 28 frequencies shown in panel a. (Adapted from Gary et al. 2018.)

mechanisms responsible for electron acceleration (shocks, turbulence, electric fields) and their subsequent transport (trapping, scattering, precipitation).

Figure 3 is a illustrative simulation. Some recent EOVSA observations of the X8.3 solar flare of 2017-Sep-10 (see Gary et al. 2018) are shown in Figure 4. The partially occulted flare, a classic flux-rope eruption seen in profile above the solar limb, was well observed by other instruments. Microwave imaging spectroscopy provides an entirely new addition to the study of such events, by providing the spatial distribution of otherwise invisible high-energy electrons as well as direct measurement of plasma parameters, including magnetic field, as a function of position and time in the radio-emitting volume.

In this example we focus on a single time, around 15:54:10 UT, shown as a "true-color" image in Figure 4f, where images at 28 frequencies are apportioned different red-green-blue weights according to their frequency. This time is during the initial, rapid rise of a flux rope seen in EUV images of the event. EOVSA shows strong radio emission below the inferred X-type reconnection point, with radio centroids located at steadily lower heights as frequency increases, shown by the shading of the central source from red to bluish-white. The EOVSA source shape also changes with frequency, from a cusp-shaped source at mid-frequencies (e.g., the 7.9 GHz image in Figure 4a), evolving toward a more loop-like shape at higher frequencies (15.9 GHz image). At frequencies below about 5 GHz, two outlier sources appear (Figure 4f) that are associated with the legs of a larger loop.

These position and morphology dependences on frequency correspond to position-dependent spectral shapes, in analogy to our simulation described above. This is demonstrated in Figure 4b-e, which are spectra measured at locations 1-4 marked in Figure 4f.

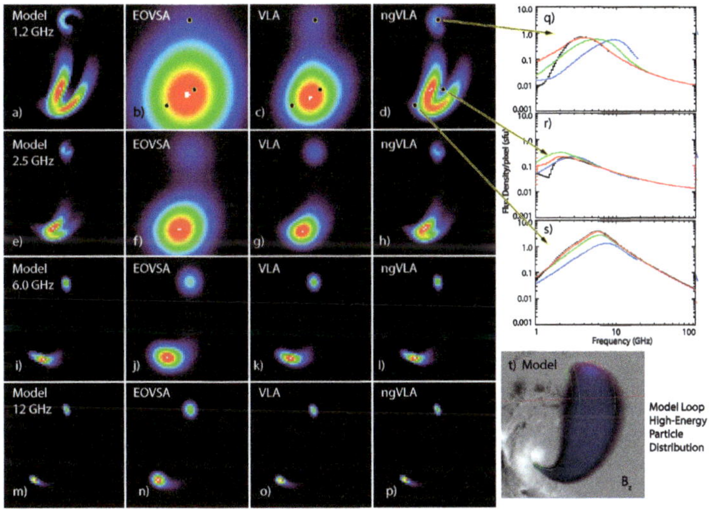

Figure 5. A simulation of the flaring loop associated with the 2017 September 04 flare, as observed by EOVSA, and its appearance as observed by the VLA and ngVLA. The left column (panels a, e, i, m) show the microwave emission calculated from the model. The next three columns shown the same emission as would be seen with the resolution of EOVSA, VLA, and ngVLA, respectively. The plots in the right column (panels q, r, s) show spectra from three spatial locations shown by the arrows, derived from the simulation for each of the four cases, model (black), EOVSA (blue), VLA (green) and ngVLA (red). Panel (t) is a rendering of the nonthermal electron distribution from the model loop, superposed on a line-of-sight magnetogram. No coronal scattering is included.

The points 1-3 located at different heights along the axis, and point 4 off to one side. The magnetic field strength inferred for each location is again indicated.

The above example thus demonstrates that, with the availability of spatially resolved microwave spectra, the diagnostic information inherent in GS emission can be exploited to provide measurements of the dynamically changing parameters in the flaring region. The ngVLA can improve on current capabilities in a few fundamental aspects. First, the ngVLA's combination of long baselines and excellent filling of smaller spacings will provide the uv coverage needed to extend imaging of solar flares to much higher spatial resolution than is currently possible. The current VLA has configurations with long spacings, but the lack of shorter spacings makes the array essentially useless for solar work beyond the C configuration (with 3 km maximum spacing). This is illustrated in Figure 5, from a simulation of the flaring loop associated with the 2017-Sep-04 flare, as observed by EOVSA. The model used to calculate the radio emission is shown in panel (t) at the lower right, where the blue shading represents the volume distribution of high-energy electrons. The first column (panels a, e, i, m) show the microwave emission at 1.2, 2.5, 6 and 12 GHz, respectively, as calculated from the model. Note the bifurcation of the loop at low frequencies, where the loop edges are bright, but not its center. This bifurcation is not captured in either EOVSA (1.2 km maximum spacing, nor the VLA C configuration (3 km maximum spacing). For the ngVLA images, spacings up to 12 km were assumed, although even longer spacings may be useful in some cases. When spectra at various points in the four data cubes are compared (panels q, r, and s), it is clear that higher resolution results in a more faithful measurement of the flux density/pixel, and features like the bifurcation of the loop become apparent. It should be noted that for the simulation in Figure 5 we have ignored scattering in the corona, which can be a limiting factor in the smallest angular size of flaring sources, as discussed further in § 6. An important strength of the ngVLA design is that (u, v)-limits can be expanded on a case-by-case basis as permitted by the inherent limits of the spatial structure in the observed source.

Second, by extending the spectral coverage towards much higher frequencies (compared with EOVSA), ngVLA will allow probing emission produced by relativistic charged particles with energies up to a few to a few dozens of MeV. This will permit quantification of spectral breaks and cutoffs and, thus, measure the highest energies to which nonthermal electrons can be accelerated. Third, having the high-frequency observations will open a new window of probing acceleration of ions. Indeed, the accelerated ions produce secondary relativistic positrons, which generate GS emission with the same intensity as electrons (provided those two populations have the same energy and angular distributions), but with the opposite circular polarization (Fleishman et al. 2013). Thus, given the magnetic field direction at the GS source, we can distinguish the positron contribution from the electron one, and in this way obtain additional constraints on acceleration of the associated ions in the flares. Finally, the greatly improved dynamic range of the images will not only provide higher-quality spectra for fitting, but will also reveal highly important, but generally much weaker, primary acceleration sources in the presence of the strong, trapped component that dominates after the first minutes of a flare.

5. Synergies with Hard X-rays and Other Data

In addition to the radio emissions already mentioned, highly energetic electrons also emit bremsstrahlung as they undergo collisions with ambient ions; this radiation shows up prominently in the HXR range (above a few keV). Bremsstrahlung is highly weighted by plasma density and so appears most strongly at the dense chromospheric footpoints of flaring loops, but coronal emission at or near acceleration sites can also be observed. Radio and HXR observations are the *only* direct diagnostics of accelerated electrons in the solar atmosphere. Since they sometimes probe different energy regimes and are sensitive to different ambient parameters (e.g., magnetic field or plasma density), they naturally complement each other in tracing and diagnosing the flare-accelerated electrons (e.g., Fleishman et al. 2016; White et al. 2011, and references therein).

Figure 2 has already shown an example of combining observations of coherent radio bursts and HXR data to trace accelerated electrons at or near the flare energy release site. Another example is given by Bain et al. (2012), who used the Nancay Radioheliograph to image a type II radio burst (another type of coherent radio emission) associated with a CME shock, while HXR images and spectra of the CME illuminated the accelerated electrons trapped within the ejection (Glesener et al. 2013). More recently, Musset et al. (2018) used measurements from RHESSI (Lin et al. 2002) to investigate diffusive trapping of electrons from keV to tens of keV scale in flare loops, and compared these results with those found for more energetic (hundreds of keV), gyrosynchrotron-emitting electrons. The use of HXRs and gyrosynchrotron emission together with 3D modeling was found to be a powerful diagnostic of the flare-accelerated electrons

(Fleishman et al. 2016; Kuroda et al. 2018). By studying the radio and HXR emissions together, a relatively full picture of the accelerated electrons was uncovered.

A variety of new HXR instrumentation is on the horizon or, in some cases, online already. The *Nuclear Spectroscopic Telescope Array* (Harrison et al. 2013), an astrophysics mission in operation since 2012, is a direct-imaging HXR telescope that performs 2-4 solar observations per year, often in coordination with other solar observing campaigns (Grefenstette et al. 2016). The solar-dedicated Spectrometer/Telescope for Imaging X-rays (STIX) instrument will launch in the coming years as part of the *Solar Orbiter (SO)* spacecraft (Krucker et al. 2016). STIX provides imaging and spectroscopy of HXRs from 4 to 150 keV with 1 keV spectral resolution. In addition, the *Focusing Optics X-ray Solar Imager (FOXSI)* is currently under consideration for launch in 2022 and would provide directly focused imaging spectroscopy of solar HXRs for the first time from a solar-dedicated spacecraft. *FOXSI* and ngVLA will be highly complementary.

6. Requirements for the ngVLA

While the ngVLA is not optimized for observing solar flares, the basic attributes of the reference design are nevertheless compelling. It provides continuous frequency coverage from 1.2 – 50 GHz and 70 – 116 GHz, and contains a multitude of antennas on baselines less than a few km. The requirement for angular resolution at radio wavelengths for flare observations is $\theta \approx 20''/\nu_{GHz}$; i.e., $1''$ at a frequency of 20 GHz, or projected baselines out to $\approx$3 km. The reason that very high resolution is not needed, in general, is that "coronal seeing" imposes limits on the usable angular resolution (Bastian 1994). However, it is likely that coronal scattering varies widely on a case-by-case basis, and as noted earlier, the continuously filled uv-plane of the ngVLA will allow the maximum spatial resolution to be tailored for each case. Of paramount importance, too, is continuous frequency coverage of the bandwidth spanned by the coherent and incoherent emissions from a solar flare. Unfortunately, the ngVLA reference instrument does not provide frequency coverage below 1.2 GHz, although the idea of commensal or standalone arrays operating at low frequencies is being explored. However, the 1.2 – 3.5 GHz coverage will still be of great interest for coherent emissions associated with magnetic energy release. The frequency coverage provided by the higher frequency bands is for observations of the incoherent gyrosynchrotron and thermal free-free emission from solar flares. In many cases, coverage up to 50 GHz will be sufficient but studies of the most energetic flares will require coverage in 70 – 116 GHz band, too (although the small field of view in this band may be problematic).

Ideally, it will be possible to observe up to 6 ngVLA frequency bands simultaneously during solar flares, depending on the science objectives, dynamic range, and image fidelity requirements. The only practical way to achieve this is through the use of subarrays. With more than 120 antennas available within the required 3+ km footprint, the array could be divided into as many as 5-6 independent subarrays, one for each frequency band, comprised of 20-25 antennas.

The time resolution required depends very much on the specific science objectives of a given observations but, loosely, observations of coherent bursts in the 1.2-3.5 GHz band require a time resolution of 10 ms or less. For observations of incoherent emissions a time resolution of 100 ms or more will be sufficient. Requirements for spectral resolution are modest: $\sim 0.1\%$ for coherent emissions and $\sim 1\%$ for incoherent emissions; i.e., more granular than needed for RFI excision.

It is important to recognize that the intense radio emissions observed on the Sun during a solar flare can introduce $10^7 - 10^9$ Jy to the system. Provisions must be made to observe and calibrate these signals without undue impact on ngVLA system electronics. This is not an insurmountable problem; the JVLA supports solar observations, as did the VLA before it, albeit in a non-optimum manner. If a signal attenuation and flux calibration strategy are designed in from the start, the ngVLA will be able to support a broad program of solar physics, including flares.

Finally, from an operational perspective, it should be noted that because flare occurrence and location cannot (yet) be reliably predicted, the ngVLA will need a flexible dynamic scheduling system for solar flare programs (similar to that currently supported by the JVLA), in order to maximize the probability of successful flare observations.

To summarize, solar flares continue to pose a number of fascinating challenges. Two have been been briefly discussed here - magnetic energy release and electron acceleration. More broadly, the ngVLA can make significant contributions to understanding the physics of flares if it supports dynamic imaging spectroscopy in all frequency bands.

Acknowledgments. This work was supported by NSF grants AST-1615807, AST-1735405, AGS-1654382, AGS-1723436, AGS-1817277 and NASA grants NNX14AK66G, NNX16AL67G, 80NSSC18K0015, NNX17AB82G, and 80NSSC18K0667.

References

Aschwanden, M. J., & Benz, A. O. 1997, ApJ, 480, 825
Bain, H. M., Krucker, S., Glesener, L., & Lin, R. P. 2012, ApJ, 750, 44
Bastian, T. S. 1994, ApJ, 426, 774
Bastian, T. S., Benz, A. O., & Gary, D. E. 1998, ARA&A, 36, 131
Bastian, T. S. et al. 2018, ASP Monograph 7, Science with a Next-Generation Very Large Array, ed. E. J. Murphy (San Francisco, CA: ASP), 87
Benz, A. O. 1985, Solar Phys., 96, 357
Butler, B. J., Campbell, D. B., de Pater, I., & Gary, D. E. 2004, New Astronomy Rev., 48, 1511
Chen, B., Bastian, T. S., White, S. M., et al. 2013, ApJ, 763, L21
Chen, B., Bastian, T. S., Shen, C., et al. 2015, Science, 350, 1238
Drake, J. F., Swisdak, M., Che, H., & Shay, M. A. 2006, Nat, 443, 553
Dulk, G. A. 1985, ARA&A, 23, 169
Fleishman, G. D. et al. 2018, ASP Monograph 7, Science with a Next-Generation Very Large Array, ed. E. J. Murphy (San Francisco, CA: ASP), 125
Fleishman, G. D., Gary, D. E., & Nita, G. M. 2003, ApJ, 593, 571
Fleishman, G. D., Nita, G. M., & Gary, D. E. 2009, ApJ, 698, L183
Fleishman, G. D., Kontar, E. P., Nita, G. M., & Gary, D. E. 2013, ApJ, 768, 190
Fleishman, G. D., Nita, G. M., Kontar, E. P., & Gary, D. E. 2016, ApJ, 826, 38
Fleishman, G. D., & Kuznetsov, A. A. 2010, ApJ, 721, 1127
Fleishman, G. D., & Mel'nikov, V. F. 1998, Physics Uspekhi, 41, 1157
Fleishman, G. D., Pal'shin, V. D., Meshalkina, N., et al. 2016, ApJ, 822, 71
Fleishman, G. D., & Toptygin, I. N. 2013, Astrophysics and Space Science Library, 388, Springer
Forbes, T. G. 1986, ApJ, 305, 553
Gary, D. E., Chen, B., Dennis, B. R., et al. 2018, ApJ, 863, 83
Gary, D. E., Fleishman, G. D., & Nita, G. M. 2013, Solar Phys., 288, 549

Gary, D. E., & Hurford, G. J. 2004, Astrophysics and Space Science Library, 314, 71
Gelfreikh, G. B. 2004, Astrophysics and Space Science Library, 314, 115
Glesener, L. & Fleishman, G. D. ArXiv e-prints 1806.00858
Glesener, L., Krucker, S., Bain, H. M., & Lin, R. P. 2013, ApJ, 779, L29
Grefenstette, B. W., Glesener, L., Krucker, S., et al. 2016, ApJ, 826, 20
Guidice, D. A., & Castelli, J. P. 1975, Solar Phys., 44, 155
Harrison, F. A., Craig, W. W., Christensen, F. E., et al. 2013, ApJ, 770, 103
Kaufmann, P., Correia, E., Costa, J. E. R., & Zodi Vaz, A. M. 1986, A&A, 157, 11
Kliem, B., Karlický, M., & Benz, A. O. 2000, A&A, 360, 715
Krucker, S., Bednarzik, M., Grimm, O., et al. 2016, Nuclear Instruments and Methods in Physics Research A, 824, 626
Kuroda, N., Gary, D. E., Wang, H., et al. 2018, ApJ, 852, 32
Lee, J. W., Gary, D. E., & Zirin, H. 1994, Solar Phys., 152, 409
Lin, R. P., Dennis, B. R., Hurford, G. J., et al. 2002, Solar Phys., 210, 3
Lysenko, A. L., Altyntsev, A. T., Meshalkina, N. S., Zhdanov, D., & Fleishman, G. D. 2018, ApJ, 856, 111
Melrose, D. B. 2017, Reviews of Modern Plasma Physics, 1, #5
Miller, J. A., Cargill, P. J., Emslie, A. G., et al. 1997, J. Geophys. Res., 102, 14631
Miller, J. A., Larosa, T. N., & Moore, R. L. 1996, ApJ, 461, 445
Musset, S., Kontar, E. P., & Vilmer, N. 2018, A&A, 610, A6
Nita, G. M., Gary, D. E., & Lee, J. 2004, ApJ, 605, 528
Raulin, J. P., Makhmutov, V. S., Kaufmann, P., et al. 2004, Solar Phys., 223, 181
Reid, H. A. S., & Ratcliffe, H. 2014, Research in Astronomy and Astrophysics, 14, 773
Rozhansky, I. V., Fleishman, G. D., & Huang, G.-L. 2008, ApJ, 681, 1688
White, S. M., Benz, A. O., Christe, S., et al. 2011, Space Sci.Rev., 159, 225

Science with a Next-Generation Very Large Array
ASP Conference Series, Monograph 7
Eric J. Murphy, ed.

Planetary Bistatic Radar

M. Brozović,[1] B. J. Butler,[2] Jean-Luc Margot,[3] Shantanu P. Naidu,[4] and T. Joseph W. Lazio[5]

[1]*Jet Propulsion Laboratory, California Institute of Technology, Pasadena, CA, USA;* `Marina.Brozovic@jpl.caltech.edu`

[2]*National Radio Astronomy Observatory, Socorro, NM; USA;* `bbutler@nrao.edu`

[3]*University of California, Los Angeles, Los Angeles, CA, USA;* `jlm@epss.ucla.edu`

[4]*Jet Propulsion Laboratory, California Institute of Technology, Pasadena, CA, USA;* `Shantanu.P.Naidu@jpl.caltech.edu`

[5]*Jet Propulsion Laboratory, California Institute of Technology, Pasadena, CA, USA;* `Joseph.Lazio@jpl.caltech.edu`

Abstract. Planetary radar observations offer the potential for probing the properties of characteristics of solid bodies throughout the inner solar system and at least as far as the orbit of Saturn. In addition to the direct scientific value, precise orbital determinations can be obtained from planetary radar observations, which are in turn valuable for mission planning or spacecraft navigation and planetary defense. The next-generation Very Large Array would not have to be equipped with a transmitter to be an important asset in the world's planetary radar infrastructure. Bistatic radar, in which one antenna transmits (e.g., Arecibo or Goldstone) and another receives, are used commonly today, with the Green Bank Telescope (GBT) serving as a receiver. The improved sensitivity of the ngVLA relative to the GBT would improve the signal-to-noise ratios on many targets and increase the accessible volume specifically for asteroids. Goldstone-ngVLA bistatic observations would have the potential of rivaling the sensitivity of Arecibo, but with much wider sky access.

1. Introduction

Planetary radar observations have been used to probe the surfaces of all of the planets with solid surfaces and many smaller bodies in the solar system (Ostro 1993, 2003), delivering information on their spins, orbital states, and surface and subsurface electrical properties and textures. Notable findings include characterizing the distribution of water at the south pole of the Moon (Stacy et al. 1997; Campbell et al. 2003), the first indications of water ice in the permanently shadowed regions at the poles of Mercury (Slade et al. 1992; Harmon et al. 1994), polar ice and anomalous surface features on Mars (Muhleman et al. 1991), establishing the icy nature of the Jovian satellites (Ostro & Pettengill 1978), and the initial characterizations of Titan's surface (Muhleman et al.

1990; Campbell et al. 2003). In multiple cases, the ground-based radar observations have served as the foundation for a subsequent space-based mission.

Radar observations are currently conducted in the S band ($\approx$ 2.3 GHz, Arecibo Observatory) and X band [$\approx$ 8.5 GHz, the Deep Space Network's Goldstone Solar System Radar (GSSR)], and future radar observations may also be conducted in the Ka band ($\sim$ 30 GHz). All of the planetary radar bands could be within the frequency coverage of the next-generation Very Large Array (ngVLA). As we discuss in more detail below (§3), the ngVLA need not be equipped with a transmitter to provide a powerful enhancement to planetary radar capabilities. Indeed, many of the results summarized previously involved *bistatic* observations in which the radar transmissions originated from one antenna and were received by a separate antenna.

Butler et al. (2004) previously considered the use of the Square Kilometre Array (SKA) as a receiver for a bistatic system. Since the time of their paper, there have been a number of developments, including multiple radar instruments on Mars orbiters, the *Cassini* radar instrument's observations of Titan, and the MESSENGER studies that confirmed earlier radar indications of polar ice at Mercury. This consideration of the ngVLA capabilities is similar to the earlier SKA consideration, but takes many of these subsequent spacecraft-based radar results into account.

We begin by motivating the scientific measurements that could be obtained from various target bodies by bistatic radar (§2), then turn to the specific benefit of the ngVLA in the context of the radar equation (§3), and conclude with a discussion on radar imaging (§4).

2. Target Bodies

In this section, we review target bodies and the science motivations for which future bistatic planetary radar observations could be relevant.

2.1. Venus

Venus is Earth's closest analog in the solar system in terms of its bulk properties, yet Venus and Earth have clearly had different evolutionary paths. There are potentially billions of Venus analogs in the Galaxy, and characterizing and understanding the differences between Venus and Earth has been given additional impetus for understanding the habitability of terrestrial-mass planets. Venus remains enigmatic on a variety of fundamental levels: The size of its core is unknown; whether the core is solid or liquid is uncertain; its atmospheric superrotation, 60× faster than the solid body, is not understood; and the atmosphere exhibits distinctive planetary-scale features that are stationary with respect to the solid body. High-precision measurements of the spin state of Venus with radar have the potential of providing key advances in all of these areas. First, a measurement of the spin precession rate ($\approx 2'' \, \mathrm{yr}^{-1}$) will yield a direct measurement of the polar moment of inertia, which is unknown. The moment of inertia provides an integral constraint on the distribution of mass in a planetary interior. Apart from bulk density, it is arguably the most important quantity needed to determine reliable models of the interior structure of Venus, including the size of its core. Second, a time history of length of day (LOD) variations at the 10 ppm level will identify the geophysical forcings responsible for spin variations, which are primarily related to transfer of angular momentum between the atmosphere and the solid planet. They will provide a crucial input to general circulation models and the key to elucidate poorly understood phenomena such as superrotation and stationary planetary-scale structures in the atmosphere.

Planetary radar provides a powerful tool for monitoring planetary spin states via observations of the "speckle displacement effect" or *radar speckle tracking* (Margot et al. 2007). Radar echoes from solid bodies exhibit spatial irregularities in the wavefront caused by the constructive and destructive interference of waves scattered by the irregular surface. The corrugations in the wavefront, i.e., speckles, are tied to the rotation of the target body. When the trajectory of the wavefront corrugations is parallel to a roughly east-west antenna baseline, echoes received at two receiving stations display a high degree of correlation. The time of day and value of the time delay at the correlation peak are directly related to the orientation and magnitude of the spin vector of the body. For typical solar system observations, the speckle size ($\sim R\lambda/D$, for a target at range R, observing wavelength λ, and diameter D) is on the order of 1 km and the high-correlation condition lasts for approximately 30 s.

The current approach using Goldstone and the Green Bank Telescope (GBT) yields instantaneous spin rate measurements at the 10 ppm level with X-band transmission from the GSSR and reception at Goldstone and the GBT. For example, with observations obtained between 2002 and 2012, the orientation of Mercury's spin axis has been measured with $5''$ precision, and measurements of the amplitude of longitude librations have revealed Mercury has a molten core (Margot et al. 2007, 2012). The accuracy of these measurements has been validated at the 1% level by independent measurements obtained by the MESSENGER spacecraft during a four-year duration (Margot et al. 2018). On-going observations include Venus, Europa, and Ganymede.

With a single pair of antennas, it is possible to obtain one measurement per day when the stringent geometry and signal-to-noise ratio (S/N) requirements are satisfied. However, measurements accumulate at a slow rate because each measurement requires simultaneous scheduling on two large radio antennas, successful transmission during the appropriate 30-second window, and successful reception at both antennas during the relevant 30-second windows. In order to fully constrain the spin axis orientation, it is imperative to secure observations at a variety of baseline orientations, which typically takes several years. An instrument such as the ngVLA would open up the possibility of securing up to 168 independent measurements in a single 20-minute session with the antennas located in the plains of San Agustin. At conjunction, the individual antenna S/N ratio and the correlation S/N ratio would exceed 100. Although the range of baseline orientations between array elements and Goldstone would remain small at any given observation epoch, the number of independent estimates of the correlation properties would improve the quality of the spin state determination by a factor of $\sqrt{N}$. Because the measurements are instantaneous, LOD variations that occur on 30-minute timescales would be detectable, which would place strong constraints on the mechanisms of angular momentum transfer.

Consequently, the ngVLA would enable (1) improved determination of the spin axis precession and therefore moment of inertia and core size; and (2) improved quantification of the amplitude of LOD variations on daily, seasonal, and secular timescales, providing strong constraints on the dynamics of the atmosphere and its interactions with the solid planet, and exploring a regime that may be common on exoplanets.

2.2. Asteroids

Radar observations of asteroids provide information on their sizes, shapes, spin states, surface properties, masses, bulk densities, orbits, and the presence of satellites (Figure 1). Recent improvements in transmitter capabilities have resulted in obtaining meter-scale spatial resolutions on various near-Earth asteroids, resolutions comparable to those obtained by spacecraft (either for fly-bys or orbiting). Thus, radar observations complement the fewer, but often more comprehensive spacecraft measurements.

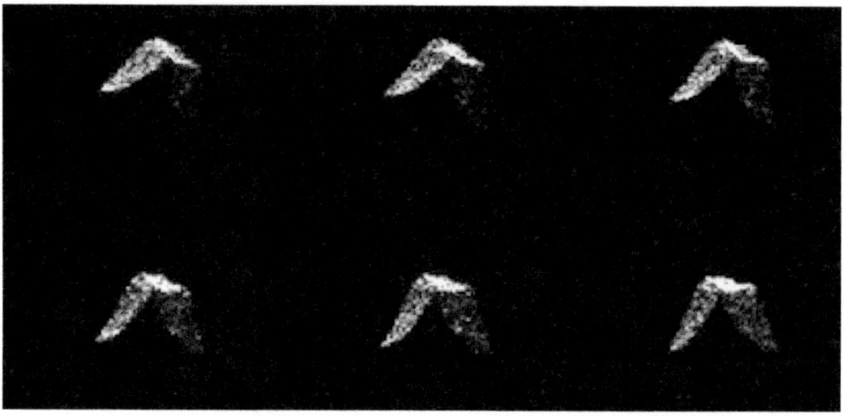

Figure 1. Radar observations of the near-Earth asteroid 2017 BQ6. Its rotation is apparent, as are the sharp, angular sides. A bright spot particularly apparent in the lower, middle panel may be a few meter-scale boulder on the surface. The sharpness of this asteroid's structure is currently unexplained. The time delay (range) increases from top to bottom, and Doppler frequency increases from left to right. (I.e., the top of the figure is closest to Earth, and the image can be considered to be a "top-down" view.) The color scale shows the echo power strength in units of standard deviation.

There have been a series of comprehensive reviews on radar observations of both near-Earth and Main Belt asteroids (Magri et al. 1999; Ostro et al. 2002; Ostro 2003; Butler et al. 2004; Benner et al. 2015; Naidu et al. 2016). We do not repeat that material here, but focus on specific aspects relative to the ngVLA.

The motivation for radar observations of asteroids is three-fold. First, asteroids represent primitive remnants of the early solar system, and their properties and orbits provide constraints on the formation and evolution of the solar system. Second, they represent targets for spacecraft (e.g., Chesley et al. 2014), for which orbital information and the presence of satellites are essential for mission planning and, for sample return missions, characterization of the surfaces is valuable. Finally, precise knowledge of their orbits is essential to assess the extent to which they might represent impact hazards to the Earth (Committee to Review Near-Earth-Object Surveys 2010), a topic that has increased in visibility over the past decade, and for which a "National Near-Earth Object Preparedness Strategy and Action Plan" has been issued (DAMIEN 2018). In particular, the orbits determined from radar observations are sufficiently precise that they can be used to assess whether a near-Earth asteroid presents any risk of colliding with the Earth over the next several decades to a century (e.g., Ostro & Giorgini 2004).

Specifically for near-Earth asteroids, bistatic radar observations can be valuable in two respects. First, for objects with close approaches to the Earth (short round-trip light travel times), it can be difficult or impossible to switch a radar facility from transmitting to receiving rapidly enough. Bistatic radar observations either simplify the observations or enable them for objects on extremely close approaches. Second, the increased sensitivity of the ngVLA would increase the range to which near-Earth asteroids could be targeted for radar observations, particularly for targets that are outside of the declination range of the Arecibo Observatory. Particularly from the perspective of planetary defense, obtaining orbits for as many near-Earth asteroids, and especially those classified as "potentially hazardous" is valuable, and the larger the volume that is accessible, the more asteroids that can be targeted. We return to this topic, in quantitative detail, in Section 3.

As quantitative estimates, we consider the improvement in range that the ngVLA might offer over the Green Bank Telescope, which is also used as the receiving element for bistatic radar. (See also Naidu et al. 2016.) If a subset of the ngVLA can be used for bistatic radar reception such that a sensitivity of *three* times that of the current GBT is obtained, it would more than *double* the accessible volume (increase the range by a factor of 30%) for near-Earth asteroid observations; if the sensitivity is *five* times that of the current GBT, it would more than *triple* the accessible volume (increase the range by a factor of 50%). Not only could Goldstone-ngVLA bistatic observations rival those of Arecibo, they would provide access to a much larger fraction of the sky.

Beyond the simple improvement in sensitivity offered by the ngVLA (§3), its antenna distribution offers the promise of improved shape modeling and spin state determinations via radar speckle tracking (Busch et al. 2010). Speckle tracking of asteroids operates in a fundamentally different regime than the case of Mercury or Venus (Margot et al. 2007). The general inability to predict the speckle trajectory in the asteroid case requires an observing configuration in which the speckle size is larger than the antenna baseline, otherwise speckles observed at different stations would not correlate. As a result, the ratio of speckle size to antenna baseline, which determines the fractional precision of the estimates, is three orders of magnitude larger for asteroids than it is for Mercury or Venus. The VLA has a dense set of antennas, but few asteroids approach the Earth sufficiently close that the VLA can be used. For example, for a 100 m object, it must be within about 10% of the lunar distance for the resulting speckle pattern to be comparable in scale to the VLA. Conversely, the VLBA has much longer baselines, allowing use of the technique to larger distances, but it has few antennas, so the number of speckle measurements that could be made is few. With a relatively dense network of antennas and antenna separations to of order 100 km, the ngVLA could be used for objects approaching to within one lunar distance (10^{-3} au), for which the number of objects is higher.

Finally, the dynamics of the Sun-Earth-Moon system allow for small asteroids to be captured into meta-stable geocentric orbits. Various predictions are that there should be a population of meter-scale "temporarily-captured orbiters" or "mini-moons," and at least one such mini-moon, 2006 RH120, has been detected (Kwiatkowski 2009; Bolin et al. 2014; Jedicke et al. 2018). The advent of future large scale-surveys, such as the Large Synoptic Survey Telescope (LSST), may result in several more being found. Due to their small size, such observations are extremely challenging, but feasible, as 2006 RH120 has been detected from Goldstone (Brozović et al. 2016). Not only do mini-moons have small radar cross sections, they are relatively close ($\approx$ 3 s round-trip light travel time). As noted above, with such short light travel times, bistatic observations would be required to avoid subjecting the transmitter to frequent power fluctuations.

2.3. Icy Satellites/Ocean Worlds

Ground-based radar observations provided some of the first clear evidence for the icy surfaces of the Galilean satellites and subsequent characterization (e.g., Campbell et al. 1977; Ostro & Pettengill 1978; Ostro et al. 1992; Black et al. 2001), demonstrating the capability to probe several meters into the surface. There have been radar observations of Saturnian satellites as well, though these are even more challenging due to Saturn's greater distance (Muhleman et al. 1990; Campbell et al. 2003; Black et al. 2007). Subsequent spacecraft investigations have provided clear evidence that at least some of the moons of Jupiter (Europa, Ganymede, Callisto) and Saturn (Enceladus, Titan), and potentially moons of Uranus (Ariel) harbor sub-surface oceans (Nimmo & Pappalardo 2016). As a consequence, the planned Europa Clipper mission would carry a radar designed to probe and constrain the thickness of the icy shell of Europa.

However, spacecraft missions are infrequent, plausibly only two missions might fly to the outer solar system in a decade, and a radar instrument might not be part of the spacecraft's payload. By contrast, ground-based radar can potentially happen essentially annually, near an outer planet's opposition, when the distance to the icy satellite is minimized.

The orbits of the Galilean satellites of Jupiter are affected by tidal interactions with Jupiter, due to their relatively small semi-major axes. Of particular interest are the tidal responses of Io and Europa as the tidal response of Io is related to the heat dissipation responsible for its active volcanism and the tidal response of Europa is related to the depth of its sub-surface ocean Tidal dissipation is parameterized by k_2/Q, where k_2 is the Love number, which is a measure of the amplitude of tidal response in the body, and Q is the quality factor or a measure of the viscous damping in the body. Both quantities are related to the properties of Jupiter's or satellites' interiors, and the ratio k_2/Q quantifies how the bulge raised by the satellite, or on the satellite, leads or trails its "precursor." The Juno mission will provide estimates of k_2, which can be combined with the radar ranging, to estimate the ratio k_2/Q. Moreover, radar ranging measurements have the advantage of being able to be carried out indefinitely, while the Juno mission is of a limited duration. (At the time of writing, the Juno prime science mission terminates in 2021 June.)

Lainey et al. (2009) used astrometric data to suggest that orbits of Io, Europa, and Ganymede have shifted due to tidal acceleration by 55 km, −125 km, and −365 km, respectively, over a period of 116 years. The highest precision astrometric measurements have a resolution of 75 km and originate from mutual occultations and eclipses.

The Arecibo planetary radar can measure a line-of-sight distance (or range) to Io with 10 km precision and distances to Europa, Ganymede, and Callisto with 1.5 km precision. However, Jupiter is only observable from Arecibo six out of every 12 years because of the constraints of the Arecibo's antenna pointing (declination range −1° to +38°). In 2015, GSSR-GBT bistatic radar demonstrated the capability to obtain ranging measurements of the Galilean moons. Both antennas are fully steerable and allow observations on a yearly basis. A range to Europa was measured with 75 km precision (Figure 2), comparable to the highest precision optical astrometry. However, ranging to Io was not possible with in this bistatic configuration due to low echo strength.

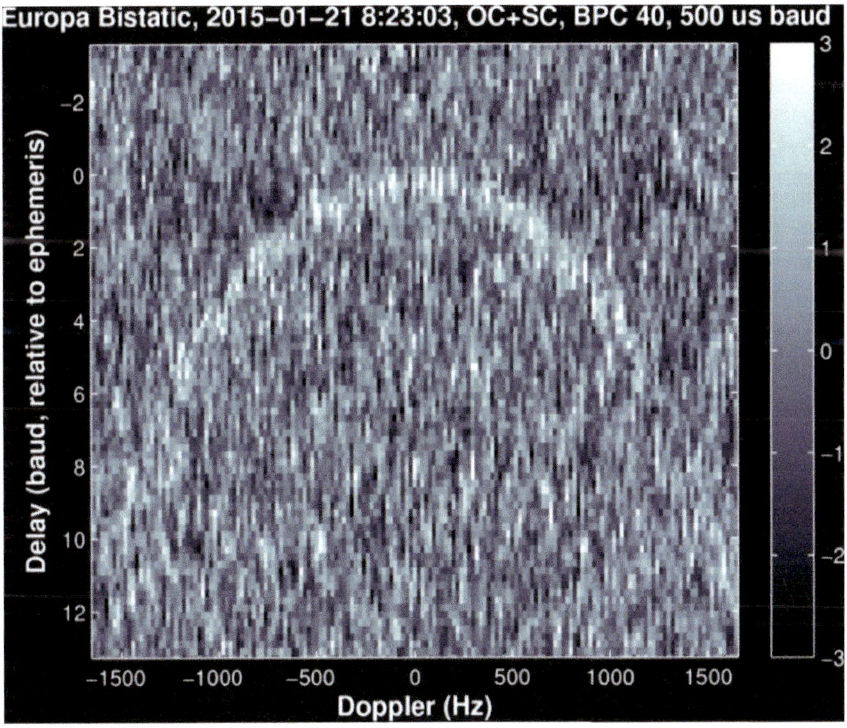

Figure 2. Goldstone Solar System Radar-Green Bank Telescope delay-Doppler image of Europa. The time delay (range) increases from top to bottom, and Doppler frequency increases from left to right. (I.e., the top of the figure is closest to Earth, and the image can be considered to be a "top-down" view.) The range resolution is 500 μs or 75 km. The color scale shows the echo power strength in units of standard deviation. The scale has been saturated at 3 units in order to enhance the echo outline. With its higher sensitivity, the ngVLA would offer higher signal-to-noise ratios and higher ranging precision.

If ngVLA achieves 3 times the sensitivity of the GBT, it would likely be able to achieve yearly ranging measurements of Europa with sub-10 km precision. Furthermore, it would possible to measure a distance to Io with 100 km precision. If a five-fold sensitivity of the ngVLA materializes, ranging precisions of 15 km–30 km for Io could be obtained. These measurements will contribute to the maintenance of highly accurate ephemerides of the Galilean satellites that would lead to continued improvements in the constraints on tidal dissipation (k_2/Q) and that could enhance, and potentially enable, future missions to these bodies (e.g., monitor Io's volcanism, explore Europa's ice shell for biosignatures).

2.4. Comets and Interstellar Objects

Much like the case for asteroids, radar observations of comets can provide information on the size, shape, and spin state of comet nuclei.

For instance, the Arecibo radar observed comet 103P/Hartley 2 shortly before NASA's EPOXI spacecraft encountered it in 2010 November. The radar observations determined that the comet nucleus has a bi-lobed shape, a result confirmed by the spacecraft (Harmon et al. 2011). Also, much like the case for asteroids, the radar observations of cometary nuclei complement the fewer, but more incisive spacecraft measurements—approximately five times as many comets have been detected by radar observations as have been visited by spacecraft.

The recent recognition of the first interstellar object, 1I/2017 U1 'Oumuamua, suggested that such objects might have extremely low optical albedos and at least this first object appeared to have a large aspect ratio, potentially in excess of 5:1 (Meech et al. 2017). The number of identified interstellar objects may increase in the future as additional wide-field surveys occur, particularly if the survey strategies explicitly account for the potential trajectories of interstellar objects. Notably, Jewitt et al. (2017) predict that there may be as many as 10^4 such objects within the orbit of Neptune at any given time. If an interstellar object did have a trajectory that took it sufficiently close to Earth to warrant radar observations, constraints on its properties would be invaluable.

3. The ngVLA and the Radar Equation

The increased sensitivity of the ngVLA would expand the set of targets for traditional bistatic delay-Doppler planetary radar. The classic radar equation is that the received power P_R is (e.g., Benner et al. 1999)

$$P_R = \frac{P_T G_T G_R \lambda^2 \sigma}{(4\pi)^3 R^4}, \tag{1}$$

where P_T is the power of the transmitter; the gains of the transmitting and receiving antennas are G_T and G_R, respectively; λ is the operational wavelength; σ is a measure of the radar cross section of the target body; and the range (distance) to the target is R. More concisely, the signal-to-noise ratio of radar observations scales as R^{-4}. This R^{-4} dependence can be understood as the product of two inverse square laws. The transmissions from the transmitting antenna to the target body suffer a R^{-2} loss by the inverse square law. By Huygen's principle, the target body re-radiates, and these emissions suffer an additional R^{-2} loss by the inverse square law.

With a fixed radar transmitter power (and antenna gain), the signal-to-noise ratio can only be improved by using increasing the gain (i.e., sensitivity) of the (bistatic) receiving element. For the ngVLA to participate in this kind of radar observation, it would have to have a "phased array" mode, in which voltages from the individual antennas are summed, after applying the appropriate time delays, so that the array appears as an effective single aperture. A phased array mode would also be valuable for observations of pulsars and for very long baseline interferometry (VLBI) imaging.

4. Radar Imaging

Existing planetary radar observations with the VLA have been used to image the radar return and make plane-of-the-sky astrometry measurements. Such *bistatic radar aperture synthesis* has produced spectacular images that convincingly demonstrated the presence of water ice at the poles of Mercury and Mars. With a factor of *five* better angular resolution, the ngVLA could produce a correspondingly better linear resolution on the surface of target bodies. Alternately, an improved angular resolution opens the possibility of producing resolved images of smaller objects or higher spatial resolution bistatic radar aperture synthesis on planets.

For example, de Pater et al. (1994) imaged 4179 Toutatis with the VLA in its A configuration. The asteroid was at a distance of 0.063 au, and the VLA's angular resolution corresponded to a linear resolution of approximately 10 km on the asteroid. They found that the asteroid showed clearly distinct residual radar features, suggestive of a bi-lobed structure, but their estimate of the separation of these features was limited by the VLA's beam size. With the improved angular resolution of the ngVLA, features with scales of about 2 km would have been distinguishable.

We thank L. Benner, M. Busch, and P. Taylor for helpful comments. This work made use of NASA's Astrophysics Data System Abstract Service. Part of this research was carried out at the Jet Propulsion Laboratory, California Institute of Technology, under a contract with the National Aeronautics and Space Administration. The National Radio Astronomy Observatory is a facility of the National Science Foundation operated under cooperative agreement by Associated Universities, Inc.

References

Benner, L. A. M., Hudson, R. S., Ostro, S. J., et al. 1999, Icarus, 139, 309; doi: 10.1006/icar.1999.6094

Benner, L. A. M., Busch, M. W., Giorgini, J. D., Taylor, P. A., & Margot, J.-L. 2015, "Radar Observations of Near-Earth and Main-Belt Asteroids," in *Asteroids IV*, eds. P. Michel, F. E. DeMeo, & W. F. Bottke (Univ. Arizona Press Tucson) p. 165; ISBN: 978-0-816-53213-1

Black, G. J., Campbell, D. B., & Nicholson, P. D. 1977, Icarus, 151, 167; doi: 10.1006/icar.2001.6616

Black, G. J., Campbell, D. B., & Carter, L. M. 2007, Icarus, 191, 702; doi: 10.1016/j.icarus.2007.06.009

Bolin, B., Jedicke, R., Granvik, M., Brown, P., Howell, E., Nolan, M. C.; Jenniskens, P., Chyba, M., Patterson, G., & Wainscoat, R. 2014, Icarus, 241, 280; doi: 10.1016/j.icarus.2014.05.026

Brozovic, M., Benner, L. A. M., Busch, M. W., Giorgini, J. D., Slade, M. A., & Lawrence, K. J. 2016, "Recent Goldstone radar observations of selected near-Earth asteroids less than 140 m in diameter," in *Asteroids: New Observations, New Models*, Proc. International Astron. Union, IAU Symposium, Volume 318, eds. S. R. Chesley, A. Morbidelli, R. Jedicke, & D. Farnocchia (Cambridge Univ. Press: Cambridge) p. 160; doi: 10.1017/S1743921315008807

Busch, M. W., Kulkarni, S. R., Brisken, W., Ostro, S. J., Benner, L. A. M., Giorgini, J. D., & Nolan, M. C. 2010, Icarus, 209, 535; doi: 10.1016/j.icarus.2010.05.002

Butler, B. J., Campbell, D. B., de Pater, I., & Gary, D. E. 2004, NewAR, 48, 1511; doi: 10.1016/j.newar.2004.09.031

Campbell, D. B., Chandler, J. F., Pettengill, G. H., & Shapiro, I. I. 1977, Science, 196, 650; doi: 10.1126/science.196.4290.650

Campbell, B. A., Campbell, D. B., Chandler, J. F., Hine, A. A., Nolan, M. C., & Perillat, P. J. 2003, Nature, 426, 137

Campbell, D. B., Black, G. J., Carter, L. M., & Ostro, S. J. 2003, Science, 302, 431; doi: 10.1126/science.1088969

Chesley, S. R., Farnocchia, D., Nolan, M. C., Vokrouhlický, D., Chodas, P. W., Milani, A., Spoto, F., Rozitis, B., Benner, L. A. M., Bottke, W. F., Busch, M. W., Emery, J. P.,

Howell, E. S., Lauretta, D. S., Margot, J.-L., & Taylor, P. A. 2014, Icarus, 235, 5; 10.1016/j.icarus.2014.02.020

Committee to Review Near-Earth-Object Surveys, 2010, *Defending Planet Earth: Near-Earth-Object Surveys and Hazard Mitigation Strategies* (National Academies Press: Washington, D.C.)

Interagency Working Group for Detecting and Mitigating the Impact of Earth-Bound Near-Earth Objects, National Science & Technology Council 2018, "National Near-Earth Object Preparedness Strategy and Action Plan"; https://www.whitehouse.gov/wp-content/uploads/2018/06/National-Near-Earth-Object-Preparedness-Strategy-and-Action-Plan-23-pages-1MB.pdf

de Pater, I., Palmer, P., Mitchell, D. L., Ostro, S. J., Yeomens, D. K., & Snyder, L. E. 2011, Icarus, 111, 489; doi: 10.1006/icar.1994.1159

Harmon, J. K., Slade, M. A., Vélez, R. A., Crespo, A., Dryer, M. J., & Johnson, J. M. 1994, Nature, 369, 213; 10.1038/369213a0

Harmon, J. K., Nolan, M. C., Howell, E. S., Giorgini, J. D., & Taylor, P. A. 2011, ApJ, 734, L2; doi: 10.1088/2041-8205/734/1/L2

Jedicke, R., Bolin, B. T., Bottke, W. F., Chyba, M., Fedorets, G., Granvik, M., Jones, L., & Urrutxua, H. 2018, Frontiers Astron. Space Sci., 5, 13; doi: 10.3389/fspas.2018.00013

Jewitt, D., Luu, J., Rajagopal, J., et al. 2017, ApJ, 850, L36

Kwiatkowski, T., Kryszczyńska, A., Polińska, M., Buckley, D. A. H., O'Donoghue, D., Charles, P. A., Crause, L., Crawford, S., Hashimoto, Y., Kniazev, A., Loaring, N., Romero Colmerero, E., Sefako, R., Still, M., & Vaisanen, P. 2009, A&A, 495, 967; doi: 10.1051/0004-6361:200810965

Lainey, V., Arlot, J.-E., Karatekin, Ö, & Van Hoolst, T. 2009, Nature, 459, 957

Magri, C., Ostro, S. J., Rosema, K. D., Thomas, M. L., Mitchell, D. L., Campbell, D. B., Chandler, J. F., Shapiro, I. I., Giorgini, J. D., & Yeomans, D. K. 1999, Icarus 140, 379

Margot, J. L., Peale, S. J., Jurgens, R. F., Slade, M. A., & Holin, I. V. 2007, Science, 316, 710

Margot, J. L., Peale, S. J., Solomon, S. C., Hauck, II, S. A., Ghigo, F. D., Jurgens, R. F., Yseboodt, M., Giorgini, J. D., Padovan, S., & Campbell, D. B. 2012, J. Geophys. Res., 117, E00L09

Margot, J. L., Hauck, S. A., Mazarico, E., Padovan, S., & Peale, S. J. 2018, "Mercury's Internal Structure," in *Mercury - The View after MESSENGER* (Cambridge University Press: Cambridge) p. 151

Meech, K. J., Weryk, R., Micheli, M., et al. 2017, Nature, doi:10.1038/nature25020

Muhleman, D. O., Butler, B. J., Grossman, A. W., & Slade, M. A. 1991, Science, 253, 1508-1513; 10.1126/science.253.5027.1508

Muhleman, D. O., Grossman, A. W., Butler, B. J., & Slade, M. A. 1990, Science, 248, 975; doi: 10.1126/science.248.4958.975

Naidu, S. P., Benner, L. A. M., Margot, J.-L., Busch, M. W., & Taylor, P. A. 2016, Astron. J., in press; arXiv:1604.01080

Nimmo, F., & Pappalardo, R. T. 2016, J. Geophys. Res. (Planets), 121, 1378; doi: 10.1002/2016JE005081

Ostro, S. J. 1993, Rev. Modern Phys., 65, 1235

Ostro, S. J. 2003, in *Encyclopedia of Physical Science and Technology* (Third Edition), ed. R. A. Meyers (Academic Press: New York) p. 295 ISBN: 978-0-12-227410-7

Ostro, S. J., & Pettengill, G. H. 1978, Icarus, 34, 268; 10.1016/0019-1035(78)90167-7

Ostro, S. J., Campbell, D. B., Simpson, R. A., Hudson, R. S., Chandler, J. F., Rosema, K. D., Shapiro, I. I., Standish, E. M., Winkler, R., & Yeomans, D. K. 1992, J. Geophys. Res., 97, 18; doi: 10.1029/92JE01992

Ostro, S. J., Hudson, R. S., Benner, L. A. M., Giorgini, J. D., Magri, C., Margot, J.-L., & Nolan, M. C. 2002, "Asteroid Radar Astronomy," in *Asteroids III*, eds. W. F. Bottke, A. Cellino, P. Paolicchi, R. P. Binzel (Univ. Arizona Press: Tucson) p. 151; ISBN: 978-0-816-52281-1

Ostro, S. J., & Giorgini, J. D. 2004, "The Role of Radar in Predicting and Preventing Asteroid and Comet Collisions with Earth," in *Mitigation of Hazardous Comets and Asteroids*

(Cambridge University Press: Cambridge) p. 38; ISBN 0521827647
Slade, M. A., Butler, B. J., & Muhleman, D. O. 1992, Science, 258, 635; 10.1126/science.258.5082.635
Smrekar, S. E., Davaille, A., & Sotin, C. 2018, Space Sci. Rev., 214, 88; doi: 10.1007/s11214-018-0518-1
Stacy, N. J. S., Campbell, D. B., & Ford, P. G. 1997, Science, 276, 1527; 10.1126/science.276.5318.1527

Science with a Next-Generation Very Large Array
ASP Conference Series, Monograph 7
Eric J. Murphy, ed.

ngVLA Observations of Coronal Magnetic Fields

G. D. Fleishman,[1] G. M. Nita,[1] S. M. White,[2] D. E. Gary,[1] and T. S. Bastian[3]

[1]*Center For Solar-Terrestrial Research, New Jersey Institute of Technology, Newark, NJ 07102, USA;* `gfleishm@njit.edu`

[2]*Space Vehicles Directorate, Air Force Research Laboratory, Kirtland AFB, NM 87123, USA;* `stephen.white.24@us.af.mil`

[3]*National Radio Astronomy Observatory, Charlottesville, VA, USA;* `tbastian@nrao.edu`

Abstract. Energy stored in the magnetic field in the solar atmosphere above active regions is a key driver of all solar activity (e.g., solar flares and coronal mass ejections), some of which can affect life on Earth. Radio observations provide a unique diagnostic of the coronal magnetic fields that make them a critical tool for the study of these phenomena, using the technique of broadband radio imaging spectropolarimetry. Observations with the ngVLA will provide unique observations of coronal magnetic fields and their evolution, key inputs and constraints for MHD numerical models of the solar atmosphere and eruptive processes, and a key link between lower layers of the solar atmosphere and the heliosphere. In doing so they will also provide practical "research to operations" guidance for space weather forecasting.

1. Introduction

Magnetic fields are important in a variety of astrophysical contexts. For example, the stored magnetic energy in an extended volume of magnetized plasma can be rapidly, catastrophically released, often accompanied by changes in the magnetic connectivity. The corresponding driving process has come to be known as magnetic reconnection, and understanding this phenomenon is central to many problems of astrophysics, such as the origin of the radio arc in the galactic center (Sofue et al. 2005), quasar superluminal sources (de Gouveia dal Pino & Lazarian 2005), the hot component of the galactic ridge X-ray emission (Tanuma et al. 1999), particle acceleration in the accretion disks of AGNs (Schopper et al. 1998), and solar-analogous problems in stellar astrophysics, such as how stellar atmospheres are heated and nonthermal particles are produced in active, highly magnetized stars (Hurley et al. 2005). Thanks to its proximity and the availability of many observing facilities across a wide range of wavelengths, the Sun is arguably the best astrophysical laboratory for obtaining clues to magnetic field structure, its evolution, and magnetic energy release.

A key area for such study is coronal magnetic fields in solar active regions. Active regions are the locations on the Sun where strong magnetic fields generated deep in the solar convection zone rise through the photosphere, often in the guise of sunspots, up into the corona. Active region magnetic field evolution and dynamics within the corona then drive many of the energetic phenomena that we observe on the Sun, converting magnetic energy into other forms.

While currently available space- and ground-based solar optical telescopes are already capable of precise measurements of the photospheric magnetic field with sub-arcsecond angular resolution and high temporal resolution using the Zeeman effect, these techniques require appropriate spectral lines and great sensitivity to be used in the corona. Next generation instruments like Daniel K. Inouye Solar Telescope (DKIST; Tritschler et al. 2016) will exploit such measurements using near-infrared coronal spectral lines (Del Zanna & DeLuca 2018), but these will only be effective above the solar limb. As we show here, radio observations are positioned to be powerful and versatile diagnostic of coronal magnetic fields using the ngVLA.

The absence of quantitative measurements of coronal magnetic fields to date has led the community to rely heavily on magnetic field extrapolations that use measurements at the photosphere as the boundary condition. For example, modern nonlinear force-free field (NLFFF) extrapolation algorithms can be used to produce models of the coronal field (e.g., De Rosa et al. 2009, 2015), but owing to the fact that the photosphere is not actually force-free, as well as other observational and theoretical limitations, such extrapolations are far from unique (De Rosa et al. 2015). Therefore, the coronal magnetic field modeling based on such extrapolations requires modification and independent verification using observational diagnostics of the magnetic field at coronal heights. Below we briefly review the available radio techniques for probing the coronal magnetic and thermal structures, and highlight synergies between the radio diagnostics and other data sets and models.

2. Sensitivity of Radio Emission to Coronal Magnetic Fields

Radio emission processes are well known to depend strongly on the magnetic field strength in the source region due to two effects: (1) the $\mathbf{v} \times \mathbf{B}$ Lorentz force causes the emitting electrons to spiral in the magnetic field to produce gyroemission, which takes the form of gyroresonance (GR) emission for thermal electrons in the hot corona of active regions, and gyrosynchrotron (GS) emission for mildly-relativistic electrons in flares; and (2) the plasma itself is birefringent, with differing indices of refraction (e.g., Fleishman & Toptygin 2013) for the ordinary (o) and extraordinary (x) magnetoionic modes, which leads to a magnetic-field-dependent polarization of each type of emission, including free-free (FF) bremsstrahlung emission (e.g., Gelfreikh 2004). Hence, both Stokes I and V polarization parameters must be observed.[1]

All three emission mechanisms, GR, GS, and FF, have been exploited in studies of radio emission from the Sun to estimate magnetic field strengths in the corona. However, such estimates based on either spatial or spectral information alone generally suffer from a large number of ambiguities that require varying degrees of assumptions to be made, or else require special conditions where the assumptions and ambiguities can be minimized. These ambiguities can be largely eliminated when both spatial and spectral information in both senses of circular polarization are available simultaneously. In this case, three-dimensional data cubes are used to provide spatially-resolved, polarized

[1]The Faraday depth of the corona is so large that Stokes Q and U are unobservable except, possibly, under special circumstances. See Allisandrakis & Chiuderi-Drago (1995).

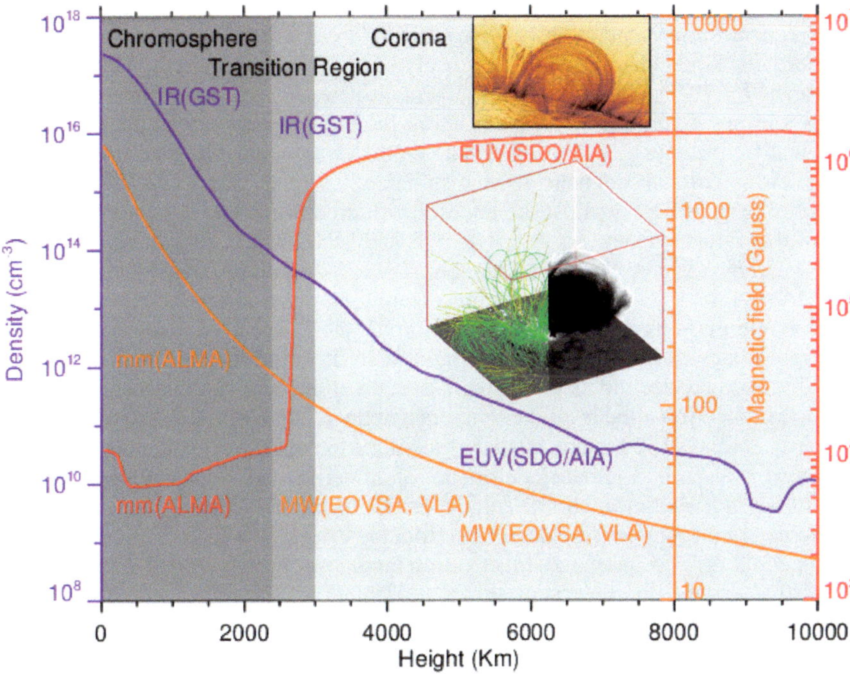

Figure 1. Schematic 1D structure of the solar atmosphere taken from a single line of sight from the MHD Bifrost model (Carlsson et al. 2016): total density (blue), temperature (red), and magnetic field (orange). The dense chromosphere and tenuous corona are separated by the relatively thin transition region-layer in which the temperature sharply rises, from thousands to millions of Kelvin. The physical parameters that can be diagnosed in each region by several ground-based and space-borne instruments, and their observational windows, are indicated in corresponding colors: GST and DKIST: Infrared (IR); EOVSA and the VLA: dm/cm wavelengths; ALMA: mm/submm wavelenghts; SDO/AIA: EUV. Insets illustrate a 2D SDO/AIA image of coronal loops, and a 3D coronal model visualized with a set of magnetic field lines (left) and electron density distribution (right).

brightness-temperature spectra over an appropriate (broad) frequency range. We refer to this as *microwave imaging spectropolarimetry.* It is precisely this capability that the ngVLA will provide, permitting the full exploitation of the magnetic field diagnostics of GR, GS, and FF emission over an unprecedentedly broad spectral band.

For active regions, the two emission mechanisms relevant for the relatively cool chromosphere and hot, magnetized corona are the GR and FF mechanisms, both of which are well understood. As described by Gary & Hurford (2004), spatially-resolved brightness temperature spectra $T_b(f)$, which can be obtained from any pixel of the multi-frequency image data cube, provide temperature as a function of frequency, $T(f) = T_b(f)$, provided the radio emission is optically thick. The optical depth of FF emission varies as $n_e^2 f^{-2} T^{-1.5}$, where n_e is the ambient electron density, so it favors dense cool regions and low frequencies. The corona is seldom optically thick to free-free absorption at frequencies $\gtrsim$3 GHz, at which point optically thick GR emission becomes relevant if the coronal magnetic field is sufficiently strong. Typically, GR emission has a significant optical depth at the lowest few harmonics s =1, 2, 3 (more seldom 4 or 5) of the gyrofrequency f_B, $f = sf_B = 2.8sB$ MHz, which allows the conversion $T(f) \Rightarrow T(sB)$; (Tun et al. 2011; Nita et al. 2011; Wang et al. 2015). Given that the temperature of the solar atmosphere drops precipitously in a narrow height range at the transition region (Figure 1), the base of the corona, the T_b spectrum likewise cuts off at those frequencies whose GR emission originates in the transition region. It is straightforward to then deduce the magnetic field strength at the base of the corona (e.g., Lee 2007) to produce so-called level-0 coronal magnetic field maps.

Wang et al. (2015) demonstrated the technique to measure coronal magnetic fields by creating a realistic 3-dimensional model of an active region, calculating the radio images from it including both GR and FF emission, and then using the level-0 technique to derive the magnetic field strength from the images. The derived magnetic field strength could then be compared directly with the known magnetic field strength in the model, to assess the differences, as shown in Figure 2c. The errors were found to be within about ±10% except in regions (blue in Figure 2c) directly over sunspots, where the opacity of the third harmonic ($s = 3$) of the gyrofrequency drops due to the small angle of the magnetic field to the line of sight, and in regions in the northwestern part of the region (orange, red, and white in Figure 2) where higher density makes the FF opacity dominant. Figure 2a,b show individual spectra at two representative points in the blue areas of panel (c), where the expected spectral signature of the $s = 3$ harmonic is missing or weak. Such points can be corrected based on the spectra [this is referred to as the level-1 technique in Wang et al. (2015)] provided the spatial resolution is sufficiently high. Figure 2d shows the magnetic field (along the cut indicated by the white dashed line in Figure 2c) from the model (black curve), from the measurements using the level-0 technique (black asterisks), and after adjustment using the level-1 (spectral shape) technique (open circles). As shown by Wang et al. (2015), to use the level-1 technique requires a high spatial resolution, which will be available using the ngVLA. In addition, the high-fidelity imaging and dynamic range of ngVLA will make it superb for exploiting such coronal magnetic field diagnostics. Differences in magnetic field strength compared to what is expected from photospheric extrapolation methods will pinpoint the location of coronal currents, which can be quite localized within an active region. Such currents are important clues to coronal heating and impending instability that may lead to flares and coronal mass ejections.

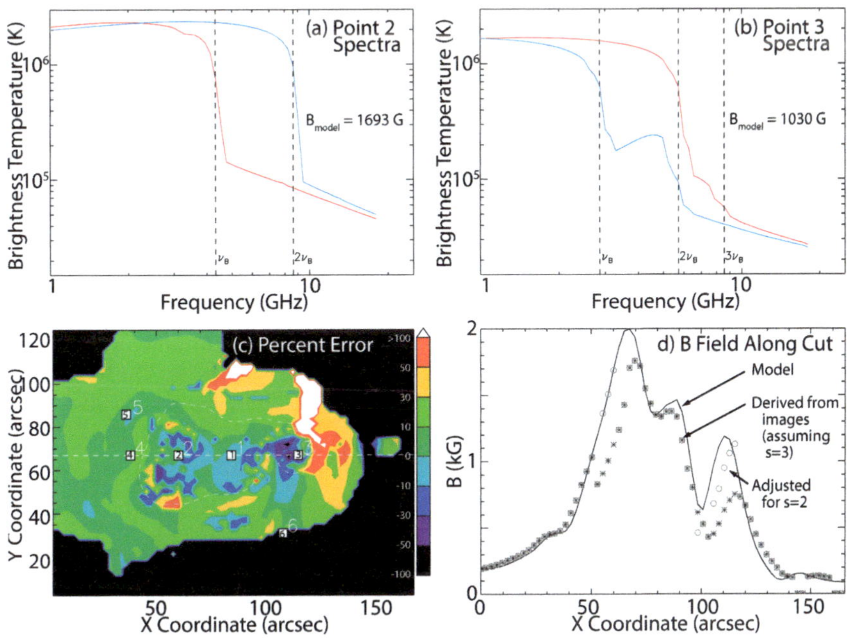

Figure 2. Results obtained from simulated images derived from an active region model. (a) Example spectra from point 2 marked in (c), in right-circular polarization (RCP, red) and left-circular polarization (LCP, blue). Only two harmonics are apparent, and the magnetic field strength derived from them is 1693 G. (b) Same as (a), for point 3 marked in (c). Note that the polarity of the field is opposite to that in panel (a). Here again two harmonics are apparent from the spectra, with a slight indication of the third harmonic in RCP. (c) Similar spectra over the entire region have been used to derive the magnetic field strength, which was then compared with the model assuming the third harmonic ($s = 3$). Plotted is the percent error of the measurements relative to the model. (d) Asterisks show the derived magnetic field along the cut indicated by the white dashed line in (c), assuming $s = 3$ compared with the model (black curve). Points that are adjusted by a factor of 1.5 to account for $s = 2$ derived from the spectra are shown as open circles. Adapted from Wang et al. (2015).

Whenever the GR emission is low, or the plasma density is particularly high, the dominant contribution to cm and mm emission comes from the FF process. The intensity of the FF emission is determined by the plasma density and temperature, while it only weakly depends on the magnetic field. In contrast, the polarization does depend on the magnetic field, primarily on its LOS component, which offers a way of probing this LOS component (Bogod & Gelfreikh 1980; Fleishman et al. 2015; Loukitcheva et al. 2017) in addition to the absolute value available from the GR diagnostics. This diagnostic will be available from the high-frequency ($\sim$ 12 – 116 GHz) ngVLA data at various heights of the upper chromosphere and TR. This diagnostics will perfectly complement the GR diagnostics of the absolute value of the magnetic field. As we show in a separate paper (Fleishman et al. 2018 ApJ, under review), the probing of two components of the magnetic field, e.g., $|B|$ and B_{los}, is exactly what is needed to dramatically improve the coronal magnetic field reconstruction.

The brightness of the optically thick thermal emission from active regions provides a direct measurement of the plasma temperature at the height range where the given radio emission is formed. Stated another way, the radio measurements offer a unique way of quantitatively studying the magneto-thermal coupling in the corona and chromosphere. For example, the GR process can be directly used to distinguish the mechanisms of coronal heating at various heights in weak-field and strong-field regions. This is different from the EUV diagnostics, which quantify the thermal structure (density and temperature), but yield only limited morphological information on the magnetic field.

3. Contribution of the ngVLA

The Sun is one of the most difficult objects to observe for a synthesis radio telescope. It has the advantage of high signal levels, but the corresponding disadvantage of needing to avoid saturation in the signal path. Further, as an imaging problem, the Sun fills the ngVLA field of view with time-varying emission on all spatial scales. An interferometer consisting of a spatially-sparse array does not sample all necessary spatial scales, and Earth-rotation synthesis cannot be used to fill in missing (u, v)-spacings due to the time-variable nature of the emission. A large number of visibility measurements is therefore needed on an appropriate distribution of spatial frequencies on short time scales; i.e., excellent instantaneous uv coverage is essential. The "Plains configuration" of ngVLA antennas, comprised of 164 antennas distributed as 94 core antennas plus 74 antennas extending out to baselines comparable to the present-day JVLA A configuration, will be an excellent match to the imaging problem posed by the Sun although solar observations will mostly use the inner few km of the array, providing $\sim 10^4$ visibilities per snapshot.

The use of 18 m antennas alone would result in a lack of short-spacing information needed to measure the emission on large spatial scales, which is particularly problematic at higher frequencies where there is relatively little contrast in the image, but the addition of a short-baseline array of 6 m antennas, plus total power antennas, will address this problem. Below 10 GHz, radio images of active regions generally have bright compact GR features that facilitate self-calibration, and very high dynamic range should be achievable, necessary to measure the fainter FF emission together with the bright GR sources in the same image. While the VLA and JVLA have produced many major solar results, they are limited in the effective spatial resolution that can be achieved: the longest baseline determines the number of resolution elements within the field of view, and reconstruction of the image generally requires about the same number of visibilities, but that must include short-spacing information. The VLA has no short-spacing information in its wider configurations, and just 350 baselines for dynamic snapshot imaging: ngVLA will be better in both these aspects by orders of magnitude, providing photometrically accurate imaging over the entire range of frequencies.

Another important consideration is that for frequencies $\gtrsim$ 20 GHz, where coronal scattering is no longer a significant factor, arcsecond and subarcsecond imaging will be possible. Both ground-based (including ALMA) and space-based solar observatories now routinely achieve subarcsecond imaging of the Sun: ngVLA will make it possible for solar microwave observations to match this resolution and resolve the fine-scale structure in the solar chromosphere that is a critical aspect of solar magnetic activity.

4. Synergies with the ngVLA

4.1. Synergies with EUV and X-ray Observations

The characteristic thermal emissions from the solar corona occur at extreme ultraviolet (EUV) and soft X-ray (SXR) wavelengths, which are absorbed by Earth's atmosphere and thus cannot be observed from the ground. A number of space missions, such as NASA's Solar Dynamics Observatory (SDO), provide high-spatial-resolution imaging at a number of EUV wavelengths that can be used to study the thermal properties of the corona. Such data complement the magnetic field studies that can be made with radio data, and combination of the two wavelength regimes can provide the full specification of density, temperature and magnetic field needed to describe the energetics of the Sun's atmosphere. Thus ngVLA data will be combined with EUV/SXR imaging obtained with future space missions.

4.2. Synergies with mm/sub-mm Observations

At frequencies above 20 GHz, GR emission is typically not expected and solar radio emission is dominated by optically-thick thermal emission from the solar chromosphere (the few thousand-km thick layer immediately above the photosphere with temperatures in the range 5000~20000 K). As noted above, the chromosphere is the region of transition between the plasma-dominated photosphere (plasma thermal pressure/magnetic pressure $\beta > 1$) and the magnetically-dominated corona ($\beta < 1$). Convective motions driven from below make the chromosphere a very dynamic layer, changing appearance on timescales of seconds. There is a great deal of interest in how magnetic fields change with height through this layer, since that controls how much magnetic flux reaches the corona, but measurements of magnetic fields are difficult at UV wavelengths (the primary emission region for chromospheric temperatures) due at least in part to the need to integrate for some time to measure weak linear polarization. In principle radio observations, which only require Stokes I and V to determine B (line of sight), can be made more rapidly and thus can "freeze" the chromospheric motion. Since the radio emission is generally optically thick above 20 GHz, there is a scaling between observing frequency and height such that higher frequencies penetrate deeper into the atmosphere. ngVLA can see most of the upper chromosphere: at 20 GHz ngVLA sees the layer with a temperature around 10000 K, while at 100 GHz emission comes from heights in the solar atmosphere where the temperature is around 7000 K. ngVLA thus complements ALMA observations at higher frequencies (Loukitcheva et al. 2017), which penetrate even deeper into the chromosphere (e.g., around 5900 K at 230 GHz). Measuring magnetic fields above an active region with both ngVLA and ALMA (Fleishman et al. 2015) simultaneously offers the prospect of tracking, at high spatial resolution in all three dimensions, the behavior of magnetic field with height through a critical part of the solar atmosphere.

4.3. Synergies with Modeling

The problem of mapping coronal magnetic fields and tracking their evolution will likely require collaboration with both IR/UV observers measuring coronal lines as well as modelers extrapolating photospheric (or chromospheric) magnetic field measurements into the corona. Radio data complement the IR/UV techniques well, since the latter are generally confined to observations off the solar disk (i.e., just above the visible limb) due to the fact that strong emission at these wavelengths from the disk itself generally overwhelms the weaker coronal lines being used to measure polarization. Use of the Hanle effect also tends to restrict these techniques to regions of weaker field. Radio observations do not suffer from this problem and can be used anywhere on the solar disk, although they have more difficulty in measuring weak fields. Extrapolation techniques are not useful for regions near the solar limb because foreshortening due to the reduced inclination of the solar surface to the line of sight severely degrades magnetic field measurements. Thus radio observations of regions on the solar disk provide the only real "ground-truth" measurements of coronal magnetic fields that can be used to validate extrapolation techniques, and hence determine whether such techniques can overcome the issue of the solar photosphere not being force-free. The next step then would be adding the chromospheric and coronal magnetic field diagnostics obtained from the radio (ngVLA and ALMA) data directly to the algorithms of nonlinear force-free field reconstructions of the magnetic field. As shown by Fleishman et al. (2018 ApJ, under review) adding such data is capable of improving all metrics of NLFFF model by at least a factor of two; in particular, to greatly improve reconstruction of the magnetic connectivity. These improved 3D magnetic reconstructions are then needed to build realistic 3D magneto-thermal models of ARs as described by Nita et al. (2018).

5. Concluding Remarks

High-resolution multi-frequency radio data is the key to attack central problems of modern solar physics. The extremely broad frequency coverage provided by the ngVLA in addition to its excellent instantaneous (u, v)-coverage, the Plains configuration plus compact array for measuring large angular scales, a total power capability, and support of polarimetry make it both ideal and wholly unique in its ability to perform high fidelity broadband spectropolarimetry of the Sun.

The Sun's magnetic field lies at the root of a number of outstanding problems in solar physics, including magnetic energy release (flares, coronal mass ejections), coronal heating and solar wind acceleration. Quantitative measurements of coronal and chromospheric magnetic fields have been lacking to date. Such measurements can be made by broadband spectropolarimetric observations of solar active regions. These can be used synergistically with a number of other next generation instruments operating at complementary wavelength regimes, including ALMA, DKIST, and space based imagers operating in EUV and SXR wavelengths.

Acknowledgments. This work was supported by NSF grants AST-1615807, AST-1735405, AST-1820613, AGS-1654382 and NASA grants NNX14AK66G, 80NSSC 18K0015, NNX17AB82G, NNX14AC87G, and 80NSSC18K0667.

References

Bogod, V. M., & Gelfreikh, G. B. 1980, Solar Phys., 67, 29

Carlsson, M., Hansteen, V. H., Gudiksen, B. V., Leenaarts, J., & De Pontieu, B. 2016, A&A, 585, A4. 1510.07581

de Gouveia dal Pino, E. M., & Lazarian, A. 2005, A&A, 441, 845

De Rosa, M. L., Schrijver, C. J., Barnes, G., Leka, K. D., Lites, B. W., Aschwanden, M. J., Amari, T., Canou, A., McTiernan, J. M., Régnier, S., Thalmann, J. K., Valori, G., Wheatland, M. S., Wiegelmann, T., Cheung, M. C. M., Conlon, P. A., Fuhrmann, M., Inhester, B., & Tadesse, T. 2009, ApJ, 696, 1780. 0902.1007

De Rosa, M. L., Wheatland, M. S., Leka, K. D., Barnes, G., Amari, T., Canou, A., Gilchrist, S. A., Thalmann, J. K., Valori, G., Wiegelmann, T., Schrijver, C. J., Malanushenko, A., Sun, X., & Régnier, S. 2015, ApJ, 811, 107. 1508.05455

Del Zanna, G., & DeLuca, E. E. 2018, ApJ, 852, 52. 1708.03626

Fleishman, G., Loukitcheva, M., & Nita, G. 2015, in Revolution in Astronomy with ALMA: The Third Year, edited by D. Iono, K. Tatematsu, A. Wootten, & L. Testi, vol. 499 of Astronomical Society of the Pacific Conference Series, 351. 1506.08395

Fleishman, G. D., & Toptygin, I. N. 2013, Cosmic Electrodynamics

Gary, D. E., & Hurford, G. J. 2004, in Astrophysics and Space Science Library, edited by D. E. Gary, & C. U. Keller, vol. 314 of Astrophysics and Space Science Library, 71

Gelfreikh, G. B. 2004, in Astrophysics and Space Science Library, edited by D. E. Gary, & C. U. Keller, vol. 314 of Astrophysics and Space Science Library, 115

Hurley, K., Boggs, S. E., Smith, D. M., Duncan, R. C., Lin, R., Zoglauer, A., Krucker, S., Hurford, G., Hudson, H., Wigger, C., Hajdas, W., Thompson, C., Mitrofanov, I., Sanin, A., Boynton, W., Fellows, C., von Kienlin, A., Lichti, G., Rau, A., & Cline, T. 2005, Nat, 434, 1098. astro-ph/0502329

Lee, J. 2007, Space Sci.Rev., 133, 73

Loukitcheva, M., White, S. M., Solanki, S. K., Fleishman, G. D., & Carlsson, M. 2017, A&A, 601, A43. 1702.06018

Nita, G. M., Fleishman, G. D., Jing, J., Lesovoi, S. V., Bogod, V. M., Yasnov, L. V., Wang, H., & Gary, D. E. 2011, ApJ, 737, 82. 1106.0262

Nita, G. M., Viall, N. M., Klimchuk, J. A., Loukitcheva, M. A., Gary, D. E., Kuznetsov, A. A., & Fleishman, G. D. 2018, ApJ, 853, 66

Schopper, R., Lesch, H., & Birk, G. T. 1998, A&A, 335, 26. astro-ph/9803329

Sofue, Y., Kigure, H., & Shibata, K. 2005, PASJ, 57, L39. astro-ph/0507568

Tanuma, S., Yokoyama, T., Kudoh, T., Matsumoto, R., Shibata, K., & Makishima, K. 1999, PASJ, 51, 161

Tritschler, A., Rimmele, T. R., Berukoff, S., Casini, R., Kuhn, J. R., Lin, H., Rast, M. P., McMullin, J. P., Schmidt, W., Wöger, F., & DKIST Team 2016, Astronomische Nachrichten, 337, 1064

Tun, S. D., Gary, D. E., & Georgoulis, M. K. 2011, ApJ, 728, 1

Wang, Z., Gary, D. E., Fleishman, G. D., & White, S. M. 2015, ApJ, 805, 93. 1503.05239

Part III

Planet Formation

Science with a Next-Generation Very Large Array
ASP Conference Series, Monograph 7
Eric J. Murphy, ed.

Resolved Substructures in Protoplanetary Disks with the ngVLA

Sean M. Andrews,[1] David J. Wilner,[1] Enrique Macías,[2]
Carlos Carrasco-González,[3] and Andrea Isella[4]

[1]*Harvard-Smithsonian Center for Astrophysics, Cambridge, MA, USA 02138;*
`sandrews@cfa.harvard.edu` , `dwilner@cfa.harvard.edu`

[2]*Department of Astronomy, Boston University, Boston, MA, USA 02215;*
`emacias@bu.edu`

[3]*Instituto de Radioastronomía y Astrofísica (IRyA-UNAM), 58089, Morelia, México;* `c.carrasco@irya.unam.mx`

[4]*Department of Physics and Astronomy, Rice University, Houston, TX, USA 77005;* `isella@rice.edu`

Abstract. Terrestrial planets and the cores of giant planets are thought to be built by the collisional agglomeration of solids spanning over 20 orders of magnitude in size within a few million years. However, there is tension between this basic picture of planet formation and standard theoretical assumptions associated with the migration of “pebbles" (~mm/cm-sized particles) from the gaseous disks and presumably much longer timescales necessary to assemble (~km-scale) “planetesimals". To confront these potential theoretical discrepancies with observational constraints, the ideal tracer of the solids concentrated in protoplanetary disk substructures is the 30 – 100 GHz continuum, which strikes the best balance in sensitivity (emission still bright), optical depth (low enough to reliably estimate densities), and angular resolution (high enough to resolve fine-scale features at disk radii as small as 1 au). With its combination of sensitivity, frequency coverage and angular resolution, the next-generation VLA will be only facility that has the capabilities to open up this new window into physics of planetesimal formation.

1. Introduction

In the “core accretion" paradigm for planet formation, terrestrial planets and giant planet cores are created by the sequential collisional agglomeration of solids over ~20 orders of magnitude in size within a few million years (e.g., Pollack et al. 1996; Raymond et al. 2014). Early in that process, the standard theoretical assumptions introduce two fundamental obstacles to that growth. The first is related to the migration of “pebbles" (~mm/cm-sized particles): as these solids decouple from the gas disk, they migrate faster than they can collide and grow (Takeuchi & Lin 2002; Brauer et al. 2007). The second is that the timescales for assembling (~km-scale) “planetesimals" is too long, given the migration and destructive impacts of their precursors (e.g., Johansen et al. 2014). The potential solution to both issues is to locally concentrate pebbles in the disk (Whipple 1972; Pinilla et al. 2012), halting their migration and slowing their relative velocities to promote rapid growth (e.g., Chiang & Youdin 2010). The requirement

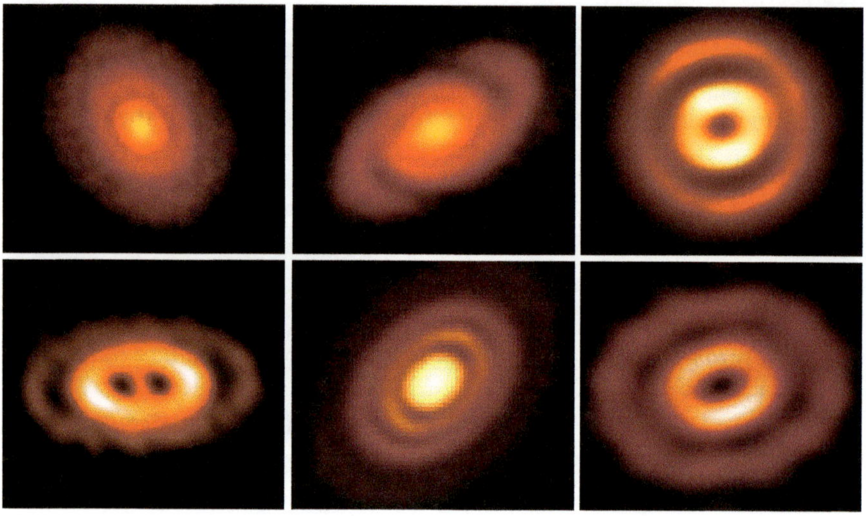

Figure 1. Examples of substructures in the ~1 mm continuum emission from nearby protoplanetary disks at scales of ~10–20 au, from the ALMA interferometer. Clockwise from top left: the V883 Ori (Cieza et al. 2016), Elias 27 (Pérez et al. 2016), HD 169142 (Fedele et al. 2017), HD 97804 (van der Plas et al. 2017), HD 163296 (Isella et al. 2016), and AA Tau (Loomis et al. 2017) disks.

to facilitate such concentrations is a gas pressure profile that is not smooth or monotonic, but rather has local maxima induced by abrupt variations in disk properties (e.g., Dzyurkevich et al. 2010; Stammler et al. 2017), dynamical effects (e.g., Baruteau et al. 2014), or fluid instabilities (e.g., Zhu et al. 2014; Flock et al. 2015).

The optimal way to directly test this hypothesis is to search for and characterize local concentrations of pebbles in protoplanetary disks. These particles emit a thermal continuum at wavelengths roughly comparable to their sizes (i.e., radio to microwave frequencies, ~10–1000 GHz); the detailed spatial distribution of this continuum emission in nearby disks can be resolved with an interferometer. If the optical depth of this emission is low, the spectrum provides constraints on the temperatures, densities, and size distribution of the constituent solids in any accessible disk region. Variations in the particle concentrations on small spatial scales would be manifested as *substructures* in the microwave continuum surface brightness morphology. The properties of those substructures provide crucial information on the mechanisms that create them and their potential for driving rapid, localized planetesimal formation.

In the past few years, the ALMA interferometer has found that such substructures on scales of $\lesssim$10–20 au (70–150 mas) are common, and perhaps ubiquitous (e.g., Zhang et al. 2016). Usually they are manifested as concentric bands or narrow rings of emission separated by pronounced depletions, or "gaps" (Isella et al. 2016; Cieza et al. 2016, 2017; Loomis et al. 2017; van der Plas et al. 2017; Cox et al. 2017; Fedele et al. 2017, 2018; Dipierro et al. 2018), although at least one disk has a spiral emission pattern (Pérez et al. 2016). Figure 1 shows a gallery of representative ALMA continuum images of disk substructures. In the two specific cases of the HL Tau and TW Hya disks, ALMA data reveal ring+gap features at even finer scales, down to (and pre-

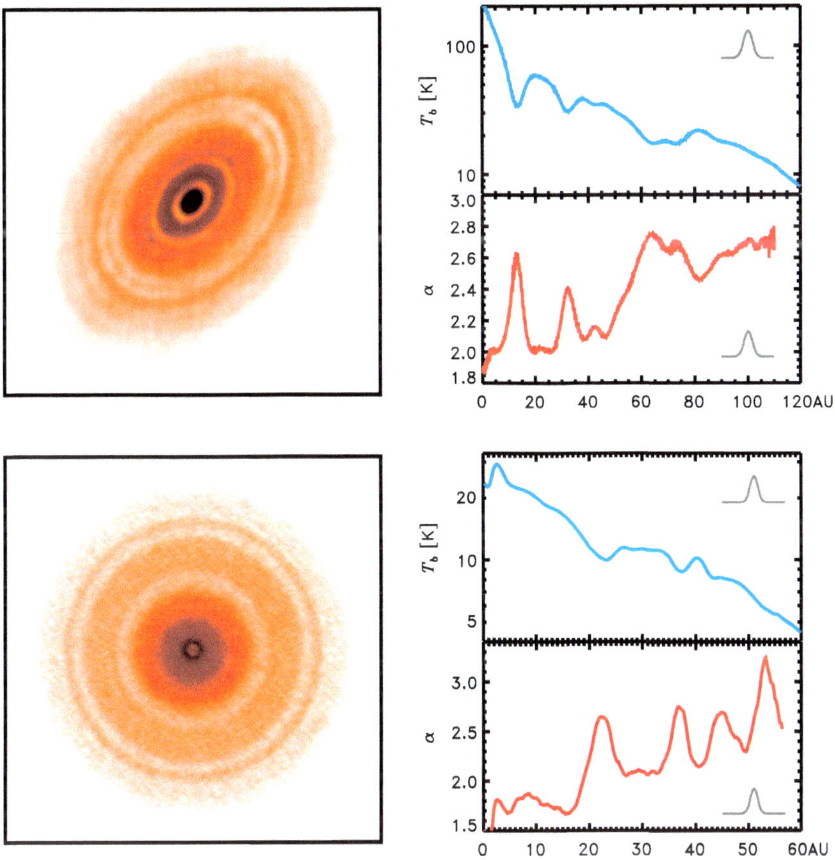

Figure 2. A more detailed look at ALMA observations of the HL Tau (*top*) and TW Hya (*bottom*) disks, based on the analyses of ALMA Partnership et al. (2015) and Huang et al. (2018). The left panels show 290 GHz continuum images. The right panels feature the azimuthally-average radial T_b (*top*) and α (340–230 GHz) profiles. The corresponding beam profiles are shown in gray. Local T_b depletions are accompanied by α enhancements. The high T_b values and low α values in the emission bands / rings are suggestive of high optical depths at these frequencies.

sumably beyond) the ~20 mas (3 and 1 au, respectively) resolution limit of the longest ALMA baseline configuration (ALMA Partnership et al. 2015; Andrews et al. 2016).

2. The Role of the ngVLA

While these ALMA discoveries have rightly ushered in a major shift in the field, they also reveal two crucial, but subtle, problems related to the interpretation of disk substructures. First, the emission substructures that have so far been observed at very high resolution appear to be largely optically thick at frequencies higher than ~200 GHz.

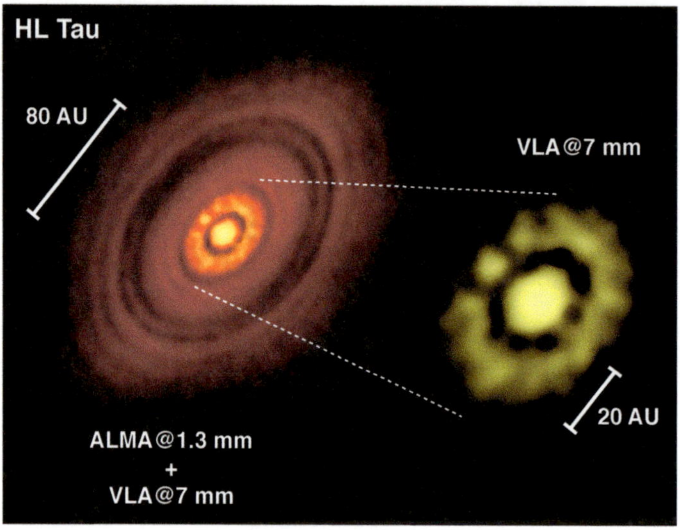

Figure 3. A comparison of the ALMA 230 GHz (*left*; ALMA Partnership et al. 2015) and VLA 43 GHz (*right*; Carrasco-González et al. 2016) continuum images for the HL Tau disk. At the lower frequency of the VLA data, the emission preferentially tracers structures in the inner disk but the observations are very expensive and only possible for the very brightest targets (and are still limited in resolution). The ngVLA will make substantially improved measurements for a large population of disks routine. (*credit: C. Carrasco-Gonzalez / Bill Saxton / NRAO / AUI / NSF*)

And second, the ring or gap features often remain unresolved (or nearly so) even for the longest ALMA baselines (scales $\lesssim$20 mas). Both of these issues are illustrated in Figure 2, which shows synthesized images along with azimuthally-averaged brightness temperature (T_b) and spectral index (α) profiles for the HL Tau (*top*) and TW Hya (*bottom*) disks for the continuum emission around 290 GHz (ALMA Partnership et al. 2015; Huang et al. 2018; see also Tsukagoshi et al. 2016). Note that the scales of the substructures are comparable to the resolutions in the T_b profiles, and also that the T_b values are similar to the expected dust temperatures at the locations where the spectral indices are near the Rayleigh-Jeans value, as would be expected for high optical depths ($\alpha \approx 2$). If this behavior ends up being common in the broader disk population, then high frequency continuum measurements will only be able to offer relatively weak limits on the densities and particle size distributions in these substructures. Moreover, all but the strongest features in the innermost disk ($\lesssim$10 au) will be inaccessible at ALMA frequencies, due to both very high optical depths and limited angular resolution.

The solution to these potential problems is to instead probe these continuum substructures at lower frequencies ($\lesssim$100 GHz), where the optical depths are correspondingly lower (a linear frequency scaling for the optical depth, $\tau_\nu \propto \nu$, is reasonable; e.g., see Ricci et al. 2010). The current VLA is too limited in sensitivity and angular resolution to optimally perform in this area, but nevertheless has produced some early results that demonstrate feasibility for the brightest available targets (e.g., Isella et al. 2014; Marino et al. 2015; Macías et al. 2016, 2017). Figure 3 illustrates perhaps the best example, for the HL Tau disk (Carrasco-González et al. 2016). Perhaps the most compelling result from these forays are the hints of substructure asymmetries that are not visible at higher frequencies due to their high optical depths.

Real progress in this line of inquiry into planetesimal formation will require the sensitivity and resolution capabilities only available with the proposed ngVLA facility. The ideal tracer of the solids concentrated in protoplanetary disk substructures is the 30–100 GHz continuum, which strikes the best balance in sensitivity (emission still bright), optical depth (low enough to reliably estimate densities), and angular resolution (high enough to resolve fine-scale features at disk radii as small as ~1 au). A modest survey could be used to understand the underlying physical mechanisms that control the prevalence, forms, scales, amplitudes, spacings, and symmetry of disk substructures and their presumably crucial roles in the planet formation process.

3. Benchmarks for Measuring Disk Substructures

The experiment design principle most relevant for probing disk substructures is ultimately related to the feasibility of detecting a microwave continuum brightness temperature contrast for a given perturbation in the local surface density of solids (or, more precisely, the associated optical depth). There is a bewildering array of potential *forms* for such perturbations – ranging from simple rings and gaps, to spirals, to azimuthally asymmetric features like vortices, and perhaps even stochastically-distributed pockets of material – depending on the various physical mechanism(s) responsible for modifying the gas pressure gradient. Aside from theoretical ideas and the preliminary work being done with ALMA, we simply do not yet know the diversity of these forms in the general disk population. So, to simplify a discussion of the tractability of studying these features, specifying the morphology is less important than the locations, spatial scales, and amplitudes of the associated particle concentrations.

The baseline temperatures T_d and optical depths τ_ν, and thereby brightness temperatures $T_b(\nu)$, should typically be higher in the inner disk (closer to the host star). As some fiducial reference points to illustrate feasibility, we consider hypothetical substructures at radii of 5 and 40 au in a disk model orbiting a ~1 $M_\odot$ host star at an age of 1 Myr (with a corresponding $L_* \approx 2\,L_\odot$). We assume that the baseline 230 GHz continuum optical depths are 10 and 1 at these locations, based roughly on models of the HL Tau and TW Hya ALMA studies referenced in the previous section, and that the optical depth spectra vary like $\tau_\nu \propto \nu^\beta$ with $\beta \approx 1$. Given a simple model of the stellar irradiation, we expect temperatures of $T_d \approx 75$ and 25 K at 5 and 40 au, respectively. These basic properties also set the expected scale of local perturbations, thought to be tied to the gas pressure scale height $H_p = c_s/\Omega$, the ratio of the sound speed to the Keplerian orbital frequency. At 5 and 40 au in this reference model, $H_p \approx 0.3$ and 2.5 au (i.e., the aspect ratio is $H_p/r \approx 0.06$). Presuming a Gaussian perturbation and a typical distance for nearby protoplanetary disks of 140 pc, we will consider substructures with FWHM sizes of $3H_p$ at 5 au (~10 mas), and ~H_p at 40 au (~40 mas), respectively.

With these assumptions and some measured or expected sensitivity and resolution metrics, we can then calculate the minimum detectable (at a given signal-to-noise ratio, SNR) fractional optical depth perturbation for a given integration time ($t_{\rm int}$). Presuming the local particle opacity does not also change (which is not necessarily the case), there is a one-to-one correspondance between optical depth and surface density perturbations. Because the base optical depths at these locations and frequencies span around

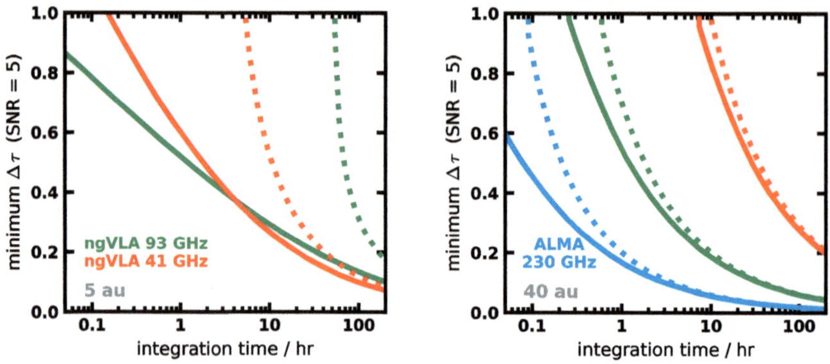

Figure 4. Illustrative examples of the minimum optical depth (fractional) contrast that can be detected at $5\,\sigma$ in a given integration time, for fiducial disk substructures at 5 (*left*) and 40 au (*right*). Solid curves are for τ_ν depletions; dotted curves are for enhancements. The blue, green, and red curves are for ALMA 230 GHz, ngVLA 93 GHz, and ngVLA 41 GHz, respectively. See the text for a detailed discussion.

unity, there is some nuance in the meaning of 'fractional optical depth perturbation' depending on whether we are considering an enhancement, $\tau'_\nu = \tau_\nu(1 + \Delta\tau_\nu)$, or a depletion, $\tau'_\nu = \tau_\nu(1 - \Delta\tau_\nu)$: we will illustrate both scenarios for clarity. However, in either case, we can quantify the sensitivity by first calculating a generic signal-to-noise ratio

$$\mathrm{SNR}_\nu = \frac{1}{\sigma_\nu}\left(\frac{t_{\rm int}}{2}\right)^{1/2}\left|T_d\,(1-e^{-\tau_\nu}) - T_d\,(1-e^{-\tau'_\nu})\right|, \tag{1}$$

where σ_ν is the rms noise level for a given beam area (in T_b units), as a function of $\Delta\tau_\nu$ and t. From that, we can then identify the smallest fractional optical depth contrast (i.e., the minimum $\Delta\tau_\nu$) that could be detected with a specific SNR threshold for a given integration time $t_{\rm int}$. Considering ALMA observations at 230 GHz (Band 6) and ngVLA measurements at 93 (W-band) and 41 GHz (Q/Ka-band), we collect estimates of σ_ν for FWHM beam diameters of 10 and 40 mas from the ALMA sensitivity calculator[1] and the metrics discussion in ngVLA Memo 17[2]. Figure 3 shows the $5\,\sigma$ (SNR = 5) optical depth "contrast curves" for the 5 and 40 au reference substructures. The solid and dotted curves show the cases of optical depth depletion and enhancement, respectively.

These contrast curves in Figure 3 are only illustrative examples, but they demonstrate the general feasibility of ngVLA observations that should play substantial roles in the study of protoplanetary disk substructures. As an approximate point of reference, gas density (thereby pressure) perturbations of ~20% induced by interactions with a ~0.1 $M_{\rm jup}$ planet (Fung et al. 2014), MHD zonal flows (Simon & Armitage 2014), or weak vortices (Goodman et al. 1987) are expected to be sufficient to trap particles in a nominal disk. Moreover, the corresponding perturbations in the local surface density of disk solids could be considerably amplified relative to the modulations in the gas densities (e.g., Paardekooper & Mellema 2006; Pinilla et al. 2012).

---

[1] `https://almascience.eso.org/proposing/sensitivity-calculator`

[2] `http://library.nrao.edu/public/memos/ngvla/NGVLA_17.pdf`

At intermediate disk radii (~tens of au; e.g., the right panel of Figure 3), where optical depths and temperatures are low but substructures should be relatively larger, there are opportunities for synergy with high frequency measurements from ALMA at matching angular resolution. For substructures at these distances from the host star, even relatively modest optical depth contrasts could be detected and resolved in a reasonable ngVLA observing time down to frequencies of ~30 GHz. For optically thicker (or warmer) cases, it would be possible to push down to 10 GHz, and thereby more directly probe concentrations of the largest accesible particle sizes (~few cm).

The real benefit of the ngVLA is more apparent at small disk radii (inside ~10 au; e.g., the left panel of Fig. 3), where optical depths and temperatures are higher and substructures should be smaller. There, the combination of long baselines and low frequencies offered by the ngVLA provide *unique* access to inner disk substructures on the smallest spatial scales. For example, ~20% density depletions at 5 au are accessible at ~30–100 GHz in ~10–30 hours of ngVLA integration, suitable for detecting the gaps in the local distribution of solids that are dynamically carved by young, super-Earth planets (e.g., Ricci et al. 2018). At the higher frequencies probed by ALMA, even relatively large density contrasts could remain hidden for even large integration times if the background optical depths are too high. Moreover, beam dilution due to limited baseline lengths creates an additional ALMA barrier; in short, there is good reason that no 230 GHz contrast curve is shown in the left panel of Figure 3.

4. Quantifying Substructures

The detectability of disk substructures is obviously fundamental, but it is not the sole value of ngVLA in this scientific context. Resolved wideband continuum measurements in the 30–100 GHz range permit more robust estimates of small-scale spatial variations in the continuum optical depth and spectral morphology (e.g., within and between the ~20 GHz-wide high-frequency ngVLA receiver bands), which in turn can provide strong constraints on the amplitude and shape of the particle size distribution and density perturbation of individual substructures. The spatial morphology of the spectrum across such a feature could reveal new insights on the hydrodynamical nature of particle traps: for example, various vortex models make different predictions about where and how strongly particles of different sizes are concentrated with respect to the local gas pressure maximum (e.g., Baruteau & Zhu 2016; Sierra et al. 2017). The spectral dependence of the brightness temperature perturbation from a substructure is itself sensitive to the underlying particle density contrast: if high concentrations of "pebbles" can be inferred, we could rule on the likelihood that fast-acting mechanisms (e.g., the streaming instability; Youdin & Goodman 2005) are indeed the dominant pathways for the formation of planetesimals. In all of these aspects, the ngVLA will be an indispensable and ground-breaking tool for these lines of work.

Finally, the unique ability of the ngVLA to probe substructure asymmetries at small disk radii creates some new opportunities in the time-domain. Long-term monitoring observations would be of extraordinarily high value for any inner disk substructures with a clear azimuthal asymmetry. The motions of such features compared to their expected Keplerian orbital rates can help illuminate the natures of the underlying source of the pressure maximum responsible for the particle trap. As a reference point, a Keplerian orbital displacement of a full 10 mas beam would take ~6 months for a structure at a radius of 5 au in a disk orbiting a solar-mass host.

5. Conclusions

The characterization of small-scale substructures in the spatial distributions of protoplanetary disk solids is the most pressing issue at the forefront of observational planet formation research. The new capabilities in high resolution wideband radio continuum imaging promised by the ngVLA will enable rapid progress in the field, delivering new insights on the particle size distributions in local concentrations of solids and offering unique access to the smallest substructure features at the locations where terrestrial planets (and most giant planets) are expected to form. In doing so, the ngVLA measurements have some natural synergy with ALMA data at higher frequencies, facilitating more robust constraints on particle size distributions and better links with resolved tracers of the molecular gas. Moreover, ngVLA observations of inner disk substructures will prove to be an important complement for tracers of inner disk material probed with the *JWST*, offering a structural template upon which to interpret those spatially unresolved infrared continuum and molecular spectra data. Overall, the ngVLA should prove to be a crucial tool for better understanding the complex metamorphosis of circumstellar disks into young planetary systems.

References

ALMA Partnership, et al. 2015, ApJ, 808, L3
Andrews, S. M., et al. 2016, ApJ, 820, L40
Baruteau, C., et al. 2014, in Protostars and Planets VI, eds. H. Beuther, R. S. Klessen, C. P. Dullemond, & Th. Henning (Univ. Arizona Press: Tucson), 667
Baruteau, C., & Zhu, Z. 2016, MNRAS, 458, 3927
Brauer, F., Dullemond, C. P., Johansen, A., Henning, Th., Klahr, H., & Natta, A. 2007, A&A, 469, 1169
Carrasco-González, C., et al. 2016, ApJ, 821, L16
Chiang, E., & Youdin, A. N. 2010, AREPS, 38, 493
Cieza, L. A., et al. 2016, Nature, 535, 258
Cieza, L. A., et al. 2017, ApJ, 851, L23
Cox., E. G., et al. 2017, ApJ, 851, 83
Dipierro, G., et al. 2018, MNRAS, 475, 5296
Dzyurkevich, N., Flock, M., Turner, N., Klahr, H., & Henning, Th. 2010, A&A, 515, 70
Fedele, D., et al. 2017, A&A, 600, 72
Fedele, D., et al. 2018, A&A, 610, 24
Flock, M., Ruge, J. P., Dzyurkevich, N., Henning, Th., Klahr, H., & Wolf, S. 2015, A&A, 574, 68
Fung, J., Shi, J.-M., & Chiang, E. 2014, ApJ, 782, 88
Goodman, J., Narayan, R., & Goldreich, P. 1987, MNRAS, 225, 695
Huang, J., et al. 2018, ApJ, 852, 122
Isella, A., Chandler, C. J., Carpenter, J. M., Pérez, L. M., & Ricci, L. 2014, ApJ, 788, 129
Isella, A., et al. 2016, Phys. Rev. Lett., 117, 251101
Johansen, A., Blum, J., Tanaka, H., Ormel, C. W., Bizarro, M., & Rickman, H. 2014, in Protostars and Planets VI, eds. H. Beuther, R. S. Klessen, C. P. Dullemond, & Th. Henning (Univ. Arizona Press: Tucson), 547
Loomis, R. A., Öberg, K. I., Andrews, S. M., & MacGregor, M. A. 2017, ApJ, 840, 23
Macías, E., et al. 2016, ApJ, 829, 1
Macías, E., et al. 2017, ApJ, 838, 97
Marino, S., et al. 2015, ApJ, 813, 76
Paardekooper, S.-J., & Mellema, G. 2006, A&A, 453, 1159
Pérez, L. M., et al. 2016, Science, 353, 1519

Pinilla, P., Birnstiel, T., Ricci, L., Dullemond, C. P., Uribe, A. L., Testi, L., & Natta, A. 2012, 538, 114
Pollack, J. B., Hubickyj, O., Bodenheimer, P., Lissauer, J. J., Podolak, M., & Greenzweig, Y. 1996, Icarus, 124, 62
Raymond, S. N., Kokubo, E., Morbidelli, A., Morishima, R., & Walsh, K. J. 2014, in Protostars and Planets VI, eds. H. Beuther, R. S. Klessen, C. P. Dullemond, & Th. Henning (Univ. Arizona Press: Tucson), 595
Ricci, L., Testi, L., Natta, A., Neri, R., Cabrit, S. & Herczeg, G. J. 2010, A&A, 512, 15
Ricci, L., Liu, S.-F., Isella, A., & Li, H. 2018, ApJ, 853, 110
Sierra, A., Lizano, S., & Barge, P. 2017, ApJ, 850, 115
Simon, J. B., & Armitage, P. J. 2014, ApJ, 784, 15
Stammler, S. M., Birnstiel, T., Panić, O., Dullemond, C. P., & Dominik, C. 2017, A&A, 600, 140
Takeuchi, T., & Lin, D. N. C. 2002, ApJ, 581, 1344
Tsukagoshi, T., et al. 2016, ApJ, 829, L35
van der Plas, G., et al. 2017, A&A, 597, 32
Whipple, F. L. 1972, in From Plasma to Planet, ed. A. Elvius, 211
Youdin, A. N., & Goodman, J. 2005, ApJ, 620, 459
Zhang, K., et al. 2016, ApJ, 818, L16
Zhu, Z., Stone, J. M., Rafikov, R. R., & Bai, X. 2014, ApJ, 785, 122

Science with a Next-Generation Very Large Array
ASP Conference Series, Monograph 7
Eric J. Murphy, ed.

Imaging Planetary Systems in the Act of Forming with the ngVLA

Luca Ricci,[1] Andrea Isella,[2] Shang-Fei Liu,[2] and Hui Li[3]

[1]*Department of Physics and Astronomy, California State University Northridge, 18111 Nordhoff Street, Northridge, CA 91330, USA;* `luca.ricci@csun.edu`

[2]*Department of Physics and Astronomy, Rice University, 6100 Main Street, Houston, TX 77005, USA*

[3]*Theoretical Division, Los Alamos National Laboratory, Los Alamos, NM 87545, USA*

Abstract. The discovery of thousands of exoplanets has shown that the birth of planets is a very efficient process in nature. Several physical mechanisms have been proposed to describe the assembly of planets in disks surrounding young stars. However, observational constraints have been sparse on account of insufficient sensitivity and resolution, especially for imaging the inner 10 – 20 au from the star, where most of planets are observed. Thanks to its unprecedented angular resolution and sensitivity at wavelengths where the emission from the circumstellar material is optically thin, the ngVLA has the potential to transform our understanding of planet formation. In this chapter, we present state-of-the-art theoretical models of planetary systems in the act of forming that demonstrate ngVLA capabilities of imaging and follow the temporal evolution of young solar system analogues up to the distance of the Orion Nebula. These images will unveil how planets form and interact with their parental disks, and will shed light on the diverse properties of exoplanetary systems.

1. Introduction

The discovery of thousands of exoplanets over the last couple of decades is a clear sign that the birth of planets is a very efficient process in nature (e.g., Burke et al. 2015). Theories invoke several possible mechanisms to describe the assembly of planets in the disks around pre-main-sequence stars, but observational constraints have been sparse on account of insufficient sensitivity and resolution. Understanding how planets form and interact with the disk is crucial also to illuminate the main characteristics of a large portion of the full planet population that is inaccessible to current and near-future observations.

Perhaps the most puzzling characteristic of the demographics of exoplanets is the broad range of observed orbital radii, which span from a fraction to several tens of au. In particular, the discovery of several "hot Jupiters" has indicated that the orbits of planets can significantly vary right after their formation (Mayor & Queloz 1995). Indeed, theory indicates that forming planets can easily migrate either towards or away from the star through exchange of angular momentum with the parent circumstellar material (Baruteau et al. 2014). Furthermore, once the parent disk has dissipated, planets can change orbital radii due to the interaction with other planets in the same system. This means that, most likely, the orbital parameters of mature exoplanets do not correspond to the initial location of their formation. Knowing where planets of different masses form in the disk is crucial to constrain the dynamical, physical and chemical history of planetary systems.

Only recently, infrared and (sub-)millimeter telescopes have achieved the angular resolution required to spatially resolve the inner regions of nearby proto-planetary disks, down to distances of $\sim 10 - 20$ au from the central star. Recent high angular resolution observations at these wavelengths resulted in the discovery of morphological features, e.g., rings and spirals, in the distribution of circumstellar gas and dust with characteristic sizes larger than 20 au (e.g., Casassus et al. 2013; ALMA Partnership 2015; Isella et al. 2016; Dipierro et al. 2018). These structures are thought to originate from gravitational perturbations by yet unseen planets in the act of forming, and provide a tool to measure planet masses and orbital radii (Jin et al. 2016; Liu et al. 2018). However, these observations also showed that the disk emission within the inner $\sim 10-30$ au is typically optically thick at sub-millimeter and shorter wavelengths (e.g., Carrasco-Gonzalez et al. 2016). This makes very difficult, if not impossible, to detect density perturbations induced by planets in the disk regions where most of the planets are expected to form. The solution to this problem is to image protoplanetary disks at wavelengths longer than about 3 mm, where even the dust emission from the innermost disk regions is optically thin (e.g., Testi et al. 2014). However, existing telescopes operating at these wavelengths such as the Jansky Vary Large Array (JVLA) and the Australia Telescope Compact Array (ATCA) do not achieve the angular resolution nor the sensitivity required to resolve the innermost disk regions.

The unprecedented angular resolution and sensitivity of a next-generation Very Large Array (ngVLA), as currently proposed, would revolutionize our understanding of planet formation by enabling routine observations of the inner $\sim 10-20$ au from the star at spatial resolutions below 1 au. ngVLA observations have the potential to reveal forming planets with masses as low as few Earth masses and unveil the dynamics of solid particles as they interact to form planetesimals. Coupled with its nearly order of magnitude increased sensitivity and angular resolution compared to the current JVLA, the ngVLA makes it possible to investigate the processes of planet formation and disk-planet interaction for large, heterogeneous samples of disks orbiting stars with different properties. Furthermore, by taking advantage of the relatively short orbital periods of planets at < 10 au from Solar-mass stars as well as by its high astrometric precision, the ngVLA can measure the motion of disk substructures as they orbit around the central star on timescales as short as a month. Finally, by imaging and monitoring the temporal evolution of young solar system analogues, the ngVLA would directly inform about the origin of our own planetary system.

In this chapter, we focus on how the ngVLA can be used to map the dust continuum emission in the inner regions of circumstellar disks and reveal the presence of young planets with different masses and orbital radii. Our main goal is to show what planets can generate disk substructures that would be directly detectable with an instrument like the ngVLA in its current proposed design.

2. ngVLA Observations of Planets in the Act of Forming

To investigate the capabilities of the ngVLA in detecting and characterizing planetary systems in the act of forming, we started by performing numerical simulations using the 2D LA-COMPASS (Los Alamos COMPutional Astrophysics Simulation Suite) bi-fluid hydrodynamic code (Li et al. 2005; Fu et al. 2014). This code, which accounts for the dynamics of gas, dust and their mutual interaction, was used to model the gravitational interaction between planets with different masses and orbital radii and gas and dust in the disk. The output of the hydrodynamic simulations were then coupled to the Monte Carlo Radiative Transfer code RADMC3D (Dullemond 2012) to generate synthetic multi-wavelength images of disks perturbed by planets, that were finally used to simulate ngVLA observations of the dust continuum at wavelengths of about 3 mm and 1 cm. As a term of comparison, we also simulated ALMA observations for the same models. This was to investigate the potential improvement allowed by the ngVLA over the observational information that can be gathered for the same disk with ALMA, as well as the possible complementarity of the two instruments, given that ALMA is particularly sensitive to small-scale disk structures at wavelengths shorter than 3 mm. The details of the method and main assumptions are described in Ricci et al. (2018), while the main results are summarized below.

2.1. Revealing Planets in the Terrestrial Planet Forming Region

ngVLA observations that achieve an angular resolution of 5 mas at a wavelength of 3 mm can detect and characterize disk structures generated by planets within 10 au from young Solar-like stars in nearby star forming regions. At a distance of 140 pc, an angular resolution of 5 mas corresponds to a spatial resolution of 0.7 au.

In particular, Figure 1 shows that the ngVLA can detect and spatially resolve strong azimuthal asymmetries in the disk generated by the interaction with planets as massive as Saturn or heavier. These pronounced asymmetries in the dust continuum emission are caused by the trapping of dust particles towards the local maximum of the gas density and pressure, the asymmetric structure of the gas being shaped by the Rossby Wave Instabilities triggered by the gap opened by the planet gravitational torques (Lovelace et al. 1999; Li et al. 2000; Lyra et al. 2009).

Furthermore, the ngVLA is capable of detecting azimuthally symmetric rings and gaps generated by planets with masses below the mass of Neptune of 17 $M_{\oplus}$. The disk models shown in Figure 1 assume a value of 10^{-5} for the α-parameter of the gas viscosity. Ricci et al. (2018) have also shown that in the case of a more viscous disk, with $\alpha = 10^{-3}$, the ngVLA would still resolve gaps opened by planets with masses as low as a few tens times the mass of the Earth.

2.2. Studying the Formation of Super-Earths

In Ricci et al. (2018), we have investigated ngVLA capabilities of detecting planet perturbations by varying the main physical parameters of the disk. For example, it is expected that the opening of a gap at the orbital radius of the planet depends on the planet mass and orbital radius, as well as on the disk viscosity and gas pressure forces, as they both act against the mechanism of gap opening by the planet's gravity (e.g., Goldreich & Tremaine 1980). Theory predics that disks characterized by low viscosity and weak gas pressure forces, the latter corresponding to low vertical pressure scale

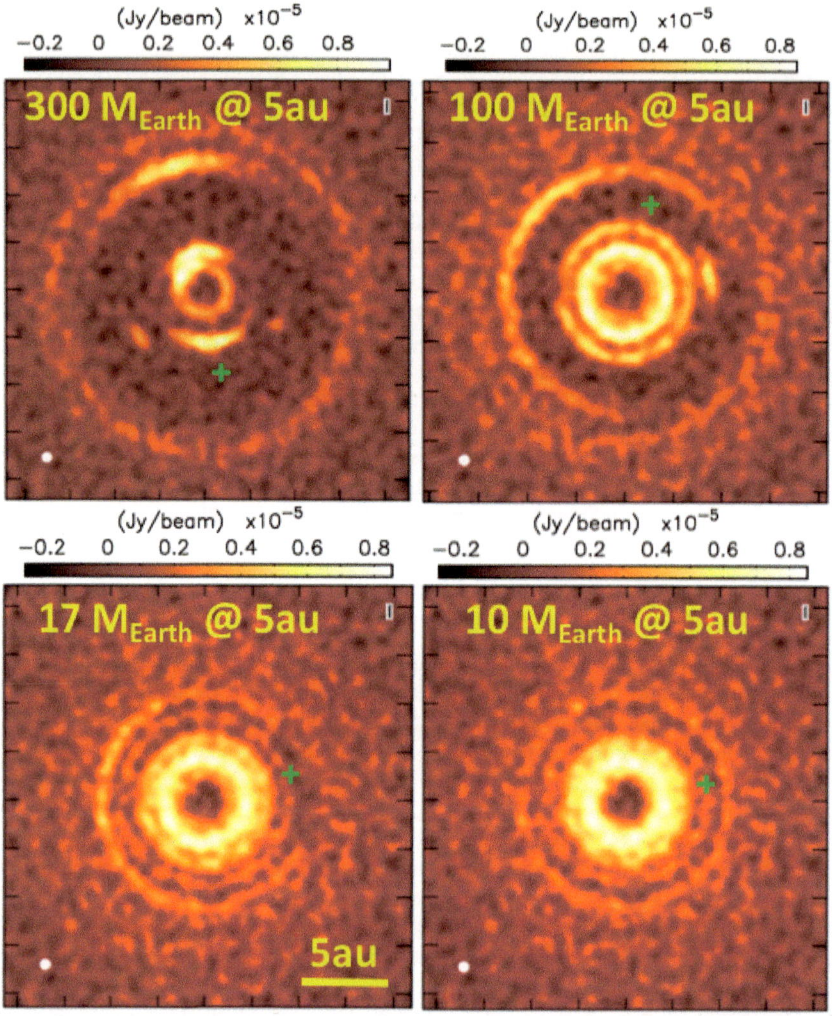

Figure 1. Simulations of ngVLA observations for the dust continuum at a wavelength of 3 mm for a disk with planets at 5 au from the central star. The planet masses are labeled in each panel, and the planet locations are indicated by the green crosses The white ellipses show the size of the synthesized beam, i.e., 5 mas, corresponding to 0.7 au at the assumed distance of 140 pc. In terms of the adopted sensitivity, the rms noise level for these observations is 0.5 μJy/beam (adapted from Ricci et al. 2018).

height, will be prone to gap opening by planets with lower mass than in a disk with higher viscosity and stronger pressure forces.

By varying the level of the viscosity parameter α and the pressure scale height h, we found that ngVLA observations could reveal the presence of planets with masses as low as $\approx 5\ M_\oplus$ at an orbital radius of 5 au. Such observations would provide direct information on the origin of super-Earths, the most common among the different types of planets detected so far, and on the environment in which they form.

2.3. Studying Planet Formation in Diverse Environments

As discussed above, ALMA observations are capable of probing the formation of planets orbiting at more than 10 – 20 au from their parent star in nearby low mass star forming regions such as, for example, Taurus and Ophiuchus. Whereas this provides a huge improvement to what has been possible in the past, these star forming regions cannot be considered as representative of star and planet formation. Most of stars and planets form indeed in massive star forming regions, where both the harsh radiation environment and the higher stellar density are expected to influence the chemistry and architecture of planetary systems. The Solar System itself was though to have formed in a much richer cluster not comparable to nearby star forming regions (Adams 2010).

Based on existing study of circumstellar disks, Ricci et al. (2018) have estimated that ngVLA observations with similar, or higher, signal-to-noise ratios, than those presented in Figure 1 are achievable for at least a couple hundreds disks within 400 pc from the Sun, a region that would encompass a large fraction of the Orion Molecular Complex, including the Orion Nebula Cluster (Robberto et al. 2013). An even larger sample could be obtained by extending these observations to disks in more massive star forming regions, such as Cepheus OB3, at 600–800 pc. Although the spatial resolution achievable at these distances would be a factor of several lower than for closer disks, ngVLA observations would still be able to clearly detect gaps opened by giant planets at orbital radii of about 5 au, and larger.

Therefore, thanks to its high angular resolution and sensitivity, the ngVLA would be capable of studying planet formation in disks in very different astrophysical environments. Besides probing environments that are more representative of star and planet formation than those sampled by the closest young regions, this investigation would allow us to quantify the effects of high stellar density and high energy radiation field on the potential that a disk has to form planets, thus providing critical inputs to theories of planet formation and early evolution.

2.4. The Case of the Young Solar Nebula

As a special case for our disk models, we investigated ngVLA capabilities of probing planetary systems in formation similar to the young Solar nebula. To this end, we performed a simulation of a disk harboring a multiple planets with masse and orbital radii as in the Solar nebula model by Walsh et al. (2011). This model, which was proposed to explain a number of observational facts in our Solar System, predicts that after a few million years from the formation of the Sun, Jupiter, Saturn, Neptune and Neptune were located at orbital radii of about 3.5, 4.5, 6 and 8 au from the early Sun, respectively.

Figure 2 shows the results of our ngVLA simulated continuum observations at a wavelength of 3 mm (left panel) and 1 cm (right). A large gap is opened by the young Jupiter and Saturn between disk radii of $\approx$ 3 and 5 au. Smaller individual gaps are opened by Uranus and Neptune, and the are both detected with the ngVLA. Inside the larger gap opened by Jupiter and Saturn, thermal emission from local concentration of dust at the Lagrangian points of the planets-star system is visible (Lyra et al. 2009).

The assumptions on the disk physical structure are similar as for the models shown in Figure 1 but with a factor of $\approx$ 10$\times$ higher gas and dust densities. These values are

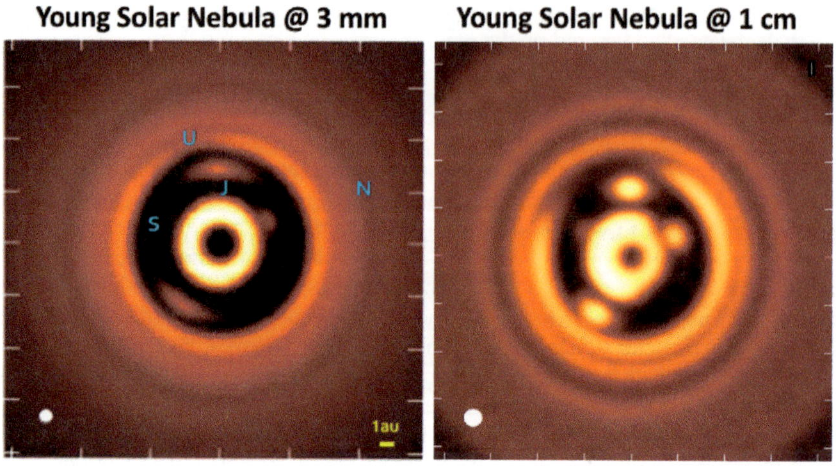

Figure 2. *Left:* Simulation of ngVLA observations for the dust continuum at a wavelength of 3 mm for a disk with planets at about 3.5 (Jupiter), 4.5 (Saturn), 6 (Uranus) and 8 au (Neptune) from the central star. The letters indicate the location of each planet. The white ellipse in the bottom left corner shows the size of the synthesized beam, i.e., 5 mas, corresponding to 0.7 au at the assumed distance of 140 pc. The rms noise level for these observations is the same as in Figure 1. The reason for the higher signal-to-noise ratio than in Figure 1 are the assumed higher gas and dust densities, a factor of $\approx 10\times$ higher than for the models in the previous figure. *Right:* Same as for left panel, but for ngVLA observations at 1 cm. In this case, the surface brightness of the disk model has been rescaled up to produce a total disk flux density of $\approx$ 1 mJy at 1 cm, similar to what estimated for the dust emission at this wavelength in a small sample of bright disks in Taurus (Rodmann et al. 2006). The size of the synthesized beam is 7 mas, or 1 au at 140 pc, and the assumed noise on the map is about 70 nJy/beam.

still within the range of the observed properties of disks in nearby star forming regions. In the case of the simulation at 1 cm, the surface brightness of the disk model has been rescaled up to produce a total disk flux density of $\approx$ 1 mJy at 1 cm, similar to what estimated for the dust emission at this wavelength in a small sample of bright disks in Taurus (Rodmann et al. 2006). This shows that the ngVLA would identify and characterize structures due to the disk-planet interaction in a Solar Nebula-like disk, at least in the brightest disks known in nearby regions.

The comparison between multi-wavelength ngVLA observations at 3 mm and 1 cm can illuminate differences in the dynamics of small solids due to the different coupling with the gas in the disk. As an example, the ngVLA map at 1 cm shown in Figure 2 displays more pronounced rings than at 3 mm, as a consequence of more efficient dust trapping of larger solids inside a gas ring. Observations of this kind would test models for the dynamics and growth of small solids (e.g., Pinilla et al. 2012), which are critical for the understanding of the formation of planetesimals (e.g., Johansen et al. 2014).

2.5. Following the Proper Motion of Disk Azimuthal Asymmetries

With the impressive collecting area and angular resolution of the ngVLA, it would be possible to follow the proper motion of the individual azimuthal structures discussed in the previous sections down to orbital radii of $\sim$ 1 au. Given the short dynamical timescales at these short separations from the central star, multi-epoch observations with the ngVLA would be able to measure these motions relatively quickly.

At the signal-to-noise ratios of the azimuthal structures shown in the left two panels in Figure 1, the angular precision of phased-referenced astrometry with the ngVLA would be of about 1 mas, and the proper motion due to the planet orbital rotation would be detected on a timespan of just a few weeks. A timespan of about 1 week would already be enough to measure the proper motion of a circumplanetary disk detected with the same signal-to-noise ratio but at an orbital radius of 1 au.

The detection of the proper motion of these features would be very important to confirm their actual location in the disk, and discard the hypothesis of background or foreground sources. Also, given the short dynamical timescales, one may expect that some of these structures may show some level of dynamical evolution over timescales that the ngVLA will be able to measure. If so, following the time evolution of the flux density and morphology of these structures would provide further constraints to theory.

3. Uniqueness to ngVLA Capabilities

While the JVLA has frequency coverage up to 50 GHz, it lacks the necessary sensitivity and angular resolution to detect the physical disk structures discussed here within $\approx$ 10 au from the central star. Observations at higher frequencies (> 50 GHz) are also important to measure the dust emission towards its brighter fluxes, therefore increasing the signal-to-noise ratio of the detected structures. In fact, our analysis shows that the 3 mm waveband is really important to be able to characterize the signposts of disk-planet interaction even for an instrument with the collecting area of the ngVLA.

ALMA is significantly more sensitive than the JVLA at detecting disk structures at small scales in young disks. According to the analysis presented in Ricci et al. (2018), ALMA can detect signatures of planets down to $\approx 20-30\ M_\oplus$ at orbital radii < 10 au for the disk parameters considered in that study. However, the gaps opened in these disk regions by planets less massive than Saturn or Jupiter are not well resolved along the radial direction by ALMA. As a consequence, inferring the planet mass would be highly uncertain. Moreover, ALMA cannot detect the multi-ring structures due to Super-Earth planets below $\approx 10-20\ M_\oplus$, as well as several of the azimuthal asymmetries predicted by disk models with low viscosity and which are visible with the ngVLA at 3 mm. Finally, at frequencies below 30 GHz, the Square Kilometre Array (SKA) is not expected to achieve enough sensitivity to study in detail structures generated by planets in disks.

References

Adams, F. C. 2010, ARA&A 48, 47

ALMA Partnership, Brogan, C. L., Perez, L. M., Hunter, T. R., Dent, W. R. F., Hales, A. S., et al. 2015, ApJ, 808, 3

Baruteau, C., Crida, A., Paardekooper, S.-J., Masset, F., et al. 2014, in Protostars & Planets VI, eds. H. Beuther, R. Klessen, C. Dullemond, & Th. Henning (Univ. Arizona Press: Tucson), p. 667

Burke, C. J., Mullally, F., Seader, S., Huber, D. et al., 2015, ApJ, 809, 8
Carrasco-Gonzalez, C., Henning, Th., Chandler, C. J., Linz, H., et al. 2016, ApJ, 821, 16
Casassus, S., van der Plas, G., Perez, S., Dent, W. R. F. et al. 2013, Nature, 493, 191
Dipierro, G., Ricci, L., Perez, L., Lodato, G., Alexander, R. D., Laibe, G., Andrews, S. et al. 2018, MNRAS, 475, 5296
Dullemond C. P., 2012, RADMC-3D: A multi-purpose radiative transfer tool (ascl:1202.015)
Fu, W., Li, H., Lubow, S., Li, S., Liang, E. 2014, ApJL, 795, 39
Goldreich, P., & Tremaine, S. 1980, ApJ, 241, 425
Jin, S., Li, S., Isella, A., Li, H. & Ji, J., 2016, ApJ, 818, 76
Johansen, A., Blum, J., Tanaka, H., Ormel, C., Bizzarro, M., & Rickman, H. 2014, in Protostars & Planets VI, eds. H. Beuther, R. Klessen, C. Dullemond, & Th. Henning (Univ. Arizona Press: Tucson), p. 547
Isella, A., Guidi, G., Testi, L., Liu, S., Li, H., et al. 2016, PhRvL, 117, 1101
Li, H., Li, S., Koller, J., Wendroff, B. B., Liska, R., Orban, C. M., Liang, E. P. T., & Lin, D. N. C. 2005, ApJ624, 1003
Li, H., Finn, J. M., Lovelace, R. V. E., & Colgate, S. A. 2000, ApJ, 533, 1023
Liu, S.-F.,Jin, S., Li, S., Isella, A. & Li, H., 2018, ApJ, 857, 87
Lovelace, R. V. E., Li, H., Colgate, S. A., & Nelson, A. F. 1999, ApJ, 513, 805
Lyra, W., Johansen, A., Klahr, H., & Piskunov, N. 2009, A&A 493, 1125
Mayor, M. & Queloz, D. 1995, Nature, 378, 355
Pinilla, P., Birnstiel, T., Ricci, L., et al. 2012, A&A 538, 114
Ricci, L., Liu, S.-F., Isella, A., & Li, Hui, 2018, ApJ, 853, 110
Robberto, M., Soderblom, D. R., Bergeron, E., Kozhurina-Platais, V., Makidon, R. B., et al. 2013, ApJS, 207, 10
Rodmann, J., Henning, Th., Chandler, C. J., Mundy, L. G., & Wilner, D. J. 2006, A&A, 446, 211
Testi, L., Birnstiel, T., Ricci, L., Andrews, S., Blum, J., et al. 2014, in Protostars & Planets VI, eds. H. Beuther, R. Klessen, C. Dullemond, & Th. Henning (Univ. Arizona Press: Tucson), p. 339
Walsh, K. J., Morbidelli, A., Raymond, S. N., O'Brien, D. P., & Mandell, A. M. 2011, Nature 475, 206

Science with a Next-Generation Very Large Array
ASP Conference Series, Monograph 7
Eric J. Murphy, ed.

Disk Winds and the Evolution of Planet-Forming Disks

I. Pascucci,[1] S. Andrews,[2] C. Chandler,[3] and A. Isella[4]

[1]*Lunar and Planetary Laboratory, Tucson, AZ, USA;* pascucci@lpl.arizona.edu

[2]*Harvard-Smithsonian Center for Astrophysics, Cambridge, MA, USA;* sandrews@cfa.harvard.edu

[3]*National Radio Astronomy Observatory, Socorro, NM, USA;* cchandle@nrao.edu

[4]*Rice University, Houston, TX, USA;* isella@rice.edu

Abstract. Disk winds are thought to play an important role in the evolution and dispersal of planet-forming disks. While high-resolution optical and infrared spectroscopy has identified several disk wind diagnostics, wind mass loss rates remain largely unconstrained mostly due to the lack of spatial resolution to measure the extent of the wind emitting region. Here, we show that the ngVLA will have the sensitivity and resolution to detect and spatially resolve the free-free emission from the fully or partially ionized component of disk winds. Hence, ngVLA observations will be critical to estimate mass loss rates and clarify the role of disk winds in the evolution and dispersal of disk mass.

1. Introduction

Disks of gas and dust around young ($\sim 1 - 10$ Myr) stars are active sites of planet formation. Hence, their evolution and dispersal directly impact which type of planetary systems can form. Accretion through the circumstellar disk is a ubiquitous phenomenon (see Hartmann et al. 2016 for a recent review) and thought to dominate disk evolution at early times (e.g., Alexander et al. 2014). However, the physical mechanism enabling accretion, and therefore global disk evolution, is as yet unclear (e.g., Turner et al. 2014).

The prevailing view for the past 40 years has been that accretion occurs because disks are viscous and transport angular momentum outward via correlated turbulent fluctuations. Magneto-rotational instability (MRI, Balbus & Hawley 1998) would produce the necessary turbulence. In this picture, viscous-dominated evolution lasts for the first few Myr of disk evolution until accretion drops below the mass loss rate from thermal disk winds driven by high-energy stellar photons, also known as photoevaporative winds. At that point photoevaporation limits the supply of gas to the inner (~ 1 au) disk which drains onto the star on the local viscous timescale, of order 100,000 years (e.g., Gorti et al. 2016). Recently, simulations including non-ideal MHD effects have shown that most of the disk is not MRI active, hence not accreting onto the star (e.g., Bai & Stone 2013). Instead, these simulations develop vigorous magneto-thermal MHD winds that extract angular momentum from the disk surface and hence enable accretion (e.g., Gressel et al. 2015). In this scenario, wind mass loss rates are comparable to mass accretion rates even at early times and disk dispersal can be rapid if the disk retains most of its magnetic flux during evolution (Bai 2016 but see also Zhu & Stone 2017). As the surface density evolution of a viscous disk is very different from that of a disk where accretion is driven by MHD winds, and surface density directly impacts planet formation, it is important to understand which physical process dominates. Detecting disk winds, spatially resolving them, and measuring wind mass loss rates is crucial to make progress in this field.

2. Disk Wind Diagnostics: The Role of cm Observations

Identifying disk winds and understanding their origin requires detecting gravitationally unbound/outflowing disk gas within a ~30 au radius from the star (e.g., Simon et al. 2017). Given the typical distance of star-forming regions (~140 pc), this requirement translates into a spatial resolution better than ~100 mas. This is why most direct evidence for disk winds relies on spatially unresolved/high-resolution (Δv ~10 km/s) optical and infrared spectroscopy of gaseous species probing the disk surface (see Ercolano & Pascucci 2017 for a recent review). Kinematic separation of different components in optical forbidden lines has demonstrated that MHD disk winds are present in the inner ~0.5 au for the majority of accreting stars while line ratios constrain the range of temperature-electron densities for the emitting gas (e.g., Simon et al. 2016). Infrared forbidden line profiles hinting to more radially extended disk winds that could be photoevaporative in nature have been also detected toward a dozen disks (e.g., Pascucci & Sterzik 2009, Sacco et al. 2012). *The missing critical parameter for a reliable computation of wind mass loss rates is a measurement of the spatial extent of the wind emitting region.*

As a fully or partially ionized disk surface emits free-free continuum radiation and H recombination lines, long mm and cm-wavelength observations can be also used to detect disk winds (Pascucci et al. 2012; Owen et al. 2013). Current radio facilities have identified candidate disk wind emission from the presence of emission in excess of the thermal dust emission and, when available, multi-wavelength cm observations have been used to exclude other possible sources of excess emission, such as gyro synchrotron non-thermal radiation and emission from very large cm-size grains or very small nanometer-size grains (e.g., Pascucci et al. 2014, Gálvan-Madrid et al. 2014). Even in these multi-wavelength studies, it remains difficult to properly separate free-free emission from a collimated fast (~100 km/s) jet (e.g., Anglada et al. 1998) and from the ionized disk surface, which hampers measuring the disk ionization fraction and the wind mass loss rate. This is illustrated in Figure 1 which reports the 3.3 cm VLA image of GM Aur (Macías et al. 2016), a star surrounded by a disk with a dust cavity (e.g., Andrews et al. 2011; Oh et al. 2016) . With a spatial resolution of $0\overset{''}{.}5$, ~ 70 au at the distance of Taurus-Auriga, an elongation perpendicular to the disk emission is barely detected. This elongation strongly suggests that a jet contributes to the excess cm emission even in disks that are thought to be substantially evolved. *Higher spatial resolution and sensitivity are necessary to separate the jet and disk wind contributions.*

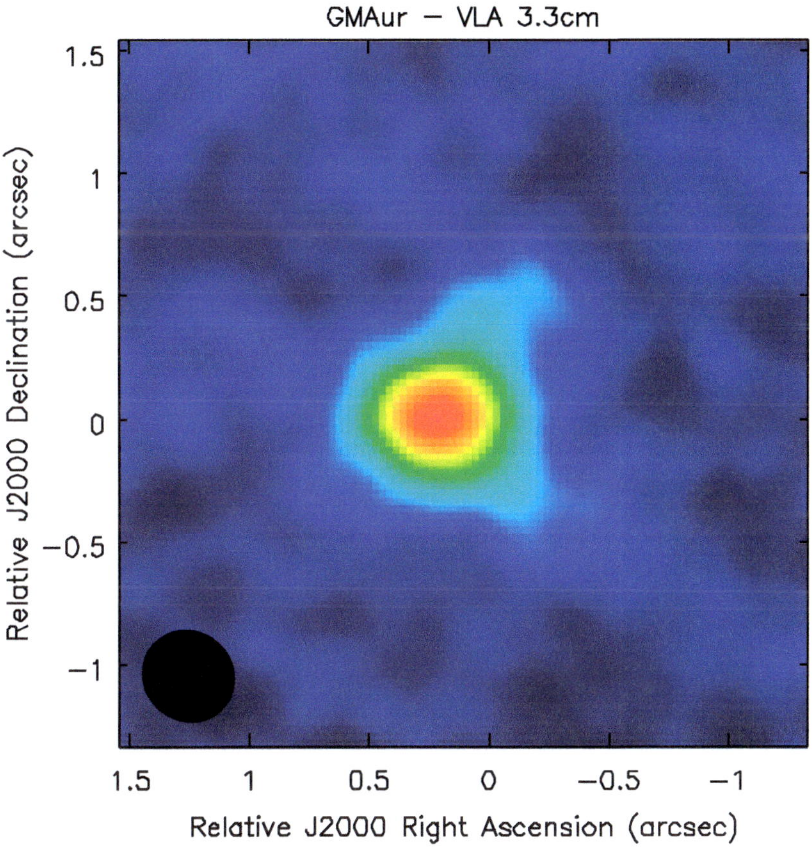

Figure 1. VLA 3.3 cm image of GM Aur by Macías et al. (2016). The north-west elongation is attributed to free-free emission from an accretion driven radio jet.

3. Connection to the Unique ngVLA Capabilities

Spatially resolving disk winds requires deep, multi-frequency radio continuum imaging at high angular resolution, and supplementary spectral imaging of H recombination lines. The continuum measurements target free-free emission from an ionized wind (and potentially jet). As dust thermal emission dominates at frequencies higher than 30 GHz, modest spectral coverage ($\Delta\nu/\nu \sim 0.8$) in the 5 – 30 GHz range is essential. For a photoevaporative wind heated and ionized solely by stellar X-rays and typical star/disk parameters the expected integrated flux density at 8 GHz is ~3μJy (Equation 3 in Pascucci et al. 2012)[1] and should arise within a ~10 au radius (Figure 4 in Owen et al.

[1]An EUV ionized layer would produce more free-free emission (Equation 2 in Pascucci et al. 2012), but observations suggest that the disk receives only a small fraction of the stellar EUV luminosity (Pascucci et al. 2014).

2013). Spatially resolving the emission in two beams and requiring a signal-to-noise of at least 5 means reaching an rms noise level of 0.3μJy/beam. Theoretical estimates for the free-free continuum (and H recombination line emission) from MHD winds have not yet been reported in the literature. However, observations tracing the jet and MHD wind of DG TauA hint at higher flux densities (~300μJy) distributed over the $\sim 0.1''$ beam of e-MERLIN (upper panel of Figure 1 in Ainsworth et al. 2013).

In a fully or partially ionized optically thin region, such as a photoevaporative wind, the H line to free-free continuum ratios increase with frequency but so does the thermal dust emission. Unless the thermal continuum and H lines can be spatially separated, the best frequency to detect H lines is around 30 GHz (lower panel of Figure 2 in Pascucci et al. 2012). At these frequencies, the integrated line flux densities from a photoevaporative wind are comparable to the free-free continuum emission (~ 3μJy) and a few percent of the total continuum (free-free + thermal dust), middle and lower panels of Figure 2 in Pascucci et al. (2012). Line widths and blueshifts with respect to the stellar velocity depend on disk inclination but typical values are ~ 10 km/s (e.g., Alexander 2008; Ercolano & Owen 2010)[2]. Such line emission is probably too weak to detect even with the ngVLA. However, if the line emission from the jet has a similar flux density as the continuum it would detectable, and be particularly important for tracing the accretion and mass-loss in embedded protostars that are obscured at optical and infrared wavelengths. With 300μJy distributed over 10 ngVLA beams obtaining a 5σ detection in a 5 km/s channel (to provide kinematic information) would take a few 10s of hours with the ngVLA; stacking multiple lines (all available in the 30 GHz band) would bring the time required down by a factor of the number of lines stacked.

4. Conclusions

In this contribution we have illustrated that the sensitivity and spatial resolution of the ngVLA can bring the study of disk winds to the next level. Although H recombination lines will remain challenging to detect, the free-free continuum emission of even a partially ionized disk surface can be separated by that of a jet and, for the first time, imaged.

The combination of collecting area and sensitivity at 30 GHz needed for this science is best matched by the ngVLA and will not be available with any other facility in the world. Although ALMA is expected to add Band-1, and perhaps array receivers, it will have the same sensitivity as the current VLA for point-source observations and the best resolution will be just over 100 mas. SKA is not currently expected to operate at frequencies as high as 30 GHz.

References

Ainsworth, R. E., Ray, T. P., Scaife, A. M. M., Greaves, J. S., Beswick, R. J. 2013, MNRAS, 436L, 64

Alexander, R. D. 2008, MNRAS, 391L, 64

Alexander, R., Pascucci, I., Andrews, S., Armitage, P., Cieza, L. 2014, in Protostars and Planets VI, 914, 475

---

[2]If H recombination lines would trace an MHD wind launched inside ~1 au line widths and blueshifts would be substantially larger (e.g., Romanova et al. 2009).

Andrews, S. M., Wilner, D. J., Espaillat, C., Hughes, A. M., Dullemond, C. P., McClure, M. K., Qi, C., Brown, J. M. 2011, ApJ, 732, 42
Anglada, G., Villuendas, E., Estalella, R., Beltrán, M. T., Rodríguez, L. F., Torrelles, José M., Curiel, S. 1998, AJ, 116, 2953
Bai, Xue-Ning & Stone, J. M. 2013, ApJ, 769, 76
Bai, Xue-Ning 2016, ApJ, 821, 80
Balbus, S. A. & Hawley, J. F. 1998, in Reviews of Modern Physics, Volume 70, Issue 1, 1-53
Ercolano, B. & Owen, J. E. 2010, MNRAS, 406, 1553
Ercolano, B. & Pascucci, I. 2017, in Royal Society Open Science, Volume 4, Issue 4
Gálvan-Madrid, R., Liu, H. B., Manara, C. F., Forbrich, J., Pascucci, I., Carrasco-Gonzalez, C., Goddi, C., Hasegawa, Y., Takami, M., Testi, L. 2014, A&A, 570L, 9
Gorti, U., Liseau, R., Sandor, Z., Clarke, C. 2016, in Space Science Reviews, Volume 205, Issue 1-4, 125-152
Gressel, O., Turner, N. J., Nelson, R. P., McNally, C. P. 2015, ApJ, 801, 84
Hartmann, L., Herczeg, G., Calvet, N. 2016, ARA&A, 54, 135
Macías, E., Anglada, G., Osorio, M., Calvet, N., Torrelles, J. M., Gómez, J. F., Espaillat, C., Lizano, S., Rodríguez, L. F., Carrasco-González, C., Zapata, L. 2016, ApJ, 829, 1
Oh, D., Hashimoto, J., Carson, J. C. et al. 2016, ApJ, 831L, 7
Owen, J. E., Scaife, A. M. M., Ercolano, B. 2013, MNRAS, 434, 3378
Pascucci, I. & Sterzik, M. 2009, ApJ, 702, 724
Pascucci, I., Gorti, U., Hollenbach, D. 2012, ApJ, 751L, 42
Pascucci, I., Ricci, L., Gorti, U., Hollenbach, D., Hendler, N. P., Brooks, K. J., Contreras, Y. 2014, ApJ, 795, 1
Romanova, M. M., Ustyugova, G. V., Koldoba, A. V., Lovelace, R. V. E. 2009, MNRAS, 399, 1802
Simon, M. N., Pascucci, I., Edwards, S., Feng, W., Gorti, U., Hollenbach, D., Rigliaco, E., Keane, J. T. 2016, ApJ, 831, 169
Sacco, G. G., Flaccomio, E., Pascucci, I., Lahuis, F., Ercolano, B., Kastner, J. H., Micela, G., Stelzer, B., Sterzik, M. 2012, ApJ, 747, 142
Simon, J. B., Bai, Xue-Ning, Flaherty, K. M., Hughes, A. M. 2017, ApJL, submitted (arXiv:1711.04770)
Turner, N. J., Fromang, S., Gammie, C., Klahr, H., Lesur, G., Wardle, M., Bai, X.-N. 2014, in Protostars and Planets VI, 914, 411
Zhu, Z. & Stone, J. M. 2017, Submitted to the AAS Journals (arXiv:1701.04627)

Science with a Next-Generation Very Large Array
ASP Conference Series, Monograph 7
Eric J. Murphy, ed.

Debris Disks: Structure and Composition

B.C. Matthews,[1,2] J.S. Greaves,[3] G. Kennedy,[4,5] L. Matrà,[6] D. Wilner,[6] and M.C. Wyatt[5]

[1]*National Research Council of Canada, Herzberg Astronomy & Astrophysics Research, Victoria, BC, Canada;* brenda.matthews@nrc-cnrc.gc.ca

[2]*University of Victoria, Department of Physics & Astronomy, Victoria, BC, Canada*

[3]*Cardiff University, Cardiff, Wales, UK*

[4]*University of Warwick, Warwick, UK*

[5]*University of Cambridge, Institute of Astronomy, Cambridge, UK*

[6]*Harvard-Smithsonian Center for Astrophysics, Cambridge, MA, USA*

Abstract. Debris disks, comprised of planetsimal belts and the dust and gas produced by their mutual collisions, are the longest-lived phase of circumstellar disks. Typically much fainter in emission than protoplanetary disks, debris disks can be found in associations in which some members still host protoplanetary disks rich in gas as well as around main sequence stars of all ages. There are even classes of debris disks around AGB stars and white dwarfs. Typically, these disks have been studied through dust emission though an increasing number of young disks are now found to host some gas which may be remnants of the protoplanetary disks or second-generation gas. The ngVLA will have a particular niche in the study of the dust population of these disks, since the size distribution of the dust can be derived from the spectral index of the spectral energy distribution from the far-infrared to the centimetre. With the longest lever arms provided by ngVLA, the size distribution can be characterized, testing models of collisional evolution. The ngVLA could also be capable of measuring the quantities of HI and OH emission associated with the disks. HI and OH are by-products of the dissociation of water, which is expected to be a primary component of outgassed material from cometary collisions. A probe of the water content of extrasolar systems has important implications for the delivery of water to terrestrial, potentially Earth-like planets.

1. Introduction

Debris disks are collisionally-dominated circumstellar disks comprised of planetesimal belts of solid bodies of 100s to 1000s of km in size. These collisionally evolve to populate circumstellar dust and gas disks that may exist co-spatially with the planetesimal bodies, but also may be subject to radiation and drag forces which act to remove small grains and gas from the systems (e.g., Hughes et al. 2018; Matthews et al. 2014; Wyatt 2008). Detectable around stars from very young ages (a few Myr) to evolved stars of many Gyr, debris disks represent the longest-lived phase of circumstellar disks. Their presence is an indication that planet formation processes in the system were successful to point of producing the parent bodies of the observed dust belt. The timescales for removal of the observed dust grains in all cases is greatly exceeded by the age of the star, meaning the dust must be continuously replenished through a collisional cascade from the largest sized objects to the dust grains.

The presence of bright debris disk emission can serve as a signpost for planetary mass bodies and orbital evolution of those bodies, which can stir the planetesimal disk and lead to enhanced collisional rates, even at very late ages. Observations of debris disks at mm and cm wavelengths have an advantage over optical scattered light imaging or thermal imaging at infrared wavelengths because the mm and cm observations probe the largest dust grains that can be detected in the disk, and they are most likely to be found near their place of creation (i.e., in the planetesimal belts), since they are not as easily removed from the systems through radiative or stellar wind forces as the smaller dust grains. They can also trace, through modeling, the underlying mass of the planetesimal belts which cannot be measured directly in extrasolar systems.

Debris disks are fundamental parts of planetary systems and many systems show evidence of multiple planetesimal belts, akin to the Solar System's own asteroid and cometary belts. Debris disk material can interact with the planets in a system, creating recognizable substructure in the disk due to resonances or secular orbital evolution. Thus, debris disks offer an opportunity to detect evidence of planets of relatively low mass at relatively large orbits (for example, a Neptune mass planet at Neptune's orbit) around stars as distant as the Sco-Cen association at 125 pc, which are not detectable by any other planet detection technique. Even in cases where no significant asymmetry is observed, the inner edge of the disk varies in width and eccentricity with the proximity and mass of the nearest planet mass, and so high resolution imaging of disks can constrain the presence and characteristics of a planet in such systems (e.g., η Corvi; Marino et al. 2018).

2. Goals and Scientific Importance

2.1. The Size Distribution of Dust in Debris Disks

One of the key components of a debris disk is the population of dust grains produced through the collisional evolution of planetesimals. The planetesimals link the debris disks to the planet formation process, but they themselves are not observable. Instead, we must glean information about the collisional evolution of the disk through measurements of the mass in dust grains produced, the distribution of the scattered light or thermal emission, and the spectral energy distribution, which reveals the temperature of the dust grains (and therefore their minimum radial separation from the star), the dominant size of dust grains present, the size distribution of those grains and, in the cases of mid-IR and other spectroscopy, the composition of the grains themselves. The clearest window into the size distribution is the spectral slope between the far-infrared and cm, which reveals whether the dust component of the disk has been generated in a collisional cascade or a more stochastic process, such as a recent collision in the disk. This latter explanation may be responsible for some of the observed bright debris disks around relatively old stars.

In the mm/submm, ALMA provides a means of measuring dust populations down to relatively low masses, dependent on our understanding of the emissivity of the grains.

With ALMA, we can also measure the spectral slope in the mm/submm through measurements in multiple bands. Due to the emissivity of individual grains, the mm data do not sample cm-scale grain sizes. Whether the spectral slope is constant or turns over in the cm also has implications for the collisional evolution of the disk. Measurement of the size distribution toward a significant number of disks (especially at young ages) will provide important insight into their (early) evolution. To sample the youngest disks, we need sensitivity to their emission in nearby star forming regions, such as Sco-Cen (d = 125 pc).

It is not the case that all observed disks present a classical Dohnanyi (1969) profile for their size distribution, which is a power law of the form $dN/da \propto a^{-q}$ where $q = 3.5$. There is in fact a range of measured values for q between 3 and 4 through numerical simulations (e.g., Pan & Sari 2005; Pan & Schlichting 2012; Gáspár et al. 2012). Draine (2006) showed that this value can be derived directly from the long wavelength slope of SEDs, so coverage throughout the range of the far-IR and cm improves our measure of q. Coverage into the cm regime could reveal whether there are any kinks in this distribution for any size scale to cm size grains.

Resolved imaging at multiple wavelengths would reveal a radially resolved spectral index in the mm/cm regime. These spectral indices can be different if the debris is fed by different types of planetesimals or different collisional processes, such as a steady state collisional cascade versus a giant planet impact in the disk (e.g. HD 111520, Draper et al., 2018, in prep). The spectral index is different for fluffy versus icy particles.

When sources are well resolved from the star, it is also possible to isolate the stellar emission from the disk, eliminating a potentially complicating factor at these wavelengths, since the stellar photosphere is no longer the dominant source of stellar emission (e.g., Cranmer et al. 2013). The chromosphere of the star produces a rise in emission in the mm and cm that can mimic the effects of a debris disk (i.e., an excess above the level of the photospheric emission extrapolated from the far-IR). Therefore, to interpret the debris disk emission properly, the resolution should be sufficient to separate the stellar emission spatially from the disk emission, which is typically $\sim 40-100$ AU, driving us to resolutions of at least $1''$, but with good sensitivity to scales out to 100s of AU.

When azimuthal structure is present in a disk, the collision rate changes as a function of both radius and azimuth in the disk. Two scenarios for producing this type of structure are i) long-term perturbations from an eccentric planet, which offsets the center of the disk from the star, and ii) capture of planetesimals into resonances during outward planet migration, which creates overdensities (clumps) of material that rotate with the planet. The disk structure resulting from these scenarios is strongly wavelength dependent (Wyatt 2006) for two reasons: the first is the aforementioned differences in radiation forces as a function of grain size, and the second is that the collision rate is now a function of location in the disk. Observations at a wide range of wavelengths, especially from the mm-cm, is diagnostic of the size distribution, where collisions are prevalent, and the likely nature of the (probably unseen) perturber.

While there are only a few detections of debris disks in continuum emission at cm wavelengths to date (e.g., Greaves et al. 2012; Ricci et al. 2012, 2015; MacGregor et al. 2016; Marshall et al. 2017), the ngVLA would significantly expand the potential for this science in the centimeter regime. The real challenge is sensitivity, and with enhanced sensitivity, nearby disks could be imaged in a range of wavelengths. In concert with

ALMA, which has now produced well resolved images of about two dozen debris disks (and counting!), these data could probe the size distribution of the solids within the disks. The effect of differing dust temperature is mitigated at longer wavelengths. The spatial distribution of grains of different sizes will be measurable by comparing resolved imaging from the ngVLA and ALMA.

There are at least 50 such systems in a feasible brightness range within 100 pc, though not all of these will show detailed structure that requires the highest spatial resolution. A typical observing strategy would be to initially image tens of systems at relatively low resolution to identify the subset that exhibit asymmetries, and then dedicate significant effort to characterizing that subset. Simulations (see Figure 1) show that we will be able to discriminate relatively modest changes in spectral slope with a combination of ALMA and ngVLA data. The lower resolution imaging will yield broad-brush measures of disk sizes as a function of wavelength, and disk-integrated mm spectral slopes to study the dust size distribution.

To address this science, we would require the measurement of flux densities over a range in wavelengths (coordinated with the mm/submm) toward a significant numbers of targets. The goal would not be to resolve each of the disks, but get a bulk measure of the flux density at each wavelength. In nearby star-forming regions, the numbers of detected debris disks number in the 10s, with 100s of potential targets in total, depending on the sensitivity limits. In the mm, debris disks around nearby stars, such as those in the 50 pc TW Hya association have flux densities in the ~1 mJy range at ALMA Band 6. There is variation, but this is a reasonable flux density expected in ALMA Band 6. Typically, we observe the flux density falling off with a β of 1, $\alpha = 3$ in the R-J limit, where $S_\nu \propto \nu^{-\alpha}$ and $\alpha = 2 + \beta$. This implies expected flux densities in the 3mm band (100 GHz) to be 10s of μJy, on the edge of observable with ALMA (2 μJy/beam rms in Band 3 requires > 1 day integration time with ALMA, but reachable in 1 hour of integration time with the ngVLA. At longer wavelengths, such as 7 mm, the flux densities will be even lower, on the order of 1 μJy/beam. This would require rms levels on the order of a fraction of a μJy/beam (0.2 μJy for a 5σ detection). Not all disks would necessarily be this faint, but reaching this level of rms would increase the number of viable disk candidates. Noise values as low as 0.2 μJy would be required, which are achievable in 5 hours with the ngVLA, but not with Band 1 with ALMA. For this science, we would not require the highest resolutions achievable with ngVLA; in fact, we will not be concerned with resolution in this project at all. Most debris disks are no larger than 40-100 AU in radius, and at 50 pc, this implies a $2 - 4''$ beam will suffice for this science without resolving the disk emission. Baselines of 0.2 km will be more than sufficient to probe these objects ($3''$ at 100 GHz, $1\rlap{.}''2$ at 40 GHz), resulting in good beam matching to the targets at 50 pc. For closer targets, a larger beam may be needed to prevent surface brightness considerations from being taken into account.

2.2. Presence and Nature of Gas

Debris disks were once thought to be essentially devoid of gas, but an increasing number of young disks around A and F stars have detected gas, typically CII or CO (for a detailed review, see Hughes et al. 2018). Around young stars particularly, the nature of the detected gas is ambiguous: is it collisionally produced like the dust, or is the gas a remnant of the putative protoplanetary disk? With the ngVLA, we can detect atomic hydrogen (HI) through its 21 cm (1.42 GHz) line in nearby debris disks. These observations will establish whether HI is created from photodestruction of water released from

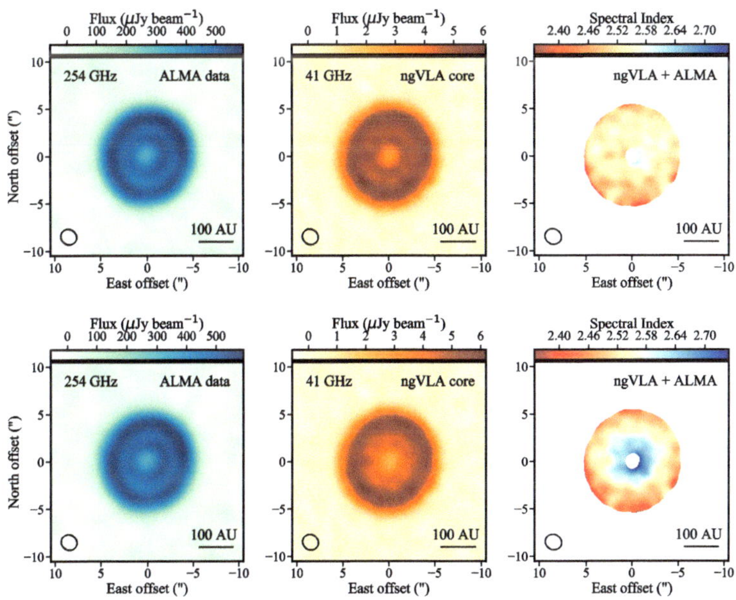

Figure 1. ALMA 1.2 mm observations (Marino et al. 2018) and ngVLA 7 mm simulations of the debris disk around HD 107146, a 100 Myr-old Solar analog (spectral type G2V) at d = 27.5 pc. The gap in the disk at a radius of 80 AU may be caused by the dynamical influence of an unseen planet. The two rows illustrate two scenarios, one where the spectral index is invariant with position (upper row), and one where the spectral index changes with radius between 20 to 170 AU by an amount that corresponds to a decrease of 0.15 in q, the power-law index of the grain size distribution (lower row). For each scenario we show: *(Left)* ALMA Band 6 image smoothed to match the $1''\!.67$ (46 AU) resolution of the ngVLA imaging simulation; *(Center)* Simulated ngVLA Band 5 image using the 114 core antennas, at 41 GHz, with 10 hr integration time, where the disk total flux density was extrapolated from mm/submm data using the observed spectral index of 2.6); *(Right)* Resolved spectral index map of the emission obtained using ALMA and the ngVLA (masked where signal-to-noise ratio falls below 20). The simulated ngVLA observations clearly recover the radial dependence in the spectral index the lower row, easily distinguishable from the simulation with no radial dependence in the upper row.

colliding exocomets, or a whether it is remnant of the protoplanetary phase of planet formation. In the exocometary scenario, the HI measurement will be compared to on-going measurements of carbon-bearing species (CI, CII , CO with *Herschel*/ALMA) to estimate the C/H ratio and water content of exocomets at the epoch where volatile delivery events are most likely to take place.

Establishing the water content of planetesimals has far reaching implications for the understanding of how and how much water is delivered to terrestrial planets. Detection of water from comet outgassing in the solar system is typically done via measurements of OH, a dissociative product of water under solar radiation. Debris disks are created by the collisional destruction of planetesimal bodies; as such, they can provide a window into the water content of planetesimals around other stars. Unlike the solar system, there is no way to observe individual planetesimals within debris disks. The importance of water for the evolution of life as we know it is well known; determining whether the fraction of water in extrasolar comets is comparable to those in the solar system would provide evidence of how much variation there is in the water delivery systems in other solar systems.

Direct detection of H_2O is prohibitively difficult due to its short photodissociation timescale, but its products OH and HI provide a means of determining how much water could be present in planetesimals. OH provides the most direct probe, since one OH molecule is produced for each H_2O molecule dissociated, but similarly to water should be readily photo-destroyed in typical debris disk environments. Therefore, neutral hydrogen (HI) produced via the first dissociation of water and the subsequent dissociation of OH is the most promising probe of the H_2O content of extrasolar planetesimals. In addition, there remains a debate on whether gas-bearing debris disks around very young systems may instead be tenuous remnants of the protoplanetary disks, which would have significant implications on the lifetime of protoplanetary disks and the formation of gas giant planets. In this scenario, HI is readily produced by photodestruction of H_2 gas due to UV and/or X-ray radiation easily penetrating through an increasingly low mass, and hence increasingly optically thin gas disk. Detection of low levels HI - as would be enabled by the ngVLA - is therefore of paramount importance to either estimate the water content of exocomets or to assess the behavior of the bulk of the gas mass in a dispersing protoplanetary disk, as well as to distinguish between these two scenarios.

Deep, high-sensitivity observations of HI at 1.42 GHz are required for detection. To optimize the latter, we require a spectral resolution equal to the expected line width (few-few tens km/s depending on system viewing geometry), and a spatial resolution of order the maximum expected disk diameter ($1 - 10''$ for typical systems between 10 and 100 pc). We do not require imaging capability; resolution of the disks would require greatly enhanced sensitivity due to the low surface brightness of the emission. For the unresolved case, the sensitivity required is 50 μJy/beam per 2 km/s channel for a $2''$ beam to be able to achieve unresolved detection of 0.08 M_{Earth} of HI produced from exocometary water in a planetary system at 50 pc, assuming a typical belt location at 50 AU distance from the star viewed face-on. Models of exocometary gas release (Matrà et al. 2015; Kral et al. 2017) predict $\sim$ 4 systems to be detectable at a significance $> 3\sigma$. If the gas (already detected in CO for 11 disks) is a tenuous remnant of the protoplanetary disk, the generally higher HI masses predicted (in the range $\sim 1 - 10\,M_{Earth}$; Kamp et al. 2007) would be detectable at high significance in systems as far as $\sim$ 100 pc.

3. Limitations of Current Instrumentation

This science case requires resolved imaging of the dust emission in the debris disk with sufficient sensitivity to measure of the spatial distribution of dust emission. To establish the dust size distribution, the disks must be resolved at multiple wavelengths. Where possible and necessary, the disks should be resolved over tens of beams to quantify the azimuthal structure at multiple wavelengths, to measure the size distribution, identify where dust is being produced in collisions, and map out the parent planetesimal locations. Frequencies from 30 – 80 GHz provide good sensitivity to thermal dust emission from large grains, the sensitivity at 30 GHz will the limiting factor in what can be imaged. An extension to 100 GHz would be useful to better define the spectral shape.

As noted above the super-sample of probable targets contains about 50 debris disk systems within 100 pc. Typical detections at 1mm are about 1-10mJy, corresponding to 10 – 100 μJy at 4 mm and 1 – 10 μJy at 1 cm. A subset of the nearest systems may not be amenable for ngVLA because their spatial extent is too large, but most have sizes from 1 – 10″. Typically disks have a relatively narrow ring structure, so the disk surface brightness scales with radius (not radius squared), and therefore the desired resolution is roughly 0.″01 – 1″, depending on the specific brightness, structure, and geometry of the system in question (which may be estimated from observations at other wavelengths or from lower resolution reconnaissance). To study the spatial structure requires several tens of beams around the disk; assuming 30 beams at a signal-to-noise ratio of 3 therefore dictates a sensitivity of 0.01 – 0.1 μJy/beam at 1 cm (this is the most stringent requirement in terms of sensitivity, imaging at shorter wavelengths to the same sensitivity would require less time).

4. Uniqueness of ngVLA

Imaging debris disks requires very high sensitivities as the disks have fractional luminosities significantly below those of proto-planetary disks (i.e., $< 10^{-2}$, down to Solar levels of 10^{-7}, which are not yet detectable; ratios of 10^{-5} are typical of detected disks to date). The masses of debris disks in dust grains are also low (several to a fraction of a lunar mass), and dust opacities drop steeply with wavelength. For this reason, very few debris disks have been detected longward of 3 mm (e.g., Ricci et al. 2012; Marshall et al. 2017). However, the increased sensitivity of the ngVLA will make it competitive for the detection of disks detected by ALMA at Bands 6 and 7 at wavelengths of 3 mm, which ALMA cannot achieve in Band 3. This will permit a probe of the cm-scale solids in the disk, which are expected to hold significantly more mass than the mm grains. Measuring the size distribution from the spectral slope in the mm/cm regime will reveal how the size distribution varies in the disk, which will permit testing of collisional models.

No other facility will be able to address this science. The Jansky VLA is not sensitive enough to probe debris disks effectively excepting the brightest nearby ones. The sensitivities are such that continuum emission from debris disks, even at $\sim$ 50 pc, could be detectable to the notional model of the ngVLA in timescales of hours at 42 and 100 GHz, which is unachievable to ALMA at these wavelengths. The SKA-1 doesn't approach these observing frequencies. The very nearby debris disks, while brighter, will have different requirements, since they are more likely to be resolved.

The ngVLA is the only observatory that can achieve the required high sensitivity to detect low HI masses in emerging planetary systems at 1.42 GHz, and this is the only way to measure the total hydrogen content in gas-bearing debris disks. In the exocometary release scenario, the ngVLA would be the only facility allowing us for the first time to probe the water content of exocomets, allowing comparison with the current-day Solar System comets and how they used to be in the epoch of water delivery to the Earth.

5. Complementarity

The measurements of the size distribution in debris disks cannot be done solely with the ngVLA, but nor can it be done by any other individual instrument. To map out the disk structure as a function of grain size requires disk observations over a wide range of wavelengths, from the optical to cm, all at a sufficient resolution to detect the disk over tens of beams. Thus, instruments that can image disks, such as ALMA, *JWST*, WFIRST, OST, and LUVOIR all have a role to play. ALMA of course will remain a highly effective instrument for the study of these disks in Bands 6 and 7, but most are be too faint to obtain effective observations out to 50 pc at 3mm and 7mm (Bands 3 and 1). The unique contribution of the ngVLA is that it is best suited to detecting the particles least affected by radiation forces at 3, 7 and 10 mm, so will yield the spatial distribution of the parent planetesimals, from which the dust detected at all other wavelengths ultimately originates.

Observations of HI with the ngVLA will be complemented by the SKA for targets in the Southern sky, completing the all-sky sample. As well, HI observations will be combined with CII + OI observations available from *Herschel* and planned with the increased sensitivity of the OST, as well as ALMA CO + CI observations to reach a complete view of the volatile composition of extrasolar planetesimals (in the exocometary scenario) or to explore how the chemical composition of a protoplanetary disk evolves as it disperses (in the primordial scenario). Finally, *JWST* will also provide valuable information on the disks' molecular content through detection of ro-vibrational transitions.

References

Cranmer, S. R., Wilner, D. J., & MacGregor, M. A. 2013, ApJ, 772, 149

Draine, B. T. 2006, in Astronomical Society of the Pacific Conference Series, Vol. 348, Astrophysics in the Far Ultraviolet: Five Years of Discovery with FUSE, ed. G. Sonneborn, H. W. Moos, & B.-G. Andersson, 58

Gáspár, A., Psaltis, D., Rieke, G. H., & Özel, F. 2012, ApJ, 754, 74

Greaves, J. S., Hales, A. S., Mason, B. S., & Matthews, B. C. 2012, MNRAS, 423, L70

Hughes, A. M., Duchene, G., & Matthews, B. 2018, ArXiv e-prints, arXiv:1802.04313

Kamp, I., Freudling, W., & Chengalur, J. N. 2007, ApJ, 660, 469

Kral, Q., Matrà, L., Wyatt, M. C., & Kennedy, G. M. 2017, MNRAS, 469, 521

MacGregor, M. A., Lawler, S. M., Wilner, D. J., et al. 2016, ApJ, 828, 113

Marino, S., Bonsor, A., Wyatt, M. C., & Kral, Q. 2018, MNRAS, arXiv:1806.01289

Marshall, J. P., Maddison, S. T., Thilliez, E., et al. 2017, MNRAS, 468, 2719

Matrà, L., Panić, O., Wyatt, M. C., & Dent, W. R. F. 2015, MNRAS, 447, 3936

Matthews, B. C., Krivov, A. V., Wyatt, M. C., Bryden, G., & Eiroa, C. 2014, Protostars and Planets VI, 521

Pan, M., & Sari, R. 2005, Icarus, 173, 342

Pan, M., & Schlichting, H. E. 2012, ApJ, 747, 113
Ricci, L., Maddison, S. T., Wilner, D., et al. 2015, ApJ, 813, 138
Ricci, L., Testi, L., Maddison, S. T., & Wilner, D. J. 2012, A&A, 539, L6
Wyatt, M. C. 2006, ApJ, 639, 1153
—. 2008, ARA&A, 46, 339

Science with a Next-Generation Very Large Array
ASP Conference Series, Monograph 7
Eric J. Murphy, ed.

Stellar Emission as a Source of Flux Bias in Debris Disks

Jacob Aaron White,[1] Jason Aufdenberg,[2] Aaron C. Boley,[3] Peter Hauschildt,[4] A. Meredith Hughes,[5] Brenda Matthews,[6] Attila Moór,[1] and David J. Wilner[7]

[1]*Konkoly Observatory, Research Centre for Astronomy and Earth Sciences, Hungarian Academy of Sciences, Konkoly-Thege Miklós út 15-17, 1121 Budapest, Hungary,* jacob.white@csfk.mta.hu

[2]*Physical Sciences Department, Embry-Riddle Aeronautical University, 600 S Clyde Morris Blvd., Daytona Beach, FL 32114, USA*

[3]*Department of Physics and Astronomy, University of British Columbia, 6224 Agricultural Rd., Vancouver, BC V6T 1T7, Canada*

[4]*Hamburger Sternwarte, Gojenbergsweg 112, 21029 Hamburg, Germany*

[5]*Department of Astronomy, Van Vleck Observatory, Wesleyan University, 96 Foss Hill Dr., Middletown, CT 06459, USA*

[6]*Herzberg Institute, National Research Council of Canada, 5071 W. Saanich Road, Victoria, BC V9E 2E7, Canada*

[7]*Harvard-Smithsonian Center for Astrophysics, 60 Garden Street, Cambridge, MA 02138, USA*

Abstract. Our understanding of stellar atmospheres and our ability to infer architectures of extrasolar planetary systems rely on understanding the emission of stars at submillimeter to centimeter wavelengths. In this chapter we describe how unconstrained stellar emission can interfere with the accurate characterization of circumstellar debris. The ngVLA is the only facility with the sensitivity that allows for the observations of a broad range of stellar spectral types in a feasible amount of time. The observations will enable the building and testing of accurate models of stellar emission, which in turn are required for evaluating both the occurrence and abundance of debris over the proposed wavelength range of the ngVLA.

1. Introduction

The next-generation Very Large Array (ngVLA), with its proposed technical specifications, will extend the search and study of debris disks to much further distances than is currently possible. The high sensitivity over the proposed wavelength range will also allow for the study of a previously difficult (and sometimes impossible) to explore size range of disk particles. For a given debris system, if the disk emission is unable to be separated spatially from its host star, an accurate model of the stellar emission is required in order to constrain the flux contributions of the star and the disk. This is particularly true for warm/hot debris systems such as exo-zodiacal dust and exo-asteroid belts.

Emission from stars over the proposed wavelength range of the ngVLA is nontrivial and generally not well-constrained (Cranmer et al. 2013), with the Sun being the most thoroughly studied star in this regime (see Liseau et al. 2015, for other G-type stars). In the atmosphere of the "quiet" Sun, the submm/cm continuum radiation is due primarily to free-free emission (Dulk 1985; Loukitcheva et al. 2004). Quiet Sun models predict a 1 mm brightness temperature (T_B) of ~ 4700 K, or ~ 80% of the Sun's photosphere T_B (Dulk 1985). The "active" Sun, with strong magnetic fields, is difficult to model because there are many contributing emission mechanisms (e.g., umbral oscillations and explosive events; Wang 2011; Wedemeyer et al. 2016). The T_B spectrum of the Sun varies significantly, with a minimum in the far-infrared/submm. This is followed by a pronounced increase in flux at mm wavelengths and very high T_B at cm wavelengths (see Figure 1).

The Sun cannot be used as a template for all stars, however, due to the differences in stellar atmospheres and magnetic activity. This is shown in Figure 1 with observations of A-type stars Sirius A, Altair, and Fomalhaut. In contrast to the Sun, the Sirius A data and corresponding PHOENIX models show that the brightness temperature continues to decrease with increasing wavelength over the observed range (White et al. 2018). Together, the Sun and Sirius A show that when considering the flux from a star, it can not be assumed that the submm/cm T_B is well represented by the photosphere's T_B. It may also not be reasonable to extend the far-infrared brightness temperature to longer wavelengths, as is often done in the literature. Further still, it can not be assumed that a given spectral type, for example an A-type star, will have the same general T_B spectrum as another A-type star. This again is seen in Figure 1 when comparing Altair, Fomalhaut, and Sirius A and is discussed further in § 2.

In this chapter, we explore how the ngVLA can be used to observe a large number of stars over the proposed wavelength range. These observations can be used to inform stellar atmosphere models (e.g., PHOENIX), which must be accurately known to study unresolved circumstellar debris.

2. Implication for Debris Disks

Poor characterizations of the radio flux from the host star in a debris system can nontrivially affect measurements of the occurrence and abundance of circumstellar debris. To see this, consider the following cases. Fomalhaut is an A3V star with a well-known extended debris ring at 140 au from the star and additional potential IR excess at much closer orbital distances (e.g., an asteroid-like belt; Acke et al. 2012). However, adopting different stellar models significantly changes the amount of this inferred excess (Su et al. 2016; White et al. 2017a). In particular, ALMA observations at 0.87 mm and 1.3 mm find a T_B of ~ 5500 K for the central emission. This is much lower than the optical photosphere temperature of 8650 K, and even lower than extrapolating far-infrared emission. Confusingly, ATCA 7 mm observations find a T_B of > 16000 K (Ricci et al. 2012), nearly ~ 200% of T_B(phot), potentially showing behavior similar to the Sun (see Figure 1). No conclusions on a potential unresolved asteroid belt can be made until the emission of the Fomalhaut star is known and subtracted from the observed flux (White et al. 2018; White 2018).

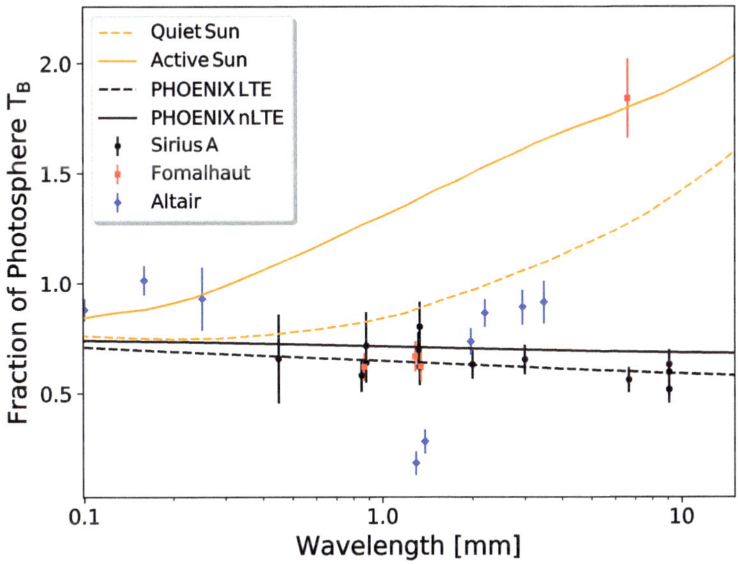

Figure 1. Temperature spectrum of the Sun, Sirius A, Fomalhaut, and Altair. The y-axis is the fractional brightness temperature (T_B), or the observed T_B normalized with the photosphere T_B of a given star. This allows for different stellar types to be easily compared on the same plot. The two orange curves represent models of the Sun at maximum activity (solid line) and minimum activity (dashed line) from Loukitcheva et al. (2004). The black points are JCMT, SMA, ALMA, and VLA observations of Sirius A (White et al. 2018) and the two black curves are PHOENIX models of Sirius A's atmosphere with a non-LTE model (solid line) and a LTE model (dashed line). The blue diamonds are *Herschel* and NOEMA observations of Altair (Thureau et al. 2014, White et al. 2019, in preparation) The red squares are ALMA and ATCA observations of Fomalhaut (Ricci et al. 2012; Su et al. 2016; White et al. 2017a; MacGregor et al. 2017).

HD 141569 is a 5 Myr-old system that contains a B9.5V star, an extended gas and dust structure, and two M dwarf companions at large separations. In 2014, unresolved 9 mm VLA observations of HD 141569 detected a higher than expected disk flux (MacGregor et al. 2016). This led to an inferred spectral index that was much shallower than average for disks, with interesting implications for the grain size distribution and dust evolution models. High resolution VLA follow-up observations in 2016, however, did not detect the expected dust structure. Instead, these observations found that the recovered emission was consistent with a point source centered on the star (White et al. 2017b). The total emission between the two semesters was also different by a factor of a few, suggesting that stellar emission, which appears to be variable, is affecting the 9 mm flux density and thus the interpretation of the dust in the system.

Another case where uncharacterized stellar emission has confounded debris disk studies is with Proxima Centauri. ALMA observations of Proxima found an unresolved excess over the "expected" stellar emission at 1.3 mm, in two different antenna configurations (Anglada et al. 2017). This excess was initially interpreted as emission from multiple debris components in the system. Follow-up analysis of the ALMA data found instead that the observations caught a very large, short lived flare (MacGregor et al. 2018). While flaring is common in M-type stars (such as Proxima), a radio flare of $\sim 1000\times$ the quiescent emission was not expected as it is nearly $10\times$ stronger than the strongest Solar flare ever detected (Krucker et al. 2013). While Proxima is indeed an M-type star, with very different emission mechanisms than the A-type stars discussed above, the ALMA observations of Proxima highlight the importance for proper characterization of stellar emission in general.

3. Uniqueness to ngVLA Capabilities

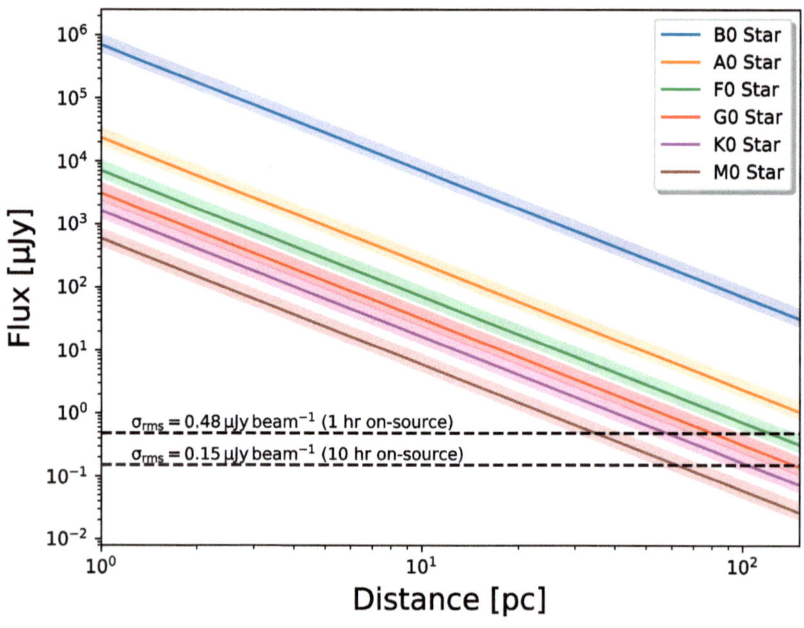

Figure 2. Estimated ngVLA Band 6 (93 GHz) flux of a given spectral type of star as a function of distance. The shaded region for each spectral type is the estimated stellar flux if the emission is best represented by a brightness temperature ranging from 75% to 150% of the photosphere temperature. The lower mass stars will likely have significant non-thermal emission due to, e.g., coronal activity and stellar magnetic fields (see orange curves in Figure 1). This non-thermal emission can lead to significantly higher emission than shown here, making these estimates lower level limits of the expected flux. The estimated $\sigma_{\rm rms}$ from 1 hr and 10 hr on-source is given by the two horizontal dashed lines (Selina et al. 2018).

The ngVLA will be the only facility capable of detecting the radio emission from a wide range of debris-poor stars over the proposed wavelength range. While the ngVLA will indeed be able to resolve some nearby stars (the 0.7 mas maximum resolution corresponds to $\sim$ 1 $D_{\odot}$ at a distance of $\sim$ 15pc), the success of this study is not dependent on resolving capabilities. Instead, this study relies on the high sensitivity provided the collecting area of the proposed 214 antennas.

Current limitations on building a complete catalog of the radio emissions of stars is largely based on the required on-source observing times. For example, ALMA observations of Sirius A highlight this issue. Sirius A is the closest A-type star to Sun at 2.6 pc and therefore is the easiest target to observe. ALMA 3 mm observations of Sirius A only required $\sim$ 1 min of on-source observing time to achieve a signal-to-noise ratio (SNR) of 15 (White et al. 2019, in preparation). To observe a Sirius-like star that is 15 pc away, it would take over 18 hr of on-source observing time. This makes observing anything but the closest and brightest debris-poor stars very difficult to do with current facilities.

In order to build a full catalog of the radio emission of stars with the ngVLA, a targeted survey approach can be adopted. The closest and brightest stars will be observed first with broad spectral coverage. With a proposed 93 GHz continumm sensitivity of $\sigma_{\rm rms}$ = 0.48 μJy beam^{-1} (Selina et al. 2018), these sources will only require on the order of minutes on-source time to get a useful SNR (e.g., >10). The observations will be used to inform stellar atmosphere modeling codes (e.g., PHOENIX; Hauschildt & Baron 1999). The feasibility of this has already been demonstrated with Sirius A White et al. (2018). The survey will then begin extending out to more distant sources, targeting a range of stellar types and properties (e.g., rapid rotators, metal rich/poor, etc.). Templates of stellar emission profiles will be made readily available so that they can be utilized in properly characterizing unresolved debris structures observed with the ngVLA.

IRAS and *Spitzer* surveys at 70 and 100 μm found that A-type stars are the most likely to host a detectable debris component (Su et al. 2006; Thureau et al. 2014). This high occurrence rate of debris makes A-type stars common targets in studies that seek to characterize debris (which in turn requires an accurate model of the star's spectrum). As can be seen in Figure 2, nearly all A-types within 150 pc will be detectable in 1 hr of observing time if their emission is characterized by > 75% of their photosphere T_B. Within 10 hr of observing time, most stars of a size and temperature greater than that of typical early G-type stars will be detectable. With current stellar number density estimates within 150 pc (e.g., Bovy 2017), this allows for hundreds of stars to be observable with the ngVLA in a reasonable amount of observing time.

The scope of this project also offers a unique opportunity for adding many new calibrators to the ngVLA's observational setup. Well-constrained stellar spectral profiles of nearby stars that exhibit no detectable variability will increase the on-sky dispersion of available flux and phase calibrators. A larger number of available calibrators can help reduce the overhead time in a given observation. If these stars were indeed used as calibrators, then it would increase the amount of data available to accomplish the overall goals of this proposed project.

4. Synergies at Other Wavelengths

A broad spectral profile is needed in order to fully characterize the structure of a given stellar atmosphere. Similarly, a broad spectral profile of a debris disk is needed in order to accurately constrain disk properties such as morphology, mass, and grain size distribution. Current observational facilities such as ALMA, NOEMA, and SMA can probe the submm-mm wavelength emission from debris disks. Future facilities such as the SKA may be able to probe the radio emission from debris disks at even longer wavelengths than the ngVLA. The "gap" between these two wavelength regimes is also under-explored in regards to stars and debris due to current sensitivity constraints, making the ngVLA a key tool in to understand the physics that occur in this regime. Only by combing data from all these facilities can we observe the full spectral profile of a star, build and test models of its emission, and use these models to studying unresolved debris disks.

5. Summary

The ngVLA will provide unprecedented sensitivity over the proposed wavelength range, creating the opportunity to observe debris disks at much further distances than currently possible. In unresolved disks, an accurate determination of the flux contribution from the disk and the star requires a well tested model of stellar emission. Current observational facilities are only able to observe a few nearby debris-poor stars, making it difficult to test models of stellar atmospheric emission. The ngVLA will have the sensitivity required to observe significantly more stars, including essentially all A-type stars within 150 pc. These observations will be absolutely essential in order to accurately constrain the frequency and abundance of circumstellar debris.

References

Acke, B., Min, M., Dominik, C., et al., 2012, A&A, 540, A125
Anglada, G., Amado, P. J., Ortiz, J. L., et al. 2017, ApJL, 850, L6
Bovy, J., 2017, MNRAS, 470, 1360
Cranmer, S. R., Wilner, D.J. and MacGregor, M.A., 2013, ApJ, 772, 149
Dulk, G. A. 1985, ARA&A, 23, 169
Hauschildt, P. H. & Baron, E. 1999, Journal of Comp. & App. Math, 109, 41
Hauschildt, P.H. and Baron, E., 2010, A&A, 509, A36
Hauschildt, P.H. and Baron, E., 2014, A&A, 566, A89
Krucker, S., Giménez de Castro, C. G., Hudson, H. S., et al., 2013, A&A Rv, 21, 58
Liseau, R., Vlemmings, W., Bayo, A., et al. 2015, A&A, 573, L4
Loukitcheva, M., Solanki, S. K., Carlsson, M. et al., 2004, A&A, 419, 747
MacGregor, M.A., Wilner, D. J., Chandler, C., et al., 2016, ApJ, 823, 79
MacGregor, M.A., Matrá, L., Kalas, P., et al., 2017, ApJ, 842, 8
MacGregor, M.A., Weinberger A.J., Wilner D.J., et al., 2018, ApJL, L2, 6
Ricci L., Testi L., Maddison S.T., Wilner D.J., 2012, A&A, 539, L6
Selina, R.J., Murphy, E.J., McKinnon, M., et al., 2018, 2018 SPIE Astronomical Telescopes and Instrumentation, arxiv:1806.08405
Su, K. Y. L., Rieke, G. H., & Stansberry, J. A. et al. 2006, ApJ, 653, 675
Su, K.Y., Rieke, G.H., Malhotra, R., et al., 2013, ApJ, 763, 118
Su, K. Y., Rieke, G. H., Defrére, D., et al., 2016, ApJ, 818, 45
Thureau, N. D., Greaves, J. S., Matthews, B. C., et al. 2014, MNRAS, 445, 2558
Wang, T., 2011, Space Sci. Rev., 158, 397
Wedemeyer, S., Bastian, T., Brajsa, R., et al., 2016, Space Sci. Rev., 200, 1
White, J. A., Boley, A. C., Dent, W. R. F., et al., 2017, MNRAS, 466, 4201
White, J. A., Boley, A. C., MacGregor M., Hughes A. M. et al. 2017, MNRAS, 474, 4500
White, J. A., Aufdenberg, J., Boley, A. C., et al., 2018, ApJ, 859, 102
White, J. A., 2018, Characterizing debris discs in the late stages of planet formation (Doctoral dissertation, University of British Columbia).

Science with a Next-Generation Very Large Array
ASP Conference Series, Monograph 7
Eric J. Murphy, ed.

Resolving the Radio Complexity of EXor and FUor-type Systems with the ngVLA

Jacob Aaron White,[1] Marc Audard,[2] Péter Ábrahám,[1] Lucas Cieza,[3] Fernando Cruz-Sáenz de Miera,[1] Michael M. Dunham,[4] Joel D. Green,[5] Manuel Güdel,[6] Nicolas Grosso,[7] Antonio Hales,[8] Lee Hartmann,[9] Kundan Kadam,[1] Joel H. Kastner,[10] Ágnes Kóspál,[1] Sebastian Perez,[11] Andreas Postel,[2] Dary Ruiz-Rodriguez,[10] Christian Rab,[12] Eduard I. Vorobyov,[13,14] and Zhaohuan Zhu[15]

[1]*Konkoly Observatory, Hungarian Academy of Sciences, Budapest, Hungary*

[2]*Department of Astronomy, University of Geneva, Versoix, Switzerland*

[3]*Facultad de Ingeniería y Ciencias, Univ. Diego Portales, Santiago, Chile*

[4]*Dept. of Physics, State University of New York at Fredonia, Fredonia, USA*

[5]*Space Telescope Science Institute, Baltimore, USA*

[6]*Department of Astrophysics, University of Vienna, Vienna, Austria*

[7]*Aix Marseille Univ, CNRS, CNES, LAM, Marseille, France*

[8]*Joint ALMA Observatory, Santiago, Chile*

[9]*Dept. of Astronomy, University of Michigan, Ann Arbor, USA*

[10]*Center for Imaging Science, Rochester Inst. of Technology, Rochester, USA*

[11]*Departamento de Astronomía, Universidad de Chile, Santiago, Chile*

[12]*Kapteyn Astronomical Inst., Univ. of Groningen, Groningen, NL*

[13]*University of Vienna, Department of Astrophysics, Vienna, Austria*

[14]*Research Inst. of Physics, Southern Federal Univ., Roston-on-Don, Russia*

[15] *Dept. of Physics and Astronomy, University of Nevada, Las Vegas, USA*

Abstract. Episodic accretion may be a common occurrence in the evolution of young pre-main sequence stars and has important implications for our understanding of star and planet formation. Many fundamental aspects of what drives the accretion physics, however, are still unknown. The ngVLA will be a key tool in understanding the nature of these events. The high spatial resolution, broad spectral coverage, and unprecedented sensitivity will allow for detailed analysis of outburst systems. The proposed frequency range of the ngVLA allows for observations of the gas, dust, and non-thermal emission from the star and disk.

1. Young Outburst Systems

Understanding the underlying mechanisms that cause the observed outbursts of young stellar objects (YSOs) is important not only for building a complete picture of stellar formation, but also for the potential implications on the planet formation process. These outbursts offer a potential solution to the *luminosity problem*, i.e., the discrepancy between the mass accretion rates inferred from protostellar luminosities (10^{-8} $M_{\odot}$ yr^{-1}) and the accretion rates predicted by cloud core collapse models (10^{-6} $M_{\odot}$ yr^{-1}). The observed outbursts may be due to *episodic accretion* of disk material onto the star, which consists of a sudden increase in the accretion rate from 10^{-8} $M_{\odot}$ yr^{-1} in quiescence up to as high as 10^{-4} $M_{\odot}$ yr^{-1} during an outburst. The relatively short $\sim 1 - 100$ yr cumulative duration of the outbursts, compared to the quiescent phase, provides a potential solution to the luminosity problem (e.g., Dunham & Vorobyov 2012).

The study of these outburst phenomena began with FU Orionis (Herbig 1966), which in 1937 showed an increase of 5 magnitudes in the V band and has since been slowly fading. Outbursts are now broadly categorized into two groups: FUors and EXors, named after their prototypes FU Orionis and EX Lupi, respectively (see Audard et al. 2014, for a recent review of these types of objects). FUors undergo large outbursts which can last 100s of years, while EXors exhibit smaller and shorter lived outbursts ($\sim$ 1 yr). Indeed, there are objects that do not fit clearly into either classification (FUor-like objects) as they can show similar spectroscopic characteristics as a FUor without showing other evidence of an outburst. This leads to the possibility that EXors and FUor represent different stages of evolution of the same object. Cieza et al. (2018) showed that EXors, for example, have lower disk masses than FUors, indicating they may represent a later stage of disk evolution. Although over the past 50 years EX/FUors have been observed with broad spectral coverage and modeled with increasing high levels of detail, there are still many open questions on their formation, frequency of occurrence, and impact on circumstellar environments.

What are the physical processes that cause these types of events? Despite being thoroughly studied, the triggering mechanisms of these outbursts are still not known. It is also not clear if the same mechanisms are responsible for both EXor and FUor-type outbursts. The ngVLA will be instrumental in addressing this issue by:

1. Accurately measuring disk masses. Knowing disk masses is crucial for understanding the FUor phenomenon, as gravitational instability may play an important role in triggering of the outbursts (e.g., Vorobyov & Basu 2005; Bae et al. 2014). The dust disks are optically thick and have a small angular size, which complicates mass estimates (e.g. Liu et al. 2017; Hales et al. 2018). However, the unprecedented angular resolution of the ngVLA, in combination with its spectral coverage at longer wavelengths where the dust can be optically thin, will enable more precise dust mass estimates. In addition, higher resolution observations of cold gas will provide further constraints on the total disk mass.

2. Dynamically constraining stellar masses. As the luminosity of EX/FUor systems can be dominated by the disk, it can be difficult to accurately determine the mass of the protostar (e.g., V883 Ori has an expected photospheric luminosity of $\sim$ 6 $L_{\odot}$ and a bolometric luminosity of $\sim$ 400 $L_{\odot}$; Cieza et al. 2016). Moreover, the inferred disk inclination affects the inferred disk luminosity and accretion rate. High spatial resolution spectroscopy with the ngVLA will allow for dynamical

constraints on both the stellar mass and inclination through the analysis of the Keplerian rotation of the gas disk (e.g., Czekala et al. 2015).

3. Determining the strength of turbulence in the inner disk. High disk turbulence has been suggested as the reason for the loss of angular momentum needed to trigger an enhanced accretion event. Flaherty et al. (2017) analyzed the DCO^+ in HD 163296's outer disk and found little turbulence, going against the theoretical predictions. This could be due to HD 163296's more evolved state or a diminished role of turbulence-driven accretion. The ngVLA will help in understanding the role of turbulence in the innermost parts of disks, where the burst is expected to be triggered, and how it can cause eruptive events.

4. Observing planet-disk-star interactions. The mass exchange between a young protoplanet and the star in the innermost parts of the disk may trigger accretion outbursts (Nayakshin & Lodato 2012). Ricci et al. (2018) estimated that the ngVLA will be able to characterize gaps and asymmetries caused by a super-Earth planet at close radii (<5 au). At those distances, it will be possible to see changes in the disk material distribution in a few years, thus making important mass transfer rate predictions in the innermost regions of the disk.

5. Assessing the impact of stellar encounters and multiplicity. Close stellar encounters either with an external star or a companion may trigger accretion outbursts (e.g., Bonnell & Bastien 1992; Pfalzner 2008). The spatial arrangement of the outburst star and the inferred intruder is sometimes not consistent with the timescales of the outburst (Liu et al. 2017), suggesting another companion in the inner unresolved disk may be present. The ngVLA will search for close-orbit companions and assess the role of companions/intruders as outburst triggers.

What are the implications of these outburst events on the origin of our Solar System and for other planetary systems? The location of snow-lines (or ice-lines) in a planet forming disk is important as an evolving snow-line can greatly impact the abundance of solid material throughout the disk and affect grain growth. ALMA observations of V883 Ori show the water snow-line at a radial distance of 42 au, in contrast with the expected distance of <5 au for a solar-like star (Cieza et al. 2016). Therefore, if these eruptive events are a common occurrence in the evolution of protoplanetary disks then their affects on the relocation of snow-lines, and how this can spur grain growth, must be considered in planet formation models.

In this chapter we explore how the proposed next generation Very Large Array (ngVLA) will be an invaluable tool for studying FUor and EXor type protostars. In § 2, we highlight key areas that still remain largely unexplored or unconstrained with current capabilities. In § 3, we show the uniqueness of the ngVLA's capabilities to study EX/FUors. In § 4, we highlight synergies at other wavelengths.

2. Exploring Accretion Mechanisms with the ngVLA

Current millimeter and radio observatories do provide valuable insight into EX/FUor systems (e.g., Kóspál et al. 2017; Liu et al. 2017; Cieza et al. 2018; Hales et al. 2018), but the major limitation in these studies is typically the resolving capabilities of the observatory (e.g., ~ 20 mas or ~ 8 au at the Orion Nebula). When the circumstellar disk around a given EX/FUor system is unresolved, or only marginally resolved, it can be difficult (if not impossible) to test many of the competing models for accretion-driven outbursts. The ngVLA will provide a major improvement in the resolving capabilities of current facilities that operate in the 1.2 – 116 GHz frequency range, where the inner disk may be transitioning to an optically thin regime.

There will indeed be many properties of EX/FUors, and circumstellar environments in general, that the ngVLA will be able to study. Some of the key areas are highlighted below.

2.1. Disk Morphology

The proposed 1000 km maximum baselines of the ngVLA will yield a resolution of 0.7 mas at 93 GHz (Selina et al. 2018). This resolution corresponds to spatial scales of 0.25 and 0.10 au for the two prototypical outburst systems, FU Orionis and EX Lupi, respectively. For EX Lupi, this spatial scale is slightly smaller than the $\sim$ 0.12 au corotation region (Sicilia-Aguilar et al. 2015) and smaller than the $\sim$ 0.4 au inner dust-free zone (Ábrahám et al. 2009). The proposed largest recoverable scales of > 20″ (Selina et al. 2018) will correspond to spatial scales of 1000s of au for FU Orionis and EX Lupi, allowing for all material with sufficient surface brightness to be detectable in a single pointing. The high level of spatial resolution will enable a detailed analysis of the morphological structure of the material in the disks.

Many of the proposed drivers of accretion in EX/FUor-type objects will have distinct morphological features that can be distinguished with the ngVLA. These features are caused by mechanisms such as gravitational instabilities, disk fragmentation, circumbinary disk interactions, and funnel flows (e.g., Armitage et al. 2001; Zhu et al. 2009; Hartmann et al. 2016). These features may be present both in the inner optically thick and the outer optically thin parts of the disks. However, current resolving capabilities make it difficult to confirm or reject the presence of any substructure in EX Lupi and FU Orionis (Liu et al. 2017; Hales et al. 2018).

The mm-cm spectral index of an EX/FUor-type disk will yield insight into the opacity and size distribution of the grains. An optically thick inner disk may be common, but the dust opacity may transition to an optically thin regime at the lower frequency range of the ngVLA. Indeed, it is likely that non-thermal emission will be significant at lower frequencies, but the high resolution capabilities should enable the thermal and non-thermal emission components to be spatially separated. Constraints on the size distribution (q) of solids can be made when a disk is optically thin. By measuring q, the grain growth, total dust mass, and mass distribution can be much more accurately constrained than is possible with current observations.

The resolution and sensitivity of the ngVLA will allow for the first tests of time-dependent evolution of disk structure. There are many mechanisms that could cause significant changes to the morphological structure of the disk on timescales that can be probed with the ngVLA. For example, if a close-in companion were present, it could open a gap in the disk or cause significant spiral structure to form (e.g. Paardekooper & Mellem 2004; Dong et al. 2015; Ricci et al. 2018). This disturbance to the disk could cause an infall of material which could further drive accretion. The movement of this material could potentially be traced over a few years. Changes over much shorter timescales would also be observable. Many disk models (e.g., the model discussed in § 3) have disk dead zones at separations of $\sim$ 0.5 au from the star. The orbital period of material at this distance would be on the order of months, and multiple observations over an ngVLA observing semester could probe morphological changes in the disk over the timescale of a single orbit. Low levels of variability in EXor/FUor-type stars have been observed at optical wavelengths (i.e., significantly lower than the outbursts themselves). If this variability is correlated with changes in the disk structure, then it could be potentially observed over the course of a single observation with the ngVLA.

2.2. Disk Chemistry

Winds and outflows are fundamental ingredients of star formation and protoplanetary disk evolution (Williams & Cieza 2011). Thus, it is not surprising that YSOs undergoing episodic accretion, such as FUors, also show high levels of outflow activity. These outflows are responsible for carrying away some of the angular momentum, allowing circumstellar matter to accrete onto the forming star. Moreover, previous studies suggest that outflows could be an effective way to disrupt the disk and envelope at early evolutionary stages (Ruíz-Rodríguez et al. 2017a,b). Accordingly, it is imperative to investigate the relation between accretion and outflows in young stellar systems, so as to disentangle the physical mechanisms that dictate the low mass star formation efficiency in turbulent clouds (Quillen et al. 2005; Krumholz et al. 2012) and efficient transport of angular momentum that permits accretion onto the star.

In addition, strong X-ray emission can be produced during FUor eruptions and rapid YSO accretion episodes more generally (Audard et al. 2014), either by accretion shocks (Grosso et al. 2010; Teets et al. 2012), magnetic reconnection (Kastner et al. 2004), or both. It has been known that X-ray radiation is a critical ionizing source in the surface disk layers at distances of $\sim 10-100$ au (Glassgold et al. 2012) and plays an important role in the dispersal of disks at pre-main sequence ages >1 Myr (Gorti et al. 2015). This is due to the large penetration depth of X-rays in the disk-envelope system, where they can ionize H_2 molecules as well as abundant atoms, such as He, C, N, and O. Hence, X-ray irradiation plays a role in the formation and position of molecular disk snow lines and dead zones, thereby ultimately determining the semimajor orbits, masses, and compositions of any resulting planets (Öberg & Bergin 2016; Cridland et al. 2017). The resulting enhanced molecular ionization leads to rich chemistry that can be probed by tracers such as HCN and HCO^+ (Kastner et al. 2008, and refs. therein).

Spatially resolved observations of outflow/disk systems with the ngVLA in the wide range of chemical tracers accessible to its Band 6 (70-116 GHz) will allow us to distinguish between chemistry induced by outflow energetics (shocks) and radiation (X-rays). Specifically, sub-arcsecond resolution mapping of lines of SO, SiO, CH_3OH, and H_2CO will provide diagnostics of shock-generated chemistry within FUor outflows. Observations of the same systems in HCO^+ will then serve to probe the presence and extent of molecular ionization, via spatial analysis of the $HCO^+/^{13}CO$ surface brightness ratio. In addition, using ^{12}CO, ^{13}CO, $C^{18}O$ line data, and adopting standard methods for correcting optical depth effects (e.g., Dunham et al. 2014; Ruíz-Rodríguez et al. 2017a,b), provides estimates of the mass and kinematic properties of the revealed outflows. In parallel, maps of HCN and CS emission will trace the dense gas within the disk and disk-outflow interface regions.

2.3. Magnetic Fields & Polarization

Some of the most promising outburst theories for EXors rely on the role of the stellar magnetic field. Armitage (2016) predicted that the outbursts could be explained by changes in the polarity and strength of the stellar magnetic fields at the kG-level. In

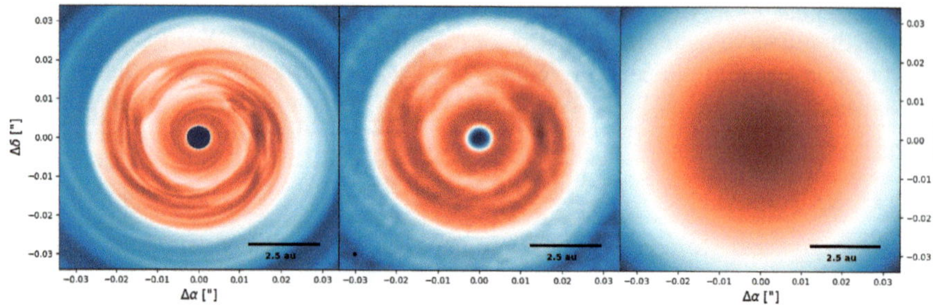

Figure 1. Simulated observations with a 10 × 10 au disk model (described in Section 3), 0.7 mas 100 GHz ngVLA observations, and 40 mas 100 GHz ALMA observations. A 2.5 au scale is given in the bottom right of each panel. The synthesized ngVLA beam is given in the bottom left of the middle panel. The simulated ngVLA observations have a 1 hr on-source $\sigma_{\rm RMS} = 0.48\ \mu$Jy beam^{-1} and 1000 km baselines.

a competing scenario, D'Angelo & Spruit (2012) proposed an instability which can lead to quasi-periodic oscillations in the inner disk and associated recurrent outbursts. This instability can occur when the accretion disk is truncated close to the corotation radius by the strong magnetic field of the star. A confirmation of the magnetic field strength would give strong support to these models, and may constrain the cause of outbursts to magnetic phenomena in young stars, a novel result. In principle, if kG-level stellar magnetic fields were present in a given system then there could be significant emission above the photospheric flux. For example, strong magnetic fields could induce synchrotron or free-free emission within a few stellar radii. The flux from this could be easily spatially separated from the inner edge of the disk and could provide constraints on the strength of the stellar magnetic fields.

The effects of magnetic fields within the disk itelf can be traced through polarization. For example, small dust grains can align with magnetic fields and cause polarization (e.g., Hull et al. 2014), allowing for another probe of the grain size distribution throughout the disk. Bright molecular lines can also be linearly polarized by magnetic fields within the disk (e.g., the Goldreich-Kylafis effect; Forbrich et al. 2008).

The large outflow jets discussed in § 2.2 can generate polarized synchrotron emission (see e.g., Galván-Madrid et al. p. 309 in this volume for a detailed overview of jets in disks). These jets have been observed in both low and high mass protostellar systems and may provide another tool in studying how accretion events can vary as a function of stellar mass. The synchrotron emission associated with these jets can lead to polarization at the lower frequency range of the ngVLA. As these jets can be much more extended from the circumstellar disk, it should be straightforward to spatially separate their emission from any disk and other polarized or non-thermal emission.

3. Uniqueness to ngVLA Capabilities

The ngVLA will be able to spatially resolve a very high level of detail in the inner regions of disks. To illustrate this, Figure 1 shows simulated observations of a detailed disk model with both the ngVLA and ALMA. The model shown is a global hydrodynamic simulation of a protoplanetary disk in the thin-disk limit, starting with the collapse of molecular cloud (Vorobyov & Basu 2015). The model used canonical values of layered disk parameters (e.g., T_{crit} = 1100K, Σ_{crit} = $100 g/cm^2$) and reproduced eruptions very similar to the observations of FUor outbursts.

To simulate this disk model with the ngVLA, we assume the disk emission is consistent with 100 GHz thermal continuum (i.e., it does not include any non-thermal effects, disk winds, or ionized emission), the system is at a distance of 150 pc, and the disk flux is normalized to 35 mJy. The highest resolution antenna configuration of 0.7 mas was used and the background noise is set to 0.48 μJy beam^{-1} (achievable in 1 hr on-source in ngVLA Band 6; Selina et al. 2018). For comparison to ALMA, a simulated observation of the same disk is included with a sensitivity equivalent to 4 hours on-source (the most extended ALMA baseline of $\sim$ 16 km corresponds to a resolution of $\sim$ 0.04 arcsec at 100 GHz).

The 10×10 au region of the disk model shows the complexity of the inner accretion disk which can be resolved by ngVLA, but not with current capabilities of ALMA. Should significant optically thick disk winds or non-thermal emission be present in a system at 100 GHz, then it could indeed make resolving the disk structure more complicated. The model in Figure 1 predicts that this inner region is very dynamic and will show detectable changes in features on the timescale of a few years. The perturbations in the disk during the outburst propagate outwards, producing spirals or concentric rings also observed in young stellar disks.

ALMA is the current state-of-the-art facility in regards to high sensitivity and resolution observations of EX/FUor-types stars. As ALMA is located in the Southern hemisphere, there are potentially many targets in the Northern hemisphere that are not easily detectable or resolved with the current Northern hemisphere observing facilities (e.g., SMA and NOEMA). Therefore, the ngVLA is uniquely positioned to provide detailed observations of a large fraction of the sky.

4. Synergies at Other Wavelengths

The frequency in which EX/FUor-types of outbursts occur is not well constrained due to the relatively small number of outburst events. Determining the frequency of these events requires large scale surveys, which is more practical with optical, near-IR, and IR telescopes. Future optical/IR survey facilities, such as LSST and WFIRST, will likely identify new outbursts and the ngVLA will be a key tool in follow-up observations and an accurate characterization of the systems. The SKA may be able to detect non-thermal emission in EX/FUors and will provide complimentary longer wavelength data.

Higher-energy observations, e.g., in the X-ray regime, can provide important complementary information on plasma heating in magnetic fields (through corona-type emission like in active stars) and the presence of *neutral* gas components in the line of sight, potentially indicating the presence of neutral winds or accretion flows (Liebhart et al. 2014). Clarifying the role of X-rays generated during massive outbursts in the evolution and dispersal of planet-forming disks and envelopes is essential to progress in connecting the first stages of planet formation to mature planetary systems.

While many of the morphological details cannot be revealed with current facilities, mm telescopes such as ALMA and NOEMA can still determine many of the bulk properties of a given EX/FUor. For example, ALMA can provide the total ~mm flux which will be useful for spectral index constraints in optically thin regimes. There are also many aspects of disk chemistry that can be measured with ALMA [e.g., CO(3-2) at 345 GHz] and future IR facilities such as JWST, that will be complimentary to the disk chemistry information obtained with the ngVLA.

The ngVLA will provide unprecedented spatial resolution and sensitivity in the 1 – 116 GHz frequency and contribute to the broad spectral coverage that is imperative to accurately models of EX/FUors and young outbursting systems.

References

Ábrahám, P., Juhász, A., Dullemond, C.P., et al., 2009, Nature, 459, 224
Armitage, P.J., Livio, M. & Pringle, J.E., 2001, MNRAS, 324, 705
Armitage, P.J., 2016, ApJL, 833, L15
Audard, M., Ábrahám, P., Dunham, M.M., et al. 2014, PPVI 387, ed H. Beuther et al.
Bae J., Hartmann L., Zhu Z., Nelson R. P., 2014, ApJ, 795, 61
Bonnell, I. and Bastien, P., 1992, ApJL, 401, L31
Cieza, L.A., Casassus, S., Tobin, J., et al., 2016, Nature, 535, 7611, 258
Cieza, L.A., Ruíz-Rodríguez, D., Perez, S., et al. 2018, MNRAS, 474, 4347
Cridland, A.J., Pudritz, R.E., Birnstiel, T., et al., 2017, MNRAS, 469, 3910
Czekala, I., Andrews, S.M., Jensen, E.L., et al., 2015, ApJ, 806, 154
D'Angelo, C.R. & Spruit, H.C., 2012, MNRAS, 420, 416
Dong, R., Zhu, Z., & Whitney, B., 2015, ApJ, 809, 93
Dunham, M.M. & Vorobyov, E.I., 2012, ApJ, 747, 52
Dunham, M.M., Arce, H.G., Mardones, D., et al.,2014, ApJ, 783, 29
Flaherty, K. M., Hughes, A. M., Rose, S. C., et al. 2017, ApJ, 843, 150
Forbrich, J., Wiesemeyer, H., Thum, C., et al., 2008, A&A, 492, 757
Galván-Madrid, R., Beltrán, M., Ginsburg A. et al. 2018, Science with a Next-Generation Very Large Array, ASP Monograph 7, ed. E. J. Murphy (San Francisco, CA: ASP), 309
Gammie C. F., 1996, ApJ, 457, 355
Glassgold, A. E., Galli, D., & Padovani, M., 2012, ApJ, 756, 157
Gorti, U., Hollenbach, D., & Dullemond, C.P., 2015, ApJ, 804, 29
Grosso, N., Hamaguchi, K., Kastner, J.H., et al., 2010, A&A, 522, A56
Hales, A.S., Pérez, S., Saito, M., et al., 2018, ApJ, 859, 111
Hartmann, L., Herczeg, G., & Calvet, N., 2016, ARA&A, 54, 135
Herbig, G.H., 1966, VA, 8, 109
Hull, C.L., Plambeck, R.L., Kwon, W., et al., 2014, ApJS, 213, 13
Kastner, J.H., Richmond, M., Grosso, N., et al., 2004, Nat, 430, 6998, 429
Kastner, J.H., Zuckerman, B., Hily-Blant, P., et al., 2008, A&A, 492, 469
Kóspál, Á., Ábrahám, P., Csengeri, T., 2017, ApJ, 843, 45
Krumholz, M.R., Klein, R.I., & McKee, C.F., 2012, ApJ, 754, 71
Liebhart, A., Güdel, M., Skinner, S.L., Green, J. 2014, A&A, 570, L11
Liu, H.B., Vorobyov, E.I., Dong, R., et al., 2017, A&A, 602, A19
Nayakshin, S. & Lodato, G., 2012, MNRAS, 426, 70
Öberg, K.I. & Bergin, E.A., 2016, ApJL, 831, L19
Paardekooper, S.J. & Mellema, G., 2004, A&A 425, L9
Pfalzner, S., 2008, A&A 492, 735
Quillen, A.C., Thorndike, S.L., Cunningham, A., et al., 2005, ApJ, 632, 941
Ricci, L., Liu, S.-F., Isella, A., & Li, H. 2018, ApJ, 853, 110
Ruíz-Rodríguez, D., Cieza, L.A., Williams, J.P., et al., 2017, MNRAS, 466, 3519
Ruíz-Rodríguez, D., Cieza, L.A., Williams, J.P., et al., 2017, MNRAS, 468, 3266
Selina, R.J., Murphy, E.J., McKinnon, M., et al., 2018, 2018 SPIE Conf., arxiv:1806.08405
Sicilia-Aguilar, A., Fang, M., Roccatagliata, V., et al., 2015, A&A, 580, A82
Teets, W.K., Weintraub, D.A., Kastner, J.H., et al., 2012, ApJ, 760, 89
Vorobyov, E.I. & Basu, S., 2005, ApJL, 633, L137
Vorobyov, E.I., Basu, S., 2015, ApJ, 805, 115
Williams, J.P. & Cieza, L.A., 2011, ARA&A, 49, 67
Zhu, Z., Hartmann, L., & Gammie, C., 2009, ApJ, 694, 1045

Science with a Next-Generation Very Large Array
ASP Conference Series, Monograph 7
Eric J. Murphy, ed.

Indirect Detection of Extrasolar Planets via Astrometry

Bryan J. Butler[1] and Brenda C. Matthews[2,3]

[1]*National Radio Astronomical Observatory, Socorro, NM, U.S.A.*

[2]*Herzberg Astronomy & Astrophysics Research Centre, National Research Council of Canada, Victoria, BC, Canada;* brenda.matthews@nrc-cnrc.gc.ca

[3]*Department of Physics & Astronomy, University of Victoria, Victoria, BC, Canada;*

Abstract. Radio wavelength astrometry of stars and other objects has a long and productive history. The use of that technique to determine whether stars have planets around them would cover a nearly unique part of the parameter space for detection of those systems. Namely, astrometric observations are most sensitive to systems with large planets in moderately wide orbits (a few to ~10 AU), because it is those systems that produce large reflex motion of the star, in a short enough measurement period (years to tens of years). In addition, astrometric observations are most sensitive to systems which are nearly face-on. Other techniques (radial velocity, or the photometric method of Kepler) are more sensitive to systems with planets in close orbits (less than ~1 AU), which are nearly edge-on. We describe here, using the *Hipparcos* and *Gaia* star catalogs, how ngVLA could use this technique on hundreds of stars, some tens of which are solar analogs, to determine whether these stars have planets orbiting them.

1. Introduction

There are now thousands of detected and confirmed exoplanets and exoplanetary systems. Through transit detections, *Kepler* has driven home the ubiquity of extrasolar planets and the diversity of extrasolar planetary systems, providing statistics of extrasolar planets in relatively close orbits around their parent stars (Borucki et al. 2010; Howell et al. 2014). The recently launched *TESS* (Transiting Exoplanet Survey Satellite) will search for planets around 2 million stars in the solar neighbourhood in the coming two years (Ricker et al. 2015). The radial velocity technique of detecting the radial velocity variations of the star due to a planet's motion around it is, like the transiting technique, limited to relatively close in planets and biased to planets which are in or near orbits viewed edge-on. In addition, these methods cannot be utilized toward young or later type stars, which exhibit variable emission from their chromospheres that preclude RV measurements to the required precision. Direct detection of exoplanets through imaging is a growth industry and a key area of current research. Optical direct imaging surveys have established that there is not a large population of giant planets on orbits of > 10 AU in young (< 100 Myr) systems (Biller & Close 2007; Nielsen 2011; Galicher et al. 2016). Older systems are not accessible to direct imaging because the thermal emission of the planets must be detected and this is only strong enough for the current generation of instrumentation for young planets.

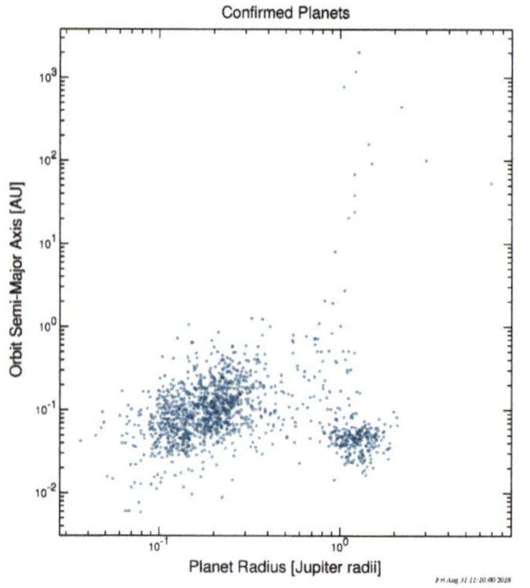

Figure 1. Known exopolanets, as of August 31, 2018 (NASA Exoplanet Archive web page).

The population of planets in systems of the age and on the scales of the outer solar system therefore remains uncharacterized. This region of the parameter space for host stars can be targeted for planet searches using the astrometric capabilities of the ngVLA. The idea of detection of extrasolar planets through astrometry in the radio is not a new one (Butler, Wootten, & Brown 2003), but ngVLA will have enhanced capabilities compared to the current VLA or ALMA due to its longer baselines. The ngVLA will have the sensitivity to detect extrasolar planets via their induced motions onto their parent stars, an "indirect detection" method similar to the RV technique. By monitoring the parent star, it is possible to infer the reflex motion of the planet. Because this technique measures plane of sky motion, it is complimentary to the RV and transit (Kepler) methods of such detections, and so probes an orthogonal parameter space. Figure 1 shows the currently confirmed exoplanets as of August 31, 2018; 3778 in total (from the NASA Exoplanet Archive web page). Note the paucity of planets in orbits similar to our own giant planets in this plot (those with orbits > 5 AU) - this is a well-known conundrum in exoplanet studies. Astrometric techniques are particularly well-suited to search for just such planets, since more massive planets in orbits further from their primary induce a larger reflex motion of that primary.

With so many planets known, what is the added impact of characterization of outer solar systems? Much of the dynamical evolution of our own solar system was driven by the outer planets. Indeed the Nice model is dependent on a violent migration event (Morbidelli et al. 2007). Even more quiescent theories of evolution depend on the ice giant migration, particularly Neptune to corral the outer planetesimal belt and stabilize the orbits of the giant planets.

There is now evidence of significant substructure at large orbital radii in protoplanetary disks (e.g., HL Tau, Elias 2-27 etc.) which suggests that formation mechanisms of planets may be active quite distant from the star, at least at early times. While our own solar system's size is typically given as the radius of the Kuiper Belt, i.e., not much beyond Neptune (Sibthorpe et al. 2018); many other debris disks (dusty circumstellar disks created by extrasolar planetesimal belts) show much larger radial extent, suggesting exoplanets may also exist at larger radii in such systems. We can search for evidence of wide orbit Jupiter and Neptune analogues using the ngVLA.

Ground based efforts to find extrasolar planets have focused in recent years in several areas, including differential astrometry, radial velocity measurements, and gravitational lensing. There are also proposals to directly image Jovian-class planets near a small number of stars. Planet detection using astrometry with ngVLA will both complement these efforts and have several important advantages:

- Astrometric searches for planets with ngVLA will use absolute (wide-angle) astrometry, avoiding the problem of solving for motions in both a target star and a background reference star. ngVLA observations will be able to tie stellar positions directly to the quasar reference frame, since both stars and quasars will be bright enough for high-accuracy astrometry with ngVLA;

- Systematic errors will be lower with ngVLA than with techniques at optical or IR wavelengths because ngVLA can observe stars 24 hours a day, providing various types of closure constraints on ngVLA astrometry. This will reduce the kinds of seasonal systematic errors that plague optical astrometry. The ngVLA at 10-100 GHz will also have less issues with weather constraints than ALMA will at higher frequencies;

- Astrometry measures the position of the star, so that the mass of the unseen planet can be easily determined. Direct detection techniques, either ground or space-based, cannot determine planetary masses, but will provide constraints on the parameters of a particular planetary system that complement those provided by astrometry; and

- Astrometric results from ngVLA will directly complement the results expected from differential astrometry with proposed ground-based IR interferometers and ALMA. ngVLA will not face the constraints on finding suitable background reference stars that are required by differential IR astrometry, so some additional stellar systems may be accessible to ngVLA compared to the optical/IR.

2. Astrometry of Planet Hosts

The orbit of any planet around its central star causes that star to undergo a reflexive motion around the star-planet barycenter. Figure 2 shows the orbit of the Sun on the planet of the sky if it were viewed from 30 light years (~9 pc). Those motions are of order milliarcseconds over periods of tens of years - it is these motions that can be astrometrically detected by ngVLA. By taking advantage of the high resolution provided by the ngVLA at even intermediate baselines, we will be able to detect this motion and thereby constrain the planetary population at large orbital radii for many solar type stars. We note that this technique has been used for decades to determine the orbits of multiple star systems (Ransom et al. 2012).

Making the usual approximation that the planet mass is small compared to the stellar mass, the stellar orbit projected on the sky is an ellipse with angular semi-major

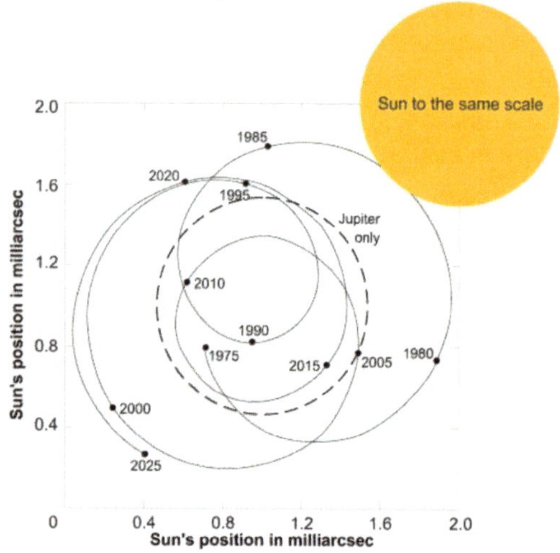

Figure 2. The wobble of the Sun on the plane of the sky, if it were viewed face on from 30 light years distance (Jones 2008).

axis θ_r (in arcsec) given by the product of the ratio of the planetary mass (M_p) to stellar mass (M_*) and the ratio of the semi-major axis of the planet in AU (a_{AU}) and the distance to the star in pc (d_{pc}):

$$\theta_r = \frac{M_p}{M_*}\frac{a_{AU}}{d_{pc}} \quad . \tag{1}$$

Astrometric resolution of an interferometer, the angular scale over which changes can be discriminated (Θ), is proportional to the intrinsic resolution, θ_{HPBW}, and inversely proportional to the signal to noise with which the stellar flux density is detected (SNR^*):

$$\Theta = \frac{\theta_{HPBW}}{2 \times SNR^*} \quad . \tag{2}$$

Note that there are systematic effects which limit this at high SNR (Reid & Honma 2014), but we are not expecting to be in that regime for these kinds of observations.

In order to detect the wobble of a star caused by an orbiting planet we must therefore calculate the induced motion on the star (meaning we must know distance to the system, the mass of the star and the assumed mass and orbit of the planet), we must know the resolution of the ngVLA, and we must know the SNR (the flux density of the star, and the sensitivity of ngVLA). We assume we can determine the distance to the system and mass of the star from information in a catalog. We also assume that we can determine effective temperature either directly or indirectly from that information. Note that at wavelengths shorter than about 2 cm, the emission from stars is mostly thermal emission from the disk of the star (see Figure 3, which shows this for the Sun). If we assume that the brightness temperature of the star is just the effective temperature,

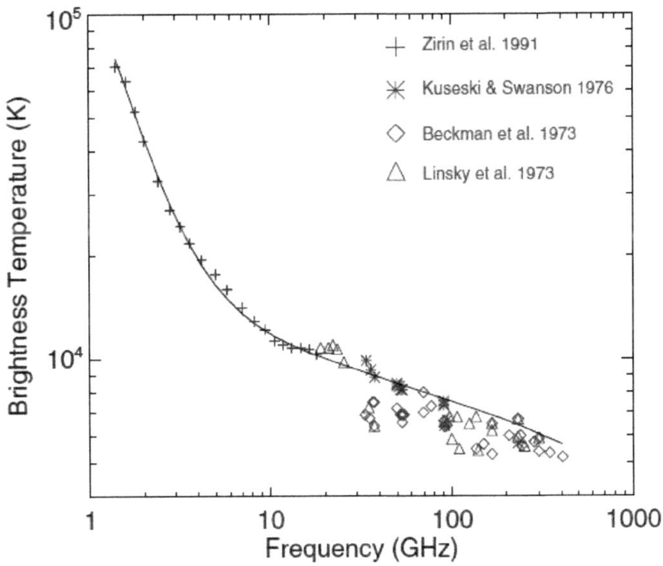

Figure 3. Figure showing the solar brightness temperature as a function of wavelength (Selhorst, Silva, & Costa 2005).

then we can calculate the flux density of the star via:

$$F_* = \frac{2 h \nu^3}{c^2} \frac{1}{e^{h \nu / k T_{eff}} - 1} \Omega \quad , \tag{3}$$

where h is the Planck constant, c is the speed of light in vacuum, k is the Boltzmann constant, T_{eff} is the effective temperature of the star, and Ω is the angular size of the star. We assume the diameter of the star can be determined from the information in the catalog. To calculate the sensitivity of ngVLA, we assume we will observe at 30 GHz, and take the noise as 1.5 μJy/min (Butler et al. 2018) and assume an on-source integration time of one hour, so final noise of ~0.2 μJy. We assume we need an SNR of 4 for each observation. We assume a maximum baseline for ngVLA of 800 km, so $\theta_{HPBW} \sim 3$ masec. For candidate planets, we assume an orbit of 5 AU (since we are interested in solar system giant planct analogs), and assume three masses: 5 times Jupiter, Jupiter, and Neptune mass.

We have access to two catalogs of stars that we can use to investigate how many of them may have planets which could be detected in this way: *Hipparcos* (Perryman et al. 1997) and *Gaia* (Gaia Collaboration et al. 2016). Both catalogs have well-determined distances to stars. The *Gaia* catalog has radii and effective temperatures for a number of stars. The *Hipparcos* catalog does not, but does contain information on spectral type and luminosity class. We use that information to infer the mass, radii, and effective temperatures using Table 15.8 in Allen's Astrophysical Quantities. For the *Gaia* catalog, we determine the mass of the star using the luminosity in the catalog, and the usual mass-luminosity relation.

For the *Hipparcos* catalog, we find 831 5*Jupiters that can be detected, 538 Jupiters, and 31 Neptunes. Since that catalog also contains values for B-V, we can classify 'solar analog' stars as those with B-V between 0.62 and 0.71, and with that classification, we can find 5*Jupiters around 43 solar analog stars, Jupiters around 26 solar analog stars, and Neptunes around 2 solar analog stars.

For the *Gaia* catalog, we find 368 5*Jupiters that can be detected, 68 Jupiters, and 3 Neptunes. That catalog does not have spectral type or luminosity class information in it, so we define a different type of 'solar candidate' which are stars that have within 30% of the radius and effective temperature of the Sun. With that classification we can find 5*Jupiters around 85 solar candidate stars, Jupiters around 10 solar candidate stars, and Neptunes around no solar candidate stars.

There is significant overlap between the two catalogs, of course, and we attribute the difference in the detectable number of stars to be due to the different methods of calculation for the two. The takeaway is that of order 1000 5*Jupiters can be detected, and 100 Jupiters can be detected, and of these some tens of the stars are similar to the Sun.

3. Finding Planets in Practice

Naturally, the detection of a planet's influence on the star requires regular observations of the star in addition to the interferometric resolution to reasonably detect a gas or ice giant akin to those in the Solar System. We will require detections of the star with a cadence that will allow us to measure the star's position 3 or 4 times during the planet's orbit. We will also have to calibrate for the stellar proper motion. *Gaia* has now measured the proper motions to high accuracy for many stars in the solar neighbourhood. Selecting out those within 100 pc, i.e., with parallaxes > 10 mas, that catalog has thousands that have proper motions that are known to better than 0.2 mas/yr.

One can readily see an observational strategy developing in which selected sub-samples of stars are routinely observed over a period of several years, and those with promising planetary like variations after 2 observations are then the focus of subsequent observational campaigns.

4. Uniqueness of ngVLA Capabilities

ALMA has been highlighted for its capability in detecting extrasolar planets through astrometry in Band 7 (Butler, Wootten, & Brown 2003). This is akin to the science that can be done at 1 cm (30 GHz) with the ngVLA, except that the ngVLA will have a much better intrinsic resolution, meaning more systems should be detectable under a planetary wobble with ngVLA than with ALMA. No other facility has such a capability.

5. Complementarity with Other Planet Search Facilities Beyond 2025

Interferometers have the resolution needed to do astrometry of nearby stellar systems to the precision needed to detect extrasolar planets. OST will not have the resolution to do astrometry or detect planets through direct imaging. A suite of upcoming telescopes has various capabilities to directly detect extrasolar planets in scattered light, namely

TMT and WFIRST, although the timelines for both are uncertain. Whether LUVOIR or HabEx are viable in 2025 is hard to foresee.

Acknowledgments. We appreciate comments from Ed Fomalont.

References

Biller, B.A., Close, L.M., 1990. ApJ, 669, L41.
Borucki, W.J., Koch, D., Basri, G., et al., 2010. Science, 327, 977.
Butler, B., Wootten, A., Brown, B., 2003. ALMA Memo 475.
Butler, B., Grammer, W., Selina, R., 2018. BAAS, 231, 342.09.
Gaia Collaboration,, Prusti, T., de Bruijne, J.H.J., 2016. A&A, 595, A1.
Galicher, R., Marois, C., Macintosh, B., et al., 2016. A&A, 594, id.A63.
Howell, S.B., Sobeck, C., Haas, M., et al., 2014. PASP, 938, 398.
Jones, B.W., 2008. Int. J. Astrobio., 7, 279.
Morbidelli, A., Tsiganis, K., Crida, A., et al., 2007. ApJ, 134, 1790.
Nielsen, E.L., 2011. PhD Thesis.
Perryman, M.A.C., Lindegren, L., Kovalevsky, J., et al., 1997. A&A, 323, L49.
Ransom, R.R., Bartel, N., Bietenholz, M.F., et al., 2012. ApJSS, 201, 6.
Reid, M.J., Honma, M., 2014. ARAA, 52, 339.
Ricker, G.R., Winn, J.N., Vanderspek, R., et al., 2015. JATIS, 1, id.014003.
Selhorst, C.L., Silva, A.V.R., Costa, J.E.R., 2005. A&A, 433, 365.
Sibthorpe, B., Kennedy, G.M., Wyatt, M.C., 2018. MNRAS, 475, 3046.

Science with a Next-Generation Very Large Array
ASP Conference Series, Monograph 7
Eric J. Murphy, ed.

Detection of Circumplanetary Disks

Zhaohuan Zhu,[1] Sean M. Andrews,[2] and Andrea Isella[3]

[1]*Department of Physics and Astronomy, University of Nevada, Las Vegas, 4505 S. Maryland Pkwy, Las Vegas, NV, 89154, USA;* `zhaohuan.zhu@unlv.edu`

[2]*Harvard-Smithsonian Center for Astrophysics, 60 Garden Street, Cambridge, MA 02138, USA;* `sandrews@cfa.harvard.edu`

[3]*Department of Physics and Astronomy, Rice University, 6100 Main Street, Houston, TX 77005, USA;* `isella@rice.edu`

Abstract. In the early phase of their formation and evolution, gaseous giant planets are expected to be surrounded by dense disks made of gas and dust generally referred as circumplanetary disks (CPDs). Similar to circumstellar disks, CPDs control the final mass of planets and are the birth places of moons, which, as in the case of Europa, might be prime locations for harboring life. The study of CPDs is fundamental for at least three reasons: (i) CPDs pinpoint the location of planets in formation, which might not be detectable by any other means, (ii) the structure of CPDs provide information on the physical processes responsible for the formation of giant planets, and (iii) the investigating CPDs will allow us to understand the formation of moons. In this study we show that the ngVLA, thanks to its unprecedented sensitivity and angular resolution, will be able to find CPDs in nearby star forming regions and measure their masses, temperatures, and radii. The combination of ngVLA and ALMA observations at shorter wavelengths should constrain the optical depth and dust properties of CPDs. Furthermore, ngVLA observations could reveal jets and winds arising from CPDs, providing key information on the planet mass accretion process. Finally, theory predicts that future 30 meter class optical/infrared telescopes, and to a more limited extent *JWST*, could detect thermal emission from CPDs. The combination of multi-wavelength observations of CPDs would constrain the thermal history of forming planets and shed light on their formation.

1. Planet and Moon Formation

During planet formation in protoplanetary disks, theoretical studies suggest that giant planets are surrounded by dense mini-disks made of gas and dust called circumplanetary disks (CPDs Lubow et al. 1999). In our solar system, the existence of CPDs in the past is supported by the coplanarity of regular satellites/moons around current giant planets. Studying CPDs is crucial for understanding satellite/moon formation (Canup & Ward 2002), and even habitability of life on moons (e.g., Europa and Enceladus). The density and temperature structure of CPDs affect the moon size and composition. If the CPD is too hot for water to condense, the moons will be made mostly of dust grains with little subsurface water ocean. The existence of CPDs can help moons to

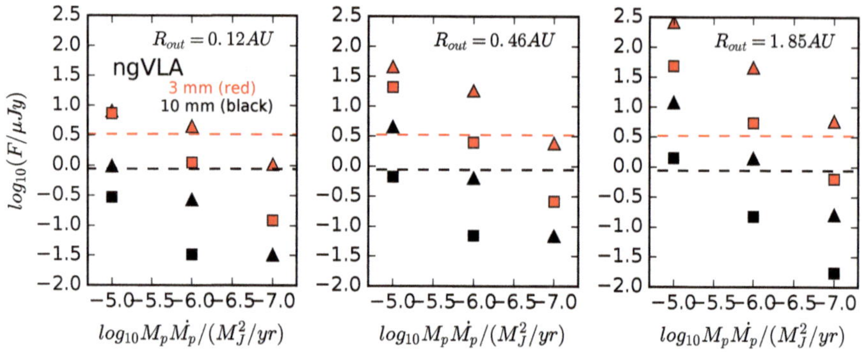

Figure 1. The flux for viscous heating dominated CPDs with different accretion rates. The panels show fluxes at 3 mm and 10 mm (red and black symbols, respectively). The red and black dashed lines (3.31 μJy/beam and 0.87 μJy/beam) indicate the 5σ level achieved by 5 hours ngVLA observation with the assumed rms noise level at these two bands (1-hour rms noise of 1.48 μJy/beam at 3 mm and 0.39 μJy/beam at 10 mm). From left to right panels, R_{out} are 0.12 AU, 0.46 AU, and 1.85 AU, which correspond to the distance of 5 AU, 20 AU, and 80 AU for a Jupiter mass planet around a solar mass star. Triangles are the cases with α=0.001 while squares are the cases with α=0.01. Adapted from Zhu et al. (2018).

move into resonance (e.g., Galilean satellites) which can lead to geological activity on moons, providing energy sources for early life.

Despite their importance for planet and moon formation and even life on moons, finding CPDs in other young stellar objects (YSOs) is challenging since these disks are a lot fainter than the central star (at near-IR) or the circumstellar disks (at radio). On the other hand, CPDs can be quite massive and potentially be detected at radio wavelengths (Isella et al. 2014; Szulágyi et al. 2018) . For all giant planets in our Solar System, the total solid mass in moons is around 0.003% of the central planet mass (Canup & Ward 2002). For Jupiter, this is 0.78 moon mass in the CPD if all the solids were in the CPD at the same time (Mosqueira & Estrada 2003). If we can detect CPDs, they not only reveal the young planets in protoplanetary disks but also provide direct constraints on the initial condition for moon formation.

2. Detection of CPDs by the ngVLA

The thermal flux from dust in CPDs depends on 1) the disk's temperature, 2) the disk's surface density, and 3) the disk's size. The disk's surface temperature can be estimated using the standard viscous accretion theory by specifying an accretion rate. To accrete a Jupiter mass within the disk's life time, the disk's accretion rate is 10^{-10} to 10^{-8} solar mass per year. The disk's surface density can be calculated by assuming an accretion viscous coefficient α. Theoretical work suggests such α coefficient is 0.001-0.01. The disk's size is roughly 1/3 of the planet's Hill sphere. For a Jupiter mass planet around a solar mass star at the closest star forming regions that is 140 pc away, the CPD radius is 0.12, 0.46, 1.85 AU (or 0.8, 3, 13 mas) when the planet is 5, 20, 80 AU away from the star. With these assumptions, Zhu et al. (2018) calculate the mm/cm flux of CPDs.

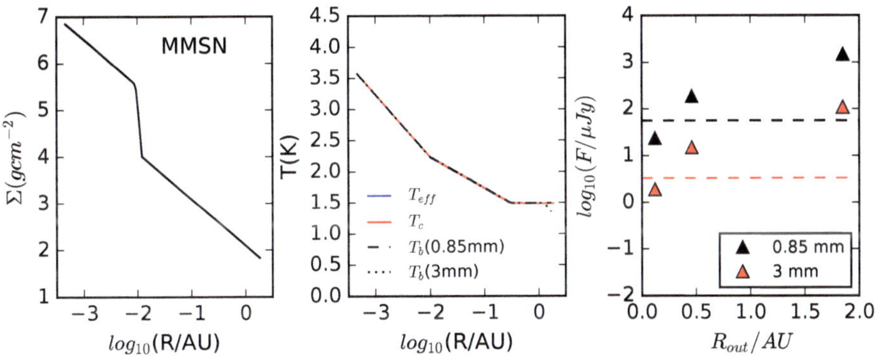

Figure 2. The minimum mass sub-nebula model and its mm/cm emission. The disk surface density and temperature (Mosqueira & Estrada 2003) are shown in the left and middle panel. The middle panel also shows the brightness temperature at 0.85 mm and 3 mm. Since the disk has a uniform temperature vertically and the disk is optically thick, the brightness temperature equals the disk midplane temperature. The integrated mm/cm emission for disks with different R_{out} are shown in the right panel. The horizontal lines are the 5-hr 5σ detection levels for ALMA and ngVLA. Adapted from Zhu et al. (2018).

If the planet is at 5 AU from the central star and is accreting at $\geq 10^{-9}$ solar mass/year, the flux from the disk is >1 μJy at 3 mm and >0.01 μJy at 1 cm based on conservative parameters. For these cases, 5 hours integration can provide 5 σ detection (Figure 1). Since the flux of CPD is roughly proportional to the planet's distance to the central star, such sensitivity can detect CPD much easier if the planet is further away from the central star.

Due to the small size of CPD, it is likely that CPD will be detected as a point source. If ngVLA can achieve 3 mas resolution, we may be able to resolve the CPD when the CPD is far away from the central star. However, such spatially resolved observation requires significantly more integration time to achieve the necessary sensitivity to image the outer part of the CPD. On the other hand, a sufficient spatial resolution may be useful for identifying gap structure in protoplanetary disk that is carved out by the planet, which will be helpful for confirming the unresolved point source detection is from the embedded planet and its CPD. Since the gap can extend from 0.5 to 1.5 planet-star distance. The gap carved out by a planet at 5 AU from the central star and 140 pc away from us can be as wide as $0''.036$. To search planets within a gap, the beam size should not be larger than the width of the gap.

Although currently we have little observational constraints on CPDs around young planets, we can use moons around Jovian planets in our solar system to study CPDs that were in our solar system. By spreading the mass of Galilean satellites into a mini disk around Jupiter, the "minimum mass sub-nebula" disk can be constructed (Lunine & Stevenson 1982; Mosqueira & Estrada 2003). Such model has a high gas surface density of few $\times 10^5$ g cm^{-2}. The disk surface density and temperature in Mosqueira & Estrada (2003) are plotted in the left and middle panel of Figure 2. The integrated the total mm/cm emission from the disk is shown in the right panel of Figure 2. Surprisingly, such a disk at 5 AU (so R_{out} ~0.12 AU) is even detectable by ALMA and ngVLA. Unfortunately ALMA's resolution is not high enough to separate CPDs from the circumstellar disks. Within one observational beam (several AU), most flux will come from the circumstellar disk around the star instead of CPDs around the planet. Thus, detecting CPDs within the background circumstellar disk is in the "background-limited" regime, where a higher resolution by a factor of f can increase the signal-to-noise ratio by a factor of f^2. Furthermore, it can separate the gap induced by the planet and the rest circumstellar disk, further decreasing the circumstellar emission around the planet. Thus, ngVLA will be crucial for separating CPDs from the circumstellar disks.

3. Radio Detection of Jets and Winds in CPDs

Most astrophysical accretion disks, from AGNs to protostars, produce jets and winds that are observable at both radio and optical wavelengths. Thus, accreting CPDs could also produce jets and winds. MHD simulations by Gressel et al. (2013) have indeed suggested that disk winds can be launched in CPDs. To estimate the centimeter free-free emission of jets launched in CPDs, Zhu et al. (2018) extrapolate the relationship for protostellar jets given in Anglada et al. (2015),

$$\frac{F_\nu d^2}{\text{mJy kpc}^2} = 0.008\left(\frac{L_{bol}}{L_\odot}\right)^{0.6} \tag{1}$$

to the planetary mass regime. If $M_p\dot{M}_p = 10^{-5}M_J^2/\text{yr}$, the accretion luminosity is $L_{bol} = 1.57\times10^{-3}L_\odot$. The estimated F_ν from the jet is 8.48 μJy at cm if d=140 pc. The flux of 8.48 μJy is detectable by the ngVLA. The detection of jet by the protoplanet will put constraints on magnetic fields around young planet and the inner CPD disk structure, which is crucial for understanding how young planets are accreting. Overall, with disk thermal emission and jets/winds, ngVLA itself provides two independent ways to detect CPD.

4. Synergy with Observations at Other Wavelengths

When CPDs are actively accreting, they can be as bright as a late M-type/early L-type brown dwarf (Quillen & Trilling 1998). Unlike the SEDs of a brown dwarf, the SEDs of accreting CPDs are a lot redder, peaking at mid-infrared. Thus, to detect CPDs in protostellar systems, we can image the young stellar system at mid-infrared, adopting high contrast imaging techniques capable of blocking most of the stellar light. With such observational strategy, a few red sources within circumstellar disks have been discovered and their photometries at infrared bands are consistent with accreting CPDs (Zhu 2015). So far, LkCa 15b is the most promising candidate for the accreting CPDs since two independent accretion indicators, Hα line and near-IR thermal emission, have been detected (Sallum et al. 2015).

In future, 30 meter class telescopes may discover many CPD candidates throughout the disk from less than 1 AU to tens of AU. Confirming them at radio wavelength bands and furthermore constraining the disk mass can only be achieved by the ngVLA. Such observations will help us to understand both planet and moon formation, and even life on moons.

References

Anglada, G., Rodríguez, L. F., & Carrasco-Gonzalez, C. 2015, Advancing Astrophysics with the Square Kilometre Array (AASKA14), 121
Canup, R. M., & Ward, W. R. 2002, AJ, 124, 3404
Gressel, O., Nelson, R. P., Turner, N. J., & Ziegler, U. 2013, ApJ, 779, 59
Isella, A., Chandler, C. J., Carpenter, J. M., Pérez, L. M., & Ricci, L. 2014, ApJ, 788, 129
Lubow, S. H., Seibert, M., & Artymowicz, P. 1999, ApJ, 526, 1001
Lunine, J. I., & Stevenson, D. J. 1982, ICARUS, 52, 14
Mosqueira, I., & Estrada, P. R. 2003, ICARUS, 163, 198
Quillen, A. C., & Trilling, D. E. 1998, ApJ, 508, 707
Sallum, S., Follette, K. B., Eisner, J. A., et al. 2015, Nat, 527, 342
Szulágyi, J., Plas, G. v. d., Meyer, M. R., et al. 2018, MNRAS, 473, 3573
Zhu, Z. 2015, ApJ, 799, 16
Zhu, Z., Andrews, S. M., & Isella, A. 2018, MNRAS, 479, 1850

Science with a Next-Generation Very Large Array
ASP Conference Series, Monograph 7
Eric J. Murphy, ed.

Dust Growth and Dust Trapping in Protoplanetary Disks

Nienke van der Marel,[1] Brenda Matthews,[1] Ruobing Dong,[2] Tilman Birnstiel,[3] and Andrea Isella[4]

[1]*Herzberg Astronomy & Astrophysics Programs, National Research Council of Canada, Victoria, BC, Canada*

[2]*Department of Physics & Astronomy, University of Victoria, Victoria, BC, Canada*

[3]*University Observatory, Faculty of Physics, Ludwig-Maximilians-Universitat Munchen, Munich, Germany*

[4]*Department of Physics & Astronomy, Rice University, Houston, TX, USA*

Abstract. ALMA has revolutionized our view of protoplanetary disks, revealing structures such as gaps, rings and asymmetries that indicate dust trapping as an important mechanism in the planet formation process. However, the high resolution images have also shown that the optically thin assumption for millimeter continuum emission may not be valid and the low values of the spectral index may be related to optical depth rather than dust growth. Longer wavelength observations are essential to properly disentangle these effects. The high sensitivity and spatial resolution of the next-generation Very Large Array (ngVLA) will open up the possibilities to spatially resolve disk continuum emission at centimeter wavelengths and beyond, which allows the study of dust growth in disks in the optically thin regime and further constrain models of planet formation.

1. Introduction

The planet formation process remains one of the major puzzles in modern-day astronomy. Planets are known to form in protoplanetary disks of gas and dust around young stars (Williams & Cieza 2011). A large amount of research has been conducted in the second half of the twentieth century, but many questions remain unanswered. In particular, planet formation is hindered by a number of growth barriers, according to dust evolution theory, while observational evidence indicates that somehow these barriers must have been overcome. The core accretion process, where the accretion of dust particles and planetesimals results in solid cores of ~Earth mass, followed by runaway gas accretion, is a promising mechanism to form the range of planets that are seen in both our Solar System and beyond (e.g., Winn & Fabrycky 2015). However, the core accretion process requires the growth of planetesimals from submicron-sized dust grains that are seen in the interstellar medium (ISM). These first steps in dust growth in protoplanetary disks, where the dust evolution is governed by the drag forces between dust and gas, have proven to be one of the most challenging parts in the planet formation process (e.g., Testi et al. 2014, and references therein).

In the last two decades, clear evidence has been found for dust grain evolution in disks beyond the grain sizes in the interstellar medium, through (interferometric) millimeter observations of disks (e.g., Beckwith & Sargent 1991; Testi et al. 2003; Andrews & Williams 2005; Natta et al. 2007; Isella et al. 2009; Ricci et al. 2010), indicating the presence of millimeter and even centimeter-sized particles in the outer regions of protoplanetary disks through multi-wavelength continuum observations and measurements of the spectral index α_{mm} (millimeter flux density $F_\nu \sim \nu^{\alpha_{mm}}$). The value of α_{mm} can provide information on the particle size in protoplanetary disks (see Testi et al. 2014, and references therein). For (sub)micrometer-sized dust, such as found in the ISM (interstellar medium), α_{mm} is typically 3.5–4.0, but when dust grows to millimeter sizes, α_{mm} is expected to decrease to 2–3 (Draine 2006; Ricci et al. 2010). When the dust emission is optically thin and in the Rayleigh-Jeans regime, the observable α_{mm} can be related to the dust opacity index $\beta = \alpha - 2$, with $\beta < 1$ for millimeter grains and larger (Natta et al. 2004). Furthermore, β is related to the dust opacity of a certain dust grain size population $n(a)$ as $\kappa_\nu \propto \nu^\beta$, where β is mostly sensitive to the maximum grain size $a_{\max}$ (Draine 2006).

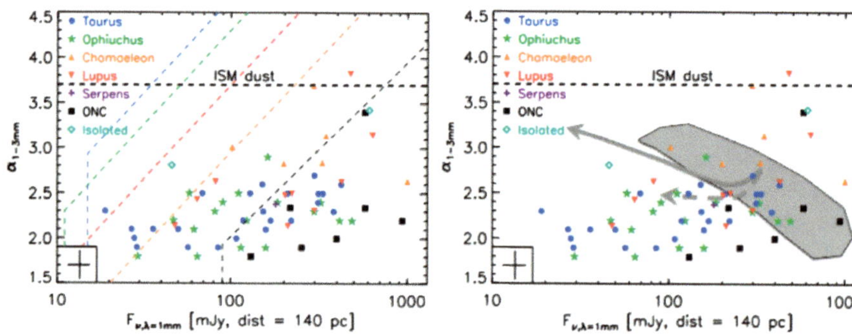

Figure 1. Spectral index α_{mm} between 1.1 and 3 mm against 1.1 mm flux density for disks in various nearby star forming regions. The dashed diagonal lines (left panel) mark the sensitivity limits of the various surveys. The grey area in the right panel illustrates the predictions of dust evolution models (Birnstiel et al. 2010), and the arrows indicate the effect of dust evolution including radial drift (solid) and pressure traps (dashed) (Pinilla et al. 2012b). Figure taken from Testi et al. (2014).

Initially, grain growth in protoplanetary disks was primarily based on unresolved or marginally resolved millimeter data in surveys of star forming regions, showing α_{mm} values well below the ISM value of 3.5–4.0 (see Figure 1), indicating millimeter-sized particles in the outer disk. For a long time, it remained unclear how these particles could be present, due to the so-called *radial drift* problem, which results in a rapid inward drift of dust particles in a disk when grown to millimeter sizes (Whipple 1972; Weidenschilling 1977). For a disk with a smooth radial density profile, both the gas surface density and temperature, and thus the pressure, decrease radially outward. This additional pressure support results in a slightly sub-Keplerian orbital velocity for the gas. In contrast, dust particles are not pressure supported and want to orbit at Keplerian speed, but as they are embedded in a sub-Keplerian gas disk, they are forced to orbit at lower speed which leads to a removal of angular momentum from the particles to the gas, causing the particles to spiral inward. The dust particles are thus experiencing a drag force by the gas, depending on its Stokes number St, which describes the coupling of the particles to the gas and depends on the dust particle size and the local gas surface density (see e.g. Brauer et al. 2008; Birnstiel et al. 2010). For small dust particles, $St \ll 1$, they are strongly coupled to the gas and do not drift inwards, but radial drift becomes significant when the particle size increases and reaches its strongest value when $St = 1$. In practice, millimeter-dust particles in the outer disk can have St values close to unity and are expected to drift inwards on time scales as short as 100 years, hence further dust growth is hindered. An additional problem is *fragmentation*: while low-velocity collisions lead to particle growth, high-velocity impacts lead to destruction according to experimental and theoretical work on dust interaction (Blum & Wurm 2000). The combination of the radial drift and fragmentation problems is also called the 'meter-size barrier' because a one-meter size object at 1 AU drifts efficiently inwards limiting further growth, and equivalently dust particles in the outer disk cannot grow beyond millimeter sizes.

A proposed solution to explain the presence of larger dust grains in disks as the start of planet formation, is a so-called *dust trap*, where dust particles are being 'trapped' in local pressure maxima in the outer disk (Whipple 1972; Klahr & Henning 1997; Rice et al. 2006; Brauer et al. 2008; Pinilla et al. 2012b). Such a pressure maximum or pressure bump can arise as a radial dust trap at the edges of dead zones (e.g., Varnière & Tagger 2006), at the edges of gas gaps cleared by planets (Zhu et al. 2012; Pinilla et al. 2012a), in zonal flows (Johansen et al. 2009; Pinilla et al. 2012b) or as azimuthal dust traps in long-lived vortices (e.g., Barge & Sommeria 1995; Klahr & Henning 1997), which can be the result of a Rossby Wave Instability of a radial pressure bump (e.g., Lovelace et al. 1999; Wolf & Klahr 2002; Lyra et al. 2009; Regály et al. 2012). Pressure maxima as a solution for dust growth beyond millimeter sizes were proposed theoretically, but without spatially resolved information on the dust grain size distribution within the disk this phenomenon remained speculative.

The *Atacama Large Millimeter/submillimeter Array* (ALMA) in Chile, which began operations in 2012, has revolutionized our view of protoplanetary disks, in particular on the existence of dust traps, starting with the discovery of the highly asymmetric millimeter-dust concentration in Oph IRS 48 (van der Marel et al. 2013). Whereas the gas and micrometer-dust are distributed along a ring the millimeter emission showed evidence for dust trapping in a vortex, thought to be generated by a Rossby Wave Instability of a radial pressure bump. Follow-up observations at centimeter wavelengths with the VLA confirmed that the azimuthal width of the dust trap decreased with wavelength (van der Marel et al. 2015a), as predicted by dust trapping models (e.g., Birnstiel et al. 2013). Oph IRS 48 is a so-called transitional disk, a disk with a cleared inner dust cavity (e.g., Espaillat et al. 2014, and references therein), which were of particular interest in dust evolution predictions for keeping millimeter-sized dust grains in the outer disk (Pinilla et al. 2012a). Other transition disks imaged with ALMA show a range of azimuthally symmetric and asymmetric dust rings (Casassus et al. 2013; Zhang et al. 2014; Pérez et al. 2014; van der Marel et al. 2015b; Pinilla et al. 2017), confirming their ring-like morphologies as seen with previous interferometers such as CARMA and the SMA (e.g., Brown et al. 2009; Isella et al. 2010; Andrews et al. 2011). The existence of radial dust traps was primarily indicated by the difference in distribution of the gas (as traced by CO isotopologues) and millimeter dust (e.g., Bruderer et al. 2014; Perez et al. 2015; van der Marel et al. 2016; Dong et al. 2017; Fedele et al. 2017) and the difference with the micrometer-sized dust grains (Garufi et al. 2013; Pinilla et al. 2015). Multi-wavelength ALMA observations between 1.3 and 0.45 millimeter provided a hint of radial trapping in the SR 21 and SR 24S disks through spatially resolved $\alpha(r)$ which decreased to <3 in the ring emission (Pinilla et al. 2015, 2017), and azimuthal changes in α_{mm} were seen in the asymmetries in HD 142527 (Casassus et al. 2015) and HD 135344B (Cazzoletti et al. 2018). Combining ALMA with VLA data at centimeter wavelengths results in further evidence for dust trapping in MWC 758 (e.g., Marino et al. 2015), but the low spatial resolution and sensitivity limits the calculation of a spatially resolved $\alpha_{mm}(r)$ map.

Another spectacular ALMA result in this context is the imaging of disks at ultra high angular resolution of ~20 mas or a few AU at the distance of nearby star forming regions. The mind-blowing image of HL Tau (ALMA Partnership et al. 2015) of the ALMA Long Baseline Campaign has revealed that even dust disks without inner cavity may exist of ring-like structures, which a remarkable similarity to the predictions of Pinilla et al. (2012b) of local pressure maxima that are required for the explanation of millimeter-sized dust in the outer disk. The HL Tau image was quickly followed by other multi-ring disks, such as TW Hya (Andrews et al. 2016) and HD 163296 (Isella et al. 2016). Explanations for the dust rings range from planets carving gaps (Lin & Papaloizou 1979), snow lines (Zhang et al. 2015), sintering (Okuzumi et al. 2016) and secular gravitational instabilities (Takahashi & Inutsuka 2016). If the dust rings are indeed caused by planets, trapping is expected to occur. Multi-wavelength observations indeed reveal radial variations of α_{mm} along the gaps and rings in HL Tau and TW Hya (Carrasco-González et al. 2016; Tsukagoshi et al. 2016), but the evidence is still marginal with the currently available data. On the other hand, evidence for radial drift is evident in observations. SMA observations already revealed evidence for a segregation between the dust and gas in the IM Lup disk (Panić & Hogerheijde 2009), with the gas outer radius being twice as large as that of the dust. Multi-wavelength continuum observations show a decrease of particle grain size with radius for a number of primordial disks (Pérez et al. 2015; Tazzari et al. 2016; Tripathi et al. 2018). Clearly, in lack of one or more dust traps in the outer disk, dust particles do drift inwards to the nearest pressure maximum. On the other hand, this implies that any extended dust disk must in fact consist of one or more pressure bumps, to keep the dust particles away from the center of the disk.

2. The Optical Depth Problem

The results described above appear very promising in explaining the presence of large dust grains in protoplanetary disks as the start of the planet formation process, but they suffer from a major problem: the spectral index α_{mm} can only be related to β in a simple linear fashion for optically thin emission. The optical depth $\tau_\nu(r)$ is defined as

$$\tau_\nu(r) = \int_{-\infty}^{\infty} \rho\kappa_\nu ds = \frac{\kappa_\nu \Sigma(r)}{\cos i} \tag{1}$$

with $\Sigma(r)$ the dust surface density at radius r and i the inclination and the emitted flux density F_ν integrated over a disk area can be computed as:

$$F_\nu = \frac{\cos i}{d^2} \int B_\nu(T(r))\left(1 - e^{-\tau_\nu(r)}\right) 2\pi r dr \tag{2}$$

with distance d, Planck function B at temperature T, which reduces to $F_\nu \propto B_\nu(T) \propto \nu^2$ for optically thick emission and $F_\nu \propto \kappa_\nu B_\nu(T) \propto \nu^{2+\beta}$ for optically thin emission

($\tau_\nu \ll 1$). This immediately illustrates that the spectral index α_{mm} is 2 (and thus $\beta \sim 0$) for optically thick emission regardless of the grain size, and a low value of β does not automatically imply large dust grains. The optical depth is generally estimated by comparing the measured brightness temperature T_b ($T_b \sim I_\nu = \frac{F_\nu}{d\Omega}$) with the solid angle $d\Omega$ and the physical dust temperature $T_{\rm dust}$ as calculated from a radiative transfer model. Typical disk observations of millimeter interferometry do show lower values for T_b than $T_{\rm dust}$, indicating $\tau_\nu \sim 0.5$ or lower, but if the optically thick emission is concentrated on size scales smaller than the resolution, optical depth effects cannot be ruled out as an explanation for the low β values based on millimeter emission (Ricci et al. 2012; Tripathi et al. 2017). This is generally measured using the filling factor f, the fraction of the disk area which is optically thick. Thus, the evidence for dust growth and dust trapping based on spectral index values from millimeter observations may not be reliable. Furthermore, if the assumption of optically thin dust emission is no longer valid, this also has potentially large consequences for disk dust mass estimates, which are generally taken as a linear relation between $M_{\rm dust}$ and integrated millimeter flux density F_ν as:

$$M_{\rm dust} = \frac{F_\nu d^2}{\kappa_\nu B_\nu(T)} \tag{3}$$

If the dust emission is optically thick, the derived disk dust mass is only a lower limit. Even in high resolution ALMA observations, the millimeter disk emission often remains unresolved (e.g., in >30% of the disks in the Lupus survey which has a spatial resolution of 0.″25 or 20 AU radius; Ansdell et al. 2018), implying that the majority of disks are much smaller than previously thought, and their emission is potentially optically thick. However, even for partially resolved extended disks the integrated flux density may be dominated by optically thick emission from either rings or from the inner part of the disk where the dust concentrates as a result of radial drift in absence of pressure maxima in the outer disk (Tripathi et al. 2018). The high spatial resolution of ALMA is a crucial factor here as high resolution observations have demonstrated that dust emission is concentrated in much narrower rings than previously thought.

3. The role of the next-generation VLA

In order to study dust growth in protoplanetary disks, spatially resolved multi-wavelength observations in the optically thin regime are crucial. The next-generation Very Large Array (ngVLA) offers high angular resolution (~10 mas) at long wavelengths (17 – 93 GHz or 3 – 18 mm) where $\tau_\nu \ll 1$. The sensitivity (~ 0.3 μJy/beam for 1 hour integration at 40 GHz) allows quality observations at high signal-to-noise, resulting in clear images of the structures in disks. The current VLA does not have the resolution or sensitivity to achieve this result in a reasonable amount of time. Of particular importance for dust growth studies is the spatially resolved spectral index β_{cm} which can be calculated across optically thin wavelengths. This opens up a large range of possibilities to answer questions on the origins and mechanisms of the structures in disks observed by ALMA.

3.1. Azimuthal Asymmetries

The azimuthal asymmetries, though to be dust trapping vortices, have gained a large interest from the community, but a lot of questions regarding their origin remain unclear. A spatially resolved $\beta_{cm}(r, \phi)$ can help to disentangle azimuthal trapping from other proposed effects to explain dust asymmetries such as eccentricities (Ataiee et al. 2014; Ragusa et al. 2017), as eccentric disks would not show an azimuthal dependence of β_{cm}. Furthermore, the measurement of $\beta_{cm}(r, \phi)$ can constrain the trapping efficiency in the disk, and in combination with estimates of the gas surface density, lead to independent constraints on the viscosity and turbulence (Birnstiel et al. 2013; Lyra & Lin 2013), which tends to diffuse the dust in the vortex. Quantitative estimates of the turbulence can further be used to estimate the time scales of dust trapping in a vortex and relate them to their observability. For AB Aur, Fuente et al. (2017) claim a decaying vortex due to the azimuthal extended emission at longer wavelengths (in contrast to dust trapping predictions), but optical depth effects need to be excluded to confirm this claim.

Another interesting consequence of trapping in a vortex is a size segregation due to the vortex's self-gravity: whereas smaller grains will be trapped in the center of the vortex, larger grains are expected to be trapped *ahead* in the azimuthal direction (Baruteau & Zhu 2016): a shift was indeed observed in the asymmetry in HD 135344B in ALMA observations, but in the opposite direction (Cazzoletti et al. 2018). In order to rule out any effects due to optically thick emission, ngVLA observations at similar angular resolution, can test the predictions of Baruteau & Zhu (2016) in a large number of disks.

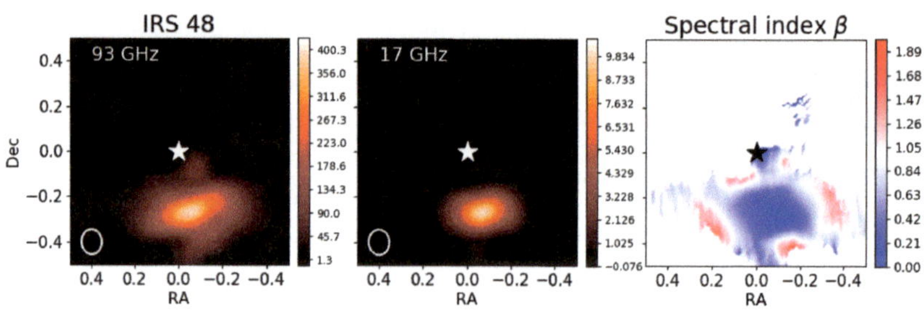

Figure 2. Simulated images of the spectral index $\beta(r, \phi)$ for the azimuthal aymmetry in Oph IRS 48 at centimeter wavelengths, as expected from ngVLA observations at 97 and 17 GHz in an 2 hour integration. Brightness units are given in μJy/beam. Oph IRS 48 is expected to show a decrease of β in the center due to dust growth.

As a demonstration, we have created intensity profiles of a highly asymmetric disks (Oph IRS 48) based on its observed morphology with ALMA and VLA and scaled to their expected centimeter flux density at 97 and 17 GHz. In Figure 2 we demonstrate that the spatially resolved β_{cm} as calculated from 3 and 18 millimeter emission can be measured with an accuracy of ~0.1 in high signal-to-noise ratio observations with the ngVLA in 2 hours per setting, using the CASA simulator and the ngVLA configuration.

3.2. Multi-Ring Disks

Spatially resolving the multi-ring systems such as HL Tau in multiple centimeter wavelengths will help to constrain their origin, which is currently completely unclear: cleared gaps by planets is a tempting explanation, but this requires (giant) planets at tens or even hundreds of AU away from the star, which are very rare in exoplanet demographics (Winn & Fabrycky 2015; Bowler 2016) and difficult to explain in reasonable time scales in planet formation theory (Helled et al. 2014). The edges of planet gaps are pressure maxima where dust is expected to be trapped; a measure of $\beta_{cm}(r)$ can immediately tell if the rings are consistent with dust traps. If not, other explanations for the origin of the rings will need to be considered.

3.3. Radial Drift

In disks without pressure maxima in the outer disk, the larger dust particles are expected to drift inwards and form a sharp outer continuum edge (Birnstiel & Andrews 2014). Although this effect has been demonstrated through comparison between the millimeter dust emission and the much more extended distribution of the gas as traced by the CO (e.g., Panić & Hogerheijde 2009; Andrews et al. 2012; de Gregorio-Monsalvo et al. 2013; Ansdell et al. 2018), a quantification of the drift and associated time scales remain unexplored due to the uncertainties introduced by the high optical depth in the inner part of the dust disk (e.g., Pérez et al. 2015; Tazzari et al. 2016; Tripathi et al. 2018). Measuring $\beta_{cm}(r)$ in optically thin wavelengths will set better constraints on the grain size distribution within the disk and will help to constrain dust evolution models in providing better estimates for drift time scales and fragmentation.

3.4. Disk Masses

Disk dust masses are generally derived from disk-integrated millimeter fluxes using the assumption of optically thin emission, but it is quite likely that this assumption is not valid, implying that many of the derived disk masses (usually calculated with the dust mass and a gas-to-dust ratio of 100) are lower limits. This has important consequences for our understanding of the dynamics in disks: gravitational instability (Boss 1997) may occur for disk masses >10% of the stellar mass, which is generally well above our current estimates of the disk mass. However, if large amounts of the millimeter emission is optically thick, these disk masses are severely underestimated and gravitational instabilities could be important for disk fragmentation, dynamics and giant planet formation than currently thought. Large disk surveys with the ngVLA of nearby star forming regions can provide us with flux density measurements of hundreds of disks, similar to current ALMA surveys. The sensitivity of the Lupus ALMA disk survey ($\sim 0.3 M_{\rm Earth}$ as 3σ detection limit; Ansdell et al. 2018) can be reached in 2 minutes per source with the ngVLA at 40 GHz, allowing full snapshot surveys in only a few hours.

4. Conclusions

Whereas ALMA has revolutionized our understanding of protoplanetary disks by revealing a large diversity of structures indicating concentrated dust growth, a better understanding and quantification of the dust growth processes can only be achieved with longer wavelength observations in the centimeter regime where the dust becomes optically thin, in particular through spectral index observations. Using the Square Kilometre Array (SKA) to derive dust spectral indices using even longer wavelengths (>5 cm) is impractical, as the dust emission becomes too faint to be detected at high spatial resolution. The ngVLA provides unprecedented possibilities to study dust growth in disks and further explore the dust trapping process.

References

ALMA Partnership, A., et al. 2015, ApJ, 808, L3
Andrews, S. M., & Williams, J. P. 2005, ApJ, 631, 1134
Andrews, S. M., et al. 2011, ApJ, 732, 42
Andrews, S. M., et al. 2012, ApJ, 744, 162
Andrews, S. M., et al. 2016, ApJ, 820, L40
Ansdell, M., et al. 2018, ApJ, 859, 21
Ataiee, S., et al. 2014, A&A, 572, A61
Barge, P., & Sommeria, J. 1995, A&A, 295, L1
Baruteau, C., & Zhu, Z. 2016, MNRAS, 458, 3927
Beckwith, S. V. W., & Sargent, A. I. 1991, ApJ, 381, 250
Birnstiel, T., Dullemond, C. P., & Brauer, F. 2010, A&A, 513, A79
Birnstiel, T., Dullemond, C. P., & Pinilla, P. 2013, A&A, 550, L8
Birnstiel, T., Andrews, S., 2014, ApJ, 780, 153
Blum, J., & Wurm, G. 2000, Icarus, 143, 138
Boss, A. P. 1997, Science, 276, 1836
Bowler, B. P. 2016, PASP, 128, 102001
Brauer, F., Dullemond, C. P., & Henning, T. 2008, A&A, 480, 859
Brown, J. M., et al. 2009, ApJ, 704, 496
Bruderer, S., et al. 2014, A&A, 562, A26
Carrasco-González, C., et al. 2016, ApJ, 821, L16
Casassus, S., et al. 2013, Nat, 493, 191
Casassus, S., et al. 2015, ApJ, 812, 126
Cazzoletti, P., et al. 2018, subm. to ApJ
de Gregorio-Monsalvo, I., et al. 2013, A&A, 557, A133
Dong, R., et al. 2017, ApJ, 836, 201
Draine, B. T. 2006, ApJ, 636, 1114
Espaillat, C., et al. 2014, Protostars and Planets VI, 497
Fedele, D., et al. 2017, A&A, 600, A72
Fuente, A., et al. 2017, ApJ, 846, L3
Garufi, A., et al. 2013, A&A, 560, A105
Helled, R., et al. 2014, Protostars and Planets VI, 643
Isella, A., Carpenter, J. M., & Sargent, A. I. 2009, ApJ, 701, 260
Isella, A., et al. 2010, ApJ, 725, 1735
Isella, A., et al. 2016, Physical Review Letters, 117, 251101
Johansen, A., Youdin, A., & Klahr, H. 2009, ApJ, 697, 1269
Klahr, H. H., & Henning, T. 1997, Icarus, 128, 213
Lin, D. N. C., & Papaloizou, J. 1979, MNRAS, 188, 191
Lovelace, R. V. E., et al. 1999, ApJ, 513, 805
Lyra, W., et al. 2009, A&A, 493, 1125
Lyra, W., & Lin, M.-K. 2013, ApJ, 775, 17
Marino, S., Perez, S., & Casassus, S. 2015, ApJ, 798, L44
Natta, A., et al. 2007, Protostars and Planets V, 767
Natta, A., et al. 2004, A&A, 424, 603
Okuzumi, S., et al. 2016, ApJ, 821, 82
Panić, O., & Hogerheijde, M. R. 2009, A&A, 508, 707
Pérez, L. M., et al. 2014, ApJ, 783, L13
Pérez, L. M., et al. 2015, ApJ, 813, 41
Perez, S., et al. 2015, ApJ, 798, 85
Pinilla, P., Benisty, M., & Birnstiel, T. 2012a, A&A, 545, A81

Pinilla, P., et al. 2012b, A&A, 538, A114
Pinilla, P., et al. 2015a, A&A, 573, A9
Pinilla, P., et al. 2015b, A&A, 584, A16
Pinilla, P., et al. 2017, ApJ, 839, 99
Ragusa, E., et al. 2017, MNRAS, 464, 1449
Regály, Z., et al. 2012, MNRAS, 419, 1701
Ricci, L., et al. 2010, A&A, 512, A15
Ricci, L., et al. 2012, A&A, 540, A6
Rice, W. K. M., et al. 2006, MNRAS, 373, 1619
Takahashi, S. Z., & Inutsuka, S.-i. 2016, AJ, 152, 184
Tazzari, M., et al. 2016, A&A, 588, A53
Testi, L., et al. 2003, A&A, 403, 323
Testi, L., et al. 2014, Protostars and Planets VI, 339
Tripathi, A., et al. 2017, ApJ, 845, 44
Tripathi, A., et al. 2018, ApJ, 861, 64
Tsukagoshi, T., et al. 2016, ApJ, 829, L35
van der Marel, N., et al. 2015a, ApJ, 810, L7
van der Marel, N., et al. 2016, A&A, 585, A58
van der Marel, N., et al. 2015b, A&A, 579, A106
van der Marel, N., et al. 2013, Science, 340, 1199
Varnière, P., & Tagger, M. 2006, A&A, 446, L13
Weidenschilling, S. J. 1977, MNRAS, 180, 57
Whipple, F. L. 1972, in From Plasma to Planet, ed. A. Elvius, 211
Williams, J. P., & Cieza, L. A. 2011, ARA&A, 49, 67
Winn, J. N., & Fabrycky, D. C. 2015, ARA&A, 53, 409
Wolf, S., & Klahr, H. 2002, ApJ, 578, L79
Zhang, K., Blake, G. A., & Bergin, E. A. 2015, ApJ, 806, L7
Zhang, K., et al. 2014, ApJ, 791, 42
Zhu, Z., et al. 2012, ApJ, 755, 6

Science with a Next-Generation Very Large Array
ASP Conference Series, Monograph 7
Eric J. Murphy, ed.

Tracing the Water Snowline in Protoplanetary Disks with the ngVLA

Ke Zhang,[1] Edwin A. Bergin,[1] Jonathan P. Williams,[2] Paola Pinilla,[3] and Sean M. Andrews[4]

[1]*University of Michigan, Ann Arbor, MI 48105, USA;* kezhang@umich.edu

[2]*Institute for Astronomy, Honolulu, HI 96822, USA*

[3]*Department of Astronomy/Steward Observatory, The University of Arizona, Tucson, AZ 85721, USA*

[4]*Harvard-Smithsonian Center for Astrophysics, 60 Garden Street, Cambridge, MA 02138, USA*

Abstract. The water snowline in protoplanetary disks is one of the most pivotal locations for planet formation: it sets a critical boundary of chemical composition in the disk and likely serves as a favorable site of planetesimal growth and planet formation. However, the water snowline at the mid-plane cannot be directly traced by water line observations as the dust emission around the water snowline is highly optically thick at these line frequencies. Alternatively, the water snowline may be traced through the sharp changes in physical properties (surface density and dust size distribution) at this transition or through the snowlines of other molecules that sublimate co-spatially with water. In this chapter, we discuss the ngVLA's capability of tracing the mid-plane water snowline through mapping sharp transitions in the continuum spectral index caused by changes of dust size distributions and by using the NH_3 snowline as a proxy. We found that with the current design of the ngVLA, a sharp transition in spectral index across the water snowline can be readily traced in relatively bright disks. A detection of the NH_3 snowline is challenging under the current design, but may still be possible if the NH_3-to-water abundance in the disk is close to that of the ISM ices. In summary, the proposed ngVLA is the only facility that can provide sufficient sensitivity and spatial resolution to trace the evolution of water snowline at the disk mid-plane and to reveal its role in accelerating the planetesimal formation and in building the architecture of planetary systems.

1. Introduction

Water is believed to be the most crucial molecule for the habitability of planetary systems (van Dishoeck et al. 2014). Its abundance and distribution in a protoplanetary disk greatly affect the atmospheric and core composition of giant planets (Öberg et al. 2011b) and the surface water content of terrestrial planets (Raymond et al. 2007).

The most critical location in the water distribution is where water vapor condenses into ices, the so-called, *water snowline*. In the Solar Nebula, the water snowline appears to be the critical boundary that separates the formation of terrestrial planets and giant planets (Hayashi 1981).

Theoretical models predict the water snowline is a pivotal location for planetary architecture, as two important physical changes occur at this transition (see Figure 1). The first one is an enhancement in the surface density of solids–as water accounts for about 50% of the total condensible mass, its condensation will significantly increase the solid mass beyond the snowline and accelerate the planet(esimal) formation (Stevenson & Lunine 1988; Pollack et al. 1996). The second change is the dust size distribution. Icy aggregates are significantly more resistant to compaction than bare silicate aggregates and thus icy aggregates can readily grow to a much larger size (Ros & Johansen 2013; Zhang et al. 2015). Once a critical decimeter-sized pebble population is formed, streaming instabilities can readily drive the creation of km-sized or larger planetesimals (Johansen et al. 2014). On the other hand, the sintering effect may break aggregates and leads to a pile-up of smaller particles in the region slightly outside the snowline (Okuzumi et al. 2016). Therefore the water snowline may play an important role in regulating the formation and chemical composition of planetesimals, and ultimately the formation of planets and their bulk composition.

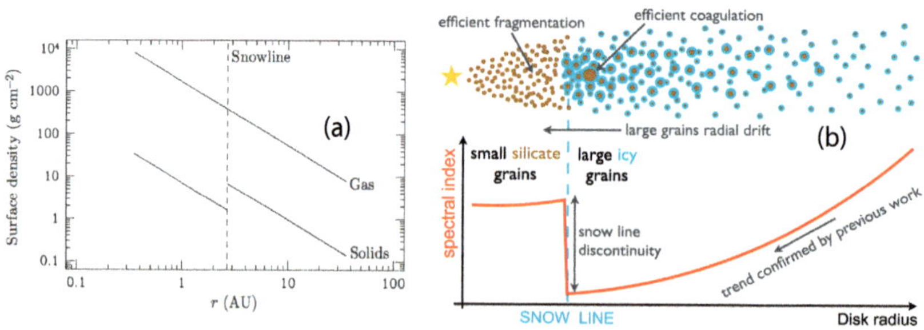

Figure 1. (a) The sharp discontinuity in the surface density of solids across the water snowline, based on Hayashi's minimum mass Solar Nebula model. The Figure is adapted from Armitage (2013). (b) An illustrative example of the sharp discontinuity in spectral index α across the water snowline, as a result of the change in dust particle size distribution. The dust emission is assumed to be optically thin. The figure is from Banzatti et al. (2015).

Despite the importance of the water snowline, it is challenging to pinpoint its location and to quantitatively study its role in planetesimal formation. First of all, planetesimal formation occurs at the dust-rich mid-plane of the disk, while water forms quite readily in disk surface layers (Bethell & Bergin 2009). Thus, water line observations can only constrain the surface snowline (Zhang et al. 2013; Blevins et al. 2016). Second, the water snowline in an accreting protoplanetary disk is around 1 – 10 AU (Kennedy & Kenyon 2008; Notsu et al. 2016, 2017). At these radii, the dust emission at relevant line frequencies of water or its isotopologues ($\geq$ 100 GHz) is largely optically thick and thereby hindering a direct detection of the snowline.

2. Tracing the Water Snowline with ngVLA

Alternatively, the water snowline maybe traced through the sharp changes of physical properties (surface density and dust size distribution) or through the snowlines of other molecules that sublimate co-spatially with water. In both approaches, observations are needed to be taken at frequencies where dust emission is optically thin and with a sufficient spatial resolution to resolve the region inside the snowline. The ngVLA is the only facility that can provide these capabilities.

2.1. Tracing the water snowline through dust property changes

A sharp change of dust properties across a water snowline may have been seen in recent ALMA observations of the V883 Ori system (Cieza et al. 2016). V883 Ori is an FU Ori type source that is undergoing a large outburst in luminosity arising from a temporary increase in the accretion rate. The luminosity outburst warms the disk and pushes the water snowline to a large radius where the dust emission is less optically thick. The 230 GHz continuum emission of the V883 Ori disk shows an intensity break due to an abrupt change in the optical depth at about 42 AU, where the disk temperature is close to the water condensation temperature. Furthermore, the spectral index across the snowline is consistent with predictions from numerical simulations that include the coagulation, fragmentation and radial drift of dust grains (Banzatti et al. 2015).

The V883 Ori study was possible thanks to its large snowline during the luminosity outburst. For the majority of protoplanetary disks, their snowlines are only at 1 – 10 AU, where the dust emission is highly optically thick at all ALMA bands. Furthermore, even in the V883 Ori case, the region inside the snowline (42 AU) appears to be optically thick at 230 GHz and thus hinders constraints on the dust size distribution. Therefore, to study the dust size change across a snowline in a large number of disks, observations need to be carried out at much lower frequencies where the dust emission is optically thin. In addition, a higher spatial resolution is needed to spatially resolve the region inside the snowline. Only the proposed ngVLA facility has sufficient sensitivity and resolution to carry out this type of observations.

To resolve a 5 AU water snowline in a disk at a distance of 100 pc, an angular resolution $\leq$50 mas is needed and a maximum recoverable scale of $\geq$500 mas is needed to compare regions inside and outside the snowline. To meet the combination of resolution and low dust optical depth, the ideal wavelengths for this experiment is at 30 GHz. The spectral index can be measured across a bandwidth of 20 GHz. For a fiducial disk, the emission is predicted to be strong and a sensitivity of 5 μJy/beam would be sufficient. Higher frequency ALMA observations at a comparable spatial resolution can provide complementary measurements on the dust temperature through optically thick emission, which helps distinguish dust property changes caused by other mechanisms. Through these spectral index observations, we can study whether dust grains grow faster in regions beyond the water snowline and quantitatively constrain the size distribution and surface density of solids across the snowline.

The ngVLA observations will provide the critical measurement of the water snowline locations in disks of different ages and stellar properties. The water snowline locations can be compared to snowlines of other abundant volatiles, such as the CO snowline measured by ALMA (Qi et al. 2013; Zhang et al. 2017). The snowlines of major volatiles set the rough boundaries of planetesimals with different elemental compositions, which further determine the bulk composition of terrestrial planets and cores of gas or ice giants (Öberg et al. 2011b). The spectral index measurements will constrain the dust size distribution and surface density changes across the snowline. These will shed crucial insight into our understanding of the planetesimal formation and the birth locations of giant planets.

2.2. Tracing the Water Snowline Through a Proxy: the NH_3 Snowline

Ammonia (NH_3) is a known major constituent of interstellar ices (~5%, Öberg et al. 2011a) and an important tracer of nitrogen chemistry in protoplanetary disks. Laboratory work of ice evaporation has shown at least a portion, or perhaps all of the NH_3 ice releases co-spatially with water (Martín-Doménech et al. 2014). Furthermore, NH_3 has inversion transitions at cm wavelengths where the dust continuum emission is less optically thick. Given these two advantages, NH_3 is the best proxy candidate to trace the water snowline in protoplanetary disks.

To resolve the NH_3 snowline, we can either spatially resolve the NH_3 emission inside its snowline or obtain high signal-to-noise spectra to constrain the gas emitting area through Keplerian velocity field in the disk as well as the centroid of the emission. As the lines are predicted to be faint (see below), the latter case is a more practical approach.

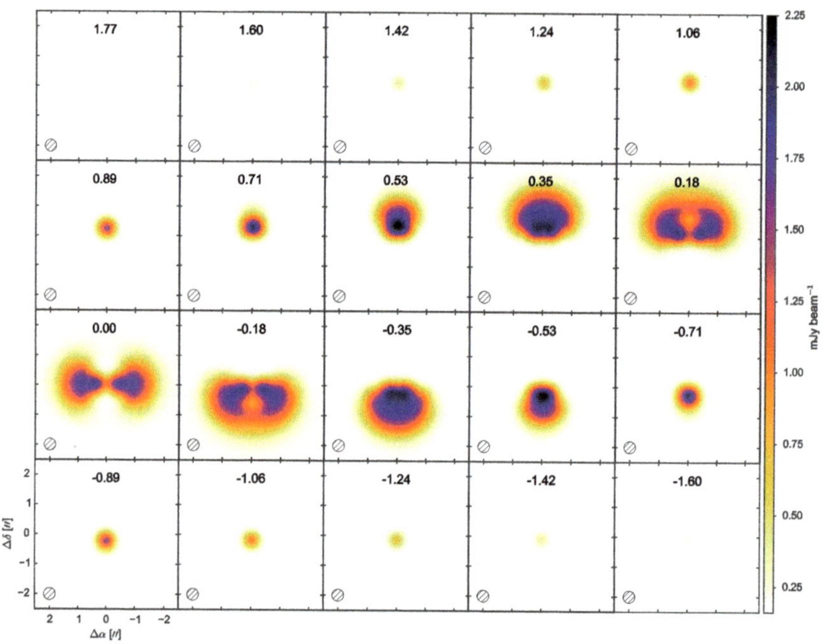

Figure 2. Emission maps as a function of velocity in a 0.″4 beam for the 1,1 transition of NH_3 in our fiducial disk model based on TW Hya (Du et al. 2015).

To explore detectability, we assume a minimal model – one that sets limits for how faint NH_3 emission would be in the most stringent case. This model assumes no NH_3 ice is provided from the ISM and NH_3 is made solely by the disk. We then use the model of TW Hya that matches water vapor emission (Du et al. 2015) to predict emission in the 1,1 and 2,2 inversion transitions around 23 GHz. This predicted channel map emission from this model is shown in Figure 1) for TW Hya. This needs to be scaled by one order of magnitude as most disk sources lie a factor of 3 more distant than TW Hya. In the future, we can perform additional calculations that assume interstellar ices are present and where sublimation would raise the gas phase abundance and hence emission levels.

For a NH_3 emission detection, we find that a 10σ detection of ammonia 1,1 transition towards TW Hya requires 0.25 mJy in 0.2 km/s channels. To survey NH_3 reliably in hundreds of disk systems we need to be able to detect this line at 150 pc at the same level, requiring 25 μJy in 0.2 km/s channels (10σ). However, since lower velocity resolution can be used for detections, a sensitivity of 120 μJy in 1.0 km/s channels (5σ) would be sufficient for a survey.

The detection of the snowline via the model provided here appears be challenging as the required sensitivity in sources at 150 pc would be 30 μJy to detect the emission from the line wings at 150 pc (using the model shown in Figure 1). Since this represents a minimal model (a maximum NH_3 abundance of $\sim 2\times10^{-6}$), any additional level of ice evaporation will increase the emission at these channels. Furthermore, we likely will need to observe disks surrounding more massive stars such as B and A spectral type. Here we still must use velocity line profiles to reach material at a few AU. Thus, if NH_3 was present in the gas at the level of 5% relative to water ice then its emission within the evaporated zone would be substantially increased, at least until the line becomes optically thick. We estimate that the line will become optically thick when the line is about a factor of 10 higher than the current model or at 300 μJy in 0.2 km s^{-1} channels.

3. Conclusion

The water snowline in protoplanetary disks is one of the most pivotal locations for planet formation: it not only sets a critical boundary of chemical composition in the disk, but may also serve as a favorable site of planetesimal growth and planet formation. The water snowline at the mid-plane cannot be directly traced by water line observations as the dust emission around the water snowline is highly optically thick at these line frequencies.

In this chapter, we discuss two alternative approaches to trace the mid-plane water snowline – by observing sharp transitions in dust properties across the snwoline or by using the NH_3 snowline as a proxy. Either approach requires observations at cm wavelengths where the dust emission is likely to be optically thin. We found that with the current design of the ngVLA ability, a sharp transition in dust property across the water snowline can be readily traced through measuring the spectral index of continuum emissions of relatively bright disks. The NH_3 snowline detection is likely to be challenging under the current design, but the NH_3 abundance can be an order of magnitude higher than our conservative model and thus still be detectable. In summary, the proposed ngVLA is the only facility that provides sufficient sensitivity and spatial resolution to trace the evolution of water snowline at the disk mid-plane and reveal its role in the formation of planetesimals and in building the architecture of planetary systems.

References

Armitage, P. J. 2013, Astrophysics of Planet Formation
Banzatti, A., Pinilla, P., Ricci, L., et al. 2015, ApJ, 815, L15
Bethell, T., & Bergin, E. 2009, Science, 326, 1675
Blevins, S. M., Pontoppidan, K. M., Banzatti, A., et al. 2016, ApJ, 818, 22
Cieza, L. A., Casassus, S., Tobin, J., et al. 2016, Nat, 535, 258
Du, F., Bergin, E. A., & Hogerheijde, M. R. 2015, ApJ, 807, L32
Hayashi, C. 1981, PTPS, 70, 35
Johansen, A., Blum, J., Tanaka, H., et al. 2014, Protostars and Planets VI, 547

Kennedy, G. M., & Kenyon, S. J. 2008, ApJ, 673, 502
Martín-Doménech, R., Muñoz Caro, G. M., Bueno, J., & Goesmann, F. 2014, A&A, 564, A8
Notsu, S., Nomura, H., Ishimoto, D., et al. 2016, ApJ, 827, 113
—. 2017, ApJ, 836, 118
Öberg, K. I., Boogert, A. C. A., Pontoppidan, K. M., et al. 2011a, ApJ, 740, 109
Öberg, K. I., Murray-Clay, R., & Bergin, E. A. 2011b, ApJ, 743, L16
Okuzumi, S., Momose, M., Sirono, S.-i., Kobayashi, H., & Tanaka, H. 2016, ApJ, 821, 82
Pollack, J. B., Hubickyj, O., Bodenheimer, P., et al. 1996, Icarus, 124, 62
Qi, C., Öberg, K. I., Wilner, D. J., et al. 2013, Science, 341, 630
Raymond, S. N., Scalo, J., & Meadows, V. S. 2007, ApJ, 669, 606
Ros, K., & Johansen, A. 2013, A&A, 552, A137
Stevenson, D. J., & Lunine, J. I. 1988, Icarus, 75, 146
van Dishoeck, E. F., Bergin, E. A., Lis, D. C., & Lunine, J. I. 2014, in Protostars and Planets VI, ed. H. Beuther, R. S. Klessen, C. P. Dullemond, & T. Henning, 835
Zhang, K., Bergin, E. A., Blake, G. A., Cleeves, L. I., & Schwarz, K. R. 2017, Nature Astronomy, 1, 0130
Zhang, K., Blake, G. A., & Bergin, E. A. 2015, ApJ, 806, L7
Zhang, K., Pontoppidan, K. M., Salyk, C., & Blake, G. A. 2013, ApJ, 766, 82

Part IV

Planetary Systems and Life

Science with a Next-Generation Very Large Array
ASP Conference Series, Monograph 7
Eric J. Murphy, ed.

Observing the Effects of Chemistry on Exoplanets and Planet Formation

Brett A. McGuire,[1] Edwin Bergin,[2] Geoffrey A. Blake,[3,4] Andrew M. Burkhardt,[5] L. Ilsedore Cleeves,[6] Ryan A. Loomis,[1] Anthony J. Remijan,[1] Christopher N. Shingledecker,[7,8] and Eric R. Willis[9]

[1]*National Radio Astronomy Observatory, Charlottesville, VA, 22903;* `bmcguire@nrao.edu`

[2]*Department of Astronomy, University of Michigan, Ann Arbor, MI 48109*

[3]*Division of Geological and Planetary Sciences, California Institute of Technology, Pasadena, CA 91125*

[4]*Division of Chemistry and Chemical Engineering, California Institute of Technology, Pasadena, CA 91125*

[5]*Harvard-Smithsonian Center for Astrophysics, Cambridge, MA 02138*

[6]*Department of Astronomy, University of Virginia, Charlottesville, VA 22904*

[7]*Max Planck Institute for Extraterrestrial Physics, Garching, Germany*

[8]*Institute for Theoretical Chemistry, University of Stuttgart, Germany*

[9]*Department of Chemistry, University of Virginia, Charlottesville, VA 22904*

Abstract. One of the primary mechanisms for inferring the dynamical history of planets in our Solar System and in exoplanetary systems is through observation of elemental ratios (i.e., C/O). The ability to effectively use these observations relies critically on a robust understanding of the chemistry and evolutionary history of the observed abundances. Significant efforts have been devoted to this area from within astrochemistry circles, and these efforts should be supported going forward by the larger exoplanetary science community. In addition, the construction of a next-generation radio interferometer will be required to test many of these predictive models in situ, while simultaneously providing the resolution necessary to pinpoint the location of planets in formation.

1. Background

The location of condensation fronts (snow lines) in protoplanetary disks (PPDs) can have substantial effects on the physical and chemical evolution of planets in the system. Indeed, many models predict that these condensation fronts are the locations where planetesimal formation begins (Ros & Johansen 2013; Schoonenberg & Ormel 2017; Drazkowska & Alibert 2017). The location of these condensation fronts is determined both by the incident radiation from the protostar, but also, in large part, by the chemical composition of the gas and ice. The desorption temperature of the bulk ice varies significantly with the chemical composition and mixing ratio (Collings et al. 2004), even of minor constituents. This can be further complicated by a variance in the sticking efficiency of the icy grains with composition (Wang et al. 2005). It is clear that the underlying chemical composition in these ices, even at low mixing ratios, can have a substantial effect on the efficiency of planet formation.

Further, the relative gas vs solid-phase inventories at snow lines govern the composition of the planets and their atmospheres. These compositions (e.g., the C/O ratio in planetary atmospheres; Öberg et al. 2011) may therefore be useful probes of the history of exoplanets. Effectively using observed exoplanet elemental ratios to test those models therefore relies heavily on our ability to accurately interpret chemical abundances measured in disks to infer the inventory of undetectable species, and how these measurements affect observed elemental ratios.

Constraining the chemical compositions in PPDs is challenging. The most complex gas-phase species yet detected in PPDs are methanol (CH_3OH; Walsh et al. 2016) and methyl cyanide (CH_3CN; Öberg et al. 2015), and more complex molecules are unlikley to be observed (Walsh et al. 2014). Further, in cold or shielded regions within disks, ices may be the major repository of C and O (Bergin et al. 2016). Direct observation of ices typically require space-based infrared facilities and a target source with a sufficiently bright background continuum *and* favorable viewing geometry. Even in massive young stellar objects with substantial ice columns, the list of firmly-detected ice species numbers less than a dozen (Boogert et al. 2015). Our knowledge of gas and ice compositions in disks must therefore rely on astrochemical models, using the detectable simple gas species as the primary constraint.

These models are not yet robust, despite substantial efforts (Garrod 2013; Herbst 2017; Wakelam et al. 2015). To reliably infer chemical compositions in PPDs, in which chemistry is closely coupled to physical conditions (temperature, density, UV flux, evolving dust populations, etc.) that vary with radius and scale height, necessitates that these models be well-constrained by observations in earlier stages of star formation, where more robust inventories can be observed. Here, we briefly discuss the state-of-the-art in progressively earlier stages of chemical evolution, and outline future directions and the technology and community support needed to enable them.

2. Chemical Evolution in PPDs ...

A number of larger scale studies of disk physical properties and chemistry have been enabled by the sensitivity and resolution of ALMA. Efforts to measure fundamental properties such as bulk mass (Ansdell et al. 2016), and larger compositional studies examining nitrogen (Guzmán et al. 2017) and deuterium (Huang et al. 2017) ratios, are now possible for larger samples sizes. With robust chemical modeling, ALMA observations of simple hydrocarbons (e.g., C_2H) can perhaps be used to infer C/O ratios indirectly (Bergin et al. 2016). As well, *JWST* and, if commissioned, the Origins Space Telescope, have significant potential to observe major ice components (e.g. H_2O, CO, CO_2, CH_3OH) toward a wide variety of targets. Yet, these infrared facilities will still largely be blind to the complex chemistry occurring within the ices, and to the combined ice/gas coupling in warmer gas. That insight is only achievable through radio observations of gas-phase species, that are sensitive to populations orders of magnitude lower in abundance, coupled with chemical modeling.

3. ... and in Complex, Star-Forming Regions

Sgr B2(N), Orion-KL, and NGC 6334I are the most chemically and physically complex stages of star formation observed and modeled (Garrod et al. 2008; Crockett et al. 2014; McGuire et al. 2017). The extreme physical environments, interplay between the gas and ice, and complex structures often result in models which can reproduce observed abundances in, at best, a qualitative fashion (Laas et al. 2011; Garrod et al. 2008; Garrod 2013). These types of physical and chemical interactions are the same that are at play in PPDs, albeit on a much larger scale. It is therefore these environments that provide a unique test environment for the similar PDR style models applied to PPDs. Improvements in these models are most likely to be achieved by testing them against a wide range of environments from quiescent clouds to massive star forming regions.

Existing facilities in the (sub-)millimeter regime, particularly ALMA, are capable of reaching the line-confusion limit in broadband line surveys of complex, star forming regions in a matter of hours. That is, the density of molecular lines in the resulting spectra is such that there is at least one spectral line every FWHM; there are no line-free channels. When line-confusion is reached, no additional information can be gained from longer integration times at the same spatial resolution. We are rapidly approaching the point at which deep drill observations with ALMA and the GBT will no longer produce new spectral lines. The end result of this limit is that no further detections of new species can be made, and thus the refinement of chemical networks will be limited to those species detectable at the confusion limit.

In the cm-wave regime, however, the most sensitive survey, the Prebiotic Interstellar Molecular Survey (PRIMOS; Neill et al. 2012) with the most sensitive facility (the GBT) toward the most line-rich source (Sgr B2) has lines in only $\sim 10-20\%$ of its channels at cm-wavelengths. Diminishing returns on sensitivity for this survey have already set in, and to increase the sensitivity substantially would require hundreds of hours of dedicated observations for every ~2 GHz window. Further, ALMA observations at higher frequencies are revealing that the region displays widely variable physical conditions within a relatively small environment, variation that is lost to single-dish GBT observations (Belloche et al. 2016). This offers a valuable opportunity to probe environment-induced chemical differentiation, but only with spatially-resolved observations which are not possible with the GBT.

While the brightest lines in Sgr B2 are readily detectable with extant facilities (e.g., the VLA and ATCA; Corby et al. 2015), the complex species are not efficiently observable, especially at high angular resolution. The only viable solution to this problem is the construction of a next-generation interferometer like the next-generation Very Large Array (ngVLA). Under the current design concept (Selina & Murphy 2017), the ngVLA is likely to not only be capable of spatially resolving chemistry in Sgr B2(N), but to reach the line-confusion limit even at cm-wavelengths, providing substantial new, complex molecular detections and their correlation with the physical conditions at work.

4. Applications to PPDs

Once refined, these models can be used to accurately interpret observations from ALMA targeting the outermost condensation fronts within disks, primarily CO snow lines (Qi et al. 2015). Application in the terrestrial planet forming region, near the H_2O snow line, will require AU-scale resolution with high sensitivity. A high-frequency ngVLA is the only facility currently in design with the required specifications. Further, while these models can then interpret results for C and O in disks, observations of the major N-bearing species, NH_3 will still be required at similar scales. An ngVLA operating at least through the 24 GHz NH_3 transition is therefore necessary.

An ngVLA design concept is needed to support broadband, high surface-brightness sensitivity capabilities at the small to modest ($0\farcs1$–$3''$) scales required to probe the effects of the widely variable physical conditions on the chemical evolution in sources like Sgr B2 and unlock chemical inventories obscured by line confusion suffered at extant facilities like ALMA. The same instrument should be capable of observing NH_3 and imaging the planet forming zone inside the water snow line in disks, requiring AU-scale (mas) resolution capabilities.

5. Chemical Evolution in Protostellar Environments

The protostellar stage of star formation spans a wide range of sources, from 'hot corinos' such as IRAS 16293 (Jørgensen et al. 2016) to chemically rich outflow sources like L1157 (Mendoza et al. 2014). The molecular outflows from active young protostars like L1157 provide a unique environment to test the immediate impact of rapid changes in temperature and density on 'pristine' molecular clouds. In these sources, distance from the protostar along the outflow corresponds directly to time, allowing modeling of temporal evolution. Hot corinos are the next stage after L1157 and before hot cores like Sgr B2(N).

The best method for constraining the chemistry in these sources is broadband spectral line surveys. These sources are relatively compact, are beam-diluted in single-dish observations, and their linewidths are typically such that high-sensitivity data are often at or approaching line confusion limits, necessitating interferometric observation (Jørgensen et al. 2016). Because they are typically moderately warm (50–150 K; Jørgensen et al. 2016; McGuire et al. 2015), complex molecules display bright features in the 1–3 mm window well-covered by ALMA, NOEMA, and the SMA.

The SOLIS project at NOEMA and IRAM, which examines a sample of solar-type protostellar sources (Ceccarelli et al. 2017), and the PILS project at ALMA, which has focused particularly on an exquisitely detailed examination of the IRAS 16293 system (Jørgensen et al. 2016), are the primary active dedicated investigations into this class of objects. Extant observational facilities are likely sufficient to provide the required insight. Ongoing and future efforts on ALMA, NOEMA, and the SMA examining a large sample of protostellar environments are critical to constraining the effects of physical environment on chemical evolution and incorporating these effects into models.

6. Chemical Evolution in Quiescent Environments

Cold (T_{ex} ~10 K), dark, starless clouds are an ideal environment in which to test the time evolution of chemical reaction networks under 'constant' physical conditions. The typical sources for exploring this stage of star and planet formation are highly extended ($\theta_s > 30''$). At these excitation temperatures, most molecules, especially those more complex than CH_3OH, will display bright emission at cm wavelengths. The best strategy for obtaining the required abundance and excitation information is therefore deep, broadband spectral line surveys at cm-wavelengths using single-dish facilities combined with chemical modeling (Cordiner et al. 2017; Kaifu et al. 2004; McGuire et al. 2018). The best extant instrument for this work is the Green Bank Telescope (GBT).

Even in a source as extended as TMC-1, there exists evidence for large chemical variation over small spatial scales (0.02–0.03 pc; Peng et al. 1998; Dickens et al. 2001), and observations of such would probe the effects of the very earliest stages of collapse on the chemistry. These observations will require high spatial resolution, high surface-brightness sensitivity, and broad spectral bandwidth to be efficient. Because of the cold temperatures, this further requires a cm-wavelength facility. While the VLA satisfies this need, it lacks the combination of surface brightness sensitivity and bandwidth necessary to conduct a sensitive line survey with any efficiency. The proposed ngVLA, in its current conceptual design, would prove more than optimal for this task.

The chemical evolution of all ranges of environments from planetary atmospheres (Arsenovic et al. 2016) to disk midplanes (Cleeves et al. 2013) to molecular clouds (Hudson et al. 2008) is impacted by radiation. The last few years have seen significant advances in modeling efforts to accurately include these effects (Shingledecker & Herbst 2018), but these efforts are young, and require substantial training on 'simple' environments like quiescent clouds where the radiation effects can be easily decoupled from dynamical physical events such as outflows and shocks. In short, efforts at the GBT and other large single-dish facilities are needed to survey quiescent chemical clouds. These efforts would be greatly bolstered by the development of broadband capabilities on the ngVLA to enable investigations of the chemical evolution upon initial gravitational collapse.

7. Summary

A combination of observational and theoretical efforts are needed that are focused on placing rigorous constraints on models of chemical evolution at each stage of the star and planet formation process. The observational requirements of these investigations necessitate a commitment by the community to support extant facilities and the development of technological advancements for the same, as well as the development of a next-generation VLA with high sensitivity and exquisite spatial resolution. The ultimate goal should be producing models that can be used to better calibrate our estimates of gas and ice compositions in PPDs, given the observations of simple gas-phase species achievable in these sources, and the construction of a radio facility capable of observationally exploiting these models *in situ* during planet formation.

References

Ansdell, M., Williams, J. P., van der Marel, N., et al. 2016, ApJ, 828, 1

Arsenovic, P., Rozanov, E., Stenke, A., et al. 2016, Journal of Atmospheric and Solar-Terrestrial Physics, 149, 180

Belloche, A., Müller, H. S. P., Garrod, R. T., & Menten, K. M. 2016, A&A, 587, A91

Bergin, E. A., Du, F., Cleeves, L. I., et al. 2016, ApJ, 831, 1

Boogert, A. C. A., Gerakines, P. A., & Whittet, D. C. B. 2015, ARAA, 53, 541

Ceccarelli, C., Caselli, P., Fontani, F., et al. 2017, ApJ, 850, 176

Cleeves, L. I., Adams, F. C., Bergin, E. A., & Visser, R. 2013, ApJ, 777, 28

Collings, M. P., Anderson, M. A., Chen, R., et al. 2004, Monthly Notices of the Royal Astronomical Society, 354, 1133

Corby, J. F., Jones, P. A., Cunningham, M. R., et al. 2015, MNRAS, 452, 3969

Cordiner, M. A., Charnley, S. B., Kisiel, Z., McGuire, B. A., & Kuan, Y.-J. 2017, ApJ, 850, 187

Crockett, N. R., Bergin, E. A., Neill, J. L., et al. 2014, ApJ, 787, 112

Dickens, J. E., Langer, W. D., & Velusamy, T. 2001, ApJ, 558, 693

Drazkowska, J., & Alibert, Y. 2017, A&A, 608, A92

Garrod, R. T. 2013, ApJ, 765, 60

Garrod, R. T., Weaver, S. L. W., & Herbst, E. 2008, ApJ, 682, 283

Guzmán, V. V., Öberg, K. I., Huang, J., Loomis, R., & Qi, C. 2017, ApJ, 836, 1

Herbst, E. 2017, International Reviews in Physical Chemistry, 36, 287

Huang, J., Öberg, K. I., Qi, C., et al. 2017, ApJ, 835, 1

Hudson, R. L., Moore, M. H., Dworkin, J. P., Martin, M. P., & Pozun, Z. D. 2008, Astrobiology, 8, 771

Jørgensen, J. K., van der Wiel, M., Coutens, A., et al. 2016, A&A, 595, A117

Kaifu, N., Ohishi, M., Kawaguchi, K., et al. 2004, PASJ, 56, 69

Laas, J. C., Garrod, R. T., Herbst, E., & Widicus Weaver, S. L. 2011, ApJ, 728, 71

McGuire, B. A., Burkhardt, A. M., Kalenskii, S. V., et al. 2018, Science, 359, 202

McGuire, B. A., Carroll, P. B., Dollhopf, N. M., et al. 2015, ApJ, 812, 1

McGuire, B. A., Shingledecker, C. N., Willis, E. R., et al. 2017, ApJL, 851, L46

Mendoza, E., Lefloch, B., Lopez-Sepulcre, A., et al. 2014, MNRAS, 445, 151

Neill, J. L., Muckle, M. T., Zaleski, D. P., et al. 2012, ApJ, 755, 153

Öberg, K. I., Guzmán, V. V., Furuya, K., et al. 2015, Nature, 520, 198

Öberg, K. I., Murray-Clay, R., & Bergin, E. A. 2011, ApJ, 743, L16

Peng, R., Langer, W. D., Velusamy, T., Kuiper, T. B. H., & Levin, S. 1998, ApJ, 497, 842

Qi, C., Öberg, K. I., Andrews, S. M., et al. 2015, ApJ, 813, 1

Ros, K., & Johansen, A. 2013, A&A, 552, A137

Schoonenberg, D., & Ormel, C. W. 2017, Astronomy & Astrophysics, 602, A21

Selina, R., & Murphy, E. J. 2017, in ngVLA Memo Series, 17

Shingledecker, C. N., & Herbst, E. 2018, Physical Chemistry Chemical Physics, 20, 5359

Wakelam, V., Loison, J. C., Herbst, E., et al. 2015, ApJS, 217, 1

Walsh, C., Millar, T. J., Nomura, H., et al. 2014, A&A, 563, A33

Walsh, C., Loomis, R. A., Öberg, K. I., et al. 2016, ApJL, 823, L10

Wang, H., Bell, R. C., Iedema, M. J., Tsekouras, A. A., & Cowin, J. P. 2005, ApJ, 620, 1027

Science with a Next-Generation Very Large Array
ASP Conference Series, Monograph 7
Eric J. Murphy, ed.

Organics in Disk Midplanes with the ngVLA

Karin I. Öberg,[1] L. Ilsedore Cleeves,[2] and Ryan Loomis[3]

[1]*Harvard-Smithsonian Center for Astrophysics, Cambridge, MA 02138;* koberg@cfa.harvard.edu

[2]*Department of Astronomy, University of Virginia, 530 McCormick Rd, Charlottesville, VA 22904;* lic3f@virginia.edu

[3]*NRAO, 520 Edgemont Rd, Charlottesville, VA 22903;* rloomis@nrao.edu

Abstract. Planets assemble in the midplanes of protoplanetary disks. The compositions of dust and gas in the disk midplane region determine the compositions of nascent planets, including their chemical hospitality to life. In this context, the distributions of volatile organic material across the planet and comet forming zones is of special interest. These are difficult to access in the disk midplane at IR and even millimeter wavelengths due to dust opacity, which can veil the midplane, low intrinsic molecular abundances due to efficient freeze-out, and, in the case of mid-sized organics, a mismatch between expected excitation temperatures and accessible line upper energy levels. At ngVLA wavelengths, the dust is optically thin, enabling observations into the planet forming disk midplane. The ngVLA also has the requisite sensitivity; using TW Hya as a case study, we show that the ngVLA will be able to map out the distributions of diagnostic organics, such as CH_3CN, in nearby protoplanetary disks.

1. Introduction

A planet's potential hospitality to life or 'habitability' depends on its bulk composition (i.e., whether it is rocky) and surface temperature (i.e., whether it can sustain liquid water). An equally important consideration, however, is whether the planet is chemically hospitable to life or chemically habitable, i.e.,whether there is access to water and soluble organic molecules on the planet's surface. Furthermore, some initial organic compositions may be more likely than others to result in a chemistry that eventually brings about an origin of life. Studies of comet compositions reveal that the Solar Nebula from which the Solar System emerged was rich in CH_3OH, CH_3CHO, CH_3CCH, HCOOH, HNCO, HC_3N and CH_3CN and other intermediate sized organic molecules (Mumma & Charnley 2011; Le Roy et al. 2015). Several of these, but especially nitriles and unsaturated hydrocarbons, have been found to be particularly well suited to form the building blocks of proteins and RNA (Powner et al. 2009; Patel et al. 2015). Understanding the prevalence of this suite of molecules in analogs to the Solar Nebula is needed to assess how often planets form in a chemical environment that is conducive to producing chemically habitable planets.

Temperate, habitable zone, planets are expected to form relatively close to their host stars, within a few au around Solar analogs, and obtain their bulk compositions from co-located disk material. Their volatile budget is set by the direct accretion of this material in the warm inner disk, and by later delivery from asteroids and comets formed at greater distances, up to 10s of au, from the star (Morbidelli et al. 2012; Raymond et al. 2014). Disk chemical compositions out to at least 30 au may thus affect the chemical habitability of nascent planets. It is important to note that we do not expect a disk to present a single composition across the inner 30 au, but rather steep chemical gradients due to radial gradients in temperature, radiation flux and density (Aikawa & Herbst 1999; Willacy & Langer 2000; Öberg et al. 2011a; Henning & Semenov 2013). With ALMA such chemical gradients are indeed observed (Figure 1). There is also not a single chemical composition at a set radius due to analogous vertical chemical gradients (Calvet et al. 1991; Aikawa & Herbst 1999), i.e.,the disk midplanes where planets and planetesimals assemble are often chemically distinct from more elevated disk layers. Most observations to date probe elevated disk layers and the community then uses models to connect these observations to midplane compostions (Öberg et al. 2015).

2. The Role of the ngVLA in Astrochemical Disk Explorations

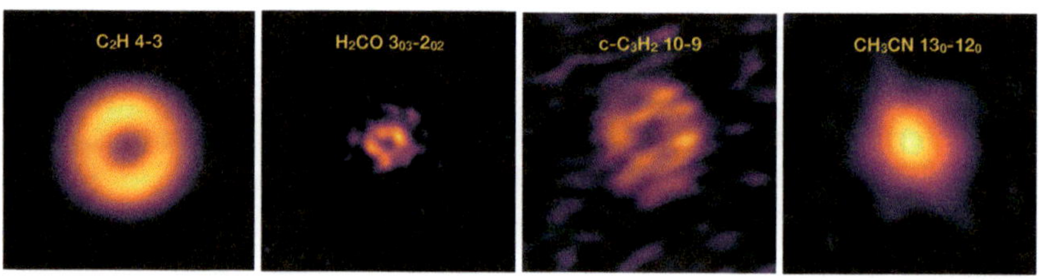

Figure 1. ALMA observations of small and mid-sized organics in the disk around nearby T Tauri star TW Hya. C_2H and c-C_3H_2 observations are from Bergin et al. (2016), H_2CO from Öberg et al. (2017) and CH_3CN from Loomis et al. (2018a). Each panel is $\sim 5'' \times 5''$ and spatial resolutions span $0\farcs4 - 0\farcs8$

The chemical composition of Solar Nebula analogs, protoplanetary disks, have been studied at sub-millimeter and millimeter wavelengths for decades, revealing a large number of 2-6 atom organic and inorganic molecules, including the nitriles CN, HCN, HNC, CH_3CN, HC_3N, hydrocarbons C_2H and c-CH_3H_2, and O-bearing organics H_2CO, CH_3OH and HCOOH (Figure 1) and (Dutrey et al. 1997; Thi et al. 2004; Öberg et al. 2011b; Chapillon et al. 2012; Qi et al. 2013b; Kastner et al. 2014; Graninger et al. 2015; Öberg et al. 2015; Walsh et al. 2016; Favre et al. 2018). It is becoming increasingly clear, however, that these observations probe the upper layers of disks, and do not reach into the planet forming midplanes (Bergner et al. 2018; Loomis et al. 2018a). First the molecular excitation temperatures in the midplane are expected to be low, and millimeter observations of mid-sized organics such as CH_3CN mainly sample the warmer more elevated molecular layers due to relatively high line upper energy levels. Second, and more seriously, high-resolution dust observations with ALMA has revealed that the dust becomes optically thick in the innermost disk (Cleeves et al. 2016; Huang et al. 2018), and the presence of sub-structure in disks may result in optically thick dust also at larger disk radii (ALMA Partnership et al. 2015; Andrews et al. 2016).

At longer wavelengths, dust opacity is less of an issue, and observations of line emission from disk midplanes becomes possible. Compared to millimeter wavelengths, such lines are intrinsically weak, however, and at least an order of magnitude increase in sensitivity is required compared to existing radio facilities, i.e.,such observations cannot be carried out with the JVLA, but would be possible with the ngVLA for a handful of key molecules in nearby disks. The ngVLA would also readily provide the resolution of a fraction of an arc second required to map out the organic content across nearby protoplanetary disks.

The ngVLA will provide access to lines of a range of mid-sized and more complex organics. For the case of disks, whose molecular emission are intrinsically weak, we only consider mid-sized organics. We suggest that the following ladders of three known disk molecules are especially advantageous to target because they present line upper energy levels that match expected disk midplane temperatures – their upper state energies range between a few and 10s of K – and cover enough of a temperature range to provide good constraints on the molecular excitation temperatures and column densities: the fundamental 1-0 and 2-1 k-ladder transitions of CH_3CN, J=1–0 to 5–4 transitions of HC_3N, and $1_{10} - 1_{11}$, $2_{11} - 2_{12}$, $3_{12} - 2_{13}$, $4_{13} - 4_{14}$ transitions of H_2CO. The same frequency range, 4–48 GHz, also presents k-ladders of CH_3OH, HCOOH and other O-bearing organics, but as explained below it is more speculative whether the ngVLA will have sufficient sensitivity to detect them.

3. Case Study: Mapping Organics in the TW Hya Protoplanetary Disk Midplane with the ngVLA

TW Hya hosts the most nearby and well-studied protoplanetary disk (Wilner et al. 2000; Kastner et al. 2002; Bergin et al. 2015; Andrews et al. 2016; Schwarz et al. 2016; Huang et al. 2018). The disk chemical composition has been constrained at millimeter and sub-millimeter wavelengths (Thi et al. 2004; Qi et al. 2008, 2013a; Kastner et al. 2014; Cleeves et al. 2015; Bergin et al. 2016; Öberg et al. 2017), revealing strong emission from CH_3CN, and the only detections to date of CH_3OH and HCOOH (Walsh et al. 2016; Favre et al. 2018; Loomis et al. 2018a). In this section we explore what investment of ngVLA time would be needed to characterize the organic chemistry in the TW Hya disk plane and extrapolate to other potential disk targets.

There are two paths to estimating line intensities at ngVLA wavelengths: predictions from disk chemistry models, and extrapolations from line emission observations at shorter wavelengths. We take the former approach, but check it against empirical extrapolations where possible. In particular, we use the disk chemistry model from Cleeves et al. (2015), which was originally optimized to reproduce the emission from small ions in TW Hya, but has also been shown to well reproduce CH_3CN emission at millimeter wavelengths (Loomis et al. 2018a).

Figure 2 shows the predicted disk integrated spectra between 1 and 50 GHz for TW Hya with a spectral resolution of 0.2 km/s. The first thing to note is the potential line richness of TW Hya. Many of these lines originate from the mid-size cometary organics that are the molecular targets of this science case. Among the strongest predicted lines are the CH_3CN k-ladders around 18 and 37 GHz. Using an empirical extrapolation from existing ALMA observations results in a factor of two higher line intensities, suggesting that the presented model spectra is conservative, at least with respect to CH_3CN. Other strong lines can be attributed to CS, C_2S, HCOOH, H_2CO,

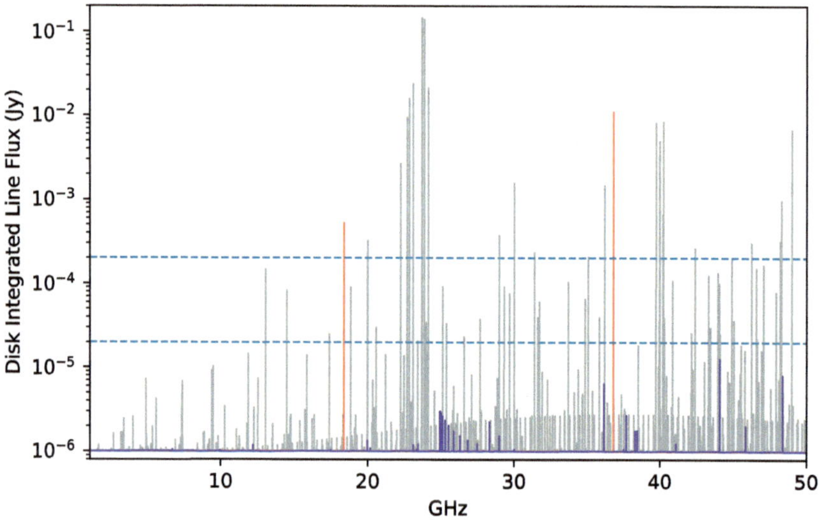

Figure 2. Predicted disk integrated spectra towards TW Hya between 1 and 50 GHz. The model spectra is generated using the disk chemistry model for TW Hya presented in Cleeves et al. (2015). The two dashed lines show the sensitivity limits for the ngVLA after 1 and 100 h integration, using the baseline sensitivity of the ngVLA 30 μJy in 10 km/s in one hour and a required spectral resolution of 0.2 km/s. CH_3CN is marked in orange and CH_3OH in blue.

NH_3, and NH_2CN (cyanimide). Detections of lines > 5σ from these species should be possible in a few hours of ngVLA time. We note that while NH_3 is not an organic, it is an especially interesting target molecule, both because it is a predicted important reservoir of nitrogen and because it could be used to measure the disk gas temperature. More complex molecules than these, as well as CH_3OH may be detectable using longer (100 h) integrations and a combination of matched filter and line stacking techniques (Loomis et al. 2018b).

The above disk integrated observations will provide important constraints on disk midplane compositions, but to characterize chemical gradients across the planet and comet forming disk regions require maps at sub-arcsecond resolution. Adopting a spatial resolution of $0.''2$, corresponding to 12 au for TW Hya's distance of 59.5 pc (Brown et al. 2016) we can estimate the sensitivity needed for mapping CH_3CN based on existing ALMA observations at $0.''8$ resolution. At this spatial resolution CH_3CN is barely resolved and almost all emission originate from within a disk radius of $0.''5$. Conservatively we assume that that emission is uniform interior of $0.''5$ (Figure 1 shows that sub-structure is likely), and the flux per beam is then 25× less than the disk integrated emission. Based on Figure 2 the strongest CH_3CN lines will have a peak flux of 10 mJy / 25 ~ 400μJy per beam, and a 5σ ngVLA map should be achievable in ~10 hours of integration.

TW Hya is unique in its proximity and therefore in its line strengths. To extend observations to a sample of disks we would need to observe disks in nearby star forming regions, such as Taurus, at distances of ~150 pc, or 2.5× farther away than TW

Hya. Based on millimeter observations, line emission from these disks are on the order of 2–5× weaker compared to TW Hya, requiring on average a factor of 10 longer integrations. To map the brightest organics in these disks would thus require ~100 h per disk.

4. Synergies and Concluding Remarks

Observations of organic molecule abundances and distributions in protoplanetary disks are key to calibrate astrochemical disk models, and to predict how often we can expect planets to form in environments that are chemically conducive for the emergence of life. The ngVLA will provide a crucial part of the puzzle by giving access to the disk midplanes where planets and comets assemble. ALMA is currently providing the complementary information on molecular distributions at scales of 10-100 a.u. in more elevated disk layers, and *JWST* will soon give access to the upper layers of the innermost disk regions. Together these three facilities will provide orthogonal constraints on the chemistry of planet forming disks and aid astrochemists in developing a complete picture of disk chemistry and the chemistry of nascent planets.

References

Aikawa, Y., & Herbst, E. 1999, A&A, 351, 233
ALMA Partnership, Brogan, C. L., Pérez, L. M., et al. 2015, ApJ, 808, L3. 1503.02649
Andrews, S. M., Wilner, D. J., Zhu, Z., Birnstiel, T., Carpenter, J. M., Pérez, L. M., Bai, X.-N., Öberg, K. I., Hughes, A. M., Isella, A., & Ricci, L. 2016, ApJ, 820, L40. 1603.09352
Bergin, E. A., Blake, G. A., Ciesla, F., Hirschmann, M. M., & Li, J. 2015, Proceedings of the National Academy of Science, 112, 8965. 1507.04756
Bergin, E. A., Du, F., Cleeves, L. I., Blake, G. A., Schwarz, K., Visser, R., & Zhang, K. 2016, ArXiv e-prints. 1609.06337
Bergner, J. B., Guzmán, V. G., Öberg, K. I., Loomis, R. A., & Pegues, J. 2018, ApJ, 857, 69
Brown, A. G. A., Vallenari, A., Prusti, T., et al. 2016, A&A, 595, A2
Calvet, N., Patino, A., Magris, G. C., & D'Alessio, P. 1991, ApJ, 380, 617
Chapillon, E., Dutrey, A., Guilloteau, S., Piétu, V., Wakelam, V., Hersant, F., Gueth, F., Henning, T., Launhardt, R., Schreyer, K., & Semenov, D. 2012, ApJ, 756, 58. 1207.2682
Cleeves, L. I., Bergin, E. A., Qi, C., Adams, F. C., & Öberg, K. I. 2015, ApJ, 799, 204. 1412.1491
Cleeves, L. I., Öberg, K. I., Wilner, D. J., Huang, J., Loomis, R. A., Andrews, S. M., & Czekala, I. 2016, ApJ, 832, 110
Dutrey, A., Guilloteau, S., & Guelin, M. 1997, A&A, 317, L55
Favre, C., Fedele, D., Semenov, D., Parfenov, S., Codella, C., Ceccarelli, C., Bergin, E. A., Chapillon, E., Testi, L., Hersant, F., Lefloch, B., Fontani, F., Blake, G. A., Cleeves, L. I., Qi, C., Schwarz, K. R., & Taquet, V. 2018, ApJ, 862, L2
Graninger, D., Öberg, K. I., Qi, C., & Kastner, J. 2015, ApJ, 807, L15. 1506.03820
Henning, T., & Semenov, D. 2013, Chemical Reviews, 113, 9016. 1310.3151
Huang, J., Andrews, S. M., Cleeves, L. I., Öberg, K. I., Wilner, D. J., Bai, X., Birnstiel, T., Carpenter, J., Hughes, A. M., Isella, A., Pérez, L. M., Ricci, L., & Zhu, Z. 2018, ApJ, 852, 122. 1801.03948
Kastner, J. H., Hily-Blant, P., Rodriguez, D. R., Punzi, K., & Forveille, T. 2014, ApJ, 793, 55. 1408.5918
Kastner, J. H., Huenemoerder, D. P., Schulz, N. S., Canizares, C. R., & Weintraub, D. A. 2002, ApJ, 567, 434. astro-ph/0111049

Le Roy, L., Altwegg, K., Balsiger, H., Berthelier, J.-J., Bieler, A., Briois, C., Calmonte, U., Combi, M. R., De Keyser, J., Dhooghe, F., Fiethe, B., Fuselier, S. A., Gasc, S., Gombosi, T. I., Hässig, M., Jäckel, A., Rubin, M., & Tzou, C.-Y. 2015, A&A, 583, A1

Loomis, R. A., Cleeves, L. I., Öberg, K. I., Aikawa, Y., Bergner, J., Furuya, K., Guzman, V. V., & Walsh, C. 2018a, ApJ, 859, 131

Loomis, R. A., Öberg, K. I., Andrews, S. M., Walsh, C., Czekala, I., Huang, J., & Rosenfeld, K. A. 2018b, AJ, 155, 182

Morbidelli, A., Lunine, J. I., O'Brien, D. P., Raymond, S. N., & Walsh, K. J. 2012, Annual Review of Earth and Planetary Sciences, 40, 251. 1208.4694

Mumma, M. J., & Charnley, S. B. 2011, ARA&A, 49, 471

Öberg, K. I., Furuya, K., Loomis, R., Aikawa, Y., Andrews, S. M., Qi, C., van Dishoeck, E. F., & Wilner, D. J. 2015, ApJ, 810, 112. 1508.07296

Öberg, K. I., Guzmán, V. V., Merchantz, C. J., Qi, C., Andrews, S. M., Cleeves, L. I., Huang, J., Loomis, R. A., Wilner, D. J., Brinch, C., & Hogerheijde, M. 2017, ApJ, 839, 43. 1704.05133

Öberg, K. I., Murray-Clay, R., & Bergin, E. A. 2011a, ApJ, 743, L16. 1110.5567

Öberg, K. I., Qi, C., Fogel, J. K. J., Bergin, E. A., Andrews, S. M., Espaillat, C., Wilner, D. J., Pascucci, I., & Kastner, J. H. 2011b, ApJ, 734, 98. 1104.1236

Patel, B. H., Percivalle, C., Ritson, D. J., Duffy, C. D., & Sutherland, J. D. 2015, NATURE CHEMISTRY, 7, 301

Powner, M. W., Gerland, B., & Sutherland, J. D. 2009, Nat, 459, 239

Qi, C., Öberg, K. I., Wilner, D. J., D'Alessio, P., Bergin, E., Andrews, S. M., Blake, G. A., Hogerheijde, M. R., & van Dishoeck, E. F. 2013a, Science, 341, 630. 1307.7439

Qi, C., Öberg, K. I., Wilner, D. J., & Rosenfeld, K. A. 2013b, ApJ, 765, L14. 1302.0251

Qi, C., Wilner, D. J., Aikawa, Y., Blake, G. A., & Hogerheijde, M. R. 2008, ApJ, 681, 1396. 0803.2753

Raymond, S. N., Kokubo, E., Morbidelli, A., Morishima, R., & Walsh, K. J. 2014, Protostars and Planets VI, 595. 1312.1689

Schwarz, K. R., Bergin, E. A., Cleeves, L. I., Blake, G. A., Zhang, K., Öberg, K. I., van Dishoeck, E. F., & Qi, C. 2016, ApJ, 823, 91. 1603.08520

Thi, W., van Zadelhoff, G., & van Dishoeck, E. F. 2004, A&A, 425, 955. arXiv:astro-ph/0406577

Walsh, C., Loomis, R. A., Öberg, K. I., Kama, M., van t Hoff, M. L. R., Millar, T. J., Aikawa, Y., Herbst, E., Widicus Weaver, S. L., & Nomura, H. 2016, ApJ, 823, L10. 1606.06492

Willacy, K., & Langer, W. D. 2000, ApJ, 544, 903

Wilner, D. J., Ho, P. T. P., Kastner, J. H., & Rodríguez, L. F. 2000, ApJ, 534, L101. astro-ph/0005019

Science with a Next-Generation Very Large Array
ASP Conference Series, Monograph 7
Eric J. Murphy, ed.

Exo-space Weather in the Era of the ngVLA[1]

Rachel A. Osten,[1,2] Michael K. Crosley,[3] and Gregg Hallinan[4]

[1]*Space Telescope Science Institute, Baltimore, MD, USA;* `osten@stsci.edu`

[2]*Center for Astrophysical Sciences, Johns Hopkins University, Baltimore, MD, USA*

[3]*Physics & Astronomy Dept., Johns Hopkins University, Baltimore, MD, USA*

[4]*California Institute of Technology, Pasadena, CA, USA*

Abstract. A forward-looking facility such as the next generation Very Large Array (ngVLA) requires forward-looking science. The ngVLA will enable stellar wind detections or robust constraints on upper limits sufficient to bridge the gap between current radio upper limits on cool stellar mass loss and current indirect estimates of stellar mass loss from astrospheric measurements. This will aid in understanding the interplay between the rotational spindown of stars and wind mass loss rates, important for an understanding not just of stellar astrophysics but the impact on planet formation and planetary dynamos. Radio observations are currently the only way to explore accelerated particles in cool stellar environments. The upgraded sensitivity of the ngVLA will enable the detection and characterization of essentially all classes of radio-active cool stars in the galaxy. This will open up new fields of study and contribute to the understanding of the stellar contribution to exo-space weather, especially with the emphasis on sensitivity above a frequency of about 10 GHz, where the transition to optically thin gyrosynchrotron emission occurs. These topics connect and extend the key science topics in the cradle of life, in an area which there is both high academic as well as popular interest, and which is expected to continue into the next several decades.

1. Introduction

With the announcement by the Kepler mission of more than two thousand confirmed transiting exoplanets, population statistics (Dressing & Charbonneau 2015) suggesting that the nearest non-transiting exoplanet in the habitable zone of its star is at a distance of 2.6 ± 0.4 pc, with the nearest transiting exoplanet in the habitable zone at a distance of $10.6^{+1.6}_{-1.8}$ pc, and recent confirmation via radial velocity techniques of a planet in the habitable zone of our nearest star (Anglada-Escudé et al. 2016), astronomy is firmly in the age of exoplanets. This is an area that grabs the public's attention as well as unites professional astronomers, planetary scientists, and astrobiologists, in the search for life. Ground- (MEarth project) and space-based (TESS, PLATO) projects have as their primary goal the discovery of even more, closer transiting exoplanets, and large ground- and space-based projects like the *James Webb Space Telescope* (*JWST*) and giant segmented mirror telescopes (E-ELT, TMT, GSMT) will characterize their atmospheres. Characterizing the atmospheres and environments of potentially habitable exoplanets will be a major focus of the astronomical community in the next ten to twenty years.

During that time, we can expect that the phase space for exoplanet demographics will be fleshed out, with major inroads into understanding the interrelationship between planet properties and stellar properties. There will be many potentially habitable worlds, and one major question will be how to move beyond that potentiality into a more informed assessment of the likelihood of seeing an *inhabited* world.

Planetary habitability is a multi-faceted issue and depends not just on the planetary parameters, but is strongly influenced by the environment set by the host star. Space weather is an important factor to consider in assessing the type of environment which may exist on a distant exoplanet. Space weather broadly posed is the impact of the star's magnetic activity: the structure of the magnetosphere itself and its interplay with close-in planetary orbits; the nonradiative heating and enhanced levels of ultraviolet-extreme ultraviolet-X-ray (UV-XEUV) radiation; a steady stellar wind; and flares, energetic particles and coronal mass ejections (CMEs) produced during magnetic eruptions. For close-in planets, orbits crossing the Alfvén surface of the stellar magnetosphere would experience severe space weather (Garaffo et al. 2016, 2017). Space weather has potentially played a role in defining habitability within our own solar system. The long-term impact of such activity was recently established in dramatic fashion by the Mars Atmosphere and Volatile EvolutioN (MAVEN) mission, which confirmed that ion loss due to solar CMEs early in Mars history likely severely depleted its atmosphere (Jakosky et al. 2015). The nonlinear dependence of a star's XEUV radiation with time will impact planetary atmospheric loss processes (Johnstone et al. 2015), and stellar activity can create false positive signatures in biomarkers (Tian et al. 2014). The star can influence the planetary environment through radiation and particles. Stellar flaring and associated coronal mass ejections, a steady stellar wind, and star-planet magnetospheric interactions are all important factors in the space weather environments of these exoplanets. Radio emission is the only wavelength regime capable of placing direct constraints on the particle environment that stars produce, and adds significant information on the steady mass loss of nearby stars.

2. Mass Loss on the Lower Half of the Main Sequence

Stellar winds affect the migration and/or evaporation of exoplanets (Lovelace et al. 2008), and are important not only for understanding stellar rotational evolution, but it also their influence on planetary dynamos (Heyner et al. 2012). At present only indirect measures of cool stellar mass loss inform these topics. Mass loss in the cool half of the HR diagram, along/near the main sequence, has been notoriously difficult to detect directly, due in part to the much lower values of mass loss here compared to other stellar environments ($\dot{M}$ of 2×10^{-14} solar masses per year for the Sun). Detection of Lyman α astrospheric absorption (Wood et al. 2004) does not detect the wind directly, but rather the bow shock created when the wind interacts with the local interstellar medium. This can only be done with high-resolution ultraviolet spectrographs in space; currently the Space Telescope Imaging Spectrograph on the *Hubble Space Telescope*, installed in 1997, is the only instrument capable of making such measurements. Non-detections of this feature do not provide upper limits to the stellar mass loss.

Cool stellar mass loss is characterized by an ionized stellar wind, whose radio flux can have a $\nu^{0.6}$ or $\nu^{-0.1}$ dependence if in the optically thick or thin regime, respectively. A direct measurement of stellar mass loss through its radio signature would be a significant leap forward not only for understanding the plasma physics of the stars themselves, but also for understanding what kind of environment those stars create. Previous attempts at a direct detection of cool stellar mass loss via radio emission have led to upper limits typically three to four orders of magnitude higher than the Sun's present-day mass loss, while indirect methods find evidence for mass loss rates comparable to or slightly higher than the Sun's present day mass loss rate (up to ~80 times solar $\dot{M}$; Fichtinger et al. 2017)

Initial work assessing the feasibility of the ngVLA's ability to detect directly the stellar winds of nearby cool, main sequence stars was done as part of the Bower et al. (2015) report of ngVLA Science Working Group 4. This brief work used analytic formulae in Drake et al. (1987), based on the expected optically thick and thin radio emission from an ionized stellar wind presented in Wright & Barlow (1975), and Panagia & Felli (1975); optically thick emission from an ionized stellar wind is expressed as

$$S_\nu = 1.6 \times 10^{11} \left(\frac{\dot{M}}{v_w}\right)^{1.33} \frac{\nu_5^{0.6} T_4^{0.1}}{d_{kpc}^2} \tag{1}$$

where S_ν is the predicted radio flux density in mJy, $\dot{M}$ is the mass loss rate in solar masses per year, v_w is the velocity of the stellar wind, ν_5 is the frequency normalized to 5 GHz, T_4 indicates the temperature of the solar wind in units of 10^4 K, and d_{kpc} is the distance in kpc. As winds from late-type stars are presumed to originate in the open field lines of the outer corona, a coronal wind temperature of 10^6 or $T_4 = 100$ is used. We expand on these calculations to different frequencies in the expected ngVLA range, and incorporate demographics of the nearest stars using catalogues of nearby stars. We use the recent continuum sensitivities quoted for the ngVLA at a variety of frequencies as posted at `http://science.nrao.edu/futures/ngvla/concepts`, extended for a 12 hour integration. The stellar escape velocity is ~600 km s^{-1} for main sequence solar-like stars and cooler, and the minimum constrainable mass loss rate for this velocity is quoted in this memo. However, we note that recent results from studies of exoplanets (Vidotto et al. 2015 and others) have found evidence for hot Jupiters embedded in slower stellar winds than would be expected based on solar scalings, with expectations of a denser stellar wind than experienced by solar system planets. With this in mind, we explore a range of assumed wind speeds, from 100 to 1200 km s^{-1}.

2.1. Increasing the Look-back Time for Studying the Young Sun

Table 1 lists nearby commonly used solar analogs within 15 pc from the Sun, and their ages. Results from high spectral resolution UV data of nearby stars where astrospheric absorption is seen suggest that there is a power-law relationship between inferred mass loss and surface X-ray flux, with a limit in surface X-ray flux of about 10^6 erg cm^{-2} s^{-1}, above which no or significantly reduced mass loss is inferred (Wood et al. 2004). This limits the ability to constrain the mass-loss history of the Sun, using analogues. In particular, the lack of detection of this feature does not provide any quantitative constraints on the mass-loss of that star. Based on this work, the most active stars appear not to have strong winds; there is evidence via astrospheric detection towards the nearby young solar analog π^1UMa (Wood et al. 2014), with age estimates between 300 and 500 million years, of a weak wind from the young Sun.

There is tension between the results returned from the astrospheric detection method and the mass loss expected from stellar rotational evolution models. Stars are expected to lose angular momentum in a wind as their rotation decreases with time; the rate of rotation period decrease can be determined from distributions of rotation periods in clusters of known ages, while MHD wind models are required to interpret the implied mass loss rates which should accompany this. Johnstone et al. (2015) performed such calculations, suggesting that the stellar mass-loss rate can be parameterized as

$$\dot{M}_\star \propto R_\star \Omega_\star^{1.33} M_\star^{-3.36} \quad . \tag{2}$$

The results suggest that the Sun at young ages should have had a mass loss rate roughly an order of magnitude higher than it does today. Fichtinger et al. (2017) presented recent results obtained from the JVLA and ALMA for a sample of nearby stars. Results are shown in Figure 1. They expanded upon the earlier spherically symmetric wind models, and showed that a conical opening angle for the wind can reduce the mass loss rate implied from a given radio flux density by about a factor of two compared to the spherically symmetric case. Their upper limits for implied mass loss are still two to three orders of magnitude above the levels of mass loss implied from the rotational evolution models, which themselves are about an order of magnitude above the mass loss rate implied by the detection of astrospheric absorption from the nearby solar analogue π^1 UMa. Results from the ngVLA will be able to make a significant leap in understanding the mass loss history of the Sun through its two order of magnitude improvement over what can be achieved at present with the JVLA. The strength of the ngVLA observations lie in the ability to provide direct constraints (either detections or upper limits) on the stellar wind mass loss; non-detections provide quantitative constraints on the mass loss rate, a marked difference from the astrospheric absorption method.

Table 1. Commonly used Solar Analogs and Potential ngVLA Mass-Loss Constraints

Name	Age (MY)	Distance (pc)	$\dot{M}$ constraint[1,2] ($\dot{M}_\odot$)
EK Dra	100	34	425
χ^1 Ori	300	8.7	55
π^1 UMa	500	14.3	116
κ^1Cet	650	9.2	60

[1] Using ngVLA sensitivity at 17 GHz, extrapolated for a 12 hour integration
[2] For $\dot{M}_\odot$=2 × $10^{-14} M_\odot$ yr^{-1} and wind speed equal to escape speed

2.2. M Dwarfs in the Solar Neighborhood

After initial skepticism about the possibility of M dwarfs hosting exoplanets, attention is fully fixed on these cool dim stars. M dwarfs are the most common type of star in our galaxy, and recent results have shown a high occurrence rate of planets around M dwarfs. This makes planets around M dwarfs one of most common modes of planet formation in our galaxy. The magnetic activity of M dwarfs can have significant influence on the chemistry and dynamics of these planets' atmospheres due to the planets' close proximity to the star, and these effects may be observable with *JWST* (Venot et

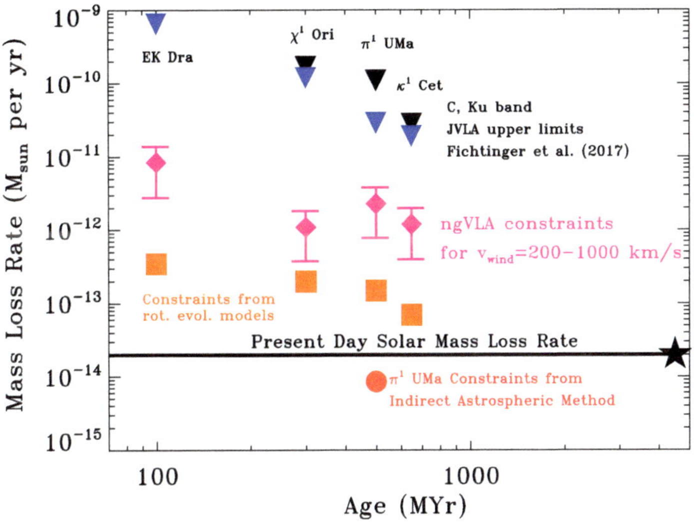

Figure 1. Summary of current mass loss constraints for nearby solar analogs, along with prospects achievable with the ngVLA. Black and blue triangles indicate upper limits from C and Ka bands obtained with the JVLA (Fichtinger et al. 2017) for spherically symmetric winds; orange squares show mass loss constraints from the rotational evolution models of Johnstone et al. (2015); and red circle shows detection from the indirect method of inferring mass loss via astrospheric absorption (Wood et al. 2014). Magenta diamonds and error bars display the grasp of the ngVLA, for wind velocities spanning 200 – 1000 km s^{-1}. The present-day solar mass loss rate is $2 \times 10^{-14}\, M_\odot\, yr^{-1}$.

al. 2016). Recent results (Garaffo et al. 2016, 2017) have shown that the stellar magnetosphere influences the inner edge of the traditional habitable zone. In the case of the iconic TRAPPIST-1 system of seven planets, magnetospheric models suggest that all but two of these planets would have orbits crossing the Alfvén surface and thus would experience severe space weather. Thus consideration of the steady stellar wind as well as time-varying flares and coronal mass ejections from M dwarfs are vital components to understanding the complex intertwining that may or may not turn a potentially habitable planet into an inhabited planet. Current studies suggest that there are about 270 M dwarfs within 10 pc (Figure 2). Several of these are already known to host exoplanets; by the time that the ngVLA is operational, at the end of the 2020s, all of these will have have been surveyed for both close-in exoplanets (via the transit and radial velocity methods), and those in more distant orbits (by coronography). Radio observations to constrain the mass loss from a steady stellar wind will be crucial measurements to add to the mix to understand the impact the star may have on exoplanet companions.

Very little is known observationally about the mass-loss of nearby M dwarfs. The nearby, magnetically active M dwarf EV Lac is one exception, having had an astrospheric detections from Lyman α absorption. The inference is for a mass loss rate of only 0.7 times the solar mass loss rate, after correcting the results of Wood et al. (2004) for the much smaller surface area of EV Lac. Wargelin & Drake (2001) presented a different method for detecting stellar winds, which makes use of an X-ray charge ex-

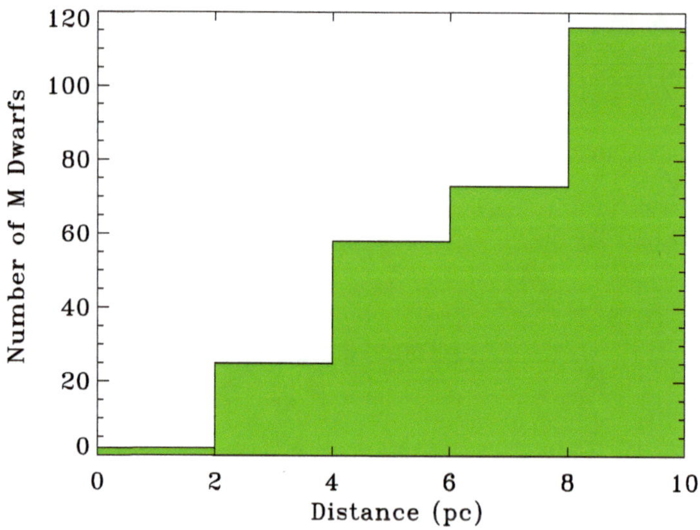

Figure 2. Number of M dwarfs as a function of distance, from the REsearch COnsortium of Nearby Stars (RECONS). There are more than about 270 M dwarfs within 10 pc; several of these are known to host exoplanets and by the time of the late 2020s all of these will have been surveyed for close-in exoplanets. An important constraint for further characterization of these exoplanets will be the extent to which the star makes its proximate environment helpful or harmful for habitability.

change halo which should exist between the highly charged ions in the stellar wind and the surrounding ISM. This method is promising; they were able to provide an upper limit to the mass loss rate of Proxima at 3×10^{-13} $M_\odot$ yr^{-1}, with approximately a factor of three uncertainty in their model. This requires the use of a space-based X-ray observatory with sufficient sensitivity and spatial resolution to resolve this emission from the stellar coronal emission, expected to be thousands of times stronger (the spatial resolution is required to discern this signature in the wings of the coronal PSF). This may be a complementary approach to the detection of radio emission of an ionized stellar wind, possible with the Lynx mission concept under study by NASA with 50 times Chandra's sensitivity and Chandra-like angular resolution. Because of the angular resolution requirement, it is most competitive for stars within about 5 pc.

The method of detecting radio emission from nearby M dwarfs will be an important contributor to understanding how the wind environment of nearby M dwarfs contributes to the habitability of orbiting exoplanets. Figure 3 shows the sensitivity of the ngVLA to detecting radio emission from an ionized stellar wind for M dwarfs within 10 pc, using similar methods as described in the above section for the mass loss history of the Sun. The green triangle depicts the constraint on mass loss rate for the nearest M dwarf, Proxima, described above, along with the estimated uncertainty in the model. The red circle shows the mass loss rate in solar masses per year for the nearby M dwarf EV Lac, as obtained from the detection of astrospheric absorption.

Sensitive, broadband observations of a small sample of the highest priority stellar systems, hosting terrestrial planets in the habitable zone, could potentially detect discrete mass ejection events. The most likely scenario is for an inactive mid-M dwarf at a distance of ~3 pc. The expected rate of large flares (radiated energy 5×10^{29} – 5×10^{32} erg) is one per week, with an estimated associated CME mass loss of 10^{-18} – -10^{-15} $M_{\odot}$/year (Osten & Wolk 2015). This will result in an increased flux of bremsstrahlung emission (optically thick in ngVLA bands) with flux densities at 30 and 80 GHz that range from 0.05-50 μJy and 0.2-200 μJy respectively, assuming equipartition between flare bolometric energy and CME kinetic energy. Large bandwidth ($\nu/\Delta\nu \sim 1$), either instantaneously or via rapid transition between observing bands, is essential to separate the CME-produced bremsstrahlung from other components of the stellar emission, such as coronal gyrosynchrotron emission. This is also aided by observations at the higher frequencies (> 50 GHz) where the bremsstrahlung will dominate over gyrosynchrotron. Observations in Stokes V can also be used to separate contaminating components from the signal. ngVLA observations will be most efficient if observing only the largest CME events (few per year per star). Such observations can be scheduled via external triggers from telescopes monitoring the stellar sample for large flares. Rapid response (< 15 minutes) is essential to this scheme.

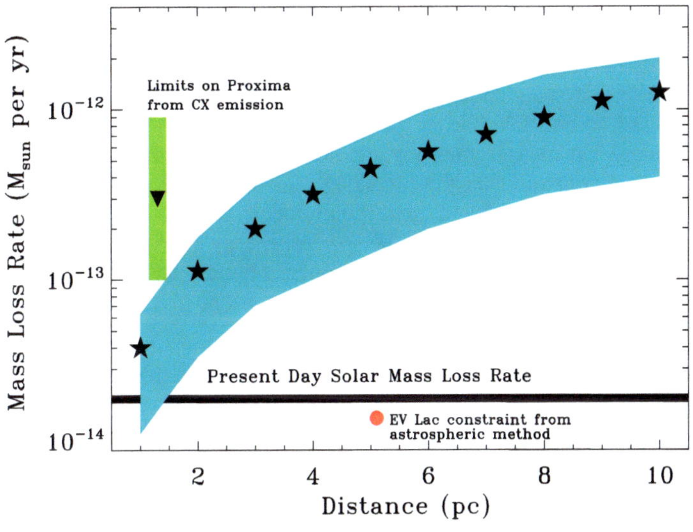

Figure 3. Regions of stellar mass loss rate which can be constrained by a 12 hour ngVLA integration for studies of radio emission from an ionized stellar wind, for M dwarfs at a range of distances from 1-10 pc. Stars indicate expected amount of optically thick emission at 28 GHz from a spherically symmetric wind with wind velocity equal to the escape velocity, and cyan shaded region gives area encompassed by a range of assumed wind velocities, from 200 – 1000 km s^{-1}. Current observational limits on M dwarf wind mass loss are indicated at the distance of the objects: the green triangle indicates the upper limit for Proxima based on limits from charge exchange emission (Wargelin & Drake 2001), along with the uncertainty in their model. The stellar wind mass loss rate implied by the detection of Lyman α astrospheric absorption towards the M dwarf EV Lac at 5.1 pc is also indicated.

2.3. Interpretation of Radio Emission

While measurements of radio emission from point sources are easy to undertake, their interpretation is complicated. The calculations in the above sub-sections have been performed under the assumption of optically thick emission, in which case the emission increases as $\nu^{0.6}$. The commonly observed gyrosynchrotron emission from radio active stars typically has a peak frequency near 10 GHz, with flat or slightly negative spectrum at higher frequencies. Indeed, the upper limits returned by Fichtinger et al. (2017) were obtained after careful removal of observed variable emission attributed to gyrosynchrotron emission. The peak frequency is a function of activity, so observations at higher frequencies are necessary to separate any wind emission from these other sources. At even higher frequencies such as those probed by ALMA, stellar chromospheric emission is detected. Thus the region between 10 and 100 GHz is ideal for searching for wind emission from nearby stars.

The stellar wind must be optically thin to nonthermal radio emission originating closer to the stellar surface. The time-averaged mass-loss from coronal mass ejections, considered a component of a stellar wind, can not be higher than a steady spherically symmetric stellar wind. Both Drake et al. (2013) and Osten & Wolk (2015) have argued for a high flare-associated transient mass loss from active stars; for the young solar analog EK Dra flare-associated CME-induced mass-loss rate of 4×10^{-11} $M_\odot$ yr^{-1}. The mass loss rate of M dwarfs is expected to vary widely as a function of magnetic activity, with flare-associated mass loss rates of 10^{-12} and 10^{-11} $M_\odot$ yr^{-1} for the active M dwarfs AD Leo and EV Lac, respectively, both at distances of about 5 pc. Constraints from ngVLA measurements would be below both of these numbers, and would provide a robust constraint on the breakdown of flare-CME connections, as implied in Crosley et al. (2018ab). Instrument requirements are a large enough bandwidth to disentangle different contributions to the spectral energy distribution and mitigate variability expected to dominate in the lower frequency bands, and circular polarization for additional constraints on ruling out nonthermal emission.

3. Particles and Fields: Contribution to Exo-Space Weather

Habitability studies are interested in the flux of accelerated particles directed outward from the stellar surface; the most energetic ones are the most impactful to an exoplanetary atmosphere. Studying the population of accelerated particles near the stellar surface can provide details of the particle spectrum. Comparison to well-studied solar events enables systematics with respect to scalings or extrapolations used for stellar space weather calculations to be investigated. Segura et al. (2010) investigated the impact on the ozone layer of a superflare from an M dwarf. The energetic particles can deplete the ozone layer of a planet in the habitable zone of an Earthlike planet around an M dwarf experiencing a superflare.

Radio observations provide the clearest signature of accelerated particles and shocks in stars arising from transient magnetic reconnection, and provide more realistic constraints on these factors than scaling by solar values (Osten & Wolk 2017). Optically thin gyrosynchrotron emission constrains the index of the accelerated particles, and also enables a constraint on magnetic field strengths in the radio-emitting source (Smith et al. 2005, Osten et al. 2016). The peak frequency is about 10 GHz, so observations at higher frequencies are necessary to disentangle optical depth effects from the interpretation.

3.1. Radio-Active Cool Stars

Variable stellar radio emission in the cool half of the HR diagram is more typical than steady flux levels, at least at microwave frequencies. This is a consequence of the dominance of emission from accelerated particles, and an indicator of the nonsteady levels of particle acceleration, varying as a function of position on the stellar disk and time. While previous studies estimating the contribution of stellar magnetic reconnection flaring from different types of active stars have used average or "typical" flux densities, the existence of extreme levels of variability (up to $1-2$ orders of magnitude) make this approach susceptible to biases. The timescales for stellar flares can vary from milliseconds, in the case of some highly circularly polarized radio bursts (Osten & Bastian 2006, 2008), to minutes to many hours or days for the most dramatic examples (Richards et al. 2003).

We examined objects characteristic of their class for several types of radio-active cool stars illustrating a large amount of variability. This included Algol, RS CVn, and BY Dra active binaries, and nearby magnetically active mid-M and late-M dwarf stars. We used a radio luminosity distribution which gives the range of radio emission expected to be observed for each type of object, under the assumption that the flux variation seen in one object is typical of the class. Figure 4 shows this time-domain information recast as a cumulative probability distribution for each object. The time-domain information for Algol, HR 1099, σ^2CrB, and UX Ari is taken from multi-year monitoring with the Green Bank Interferometer (Richards et al. 2003). The M dwarf distribution comes from extensive 8 GHz light curves of the nearby M dwarf EV Lac (Osten et al. 2005, Osten et al. 2019, in preparation), while the ultracool dwarf LP349-25 was used for the distribution of radio luminosities for ultracool dwarfs (Ngoc et al. 2007, Osten et al. 2009). For young stellar objects, the radio luminosity distribution was obtained from the 556 sources detected in the deep centimeter wavelength catalog of the Orion Nebula Cluster (Forbrich et al. 2016). For this sample, objects with negative spectral index were removed as potential contaminating extragalactic sources, and the distribution of radio luminosities for the remaining 478 sources was computed.

We used demographic information and space densities (listed in Table 2) to come up with a likelihood of observing a given radio luminosity level for types of stars at a given distance. This is more probabilistic and represents reality better than taking the largest radio luminosity of a given type, assuming that that is characteristic level of radio emission, and projecting that to find the farthest distance to which the system is sensible, to arrive at source numbers. An rms for given frequency and integration time determines the distance to which an observation is sensitive, for a source of a particular radio luminosity,

$$L_R = F_{3\sigma} 4\pi d_{\text{sens}}^2 \tag{3}$$

with L_R the radio luminosity, $F_{3\sigma}$ the flux density density sensitivity, and d_{sens} the resulting sensitivity distance. Then, the number of stars is

$$\#Stars = \left(\frac{f_{\text{sky}}}{4\pi}\right) \frac{4\pi n_{\text{obj}}}{3} \left(\frac{L_R}{4\pi F_{3\sigma}}\right)^{3/2} P(L_R) \tag{4}$$

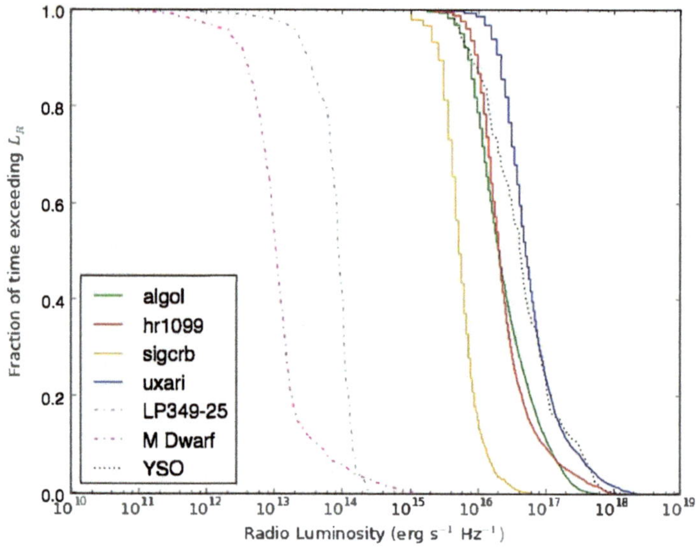

Figure 4. Cumulative distributions of 8 GHz cm radio luminosities for different categories of variable cool stellar objects: red, blue, yellow, and green histograms are active binary systems (Algol and RS CVn-type systems); the purple dashed histogram is the result of several multi-wavelength campaigns of the M dwarf flare star EV Lac (Osten et al. 2005, Osten et al. 2019, in prearation.); grey red dotted histogram is the result of several monitoring observations of the ultracool dwarf LP349-25 (Ngoc et al. 2007, Osten et al. 2009); gray dotted histogram is the results of a deep centimeter-wavelength catalog of the Orion Nebula Cluster (Forbrich et al. 2016). Converting these to probability distributions enables a more realistic assessment of the likelihood of observing a given radio luminosity, and hence detectability of these types of radio sources, than averages or maximum luminosities.

where $f_{\rm sky}$ is the fraction of the sky visible from the ngVLA site, $\approx$ 10.3 steradian (Condon et al. 1998), $n_{\rm obj}$ is the space density of the type of star, L_R the radio luminosity, $F_{3\sigma}$ the sensitivity threshold, and $P(L_R)$ the probability of the stellar object having that luminosity. We realize that the assumption of spherical geometry will break down at the scale of the Milky Way thick disk, or ~1 kpc, but this treatment gives a simplified approach to the maximum distance detectable. Figure 5 plots the number of stars detectable with the ngVLA as a function of radio luminosity, using the radio luminosity distributions shown in Figure 4. Only luminosity ranges available from the data shown in Figure 4 are used, and inflections in the radio luminosity distribution show up as departures from the $\propto L_R^{3/2}$ trend of number of stars as a function of radio luminosity in Figure 5. The active binary systems have larger intrinsic radio luminosities than the M dwarf classes.

Figure 6 shows the number of stars detectable as a function of distance. Here the distance at which the trend for each type of star flattens indicates the distance at which the smallest radio luminosity present in Figure 4 is no longer detected. All stars are detectable at small distances, and the number scales as distance^3. Eventually only the brighter stars are able to be detected at farther distances, so the rate of additional star

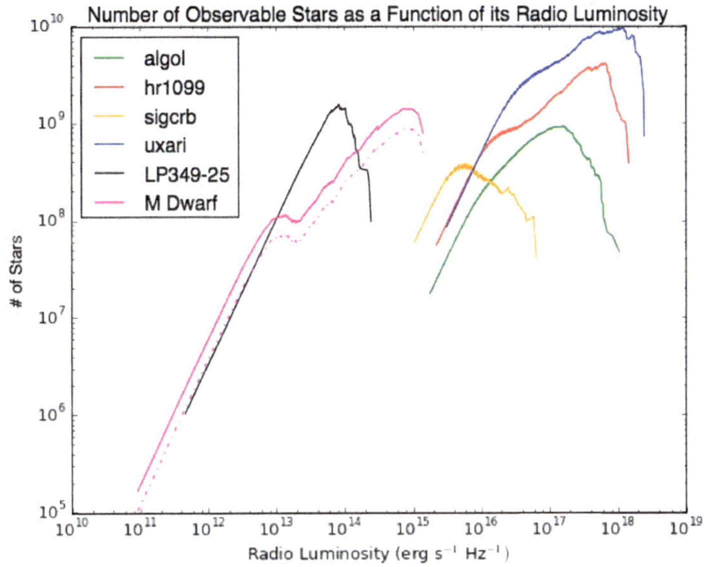

Figure 5. Number of stars detectable with the ngVLA as a function of radio luminosity, using the radio luminosity distributions shown in Figure4, for a . five minute integration at a frequency of 10 GHz.

counts decreases until no more stars would be detected at the largest distances. The calculation for M dwarfs used two different space density values as listed in Table 2.

Table 2. Categories of radio-active cool stars and references to their space densities.

Category	Space Density stars pc^{-3}	Reference
active binaries	3.7×10^{-5}	Favata et al. (1995)
early-mid M dwarfs	0.08	Reid et al. (2007)
	0.05	Reid et al. (2008)
ultracool dwarfs	0.013	Reid et al. (2008)

Our approach for YSOs differed slightly. Since YSOs are found in clusters at discrete distances, for this population, we estimated the fraction of sources which would be detectable in a cluster similar to the ONC, but placed further away. For these purposes we used the ngVLA's continuum rms in one hour at 8 GHz to set the percent of sources detected for a 3σ detection. Figure 7 displays the result of this calculation, and demonstrates that 100% of the sources could be recovered in clusters at 2 kpc distance, sufficient to characterize many of the typical radio-emitting sources in star forming regions located in the nearest parts of the Sagittarius and Perseus arms of our galaxy.

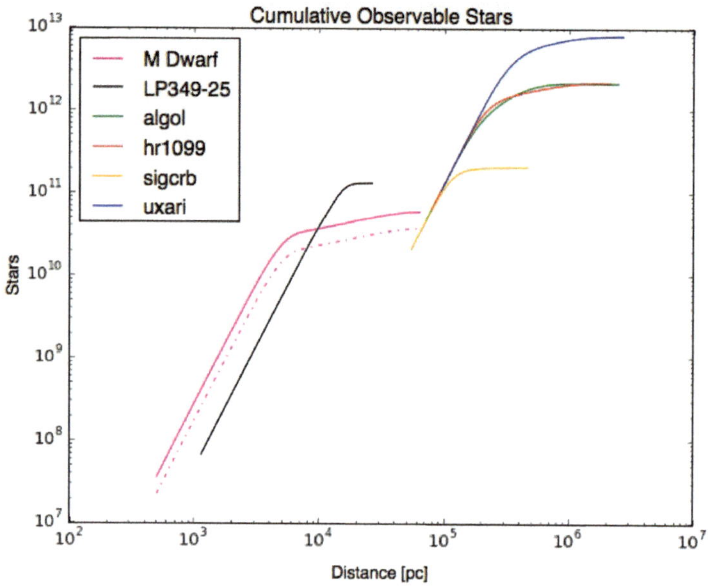

Figure 6. Cumulative number of stars observable as a function of distance.

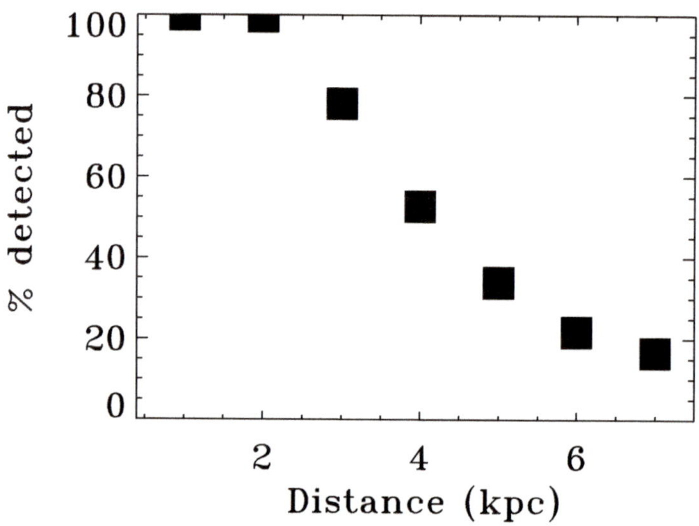

Figure 7. Percentage of the 478 radio sources detected in the deep exposure of the Orion Nebula Cluster reported by Forbrich et al. (2016), for clusters at various distances. Sensitivities appropriate to a one hour 8 GHz observation with the ngVLA are used.

3.2. Constraining Particle Distributions and Magnetic Fields in Magnetic Reconnection Flares

Radio observations provide a unique way to characterize the nature of the accelerated electron population near the star. This is important for understanding the radiation and particle environment in which close-in exoplanets are situated.

An aspect of star-planet interactions which can be probed with radio observations is the magnetospheric interactions of a magnetized close-in planet with its host star. Similar to the idea that the proximity of two stellar magnetospheres due to close passage can cause periodic episodes of magnetic reconnection (as in Massi et al. 2006), recent results (Pilliterri et al. 2014) have suggested a triggering mechanism for regularly recurring stellar flares on stars hosting close-in exoplanets (hot Jupiters). For radio wavelength observations, the time-dependent response of radio emission reveals the changing nature of the magnetic field strength and number and distribution of accelerated particles. As stellar radio emission typically has a peak frequency near 10 GHz, a wide bandwidth system spanning this range can probe the optically thick and thin conditions, diagnose the changing conditions during the course of a magnetic reconnection flare associated with close passage of an exoplanet, and deduce the nature of the accelerated particle population through measurements of spectral indices from confirmed optically thin emission. This would provide constrains on the accelerated particle population of close-in exoplanets unavailable from any other observational method; such a constraint is necessary to perform detailed modeling of the atmospheres of such exoplanets, due to the influence of accelerated particles in affecting the chemical reactions in terrestrial planet atmospheres (Jackman et al. 1990).

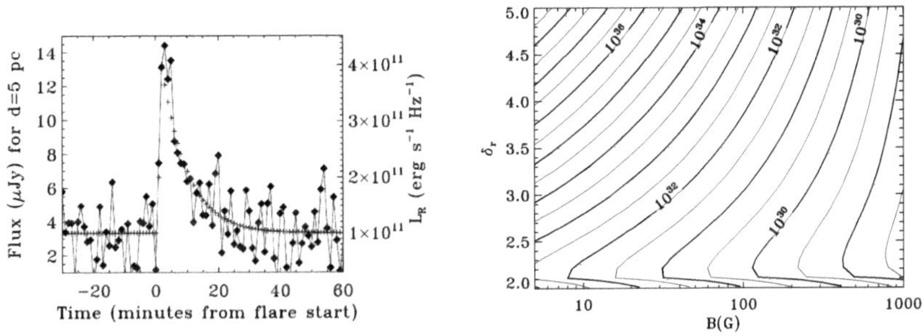

Figure 8. (*Left:*) Light curve of a short duration (lasting 5 minutes), small enhancement (factor of three increase above quiescence) flare on a nearby (5 pc) star. Right y-axis gives intrinsic radio luminosity, left y-axis gives the estimated radio flux density at 17 GHz. Noise is estimated in each 1 minute bin by offsetting the flux by a random number times the expected rms values. (*Right:*) Contour plot of δ_r, the index of accelerated particles, and B, the magnetic field strength in the radio-emitting source, for the radio flare seen above.

Smith et al. (2005) and Osten et al. (2016) laid out the basic framework for such a measurement, which requires multi-wavelength observations of the radio flare and its counterpart at higher energies. The uniqueness of the measurement lies in the difficulty in determining the characteristics of accelerated particles in nondegenerate environments in the cool half of the HR diagram. The usual technique for solar physicists is to deconvolve hard X-ray spectra exhibiting optically thin measurements of nonthermal bremsstrahlung emission to determine the power-law index of the accelerated electrons and source size. Because of astrophysical detector sensitivity coupled with typically hot thermal flaring plasma, this observational method has not yielded results in stellar cases. Radio constraints on accelerated particle characteristics are therefore the only measurement technique that will yield results for planet-hosting stars in the solar neighborhood.

The integrated radio flare energy provides a constraint on the accelerated particle kinetic energy, under the assumption of optically thin emission. The radio light curve provides constraints on both the index of the distribution of accelerated particles, as well as the magnetic field strength in the radio-emitting source. The contour plot of δ_r, the index of accelerated particles, versus the magnetic field strength in the radio-emitting source as a function of kinetic energy in the accelerated particles in a stellar superflare reported in Osten et al. (2016) was for an event with a peak radio luminosity at 15 GHz of 5×10^{16} erg s^{-1} Hz^{-1}. Based on Figure 5, there are numerous M dwarfs detectable with the ngVLA at the lower radio luminosity end, near 10^{11-12} erg s^{-1} Hz^{-1} for which similar analyses could be performed. Using the same methodology, we worked out the sensitivity to a radio flare with a peak flare enhancement of only 3 times this base level, for a flare lasting only 5 minutes total duration. Such an example is shown in the left part of Figure 8. These types of observations would be most sensitive to the nearest M dwarfs, in order to study time-variable emission over very short timescales (minutes). Quiescent radio luminosities as low as 10^{11} erg s^{-1} Hz^{-1} (at the limit of current detection capabilities; Osten et al. 2015) would be detectable in short integrations, allowing for study of small enhancement events only factors of a few larger. The kinetic energies probed here are moderate to large sized solar flares, and probe magnetic field strengths in the tens to hundreds of Gauss level.

A wide bandwidth system spanning the 10 – 20 GHz range can probe the optically thick and thin conditions, diagnose the changing conditions during the course of a magnetic reconnection flare associated with the close passage of an exoplanet, and deduce the nature of the accelerated particle population through measurements of spectral indices from confirmed optically thin emission. Operation of multiple subarrays would enable the frequency coverage available simultaneously to be significantly expanded. This would provide constraints on the accelerated particle population of close-in exoplanets unavailable from any other observational method; such a constraint is necessary to perform detailed modeling of the atmospheres of such exoplanets, due to the influence of accelerated particles in affecting the chemical reactions in terrestrial planet atmospheres.

4. Summary

The influence that stars have on their near environments is a timely research topic now, and is expected only to grow in the future as exoplanet discoveries grow. While we expect that new discoveries in this area in the next decade will fill in some areas where we currently lack insight, there will remain gaps in our knowledge which can only be filled by a next generation sensitive radio telescope operating in the deka-GHz range. Observations with the SKA at lower frequencies will be sensitive to magnetic and plasma emissions from stars, and will be able to study magnetic reconnection flaring and associated radio emission variability. The emphasis on lower frequencies with the SKA, however, means that the studies discussed here that are achievable with the ngVLA will not be possible to undertake with SKA observations due to optical depth effects and frequency dependence of the emission.

References

Anglada-Escudé, G. et al. 2016 Nature, 536, 437
Bower, G. et al. 2015 arXiv 1510.06432
Condon, J. et al. 1998 AJ 115, 1693
Crosley, M. K. & Osten, R. A. 2018a ApJ 856, 39
Crosley, M. K. & Osten, R. A. 2018b ApJ, submitted
Drake, S. A. et al. 1987 AJ 94, 1280
Drake, J. et al. 2013 ApJ 764, 170
Dressing, C. & Charbonneau, D. 2015 ApJ 807, 45
Favata, F. et al. 1995 A&A 298, 482
Fichtinger, B. et al. 2017 A&A 599, 127
Forbrich, J. et al. 2016 ApJ 822, 93
Garraffo, C. et al. 2016 ApJ 833, L4
Garraffo, C. et al. 2017 ApJ 843, L33
Heyner, D. et al. 2012 ApJ 750, 133
Jackman, C. H. et al. 1990 JGR 95, 7417
Jakosky, B. et al. 2015 Science 350, 0210
Johnstone, C. P. et al. 2015 A&A 577, 28
Lovelace, R. et al. 2008 MNRAS 389, 1233
Massi, M. et al. 2006 A&A 453, 959
Ngoc, P. et al. 2007 ApJ 658, 553
Osten, R. A. & Bastian, T. 2006 ApJ 637, 1016
Osten, R. A. & Bastian, T. 2008 ApJ 674, 1078
Osten, R. A. et al. 2005 ApJ 621, 398
Osten, R. A. et al. 2009 ApJ 700, 1750
Osten, R. A. et al. 2016 ApJ 832, 174
Osten, R. A. & Wolk, S. J. 2015 ApJ 809, 79
Osten, R. A. & Wolk, S. J. 2017 IAUS 328, 243
Panagia, N. & Felli, M. 1975 A&A 39, 1
Pilliterri, I. et al. 2014, ApJ 785, 145
Reid, I. N. et al. 2007AJ 133, 2825
Reid, I. N. et al. 2008 AJ 136, 1290
Richards, M. et al. 2003 ApJS 147, 337
Segura, A. et al. 2010 AsBio 10, 751
Smith, K. et al. 2005 A&A 436, 241
Tian, F. et al. 2014 E&PSL 385, 22
Venot, O. et al. 2016 ApJ 830, 77
Vidotto, A. et al. 2015 MNRAS 449, 4117
Wargelin, B. J. & Drake, J. 2001 ApJ 546, L57
Wood, B. E. et al. 2004 AdSpR 34, 66
Wood, B. E. et al. 2014 ApJ 781, L33
Wright, A. E. & Barlow, M. J. 1975 MNRAS 170, 41

Science with a Next-Generation Very Large Array
ASP Conference Series, Monograph 7
Eric J. Murphy, ed.

Prebiotic Molecules

Brett A. McGuire,[1] P. Brandon Carroll,[2] and Robin T. Garrod[3]

[1]*National Radio Astronomy Observatory, Charlottesville, VA, 22903;* `bmcguire@nrao.edu`

[2]*Harvard-Smithsonian Center for Astrophysics, Cambridge, MA, 02138;* `paul.carroll@cfa.harvard.edu`

[3]*University of Virginia, Charlottesville, VA, 22904;* `rgarrod@virginia.edu`

Abstract. We aim to greatly expand our understanding of the molecular inventory of prebiotic molecules and their chemical evolution by leveraging the ngVLA's unique capabilities to detect new complex organic molecules. Informed by state-of-the-art chemical models, we will search for N, O, and S-bearing small aromatic molecules, direct amino acid precursors, biogenic species such as sugars, and chiral molecules well beyond the reach of modern facilities.

1. Introduction

Extraterrestrial amino acids, the chemical building blocks of the biopolymers that comprise life as we know it on Earth, are present in meteoritic samples. More recently, glycine (NH_2CH_2COOH), the simplest amino acid, was detected by the Rosetta mission in comet 67P. Despite these exciting discoveries, our understanding of the chemical and physical pathways to the formation of (pre)biotic molecules is woefully incomplete. This is largely because our knowledge of chemical inventories during the different stages of star and planet formation is incomplete. It is therefore imperative to solidify our accounting of the chemical inventories, especially of critical yet low-abundance species, in key regions and to use this knowledge to inform, expand, and constrain chemical models of these reactions. This is followed naturally by a requirement to understand the spatial distribution and temporal evolution of this inventory. Here, we briefly outline a handful of particularly-impactful use cases in which the ngVLA will drive the field forward.

2. Expanding Chemical Inventories: Pushing cm-Wave Chemical Complexity to the Confusion Limit

Existing facilities in the (sub-)millimeter regime, particularly ALMA, are capable of reaching the line-confusion limit in molecular line surveys of the richest interstellar sources in a matter of hours. That is, the density of molecular lines in the resulting spectra is such that there is at least one spectral line every FWHM; there are no line-free channels. When line-confusion is reached, no additional information can be gained

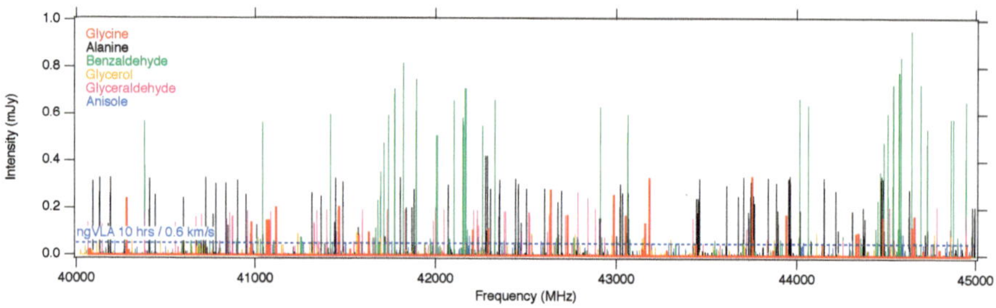

Figure 1. Simulation of six complex organic molecules in a 5″ source at T_{ex} = 100 K, ΔV = 3 km s^{-1}, and N_T = 5 × 10^{13} cm $^{-2}$. The simulation assumes the source completely fills the beam. Glycine transitions have been shown in **red** for emphasis. The approximate noise level of a 10 hour ngVLA integration with 0.6 km s^{-1} resolution is shown as a dashed **blue** line.

from longer integration times at the same spatial resolution. We are rapidly approaching the point at which deep drill observations with ALMA and the GBT will no longer produce new spectral lines in the richest sources.

In the cm-wave regime, however, the most sensitive survey (PRIMOS) with the most sensitive facility (the GBT) toward the most line-rich source (Sgr B2) has lines in only ~10-20% of its channels. Diminishing returns on sensitivity for this survey have already set in with an rms of ~2.75 mJy/beam (in ~0.2 km/s channels) - to increase the sensitivity substantially would require hundreds of hours of dedicated observations for every ~2 GHz window. With the higher sensitivity of the ngVLA, at interferometric resolutions better matched to the size of the cores in this source (~1–5″), we can reach the line-confusion limit in this survey, which we project will set in around 100 μJy/beam rms. These rms values are reachable with the ngVLA, on appropriate spatial scales, in 4 – 10 hours.

Observation of a substantial number of predicted, but as yet undetected, complex prebiotic species are needed to truly understand chemical evolution toward glycine and other biogenic molecules. State-of-the-art models predict these molecules will display emission lines with intensities that are easily detectable with the ngVLA, but well below the current detectability thresholds of ALMA, GBT, IRAM, etc. Figure 1 shows simulations of a representative set of the types of molecules whose discovery will be enabled by the ngVLA: N, O, and S-bearing small aromatic molecules, direct amino acid precursors, biogenic species such as sugars, and chiral molecules.

3. Grasping for Truly Biogenic Species: Glycine and Glyceraldehyde

The detection and characterization of glycine and glyceraldehyde (the simplest sugar) would be transformational for the field. Using state-of-the-art chemical models combined with observationally-constrained physical conditions and temperature-density profiles, we have simulated the expected intensity of the rotational transitions of these molecules toward two high-profile targets: Sgr B2(N) and IRAS 16293.

For glyceraldehyde, we predict intensities of 100–200 μJy/beam toward Sgr B2(N), which puts it within reach for the ngVLA with a dedicated observing session, although many lines will be at or around the confusion limit. By contrast, we expect that the emission in IRAS 16293 is likely to be too weak. For glycine, intensities at the lower end of the ngVLA frequency range (<10 GHz) are unlikely to be sufficient for detection, even with the ngVLA, while intensities at mm-wave frequencies are at or below the current line-confusion limit. In the heart of the ngVLA range (10 – 50 GHz), however, our models predict line intensities of order $\sim$500 μJy/beam in Sgr B2(N). This offers substantial room for variance in the actual abundance while still remaining detectable, but only with the sensitivity and spatial resolution of the ngVLA.

In IRAS 16293, a population of glycine at the abundances predicted by our chemical model is likely beyond the reach of even the ngVLA. However, the cold chemistry in this region is not as well-matched to the hot core model of Garrod (2013) used here, and a more optimistic abundance estimate, based on the detected abundance of glycine in comet 67P by Altwegg et al. (2016), indicates transitions of up to $\sim$1 mJy/beam (at $0\farcs2$ resolution) are possible. These values are only realistically achievable with the ngVLA.

4. Explorations of Interstellar Chirality

Chiral molecules, that is, molecules whose mirror image is not identical to the original, are central to biological function. Molecular chirality has a profound effect on the structure and function of biological molecules, as a result of nature's use of only one of the mirror images in biological processes (a concept known as homochirality). There is no energetic basis for the dominance in life of one handedness of a chiral molecule over another. Rather, a plausible explanation is that the slight primordial excess of one handedness was inherited from the nascent molecular inventory and subsequently enhanced and enriched catalytically by life. Material in molecular clouds from which planetary systems form is processed through circumstellar disks, and can subsequently be incorporated into planet(esimal)s. Thus, a primordial excess found in the parent molecular cloud may be inherited by the fledgling system. The detection of chiral molecules toward molecular clouds is therefore key to advancing our understanding of this process. Chiral molecules, like other complex species detected earlier, are necessarily large, with propylene oxide, the only detected chiral species to date (McGuire & Carroll et al. 2016), being perhaps the only example simple enough for detection with existing facilities. The ngVLA will provide the sensitivity and angular resolution required to detect additional, biologically-relevant chiral species, such as glyceraldehyde.

Indeed, one possible route to generate a chiral excess is through UV-driven photodissociation of chiral molecules by an excess of left or right circularly polarized light. The ability not only to detect, but to image the abundance of chiral species at spatial scales commensurate with observations of circularly polarized light toward star-forming regions would be an immense leap forward. Using known, polarization-dependent photodissociation cross sections from laboratory studies, these observations would enable quantitative estimates of potential UV-driven excess. While such studies are well beyond the capability of existing observatories, they would be achievable with the ngVLA.

5. Uniqueness to ngVLA Capabilities

As discussed above, the cm-wave regime is an under-explored, but extraordinarily-rich wavelength regime for prebiotic chemical studies. Existing facilities in this wavelength range, particularly the GBT and VLA, have produced stunning scientific results, but they have nearly reached the limits of their capabilities in this area, particularly in their sensitivity. The GBT beam is poorly matched to compact sources and suffers incredibly from beam dilution, while the VLA lacks the raw collecting area. Finally, the Band 1 receivers at ALMA will cover only a portion of the critical 10–50 GHz range necessary to enable this science, and at far worse brightness sensitivity. The ngVLA is the only facility, extant or planned, that can unlock the specific science goals outlined above at these frequencies.

6. Synergies at Other Wavelengths

The complex chemistry described here, which is detected and probed through gas-phase observations, has its genesis in the icy mantles of dust grains. As a protostar turns on and warms the surrounding medium, or through the passage of shocks, these ices and the complex molecules that formed within them, are liberated into the gas phase where they are detected. Observational studies of these ices are extraordinarily limited, and no species more complex than methanol (CH_3OH) has been detected. This will change with the launch of the *JWST*, which will enable the study of a far larger sample of molecular ices, with far greater complexity than is currently possible. The interpretation of these observations, and their synergy with gas-phase molecular inventories and observations, will rely upon the types of ngVLA observations outlined here.

References

Altwegg, K, Balsiger, H., Bar-Nun, A., Bertheller, J.J., Bieler, A. et al. 2016 Science Advances 2, e1600285

Garrod, R. T. 2013 ApJ 765, 60

McGuire, B.A. & Carroll, P.B. et al. 2016 Science 352, 1449.

Science with a Next-Generation Very Large Array
ASP Conference Series, Monograph 7
Eric J. Murphy, ed.

Complex Organic Molecules in Hot Molecular Cores/Corinos: Physics and Chemistry

Maria T. Beltrán and Víctor M. Rivilla

INAF-Osservatorio Astrofisico di Arcetri, Largo E. Fermi 5, I-50125 Firenze, Italy; mbeltran@arcetri.astro.it, rivilla@arcetri.astro.it

Abstract. Hot molecular cores and hot corinos are the richest reservoirs of complex organic molecules in the Galaxy, including key species for prebiotic processes. The superb capabilities of the ngVLA will allow us not only to detect new heavy complex organic species, but also to spatially resolve their emission from the envelope down to the accretion disks embedded in the hot core and the hot corino and to chemically and physically characterize the region where the emission originates.

1. Introduction

Hot molecular cores (HMCs), the cradles of massive stars, are the most chemically rich sources in the Galaxy (Bisschop et al. 2007; Belloche et al. 2013; Rivilla et al. 2017). The typical masses of these cores (few hundreds of $M_{\odot}$) make them the most important reservoirs of complex organic molecules (COMs), including key species for prebiotic processes. This rich chemistry is thought to be the result of the evaporation of dust grain mantles by the strong radiation of the deeply embedded early-type star(s). Our own Sun may have been born in a high-mass star-forming region (Adams 2010), so our Earth may have inherited the primordial chemical composition of its parental hot core region, as suggested by recent studies of oxygen and sulfur chemistry in comets (Taquet et al. 2016; Drozdovskaya et al. 2018).

The immediate surroundings of high-mass protostars are not the only regions rich in COMs. In recent years, the environments of some low-mass protostars have also revealed a very complex chemistry, with the presence of large molecules with abundances similar to those found towards high-mass protostars. These low-mass cores have been called hot corinos (Bottinelli et al. 2004). Although the mechanism responsible for the emission of COMs would be the increase of temperature in the inner envelope, like for the high-mass hot cores, it is not clear what is producing such an enhancement: shocks associated with infall or with outflows (Ceccarelli et al. 1996; Charnley & Rodgers 2005). The fact that hot corinos are associated with low-mass young stellar objects allows us to study directly the environment where solar-analogs are now forming.

In this chapter, we discuss how the ngVLA can help us to study the emission of heavy COMs in both low- and high-mass star-forming regions. The emission of COMs is important not only because it allows us to understand how chemistry may have developed to eventually form life in our Earth, but also because COMs are a powerful tool for studying the physical properties and kinematics of the dense regions very close to the central protostars.

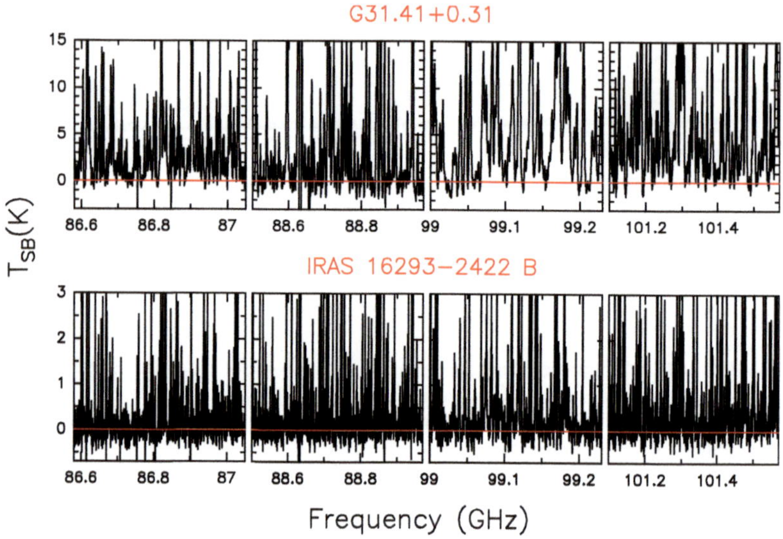

Figure 1. ALMA 3 mm spectra at ~1″ angular resolution of the hot molecular core G31.41+0.31 (*upper panels*) and the hot corino IRAS 16293-2422 B (*lower panels*) (V. M. Rivilla, private communication). Red lines indicate T_{SB}=0 K. The 3σ noise level at each panel falls inside the red line. The (sub)millimeter spectra of chemically rich hot cores/corinos like these two examples are full of molecular lines, which complicates the detection of very complex molecules with weak emission due to line confusion limit and blending problems.

2. COMs at Centimeter Wavelengths

COMs in high- and low-mass star-forming regions can be observed at (sub)millimeter wavelengths but are severely affected by line blending, especially in turbulent regions where the line widths are broad. In fact, many ALMA observations of the most chemically rich regions are essentially at the line confusion limit (Figure 1) – that is, no additional molecules can be detected that emit below that threshold regardless of the sensitivity (rms) of the spectrum. Moving to the centimeter-wavelength window, the number of spectral lines in a selected spectral portion is significantly lower (see Figure 2) with respect to the millimeter window, and hence, the lines are less blended. Moreover, at millimeter wavelengths, the dust opacity is commonly very high, and may produce strong line absorption, preventing in some cases the detection of the line emission itself, while at centimeter wavelengths the dust emission is usually optically thin.

The study of COMs at centimeter wavelengths is suited to:

- **Detect heavy complex organic molecules in star forming-regions**. The PRIMOS survey, the GBT Legacy Survey of Prebiotic Molecules Toward Sgr B2N (Neill et al. 2012; Loomis et al. 2013; Zaleski et al. 2013), has detected in the last few years many COMs in this range in the massive HMC Sgr B2N in the Galactic Center. Some examples are: propenal (CH_2CHCHO), propanal (CH_3CH_2CHO), simple aldehyde sugars like glycolaldehyde (CH_2OHCHO), the first keto ring molecule detected in a interstellar cloud, cyclopropenone (c−H_2C_3O), keten-

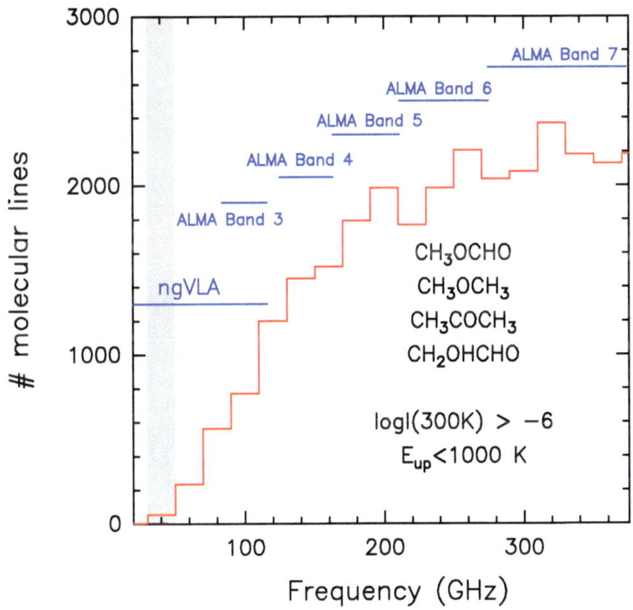

Figure 2. Distribution of the number of molecular lines as a function of frequency of four O-bearing COMs that are abundant in hot cores/corinos: methyl formate (CH_3OCHO), dimethyl ether (CH_3OCH_3), acetone (CH_3COCH_3), and glycolaldehyde (CH_2OHCHO). We have selected molecular transitions with log I(300 K) > −6 and E_{up} < 1000 K from the JPL molecular catalogue. The frequency range shown in Figures 3 and 4, 30–50 GHz, is indicated with a grey band. The spectral coverage of the ngVLA and ALMA Bands 3, 4, 5, 6 and 7 is shown.

imine (CH_2CNH), ethanimine (CH_3CHNH), cyanomethanimine (HNCHCN) (see Figure 3), and acetamide (CH_3CONH_2), the largest interstellar molecule detected with a peptide bond, key for prebiotic chemistry. Some of these species are, in fact, not suitable for observations at higher frequencies (millimeter and submillimeter regime). In general, the heavier the molecule, the more shifted to centimeter wavelengths its rotational spectrum is. In fact, these large molecules, with more than 8 atoms, have a considerable number of transitions at frequencies < 50 GHz.

- **Study circumstellar disks surrounding the protostars**. Heavy COMs are in general less abundant than typical HMC tracers, such as methyl cyanide, and therefore, their emission is expected to be optically thinner. The combination of low dust opacity at centimeter wavelengths and of low line opacity of these large and low abundant COMs, makes these species suitable to study the kinematics of high-mass star-forming regions – in particular of possible accretion disks – in a very pristine way. This probe is very important to understand better the formation of OB-type stars, for which the existence of true accretion disks is still under debate (e.g., Beltrán & de Wit 2016). The kinematics of the gas close to the central protostar should tell us whether high-mass stars have Keplerian

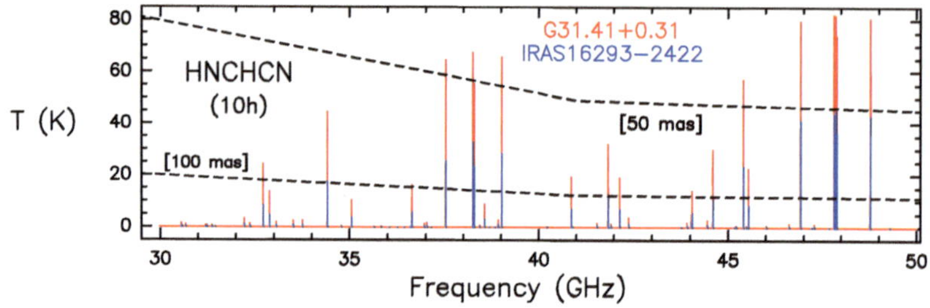

Figure 3. Synthetic spectra obtained with MADCUBA of cyanomethanimine, HNCHCN, for G31.41+0.31 hot molecular core (red) and IRAS16293-2422 B hot corino (blue). We have assumed an excitation temperature of 100 K, and column densities of 3×10^{17} cm^{-2} and 3×10^{16} cm^{-2} for G31.41+0.31 and IRAS 16293–2422, respectively. The line widths considered are 5.0 km s^{-1} and 1.5 km s^{-1} for G31.41+0.31 and IRAS 16293–2422, respectively, consistent with those found at milllimeter wavelengths (Rivilla et al. 2017; Jørgensen et al. 2016). We have overplotted with dashed lines the 3σ noise level for an integration time of 10 hr with two different angular resolutions, assuming a spectral resolution of 1 km s^{-1}, according to the specifications of the ngVLA.

circumstellar disks, like those observed towards lower-mass counterparts (e.g., L1527 IRS: Tobin et al. 2012; Ohashi et al. 2014; HD 163296: de Gregorio-Monsalvo et al. 2013), or whether they are surrounded by rotating structures undergoing solid-body rotation.

3. Measurements Required

The ngVLA will provide an unprecedented sensitivity and spatial resolution in the centimeter wavelength range that will allow us to detect new complex species and spatially resolve the emission of heavy COMs, e.g., those previously detected with the GBT (resolution > 16″). By targeting two of the most chemically rich star-forming regions, a high-mass hot molecular core (G31.41+0.31: Beltrán et al. 2009; Rivilla et al. 2017) and a low-mass hot corino (IRAS 16293-2422: Jørgensen et al. 2016; Martín-Doménech et al. 2017), it will be possible to detect prebiotic species, such as HNCHCN (see Figure 3), for the first time outside the Galactic Center. In addition, the high-angular resolution provided by the ngVLA will allow us to map the distribution of heavy COMs from the envelope down to the accretion disks embedded in the hot core and hot corino, respectively, and characterize the region where the emission originates. It is important to understand whether the emission comes from the inner circumstellar disk where planets could form, or whether it originates in an outer region of the disk/envelope, where comets reside.

In particular, the ngVLA capabilities would allow us to:

- **Detect heavy (prebiotic) COMs, for the first time outside the Galactic Center**. First detections of new species in HMCs have been carried out mainly towards Sgr B2 (see Belloche et al. 2008; Zaleski et al. 2013). This complex, located in the Central Molecular Zone, has been observed at different wavelengths and as a result, the first detections of heavy COMs in the Galaxy, such as ethylene glycol or ethyl formate (10 and 11 atoms, respectively), have been obtained there (e.g., Hollis et al. 2000, 2002). The cores in Sgr B2, however, are not typical HMCs. The current star formation rate of Sgr B2 qualifies it as a mini-starburst region and its location close to the Galactic Center makes the physical conditions in that environment quite extreme, which could have significant consequences on the chemistry. Therefore, the chemistry in Sgr B2 might not be representative of that typical of HMCs in the disk of the Galaxy. In addition, the presence of several velocity components along the line of sight makes line identification more difficult. Therefore, to study the chemistry of COMs and the environments where they form, it is important to detect heavy COMs around more typical star-forming regions. The two star-forming cores G31.41+0.31 and IRAS 16293−2422 are among the best candidates to detect new molecular species in high- and low-mass star-forming regions, respectively, outside the Galactic Center because of their chemical richness (see Figure 1). They, thus, will serve as templates for future observations of a range of sources with different stellar masses. In particular, large molecules with > 9 atoms have been already detected in both regions (Jørgensen et al. 2016; Rivilla et al. 2017). However, these two sources are not the only ones showing a chemically rich spectrum. HMCs are the cradles of OB stars, and it is therefore expected all high-mass (proto) stars to be associated with them (Beltrán 2010). Regarding the low-mass regime, with the advent of ALMA, the number of known hot corinos is increasing

To detect new species, high sensitivity is more important than high-angular resolution. In principle, it would not even be necessary to resolve the emission. In practice, however, it is important to resolve it, at least slightly, to study the spatial distribution of the COMs and check whether the emission of different species is spatially correlated, i. e., coming from the same region.

We have used the example of the prebiotic species cyanomethanimine (HNCHCN, see Figure 3), only detected so far by the GBT in Sgr B2N within the PRIMOS Project (Zaleski et al. 2013). To estimate the sensitivity (rms) required to detect emission from HNCHCN, we have modeled its emission assuming LTE conditions, an excitation of 100 K, typical of hot cores and hot corinos, and column densities of 3×10^{17} cm^{-2} and 3×10^{16} cm^{-2} for G31.41+0.31 and IRAS 16293−2422, respectively (see caption of Figure 3). In this figure, we have plotted the 3σ noise level for an angular resolution of 50 mas and 100 mas and an integration time of 10 hr. As seen in the figure, at 50 mas angular resolution, only emission from the high-mass hot molecular core G31.41+0.31 would be detectable. Since IRAS 16293−2422 is located at only 120 pc, however, the spatial resolution achieved even with 100 mas (∼12 au), is already very high.

- **Study of the physical properties and kinematics of the circumstellar disks.** The distance to the solar-type protostar IRAS 16293-2422 is 120 pc. Therefore, with a spatial resolution of ∼5–10 au, achieved with an angular resolution of 50–100 mas, we should be able to study the kinematics and the chemistry of the inner disk region where planets could form. On the other hand, the O-type (proto)star G31.41+0.31 is located much farther away, at 7.9 kpc. By observing the core with an angular resolution of 35 mas (a good compromise between angular resolution and sensitivity), the spatial scales achieved would be ∼280 au, typical of disks around intermediate-mass protostars (e.g., Beltrán & de Wit 2016). Therefore, if

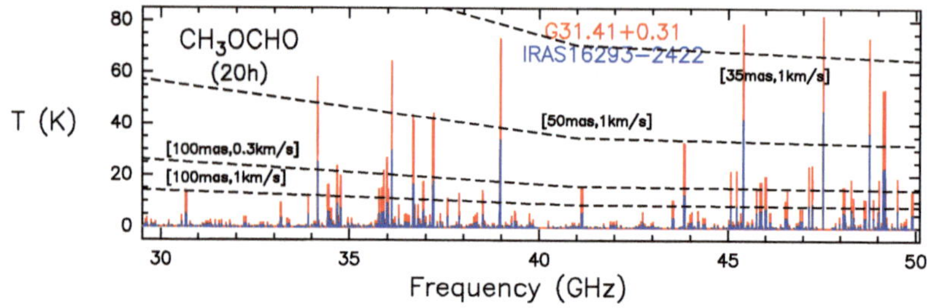

Figure 4. Synthetic spectra obtained with MADCUBA of methyl formate, CH_3OCHO, for G31.41+0.31 hot molecular core (red) and IRAS16293-2422 B hot corino (blue). We have assumed an excitation temperature of 100 K. The column densities for G31.41+0.31 and IRAS 16293−2422 are 5×10^{18} cm^{-2} and 5×10^{17} cm^{-2}, respectively, from Rivilla et al. (2017a) and Jorgensen et al. (2012). The linewidths considered are 5.0 km s^{-1} and 1.5 km s^{-1} for G31.41+0.31 and IRAS 16293−2422, respectively. We have overplotted with dashed lines the 3σ noise level of an integration time of 20 hr with different combinations of angular and spectral resolutions, indicated with labels, according to the specifications of the ngVLA.

a circumstellar disk is surrounding this massive star, the ngVLA should be able to detect it. Note that circumstellar disks have been detected up to late O-type stars (e.g., Johnston et al. 2015), but the luminosity of G31.41+0.31 ($3 \times 10^5 L_{\odot}$) indicates that this HMC is hosting O6–O7 (proto)stars. Therefore, detecting such a disk would be crucial to confirm that stars of all masses form by disk-mediated accretion.

To study the physics and kinematics of the disks, we have chosen the well-known high-density tracer methyl formate (CH_3OCHO, see Figure 4). Methyl formate is a complex (8 atoms) molecule, which is a good thermometer and has been clearly detected in both regions with observations at much lower angular resolution (Jørgensen et al. 2012; Beltrán et al. 2018). To estimate the required sensitivity, we have modeled the methyl formate emission assuming LTE conditions, an excitation of 100 K, typical of hot cores and hot corinos, and a column density of 5×10^{18} cm^{-2} and 5×10^{17} cm^{-2} for G31.41+0.31 and IRAS 16293−2422, respectively (see caption of Figure 4), consistent with what found at millimeter wavelengths in the inner regions of hot cores/corinos (Jørgensen et al. 2012; Rivilla et al. 2017), and line widths of 5 km s^{-1} and 1.5 km s^{-1}. The rms noise has been estimated for different angular and spectral resolutions. As seen in Figure 4, the methyl formate emission can be observed at high enough spatial resolution to study properly the kinematics of the circumstellar disks: up to ~5 au for IRAS 16293−2422 and up to ~280 au for G31.41+0.31.

4. Synergy with Other Instruments

Complex organic molecules can be observed from centimeter to millimeter wavelengths. As mentioned before, to detect new species, high sensitivity is more important than high-angular resolution. Therefore, it is not surprising that some of the most recent detections have been done with a single-dish telescope, the GBT, within the PRIMOS Project (Zaleski et al. 2013). However, although single-dish observations are a good starting point to detect new species, high-sensitivity interferometric observations are absolutely needed. On the other hand, because the emission of heavy COMs can be very weak and compact, it is not unusual that species previously not detected with a single dish, because of beam dilution, and easily detected with an interferometer. On the other hand, it is important to resolve the emission of COMs to study the spatial distribution of the COMs and check whether the emission of different species is spatially correlated, i.e., coming from the same region. This can help to constrain better the formation routes of different species and to distinguish between different chemical models (e.g., gas phase formation vs. grain surface).

Regarding the millimeter regime, COMs have been detected with facilities such as IRAM NOEMA and ALMA, although as explain before, for chemically rich sources, the millimeter observations can suffer from line confusion limit and line blending, which might difficult the identification of new species. In any case, millimeter and centimeter observations are complementary. By observing COMs at millimeter wavelengths with ALMA or NOEMA, we should be able to trace high-energy transitions of the same molecules and obtain a complete picture of the emission of such heavy species. This synergy will be very important to constrain better the formation routes of heavy COMs.

Acknowledgments. V.M.R. is funded by the European Union's Horizon 2020 research and innovation programme under the Marie Skłodowska-Curie grant agreement No 664931.

References

Adams, F. C. 2010, ARA&A, 48, 47
Belloche, A., Menten, K. M., Comito, C. et al. 2008, A&A, 482, 179
Belloche, A., Müller, H. S. P., Menten, K. M. et al. 2013, A&A, 559, A47
Beltrán, M. T. in Computational Star Formation, Proceedings of the International Astronomical Union, IAU Symp. 270, 33
Beltrán, M. T., Codella, C., Viti, S. et al. 2009, ApJ, 690, L93
Beltrán, M. T., Cesaroni, R., Rivilla, V. M. et al. 2018, A&A, arXiv:1803.05300
Beltrán, M. T., & de Wit, W. J. 2016, A&ARv, 24, 6
Bisschop, S. E., Jørgensen, J. K., van Dishoeck, E. F. et al. 2007, A&A, 465, 913
Bottinelli, S., Ceccarelli, C., Neri, R. 2004, ApJ, 617, L69
Ceccarelli, C., Hollenbach, D. J., Tielens, A. G. G. M. 1996, ApJ, 471, 400
Charnley, S., & Rodgers, S. D. 2005, Highlights of Astronomy, 13, 511
de Gregorio-Monsalvo, I., Ménard, F., Dent, W. et al. 2013, A&A, 557, A133
Drozdovskaya, M. N., van Dishoeck, E. F., Jørgensen, J. K. et al. 2018, MNRAS, 476, 4949
Hollis, J. M., Lovas, F. J., Jewell, P. R. 2000, ApJ, 540, L107
Hollis , J. M., Lovas, F. J., Jewell, P. R., Coudert, L. H. 2002, ApJ, 571, L59
Johnston, K. G., Robitaille, T. P., Beuther, H. et al. 2015, ApJ, 813, L19
Jørgensen, J. K., Favre, C., Bisschop, S. E. 2012, ApJ, 757, L4
Jørgensen, J. K., van der Wiel, M. H. D. Coutens, A. et al. 2016, A&A, 595, A117
Loomis, R. A., Zaleski, D.łP., Steber, A. L. et al. 2013, ApJ, 765, L9
Martín-Doménech, R., Rivilla, V. M., Jiménez-Serra, I. et al. 2017, MNRAS, 469, 2230
Neill, J. L., Muckle, M. T., Zaleski, D. P. et al. 2017, ApJ, 755, 153
Ohashi, N., Saigo, K., Aso, Y. et al. 2014, ApJ, 796, 131
Rivilla, V. M., Beltrán, M. T., Cesaroni, R. et al. 2017, A&A, 598, A59
Taquet, V., Furuya, K., Walsh, C., van Dishoeck, E. F. 216, MNRAS, 462, 99
Tobin, J. J., Hartmann, L., Chiang, H.-F. et al. 2012, Nature, 492, 83
Zaleski, D. P., Seifert, N. A., Steber, A. L. et al. 2013, ApJ, 765, L10

Science with a Next-Generation Very Large Array
ASP Conference Series, Monograph 7
Eric J. Murphy, ed.

SETI Searches for Evidence of Intelligent Life in the Galaxy

Steve Croft,[1] Andrew P. V. Siemion,[1,2,3] James M. Cordes,[4] Ian S. Morrison,[5] Zsolt Paragi,[6] and Jill Tarter[3]

[1]*Department of Astronomy, University of California, Berkeley, 501 Campbell Hall #3411, Berkeley, CA 94720, USA*

[2]*Department of Astrophysics/IMAPP, Radboud University, P.O. Box 9010, NL-6500 GL Nijmegen, The Netherlands*

[3]*SETI Institute, Mountain View, CA 94043, USA*

[4]*Cornell Center for Astrophysics and Planetary Science and Department of Astronomy, Cornell University, Ithaca, NY 14853, USA*

[5]*Curtin Institute of Radio Astronomy, Curtin University, Perth, Australia*

[6]*Joint Institute for VLBI ERIC (JIVE), Postbus 2, NL-7990 AA Dwingeloo, The Netherlands*

Abstract. Radio SETI experiments aim to test the hypothesis that extraterrestrial civilizations emit detectable signals from communication, propulsion, or other technologies. The unprecedented capabilities of next generation radio telescopes, including the ngVLA, will allow us to probe hitherto unexplored regions of parameter space, thereby placing meaningful limits on the prevalence of technological civilizations in the Universe (or, if we are fortunate, making one of the most significant discoveries in the history of science). The ngVLA provides critical capabilities in the 10 – 100 GHz range, and will be a valuable complement to the SKA in the southern hemisphere, as well as surveying the sky at frequencies underexplored by previous SETI experiments.

1. Description of the Problem

Almost all stars are now thought to have planets, and a substantial fraction (Kasting et al. 2014; Dressing & Charbonneau 2015) have a planet in the "habitable zone" where liquid water may exist on the surface. The local Universe is thus a target-rich environment to search for signs that life may have taken hold elsewhere in our Galaxy. This can be accomplished either by looking for atmospheric biosignatures (using the next generation of space-based or extremely large ground-based optical telescopes), or by looking for advanced life by searching for signatures of technology (Tarter 2001). The latter will be an area in which the ngVLA will excel. The search for extraterrestrial intelligence (SETI) is also complementary to other proposed "cradle of life" studies to be undertaken by the ngVLA, including studies of star and planet formation, and the formation of complex molecules (e.g., see Isella et al. 2015).

The frequency coverage of the ngVLA is very complementary to the bands to be covered by the SKA (Siemion et al. 2013), allowing a comprehensive SETI search to be undertaken over three decades of radio frequency. These new instruments will also provide the sensitivity required to detect analogs of Earthbound transmitters such as aircraft radar from the distance of several nearby stars, in addition to more powerful (or directional) transmitters from across the Galaxy.

2. Scientific Importance

"Are we alone?" is one of the most profound human questions. Yet at the turn of the seventh decade (Cocconi & Morrison 1959; Drake 1961) of SETI searches, there is, as yet, no evidence that intelligent life has arisen elsewhere in our Universe. One might infer that technological life beyond Earth is rare, or even non-existent, but in reality such a small volume of parameter space has been explored to date that this conclusion would be premature. The search for extrasolar planets provides an instructive example. As late as the 1980s, searching for planets orbiting other stars was a fruitless task, but as new techniques and powerful new instruments became available, the field has exploded (e.g., Borucki et al. 2010). We now understand that the Galaxy is replete with planets of a huge variety of sizes, temperatures, and compositions, including many similar in character to our own.

3. Astronomical Impact

The detection of the first exoplanets opened an entirely new field of research. Aside from the cultural impact of a SETI detection, the dawning of the age of observational astrobiology will mark one of the most significant milestones in the history of astronomy. Even the detection of a simple beacon, indicative of the presence of another technological civilization in our Galaxy, would enable us to begin to understand the processes by which the Universe gives rise to life, and intelligence, launching a new Copernican Revolution by which we more fully understand our place in the cosmos. Were a signal to contain information that we were able to comprehend, we could potentially understand what awaits us as a species, or begin to investigate "the archaeology of the future" (Morrison 1997).

4. Anticipated Results

Radio SETI experiments aim to test the hypothesis that extraterrestrial civilizations emit detectable signals from communication, propulsion, or other technologies. SETI attempts to constrain the terms in the Drake Equation (Drake 1961; Worden et al. 2017); in particular, given that we now know that habitable planets are common, can we attempt to constrain the fraction of such planets that develop life, intelligence, and technology, and the lifetime over which signs of such technology are detectable? We do not know, a priori, the frequency spectrum, luminosity function, duty cycle, or antenna gain and beaming fraction of ETI transmissions. To have the best chance of success, therefore, a SETI survey must attempt to explore a large region of parameter space in frequency (coverage and resolution), time (resolution, cadence, and duration), and sensitivity. While it is challenging to qualitatively compare one SETI survey to another, increases in computational power and survey speed are enabling modern SETI searches that far exceed the capabilities of their predecessors (Enriquez et al. 2017).

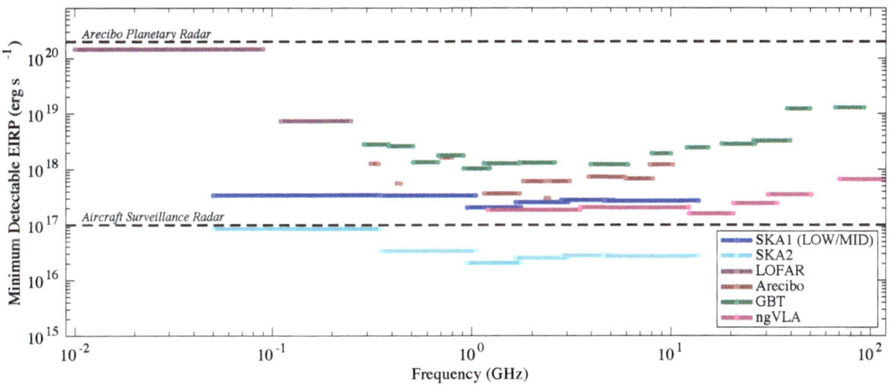

Figure 1. Sensitivity of the the ngVLA to narrow-band transmitters at 15 pc, as compared with other facilities actively performing SETI searches over the same band (following Siemion et al. 2013). Here we assume a significance threshold $\sigma = 15$, bandwidth $\Delta b = 0.5$ Hz and integration time $t = 10$ minutes. A transmitter is detectable if its EIRP is above the curve for a given telescope. Thus a transmitter with an EIRP of 2×10^{20} ergs/sec (planetary radar) is detectable with all of the telescopes shown. A transmitter with an EIRP of 1×10^{17} ergs/sec (airport radar) is only detectable with the realization of the SKA2 at 15 pc, but could also be detected by the ngVLA for stars closer than $\sim$ 10 pc. Sensitivities for the Green Bank Telescope (GBT), Arecibo and LOFAR were taken from those facilities observing guides. For LOFAR we assume only core stations are used. The ngVLA sensitivity was taken from Selina et al. (2018).

As a fast survey instrument with high sensitivity and wide frequency coverage, the ngVLA will provide powerful constraints on the space density of technologically advanced civilizations in our Galaxy, if not actually directly detecting their existence.

5. Limitations of Current Astronomical Instrumentation

Although the power of extraterrestrial transmissions may be considerably higher than those possible with human technologies, it is nevertheless useful to use our own transmission capabilities as a guide. Our most powerful beamed transmitter, the Arecibo telescope when used as a planetary radar, has an equivalent isotropic radiated power (EIRP) of 2×10^{20} erg/s, and would be detectable by the ngVLA at distances of kiloparsecs. ETI signals comparable to our most powerful wide-angle transmitters, such as airport radars (EIRP $\sim 10^{17}$ erg/s) would be detectable from planets orbiting the nearest few hundred stars to the Sun. The ngVLA will complement SKA1-LOW and SKA1-MID as the only facilities with the capability to detect "leakage" transmissions from omnidirectional transmitters with power close to the brightest transmitters on Earth (Figure 1).

6. Connection to Unique ngVLA Capabilities

The unprecedented capabilities of next generation radio telescopes, including the ngVLA, will allow us to probe hitherto unexplored regions of parameter space (Ekers et al. 2002), placing meaningful limits on the prevalence of technological civilizations in the Universe, or, if we are fortunate, making one of the most significant discoveries in the history of science.

The ngVLA provides critical capabilities in the 10 – 100 GHz range, a region of the spectrum used by many human technologies, to survey the sky at sensitivities unmatched by other facilities. At frequencies where the ngVLA and SKA overlap, the ngVLA will primarily see the northern hemisphere and the SKA the southern.

Additionally, ngVLA combines high sensitivity with very high resolution (of order 10 mas). With such high resolution, the motion of a putative transmitter on a planet or spacecraft, orbiting its star at a distance of ~AU, could be discerned astrometrically even for stars at distances of ~ 100 pc. Measuring the motion of such a transmitter would be a powerful additional tool for distinguishing artificial transmitters in orbit around nearby stars.

7. Experimental Layout

SETI searches have two key goals:

1. Find signals that appear artificial (e.g., narrow band, modulated, pulsed, or otherwise obviously "unnatural").

2. Localize such signals at a specific right ascension and declination [to eliminate the possibility that they are radio frequency interference (RFI) from transmitters close to the telescope or from Earth-orbiting satellites].

Searches can either be performed using baseband voltage data, or, after an FFT operation, spectrograms or "waterfall plots" of intensity as a function of frequency and time. Classical SETI algorithms have typically focused on the search for narrow-band signals, but increased computing power and more sophisticated algorithms (including machine learning approaches) will enable the search for increasingly complex signals, as well as better classification of confounding RFI.

We envisage an ngVLA SETI search operating commensally with most primary science users of the array, but in addition a smaller amount (tens to hundreds of hours per year) of TAC-allocated time for pointed observations in support of time-critical or other targets of interest. Any star within the instantaneous field of view of the ngVLA is a potential SETI target, and as such a commensal search with the ngVLA would enable stringent limits on ETI transmissions for millions of stars, in addition to limits at higher EIRP for more distant targets such as galaxies. In addition to stars and galaxies, solar system objects have in the past also been targets for SETI searches (e.g., Freitas & Valdes 1980; Enriquez et al. 2018; Tingay et al. 2018); commensal or targeted searches of such objects with the ngVLA could easily detect transmitters with EIRPs measured in milliwatts.

The proposed observations require that the ngVLA have the capability to deliver time domain voltage data from every antenna to dedicated signal processing hardware to be installed as either a user or facility instrument. Similar to user instruments operating at Arecibo, Green Bank, and Parkes, such a system would run in a commensal mode, capturing or processing voltage data alongside non-SETI observations, in addition to standalone observations of SETI targets. Such equipment would consist of ~ 100 PB storage and 10 PFLOPS compute (assuming Moore's Law extrapolation from current equipment) enabling digitization of the entire ngVLA bandwidth and be capable of forming and searching both incoherent and coherent (phased) beams.

As well as operating commensally with primary science programs, flexible sub-arraying capabilities would allow some of the antennas to undertake a dedicated SETI search simultaneously with other users of the array (albeit at reduced sensitivity). When operating in a commensal or survey mode, the primary metric of interest is survey speed, which favors an LNSD (large number of small diameter dishes) design.

8. Complementarity

The ability to access baseband data is an important capability not just for SETI, but for fast transient applications such as pulsar and FRB studies (e.g., see Law et al. 2018, this volume p. 773). Baseband products can also be used for spectral line or other studies, even regenerating higher-level data products to custom specifications from archival data long after the observations are taken (assuming these data are stored).

The proposed search will also provide a superb dynamic catalog of RFI, classified using machine learning algorithms, publicly accessible to other science users. To avoid inadvertent flagging of signals of interest to SETI, the ability to bypass any automatic RFI excision built in to the system is important.

By 2025, Breakthrough Listen (Worden et al. 2017) will have completed an unprecedented 10-year SETI survey. Even if Listen, or other surveys such as those that are ongoing at the Allen Telescope Array (Harp et al. 2016), do not confirm the existence of technological civilizations beyond Earth, they will constrain the parameter space in which such civilizations may exist, motivating a push to higher sensitivities and different frequencies. In the event of an ETI detection, or of the detection of biosignatures or other contingent results from *JWST* (Greene et al. 2016), *TESS* (Ricker et al. 2015), or other planet finding and characterization experiments, the ngVLA will provide critical capabilities for follow-up observations.

Acknowledgments. Funding for Breakthrough Listen research is provided by the Breakthrough Prize Foundation.

References

Borucki, W. J., Koch, D., Basri, G., et al. 2010, Science, 327, 977
Cocconi, G., & Morrison, P. 1959, Nat, 184, 844
Drake, F. D. 1961, Physics Today, 14
Dressing, C. D., & Charbonneau, D. 2015, ApJ, 807, 45
Ekers, R., Morrison, P., Group, S. S. . T. W., & Institute, S. 2002, SETI 2020: A Roadmap for the Search for Extraterrestrial Intelligence (SETI Press)
Enriquez, J. E., Siemion, A., Price, D. C., et al. 2017, ApJ, submitted
Enriquez, J. E., Siemion, A., Lazio, T. J. W., et al. 2018, Research Notes of the American Astronomical Society, 2, 9
Freitas, R. A., & Valdes, F. 1980, Icarus, 42, 442
Greene, T. P., Line, M. R., Montero, C., et al. 2016, ApJ, 817, 17
Harp, G. R., Richards, J., Tarter, J. C., et al. 2016, AJ, 152, 181

Isella, A., Hull, C. L. H., Moullet, A., et al. 2015, ArXiv e-prints, arXiv:1510.06444
Kasting, J., Kopparapu, R., Ramirez, R., & Harman, C. 2014, Proceedings of the National Academy of Sciences of the United States of America, 111, 12641, cited By 32
Law, C. J., Bower, G. C., Burke-Spolaor, S., Butler, B. J., Demorest, P., Lazio, T. J. W., & Linford, J. D. 2018, Science with a Next-Generation Very Large Array, ASP Monograph 7, ed. E. J. Murphy (San Francisco, CA: ASP), 773
Morrison, P. 1997, Nothing Is Too Wonderful to Be True, Masters of Modern Physics (American Inst. of Physics)
Ricker, G. R., Winn, J. N., Vanderspek, R., et al. 2015, Journal of Astronomical Telescopes, Instruments, and Systems, 1, 014003
Selina, R. J., Murphy, E. J., McKinnon, M., et al. 2018, in Society of Photo-Optical Instrumentation Engineers (SPIE) Conference Series, Vol. 10700, Society of Photo-Optical Instrumentation Engineers (SPIE) Conference Series, 107001O
Siemion, A. P. V., Demorest, P., Korpela, E., et al. 2013, ApJ, 767, 94
Tarter, J. 2001, ARA&A, 39, 511
Tingay, S. J., Kaplan, D. L., Lenc, E., et al. 2018, ApJ, submitted
Worden, S. P., Drew, J., Siemion, A., et al. 2017, Acta Astronautica, 139, 98

Part V

Stellar Systems and Formation

Science with a Next-Generation Very Large Array
ASP Conference Series, Monograph 7
Eric J. Murphy, ed.

Stellar Activity on Red Giant and Supergiant Stars: Mass Loss and the Evolution of the Stellar Dynamo.

Graham M. Harper

Center for Astrophysics and Space Astronomy, University of Colorado Boulder, CO, USA; graham.harper@colorado.edu

Abstract. In this Chapter we examine the role of the ngVLA to further our understanding of the different manifestations of convective or turbulence-driven stellar activity on red giant and supergiant stars. The combination of high spatial resolution and high sensitivity will enable the ngVLA to significantly improve our understanding of the processes that dissipate energy in the extended atmospheres of cool evolved stars, and drive ubiquitous stellar outflows. The high spatial resolution will enable us to image the surfaces of nearby red supergiants, and to measure the atmospheric extent of red giants. Multi-frequency observations will permit thermal continuum tomography on the largest angular diameter stars, providing key empirical data to test theoretical models. The complementary frequencies and similar spatial resolutions of the ngVLA and ALMA will be a powerful synergy.

1. Context

Cool stars have vigorous sub-photospheric convection zones that lead to turbulent motions in their photospheres, and in the layers above. The interaction of surface magnetic fields (e.g., Auriére et al. 2015) and dynamic ionized plasma leads to complex phenomena including starspots, active regions, chromospheres, and stellar outflows. These manifestations of *Stellar Activity* (Schrijver & Zwaan 2000) are not well understood but have important astrophysical consequences, ranging from the effects of mass loss on galactic structure and chemical evolution, the nature of space-weather for exo-planets orbiting red giants, to the effects of cool supergiant mass-loss history on the interpretation of early-time supernovae spectra (e.g., Dessart et al. 2017). Theoretical models have not been able to reproduce observational signatures of stellar activity, and progress is being led by state-of-the-art observations made at the highest spatial resolution and with the greatest sensitivity.

Thermal radio continuum observations are powerful diagnostics in this context because the atomic cross-sections are very accurately known (Hummer 1988) and the emission source term is the Planck function, which is linear in electron temperature. For example, resolved VLA images provided a ground breaking discovery and new insights into red supergiant (RSG) outflows with the finding by Lim et al. (1998) that Betelgeuse's extended atmosphere is cooler and much less ionized than predicted by leading theoretical and semi-empirical models of the time (see Hartmann & Avrett 1984; Harper et al. 2001).

However, red giants and supergiants are not strong radio emitters and have relatively small angular sizes so that future progress requires greater sensitivity *and* higher spatial resolution. The ngVLA will be able to provide both of these.

2. Mass Loss

Red giants and RSGs return mass that is enriched by nuclear processing back into the interstellar medium, supplying giant molecular clouds that may host the next generation of star formation, and leading to new stars and their exo-planets. The rate at which mass is returned by cool stars is, to a first approximation, proportional to a star's surface area and the inverse cube of the surface escape speed (Holzer et al. 1983). So low surface-gravity cool evolved stars with radii $10-10^3 R_\odot$ currently inject more processed material than their main-sequence progenitors. However, the mechanisms that drive mass-loss for K through mid-M giants and the yellow and RSGs are very poorly understood: They have too little dust or molecules in their outflows for radiation driven winds; they have too little photospheric variability or pulsation to have acoustic and shock driven winds (e.g., Arroyo-Torres et al. 2015), and they are too cool for a Parker-like thermal winds. Some form of magnetic activity, or a combination of processes, is likely to be the cause of these ubiquitous mass outflows (Lamers & Cassinelli 1999). Clearly a quantitative physical model for mass loss is lacking for these stars, and even for the later spectral-type pulsating asymptotic giant branch (AGB) stars the role of magnetic fields in driving and shaping the outflows is not understood.

2.1. What Empirical Data do we Need to Make Progress?

Outflows from cool evolved stars have terminal wind speeds, v_∞, that are typically a small fraction of the surface escape speed, v_{esc}. This means that **most** of the energy that goes into driving the outflows, proportional to $v_{esc}^2 + v_\infty^2$, goes into overcoming the star's gravitational potential, i.e., $\propto v_{esc}^2$. Therefore, *the optimum region to study mass loss mechanisms is within the first few stellar radii where most of the energy goes into the wind and the thermodynamic signatures of the outflow driving mechanisms will be most apparent.*

Nearby red giants have photospheric angular diameters, θ_*, of 10 – 20 mas, and the two nearest RSGs (Betelgeuse: M2 Iab, and Antares: M1.5 Iab) have $\theta_* \simeq 42 - 44$ mas (Ohnaka et al. 2013; Montargès et al. 2016). The luminous Herschel's Garnet Star (μ Cep: M2 Ia) has $\theta_* = 14.1$ mas (Perrin et al. 2005), the next nearest luminosity class Iab-Ib RSG (CE Tau) has $\theta_* \simeq 10$ mas (Cruzalèbes et al. 2013), and α^1 Her (M5 Ib-II) has $\theta_* = 33$ mas (Benson et al. 1993). Importantly, at 50 GHz, the atmospheres of RSGs have angular extents about twice that of their photospheres, and the ngVLA, with baselines of 200 km, would provide a resolution of better than 10 mas. This would match the highest ALMA spatial resolutions, for example, as shown in Figure 1.

This figure shows the 338 GHz (0.89 mm) thermal continuum image of the RSG Betelgeuse obtained with ALMA with the longest (16 km) baselines, providing a spatial resolution of 15 mas (O'Gorman et al. 2017). At this wavelength the atmosphere of the star has an angular diameter of ~ 55 mas, and this image shows the non-uniform heating in the atmosphere that might be related to giant convection cells or regions of enhanced magnetic fields. The thermal cm-continuum opacity $\kappa_\lambda \propto \lambda^{2.1}$, so observations at multiple wavelengths map out the temperature and ion density distribution as a function of radius, providing vital clues to the processes levitating and heating

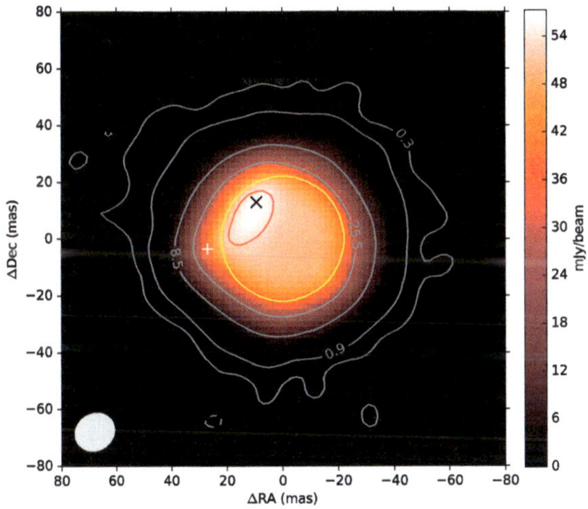

Figure 1. 338 GHz ALMA map of Betelgeuse with a half-power beam width of $\simeq$ 15 mas (shown lower left), from O'Gorman et al. (2017). The yellow circle is the photospheric angular diameter measured with VLTI-PIONIER in H-band ($\simeq 1.6\mu$m) by Montargès et al. (2016). The larger angular extent seen at ALMA frequencies reveals the lower chromosphere and temperature minimum region. In contrast, the frequencies probed by the VLA sample the more extended chromosphere and wind acceleration region. In this image localized excess emission is marked with crosses. The ngVLA would be able to image the extended regions with the same spatial resolution as ALMA, allowing us to build a tomographic map from the upper-photosphere out into the region where the outflow is initiated. Credit: O'Gorman et al., A&A, Vol. 602, p. L11, 2017, reproduced with permission ©ESO.

the stellar outflow. This requires that the ngVLA 200 km baselines support a range of observing frequencies.

3. Dynamos in Cool Evolved Stars

Most studies of solar-type $\alpha\Omega$ dynamos have focused on main-sequence stars that explore a relatively small parameter space of radius and mass, with a wide range of rotation rates (e.g., Brun & Browning 2017). Red giants, however, provide a different challenge for dynamo theories because they typically have low rotation rates but a much greater range of radii and convection zone depths. It is an open question as to whether an $\alpha\Omega$ dynamo is maintained on the red giant branch or whether a turbulent dynamo (e.g., Durney et al. 1993) is responsible for the generation of magnetic fields that power the ultraviolet (UV) and X-ray emission, and drive stellar outflows. Another important application for the long baselines of the ngVLA is to measure the extent of the magnetically heated chromosphere of red giant stars. Spatial resolution breaks the *spatial impasse* that limits what can be learned from disk integrated flux diagnostics alone (e.g., Harper et al. 2013). There are some recent clues that red giants show signs of magnetic cycles (Sennhauser & Berdyugina 2011) and non-uniform optical brightness distributions (Richichi et al. 2018), indications that magnetic phenomena might be important in creating starspots and active regions. High spatial resolution ngVLA observations would reveal the radio size and uniformity of the chromosphere for the nearest red giant stars. This information would help constrain the nature of stellar dynamos, and thus the shape of magnetic fields that drive outflows from red giants. Such radio observations might also help determine whether low-activity red giants, so called basal-flux stars, are heated by acoustic shocks (Pérez Martínez et al. 2014). These would have compact chromospheres, but if they too are heated by magnetic fields (Judge & Carpenter 1998) they may have more extended atmospheres. There are 12 K and M non-Mira red giants with limb-darkened angular diameters >10 mas obtained with the Mark III Stellar Optical Interferometer (Mozurkewich et al. 2003) that would be resolved at 50 GHz with 200 km baselines. These stars are given in Table 1 along with the MK spectral-types from Keenan & McNeil (1989). The atmospheric electron density scale-heights are expected to be small compared to the stellar radii, so that the radio specific intensity distribution is, to first order, a top-hat and thus relatively easy to interpret. A sweep through radio continuum frequencies provides an opportunity to build a tomographic map of the mean atmospheric temperature profile: ngVLA observations would sample the key chromosphere and wind regions, while synergistic ALMA observations would probe the temperature minimum and upper photosphere (see Figures 1 and 2 of O'Gorman et al. 2013).

Table 1. Stellar properties for red giants with θ_* >10 mas (Mozurkewich et al. 2003).

HR	Name	Spectral-Type	Ang. Diam. (mas)	T_{eff} (K)
5340	α Boo	K1.5 III	21.37 ± 0.25	4226 ± 53
5563	β Umi	K4- III	10.30 ± 0.10	3849 ± 47
1457	α Tau	K5+ III	21.10 ± 0.21	3871 ± 48
337	β And	M0+ IIIa	13.75 ± 0.14	3763 ± 46
6056	δ Oph	M0.5 III	10.47 ± 0.12	3721 ± 47
911	α Cet	M1.5 IIIa	13.24 ± 0.26	3578 ± 53
2216	η Gem	M2 IIIa	11.79 ± 0.12	3462 ± 43
8775	β Peg	M2.5 II-III	17.98 ± 0.18	3448 ± 42
2286	μ Gem	M3 IIIab	15.12 ± 0.15	3483 ± 43
4910	δ Vir	M3+ III	10.71 ± 0.11	3602 ± 44
921	ρ Per	M4+ IIIa	16.56 ± 0.17	3281 ± 40
6146	g Her	M6- III	19.09 ± 0.19	3008 ± 37

4. Synergies with other Observatories

Free-free radio continuum diagnostics *directly* relate to the thermodynamic state of the atmospheres, and spatially-resolved multi-wavelength observations provide measures of the temperature and electron (ion) density scale-heights. The radio therefore provides the physical context with which to interpret optical and IR images made a similar high spatial resolutions. The 10 mas spatial resolution obtainable with the ngVLA complements the spectro-interferometry and imaging currently being made with the *Very Large Telescope*, (e.g., Kervella et al. 2016; Montargès et al. 2017).

Another ngVLA synergy is that with UV spectroscopy. UV emission line fluxes and radio optical depths are proportional to the emission measure, i.e.,

$$F_{UV} \propto \int n_e n_H \, dR \, dA \qquad \tau_{radio} \propto \int n_e^2 \, dR$$

where n_e and n_H are the electron and hydrogen densities, respectively. dR and dA are the radial and area integral elements, respectively. Since $n_e \propto n_H$, then

$$F_{UV} \propto \tau_{radio} \, dA.$$

The combination of radio and UV flux data, especially when spatially resolved, i.e., when dA is measured, provides very powerful constraints on inhomogeneous atmospheric structures because of the very different temperature sensitivities of radio and collisionally excited UV line emission.

5. Astrophysical Impact

One of the major goals of this research is to be able to build a quantitative predictive model of atmospheric structure and mass loss based on underlying physical principles. Such a model could be used to improve stellar and galactic chemical evolution calculations, and predict the mass-loss histories of RSG which would in turn help to interpret early-time supernova spectra (Moriya et al. 2017). Today we are very far from such a model. The observational constraints that the ngVLA will provide on red giant and RSG extended atmospheres, where the winds accelerate, would be a major step forward. ngVLA would provide the basic measurements of size, electron density scale-height, and degree of uniformity that any theoretical model must satisfy.

6. Uniqueness of the ngVLA Capabilities

The power of the ngVLA lies in its high-spatial resolution *and* high sensitivity which, when combined, provide the ability to probe the spatial scales of the extended near-star atmospheres of evolved stars. This is the zone where the physics that controls different aspects of stellar activity will reveal itself.

Acknowledgments. GMH thanks CU-CASA for infrastructure resources to support ongoing stellar astrophysics radio research. This research has made use of NASA's Astrophysics Data System Bibliographic Services and the VizieR catalogue access tool, CDS, Strasbourg, France (Ochsenbein et al. 2000).

References

Arroyo-Torres, B., Wittkowski, M., Chiavassa, A., et al. 2015, A&A, 575, 50

Auriére, M., Konstantinova-Antova, R., Charbonnel, C., et al. 2015, A&A, 574, 90

Benson, J. A., Dyck, H. M., Ridgway, S. T., Dixon, D. J., Mason, W. L., & Howell, R. R. 1993, AJ, 105, 736

Brun, A. S. & Browning, M. K. 2017, Living Reviews in Solar Physics, 14, 4

Cruzalèbes, P., Jorrisen, A., Rabbia, Y. et al. 2012, MNRAS, 434, 437

Dessart, L., Hillier, D. J., & Audit, E. 2017, A&A, 605, A83

Durney, B. R., De Young, D. S., & Roxburgh, I. W. 1993, Solar Phys., 145, 207
Harper, G. M., Brown, A., & Lim, J. 2001, ApJ, 551, 1073
Harper, G. M., O'Riain, N., & Ayres, T. R. 2013, MNRAS, 428, 2064
Hartmann, L. & Avrett, E. 1984, ApJ, 284, 238
Holzer, T. E., Flå, T., & Leer, E. 1983, ApJ, 275, 808
Hummer, D. G. 1988, ApJ, 327, 477
Judge, P. G. & Carpenter, K. G. 1998, ApJ, 494, 828
Keenan, P. C. & McNeil, R. C. 1989, ApJS, 71, 245
Kervella, P., Lagadec, E., Montargès, M., et al. 2016, A&A, 585, A28
Lamers, H. J. G. L. M. & Cassinelli, J. 1999, *Introduction to Stellar Winds*, CUP, Cambridge UK
Lim, J., Carili, C. L., White, S. M., Beasley, A. J., & Marson, R. G. 1998, Nat, 392, 575
Moriya, T. J., Yoon, S.-C., Graäfener, G., & Blinnikov, S. I. 2017, MNRAS, 469, L108
Pérez Martínez, M. I., Schröder, K.-P., & Hauschidlt, P. 2014, MNRAS, 445, 270
Montargès, M., Kervella, P., Perrin, G., et al. 2016, A&A, 588, A130
Montargès, M., Chiavassa, A., Kervella, P., Ridway, S. T., Perrin, G., Le Bouquin, J.-B., & Lacour, S. 2017, A&A, 605, 108
Mozurkewich, D., Armstrong, J. T., Hindsley, R. B. et al. 2003, AJ, 126, 2502
Ochsenbein, F., Bauer, P., & Marcout, J. 2000, A&AS, 143, 23
Ohnaka, K., Hofmann, K.-H., Schertl, D., et al. 2013, A&A, 555, 183
O'Gorman, E., Harper, G. M., Brown, A., Drake, S., & Richards, A. M. S. 2013, AJ, 146, 98
O'Gorman, E., Kervella, P., Harper, G. M., Richards, A. M. S., Decin, L., Montargès, M., & McDonald, I. 2017, A&A, 602, L10
Perrin, G., Ridgway, S. T., Verhoelst, T., et al. 2005, A&A, 436, 317
Richichi, A., Sharma, S., Pandey, A. K., Sinha, T., & Norharizan, M. D. 2018, New Astronomy, 59, 28
Schrijver, C. J. & Zwaan, C. 2000, *Solar and Stellar Magnetic Activity*, CUP, New York USA
Sennhauser, C. & Berdyugina, S. V. 2011, A&A, 529, 100

Science with a Next-Generation Very Large Array
ASP Conference Series, Monograph 7
Eric J. Murphy, ed.

ngVLA Studies of Classical Novae

Justin D. Linford,[1,2] Laura Chomiuk,[3] and Michael P. Rupen[4]

[1]*Department of Physics, The George Washington University, Washington, DC 20052, USA;* `jlinford@gwu.edu`

[2]*Astronomy, Physics, and Statistics Institute of Sciences (APSIS), 725 21st St NW, Washington, DC 20052, USA*

[3]*Center for Data Intensive and Time Domain Astronomy, Department of Physics and Astronomy, Michigan State University, 567 Wilson Road, East Lansing, MI 48824, USA;* `chomiuk@pa.msu.edu`

[4]*Herzberg Astronomy and Astrophysics, National Research Council, Penticton, BC, Canada;* `michael.rupen@nrc-cnrc.gc.ca`

Abstract. Observations with modern radio telescopes have revealed that classical novae are far from the simple, spherically symmetric events they were once assumed to be. It is now understood that novae provide excellent laboratories to study several astrophysical properties including binary interactions, stellar outflows, and shock physics. The ngVLA will provide unprecedented opportunities to study these events. It will enable us to observe more distant and fainter novae than we can today. It will allow us to simultaneously resolve both the thermal and non-thermal components in the ejecta. Finally, monitoring novae with the ngVLA will reveal the evolution of the ejecta in better detail than is possible with any current instrument.

1. Introduction

A nova outburst results when sufficient mass accretes onto the surface of a white dwarf from a companion star, triggering a thermonuclear explosion. In classical novae the bulk of the emission comes from the warm, expanding ejecta – 10^{-7} – 10^{-3} $M_\odot$ expanding at a few thousand km/s. The prevailing theories assume that the explosion occurs as a single, spherically symmetric ejection event and predict a simple relationship between the white dwarf mass, the accretion rate, and the ejecta mass and energetics of the explosion (e.g., Yaron et al. 2005). However, observations with modern instruments indicate that nova eruptions are far from simple. There is now evidence for multiple ejection events (e.g., Nelson et al. 2014), common envelopes (e.g., Chomiuk et al. 2014), non-spherical geometry (e.g., Slavin et al. 1995; Linford et al. 2015), and even jet-like structures in the ejecta (e.g., Rupen et al. 2008).

One of the most surprising discoveries of the *Fermi Gamma-ray Space Telescope* was that novae produce GeV γ-ray emission. While MeV emission from novae was predicted (and yet never observed; Hernanz 2013), GeV emission requires a population of accelerated particles that was not expected to be present in nova explosions. The first nova detected by *Fermi*/LAT was V407 Cyg, which had a Mira giant companion

(Abdo et al. 2010). The nova was therefore embedded in the dense wind of its companion star, providing an ideal environment for producing strong shocks when the fast nova ejecta slammed into the slow wind material. However, even novae with main sequence companions are capable of producing detectable γ-rays (e.g., Ackermann et al. 2014; Cheung et al. 2016), which is surprising given the negligible stellar winds leading to low density circumstellar material surrounding these binaries. Recent results from studies at radio wavelengths demonstrate of the importance of shocks in novae, including detection of γ-ray producing shocks in several sources (e.g., Chomiuk et al. 2014; Weston et al. 2016a,b), and the realization that multiple, long-lived outflows are much more common than previously assumed in these explosions (e.g., Nelson et al. 2014). Here we illustrate how the next generation Very Large Array will contribute to the study of classical novae, with an emphasis on characterizing the shocks that develop early in the evolution of these complex explosions.

2. The Case of V959 Mon

The 2012 nova V959 Mon was one of the first classical novae with a main sequence companion detected by *Fermi*. In fact, the nova was discovered by *Fermi* as it was too close to the Sun for optical observations. Chomiuk et al. (2014) were able to directly image the shocked regions with the Very Long Baseline Array (VLBA) and European VLBI Network (EVN). As shown in Figure 1, the synchrotron-emitting compact regions did not align with the bipolar structure of the expanding thermal gas, as imaged by the Karl G. Jansky Very Large Array (VLA). Instead, the combined VLBA, EVN, and VLA observations revealed that the compact components were moving on a path roughly 45° to the VLA structure. High-resolution observations with the VLA separated by more than a year also revealed an apparent 90° flip in the orientation of the bipolar outflow (Figure 2). Chomiuk et al. (2014) concluded that the nova ejecta had two distinct components: a slow flow in the equatorial plane of the binary with a velocity of $\sim$ 480 km s^{-1}, and a faster flow in the polar direction with a velocity of $\sim$ 1200 km s^{-1}. Analysis of high-resolution VLA images of these two outflows led to a distance estimate of 1.4 ± 0.4 kpc (Linford et al. 2015). The interaction between these flows led to the shocks seen in the VLBI observations, and inferred from the γ-ray detections.

In our model (Figure 3), the initial ejection of material from the nova is relatively slow (a few hundred km s^{-1}). This slow ejecta forms a common envelope around the binary and angular momentum is transferred such that a dense torus forms in the orbital plane (Porter et al. 1998). Later, a fast wind develops and is confined by the torus to be ejected perpendicular to the orbital plane. This fast flow dominated the VLA images during the first year of observing. The slow equatorial flow in the torus remained optically thick much longer, and dominated the VLA images at late times. The collision of the fast flow with the slower torus leads to shocks, creating a population of accelerated particles (Chomiuk et al. 2014). In V959 Mon, these accelerated particles were detected via radio synchrotron emission (Chomiuk et al. 2014). There is new evidence that the shocks are also at least partially responsible for the optical emission. Li et al. (2017) found a correlation between the optical and γ-ray light curves for V5856 Sgr that indicated the optical emission could be reprocessed shock emission.

In addition to uncovering the shocks, observations of V959 Mon in both the radio (Linford et al. 2015) and X-rays (Nelson et al. 2019, in preparation) suggest that the initial expansion stalled during the first month after eruption. In the radio, the measured

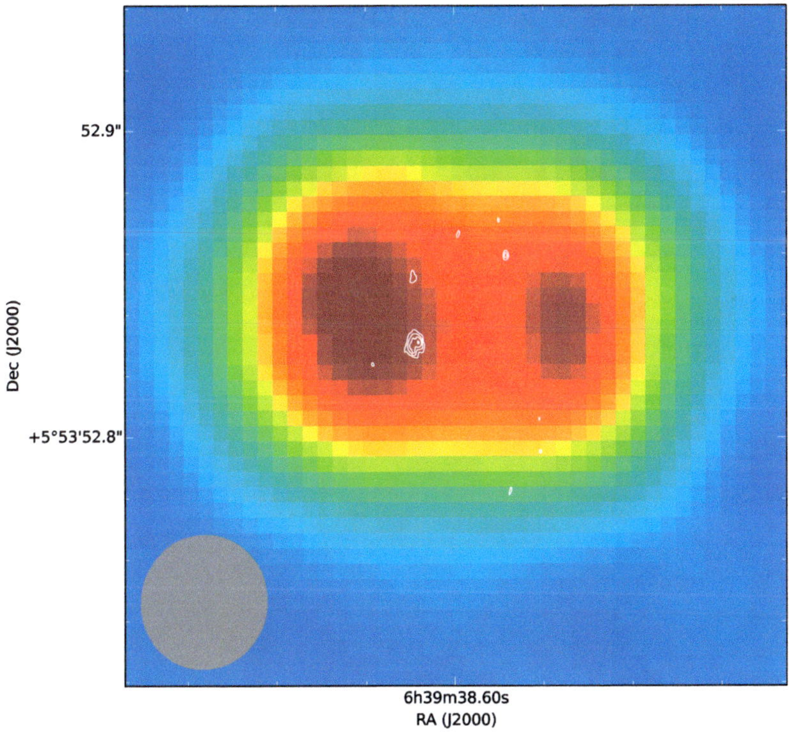

Figure 1. Imaging of the γ-ray detected nova V959 Mon (2012), with compact regions of synchrotron emission (white contours) superimposed on the thermal ejecta (color scale). The synchrotron contours were imaged by the VLBA at 5 GHz 106 days after discovery, while the morphology of the thermal ejecta was revealed by the Jansky VLA at 36.5 GHz, 126 days after discovery. The gray ellipse in the lower left corner represents the 36.5 GHz VLA restoring beam. Adapted from Chomiuk et al. (2014).

angular expansion rate of the bipolar material is inconsistent with expansion starting at the time of eruption. Instead, the expanding material appears to have been ejected ~ 25 days after the initial eruption. In the X-rays, there is an observed drop in column density towards the X-ray emitting region that is best described by a shell of material that was ejected ~ 30 days after the nova was initially detected.

3. Impact of the ngVLA: Sensitivity

The radio emission from classical novae provides information on the amount of mass ejected from the surface of the white dwarf. Because thermal bremsstrahlung dominates the radio emission from the ejecta, radio observations trace the bulk of the ejecta more simply and more accurately than observations at other wavelengths (Seaquist & Bode 2008). For the first 100–200 days after eruption, the ejecta are optically thick at radio frequencies and the flux density is proportional to the electron temperature and the

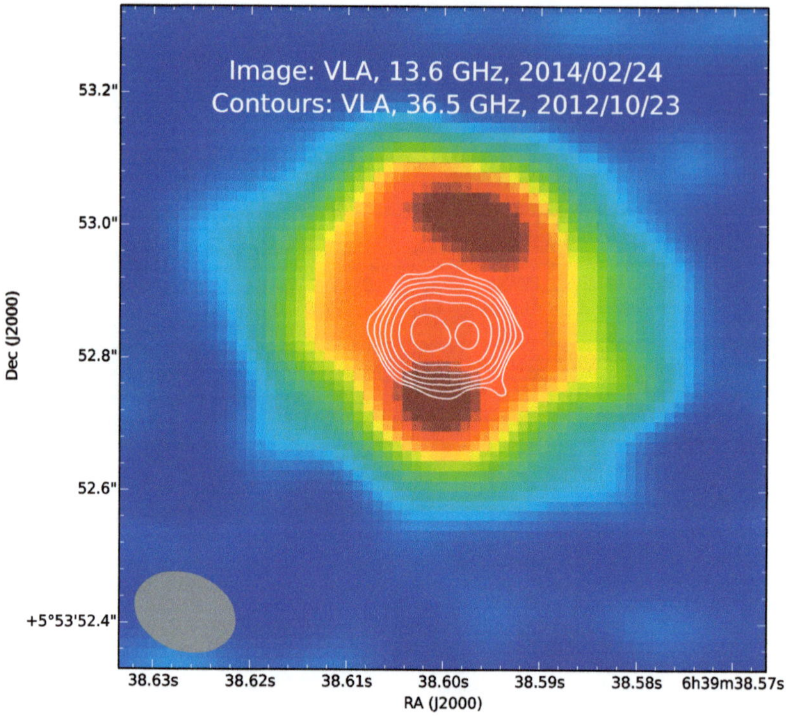

Figure 2. The evolution of V959 Mon over the two years following nova explosion, as imaged by the Jansky VLA which is sensitive to the thermal ejecta. The plane of the binary system is oriented in the N-S (up-down) direction. **White Contours** show the 36.5 GHz image of V959 Mon 126 days after discovery, which is dominated by a fast bipolar outflow perpendicular to the orbital plane. **Color** shows the 13.6 GHz image of V959 Mon 615 days after discovery, which is dominated by the slowly expanding torus oriented along the orbital plane. The Gray ellipse in the lower left represents the 13.6 GHz VLA restoring beam. Adapted from Chomiuk et al. (2014).

projected area. This means that the sample of radio-detected novae is biased, since novae with small ejecta masses can only be detected if they are nearby (i.e., within a few kpc). The increased sensitivity of the ngVLA will remove this bias by allowing the study of the full range of novae, with observations of even low-luminosity sources out to the Galactic Center.

The recent discovery of γ-ray emission from classical novae renewed interest in understanding the causes and physics of shocks in classical novae. One of the best tracers of shocks is synchrotron emission, which results in both steeply-falling radio spectra and high radio brightness temperatures ($\gtrsim 10^5$ K). Spectral index measurements with the VLA have been moderately successful in identifying synchrotron emission early in a nova's evolution (e.g., Linford et al. 2017). Another, more subtle indicator of shocks is the presence of strong radio emission at early times. The source size may be inferred from its age and optical expansion velocities; if the radio flux density corresponds to brightness temperatures above 10^5 K, there must be at least some contribution from synchrotron emission (e.g., Weston et al. 2016a,b). This analysis

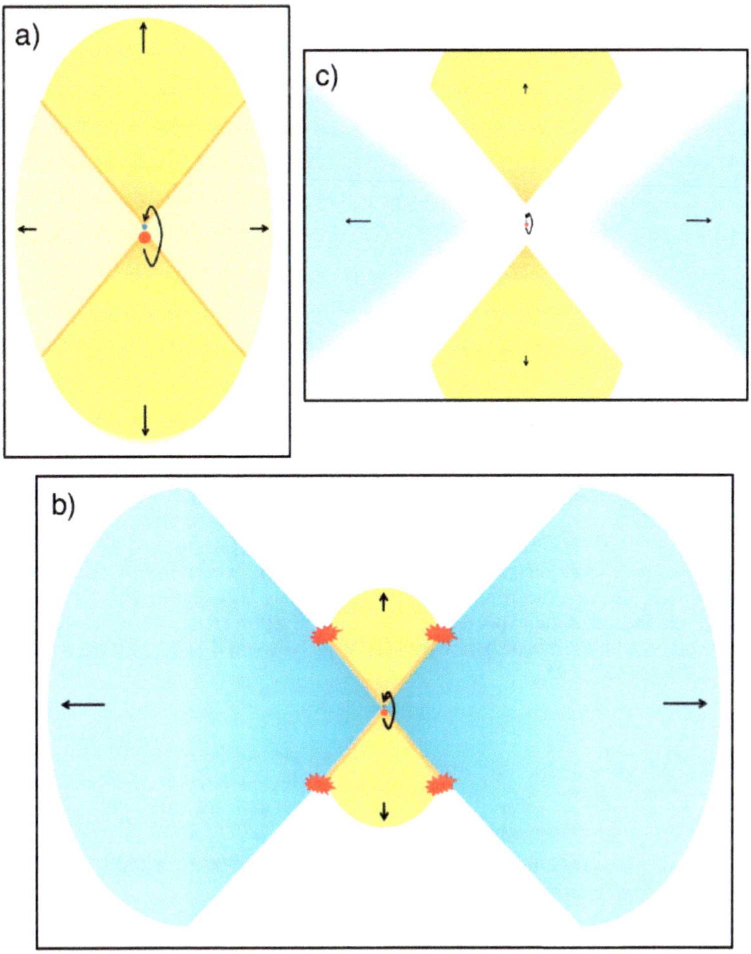

Figure 3. A cartoon picture to explain the evolution of V959 Mon. a) At early times, the ejecta interacts with the binary and a dense torus is formed (darker yellow, oriented vertically). b) Later, a fast wind (blue) shocks against the torus leading to synchrotron and γ-ray emission (red regions). This phase corresponds to the image of V959 Mon in Figure 1. c) At late times, the wind material is too diffuse to detect and the dense torus dominates the radio emission. Adapted from Chomiuk et al. (2014).

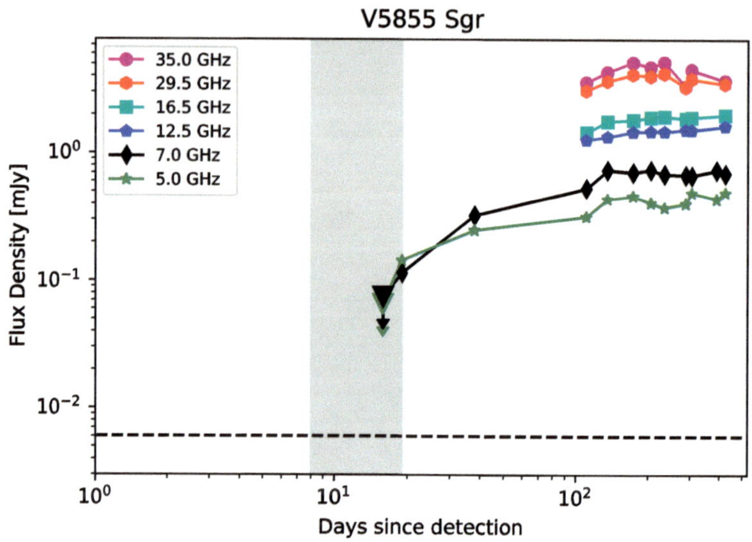

Figure 4. The early radio light curve of V5855 Sgr. Note the sudden transition from non-detection to detection at C-band (4 – 8 GHz). Also, note that for the first detection, the 5 GHz flux density is higher than the 7 GHz flux density, which is opposite of what is expected for a source emitting via thermal bremsstrahlung. The gray region indicates the time range while the nova was detected by *Fermi*. The dashed line indicates the expected 3σ rms noise level of ~6 μJy/beam for 10 minute observations with 2 GHz of bandwidth centered between 4 and 12 GHz on the next generation VLA.

rests on deep low-frequency (<7 GHz) observations early in the outburst, which have been rather sparse. Even where there are such observations, one cannot tell how long the synchrotron (shock) emission persists, without more direct indicators, since the increasing size of the remnant makes lower brightness limits from the radio flux density less constraining as the source expands.

The 2016 eruption of nova V5855 Sgr (TCP J18102829-2929590; CBET 4332) provided an intriguing test case. The first observation was a non-detection, while observations three days later showed strong emission. Moreover, the spectral index in the second epoch was characteristic of optically thin synchrotron emission, with $S_\nu \propto \nu^{-0.7}$. This shows that early synchrotron emission can appear quite suddenly. By the third observation, 19 days after the second epoch, the nova appeared to be emitting via optically thick thermal bremsstrahlung and no clear evidence for synchrotron emission remained. Due to the sparse sampling of the light curve, it is not known how long the synchrotron emission persisted or if it showed any correlation with the γ-ray or X-ray emission.

With its high sensitivity, the ngVLA will be capable of uncovering synchrotron emission from less luminous and more distant novae. Shorter snapshot observations could be used to monitor novae at a high cadence (perhaps even daily) to look for the emergence and disappearance of synchrotron emission early in their evolution. This

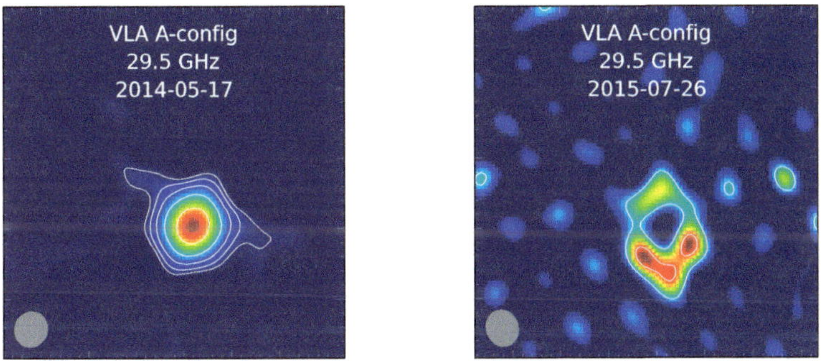

Figure 5. (*Left:*) VLA A configuration image of V339 Del taken 2014-05-17. (*Right:*) VLA A configuration image of V339 Del taken 2015-07-26. Note the appearance of a circular structure in the later image. Due to the VLA configuration schedule, there was no way of resolving the the ejecta in between the A configurations. Over a year's worth of evolution must therefore be inferred from 2 snapshots. The next generation VLA, with high angular resolution year-round, will for the first time allow imaging of the *entire* evolution of a nova.

will provide detailed information about the development and evolution of shocks in the nova ejecta, and allow direct comparison with high-cadence observations with X-ray and γ-ray instruments.

4. Impact of the ngVLA: Imaging

High resolution ngVLA images will allow us to answer fundamental questions about novae. With a diversity of baseline lengths extending to 1000 km, every ngVLA observation of a nova has the potential to reveal structure like that imaged in Figure 1 – simultaneously tracing both $\sim 10^4$ K thermal ejecta and compact synchrotron structure.

What are the shapes of the expanding ejecta and how do they evolve?

It is still unclear whether nova ejecta are dominated by a spherical shell, a ring, or a bipolar structure (e.g., Slavin et al. 1995; Sokoloski et al. 2013). It is also unclear if nova explosions share a common ejecta morphology and shaping physics, or if their structures are shaped by a diversity of processes. ngVLA imaging will enable the search for evidence of a dense equatorial torus which could lead to the bipolar structure seen in V959 Mon (e.g., Ribeiro et al. 2009). It is unknown exactly when this torus forms (or if it is already present prior to the eruption), so early images with higher angular resolution than the VLA allows are vital. Understanding the evolving shape of a nova remnant is essential to interpreting observations at all wavelengths, from radio to γ-rays, and to testing the two-flow model.

The transition from the radio emission being dominated by the fast wind to being dominated by the slowly expanding torus has yet to be definitively captured, although Healy et al. (2017) reported a possible example in eMERLIN observations of V959 Mon. In many novae the observations constrain only one of these phases. In the Fermi-detected nova V339 Del, for example, there is evidence for the torus at late times, but no strong imaging evidence for the faster bipolar outflow (see Figure 5). This is due in large part to the VLA only being in A configuration for roughly one quarter of the time. Even in A configuration, the VLA only has sufficient angular resolution to resolve the ejecta in the nearest novae. Having high sensitivity and high angular resolution of the ngVLA for every observation will capture the evolution from one phase to the other in detail, in a single source, and show directly whether our hypothesized ejecta morphology is universal.

Does the ejecta begin to expand immediately after eruption, or does the initial expansion stall?

As discussed in Section 2, V959 Mon showed evidence for a stall in the expansion of the ejecta during the first month after eruption. With current radio instruments, it is only possible to detect such a stall phase several months after the eruption and only for relatively nearby novae (within $\sim$ 4 kpc). The ngVLA will have an order-of-magnitude better resolution and higher sensitivity than the VLA, and the resulting imaging should show both whether this such stalls are universal, and when and how the expansion resumes. The exact timing of this apparent stall can differentiate between a common envelope and a confinement phase (in which the nova ejecta are inhibited by external material present prior to the eruption).

What is the energy budget of novae?

Combining ngVLA images of the expanding ejecta with velocity measurements from optical spectroscopy will allow for the application of expansion parallax techniques similar to those used for V959 Mon (Linford et al. 2015). Knowing the distances to the novae will help determine whether only nearby novae are detected by *Fermi* (e.g., Morris et al. 2017; Franckowiak et al. 2018), or if there is a large range in their γ-ray luminosities. The γ-ray luminosities are also extremely valuable for understanding the energetics of these sources and determining whether the γ-ray production mechanism is leptonic or hadronic (e.g., Metzger et al. 2015; Vlasov et al. 2016).

Where is the early synchrotron radiation?

In addition to the expanding thermal ejecta, there has been evidence for early synchrotron emission in novae (e.g., Taylor et al. 1982; Weston et al. 2016a,b; Linford et al. 2017). Presumably, this early non-thermal emission is related to the shocks that produce the γ-ray emission, but the relation is not clear (e.g., Vlasov et al. 2016). For V959 Mon, For V959 Mon, observing both thermal and non-thermal components required a complicated campaign with multiple telescopes (VLA, EVN, VLBA, & eMERLIN). The higher sensitivity and longer baselines of the ngVLA will enable detection of synchrotron emission with higher confidence and earlier in the evolution of the nova than current capabilities allow. The wide frequency range of the ngVLA will also allowing both the thermal and non-thermal components, simultaneously, and with a single telescope!

5. Conclusions

The ngVLA will be an amazing instrument for studying classical novae. The increased sensitivity will enable detection of dimmer and more distant novae. Fast wideband monitoring will allow probes of synchrotron emission from shocks at early times. The large range in baselines will enable simultaneous imaging of both the diffuse thermal ejecta and the regions of shock emission. Finally, with high angular resolution at all times, every radio light curve data point will also be a frame in a movie tracing the evolution of the ejecta from shortly after the explosion to its final dispersal into the interstellar medium. These movies will provide unprecedented opportunities to study these dynamic and enigmatic sources.

Acknowledgments. We are grateful for the pioneering work of R. M. Hjellming using the original VLA.

References

Abdo, A. A., Ackermann, M., Ajello, M., et al. 2010, Science, 329, 817
Ackermann, M., Ajello, M., Albert, A., et al. 2014, Science, 345, 554
Cheung, C. C., Jeanm P., Shore, S. N., et al. 2016, ApJ, 761, 173
Chomiuk, L., Krauss, M. I., Rupen, M. P., et al. 2012, ApJ, 761, 173
Chomiuk, L., Linford, J. D., Yang, J., et al. 2014, Nature, 514, 339
Franckowiak, A., Jean, P., Wood, M., Cheung, C. C., & Buson, S. 2018, A&A, 609, 120
Healy, F., O'Brien, T. J., Breswick, R., Avison, A., & Argo, M. K. 2017, MNRAS, 469, 3976
Hernanz, M. 2013, arXiv:1301.1660
Li, K.-L., Metzger, D. D., Chomiuk, L., et al. 2017, Nature Astronomy, 1, 697
Linford, J. D., Ribeiro, V. A. R. M., Chomiuk, L., et al. 2015, ApJ, 805, 136
Linford, J. D., Chomiuk, L., Nelson, T., et al. 2017, ApJ, 842, 73
Metzger, B. D., Finzell, T., Vurn, I., Hascoët, R., Beloborodov, A. M., & Chomiuk, L. 2015, MNRAS, 450, 2739
Morris, P. J., Cotter, G., Browm, A. M., & Chadwick, P. M. 2017, MNRAS, 465, 1218
Mukai, K. 2017, PASP, 129, 2001
Nelson, T., Chomiuk, L., Roy, N., et al. 2014, ApJ, 785, 78
Porter, J. M., O'Brien, T, J., &Bode, M. F. 1998, MNRAS, 296, 943
Ribeiro, V. A. R. M, Bode, M. F., Darnley, M. J., et al. 2009, ApJ, 703, 1955
Rupen, M. P., Mioduszewski, A. J., & Sokoloski, J. L. 2008, ApJ, 688, 559
Seaquist, E. R., & Bode, M. F. 2008 in Classical Novae, 2nd Edition. Cambridge Astrophysics Series, No. 43, Cambridge: Cambridge University Press, Ed. M. F. Bode & A. Evans, 141
Slavin, A. J., O'Brien, T. J., & Dunlop, J. S. 1995, MNRAS, 276, 353
Sokoloski, J. L., Rupen, M. P., & Mioduszewski, A. J. 2008, ApJ, 685, L137
Sokoloski, J. L., Cortts, A. P. S., Lawrence, S., & Uthas, H. 2013, ApJ, 770, L33
Sokoloski, J. L., Lawrence, S., Crotts, A. P. S., & Mukai, K. 2017, APCS2016, arXiv:1702.05898
Taylor, A. R., Seaquist, E. R., Hollis, J. M., & Pottasch, S. R. 1987, A&A, 183, 38
Vlasov, A., Vurm, I., & Metzger, B. D. 2016, MNRAS, 463, 394
Weston, J. H. S., Sokoloski, J. L., Metzger, B. D., et al. 2016a, MNRAS, 457, 887
Weston, J. H. S., Sokoloski, J. L., Chomiuk, L., et al. 2016b, MNRAS, 460, 2687
Yaron, O., Prialnik, D., Shara, M. M., & Kovetz, A. 2005, ApJ, 623, 398

Science with a Next-Generation Very Large Array
ASP Conference Series, Monograph 7
Eric J. Murphy, ed.

Evolved Stars

Lynn D. Matthews[1] and Mark J Claussen[2]

[1]*Massachusetts Institute of Technology Haystack Observatory, Westford, MA, 01886 USA;* lmatthew@mit.edu

[2]*National Radio Astronomy Observatory, P.O. Box O, Socorro, NM, 87801 USA;* mclaussen@nrao.edu

Abstract. This chapter reviews some of the expected contributions of the ngVLA to the understanding of the late evolutionary stages of low-to-intermediate mass stars, including asymptotic giant branch (AGB) stars, post-AGB stars, and pre-planetary nebulae. Such objects represent the ultimate fate of stars like the Sun, and the stellar matter they lose to their immediate vicinity contributes significantly to the chemical enrichment of galaxies. Topics addressed in this chapter include continuum imaging of radio photospheres, studies of circumstellar envelopes in both thermal and nonthermal lines, and the investigation of the transition stages from the AGB to planetary nebulae using radio wavelength diagnostics. The authors gratefully acknowledge contributions to the content of this chapter from members of the evolved star community.

1. Introduction

Evolution onto the asymptotic giant branch (AGB) marks the final thermonuclear burning stage in the lives of low-to-intermediate mass stars ($0.8 \lesssim M_* \lesssim 8\ M_\odot$). The AGB is characterized by a dramatic increase in luminosity (factors of 10^4 or more) and significant mass-loss ($\dot{M} \sim 10^{-8}$ to $10^{-4} M_\odot\ \mathrm{yr}^{-1}$) through cool, low-velocity ($\sim$10 km s^{-1}) winds. Over the course of roughly one million years, this mass loss leads to the formation of extensive circumstellar envelopes (CSEs) of gas and dust that may span more than a parsec in size.

Knowledge of the mass-loss histories of AGB stars is fundamental to understanding the complete life-cycle of our Galaxy. The winds of AGB stars are rich in heavy elements and molecules, sowing the seeds for future generations of star and planet formation. Indeed, mass loss during the AGB is believed to be responsible for ~50% of the chemical enrichment in the Galaxy (e.g., Van Eck et al. 2001). However, many aspects of the lives of AGB stars remain poorly understood, including the wind driving mechanisms, the geometry of AGB mass loss, and the evolutionary paths for stars of different initial masses (see Höfner & Olofsson 2018).

Radio observations play a crucial role in understanding stellar evolution on the AGB and beyond. Continuum radiation is emitted from the optically thick radio photospheres of AGB stars, as well as from the jets and torii that may appear during subsequent evolution to the post-AGB and pre-planetary nebula (pPN) stages. The cool, extended atmospheres and CSEs of evolved stars also give rise to numerous spectral lines at cm and mm wavelengths, offering powerful diagnostics of the chemistry, temperature, and density structure of the CSE, as well as wind outflow speeds and other kinematic information. Studies of radio lines supply crucial insight into the stellar mass-loss history and probe how evolved, mass-losing stars interact with their interstellar environments and shape the composition and small-scale structure of the ISM. The specific properties of CSEs gleaned from radio observations also supply insight into their role in the subsequent evolution into planetary nebulae (PNe) and Type Ia supernovae.

The ngVLA will provide enormous gains in our understanding of the end stages of the lives of Sun-like stars. Its exquisite sensitivity will permit imaging observations of large numbers of spectral lines with high angular and spectral resolution, yielding the most detailed and comprehensive maps to date of CSEs from the smallest to the largest scales. The ultrawide bandwidths will make it possible to observe simultaneously both lines and the (relatively weak) stellar continuum emission, providing precise astrometric registration required for the testing of sophisticated radiative transfer and hydrodynamical modeling of AGB star atmospheres and envelopes.

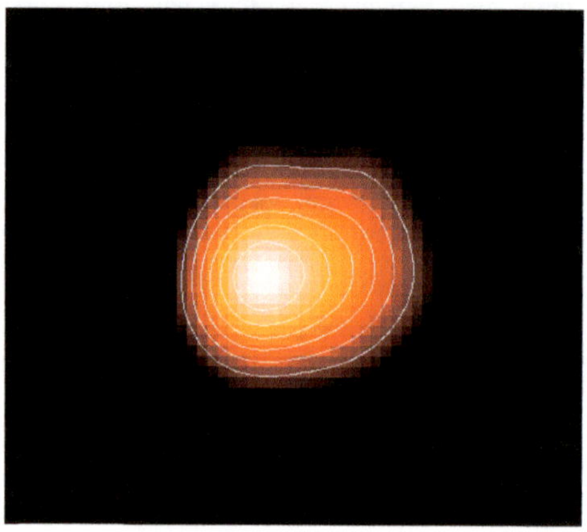

Figure 1. The radio photosphere of R Leo imaged with the VLA at 7 mm (Matthews et al. 2018). The star shows a non-spherical shape and evidence for a non-uniform surface. The image was produced using a sparse model reconstruction algorithm, enabling super-resolution of ~0.75 times the dirty beam FWHM of ~38 mas. The peak flux density is ~4 mJy beam^{-1} and the image is ~150 mas on a side. The ngVLA will enable comparable resolution on AGB stars at distances of up to 1 kpc as well as the resolution of finer surface details on nearby stars.

2. Continuum Observations of Radio Photospheres

AGB stars emit optically thick continuum radiation at cm and (sub)mm wavelengths from a region at $\sim 2R_\star$ known as the "radio photosphere" (Reid & Menten 1997; see also Harper 2018, this volume p. 265). In a typical oxygen-rich AGB star, the radio photosphere has a diameter of ~4 AU and lies near the outskirts of the so-called molecular layer or "MOLsphere" (Tsuji 2000, 2001), but interior to the dust-formation zone and maser-emitting regions (see Section 3.2). Thus the radio photosphere samples an atmospheric region impacted by pulsation, convection, shocks, and other key processes responsible for the transport of material from the stellar surface to the outflowing wind.

Presently it is possible to resolve the radio surfaces of the nearest AGB stars ($d \lesssim 150$ pc) at $\lambda \lesssim 1$ cm using the VLA and ALMA in their most extended configurations. Such observations have led to tantalizing evidence for non-uniform radio surfaces and non-spherical photospheric shapes that evolve over time, most likely as a result of pulsation and convective effects (Reid & Menten 2007; Matthews et al. 2015b, 2018; Vlemmings et al. 2017; Figure 1). The ngVLA is expected to provide an angular resolution of ~26 mas at 8 GHz and ~2 mas at 93 GHz[1]; thus at shorter wavelengths, the ngVLA will be able to resolve the surfaces of AGB stars to distances beyond a kpc, making it possible to assess the ubiquity of spots and other surface features and measure their brightness temperatures and temporal evolution. For nearby stars ($d \lesssim 150$ pc), it will be possible to resolve radio photospheres over nearly two decades in frequency, effectively providing "tomography" of the atmosphere (see Lim et al. 1998; O'Gorman et al. 2015; Matthews et al. 2015b). The ngVLA will be complementary to ALMA for such work, as cm bands probe higher layers of the atmosphere than can be studied with ALMA and also suffer negligible contamination from line emission and dust. The further addition of contemporaneous imaging with optical/infrared interferometers such as the Center for High Angular Resolution Astronomy (CHARA) Array and Very Large Telescope Interferometer (VLTI) would enable the most stringent tests to date of sophisticated 3D models of AGB star atmospheres (e.g., Freytag et al. 2017).

Sensitivities of $\sim 0.4\,\mu$Jy beam^{-1} in one hour at $\lambda \approx 7$ mm will also permit the 5σ detection of unresolved radio photospheres to distances of at least 5 kpc. Over the course of roughly a year, the radio light curve observations of Reid & Menten (1997) found levels of variability in a small sample of AGB stars at 8 GHz to be $\lesssim 15\%$, implying that shock speeds within the radio photosphere are limited to $\lesssim 5$ km s^{-1}. But at 7 mm, evidence is now seen for larger flux density changes in data taken several years apart (Matthews et al. 2015b, 2018). The ngVLA will enable for the first time studies of the variability of radio photospheres over both short (intra-pulsation cycle) and long (several years) timescales for a statistical samples of stars. Such data will be extremely powerful for understanding the roles of shocks in governing the physics of AGB star atmospheres, as well as for gathering for the first time statistics on the frequency of possible eruptive or flaring activity in AGB stars. Evidence for the latter phenomena has already been seen at other wavelengths, including the ultraviolet (Sahai et al. 2011b, 2015) and in X-rays (Karovska et al. 2005). Such events may be triggered by accretion activity linked with a companion, chromospheric-like activity, or other still unknown mechanisms.

3. Spectral Line Studies of the CSEs of AGB Stars

3.1. Molecular Lines (Thermal)

The rich chemistry and cool temperatures of the extended atmospheres and CSEs of AGB stars give rise to emission from a multitude of molecular lines at cm and mm wavelengths, predominantly through rotational transitions (e.g., Turner & Ziurys 1988).

---

[1]Throughout this chapter we assume a maximum baseline length of ~300 km (Selina & Murphy 2017).

Together these various line transitions provide important diagnostic information about the range of temperature and density of the CSE and its kinematics. However, with the exception of masing lines (see Section 3.2), the bulk of these lines are quite weak, requiring excellent sensitivity for their identification and mapping, particularly at the high angular resolutions needed to spatially resolve their distribution and kinematics within the CSE.

3.1.1. Astrochemistry and Searches for New Circumstellar Molecules

The leap in sensitivity of the ngVLA is expected to lead to the discovery of numerous additional molecular species in the envelopes of AGB stars. These are likely to include pre-biotic molecules, carbon chains, and other complex molecules suspected of having their origin in AGB star winds (e.g., Ziurys 2006). The frequency range covered by the ngVLA will have advantages over ALMA for searching for many complex molecules in CSEs, both because heavier molecules have lower frequency rotational transitions, and because line blending below ~100 GHz is much less severe than at higher frequencies. Even for commonly observed molecular transitions such as the CO $J = 1 - 0$ line, the ngVLA will provide important new statistics by enabling detection of late-type stars to distances of up to ~10 kpc, including within the Galactic Center Region.

3.1.2. Spiral Patterns in the Mass Ejecta of AGB Stars

Binary motion can generate spiral patterns in the CSEs of AGB stars, and modeling of the spiral pattern can be used to set constraints on orbital parameters (e.g., period, eccentricity, and inclination; see Mastrodemos & Morris 1999; Kim & Taam 2012a, b, c). Interferometric molecular-line imaging of CO lines has detected such patterns in a small number of objects (e.g., Maercker et al. 2012, Kim et al. 2017). For the carbon star CIT 6, the spiral pattern was found via VLA imaging of the HC_3N(4-3) at 36.3 GHz (Claussen et al. 2011). However, the unprecedented sensitivity of ngVLA will enable searches for such patterns in a statistical sample of AGB stars.

3.2. Non-Thermal Lines (Masers)

Molecular maser emission is frequently detected in the CSEs of AGB stars (and in young PNe; see below). The most widely observed masers in circumstellar environments include transitions of OH (1.6 GHz), H_2O (22 GHz), SiO v=0,1,2,3, J=1−0 (43 GHz), and SiO v=1, J=2−1 (86 GHz) (e.g., Alcolea 2004). These various transitions require different combinations of temperature and density to excite, hence observations of multiple transitions within a single star can be used to probe material from different regions in the CSE, ranging from a few AU from the star for SiO masers to r ~1000 AU for OH lines.

The strongest circumstellar masers show brightness temperatures of many millions of K, and the intensity and high excitation of some of these lines have enabled exquisitely high resolution mapping of CSEs using Very Long Baseline Interferometry (VLBI; e.g., Cotton et al. 2004; Gonidakis et al. 2013; Desmurs et al. 2014; Figure 2). However, there are various limitations of VLBI studies, including losses of up to ~50% of the flux density (e.g., Desmurs et al. 2017) and the difficulty of phase referencing (and thus, astrometry) at frequencies above ~20 GHz (Beasley & Conway 1995). The ngVLA will be able to systematically observe circumstellar masers with excellent angular resolutions (~10 mas at 22 GHz). But in contrast to current VLBI arrays, it will have much higher sensitivity, negligible flux loss, and excellent instantaneous (u, v)-coverage. The high signal-to-noise ratios (SNR) that will be achieved will additionally enable highly accurate positioning of maser lines (in principle, the achievable astrometric precision is approximately the angular resolution divided by the SNR). And crucially, the ultra-wide bandwidths of the ngVLA will permit the routine simultaneous detection of multiple maser transitions, along with the stellar photosphere in the continuum, allowing placement of the distributions of the different masers in the context of the inner circumstellar layers. In nearby stars, it will also be possible to study in detail the movements of maser spots and the relation between the structure of different maser species during the course of the stellar pulsation. These types of data will be vital for identifying the maser pumping mechanisms and providing quantitative constraints on models of AGB star atmospheres (e.g., Gray et al. 2009, 2016).

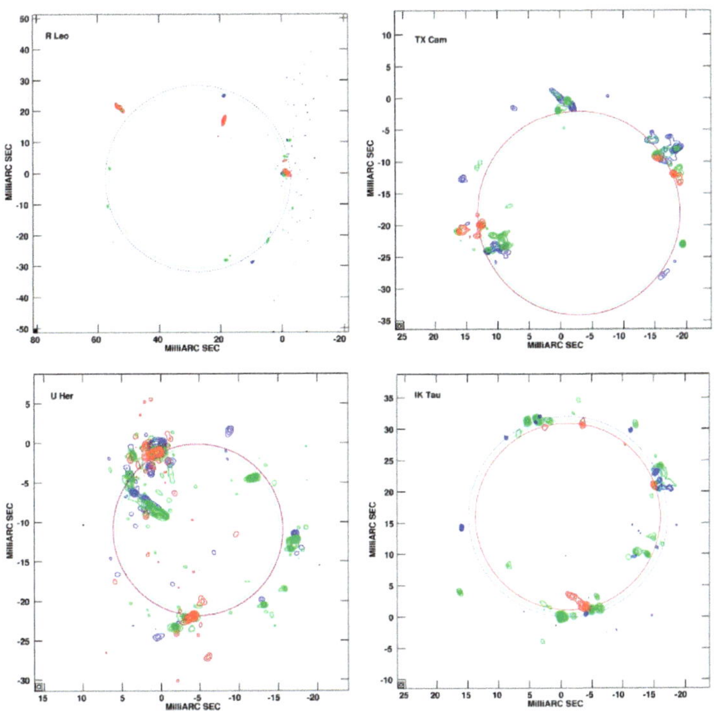

Figure 2. Very Long Baseline Array (VLBA) maps of SiO $v = 1$ (in blue), v=2 (in green), and v=3 (in red), J=1−0 maser lines from R Leo, TX Cam, U Her, and IK Tau (Desmurs et al. 2014). The relative positions of the maser spots of the different lines should provide clues to understanding the SiO pumping, but presently the relative astrometry is very uncertain.

The ngVLA performance will also enable systematic high-resolution mapping of relatively weak maser lines ($T_b \lesssim 10^4$ K), whose properties are not yet well understood and for which very little observational information is currently available. Examples include the SiO v=0 masers (useful to study intermediate shells of CSEs; Bolboltz & Claussen 2004); ^{29}SiO v=0,1, ^{28}SiO v=2, J=2−1, and ^{28}SiO v=3 masers (which are important for the study of SiO maser pumping; Soria-Ruiz et al. 2005, Desmurs et al. 2014), and HCN J=1−0 masers at 88.6 GHz (which may be the best way to probe the very inner regions in C-rich CSEs; Lucas et al. 1988, Menten et al. 2018).

The study of circumstellar OH masers will benefit from the ngVLA's excellent angular resolution even at longer wavelengths (18 cm), along with a factor of several gain in line sensitivity over the present VLA. This will make OH masers around late-type stars accessible throughout the Galaxy and enable new constraints on brightness temperature and maser gain models. In addition to surveys, follow-up imaging studies with a combination of high sensitivity, high angular resolution, high spectral resolution ($\lesssim$0.07 km s^{-1}), and full Stokes parameters will enable separation of individual maser components and detection of the often present signatures of strong linear or circular polarization that are otherwise washed out by insufficient spatial and velocity resolution (e.g., Wolak et al. 2012, 2013).

A special category of OH maser sources is the OH/IR stars, a subclass of evolved stars believed to be in the final stages of their evolution on the AGB. Their dusty winds and exceptionally high mass-loss rates (up to a few times 10^{-4} $M_\odot$ yr^{-1}) render them optically thick at visible wavelengths, and they generally emit a strong characteristic "doubled-peaked" 1612 MHz OH maser emission profile (e.g., Engels 1985). To test the predictions of evolutionary models, in particular the role of metallicity, observational properties of OH/IR stars during their obscured phase *and* reliable distances are needed. For OH/IR stars, the latter can be derived via the so-called "phase-lag" method (Engels et al. 2015; Etoka et al. 2017), using a combination of: (i) the linear diameter of the shell as obtained from the light curve of the varying OH maser peaks; and (ii) an angular diameter obtained from an interferometric map of the OH emission. In order to get a precise angular diameter determination it is vital to map the faint "inter-peak" emission from which the full extent of the shell and its geometry can be inferred. While challenging with current instruments, the enhanced sensitivity and resolution of the ngVLA will facilitate such measurements for larger samples of OH/IR stars. For the nearest stars ($d \lesssim$200 pc) it should become possible to do direct comparisons of such distances with OH maser parallax measurements.

3.3. The H I 21-cm Line

The CSEs surrounding AGB stars can become enormously extended, in some cases spanning more than a parsec (e.g., Villaver et al. 2012). As densities drop with increasing distance from the star, molecules, including CO, become dissociated by the interstellar radiation field. Molecular lines therefore cannot probe the outer regions of the CSE. However, neutral atomic hydrogen (H I) persists in this environment, and recent studies have demonstrated that its 21 cm line emission provides a powerful probe of outer CSE (representing $\gtrsim 10^5$ yr of mass loss history), as well as the interfaces between AGB stars and their environments (e.g., Figure 3). While FIR and/or far-ultraviolet emission can sometimes supply complementary information on the CSE outskirts, only H I studies supply *kinematic* information on the extended gas (e.g., Matthews et al. 2008).

Surveys carried out with the Nançay Radio Telescope have now identified dozens of candidate H I shells around nearby AGB stars (Gérard & Le Bertre 2006; Gérard et al. 2011 and in prep.). However, follow-up interferometric observations are critical for studying the detailed spatial and kinematic structure of the H I envelopes and for disentangling possible contamination from line-of-sight emission. Only a handful of

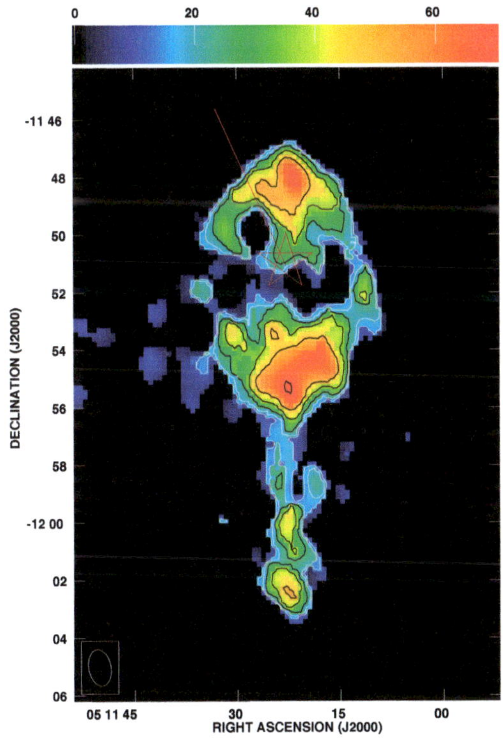

Figure 3. VLA H I total intensity image of the AGB star RX Lep (Matthews et al. 2013). The intensity scale units are Jy beam^{-1} m s^{-1}. A star symbol denotes the stellar position and the direction of space motion is shown by a red line. The projected extent of the H I shell and tail in the north-south direction is ~0.7 pc.

AGB stars have been imaged in the H I line with the VLA to date, owing to the challenging and time-consuming nature of these observations (e.g., Matthews & Reid 2007; Matthews et al. 2008, 2013). However, the ngVLA will provide substantial gains in sensitivity, survey speed, and in the mitigation of background confusion for circumstellar H I observations.

The quasi-random distribution of the individual antennas in the ngVLA core will considerably improve the (u, v)-coverage compared with the existing VLA and reduce the periodic sidelobe effects caused by regularly spaced antennas, thereby helping to mitigate the problems of contamination from interstellar H I emission along the line-of-sight. (cf. Matthews et al. 2011, 2013). The addition of a large ($\gtrsim$45 m) single-dish antenna to the array core (Frayer 2017) would also significantly enhance the ability to subtract off line-of-sight emission and to image extremely low surface brightness H I emission from H I shells, tails, and other structures spanning tens of arcminutes (see e.g., Matthews et al. 2013, 2015a).

Even in the nearest AGB stars, circumstellar H I emission is typically weak with narrow linewidths ($\sim$10 km s^{-1}), requiring good spectral resolution ($\lesssim$1 km s^{-1}) to measure velocity gradients and other kinematic features (e.g., Matthews et al. 2008, 2011, 2013). For sightlines or velocity ranges that are not confusion limited, the sensitivity gained by the factor of $\sim$4 increase in collecting area of the ngVLA will enable expansion of the study of circumstellar H I to significantly larger samples of stars, Finally, the ngVLA core configuration is expected to have baselines up to 1.55 km, which will provide $\sim$50% higher spatial resolution than the current VLA D configuration for mapping the complex kinematics of the emission, including cases where the H I emission may trace the extension of axisymmetric outflow structures seen at smaller radii in molecular lines (Hoai et al. 2014).

4. Evolution Past the AGB

4.1. Understanding the AGB→PN Transition

The mass loss characteristic of evolved stars increases with evolutionary stage (Balick & Frank 2002, Bujarrabal et al. 2001). During the pPN stage, the star ejects a significant fraction of its initial mass ($\sim$ 1 $M_\odot$) via a "superwind" within only a few thousand years (e.g., Lewis 2001). During this phase, the core of the star becomes visible and illuminates (and will later ionize) a very massive nebula, forming a PN, whose central stellar core will then evolve into a white dwarf (WD). However, very few details are known about how these events unfold.

Observations of molecular lines at cm and mm wavelengths are a powerful tool for studying the AGB/PN transition, including the disks, torii, and outflows that commonly appear during this stage—as well as the central stars themselves (e.g., Alcolea et al. 2007; Bujarrabal et al. 2013; Sahai et al. 2017). The ngVLA will complement studies performed with ALMA by extending the range of molecular transitions that can be observed with exquisite sensitivity and spatial resolution to include such lines as ^{12}CO & ^{13}CO J=1−0 (110-115 GHz), HCN J=1−0 (89 GHz), SiO v=1, J=1−0 & J=2−1 (43 and 87 GHz), CS $J = 1-0$ & $2-1$ (49 and 98 GHz). At these frequencies, the ngVLA is expected to achieve angular resolutions of $\sim$2–5 mas—of order ten times better than present ALMA resolutions. At the same time, the ngVLA's continuum sensitivity will allow probing the dust component of the central torii in a statistical sample of post-AGB objects, determine their cm-mm spectral energy distributions, and constrain the grain sizes and dust masses. Combined, this information will address key questions about the geometry and origin of these structures (e.g., Bondi-Hoyle accretion vs. common-envelope ejection).

Even at very high spatial and spectral resolutions, the impressive sensitivity of the ngVLA will allow maps of lines arising from hot inner regions, where brightness temperatures of several hundred K are expected. The innermost layers of pPNe also develop very compact H II regions, resulting from the evolution of the central star to a WD and the ensuing ionization of the surrounding material. Therefore, those regions trace the last phases of the nebular ejection. Their continuum emission is opaque or nearly opaque and very intense at most ngVLA frequencies (T_b>1000 K); recombination lines with T_b $\sim$10% of the continuum (i.e., >100 K) are also expected (e.g., Kwok & Bignell 1984; Sánchez Contreras et al. 2017). The ngVLA will therefore enable studies with high accuracy of both the neutral and the ionized gas from very inner regions of pPNe,

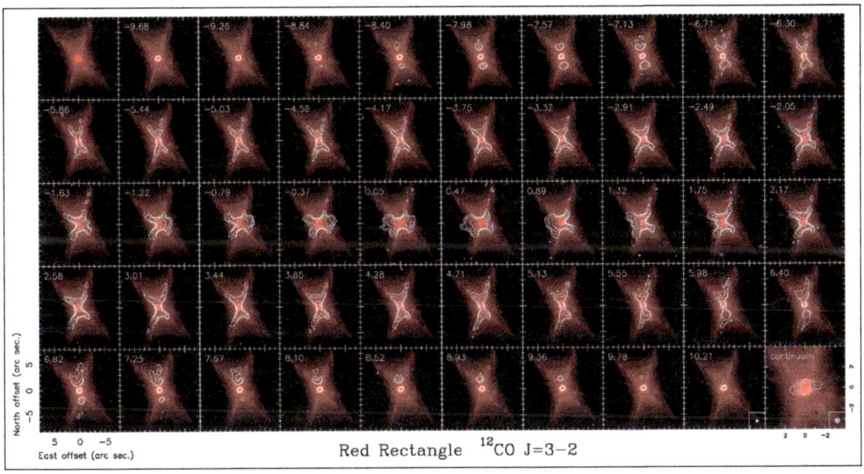

Figure 4. ALMA observations of ^{12}CO J=3–2 emission from the pPN object the Red Rectangle (Bujarrabal et al. 2013; contours). A *Hubble Space Telescope* optical image is also shown in color. The equatorial feature at the central velocities is a rotating disk; the X-shaped features come from more diffuse gas in expansion. It is thought that rotating disks in the centers of pPNe can account for their characteristic bipolar shapes, but only very high angular resolution allows mapping the relevant components.

where vital clues to understand the late evolutionary phases of stars are most likely hidden.

4.2. The Role of Binarity in Post-AGB Evolution

Binary companions are thought to play a significant role in the late evolutionary stages of Sun-like stars. For example, binary interactions are widely believed to underlie the formation of most PNe and may hold the key to the resolution of a long-standing puzzle, namely that although PNe evolve from AGB stars, whose CSEs typically appear spherical with relatively, slow, isotropic expansion ($V_{\rm exp} \sim 5 - 15\,{\rm km\,s^{-1}}$), the vast majority of PNe and pPNe exhibit axisymmetric structures, with a variety of elliptical, bipolar, and multipolar morphologies, as well as fast, collimated outflows ($V_{\rm exp} \gtrsim 50 - 100\,{\rm km\,s^{-1}}$; e.g., Figure 4).

About half of post-AGB objects are known or likely binaries and show prominant disks with little or no extended nebulosity. These have been dubbed "dpAGB" objects, and their dusty disks are remarkably similar to those of pPNe in harboring substantial masses of very large (mm-sized) dust grains (e.g., Sahai et al. 2011a). This supports the idea that the formation of dense "waists" in post-AGB objects is intimately linked to binarity. Surveys of dpAGBs and pPNe with the ngVLA to measure the cm-wavelength SEDs of a large population of these targets will allow the characterization of the dust disks in both classes of objects and new insights into the role of binarity.

4.3. Water Fountain Sources

Water fountain sources (WFs) are thought to be one of the first manifestations of highly collimated outflows in the very short transition from the AGB to PNe (see, e.g., Claussen et al. 2007). Simultaneous ngVLA observations of the continuum and maser emission from WFs in full polarization mode at frequencies of 1.6 GHz (OH masers), 22 GHz (H_2O masers), and 43 GHz (SiO masers) are expected to significantly advance our understanding of these sources and hence this important but short-lived evolutionary stage of Sun-like stars. Observations with the ngVLA of optically thin thermal lines such as HCN, $H^{13}CN$ $J = 1-0$, CS $J = 1-0$ & $2-1$, and SO $3,2-2,1$ are also expected to provide unique insights (Sahai et al. 2017).

Studies of cm wavelength continuum from WFs will enable the detection and study of the expected thermal radio jets and their powering sources. Even though thermal free-free emission produced by photoionization of the gas in the parental CSE is not expected to be significant [the central star(s) are too cold], thermal free-free emission from shock-ionized gas could be present. A similiar phenomenon is observed in low-mass young stellar objects, where ionization of the gas in the local environment may occur through strong shocks of the wind against the ambient gas (e.g., Rodríguez et al. 1999). Radio jets have been detected towards the objects in the pPNe stage, but it will be crucial to explore of the youngest phase of these jets (i.e., during the WF phase) in order to fully understand their underlying driving mechanism.

From the formulation given in Torrelles et al. (1985) and Curiel et al. (1989), it is estimated that a wind with a mass-loss rate of 10^{-7} $M_\odot$ yr^{-1} and a velocity of 500 km s^{-1} can produce via shocks thermal free-free emission at a level of $\sim$ 1 μJy at $\lambda \sim$1 cm for a source at $d \approx$10 kpc. This emission can be detected with the ngVLA with SNR$\approx$10 for an unresolved source (assuming rms noise 0.1 μJy beam^{-1} after 10 hours of integration). By studying the radio continuum emission at different frequencies with the ngVLA, together with the kinematics of masers in the outflow, it will be possible to obtain basic parameters of the central source of the WFs (e.g., mass-loss rate and wind velocity) and to understand the physical relationship between the masers and the continuum (e.g., Orosz et al. 2018).

Observations of the maser emission from WFs with the ngVLA over multiple epochs will also enable measurements of proper motions and enable the derivation of 3D space velocities. In particular, this will be possible for 22 GHz H_2O masers, where individual maser clumps have been observed to be moving at velocities of $\sim$100 km s^{-1} on time scales of months for sources at 10 kpc distance. Because the continuum emission will be observed simultaneously, it will be possible to derive relative positions of the masers with respect to the central source with high precision. This is mandatory to enable detailed modeling of the motions of the gas, from which one may then extract information on the physical properties of the envelope created during the AGB phase (e.g., its gas density and mass; Orosz et al. 2018). As the mass-loss rate climbs to $\sim 10^{-5} - 10^{-4} M_\odot$ yr^{-1} during the superwind phase, it will also be possible to test WF models by searching for evidence of deceleration of the jets using the spatio-kinematics of the masers (Imai et al. 2005).

Finally, in addition to allowing measurements of proper motions, observations of H_2O masers in WFs with the ngVLA will provide new information on the relationship between the magnetic field and velocity vectors through polarization measurements of the masers and their 3D velocities. Similar techniques have already been applied to the studies of YSOs (e.g., Goddi et al. 2017; see also Hunter et al. 2018, this volume p. 321). This will be invaluable for testing the role of magnetic fields in jet launching.

Acknowledgments. The authors gratefully acknowledge contributions to this Chapter from the following individuals: Javier Alcolea; Valentín Bujarrabal; Dieter Engels; Sandra Etoka; Eric Gérard; José-Francisco Gomez; Hiroshi Imai; Thibaut Le Bertre; Anita M. S. Richards; Raghvendra Sahai; José-Maria Torrelles.

References

Alcolea, J. 2004, in Proceedings of the 7th Symposium of the European VLBI Network, ed. by R. Bachiller, F. Colomer, J.-F. Desmurs, and P. de Vicente, Observatorio Astronomico Nacional of Spain, 169

Alcolea, J., Neri, R., & Bujarrabal, V. 2007, A&A, 468, L41

Balick, B., & Frank, A. 2002, ARA&A, 40, 439

Beasley, A. J. & Conway, J. E. 1995, in Very Long Baseline Interferometry and the VLBA, ASP Conf. Series, Vol. 82, ed. by J. A. Zensus, P. J. Diamond, and P. J. Napier, (San Francisco: ASP), 327

Boboltz, D. A., & Claussen, M. J. 2004, ApJ, 608, 480

Bujarrabal, V., Castro-Carrizo, A., Alcolea, J., & Sánchez Contreras, C. 2001, A&A, 377, 868

Bujarrabal, V., Castro-Carrizo, A., Alcolea, J., Van Winckel, H., & Sánchez Contreras, C., Santander-García, M., Neri, R. & Lucas, R. 2013, A&A, 557, L11

Claussen, M. J, Sahai, R., & Morris, M. 2007, Asymmetrical Planetary Nebulae IV, `http://www.iac.es/proyect/apn4`

Claussen, M. J., Sjouwerman, L. O., Rupen, M. P., Olofsson, H., Schöier, F. L., Bergman, P., & Knapp, G. R. 2011, ApJ, 739, 5

Cotton, W. D, Mennesson, B., Diamond, P. J., et al. 2004, A&A, 414, 275

Curiel, S., Rodríguez, L. F., Bohigas, J., Roth, M., Canto, J., & Torrelles, J. M. 1989, ApL&C, 27, 299

Desmurs, J.-F., Bujarrabal, V., Lindqvist, M., et al. 2014, A&A, 565, A127

Desmurs, J.-F., Alcolea, J., Bujarrabal, V., Colomer, F., & Soria-Ruiz, R. 2017, in Astrophysical Masers: Unlocking the Mysteries of the Universe, ed. by A. Tarchi, M. J. Reid, & P. Castangia, IAU Symp. 336, in press (arXiv:1712.07380)

Engels, D. 1985, in The Milky Way Galaxy, IAU Symposium No. 106, ed. by H. van Woerden, R. J. Allen, and W. B. Burton (Dordrecht: Reidel), 131

Engels, D., Etoka, S., Gérard, E., & Richards, A. 2015, in Why Galaxies Care about AGB Stars III: A CLoser Look in Space and Time, ASP Conf. Series, Vol. 497, ed. by F. Kerschbaum, R. F. Wing, and J. Hron (San Francisco: ASP), 473

Etoka, S., Engels, D., Gérard, E., & Richards, A. M. S. 2017, in Astrophysical Masers: Unlocking the Mysteries of the Universe, IAU Symposium No. 336, ed. by A. Tarchi, M. J. Reid, & P. Castangia, in press (arXiv:1711.09694)

Frayer, D. T. 2017, ngVLA Memo #14

Freytag, B., Liljegren, S., & Höfner, S. 2017, A&A, 600, A137

Gérard, E. & Le Bertre, T. 2006, AJ, 132, 2566

Gérard, E., Le Bertre, T., & Libert, Y. 2011, SF2A-2011, ed. by G. Alecian et al., 419

Goddi, C., Surcis, G., Moscadelli, L., Imai, H., Vlemmings, W. H. T., van Langevelde, H. J., & Sanna, A. 2017, A&A, 597, A43

Gonidakis, I., Diamond, P. J., & Kemball, A. J. 2013, MNRAS, 433, 3133

Gray, M., Baudry, A., Richards, A. M. S., Humphreys, E. M. L., Sobolev, A. M., & Yates, J. A. 2016, MNRAS, 456, 374

Gray, M., Wittkowski, M., Scholz, M., Humphreys, E. M. L., Ohnaka, K., & Boboltz, D. 2009, MNRAS, 394, 51

Harper, G. M. 2018, Science with a Next-Generation Very Large Array, ASP Monograph 7, ed. E. J. Murphy (San Francisco, CA: ASP), 265

Hoai, D. T., Matthews, L. D., Winters, J. M., Nhung, P. T., Gérard, E., Libert, Y., & Le Bertre, T. 2014, A&A, 565, 54
Höfner, S. & Olofsson, H. 2018, A&ARv, 26, 1
Hunter, T. R., Brogan, C. L., Bartkiewicz, A., Chibueze, J. O., Cyganowski, C. J. et al. 2018, Science with a Next-Generation Very Large Array, ASP Monograph 7, ed. E. J. Murphy (San Francisco, CA: ASP), 321
Imai, H., Nakashima, J.-i., Diamond, P. J., Miyazaki, A., & Deguchi, S. 2005, ApJL, 622, L125
Karovska, M., Schlegel, E., Hack, W., Raymond, J. C., & Wood, B. E. 2005, ApJL, 623, L137
Kim, H. & Taam, R. E. 2012a, ApJ, 744, 136
Kim, H. & Taam, R. E. 2012b, ApJ, 759, 59
Kim, H. & Taam, R. E. 2012c, ApJL, 759, L22
Kim, H., Trejo, A., Liu, S.-Y., et al. 2017, NatAs, 1, 60
Kwok, S. & Bignell, R. C. 1984, ApJ, 276, 544
Lewis, B. M. 2001, AJ, 121, 426
Lim, J., Carilli, C. L., White, S. M., Beasley, A. J., & Marson, R. G. 1998, Nature, 392, 575
Lucas, R., Omont, A., & Guilloteau, S. 1988, A&A, 194, 230
Maercker, M., Mohamed, S., Vlemmings, W. H. T., et al. 2012, Nature, 490, 232
Mastrodemos, N. & Morris, M. 1999, ApJ 523, 357
Matthews, L. D., Gérard, E., & Le Bertre, T. 2015a, MNRAS, 449, 220
Matthews, L. D., Le Bertre, T., Gérard, E., & Johnson, M. C. 2013, AJ 145, 97
Matthews, L. D., Libert, Y., Gérard, E., Le Bertre, T., & Reid, M. J. 2008, ApJ, 684, 603
Matthews, L. D., Libert, Y., Gérard, E., Johnson, M. C., & Dame, T. M. 2011, AJ, 141, 60
Matthews, L. D. & Reid, M. J. 2007, AJ, 133, 2291
Matthews, L. D., Reid, M. J., & Menten, K. M. 2015b, ApJ, 808, 36
Matthews, L. D., Reid, M. J., Menten, K. M., & Akiyama, K. 2018, AJ, 156, 15
Menten, K. M., Wyrowski, F., Keller, D., & Kamiński, T. 2018, A&A, 613, 49
O'Gorman, E., Harper, G. M., Guinan, E. F., Richards, A. M. S., Vlemmings, W., & Wasatonic, R. 2015, A&A, 580, A101
Orosz, G., Gomez, J. F., Imai, H., et al. 2018, MNRAS Letters, in press (arXiv:1809.07505)
Reid, M. J. & Menten, K. M. 1997, ApJ, 476, 327
Reid, M. J. & Menten, K. M. 2007, ApJ, 671, 2068
Rodríguez, L., F., Anglada, G., & Curiel, S. 1999, ApJS, 125, 427
Sahai, R., Claussen, M. J, Schnee, S., Morris, M. R., & Sánchez Contreras, C. 2011a, ApJL, 739, L3
Sahai, R., Neill, J. D., Gil de Paz, A., & Sánchez Contreras, C. 2011b, ApJL, 740, L39
Sahai, R., Sanz-Forcada, J., Sánchez Contreras, C., & Stute, M. 2015, ApJ, 810, 77
Sahai, R., Vlemmings, W. H. T., Gledhill, T., et al. 2017, ApJL, 835, L13
Sánchez Contreras, C., Báez-Rubio, A., Alcolea, J., Bujarrabal, V., & Martín-Pintado, J. 2017, A&A, 603, A67
Selina, R. & Murphy, E. J. 2017, ngVLA Memo #17
Soria-Ruiz, R., Colomer, F., Alcolea, J., et al. 2005, A&A, 432, L39
Torrelles, J. M., Ho, P. T. P., Rodríguez, L. F., & Canto, J. 1985, ApJ, 288, 595
Tsuji, T. 2000, ApJ, 540, L99
Tsuji, T. 2001, in Galaxies and their Constituents at the Highest Angular Resolutions, IAU Symposium 205, ed. by R. T. Schilizzi, (San Francisco: ASP), 316
Turner, B. E. & Ziurys, L. M. 1988, in Galactic and Extragalactic Radio Astronomy, 2nd Edition, ed. by G. L. Verschuur and K. I. Kellermann, (Berlin: Springer-Verlag), 200
Van Eck, S., Goriely, S., Jorissen, A., & Plez, B. 2001, Nature, 412, 793
Villaver, E., Manchado, A., & García-Segura, G. 2012, ApJ, 748, 94
Vlemmings, W., Khouri, T., O'Gorman, E., et al. 2017, NatAs, 1, 848
Wolak, P., Szymczak, M., & Gérard, E. 2013, MNRAS 430, 2499
Wolak, P., Szymczak, M., & Gérard, E. 2012, A&A 537, A5
Ziurys, L. M. 2006, PNAS, 103, 122274

Science with a Next-Generation Very Large Array
ASP Conference Series, Monograph 7
Eric J. Murphy, ed.

Detecting Infall in High-Mass Protostellar Objects

Maria T. Beltrán,[1] Qizhou Zhang,[2] and Víctor M. Rivilla[1]

[1]*INAF-Osservatorio Astrofisico di Arcetri, Largo E. Fermi 5, I-50125 Firenze, Italy;* `mbeltran@arcetri.astro.it`, `rivilla@arcetri.astro.it`

[2]*Harvard-Smithsonian Center for Astrophysics, 60 Garden St. Cambridge, MA 02138, USA;* `qzhang@cfa.harvard.edu`

Abstract. Red-shifted molecular absorption against a bright background continuum source is the clearest way of diagnosing infall and accretion in high-mass star-forming regions. Up to now, red-shifted absorption at centimeter wavelengths has been only detected toward the brightest ultracompact and hyper compact (UC/HC) HII regions with limited angular resolution and sensitivity. The high sensitivity and angular resolution of the ngVLA will allow us to detect several NH_3 inversion transitions with a wide range of excitation energies (~20 K to 2000 K) in red-shifted absorption toward high-mass accretion disk candidates with luminosities $> 10^3\,L_\odot$, even for the faintest UC/HC HII regions. In particular, the ngVLA will detect infall toward massive stars shortly after the onset of an HII region at an angular resolution of 20 mas (~100 au at the typical distances of high-mass protostars).

1. Introduction: the Formation of High-Mass Stars

The role of accretion disks in the formation of low-mass stars has been well assessed by means of high angular resolution observations at various wavelengths. These findings confirm the prediction that conservation of angular momentum during the collapse leading to the formation of a star is bound to produce flattening and rotation of the collapsing core. Therefore, the existence of such disks around low-mass young stellar objects (YSOs) strongly supports the scenario where the formation of *low-mass* stars proceeds through contraction of a dense molecular clump and subsequent inside-out collapse with accretion onto a protostellar core. In synthesis, one can say that *low-mass stars form through accretion.*

What about *high-mass* stars? The formation process of high-mass stars has puzzled the astrophysical community for decades because of the apparent stellar mass limit for spherical accretion. Beyond this limit, theory predicted that it is impossible to continue accreting material because the stellar wind and the radiation pressure from the newly-formed early-type star would stop the infall (e.g., Kahn 1974; Yorke & Krügel 1977; Wolfire & Casinelli 1987). In the 1990s, different theoretical scenarios proposed non-spherical accretion as a possible solution for the formation of OB-type stars (Nakano 1989; Jijina & Adams 1996), and in recent years, theoretical ideas and simulations appear to have converged to a disk-mediated accretion scenario (e.g., Krumholz et al. 2009; Kuiper et al. 2010; Klassen et al. 2016). In fact, competing theories that propose very different high-mass star-formation mechanisms, such as models suggesting that massive star-formation is initiated by the monolithic collapse of a turbulent molecular core (McKee & Tan 2002), or those based on competitive accretion (Bonnell & Bate 2006), all predict the existence of circumstellar accretion disks through which the material is channeled onto the forming star. With this in mind, it is clear that the detection of infall in circumstellar disks represents a crucial test to prove the accretion scenario also applies in the case of high-mass stars.

At present, several authors have reported on detections of disks around high-mass YSOs (see Beltrán & de Wit 2016, for a review). In most cases, the existence of such a disk is also supported by the detection of a bipolar jet/outflow along the apparent rotational axis of the disk. It must be noted that some of these disks resemble massive toroids rotating about a cluster of high-mass stars, rather than geometrically thin circumstellar disks. In any case, no matter whether the disk is circumstellar or "circum-cluster", these findings strongly support the accretion scenario also for the high-mass regime.

Notwithstanding these important results, the presence of disks rotating about high-mass stars is not sufficient by itself to prove unambiguously the accretion model: what is needed is iron-clad evidence of *infall*. Such evidence is very difficult to find, as the free-fall velocity becomes significant only very close to the accreting star, i.e., over a region of a few 0.01 pc (~2000 au), which is very difficult to access and disentangle from the surrounding quiescent or rotating material. In this chapter we discuss how to characterize the infall of material in a sample of 36 high-mass accretion disk candidates covering a broad range of luminosities, from $10^3\,L_\odot$ to $10^6\,L_\odot$, compiled by Beltrán & de Wit (2016) with the ngVLA.

2. Tracing Infall

Red-shifted molecular absorption against a bright background continuum source is possibly the clearest way of diagnosing infall and accretion. For this, it is necessary to observe high-density tracers with excitation energies below the brightness temperature of the central continuum source, and observe them especially with interferometers (because the emission does not suffer from beam dilution). The first studies using this technique were conducted at centimeter wavelengths with the VLA by observing NH_3 in absorption against the bright ultracompact (UC) HII region G10.62−0.4 (Ho & Haschick 1986; Keto et al. 1988). Since, infall has been detected toward other ultracompact/hypercompact (UC/HC) HII regions: e.g., W51 (Zhang & Ho 1997), G24.78+0.08 (Beltrán et al. 2006a), NGC 7538 IRS1 (Goddi et al. 2015), also observed in NH_3 with the VLA or the upgraded Jansky VLA. Ammonia is the best species to study infall in dense cores because it is possible to trace excitation up to temperatures of ~2000 K by observing its inversion transitions within a relatively narrow frequency range at centimeter wavelengths, 20 – 40 GHz, which cannot be observed with (sub)millimeter interferometers such as SMA, IRAM NOEMA, or ALMA. This allows us to conduct the tomography of the infall in the core. In addition, in case the main line is optically thick, one can always use the hyperfine satellite lines, which are usually optically thin, to study the absorption.

Red-shifted absorption at high-angular resolution has also been observed against bright dust continuum sources at millimeter wavelengths (e.g., Zapata et al. 2008; Girart et al. 2009; Beltrán et al. 2018). For high-mass protostars, however, dust emission from envelopes and disks becomes so optically thick that deep probes of infall motion are prevented. In addition, the brightness temperatures of the UC/HC HII regions are typically 10^3–10^4 K, much higher than those of the dust continuum sources (a few hundreds of K). This bright background enables detection of line absorption with much higher signal-to-noise ratio, as well as probing infall closer to the central protostar with higher energy molecular line transitions.

What can we learn from the gravitational collapse by observing red-shifted absorption? We can estimate the infall rate $\dot{M}_{\rm inf}$ of material in the core. Note that red-shifted absorption profiles trace infalling material in the inner part of the core or in the disk, but not actual accretion onto the central star. Following Beltrán et al. (2006a), the infall rate in a solid angle Ω can be estimated from the expression $\dot{M}_{\rm inf} = (\Omega/4\pi)\,2\pi\, m_{\rm H_2}\, N\, R_0^{1/2}\, R^{1/2}\, V_{\rm inf}$, where N is the gas column density, R_0 is the radius of the continuum source, $V_{\rm inf}$ is the infall velocity, and R is the radius at which $V_{\rm inf}$ is measured. The main caveat of this method is the uncertainty on R. In fact, the radius at which $V_{\rm inf}$ is associated is not known, and usually the size of the infalling core is given as an upper limit. Red-shifted absorption can also help us to constrain star-formation theories by studying whether infall spins down or spins up. In fact, the advantage of tracing a wide range of excitation temperatures (~20 K to 2000 K) with several transitions of NH_3 is that by measuring the velocity of the absorption feature for each transition it is possible to study whether the infall velocity (assumed to be the difference between the velocity of the red-shifted absorption dip and the systemic LSR velocity) changes with the energy of the line. If the infall velocity increases with the energy of the transition (spins up), this would suggest that the infall is accelerating towards the center of the core, consistent with gravitational collapse (Beltrán et al. 2018). On the other hand, if the infall velocity decreases (spins down), it could suggest that magnetic braking is acting and removing angular momentum (Basu & Mouschovias 1994). This would suggest that the gravitational collapse of the core is controlled by magnetic fields (Girart et al. 2009). In addition, a detailed modeling of the NH_3 line emission and of the spectral energy distribution with an infalling envelope model can give us information on the mass of the central star, its age and luminosity, and the mass accretion rate, similar to what Osorio et al. (2009) have done for the hot molecular core G31.41+0.31.

3. Uniqueness to ngVLA Capabilities

Up to now, red-shifted absorption at centimeter wavelengths has been only detected toward the brightest UC/HC HII regions and with limited angular resolution and sensitivity using the VLA. These limits mean that, in many cases, the absorption has not been spatially resolved and only the lower excitation energy transitions, which are probably tracing the outer regions of the massive envelope, have been observed. Therefore, to trace infall in high-mass cores, with luminosities ranging from $10^3\,L_\odot$ up to $10^6\,L_\odot$, high sensitivity and high-angular resolution are essential. High sensitivity is needed to trace high-excitation transitions of high-density tracers, which are those tracing the material closest to the protostar. On the other hand, high-angular resolution would allow us to resolve spatially the absorption, even for distant objects. A good tracer of infall is NH_3 because, as already mentioned, multi transitions at different energies can be observed in red-shifted absorption against bright ionized sources.

The high sensitivity of the ngVLA will allow us to detect NH_3 inversion transitions in red-shifted absorption toward all the sources of the sample of high-mass accretion

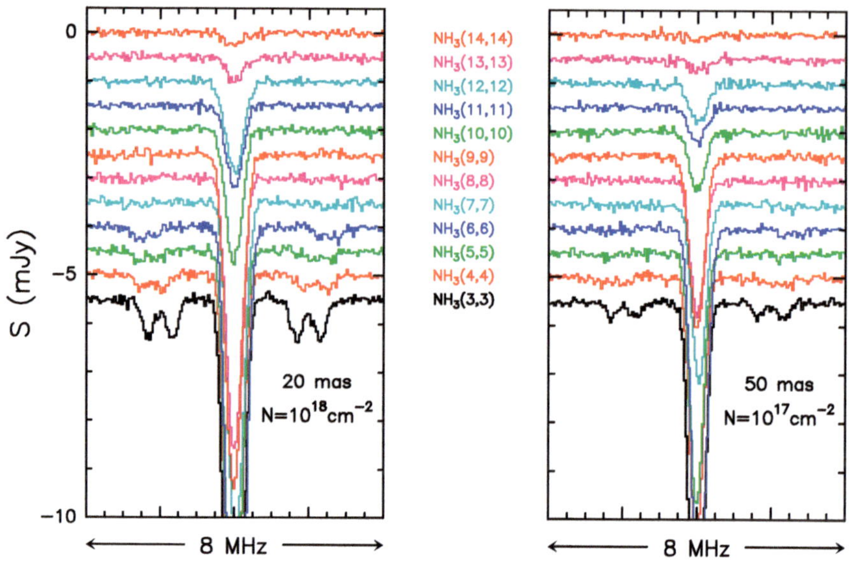

Figure 1. Synthetic spectrum of NH_3 obtained with MADCUBA, assuming LTE conditions. Different transitions, from the (3,3) inversion transition to the (14,14) one, are shown with different colors. For display purposes, the transitions have been shifted by 0.5 mJy in the y-axis. The left and right panels show two cases for different angular resolutions and column densities of NH_3: 20 mas and 10^{18} cm^{-2}, and 50 mas and 10^{17} cm^{-2}, respectively. The linewidth considered is 5 km s^{-1}. The brightness temperature of the UC/HC HII region is 10^4 K. We have considered a noise level of 45 μJy, which is the rms achieved for an integration time of 10 hr and a spectral resolution of 1 km s^{-1}, according to the specifications of the ngVLA.

disk candidates with luminosities ranging from $10^3\,L_\odot$ up to $10^6\,L_\odot$ compiled by Beltrán & de Wit (2016), including the faintest ionized ones. In particular, the ngVLA will detect infall toward massive stars shortly after the onset of an HII region, and at a maximum angular resolution of 20 mas. Figure 1 shows the synthetic spectrum of NH_3, from the (3,3) inversion transition to the (14,14) one, obtained with MADCUBA (Madrid Data Cube Analysis on ImageJ), which is a software developed in the Center of Astrobiology (Madrid, INTA-CSIC) to visualize and analyze single spectra and datacubes (Rivilla et al. 2016; Martín et al. 2019, in prep.). The spectra have been modeled for two different angular resolutions, 20 mas and 50 mas with different column densities of NH_3, 10^{18} cm^{-2} and 10^{17} cm^{-2}, respectively. As seen in the figure, an rms of ~45 μJy/beam, achievable after a 10 hr integration per source, would allow us to detect the main lines of the weakest inversion transitions of ammonia, up to the NH_3 (14, 14) transition, and the satellite lines of the strongest ones, for NH_3 column densities consistent with (or even lower than) those found around high-mass young stellar cores (e.g., Beltrán et al. 2006a; Goddi et al. 2015). Spectral resolution should not be a problem for the ngVLA, considering that a modest channel spacing of ~1 km/s is less than the typical line widths of high-mass protostellar cores of a few km/s. Regarding the angular resolution, 20 mas corresponds to 100 au at a distance of 5 kpc, which is the typical distance of O-type young stars (e.g., Beltrán et al. 2006b). Typical sizes of accretion disks around low-mass stars are of the same order, ~100 au. Note that the highest angular resolution provided by VLA at the NH_3 transition frequencies is about 50 mas. Given the high mass of these cores, they often fragment and form multiple protostars. Therefore, the high-angular resolution of ngVLA will resolve the emission and trace the accretion flows around individual protostars.

The milli-arcsecond angular resolution achieved with the ngVLA and the wide range of excitation energies covered by the NH_3 inversion transitions (from ~20 K to 2000 K) will allow us to trace the kinematics of circumstellar gas from the envelope down to the circumstellar disk, and infer the properties of the accretion to better constrain theories of massive star formation. In fact, although the two competing theories on massive star-formation propose a disk-mediated accretion scenario, the properties of such disks are expected to be different. In fact, while in the turbulent core collapse case one would expect the material to be incorporated onto the central protostar though a true accretion disk in Keplerian motion with a size of a few hundreds of au, in the competitive accretion, the disks could be asymetric, truncated and very small (< 100 au) due to tidal interactions (Cesaroni et al. 2006) in the crowded star-forming cores (according to this model, high-mass stars always form in clusters). Alternatively, accretion could proceed to the central protostar through filaments rather than through a disk (Goddi et al. 2018).

4. Synergy at Other Wavelengths

The study of infall at centimeter wavelengths complements: *i)* ALMA or NOEMA observations of molecular outflows aimed at estimating the mass outflow rate, which is related to the mass accretion rate onto the protostar as $\dot{M}_{\rm acc} = \dot{M}_{\rm out}/6$ (Tomisaka 1998; Shu et al. 1999), and *ii) JWST* observations of Brγ lines aimed at determining the mass accretion rate in the disk. By comparing the mass infall rate obtained at centimeter wavelengths to the mass accretion rate one could determine concretely whether the former are much higher than the latter, as suggested by recent studies (Beltrán & de Wit 2016). Such a mismatch, which could indicate pile-up of material in the disk and the existence of episodic accretion, was recently observed in the high-mass stellar object S255IR (Caratti o Garatti et al. 2017).

5. Conclusions

The superb sensitivity and angular resolution provided by the ngVLA will allow us to study the gravitational collapse in high-mass star-forming sources of all luminosities (from $10^3\,L_\odot$ to $10^6\,L_\odot$). The detection of red-shifted absorption in several inversion transitions of NH_3 with different excitation energies will allow us to trace the material down to disk scales and to derive the properties of the collapse and of the accretion disks that will help us to better constrain theories of high-mass star formation. In particular, by studying the infall velocity field, i.e., whether infall spins down or spins up, we will infer what controls the gravitational collapse in high-mass star-forming regions.

Acknowledgments. V.M.R. is funded by the European Union's Horizon 2020 research and innovation programme under the Marie Skłodowska-Curie grant agreement No 664931.

References

Basu, S., & Mouschovias, T. Ch. 1994, ApJ, 432, 720
Beltrán, M. T., & de Wit, W. J. 2016, A&ARv, 24, 6
Beltrán, M. T., Cesaroni, R., Codella, C., Testi., L. et al. 2006a, Nature, 443, 427
Beltrán, M. T., Brand, J., Cesaroni, R. et al. 2006b, A&A, 447, 221
Beltrán, M. T., Cesaroni, R., Rivilla, V. M. et al. 2018, A&A, arXiv:1803.05300
Bonnell, I. A., & Bate, M. R. 2006, MNRAS, 370, 488
Caratti o Garatti, A., Stecklum, B., Garcia López, R. et al. 2017, Nature Physics, 13, 276
Cesaroni, R., Galli, D., Lodato, G. et al. 200, Nature, 444, 703
Girart, J. M., Beltrán, M. T., Zhang, Q. et al. 2009, Science, 324, 1408
Goddi, C., Zhang, Q., & Moscadelli, L. 2015, A&A, 573, A108
Goddi, C., Ginsburg, A., Maud, L. et al. 2018, arXiv:1805.05364
Kahn, F. 1974, A&A, 37, 149
Keto, E. R., Ho, P. T. P., & Haschick, A. D. 1988, ApJ, 324, 920
Klassen, M., Pudritz, R. E., Kuiper, R., Peters, Th., Banerjee, R. 2016, ApJ, 823, 28
Krumholz, M. R., Klein, R. I., McKee, C. F., Offner, S. S. R., & Cunningham, A. J. 2009, Science, 323, 754
Kuiper, R., Klahr, H., Beuther, H., & Henning, T. 2010, ApJ, 722, 1556826, 161
Ho, P. T. P., & Haschick, A. D. 1986, ApJ, 304, 501
Jijina, J., & Adams, F. C. 1996, ApJ, 462, 874
McKee, C. F., & Tan, J. C. 2002, Nature, 416, 59
Nakano, T. 1989, ApJ, 345, 464
Osorio, M., Anglada, G., Lizano, S., D'Alessio, P. 2009, ApJ, 694, 29
Rivilla, V. M., Fontani, F., Beltrán, M. T. et al. 2016, ApJ,
Shu, F., Allen, A., Shang, H. et al. 1999, NATO Adv. Sci. Institutes Ser. C., Lada, C. & Kylafis, N. (ed) NATO Advanced Science Institutes (ASI) Series C, 540, 193
Tomisaka, K. 1998, ApJL, 502, 163
Wolfire, M, & Cassinelli, J. 1987, ApJ, 319, 850
Yorke, H. W., & Krügel, E. 1977, A&A, 54, 183
Zapata, L., Palau, A., Ho, P. T. P. et al. 2008, A&A, 479, L25
Zhang, Q., & Ho, P. T. P. 1997, ApJ, 488, 241

Science with a Next-Generation Very Large Array
ASP Conference Series, Monograph 7
Eric J. Murphy, ed.

Star-Forming Filaments and Cores on a Galactic Scale

James Di Francesco,[1] Jared Keown,[2] Rachel Friesen,[3] Tyler Bourke,[4] and Paola Caselli[5]

[1]*National Research Council of Canada, Victoria, BC, Canada;* `james.difrancesco@nrc-cnrc.gc.ca`

[2]*University of Victoria, Victoria, BC, Canada;* `jkeown@uvic.ca`

[3]*National Radio Astronomy Observatory, Charlottesville, VA, U.S.A.;* `rfriesen@nrao.edu`

[4]*Square Kilometre Array Organization, Macclesfield, UK;* `t.bourke@skatelescope.org`

[5]*Max Planck Institute for Extraterrestrial Physics, Garching, Germany;* `caselli@mpe.mpg.de`

Abstract. Continuum observations of molecular clouds have revealed a surprising amount of substructure in the form of filaments of a few pc length and cores of ~0.1 pc diameter. Understanding the evolution of these substructures towards star formation requires the kinematic and dynamical insights provided uniquely by sensitive line observations at high angular and spectral resolution. In this chapter, we describe how an ngVLA can probe effectively the dynamics of filaments and cores in nearby star-forming molecular clouds using the NH_3 rotation-inversion transitions at 24 GHz. Such emission has been proven to trace well the high column density environments of star-forming cores and filaments but higher-resolution observations are needed to reveal important details of how dense gas is flowing within and onto these substructures. In particular, we describe how 150 × 18 m antennas with a maximum baseline of 1 km can be used to map sensitively NH_3 emission across high column density locations in clouds in roughly an order of magnitude less time than with the current Jansky VLA.

1. Introduction

Stars form out of molecular cloud gas, when dense pockets (i.e., "cores") become gravitationally unstable and collapse (see Di Francesco et al. 2007 for a review). Recent far-infrared/submillimeter continuum observations (e.g., from *Herschel* or the JCMT) of Galactic clouds at distances < 3 kpc have revealed close connections between the detailed substructures of clouds and star formation. First, molecular clouds are suffused with filaments, parsecs-long substructures of ~0.1 pc width, regardless of their star-forming activity (André et al. 2010; Ward-Thompson et al. 2010). Second, clouds can have hundreds of cores, many of which appear bound and hence likely to form stars in the future (Könyves et al. 2010). Third, core formation, and hence star formation, appears to be most efficient within supercritical filaments above a given column density threshold equivalent to $A_V \sim 7$ magnitudes (André et al. 2010, 2014). Indeed, the

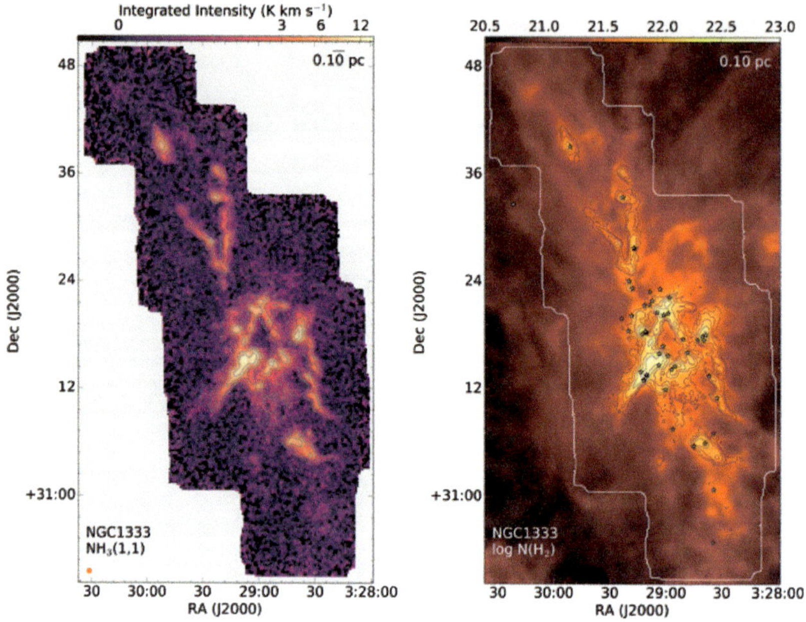

Figure 1. (*Left:*) integrated intensity map of the NH_3 (1, 1) line for the NGC 1333 region obtained by the Green Bank Ammonia Survey (GAS; Friesen et al. 2017). Contours are drawn at [3, 6, 12, 24, ...]-σ, where 1 $\sigma \approx 1.0$ K km s^{-1} was estimated from emission-free pixels. Beam size and scale bar are shown in the bottom left and top right corners, respectively. (*Right:*) the H_2 column density, log10 [N(H_2) cm^{-2}], derived from SED fitting of *Herschel* submillimeter dust continuum data (A. Singh et al. 2018, in preparation). Contours show the NH_3 (1,1) integrated intensity, as in the left panel. White contours show the GAS map extent. Stars show the locations of Class 0/I and flat-spectrum protostars (Dunham et al. 2015).

resulting core mass function strikingly resembles the initial stellar mass function, suggesting the process that forms cores also ultimately determines stellar masses (Könyves et al. 2015). Finally, high-mass star and cluster formation occur most efficiently where supercritical filaments appear to intersect (Schneider et al. 2012).

Understanding the relationships between filamentary substructures and star formation requires kinematic and dynamical insights. For example, how does gas in clouds assemble into filaments and cores? Also, how will the gas in these substructures evolve? Is there further coherent substructure within filaments (e.g., "fibers"; Hacar et al. 2013, 2017). The continuum observations that identified filaments in molecular clouds do not themselves have the ability to trace their kinematics and dynamics. Instead, observations of line emission are essential to determine how mass flows within filaments and cores and whether or not such substructures are stable. More specifically, to understand the structure and dynamical evolution of molecular clouds in general, large-scale surveys of lines must be performed at sensitivities and resolutions not possible with currently available observatories.

The NH_3 rotation-inversion transitions at 24 GHz [e.g., (1,1), (2,2), (3,3), etc.] can probe effectively the kinematics and dynamics of star-forming substructures in nearby molecular clouds. NH_3 emission has been shown repeatedly to trace best the loca-

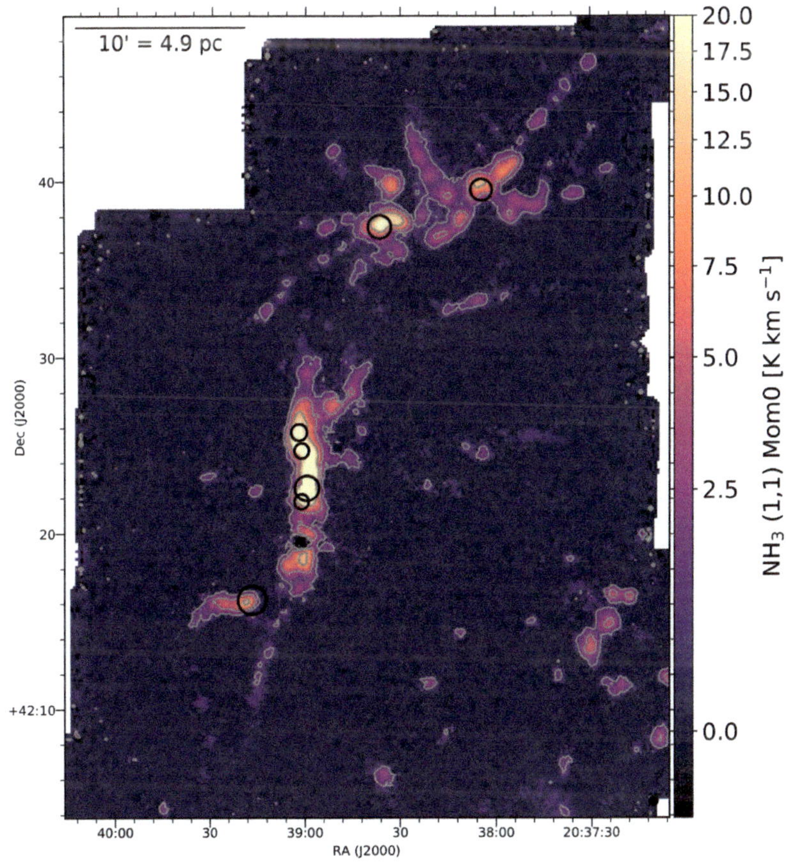

Figure 2. Integrated intensity map of the NH_3 (1,1) line for the Cygnus X North region, including DR21, obtained by the KEYSTONE survey (Keown et al. 2018). Colors range from -1 K km s^{-1} (black) to 10 K km s^{-1} (yellow). Contours are at 1.0, 3.5, and 10 K km s^{-1}. Circles show the locations of H_2O maser emission simultaneously detected during the NH_3 observations. The size of the circle indicates the relative brightness of the maser emission.

tions of significant column density revealed by continuum emission from dust (e.g., Benson & Myers 1989). This behavior follows partly because the NH_3 transitions are excited in moderately dense gas; e.g., the critical densities of (1,1) and (2,2) at 10 K are 10^{3-4} cm^{-3} (Ho & Townes 1983). In addition, these NH_3 transitions have hyperfine structure that spreads out their overall optical depths over numerous components, enabling better probes of all dense gas along the line of sight (Crapsi et al. 2007). In contrast, other molecules have emission that is too optically thick to sample such environments (e.g., ^{12}CO) or have been themselves too drastically depleted in cold dense gas to be effective probes (e.g., $C^{18}O$; see Di Francesco et al. 2007). Even other less-abundant "dense gas tracers" like HCN or CN emission may yet suffer from optical depth and depletion issues on the scales of cores and filaments. Furthermore, the hyperfine structure of the NH_3 transitions further allows, through simultaneous fitting, direct determinations of excitation temperature and opacity, and hence column density. Finally, these NH_3 transitions can directly provide the gas kinetic temperature along the line of sight via their ratios (Walmsley & Ungerechts 1983), unlike those of other nitrogen-based molecules (e.g., N_2H^+).

Given the utility of NH_3 emission, it is being widely observed to trace the kinematics and dynamics of moderately dense gas in molecular clouds, especially in filaments and cores. These programs are being largely run from the Green Bank Telescope (GBT) using its unique K-band Focal Plane Array (KFPA) instrument to map many square degrees of sky in nearby molecular clouds, Giant Molecular Clouds, and large swaths of the Galactic Plane. Figures 1 and 2 show examples of recent wide-field GBT KFPA observations of NH_3 (1,1) emission toward star-forming molecular clouds. Figure 1 (left) shows the integrated intensities of NH_3 (1,1) emission toward the NGC 1333 star-forming region of the Perseus molecular cloud from the Green Bank Ammonia Survey (GAS) with the KFPA (Friesen et al. 2017), while Figure 1 (right) shows the H_2 column densities of the same region derived from continuum data obtained by *Herschel* (Singh et al. 2018). Figure 2 shows the integrated intensity of NH_3 (1,1) emission toward the Cygnus X North region, including DR21, from the K-band Examinations of Young STellar Object Natal Environments (KEYSTONE) survey (Keown et al. 2018). In both cases, widespread NH_3 emission indicative of dense, star-forming gas is seen, much of it in filaments.

Such wide-field NH_3 mapping surveys are already changing our understanding of the stability of dense cores and hierarchical structures - and hence, their ability to collapse and form stars, disperse without collapse, fragment, and accrete additional mass. For example, Kirk et al. (2017) combined GAS and JCMT Gould Belt Legacy Survey data to estimate the virial states of dense cores within the Orion A molecular cloud. They found that none of the dense cores are sufficiently massive to be bound when considering only the balance between self-gravity and the thermal and non-thermal motions present in the dense gas, in contrast to analyses which include only thermal gas motions. Instead, most of the dense cores are pressure-confined by the additional pressure binding imposed by the weight of the ambient molecular cloud material and additional smaller pressure term. Over larger spatial scales, Keown et al. (2017) showed that NH_3-identified structures in Cepheus are gravitationally dominated, yet may be in or near a state of virial equilibrium. Filamentary structures in Cepheus have virial parameters $\ll 2$, such that they should be gravitationally unstable unless there is significant support by magnetic fields. This result is consistent with a similar analysis of the more massive, more actively star-forming Serpens South region (Friesen et al. 2016), and hints that large-scale gravitational instability is a general property of the dense gas structure in star-forming environments.

Though wide-field observations such as those in Figures 1 and 2 are enabling significant advances to be made about understanding star formation, it remains challenging to recover from such data important details of the kinematics and dynamics of the dense gas. For context, the GBT's relatively low angular resolution, i.e., 33″ at 24 GHz, is equivalent to 0.07 pc at the 420 pc distance of the Orion molecular cloud complex, but cores and filaments each have characteristic widths of 0.1 pc (Di Francesco et al. 2007; Arzoumanian et al. 2011). Hence, the GBT observations do not resolve the gas kinematics and dynamics of such structures in clouds much more distant that Orion. Indeed, higher-resolution observations are critical for such targets, to probe for further substructure and allow details of mass flow and dynamical stability to be recovered. Though the Jansky VLA can be currently used to observe NH_3 emission at 24 GHz, its limited point-source sensitivity to line emission, its relatively small field-of-view, and its insensitivity to moderate spatial scales, qualities effectively hardwired since the beginning of the VLA's operations in 1980, make it challenging to map the internal details of multi-pc long filaments and numerous cores. Instead, a next generation Very Large Array (ngVLA) with improved sensitivities and wider fields-of-view will enable acquisition of the data needed to probe best the kinematics and dynamics of dense star-forming gas. With a flexible ngVLA, targets could include large samples of targeted individual pre-/protostellar cores and their host filaments or wide-field mapping across star-forming molecular clouds.

2. NH_3 Observing with ngVLA

This science is uniquely addressed with sensitive observations of low surface brightness emission at high spectral and spatial resolution in K-band. The current ngVLA design will enable acquisition of such data through two important features: i) a large amount of surface area over relatively small baselines to obtain high sensitivity to compact sources, and ii) a Short Baseline Array (SBA) of antennas in a compact fixed configuration that provides high sensitivity to extended sources. As stated above, the Jansky VLA does not have the sensitivities or field-of-view in K-band for wide-field imaging of NH_3 emission (see also below). The GBT is useful for locating NH_3 emission but it does not have the intrinsic angular resolution to resolve cores and filaments beyond Orion, e.g., in several key Giant Molecular Clouds like Cygnus X (see Figure 2). At present, no other single-dish radio telescope on the planet is as optimally equipped for wide-field K-band observing than the GBT. The SKA's current specifications on sensitivity only go up to 15 GHz. While there is a specification for aperture efficiency at 20 GHz, the SKA antennas are not being optimized for performance above 15 GHz. (We note that design work indicates that performance of the SKA antennas will be good at 25 GHz but their actual performance remains to be determined.) ALMA can observe emission from N_2H^+, a nitrogen-based molecule chemically similar to NH_3 that can also trace dense gas, but the critical density of its lowest rotational transition is 1-2 orders of magnitude higher than those of the lower NH_3 transitions, suggesting it is less able to map typically lower density filaments or the transition regions between dense cores and the filaments within which they are embedded. Moreover, N_2H^+ cannot be used to determine kinetic gas temperatures directly as NH_3 can. (Note that an ngVLA with frequency coverage up to 116 GHz will be able to detect N_2H^+ 1-0 emission at 93

GHz at higher sensitivity than ALMA, allowing sensitive probes of compact dense gas in cores.)

No other facilities are planned to sample K-band frequencies with the sensitivities and resolutions of the ngVLA. Synergies may exist with future far-infrared or submillimeter space missions like SPICA and OST, if such facilities have wide-field continuum imaging capabilities. In particular, the potential addition of polarizers on these telescopes would provide high-resolution information on magnetic field morphologies, from which ngVLA line data could allow magnetic field strengths to be determined using the Davis-Chandrasekhar-Fermi method. Though cores and filaments emit largely at far-infrared and submillimeter wavelengths, maps of their internal structures at resolutions higher than *Herschel* or the JCMT, i.e., 1-3″, could arise from extinction mapping of molecular cloud material using deep wide-field near-infrared observations from next-generation facilities such as TMT/E-ELT or LSST. Complementarity with kinematic information on such scales is therefore vital.

There are hundreds of known molecular clouds, from those situated relatively near the Sun (0.1 – 0.5 kpc distance) to those largely confined to the Galactic Plane (~10 kpc distance), so there are several targets available from any potential ngVLA site. The targets, however, are generally within 30 degrees of the Galactic Plane. Dense gas is found within molecular clouds in relatively isolated locations, a small percentage of total cloud surface areas. Nominally, such locations exhibit extended NH_3 emission over ~10′ × 10′ scales in single-dish maps. The typical peak brightnesses of NH_3 (1,1) emission are 1-3 K (T_{mb}) in a 33″ beam. Assuming no beam dilution, those brightnesses translate to 7.5 – 25 mJy beam^{-1} for a "standard resolution" ~4″ beam (see below). Detection of such emission at an signal-to-noise radio of ~5, the minimum needed for high-quality line fitting, requires per channel sensitivities of ~5 mJy beam^{-1}.

3. Example ngVLA Mosaic Observations

To determine examples of reasonable observations for an ngVLA, we first assume a standard target resolution of 4″ FWHM, similar to that provided by the D configuration of the Jansky VLA. Going to higher resolution has a profound impact on surface brightness sensitivity, so there are tradeoffs expected between sensitivity and field-of-view. Of course, more distant clouds for which higher resolution may be most beneficial will be intrinsically more compact and require fewer pointings per mosaic. Single-dish data show NH_3 emission in dense gas regions of nearby molecular clouds to be extended over ~10′ × 10′ fields, leading to a mapped image size of ~100 arcmin^2. In addition, the target maximum angular scale is ~30″. The important addition of an SBA to the ngVLA will enable low surface brightness emission to be detected over larger scales. Here, however, we focus on ngVLA observations without an SBA, to provide a more straightforward comparison with the current Jansky VLA.

We expect that a target sensitivity of ~0.5 K in channels 0.05 km s^{-1} wide over a mosaic of 10′ × 10′ provides a reasonable minimum goal for ngVLA observations of dense gas in star-forming regions. Such high spectral resolution is necessary to resolve NH_3 lines at low temperatures, e.g., the thermal line width of NH_3 at 10 K is 0.07 km s^{-1}. According to the VLA Exposure Calculator[1], 24 GHz can be observed

[1]https://obs.vla.nrao.edu/ect/

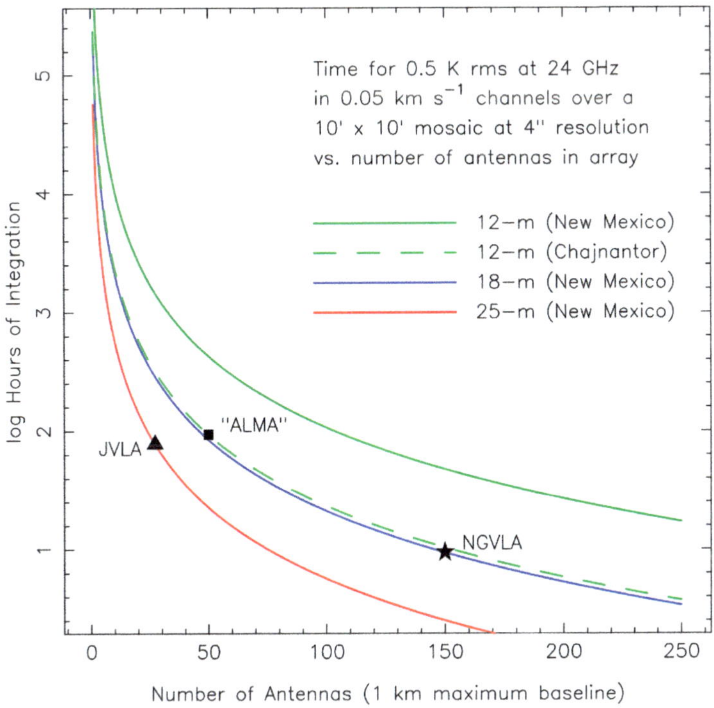

Figure 3. Hours of integration required to reach a 1 σ rms = 0.5 K at 24 GHz in a 0.05 km s^{-1} channel over a $10' \times 10'$ mosaic at $4''$ FWHM resolution given the number of antennas in an array. Note that this resolution requirement means the number of antennas within about ~1 km maximum baseline. The number of hours is obtained using the current Jansky VLA Exposure Calculator and scaling the integration time by the D^2N^4 scaling relation for mapping speed. (D is the antenna diameter and N is the number of antennas.) The solid red line shows the number of 25 m antennas within 1 km needed to reach this target sensitivity in New Mexico (i.e., assuming Jansky VLA site conditions). The black triangle indicates the current Jansky VLA of 27 antennas. The solid green line shows the number of 12 m diameter antennas within 1 km needed in New Mexico, while the dashed green line shows the number of 12 m ALMA-like antennas required at the Llano de Chajnantor in Chile which has superior site conditions. For reference, the black square indicates the 50 × 12 m antennas of ALMA, though ALMA has no plans for K-band receivers for the foreseeable future. The solid blue line shows the number of 18 m antennas within 1 km needed to reach the target sensitivity in New Mexico, with a star indicating the reasonable number (150) needed in a compact distribution of ngVLA antennas.

in a single pointing to 0.5 K sensitivity in 0.05 km s^{-1} channels with 25 antennas in D configuration (i.e., ~ 3 – 4″ FWHM resolution) in ~1.2 hours of winter time on source or ~2.3 hours with overheads. (Baselines of up to 4 km are needed to reach 1″ resolution.) A ~10′ × 10′ field of dense gas, however, requires ~126 pointings for Nyquist sampling in a mosaic. Indeed, a mosaic may benefit from a $\sqrt{2}$ increase in sensitivity from overlapping pointings, equivalent to a savings of time by a factor of 2. Accordingly, the Jansky VLA can reach the target sensitivity over a 10′ × 10′ mosaic in approximately 78 hours. Though tractable for a small cloud of 100 arcmin^2 size or less, mapping the NH_3 emission of larger clouds with the Jansky VLA would take numerous such mosaics to cover their entirety, and hence would require a prohibitively large amount of observing time. For example, the mapped region of Cygnus X shown in Figure 2 consists of eleven such 10′ × 10′ fields.

Figure 3 shows the number of hours needed to reach the target sensitivity of 0.5 K in 0.05 km s^{-1} channels over a 10′ × 10′ mosaic for various numbers of antennas of different size. The trends follow the well-known D^2N^4 scaling relation for mapping speed, where D and N are the diameter and number of the antennas, respectively. The hours needed for the Jansky VLA of 27 antennas of 25 m diameter is indicated, as is the impact on observing time given a lesser or greater number of such antennas. Also shown in Figure 3 are curves indicating the observing times needed to reach the target sensitivity for 18 m and 12 m diameter antennas in New Mexico, again following the D^2N^4 scaling relation. For further comparison, a curve is also shown in Figure 3 for ALMA-like 12 m antennas on the Llano de Chajnantor in Chile, which has site conditions superior to those in New Mexico (e.g., a 50% better atmospheric opacity at 24 GHz is assumed). ALMA has 50 × 12 m antennas and the specific hours of integration needed for it to reach the target sensitivity is indicated in Figure 3. Note, however, that there are no plans to install K-band receivers on ALMA for the foreseeable future. Also, the sensitivity performance of ALMA equipped with K-band receivers would be merely comparable to that of the Jansky VLA.

Figure 3 indicates that ~150 × 18 m antennas within 1 km distance would be sufficient to improve the observing speed of observing such a mosaic to the target sensitivity by roughly an order of magnitude, i.e., 9.4 hours. In contrast, hundreds of 12 m antennas in New Mexico would be needed to reach a similar performance improvement. Note that for the ngVLA curves in Figure 3 we assumed current Jansky VLA receiver and antenna performances and did not include further sensitivity improvements from receiver upgrades or better dish surfaces.

We conclude that 150 antennas with a maximum baseline of ~1 km are required to fulfill this science case. Such antenna numbers are needed as an absolute minimum to enable roughly an order of magnitude improvement over the current Jansky VLA. Compact placement of ngVLA antennas is critical for the ability to image low surface brightness emission from dense gas. For context, a circular area of 1-km diameter has the same surface area as ~3100 antennas of 18 m diameter, so the required configuration will be tight but doable. At present, however, the ngVLA main array concept includes 214 × 18 m antennas of which only ~44% (i.e., 94) would be located within 1.3 km and another ~35% located within 36 km. Hence, it may be worthwhile for the inner ngVLA antennas to have some limited reconfigurability for low surface brightness imaging. Note that the inclusion of an SBA in the ngVLA concept will provide complementary sensitivity to low surface brightness emission on large scales. Indeed, a subset of a few SBA antennas equipped with K-band focal plane arrays to obtain total power observations efficiently would be very beneficial for this science. Detailed simulations of ngVLA observations of dense gas structures, including inputs from various SBA designs, are urgently needed.

Acknowledgments. JDF and JK acknowledge the support of a Discovery Grant by the Natural Science and Engineering Council of Canada.

References

André, Ph., Di Francesco, J., Ward-Thompson, D., et al. 2014, in Protostars and Planets VI, ed. H. Beuther, R. S. Klessen, C. P. Dullemond, & T. Henning, (University of Arizona Press, Tucson), p. 27-51
André, Ph., Men'shchikov, A., Bontemps, S., et al. 2010, A&A, 518, L102
Arzoumanian, D., André, Ph., Didelon, P., et al. 2011, A&A, 529, L6
Benson, P., & Myers, P. C. 1989, ApJS, 71, 89
Crapsi, A., Caselli, P., Walmsley, M., et al. 2007, A&A, 470, 221
Di Francesco, J., Evans, N. J., II., Caselli, P., et al. in Protostars and Planets V, ed. B. Reipurth, D. Jewitt & K. Keil, (University of Arizona Press, Tucson), p. 17
Dunham, M. M., Allen, L. E., Evans, N. J., II, et al. 2015, ApJS, 220, 11
Friesen, R. K., Pineda, J. E., Rosolowsky, E., et al. 2017, ApJ, 843, 63
Friesen, R. K., Bourke, T. L., Di Francesco, J., et al. 2016, ApJ, 833, 204
Hacar, A., Tafalla, M., & Alves, J. 2017, A&A, 606, 24
Hacar, A., Tafalla, M., Kauffmann, J., & Kovács, A. 2013, A&A, 554, 55
Ho, P. T. P., & Townes, C. H. 1983, ARA&A, 21, 239
Keown, J., Di Francesco, J., Rosolowsky, E., et al. 2018, in preparation
Keown, J., Di Francesco, J., Kirk, H., et al. 2017, ApJ, 850, 3
Kirk, H., Friesen, R. K., Pineda, J. E., et al. 2017, ApJ, 846, 144
Könyves, V., André, Ph., Men'shchikov, A. et al. 2015, A&A, 584, A91
Könyves, V., André, Ph., Men'shchikov, A., et al. 2010, A&A, 518, L106
Schneider, N., Csengeri, T., Hennemann, M., et al. 2012, A&A, 540, L11
Singh, A., et al. 2018, in preparation
Walmsley, M., & Ungerechts, H. 1983, A&A, 122, 164
Ward-Thompson, D., Kirk, J., André, Ph., et al. 2010, A&A, 518, L92

Science with a Next-Generation Very Large Array
ASP Conference Series, Monograph 7
Eric J. Murphy, ed.

Resolving the Structure and Kinematics of the Youngest HII Regions and Radio Jets from Young Stellar Objects

Roberto Galván-Madrid,[1] Maite Beltrán,[2] Adam Ginsburg,[3] Carlos Carrasco-González,[1] Hauyu Baobab Liu,[4] Luis F. Rodríguez,[1] and Stanley Kurtz[1]

[1]*Instituto de Radioastronomía y Astrofísica, Universidad Nacional Autónoma de México, Apdo. Postal 72-3 (Xangari), Morelia, Michoacán 58089, Mexico.;* `r.galvan@irya.unam.mx`

[2]*INAF-Osservatorio Astrofisico di Arcetri, Largo E. Fermi 5, I-50125 Firenze, Italy.*

[3]*National Radio Astronomy Observatory, 1003 Lopezville Rd., Socorro, NM 87801, USA.*

[4]*European Southern Observatory, Karl-Schwarzschild-Straße 2, D-85748 Garching bei München, Germany.*

Abstract. In this contribution we explore the new science that a next generation Very Large Array (ngVLA) would be able to perform on the topics of the youngest HII regions and (proto)stellar jets. Free-free continuum and radio recombination line (RRL) emission are often the only way of peering into the dense envelopes surrounding (proto)stars of all masses, and trace their initial feedback in the form of 'radio jets', 'hypercompact HII regions', or photoevaporating, partially-ionized flows. Properly disentangling free-free from dust emission is also mandatory in studies of protoplanetary and accretion disks. Current VLA research has reached an impasse in which a population of faint ionized radio sources, probably corresponding to the above mentioned objects, is detected, but their nature is mostly unknown. The ngVLA would allow us to resolve the density structure and kinematics of such sources, revolutionizing our knowledge of star formation across the entire stellar-mass spectrum.

1. Introduction

1.1. Free-Free and Recombination Line Emission from High-Mass Young Stellar Objects

Thermal free-free emission from ionized gas is a ubiquitous signpost of the physical processes associated with star formation. Such emission is complementary to and as important as that from dust and cold gas. Massive (proto)stars can produce enough Lyman continuum photons to significantly ionize their own envelopes. The VLA was key to the discovery and understanding of ultracompact (UC) and hypercompact (HC) HII regions: the small (< 0.1 pc ≈ 20000 au) pockets of ionized gas associated with newborn massive ($M_\star > 15\ M_\odot$) stars (Dreher & Welch 1981; Ho & Haschick 1981; Kurtz et al. 1994; De Pree et al. 2014; Kalcheva et al. 2018). The nature of the fainter

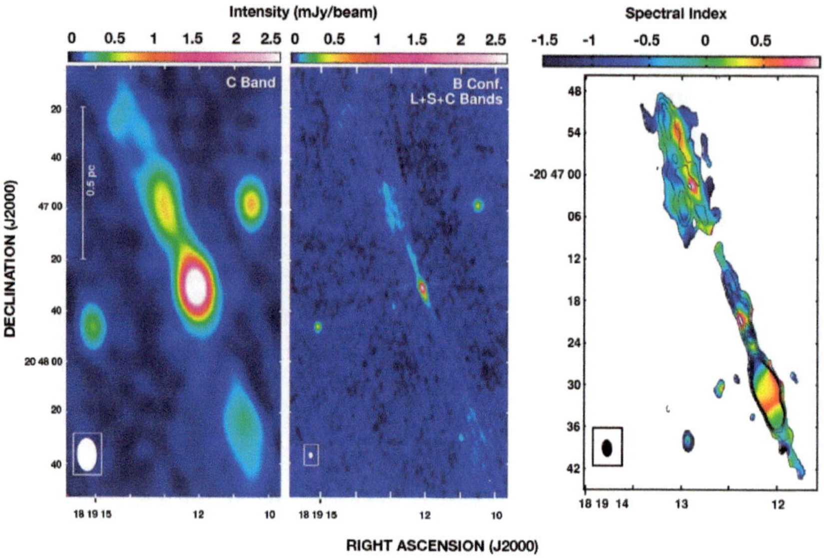

Figure 1. The radio jet from IRAS 18162–2048, which powers Herbig-Haro objects 80 and 81. This object is the prototypical example of a radio jet from a high-mass YSO, well studied thanks to its proximity (1.7 kpc), brightness, and anomalously large (pc) extension. The *left* and *central* panels show low and high (arcsecond) angular resolution views of the thermal jet core and inner jet knots. The *right* panel shows the spectral index derived from multifrequency synthesis imaging over the *L*, *S* and *C* bands. Positive values at the core and jet axis mark the thermal free-free component, while negative spectral indices pinpoint to non-thermal emission. Taken from Rodríguez-Kamenetzky et al. (2017). The ngVLA will allow us to perform this type of studies in statistically significant samples of radio jets from both nearby low-mass YSOs and high-mass YSOs at kpc distances.

ionized sources detected by the VLA in massive star formation regions (Rosero et al. 2016; Moscadelli et al. 2016) is not clear, however, since they are barely detected (flux densities $\lesssim 100\ \mu$Jy at $\nu > 30$ GHz) and mostly unresolved (sizes < 100 mas, or < 500 au at 5 kpc).

Competing theoretical models suggest that these faint free-free sources could either be: i) gravitationally trapped HII regions (Keto 2003), expected to exist in the late stages of accretion in the formation of stars more massive than 20 to 30 $M_\odot$; ii) outflow-confined HII regions (Tanaka et al. 2016), expected to form if there is a photoionized disk-wind outflow; or iii) radio jets similar to those found in their lower-mass counterparts (Beltrán et al. 2016; Purser et al. 2016). Figure 1 shows the iconic HH 80–81 radio jet, one of the brightest known, that is powered by the $\sim 10\ M_\odot$ protostar IRAS 18162–2048 (Girart et al. 2017). A significant fraction of the faint radio sources in massive star formation regions could be stellar winds from very young (Carrasco-González et al. 2015) or unembedded massive stars (Dzib et al. 2013), or even colliding-wind binaries with synchrotron excesses at $\nu < 4$ GHz (Williams et al. 1990; Pittard & Dougherty 2006; Ginsburg et al. 2016). Figure 2 illustrates the differences between the relatively

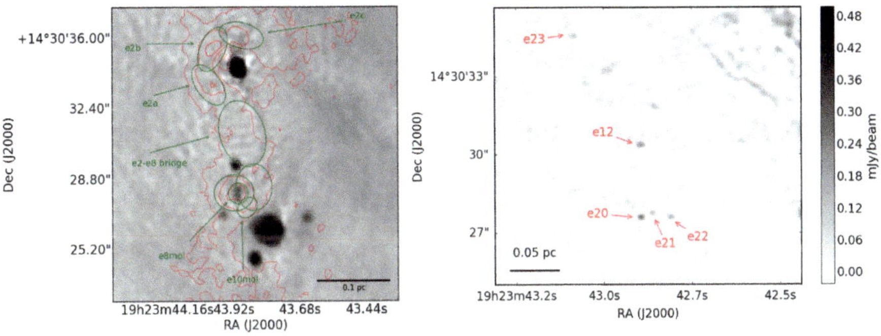

Figure 2. *Left:* Examples of bright hypercompact (HC) H II regions: the well-known e2/e8 system in the massive cluster forming region W51A. Gray-scale shows the free-free 2-cm continuum observed at 0."3 resolution. Contours show their surrounding molecular (formaldehyde) emission. Current facilities such as the VLA can map the continuum of all H II regions and the recombination line emission of some of these with $S_{\rm cm} \sim 10 - 100$ mJy. *Right:* Examples of faint ionized sources that are candidate HC H IIs, radio jets, or stellar-wind systems. Their flux densities $S_{\rm cm} \lesssim 0.1 - 1$ mJy and sizes $l < 100$ mas require the ngVLA to be characterized. Taken from Ginsburg et al. (2016).

bright HC H IIs in W51A, whose nature as photoionized regions around young massive stars is well established, and the compact, fainter sources that are candidates to be HC H IIs, thermal radio jets, or stellar-wind systems.

The high angular resolution and sensitivity of the ngVLA in both cm continuum and Radio Recombination Lines (RRLs) would open a new window to perform systematically detailed studies of the density and velocity distribution of the ionized gas present in the formation of massive stars. In Sections 2 and 3, we give more detailed sensitivity estimates for this scientific case. Regardless of whether or not current theoretical models are correct in detail, free-free emission is the only way of tracing the stellar content in the deeply embedded regions where the formation of massive stars occurs.

The next generation VLA will allow us to finally resolve the structure and kinematics of ionized material, the closest gaseous component to the (proto)star, in both low- and high-mass star formation regions.

1.2. Free-Free and Recombination Line Emission from Low-Mass Young Stellar Objects

The VLA was the most important tool for the discovery of and understanding that low-mass Young Stellar Objects (YSOs) have at their centers collimated jets of free-free emission, known as thermal radio jets, powering the observed molecular outflows and Herbig-Haro systems at larger scales (Rodríguez et al. 1994; Anglada 1995; Rodríguez 1997; Beltrán et al. 2001; Liu et al. 2014; Tychoniec et al. 2018). This finding was an important confirmation of a key prediction of star formation theories (Shu et al. 1994). Low-mass YSOs do not emit significant EUV photons from their photospheres. Thus, ionization produced by photons arising from the shock of the – mostly – neutral stellar wind against the surrounding high-density gas was established as a viable mechanism (Torrelles et al. 1985; Curiel et al. 1987; Anglada et al. 1998). Figure 3 shows a compilation of measurements of bolometric and radio luminosities of sources interpreted as being radio jets, compared to the theoretical expectations for their 'jet vs HII region' nature.

In the most embedded and youngest low-mass YSOs (Class 0 and I, or low-mass protostars), the radio jets are the only way of peering into the dense envelopes and tracing the ejection processes close (< 10 au) to the central objects. Even with the $\sim$ 10 μJy/beam sensitivity and $\sim$ 10 au resolution currently achievable with the VLA for the nearest low-mass star forming regions, however, our understanding of ejection phenomena in low-mas YSOs is far from complete. Specifically, the relatively bright radio jets from Class 0 and I YSOs are barely resolved in the continuum, and their kinematics are unknown due to the practical impossibility of detecting their RRL emission. Furthermore, jet emission in the more evolved Class II YSOs – the protoplanetary disks, now one of the most active fields in astronomy – is far weaker and hardly detectable by the VLA in the most optimistic cases (Galván-Madrid et al. 2014; Macías et al. 2016). For Class II (primordial disks) and Class III (remnant disks) YSOs, other physical processes become observable. Disk photoevaporation, one of the main candidates to cause protoplanetary disk dispersal, may be detected through faint free-free and RRL emission (Pascucci et al. 2012; Owen et al. 2013). Also, unresolved non-thermal (gyro)synchrotron emission from active magnetospheres is often detected with the VLA in some of these more-evolved YSOs (Forbrich & Wolk 2013; Liu et al. 2014; Rivilla et al. 2015). The ngVLA will enable us to resolve, for the first time, the structure and kinematics of the free-free emission from YSOs at all proto-stellar and young-stellar evolutionary stages from Class 0 to Class III, as well as disentangle it from other 'contaminating' processes such as (gyro)synchrotron emission. A full characterization of the free-free emission from YSOs at frequencies $\nu > 30$ GHz is also necessary to separate it from dust emission. This disentangling is a mandatory issue to address in studies of planet formation, which need to get accurate measurements of the mass distribution of disks around low-mass YSOs.

2. Scientific Capabilities with Continuum Observations

We estimate here the scientific outcome of deep (rms noise 0.5 μJy/beam) continuum total intensity observations from 1 to 50 GHz with an angular resolution of 30 mas or better. These values are about the ngVLA 1-hr sensitivity and resolution quoted in the project overview by Carilli et al. (2015). All the science cases described below require such wide frequency coverage to image the respective free-free emission and disentangle it from other 'contaminating' emission mechanisms. At $\nu < 4$ GHz, there could be a contribution from (gyro)synchrotron emission that would need to be properly subtracted (see Hull et al. 2018, this volume p. 357). At $\nu > 50$ GHz dust emission starts to dominate in some objects. This transition region also needs to be covered to disentangle properly the competing emission mechanisms (Isella et al. 2015).

A wide field-of-view and the ability to mosaic are needed to make studies over statistically significant samples, except for highly clustered regions (e.g., Forbrich et al. 2016). A recent VLA survey of YSOs in a nearby star formation region needed 134 pointings to create a Nyquist-sampled mosaic of an area equivalent to 0.5 deg^2 (Hauyu Baobab Liu, personal communication). Under the assumption of 18-m dishes, the ngVLA would need about half the number of pointings than the current VLA. At the sensitivities considered in the following Sections, a full survey of the entire YSO

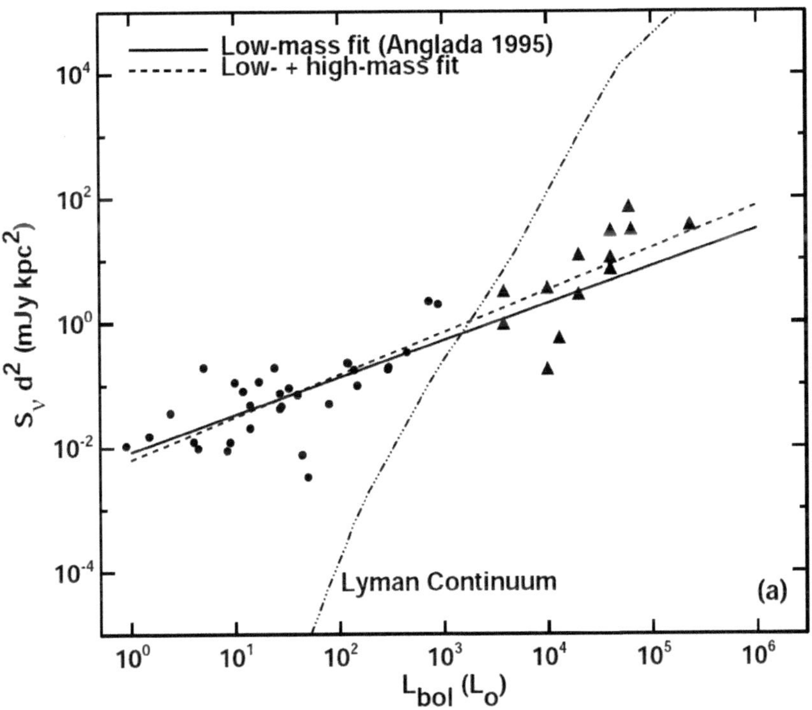

Figure 3. Empirical correlations between the YSO bolometric luminosity $L_{\rm bol}$ and the cm radio luminosity $S\,d^2$. The plotted points and fits correspond to objects best interpreted as radio jets, from the low- to high-luminosity regime. The steeper dash-dotted line shows the expected radio luminosity for H ii regions produced by a ZAMS star of the corresponding $L_{\rm bol}$. HC and UC H iis are observed to populate this locus for $L_{\rm bol} > 10^3\ L_\odot$ (not plotted). The nature of radio objects close to the intersection – of which exact value is somewhat model dependent – is particularly difficult to determine (jet vs H ii region). These objects at $\sim 10^3 - 10^4\ L_\odot$ are highly relevant for understanding the formation of intermediate- and high-mass stars. The ngVLA will allow us to perform systematically resolved studies of radio jets in nearby star formation regions down to the brown-dwarf regime, and of HC H iis at kpc distances down to $\lesssim 10^3\ L_\odot$. Taken from Carrasco-González (2010). See also the recent review by Anglada et al. (2018).

population ($\sim 10^3$ sources) in a nearby star-forming cloud could be completed with a time investment of tens of hours.

2.1. Radio Jets from Low-Mass YSOs

We use the empirical correlations between the radio luminosity of YSOs and their bolometric luminosity (Anglada 1995; Shirley et al. 2007; Anglada et al. 2015), which appear to hold even in the brown dwarf regime (Morata et al. 2015). We also assume a distance of 140 pc throughout this section, characteristic of nearby low-mass star formation regions such as Taurus and Ophiuchus (Ortiz-León et al. 2017). For the lowest-luminosity case, $L_{\rm bol} = 0.01\ L_\odot$, this relation gives a 3.6 cm flux density $F_{3.6\rm cm} = 10\ \mu$Jy, which is only detectable with deep VLA integrations. The jet is comprised of a jet core and extended emission typically in the form of jet knots (see Figure 1 for one of the few well studied cases). The jet core is usually the brightest and we consider it to have ~ 50% of the total jet emission, ($F_{3.6\rm cm,core} = 5\ \mu$Jy). Such emission would be detectable easily with the ngVLA at a signal-to-noise ratio (S/N) ~ 10 in ~ 1 hr integrations (Carilli et al. 2015).

Depending on its driving mechanism, evolutionary stage, and mass, the jet core emission can be spread through very different scales (Frank et al. 2014), from magnetospheric (diameter ~ 0.1 au = 0.7 mas at 140 pc) in the X-wind model to disk-wind scales (~ 10 au = 70 mas). The ngVLA with an angular resolution of 5-20 mas would be able to distinguish between jet launching models, detecting unresolved or barely resolved cores for the former scenario and resolved ones for the latter. It is also possible that some jets are collimated externally through ambient pressure or magnetic fields, in which case the emission would extend to tens of au (Albertazzi et al. 2014) and be more easily resolved. For an unresolved X-wind jet core, most of the flux density would be inside the beam, so it would be easily detected. For an resolved disk-wind jet core, sensitivity to different parts of the emission is model dependent, but assuming a conservative electron temperature $T_e = 5000$ K, then $T_B = 50 - 5000$ K for $\tau = 0.01 - 1$, so the fainter optically-thin emission could be detected at S/N ~ $50/7$ =~ 7 for 1 hr integration at 10 GHz (Carilli et al. 2015).

2.2. Photoevaporating Flows from Class II YSOs

During the Class II stage, free-free emission from the radio jet becomes weaker due to the much lower accretion rates compared to the Class 0 and I stages. Free-free emission from disk photoevaporation by EUV photons could then take over as the principal continuum emission (Pascucci et al. 2014; Galván-Madrid et al. 2014; Macías et al. 2016). The possibility of detecting and imaging this emission is highly relevant for understanding the mechanisms of protoplanetary disk dispersal (Alexander et al. 2014). Pascucci et al. (2012) applied the photoevaporation-flow model of Hollenbach et al. (1994) to estimate the free-free continuum and hydrogen recombination line emission due to EUV photoevaporation. Using their prescription, disks with an ionizing photon luminosity as low as $\Phi = 10^{40}\ \rm s^{-1}$ at 140 pc would have centimetric flux densities (almost constant with frequency) $S_{\rm cm} \sim 4\ \mu$Jy. Therefore, EUV photoevaporation happening in low-mass protoplanetary disks could be easily detectable with the ngVLA at S/N ~ 10 in 1 hour integrations (Carilli et al. 2015).

2.3. Faint Ionized Sources in High-Mass Star Formation

The VLA has been the main tool for discovering and characterizing the small pockets of ionized gas around newly formed O-stars: the so-called ultracompact (UC) and hypercompact (HC) HII regions. In the following, we assume ionization equilibrium – something debatable in the faintest sources that the ngVLA will help to understand – and use the observational calibration for main-sequence OB stars of Vacca et al. (1996). We also assume a typical distance of 5 kpc for massive star formation regions throughout this section. A B0 main sequence star of 20 $M_\odot$ has $L_{\rm bol} = 10^{4.8}\ L_\odot$ and Lyman continuum photon rate $N_{\rm Ly} = 10^{48.1}\ \rm s^{-1}$, which translates into a 8 GHz optically-thin flux density ≈ 480 mJy (quite constant across frequencies since the dependency for optically thin emission is $\nu^{-0.1}$). This emission level indeed defined the target sensitivities for the past few decades (Wood & Churchwell 1989; Kurtz et al. 1994; Urquhart et al. 2009).

Extrapolating the empirical calibration used above to lower masses, the HC HII of an intermediate-mass star with $L_{\rm bol} = 10^3\ L_\odot$ has an 8 GHz flux density of 120 μJy (e.g., Cesaroni et al. 2015). The VLA is currently allowing observers to detect, but not to resolve, a population of radio sources in massive star formation regions with centimeter flux densities $\lesssim 100\,\mu$Jy (e.g., Rosero et al. 2016; Moscadelli et al. 2016). Their interpretation as faint HC HIIs, however, is not unique, and many of them can be as well interpreted as radio jets from high-mass YSOs. Extrapolating the radio-jet flux to bolometric luminosity correlation to high masses gives jet centimeter flux densities that are below HC HII flux densities for all bolometric luminosities above $L_{\rm bol} \sim 10^3\ L_\odot$ (see Figure 3, we note that the exact intersection is model dependent). Radio jets, however, are often assumed to have ionization fractions ~ 0.1 for low-mass YSOs, whereas UC and HC HIIs are considered to be fully ionized. The theoretically expected quenching of HC HIIs due to their surrounding neutral accretion flows (Walmsley 1995; Peters et al. 2010) and a higher ionization fraction in jets from intermediate- and high-mass YSOs could easily invert the relative radio brightnesses of the two different physical scenarios. Observationally, there is evidence that a jet nature may be the correct interpretation for the faint ionized sources of some YSOs with luminosities up to $L_{\rm bol} \sim 10^4\ L_\odot$ (Rosero et al. 2016; Moscadelli et al. 2016; Purser et al. 2016). At these luminosities there could be a transition to HII-region-dominated emission (Guzmán et al. 2014, 2016).

The ngVLA is needed to understand the nature of this population of objects with cm flux densities $S_{\rm cm} \lesssim 100\ \mu$Jy and sizes $l \lesssim 100$ mas (see Figure 2). Assuming a conservative electron temperature $T_e = 5000$ K, then $T_B = 50-5000$ K for $\tau = 0.01-1$, so a 1 hour integration with an rms noise $\sim 0.5\ \mu$Jy/beam ~ 10 K would suffice to map very optically thin emission at 5 – 30 mas resolution, which would correspond to 25 – 150 au at 5 kpc.

3. Scientific Capabilities with Recombination Line Observations

We estimate here the scientific outcome of deep (rms noise of 100 μJy/beam in channels 5 km/s wide) total intensity observations of RRLs from H78α (5.0 GHz) to H40α (99.0 GHz). The desired maximum angular resolution is 100 mas for unresolved detections, 30 mas for resolved detections, or slightly better for the few brightest cases. A hundred milliarcseconds correspond to 14 au for objects in nearby (140 pc) low-mass star forming regions, or to 500 au at 5 kpc.

Continuum imaging is not enough to model and fully interpret ionized sources in star formation. The kinematic information inferred from hydrogen recombination lines is also needed. Recombination lines in the cm radio and (sub)mm are weak compared to the often used optical and infrared lines such as Lyα, Hα, and Brγ. They are unobscured, however, which makes RRLs a unique tool for revealing the kinematics in the embedded regions where star formation occurs (Martín-Pintado et al. 1989; Peters et al. 2012; Jiménez-Serra et al. 2013; Plambeck et al. 2013; Zhang et al. 2017).

The VLA, ALMA, and other interferometers have been able to detect systematically RRLs from the UCs and HC HIIs produced by O and early-B type stars (Keto et al. 2008), but not for lower masses. The resolved kinematics of ionized gas in star formation is almost unexplored except for the brighter UC HIIs and some HC HIIs (e.g., Garay et al. 1986; Liu et al. 2012). Also, only a couple of detections of RRL emission from bright radio jets have been reported (Jiménez-Serra et al. 2011, 2013), probably with significant contribution from an underlying Hɪɪ region. Attempts to detect RRLs in iconic objects such as the jet of IRAS 18162–2048 (Figure 1 shows its continuum emission) have produced negative results (Roberto Galván-Madrid, unpublished). Indeed, RRLs are more easily detected in HC Hɪɪs than in radio jets because the former tend to be brighter and their line widths are one order of magnitude smaller. Even 'broad linewidth' HC Hɪɪs have $\Delta V \sim 30 - 50$ km s^{-1}, whereas jet motions appear to have ΔV of several hundreds of km s^{-1}, as inferred from their proper motions (Curiel et al. 2006; Rodríguez-Kamenetzky et al. 2016; Guzmán et al. 2016) and a few tentative RRL detections (Jiménez-Serra et al. 2011). Therefore, ngVLA measurements of RRL line widths alone, even if angularly unresolved, would serve as a discriminator for competing models of the interpretation of faint ionized sources in massive star formation.

Lines such as H51α are detectable for $L_{\rm bol} = 10^5~L_\odot$ objects even with the sensitivities of the VLA before its upgrade (e.g., Galván-Madrid et al. 2009). In the following, we consider the detectability of RRL emission from fainter sources at kpc distances with the ngVLA. The LTE line-to-continuum ratio $S_{\rm line}\Delta V/S_{\rm cont}$ goes as $\nu^{1.1}$ (Gordon & Sorochenko 2002). Observationally, it has been measured to be $S_{\rm line}\Delta V/S_{\rm cont}$ = $30 - 100$ km s^{-1} for the H30α line (231.9 GHz) for HC Hɪɪs with $L_{\rm bol} \sim 10^5 L_\odot$ (Keto et al. 2008). We consider $S_{\rm line}\Delta V/S_{\rm cont} = 50$ km s^{-1} for the H30α line and $\Delta V = 25$ km s^{-1} for all lines. This estimate translates into $S_{\rm line}\Delta V/S_{\rm cont} \sim 17.6$ km s^{-1} for H41α (90.0 GHz), 8.9 km s^{-1} for H51α (48.1 GHz), and 0.8 km s^{-1} for H106α (5.4 GHz). We further consider a HC Hɪɪ region with a 48 GHz continuum flux density $S_{\rm 48GHz} = 1$ mJy, such as the ones detected in $L_{\rm bol} \sim 10^4~L_\odot$ YSOs by van der Tak & Menten (2005). This HC Hɪɪ would have a velocity-integrated H51α flux $S_{\rm line}\Delta V = 8.9$ mJy km s^{-1}. At this frequency, the expected ngVLA noise in a 5 km s^{-1} channel for a 10 hour integration is $\sim 32~\mu$Jy/beam (Carilli et al. 2015). For the assumed FWHM linewidth of 25 km s^{-1}, the line peak would be $S_{\rm peak} \sim 334~\mu$Jy, detectable at S/N $\sim$ 10.

The above discussion applies to beam-matched or angularly unresolved observations with a HPBW $\sim$ 100 mas. Observations at higher angular resolution would need to aim at brighter objects or integrate longer. The S/N of individual line detections for these faint Hɪɪ regions would go up by factors $\sim$ 2 for the H41α line at 90 GHz, so resolved RRL mapping would be less expensive using high-frequency RRLs. Lines in the frequency range $50 < \nu < 100$ GHz could be the right compromise between high brightness and the absence of contamination from molecular lines compared to RRLs at ALMA frequencies. Therefore, RRL observations are an important motivation for the ngVLA to operate at frequencies $\nu > 50$ GHz.

Non-detections at frequencies $\nu < 20$ GHz or in fainter objects could be solved by taking advantage of stacking of lines from neighboring quantum numbers. The VLA Ka and Q bands ($36 - 50$ GHz) harbor 12 α hydrogen RRLs (H62α to H51α) that could be stacked together in a given object to decrease the noise by factors of $\sim 3 - 4$. This opportunity highlights the importance of having a large simultaneous frequency coverage.

Although the previous discussion considered a HC Hɪɪ from a massive (proto)star at kpc distances, it also applies to radio jets from low-mass YSOs at the canonical distance of 140 pc. These radio jets have cm flux densities $S_{\rm cm} \sim 1$ mJy for luminosities $L_{\rm bol} > 10~L_\odot$ (see Figure 3). Their RRL emission could be spread through several hundred km s^{-1}. Assuming that the radio-jet line width is $\times$10 larger than for HC Hɪɪs, the S/N would decrease by factors of $\sim 10^{0.5} \sim 3$ if the velocity binning increases from $5 - 50$ km s^{-1}.

In conclusion, although more expensive than continuum observations, kinematical imaging of hydrogen recombination line emission is feasible with the ngVLA. Such observations would open for the first time a window into understanding the kinematics of ionized gas in the formation of both low- and high-mass stars.

4. Peering into the Innermost Regions in the Formation of Stars of all Masses

From the previous sections, it can be concluded that the ngVLA will allow us to perform, for the first time, studies of statistically significant samples of the resolved structure and kinematics of the ionized gas associated with stars in formation and young stars of all masses. This ionized gas is often the closest gaseous component to the (proto)star itself. Table 1 summarizes the basic continuum properties of the objects that we have considered in detail among the anticipated 'radio zoo'.

Object	Typical distance [kpc]	8 GHz Flux Density [μJy]	Size [mas (au)]
Faint HC H II or massive radio jet	5	< 100	< 100 (< 500)
Very low luminosity radio jet	0.14	5	0.7 – 70 (0.1 – 10)
Photoevaporation flow from low-mass disk	0.14	4	7 – 350 (1 – 50)

Table 1. Free-free continuum properties of the considered objects.

5. Uniqueness of ngVLA Capabilities

Only the VLA and a future ngVLA can observe the free-free continuum emission associated with star formation in the range of frequencies where it dominates (4 – 50 GHz). At frequencies < 4 GHz and > 50 GHz, (gyro)synchrotron and dust emission, respectively, can be a significant 'contaminant', or even dominate the observed flux density. Having a telescope with frequency coverage beyond the overlapping ranges is needed to model properly and subtract the contribution to the emission from those other mechanisms. Conversely, properly subtracting the free-free contribution will be mandatory in the much anticipated studies of dust emission in protoplanetary disks.

Similarly, hydrogen radio recombination lines from about 20 to 100 GHz have the right compromise between brightness (they are too faint at lower frequencies) and the absence of confusion with the rich molecular forest seen at higher (ALMA) frequencies.

A field-of-view of at least a few arcminutes size, as given by the proposed 18-m dishes, is needed to perform surveys of large numbers of sources in areas of the order of a square degree in reasonably long integrations of tens of hours.

Although not discussed at length in this paper (see Hull et al. 2018, this volume p. 357), the capability to image in full polarization mode is highly desirable, since a few radio jets are known to have synchrotron emission beyond the thermal core (Carrasco-González et al. 2010), and that might be the rule rather than the exception.

Finally, a large instantaneous bandwidth (of the order of the central frequency where possible) would uniquely open the time-domain window in a systematic way. Radio jets (Liu et al. 2014) and some HC HIIs (De Pree et al. 2014) are known to be variable in timescales from days to years, a phenomenon with important implications to the properties of accretion in YSOs (Galván-Madrid et al. 2011; Hunter et al. 2017).

6. Synergies at other Wavelengths

The ngVLA would naturally complement and enhance the star formation research done with ALMA and the SKA. The studies of free-free and recombination line emission proposed in this chapter will consistently be put in the context of ALMA observations of the corresponding dust and molecular line emission. All tracers are needed to have a comprehensive physical picture. SKA observations at lower-frequencies would greatly help to disentangle non-thermal emission mechanisms. The proposed ngVLA research would also benefit observations with the next generation of ground- and space-based near- and mid-IR telescopes, such as the TMT, GMT, E-ELT, *JWST*, and WFIRST, since those observations are the most efficient to discover and classify YSOs. Finally, high-angular resolution mid-IR and far-IR observations, such as the ones provided by SOFIA and other future facilities, are in high demand to pinpoint the sources of luminosity at resolutions that are not too coarse compared with instruments like ngVLA, ALMA, and *JWST*.

Acknowledgments. The National Radio Astronomy Observatory is a facility of the National Science Foundation operated under cooperative agreement by Associated Universities, Inc. RGM acknowledges support from UNAM-PAPIIT program IA102817.

References

Albertazzi, B. 2014, Science, 346, 325
Alexander, R., Pascucci, I., Andrews, S., Armitage, P., & Cieza, L. 2014 Protostars and Planets VI ed H. Beuther et al (Tucson, AZ: Univ. Arizona Press) 475
Anglada, Guillem 1995, RMxAC, 1, 67
Anglada, G., Rodríguez, L. F., Carrasco-González, C. Advancing Astrophysics with the Square Kilometre Array (AASKA14) 121
Anglada, G., Rodríguez, L. F., Carrasco-González, C. 2018, A&ARv, 26, 3
Anglada, Guillem, Villuendas, Eva, Estalella, Robert, Beltrán, Maria T., Rodríguez, Luis F., Torrelles, José M., Curiel, Salvador 1998, AJ, 116, 2953
Beltrán, M. T., Cesaroni, R., Moscadelli, L., Sánchez-Monge, A., Hirota, T., Kumar, M. S. N. 2016, A&A, 593A, 49
Beltrán, Maria T., Estalella, Robert, Anglada, Guillem, Rodríguez, Luis F., Torrelles, José M. 2001, AJ, 121, 1556
Carrasco-González, Carlos. 2010, PhD Thesis, Universidad de Granada.
Carrasco-González, Carlos, Rodríguez, Luis F., Anglada, Guillem, Martí, Josep, Torrelles, José M., Osorio, Mayra 2010, Science, 330, 1209
Carrasco-González, C., et al. 2015, Science, 348, 114
Carilli, C. L., et al. 2015, arXiv:1510.06438
Cesaroni, R., Pestalozzi, M., Beltrán, M. T., Hoare, M. G., Molinari, S., Olmi, L., Smith, M. D., Stringfellow, G. S., Testi, L., Thompson, M. A. 2015, A&A, 579A, 71
Curiel, Salvador, Cantó, Jorge, Rodríguez, Luis F. 1987, RMxAA, 14, 595

Curiel, Salvador, et al. 2006, ApJ, 638, 878

De Pree, C. G., Peters, T., Mac Low, M.-M., Wilner, D. J., Goss, W. M., Galván-Madrid, R., Keto, E. R., Klessen, R. S., Monsrud, A. 2014, ApJ, 781, L36

Dreher, J. W., Welch, W. J. 1981, ApJ, 245, 857

Dzib, S. A., Rodríguez-Garza, C. B., Rodríguez, L. F., Kurtz, S. E., Loinard, L., Zapata, L. A., Lizano, S. 2013, ApJ, 772, 151

Forbrich, J., Wolk, S. J. 2013, A&A, 551A, 56

Forbrich, J., Rivilla, V. M., Menten, K. M., Reid, M. J., Chandler, C. J., Rau, U., Bhatnagar, S., Wolk, S. J., Meingast, S. 2016, ApJ, 822, 93

Frank, A., et al. 2014 Protostars and Planets VI ed H. Beuther et al (Tucson, AZ: Univ. Arizona Press) 451

Galván-Madrid, Roberto, Keto, Eric, Zhang, Qizhou, Kurtz, Stan, Rodríguez, Luis F., Ho, Paul T. P. 2009, ApJ, 706, 1036

Galván-Madrid, R., Liu, H. B., Manara, C. F., Forbrich, J., Pascucci, I., Carrasco-González, C., Goddi, C., Hasegawa, Y., Takami, M., Testi, L. 2014, A&A, 570, L9

Galván-Madrid, R., Peters, T., Keto, E. R., Mac Low, M.-M., Banerjee, R., Klessen, R. S. 2011, MNRAS, 416, 1033

Garay, Guido, Rodríguez, Luis F., van Gorkom, J. H. 1986, ApJ, 309, 553

Ginsburg, Adam, et al. 2016, A&A, 595A, 27

Girart, J. M., et al. 2017, ApJ, 847, 58

Gordon, M. A. & Sorochenko, P. N. 2002, Radio Recombination Lines: Their Physics and Astronomical Applications (Dordrecht: Kluwer)

Guzmán, A. E., Garay, G., Rodríguez, L. F., Moran, J., Brooks, K. J., Bronfman, L., Nyman, L.-A., Sanhueza, P., Mardones, D. 2014, ApJ, 796, 117

Guzmán, Andrés E., Garay, Guido, Rodríguez, Luis F., Contreras, Yanett, Dougados, Catherine, Cabrit, Sylvie 2016, ApJ, 826, 208

Hollenbach, David, Johnstone, Doug, Lizano, Susana, Shu, Frank 1994, ApJ, 428, 654

Ho, P. T. P., Haschick, A. D. 1981, ApJ, 248, 622

Hull, C. L. H., Carrasco-González, C., Williams, P. K. G., et al. 2018, Science with a Next-Generation Very Large Array, ASP Monograph 7, ed. E. J. Murphy (San Francisco, CA: ASP), 357

Hunter, T. R., Brogan, C. L., MacLeod, G., Cyganowski, C. J., Chandler, C. J., Chibueze, J. O., Friesen, R., Indebetouw, R., Thesner, C., Young, K. H. 2017, ApJ, 837, L29

Isella, Andrea, et al. 2015, arXiv:1510.06444

Jiménez-Serra, I., Martín-Pintado, J., Báez-Rubio, A., Patel, N., Thum, C. 2011, ApJ, 732, L27

Jiménez-Serra, I., Báez-Rubio, A., Rivilla, V. M., Martín-Pintado, J., Zhang, Q., Dierickx, M., Patel, N. 2013, ApJ, 764, L4

Kalcheva, I. E., Hoare, M. G., Urquhart, J. S., Kurtz, S., Lumsden, S. L., Purcell, C. R., Zijlstra, A. A. 2018, arXiv:1803.09334

Keto, Eric. 2003, ApJ, 599, 1196

Keto, Eric, Zhang, Qizhou, Kurtz, Stanley 2008, ApJ, 672, 423

Kurtz, S., Churchwell, E., Wood, D. O. S. 1994, ApJS, 91, 659

Liu, Hauyu Baobab, et al. 2014, ApJ, 780, 155

Liu, Hauyu Baobab, Jiménez Serra, Izaskun, Ho, Paul T. P., Chen, Huei-Ru, Zhang, Qizhou, Li, Zhi-Yun 2012, ApJ, 756, 10

Macías, Enrique, et al. 2016, ApJ, 829, 1

Martín-Pintado, J., Bachiller, R., Thum, C., Walmsley, M. 1989, A&A, 215, L13

Morata, Oscar, et al. 2015, ApJ, 807, 55

Moscadelli, L., Sánchez-Monge, A., Goddi, C., Li, J. J., Sanna, A., Cesaroni, R., Pestalozzi, M., Molinari, S., Reid, M. J.
2016, A&A, 585A, 71

Ortiz-León, Gisela, et al. 2017, ApJ, 834, 141O

Owen, James E., Scaife, Anna M. M., Ercolano, Barbara 2013, MNRAS, 434, 3378

Pascucci, I., Gorti, U., Hollenbach, D. 2012, ApJ, 751, L42

Pascucci, I., Ricci, L., Gorti, U., Hollenbach, D., Hendler, N. P., Brooks, K. J., Contreras, Y. 2014, ApJ, 795, 1
Plambeck, R. L., et al. 2013, ApJ, 765, 40
Peters, Thomas, Banerjee, Robi, Klessen, Ralf S., Mac Low, Mordecai-Mark, Galván-Madrid, Roberto, Keto, Eric R. 2010, ApJ, 711, 1017
Peters, Thomas, Longmore, Steven N., Dullemond, Cornelis P. 2012, MNRAS, 425, 2352
Pittard, J. M., Dougherty, S. M. 2006, MNRAS, 372, 801
Purser, S. J. D., Lumsden, S. L., Hoare, M. G., Urquhart, J. S., Cunningham, N., Purcell, C. R., Brooks, K. J., Garay, G., Guzmán, A. E., Voronkov, M. A. 2016, MNRAS, 460, 1039
Rivilla, V. M., Chandler, C. J., Sanz-Forcada, J., Jiménez-Serra, I., Forbrich, J., Martín-Pintado, J. 2015, ApJ, 808, 146
Rodríguez, L. F. 1997, in IAU Symp. 182, Herbig-Haro Flows and the Birth of Stars, ed B. Reipurth and C. Bertout (Cambridge: Cambridge Univ. Press) 83
Rodríguez, Luis F., Garay, Guido, Curiel, Salvador, Ramirez, Solange, Torrelles, José M., Gómez, Yolanda, Velazquez, Arturo 1994, ApJ, 430, L65
Rodríguez-Kamenetzky, Adriana, Carrasco-González, Carlos, Araudo, Anabella, Torrelles, José M., Anglada, Guillem, Martí, Josep, Rodríguez, Luis F., Valotto, Carlos 2016, ApJ, 818, 27
Rodríguez-Kamenetzky, Adriana, Carrasco-González, Carlos, Araudo, Anabella, Romero, Gustavo E., Torrelles, José M., Rodríguez, Luis F., Anglada, Guillem, Martí, Josep, Perucho, Manel, Valotto, Carlos 2017, ApJ, 851, 16
Rosero, Viviana, et al. 2016, ApJS, 227, 25
Shirley, Yancy L., Claussen, Mark J., Bourke, Tyler L., Young, Chadwick H., Blake, Geoffrey A. 2007, ApJ, 667, 329
Shu, Frank, Najita, Joan, Ostriker, Eve, Wilkin, Frank, Ruden, Steven, Lizano, Susana. 1994, ApJ, 429, 781
Tanaka, Kei E. I., Tan, Jonathan C., Zhang, Yichen 2016, ApJ, 818, 52
Torrelles, J. M., Ho, P. T. P., Rodríguez, L. F., Cantó, J. 1985, ApJ, 288, 595
Tychoniec, Ł, et al. 2018, arXiv:1806.02434
Urquhart, J. S., Hoare, M. G., Purcell, C. R., Lumsden, S. L., Oudmaijer, R. D., Moore, T. J. T., Busfield, A. L., Mottram, J. C., Davies, B. 2009, A&A, 501, 539
van der Tak, F. F. S., Menten, K. M. 2005, A&A, 437, 947
Vacca, William D., Garmany, Catharine D., Shull, J. Michael 1996, ApJ, 460, 914
Walmsley, M. 1995, RMxAC, 1, 137
Williams, P. M., van der Hucht, K. A., Pollock, A. M. T., Florkowski, D. R., van der Woerd, H., Wamsteker, W. M. 1990, MNRAS, 243, 662
Wood, Douglas O. S., Churchwell, Ed 1989, ApJS, 69, 831
Zhang, Qizhou, Claus, Brian, Watson, Linda, Moran, James 2017, ApJ, 837, 53

Science with a Next-Generation Very Large Array
ASP Conference Series, Monograph 7
Eric J. Murphy, ed.

Understanding Massive Star Formation through Maser Imaging

Todd R. Hunter,[1] Crystal L. Brogan,[1] Anna Bartkiewicz,[2]
James O. Chibueze,[3,4,5] Claudia J. Cyganowski,[6] Tomoya Hirota,[7]
Gordon C. MacLeod,[8] Alberto Sanna,[9] and José-Maria Torrelles[10]

[1]*NRAO, Charlottesville, VA, USA;* `thunter@nrao.edu, cbrogan@nrao.edu`

[2]*Nicolaus Copernicus University, Torun, Poland;* `annan@astro.umk.pl`

[3]*Department of Physics and Astronomy, University of Nigeria, Nsukka, Nigeria*

[4]*SKA South Africa, Cape Town, South Africa;* `james.chibueze@gmail.com`

[5]*Centre for Space Research, Physics Department, North-West University, Potchefstroom, South Africa*

[6]*SUPA, School of Physics and Astronomy, University of St. Andrews, St. Andrews, UK;* `cc243@st-andrews.ac.uk`

[7]*National Astronomical Observatory of Japan, Mitaka, Tokyo, Japan;* `tomoya.hirota@nao.ac.jp`

[8]*Hartebeesthoek Radio Astronomy Observatory, Krugersdorp, South Africa;* `gord@hartrao.ac.za`

[9]*Max-Planck-Institut für Radioastronomie, Bonn, Germany;* `asanna@mpifr-bonn.mpg.de`

[10]*Institut de Ciéncies de l'Espai, Barcelona, Spain;* `chema.torrelles@ice.cat`

Abstract. Imaging the bright maser emission produced by several molecular species at centimeter wavelengths is an essential tool for understanding the process of massive star formation because it provides a way to probe the kinematics of dense molecular gas at high angular resolution. Unimpeded by the high dust optical depths that affect shorter wavelength observations, the high brightness temperature of these emission lines offers a way to resolve accretion and outflow motions down to scales as fine as ~1-10 au in deeply embedded Galactic star-forming regions, and at sub-pc scales in nearby galaxies. The next generation Very Large Array will provide the capabilities needed to fully exploit these powerful tracers.

1. Introduction: The Scientific Importance of Masers

The process of star formation leads to the concentration of molecular gas to high densities in molecular cloud cores. The potential energy released by gravitational collapse and accretion onto the central protostars heats and excites the surrounding material through infrared radiation and high velocity bipolar outflows. Both of these feedback mechanisms (radiative and mechanical) naturally produce population inversions between specific pairs of energy levels in several abundant molecules, including H_2O, CH_3OH, OH, NH_3, SiO, and H_2CO. The resulting non-thermal maser emission in the corresponding spectral transitions provides a beacon whose brightness temperature far exceeds the more commonly-excited thermal emission lines. Consequently, maser lines at centimeter wavelengths have traditionally provided a powerful probe of star formation through single-dish surveys and interferometric imaging. In general, they trace hot, dense molecular gas, revealing the kinematics of star-forming material within a few 1000 au from very young stars, including accretion disks and their associated jets, as well as shocks in the outflow lobes where the jets impact ambient gas. Masers are generally more prevalent in regions surrounding massive protostars, due to their higher luminosities and more energetic outflows. Furthermore, masers are sensitive indicators of sudden changes in the pumping conditions near the protostars, and, recently, it has been recognized that maser flares, in those lines which are pumped by infrared photons, can be directly associated with bursts of accretion onto the stars. In this context, maser emission provides a unique tool for probing how massive stars form, allowing us to reconstruct the gas dynamics in the vicinity of young stars with tens of Solar masses, as well as to study the accretion process in the time domain.

2. Description of the Problem and Limits of Current Facilities

With the advent of the Atacama Large Millimeter/submillimeter Array (ALMA), imaging thermal lines at high resolution has become easier, and recent results have begun to place previous and ongoing maser studies into better physical context (see, e.g., Orion Source I; Plambeck & Wright 2016; Hirota et al. 2017). At the distances of more typical massive star-forming regions, however, the brightness temperature sensitivity of ALMA is still not sufficient to trace the accretion flow and accompanying jet structures that surround massive protostars, because of the high angular and spectral resolution required. Moreover, at the shorter wavelengths of ALMA, the combination of molecular line confusion and high dust opacity toward the hot cores in protoclusters will often block the most interesting details from ALMA's view. In contrast, the centimeter maser transitions propagate unobscured from the innermost regions, providing a strong signal for self-calibration, and thus enabling high dynamic range imaging on long baselines.

Unfortunately, the angular resolution of the VLA is insufficient to resolve the details of accreting gas, particularly in the 6 GHz band where the beamsize is limited to $\sim 0\farcs3$. In the handful of nearest examples of massive star formation ($d \sim 1$ kpc), this resolution corresponds to 300 au (e.g., Brogan et al. 2016). However, in the majority of star-forming sites across the Milky Way at several kpc from the Sun, it exceeds 1000 au, which is often more than the separation of protostars in the centers of massive protoclusters (e.g., Palau et al. 2013). Thus, an order of magnitude improvement in angular resolution (requiring at least $\sim$300 km baselines) is needed to resolve the spatial morphology and kinematics of disks, or other accretion structures, at scales of 10 – 100 au in a large sample of massive protostars. Such a resolution would also enable three dimensional measurements of gas velocity via multi-epoch proper motions. For instance, with an angular resolution of 10 mas in the bright H_2O maser line, it is possible to determine the maser positions with an accuracy better than 0.1 mas (assuming a signal-to-noise ratio > 100), and then to measure proper motions of order 10 km s^{-1} in a few months only, for sources located up to several kpc distance. Furthermore, proper motion measurements of different maser species toward the same region have the potential to trace simultaneously the complementary kinematic structures around a young star, providing a unique picture of the gas dynamics locally (Sanna et al. 2010; Goddi et al. 2011). These measurements are currently conducted by means of Very Long Baseline Interferometry (VLBI) observations, but with the small number of available antennas their sensitivity is limited to non-thermal processes exceeding brightness temperatures of $\sim 10^7$ K (e.g., Matsumoto et al. 2014; Bartkiewicz et al. 2009). An example of proper motion observations at these scales, using three different VLBI facilities, is shown in Figure 1 (Burns et al. 2017).

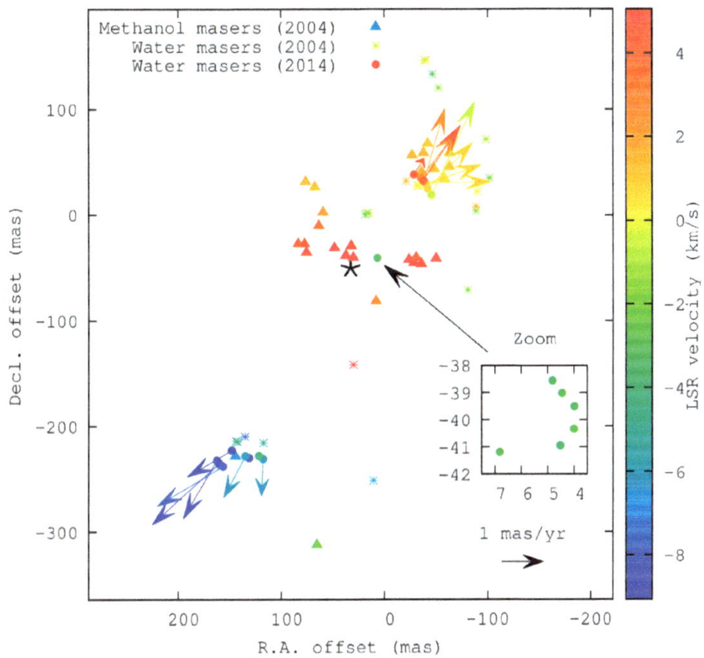

Figure 1. Central region of the dominant member (MM1) of the massive proto-cluster AFGL 5142 (figure extracted from Burns et al. 2017). Combined view of 22 GHz water masers (filled circles) observed with VERA in 2010, 22 GHz water masers (asterisk) observed with the VLBA in 2004 (Goddi & Moscadelli 2006) and 6.7 GHz methanol masers (triangles) observed with the EVN in 2004 (Goddi et al. 2007). The inset shows the trajectory of maser feature A, moving in a clockwise fashion. The trajectory of feature A and proper motions of other masers are all converted to the YSO frame. The black asterisk symbol indicates the approximate origin of the episodic ejections, estimated from least-squares fitting of ellipses to the VERA maser data.

While current VLBI facilities (VLBA, EVN, eMERLIN, KVN, VERA, and LBA) have the requisite angular resolution to trace maser proper motions accurately, studies at these scales currently suffer from poor surface brightness sensitivity, which affects the science in two key ways. First, only the brightest maser spots can be detected, reducing the fidelity with which kinematic structures can be delineated in a single epoch, and reducing the number of potential spots that will persist over multiple epochs (used for proper motion studies). Second, the thermal radio continuum emission which arises in the immediate vicinity of massive protostars, with flux densities of < 1 mJy typically (Cyganowski et al. 2011), cannot be observed with the masers simultaneously, leading to (relative) positional uncertainties between the protostellar and maser components. The resulting ambiguity of the dynamical center severely hinders the interpretation of multi-epoch proper motion measurements, which are essential to understand the mass, momentum, and kinetic energy of the inner jet where it transitions into a bipolar molecular outflow.

Studying additional objects at scales of 10-100 au in a comprehensive list of maser lines, and with sufficient brightness sensitivity to image simultaneously the associated thermal free-free continuum emission, will be an important task for the next generation Very Large Array (ngVLA); see the accompanying chapter on Jets from YSOs (Galván-Madrid et al. 2018, this volume p. 309). These studies will test and expand our current picture of massive star formation into the broader context of the Milky Way. Moreover, with 300 km baselines, it will also be possible to trace molecular gas structures down to 0.2 pc scales in the star-forming clouds in nearby extragalactic nuclei (e.g., Maffei 2 / IC 342 / M82) in the bright water maser and Class I methanol maser lines, enabling the study of individual massive protoclusters in these objects for the first time.

3. Astronomical Impact

While each maser transition offers a unique view into particular phenomena of massive star formation, masers can be broadly classified into major categories. The Class II CH_3OH maser lines, primarily at 6.7 GHz, 12.2 GHz, and 19.9 GHz, trace hot molecular gas that is (at least) moderately close to the youngest massive protostars ($\lesssim$ 1000 au), such that they can provide sufficiently intense mid-infrared emission to pump the maser transitions (e.g., Moscadelli et al. 2011; Bartkiewicz et al. 2014). The light curves of this maser species also show intriguing properties. Quasi-periodic flares in one or more Class II CH_3OH maser lines (120 – 500 days) have been observed in about a dozen objects (e.g., Goedhart et al. 2014); in one case, the 4.83 GHz H_2CO maser also shows correlated flaring (Araya et al. 2010). Recently, two spectacular accretion outbursts in massive protostars have been accompanied by strong flaring of these lines, S 255 NIRS3 (Caratti o Garatti et al. 2017; Moscadelli et al. 2017) and NGC 6334I-MM1 (Hunter et al. 2017, 2018; MacLeod et al. 2018), supporting the idea that maser flares might be caused by a variable accretion rate onto the central protostar. These extraordinary events led to the formation of the international Maser Monitoring Organization (M2O), with the goal of detecting and reporting future maser flares so that interferometers can be alerted to study the accretion event while it is still underway. Such an accretion event is also expected to yield variation in the continuum emission from the thermal jet (Cesaroni et al. 2018) and/or the hypercompact HII region (Brogan et al. 2018). Since both of these phenomena are powered by the protostar, the ability of ngVLA to perform simultaneous observations of the continuum and the masers will enable direct measurements of the correlations between them, yielding important constraints on the physics of the accretion mechanism.

The Class II CH_3OH maser lines (Sobolev et al. 1997; Cragg et al. 2005), along with the 1.6 GHz ground state OH lines and several excited state OH lines which are radiatively pumped (at 4.66 GHz, 4.75 GHz, 4.765 GHz, 6.030 GHz, and 6.035 GHz), are also seen to trace the ionization front of ultracompact HII regions (e.g., Fish & Reid

2007), which are powered by the more evolved massive protostars and Zero-Age Main Sequence (ZAMS) OB stars. Although excited OH lines are generally considered rare, a recent unbiased survey found that the 6.035 GHz line is detected toward nearly 30% of Class II CH_3OH masers and with a similar distribution in Galactic latitude (Avison et al. 2016). A similar detection rate is seen in survey of the 4.765 GHz line (Dodson & Ellingsen 2002). A simple explanation is these excited OH masers always occur in the same objects that power Class II masers, but simply have a correspondingly shorter mean lifetime or duty cycle, perhaps reflecting how long they remain above current sensitivity levels following each successive accretion outburst. In rare cases, the main line OH masers can also trace outflow motion (e.g., W75N and W3OH-TW Fish et al. 2011; Argon et al. 2003).

The strong water maser line at 22 GHz also traces gas close to massive and intermediate mass protostars. Often these lines span a broad (LSR) velocity range, of several tens of km s^{-1}, about the systemic velocity of the young stars, particularly compared to both classes of methanol masers ($\lesssim 10\,km\,s^{-1}$). In some cases, water masers clearly arise from gas in the first few hundred au of the jet, such as in Cepheus A (e.g., Torrelles et al. 2011; Chibueze et al. 2012), or in bow shocks somewhat further out (e.g., Sanna et al. 2012; Burns et al. 2016). With continent-scale baselines, proper motion studies of these masers reveal the 3D velocities and orientations of collimated jets and/or wide-angle winds in the inner few 1000 au from the central protostars (e.g., Torrelles et al. 2001, 2003, 2014; Moscadelli et al. 2007; Sanna et al. 2010; Burns et al. 2017). When these studies are combined with high-resolution radio continuum observations of radio thermal jets, they can allow us to quantify the outflow energetics directly produced by the star formation process (e.g., Moscadelli et al. 2016; Sanna et al. 2016), as opposed to estimates of the molecular outflow energetics that are attainable on scales greater than 0.1 pc (typically through CO isotopologues). Long-term monitoring studies demonstrate that water masers are also highly variable (e.g., Felli et al. 2007), and their primary pumping mechanism is not believed to be radiative but collisional. Thus, since water masers are fundamentally produced in specific ranges of gas density and temperature within shocked gas layers, they are likely to trace different types of coherent motions at different stages of protostellar evolution. This is the case, for instance, of the star-forming region W75 N, where the 22 GHz masers (and the radio continuum) show different spatial distributions around two distinct young stars at different evolutionary stages (Torrelles et al. 2003; Carrasco-González et al. 2015). The 22 GHz line is also unique in exhibiting the 'superburst' phenomenon, in which brief flares reach 10^5 Jy or more. It has happened in only a few objects including Orion KL (Hirota et al. 2014, and references therein) and G25.65+1.05 (Lekht et al. 2018), but it has repeated in both, and appears to be due to interaction of the jet with high density clumps in the ambient gas, but within a few thousand au of the central protostar. In addition to Galactic studies, the detection and imaging of water masers in nearby star-forming galaxies provides a powerful probe of optically-obscured areas of star formation like the overlap region of the Antennae (Brogan et al. 2010). Additional uses of extragalactic masers are described in the accompanying chapter on Megamaser Cosmology (Braatz et al. 2018, this volume p. 821).

In contrast to H_2O masers and Class II CH_3OH masers, the Class I CH_3OH maser lines [primarily at 25 GHz, 36 GHz, 37 GHz, 44 GHz, and 95 GHz, but see Müller et al. (2004) for a more complete list] typically arise from collisionally-excited gas located much further from the protostar ($\sim 0.1 - 0.5$ pc) where the bipolar molecular outflow lobes impact ambient gas. VLA surveys of these masers show that they often coincide with 4.5 μm emission that traces shocked gas in active outflows (Towner et al. 2017; Cyganowski et al. 2012, 2009). A similar maser phenomenon occurs in the ortho-NH_3 (3,3) and (6,6) lines at 23.87 GHz and 25.06 GHz (Brogan et al. 2011; Kraemer & Jackson 1995). Unfortunately, the VLA beam size is insufficient to resolve the structure of individual Class I maser features but VLBI observations resolve out most of the emission. Furthermore, the sensitivity of VLBI arrays is insufficient to map weak maser features detected by the VLA and single dishes (e.g. Matsumoto et al. 2014). We note that a portion of Class II maser emission also tends to be resolved out on VLBI scales (Pandian et al. 2011; Bartkiewicz et al. 2009; Minier et al. 2002). Thus, the ngVLA will provide a unique tool to study the spatial and velocity structures of both types of methanol masers.

The Class I CH_3OH masers are particularly abundant in the Galactic Center star-forming clouds (McEwen et al. 2016), and have recently been detected in starburst galaxies (Ellingsen et al. 2017; McCarthy et al. 2017). With ngVLA sensitivity and resolution, these masers could be used to probe such star-forming clouds in nearby galaxies (including Maffei 2 / IC 342 / M82).

Many maser lines are significantly polarized, thus it is important to observe them with full Stokes products in order to obtain the highest fidelity imaging. Furthermore, recent full polarization images of various Class I and Class II methanol maser lines (e.g., Surcis et al. 2013; Dall'Olio et al. 2017; Momjian & Sarma 2017) and the 22 GHz water line (e.g., Surcis et al. 2011; Goddi et al. 2017) have been used successfully to measure the magnetic field toward massive protostars via Zeeman splitting. These maser lines thus offer the potential to help us understand the degree to which magnetic fields influence or control accretion and outflows in massive protostars by providing very valuable information of magnetic field and velocity vectors in the same maser features (Sanna et al. 2015; Goddi et al. 2017).

Finally, maser emission from the vibrationally-excited levels of SiO offers a powerful (though rare) probe of the innermost hot gas surrounding massive protostars. For example, in one spectacular nearby case (Orion KL), movies of the vibrationally-excited SiO J=1-0, v=0 and v=1 transitions at 43 GHz have revealed a complicated structure of disk rotation and outflow (Matthews et al. 2010). Additional massive protostars (at greater distances) have recently been detected in these lines (Cordiner et al. 2016; Ginsburg et al. 2015; Zapata et al. 2009). The increased sensitivity of the ngVLA will no doubt yield further detections and enable new detailed images of the inner accretion structures in these objects. Additional uses of all maser species is described in the accompanying chapter on Evolved Stars (Matthews et al. 2018, this volume p. 281).

4. Connection to Unique ngVLA Capabilities

With its proposed frequency span, the ngVLA will uniquely provide access to all of the most important maser transitions from OH at 1.6 GHz to CH_3OH at 109 GHz. While the Long Baseline Observatory's Very Long Baseline Array (VLBA) also covers most of this frequency span, it lacks the sensitivity at the critical baseline lengths of up to 300 km to image simultaneously the continuum emission. The ngVLA will provide the required balance between angular resolution and brightness sensitivity, filling an important gap in existing capabilities. With the ability to image the non-thermal and thermal processes simultaneously, it will finally be possible to link the studies of the 3D gas dynamics (using the masers) with studies of the physical conditions of the ionized and molecular gas (using the continuum and strong, compact thermal lines like ammonia, respectively) at the same spatial resolution. Also, the ability to acquire high-fidelity images of all of these maser lines in just a few tunings will promote more uniform surveys of massive protostars as well as enable rapid monitoring of protostars currently undergoing an accretion outburst. Furthermore, the broader bandwidth receivers will provide more robust measurements of the spectral index of the continuum emission by promoting the ability to obtain all the necessary observations at a common epoch. Finally, the high spectral resolution and full polarization capability of the ngVLA will allow measurements of the magnetic field in the masing molecular gas via the Zeeman effect in methanol and water, which is a fundamental quantity for understanding the physics of star formation.

5. Experimental Layout

In addition to studies of individual high mass protostellar objects (HMPOs), such as those currently undergoing an outburst, one can foresee a large ngVLA project to perform multi-epoch imaging of a significant sample of (several dozen) high-mass star-forming regions (HMSFRs). These HMSFRs typically contain clusters of massive protostars in diverse evolutionary states, a phenomenon that is often termed a "proto-Trapezium" (Megeath et al. 2005) or a "protocluster" (Minier et al. 2005). The following project would allow us: (1) to take a census of the young stellar objects across a broad mass range in these protoclusters, directly measuring the initial-mass-function; (2) to measure the kinematics of the gas undergoing maser emission, which will trace the flow of gas near the most massive stars; and (3) to study the way massive stars feed energy back into the protocluster environment and influence the formation of Solar-mass stars (and vice versa), by combining points (1) and (2).

Most protoclusters will be easily encompassed by a single pointing of the ngVLA in the lower bands, in which the primary beam full width half power (FWHP) is envisioned to be $\sim 10'$ (15 pc at 5 kpc) at 6.7 GHz. The largest ones may require a few pointings at the highest band where the beam is only $\sim 40''$ (1 pc at 5 kpc) at 100 GHz. Given that protostars at different stages of evolution may excite different masers, imaging these fields in all the ngVLA bands, and using all available antennas with full polarization products will efficiently identify all the massive protostars via their continuum emission (ionized gas and/or dust) and various maser lines. To illustrate the level of sensitivity required, we consider the continuum emission from jets which are closely linked to maser activity (see §3). Furthermore, we consider specifically jets from intermediate-mass protostars in order to define a sensitivity that will probe a broad range of massive protostars. The emission from individual clumps in jets observed by the VLA toward intermediate-mass protostars is approximately 1 mJy at 2 – 20 GHz in nearby (400 pc) regions like Serpens (Rodríguez-Kamenetzky et al. 2016), which translates to only 6 μJy for similar examples that are expected to populate more massive protoclusters at 5 kpc. In order to measure the SED of such an object, we require a $\approx 6\sigma$ detection in each 1 GHz subband. With the current VLA, we would need 44 hr of VLA observing time to reach 1 μJy rms in X-band alone. With the increased effective collecting area of the ngVLA, this time requirement would drop to $\lesssim 1$ hr per band, meaning that a multi-band survey of many fields would become feasible. Also, with this sensitivity, young lower-mass T Tauri stars, which are chromospherically active stars associated with highly-variable faint (synchrotron) radio continuum emission, will be also detected in the lower frequency bands (e.g., C-band), providing information about the low-mass population (Forbrich et al. 2017).

The available correlator channels will be distributed accordingly among low resolution continuum windows and up to 20 high-resolution spectral line windows in each of the 6 bands. Due to the 300 km baselines, we note that the maximum channel width for a 24 GHz (K-band) continuum observation is 0.5 MHz in order to limit bandwidth smearing to less than 2% across the FWHP of the primary beam, which means that ~28000 dual-polarization channels will be needed for the 14 GHz instantaneous bandwidth available in that band. In addition, the high-resolution line windows will require between 256 channels and 1024 channels each in order to provide Hanning-smoothed spectra with 0.1 km s^{-1} resolution across a velocity width of 12 km s^{-1} to 50 km s^{-1}. The broadest lines (primarily water) will require 4096 channels to cover ~200 km s^{-1}, so an additional ~20000 full-polarization channels are also needed for the maser lines. Thus, the number of correlator resources needed for the line observations is comparable to that needed for the continuum. A list of maser lines detectable in each band is summarized in Table 1.

Table 1. Detected maser lines in each ngVLA band; see Table 1 of Menten (2007) for further details

band	Range (GHz)	Species and line frequencies (GHz)	# lines
1	1.2 - 3.5	ground-state OH (1.612, 1.665, 1.667, 1.720)	4
2	3.5 - 12.3	excited OH (4.66, 4.75, 4.765, 6.031, 6.035, 6.049)	6
		CH_3OH Class I (9.936)	1
		CH_3OH Class II (6.668, 12.18)	2
		ortho-H_2CO (4.83)	1
3	12.3 - 20.5	CH_3OH Class II (19.97)	1
		excited OH (13.441, Baudry et al. 1981; Caswell 2004)	1
4	20.5 - 34	H_2O (22.235)	1
		CH_3OH J_2-J_1-series Class I (24.9-30.3)	15
		CH_3OH Class I (23.445, Voronkov et al. 2011)	1
		CH_3OH Class II (23.12, Cragg et al. 2004)	1
		ortho-NH_3 (3,3) (6,6) (9,9) (12,12)	4
		other NH_3 inversion lines (thermal and/or maser)	~dozen
		excited OH (23.8, Baudry et al. 1981)	1
5	30.5 - 50.5	Class I CH_3OH (36.169, 44.069)	2
		Class II CH_3OH (37.7, 38.29, 38.45)	3
		SiO 1-0 v=1,2 (43.12, 42.82)	2
6	70 - 116	CH_3OH Class I (84.521, 95.169, 104.3)	3
		CH_3OH Class II (76.2, 76.5, 85.5, 86.6, 104.1, 107, 108.8)	7
		SiO 2-1 v=1,2 (86.24, 85.64)	2

With the calibrated (u, v)-data, we will first construct multi-band continuum images with a matched resolution of $0\farcs05$ in the range $5-100$ GHz, using all the baselines (~360 km) at 5 GHz, and tapering the data to 60 km at 30 GHz and 18 km at 100 GHz. The photometry from these images will provide the spectral energy distributions of all the individual protostars (or compact binaries) on scales of 250 au $\times$ $(d/5$ kpc$)$, giving an immediate census of the protocluster. At these scales, images of thermal line emission from tracers of warm dense gas can also be attempted. Finer details of the individual objects can then be pursued by imaging the continuum using the longer baselines in the higher frequency bands, for instance, as fine as 0.″01 at 22 GHz [50 au $\times(d/5\mathrm{kpc})$]. This step will allow us to distinguish between jets (linearly-shaped emission), ultra-compact and hypercompact HII regions (more symmetric emission), and T Tauri stars (non-thermal point-like emission). At this scale, in the higher frequency bands, disk-like accretion structures may be visible in the (still) optically-thin dust emission. Next, we can image the full field in the individual maser lines at moderate spectral resolution ($\sim$1 km s^{-1}) and in both polarizations, tabulating which transitions are associated with each object. In some cases, this process will include identifying the angular scale at which the emission becomes resolved in order to maintain sensitivity to the weakest features. We can then make spatially smaller cubes of each maser line, imaging several smaller sub-fields (i.e., those containing all the strong sources) taking advantage of the highest angular and spectral resolution ($\sim$0.1 km s^{-1}) in order to analyze the maser gas kinematics associated with each protostar. In many cases, we would make these cubes with full polarization in order to measure the magnetic fields. Finally, the strongest compact features can then be used to measure proper motions over several months or years.

On the one hand, such an ambitious project would require the efforts of many students and researchers at many international institutions, but on the other hand, it will foster new synergies between astronomers with diverse expertise, and provide a global picture of the youngest star-forming sites across the Milky Way.

6. Complementarity with Respect to Existing and Planned (>2025) Racilities

The science described here will share many synergies with other future facilities. Of course, ALMA will likely be the closest complementary facility as it can observe the dust emission and strong thermal lines at comparable angular resolutions. The Square Kilometer Array (SKA) will provide access to high resolution imaging of the lowest frequency masers, such as OH at 1612 MHz, 1665 MHz, 1667 MHz and 1720 MHz, as well as covering the southern Galactic plane in the additional maser lines that will be accessible in its highest frequency bands including Class II methanol masers. Toward many target regions, imaging at mid-infrared wavelengths by the Thirty Meter Telescope (TMT) and the European Extremely Large Telescope (ELT) will be quite informative as these facilities can potentially deliver 0.″05-0.″2 resolution at 7-28 μm which can probe through a portion of the typically high column of extinction. The *James Webb Space Telescope* can also provide sensitive integral field images for regions where saturation is not prohibitive. In both cases, the short wavelengths will delineate outflow cavities on larger scales, while the longer wavelengths (when combined with ALMA observations) will enable measurements of the luminosities of the individual massive protostars. Finally, future X-ray space telescopes such as Lynx (Vikhlinin 2018) can potentially provide crucial measurements of transient hot gas following accretion outbursts in massive protostars, as well as study the variable emission from the lower mass members of the protocluster (see, e.g., Montes et al. 2015; Townsley et al. 2014).

Acknowledgments. The National Radio Astronomy Observatory is a facility of the National Science Foundation operated under agreement by Associated Universities, Inc.

References

Araya, E. et al. 2010, ApJ, 717, 133
Argon, A. L., Reid, M. J., & Menten, K. M. 2003, ApJ, 593, 925
Avison, A., Quinn, L. J., Fuller, G. A., et al. 2016, MNRAS, 461, 136
Bartkiewicz, A., Szymczak, M., & van Langevelde, H. J. 2014, A&A, 564, A110
Bartkiewicz, A., Szymczak, M., van Langevelde, H. J., Richards, A. M. S., & Pihlström, Y. M. 2009, A&A, 502, 155
Baudry, A., Walmsley, C. M., Winnberg, A., & Wilson, T. L. 1981, A&A, 102, 287
Braatz, J., et al. 2018, Science with a Next-Generation Very Large Array, ASP Monograph 7, ed. E. J. Murphy (San Francisco, CA: ASP), 821
Brogan, C.L., et al. 2018, in Proc. of IAU Symp. 336, ed. A. Tarchi, M.J. Reid & P. Castangia, Cambridge University Press
Brogan, C.L., et al. 2016, ApJ, 832, 187
Brogan, C. L., Hunter, T. R., Cyganowski, C. J., et al. 2011, ApJ, 739, L16
Brogan, C., Johnson, K., & Darling, J. 2010, ApJ, 716, L51
Burns, R. A., Handa, T., Imai, H., et al. 2017, MNRAS, 467, 2367
Burns, R. A., Handa, T., Nagayama, T., Sunada, K., & Omodaka, T. 2016, MNRAS, 460, 283
Caratti o Garatti, A., Stecklum, B., Garcia Lopez, R., et al. 2017, Nature Physics, 13, 276
Carrasco-González, C., Torrelles, J. M., Cantó, J., et al. 2015, Science, 348, 114
Caswell, J. L. 2004, MNRAS, 352, 101
Cesaroni, R., Moscadelli, L., Neri, R., et al. 2018, A&A, 612, 103, arXiv:1802.04228
Chibueze, J. O., Imai, H., Tafoya, D., et al. 2012, ApJ, 748, 146
Cordiner, M. A., Boogert, A. C. A., Charnley, S. B., et al. 2016, ApJ, 828, 51
Cragg, D. M., Sobolev, A. M., & Godfrey, P. D. 2005, MNRAS, 360, 533
Cragg, D. M., Sobolev, A. M., Caswell, J. L., Ellingsen, S. P., & Godfrey, P. D. 2004, MNRAS, 351, 1327
Cyganowski, C. J., Brogan, C. L., Hunter, T. R., et al. 2012, ApJ, 760, L20
Cyganowski, C. J., Brogan, C. L., Hunter, T. R., & Churchwell, E. 2011, ApJ, 743, 56
Cyganowski, C.J., et al. 2009, ApJ, 702, 1615
Dall'Olio, D., Vlemmings, W. H. T., Surcis, G., et al. 2017, A&A, 607, A111
Dodson, R. G., & Ellingsen, S. P. 2002, MNRAS, 333, 307
Ellingsen, S. P., Chen, X., Breen, S. L., & Qiao, H.-H. 2017, MNRAS, 472, 604
Felli, M., Brand, J., Cesaroni, R., et al. 2007, A&A, 476, 373
Fish, V. L., Gray, M., Goss, W. M., & Richards, A. M. S. 2011, MNRAS, 417, 555
Fish, V. L., & Reid, M. J. 2007, ApJ, 670, 1159
Forbrich, J., Reid, M. J., Menten, K. M., et al. 2017, ApJ, 844, 109
Galván-Madrid, R., et al. 2018, Science with a Next-Generation Very Large Array, ASP Monograph 7, ed. E. J. Murphy (San Francisco, CA: ASP), 309
Ginsburg, A., Walsh, A., Henkel, C., et al. 2015, A&A, 584, L7
Goddi, C., et al. 2017, A&A, 597, 43
Goddi, C., Moscadelli, L., & Sanna, A. 2011, A&A, 535, L8
Goddi, C., Moscadelli, L., Sanna, A., Cesaroni, R., & Minier, V. 2007, A&A, 461, 1027
Goddi, C., & Moscadelli, L. 2006, A&A, 447, 577
Goedhart, S., et al. 2014, MNRAS, 437, 1808
Hirota, T., Machida, M. N., Matsushita, Y., et al. 2017, Nature Astronomy, 1, 0146
Hirota, T., Tsuboi, M., Kurono, Y., et al. 2014, PASJ, 66, 106
Hunter, T. R., Brogan, C. L., et al. 2018, ApJ, 854, 170
Hunter, T. R., Brogan, C. L., MacLeod, G. C., et al. 2017, ApJ, 837, L29
Kraemer, K. E., & Jackson, J. M. 1995, ApJ, 439, L9
Lekht, E. E., et al. 2018, Astronomy Reports, 62, 213
MacLeod, G. C., et al. 2018, MNRAS, 478, 1077, arXiv: 1804.05308
Matsumoto, N., Hirota, T., Sugiyama, K., et al. 2014, ApJ, 789, L1
Matthews, L., et al. 2018, Science with a Next-Generation Very Large Array, ASP Monograph 7, ed. E. J. Murphy (San Francisco, CA: ASP), 281

Matthews, L., et al. 2010, ApJ, 708, 80
McCarthy, T. P., Ellingsen, S. P., Chen, X., et al. 2017, ApJ, 846, 156
McEwen, B. C., Sjouwerman, L. O., & Pihlström, Y. M. 2016, ApJ, 832, 129
Megeath, S. T., Wilson, T. L., & Corbin, M. R. 2005, ApJ, 622, L141
Menten, K. M. 2007, Astrophysical Masers and their Environments, IAU Symposium 242, 496
Minier, V., Burton, M. G., Hill, T., et al. 2005, A&A, 429, 945
Minier, V., Booth, R. S., & Conway, J. E. 2002, A&A, 383, 614
Momjian, E. & Sarma, A. 2017, ApJ, 834, 168
Montes, V. A., Hofner, P., Anderson, C., & Rosero, V. 2015, ApJS, 219, 41
Moscadelli, L., Sanna, A., Goddi, C., et al. 2017, A&A, 600, L8
Moscadelli, L., Sánchez-Monge, Á., Goddi, C., et al. 2016, A&A, 585, A71
Moscadelli, L., Sanna, A., & Goddi, C. 2011, A&A, 536, A38
Moscadelli, L., Goddi, C., Cesaroni, R., Beltrán, M. T., & Furuya, R. S. 2007, A&A, 472, 867
Müller, H. S. P., Menten, K. M., & Mäder, H. 2004, A&A, 428, 1019
Palau, A., Fuente, A., Girart, J. M., et al. 2013, ApJ, 762, 120
Pandian, J. D., Momjian, E., Xu, Y., Menten, K. M., & Goldsmith, P. F. 2011, ApJ, 730, 55
Plambeck, R. L., & Wright, M. C. H. 2016, ApJ, 833, 219
Rodríguez-Kamenetzky, A., Carrasco-González, C., Araudo, A., et al. 2016, ApJ, 818, 27
Sanna, A., Moscadelli, L., Cesaroni, R., et al. 2016, A&A, 596, L2
Sanna, A., Surcis, G., Moscadelli, L., et al. 2015, A&A, 583, L3
Sanna, A., Reid, M. J., Carrasco-González, C., et al. 2012, ApJ, 745, 191
Sanna, A., Moscadelli, L., Cesaroni, R., et al. 2010, A&A, 517, A78
Sobolev, A. M., Cragg, D. M., & Godfrey, P. D. 1997, A&A, 324, 211
Surcis, G., Vlemmings, W. H. T., van Langevelde, H. J., et al. 2013, A&A, 556, A73
Surcis, G., Vlemmings, W. H. T., Curiel, S., et al. 2011, A&A, 527, A48
Torrelles, J. M., Trinidad, M. A., Curiel, S., et al. 2014, MNRAS, 437, 3803
Torrelles, J. M., Patel, N. A., Curiel, S., et al. 2011, MNRAS, 410, 627
Torrelles, J. M., Patel, N. A., Anglada, G., et al. 2003, ApJ, 598, L115
Torrelles, J. M., Patel, N. A., Gómez, J. F., et al. 2001, Nat, 411, 277
Towner, A. P. M., Brogan, C. L., Hunter, T. R., et al. 2017, ApJS, 230, 22
Townsley, L. K., Broos, P. S., Garmire, G. P., et al. 2014, ApJS, 213, 1
Vikhlinin, A. 2018, American Astronomical Society Meeting Abstracts #231, #103.04
Voronkov, M. A., Walsh, A. J., Caswell, J. L., et al. 2011, MNRAS, 413, 2339
Zapata, L. A., Menten, K., Reid, M., & Beuther, H. 2009, ApJ, 691, 332

Science with a Next-Generation Very Large Array
ASP Conference Series, Monograph 7
Eric J. Murphy, ed.

New Frontiers in Protostellar Multiplicity with the ngVLA

John J. Tobin,[1] Patrick Sheehan,[1] and Doug Johnstone

[1]*University of Oklahoma,* `jjtobin@ou.edu`

[2]*NRC-Herzberg*

Abstract. The ngVLA will enable significant advances in our understanding of the formation and evolution of multiple star systems in the protostellar phase, building upon the breakthroughs enabled by the VLA. The high-sensitivity and resolution at 3 mm wavelengths and longer will enable closer multiple systems to be discovered in the nearby star forming regions. The ngVLA is incredibly important for multiplicity studies because dust opacity at short wavelengths (< 3 mm) can hide multiplicity and the long wavelengths are needed to reveal forming multiples in the youngest systems. The samples sizes can be expanded to encompass star forming regions at distances of at least 1.5 kpc, enabling statistical studies that are on par with studies of field star multiplicity. We verify the capability of the ngVLA to detect and resolve multiple star systems at distances out to 1.5 kpc using empirical examples of systems detected by the VLA and scaling them to greater distances. We also use radiative transfer models and simulations to verify that the ngVLA can resolve close binary systems from their dust emission at these distances. The ngVLA will also have excellent imaging capability and the circum-multiple environments can also be examined in great detail.

1. Introduction

Star formation occurs as a consequence of dense gas clouds collapsing under their own gravity, once the gravitational force is able to overcome sources of support (e.g., thermal pressure, magnetic fields, turbulence; McKee & Ostriker 2007). The star formation process frequently results in the formation of two or more stars that comprise a gravitationally bound system, given that nearly half of Sun-like stars (in terms of stellar mass) are found in binary or higher-order multiple systems (Duquennoy & Mayor 1991; Raghavan et al. 2010). The degree of stellar multiplicity strongly depends on stellar mass. Stars more massive than the Sun have a higher fraction of multiplicity and stars less massive than the Sun have a lower degree of multiplicity, but still with a multiplicity fraction upwards of 25%; see Duchêne & Kraus (2013) for a recent review. Thus, multiple star formation is a common outcome of the star formation process for all stellar masses and a comprehensive understanding of star formation must account for multiplicity.

Recent surveys in the infrared, millimeter, and centimeter have also shown a high degree of multiplicity in the protostar phase (Connelley et al. 2008; Looney et al. 2000). Both the youngest protostars (Class 0 sources, André et al. 1993) and more evolved protostars (Class I), show a higher degree of multiplicity than field stars (e.g., Reipurth et al. 2004; Chen et al. 2013; Tobin et al. 2016b). The largest mass reservoir is available during the protostellar phase, making it the most promising epoch for companion star formation to occur (Tohline 2002). Thus, the distribution of field star multiplicity is likely derived from the primordial distribution of companions that form during the protostar phase. While the mechanisms of multiple star formation are still uncertain, the distribution of separations in the protostellar phase may reveal the signature of their formation process. The peak of the companion separation distribution for field solar-type stars is ~50 AU (Raghavan et al. 2010), but the formation route for these systems cannot be determined from evolved stellar populations alone because they have undergone Myr to Gyr of dynamical evolution. Moreover, not all systems that form as multiples may remain multiple throughout their lives, but they may still have formed in the presence of one or more companion stars (Sadavoy & Stahler 2017). Some companions can be ejected through dynamical interactions (e.g., Reipurth & Mikkola 2012) or become unbound after the dispersal of the star forming core. This means that the formation of nascent planetary systems may have been influenced by companion stars, even if they are no longer bound to the system. Thus, to reveal the origins of stellar multiplicity and its effects on proto-planetary systems, multiplicity must be characterized during the earliest stage of the star formation process.

There are two favored routes to explain the formation of multiple star systems: disk fragmentation due to gravitational instability (e.g., Stamatellos & Whitworth 2009; Kratter et al. 2010) and turbulent fragmentation within the molecular cloud (e.g., Padoan & Nordlund 2002; Offner et al. 2010). Disk fragmentation will preferentially result in the formation of close (<500AU) multiple star systems and requires a large (R_{disk} ~50 AU) and massive rotationally-supported disk to have formed around the primary star. Turbulent fragmentation can result in the formation of both wide and close multiple systems. In this scenario, the initial protostars form with separations ~1000 AU, and depending on their relative motions and masses they can migrate inward to radii ~100 AU (Offner et al. 2010), remain at wide radii, or drift further apart. However, the expected trends can only be revealed statistically, requiring large samples of protostars to be observed with spatial resolution better than 50 AU (the average field star separation). It is also important to point out that rotation of the protostellar cloud itself has also been suggested as a mechanism to form multiple star systems (Burkert & Bodenheimer 1993), but current observational evidence points more toward turbulence for the formation of wide companions (Lee et al. 2016).

A key difference in examining the formation of multiple stars at radio/millimeter wavelengths versus optical/infrared is that direct emission from the protostellar photosphere is not being detected. Instead, the dust emission from the individual circumstellar disks and/or free-free emission from the base of the protostellar jet are being probed. Thus, the observations are tracing fragmentation, but cannot directly confirm the protostellar nature of the multiple observed sources. Tobin et al. (2016b) examined the nature of the emission detected and concluded that multiple sources of emission very likely correspond to true multiplicity, but we note, however, that the possibility of a small percentage of false positives cannot be excluded. Nevertheless, observations at millimeter/centimeter wavelengths are generally the only available tool to study multiplicity toward such deeply embedded objects.

The multiplicity statistics of field stars have the benefit of large samples and are not subject to the same limitations of protostellar multiplicity studies. Nonetheless, it is useful to use the field multiplicity studies as a baseline for the survey requirements of future protostellar multiplicity studies. The most recent compendium of field solar-type star multiplicity had a sample of 454 stars in a volume limited sample, finding a total of 259 companions with an average companion separation of 50 AU in an approximate Gaussian distribution (Raghavan et al. 2010). Assuming that the field multiplicity distribution represents the distribution of protostellar multiplicity (it probably does not), we can use this distribution in order to estimate how many protostellar multiples would need to be detected in order to obtain the same level of statistical accuracy. Restricting the relevant parameter space to a separation range between 1 AU and 1000 AU, the Raghavan et al. (2010) sample contains ~126 companion stars within this range. It is necessary to limit the parameter space in this regard because scales less than 1 AU will likely be difficult to examine for protostars and scales greater than 1000 AU will be dominated by clustering and not physical association. In order to observe 126 protostellar companions to match the sample size, at least 630 protostars must be observed if the average companion frequency of protostars is in this range is 20% (Tobin et al. 2016b). This would require that all protostars within the Gould Belt be observed, sampling over different star formation conditions. If the youngest (Class 0) protostars are required for such characterization, then there are much less than 630 in the entire Gould Belt (Dunham et al. 2014). Surveys of more distant, massive star forming regions are therefore required to observe this number of protostars in a single region.

2. Setting the Stage: The VANDAM Survey

A first large, systematic survey to examine protostellar multiplicity in an entire star forming region was carried out by the VLA Nascent Disk and Multiplicity (VANDAM) Survey (Tobin et al. 2015). This survey was carried out toward the Perseus star forming region, observing 45 Class 0 protostars (including possible objects in transition; Class 0/I), 37 Class I protostars, and 12 Class II young stars. All protostars were observed at a wavelength of 9 mm (33 GHz), at a uniform sensitivity of ~9 μJy, and at a uniform resolution of 15 AU ($0\farcs065$). From these data, we identified 18 systems that were multiple with companion separations <500 AU; 16 of these companions were new detections by the VANDAM survey. We also detected a number of systems with separations >500 AU as well as many hierarchical multiple systems. Some multiple systems had separations as small as ~19 AU, near the limit of our spatial resolution. Figure 1 shows an example of a very close multiple system. Figure 2 then shows an example of a hierarchical system toward L1448 IRS3 where there is a triple, a binary, and a single system all within 5000 AU. The triple was revealed by ALMA to have a circum-multiple disk, strongly indicating that the system formed via disk fragmentation (Tobin et al. 2016a), and the system as a whole in Figure 2 illustrates how fragmentation is a multi-scale process.

The separations of all companion stars detected in the VANDAM survey, are shown as a histogram in Figure 3. Two features are obvious: 1) there appears to be two peaks, one at ~75 AU and another >1000 AU, and 2) the separation distribution is in excess of the field, except for separations between ~300 AU and ~1000 AU. It is argued in Tobin et al. (2016b) that this bimodality results from both disk fragmentation (~75 AU part) and turbulent fragmentation (>1000 AU part) happening to produce the observed distribution. There are a lower number of detected companions between ~300 AU and ~1000 AU, indicating that neither formation mechanism is efficient at these scales.

While this survey of Perseus represents a major advance, it was limited in terms of the number of sources. Considering uncertainties, the distribution could be consistent

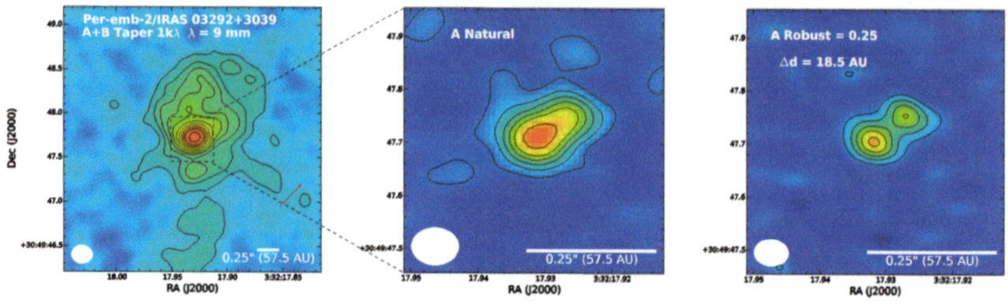

Figure 1. Images of Per-emb-2 (IRAS 03292+3039) at 9 mm from the VLA at increasing resolution from left to right. The left panel, with the lowest resolution and most sensitivity to extended structure, shows significant/structured emission surrounding a bright source that we interpret as the position of the main protostar; the middle and right panels zoom-in on the region outlined with a dashed box. The middle panels with higher resolution have resolved-out the extended structure and only detect the bright peak at the position of the protostar; however, the source appears extended at this resolution. The highest resolution image in the right panel shows that the source is resolved into two sources separated by 18.5 AU. The contours in each panel are [-6, -3, 3, 6, 9, 12, 15, 20, 25, 30, 35, 40, 50, 60, 70, 80, 90, 100, 150] $\times\, \sigma$, where $\sigma = 7.3\ \mu$Jy, 9.6 μJy, 11.9 μJy from left to right at 9 mm.

with a flat separation distribution (when considering the logarithm of the separation). Thus, it is clear that greater numbers are required and more than one star forming region needs to be examined in order to understand if this is a common distribution or an outlier. However, Perseus still offers an excellent template for studies of multiplicity to larger samples and more distant star forming regions. It remains to be seen if the separation distribution observed toward Perseus is 'universal,' and its implications for the formation mechanisms of multiple stars are just beginning to be understood (Tobin et al. 2016a; Lee et al. 2016; Offner et al. 2016).

Expanding on the discussion at the end of the §1, it will be important to determine if the observed bimodality is statistically significant. One way to do this is to obtain enough statistics such that the uncertainties in the individual bins of Figure 3 are statistically distinct. Assuming the same distribution of companion separations in Perseus, the bin at 1.875 log(AU) will be inconsistent with the bin at 2.375 log(AU) at the 4σ level when the sample size reaches ~500. Thus, samples in excess of 500 protostars are needed to statistically distinguish between bimodal and log-flat distributions.

A challenge to observing larger numbers of protostars is that the most populous star forming regions are more distant. Orion is ~400 pc away and the VLA offers a best spatial resolution of ~30 AU at 8 mm, this is just a bit better that the typical separation of field solar-type stars and there are likely to be more multiple systems lurking at separations below 30 AU. Perseus has at least three systems with separations ~20 AU (Figures 1 and 3). Another limiting issue in Orion is that sources are 3× fainter due to the increased distance, so observing with the same sensitivity requires 9× more observing time as compared to Perseus. It quickly becomes impractical to observe even more distant star forming regions with sensitivity to protostars having typical luminosities (i.e., $1 - 3\ L_{\odot}$; Fischer et al. 2017).

Finally, it is important to highlight that in the VANDAM survey, the spatial association of young stars was observed and not boundedness, which could not be evaluated

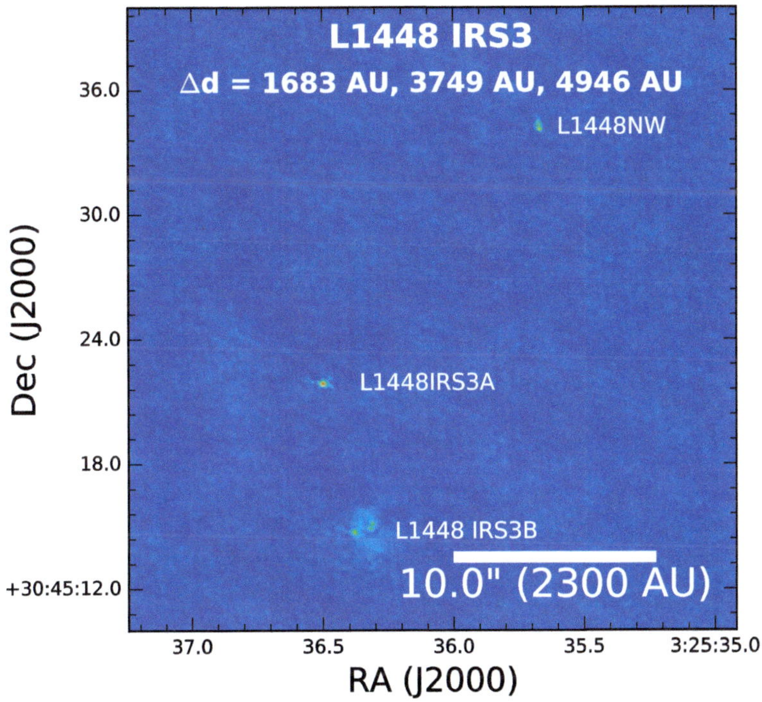

Figure 2. Image of the L1448-N or IRS3 region at 9 mm from the VLA. L1448 IRS3B is a close triple, L1448NW (L1448 IRS3C) is a binary, and L1448 IRS3A is single at the limit of our resolution. Both L1448 IRS3B and L1448NW are Class 0 sources while L1448 IRS3A is a Class I system. Separations written inside the figure are relative to L1448 IRS3B.

from the data at hand. Many (or most) of the companions shown in Figure 2 at separations > 1000 AU do not likely reflect bound systems. Some could be line of sight associations (more frequent with increasing separation), or when they finish accreting material, outflows may have removed enough mass such that they are unbound. The closer companions (< 500 AU) have a higher likelihood of being bound, but even so, interactions may alter or destroy some of these multiple systems. Thus, the observed companion separation distribution in the protostellar phase may reflect the initial companion separation distribution and it will dynamically evolve over Myr to Gyr toward toward the field star separation distribution.

The ngVLA has excellent potential to evaluate the boundedness of systems with <50 AU separations. The angular resolution in the nearby regions will uncover closer multiple systems, and the positional accuracy granted by the increased angular resolution will enable the orbital motion of <50 AU companions to be measured on time scales of ~5 years. Such analysis can incorporate previous VLA data as needed to es-

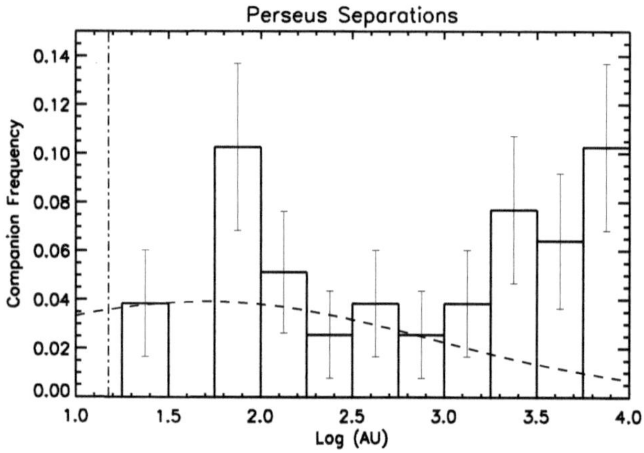

Figure 3. Histogram of companion star frequency versus separation for the entire sample of multiple sources in Perseus. The dashed curve is the Gaussian fit to the field star separation distribution from Raghavan et al. (2010). The vertical dot-dashed line corresponds to the approximate resolution limit of the VLA in A-configuration (~15 AU) toward Perseus at a distance of 230 pc.

tablish likely orbital solutions with an extended time baseline. Such measurements of binary/multiple orbits will also offer constraints on the protostar masses.

3. Examining Close Multiplicity Studies out to ≥1.5 kpc

The ngVLA offers a number of features that overcome the limitations of the current VLA. The full array will provide an angular resolution of $0\farcs01$ at 7.3 mm and $0\farcs014$ at 1.1 cm (ngVLA Memo #17). These two wavelength bands are ideal for examining protostellar multiplicity because they probe both dust and free-free emission. The addition of free-free emission to the dust emission makes the protostars themselves stand out, enhancing their detectability. The dramatically increased resolution of the ngVLA will enable the range of separations to be expanded, with better sensitivity to close companions. For Perseus (d = 230 pc), we will be able to look for companions down to 2.3 AU separations. Toward Orion (d = 400 pc), the separation limit will be pushed down to 4 AU, and in more distant, more populous star forming regions (e.g., Cygnus-X at 1500 pc) the separation limit will be 15 AU. The ngVLA will enable us to examine separations in these distant Galactic star forming regions down to the same scale that we are currently able to do in Perseus!

Angular resolution improvements alone, however, are not sufficient to address questions of protostellar multiplicity, because these studies require statistics and not a few case studies. With the drastic increase in the number of antennas, and hence the collecting area, projected for the ngVLA, it is important to consider the detectability of protostellar multiple systems in progressively more distant star forming regions. Using IRAS 03292+3039 from Perseus (Figure 1) as a case study, current VLA observations detect sources A and B with peak brightnesses of 170 μJy beam^{-1} and 130 μJy beam^{-1}, respectively. If a similar system were located at the distance of Cygnus-X, their respective flux densities would be 4 μJy and 3 μJy. These flux densities are completely impractical to observe at 30 GHz for the current VLA; reaching a signal-to-noise ratio (S/N) of 10 with the VLA would take 10 days! However, the ngVLA can reach a sensitivity of 0.26 μJy beam^{-1} (S/N$\sim$10) in 1 hour on-source. Thus, detecting such faint and close companions in massive, distant star forming regions will be routine with the ngVLA because the increased sensitivity.

We have concentrated on the capabilities of the ngVLA at $\sim$7.3 mm because of the ability to directly compare with data from the VANDAM surveys. However, the 3 mm band of the ngVLA also holds significant potential to enable studies of multiplicity out to star forming regions even more distant than Cygnus-X. At 3 mm, most emission will be from pure dust emission, and the brightness of this emission increases steeply with decreasing wavelength $\lambda^{-(2+\beta)}$ (increasing with frequency $\nu^{2+\beta}$), where β is the dust opacity spectral index which is typically observed to be between 0 and 2 in disks (e.g., Ricci et al. 2010). IRAS 03292+3039 was also observed with ALMA at 3 mm using 14 km baselines, obtaining nearly the highest resolution offered by the facility at 3 mm. The 3 mm flux densities of the two companions were 1200 μJy and 1050 μJy, respectively. Scaling these to the distance of Cygnus-X they could be detected with S/N > 20 with the ngVLA in 1 hr on-source with a sensitivity of 0.86 μJy beam^{-1}. At an even greater distance of $\sim$3 kpc, they could be detected with a S/N=8 in 1 hr as well! Thus, the increased sensitivity of the ngVLA, coupled with the higher angular resolution will radically enhance the study of multiple star formation. Furthermore, the addition of a 3 mm band to the ngVLA would be particularly advantageous for the detection of multiple protostars in even more distant regions than possible with the 7.3 mm band due to the better resolution and the fact that dust emission increases steeply with decreasing wavelength.

Obtaining sensitive images will still take a few hours of time with overheads toward the more distant star forming regions, but star forming regions at greater distances occupy less area of the sky in terms of their solid angle. The wide field of view offered by the ngVLA (a factor of 2 increase in solid angle over the VLA), coupled with the increased sensitivity, means that many more protostars can be captured in a single pointing than in the nearby star forming regions, this will somewhat offset the need for longer integrations times because more sources will be observed in a single observation.

While ALMA of course has unmatched sensitivity to dust emission at shorter wavelengths, the shorter wavelength becomes a disadvantage when attempting to study the youngest protostars. On scales less than $\sim$100 AU, the dense inner envelope or disks of protostars can be opaque at wavelengths as long as 1.3 mm and 3 mm, hiding small-scale multiplicity (see Tobin et al. 2018, this volume p. 345) Moreover, at higher frequencies the field of view of ALMA becomes small, making it unable to simultaneously observe as many protostars at high resolution. Thus the ngVLA capability of $0\farcs01$ or better resolution at wavelengths between 2 cm and 3 mm is an absolutely unique and critical capability to examine the formation of multiple stars during the early stages of protostellar evolution.

Because detection of dust emission from the small circumstellar disks in multiple systems is the most well-defined route to detection, it is important to further demonstrate their detectability. We ran a few radiative transfer models of protostars with small disks in binary star systems and simulated their observation with the full ngVLA using the CASA SIMOBSERVE task. The modeling is described in more detail in Tobin et al.

(2018, this volume p. 345). We show the results of our simulations in Figure 4, where we have simulated the observation of two protostars, separated by 5 AU, each with a 1 AU radius and 0.001 $M_\odot$ disk. These protostars are simulated at a distance of 400 pc, and they can be well-detected by the ngVLA with S/N = 20 in 1 hr. We also computed a model for a 15 AU separation binary system at a distance of 1.5 kpc and each disk having a radius of 3 AU and a mass of 0.01 $M_\odot$. The 1.5 kpc binary can be detected by the ngVLA with S/N = 10 in 1 hr. Thus, the ngVLA will enable the detection of extremely close binary protostars with separations as small as 3 AU (at 230 pc) from their dust emission alone, under the assumption that circumstellar disks have radii of order their separation/3 (Artymowicz & Lubow 1994). However, if each component has some free-free emission, in addition to the dust, they can likely be resolved at closer separations and their disks would not need to be as massive.

In addition to the ngVLA's ability to examine extremely close companions, the imaging capabilities of the ngVLA are superb enough to enable circum-multiple environments to be characterized. Using an ALMA image at 0.87 mm of L1448 IRS3B system, a triple system with a surrounding circum-multiple disk with spiral structure (Tobin et al. 2016a), we scaled the surface brightness assuming optically thin dust emission and β=1, implying a flux density scaling as λ^{-3} (see discussion in preceding paragraphs). Then we simulated a 1 hour observation at 3 mm, and a 2 hour observation at 7.3 mm using the 168 antenna 'Plains array,' keeping the distance at 230 pc. We show the results of modeling in Figure 5; the structure observed at 0.87 mm is fully recovered with high fidelity at 3 mm and the image is also well-recovered at 7.3 mm but with lower S/N due to the fainter dust emission. Thus, the ngVLA will also enable the imaging of circum-multiple environments simultaneously while probing for multiplicity. At 3 mm and 7.3 mm, the sensitivity of the ngVLA will be far superior to that of ALMA at these wavelengths and have higher angular resolution even with only the 'Plains' array. The circum-multiple emission is more likely to be optically thin at these wavelengths as compared to shorter wavelengths with ALMA.

The ngVLA will open up three exciting regions of parameter space in protostellar multiplicity studies. It will conduct high-S/N imaging toward nearby star forming regions (i.e., Perseus, Taurus) with enough angular resolution to search for companion protostars that have ~5× smaller separation than can be examined with the VLA. The ngVLA will also enable studies of protostellar multiplicity to obtain far greater statistics through the simultaneous observation of larger numbers of protostars in more distant regions (e.g., Cygnus-X) with the same or better sensitivity than is currently possible toward the nearby regions. Finally, the ngVLA can examine the environments around multiple star systems with high-S/N, with best results at the shortest wavelengths.

4. Synergies with Other Facilities

While the ngVLA will be able to identify very close multiple systems perhaps out to 3 kpc, ALMA will be very useful in characterizing the immediate surroundings of multiple star systems. The brighter dust emission at shorter wavelengths may make ALMA an ideal complement to very high resolution studies with the ngVLA, both in the dust continuum and molecular lines, provided that the dust emission is not too optically thick. ALMA will be able to provide access to the necessary molecular lines to characterize the immediate environs around proto-multiple systems. This is because many of the most abundant, disk-tracing molecules have their strongest rotation transitions at

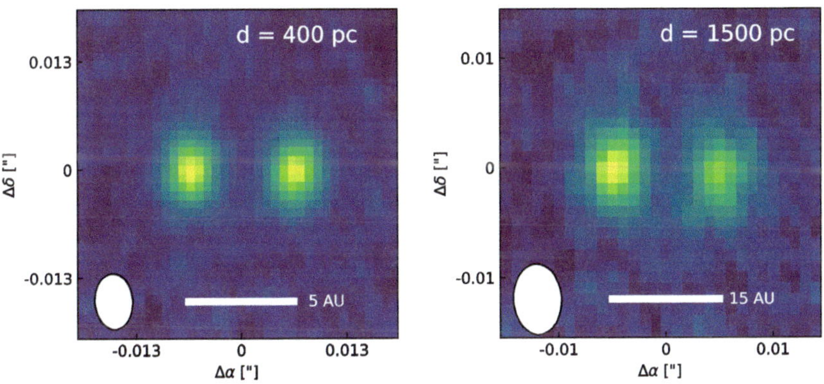

Figure 4. Synthetic ngVLA observations of binary systems comprised of emission from compact circumstellar disks. The left panel is at a distance of 400 pc having R_{disk} = 1 AU and M_{disk} = 0.001 $M_{\odot}$, and the right pane is at a distance of 1.5 kpc R_{disk} = 3 AU and M_{disk} = 0.01 $M_{\odot}$. Thus, these models demonstrate the feasibility of detecting small disks in multiple star systems at close separations.

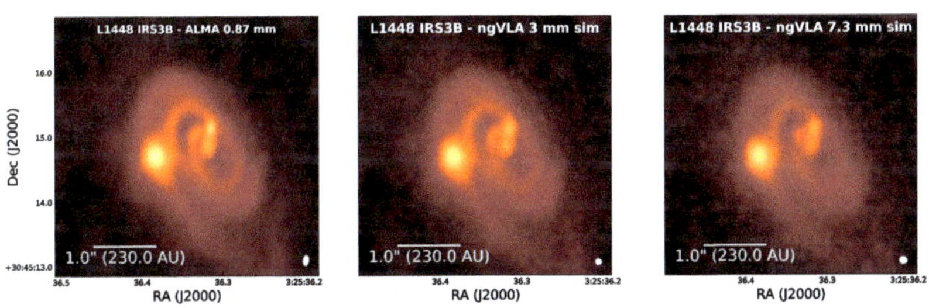

Figure 5. Synthetic ngVLA observations of the triple system L1448 IRS3B using only the 168 antenna 'Plains' array. These simulations used a scaled ALMA 0.87 mm image as the model (left) and utilize the maximum bandwidth available for the respective ngVLA bands, 40 GHz (3 mm; middle) and 20 GHz (7.3 mm; right). The 3 mm simulation is one hour on-source and is imaged using Briggs weighting, robust=0.5, and a (u, v)-taper at 2000 kλ; the 7.3 mm simulation is two hours on-source and the rest of the imaging parameters are the same as the 3 mm.

wavelengths shorter than 3 mm; the ngVLA will only be able to access the ($J = 1 \rightarrow 0$) transitions for most molecules, and these transitions do not generally have the brightest emission in the warm (>20 K) regions in the immediate vicinity of protostars. Furthermore, ALMA dust continuum observations can detect circum-multiple disks around the protostar that might have column densities too low to detect with the ngVLA, but as shown in Figure 5 the ngVLA will have this capability as well.

The more distant star forming regions have a disadvantage relative to the nearby ones in that their protostellar content is not as well-characterized due to the low resolution (>1″) of previous mid-to-far-infrared surveys. The *James Webb Space Telescope* will undoubtedly survey numerous massive star forming regions and infrared dark clouds (IRDCs) that harbor significantly more young stars than the nearby regions at wavelengths between 10 and 28 μm. Thus, by the time the ngVLA is conducting early science, the protostellar content of more distant star forming regions is likely to be much more well-characterized, enabling the multiplicity results obtained by the ngVLA to be put into a similar context as the results toward nearby star forming regions.

5. Summary

The ngVLA will open a new window into the study of multiplicity during the protostellar phase, providing a much clearer picture of where most companion stars are forming. Thus, the ngVLA will improve our understanding of just how many stars and proto-planetary disks begin their lives initially as part of a multiple system and how multiplicity affects the evolution of planetary systems, whether or not the system remains a multiple in its main sequence life. The increased resolution of the ngVLA will enable both closer multiples to be detected than previously possible for protostellar systems, and coupled with the increased sensitivity, larger numbers of multiple star systems can be observed in order to obtain statistics that equal or surpass that of the field solar-type stars.

References

André, P., Ward-Thompson, D., & Barsony, M. 1993, ApJ, 406, 122
Artymowicz, P., & Lubow, S. H. 1994, ApJ, 421, 651
Burkert, A., & Bodenheimer, P. 1993, MNRAS, 264, 798
Chen, X., Arce, H. G., Zhang, Q., et al. 2013, ApJ, 768, 110
Connelley, M. S., Reipurth, B., & Tokunaga, A. T. 2008, AJ, 135, 2526
Duchêne, G., & Kraus, A. 2013, ARA&A, 51, 269
Dunham, M. M., Stutz, A. M., Allen, L. E., et al. 2014, Protostars and Planets VI, 195
Duquennoy, A., & Mayor, M. 1991, A&A, 248, 485
Fischer, W. J., Megeath, S. T., Furlan, E., et al. 2017, ApJ, 840, 69
Kratter, K. M., Matzner, C. D., Krumholz, M. R., & Klein, R. I. 2010, ApJ, 708, 1585
Lee, K. I., Dunham, M. M., Myers, P. C., et al. 2016, ApJ, 820, L2
Looney, L. W., Mundy, L. G., & Welch, W. J. 2000, ApJ, 529, 477
McKee, C. F., & Ostriker, E. C. 2007, ARA&A, 45, 565
Offner, S. S. R., Dunham, M. M., Lee, K. I., Arce, H. G., & Fielding, D. B. 2016, ApJ, 827, L11
Offner, S. S. R., Kratter, K. M., Matzner, C. D., Krumholz, M. R., & Klein, R. I. 2010, ApJ, 725, 1485
Padoan, P., & Nordlund, Å. 2002, ApJ, 576, 870
Raghavan, D., McAlister, H. A., Henry, T. J., et al. 2010, ApJS, 190, 1

Reipurth, B., & Mikkola, S. 2012, Nat, 492, 221
Reipurth, B., Rodríguez, L. F., Anglada, G., & Bally, J. 2004, AJ, 127, 1736
Ricci, L., Testi, L., Natta, A., et al. 2010, A&A, 512, A15
Sadavoy, S. I., & Stahler, S. W. 2017, MNRAS, 469, 3881
Stamatellos, D., & Whitworth, A. P. 2009, MNRAS, 392, 413
Tobin, J. J., Dunham, M. M., Looney, L. W., et al. 2015, ApJ, 798, 61
Tobin, J. J., Kratter, K. M., Persson, M. V., et al. 2016a, Nat, 538, 483
Tobin, J. J., Looney, L. W., Li, Z.-Y., et al. 2016b, ApJ, 818, 73
Tohline, J. E. 2002, ARA&A, 40, 349

Science with a Next-Generation Very Large Array
ASP Conference Series, Monograph 7
Eric J. Murphy, ed.

Exploring Protostellar Disk Formation with the ngVLA

John J. Tobin,[1] Patrick Sheehan,[1] Doug Johnstone,[2] and Rajeeb Sharma[1]

[1]*University of Oklahoma,* `jjtobin@ou.edu`

[2]*NRC-Herzberg*

Abstract. The formation and evolution of disks early in the protostellar phase is an area of study in which the ngVLA is poised to make significant breakthroughs. The high-sensitivity and resolution at wavelengths of 3 mm and longer will enable forming disks to be examined with unprecedented detail. The need to observe dust emission at wavelengths of 3 mm and longer is motivated by the fact that dust emission at these wavelengths is more likely to be optically thin, which is essential to understanding the structure of these disks. We explore the feasibility of detecting and resolving protostellar disks with a variety of radii, masses, and distances, out to distances as large as 1.5 kpc using radiative transfer models and simulations with the proposed ngVLA configuration. We also examine the potential for the ngVLA to enable studies of grain growth and radial migration of dust particles early in the protostellar phase with the broad multi-wavelength coverage. Studies of grain growth will require wavelength coverage extending at least to ~4 cm to characterize and quantify the location and intensity of free-free emission, which is expected to be generated at <10 AU scales.

1. Introduction

The formation of disks occurs as a natural consequence of angular momentum conservation during the star formation process. As such, proto-planetary disks are found nearly ubiquitously toward pre-main sequence stars, with higher fractions of disks found toward members of clusters/associations having younger collective ages (Hernández et al. 2008). The origins of proto-planetary disks can be traced to the disks that form during the early phase of the star formation process, around Class 0 protostars. Class 0 protostars are characterized by a dense envelope of infalling material that feeds the protostellar disk. As the envelope dissipates due to continued accretion and erosion of the envelope by protostellar outflows (e.g., Frank et al. 2014), it becomes a Class I protostar. The Class 0 phase is expected to last ~150 kyr, while the combined Class 0 and Class I phases are expected to last ~0.5 Myr (Dunham et al. 2014); these ages estimates are based on the assumption that pre-main sequence stars hosting proto-planetary disks (objects in the Class II phase) have an average age of ~2 Myr. These young, embedded disks are often collectively referred to as protostellar disks.

Protostellar disks are important for the formation and mass assembly of the star, as well as the planet formation process. Most of the mass that is accreted onto a star must pass through the disk, and these protostellar disks are the initial conditions for disk evolution into a proto-planetary disk. The radius and mass of the forming disk can be regulated by the angular momentum of the infalling material, and magnetic fields can remove angular momentum efficiently in the absence of dissipative processes (e.g., Li et al. 2014). As such, the structure of these protostellar disks is expected to be connected to their formation conditions, and clear detections of rotationally supported disks have been found toward Class 0 and I protostars (Tobin et al. 2012; Harsono et al. 2014). The disks may also increase in radius later via viscous evolution and the outward transport of angular momentum (e.g., Manara et al. 2016). Furthermore, if gravitational instability in disks is a viable mechanism for angular momentum transport, the formation of multiple star systems, and possibly giant planet formation, the disks in the protostellar phase are those most likely to have the requisite conditions (Kratter et al. 2010; Tobin et al. 2016). Finally, the growth of solid material can begin during the protostellar phase and this may catalyze the later formation of planets via the core accretion process if the protostellar disk is able to efficiently grow dust grains to pebble/rock sizes in order to promote planetesimal growth.

Typical proto-planetary disks have masses of ~ 0.005 $M_{\odot}$, radii of ~50 AU, and surface density profiles $\propto R^{-1}$, meaning that most of the disk mass is at large radii (Andrews et al. 2013; Ansdell et al. 2016). Proto-planetary disks tend to be significantly lower in mass than the protostellar disks, which can have masses > 0.1 $M_{\odot}$ (Tobin et al. 2015; Jørgensen et al. 2009; Tychoniec et al. 2018). The typical masses, radii, and/or surface density profiles for protostellar disks are still poorly constrained, and the ngVLA has significant potential to unlock some of these characteristics.

2. Resolving and Characterizing Youngest Disks

Observational studies of young disks have only just begun to reach samples larger than 10 – 20 systems (Tobin et al. 2015; Segura-Cox et al. 2016), and young protostar systems (Class 0 and Class I phases) are inherently more rare than pre-main sequence stars hosting proto-planetary disks. This is because the lifetime of the protostar phase is expected to be at least ~4× shorter than the lifetime of a PMS star with a disk (Dunham et al. 2014). Specifically, the *Spitzer cores2disks* survey along with the entire *Spitzer* Gould Belt survey contain only ~200[1] Class 0 and I protostars. Thus, to observe large numbers of protostellar disks, more populous star forming regions need to be observed and these are only present at distances ≥ 400 pc. For example, Orion alone contains ~315 Class 0 and I protostars. The need to examine protostellar disks at 100s of pc distances makes their characterization difficult to achieve with the same mass sensitivity and spatial resolution as the more nearby proto-planetary disk systems. Moreover, high mass sensitivity is particularly difficult at long wavelengths where the dust emission is intrinsically fainter [$F_{\lambda} \propto \lambda^{-(2+\beta)}$, where β is the dust opacity spectral index]. Figure 1 shows an example radio spectrum of a protostar from the submillimeter to centimeter wavelengths which illustrates the expected spectral properties of protostar systems as a function of wavelength. Protostars often exhibit free-free emission that dominates at wavelengths >1 cm, but can can contribute to the overall flux density at 7.5 mm. Thus, multi-wavelength observations are important to measure and remove the contribution of free-free emission to the flux density at wavelengths where we aim to trace dust emission. We will discuss this aspect further in §4. For the sake of discussion, throughout this chapter we will use the dust opacity normalization of 0.899 cm^2 g^{-1} (dust only) at

[1]This number corresponds to sources with the most firm characterization based on bolometric temperature.

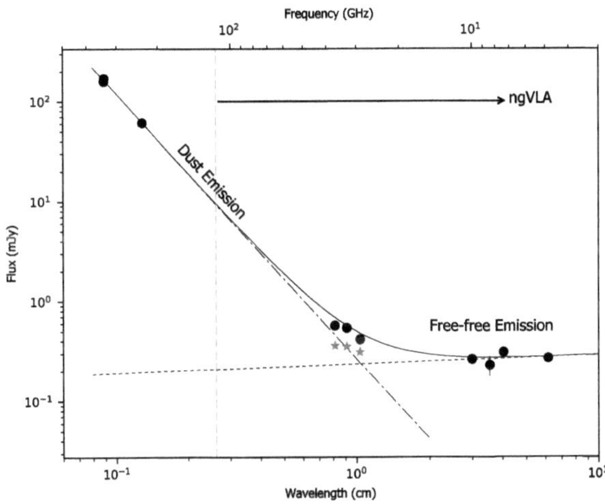

Figure 1. Spectrum from millimeter to centimeter wavelengths of a typical protostar in Orion. The emission from dust dominates at short wavelengths and free-free dominates at long wavelengths. The ngVLA will fully probe the transition from dust emission to free-free. The 8 mm to 1 cm bands have some free-free emission that is typically removed in current studies using power-law fits to the long wavelength emission. The ngVLA will enable improvements by spatially resolving the free-free emitting region(s). The points at 0.87 mm and 1.3 mm are from ALMA, the points at all longer wavelengths are from the VLA; the green star points are the flux densities between 8 mm and 1 cm with the extrapolated free-free contribution subtracted. Error bars are plotted with each point, but in most cases the statistical uncertainties are smaller than the points plotted.

1.3 mm from (Ossenkopf & Henning 1994), assuming β=1 for extrapolation to longer wavelengths (0.155 cm^2 g^{-1} at 7.5 mm).

The advent of the Karl G. Jansky Very Large Array (VLA) with its enhanced continuum sensitivity has enabled surveys of protostellar disks at wavelengths between 6.7 mm and 1 cm. Dust emission is more optically thin than at shorter wavelengths and thus the emission better traces column density due to the power-law decrease of dust opacity with increasing wavelength (Hildebrand 1983). Two VLA Nascent Disk and Multiplicity (VANDAM) surveys have been carried out with the VLA toward the Perseus and Orion star forming regions, observing a total of over 200 protostar systems in each region. The VANDAM surveys took ~600 hours of time on the VLA. These surveys were conducted in Ka-band at a central wavelength of 9 mm, while observing at the highest possible angular resolutions, translating to spatial resolutions of ~15 AU and ~30 AU in Perseus and Orion, respectively. Example images of disks observed with the VLA are shown in Figure 2. Also note that the VANDAM: Perseus survey observed the entire sample at 4.1 cm and 6.4 cm, but we focus on the dust emission at 5.9 mm to 10 mm in this chapter. This corresponds to the proposed ngVLA band that is most similar to the wavelengths used in VANDAM and is centered at ~7.5 mm.

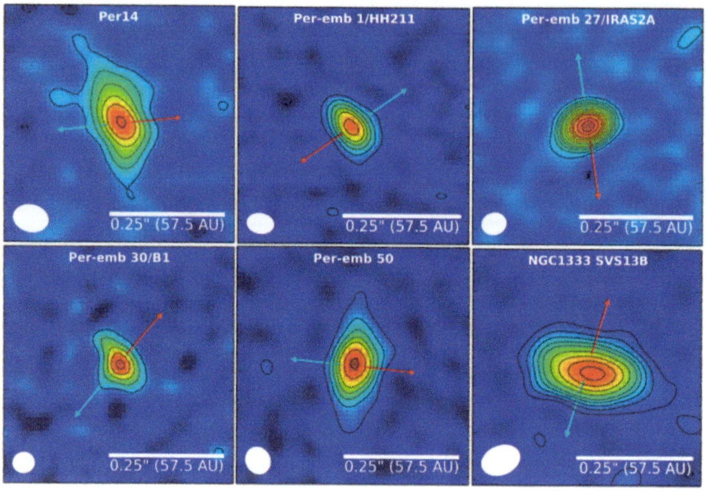

Figure 2. Images of some well-resolved disks from the VANDAM: Perseus survey at 8 mm, from Segura-Cox et al. (2016). The contours in each panel are [-6, -3, 3, 6, 9, 12, 15, 20, 25, 30, 35, 40, 50, 60, 70, 80, 90, 100, 150] $\times \sigma$, where σ = 11 μJy at 8 mm.

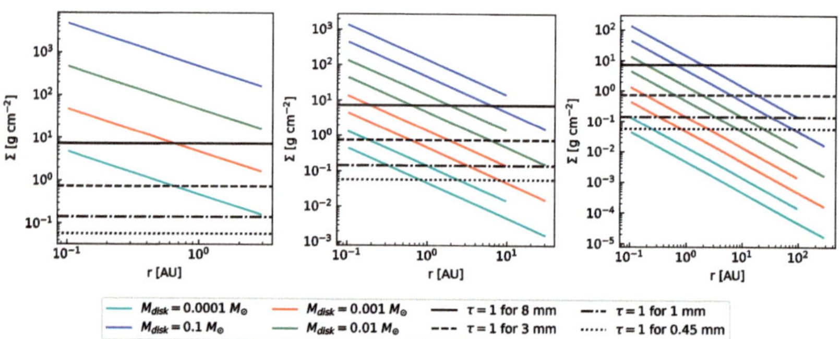

Figure 3. Surface density profiles of modeled disks versus radius. Note that the range of radius plotted increases from left to right; the 3 AU disks are plotted in the left panel, the 10 AU and 30 AU disks are plotted in the middle panel, and the 100 AU and 300 AU disks are plotted in the right panel. Horizontal lines are plotted that denote the surface density corresponding to τ = 1 at 7.5 mm (solid), 3 mm (dashed), 1 mm (dash-dot), and 0.45 mm (dotted). Observations at 7.5 mm will not be immune to dust opacity effects, but they can probe surface densities 10$\times$ larger without becoming optically thick. The dust mass opacities calculated are 0.15, 1.35, 7.03, 17.43 cm^2 g^{-1} at wavelengths of 7.5, 3, 1, and 0.45 mm, respectively.

The VANDAM: Perseus survey detected candidate disks, meaning that extended dust continuum emission was resolved, toward only 12/42 Class 0 protostars and even fewer toward Class I protostars 5/37 (Segura-Cox et al. 2018). We show examples of some of the well-resolved disks detected in the VANDAM: Perseus survey in Figure 2 (Segura-Cox et al. 2016). The disks in Perseus were further characterized by Tychoniec et al. (2018), focusing on the integrated dust emission at 9 mm toward all sources whether or not they were resolved. Tychoniec et al. (2018) found that Class 0 protostars tend to have more mass within radii <100 AU relative to the Class I protostars. Furthermore, both Class 0 and Class I protostars had higher masses than the more-evolved proto-planetary disks (Class II sources). The low percentage of well-resolved disks toward Class 0 and I protostars may be due to both the limited spatial resolution and sensitivity of the VLA. The sensitivity may be especially limiting for Class I protostars as, due to their disk evolution, they have less overall mass (thus lower flux densities). The ability to observe at wavelengths of ~9 mm is extremely important for examining protostellar disks, especially around Class 0 targets. If they are compact (R< 50 AU) and massive (> 0.01 $M_\odot$), they are likely optically thick at shorter wavelengths, but remain optically thin at most radii at 9 mm. We show radial surface density profiles for model protostellar disks and their corresponding radii where $\tau = 1$ in Figure 3, thereby underscoring that long wavelength observations of dust emission are essential to resolve and quantify the majority of the mass within protostellar disks. The opacity of disks at different wavelengths depending on mass and radius will be discussed in more detail in the following section.

Despite the advances enabled by the VLA surveys, they are not without limitations. Due to the increased distance to Orion, the VANDAM: Orion survey had 2.2× less mass sensitivity than the Perseus survey, despite spending 2× more time on source. The VANDAM: Orion survey used 240 hours in the VLA A-configuration to observe 100 Class 0 protostars, accounting for ~50% of the available array time at high-frequencies. Thus, high-sensitivity at ~8 mm wavelengths toward large protostar samples at a distance of ≥400 pc requires a significant utilization of available observing time for the current VLA; the time required quickly becomes unrealistic for protostars at much greater distances.

The ngVLA is needed to further advance studies of disks because ALMA will also not be as sensitive to long wavelength dust emission as the ngVLA. Using ALMA at 3 mm, 1 hour of integration will achieve a sensitivity of 15 μJy. Assuming a distance of 230 pc, this translates to a 1σ mass sensitivity of $1.6\times10^{-4}\ M_\odot$. Therefore, ALMA Band 1 (~7.5 mm) will only be about as sensitive as the current VLA. The ngVLA will reach 1σ mass sensitivities of $5.0\times10^{-5}\ M_\odot$ and $8.2\times10^{-6}\ M_\odot$ at wavelengths of 7.5 mm and 3 mm, respectively. Thus, the ngVLA will be vastly superior to ALMA at these long wavelengths for the characterization of dust emission from protostellar disks. However, these mass estimates assume that all emission is within one beam. We will conduct a more realistic assessment of the ngVLA's ability to detect resolved disks in the following section.

3. Detectability of Protostellar Disks with the ngVLA

The VANDAM: Perseus survey was able to detect disks with total masses > 0.025 $M_\odot$ (gas+dust, assuming a gas-to-dust ratio of 100:1 and a 5σ detection limit) and resolve disks with observed radii >10 AU at a wavelength of 8 mm (Segura-Cox et al. 2018).

However, most of the resolved disks have masses > 0.1 $M_\odot$. The sensitivity of the ngVLA will enable a ~10× leap in both resolution and sensitivity. In order to quantitatively characterize the ability of the ngVLA to detect and resolve disks toward both nearby and distant star forming regions, we have computed a small suite of radiative transfer models at wavelengths of 7.5 mm and 3 mm, varying luminosity (1.0, 10.0, 100.0 $L_\odot$), disk radius (3.0, 10.0, 30.0, 100.0, 300.0 AU), and disk mass (10^{-4}, 10^{-3}, 10^{-2}, 10^{-1} $M_\odot$). The disks are assumed to have a radial surface density profile proportional to R^{-1}, scale height (H/R) of 0.1 at 1 AU, and flaring $H \propto R^{1.15}$; the disk density profiles are truncated at the specified radii and do not exponentially drop-off. All models have a surrounding envelope with a mass of 0.1 $M_\odot$ and a radius of 1500 AU. The dust opacities we use are derived from Woitke et al. (2016) and have a maximum particle size of 1 mm, yielding dust mass opacities of 0.15 cm^2 g^{-1} and 1.35 cm^2 g^{-1} at 7.5 mm and 3 mm, respectively. Radiative transfer was computed with the RADMC3D code (Dullemond et al. 2012), and model images were generated at three distances (230, 400, and 1500 pc) and three viewing geometries (i = 25, 45, and 75°). We simulated ngVLA observations of these models with the CASA *simobserve* task using the full ngVLA configuration, assuming 1 hour on source and the estimated noise of 0.26 μJy from this same integration time (ngVLA Memo #17).

A subset of these models are shown in Figure 4 at a wavelength of 7.5 mm, distances of 230 pc, 400 pc, and 1.5 kpc, and at an inclination of 75°. At distances of 230 pc and 400 pc, disks with masses of > 0.001 $M_\odot$ toward protostars with luminosities of 10 $L_\odot$ can be detected and resolved. We note, however, that at a mass of 0.001 $M_\odot$, the 300 AU radius disk is not as well detected due to the mass being spread over a larger disk size. We also find that disks with radii of 3 and 10 AU can also be detected for a disk mass as small as 0.0001 $M_\odot$.

At a distance of 1.5 kpc and the same luminosity, disks with masses of 0.1 $M_\odot$ can be detected and resolved at all radii, and at a mass of 0.01 $M_\odot$ the disks with $R_{disk} \leq 100$ AU can still be detected. Disks with masses much below 0.01 $M_\odot$ are not well-detected in a 1 hour observation. While optimal imaging parameters will depend on the radii and mass of each disk, the visibility data themselves can be utilized to fit the disk radii with greater accuracy from the images alone, given the possible limitations of maximum recoverable scale and surface brightness. The ngVLA will enable the physical structure of protostellar disks with radii as small as ~3 AU to be characterized in the nearby star forming regions. The ngVLA is the only facility that will be able to do this given the necessity of very high angular resolution and long-wavelengths; this will be a major advance in the capability of examining small-scale structures in protostellar systems.

Furthermore, observations at 3 mm (not shown for brevity) enable the detection of all the disks detected at 7.5 mm (Figure 4), but lower mass disks will be able to be detected as well given the factor of ~6 - 16× increase in flux density going from 7.5 mm to 3 mm (depending on dust opacity spectral index). The 3 mm band will be especially effective for the detection of disks in star forming regions out to 1.5 kpc, as disk masses of 0.001 $M_\odot$ can be detected more robustly. Thus, both the 7.5 mm and 3 mm ngVLA bands will be essential to further characterizing disks in the protostellar phase, and the full ngVLA will enable disks to be detected and resolved out to star forming regions at distances of at least 1.5 kpc. This capability will greatly expand the number of disks that can be observed compared with the current VLA by enabling significantly smaller radii and masses to be detected and resolved. By including all star forming regions out to

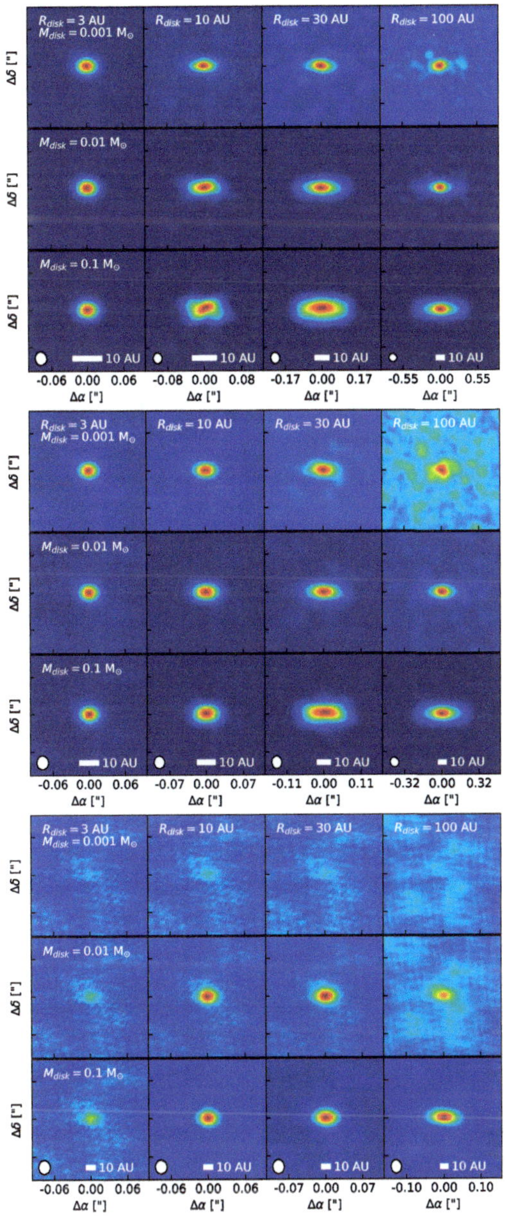

Figure 4. Model images of protostellar disks at 7.5 mm computed for a three distances 230 pc (top), 300 pc (middle), and 1500 pc (bottom). For each distance we have generated synthetic observations for a variety of radii and disk masses and run through a simulated observation of 1 hr with the full ngVLA configuration. (u, v)-tapering has been applied to the disks with $R_{disk} \geq 30$ AU at the level of $R_{disk}/5.0$ translated to angular units.

$\sim$1.5 kpc (e.g., Cygnus-X; Kryukova et al. 2014), more than 1000 additional protostars and their disks will be accessible to the ngVLA. Moreover, the 100s of protostars star-forming regions at distances of 400 pc or less (Dunham et al. 2014) will be able to have their disks resolved, where only unresolved observations were possible with the current VLA.

Optical depth is another significant advantage of the ngVLA over other millimeter interferometers like ALMA that primarily operate at wavelengths shorter than 3 mm. The surface density profiles for the modeled disks are shown in Figure 3 with horizontal lines marking the surface density corresponding to $\tau = 1$ for wavelengths between 7.5 mm and 0.45 mm. Due to the fixed grid of masses and varying radii, the surface density profiles are larger for smaller disk radii. However, it can be seen that 7.5 mm wavelengths can trace about 10× higher surface densities than at 1 mm before becoming optically thick. The opacity can be significant even at 7.5 mm, particularly for disks with $M_{disk} = 0.1\ M_{\odot}$. For the 100 and 300 AU radius and $M_{disk} = 0.1\ M_{\odot}$ disks, only the inner few AU are optically thick at 7.5 mm, while the entire 100 AU disk is optically thick at 1 mm. The disks with radii between 10 and 30 AU and $M_{disk} = 0.1\ M_{\odot}$ are optically thick out to $\sim$10 AU in both cases, but lower mass disks in this range of radii are only optically thick in their inner few AU. However, these disks are completely optically thick at 1 mm and 0.45 mm until the disk masses are $\sim$ 0.0001 $M_{\odot}$. Note that τ in Figure 3 is calculated for an inclination of 0° (face-on), and intermediate to edge-on inclinations will have substantially more opacity for a given surface density profile. Therefore, we can conclude that 7.5 mm wavelengths are not necessarily immune to issues of opacity, but they are significantly less affected than 1 mm and 0.45 mm. This will enable ngVLA to significantly improve on the characterization of protostellar disks, while the ability of ALMA to probe their structure is more limited.

The ability to detect these low-mass, small radius disks is also extremely important for the characterization of multiple star systems from the emission of their circumstellar disks. This capability is described in more detail in Tobin et al. (2018, this volume p. 333). Furthermore, while we do not specifically address the likelihood of substructure in protostellar disks, growing numbers of proto-planetary disks exhibit a variety of substructure when observed with sufficiently high resolution in the dust continuum (ALMA Partnership et al. 2015; Andrews et al. 2016; Sheehan & Eisner 2018, Andrews et al. 2018 in prep.). Therefore, it is entirely possible that protostellar disks themselves will also not have smooth structures. The higher-resolution at longer wavelengths of the ngVLA will be crucial for characterizing whether or not protostellar disks exhibit significant substructure. This is because, as discussed previously, the more massive protostellar disks are largely opaque at shorter wavelengths and the longer wavelengths with lower opacity are more ideal to reveal substructure if present.

4. Characterizing Early Grain Growth

The ability to detect and resolve disks with a variety of radii in the nearby star forming regions makes it possible to characterize the spectral index of dust emission both in the limit of integrated flux densities and spatially-resolved emission throughout the disks. With a shortest wavelength of 3 mm, the ngVLA can characterize the spectral index of dust emission at least between 3 mm and 1 cm. The sensitivity of the >1 cm wavelength ngVLA bands will also have the capability of detecting and resolving dust emission. This capability is incredibly important because the rate of growth for solids in disks can be examined as a function of evolution through the protostellar phase. However, at wavelengths >1 cm dust emission becomes more difficult to detect due to the power-law decrease in flux density of dust emission at longer wavelengths, thus the ngVLA is crucial to enable the detection of dust emission at wavelengths >1 cm because the current VLA does not have the needed angular resolution or sensitivity.

The ability to measure the spectral index of dust emission over a wide range of wavelengths is important because it is connected to both the maximum dust grain sizes and the slope of the assumed power-law dust size distribution; dust grains are most efficiently detected when observed at a wavelength comparable to their size. Through radiative transfer modeling, using the same techniques that generated the images in Figure 4, the dust emission over a wide range of wavelengths can be used to determine the maximum particle sizes. Such studies are currently possible toward a few select systems, but the limited sensitivity of the VLA makes detection of dust emission out to large disk radii and low disk masses challenging. Furthermore, the low angular resolution of the current VLA at long wavelengths means that characterizing dust emission a wavelengths of several cm must use integrated flux densities. Resolved studies of disk spectral indices have been done for a few of the most nearby disk-hosting PMS stars (Pérez et al. 2012, 2015). Doing such modeling for protostars spanning the full range of evolution will enable the rate of grain growth to be understood, and we will learn whether or not this is a linear/universal process, or if it varies strongly between protostars. However, in order to accomplish this, greater sensitivity and resolution than are possible with the VLA are needed to sufficiently resolve the inner regions of protostellar disks and accurately separate free-free emission from dust emission.

Separating the free-free jet emission from dust emission is incredibly important, especially for examining dust emission at >1 cm (Figure 1). The resolution of the ngVLA will enable the dust emission to be resolved from the free-free emission out to wavelengths of several cm. However, the disk needs to be well-resolved for this to be possible. Compact free-free emission is expected to originate from the inner ~10 AU (Anglada et al. 1998) and the ngVLA will have a spatial resolution of ~11.5 AU in the 3.75 cm (8 GHz) band for protostars at a distance of 230 pc. Thus, the emission from the two distinct processes can be separated spatially with multi-wavelength observations, and the region containing free-free emission can be isolated and removed from the analysis. This is not possible with the angular resolution of the current VLA and will be an unique capability of the ngVLA. We also note that free-free emission can also originate from a disk wind, but without the ngVLA we cannot fully characterize the nature of free-free emission to properly account for it. Finally, free-free emission is can still contribute at wavelengths <1 cm (Figure 1), and the higher angular resolution at shorter wavelengths will help further characterize its nature.

ALMA can add to the characterization of dust growth with its equivalent angular resolution at shorter wavelengths. While small disks and the inner radii of large disks will be optically thick (Figure 3), ALMA observations should be more sensitive to the outer regions of protostellar disks where the concentration of large grains is expected to be lower. This is because grain growth is slower at larger radii, having lower densities, and the radial drift process is expected to move large particles to smaller radii (Weidenschilling 1977). Therefore, in the regions where the dust emission is not optically thick, short wavelength observations from ALMA will be helpful to characterize dust growth at all disk radii, especially if the dust emission is more compact at longer wavelengths (Pérez et al. 2012, 2015; Segura-Cox et al. 2016).

Furthermore, the ngVLA may not be able to probe the transition from the envelope to disk as well as ALMA, because the envelope will have significantly lower column density and thereby low surface brightness. On top of that, if the envelopes have mostly small dust grains < 10 μm, the dust emission drops much faster in wavelength than it does in the disk due to the smaller grains. Thus, ALMA will enable the connection of the disk to the infalling/rotating envelope to be better understood through both the observations of shorter wavelength dust and molecular lines tracing envelope and disk kinematics. The ngVLA then will be the ideal instrument for examining the presence and structure of the disks around protostars to smaller radii. This will complement ALMA's ability to examine the structure of proto-planetary disks around more evolved young stars.

5. Uniqueness of ngVLA Capabilities

The ngVLA will have extraordinary capabilities to examine disks toward protostars throughout the Class 0 and I phases. It is crucial to examine the disks at long wavelengths where the dust emission is optically thin over a large range of radii to probe nearly the entire disk mass reservoir. Moreover, the ngVLA will be able to fully resolve young disks with radii as small as ~3 AU in the nearby star forming regions. While ALMA can reach such spatial scales at short wavelengths, the disks around protostars are likely to be optically thick at such small radii (Figure 3), making it impossible for ALMA to examine the internal structure of their protostellar disks in many cases.

Due to the optical depth of disks (Figure 3), the radial distribution of dust particle sizes cannot be explored with short wavelength data from ALMA. Longer wavelength observations and ~10 mas resolution are crucial to map the radial distribution of dust emission. While ALMA has 3 mm receivers and is expected to have ~1 cm receivers (Band 1) in the future, the best angular resolution at 3 mm is ~0.″05, and ~0.″14 at 1 cm. Thus, ALMA at 3 mm will only have the spatial resolution that VANDAM had with the VLA toward Perseus at 8 mm, and ALMA will have lower angular resolution at 1 cm; ALMA will not be able to improve upon the resolution of the VANDAM survey at wavelengths where dust emission is less affected by opacity. Furthermore, ALMA will be significantly less sensitive than the ngVLA at these same wavelengths; see §2. In summary, the ngVLA will be superior to ALMA in examining protostellar disks at wavelengths of 3 mm and longer.

Lastly, the ngVLA will be able to examine the vertical settling of large dust grains for edge-on disks. In nearby star forming regions, (e.g., Taurus, Oph at 140 pc), edge-on disks can be observed to determine how settled the disk midplanes are. The disk midplanes can be examined from 3 mm to 1 cm to see if there is variation with wavelength in the vertical extent of the dust emission. If the vertical structure of the dust emission is approximately H/R ~0.1, then at a radius of 14 AU the vertical height of the dust at 7.5 mm can be resolved (assuming 0.″01 resolution at 140 pc). This study also cannot be done with ALMA at shorter wavelengths because the disk midplanes are opaque (Lee et al. 2016). Through modeling, it can be determined if the dust emission is more concentrated in the disk midplanes going to longer wavelengths, consistent with expectations of dust settling (D'Alessio et al. 2006) and low levels of turbulence in the disks (e.g., Flaherty et al. 2017).

6. Summary

The ngVLA will enable a new revolution in the study of protostellar disks with the ability to probe the spatial structure of protostellar disks, where many of these disks cannot be resolved with current instrumentation. The ngVLA will enable us to resolve dust emission from disks that are as small as 3 AU in the nearby star forming regions and examine the full spatial extents of protostellar disks using emission from optically thin dust. Together, this information will enable us to determine the spectral index and thereby the sizes of the emitting dust grains to be characterized as a function of radius, and perhaps height for edge-on disks. The ngVLA will enable the structure and early grain growth of protostellar disks to be better understood, in addition to how quickly the conditions for planet formation are established in disks.

References

ALMA Partnership, Brogan, C. L., Pérez, L. M., et al. 2015, ApJ, 808, L3
Andrews, S. M., Rosenfeld, K. A., Kraus, A. L., & Wilner, D. J. 2013, ApJ, 771, 129
Andrews, S. M., Wilner, D. J., Zhu, Z., et al. 2016, ApJ, 820, L40
Anglada, G., Villuendas, E., Estalella, R., et al. 1998, AJ, 116, 2953
Ansdell, M., Williams, J. P., van der Marel, N., et al. 2016, ApJ, 828, 46
D'Alessio, P., Calvet, N., Hartmann, L., Franco-Hernández, R., & Servín, H. 2006, ApJ, 638, 314
Dullemond, C. P., Juhasz, A., Pohl, A., et al. 2012, RADMC-3D: A multi-purpose radiative transfer tool, Astrophysics Source Code Library, ascl:1202.015
Dunham, M. M., Stutz, A. M., Allen, L. E., et al. 2014, Protostars and Planets VI, 195
Frank, A., Ray, T. P., Cabrit, S., et al. 2014, ArXiv e-prints, arXiv:1402.3553
Harsono, D., Jørgensen, J. K., van Dishoeck, E. F., et al. 2014, A&A, 562, A77
Hernández, J., Hartmann, L., Calvet, N., et al. 2008, ApJ, 686, 1195
Hildebrand, R. H. 1983, QJRAS, 24, 267
Jørgensen, J. K., van Dishoeck, E. F., Visser, R., et al. 2009, A&A, 507, 861
Kratter, K. M., Matzner, C. D., Krumholz, M. R., & Klein, R. I. 2010, ApJ, 708, 1585
Kryukova, E., Megeath, S. T., Hora, J. L., et al. 2014, AJ, 148, 11
Lee, K. I., Dunham, M. M., Myers, P. C., et al. 2016, ApJ, 820, L2
Li, Z.-Y., Banerjee, R., Pudritz, R. E., et al. 2014, Protostars and Planets VI, 173
Manara, C. F., Rosotti, G., Testi, L., et al. 2016, A&A, 591, L3
Ossenkopf, V., & Henning, T. 1994, A&A, 291, 943
Pérez, L. M., Carpenter, J. M., Chandler, C. J., et al. 2012, ApJ, 760, L17
Pérez, L. M., Chandler, C. J., Isella, A., et al. 2015, ApJ, 813, 41
Segura-Cox, D. M., Harris, R. J., Tobin, J. J., et al. 2016, ApJ, 817, L14
Sheehan, P. D., & Eisner, J. A. 2018, ApJ, 857, 18
Tobin, J. J., Hartmann, L., Chiang, H.-F., et al. 2012, Nat, 492, 83
Tobin, J. J., Dunham, M. M., Looney, L. W., et al. 2015, ApJ, 798, 61
Tobin, J. J., Kratter, K. M., Persson, M. V., et al. 2016, Nat, 538, 483
Tobin, J. J., Sheehan, P., & Johnstone, D. 2018, Science with a Next-Generation Very Large Array, ASP Monograph 7, ed. E. J. Murphy (San Francisco, CA: ASP), 333
Tychoniec, Ł., Tobin, J. J., Karska, A., et al. 2018, ArXiv e-prints, arXiv:1806.02434
Weidenschilling, S. J. 1977, MNRAS, 180, 57
Woitke, P., Min, M., Pinte, C., et al. 2016, A&A, 586, A103

Science with a Next-Generation Very Large Array
ASP Conference Series, Monograph 7
Eric J. Murphy, ed.

Magnetic Fields in Forming Stars with the ngVLA

Charles L. H. Hull,[1,2,3] Carlos Carrasco-González,[4] Peter K. G. Williams,[5] Josep M. Girart,[6] Timothy Robishaw,[7] Roberto Galván-Madrid,[4] and Tyler Bourke[8]

[1]*National Astronomical Observatory of Japan, NAOJ Chile Observatory, Alonso de Córdova 3788, Office 61B, Vitacura 763 0422, Santiago, Chile;* `chat.hull@nao.ac.jp`

[2]*Joint ALMA Observatory, Alonso de Córdova 3107, Vitacura 763 0355, Santiago, Chile*

[3]*NAOJ Fellow*

[4]*Instituto de Radioastronomía y Astrofísica (IRyA-UNAM), Universidad Nacional Autónoma de México, Campus Morelia, Apartado Postal 3-72, 58090 Morelia, Michoacán, Mexico*

[5] *Harvard-Smithsonian Center for Astrophysics, 60 Garden St, MS 20, Cambridge, MA 02138, USA*

[6]*Institut de Ciències de l'Espai (CSIC-IEEC), Campus UAB, Carrer de Can Magrans S/N, E-08193 Cerdanyola del Vallès, Catalonia, Spain*

[7]*Dominion Radio Astrophysical Observatory, National Research Council Canada, PO Box 248, Penticton, BC, V2A 6J9, Canada;*

[8]*Square Kilometre Array Organization, Jodrell Bank Observatory, Lower Withington, Macclesfield, Cheshire SK11 9DL, UK;*

Abstract. The magnetic field plays an important role in every stage of the star-formation process from the collapse of the initial protostellar core to the star's arrival on the main sequence. Consequently, the goal of this article is to explore a wide range of magnetic phenomena that can be investigated using the polarization capabilities of the next generation Very Large Array (ngVLA). These include: (1) magnetic fields in protostellar cores via polarized emission from aligned dust grains, including in regions optically thick at wavelengths observable by the Atacama Large Millimeter/submillimeter Array (ALMA); (2) magnetic fields in both protostellar cores and molecular outflows via spectral-line polarization from the Zeeman and Goldreich-Kylafis effects; (3) magnetic fields in protostellar jets via polarized synchrotron emission; and (4) gyrosynchrotron emission from magnetospheres around low-mass stars.

1. Scientific Importance & Anticipated Results

1.1. Dust polarization in Low-Mass Protostellar Objects

Mapping the morphology of magnetic fields in the cores and envelopes surrounding young, low-mass protostars is critical to further our understanding of how magnetic fields affect the star-formation process at early times. Under most circumstances, and for the typical grain sizes found in the ISM ($\ll$100 μm), elongated dust grains are thought to align themselves with their long axes perpendicular to magnetic field lines (e.g., Hildebrand 1988; Lazarian 2007; Hoang & Lazarian 2009; Andersson et al. 2015), resulting in thermal radiation from the grains that is polarized perpendicular to the magnetic field (see Heiles et al. 1993 for a description of not only linear dust polarization, but also the other polarization mechanisms that are mentioned in the following sections). Thus, this type of observation can be used to infer the magnetic field morphology in the plane of the sky.

Polarized thermal dust emission from both low- and high-mass protostellar objects has been studied extensively at millimeter wavelengths at 1–2″ resolution by the Berkeley Illinois Maryland Association millimeter array (BIMA; e.g., Rao et al. 1998; Girart et al. 1999; Lai et al. 2001; Matthews et al. 2005; Cortes & Crutcher 2006; Kwon et al. 2006), the Submillimeter Array (SMA; e.g., Girart et al. 2006, 2009; Tang et al. 2009; Alves et al. 2011; Zhang et al. 2014), the Combined Array for Research in Millimeter-wave Astronomy (CARMA; e.g., Hull et al. 2013; Hughes et al. 2013; Hull et al. 2014; Stephens et al. 2014; Segura-Cox et al. 2015), and ALMA (e.g., Cortes et al. 2016; Hull et al. 2017a,b; Cox et al. 2018; Maury et al. 2018). With the upgraded Karl G. Jansky Very Large Array (VLA), it has recently become possible to begin performing dust polarization studies at centimeter wavelengths, using the highest-frequency bands available: K, Ka, and Q bands (18–50 GHz). Both Cox et al. (2015, see Figure 1) and Liu et al. (2016) studied the polarized dust emission at 37 and 43 GHz from NGC 1333-IRAS 4A, one of the first sources to be imaged in full polarization and high resolution at submillimeter wavelengths using the SMA (Girart et al. 2006). More recently, Liu et al. (2018) have studied the 40–48 GHz polarization toward the Class 0 protostar IRAS 16293, another bright source that was an early target for millimeter-wave polarization observations (Rao et al. 2009, 2014).

For the VLA, NGC 1333-IRAS 4A and IRAS 16293 are two of the very few sources with polarized dust emission bright enough that they can be imaged within a few hours. A typical protostellar core may be $\sim 10\times$ fainter than IRAS 4A or IRAS 16293. In order to probe the magnetic field in the innermost $\sim$ 1000–100 au regions (which can be resolved with the 1000–10 mas scales recoverable with the ngVLA) of these more “normal” objects, we need the sensitivity and the resolution of the ngVLA at the $\sim$ 30–90 GHz wavelengths where there is still appreciable dust emission, but where material is still optically thin even in the densest inner regions. For example, the peak flux of IRAS 4A at 30 GHz as detected by Cox et al. (2015) is $\sim 3000\,\mu$Jy in an 0.25″ synthesized beam. In <1 hr of integration time, the ngVLA will be able to detect 1% polarized dust emission at a frequency of 30 GHz at a level $> 3\sigma$ in objects $10\times$ fainter than IRAS 4A.

1.1.1. Polarization in Protoplanetary Disks

Regarding protoplanetary disk polarization at long wavelengths: a notable non-detection is HL Tau—one of the brightest known protoplanetary disks—where exceptionally

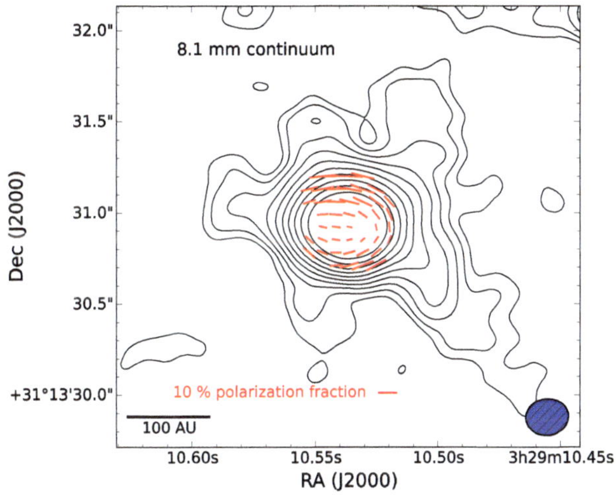

Figure 1. VLA 8 mm polarization observations of the embedded Class 0 protostellar source NGC 1333-IRAS 4A (Cox et al. 2018). Contours show total intensity (Stokes *I*) emission; red line segments show the inferred magnetic field, rotated by 90° relative to the dust polarization. IRAS 4A is one of the brightest low-mass protostellar sources known; typical sources are not bright or polarized enough for dust polarization to be detected toward them with the VLA.

deep 7 mm VLA observations by Carrasco-González et al. (2016) were unable to detect dust polarization, despite having an rms noise of $\sim 3\,\mu$Jy beam^{-1} (after integrating for nearly 15 hrs on source), simply because the dust emission is too faint at those long wavelengths. Polarization fractions of $\sim 1\%$ are expected in protostellar disks; thus, an order-of-magnitude improvement in sensitivity is needed in order to detect polarization in these sources.

HL Tau is $\sim 10\times$ brighter than a "typical" extended protoplanetary disk: for example, the subsample of transition disks in the ALMA survey of ρ Oph by Cox et al. (2017) has a median integrated flux of ~ 260 mJy at $850\,\mu$m, whereas HL Tau has an integrated flux of brightness of ~ 2.4 Jy at the same wavelength (Stephens et al. 2017). Assuming this approximate factor of 10 ratio holds at longer wavelengths, a typical large disk would have a peak 3 mm flux of ~ 3 mJy in a $\sim 0\farcs5$ beam, $10\times$ lower than the value reported in Stephens et al. (2017). In < 1 hr of integration time, the ngVLA will be able to detect 1% polarized dust emission at a wavelength of 3 mm at a level $> 3\sigma$ in such an object, enabling studies of long-wavelength polarization in a statistically significant sample of many of the famous, extended disks present in a wide range of star-forming regions. These studies will also allow us to characterize the polarization in regions optically thick at ALMA wavelengths (e.g., in HL Tau; ALMA Partnership et al. 2015; Carrasco-González et al. 2016; Jin et al. 2016); will enable us to characterize the polarization properties of grains by observing the polarization spectrum—i.e., the source polarization as a function of wavelength—from infrared to (sub)millimeter to centimeter wavelengths; and will allow us to better understand the physics behind the transition from magnetic alignment of dust grains to other polarization mechanisms (see below), which are thought to be dependent on dust-grain growth.

Recent ALMA polarization observations toward several protoplanetary disks have revealed different mechanisms causing the polarization: self-scattering at 870 μm (observational: Kataoka et al. 2016; Stephens et al. 2017; Lee et al. 2018; Girart et al. 2018; Hull et al. 2018; theoretical: Kataoka et al. 2015; Yang et al. 2017), alignment with the dust-emission gradient at 3 mm (observational: Kataoka et al. 2017; theoretical: Tazaki et al. 2017), and a seemingly linear combination of the two at 1.3 mm (Stephens et al. 2017). Recent (sub)millimeter ALMA polarization observations of younger, more embedded Class 0/I protostars have even shown hints of scattering in the innermost, ~ 100 au regions of a few sources (Cox et al. 2018; Harris et al. 2018; Sadavoy et al. 2018). However, self-scattering is a highly wavelength-dependent phenomenon, and it is maximum around a wavelength of $2\pi a_{\rm max}$, where $a_{\rm max}$ is the maximum grain size. Thus, at long wavelengths far from this value (i.e., those accessible to the ngVLA), it may be possible that grains aligned with the magnetic field dominate the dust polarization.

Considering that ALMA will eventually have Band 1 (~ 40 GHz), and that it already has a functioning polarization system at Band 3 (~ 100 GHz), lower-frequency K and Ka band (18–40 GHz) observations with the ngVLA would be complementary, and possibly unique to the instrument. Also, without considering the unknown polarization efficiency as a function of wavelength, 18–40 GHz observations are easier than higher Q band (40–50 GHz) observations at the VLA site due to the weather constraints.

1.2. Spectral-Line polarization: Zeeman and Goldreich-Kylafis effects

Molecular and atomic lines are sensitive to magnetic fields, which cause their spectral levels to split into magnetic sub-levels. When threaded by a magnetic field, atomic hydrogen and molecules with a strong magnetic dipole moment will have the degeneracy in magnetic sub-levels lifted for states with non-zero angular momentum. Examples lying within the nominal 1.2–116 GHz frequency range of the ngVLA include lines from H at 1.42 GHz; OH at 1.61–1.72 GHz & 6.01–6.05 GHz; CH_3OH at 6.7, 36, & 44 GHz; C_4H at 9.5 & 19.0 GHz; CCS at 11, 22, 33, & 45 GHz; C_2H at 87 GHz; SO at 13, 30, 63, 86, & 99 GHz; and CN at 113.1–113.6 GHz. This lifting of the degeneracy will split the radio frequency transitions into a number of linearly and elliptically polarized components slightly separated in frequency. This is known as the Zeeman effect (Troland & Heiles 1986; Heiles et al. 1993; Crutcher 1999). Measuring this effect is the only way to directly measure the magnetic field strength (yielding either the line-of-sight field strength for Zeeman-broadened transitions, or the entire field strength for transitions that are completely split, (e.g., Galactic OH masers[1]). Directly measuring the magnetic field strength is essential to achieve a more robust understanding of the dynamical relevance of the magnetic fields before and during the collapse of the star-forming core. Galactic OH masers can also be used to map out the large-scale magnetic field in the Milky Way since the star-forming regions in which they're found seem to retain information about the large-scale field in their vicinity (Davies 1974; Reid & Silverstein 1990). The ngVLA will play a critical role in extending Zeeman measurements beyond our Galaxy by detecting magnetic fields from extragalactic OH masers

[1] Linear and circular polarization from maser emission of centimeter and millimeter lines of SiO, CH_3OH, H_2O, and OH would allow us to probe the dense environment around young stellar objects and the circumstellar material around evolved stars.

in nearby galaxies like M33 and M31 and from OH megamasers in distant starbursting galaxies (Robishaw et al. 2008).

For other types of molecules, linear polarization can arise whenever an anisotropy in the radiation field yields a non-LTE population of magnetic sub-levels. This is the so-called Goldreich-Kylafis (G-K) effect, which traces the plane-of-sky magnetic field orientation (Goldreich & Kylafis 1981, 1982; Kylafis 1983; Deguchi & Watson 1984). This effect is expected to be easiest to observe where the lines have an optical depth of ~ 1; when the ratio of the collision rate to the radiative transition rate (i.e., the spontaneous emission rate) is also ~ 1; and for the lowest rotational transitions of simple molecules such as CO, CS, HCN, SiO, or HCO^+. So far, the G-K effect has been measured mostly around molecular outflows (e.g., Girart et al. 1999; Lai et al. 2003; Cortes et al. 2005; Vlemmings et al. 2012; Girart et al. 2012; Ching et al. 2016). This suggests that G-K observations with the ngVLA at 7 mm (SiO, CS) or 3 mm (CO, SiO, HCO^+) will provide new insight into the properties of the putative helical magnetic field near the launching point of bipolar outflows around young stellar objects, where a significant fraction of the outflow appears to be molecular. Molecular outflows from protostellar objects tend to have a parabolic, multi-layered, "onion-skin" type structure, with the lowest/highest-velocity material located in the outermost/innermost shells. Therefore, measuring the polarization at different velocities enables us to reconstruct the 3-dimensional magnetic field morphology in the source.

Polarization from thermal lines requires detection of emission with a high signal-to-noise ratio at high spectral resolution. The layout of the ngVLA (including the high-density central core of antennas) will yield excellent brightness sensitivity over a range of spatial scales relevant to star formation (approximately 10–1000 mas). The types of sources where the ngVLA would have a clear science impact detecting the Zeeman splitting and the G-K effect include the inner regions of various types of star-forming cores (e.g., warm cores, hot corinos, hot cores), proto/circumstellar envelopes, the launching regions of molecular outflows and jets, and disks.

1.3. Synchrotron Emission from Protostellar Jets

Jets play an important role in the star-formation process. They are believed to regulate accretion from the disk onto the protostar at the earliest (Class 0/I) stages of star formation. However, jets are not only important in the field of the star formation—protostellar jets are usually considered a less energetic manifestation of the powerful relativistic jets driven by supermassive black holes at the center of active galaxies or stellar-mass black holes in our own Galaxy (e.g., de Gouveia Dal Pino 2005). Despite their importance, we still do not know how protostellar jets are launched and collimated. In the case of relativistic jets, observations of polarized synchrotron emission suggest that a helical configuration of the magnetic field is responsible for the collimation of the jet (e.g., Lyutikov et al. 2005; Gómez et al. 2008). In the case of protostellar jets, theoretical models also point to a fundamental role of the magnetic field, which is thought to have a helical configuration (e.g., Shu et al. 1994); helical magnetic fields are a consequence of the rotation of the protostar-disk system. As for the collimation of jets, they may be collimated internally. However, other collimation mechanisms by external agents, such as ordered magnetic fields present in the protostar's neighborhood, have also been proposed (e.g., Albertazzi et al. 2014); some observations are consistent with external collimation mechanisms (e.g., Carrasco-González et al. 2015).

The magnetic field is extremely difficult to study in the case of protostellar jets. Jets have been observed in great detail at radio wavelengths, although they emit mainly continuum thermal radiation (e.g., Anglada 1996), which contains no information about the magnetic field. However, in the last decades, very sensitive observations have shown that synchrotron emission can also be present in some protostellar jets (see Carrasco-González et al. 2010 and references therein). While the material in the jet moves at velocities of a few $\times$ 100 km s^{-1}, it has been shown that under some conditions, a population of particles can be accelerated up to relativistic velocities in strong shocks against the ambient medium (e.g., Rodríguez-Kamenetzky et al. 2016). Under these conditions, these accelerated particles can emit synchrotron emission, thus allowing the study of the magnetic field in the jet via synchrotron polarization. However, synchrotron emission in these objects is extremely weak, with intensities on the order of only a few $\times$ 10 μJy in Stokes I. At the beginning of the 2000s, we only knew of a handful of bright objects that show hints of synchrotron emission (i.e., negative spectral indices at centimeter wavelengths). In the last few years, more candidates have been detected as a result of the recently improved sensitivity in the ATCA and VLA (e.g., Purser et al. 2016; Osorio et al. 2017). Yet, it has only been possible to detect linearly polarized emission in a single object, HH 80-81 (Carrasco-González et al. 2010, see Figure 2). Even though HH 80-81 is the brightest known protostellar jet, this unique detection at a single wavelength (6 cm) was only possible with significant observational effort (15 hrs of observations with the pre-upgrade VLA). Therefore, given the sensitivity of current instruments, it is still prohibitive to extend this study to multiple wavelengths in a sample of several protostellar jets.

In order to infer the 3-dimensional configuration of the magnetic field in the jet it is necessary to study the properties of the polarization of the synchrotron emission at several wavelengths. For example, for a helical magnetic field, we expect to observe gradients of the polarization fraction and the rotation measure (RM) across the jet width (Lyutikov et al. 2005). This implies the necessity to observe with high sensitivity to detect very low polarization fractions in intrinsically weak emission, and with high angular resolution to resolve the jet width (a few $\times$ 10 milliarcseconds). In principle, these requirements could be reached by the VLA. However, protostellar jets are very dense (electron densities of the order of a few $\times$ 1000 cm^{-3}), and thus large RMs are expected. High sensitivity is currently achieved by averaging large frequency ranges. However, this introduces depolarization across the bandwidth when large RMs are present, rendering the polarized signal impossible to detect. It is therefore necessary to obtain high sensitivities even in narrow frequency ranges, which can only be achieved by increasing the observation time and/or the number of antennas in the interferometer.

Finally, it is not only desirable to detect polarization at several frequencies, but also at very different physical scales. The jet is expected to show complex morphology: extended diffuse emission, bow shocks, extended lobes, highly collimated parts, and compact internal shocks (e.g., Rodríguez-Kamenetzky et al. 2017). We also expect to see changes in the density across and along the jet. All of these characteristics imply dramatic changes of the properties of the emission (e.g., RM, brightness, and polarization fraction) at different scales. These changes can actually be used to obtain information about the jet's physical structure by, for example, using new wide-band spectropolarimetry modeling techniques that have been successfully applied to very bright, but unresolved, radio galaxies (e.g., O'Sullivan et al. 2017; Pasetto et al. 2018). While this type of study would be very difficult to perform in distant extragalactic jets, in nearby protostellar jets it could be performed simply by being sensitive simultane-

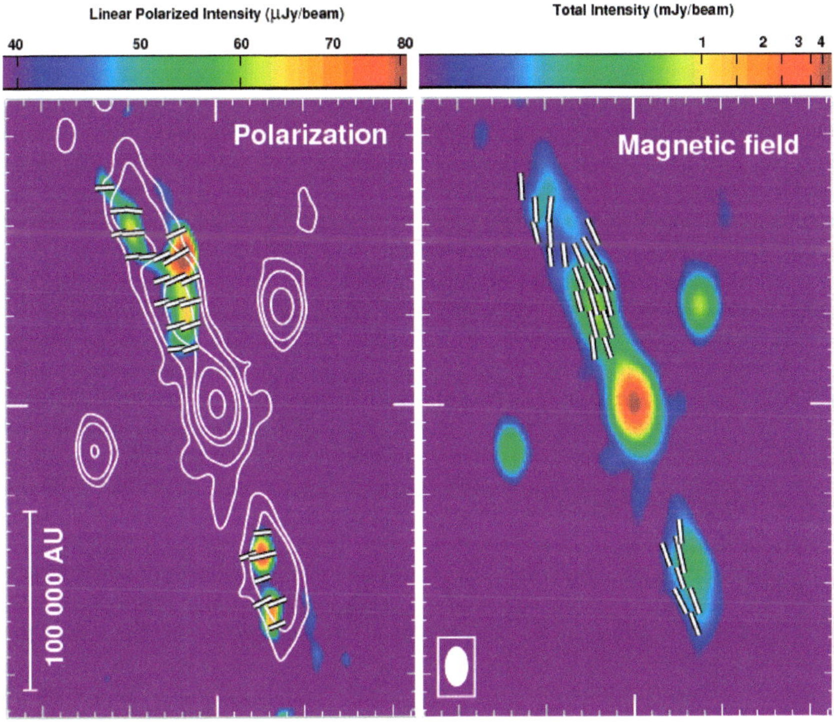

Figure 2. Observations of linearly polarized emission at 6 cm in the protostellar jet HH 80-81 (Carrasco-González et al. 2010). This is the first and only time that polarized emission has been detected in a protostellar jet. This result demonstrated that detecting polarized synchrotron emission is possible in protostellar jets, and allowed the study of the magnetic field in this object using techniques similar to those used in relativistic jets. However, the weakness of the emission and the likelihood of large rotation measures make it extremely difficult to extend this kind of study to other wavelengths and/or protostellar jets using current instruments. (*Left:*) Color scale is the linearly polarized emission at 6 cm. Line segments indicate the polarization orientation of the electric field. Contours are the total 6 cm emission from the protostellar jet. (*Right:*) White line segments indicate the orientation of the inferred (projected on the plane of the sky) magnetic field obtained from the polarization measurements. Color scale is the total 6 cm emission of the jet. The apparent magnetic field is parallel to the direction of the jet. This is consistent with a helical magnetic field configuration. However, in order to to infer the 3-dimensional configuration of the magnetic field it is necessary to study the polarization properties at several wavelengths.

ously to scales from a few milliarcseconds to several arcseconds. This will be easily achieved by an interferometer such as the ngVLA, with a large number of antennas spread over a few thousand kilometers.

1.4. Gyrosynchrotron from Active Magnetospheres Around Young Stellar Objects

Gyrosynchrotron radio emission from active magnetospheres is present in low-mass young stellar objects (YSOs) (e.g., Andre 1996) as well as their more evolved counterparts (Güdel & Benz 1993), including our Sun (Nita et al. 2004). Studying this emission in pre-main-sequence stars sheds light on their magnetic activity and the overall "space weather" that they drive, which has important effects on the chemistry and evolution of the surrounding protoplanetary and debris disks (e.g., Wilner et al. 2000).

The characterization of magnetospheric radio emission is also necessary for the proper analysis of YSO dust emission, because the two processes lead to emission at overlapping radio frequencies—in particular, the range of frequencies where the ngVLA will be most powerful. Non-thermal gyrosynchrotron emission can be distinguished from thermal emission mainly through its variability on timescales from minutes to months (Forbrich et al. 2015), its generally (but not always) negative spectral index (Garay et al. 1996), circular polarization fraction of a few to $\sim$20% (Andre 1996), and its brightness temperatures $\gg 10^4$ K (Ortiz-León et al. 2017). Even the upgraded VLA, however, has difficulty achieving the performance needed to fully understand YSO radio emission at centimeter wavelengths. In a study using the VLA, Liu et al. (2014) characterized the radio flux, variability, and circular polarization of YSOs from Class 0 to Class III, achieving a noise of $\sim 20\,\mu$Jy using observing blocks of a few hours and averaging over a few GHz of bandwidth. The results were consistent with earlier findings (e.g., Gibb 1999) showing that centimeter radio emission from YSOs is: (1) dominated by free-free emission from collimated winds or radio jets in the early Class 0 and I stages (Anglada et al. 2015); (2) difficult to detect in Class IIs (Galván-Madrid et al. 2014); and (3) detectable in some but not all Class III YSOs (Forbrich et al. 2007; Dzib et al. 2013). The emission from Class IIIs appears to be dominated by gyrosynchrotron radiation, but in Class IIs there are individual case studies pointing toward either free-free or gyrosynchrotron (Macías et al. 2016; Galván-Madrid et al. 2014).

The high sensitivity, wide frequency coverage, and angular resolution of the ngVLA would transform studies of both thermal and non-thermal YSO radio emission, allowing robust disentanglement of the two processes in snapshot observations as short as a few seconds. Scaling the nominal sensitivities reported by Carilli et al. (2015), the ngVLA could obtain full-polarization measurements with noise levels of $\sim 10\ \mu$Jy in 5-second integrations. Moreover, the large simultaneous bandwidth would allow full characterization of the non-thermal radio SEDs and their overlap with free-free or dust thermal emission truly simultaneously, something that has not been possible with the VLA. It is quite likely that the non-thermal SEDs of YSOs have complexities due to multiple magnetospheric components, binary interactions (Salter et al. 2010), and other phenomena yet to be discovered.

The same attributes of the ngVLA that will make it a transformative tool for studies of YSO magnetospheres will produce similar gains in studies focusing on other objects with comparable magnetic properties. In particular, the magnetospheres of fully-convective YSOs can be dipolar and rotation-dominated, with strong commonalities to those found in very-low-mass stars and brown dwarfs, magnetic chemically peculiar (MCP) stars, and even the Solar System giant planets (Donati & Landstreet 2009; Schrijver 2009). The magnetism of these objects is quite different from that of the Sun, being driven by a fully-convective dynamo and resulting in large-scale current systems, coherent maser emission, and potentially trapped particle radiation belts (Hallinan et al.

2006; Williams 2017). The resulting radio emission is generally quite broadband and highly polarized (e.g., Williams et al. 2015), and current studies are often sensitivity-limited; for example, the VLA is unable to detect even the 2 pc-distant ultracool dwarf binary Luhman 16AB (Osten et al. 2015). The ngVLA will therefore enable an entirely new class of time-resolved, full-polarization studies of magnetospheric activity in YSOs and beyond.

2. Measurements Required

Continuum polarization observations are relatively straightforward, generally requiring a single-pointing per source, integrating down to the point where a ~ 1–3% polarization signal can be detected at the $> 3\sigma$ confidence level across the desired region of the source. Spectral line polarization requires high spectral resolution, where the rule of thumb is ~ 6 channels across the full-width-at-half-max (FWHM) of the line. This is especially important for Zeeman observations of masers, where a velocity resolution of < 0.1 km/s is required. Polarization levels can be low, on the order of a few $\times\ 0.1\%$, so an accurate characterization of the instrumental polarization is required, down to a level of $\sim 0.1\%$, similar to what is achievable with the ALMA polarization system. The more compact sources (disks, protostellar envelopes, etc.) will generally fill only a fraction of the primary beam; however, for more extended sources, and in the event that there is more than one source (e.g., a few disks or protostars) in one field of view, either separate pointings and/or mosaics will have to be performed, or the wide-field polarization characteristics will have to be well characterized in order to correct for off-axis effects (see, e.g., work done by Jagannathan et al. 2017, 2018). Other standard polarization-specific calibration requirements involve solving for leakages (D-terms) and the XY (or RL) phase of the reference antenna.

The design of the ngVLA antennas and feeds should be informed by careful consideration of polarimetric goals – unfortunately, there is always a polarimetric sacrifice to be made for any given design. Linearly polarized radiation (Stokes Q and U) – especially weak polarization – is best measured via the cross-correlation of the orthogonal outputs from native circular feeds (with a penalty in measuring Stokes V); likewise, circularly polarized radiation is best measured by cross-correlating the outputs from orthogonal native linear feeds (with a penalty in measuring Stokes Q). New interferometers, except for the VLA, are employing native dual-linear feeds. While these feeds can be converted to dual-circular by means of hybrids or dielectric quarter-wave phase shifters, such devices provide frequency-dependent polarization, operate over narrow bandwidths, and can increase not only the system temperature, but also the complexity and the cost of the receiver design.

An altitude-azimuth (alt-az) mount is ideal for polarization measurements because it causes a linearly polarized astronomical source to rotate relative to the feed as a source is tracked. This rotation greatly reduces systematic effects and makes calibration much easier. The Australian Square Kilometre Array Pathfinder (ASKAP) interferometer has been designed with dishes on alt-az mounts, but the designers have opted to rotate their entire dish structure with the sky—mimicking the behavior of an equatorial mount—in order to simplify wide-field polarimetric imaging. This is even better than a standard alt-az mount because the polarization rotation can be done instantaneously, without having to wait for the angle to change while tracking.

A final polarization consideration involves the choice to use an off-axis Gregorian secondary-focus design for the dishes. When it comes to measuring polarization, symmetry is better: sidelobes and spillover radiation tend to be polarized by an amount that increases with asymmetry. Moreover, ground radiation is highly polarized.

3. Uniqueness to ngVLA Capabilities

The centimeter-wave/$\sim$ 30 GHz capabilities of the ngVLA are unique in that the other extant telescopes that can currently perform such observations—namely, the VLA—doesn't have the sensitivity to perform the type of observations we require. In the case of future ALMA Band 1 observations at $\sim$ 1 cm wavelengths, the ngVLA's resolution and sensitivity will be much higher, enabling the studies we discuss above. Lower frequency telescopes probe physical phenomena that are completely different from the various polarization mechanisms we discuss above.

4. Synergies at Other Wavelengths

For continuum science, as an interferometer the ngVLA will fill a gap between the Square Kilometre Array (SKA)—which will address many different topics related to magnetic fields and polarization—and ALMA, which will look at many of the same objects, but at high frequencies where the material in the regions of interest (i.e., the inner parts of protoplanetary disks) is sometimes optically thick. The ngVLA will perform extremely high-resolution, high-sensitivity, targeted polarization science at the same time that the Origins Space Telescope (OST), if funded and constructed, is performing surveys of continuum and spectral-line polarization toward entire star-forming clouds across the Galaxy. The ngVLA will perform long-wavelength follow-up of observations from the myriad telescopes that are currently available (or are soon coming online) and have polarization capabilities at far-infrared and (sub)millimeter wavelengths, including ALMA, SOFIA (HAWC+ polarimeter), the James Clark Maxwell Telescope (JCMT; POL2 polarimeter), the Large Millimeter Telescope (LMT; TolTEC polarimeter), and the BLAST-TNG balloon-borne experiment.

Acknowledgments. The authors thank Carl Heiles for his help in clarifying the implications of the ngVLA receiver design on the telescope's polarimetric capabilities. C.L.H.H acknowledges the support of the JSPS KAKENHI grant 18K13586. J.M.G. is supported by the MINECO (Spain) AYA2017-84390-C2 grant. R.G.-M. acknowledges support from UNAM-PAPIIT program IA102817. P.K.G.W. acknowledges support for this work from the National Science Foundation through Grant AST-161477. The National Radio Astronomy Observatory is a facility of the National Science Foundation operated under cooperative agreement by Associated Universities, Inc.

References

Albertazzi, B., Ciardi, A., Nakatsutsumi, M., et al. 2014, Science, 346, 325
ALMA Partnership, Brogan, C. L., Pérez, L. M., et al. 2015, ApJ, 808, L3
Alves, F. O., Girart, J. M., Lai, S.-P., Rao, R., & Zhang, Q. 2011, ApJ, 726, 63
Andersson, B.-G., Lazarian, A., & Vaillancourt, J. E. 2015, ARA&A, 53, 501
Andre, P. 1996, in Astronomical Society of the Pacific Conference Series, Vol. 93, Radio Emission from the Stars and the Sun, ed. A. R. Taylor & J. M. Paredes, 273–284

Anglada, G. 1996, in Astronomical Society of the Pacific Conference Series, Vol. 93, Radio Emission from the Stars and the Sun, ed. A. R. Taylor & J. M. Paredes, 3–14
Anglada, G., Rodríguez, L. F., & Carrasco-Gonzalez, C. 2015, Advancing Astrophysics with the Square Kilometre Array (AASKA14), 121
Carilli, C. L., McKinnon, M., Ott, J., et al. 2015, ArXiv e-prints, arXiv:1510.06438
Carrasco-González, C., Rodríguez, L. F., Anglada, G., et al. 2010, Science, 330, 1209
Carrasco-González, C., Torrelles, J. M., Cantó, J., et al. 2015, Science, 348, 114
Carrasco-González, C., Henning, T., Chandler, C. J., et al. 2016, ApJ, 821, L16
Ching, T.-C., Lai, S.-P., Zhang, Q., et al. 2016, ApJ, 819, 159
Cortes, P., & Crutcher, R. M. 2006, ApJ, 639, 965
Cortes, P. C., Crutcher, R. M., & Watson, W. D. 2005, ApJ, 628, 780
Cortes, P. C., Girart, J. M., Hull, C. L. H., et al. 2016, ApJ, 825, L15
Cox, E. G., Harris, R. J., Looney, L. W., et al. 2018, ArXiv e-prints, arXiv:1802.00449
—. 2015, ApJ, 814, L28
—. 2017, ApJ, 851, 83
Crutcher, R. M. 1999, ApJ, 520, 706
Davies, R. D. 1974, in IAU Symp., Vol. 60, Galactic Radio Astronomy, ed. F. J. Kerr & S. C. Simonson (Dordrecht: Reidel), 275
de Gouveia Dal Pino, E. M. 2005, Advances in Space Research, 35, 908
Deguchi, S., & Watson, W. D. 1984, ApJ, 285, 126
Donati, J. F., & Landstreet, J. D. 2009, ARA&A, 47, 333
Dzib, S. A., Loinard, L., Mioduszewski, A. J., et al. 2013, ApJ, 775, 63
Forbrich, J., Rodríguez, L. F., Palau, A., et al. 2015, ApJ, 814, 15
Forbrich, J., Preibisch, T., Menten, K. M., et al. 2007, A&A, 464, 1003
Galván-Madrid, R., Liu, H. B., Manara, C. F., et al. 2014, A&A, 570, L9
Garay, G., Ramirez, S., Rodriguez, L. F., Curiel, S., & Torrelles, J. M. 1996, ApJ, 459, 193
Gibb, A. G. 1999, MNRAS, 304, 1
Girart, J. M., Beltrán, M. T., Zhang, Q., Rao, R., & Estalella, R. 2009, Science, 324, 1408
Girart, J. M., Crutcher, R. M., & Rao, R. 1999, ApJ, 525, L109
Girart, J. M., Patel, N., Vlemmings, W. H. T., & Rao, R. 2012, ApJ, 751, L20
Girart, J. M., Rao, R., & Marrone, D. P. 2006, Science, 313, 812
Girart, J. M., Fernandez-Lopez, M., Li, Z.-Y., et al. 2018, ArXiv e-prints, arXiv:1803.06165
Goldreich, P., & Kylafis, N. D. 1981, ApJ, 243, L75
—. 1982, ApJ, 253, 606
Gómez, J. L., Marscher, A. P., Jorstad, S. G., Agudo, I., & Roca-Sogorb, M. 2008, ApJ, 681, L69
Güdel, M., & Benz, A. O. 1993, ApJ, 405, L63
Hallinan, G., Antonova, A., Doyle, J. G., et al. 2006, ApJ, 653, 690
Harris, R. J., Cox, E. G., Looney, L. W., et al. 2018, ArXiv e-prints, arXiv:1805.08792
Heiles, C., Goodman, A. A., McKee, C. F., & Zweibel, E. G. 1993, in Protostars and Planets III, ed. E. H. Levy & J. I. Lunine, 279–326
Hildebrand, R. H. 1988, QJRAS, 29, 327
Hoang, T., & Lazarian, A. 2009, ApJ, 697, 1316
Hughes, A. M., Hull, C. L. H., Wilner, D. J., & Plambeck, R. L. 2013, AJ, 145, 115
Hull, C. L. H., Plambeck, R. L., Bolatto, A. D., et al. 2013, ApJ, 768, 159
Hull, C. L. H., Plambeck, R. L., Kwon, W., et al. 2014, ApJS, 213, 13
Hull, C. L. H., Mocz, P., Burkhart, B., et al. 2017a, ApJ, 842, L9
Hull, C. L. H., Girart, J. M., Tychoniec, Ł., et al. 2017b, ApJ, 847, 92
Hull, C. L. H., Yang, H., Li, Z.-Y., et al. 2018, ArXiv e-prints, arXiv:1804.06269
Jagannathan, P., Bhatnagar, S., Brisken, W., & Taylor, A. R. 2018, AJ, 155, 3
Jagannathan, P., Bhatnagar, S., Rau, U., & Taylor, A. R. 2017, AJ, 154, 56
Jin, S., Li, S., Isella, A., Li, H., & Ji, J. 2016, ApJ, 818, 76
Kataoka, A., Tsukagoshi, T., Pohl, A., et al. 2017, ApJ, 844, L5
Kataoka, A., Muto, T., Momose, M., et al. 2015, ApJ, 809, 78
Kataoka, A., Tsukagoshi, T., Momose, M., et al. 2016, ApJ, 831, L12

Kwon, W., Looney, L. W., Crutcher, R. M., & Kirk, J. M. 2006, ApJ, 653, 1358
Kylafis, N. D. 1983, ApJ, 267, 137
Lai, S.-P., Crutcher, R. M., Girart, J. M., & Rao, R. 2001, ApJ, 561, 864
Lai, S.-P., Girart, J. M., & Crutcher, R. M. 2003, ApJ, 598, 392
Lazarian, A. 2007, J. Quant. Spec. Radiat. Transf., 106, 225
Lee, C.-F., Li, Z.-Y., Ching, T.-C., Lai, S.-P., & Yang, H. 2018, ApJ, 854, 56
Liu, H. B., Hasegawa, Y., Ching, T.-C., et al. 2018, ArXiv e-prints, arXiv:1805.02012
Liu, H. B., Galván-Madrid, R., Forbrich, J., et al. 2014, ApJ, 780, 155 5.
Liu, H. B., Lai, S.-P., Hasegawa, Y., et al. 2016, ApJ, 821, 41
Lyutikov, M., Pariev, V. I., & Gabuzda, D. C. 2005, MNRAS, 360, 869
Macías, E., Anglada, G., Osorio, M., et al. 2016, ApJ, 829, 1
Matthews, B. C., Lai, S.-P., Crutcher, R. M., & Wilson, C. D. 2005, ApJ, 626, 959
Maury, A. J., Girart, J. M., Zhang, Q., et al. 2018, MNRAS, arXiv:1803.00028
Nita, G. M., Gary, D. E., & Lee, J. 2004, ApJ, 605, 528
Ortiz-León, G. N., Loinard, L., Kounkel, M. A., et al. 2017, ApJ, 834, 141
Osorio, M., Díaz-Rodríguez, A. K., Anglada, G., et al. 2017, ApJ, 840, 36
Osten, R. A., Melis, C., Stelzer, B., et al. 2015, ApJ, 805, L3
O'Sullivan, S. P., Purcell, C. R., Anderson, C. S., et al. 2017, MNRAS, 469, 4034
Pasetto, A., Carrasco-González, C., O'Sullivan, S., et al. 2018, ArXiv e-prints, arXiv:1801.09731
Purser, S. J. D., Lumsden, S. L., Hoare, M. G., et al. 2016, MNRAS, 460, 1039
Rao, R., Crutcher, R. M., Plambeck, R. L., & Wright, M. C. H. 1998, ApJ, 502, L75
Rao, R., Girart, J. M., Lai, S.-P., & Marrone, D. P. 2014, ApJ, 780, L6
Rao, R., Girart, J. M., Marrone, D. P., Lai, S.-P., & Schnee, S. 2009, ApJ, 707, 921
Reid, M. J., & Silverstein, E. M. 1990, ApJ, 361, 483
Robishaw, T., Quataert, E., & Heiles, C. 2008, ApJ, 680, 981
Rodríguez-Kamenetzky, A., Carrasco-González, C., Araudo, A., et al. 2016, ApJ, 818, 27
—. 2017, ApJ, 851, 16
Sadavoy, S. I., Myers, P. C., Stephens, I. W., et al. 2018, ArXiv e-prints, arXiv:1804.05968
Salter, D. M., Kóspál, Á., Getman, K. V., et al. 2010, A&A, 521, A32
Schrijver, C. J. 2009, ApJ, 699, L148
Segura-Cox, D. M., Looney, L. W., Stephens, I. W., et al. 2015, ApJ, 798, L2
Shu, F., Najita, J., Ostriker, E., et al. 1994, ApJ, 429, 781
Stephens, I. W., Looney, L. W., Kwon, W., et al. 2014, Nat, 514, 597
Stephens, I. W., Yang, H., Li, Z.-Y., et al. 2017, ApJ, 851, 55
Tang, Y.-W., Ho, P. T. P., Girart, J. M., et al. 2009, ApJ, 695, 1399
Tazaki, R., Lazarian, A., & Nomura, H. 2017, ApJ, 839, 56
Troland, T. H., & Heiles, C. 1986, ApJ, 301, 339
Vlemmings, W. H. T., Ramstedt, S., Rao, R., & Maercker, M. 2012, A&A, 540, L3
Williams, P. K. G. 2017, in Handbook of Exoplanets, ed. H. J. Deeg & J. A. Belmonte (Springer Verlag). http://adsabs.harvard.edu/abs/2017arxiv170704264
Williams, P. K. G., Berger, E., Irwin, J., Berta-Thompson, Z. K., & Charbonneau, D. 2015, ApJ, 799, 192
Wilner, D. J., Ho, P. T. P., Kastner, J. H., & Rodríguez, L. F. 2000, ApJ, 534, L101
Yang, H., Li, Z.-Y., Looney, L. W., Girart, J. M., & Stephens, I. W. 2017, MNRAS, 472, 373
Zhang, Q., Qiu, K., Girart, J. M., et al. 2014, The Astrophysical Journal, 792, 116. http://stacks.iop.org/0004-637X/792/i=2/a=116

Science with a Next-Generation Very Large Array
ASP Conference Series, Monograph 7
Eric J. Murphy, ed.

Imaging Stellar Radio Photospheres with the Next Generation Very Large Array

C. L. Carilli,[1] B. Butler,[1] K. Golap,[1] M. T. Carilli,[2] and S. M. White[3]

[1] *National Radio Astronomy Observatory, Charlottesville, VA, 22903;* ccarilli@nrao.edu

[2] *University of Colorado, Boulder, CO*

[3] *Air Force Research Lab, Albuquerque, NM*

Abstract. We perform simulations of the capabilities of the next generation Very Large Array to image stellar radio photospheres. For very large (in angle) stars, such as red supergiants within a few hundred parsecs, good imaging fidelity results can be obtained on radio photospheric structures at 38 GHz employing standard techniques, such as disk model fitting and subtraction, with hundreds of resolution elements over the star, even with just the ngVLA-classic baselines to 1000 km. Using the ngVLA Rev B plus long baseline configuration (with baselines out to 9000 km, August 2018), we find for main sequence stars within ~ 10 pc, the photospheres can be easily resolved at 85 GHz, with accurate measures of the mean brightness and size, and possibly imaging large surface structures, as might occur on e.g., active M dwarf stars. For more distant main sequence stars, we find that measurements of sizes and brightnesses can be made using disk model fitting to the (u, v)-data down to stellar diameters ~ 0.4 mas in a few hours. This size would include M0 V stars to a distance of 15 pc, A0 V stars to 60 pc, and Red Giants to 2.4 kpc. Based on the *Hipparcos* catalog, we estimate that there are at least 10,000 stars that will be resolved by the ngVLA. While the vast majority of these (95%) are giants or supergiants, there are still over 500 main sequence stars that can be resolved, with ~ 50 to 150 in each spectral type (besides O stars). Note that these are lower limits, since radio photospheres can be larger than optical, and the *Hipparcos* catalog might not be complete. Our initial look into the *Gaia* catalog suggests these numbers might be pessimistic by a factor few.

1. Introduction

The field of stellar atmospheres, and atmospheric activity, has taken on new relevance in the context of the search for habitable planets, due to the realization of the dramatic effect 'space weather' can have on the development of life (Osten et al. 2018; Trigilio et al. 2018). Observations at radio wavelengths provide some of the most powerful probes of exo-space weather, including: (i) high brightness coherent processes, such as masers associated with outflows in forming stars, (ii) non-thermal coherent and incoherent gyro-synchrotron emission due to magnetic activity in stellar coronae, and (iii) thermal emission from extreme mass loss events, such as novae (Gudel 2002; Linford et al. 2018, this volume p. 271 and 265).

In this article, we consider the more quiescent properties of stellar radio photospheres, namely, the thermal emission from the $\tau \sim 1$ surface of the photospheres of stars of different types. The size of stars, as measured by their stellar radio photospheres, is a strong function of wavelength, since the free-free optical depth drops quickly with frequency ($\propto \nu^{-2.1}$). Hence, radio observations over a range in frequencies provide a tomographic image of the density and temperature structure of the transition from the optical photosphere to the chromosphere. For giant stars, the radio photosphere will be larger, and typically cooler, than the optical photosphere at long wavelengths, and approach the optical photospheric size and temperature at shorter wavelengths. For instance, in red super-giants, the radio photospheres at $\sim$ 6 cm are typically 7× larger, and 3× cooler, than the optical photosphere, while at $\sim$ 1 mm the radio and optical photospheric temperatures and sizes are comparable to within $\leq 30\%$ (Lim et al. 1998; O'Gorman et al. 2015, 2017; Harper et al. 2013). For main sequence stars, there should be closer correspondance between radio and optical photospheric sizes, although even in these cases, at cm and mm wavelengths the stellar atmospheres can become optically thick in the chromosphere, which would enhance temperatures by a few thousand Kelvin (e.g., the Sun has a brightness temperature at 100 GHz of 7300 K; White et al. 2017).

Of particular interest is the question of the end-of-life mass loss in red giants and supergiants through strong winds. Current observations suggest elevated atmospheres, with substantial large-scale temperature structure, heated by sound or Alfven waves driven by large convective cells bubbling-up from the stellar interior (Lim et al. 1998; O'Gorman et al. 2015, 2017; Harper 2018; Matthews & Claussen 2018). For Solar-type and lower mass main sequence stars, recent interest has focused on atmospheric conditions that can drive strong winds, leading to star-planet magnetic interactions, possibly triggering auroral radio emission (Trigilio et al. 2018; Fichtinger et al. 2017; Hallinan et al. 2015).

To date, direct imaging of stellar radio photospheres has been limited to a handful of the largest, closest stars, such as red giants, super-giants, and AGB stars (Lim et al. 1998; Vlemmings et al. 2017; O'Gorman et al. 2013; Matthews & Claussen 2015). Likewise, detection of thermal emission from nearby solar-type main sequence stars is just possible with deep observations with the VLA and ATCA (Villadsen et al. 2014; O'Gorman et al. 2017), but imaging is certainly impossible with current facilities. The situation is similar in the optical and near-IR, with a moderate number of large, hot stars resolved through optical interferometry or direct imaging with *HST* (Boyajian et al. 2015; Monnier et al. 2007; Gilliland & Dupree 1996; Mozurkewich et al. 2003). For optical interferometers, structure is typically inferred through model fitting of interferometric closure quantities (Monnier 2003; Defrere et al. 2018), although hybrid closure-imaging approaches are being explored with the new VLTi (Montarge et al. 2018).

The 'next generation Very Large Array' (ngVLA) opens a new window on stellar radio photospheres, through sub-milliarcsecond resolution at up to 90 GHz with sub-μJy sensitivity (McKinnon et al. 2016; Carilli et al. 2015; Selina & Murphy 2017). The ngVLA rms brightness sensitivity in one hour at 85 GHz and 0.4 mas resolution is $\sim$ 1000 K. There are thousands of stars with sizes larger than 0.4 mas, with brightness temperatures above a few thousand Kelvin at 90 GHz. These include nearby main sequence stars, and more distant giants and supergiants. We investigate the ability of the ngVLA to provide high fidelity images of the radio photospheres of larger (in angle) stars, and to determine sizes and flux densities for sub-mas stars.

2. Configuration, Model, and Process

We employ the newly accepted ngVLA configuration (August 2018). This configuration includes the original ngVLA Rev B configuration distributed across the U.S. Southwest and Mexico. The Rev B configuration includes 214 antennas of 18 meter diameter, to baselines up to 1000 km (Carilli & Erickson 2018; Selina & Murphy 2017). The new ngVLA configuration includes 30 more antennas distributed to baselines of 9000 km, including clusters of two to three antennas at sites such as Hawaii and St. Croix. For image simulations, the task SIMOBSERVE within CASA was used to (u, v)-data sets. Instructions on how this is done are found on the ngVLA web page.

The ngVLA has a highly centrally condensed antenna distribution. The naturally weighted beam for this centrally condensed distribution leads to a very non-Gaussian PSF, with two skirts extending to 10× the nominal full resolution of the longest baselines at the 50% to 20% level, and 100× the full resolution at the 20% to 10% level, for just the Rev B array alone. This PSF problem becomes an order of magnitude more severe when including the longer baselines, in terms of the relative angular size of the PSF 'spike' corresponding to the longest baselines vs. the skirts. These prominent skirts represent the major challenge when trying to image complex structure. The imaging process entails a balance between sensitivity and synthesized beam shape, using visibility tapering and Briggs robust weighting (Carilli et al. 2018; Cotton & Condon 2017; Carilli 2017, 2016). In all cases we employed the CLEAN algorithm with Briggs weighting, and adjusted the robust parameter (R) and (u, v)-taper to give a reasonable synthesized beam and noise performance.

We employ a template stellar disk model that includes a few large cold and hot spots at the ±10% level (Figure1a) . Such spots have been seen on the radio photospheres of giant and supergiant stars. Such large structures are not present on most main sequence stars, but they provide some 'test structures' that exercise the imaging algorithms at high resolution. Moreover, active stars, such as M-dwarves, may have large temperature structures, such as giant sunspot regions. We then adjust the star size and mean brightness temperature according to the stellar types and assumed distance. We emphasize that the models below are not meant to be accurate reproductions of a specific star, just representative.

The new long baselines extend from 1000 km to 9000 km. The longest baselines that include correlation to the ngVLA core are about 5000 km (e.g., Socorro to Hawaii). Beyond that, only outer stations are involved in the cross correlation, and we will show that the sensitivity drops dramatically beyond 5000 km baselines, as expected.

3. Results

3.1. Betelgeuse (α Orionis) at 38 GHz

Betelgeuse (α Orionis, J0555+0724, spectral type M1 Ia, Mass = $11.7\,M_{\odot}$), is a red supergiant at a distance of 222 pc. For the Betelgeuse-like model, we adopt parameters for the radio photosphere guided by existing observations at 22 GHz to 43 GHz from the VLA (Lim et al. 1998; O'Gorman et al. 2017). The model has a diameter or 100 mas and a total flux density of 21.7 mJy at 1 cm wavelength. The implied mean brightness temperature is about 3000 K. The model includes large cold and hot spots with Gaussian profiles, with sizes (FWHM) up to 20% of the stellar diameter, and peak deviations from the mean of ±12%, qualitatively consistent with radio imaging of Betelgeuse.

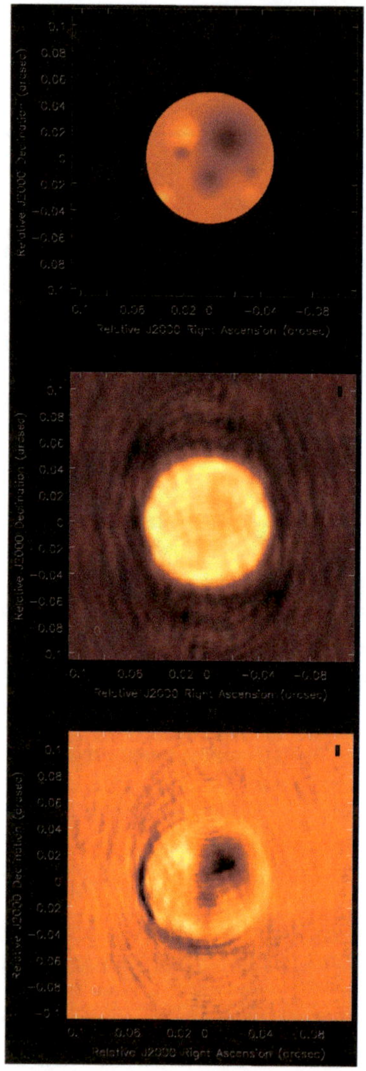

Figure 1. Images of Betelgeuse at 38 GHz. (*Top:*) the input model, including cold and hot spots of peak magnitude ~ ±12%. (*Middle:*) The ngVLA image at 6.5mas × 4.0mas resolution, for a 4 hour synthesis. The rms noise is 0.51 μJy beam^{-1}, and the peak surface brightness is 0.1 mJy beam^{-1}. (*Bottom:*) Image with the ngVLA, after fitting for a disk model, and subtracting the model from the visibilities. The rms noise is 0.25 μJy beam^{-1}, and the peak surface brightness is 7.8 μJy beam^{-1}.

The model is shown in Figure 1a. We simulated a 4 hour observation of this model with the ngVLA at 38 GHz (Figure 1b). For imaging, we adopt Briggs weighting with $R = -0.5$, and we use a multiscale clean with a loop gain of 0.03 to mitigate the CLEAN instability. The resulting image has an rms noise of 0.6 μJy beam^{-1}, and the synthesized beam has a FWHM = 6.5 mas × 4.0 mas. The disk is imaged at high signal to noise, however, the hot and cold spots are only marginally perceptible. There is a depression at the position of the largest cold spot of ~ 15%, but the overall impression is that of an edge-brightened disk. This result is a well known issue with interferometric imaging of objects with relatively flat brightness distributions and hard edges, such as the disks of planets in our Solar system (de Pater et al. 2016). The result is due to the natural tendency for an interferometer to enhance edges and sharp gradients in images, and the general CLEAN instability for diffuse sources (Cornwell 2008).

In the case of Solar system planets, the standard analysis technique is to perform disk model fitting to the visibilities, then subtract the disk model from the (u, v)-data, to look for under-lying structures (de Pater et al. 2016). We have performed such an analysis on the Betelgeuse simulation. The results from the (u, v)-model fit are listed in Table 1. The residual image after model subtraction from the visibilities is shown in Figure 1c. The model fit provides an accurate realization of the mean disk, both in terms of size and total flux density, to better than 1%. The residual image shows the more prominent hot and cold spots, with magnitudes within 20% of the expected deviations. One artifact of the model and fitting process is that the largest cold spot is significant enough to perturb the central position of the fit. Even a ~ 0.5% position shift (relative to the stellar diameter), will lead to sharp positive/negative edges in the residual image.

Table 1. (u, v) Fitting Results

Star	Max Baseline ($\times 10^8 \lambda$)	S_{mod} (mJy)	D_{mod} (mas)	S_{fit} (mJy)	D_{fit} (mas)
α Orionis 38 GHz	1.3	21.7	100	21.6±0.2	100.6±0.5
α Can Maj 85 GHz	3.0	1.63	6.0	1.6295±0.0002	6.039±0.005×6.042±0.003
α Can Maj 85 GHz	–	1.63	6.0	1.629±0.0002	6.040±0.005×6.043±0.003
θ Leonis 85 GHz	3.0	0.0255	0.723	0.0251±0.00077	0.58±0.41×0.61±0.12
θ Leonis 85 GHz	–	0.0255	0.723	0.0251±0.00075	0.56±0.17×0.63±0.08
A2 V 85 GHz at 100 pc	–	0.0064	0.36	0.0063±0.0004	0.45±0.14×0.33±0.10

3.2. Sirius (α Canis Majoris) at 85 GHz

Sirius is the brightest star in the sky in the optical (α Canis Majoris, J0645-1642), at a distance of 2.6 pc from Earth. It is a binary system, with the brighter star being a hot, 2 $M_\odot$, main sequence star of spectral type A0, and the second star being a 1 $M_\odot$ white dwarf. We focus on the A0 star.

For the radio model of a Sirius-like star, and subsequent analyses below, we adopt the parameters of the optical photosphere, in terms of diameter and brightness temperature, since the radio and optical photospheres for hot main sequence stars at 85 GHz are likely to be similar at the ~ 30% level, in size and brightness (White et al. 2018).

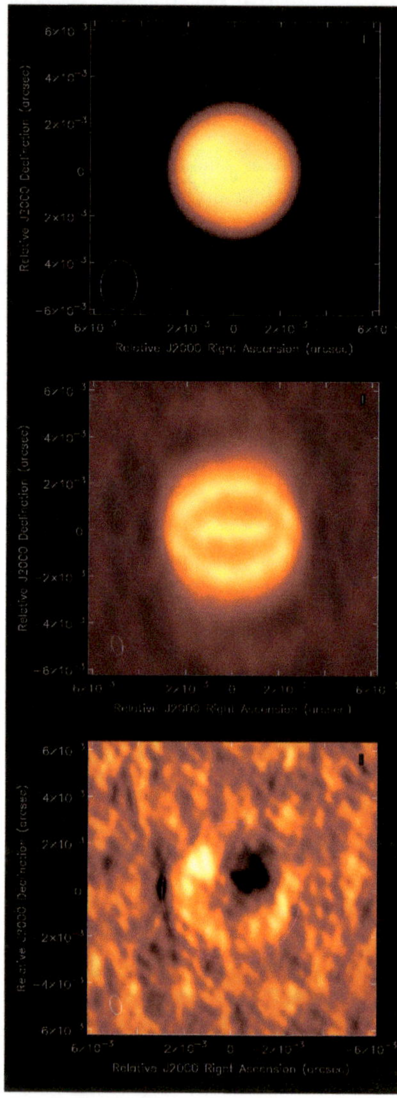

Figure 2. (*Top:*) Image of Sirius with the ngVLA for a 16 hour synthesis, made using (u, v)-weighting appropriate for the original Rev B configuration (1000 km maximum baseline). The resolution is 2.1 mas × 1.6mas and the rms = $0.6\,\mu$Jy beam^{-1}. (*Middle:*) Image of Sirius with (u, v)-weighting that incorporates the longer baselines. The resolution is 0.85mas × 0.47mas and the rms = $0.5\,\mu$Jy beam^{-1}. (*Bottom:*) Image of Sirius with the same resolution as (b), after fitting for a disk model and subtracting the model from the (u, v)-data. The pink circle shows roughly the position of the stellar disk. The peak on the image is 3.5μJy beam^{-1}. (Note: please refer to Figure 1a for the model structure, scaled accordingly).

We feel this assumption is conservative, in that the radio photospheres could be larger. In this model, the angular diameter is 6 mas and the total flux density at 85 GHz is 1.63 mJy. The mean brightness temperature is about 9000 K. We incorporate the same stellar structure as in Section 3.1 (hot and cold spots). Such structures are unlikely in hot main sequence stars, but they allow us to explore the imaging capabilities of the array at the highest resolutions.

We simulate a 16 hour observation at 85 GHz with the Rev B configuration plus the long baselines. We first image with a taper appropriate to the scales of the Rev B baselines only: $R = -1.3$, a cell size of 0.1 mas, and an outer (u, v)-taper of 1.4 mas. The resulting image is shown in Figure 2a. The beam FWHM is 2.1 mas × 1.6 mas, and the rms is 0.6 μJy beam^{-1} (the Naturally weighted theoretical noise is 0.2 μJy beam^{-1}). The resulting image shows a resolved star, but the resolution is inadquate to isolate any region clearly. There may be a depression at the location of the biggest cold spot, but again, the image is not conclusive.

We then image with a taper more appropriate for the longer baselines: $R = -1.3$, cell size of 0.05 mas, and an outer (u, v)-taper of 0.35 mas. The resulting synthesized beam FWHM is 0.85 mas × 0.47 mas, and the image noise is about 0.5 μJy beam^{-1}. Figure 2b shows the resulting image. The structure is clearly dominated by imaging artifacts.

We perform a similar disk (u, v)-model fitting and subtraction as in Section 3.1. The resulting image of the residuals is show in Figure 2c, using the same weighting as is used in 2b. There is a reasonable indication of both the primary cold and hot spots, at levels within 30% of the expected values relative to the mean disk brightness, and even marginal evidence for the next lowest cold spot.

The results for the disk fitting to the visibilities are given in Table 1. We performed two fits: one using baselines from the original Rev B configuration out to 1000 km ($3 \times 10^8 \lambda$), and the second using all the long baselines (out to 9000 km). In this case, the results are much the same, since the star is 6 mas in diameter, and the resolution of even the Rev B baselines is a factor few smaller than this.

3.3. θ Leonis at 85 GHz

We next consider a model of a hot main sequence star at a distance 20 times that of Sirius, and determine how well the stellar properties can be determined in a one hour synthesis, appropriate for surveys of large numbers of stars. We adopt the parameters comparable to those of θ Leonis, a 2.5 $M_\odot$ A2 V star at a distance of 51 pc (J1114+1525). The model in this case has a total flux density at 85 GHz of 25.5 μJy, and a diameter of 0.723 mas. We simulate a 1 hour observation at 85 GHz using the Rev B plus long baseline configuration. The resulting image is shown in Figure 3. We use $R = -1.0$ and an outer (u, v)-taper of 0.35 mas. The beam FWHM = 0.84 mas × 0.62 mas, and the noise is about 2.0 μJy beam^{-1}. The resulting image is shown in Figure 3a.

More relevant than direct imaging in this case is (u, v)-model fitting. Table 1 shows disk modeling fitting results, again using baselines appropriate for the Rev B configuration only, and then including the long baselines. In this case, the total flux density is well fit by both short and long baseline fits, but the errors on the diameters are significantly lower when including the longer baselines.

Figure 4 shows a radial binning of the Real part of the visibilities vs (u, v)-distance, plus the model visibilities. The Rev B only configuration does not reach the first null in the visibility function. The longer baselines sample the first three nulls at reasonable

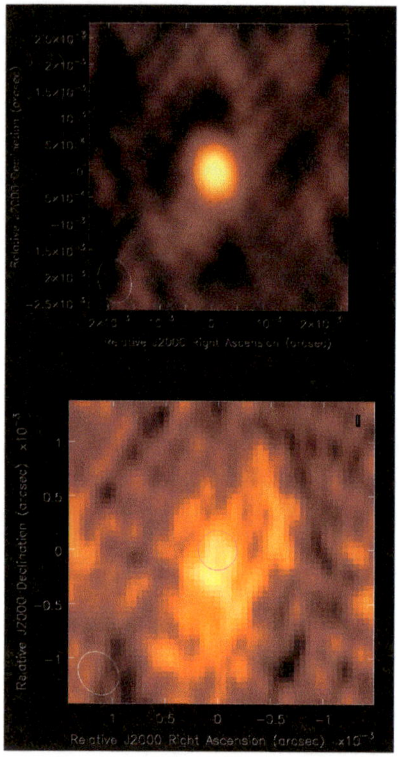

Figure 3. (*Top:*) Image of θ Leonis with the Rev B plus long baseline configuration of August, 2018, for a one hour synthesis. The resolution is 0.84 mas $\times$ 0.62 mas and the rms = $2\,\mu$Jy beam^{-1}. The peak on the image is $25\,\mu$Jy beam^{-1}. (*Bottom:*) Image of an A2 V star at a distance of 10 pc using the Rev B plus long baseline configuration of August, 2018, for a 4 hour synthesis. The resolution is 0.43 mas $\times$ 0.36 mas and the rms = $0.7\,\mu$Jy beam^{-1}. The peak in the field is $3.6\,\mu$Jy beam^{-1}. The circle shows the model photospher position and size.

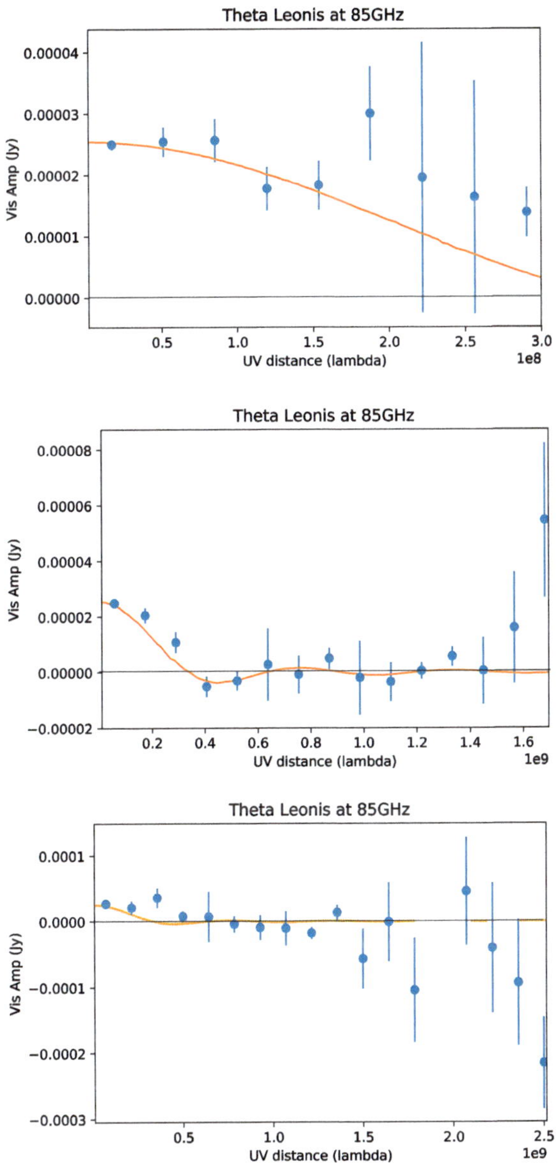

Figure 4. Azimuthally averaged real part of the visibility curve for θ Leonis at 85 GHz. Points are the measurements for a 4 hour synthesis with the ngVLA-classic baselines to 1000 km (top) and the Rev B plus long baseline configuration truncated at about 5000 km (middle), and at 9000 km (bottom). Beyond about 5000 km, the measurements become very noisy since the large number of antennas in the Core and Plains no longer contribute.

signal to noise, out to $1.4 \times 10^9\lambda$ (= 5000 km). Beyond that, the sensitivity drops dramatically. This sensitivity drop is due to the fact that only the outer antennas are involved in the cross correlation, and the Core and Plains antennas no longer contribute.

3.4. An A2 V Star at 100 pc: Limiting Case

To test the limits of the array, we model a hot main sequence star as above (A2 V star), but now at a distance of 100 pc, or 40 times further than Sirius. We simulate a 4 hour synthesis at 85 GHz. The total flux density in the model is 6.4 μJy, and the stellar diameter is 0.36 mas. We make an image using $R = -1.0$, taper = 0.15 mas, and a cell size of 0.05 mas. The resulting resolution is 0.4 mas and rms noise is 0.70 μJy (vs. 0.36 μJy Naturally weighted theoretical noise). The resulting image is shown in Figure3b. There is a fuzzy source at the expected position, but determining the size and total flux density from such an image is problematic, and disk fitting to the visibilities is the only viable method.

The results for disk fitting to the visibilities are again given in Table 1 using the full array. We find that the flux density is well determined ($\geq 10\sigma$), and the diameter is fit to $\sim 3\sigma$. We consider this the limiting case for determining stellar properties with the ngVLA.

Table 2. *Hipparcos* Stars > 0.4 mas, $\delta > -40°$

	I	II	III	IV	V	Total
O	6	1	0	0	2	9
B	114	77	81	55	61	388
A	35	32	105	118	61	351
F	55	47	148	245	159	654
G	67	106	1120	351	107	1751
K	55	117	5351	227	104	5854
M	27	32	1039	2	44	1144
Total	359	412	7884	998	538	10151

3.5. Number of Stars

Based on the simulations above, we adopt an extreme limiting case for determining stellar parameters of: a 1σ limit in 1hr of 1.4 μJy beam^{-1} at 85 GHz and 0.4 mas stellar size. This limit imples an rms brightness temperature limit of 1500 K. M-type stars have optical photospheric temperatures of $\sim$ 3000 to 4000 K, which would be at the limit of detection at this resolution. However, as mentioned above, at cm and mm wavelengths, the stellar atmospheres of main sequence stars can become optically thick in the chromosphere, which would enhance temperatures by a few thousand Kelvin. A size of 0.4 mas corresponds to the diameter of an M0 V star (optical radius of $0.6R_\odot$) at a distance of 15 pc. For A0 V stars (radius of $2.5R_\odot$), the distance is 60 pc. Red giant stars (radii of order $100R_\odot$), could be resolved to 2.5 kpc.

The *Hipparcos* catalog has a detailed listing of stellar spectral type, luminosity class, and distance (Perryman et al. 1997). From these quantities, we derive angular sizes using standard relationships for stellar optical photospheric diameters from

Allen's Astrophysical Quantities (Table 15.8). We then select for stars above −40° declination, with angular diameters > 0.4 mas. The results as a function of spectral type and luminosity class, are given in Table 2. The total number of stars evaluated was 32886 (meaning stars with fully specified spectral type and luminosity class, not Carbon stars, and above −40° declination [1]). Of these, 10151 are larger than 0.4 mas. The vast majority of the resolved stars are in the Giant or Supergiant class (9613 or 95% in class IV or lower). Yet there are still more than 500 main sequence stars (class V), that could be resolved by the ngVLA, with ~ 50 to 150 in each spectral type, except for the O stars.

For those interested in making direct comparisons to the Sun, there are 43 'Solar-type' stars ($B-V$ color index between 0.65 and 0.67), and 145 'Solar-analog' stars ($B-V$ color index between 0.62 and 0.71), that could be resolved by the ngVLA.

We note that these are lower limits, since the radio photospheres can be substantially larger than the optical photospheres, in particular in Giants and Supergiants, or hotter, in the case of main sequence stars. Additionally, the *Hipparcos* catalog is much smaller than the new *Gaia* survey (Andrae et al. 2018). However, by definition, the larger (in angle) stars of a given type/class will also be closer, and hence brighter and more likely to be in both catalogs.

For completeness, we have also searched the *Gaia* catalog. Of the almost 1.7 billion stars in that catalog, there are 77 million which have both effective temperature and radius cataloged, of which 54 million are above −40° declination. Based on the radii and distances in the catalog, we find that there are about 40,000 stars larger than 0.4 mas and $\delta \geq -40°$. Of these, we estimate that 246 are "solar analogs", meaning with tabulated radius and temperature within 30% Solar. The *Gaia* catalog does not contain information about spectral type or luminosity class, so no table similar to Table 2 for the *Hipparcos* catalog can be made, nor does it contain values for $B-V$, so we cannot define 'solar-type' stars in quite the same way as for *Hipparcos*.

4. Conclusions

The ngVLA will transform the field of stellar radio photospheres, allowing for imaging and parameter estimation of thousands of main sequence and giant stars. Such studies are directly relevant to determining the structure of stars in the key transition region from the optical photosphere to the chromosphere – the regions powering exo-space weather phenomena.

For very large (in angle) stars, such as red supergiants within a few hundred parsecs, good fidelity results can be obtained, with many resolution elements across the star. Even the closer main sequence stars (within ten parsecs or so), can be resolved and imaged, although with reduced image fidelity. In these cases, large scale temperature structures are not expected on the radio photospheres, and the primary physical parameters of diameter and brightness will be well constrained.

Using the *Hipparcos* catalog measurements of stellar spectral type, luminosity class, and distance for over 100,000 stars, we estimate that there are at least 10,000 stars that will be resolved by the ngVLA. While the vast majority of these (95%) are

[1]The 32886 was out of a total number of stars in the catalog of 118218, of which 91488 are above −40° declination

giants or supergiants, there are still over 500 main sequence stars that can be resolved, with ~ 50 to 150 of each spectral type (besides O stars). Note that these are lower limits, since radio photospheres can be larger than optical, and the *Hipparcos* catalog might not be complete. An initial investigation of the *Gaia* catalog suggests these numbers may be pessimistic by a factor of a few.

Acknowledgments. The National Radio Astronomy Observatory is a facility of the National Science Foundation operated under cooperative agreement by Associated Universities, Inc.

References

Andrae, R., Fouesneau, M., Creevey, O. et al. 2018, A& A, in press
Boyajian, T., von Braun, K., Feiden, G. A., et al. 2015, MNRAS, 447, 846
Carilli, C. & Erickson, A. 2018, ngVLA Memo No. 41
Carilli, C. et al. 2018, ngVLA Memo No. 35
Carilli, C. 2017, ngVLA Memo No. 16
Carilli, C. 2016, ngVLA Memo No. 12
Carilli, C. et al. 2015, ngVLA Memo No. 5
Cornwell, T. 2008, IEEE Journal of Selected Topics in Signal Processing, 2, 793
Cotton, W. & Condon, J. 2017, ngVLA Memo No. 30
Defrere, D., Absil, O., Berger, J.-P., et al. 2018, Experimental Astronomy, in press
Fichtinger, B., Gudel, M., Mutel, R. et al,. 2017, A & A, 599, 127
Gilliland , R. & Dupree, A. 1996, ApJ, 436, L29
Gudel, M 2002, ARAA, 40, 217
Hallinan, G., Littlefair, S., Cotter, G. et al. 2015, Nature, 523, 568
Harper, G. 2018, Science with a Next-Generation Very Large Array, ASP Monograph 7, ed. E. J. Murphy (San Francisco, CA: ASP), 265
Harper, G., O'Riain, N., Ayres, T. 2013, MNRAS, 428, 2064
Lim, J., Carilli, C., White, S., Beasley, A., Marson, R. 1998, Nature, 392, 575
Linford, J., Chomiuk, L., & Rupen, M. 2018, Science with a Next-Generation Very Large Array, ASP Monograph 7, ed. E. J. Murphy (San Francisco, CA: ASP), 271
Matthews, L. & Claussen 2018, Science with a Next-Generation Very Large Array, ASP Monograph 7, ed. E. J. Murphy (San Francisco, CA: ASP), 281
Matthews, L., Reid, M., Menten, K. 2015, ApJ, 808, 36
McKinnon, M. et al. 2016, SPIE, 9906, 27
Monnier, J. et al. 2007, Science, 317, 342
Monnier, J. 2003, Reports on Progress in Physics, 66, 789
Montarge, M., Norris, R., Chiavassa, A. et al. 2018, A& A, 614, 12
Mozurkewich, D., Armstrong, J., Hindsley, R. et al. 2003, AJ, 126, 2502
Ochsenbein, F., Bischoff, M. & Egret, D. 1981, A& A Supp, 43, 259
O'Gorman E. et al. 2013, AJ, 146, 98
O'Gorman, E. et al. 2015, A& A, 580, 101
O'Gorman, E. et al. 2017, A& A, 602, L10
Osten, R., Crosley, M., Gudel, M. et al. 2018, Report for National Academy of Science Committee on Exoplanet Science Strategy
de Pater, I., Sault, R., Butler, B. et al. 2016, Science, 352, 1198
Perryman M.A.C., Lindegren L., Kovalevsky J. et al. 1997, A& A, 323, L49
Selina, R. & Murphy, E. J. 2017, ngVLA Memo. No. 17
Trigilio, C., Umana, G., Cavallaro, F. et al. 2018, MNRAS, in press
Villadsen, J., Hallinan, G., Bourke, S. et al. 2014, ApJ, 788, 112
Vlemmings, W., Khouri, T., O'Gorman, E. et al. 2017, NatAs, 1, 848
White, J., Aufdenberg, J., Boley, A. et al. 2018, ApJ, 859, 102
White, S.M., Iwai, K., Phillips, N. et al. 2017, Solar Phys, 292, 88

Science with a Next-Generation Very Large Array
ASP Conference Series, Monograph 7
Eric J. Murphy, ed.

Deuteration in Starless and Protostellar Cores

Rachel K. Friesen,[1] Maria T. Beltrán,[2] Paola Caselli,[3] and Robin T. Garrod[4]

[1]*National Radio Astronomy Observatory, Charlottesville, VA, USA;* rfriesen@nrao.edu

[2]*INAF-Osservatorio Astrofisico di Arcetri, Largo E. Fermi 5, I-50125 Firenze, Italy;* mbeltran@arcetri.astro.it

[3]*Max-Planck-Institut für extraterrestrische Physik, Garching, Germany;* caselli@mpe.mpg.de

[4]*University of Virginia, Charlottesville, VA, USA;* rgarrod@virginia.edu

Abstract. In dense starless and protostellar cores, the relative abundance of deuterated species to their non-deuterated counterparts can become orders of magnitude greater than in the local interstellar medium. This enhancement proceeds through multiple pathways in the gas phase and on dust grains, where the chemistry is strongly dependent on the physical conditions. In this article, we discuss how sensitive, high resolution observations with the ngVLA of emission from deuterated molecules will trace both the dense gas structure and kinematics on the compact physical scales required to track the gravitational collapse of star-forming cores and the subsequent formation of young protostars and circumstellar accretion regions. Simultaneously, such observations will play a critical role in tracing the chemical history throughout the various phases of star and planet formation. Many low-J transitions of key deuterated species, along with their undeuterated counterparts, lie within the 60 – 110 GHz frequency window, the lower end of which is largely unavailable with current facilities and instrumentation. The combination of sensitivity and angular resolution provided only by the ngVLA will enable unparalleled detailed studies of the physics and chemistry of the earliest stages of star formation.

1. Introduction

In the local interstellar medium, the relative abundance of deuterium to hydrogen is $\sim 10^{-5}$ (Oliveira et al. 2003). In cold, dense molecular cores, however, observations have revealed abundances of deuterated molecules relative to their hydrogen-containing counterparts (i.e., DCO^+ and HCO^+; N_2H^+ and N_2D^+) that are orders of magnitude greater than the elemental abundance of D/H (Wilson et al. 1973). Furthermore, doubly- and triply-deuterated species have been detected toward both starless and protostellar cores (i.e., NH_3 and ND_2H, ND_3; Roueff et al. 2000; Loinard et al. 2001; van der Tak et al. 2002; Lis et al. 2002).

In cold molecular clouds where HD is the primary reservoir of deuterium, enhancement of deuterated species can occur through the exothermic chemical reaction

$$H_3^+ + HD \rightleftharpoons H_2D^+ + H_2 + \Delta E \tag{1}$$

where $\Delta E = 232$ K (Millar et al. 1989). At very low temperatures ($T \lesssim 20$ K) the forward reaction dominates due to the small energy barrier (Millar et al. 1989; Roberts et al. 2003; Walmsley et al. 2004). Other deuterated molecules, like N_2D^+, then form through ion-molecule reactions between H_2D^+ and species such as N_2. At the high densities characteristic of the interiors of highly-evolved starless cores ($n \gtrsim 10^5$ cm^{-3}), CO and other neutral species begin to deplete from the gas phase through freeze-out onto dust grains. This reduces the destruction rate of the deuterated ions thus formed (Caselli et al. 1999), further enhancing their abundances relative to their non-deuterated counterparts.

Deuterated molecules - particularly of N-bearing species, due to the depletion of C-bearing molecules from the gas phase - are thus selective tracers of the coldest, densest gas in molecular clouds and star-forming cores. In these cold regions, observations of the low-J transitions of deuterated molecules and their non-deuterated counterparts are particularly important to determine independently their excitation conditions and column densities, and hence the deuterium fractionation as a function of radius across a core. In addition, many of these species are also predicted to distinguish well between disks and pseudodisks in magnetised models of core collapse (Hincelin et al. 2016), and are thus critical to distinguish rotationally-supported disks from collapsing pseudodisks, and to measure infall.

In warmer regions, such as the hot core or hot corino surrounding young protostars, several mechanisms are possible to enhance the deuterium fractionation of the gas. Detections of deuterated forms of methanol and formaldehyde suggest that deuterium fractionation may also occur on the surface of dust grains (Aikawa et al. 2001; Willacy 2007). Observations of circumstellar disks and the Orion bar also suggest that enhanced deuterium fractions may also be produced in the gas phase with CH_2D^+ as the catalyst instead of H_2D^+ (Parise et al. 2009; Öberg et al. 2012). Observations of CH_2D^+ are critical to distinguish between these mechanisms, but have been extremely limited by telescope sensitivity and receiver availability, with only a single tentative detection of the $1_{01} - 0_{00}$ (para) feature at 278 GHz (Roueff et al. 2013).

2. Scientific Importance

The scientific importance of observing low-level transitions of deuterated molecules with high resolution and sensitivity in dense starless and protostellar cores is twofold: first, to understand the chemical evolution from cloud to core to stellar system, and to shed light on the origin of chemical species in our own Solar System; and second, to better trace the physical structure and kinematics of dense cores on small, ~ tens of au scales as they collapse under self-gravity to form protostars, circumstellar accretion regions, and protoplanetary disks.

In dense cores, it is not yet clear what species are the best tracers of the cold molecular gas prior to the formation of a protostar. In some chemical models, all heavy elements are expected to deplete onto dust grains, leaving only H_3^+ and its deuterated isotopologues in the gas phase (Walmsley et al. 2004). Despite both H_2D^+ and D_2H^+ being observed toward multiple star-forming regions with single dish resolution (Stark et al. 1999; Caselli et al. 2003; Stark et al. 2004; Vastel et al. 2004; Friesen et al. 2010), H_2D^+ has only been detected with ALMA toward a single, low-mass prestellar core (Friesen et al. 2014).

Table 1. Prominent transitions of deuterated molecules between 60 and 115 GHz

Molecule	**Transition**	**Rest Frequency** (GHz)
o-CH_2D^+	$1_{1,0}$ - $1_{1,1}$	67.273
p-NHD_2	$1_{1,1}$ - $1_{0,1}$	67.842
$D^{13}CO^+$	1 - 0	70.733
DCO^+	1 - 0	72.0393
$D^{13}CN$	1 - 0	71.175
DCN	1 - 0	72.415
CCD	1 - 0	72.108
$DN^{13}C$	1 - 0	73.368
DNC	1 - 0	76.306
N_2D^+	1 - 0	77.109
HDO	$1_{1,0}$ - $1_{1,1}$	80.6
o-NH_2D	$1_{1,1}$ - $1_{0,1}$	85.928
p-NH_2D	$1_{1,1}$ - $1_{0,1}$	110.15
CH_2DOH	*Multiple*	67 - 95

Detections of N_2D^+ and (singly- and doubly-) deuterated NH_3 toward evolved cores, however, imply that nitrogen-bearing species may remain abundant in these regions (see Figure 1). In these cold regions, the ground-level rotational transitions of these species are the most sensitive probe of molecular abundances. Observations are needed to test and refine astrochemical models. The rest frequencies of these transitions for many key deuterated species lie below 80 GHz, however, and are not accessible by most current single-dish telescopes or interferometers (the IRAM 30 m Telescope and NOEMA interferometer can now observe at frequencies as low as 71 GHz, but with significantly lower resolution and sensitivity than will be available with the ngVLA). Some of these transitions are listed in Table 1. For many of these species, their non-deuterated counterparts also have observable transitions in this window, enabling direct measurements of the deuterium fractionation as a function of species. This band contains the low excitation lines of multiple deuterated species of interest in both starless cores and their more evolved, protostellar counterparts, observations of which are crucial to determining the relative importance of gas phase and grain surface reactions to the overall deuterium fractionation history of star-forming cores. In particular, this window contains the $J = 1-0$ transitions of both N_2H^+ and N_2D^+, whose isolated hyperfine components enable high-precision measurements of the kinematics of the densest gas in star-forming cores.

Once formed, young protostars heat their environments, forming 'hot cores' ('hot corinos' in low mass objects) with sizes corresponding to where the heated gas reaches the sublimation temperature of the icy grain mantles ($\sim$ 100 K). Many complex organic molecules are thought to form on warm dust grains ($\sim 30-40$ K), and are returned to the gas phase within the hot core. Alternatively, some species may form in the gas phase from simpler molecules also released from the grains. Deuterated species are excellent tracers of whether a particular species forms in the gas phase or on dust grains. Mea-

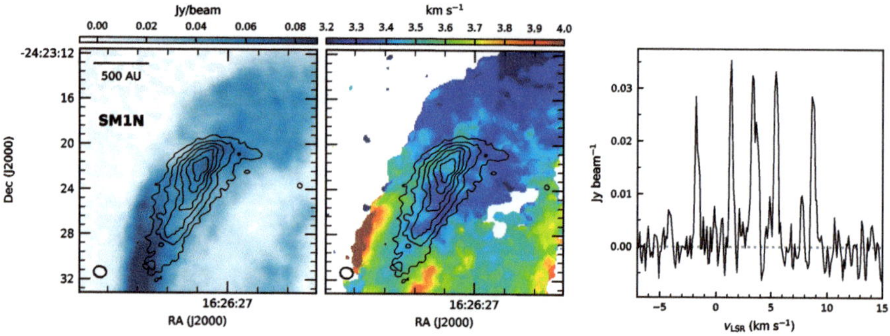

Figure 1. NH_2D $1_{1,1} - 1_{0,1}$ integrated intensity (left), first moment (middle), and peak spectrum (right) observed with ALMA at 1″ angular resolution toward the dense, star-forming SM1N core in the Ophiuchus molecular cloud (Friesen et al. 2019, in preparation). Black contours show 1 mm continuum emission at similar resolution. The NH_2D transition is just available at the lower edge of ALMA's Band 3. The hyperfine line strengths indicate that the emission is optically thick at the continuum peak, emphasizing the need to target these types of objects with denser gas-tracing deuterated species and rarer isotopologues. In addition, the extended nature of the potentially infalling gas around a young protostar requires both high resolution as well as a compact array component to recover larger-scale structures.

surements of the relative abundances of species like HCO^+, N_2H^+, NH_3, and CH_3OH and their deuterated forms in these environments are needed to identify the dominant deuterium fractionation routes in the gas-phase and on grains. These lines will provide detailed information on the region where ices are recently vaporised, testing both the kinematics and ice fractionation. Identification of deuterated methanol lines at higher frequencies in these objects is challenging due to the forest of emission lines present; at 60 − 80 GHz, however, the sparser line distribution allows the easy identification of several CH_2DOH lines within the band and an accurate measure of the deuterium fraction. Furthermore, only the ngVLA can enable sensitive, high resolution observations of the low energy transition of CH_2D^+ at 67 GHz to test directly its importance to deuterium fractionation in protostellar cores and disks.

3. Observing Deuterated Species with the ngVLA

The ngVLA is best suited to detect and map the emission between 60 - 110 GHz from deuterated species at resolutions of a few tens of au toward dense molecular cores in nearby star-forming regions in the Galaxy. The ngVLA's sensitivity in this regime will enable detailed maps of the distribution of multiple deuterated species (simultaneously with their non-deuterated counterparts in some cases) across starless and protostellar cores with resolution down to only a few tens of au. With this sensitivity and resolution, we can trace the deuteration trail from starless cores to protostellar envelopes, and further to the evolution of protoplanetary disks.

Evolved starless cores with large contrast between their central and mean density are the ideal targets for this study. Furthermore, cores on the cusp of star formation are the best targets to investigate collapse dynamics, since at earlier stages, the infall velocity has yet to reach the supersonic values predicted in hydrodynamic models, while at later times the collapse motions may be masked by strong outflows.

For these objects, sensitive imaging at $\sim$ 65 GHz to 90 GHz with angular resolution $\lesssim 1''$ will provide physical resolutions of $\sim$ 100 au–500 au in the nearest star-forming regions. This resolution is needed to resolve the peak of the density distribution of typical evolved starless cores, to study the chemical composition of the innermost regions (where stellar systems will form), and test current theories of complete freeze-out of elements heavier than He. Within the cores, the turbulent motions of the gas are largely dissipated, and the lines are thus expected to be narrow. Kinematics on these size scales are critical to distinguish between rotationally supported disks and pseudodisks at young ages (Harsono et al. 2015), and to track the increase in infall velocity at small scales predicted by hydrodynamic models (e.g., Foster & Chevalier 1993).

We have calculated the line strengths for a number of deuterated species and their non-deuterated counterparts in a starless core based on density, temperature and abundance models derived for the Ophiuchus H-MM1 core by Harju et al. (2017), assuming 0.25 km s^{-1} velocity dispersion, $1''$ angular resolution, and a distance of 250 pc. Figure 2 shows the resulting line brightnesses of NH_2D, N_2H^+, N_2D^+, DCO^+ and HCO^+ transitions from Table 1. Under the assumption that the deuterium fractionation levels are close to those currently measured toward starless cores, a sensitivity of 2 mJy beam^{-1} will provide good signal-to-noise ratio detections for most species, as well as for the hyperfine structure of the NH_2D lines, enabling precise measurements of line excitation and opacity needed for abundance analysis. We therefore require a sensitivity of 2 mJy beam^{-1} (0.4 K) within a velocity resolution of 0.1 km s^{-1}. Ideally, a flexible spectral setup will allow the targeting of multiple lines within a single observation.

In protostellar cores, higher angular resolution is needed to resolve the hot core or hot corino, where the bulk of the deuterated organic species are expected to be found, and to determine the critical processes driving the enhanced deuterium fractionation in these environments. The physical hot core/corino radii range from several tens of au for low-mass sources to $\sim$ 1000 au for high mass sources. An angular resolution of $0''.2$ will match the predicted corino size for nearby sources (spatial resolution of 50 au at a distance of 250 pc), and resolve the hot core in more distant, high mass sources (spatial resolution of 400 au at a distance of 2 kpc). Lines will be broader than in the cold cores, allowing a larger velocity resolution of 1.5 km s^{-1}. Simulations of CH_2DOH in a $1''$ beam toward a typical massive hot core suggest that a brightness sensitivity of 0.3 K will allow the detection of multiple lines within the band (Beltrán et al. 2015). In a $0''.2$ beam, the equivalent flux density sensitivity thus required is 70 μJy beam^{-1}. Spatial differentiation between species such as N_2H^+ and N_2D^+, and variations in the deuterium fractionation of N_2D^+ between sources, further test chemical evolution models (Tobin et al. 2013) and may provide an elusive age indicator for the youngest protostars (Emprechtinger et al. 2009; Fontani et al. 2011).

At the very small physical scales probed here, we are targeting very high density regimes. Some lines may be optically thick in starless cores, while dust opacities may be high in protostellar cores. Rarer isotopologues of the targeted species are present within the proposed frequency band (see Table 1) and, in case dust opacity is an issue, heavier and more complex deuterated molecules at lower frequencies may also be targeted.

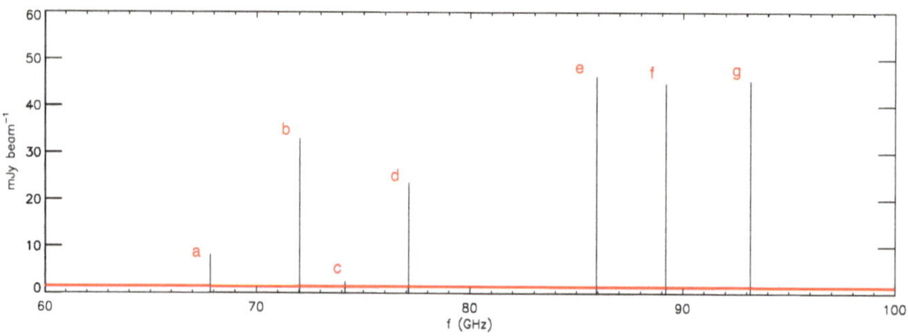

Figure 2. Predicted line brightnesses in a 1″ beam toward an evolved, starless core based on a chemical model by Harju et al. (2017). Lines include a) p-NHD_2 ($1_{1,1} - 1_{0,1}$), b) DCO^+ ($1 - 0$), c) NH_2D ($2_{1,2} - 2_{0,2}$), d) N_2D^+ ($1 - 0$), e) o-NH_2D ($1_{1,1} - 1_{0,1}$), f) HCO^+ ($1 - 0$), g) N_2H^+ (1-0). The red line shows ~ 2 mJy beam^{-1}.

4. Uniqueness to ngVLA Capabilities

Historically, very few telescopes (neither interferometer nor single dish) have had detectors working in this frequency band, leading to a major dearth of data on these low-level transitions of important deuterium-bearing species in star-forming cores. Ideally, a frequency range of $\sim$ 60-110 GHz would allow measurements of the critical species discussed here. In particular, no other telescope can observe both the N_2D^+ and N_2H^+ $J = 1 - 0$ lines at 77 GHz and 93 GHz, respectively, with the sensivity and resolution required for deuterium studies in starless and protostellar cores. As of 2017, an upgrade of the NOEMA interferometer allows observations between 71 GHz and 80 GHz, but with orders of magnitude less sensitivity and imaging capability as the ngVLA. Furthermore, the frequency band between 60 GHz and 71 GHz is not covered in whole by any other telescope at this time, such that the 67 GHz CH_2D^+ line is currently unobservable with current facilities.

In the future, Band 2 and/or Band 2/3 receivers will likely be developed for ALMA that cover the low-level transitions of deuterated species discussed here. ALMA observations, however, will lack the needed sensitivity to map these species toward dense cores at resolutions of a few tens of au in a typical, nearby star-forming cloud. No other planned facilities combine the required sensitivity and resolution of the ngVLA.

The sensitivity and high angular resolution of the ngVLA are critical to mapping the distribution of deuterated species across starless and protostellar cores. In particular, observations with the ngVLA will be able to target, for the first time, the chemical transition at only a few tens of au between hot core and cold envelope, where we expect a transition between the dominant chemical pathways to enhanced deuterium fractionation.

Infrared absorption observations with *JWST* and ELTs targeting young protostars will enable detailed analysis of the ice composition on dust grains, providing complementary information to the gas-phase observations enabled by the ngVLA. In combination, this will allow rigorous tests of astrochemical models of grain-surface and gas-phase deuterium fractionation, crucial to exploit fully the diagnostic power of deuterated molecules and their chemical history throughout the various phases of star and planet formation.

References

Aikawa, Y., Ohashi, N., Inutsuka, S.-i., Herbst, E., & Takakuwa, S. 2001, ApJ, 552, 639
Beltrán, M. T., Bianchi, E., Brand, J., et al. 2015, ArXiv e-prints, arXiv:1509.02702
Caselli, P., van der Tak, F. F. S., Ceccarelli, C., & Bacmann, A. 2003, A&A, 403, L37
Caselli, P., Walmsley, C. M., Tafalla, M., Dore, L., & Myers, P. C. 1999, ApJ, 523, L165
Emprechtinger, M., Caselli, P., Volgenau, N. H., Stutzki, J., & Wiedner, M. C. 2009, A&A, 493, 89
Fontani, F., Palau, A., Caselli, P., et al. 2011, A&A, 529, L7
Friesen, R. K., Di Francesco, J., Bourke, T. L., et al. 2014, ApJ, 797, 27
Friesen, R. K., Di Francesco, J., Myers, P. C., et al. 2010, ApJ, 718, 666
Harju, J., Daniel, F., Sipilä, O., et al. 2017, A&A, 600, A61
Hincelin, U., Commerçon, B., Wakelam, V., et al. 2016, ApJ, 822, 12
Lis, D. C., Roueff, E., Gerin, M., et al. 2002, ApJ, 571, L55
Loinard, L., Castets, A., Ceccarelli, C., Caux, E., & Tielens, A. G. G. M. 2001, ApJ, 552, L163
Millar, T. J., Bennett, A., & Herbst, E. 1989, ApJ, 340, 906
Öberg, K. I., Qi, C., Wilner, D. J., & Hogerheijde, M. R. 2012, ApJ, 749, 162
Oliveira, C. M., Hébrard, G., Howk, J. C., et al. 2003, ApJ, 587, 235
Parise, B., Leurini, S., Schilke, P., et al. 2009, A&A, 508, 737
Roberts, H., Herbst, E., & Millar, T. J. 2003, ApJ, 591, L41
Roueff, E., Gerin, M., Lis, D. C., et al. 2013, Journal of Physical Chemistry A, 117, 9959
Roueff, E., Tiné, S., Coudert, L. H., et al. 2000, A&A, 354, L63
Stark, R., van der Tak, F. F. S., & van Dishoeck, E. F. 1999, ApJ, 521, L67
Stark, R., Sandell, G., Beck, S. C., et al. 2004, ApJ, 608, 341
Tobin, J. J., Bergin, E. A., Hartmann, L., et al. 2013, ApJ, 765, 18
van der Tak, F. F. S., Schilke, P., Müller, H. S. P., et al. 2002, A&A, 388, L53
Vastel, C., Phillips, T. G., & Yoshida, H. 2004, ApJ, 606, L127
Walmsley, C. M., Flower, D. R., & Pineau des Forêts, G. 2004, A&A, 418, 1035
Willacy, K. 2007, ApJ, 660, 441
Wilson, R. W., Penzias, A. A., Jefferts, K. B., & Solomon, P. M. 1973, ApJ, 179, L107

Science with a Next-Generation Very Large Array
ASP Conference Series, Monograph 7
Eric J. Murphy, ed.

Massive and Evolved Stars with the ngVLA

Thomas J. Maccarone,[1] Saida Caballero-Nieves,[2] Nathan Smith,[3] and Nora Lützgendorf[4]

[1]*Texas Tech University, Lubbock, TX, USA;* `Thomas.Maccarone@ttu.edu`

[2]*Florida Institute of Technology, Melbourne, FL, USA*

[3]*University of Arizona, Tucson, AZ, USA*

[4]*Space Telescope Science Institute, Baltimore, MD, USA*

Abstract. The next generation Very Large Array will have excellent sensitivity for detecting the thermal emission from massive stars and from red giants. This will allow direct imaging of the winds for a large number of hot massive stars. It will also allow using the radio emission for the massive stars as a way to detect stars to allow high resolution measurements can be made, even with large extinction. A few examples of the utility of the high resolution measurements are given: dynamics of globular clusters with red giants, detection of intermediate mass stripped stars in binaries, and measurement of masses of stars in massive binaries.

1. Description of the Problem

Winds of hot massive stars are strong, spatially extended sources of free-free emission in the radio band. The acceleration of stellar winds is often probed via optical and ultraviolet emission lines, but is rarely directly resolved. Stellar winds from many classes of stars allow us to have bright radio sources that can be used as dynamical tracers, both for measuring the masses of massive stars in binaries and for measuring the gravitational potential fields in globular clusters, and should also provide a check on orbits around the Galactic Center. Estimates of the masses of the most massive stars have been extremely difficult in the past, with luminosities and single star evolutionary tracks and/or the Eddington limit used for mass estimates, rather than dynamical measurements typically used as a technique. By providing a tool for obtaining precise astrometry and high dynamic range, even in highly obscurred parts of the sky, radio measurements of stellar masses will provide crucial information for understanding massive stars.

With globular clusters, it is quite controversial about whether the clusters contain intermediate mass black holes. Radial velocity measurements and proper motions of stars in the optical band are a valuable tool for probing the inner dynamics of star clusters, but are complicated by hard-to-constrain crowding effects. Using radio tracers of stellar dynamics will allow a complementary test of whether there are intermediate mass black holes in clusters, and will also allow measurements of the proper motions of the most extincted clusters.

Observations with the next generation Very Large Array (ngVLA) should also be able to resolve stellar winds directly, and provide direct imaging constraints on the clumpiness and acceleration of stellar winds. They should also be able, through direct imaging, to determine whether stellar winds are anisotropic, as may be expected for rapidly rotating stars.

2. Scientific Importance and Astronomical Impact

2.1. Understanding Massive Star Winds

The winds from massive stars affect the whole evolution of the stars as well as the properties of the supernovae and massive remnants produced later in their lifetimes (see e.g., Smith 2014, for a review). Understanding the stellar wind mass loss rates, and how the winds are accelerated, is thus an important process for interpreting supernovae, understanding the formation of compact objects, and understanding the deposition of metals into the Universe. Furthermore, the kinetic power of the winds themselves can sometimes be important for powering superbubbles.

Direct imaging of stellar winds has been done for some low mass stars, but never for massive stars. First, one can study, in detail, the spatial scale on which clumping in stellar winds takes place. It can be well established from the difference between line strengths in P Cygni profiles, which scale linearly with density, and free-free and recombination line tracers, which scale as density squared, that significant clumping must be present (Abbott et al. 1981; Smith 2014). For massive stars, the clumping factors are typically ~ 5 (Repolust et al. 2004; Markova 2004). Analysis which ignores clumping overestimates the mass loss rates by factors of ~ 3 in most cases, but in some extreme cases, overestimates of a factor of 10 have been suggested (Fullerton et al. 2006). Similar clumping factors have been suggested for cool evolved stars(Harper 2010).

Theoretical considerations suggest that the clumping is taking place primarily in the inner regions of the stellar wind, but at the present time, there is no direct imaging evidence of this. The spatial scale for the clumping should be smaller than the size scale of the star, but with the ngVLA Long Baseline Array (LBA; see Selina et al., 2018, this volume p. 15), this may be resolvable. Regardless, it should be straightforward to determine whether the clumping is taking place preferentially in the inner region of the wind.

Additionally, studies can be made of the departures from isotropy of stellar winds. Be star winds show a variety of lines of evidence for being disk-like. Understanding these stars' winds is essential for developing a universal theory of mass loss, since it is generally believed that stellar winds from rapidly rotating stars are launched through a fundamentally different mechanism than those from other stars (Araya et al. 2018).

The core figure of merit for stellar wind observations would be the ability to make good measurements of the flux density per beam in each beam in reasonable exposure times. As a conservative flux estimate, we start with the equations from (Wright & Barlow 1975) for a constant velocity, smooth stellar wind, using the form from Güdel (2002):

$$R_{thick} = 8 \times 10^{28} \left(\frac{\dot{M}}{v}\right)^{2/3} T^{-0.45} \nu^{-0.7} \tag{1}$$

where R_{thick} is the radius of the optically thick part of the stellar wind, $\dot{M}$ is the mass loss rate in $M_\odot$ yr^{-1}, T is the wind temperature in K, v is the wind velocity in km/s, and ν is the frequency in Hz at which the observations are made.

Then, the flux density, S_ν in mJy from the stellar wind is given by:

$$S_\nu = 9 \times 10^{10} \left(\frac{\dot{M}}{v} \right)^{4/3} T^{0.1} d_{\rm pc}^{-2} \nu^{0.6} \tag{2}$$

if R_{thick} is greater than the stellar radius and

$$S_\nu = 5 \times 10^{39} \left(\frac{\dot{M}}{v} \right)^{2} T^{-0.35} R_*^{-1} d_{\rm pc}^{-2} \nu^{-0.1} \tag{3}$$

if R_{thick} is smaller than the stellar radius, R_*, with d the distance to the star in pc.

For a massive star with a reasonably high mass loss rate, we can take $\dot{M} = 10^{-5} M_\odot$ yr $^{-1}$, v = 1000 km/s, T = 10^4 K and ν = 15 GHz. We find that the wind will be optically thick out to about 30 AU, considerably bigger than the stellar radius of massive stars. Next we can take d = 1 kpc, and we find that S_ν = 6 mJy for such a star, with an angular size scale of 30 mas. With 1000 km baselines, the angular resolution at 15 GHz would be 4 mas, allowing for about 50 beams across the source. If one assumes homogeneous surface brightness, this would yield 100 μJy/beam flux densities, meaning that $\sim$ 100σ detections would be possible in each beam. Having so many well-detected beams would allow for careful studies of source structure, source variability, and source spectra, by making similar measurements at other frequencies. This would thus allow, in turn, for many stars, the decomposition of the radio emission into components from the thermal part of the wind, synchrotron emission from shocks (which should mostly contribute at lower frequencies), and from the stellar photosphere (which should contribute most at higher frequencies — see Carilli et al. 2018, this volume p. 369).

2.2. Understanding Massive Star Masses and Orbits

It is expected that the most massive stars should be found predominantly in binary (or other multiple) star systems with other massive stars, unless they have undergone mergers (Sana et al. 2012). Still, with the most massive stars, binary companions can be emitting a small fraction of the total starlight in the system, and hence can be extremely difficult to detect. Additionally, the radial velocity wobbles in such systems can also be very small. Precise astrometric measurements of the stars can provide an alternative means of establishing binarity in many cases, and, in the best cases, of estimating the masses of the stars.

The highest mass stars are expected to be about $\sim$ 300 $M_\odot$ or more (Crowther et al. 2010, 2016), and they show mass loss rates of $2 - 5 \times 10^{-5} M_\odot$/yr, despite being in the Large Magellanic Cloud, where low metallicity might be expected to suppress mass loss(Crowther et al. 2010). A small number of similar stars have been seen in the Milky Way, but extinction has generally made it difficult to identify such stars in our own Galaxy; the Magellanic Cloud stars themselves obviously cannot be measured with the ngVLA, but these similar stars can be identified in Galactic radio plane surveys, and studied carefully with additional follow-up. Conservatively assuming a typical 10 kpc distance, along with $\dot{M} = 2 \times 10^{-5} M_\odot$/yr, v_W of 2500 km/s, we can expect radio flux densities of about 50 μJy for these stars at 15 GHz, and angular sizes of about 3 mas (and perhaps a bit smaller if the winds are significantly hotter).

These objects are thus well-suited for astrometric work with the ngVLA. Positional measurements with precision of about 50μarcsec should be reached in about one hour of observations, and with a set of longer observations, it should be possible to make geometric parallax measurements for these stars. Furthermore, if there are binary companions with even 10% of the flux of the primary stars, they should be detectable as well. In such a case, following the proper motions of both stars gives a "visual binary" type mass estimate for the two stars.

To establish that a system is a binary without detecting its companion star, one needs to make a measurement of the acceleration of the motion of the object. As a rule-of-thumb, let us consider a binary with separation of 100 AU (a fairly typical value in star-forming regions; Griffiths et al. 2018). Such a binary, if the total system mass is 100 $M_\odot$ will have an orbital period of about 100 years. If the two stars are 10 $M_\odot$ and 90$M_\odot$, then the more massive star will have a reflex motion of about 10 AU, which will correspond to a total variation in position of 1 mas. Over the course of 10 years of observations, it will move through 12° of its orbit, meaning that the deviation from a pure straight line will be about 20 μas. The $\approx$ 0.1 AU/year velocity of the star's motion, corresponding to 0.5 km/s, is beyond present capabilities for radial velocity precision for massive stars, and is likely to remain that way – thus even if such a star can be detected in the infrared or optical bands with high signal-to-noise, its binarity can be studied only astrometrically. For a star like the one discussed in the preceding paragraph, spending about 20 hours per year on obtaining positions accurate to 10 μas each would yield a data set sufficient for ruling out a straight-line fit. For more nearby star-forming regions, of course, such measurements could be made more quickly. For supergiant stars where the stellar photosphere may exceed the brightness of the stellar wind, such measurements could be made even more easily using the photospheric emission.

2.3. Detecting Intermediate Mass Stripped Stars

Stripped stars produced by binary stellar evolution are likely to be quite common in the Galaxy due to binary interactions, and to be an important source of ionizing radiation, due to their exceptionally hot temperatures (Götberg et al. 2018). Despite their importance, they can be difficult to detect when they have a massive binary companion, because of the higher luminosities (especially in the less extincted red bands) of their companion stars. The stellar winds from these stars can make them detectable in the radio. Taking numbers typical for these stars (Vink 2017), $\dot{M}$=$10^{-7}M_\odot$/yr, v = 2000 km/s, T=50000 K, d = 2 kpc, and an observing frequency of 30 GHz, the expected radius of the optically thick part of the stellar wind is 0.2 AU (meaning that these objects would appear to be point sources with inverted radio spectra unless they are very nearby), and the expected flux density is 0.2μJy. In such systems, then, either two objects will be seen as a resolved binary, or the stellar winds are likely to be interacting such that synchrotron emission will be seen, and the systems will be identifiable as colliding wind binaries.

2.4. Proper Motions of Globular Cluster and Galactic Center Stars

The search for intermediate mass black holes in globular clusters has been a long-running source of controversy. A few approaches are typically used — searches for accretion (Bahcall & Ostriker 1975; Maccarone 2004; Wrobel et al. 2018, this volume p. 743), searches for radial velocity excesses toward the center of the cluster (Noyola et al. 2008), and searches for proper motion velocity excesses toward the center of the cluster (Watkins et al. 2015). All of these approaches have caveats. E.g., searches for evidence of accretion have been upper limits in current work, and only in a few cases is the gas content of the cluster well-established. With optical measurements, there are always potential problems associated with crowding in the inner regions of the clusters. Crowding problem will, in fact, be *worse* with *JWST* than with *HST* in the modes with large fields of view, but can be ameliorated with aperture masking. For radial velocities there can be additional complications due to the effects of rotation and binarity of the stars measured. Radio-based proper motions are ideal for solving this problem, so long as a sufficiently abundant set of dynamical tracers can be detected in the radio band.

Detection of photospheric emission from globular cluster stars in the radio will require prohibitive exposure times. Attempts to make radio measurements of globular cluster star motions thus rely on the use of radio-bright tracers. At the present time, pulsars and X-ray binaries may provide such tracers for a subset of clusters. These can be useful for bulk proper motion measurements and, in M4, for geometric parallax measurements. No cluster has a large enough number of detected pulsars and X-ray binaries to make useful radial velocity dispersion measurements.

It may become possible with ngVLA to make sure measurements with pulsars in Terzan 5, but ideally, we would be able to obtain larger samples of objects which are bright enough in radio to allow proper motion measurements. In fact, the stellar winds from red giants should provide such tracers. For a wind mass loss rate of $10^{-8} M_{\odot}$/yr (Meszaros et al. 2008), a wind speed of 10 km/s, and a distance of 4 kpc, the expected radio flux density is 11 μJy, while the flux density would be 4 μJy for $2 \times 10^{-9} M_{\odot}$/yr (Cohen 1976), and clumping in the winds may make the radio fluxes brighter than these values.

To this sample of objects we can add the millisecond pulsars and X-ray binaries, which will generally be significantly brighter than 11 μJy. With such a flux density, using only the ngVLA LBA, with angular resolution of about 1.5 mas (approximately the same as the expected angular size of the optically thick part of the stellar wind), each one-hour measurement will have a positional accuracy of about 150 μas. Over a 2-year time baseline, then, the proper motion accuracy of each star would be obtained to a precision of about 0.8 km/s. As a result, then, the precision of the velocity dispersion within a bin would be set by the sampling of the velocity distribution, rather than by the precision of the measurements, unless very large numbers of stars were detected.

The uncertainty on the one-dimensional velocity dispersion from a set of Gaussian-distributed measurements will be $\sigma_v \sqrt{1/(2N)}$, where σ_v is the velocity dispersion of the sample and N is the number of stars measured. With two dimensions measured, there will be an extra $\sqrt{2}$ term, meaning that $\sigma_{v,m}$ for a bin, the measured velocity dispersion within that bin, will have a precision of $\frac{1}{2}\sqrt{1/N}$. Thus, with 20 stars per bin, 30% changes in the velocity dispersion will be measurable at the 3σ level. The massive globular clusters in the Galaxy tend to contain a few hundred to a few thousand bright red giant stars, meaning that the velocity dispersions should be measurable to $\sim$ 5% for these clusters, providing significantly better tests of the IMBH hypothesis than can currently be achieved. For some of the less massive clusters it will, at least, be possible to use the few brightest red giants to measure the clusters proper motions, and often geometric parallaxes, even for the case of heavily extincted clusters where *Gaia* is not effective.

3. Connection to Unique ngVLA Capabilities

The interest in resolved emission at high frequencies on angular scales of ~ 10−100 mas is fundamentally ngVLA science. At the present time, only eMERLIN covers a similar angular resolution scale, but eMERLIN has much worse sensitivity than the ngVLA will, is located far North so that a much smaller fraction of the Galactic Plane is observable, and is in a location where the prevailing weather conditions make an expansion beyond 22 GHz unlikely to occur (or to provide much useful data). For the astrometric work on massive stars, ngVLA allows seeing through large extinction and looking at the Galactic massive star population in a way that cannot be done easily in optical and infrared. For the astrometric work on giants in globular clusters, the ngVLA measurement allow avoidance of the crowding and the highest possible angular resolution through the use of the ngVLA LBA. Without the continental scale baselines provided by the ngVLA LBA, the proper motion precision could still be obtained by achieving higher signal-to-noise measurements, but it would be more difficult to ensure that there was no crowding affecting the measurements.

Acknowledgments. We thank Selma de Mink for suggestions which helped start this project, and Jay Strader, Eva Noyola and Laura Watkins for discussions of globular cluster dynamics.

References

Abbott D.C., Bieging J.H., Churchwell E., 1981, ApJ, 250, 645
Araya I., Curé M., ud-Doula A., Santillán A., Cidale L., 2018, MNRAS, 477, 755
Bahcall J.N., Ostriker J.P., 1975, Nature, 256, 23
Carilli, C. L., Butler, B., Golap, K., Carilli, M. T. et al. 2018, Science with a Next-Generation Very Large Array, ASP Monograph 7, ed. E. J. Murphy (San Francisco, CA: ASP), 369
Cohen J.G., 1976, ApJ, 203, 587
Crowther P.A., Schnurr O., Hirschi R., Yusof N., Parker R.J., Goodwin S.P., Kassim H.A., 2010, MNRAS, 408, 731
Crowther P. et al., 2016, MNRAS, 458, 624
Fullerton A.W., Massa D.L., Prinja R.K., 2006, ApJ, 637, 1025
Götberg Y., de Mink S.E., Groh J.H., Kupfer T., Crowther P.A., Zapartas E., Renzo M., 2018, A&A, 615, A78
Griffiths D.W., Goodwin S.P., Caballero-Nieves S.M., 2018, MNRAS, 476, 2493
Güdel 2002, ARA&A, 40, 217
Harper G.M., 2010, ApJ, 720, 1767
Maccarone T.J., 2004, MNRAS, 351, 1049
Markova N., Puls J., Markov H., 2004, A&A, 413, 693
Meszaros S., Dupree A.K., Szentgyorgyi, A., 2008, AJ, 135, 1117
Noyola E. Gebhardt K., Bergmann M., 2008, ApJ, 676, 1008
Repolust T. Puls J, Herrero A., 2004, A&A, 415, 349
Sana H., et al., 2012, Science, 337, 444
Selina, R. J., Murphy, E. J., McKinnon, M., et al. 2018, Science with a Next-Generation Very Large Array, ASP Monograph 7, ed. E. Murphy (San Francisco, CA: ASP), 15
Smith N., 2014, ARA&A, 52, 487
Vink J., 2017, A&A, 607L, 8
Watkins L.L., van der Marel R.P., Bellini A., Anderson J., 2015, ApJ, 803, 29
Wright A.E., Barlow M.J., 1975, MNRAS, 170, 41
Wrobel, J. M., Miller-Jones, J. C. A., Nyland, K. E., & Maccarone, T. J. 2018, Science with a Next-Generation Very Large Array, ASP Monograph 7, ed. E. J. Murphy (San Francisco, CA: ASP), 743

Science with a Next-Generation Very Large Array
ASP Conference Series, Monograph 7
Eric J. Murphy, ed.

Planetary Nebulae

Joel H. Kastner,[1] Albert Zijlstra,[2] Bruce Balick,[3] and Raghvendra Sahai[4]

[1]*Rochester Institute of Technology, Rochester, NY, USA;* `jhk@cis.rit.edu`

[2]*University of Manchester, Manchester, UK;* `a.zijlstra@manchester.ac.uk`

[3]*University of Washington, Seattle, WA, USA;* `balick@uw.edu`

[4]*JPL, Pasadena, CA, USA;* `raghvendra.sahai@jpl.nasa.gov`

Abstract. Planetary nebulae (PNe) represent the near endpoints of evolution for stars of initial mass ~1–8 $M_\odot$, wherein the envelope of an asymptotic giant branch (AGB) star becomes photodissociated and ionized by high-energy radiation from a newly emerging white dwarf that was the progenitor star's core. It is increasingly evident that most PNe are descended from binary systems. PNe hence provide unique insight into a diverse range of astrophysical phenomena, including the influence of companion stars on the late stages of stellar evolution; stellar wind interactions and shocks; the physics and chemistry of photoionized plasmas and photon-dominated regions (PDRs); and enrichment of the ISM in the products of intermediate-mass stellar nucleosynthesis. We describe specific examples of the potential impact of the ngVLA in each of these areas.

1. Introduction

Planetary nebulae (PNe) serve as textbook examples of astrophysical plasmas and shock processes and provide essential tests of theories of stellar evolution and the origin and enrichment of the heavy elements in the universe (e.g., Kwok 2000). Classically, a PN results from the ejection and ionization of the envelope of an asymptotic giant branch (AGB) star — the descendant of a progenitor of initial mass ~1–8 $M_\odot$ (see chapter by Matthews & Claussen 2018, in this volume p. 281) — wherein the source of ionizing (UV) photons is a newly emerging white dwarf that was the progenitor star's core. PNe also exhibit a dazzling variety of aspherical morphologies: elliptical; bipolar or multipolar; highly point-symmetric; chaotic and clumpy (Sahai et al. 2011). The physical mechanisms responsible for this PN morphological menagerie have been the subject of intense interest and hot debate among PN researchers for well over two decades (e.g., Balick & Frank 2002; Akashi et al. 2018, and references therein). The PN shaping problem is multifaceted, with connections to (and implications for) a wide variety of astrophysical systems. Areas of particular importance are binary star astrophysics; stellar wind interactions and shocks; the physics and chemistry of photoionized plasmas and photon-dominated regions (PDRs); and enrichment of the ISM in the products of intermediate-mass stellar nucleosynthesis. The study of PNe with the ngVLA will have major impact in all of these areas. Here, we offer a few prime examples.

2. Recombination Lines: Nebular Dynamics and Abundances of Light Elements

The radio spectra of planetary nebulae include a large number of bound-bound recombination lines. The brightest lines are those of the H$n\alpha$ series. The helium lines are a factor of 10 fainter, and the carbon lines fainter still. Beyond this, the lines from the different elements merge. These radio recombination lines provide a powerful diagnostic of PN internal regions that is not affected by extinction, and is much less temperature sensitive than optical forbidden lines. Radio lines provide a 2-D map of PN internal dynamics (from the line profiles), the local electron temperatures (from the line to continuum ratios), and the helium abundance (from the helium to hydrogen line ratios). The power of these methods was shown by Roelfsema et al. (1991) for the planetary nebula NGC 7027.

For an *HST*-like resolution of $0\farcs1$, the ngVLA has a line sensitivity (1σ) in 1 hour of $\sim 100\,\mu$Jy beam^{-1}, assuming a linewidth of 10 km s^{-1} that is appropriate for PNe like NGC 7027 (Figure 2). The line to continuum ratio is typically 10 per cent, increasing as $\nu^{1.1}$. Young planetary nebulae have brightness temperatures of $T_b = 10^3$ K at 5 GHz, with a continuum flux of around 5 mJy beam^{-1}, and ~50% half have $T_b \gtrsim 10^2$ K (van de Steene & Zijlstra 1995). Hence ngVLA mapping of radio recombination lines would be feasible even at this extreme resolution. In fainter regions, lower angular resolution can be used, and multiple recombination lines within the observed frequency band can be summed to increase signal-to-noise. Higher angular resolution would not be useful; for thermal lines, most collecting area should be within the inner 100 km.

The continuum flux declines only as $\nu^{-0.1}$ towards higher frequencies, while the recombination lines increase in strength and the ngVLA sensitivity at constant resolution is fairly constant. The recombination lines can thus be mapped over the full frequency range available to the ngVLA, which spans transitions from H40α to well over H120α, with reduced sensitivity above 50 GHz due to the higher angular resolution. These observations would provide unique detail concerning the 2-D velocity fields of PNe, while the velocity-depth profile can be traced by observing at frequencies where the nebula becomes optically thick. Deviations from LTE can be measured; at low n, the H$n\alpha$ lines are sensitive to laser amplification within circumstellar disks at the cores of PNe that are of particular interest for shaping studies (Aleman et al. 2018, see §4).

3. Precise Expansion Velocities and Distances from Nebular Proper Motions

The fundamental parameters of PNe — such as emission line and continuum luminosities and linear dimensions — are generally poorly determined, thanks to the notorious difficulty in ascertaining PN distances. Even in the era of *Gaia*, trigonometric parallax distances to all but the nearest and brightest PNe will likely remain uncertain, given the challenges inherent in obtaining precise astrometry for faint and/or heavily optically obscured central stars or diffuse PN emission. The ngVLA will enable the broad application of the expansion parallax distance determination technique, which was pioneered via VLA and HST studies of relatively nearby PNe and compact H II regions (e.g., Masson 1989; Acord et al. 1998; Palen et al. 2002; Guzmán et al. 2006), to the hundreds of Galactic PNe recently discovered via, e.g., large-area Hα surveys (Frew et al. 2013; Sabin et al. 2014). With angular resolution a factor ~10 times better than that of the seminal Masson (1989) study of NGC 7027 (d = 890 pc) over the same (5–15

GHz) frequency range, the ngVLA can provide expansion parallax distances accurate to ~10% for PNe out to ~10 kpc, with the potential for much higher precision distance determinations for objects in the solar neighborhood.

Conversely, the precise measurement of proper motions of nebular gas for the significant subset of nearby PNe that have bright central stars — and, hence, will have well-determined *Gaia* distances — offers an entirely new approach to modeling PN bulk gas expansion and the short-timescale evolution of ionization fronts. A planetary nebula is shaped as faster and highly directed stellar outflows from the PN core and/or a binary companion (see § 4) supersonically overtake the much more massive, isotropic, slow winds deposited over thousands of years as the star ascended the AGB. The geometry and momentum flux of the fast, collimated wind evolves as the central star of a planetary nebulae (CSPN) traverses the H-R diagram from the tip of the AGB (T_{eff} ~ 3500 K) to the white dwarf turnoff (~200 kK) at constant luminosity ($L \sim 10^{3.6} L_{\odot}$). The fast-slow wind interactions produce shock interfaces of varying shapes that expand at Mach 3–10 (30–100 km s^{-1}) into the ambient AGB gas upstream. While the shocks themselves are only detectable, and only under specific conditions, via X-rays (Kastner et al. 2012), the sharp (spatially unresolved) outer edges of a bright rim of wind-displaced and compressed slow gas mark their locations. The movements of this rim can be traced via observations with geometrically stable imaging instruments with ≤100 mas spatial resolution (e.g., Schönberner et al. 2018) — a requirement readily satisfied by the ngVLA. Indeed, the long-baseline synthesis imaging enabled by the ngVLA will be the tool of choice for tracking the progress of shock interfaces after HST has reached mission end.

The changes of rim shape measured via multi-epoch ngVLA imaging, coupled with ever-improving hydro simulations of the rim formation process and new model atmosphere calculations of the evolving CSPN, will allow us to retrace the histories of the stellar winds in a strategically selected sample of bright, nearby targets. For these objects, we can obtain both the tangential expansion speeds and radial velocities (via radio or optical recombination line profile modeling; § 2). This will effectively eliminate a major source of uncertainty in ascertaining the three-dimensional structures of PNe, leading to radical advances in our understanding of PN shaping. Such ngVLA studies can be carried out at wavelengths 2 to 4 cm in all but the worst of local weather. In contrast, adaptive optics (AO) and interferometric near-IR imaging aimed at the same science is affected by a spatially complex mixture of scattering small dust and emitting cool dust that is difficult to disentangle, and the PSFs of optical/near-IR AO instruments tend to be too sensitive to atmospheric conditions for purposes of precise measurements of proper motions of complicated nebular patterns over few-year temporal baselines.

4. Nebular Structures and Substructures: Diagnosing Close Binary Interactions

As noted by Matthews & Claussen (2018, this volume p. 281), there is now broad agreement that the transformation from a quasi-spherical wind during the progenitor star's AGB phase to nonspherical or highly collimated outflows during post-AGB, pre-PN, and PN phases can be traced to the influence of a companion to the mass-losing central star (Frank et al. 2018; Akashi et al. 2018, and references therein). This consensus was forged on the basis of two decades of (mostly *HST*) high-resolution imaging which, when placed in the context of detailed simulations of PN morphological evolution, has shown convincingly that PNe are initially shaped, during late AGB and post-AGB (pre-PN) stages, from the inside out by fast, collimated outflows (Sahai & Trauger 1998; Huarte-Espinosa et al. 2012). Such collimated outflows are a natural consequence of interacting binary systems, whether via common envelope (CE) evolution or jets associated with a companion star's accretion disk. Under either scenario, the highly directed outflows power their way through through the far more massive but more spherically symmetric remains of slow winds previously produced from the parent AGB star. After a few centuries these jets produce the observed pairs of lobes that characterize bipolar/multipolar pre-PNe and PNe. The sculpting action of jets may also be responsible for weaker departures from sphericity apparent in the vast majority of (elliptical) PNe (Soker 1990; Huarte-Espinosa et al. 2012).

This realization is forcing a paradigm shift in the goals and strategies of observational programs that can be undertaken over the upcoming decade with radio and IR telescopes around the world and in orbit. The effects of the formation of accretion disks and common envelope evolution are arguably more readily studied in pre-PNe and PNe than in SNe Ia, novae, and cataclysmic binaries; compared to these exciting but highly transient sources, the binary system outflows presented by PNe and their progenitors are closer, far more numerous, and slower to unfold. Spectroscopic instruments, such as the ngVLA, that combine very high spatial precision and larger collecting area with geometric stability can exploit the resulting opportunities to trace wind-sculpted density structures over a wide range of size scales, as well as monitor small-scale structural evolution over long temporal baselines, for significant samples of pre-PNe and PNe.

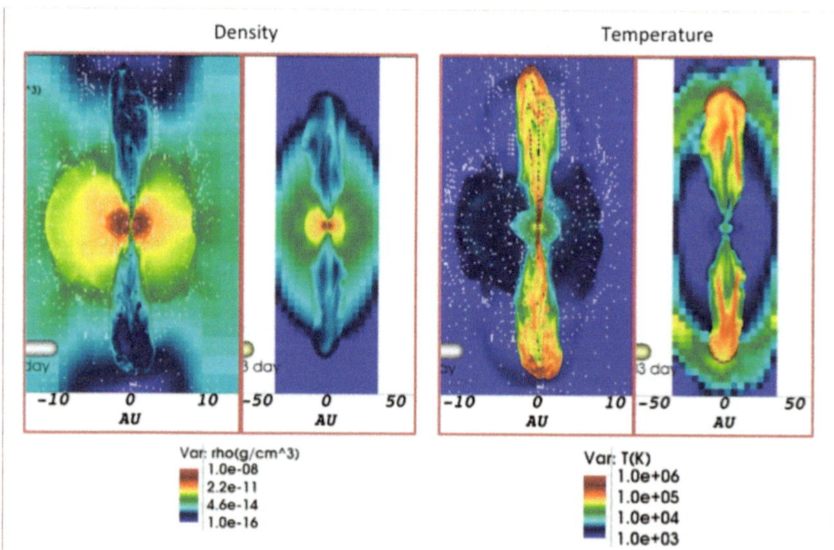

Figure 1. Results of 3-D adaptive mesh refinement simulations of a high-momentum fast wind being driven into the ejecta resulting from the onset of a binary star common envelope event. The left and right panels illustrate the distributions of density and temperature, respectively, at t = 780 days (left subpanels) and t = 1600 days (right subpanels) after initiation of the fast wind. Note the change of scale between the two times as the nebular lobes expand. Figure from Frank et al. (2018).

The sensitivity and resolving power of the ngVLA will finally provide the data needed to inform and test AGB and post-AGB ejecta (hence PN) shaping models. Most such models invoke Roche lobe overflow resulting in companion accretion disk and jet formation (Akashi et al. 2018) or wind interactions resulting from the sudden onset of the CE phase of a close (AGB + MS star) binary system (Frank et al. 2018). A simulation of the collimated outflows emerging within a few years of the onset of CE is presented in Figure 1. The characteristic size scales of the outflow collimator (i.e., the CE ejecta) and the wind interaction region are ~10–50 AU, requiring resolution of order 10 mas given typical post-AGB and PNe distances (a few kpc). The ngVLA will have unparalleled capability to probe such size scales via 10–100 GHz imaging. The collimating structures (i.e., disks and tori) can be traced in thermal continuum emission (i.e., Bremsstrahlung and/or emission from large dust grains) as well as CO isotopologues and dense gas tracers (CS, HCN, HCO^+). Subarcsecond low-T_{ex} molecular line imaging with the ngVLA will complement ALMA studies in distinguishing kinematically between Keplerian disks, expanding tori, and outflow structures (e.g., Bujarrabal et al. 2018). The potent combination of ngVLA and ALMA will hence represent the tool of choice for studies of AGB or post-AGB stars that are the suspected sites of recent CE events, or the cores of young PNe whose complex large-scale morphologies, evidence for jets, etc., makes them the suspected hosts of interacting binaries.

5. Central Stars: Binaries and Winds

Given the importance of central binaries in PN shaping, concerted observational effort has been devoted to searches for companions to PN central stars in recent years (via, e.g., central star photometric monitoring), resulted in numerous detections (Hillwig 2018). However, such optical survey work remains limited to relatively bright and nearby central stars. Meanwhile, roughly 50% of nearby PNe show an X-ray source at the cental star, and most of these point-like X-ray sources appear too hard to be due to shock processes (Kastner et al. 2012). The most likely explanation is that a low-mass companion star has accreted material from the primary during AGB mass loss and subsequent PN ejection, and has been spun up; the resulting fast rotation reactivates the secondary's internal magnetic dynamo, leading to coronal emission. This model predicts coronal radio emission as well, if such stars follow the Guedel-Benz relation between X-ray luminosity and radio luminosity density (e.g., Forbrich et al. 2011). Indeed, pilot e-Merlin observations have detected radio continuum emission from LoTr 5 — a PN (at d = 500 pc) known to harbor a central binary that is a luminous, point-like X-ray source — at a level consistent with the flux density predicted from the X-rays. This demonstrates the potential of the ngVLA, which can be used to efficiently survey all nearby PNe to search for central star companions.

The mass-losing central stars of planetary nebulae themselves have not yet been detected in the radio. They are expected to show weak emission from stellar winds, in addition to chromospheric emission from spun-up companions. Stellar winds are the largest uncertainty in the evolution of PNe: they remove some of the remaining hydrogen envelope and thereby speed up the evolution towards higher temperatures. These winds will finally be measurable with the ngVLA. Using the equations in Wright & Barlow (1975) for typical central star wind parameters (a mass loss rate of $10^{-8}\,M_\odot\,yr^{-1}$ and wind speed of $v_w = 10^3\,km\,s^{-1}$), and assuming a distance of 1 kpc, the predicted flux density at 5 GHz is 0.2 μJy, increasing as $\nu^{0.7}$. Such wind emission becomes detectable at 40 GHz; PNe at smaller distances would be detectable at lower frequencies or mass loss rates. The aforementioned survey of PN central stars hence would not only have the potential to directly detect low-mass binary companions, but would resolve such companions from primaries that are themselves wind radio emission sources at separations as small as ~100 AU. Such capabilities will be far superior to those of other (e.g., Chandra X-ray) central star wind and companion detection techniques.

6. Understanding Physical and Chemical Processes in Irradiated Molecular Gas

Though best known as $\sim 10^4$ K optical emission line sources, a subset of PNe retain significant masses of cold (< 100 K), dense ($\sim 10^4 - 10^6$ cm^{-3}) molecule-rich AGB ejecta (e.g., Huggins et al. 2005). This molecular gas is irradiated from within by UV from hot (30–200 kK) PN central stars. A majority of PNe also display point-like X-ray emission from their central stars (or companions) and/or diffuse X-ray emission from rarefied, shock-heated ($T > 10^6$ K) plasmas (Kastner et al. 2012). Such extremes of physical conditions within individual, readily resolvable objects with well-defined molecular gas irradiation geometries make PNe particularly fertile ground for the study of radiation-driven molecular gas heating and chemistry. The lessons learned can be applied to PDR environments ranging in scale from individual protoplanetary disks orbiting T Tauri stars to H II regions associated with massive young star clusters.

In PNe, as in any PDR environment, FUV and EUV photons are readily absorbed by neutral H and dust (respectively) in molecular surface layers, strongly influencing the heating of the gas in these layers. X-rays penetrate to larger depths, where they can ionize H_2 molecules as well as abundant atoms, such as He, N, and O, that have relatively high ionization potentials or tightly bound inner-shell electrons; the presence of highly reactive and/or abundant ions (e.g., H_3^+, He^+, N^+) can then lead to rich molecular chemistries via a miriad of reaction pathways. The molecules HCO^+, N_2H^+, HCN, HNC, CN, and C_2H, along with radio tracers of H_2 density (i.e., CO and its isotopologues), represent particularly potent diagnostics of the effects and relative penetration depths of FUV/EUV and X-ray photons. The potential of PNe in this regard is apparent from the single-dish and interferometric molecular line surveys that have been carried out thus far for a relatively small number of objects (e.g., Bachiller et al. 1997; Schmidt & Ziurys 2017a,b; Bublitz et al. 2018, see Figure 2).

The leap in frequency coverage of the ngVLA will unlock this potential: all of the transitions in Figure 2 will be accessible to the ngVLA's Band 6 (70–116 GHz) at sensitivities and spatial resolution exceeding that of ALMA by factors of ~5 and ~10, respectively. Thus, the ngVLA will greatly expand the number of PNe that can be surveyed in the aforementioned species; molecular line surveys of objects analogous to the well-studied, molecule-rich PNe in the solar neighborhood (e.g., the Helix, Ring, Dumbbell, NGC 7027, etc.) can be undertaken for objects as distant as the Galactic Bulge. Furthermore, the ngVLA will facilitate spatially resolved studies of the same species in many dozens of PNe for which only single-dish molecular line data exist and/or only CO or (near-IR) H_2 emission has been detected thus far (Kastner et al. 1996; Huggins et al. 2005). We can thereby measure the spatial gradients of temperature- and ionization-sensitive molecular line ratios (e.g., HNC/HCN, CN/HCN, HCO^+/CO) as functions of distance from UV- and X-ray-luminous PN central stars so as to constrain molecular gas ionization and photodissociation rates and inform models of UV- and

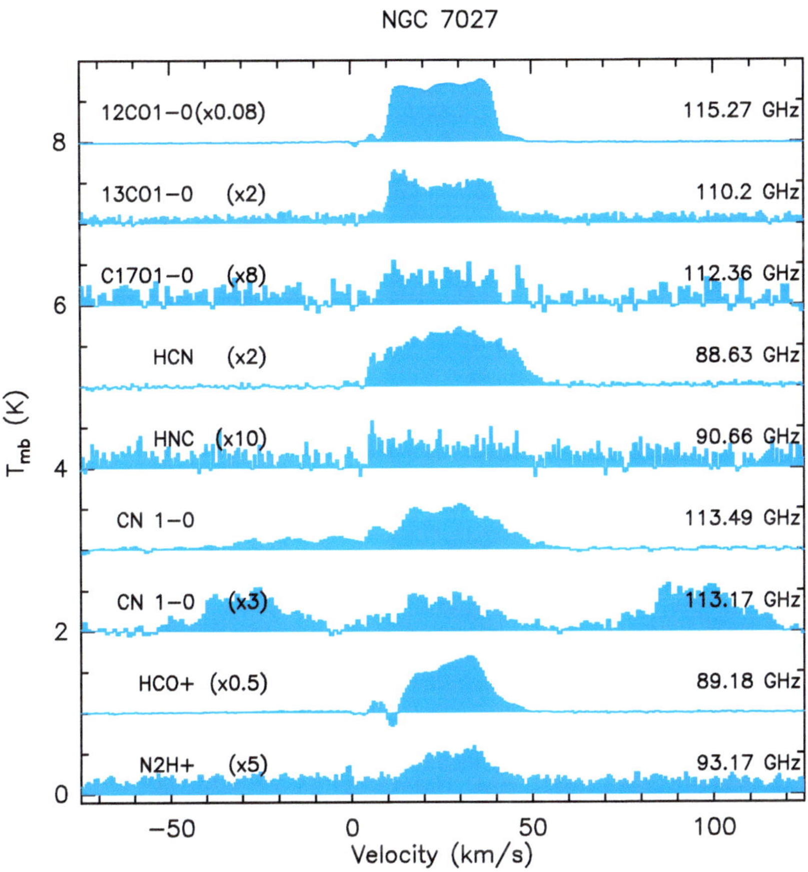

Figure 2. Selected spectra from a recent single-dish (IRAM 30 m telescope) molecular line survey of planetary nebula "poster child" NGC 7027 (adapted from Bublitz et al. 2018). All of these transitions will be accessible to the ngVLA's Band 6 (70–116 GHz).

X-ray-driven molecular chemistry in H II region PDRs and protoplanetary disks (e.g., Graninger et al. 2014; Cleeves et al. 2015).

Finally, the exquisite sensitivity of the ngVLA will be especially conducive to observations of isotopologues of CN and HCN in AGB stars, pre-PNe, and PNe aimed at measurements of the isotopic ratios $^{12}C/^{13}C$ and $^{14}N/^{15}N$ (Lee et al. 2013). Ascertaining the range of these ratios in major sources of ISM C and N enrichment, like PNe, is key to addressing longstanding questions concerning Galactic chemical evolution and the origins of pre-solar grains (e.g., Hily-Blant et al. 2017; Ziurys et al. 2017, and references therein).

References

Acord, J. M., Churchwell, E., & Wood, D. O. S. 1998, ApJ, 495, L107
Akashi, M., Bear, E., & Soker, N. 2018, MNRAS, 475, 4794
Aleman, I., Exter, K., Ueta, T., et al. 2018, MNRAS, 477, 4499
Bachiller, R., Forveille, T., Huggins, P. J., & Cox, P. 1997, A&A, 324, 1123
Bublitz, J., Kastner, J.H., Santander-Garcia, M., et al. 2018, in preparation
Bujarrabal, V., Alcolea, J., Mikołajewska, J., Castro-Carrizo, A., & Ramstedt, S. 2018, A&A, 616, L3
Balick, B., & Frank, A., 2002, ARAA, 40, 439
Cleeves, L. I., Bergin, E. A., Qi, C., Adams, F. C., & Öberg, K. I. 2015, ApJ, 799, 204
Frank, A., Chen, Z., Reichardt, T., et al. 2018, in Proceedings of "Asymmetrical Planetary Nebular VII" (special issue of *Galaxies*), in press; arXiv:1807.05925
Frew, D. J., Bojičić, I. S., & Parker, Q. A. 2013, MNRAS, 431, 2
Forbrich, J., Wolk, S. J., Güdel, M., et al. 2011, 16th Cambridge Workshop on Cool Stars, Stellar Systems, and the Sun, 448, 455
Graninger, D. M., Herbst, E., Öberg, K. I., & Vasyunin, A. I. 2014, ApJ, 787, 74
Guzmán, L., Gómez, Y., & Rodríguez, L. F. 2006, RMxAA, 42, 127
Hillwig, T. 2018, Galaxies, 6, 85
Hily-Blant, P., Magalhaes, V., Kastner, J., et al. 2017, A&A, 603, L6
Huarte-Espinosa, M., Frank, A., Balick, B., et al. 2012, MNRAS, 424, 2055
Huggins, P. J., Bachiller, R., Planesas, P., Forveille, T., & Cox, P. 2005, ApJS, 160, 272
Kastner, J. H., Weintraub, D. A., Gatley, I., Merrill, K. M., & Probst, R. G. 1996, ApJ, 462, 777
Kastner, J. H., Montez, R., Jr., Balick, B., et al. 2012, AJ, 144, 58
Kwok, S. 2000, *The Origin and Evolution of Planetary Nebulae*. New York: Cambridge University Press (Cambridge Astrophysics Series; 33)
Lee, C.-F., Yang, C.-H., Sahai, R., & Sánchez Contreras, C. 2013, ApJ, 770, 153
Masson, C. R. 1989, ApJ, 336, 294
Matthews, L. & Claussen, C. 2018, Science with a Next-Generation Very Large Array, ASP Monograph 7, ed. E. J. Murphy (San Francisco, CA: ASP), 281
Palen, S., Balick, B., Hajian, A. R., et al. 2002, AJ, 123, 2666
Roelfsema, P. R., Goss, W. M., Pottasch, S. R., & Zijlstra, A. 1991, A&A, 251, 611
Sabin, L., Parker, Q. A., Corradi, R. L. M., et al. 2014, MNRAS, 443, 3388
Sahai, R., & Trauger, J. T. 1998, AJ, 116, 1357
Sahai, R., Morris, M., & Villar 2011, AJ, 141, 134
Schönberner, D, Balick, B., & Jacob, R. 2018, A&A 609, A126.
Schmidt, D. R., & Ziurys, L. M. 2017a, ApJ, 835, 79
Schmidt, D. R., & Ziurys, L. M. 2017b, ApJ, 850, 123
Soker, N. 1990, AJ, 99, 1869
van de Steene, G. C., & Zijlstra, A. A. 1995, A&A, 293, 541
Wright, A. E., & Barlow, M. J. 1975, MNRAS, 170, 41
Ziurys, L. M., Woolf, N. J., Schmidt, D. R., Zack, L. N., & Zega, T. 2017, 80th Annual Meeting of the Meteoritical Society, 1987, 6408

Science with a Next-Generation Very Large Array
ASP Conference Series, Monograph 7
Eric J. Murphy, ed.

Probing Strong Binary Interactions and Accretion in AGB Stars with the ngVLA

Raghvendra Sahai

JPL, Pasadena, CA, USA; `raghvendra.sahai@jpl.nasa.gov`

Abstract. Understanding strong binary interactions is of wide astrophysical importance, and the deaths of most stars in the Universe that evolve in a Hubble time could be fundamentally affected by such interactions. These stars end their lives, evolving from Asymptotic Giant Branch (AGB) stars with extensive mass-loss into planetary nebulae with a spectacular array of morphologies. Binarity, and the associated formation of accretion disks (that drive collimated, fast jets) during the very late AGB or early post-AGB phase is believed to produce this dramatic morphological transformation. But the evidence for binarity and accretion during the AGB phase has been hard to obtain due to observational limitations. However, recent observations at UV and X-ray wavelengths have broken thru the observational barrier – our studies using *GALEX* reveal a candidate population of AGB stars, generally with strongly-variable far-ultraviolet (FUV) emission (fuvAGB stars), and our follow-up studies with *XMM-Newton*, *Chandra*, and *HST* of a few key objects supports our hypothesis that these objects have companions that are actively accreting material from the primary. The most prominent fuvAGB star has been detected with the VLA, showing the presence of variable non-thermal emission. The ngVLA, with its unprecedented sensitivity, is needed to survey a statistical sample of fuvAGB stars over the ~3-90 GHz range to search for and characterize the nature of the radio emission from fuvAGB stars and test our binarity+accretion hypothesis. Such a survey will distinguish between binarity/accretion-related radio emission that is expected to have both thermal and non-thermal components and display significantly time-variable on short time-scales (minutes to weeks), and single-star chromospheric emission from the primary that is expected to be thermal, possibly with time-variability, but only on long time-scales (many months to a year).

1. Introduction – Binarity and the AGB to PN Transition

A major astrophysical accomplishment of the 20th century was the success of models of the structure and evolution of stars of all masses from adolescence to retirement. The challenges continuing into the 21st century are now stellar birth, stellar death, and the impact of binary interactions on stellar evolution. Binary interactions in particular affect our understanding of stars in general, since such interactions affect the observable qualities of stars and stellar populations.

A glaring example of the effect of binary interactions on stellar evolution are planetary nebulae (PNe) that represent the bright end-stage of most stars in the Universe that evolve in a Hubble time (i.e., those with main-sequence masses of $1 - 8\ M_{\odot}$). PNe have wide-ranging astrophysical importance, covering diverse topics such as mass-loss and its effect on stellar evolution, the chemical evolution of the ISM, the cosmological distance ladder, astrophysical jets, common-envelope evolution, and intermediate-luminosity transients.

A long-standing puzzle for PNe formation is that while modern imaging surveys indicate that the vast majority of PNe deviate strongly from spherical symmetry (e.g., Schwarz, Corradi & Melnick 1992, Manchado et al. 1996, Sahai & Trauger 1998 [ST98], Sahai et al. 2011a, Stanghellini et al. 2016), the progenitors of PNe, AGB stars, have spherically-symmetric circumstellar envelopes (CSEs) resulting from mass-loss. Spherical mass-loss on the AGB implictly implies that potential physical causes of asymmetry present during the main-sequence phase (e.g., stellar rotation, which is very small and can be ignored at ages $> few \times 10^8$ yr) are not significant in affecting mass-loss on the AGB and beyond. Binarity provides a source of angular momentum, as well as a preferred axis to a stellar system, and is now widely believed to dramatically affect the evolutionary transition from the AGB to the planetary nebula (PN) phase (e.g., Balick & Frank 2002). ST98 proposed that highly-collimated, fast jet-like outflows at the late-AGB or pre-PN (PPN) phase sculpt the AGB mass-loss envelopes from the inside-out, producing the observed aspherical morphologies of PPNe (Sahai et al. 2007) and PNe. The engines that drive these outflows must reside in accretion disks that can be produced as a result of binarity and resulting accretion modes due to, e.g., Bondi-Hoyle or Roche-lobe accretion or during common-envelope evolution (e.g., Blackman & Lucchini 2014).

If binarity plays a significant role in the PN population, one may question whether the PN population is representative of all $1-8\,M_{\odot}$ stars, and one may need to significantly revise how PN are used as problems of stellar and chemical evolution (e.g., De Marco & Izzard 2017, Kwitter et al. 2014, Akashi & Soker 2013, Ciardullo et al. 2002). Hence, obtaining observational constraints on binarity and its effects in the general population of AGB stars is vital.

1.1. Finding Binaries in AGB Stars – An Observational Challenge

Until recently, observational evidence of binarity in AGB stars had been sorely lacking simply because AGB stars are very luminous and variable, invalidating standard techniques for binary detection (e.g., radial-velocity and photometric variations due to a companion star, direct imaging.) An innovative technique to search for FUV and NUV emission in AGB stars with *GALEX* has now provided a large candidate sub-class of AGB stars with close binary companions (Sahai et al. 2008 [Setal08], Sahai et al. 2011b [Setal11], Sahai et al. 2016). In this sub-class, for objects with emission in the FUV (1344-1786Å) (fuvAGB stars), the observed FUV flux densities are typically a factor $> 10^6$ larger than expected for the primary's photospheric emission (Setal08), and show strong variability (Figure 1).

It should be noted that fuvAGB stars in general (based on their optical spectra), do not belong to the well-studied class of symbiotic stars (red giant stars with white-dwarf [WD] companions) and have never been classified as such. So if the compact companions in fuvAGB star systems are WDs, then these must be quite cool ($T_{eff} \lesssim 20,000$ K) (Sahai et al. 2015 [Setal15]).

1.2. UV & X-Ray Emission: Binarity+Accretion and/or Chromospheres?

The FUV source in fuvAGB stars has been hypothesized to be likely dominated by emission due to variable accretion activity associated with a close companion (Setal08, Setal11, Ortiz & Guerrero 2016). Small X-ray surveys using *XMM-Newton* and *Chan-*

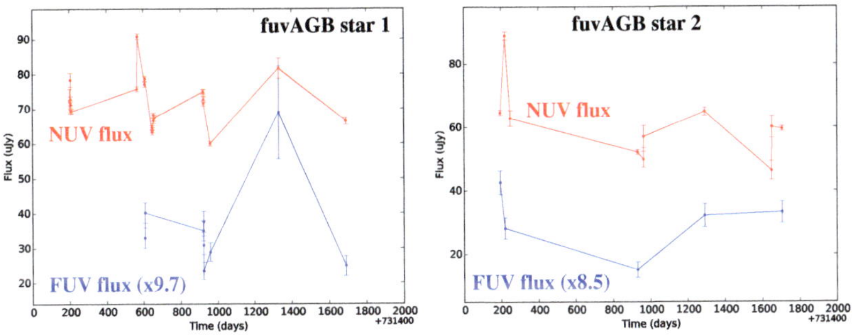

Figure 1. Two fuvAGB stars observed with *GALEX* showing strong variability in their FUV (blue) and NUV (red) emission (the FUV flux densities have been scaled up by arbitrary factors.)

dra support this hypothesis, finding X-ray emission in about 50% of fuvAGB stars. The X-ray emission is characterised by relatively high luminosities $Lx \sim (0.002 - 0.11)\, L_{\odot}$, and very high plasma temperatures $Tx \sim (35 - 160) \times 10^6$ K (Figure 2, adapted from Setal15). Amongst fuvAGB stars, objects with large FUV/NUV ratios, $R_{fuv/nuv} > 0.2$, have a much higher probability of being detected in X-ray emission (Sahai et al. 2016), and are almost certainly binaries with accretion activity powering the high-energy emission. The UV and X-ray emission is variable (Figure 3) with both periodic and stochastic components, signatures of active, ongoing accretion (Setal15).

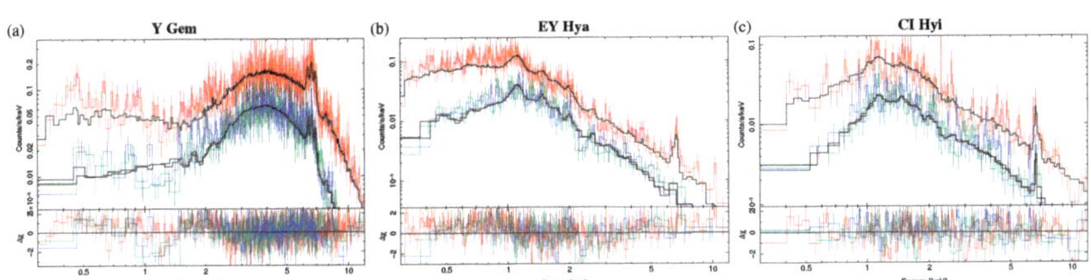

Figure 2. X-ray spectra fuvAGB stars Y Gem, EY Hya and CI Hyi using XMM/EPIC detectors pn (*red*), MOS1 (*blue*), MOS2 (*green*), together with APEC model fits (*black*); bottom panels show residuals.

A recent STIS spectroscopic study of the prototype high FUV/NUV ratio star, Y Gem, shows the presence of flickering on time-scales of $\lesssim$20 s and high-velocity infall and outflows, and thus directly supports the binary/accretion hypothesis (Sahai et al. 2018). Setal11 discuss five plausible models for the UV emission from Y Gem and conclude that the most plausible one is line and/or continuum emission associated with variable accretion activity and a disk around a companion star – the accretion shock may reside on (a) the disk or (b) the stellar surface of the companion (as, e.g., in T Tauri stars: Calvet & Gullbring 1998). Further support for this model comes from Y Gem's unusually narrow CO $J = 2 - 1$ line profile (Setal11) that suggests the possibility of a large, orbiting reservoir of molecular material in this object that may be associated with the central accretion disk.

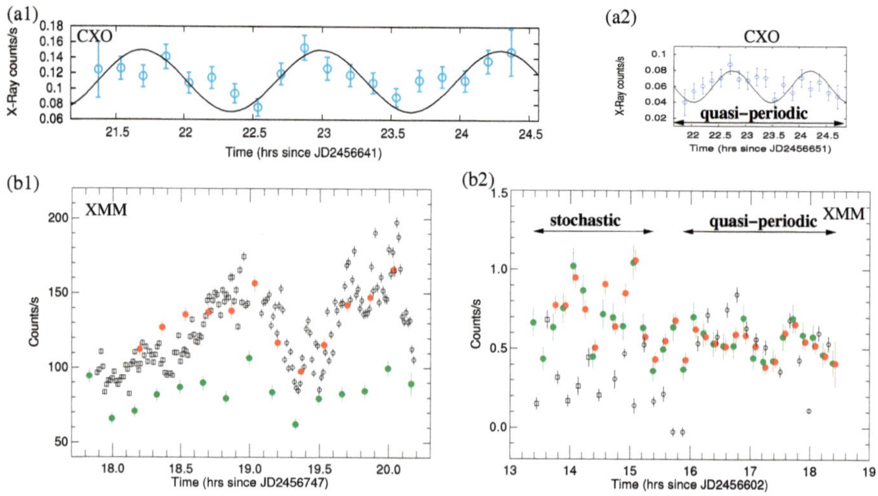

(1) Y Gem **(2) EY Hya**

Figure 3. X-ray and UV variability in the fuvAGB stars Y Gem and EY Hya. (*a*) CXO (ACIS-S), and (*b*) XMM (EPIC=pn+MOS1+MOS2: red, MOS=MOS1+MOS2: green, UVM2: black squares, UVW2: black circles). The EPIC, MOS and UVW2 data have been respectively re-scaled as follows: 80, 130, 2.5 (Y Gem), and 2, 5, 0.8 (EY Hya). A sinusoidal fit (by eye) with periods, P=1.35 and 1.45 hr are shown for the X-ray light curves of Y Gem and EY Hya in panels *b1* & *b2*.

This object (and objects like it) may represent the earliest phases of an AGB star with a growing accretion disk which will ultimately produce collimated jets that are now believed to sculpt the round circumstellar envelopes of AGB stars into bipolar planetary nebulae. Following the discovery of Y Gem's remarkable UV emission with this technique (Setal11), we obtained multifrequency (5.5, 22, & 30.5 GHz) observations of Y Gem, detecting the source at all 3 frequencies. These data suggest that we are probing an ionized accretion disk with magnetospheric accretion in this object. New VLA observations of Y Gem, covering a more extensive wavelength space, as well as multiple epochs, have been obtained and are being analyzed in order to test this scenario.

Objects with little or no FUV emission, i.e., with $R_{fuv/nuv} \lesssim 0.1$, which dominate the population of UV-emitting AGB stars, the UV emission may have a different source. From an analysis of the NUV emission in 179 AGB stars, Montez et al. (2017) argue that the origin of the *GALEX*-detected UV emission is intrinsic to the AGB star (chromospheric & photospheric emission), and is unrelated to binarity. Ortiz & Guerrero (2016), from a study of a volume-limited sample (<0.5 kpc) of 58 AGB stars, conclude that the detection of NUV emission with a very large observed-to-predicted ratio, $Q_{NUV} > 20$ (Ortiz & Guerrero 2016), is evidence for binarity in these objects.

2. Radio Observations

Our current working hypothesis for UV emission from AGB stars is that objects with a close companion produce high FUV/NUV ratios at least some of the time (since accretion can be variable) and/or a very large NUV excess, whereas single AGB stars (or those with large binary separations) always show low FUV/NUV ratios, as the emission in these is chromospheric. High-sensivity radio observations offer an unprecedented opportunity for testing the above hypothesis and leading to an understanding of binarity and binary interaction in AGB stars by observing the radio emission from AGB stars with UV emission. We describe below our VLA study of Y Gem (Sahai & Claussen 2018, *in prep*) that motivates and demonstrates this concept.

Our VLA A-array observation in December 2012 of Y Gem revealed an unresolved radio source with flux densities of 0.176 ± 0.02 mJy, 1.75 ± 0.20 mJy, and 3.62 ± 0.31 mJy at 5.5, 22 and 30.5 GHz, respectively (Figure 4b). These flux densities are far larger (factor ~100) than that expected from photospheric emission. The spectral-index, α, is ~ 2 in the (22–30.5) GHz, and ~ 1.65 in the (5.5–30.5) GHz range, arguing against the emission arising in an ionized outflow, which would result in $\alpha \sim 0.6$. Interpreting the (22–30.5) GHz SED as optically-thick emission from ionized gas (component 1), we find that there is a significant excess above such emission at 5.5 GHz, clearly requiring an additional source (component 2) at the low-frequency end – a plausible mechanism for this is gyrosynchrotron radiation. Modeling of component 1 shows that it arises in a region of size $\theta < 0\farcs03$ (17.5 AU at 580 pc), with an average electron density, $n_e > 8 \times 10^6\,\mathrm{cm}^{-3}$.

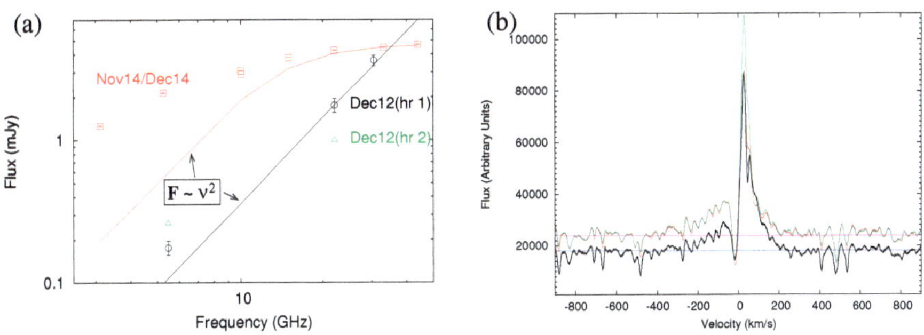

Figure 4. (a) Variable radio emission (VLA data) from Y Gem. Free-free emission components have been fitted to the data at K (22 GHz) band and higher frequencies. The lower-frequency data shows significant excess emission above these components at both epochs, implying the presence of a variable component that may be due to gyrosynchrotron radiation from material in an accretion flow. (b) Y Gem's Hα line profile at 3 epochs 1 Apr 2012 (red), 4 Apr 2012 (green), and 11 May 2012 (black) showing a P-Cygni type profile with broad wings, signature of a fast outflow – the profile is time-variable.

Plasma trapped in a magnetic field in the vicinity of the primary AGB star and/or the accretion disk may be the source of the non-thermal radio emission (component 2) from Y Gem – the presence of such plasma is revealed by the detection of strong ($L_x \sim 0.049\,L_\odot$) X-ray emission from Y Gem with *Chandra* (Setal15). A model fit to the spectrum shows the presence of plasma with a very high temperature, $T_x \sim 5.4$ keV, suggesting that the X-rays come from coronal emission due to plasma confined by a strong magnetic field.

Y Gem's UV and X-ray variability indicates episodic accretion (a) onto the disk, from matter ejected by the AGB star (e.g., Mastrodemos & Morris 1998), or (b) onto the companion from the inner disk region. If the mechanism is type *a* then one might expect shock and associated thermal emission, and if type *b* and if the companion has a strong magnetic field, then one would expect gyroresonance or gyrosynchrotron emission at radio wavelengths. Y Gem's radio flux densities are consistent with those observed in stellar sources with known accretion activity:

For example, in the type *a* source Mira [a well-studied AGB star which shows variable UV line emission, and is a known (symbiotic) binary], the compact companion accretes matter from the primary's wind. The radio flux densities of Mira (scaled to Y Gem's distance and factor 43 higher FUV flux, at its peak UV-emission state) of 0.38, 1.2 & 1.9 mJy at 8.5, 22.5, & 43.3 GHz, are comparable to those of Y Gem. The UV variability of Mira results from changes in the accretion rate onto the companion (Matthews & Karovska 2006).

We note that Y Gem is not a symbiotic star, and also consider type *b* sources, namely other radio-variable stellar sources with accretion flows. Y Gem's radio flux densities are also consistent with emission from such a flow. An example is the non-thermal (likely) gyrosynchrotron emission source observed in T Tau S at 2–3.6 cm. This emission, which may be due to a scaled-up solar-like flare or accretion-related (Skinner & Brown 1994), would produce a 3.6 cm flux density of about 0.3 mJy at Y Gem's distance.

3. The Role of ngVLA

The discovery of the large and variable UV and X-ray flux densities, and subsequent detection of radio emission from Y Gem supports our model with episodic accretion in a binary system for the high energy emission from fuvAGB stars. Although X-ray observations (with *XMM-Newton* and *Chandra*) and UV observations (with *HST*) are valuable probes of binary-associated accretion activity in AGB stars, our studies show that these are sensivity-limited to the brightest few ($\lesssim$ 10) objects in the full sample.

The ngVLA's sensitivity is needed to carry out a survey of a statistical sample of fuvAGB stars over the ~3-90 GHz range (see below), and to probe their short and long-term variability. Such a survey will enable us to search for and investigate the nature of radio emission from fuvAGB stars. Accretion-related radio emission is expected to have both thermal and non-thermal components and display significantly time-variable on short time-scales (minutes to weeks), whereas chromospheric emission is expected to be thermal, possibly with time-variability, but only on long time-scales (many months to a year).

The long-term UV variability time-scale in fuvAGB stars ($\sim 10^7$ sec in Y Gem) is likely related to the semi-regular variability of the primary (the optical light curve of Y Gem (Figure 4b) shows 3 periodicities at 280 d, 158 d and 172 d) due to its pulsations producing variations in the accretion mass-flux. We find variability on a medium time-scale (of a few days) in the Hα profile of Y Gem suggesting variations in the inner regions (few×100 AU) of a fast outflow (e.g., a jet powered by the accretion disk). The shortest variability timescale of $\sim$ 20 s is due to the flickering phenomenon that characterizes active accretion disks. The wide frequency coverage provied by the ngVLA will enable detailed modeling the radio emission, and decomposition into thermal and non-thermal components.

Using the ngVLA's predicted continuum performance as given by Selina et al. (2018), we find that with 15 min integration time per source, we can achieve a 5σ sensitivity of $\lesssim 2.5\,\mu$Jy/beam ($4.8\,\mu$Jy/beam) for any ngVLA band in the 8-41 GHz (90 GHz) frequency range. Since this is a factor > 100 lower than Y Gem's flux density at > 10 GHz, we can carry out a large multi-band survey of several 100 AGB stars with UV emission with a modest expenditure of ngVLA time. Such a project cannot be done with any other facility: it is prohibitive in terms of time for the VLA even for the mere detection of AGB stars at the few μJy level, and the VLA certainly lacks the sensitivity to probe the short-term (i.e., over minutes to hours) variability of these objects.

The brightest (say) 20 of these would then define a key subsample for detailed follow-up time-monitoring studies. Since the ngVLA covers the low-J lines of several molecular species that are typically detected in the winds from AGB stars (e.g., CO, ^{13}CO, HCN, SiO, CS), a pilot survey of these lines in the key subsample, together with ALMA observations of high-J lines, would help to probe the gas mass-loss rates and kinematics of the inner regions of outflows, and the presence of torii and/or disks in these stars resulting from the binary interaction (e.g., Sahai et al. 2017a,b).

Acknowledgments. RS gratefully acknowledges the contributions of his colleagues Carmen Sánchez Contreras, Jorge Sanz-Forcada, Mark Claussen, and C. Muthumariappan to the UV, X-ray, optical and radio studies described here, and discussions with Orsola de Marco, Eric Blackman, Albert Zijlstra, Joel Kastner, and Bruce Balick (in no particular order) during numerous proposal preparations, that have been helpful in motivating the study described here.

References

Akashi, M., & Soker, N. 2013, MNRAS, 436, 1961
Balick, B. & Frank, A. 2002, ARA&A, 40, 439
Blackman, E. G., & Lucchini, S. 2014, MNRAS, 440, L16
Calvet, N., & Gullbring, E. 1998, ApJ, 509, 802
Ciardullo, R., Feldmeier, J. J., Jacoby, G. H., et al. 2002, ApJ, 577, 31
De Marco, O., & Izzard, R. G. 2017, PASA, 34, e001
Kwitter, K. B., Méndez, R. H., Peña, M., et al. 2014, RMxAA, 50, 203
Manchado, A., Guerrero, M. A., Stanghellini, L., & Serra-Ricart, M. 1996, The IAC morphological catalog of northern Galactic planetary nebulae, Publisher: LaLaguna, Spain: Instituto de Astrofisica de Canarias (IAC)
Mastrodemos, M. & Morris, M. 1998, ApJ 497, 303
Matthews, L.K., & Karovska, M. 2006, ApJ, 637, L49
Montez, R., Jr., Ramstedt, S., Kastner, J. H., Vlemmings, W., & Sanchez, E. 2017, ApJ, 841, 33
Ortiz, R., & Guerrero, M. A. 2016, MNRAS, 461, 3036
Sahai, R., Morris, M., Sánchez Contreras, C., & Claussen, M. 2007a, AJ, 134, 2200
Sahai, R. & Trauger, J.T. 1998, AJ, 116, 1357
Sahai, R., Morris, M., Sánchez Contreras, C., & Claussen, M. 2007, AJ, 134, 2200
Sahai. R., Findeisen, K., Gil de Paz, A., Sánchez Contreras, C. 2008, ApJ, 689, 1274
Sahai, R., Morris, M. R., & Villar, G. G. 2011a, AJ, 141, 134
Sahai, R., Neill, J. D., Gil de Paz, A., & Sánchez Contreras, C. 2011b, ApJ, 740, L39
Sahai, R., Sanz-Forcada, J., Sánchez Contreras, C., & Stute, M. 2015, ApJ, 810, 77
Sahai, R., Sanz-Forcada, J., & Sánchez Contreras, C. 2016, Journal of Physics Conference Series, 728, 042003
Sahai, R., Vlemmings, W. H. T., Gledhill, T., et al. 2017a, ApJ, 835, L13

Sahai, R., Vlemmings, W. H. T., & Nyman, L.-Å. 2017b, ApJ, 841, 110
Sahai, R., Sánchez Contreras, C., Mangan, A. S., et al. 2018, ApJ, 860, 105
Schwarz, H.E., Corradi, R.L.M. & Melnick, J. 1992, A&A Suppl.Ser., 96, 23
Selina, R. J., Murphy, E. J., McKinnon, M., et al. 2018, Society of Photo-Optical Instrumentation Engineers (SPIE) Conference Series, 10700, 107001O
Skinner, S.L. & Brown, A. AJ, 1994, 107, 1461
Stanghellini, L., Shaw, R. A., & Villaver, E. 2016, ApJ, 830, 33

Science with a Next-Generation Very Large Array
ASP Conference Series, Monograph 7
Eric J. Murphy, ed.

A Six-Dimensional Tomographic View of Galactic Star-Formation

Laurent Loinard[1] and Mark J. Reid[2]

[1]*Centro de Radioastronomía y Astrofísica, Universidad Nacional Autónoma de México, Morelia, Michoacán, México* `l.loinard@irya.unam.mx`

[2]*Harvard-Smithsonian Center for Astrophysics, Cambridge, Massachusetts, USA* `reid@cfa.harvard.edu`

Abstract. Various sign-posts of recent star-formation activity, such as water and methanol maser emission or magnetically active low-mass young stars, can be detected with Very Long Baseline Interferometry (VLBI) radio arrays. The extremely accurate astrometry already attainable with VLBI instruments implies that the trigonometric parallax and the proper motion of these objects can be measured to better than 1% for sources within a few hundred parsecs of the Sun, and better than 10% for objects at a few kiloparsecs. An ngVLA with baselines extending to several thousand km will have a sensitivity more than one order of magnitude better than current VLBI instruments, and will enable such highly accurate astrometric measurements to be performed throughout the Milky Way. This will provide a full six-dimensional view (three spatial and three velocity coordinates) of star-formation in the Galactic disk, and have a transformative impact on our understanding of both star-formation processes and Galactic structure.

1. Introduction

Even when considering only the Milky Way galaxy, star-formation involves a wide variety of relevant spatial scales. Individual protostars accrete through disks that are 10–100 AU across, from protostellar envelopes whose radii are typically 10,000 AU. In addition, stars typically do not form in isolation. Instead, star-formation involves molecular cores, filaments, clouds, and complexes with scales between a tenth and several hundred parsecs. Finally, star-formation in the Galactic disk is organized into a coherent spiral pattern, where individual spiral arm segments can be traced out over tens of kiloparsecs. On all of these scales, many questions regarding the physical processes underlying star-formation remain unanswered. Let us provide just three examples, involving widely different spatial scales.

- While a paradigm exists for the formation of isolated low-mass stars (Shu, Adams & Lizano 1987), the formation of multiple stellar systems remains poorly understood. The two leading (and not necessarily mutually exclusive) contending theories are turbulent and disk fragmentation (Padoan et al. 2007; Adams et al. 1989). These two theories make different predictions on the distribution of separations in the resulting multiple systems that can be tested through high angular resolution observations (e.g., Tobin et al. 2016a,b). Tight multiple systems (with sizes of a fraction of an AU) are particularly interesting in this context because their orbital periods are short, so they can be fully characterized within a reasonable amount of time.

- Parsec-scale filamentary structures are now known to be ubiquitous in regions of star-formation (André et al. 2017), but their exact impact on the star-formation process remains unclear. It has been argued that their fragmentation might lead to the formation of dense cores and that gas motion along the filaments could be important to sustain the accretion onto the protostars contained in these cores (Lu et al. 2018). To unambiguously test this possibility, it is important to establish the exact orientation of the filaments, the kinematics of the gas motion, as well as the location and kinematics of the young stars embedded within the filaments.

- Spiral structure is a very common feature in disk galaxies, and it is well established that star-formation (in spiral galaxies) occurs mainly along the spiral arms. Yet, the very origin of the spiral structure and the exact relationship between spiral structure and star-formation remain elusive (see Dobbs et al. 2014 for a recent discussion). A popular scenario, the spiral density wave theory of Lin & Shu (1964) combined with an analysis of the gas response to the resulting stellar potential (Roberts 1969), leads to the prediction of organized *streaming motions* across the spiral arms that should be detectable for both the gaseous and the stellar components. However, detailed numerical simulations do not appear to support this theory. Thus, the unambiguous detection of streaming motions (or stringent upper limits on their non existence) would place important constraints on the mechanisms relating spiral structure and star-formation. The necessary observations would have to constrain the exact location and structure of the spiral arms as well as the kinematics of the young stars and the associated ISM.

These three examples illustrate the need for both accurate distances to young stars (to establish the orientation of filaments in star-forming regions or the location of spiral arm features in the Milky Way) and accurate measurements of their 3-dimensional velocity vectors, at scales ranging from individual stellar systems, to specific star-forming regions, to the entire Galactic disk. Formally, this means constraining the distribution function $f(\mathbf{x}, \dot{\mathbf{x}})$ for a sufficiently large sample of young stars, and can be achieved through a combination of dedicated spectroscopic and astrometric observations. The spectroscopic observations are necessary to provide the velocity component projected along the line of sight and typically require sensitive optical or infrared observations. The remaining 5 dimensions (3 spatial coordinates, and two velocity components) require accurate positions, trigonometric parallaxes, and proper motions. The *Gaia* mission will be transformative in providing this type of information for many different populations of astronomical objects, including some young stars. However, given the dust opacity associated with individual star-forming regions and the Galactic plane as a whole, *Gaia* will be limited to young stars that are both nearby and not deeply enshrouded in dust. Very Long Baseline Interferometry (VLBI) observations at radio wavelengths are immune to dust obscuration and can already deliver trigonometric parallaxes with an accuracy as good as 10 μas, and proper motions accurate to better than a few μas yr^{-1} (Reid & Honma 2014). Taking advantage of this, VLBI observations have been used to study star-forming regions in the Milky Way on a variety of scales[1] as we will describe in §2. We will then proceed, in §§3 and 4, to discuss the impact that baselines in thousands of km on the ngVLA would have on the study of Galactic star-formation, as well the synergies with other existing and planned facilities.

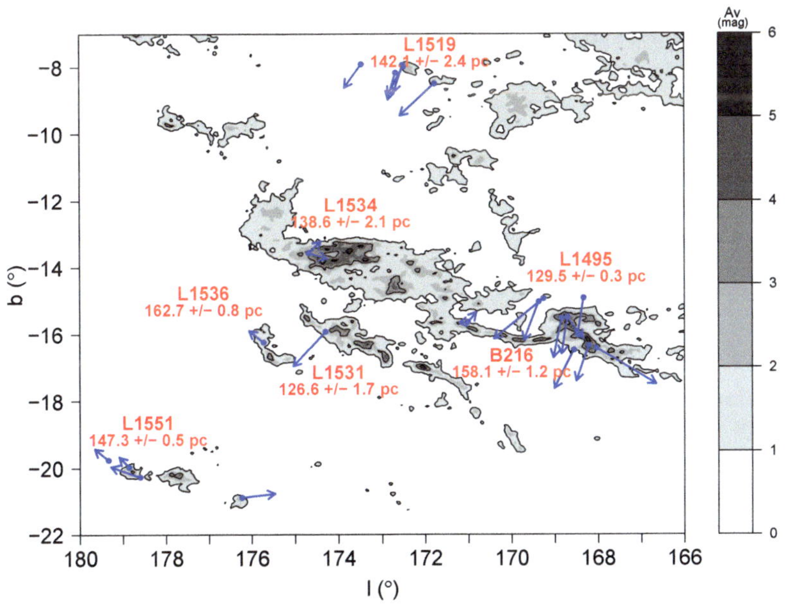

Figure 1. Three-dimensional structure of the Taurus star-forming regions (adapted from Galli et al. 2018). The underlying grey-scale image shows the extinction map derived by Dobashi et al. (2005). The different dust clouds in the region are indicated, together with their distance measured though VLBI astrometric observations. The blue arrows correspond to the proper motions (also determined from VLBI observations) expressed in the LSR.

2. Existing VLBI Observations

2.1. Structure and Kinematics of Nearby Regions

At the distance of the nearest star-forming regions (Taurus, Ophiuchus, Orion, etc.; 100 pc < d < 500 pc), VLBI observations of numerous magnetically-active young stars have enabled the measurement of their distance and velocity to an accuracy of about 1%. For instance, the distance to the Orion Nebula Cluster (ONC; the nearest region of high-mass star-formation) has been refined to 388 ± 5 pc through a series of VLBI measurements of an increasingly representative sample of individual young stars (Sandstrom et al. 2007; Hirota et al. 2007; Menten et al. 2007; Kim et al. 2008; Kounkel

[1]Other applications include extragalactic proper motions, as well as astrometry of pulsars, compact objects, and evolved stars but they will not be discussed here. See Reid & Honma (2014) for a review.

et al. 2017). An analysis of the *Gaia* DR2 results by Kounkel et al. (2018) on a different subset of young stars in the ONC yields a fully consistent distance determination of 389 ± 3 pc.

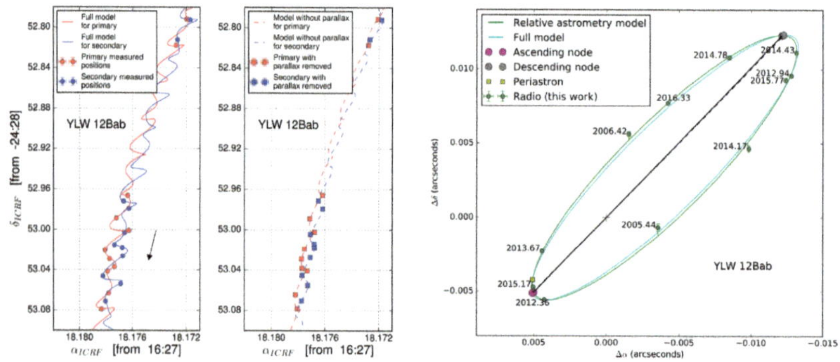

Figure 2. VLBI resolved observations of the tight binary system YLW12B in Ophiuchus (from Ortiz-León et al. 2017a). The leftmost panel shows the positions of the two stars in the systems (as blue and red dots, respectively) measured a different epochs, together with fits to these positions that combine trigonometric parallaxes, systemic, and orbital proper motions. The central panel shows the same positions and fits but with the parallax contribution subtracted. The rightmost panel shows the relative orbit between the two stars in the system with the best Keplerian fit.

Nearby star-forming regions are typically a few to several tens of parsecs across, so an accuracy of order 1% on the distance to individual stars is largely sufficient to resolve their depth along the line of sight. As a consequence, VLBI observations have been able to reconstruct the 3D spatial structure of several of these regions. The example of Taurus, where the depth (~30 pc) is comparable to the extent of the complex on the plane of the sky, is illustrated in Figure 1. A combination of VLBI astrometry and molecular gas spectroscopy can then be used to reconstruct the six-dimensional (space-velocity) structure of the complexes. Preliminary result of this type have been presented by Kounkel et al. (2017) who showed that two sub-regions along the Orion A complex (the ONC and Lynds 1641) are approaching each other both in the plane of the sky (as indicated by their relative proper motions) and along the line of sight. The latter conclusion stems from the combined facts that (i) the ONC (at 389 pc) is significantly nearer than L 1641 (at 430 pc), and (ii) the radial velocity of the molecular gas associated with the ONC is less blue-shifted than that of L 1641.

2.2. Multiple Stellar Systems

An interesting by-product of the intensive VLBI observations that led to the accurate distances to nearby star-forming regions mentioned above, has been the identification and spatial resolution of a significant number of very tight binary systems (Ortiz-León et al. 2017a,b, Kounkel et al. 2017). As an illustration, Figure 2 shows the case of the YLW12B system in Ophiuchus, which is resolved into two stars in an orbit with a semi-major axis of 12.6 mas (1.73 AU) and an orbital period of 1.42 yr. This tight binary is itself orbited by a third star on a wider orbit, making the entire system a hierarchical triple. Modelling the measured positions with a combination of parallax, systemic,

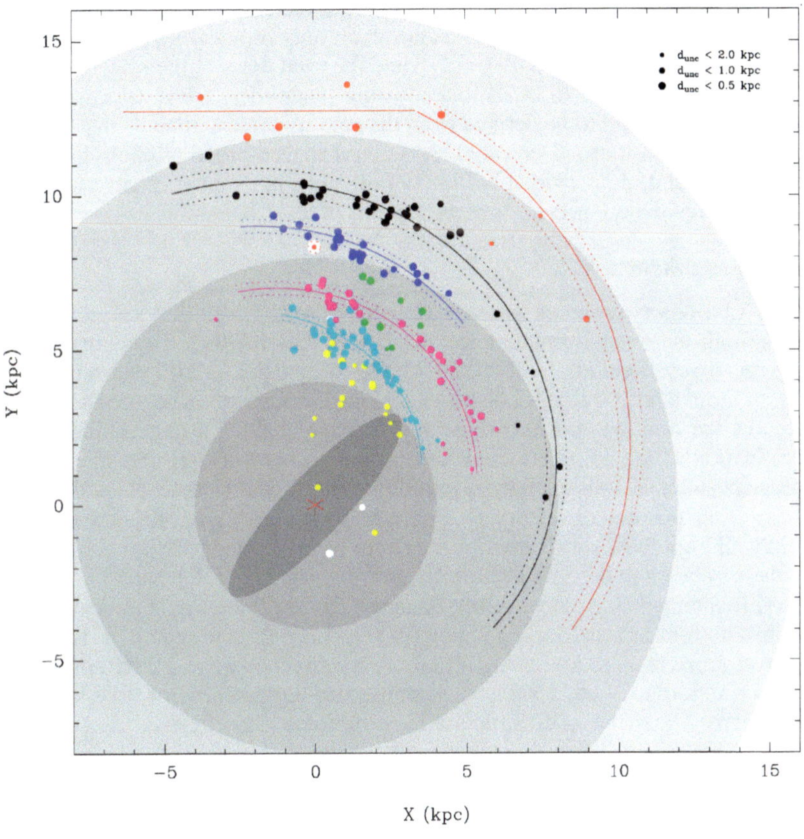

Figure 3. Location of high-mass star-forming regions in the Milky way with VLBI-measured distances (from Reid et al. 2018, in prep.) overlaid on a face-on sketch of the Galaxy. Different symbols and colors show different arm segments.

and orbital motions results in extremely accurate mass determinations for the individual stars ($1.397 \pm 0.019\, M_\odot$ and $1.258 \pm 0.006\, M_\odot$). Such observations, obtained for a statistical sample of young multiple systems, provide important constraints for theoretical models of early stellar evolution, and inform models of multiple star-formation on the observed distribution of system architectures.

It is important to appreciate that two distinct properties of VLBI observations came into play to enable these results. On the one hand, VLBI observations provide coordinates for the individual stars relative to a fixed quasar –and not just relative to one another. This registration of the positions relative to a fixed reference is what enables individual masses to be measured, rather than only the total mass of the system. Second, the very high angular resolution of VLBI images (of order 1 mas) enabled the resolution of the different stars in the systems. This combination of VLBI image properties are not available at any other wavelengths. Stellar mass determination based on orbital motion modelling has been obtained for increasingly tight systems from optical or near-infrared ground-based observations using techniques like adaptive optics and aperture masking. Those observations, however, only provide the relative positions of the stars in a given system so they only yield the total mass of the systems. Individual stellar masses then have to be estimated using model-dependent assumptions on the mass ratio (which tend to be derived from the very theoretical models that the observations intend to constrain). Dedicated astrometric spaced-based missions such as *Gaia*, on the other hand, do provide absolute coordinates, but cannot resolve tight systems (the angular resolution of *Gaia* is several tenths of an arcsecond).

2.3. Galatic Structure

Using VLBI observations of water and methanol maser emission associated with regions of high-mass star-formation, it is possible to measure their trigonometric parallax to high accuracy. Reid et al. (2009, 2014) and Honma et al. (2012) have obtained distances to more than 100 regions distributed along Galactic spiral segments within about 10 kpc of the Sun (Figure 3). These results have enabled a huge refinement in the determination of the spiral structure of our Galaxy, and of the fundamental Galactic parameters, such as a distance from the Sun to the Galactic center R_0 and the circular velocity at the position of the Sun θ_0. An interesting conclusion of this work is that essentially all high-mass star-formation appear to be associated with know spiral features –i.e. there appears to be little or no inter-arm high-mass star-formation activity in our Galaxy. In a remarkable recent result, Sanna et al. (2017) were able to measure the distance to a high-mass star-forming region (G007.47+00.05) located on the far side of the Galaxy at more than 20 kpc using VLBI observations of water maser emission. This result is accurate to about 15%, and demonstrates the potential of very long baseline observations to map the spiral structure over the entire Galactic disk.

3. The Impact of the ngVLA

The examples shown in §2 demonstrate that VLBI astrometry has already provided fundamental constraints on star-formation processes at scales ranging from a fraction of an AU to tens of kiloparsecs. Continental-size baselines on the ngVLA will result in a sensitivity more than one order of magnitude better than that of existing VLBI facilities such as the VLBA or the EVN, dramatically increasing the number of detectable sources, both as targets and calibrator. At frequencies above $\sim$ 5 GHz, the astrometric accuracy of VLBI observations for a given maximum baseline length is limited by residual delay errors which scale linearly with the separation between the target and the calibrator. The densification of calibrators resulting from the increased sensitivity of the ngVLA will therefore directly translate to an increase in astrometric accuracy, which will become as good as 1 μas for trigonometric parallax measurements (Reid et al. 2018 in this volume, p. 523). In other words, **the astrometry delivered by the ngVLA equipped with baselines of thousands of km will surpass that of *Gaia* by one order of magnitude**. The combination of this and the detectability of weaker target sources will have several very important consequences.

- In nearby regions, the ngVLA will deliver distance and velocity measurements accurate to better than 0.1%. In addition, the number of detectable targets will increase very significantly. At the moment, only about 10–20% of young stars in a given region is detectable with VLBI arrays. Given the radio luminosity function of young stars (Ortiz-León et al. 2017a), the ngVLA will be able to detect the majority of sources. This will provide a full six-dimensional description of each individual region, that will be directly comparable with numerical simulation of the formation and evolution of stellar clusters. It will also increase the number of multiple stellar systems for which the architecture and extremely accurate individual masses will become available. As explained in §2.2, VLBI observation is currently the only technique that can provide the combination of resolved images and coordinates registered to a fixed reference necessary to measure individual masses for systems tighter than several tenths of an arcsecond.

- The magnetically active young stars mentioned in §2.1 are currently only detectable up to distances of about 500 pc. With the ngVLA, they will be detectable in the nearest spiral arm segments. Since such active stars are typically low-mass, they are much more numerous than the maser sources associated with high-mass protostars. As a consequence, their detection in nearby spiral arm segments will enormously improve the way that spiral arms can traced. This will make it possible to map in detail the distribution of star-formation activity and the kinematics of the associated young stars along and across spiral arm segments. Combined with observations of the molecular ISM component, this will provide fundamental constraints on the star-formation mechanisms in spiral galaxies and their relationship with spiral structure.

- The ngVLA will be able to detect and measure accurate distance to maser associated with high-mass star-forming regions throughout the Milky Way Galaxy. The measurement reported by Sanna et al. (2017) of a parallax for a water maser at more than 20 kpc was only possible because of the extraordinary brightness of that specific maser. With the ngVLA, however, such measurements will become possible for a large population of masers on the far side of the Milky Way. This will permit the delineation of the entire spiral structure of the Milky Way, with fundamental implications for the analysis of star-formation on the scale of the entire Galaxy.

In addition to the purely astrometric results mentioned here, long baselines on the ngVLA will have the potential to reveal interesting physics related to stellar magnetic activity. For instance there is evidence that tight binary systems are more frequently radio-active than single stars (or wider binaries) and that the exact location of the radio emission centroid is modulated by the orbital motion of the system (e.g., Massi et al. 2002; Torres et al. 2012). The increased sensitivity and astrometric accuracy of the ngVLA will provide statistically relevant constraints on these and a variety of other phenomena that might prove relevant to exo-space wheather.

4. Synergies at Other Wavelengths and with Other Instruments

As detailed above, the ngVLA equipped with very long baselines will provide a tomographic six-dimensional view of star-formation in our Galaxy, providing information on scales from 1 AU to tens of kpc. This will be highly complementary of the *Gaia* satellite to reveal the dust obscured Galactic plane and dust-enshrouded nearby star-formation. In addition, since many ingredients enter the process of star-formation, observations at multiple wavelengths are necessary to make progress. The ngVLA itself, particularly combined with ALMA, large single dish telescopes, and future infrared space observatories, will enable the mapping of dust, molecular, atomic, and ionized gas at high spatial resolution in individual protostars and extended star-forming regions. Such observations might also reveal the role of magnetic fields both at large and small scales. The study of the physical mechanisms leading to continuum radio emission in low-mass young stars and their relationship to binarity and other stellar properties will require low frequency observations from the ngVLA itself, but also from the SKA. Finally, X-ray telescopes will be useful both to probe ISM distributions and stellar activity.

Acknowledgments. L.L. acknowledges the financial support of DGAPA, UNAM (project IN112417), and CONACyT, México.

References

Adams, F. C., Ruden, S. P., & Shu, F. H. 1989, ApJ, 347, 959
André, P., Men'shchikov, A., Bontemps, S., et al. 2010, A&A, 518, L102
Dobashi, K., Uehara, H., Kandori, R., et al. 2005, PASJ, 57, S1
Dobbs, C. L. 2014, Setting the scene for *Gaia* and LAMOST, 298, 221
Gaia Collaboration, Prusti, T., de Bruijne, J. H. J., et al. 2016, A&A, 595, A1
Galli, P. A. B., Loinard, L., Ortiz-Léon, G. N., et al. 2018, ApJ, 859, 33
Hirota, T., Bushimata, T., Choi, Y. K., et al. 2007, PASJ, 59, 897
Honma, M., Nagayama, T., Ando, K., et al. 2012, PASJ, 64, 136
Kim, M. K., Hirota, T., Honma, M., et al. 2008, PASJ, 60, 991
Kounkel, M., Covey, K., Suárez, G., et al. 2018, AJ, 156, 84
Kounkel, M., Hartmann, L., Loinard, L., et al. 2017, ApJ, 834, 142
Lin, C. C., & Shu, F. H. 1964, ApJ, 140, 646
Lu, X., Zhang, Q., Liu, H. B., et al. 2018, ApJ, 855, 9
Massi, M., Menten, K., & Neidhöfer, J. 2002, A&A, 382, 152
Menten, K. M., Reid, M. J., Forbrich, J., & Brunthaler, A. 2007, A&A, 474, 515
Ortiz-León, G. N., Dzib, S. A., Kounkel, M. A., et al. 2017, ApJ, 834, 143
Ortiz-León, G. N., Loinard, L., Kounkel, M. A., et al. 2017, ApJ, 834, 141
Reid, M. J., Menten, K. M., Brunthaler, A., et al. 2014, ApJ, 783, 130
Reid, M. J., Menten, K. M., Zheng, X. W., et al. 2009, ApJ, 700, 137
Reid, M. J., & Honma, M. 2014, ARA&A, 52, 339
Reid, M.J., Loinard, L., Maccarone, T.J., 2018, Science with a Next-Generation Very Large Array, ASP Monograph 7, ed. E. J. Murphy (San Francisco, CA: ASP), 523
Padoan, P., Nordlund, Å., Kritsuk, A. G., Norman, M. L., & Li, P. S. 2007, ApJ, 661, 972
Sandstrom, K. M., Peek, J. E. G., Bower, G. C., et al. 2007, ApJ, 667, 1161
Reid, M., Loinard, L., & Maccarone, T. 2018, Science with a Next-Generation Very Large Array, ASP Monograph 7, ed. E. J. Murphy (San Francisco, CA: ASP), 523
Roberts, W. W. 1969, ApJ, 158, 123
Sanna, A., Reid, M. J., Dame, T. M., Menten, K. M., & Brunthaler, A. 2017, Science, 358, 227
Shu, F. H., Adams, F. C., & Lizano, S. 1987, ARA&A, 25, 23
Tobin, J. J., Kratter, K. M., Persson, M. V., et al. 2016, Nat, 538, 483
Tobin, J. J., Looney, L. W., Li, Z.-Y., et al. 2016, ApJ, 818, 73
Torres, R. M., Loinard, L., Mioduszewski, A. J., et al. 2012, ApJ, 747, 18

Part VI

Galaxy Ecosystems

Science with a Next-Generation Very Large Array
ASP Conference Series, Monograph 7
Eric J. Murphy, ed.

Radio Continuum Emission from Galaxies: An Accounting of Energetic Processes

Eric J. Murphy,[1] James J. Condon,[1] Antxon Alberdi,[2] Loreto Barcos-Muñozarcos,[3,1] Robert J. Beswick,[4] Elias Brinks,[5] Dillon Dong,[6] Aaron S. Evans,[7,1] Kelsey E. Johnson,[7,1] Robert C. Kennicutt Jr.,[8] Sean T. Linden,[7,1] Tom W. B. Muxlow,[4] Miguel Pérez-Torres,[2] Eva Schinnerer,[9] Mark T. Sargent,[10] Fatemeh S. Tabatabaei,[2,11] and Jean L. Turner[12]

[1]*National Radio Astronomy Observatory, Charlottesville, VA 22903;* `emurphy@nrao.edu`

[2]*Instituto de Astrofisica de Andalucía (IAA, CSIC); Glorieta de la Astronomía s/n, 18008-Granada, Spain*

[3]*Joint ALMA Observatory, Alonso de Córdova 3107, Vitacura, Santiago, Chile*

[4]*Jodrell Bank Centre for Astrophysics/e-MERLIN, The University of Manchester, Manchester, M13 9PL, UK*

[5]*School of Physics, Astronomy and Mathematics, University of Hertfordshire, Hatfield AL10 9AB, UK*

[6]*California Institute of Technology, MC 100-22, Pasadena, CA 91125, USA*

[7]*Department of Astronomy, University of Virginia, 3530 McCormick Road,Charlottesville, VA 22904, USA*

[8]*Institute of Astronomy, University of Cambridge, Madingley Road, Cambridge CB3 0HA, UK*

[9]*Max Planck Institut für Astronomie, Königstuhl 17, Heidelberg D-69117, Germany*

[10]*Astronomy Centre, Department of Physics and Astronomy, University of Sussex, Brighton BN1 9QH, UK*

[11]*Departamento de Astrofísica, Universidad de La Laguna, E-38206 La Laguna, Spain*

[12]*Department of Physics and Astronomy, UCLA, Los Angeles, CA 90095, USA*

Abstract. Radio continuum observations have proven to be a workhorse in our understanding of the star formation process (i.e., stellar birth and death) from galaxies both in the nearby universe and out to the highest redshifts. In this article we focus on how the ngVLA will transform our understanding of star formation by enabling one to map and decompose the radio continuum emission from large, heterogeneous samples of nearby galaxies on $\gtrsim$ 10 pc scales to conduct a proper accounting of the energetic processes powering it. At the discussed sensitivity and angular resolution, the ngVLA will simultaneously be able to create maps of current star formation activity at ~100 pc scales, as well as detect and characterize (e.g., size, spectral shape, density, etc.) discrete HII regions and supernova remnants on 10 pc scales in galaxies out to the distance of the Virgo cluster. Their properties can then be used to see how they relate to the local and global ISM and star formation conditions. Such investigations are essential for understanding the astrophysics of high-z measurements of galaxies, allowing for proper modeling of galaxy formation and evolution.

1. Introduction

Radio continuum observations have proven to be a workhorse in our understanding of the star formation process (i.e., stellar birth and death) from galaxies both in the nearby universe and out to the highest redshifts. A next-generation Very Large Array (ngVLA), as currently proposed, would revolutionize our understanding of the emission mechanisms that power the radio continuum emission in and around galaxies by enabling the routine construction of $\sim 1.2 - 116$ GHz radio spectral maps. Coupled with its nearly order of magnitude increased sensitivity compared to the current Jansky VLA, the ngVLA makes it possible to use this frequency window to investigate the distinct physical processes linked to stellar birth and death for large, heterogeneous samples of galaxies for the first time. Furthermore, by delivering such a finely sampled spectrum over this entire frequency range with a single instrument will allow robust separation of the various emission components, which is currently the main challenge for multi-frequency radio studies. Each observation will provide enough sensitivity and spectral coverage to robustly decompose and accurately quantify the individual energetic components powering the radio continuum, thus providing unique information on the non-thermal plasma, ionized gas, and cold dust content in the disks and halos of galaxies.

In this chapter we focus on how the ngVLA can be used to map and decompose the radio continuum emission from large heterogeneous samples of nearby galaxies on $\gtrsim 10$ pc scales to conduct a proper accounting of their energetics. In doing so, we will be able to determine how star formation, AGN, and physical conditions in the ISM give rise to varying contributions of:

- non-thermal synchrotron emission powered by accelerated CR electrons/positrons.
- free-free emission from young massive star-forming (HII) regions.
- anomalous microwave emission (AME), which is a dominant, but completely unconstrained, foreground in CMB experiments.
- cold, thermal dust emission that accounts for most of the dust and total mass content of the ISM in galaxies.

2. The Energetic Processes Powering Radio Continuum Emission

Radio continuum emission from galaxies covering $\sim 1.2 - 116$ GHz is powered by an eclectic mix of physical emission processes, each providing completely independent in-

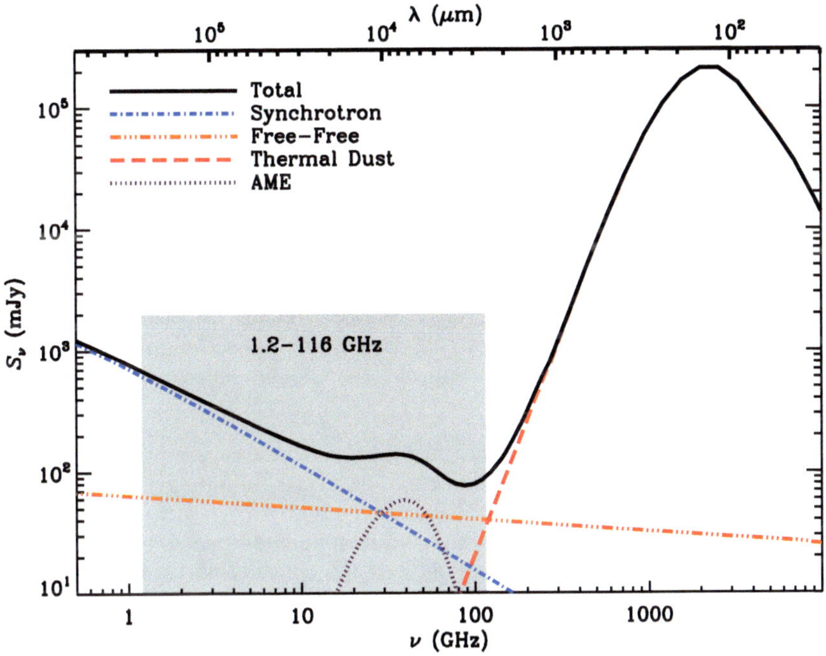

Figure 1. A model spectrum illustrating the various emission processes (non-thermal synchrotron, free-free, AME, thermal dust) that contribute to the observed radio frequency range to be covered by the ngVLA. Only in the proposed ngVLA frequency range (1.2–116 GHz, highlighted) do all major continuum emission mechanisms contribute at similar levels, making this range uniquely well-suited to next-generation continuum studies.

formation on the star formation and ISM properties of galaxies (see Figure 1). As such, it provides a unique window into the ISM and the process of star formation in galaxies. These processes include non-thermal synchrotron, free-free (thermal bremsstrahlung), anomalous microwave, and thermal dust emission that are directly related to the various phases of the ISM and provide a comprehensive picture of how galaxies convert their gas into stars. Each of these emission components, described in detail below, are of low surface brightness in the $\sim 30-116$ GHz frequency range, and therefore difficult to map in a spatially resolved manner at $\sim 10-100$ pc scales in the general ISM of nearby galaxies using existing facilities. Consequently, our current knowledge about the emissions processes over this frequency range is limited to the brightest star-forming regions/nuclei in the most nearby sources (e.g., Clemens et al. 2008, 2010; Leroy et al. 2011; Murphy 2013; Murphy et al. 2015; Barcos-Muñoz et al. 2015), providing no information on how the situation may change in drastically different ISM conditions (e.g., Tabatabaei et al. 2018) that may be more representative of those in high-redshift galaxies, where we have to rely on globally integrated measurements.

- **Non-Thermal Synchrotron Emission:** At ~GHz frequencies, radio emission from galaxies is dominated by non-thermal synchrotron emission resulting, indirectly, from star formation. Stars more massive than $\sim 8\,M_\odot$ end their lives as core-collapse supernovae, whose remnants are thought to be the primary accelerators of cosmic-ray (CR) electrons, giving rise to the diffuse synchrotron emission observed from star-forming galaxies. Thus, the synchrotron emission observed from galaxies provides a direct probe of the still barely understood relativistic (magnetic field + CRs) component of the ISM. As illustrated in Figure 1, the synchrotron component has a steep spectral index, typically scaling as $S_\nu \propto \nu^{-0.83}$ with a measured rms scatter of 0.13 (Niklas et al. 1997). By covering a frequency range spanning 1.2 – 116 GHz, the ngVLA will be sensitive to CR electrons spanning an order of magnitude in energy (i.e., ~ 1 – 30 GeV), including the population that may drive a dynamically-important CR-pressure term in galaxies (e.g., Socrates et al. 2008). Further, the ngVLA will allow for investigations of the radio spectra of galaxies that are not self-similar across a range of physical scales as the injection of fresh CRs are able to modify the spectra locally.

- **Free-Free Emission:** The same massive stars whose supernovae are directly tied to the production of synchrotron emission in star-forming galaxy disks are also responsible during their lifetime for the creation of HII regions. The ionized gas produces free-free emission, which is directly proportional to the production rate of ionizing (Lyman continuum) photons and optically-thin at radio frequencies. In contrast to optical recombination line emission, no hard-to-estimate attenuation term is required to link the free-free emission to ionizing photon rates, making it an ideal, and perhaps the most reliable, measure of the current star formation in galaxies. Unlike non-thermal synchrotron emission, free-free emission has a relatively flat spectral index, scaling as $S_\nu \propto \nu^{-0.1}$. Globally, free-free emission begins to dominate the total radio emission in normal star-forming galaxies at $\gtrsim$30 GHz (e.g., Condon 1992; Murphy et al. 2012), exactly the frequency range for which the ngVLA will be delivering an order of magnitude improvement compared to any current or planned facility.

- **Thermal Dust Emission:** At frequencies $\gtrsim$100 GHz, (cold) thermal dust emission on the Rayleigh-Jeans portion of the galaxy far-infrared/sub-millimeter spectral energy distribution can begin to take over as the dominant emission component for regions within normal star-forming galaxies. This in turn provides a secure handle on the cold dust content in galaxies, which dominates the total dust mass. For a fixed gas-to-dust ratio, the total dust mass can be used to infer a total ISM mass (e.g., Heiles et al. 1988; Dame et al. 2001; Scoville et al. 2016). Given the large instantaneous bandwidth offered by the ngVLA, approximately an order of magnitude increase in mapping speed at 100 GHz compared to ALMA (Carilli 2017), such observations will simultaneously provide access to the $J = 1 \rightarrow 0$ line of CO revealing the molecular gas fraction for entire disks of nearby galaxies. Alternatively, combining HI observations (also obtained with the ngVLA) with $J = 1 \rightarrow 0$ CO maps, one can instead use the thermal dust emission to measure the spatially varying gas-to-dust ratio directly.

- **Anomalous Microwave Emission (AME):** In addition to the standard Galactic foreground components described above, an unknown component has been found to dominate over these at microwave frequencies between ~ 10 – 90 GHz, and is seemingly correlated with 100 μm thermal dust emission. Cosmic microwave background (CMB) experiments were the first to discover the presence of AME

(Kogut et al. 1996; Leitch et al. 1997), whose origin still remains unknown (see Dickinson et al. 2018, for a review). Its presence as a foreground is problematic as the precise characterization and separation of foregrounds remains a major challenge for current and future CMB experiments (e.g., BICEP2/Keck Collaboration et al. 2015; Planck Collaboration et al. 2016). At present, the most widely accepted explanation for AME is the spinning dust model (Erickson 1957; Draine & Lazarian 1998; Planck Collaboration et al. 2011; Hensley & Draine 2017) in which rapidly rotating very small grains, having a nonzero electric dipole moment, produce the observed microwave emission. The increased sensitivity and mapping speed of the ngVLA will allow for an unprecedented investigation into the origin and prominence of this emission component both within our own galaxy and others, ultimately helping to improve upon the precision of future CMB experiments.

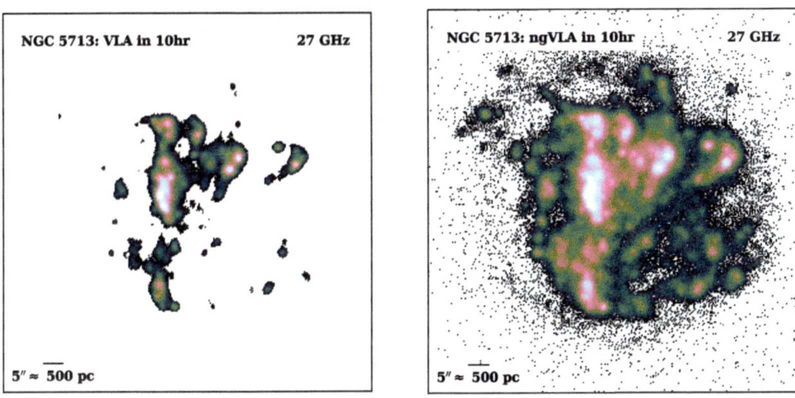

Figure 2. Both panels show a model 27 GHz free-free emission image of NGC 5713 ($d_L \approx 21.4$ Mpc, SFR $\approx 4\,M_\odot\,\mathrm{yr}^{-1}$) based on an existing H$\alpha$ narrow band image at its native ($\approx 1''$) angular resolution taken from SINGS (Kennicutt et al. 2003). The left and right panels indicate the emission that would be detected at the 3σ level after a 10 hr on-source integration time using the current VLA in C-configuration ($1\sigma \approx 1.5\,\mu\mathrm{Jy\,bm}^{-1}$) and the ngVLA with a $1''$ sculpted beam ($1\sigma \approx 0.17\,\mu\mathrm{Jy\,bm}^{-1}$). In a relatively modest integration, the ngVLA is able to recover a significant fraction of star formation activity that is completely missed by the VLA.

3. Robust Mapping of Star Formation within Nearby Galaxies on a Range of Physical Scales

For a proper decomposition of the radio continuum emission into its component parts, one needs to have spectral coverage at frequencies low enough (i.e., < 10 GHz) to be dominated by the non-thermal, steep spectrum component, having a spectral index of -0.83 (Condon 1992; Niklas et al. 1997; Murphy et al. 2011) and at frequencies high enough (i.e., > 50 GHz) where the emission becomes completely dominated by thermal emission, having a spectral index of -0.1. Given the potential for a significant contribution from AME (e.g., Murphy et al. 2010, 2018; Scaife et al. 2010; Hensley et al. 2015), peaking at frequencies $\sim$ 20 – 40 GHz, coarse coverage spanning that spectral region is critical to account for such a feature. To date, the shape of the AME feature is largely unconstrained, and has not been carefully measured in the ISM of extragalactic sources. This is largely due to the insensitivity of current facilities to conduct a proper search for AME in nearby galaxies and map the feature with enough frequency resolution to provide useful constraints on its shape.

Broadband imaging spanning the full ngVLA frequency range of 1.2 – 116 GHz will therefore be extremely powerful to properly decompose radio continuum emission from galaxies into its constituent parts. Additionally having frequency coverage below 8 GHz provides sensitivity to free-free absorption, which is common in nearby luminous infrared galaxies (e.g., Condon et al. 1991; Clemens et al. 2010; Barcos-Muñoz et al. 2015). For individual (ultra compact) HII regions, the turnover frequency can be as high as $\approx$20 GHz (Murphy et al. 2010). A fundamental goal of the ngVLA will be to produce star formation maps for a large, heterogenous samples of nearby galaxies at $\approx 1''$ resolution. This will deliver Hα-like images that optical astronomers have relied on heavily for decades without having the additional complications of extinction and contamination by nearby [NII] emission, which make such images challenging to interpret.

By achieving arcsecond-like resolution that is commensurate with ground-based optical facilities, the ngVLA will be able to probe $\approx$100 pc scales out to the distance of Virgo (the nearest massive cluster at $d \approx 16.6$ Mpc), which are the typical sizes of giant molecular clouds (GMCs) and giant HII regions. At an rms sensitivity of 0.15 μJy bm^{-1} at 27 GHz, such radio maps will reach a sensitivity espressed in terms of star formation rate (SFR) density of $\approx 0.005\, M_{\odot}\,\mathrm{yr}^{-1}\mathrm{kpc}^{-2}$, matching the sensitivity of extremely deep Hα images such as those included in the Local Volume Legacy survey (Kennicutt et al. 2008). An example of this is illustrated in Figure 2 where an existing Hα narrow band image taken from SINGS (Kennicutt et al. 2003) was used to create a model 27 GHz free-free emission map at $1''$ resolution for the nearby star-forming galaxy NGC 5713. With a 10 hr integration the ngVLA will be able to map the entire disk of NGC 5713 down to an rms of $\approx$0.15 μJy bm^{-1} ($\approx$35 mK). A comparison of what can currently be delivered with the VLA for the same integration time is also shown, indicating that only the brightest star-forming peaks are able to be detected. To make a map to the same depth using the current VLA would take $\gtrsim$800 hr! This is the same amount of time it would take to roughly survey $\gtrsim$80 galaxies with the ngVLA.

Using the same data, but applying a different imaging weighing scheme to create finer resolution maps (i.e., $0\farcs1$, or even higher for brighter systems), similar multi-frequency radio continuum analyses can be performed for discrete HII regions and supernova remnants (SNRs) to complement high-resolution, space-based optical/NIR observations (e.g., *HST*, *JWST*, WFIRST, etc.). At an angular resolution of $0\farcs1$, the data would sample $\approx$10 pc scales in galaxies out to the distance of Virgo to resolve and characterize (e.g., size, spectral shape, density, etc.) discrete HII regions and SNRs with a sensitivity to diffuse free-free emission corresponding to a SFR density of $\approx 0.5\, M_{\odot}\,\mathrm{yr}^{-1}\mathrm{kpc}^{-2}$.

This is a transformational step for studies of star formation in the local universe covering a large, heterogeneous set of astrophysical conditions. This statement is independent of the fact that with such observations using the ngVLA, having its wide-bandwidth, a number of RRL's will come for free (e.g., see Balser et al. 2018, in this volume p. 431). The detection of such lines (individually or through stacking), coupled with the observed continuum emission, can be used to quantify physical conditions for the HII regions such as electron temperature. It is without question that the ngVLA will make radio observations a critical component for investigations carried out by the entire astronomical community studying star formation and the ISM of nearby galaxies.

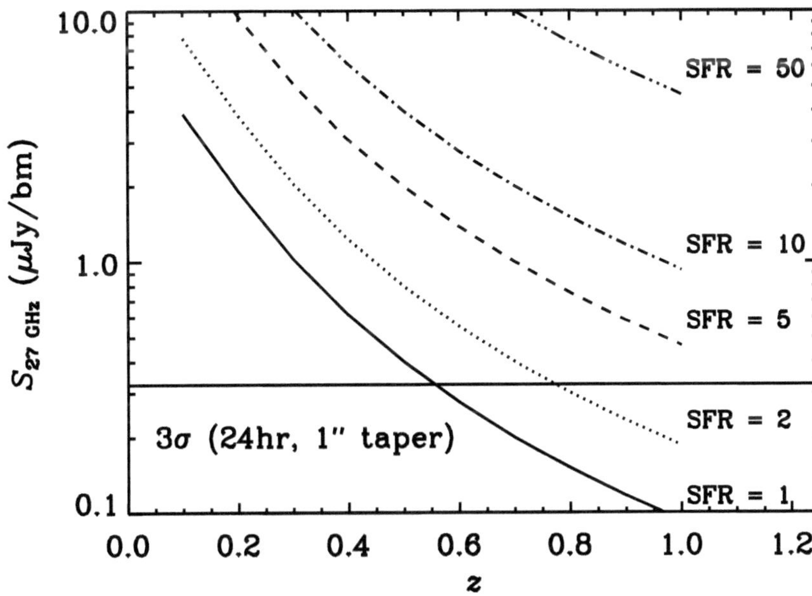

Figure 3. Telescope sensitivity in μJy bm^{-1} at 27 GHz plotted against redshift indicating the expected brightness of a 4 kpc disk galaxy forming stars at a rate of 1, 2, 5, 10, and 50 $M_\odot$ yr^{-1}. Given the current sensitivity specifications, by tapering to a 1$''$ synthesized beam, the ngVLA will have enough brightness temperature sensitivity to resolve a Milky Way like galaxy forming stars at a rate of a $\sim 3\,M_\odot$ yr^{-1} out to $z \sim 1$ after a 24 hr integration. Using the existing VLA, the detection of such a galaxy would take $\gtrsim$2000 hr!

4. A Robust Star Formation Indicator at All Redshifts

While deep field surveys aimed at studying the cosmic star formation activity at high redshifts is the focus of another chapter in this volume (i.e., Barger et al. 2018, p. 551), we briefly describe what the ngVLA can do for studies of star formation out to moderate redshifts (i.e., $z \lesssim 1$). With the increased collecting area and bandwidth of the ngVLA, one would be able to measure and resolve rest-frame 60 GHz emission from a Milky Way like galaxy out to $z \sim 1$ forming stars at rates of $\sim 3\,M_\odot$ yr^{-1} after a 24 hr integration. This is illustrated in Figure 3 which shows the observed 27 GHz brightness of 4 kpc diameter disk galaxies with a range of SFRs. We note that a choice of 4 kpc for a typical disk diameter may be slightly conservative, given that the 10 GHz sizes found for μJy sources at $z \sim 1$ in deep VLA observations of GOODS-N are reported to be $\sim$ 1.2 kpc (Murphy et al. 2017). The $3\,\sigma$ 27 GHz rms of the ngVLA tapered to a 1″ synthesized beam is based on the updated sensitivities given in (Selina & Murphy 2017)[1]. Accordingly, such observations provide highly robust extinction-free measurements of SFRs for comparison with other optical/UV diagnostics to better understand how extinction on both galactic scales and within individual star-forming regions, evolves with redshift. By coupling these higher frequency observations with those at lower radio frequencies one can accurately measure radio spectral indices as a function of redshift to better characterize thermal and non-thermal energetics independently. This therefore enables one to determine if these physically distinct components remain in rough equilibrium with one another or if there are changes which in turn affects the global star formation activity as a function of lookback time.

5. Uniqueness of ngVLA Capabilities

While the VLA has frequency coverage up to 50 GHz, it lacks the higher frequencies ($\gtrsim$ 50 GHz) necessary to measure the location where the radio spectrum is completely dominated by free-free emission and/or properly constrain the spectral shape of AME. This is required to robustly account for the energetic contributions of each emission component to the total radio continuum emission of star-forming galaxies. Furthermore, at the top end of the ngVLA frequency range, cold thermal dust emission associated with the molecular ISM may also be detectable at low surface brightness levels, which can be characterized by a rising spectrum. Perhaps most importantly, the lack of sensitivity by the VLA and other telescopes (e.g., ALMA, NOEMA, etc.) makes such continuum mapping for entire galaxies, rather than just the brightest H II regions within them, impossible, requiring the factor of $\gtrsim 10\times$ improvement in sensitivity afforded by the ngVLA. It is only with the ngVLA that the full potential of the radio continuum spectrum can be used as a tool to properly constrain the various energetic processes that power its emission for a broad range of heterogeneous conditions in galaxies.

6. Synergies at Other Wavelengths

Nearby galaxies provide our only laboratory for understanding the detailed physics of star formation and AGN activity across much larger ranges of physical conditions than offered by our own Milky Way. They are the workhorses for testing and applying physical models of star formation and their associated feedback processes that are used in galaxy evolution models to explain statistical observations for large populations of galaxies at high redshifts, for which it is impossible to conduct the detailed astrophysical experiments described here. Consequently, the transformational set of observations enabled by the ngVLA, and discussed in this chapter, will illuminate the relation between thermal and non-thermal energetic processes in the ISM and be highly synergistic with observations from shorter wavelength ground- and space-based telescopes that have access to large numbers of other diagnostic (line and continuum imaging) that may be difficult to interpret due to extinction at both low and high redshift. The same can be said for synergy with longer wavelength, far-infrared telescopes that will provide access to dust continuum and fine structure line emission that can be used to

[1] `http://ngvla.nrao.edu/page/refdesign`

characterize the cold/warm neutral phase of the ISM, but lack the angular resolution necessary to study discrete ($\gtrsim$ 10 pc) star-forming regions in large samples of galaxies.

References

Barcos-Muñoz, L., Leroy, A. K., Evans, A. S., et al. 2015, ApJ, 799, 10
Barger, A., Kohno, K., Murphy, E. J., et al., 2018, ASP Monograph 7, Science with a Next-Generation VLA, ed. E. J. Murphy (San Francisco, CA: ASP), 551
Balser, D. S., Anderson, L. D., Bania, T. M., et al. 2018,, ASP Monograph 7, Science with a Next-Generation VLA, ed. E. J. Murphy (San Francisco, CA: ASP), 431
BICEP2/Keck Collaboration, Planck Collaboration, Ade, P. A. R., et al. 2015, Physical Review Letters, 114, 101301
Carilli, C., 2017, ngVLA Memo #12
Clemens, M. S., Vega, O., Bressan, A., et al. 2008, A&A, 477, 95
Clemens, M. S., Scaife, A., Vega, O., & Bressan, A. 2010, MNRAS, 405, 887
Condon, J. J., Huang, Z.-P., Yin, Q. F., & Thuan, T. X. 1991, ApJ, 378, 65
Condon, J. J. 1992, ARA&A, 30, 575
Dame, T. M., Hartmann, D., & Thaddeus, P. 2001, ApJ, 547, 792
Dickinson, C., Ali-Haïmoud, Y., Barr, A., et al. 2018, NewAR, 80, 1
Draine, B. T., & Lazarian, A. 1998, ApJ, 494, L19
Erickson, W. C. 1957, ApJ, 126, 480
Heiles, C., Reach, W. T., & Koo, B.-C. 1988, ApJ, 332, 313
Hensley, B., Murphy, E. J., & Staguhn, J. 2015, MNRAS, 449, 809
Hensley, B. S., & Draine, B. T. 2017, ApJ, 836, 179
Kennicutt, Jr., R. C., Armus, L., Bendo, G., et al. 2003, PASP, 115, 928
Kennicutt, R. C., Jr., Lee, J. C., Funes, J. G., et al. 2008, ApJS, 178, 247-279
Kogut, A., Banday, A. J., Bennett, C. L., et al. 1996, ApJ, 460, 1
Leroy, A. K., Evans, A. S., Momjian, E., et al. 2011, ApJ, 739, L25
Leitch, E. M., Readhead, A. C. S., Pearson, T. J., & Myers, S. T. 1997, ApJ, 486, L23
Murphy, E. J., Helou, G., Condon, J. J., et al. 2010a, ApJ, 709, L108
Murphy, E. J., Condon, J. J., Schinnerer, E., et al. 2011, ApJ, 737, 67
Murphy, E. J., Bremseth, J., Mason, B. S., et al. 2012, ApJ, 761, 97
Murphy, E. J. 2013, ApJ, 777, 58
Murphy, E. J., Dong, D., Leroy, A. K., et al. 2015, ApJ, 813, 118
Murphy, E. J., Momjian, E., Condon, J. J., et al. 2017, ApJ, 839, 35
Murphy, E. J., Linden, S. T., Dong, D., et al. 2018, ApJ, 862, 20
Murphy, T., Cohen, M., Ekers, R. D., et al. 2010b, MNRAS, 405, 1560
Niklas, S., Klein, U., & Wielebinski, R. 1997, A&A, 322, 19
Planck Collaboration, Ade, P. A. R., Aghanim, N., et al. 2011, A&A, 536, A20
Planck Collaboration, Adam, R., Ade, P. A. R., et al. 2016, A&A, 586, A133
Scaife, A. M. M., Nikolic, B., Green, D. A., et al. 2010, MNRAS, 406, L45
Scoville, N., Sheth, K., Aussel, H., et al. 2016, ApJ, 820, 83
Selina, R. and Murphy, E. J., 2017, ngVLA Memo #17
Socrates, A., Davis, S. W., & Ramirez-Ruiz, E. 2008, ApJ, 687, 202-215
Tabatabaei, F. S., Minguez, P., Prieto, M. A., & Fernández-Ontiveros, J. A. 2018, Nature Astronomy, 2, 83

Science with a Next-Generation Very Large Array
ASP Conference Series, Monograph 7
Eric J. Murphy, ed.

Radio Recombination Lines from H II Regions

Dana S. Balser,[1] L. D. Anderson,[2] T. M. Bania,[3] John M. Dickey,[4] D. Anish Roshi,[1] Trey V. Wenger,[5] and T. L. Wilson[6]

[1]*National Radio Astronomy Observatory, Charlottesville, VA, USA;* `dbalser@nrao.edu`, `aroshi@nrao.edu`

[2]*West Virginia University, Morgantown, WV, USA;* `loren.anderson@mail.wvu.edu`

[3]*Boston University, Boston, MA, USA;* `bania@bu.edu`

[4]*University of Tasmania, Hobart, TAS, Australia;* `John.Dickey@utas.edu.au`

[5]*University of Virginia, Charlottesville, VA, USA;* `tvw2pu@virginia.edu`

[6]*Max-Planck-Institut für Radioastronomie, Bonn, Germany;* `twilson@mpifr-bonn.mpg.de`

Abstract. The ngVLA will create a Galaxy-wide, volume-limited sample of H II regions; solve some long standing problems in the physics of H II regions; and provide an extinction-free star formation tracer in nearby galaxies.

1. Introduction

Here we discuss three scientific areas that could be revolutionized by ngVLA observations of radio recombination line (RRL) and continuum emission from H II regions: Galactic structure, H II region physics, and star formation in nearby galaxies.

H II regions are zones of ionized gas surrounding recently-formed high-mass OB-type stars. They are the archtypical tracer of spiral structure. Since they are young (< 10 Myr), their chemical abundances represent Galactic abundances today, and reveal the effects of billions of years of Galactic chemical evolution. H II regions are often part of a high-mass star formation complex that consists of a thin neutral photodissociation region (PDR) between a hot ($\sim 10^4$ K) H II region and dense, cool ($\sim 10 - 100$ K) molecular material from which the stars were formed.

H II regions are the brightest objects in the Milky Way at infrared and mm wavelengths and can be detected across the Galaxy. They are characterized in the infrared by $\sim 20\,\mu$m emission from stochastically heated small dust grains mixed with the ionized gas surrounded by $\sim 10\,\mu$m emission from polycyclic aromatic hydrocarbon (PAH) molecules located in the PDR (Anderson et al. 2014). The ionized gas produces bright, free-free (thermal) radio continuum emission that is coincident with the $\sim 20\,\mu$m infrared emission. When free electrons recombine and cascade to the ground state they produce recombination line emission which is found at radio wavelengths for the higher principal quantum number levels ($n > 50$). Currently, only emission from hydrogen and helium is detected in HII regions at radio wavelengths, but heavier elements (e.g., carbon) are detected in the denser, cooler PDRs (Wenger et al. 2013).

RRLs are an excellent, extinction-free diagnostic of HII regions. Unlike radio continuum emission, they uniquely probe the thermal emission and therefore unambiguously identify an HII region. The RRL parameters alone provide information about HII region kinematics (Anderson et al. 2012), turbulent motions (Roshi 2007), and He/H abundance ratios (Balser 2006). Measuring the RRL and continuum emission together allows the calculation of many physical HII region properties: electron (thermal) temperature, rms electron density, the H-ionizing stellar luminosity (number of OB-type stars), etc. (Gordon & Sorochenko 2009).

The main limitation of using RRLs to probe Galactic structure and star formation compared to similar tracers at optical or infrared wavelengths is that they have much lower intensities. This is particularly an issue in external galaxies. For example, RRL emission has been primarily detected from nearby starburst galaxies powered by massive young star clusters. Recently, however, RRL emission was detected in the normal galaxies M51 and NGC 628 at centimeter wavelengths (Luisi et al. 2018). Moreover, for some radio facilities the spatial resolution is not sufficient to resolve individual HII regions, which limits the analysis. The ngVLA will provide both the sensitivity and spatial resolution to remedy these problems and place RRLs as an HII region diagnostic on par with, for example, Hα.

2. Galactic Structure

The ngVLA is necessary to create a Galaxy-wide, volume-limited sample of HII regions. Since radio waves penetrate the dust located within the Galactic disk, an HII region map of the entire Milky Way is possible at these wavelengths. This is needed to understand the global properties of the Milky Way, and to compare its star formation to that of external galaxies. There is strong evidence that we have only discovered about 25% of the HII regions in the Milky Way, which impedes our understanding of Galactic structure and Galactic chemical evolution (Anderson et al. 2014).

The implications for Galactic structure studies using RRLs were recognized shortly after their discovery in 1965 and prompted large-scale HII region surveys (Wilson et al. 1970; Reifenstein et al. 1970). RRLs measure the systemic velocity of an HII region and therefore their motion around the center of the Galaxy. Using a rotation curve model together with the velocity yields the Heliocentric distance of the HII region. Distances computed in this way are known as “kinematic distances.” Improvements in receiver performance resulted in follow-up surveys in the 1980s that increased the number of sources (Wink et al. 1983; Caswell & Haynes 1987; Lockman 1989). These RRL surveys, however, were limited in sensitivity and had inaccurate distance determinations. A third generation of Galactic HII region discovery surveys (HRDS) is almost complete using primarily the Green Bank Telescope (GBT) in the Northern sky (Bania et al. 2010) and the Australia Telescope Compact Array (ATCA) in the Southern sky (Brown et al. 2017). The HRDS has doubled the number of known HII regions. It has the sensitivity to detect all nebulae ionized by an O-type star out to a distance of 20 kpc from the Sun. This has primarily been achieved by the flexibility of spectrometers that can simultaneously observe multiple RRLs which can be averaged together to increase the signal-to-noise ratio (SNR). Maser parallax distances together with better models of Galactic rotation have improved our understanding of kinematic distance determinations (Reid et al. 2014; Wenger et al. 2018).

These new surveys have made several discoveries about the morphological and chemical structure of the Milky Way. *Cold H I gas and molecular clouds are often used to define spiral structure, but such tracers are not always associated with high-mass star formation* (Anderson et al. 2009). Therefore, Koo et al. (2017) used HI surveys together with the HRDS to characterize the spiral structure in the outer Galaxy and found that a four-arm spiral model with pitch angle of 12° was a good fit to the data. Yet the number of HII regions sampling the outer-most spiral arms is limited. For example, deep RRL observations with the GBT were only able to detect a handful of HII regions in the outer Scutum-Centaurus arm (Armentrout et al. 2017). [N.B., the RRL spectral sensitivity of the Jansky Very Large Array (VLA) is comparable to the GBT.]

For HII regions in thermal equilibrium the nebular electron temperature is determined by the abundance of the coolants (O, N, and other heavy elements). Balser et al. (2015) used this to derive electron temperatures for a sub-set of the HRDS HII regions and used them to derive the nebular [O/H] abundances. Figure 1 shows the resulting metallicity map and reveals the chemical structure of the Milky Way disk. The radial metallicity gradient is obvious from this map and has been detected before, but there is also azimuthal structure that was revealed for the first time. This may be due to radial mixing from the Galactic Bar whose major axis is aligned toward an azimuth of about 30°. The sample is limited, however, since only the most accurate RRL and continuum data is selected. More sources are required to better characterize this chemical structure and therefore constrain the physical mechanisms that might be in play. The ngVLA will add thousands of sources to Figure 1 (see below) and reveal any detailed metallicity structure across the Galactic disk.

The *WISE* catalog of Galactic HII regions contains about 8000 objects identified by their mid-infrared and radio properties as HII region (Anderson et al. 2014). Currently only $\sim$ 2000 sources in this catalog are known HII regions with RRL or Hα emission. Of the remaining sources $\sim$ 2000 have spatially coincident radio continuum emission and are very likely HII regions, and $\sim$ 4000 are radio quiet at the sensitivity limits of existing radio continuum surveys. There is some evidence that the radio quiet sources in the *WISE* HII region catalog are bona fide HII regions (Armentrout et al. 2018, in prep.). Because of their extreme luminosities, HII regions are essential for computing extragalactic star formation rates in combination with infrared and other radio data. The HRDS, when complete, will be the first survey to detect all HII regions created by at least one O-type star, but there will remain thousands of likely HII regions whose RRL emission cannot be detected with extant facilities.

The ngVLA will have the sensitivity to create a Galaxy-wide, volume-limited sample of HII regions. The optimal frequency range is 4-12 GHz (ngVLA band 2), where classical HII regions are optically thin, in LTE, and many adjacent transitions can be averaged to increase the SNR. The HRDS spectral sensitivity at these frequencies is $\sim$ 1 mJy. To detect all sources in the *WISE* Galactic HII region catalog requires a spectral sensitivity of $\sim 3\,\mu$Jy beam^{-1} (Anderson et al. 2014). This assumes a radio continuum flux density of 0.2 mJy at 21 cm, a SNR of 5, an optically thin nebula, and a line-to-continuum ratio of 0.1 at 8 GHz (e.g., Brown et al. 2017). Using the ngVLA spectral

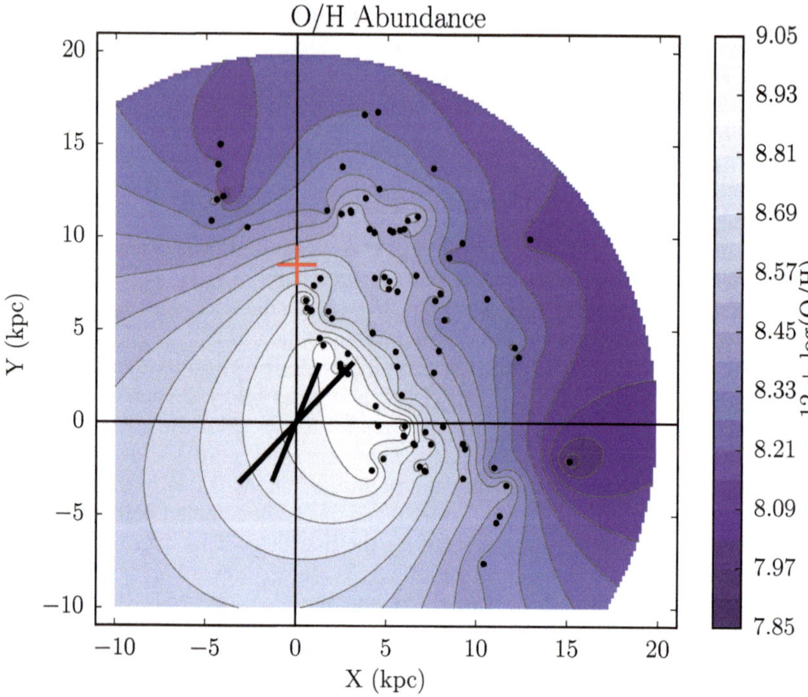

Figure 1. Face-on Galactic [O/H] abundance ratio image using Kriging to interpolate between the discrete H II region values taken from Balser et al. (2015). The points indicate the location of the discrete H II regions. The solid lines intersect at the Galactic Center. The red lines mark the location of the Sun. The thick lines correspond to the central locii of the putative "short" and "long" bars.

sensitivity from Selina & Murphy (2017) for a 50 mas beam[1], the weakest sources in the *WISE* catalog would require 15 hr of integration time. The ngVLA sensitivity will be less for sources that are resolved by the longest baselines. At a distance of 20 kpc an H II region with a physical size of 1 pc has an angular size of 10″. Since the ngVLA core of antennas will contain a large fraction of the collecting area the sensitivity should not be significantly reduced. In principle, the GBT could fill in the missing flux density. Phased array feeds in the central ngVLA core would increase the efficiency. Such a survey would complement MeerGAL, a high frequency (~ 10 GHz) survey of the Galactic plane with MeerKat in the Southern sky.

A Galactic H II region pointed survey toward *WISE* H II region candidates may be possible with the *James Webb Space Telescope* (*JWST*), which will be sensitive to wavelengths from 0.6 – 28 μm, using collisonally excited lines (CELs) in the mid-infrared. RRLs provide a more direct measure of the ionizing flux, but CELs probe the hardness of the radiation field. Most optical and near-infrared studies, however, trace the kine-

[1]The ngVLA spectral rms is 37.3 μJy beam^{-1} for a beam size of 50 mas at 8 GHz with an integration time of 1 hr and a FHWM line width of 10 km s^{-1}.

matic and chemical structure of stars. Recent surveys such as APOGEE, GAIA-ESO, LAMOS, and RAVE are sampling hundreds of thousands of stars and will soon have accurate distances from *Gaia*. But these surveys do not probe the gas component of the Milky Way or very far into the disk due to extinction by dust. Understanding the morphological and chemical structure of both the stars **and** the gas is critical to constrain Galactic formation and evolution models (e.g., Kubryk et al. 2013).

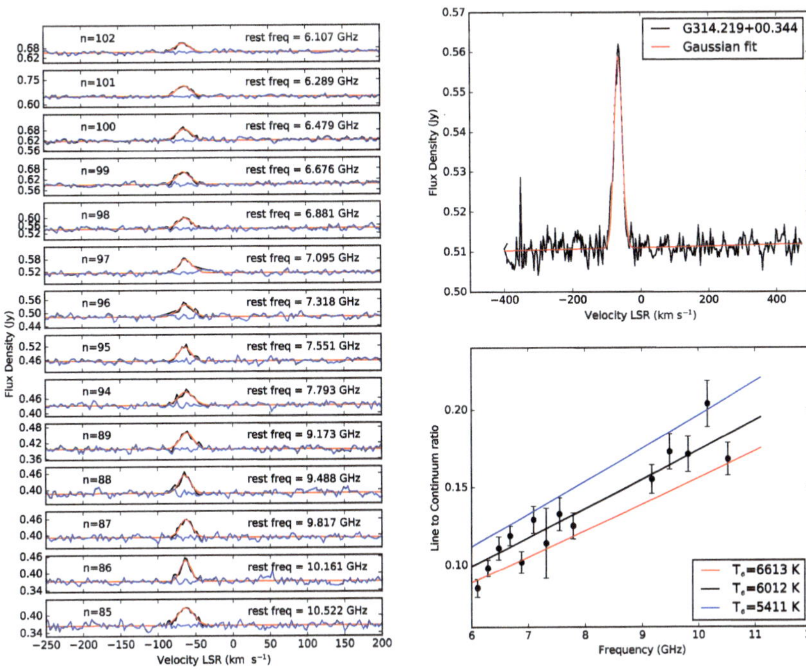

Figure 2. RRL spectra for G314.219+00.344 using the ATCA taken from Brown et al. (2017). Individual spectra (left) and an averaged spectrum (upper-right) are shown where the red curves are Gaussian fits to the data. The lower-right panel plots the line-to-continuum ratio as a function of frequency.

3. HII Region Physics

The ngVLA could solve some long standing problems in the physics of HII regions including, for example, constraining the collision rates from free electrons, characterizing temperature fluctuations, and understanding the role of magnetic fields within HII regions.

Recombination lines from hydrogen and helium at radio wavelengths are strongly affected by collisions from free electrons and therefore accurate collision rates are important for predicting the line opacity. Under some physical conditions the opacities can vary by as much as a factor of five depending on the rates used (F. Guzmán, private communication; also see Guzmán et al. 2016). Accurate measurements of the RRL intensity ratio over a frequency range wide enough for the opacity to vary significantly would provide vital observational constraints for the collision rates. For example, the flexible spectrometer on the ATCA allows RRLs to be observed simultaneously from 6-12 GHz so that we can begin to constrain the atomic physics (see Figure 2). We are working on a detailed strategy to detect each RRL transition with a high SNR over a variety of physical conditions. This will require the sensitivity of the ngVLA.

One of the key problems in the physics of ionized nebulae is that abundances derived from recombination lines (RLs) are significantly larger, by as much as a factor of two, than those derived from CELs (Garcá-Rojas & Esteban 2007). These differences can be explained by temperature fluctuations in the ionized gas. The concept of temperature fluctuations was developed by Peimbert (1967) to explain how electron temperatures derived from RLs were different than those derived from CELs. Temperatures are also expected to increase at the HII region boundary because of photon hardening. That is, photons with energies of 20 eV have a smaller absorption cross section so they can reach the nebular boundary where they are absorbed and thus raise the electron temperature. Wilson et al. (2015) suggested that optical observations produced different electron temperature profiles in the Orion nebula compared with RRLs due to the effects of dust. The radio observations were performed with the GBT instead of the VLA because Orion is extended, but to probe small-scale structure requires better spatial resolution. The smallest scales over which the temperatures vary is not known. Therefore to measure both the small and large-scale temperature variations in HII regions requires calculating the radio line-to-continuum ratio over a wide range of spatial scales with high sensitivity. This may be possible by combing data from the VLA and GBT, but the systematic errors using this method may be too large. The internal calibration with the ngVLA would provide a more accurate measure of the electron temperature over a wider range of spatial scales with significantly better sensitivity.

Magnetic fields may play an important role in the formation of stars and the Zeeman effect is the only method that directly measures the *in-situ* magnetic field strength, but it is a challenging experiment. Most of the work thus far has probed molecular (e.g., CN) or neutral (e.g, HI) components of high-mass star formation regions (Crutcher et al. 2010). To our knowledge there has never been a Zeeman detection from RRLs (e.g., Troland & Heiles 1977), so an ngVLA detection would be the first. Such a measurement is difficult since the hydrogen RRL width ($\sim$ 25 km s^{-1}) is broader than the Zeeman splitting and therefore a very high SNR is required. High spatial resolution is necessary since magnetic field variations have been detected and will be averaged over a large beam. Since many bright HII regions are extended, however, the larger spatial scales, or shorter baselines for an interferometer, must be sampled.

We estimate the sensitivity of the ngVLA for Zeeman measurements in HII regions by assuming a hydrogen RRL intensity of 50 mJy beam^{-1} and a line width of 30 km s^{-1} observed with a resolution of $1''$ at 8 GHz. Using the ngVLA spectral sensitivity from Selina & Murphy (2017) for a 1 arcsec beam and Equation 2 from Troland & Heiles (1982), we estimate a $3\,\sigma$ limiting line-of-sight magnetic field strength of 295 μG in 1 hr of integration time. This assumes we are able to average 20 RRLs with six spectral channels across the FWHM line width.

4. Star Formation in Nearby Galaxies

Using RRLs the ngVLA will provide an extinction-free star formation tracer in nearby galaxies that is comparable to Hα for the sensitivity and spatial resolution afforded by existing optical instruments. That is particularly important in extreme environments with clusters of massive stars since star formation may proceed differently in these objects. These environments are, moreover, often opaque at ultraviolet, optical, or near infrared wavelengths due to dust.

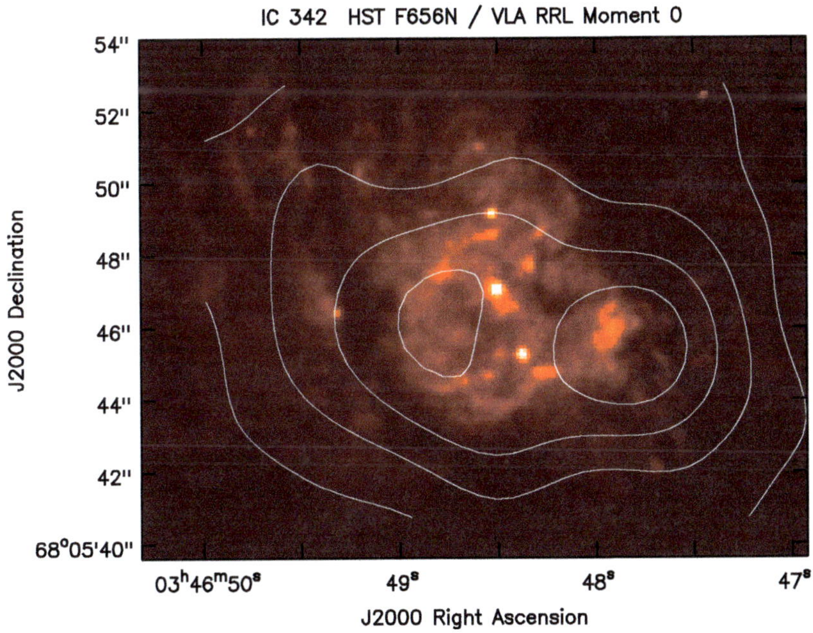

Figure 3. Star formation diagnostics in the IC 342 nuclear region taken from Balser et al. (2017). Shown in color is the *HST* Hα (F656N) image together with the VLA Ka-band RRL-integrated intensity map in contours. The contour levels are 0.2, 0.4, 0.6, and 0.8 times the peak intensity (0.1324 Jy beam^{-1} km s^{-1}). The VLA synthesized beam size is $4.''5 \times 4.''5$.

Radio continuum measurements are excellent probes of these regions since they can penetrate the gas and dust that surrounds the young star clusters. Observations at multiple frequencies measure the thermal emission associated with HII regions and non-thermal emission related to supernova remnants (e.g., Turner & Ho 1983; Johnson et al. 2004; Murphy et al. 2018, this volume p. 421). Radio recombination lines are fainter than radio (e.g., Turner & Ho 1983; Johnson et al. 2004; Murphy et al. 2018). continuum emission, but they directly measure the thermal emission and yield important information about the dynamics of these star clusters. Anantharamaiah et al. (2000) combined both RRL and continuum data at several different frequencies to constrain models of these young massive star clusters. Technical improvements provided by the VLA increased the sensitivity of such work by almost an order of magnitude (Kepley et al. 2011), and enabled RRL studies of less massive star clusters. For example, Balser et al. (2017) detected RRL emission in IC 342 for the first time and revealed thermal emission to the east and west of the nuclear star cluster that was associated with giant molecular clouds (see Figure 3). The best fit model is a collection of many hundreds of compact (~ 0.1 pc) HII regions ionized by an equivalent of ~ 2000 O6 stars. Even with the increased sensitivity of the VLA these observations are challenging and we do not detect RRL emission in gas located beyond the most central regions. Better sensitivity over many spatial scales is required. Also, Anantharamaiah et al. (2000) showed that because of free-free opacity effects the RRL are brighter at higher frequencies and therefore it is important to observe RRLs over a large range of frequencies (e.g., 1-100 GHz).

The ngVLA has the required sensitivity spanning a wide range of spatial and frequency scales. Ideally, we want to resolve individual HII regions but also have the sensitivity to detect extended emission (e.g., the diffuse ionized gas). So we need to probe scales from $\sim 1 - 1000$ pc in galaxies with distances between $1 - 20$ Mpc, or angular scales $0\farcs01 - 200''$. To estimate the required spectral sensitivity we use a model HII region. We assume a homogeneous sphere ionized by an O5 star at a distance of 2 Mpc with an electron density and temperature of 1,000 $\mathrm{cm^{-3}}$ and 8,000 K, respectively. Using the stellar models of Martins et al. (2005) the continuum flux density at a frequency of 5 GHz is 0.1 mJy (see appendix G.1 in Balser 1995). This continuum flux density estimate is similar to the values found in compact sources in nearby galaxies (e.g., Tsai et al. 2006). Assuming a line-to-continuum ratio of 0.1 yields a RRL intensity of 10 μJy. Using the ngVLA sensitivity from Selina & Murphy (2017) for a 50 mas beam, we estimate an integration time of 35 hr assuming a SNR of 5, a spectral resolution of 5 $\mathrm{km\,s^{-1}}$, and averaging 20 RRLs. This is over two orders of magnitude more sensitive than our VLA RRL observations. For an optically thin, unresolved HII region the results are not very sensitive to the electron density, electron temperature, or frequency. Moreover, the flux density only increases by approximately a factor of two for an O3 star relative to an O6 star. The integration time can therefore be approximated as $t_{\rm int} \approx 35 \times (d/2\,\mathrm{Mpc})^4$ hr for detecting RRL emission from HII regions ionized by a single star, where D is the distance.

The RRL data probe the ionized gas and will complement observations that sample the neutral (HI; ngVLA) and molecular (CO; ALMA, ngVLA) components to characterize these star formation complexes. Detection of submillimeter H and He RRLs with ALMA trace the hardness of the radiation field, or indirectly the high-mass end of the initial mass function, in nearby massive young star clusters (e.g., see Scoville & Murchikova 2013). Similar observations with the ngVLA are possible at 3 mm assuming the ngVLA is about five times more sensitive than ALMA (A. Bolatto, private comm.). In principle the *JWST* Paα and ngVLA RRL intensity ratio could directly measure the dust extinction as a function of velocity.

5. Summary

The ngVLA will allow for sensitive observations of RRLs and impact the following areas:

- *Galactic Structure:* The ngVLA will have the sensitivity to create a Galaxy-wide, volume-limited sample of HII regions for the first time. This is needed to determine the global star formation properties of the Milky Way, and to better characterize both the morphological and chemical structure of the Galactic disk.

- *HII Region Physics:* The ngVLA could solve several long standing problems in HII region physics. These include discriminating between different models of the collision rates from free electrons that determine the opacity, characterizing temperature fluctuations which effect abundance determinations, and directly measuring the magnetic field strength within HII regions.

- *Star Formation in Nearby Galaxies:* The ngVLA will have both the necessary sensitivity and spatial resolution to probe star formation in a statistically significant sample of nearby galaxies. RRLs provide an extinction-free tracer of the physical conditions and dynamics of the ionized gas.

References

Anantharamaiah, K. R., Viallefond, F., Mohan, N. R. et al. 2000, ApJ, 537, 613
Anderson, L. D., Bania, T. M., Balser, D. S., & Rood, R. T. 2012, ApJ, 754, 62
Anderson, L. D., Bania, T. M., Balser, D. S., et al. 2014, ApJS, 212, 1
Anderson, L. D., Bania, T. M., Jackson et al. 2009, ApJS, 181, 255
Armentrout, W. P., Anderson, L. D., Balser, D. S. et al. 2017, ApJ, 841, 121
Balser, D. S. 1995, PhD thesis
Balser, D. S. 2006, AJ, 132, 2326
Balser, D. S., Roshi, D. A., Jeyakumar, S. et al. 2016, ApJ, 816, 22
Balser, D. S., Wenger, T. V., Anderson, L. D., & Bania, T. M. 2015, ApJ, 806, 199
Balser, D. S., Wenger, T. V., Goss, W. M., Johnson, K. E., & Kepley, A. A., 2017, ApJ, 844, 73
Bania, T. M., Anderson, L. D., Balser, D. S., & Rood, R. T. 2010, ApJ, 718, L106
Brown, C., Jordan, C., Dickey, J. M. 2017, AJ, 154, 23
Caswell, J. L., & Haynes, R. F. 1987, A&A, 171, 261
Crutcher, R. M., Wandelt, B., Heiles, C. et al. 2010, ApJ, 725, 466
Garcá-Rojas, J., & Esteban, C. 2007, ApJ, 670, 457.
Gordon, M. A., & Sorochenko, R. L. 2009, ASSL, Vol. 282, Radio Recombination Lines
Guzmán, F., Badnell, N. R., Williamns, R. J. R. et al. 2016, MNRAS, 459, 3498
Johnson, K. E., Indebetouw, R., Watson, C., & Kobulnicky, H. A. 2004, AJ, 128, 610
Kepley, A. A., Chomiuk, L., Johnson, K. E. et al. 2011, ApJ, 739, L24
Koo, B.-C., Park, G., Kim, W.-T. et al. 2017, PASP, 129, 94102
Kubryk, M., Prantzos, N., & Athanassoula, E. 2013, MNRAS, 436, 1479
Lockman, F. J. 1989, ApJS, 71, 469
Luisi, M., Anderson, L. D., Bania, T. M. et al. 2018, PASP, in press (ArXiv:1805.11460)
Martins, F., Schaerer, D., & Hillier, D. J. 2005, A&A, 436, 1049
Murphy, E. J., Condon, J. J., Alberdi, A., et al. 2018, 2018, ASP Monograph 7, Science with a Next-Generation Very Large Array, ed. E. J. Murphy (San Francisco, CA: ASP), 421
Peimbert, M. 1967, ApJ, 150, 825
Reid, M. J., Menten, K. M., Brunthaler, A., et al. 2014, ApJ, 783, 130
Reifenstein, E. C., III, Wilson, T. L., Burke, B. F. et al. 1970, A&A, 4, 357
Roshi, D. A. 2007, ApJ, 658, L41
Scoville, N., & Murchikova, L. 2013, ApJ, 779, 75
Selina, R., & Murphy, E. J. 2017, ngVLA Memo 17.
Troland, T. H. & Heiles, C. 1977, ApJ, 214, 703
Troland, T. H. & Heiles, C. 1982, ApJ, 252, 179
Tsai, C.-W., Turner, J. L., Beck, S. C. et al. 2006, AJ, 132, 2383
Turner, J.L. & Ho, P.T.P. 1983, ApJ, 268, L79
Wenger, T. V., Balser, D. S., Anderson, L. D. & Bania, T. M. 2018, ApJ, 856, 52
Wenger, T. V., Bania, T. M., Balser, D. S., & Anderson, L. D. 2013, ApJ, 764, 34
Wilson, T. L., Bania, T. M., & Balser, D. S. 2015, ApJ, 812, 45
Wilson, T. L., Mezger, P. G., Gardner, F. F., & Milne, D. K. 1970, A&A, 6, 364
Wink, J. E., Wilson, T. L., & Bieging, J. H. 1983, A&A, 127, 211

Science with a Next-Generation Very Large Array
ASP Conference Series, Monograph 7
Eric J. Murphy, ed.

How Do Cold Gas Outflows Shape Galaxies?

Alberto D. Bolatto,[1] Lee Armus,[2] Adam K. Leroy,[3] Sylvain Veilleux,[1] Fabian Walter,[4] and Richard Mushotzky[1]

[1]*University of Maryland, Department of Astronomy, College Park, MD 20742, USA;* `bolatto@astro.umd.edu`, `veilleux@astro.umd.edu`, `richard@astro.umd.edu`

[2]*Infrared Processing and Analysis Center, California Institute of Technology, 1200 E. California Boulevard, Pasadena, CA 91125, USA;* `lee@ipac.caltech.edu`

[3]*Ohio State University, Department of Astronomy, Columbus, OH 43210, USA;* `leroy.42@osu.edu`

[4]*Max Planck Institute for Astronomy, Heidelberg, Germany;* `walter@mpia.de`

Abstract. The ngVLA will obtain breakthrough observations of the cold phases (molecular and atomic) of galactic outflows. These observations will be key to study the driving mechanisms of cold winds, the fate of the gas, the mass-loss rates, the conditions that trigger cold outflows, the fraction of gas that escapes galaxies, the effects of winds on suppressing star formation and slowing black hole growth, the statistical properties of outflows, and the relation between large scale outflows and ultra-fast outflows and broad absorption line outflows. Finally, we highlight the ngVLA complementarity with existing and planned facilities, such as ALMA/SKA, OST/SPICA, and Lynx/ATHENA.

1. The Importance of Galaxy-scale Winds

It has become increasingly clear over the last two decades that it is impossible to understand how galaxies form and evolve through cosmic time without a deeper understanding of how star formation and black hole accretion affect the growth of galaxies and the state of their gas reservoirs. The failure of both large-scale cosmological simulations and detailed physical simulations of individual galaxies to reproduce the observed characteristics of the galaxy population has driven home the point - understanding how feedback works is probably the key open question in galaxy evolution. This is intimately linked to one of the fundamental questions identified in the *"New Worlds, New Horizons"* decadal report: how do baryons cycle in and out of galaxies?

Galactic winds are thought to shape the galaxy mass function, heat the circumgalactic medium, play a critical role in quenching star formation, and pollute the intergalactic medium with heavy elements (e.g., Veilleux et al. 2005). Galaxies are not closed systems, and winds are necessary to explain chemical evolution, as well as playing a key role in shaping the disk-halo interface. The fastest and most energetic winds arise due to feedback from active galactic nuclei (AGN) or strong starbursts, while less powerful, and more localized, galactic fountains due to clustered star formation recycle material between the disk and the halo even in less active galaxies. On smaller scales, super star clusters can also drive outflows that shape their final stellar mass and affect their environment (e.g., Herrera & Boulanger 2017).

Detailed studies of nearby systems have frequently focused on the warm or hot phases of galactic winds, which are often visible in X-rays or optical emission lines. The manifestation of these hot winds can often be quite spectacular, consisting of huge, bipolar nebulae (e.g., Strickland et al. 2004), but they typically carry very little mass $M_{Xray} \sim 10^6\,M_\odot$ (e.g., Lehnert et al. 1999). One of the important effects of winds is to shut down star formation by removal of gas, but this hot gas is very low density compared to the cold gas out of which young stars will ultimately form. But galactic winds are multi-phase phenomena (Figure 1). When present, colder phases, containing neutral atomic and dense molecular gas, can dominate the mass budget of the outflow (Walter et al. 2002; Rupke et al. 2005; Feruglio et al. 2010; Alatalo et al. 2011; Rupke & Veilleux 2013).

Until recently, observations of the cooler phases of galactic outflows have been hindered by a lack of sensitivity and/or spatial resolution to properly image the low surface brightness wind and unambiguously connect it to the processes in the disk that power the outflow. With *Herschel*, it has been possible to detect fast, dense outflows in a number of local ultra-luminous infrared and starburst galaxies (e.g., Sturm et al. 2011; Veilleux et al. 2013). In some cases the mass outflow rates are comparable to, or larger than, the star formation rates of the galaxies themselves, suggesting a significant impact on the lifetime of the active phase. However, with *Herschel* it was only possible to detect the nearest, brightest sources, and the winds were nearly always unresolved. ALMA and the IRAM Plateau de Bure Interferometer are currently making important advances in finding and imaging molecular outflows (Bolatto et al. 2013a; Combes et al. 2013; Cicone et al. 2014; Sakamoto et al. 2014; García-Burillo et al. 2014; Zschaechner et al. 2016; Veilleux et al. 2017), but the resolution and surface brightness sensitivity requirements make identifying and characterizing molecular winds challenging with present-day facilities.

2. The Need for the ngVLA

The ngVLA will have the sensitivity and frequency coverage to enable a giant leap in our understanding of the cold phases of galactic winds. The ngVLA will be able to: 1) identify molecular and neutral atomic winds using sensitive, high resolution imaging, 2) characterize their mass, outflow rate, and physical state using molecular multi-line spectroscopy and 21 cm emission, 3) diagnose their fate by imaging their extent with high surface brightness sensitivity across multiple phases and providing detailed kinematics, and 4) study their physical launching mechanisms.

Imaging galactic outflows requires the combination of high spectral sensitivity, resolution, and the ability to recover a wide range of spatial scales – all things for which the ngVLA will excel. Close to the source, typical column densities may correspond to optical extinction (A_V) of a few magnitudes, rapidly decaying with distance from the source as the outflow expands and gets diluted. The hydrogen column density corresponding to $A_V \sim 1$ is N(H) $= 1.9 \times 10^{21}\,\mathrm{cm}^{-1}$ (Bohlin et al. 1978; Rachford et al. 2009). At $\lambda = 21$ cm the ngVLA is capable of attaining an atomic hydrogen column density rms of N(H) $\sim 5.2 \times 10^{20}\,\mathrm{cm}^{-1}$ in $10\,\mathrm{km\,s}^{-1}$ in one hour, at a resolution of $1''$. Therefore assuming a linewidth of 30 km/s and an integration time of 24 hours, it can detect and image atomic outflows down to N(H) $\sim 3 \times 10^{20}\,\mathrm{cm}^{-1}$ at 5σ,

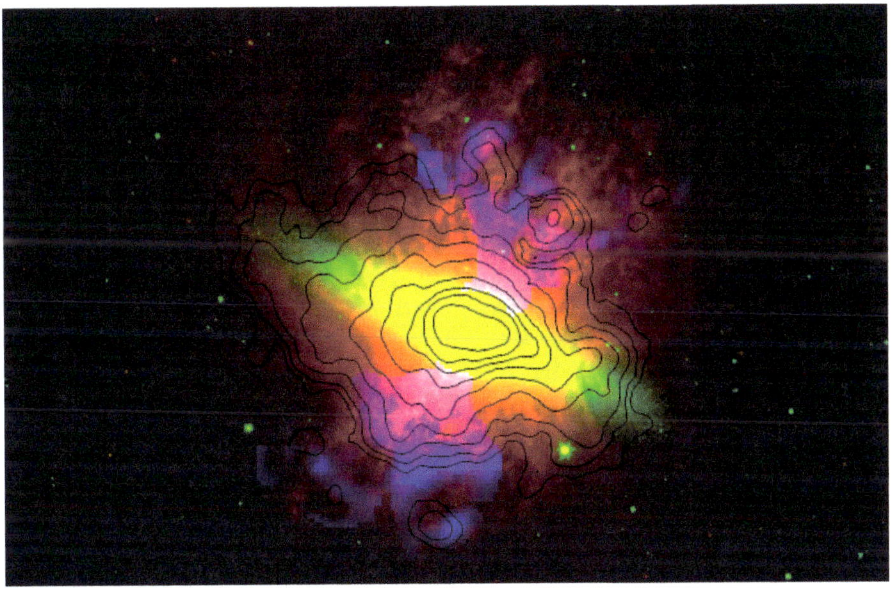

Figure 1. The cooler phases of the starburst-driven wind in the prototypical starburst M 82. The red, green, and blue colors correspond respectively to PAH emission (from IRAC 8.0 μm), the stellar component (from IRAC 3.6 μm), both from Kennicutt et al. (2003), and HI emission kinematically identified to be part of the outflow (from Leroy et al. 2015) imaged at 24″ resolution. The black contours show the CO emission (in this case CO 2 − 1) at 20″ resolution obtained by the IRAM 30m telescope. The contours correspond to both the disk and the outflow emission (also from Leroy et al. 2015).

which corresponds to $A_V \sim 0.1$ mag in $30\,\mathrm{km\,s^{-1}}$ (equivalent to a surface density of $\Sigma_{\mathrm{HI}} \sim 1.5\,M_\odot\,\mathrm{pc}^{-2}$, or a mass $\mathrm{M(HI)} \sim 2.5 \times 10^4\,M_\odot$ at $d = 30\,\mathrm{Mpc}$).

The molecular content of outflows may decrease faster than their atomic content, as molecular cloudlets get shattered and dissociated (see, for example, Leroy et al. 2015). At $\lambda = 3$ mm the ngVLA has a line sensitivity of 10 mK in 10 km/s at a resolution of 1″. Assuming a CO 1 − 0 emission with linewidth of 30 km/s, and a CO-to-H_2 factor of $X_{\mathrm{CO}} \sim 0.5 \times 10^{20}\,\mathrm{cm^{-2}\,(K\,km\,s^{-1})^{-1}}$ (intermediate between Milky Way GMCs and optically thin emission, Bolatto et al. 2013a,b), the ngVLA will be capable of detecting molecular outflows down to $A_V \sim 0.01$ mag at 5σ (equivalent to $\Sigma_{\mathrm{H2}} \sim 0.15\,M_\odot\,\mathrm{pc}^{-2}$, or $\mathrm{M(H_2)} \sim 2.5 \times 10^3\,M_\odot$ at $d = 30\,\mathrm{Mpc}$) in 24 hours at 1″ resolution. ALMA is capable of comparable Rayleigh-Jeans brightness temperature sensitivities in CO $J = 3 - 2$. However, the factor of $\sim$ 27 higher critical density required to excite the $J = 3 - 2$ transition compared to $J = 1 - 0$ means that only very dense molecular gas will be bright in the CO $J = 3 - 2$ line. As we emphasize below, the combination of the ngVLA and ALMA provides a powerful set of tools for studying a wide range of density and excitation conditions in molecular outflows.

Some of the key open questions that ngVLA observations will help answer are:

1. *What are the driving mechanisms of cold winds?* Mechanical feedback from supernovae (e.g., Fujita et al. 2009), radiation pressure (e.g., Murray et al. 2011),

cosmic ray pressure gradients (e.g., Uhlig et al. 2012), entrainment facilitated by Kelvin-Helmholtz instabilities (e.g., Heckman et al. 2000), and direct driving by interaction with AGN jets (e.g., Wagner & Bicknell 2011) have all been proposed as ways to inject momentum in the gas. It remains unclear, however, how they combine, and which one if any dominates (Hopkins et al. 2012; Muratov et al. 2015). A combination of high-resolution and sensitivity observations of the cold phases as they are ejected are key to solve this problem. For example, it appears that radiation pressure is insufficient to explain the high-resolution properties of the NGC 253 molecular outflow (Walter et al. 2017).

2. *What is the fate of the launched gas? Is molecular gas reformed in the wind? Does expelled gas change phases?* Imparting momentum to molecular cloudlets without destroying them has proven difficult in numerical simulations, while at the same time it is expected that part of the hot phase outflow may cool and reform a cold phase (e.g., Thompson et al. 2016). Observations of the outflow in M 82 strongly suggest that there is conversion of molecular into neutral atomic gas as the outflowing gas progresses away from the galaxy (Leroy et al. 2015). High sensitivity, resolved imaging in neutral atomic gas and molecular tracers is needed to answer these questions.

3. *What are the wind mass loss rates? What is the mass-loss to star-formation-rate ratio (the mass loading parameter)? Are mass-loading parameters and mass-loss rates as high as predicted for low-mass galaxies?* These parameters are key inputs to cosmological simulations, necessary to understand the precise effects of feedback on galaxy growth. Current estimates of the efficiency with which momentum is imparted to the different gas phases, implemented as sub-grid recipes in physical galaxy simulations, suggest mass loading parameters of order $\eta \sim 10$ for massive galaxies, and as high as $\eta \sim 100$ for dwarf galaxies (Muratov et al. 2015). While higher mass galaxies are expected to retain most of their metals in their circum-galactic environment, dwarf galaxies should heavily pollute the IGM (Muratov et al. 2017). Such values of η appear to be necessary to reproduce galaxy properties in cosmological simulations. It remains unclear, however, how to precisely attain such high mass loading efficiencies in detailed simulations Kim & Ostriker (2018). Most of the mass loss is due to the denser, cold phases, likely dominated by the neutral atomic and molecular gas that is directly observable with the ngVLA. The key to measuring accurate molecular mass loss rates is an understanding of the density structure in the wind, critically dependent on observations of several molecular species in the outflowing gas. These observations require extremely high sensitivity, wide bandwidth, and the ability to spatially and kinematically map the emission from the outflow.

4. *What are the conditions triggering cool outflows?* Observations suggest the existence of a star formation surface density threshold for launching large outflows (e.g., Newman et al. 2012), and a similar threshold in luminosity appears to exist for AGN-launched outflows (e.g., Veilleux et al. 2013; Wylezalek et al. 2017). Lower velocity galactic fountains and even radiation pressure-driven outflows, can also occur over extended areas of disks. Systematic demographics of the different properties of the multiphase outflows and their hosts, collected through a combination of radio and other multi-wavelength observations (see below), will provide valuable information about the conditions and triggers of outflows.

5. *What fraction of the outflowing gas reaches escape velocity versus falling back and being re-accreted/recycled?* Detailed simulations suggest that winds and fountains go through phases, dominated alternatively by outflow and inflow (Kim & Ostriker 2018). Observations show velocity gradients in the expelled gas, that can be interpreted in some cases as acceleration (Walter et al. 2017), or deceleration (Martini et al. 2018). Generally, recycled material may play an important role in lengthening the gas depletion timescale of galaxies (Davé et al. 2011), feeding galaxies at late cosmic times, and allowing for the exchange of processed gas between galaxies (Anglés-Alcázar et al. 2017b). Understanding the fraction of escaping gas and the relative amounts of expelled and recycled material, requires high sensitivity and high angular resolution observations of molecular and atomic gas tracers in galaxies and their circum-galactic environments.

6. *Are winds only effective at suppressing the formation of stars (negative feedback), or can they also trigger star formation in galaxies?* Although outflows driven by star formation or AGN are frequently invoked as mechanisms to expel gas and/or quench star formation (e.g., Alatalo et al. 2015), star formation can also be enhanced by compressive turbulence driven by the mechanical energy input of the wind (e.g., van Breugel et al. 1985; Croft et al. 2006). These positive feedback processes can also have an important effect on galaxy evolution (Silk 2013). Multi-wavelength radio techniques are particularly well suited to studying this problem, since they can be used to simultaneously image the outflowing gas and measure the gas kinematics, as well as derive the star formation rate (via the thermal free-free radio continuum emission) on sub-kpc scales.

7. *How do winds affect the growth of black holes?* Simulations suggest that black hole growth, particularly at early times, is limited by stellar feedback, which expels gas from galactic nuclei, limiting accretion. As a consequence, black holes can be undermassive in low-mass galaxies with respect to their high-mass counterparts, causing them to fall below the $M_{BH} - \sigma_{halo}$ relation (e.g., Anglés-Alcázar et al. 2017a). The ability of the ngVLA to resolve and measure gas content and kinematics in the innermost regions of galaxies, is key to quantifying the effects of winds on the growth of supermassive black-holes.

8. *What are the statistical properties of cold winds in the universe? Are these a rare phenomenon confined to AGN and starbursts, or are they a general feature of galaxies? What is the redshift evolution of starburst- and quasar-driven winds?* Observations suggest that galactic winds are ubiquitous at high redshift (e.g., Newman et al. 2012). We know very little about their cold components, except in a handful of spectacular examples (e.g., Maiolino et al. 2012). Fast outflows are seen in powerful IR galaxies in the local Universe (Veilleux et al. 2013), but studies have been limited to a handful of the brightest, most energetic sources. Large, sensitive, systematic surveys of molecular outflows reaching out to epochs where most galaxies are rapidly evolving are necessary to establish the importance of outflows in a cosmological context.

9. *What is the relation between the large scale cold outflows and the AGN that seem to trigger them? Are the ultra-fast outflows (UFOs) seen in the X-rays on sub-parsec scale the ultimate drivers of the most powerful cold outflows, or are the more common but slower soft X-ray and ultraviolet (UV) warm absorbers and*

UV broad absorption line (BAL) outflows seen on larger scale than the UFOs a better predictor of these cold outflows? How is the energy in these nuclear winds transferred to drive the galaxy-scale cold outflows? The kinetic power that is available from an AGN wind scales with v_{out}^3, rising quite rapidly with outflow velocity. Thus the fastest outflows are the ones that can potentially produce the largest feedback effects, provided they couple with the ISM effectively. X-ray observations identify a type of outflow that can reach modestly relativistic speeds $v_{out} \sim 0.03\,c - 0.3\,c$, consisting of a highly ionized flow originating from the accretion disk itself with column densities as high as 10^{24} cm^{-2} (Reeves et al. 2003; Tombesi et al. 2010). In theory, these very fast accretion-disk winds can drive shocks into the host galaxy ISM and create shock-driven over-pressurized bubbles that give rise to the large-scale outflows observed in ionized, neutral, and molecular gas (Faucher-Giguère & Quataert 2012; Tombesi et al. 2015). Testing these models, and directly linking fast accretion disk outflows with galactic winds requires high-quality, velocity-resolved imaging of the molecular and neutral atomic gas at high spatial resolution in the central regions of AGN with identified X-ray UFOs. A more complete survey of AGN will help relate the cold outflows to the more prevalent warm absorbers (Crenshaw & Kraemer 2012) and BAL outflows (Gibson et al. 2009). The prospects are good that ngVLA and future UV-optical facilities will be able to spatially resolve the regions where the BAL outflows interact with the ISM of the AGN host galaxies (Moe et al. 2009; Bautista et al. 2010; Dunn et al. 2010).

3. The Unique Discovery Space of the ngVLA

The ngVLA has a unique part of the discovery space to fill, of surface brightness sensitivity at high enough resolution to spatially resolve outflows close in and trace them far out from galaxy centers while at the same time studying the launching mechanisms and compositions.

Obtaining these measurements requires imaging and kinematics of the cool molecular and atomic component. This requires high surface brightness sensitivity on physical scales of 20 – 100 pc out to 100 Mpc, with good imaging dynamic range (frequently the faint wind emission near bright line or continuum emission from the galaxy/AGN). A distance of 100 Mpc encloses a volume big enough to find several examples of (ultra-)luminous infrared galaxies and powerful AGN. Physical resolution of 20–100 pc (~ 20 pc is the width of the streamers in NGC 253) is necessary to study the base of the outflows and their launching mechanisms, while high surface brightness sensitivity allows us to follow the outflow further out and search for accelerations/velocity gradients that constrain the final velocity. Even higher resolution is useful to study the disk/nucleus next to the outflow (usually much brighter), and investigate the connections to energy sources such as super star clusters, collections of supernova remnants, or SMBH accretion disks. To find and trace the extent and composition of winds it is necessary to image the CO, HI, and OH lines along with the continuum, as well as a host of tracers of local physical conditions. To characterize the physical state of winds and their mass loss, CO isotopologues, shock tracers, probes of the diverse excitation mechanisms (IR/UV/CR/X-ray) and tracers of the density distribution are key (e.g., Lindberg et al. 2016; Walter et al. 2017).

4. Complementarity with Existing and Planned Facilities

Three of the NASA flagship mission concepts under study have significant complementarity for the study of galaxy outflows, which are naturally multi-phase, multi-wavelength phenomena. In the following we focus on the complementary capabilities brought by the Infrared Origins Space Telescope (OST), the X-ray Lynx Telescope, and the Large UV/Optical/Infrared Surveyor (LUVOIR), while also discussing the Square Kilometre Array (SKA) and the Atacama Large Millimeter/submillimeter Array (ALMA).

With a huge leap in line and surface brightness sensitivity over all previous and planned IR missions, high spatial resolution, and the ability to quickly map large areas of the sky to generate unbiased spectroscopic samples, the OST will be able to find cool outflows spanning from the present day to over 90% of the age of the Universe. These targets will be prime targets for detailed follow-up study with the ngVLA at high angular and spectral resolution. For example, the Medium Resolution Survey Spectrometer (MRSS) on OST can detect columns of extra-planar neutral gas with $N_{HI} \sim 10^{20}\,\mathrm{cm}^{-2}$ and $n_H \sim 1\,\mathrm{cm}^{-3}$ via [CII] at sub-kpc resolution out to 30 Mpc in 10 minutes, and fine-structure lines in fast moving galactic winds down to $\sim 2 \times 10^{-21}\,\mathrm{W\,m}^{-2}$ in one hour, mapping nearby galaxies and detecting sources well into the epoch of re-ionization. For example, outflow signatures such as P-Cygni profiles (blueshifted absorption and redshifted emission) are easily detectable in the far-infrared molecular OH feature at $\lambda_{rest} = 79\,\mu\mathrm{m}$ from a source like Mrk 231 out to $z \sim 5$ in an hour, making surveys of large numbers of galaxies feasible in the far-infrared for the first time. Through blind, low-resolution spectroscopic surveys and detailed studies of nearby starbursts and AGN, OST working together with the ngVLA, will deliver a complete picture of feedback on the cold and warm atomic and molecular ISM in rapidly evolving galaxies. The proposed JAXA/ESA mission SPICA, recently selected as one of three candidate missions to be further studied for ESA's M5 mission call, will also be an extremely capable observatory for studying galactic outflows in the infrared (González-Alfonso et al. 2017), although it will have significantly less collecting area than OST and a narrower wavelength coverage.

While OST searches for cool outflows, Lynx will characterize the highly ionized $10^6 - 10^7$ K plasma phase (although measurements in absorption can also measure cold phases). The crucial capabilities of the Lynx telescope for galactic winds are high angular resolution ($\sim 0\farcs5$), high spectral resolution of ~ 2 eV FWHM independent of energy (corresponding to $R \sim 700$ at the very strong OVII line, and high signal-to-noise imaging over a field of view of $\sim 5'$. Lynx also has a grating with energy-resolution of 0.3 eV, capable of resolving the kinematics of the hot wind phases, and producing density diagnostics for the soft X-ray emitting winds. This combination of parameters allows measurement of dynamics (velocities of $\sim 200\,\mathrm{km\,s}^{-1}$), chemical composition (through lines from C, N, O, Ne, Mg, Si, S, Fe, and Ni), and diagnostics of ionization state. This allows Lynx to study the narrow emission-line regions of hot plasma bubbles inflated by AGN, as well as measuring metallicities outside the cores of massive clusters out to a redshift $z \sim 3$, and determining the mechanisms of gas ionization. In dusty winds Lynx can also distinguish dust features in absorption as the binding energy of atoms in grains is ~ 1 eV, allowing it to detect whether oxygen is bound in grains or in the gas phase. Lynx's planned sensitivity will allow the study of M 82-like starbursts at moderate redshifts, the most luminous starbursts at $z \sim 2 - 3$, and lensed objects at $z \sim 4$. In addition Lynx's high sensitivity and resolution in the Fe K band will allow a detailed analysis of the ultrafast outflows, determining their energetics and geometry over a wide range of redshifts, luminosities and black hole masses. Similarly, ESA's Advanced Telescope for High Energy Astrophysics (ATHENA) to be launched in the early 2030s, will also be able to detect and measure the hot gas associated with galactic outflows. However, ATHENA's significantly lower spatial and spectral resolution compared to Lynx ($5''$ and 2.5 eV compared to $1''$ and 0.3 eV, respectively) will make it better suited to studying outflows from powerful AGN, and likely limit its usefulness for low-velocity, starburst-driven winds.

LUVOIR will bring a leap in sensitive high-resolution imaging and spectroscopy across the full UV/OIR ($0.1 - 3\,\mu$m) bandpass, with the capability to resolve structures on scales of 100 parsecs or smaller in all galaxies at $z \leq 4$. The excellent match in resolution between LUVOIR and ngVLA will allow for direct comparisons between the warm and cool phases of the outflows and their impact on these galaxies. In particular, LUVOIR will provide spatially resolved maps of the UV BAL outflows in quasars to directly probe the mechanisms involved in driving the most extreme cold outflows. The 50–100 times larger UV spectroscopic sensitivity of LUVOIR compared with *HST* will bring a hundredfold more UV-bright quasars within its reach. It will also be able to use the much more numerous UV-bright galaxies as background sources. With this dramatic gain in sensitivity, LUVOIR will be able to efficiently probe gas and metals outside of galaxies, and the complex interplay between accretion flows onto galaxies and gas outflowing from the galaxies. These results will nicely complement the low surface brightness maps of galaxies and their halos produced by ngVLA.

The mid-frequency SKA has significant overlap with the low-frequency capabilities of the ngVLA, and so it will be able to perform HI and low-frequency OH studies of outflowing gas. The SKA phase 1 is comparable to the ngVLA, with a somewhat smaller collecting area but without the high-frequency capabilities. The precise sensitivity comparison will depend on decisions about array configurations, and the upper frequency cutoff of the SKA. However, it is clear that much of the molecular and high-resolution science associated with the outflows will be out of reach of even a phase 2 mid-frequency SKA. With a full mid-frequency SKA built, the complementarity with the ngVLA will be excellent, with the former focusing on the low frequency science while the ngVLA observes at higher frequencies ($\nu > 10 - 20$ GHz).

Compared to the the ngVLA, ALMA, which currently drives much of the research on high-resolution studies of molecular outflows in galaxies, will have significantly lower line sensitivity in the region of frequency overlap (ALMA bands 1 to 3). But because of its extensive high frequency coverage the combination of ALMA and ngVLA will be extremely complementary, particularly for studies of and launching mechanisms and chemistry and excitation in outflows.

Acknowledgments. ADB acknowledges support from NSF-AST 1412419.

References

Alatalo, K., Blitz, L., Young, L. M., et al. 2011, ApJ, 735, 88
Alatalo, K., Lacy, M., Lanz, L., et al. 2015, ApJ, 798, 31
Anglés-Alcázar, D., Faucher-Giguère, C.-A., Quataert, E., et al. 2017, MNRAS, 472, L109
Anglés-Alcázar, D., Faucher-Giguère, C.-A., Kereš, D., et al. 2017, MNRAS, 470, 4698
Bautista, M. A., Dunn, J. P., Arav, N., Korista, K. T., Moe, M., & Benn, C. 2010, ApJ,713, 553
Bohlin, R. C., Savage, B. D., & Drake, J. F. 1978, ApJ, 224, 132
Bolatto, A. D., Wolfire, M., & Leroy, A. K. 2013, ARA&A, 51, 207

Bolatto, A. D., Warren, S. R., Leroy, A. K., et al. 2013, Nat, 499, 450
van Breugel, W., Filippenko, A. V., Heckman, T., & Miley, G. 1985, ApJ, 293, 83
Cicone, C., Maiolino, R., Sturm, E., et al. 2014, A&A, 562, A21
Combes, F., García-Burillo, S., Casasola, V., et al. 2013, A&A, 558, A124
Crenshaw, D. M., & Kraemer, S. B. 2012, ApJ, 753, 75
Croft, S., van Breugel, W., de Vries, W., et al. 2006, ApJ, 647, 1040
Davé, R., Oppenheimer, B. D., & Finlator, K. 2011, MNRAS, 415, 11
Dunn, J. P., Bautista, M., Arav, N., et al. 2010, ApJ, 709, 611
Faucher-Giguère, C.-A., & Quataert, E. 2012, MNRAS, 425, 605
Feruglio, C., Maiolino, R., Piconcelli, E., et al. 2010, A&A, 518, L155
Fujita, A., Martin, C. L., Mac Low, M.-M., New, K. C. B., & Weaver, R. 2009, ApJ, 698, 693
García-Burillo, S., Combes, F., Usero, A., et al. 2014, A&A, 567, A125
Gibson, R. R., Jiang, L., Brandt, W. N., et al. 2009, ApJ, 692, 758
González-Alfonso, E., Armus, L., Carrera, F. J., et al. 2017, PASA, 34, e054
Heckman, T. M., Lehnert, M. D., Strickland, D. K., & Armus, L. 2000, ApJS, 129, 493
Herrera, C. N., & Boulanger, F. 2017, A&A, 600, A139
Hopkins, P. F., Quataert, E., & Murray, N. 2012, MNRAS, 421, 3522
Kennicutt, R. C., Jr., Armus, L., Bendo, G., et al. 2003, PASP, 115, 928
Kim, C.-G., & Ostriker, E. C. 2018, ApJ, 853, 173
Lehnert, M. D., Heckman, T. M., & Weaver, K. A. 1999, ApJ, 523, 575
Leroy, A. K., Walter, F., Martini, P., et al. 2015, ApJ, 814, 83
Lindberg, J. E., Aalto, S., Muller, S., et al. 2016, A&A, 587, A15
Maiolino, R., Gallerani, S., Neri, R., et al. 2012, MNRAS, 425, L66
Martini, P., Leroy, A. K., Mangum, J. G., et al. 2018, arXiv:1802.04359
Moe, M., Arav, N., Bautista, M. A., & Korista, K. T. 2009, ApJ, 706, 525
Muratov, A. L., Kereš, D., Faucher-Giguère, C.-A., et al. 2015, MNRAS, 454, 2691
Muratov, A. L., Kereš, D., Faucher-Giguère, C.-A., et al. 2017, MNRAS, 468, 4170
Murray, N., Ménard, B., & Thompson, T. A. 2011, ApJ, 735, 66
Newman, S. F., Genzel, R., Förster-Schreiber, N. M., et al. 2012, ApJ, 761, 43
Rachford, B. L., Snow, T. P., Destree, J. D., et al. 2009, ApJS, 180, 125
Reeves, J. N., O'Brien, P. T., & Ward, M. J. 2003, ApJ, 593, L65
Rupke, D. S. N., & Veilleux, S. 2013, ApJ, 768, 75
Rupke, D. S., Veilleux, S., & Sanders, D. B. 2005, ApJS, 160, 115
Sakamoto, K., Aalto, S., Combes, F., Evans, A., & Peck, A. 2014, ApJ, 797, 90
Silk, J. 2013, ApJ, 772, 112
Sturm, E., González-Alfonso, E., Veilleux, S., et al. 2011, ApJ, 733, L16
Strickland, D. K., Heckman, T. M., Colbert, E. J. M., Hoopes, C. G., & Weaver, K. A. 2004, ApJS, 151, 193
Thompson, T. A., Quataert, E., Zhang, D., & Weinberg, D. H. 2016, MNRAS, 455, 1830
Tombesi, F., Cappi, M., Reeves, J. N., et al. 2010, A&A, 521, A57
Tombesi, F., Meléndez, M., Veilleux, S., et al. 2015, Nat, 519, 436
Uhlig, M., Pfrommer, C., Sharma, M., et al. 2012, MNRAS, 423, 2374
Veilleux, S., Cecil, G., & Bland-Hawthorn, J. 2005, ARA&A, 43, 769
Veilleux, S., Meléndez, M., Sturm, E., et al. 2013, ApJ, 776, 27
Veilleux, S., Bolatto, A., Tombesi, F., et al. 2017, ApJ, 843, 18
Wagner, A. Y., & Bicknell, G. V. 2011, ApJ, 728, 29
Walter, F., Weiss, A., & Scoville, N. 2002, ApJ, 580, L21
Walter, F., Bolatto, A. D., Leroy, A. K., et al. 2017, ApJ, 835, 265
Wylezalek, D., Schnorr Müller, A., Zakamska, N. L., et al. 2017, MNRAS, 467, 2612
Zschaechner, L. K., Walter, F., Bolatto, A., et al. 2016, ApJ, 832, 142

Science with a Next-Generation Very Large Array
ASP Conference Series, Monograph 7
Eric J. Murphy, ed.

Accreting Supermassive Black Holes in Nearby Low-Mass Galaxies

Kristina Nyland[1] and Katherine Alatalo[2]

[1]*National Radio Astronomy Observatory, Charlottesville, VA, 22903;* `knyland@nrao.edu`

[2]*Space Telescope Science Institute, 3700 San Martin Dr., Baltimore, MD 21218, USA;* `kalatalo@stsci.edu`

Abstract. The ngVLA will facilitate deep surveys capable of detecting the faint and compact signatures of accreting supermassive black holes (SMBHs) with masses below one million solar-masses hosted by low-mass ($< 10^9\, M_\odot$) galaxies. This will provide important new insights on both the origins of supermassive black holes and the possible impact of active galactic nucleus-driven feedback in a currently unexplored mass regime.

1. Introduction and Motivation

It is now well-established that supermassive black holes (SMBHs) with masses ranging from $10^6 - 10^{10}\, M_\odot$ commonly reside in the nuclei of galaxies. A number of lines of evidence, including the observed scaling relations between SMBHs and their hosts, further suggest that the formation and growth of SMBHs and their hosts are inextricably linked (Kormendy & Ho 2013; Heckman & Best 2014). Despite their prevalence and importance to our understanding of galaxy evolution, the origin of supermassive SMBHs at high redshift (e.g., hierarchical merging vs. direct gas collapse) and their formation efficiency remain an open area of research (Volonteri 2010; Bellovary et al. 2011; Shirakata et al. 2016; Latif & Ferrara 2016). Thus, a key observational parameter for constraining the formation of SMBHs is the mass distribution of so-called SMBH "seeds" (Greene et al. 2012; Reines & Volonteri 2015; Mezcua et al. 2016).

1.1. Insights on SMBH Formation from Nearby Low-Mass Galaxies

Knowledge of the shape of the SMBH seed mass distribution at high redshift ($z > 6$) would provide strong constraints on the dominant SMBH formation channel, thus elucidating a key missing element in our understanding of SMBH-galaxy co-evolution. While direct mass measurements of SMBH seeds in the early universe are not yet feasible (Reines & Comastri 2016; Volonteri et al. 2017), the detection of their radiative signatures provides information on the SMBH occupation fraction as well as indirect SMBH mass constraints.

Thus far, studies of high-z SMBHs have been limited to luminous quasars, which may have formed via atypical processes to build the massive SMBHs needed to power their central engines at such early cosmic epochs (e.g., Mortlock et al. 2011). The population of accreting SMBHs with $M_{\rm BH} < 10^6\, M_\odot$ in the local Universe residing in lower-mass and dwarf galaxies offers an alternative means for studying SMBH seed formation since these systems have not experienced substantial merger or accretion driven growth (e.g., Mezcua 2017).

1.2. SMBH-driven Feedback in Low-Mass Galaxies?

In addition to their importance for constraining SMBH seed formation and growth scenarios, accreting SMBHs hosted by low-mass galaxies may also impact their hosts through energetic feedback. While supernova-driven feedback is believed to be the dominant regulatory mechanism associated with low-mass galaxies (e.g., Martin-Navarro & Mezcua 2018), recent observational and theoretical evidence suggests that active galactic nuclei (AGNs) powered by the sub-million-solar-mass SMBHs residing in their nuclei may also be capable of producing significant energetic feedback. From a theoretical standpoint, analytical models suggest that AGN feedback in low-mass galaxies may provide an efficient mechanism for the displacement of gas from the host galaxy (Silk 2017; Dashyan et al. 2018), though this possibility remains controversial (e.g., Trebitsch et al. 2018). This scenario may be most plausible in galaxies with stellar masses in the range of $10^7 \lesssim M_* \lesssim 10^9\,M_\odot$ when preceded by substantial supernova feedback capable of rarefying the ISM, thus making it more susceptible to disruption via AGN feedback (Prieto et al. 2017; Hartwig et al. 2018).

Observational evidence for AGNs hosted by low-mass galaxies has become increasingly common over the past decade (Barth et al. 2008; Greene & Ho 2007; Reines et al. 2013; Moran et al. 2014; Lemons et al. 2015; Sartori et al. 2015; Nucita et al. 2017; Mezcua et al. 2018), though the identification of AGN-driven feedback in low-mass and dwarf galaxies has remained challenging due to the resolution and sensitivity limitations of existing telescopes. However, recent spatially-resolved optical emission line studies have provided tentative support for AGN feedback in the low-mass-galaxy regime (e.g., Penny et al. 2018).

At radio frequencies, the identification of dwarf galaxies harboring jetted AGNs with the potential to impart feedback on their hosts is also challenging with current instruments such as the Karl G. Jansky Very Large Array (VLA; Nyland et al. 2016; Padovani 2016), which lack adequate collecting area and angular resolution to detect their faint, compact signatures. Despite their importance for probing the impact of jet-driven feedback physics in the low-mass regime, only a handful of candidate jetted AGNs hosted by low-mass galaxies with $M_{\rm BH} < 10^6\,M_\odot$ are known, including NGC 4395 (Wrobel & Ho 2006), Henize 2-10 (Reines et al. 2011, 2013), and NGC 404 (Nyland et al. 2017).

1.3. NGC 404: A Case Study

In Figure 1, we illustrate the multiwavelength nuclear properties of the nearby dwarf galaxy NGC 404, which harbors a nuclear radio jet with an extent of ~10 pc characterized by a steep radio spectral index from 1 to 18 GHz ($\alpha \sim -1$; Nyland et al. 2017). The spatial coincidence of the radio source, hard X-ray source, dynamical center of NGC 404, and the rotating circumnuclear CO(2–1) disk are most consistent with a confined radio jet launched by the central low-mass ($M_{\rm BH} < 10^5\,M_\odot$ from stellar dynamical modeling; Nguyen et al. 2017), low-Eddington-ratio ($L_{\rm bol}/L_{\rm Edd} \sim 10^{-6}$; Nyland et al. 2012) SMBH that has been disrupted by the ambient interstellar medium (ISM). Multiwavelength evidence for shock excitation supports the possibility of an interaction between the confined jet and the ISM. This includes the detection of extended Hα and

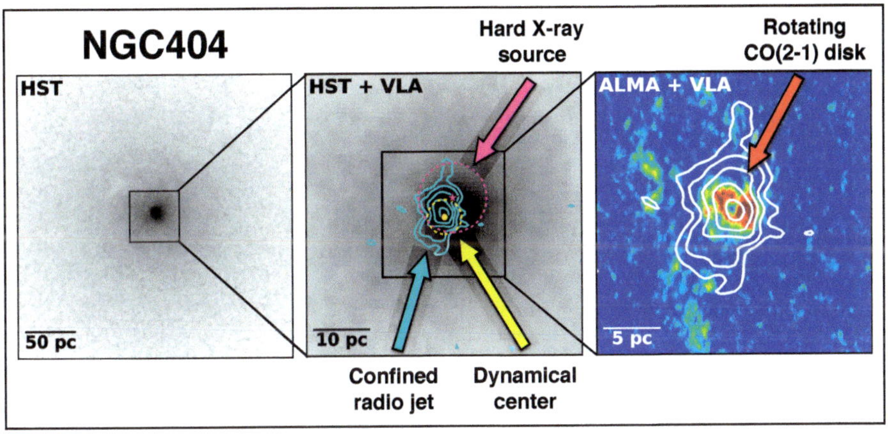

Figure 1. *HST*, VLA, *Chandra* and ALMA data identify a confined radio jet possibly interacting with the ambient ISM in the nucleus of the dwarf galaxy NGC 404. Adapted from Nyland et al. (2017).

soft X-ray emission from *HST* and *Chandra*, respectively, as well as strong [FeII] at 26 μm and rotational H_2 emission in archival *Spitzer* spectroscopic data (Nyland et al. 2017).

The nucleus of NGC 404 therefore offers a unique local laboratory for directly constraining the energetic impact on the ambient ISM by a jetted AGN hosted by a dwarf galaxy. However, the identification of additional jetted AGNs in low-mass galaxies will ultimately be needed to place NGC 404 in the broader context of galaxy evolution. Given the limited sensitivity and resolution of current radio telescopes, building a larger sample of jetted low-mass AGNs must await the availability of next generation instruments such as the next generation Very Large Array (ngVLA).

2. Prospects for the ngVLA

2.1. Demographics and Energetics of a Hidden SMBH Population

The ngVLA, with its roughly order of magnitude increase in sensitivity and angular resolution compared to the VLA, will greatly improve our ability to study compact radio sources in the nuclei of low-mass galaxies associated with SMBH accretion. The local volume of nearby galaxies within ~10 Mpc encompasses 869 galaxies, about 75% of which are dwarf galaxies (Karachentsev et al. 2013). Recent studies (e.g., Mezcua et al. 2016) have suggested that accreting SMBHs with masses in the range of $10^3 \lesssim M_{\rm BH} \lesssim 10^6$ M$_\odot$ may commonly reside in the nuclei of nearby low-mass galaxies, thus motivating deep searches for the radio continuum signatures of this population. However, identifying SMBHs in this population of galaxies is inherently difficult due to their expected low masses ($\rm M_{BH} < 10^6$ $M_\odot$) and faint accretion signatures.

The identification of NGC 404 analogs in other dwarf or low-mass galaxies would offer new insights into the occupation fraction of SMBHs analogous to the SMBH seeds that formed at high redshift, thus profoundly impacting our understanding of the origin of SMBHs. Deep, high-angular-resolution observations with the ngVLA will both help

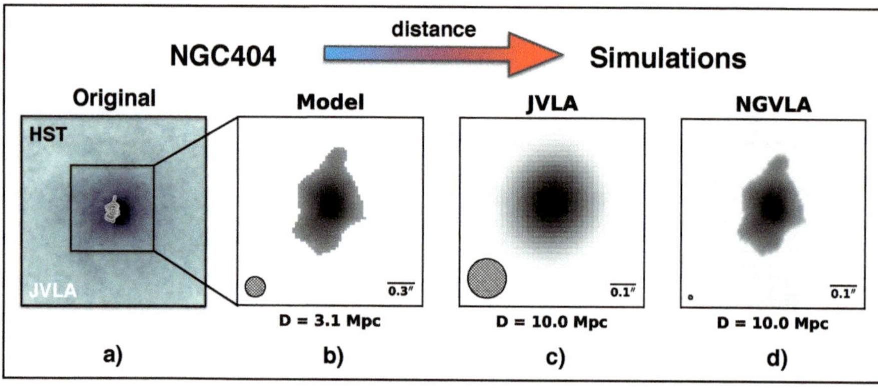

Figure 2. Simulated VLA and ngVLA images of an analog to the jetted AGN hosted by the nearby dwarf galaxy NGC 404 as it would appear if it were $\sim 3\times$ more distant. **a)** An *HST* optical image is shown in the background colorscale and the VLA *Ku*-band (15 GHz) radio contours are overlaid in white. The angular resolution of the radio data is $\theta_{\rm FWHM} = 0\farcs20$ and the extent of the radio jets from end to end is 1.13″ (17 pc). The NGC 404 VLA data were originally published in Nyland et al. (2017). **b)** Model radio image of NGC 404 based on the original data shown in panel a) and shown at the true source distance of d = 3.1 Mpc. **c)** Simulated VLA A-configuration map of an NGC 404 analog shifted to a distance of d = 10.0 Mpc at 15 GHz with $\theta_{\rm FWHM} = 0\farcs15$. **d)** Simulated map of an NGC 404 analog at d = 10.0 Mpc as it would appear if imaged with the ngVLA at 15 GHz. The angular resolution is $\theta_{\rm FWHM} = 0\farcs014$. Adapted from Nyland et al. (2018).

constrain the SMBH seed mass distribution[1] and also provide new constraints on the energetic impact of AGN feedback associated with SMBHs with masses below one million solar-masses. These new insights will provide crucial tests of our understanding of SMBH-galaxy co-evolution.

2.2. Imaging Simulations

We present ngVLA and VLA simulations of a more distant analog to the confined jet with an extent of ~10 pc in the center of NGC 404 in Figure 2. To produce the input model, we scaled the image pixel size such that it corresponds to an NGC 404 analog located at 10 Mpc, or $\sim 3\times$ further than the true distance to NGC 404. For the ngVLA simulations, we used the ngVLA configuration with 300 antennas and maximum baselines of $B_{\rm max} \sim 513$ km as shown in Nyland et al. (2018). The VLA simulations were performed in the A-configuration, which has $B_{\rm max} = 36.4$ km.

Figure 2 illustrates the inability of the VLA to spatially resolve the morphology of a 10 pc jet at a distance of 10 Mpc in the most extended A configuration. This is in contrast to the simulated ngVLA observations, which successfully resolve the extended structure of a ~10-pc-scale jet at a distance of 10 Mpc at 15 GHz. As discussed in detail in Nyland et al. (2018), both the VLA in the A configuration and the ngVLA

[1]Additional considerations on the topic of constraining SMBH seed formation with the ngVLA are provided in Plotkin & Reines (2018, this volume p. 719).

would be able to detect emission from an NGC 404 analog at a distance of 10 Mpc in a reasonable amount of on-source integration time ($\lesssim$ 15 hr). However, only the ngVLA would be able to spatially resolve the emission. Given the low-luminosity of NGC 404 ($L_{15\,\mathrm{GHz}} \sim 3.4 \times 10^{17}$ W Hz^{-1}), spatially resolving the morphology of the source is crucial to distinguish its origin from nuclear star formation (Nyland et al. 2017).

3. Summary

In addition to placing constraints on the basic physical properties (e.g., extent, luminosity, and energy) of jetted AGNs in nearby low-mass galaxies, observations of the molecular ISM as traced by the CO(1–0) line at 115 GHz will probe the amount of energy transferred to the ISM and, in conjunction with supporting multiwavelength ancillary data from telescopes such as ALMA and the *James Webb Space Telescope*, its impact on the ambient star formation efficiency. Thus, the unprecedented capabilities of the ngVLA will enable new advancements in our understanding of SMBH formation and the impact of jet-ISM feedback driven by SMBH seed analogs in the local Universe.

References

Barth, A. J., Greene, J. E., & Ho, L. C. 2008, AJ, 136, 1179
Bellovary, J., Volonteri, M., Governato, F., et al. 2011, ApJ, 742, 13
Dashyan, G., Silk, J., Mamon, G. A., Dubois, Y., & Hartwig, T. 2018, MNRAS, 473, 5698
Greene, J. E., & Ho, L. C. 2007, ApJ, 670, 92
Greene, J. E., Zakamska, N. L., & Smith, P. S. 2012, ApJ, 746, 86
Hartwig, T., Volonteri, M., & Dashyan, G. 2018, MNRAS, 476, 2288
Heckman, T. M., & Best, P. N. 2014, ARA&A, 52, 589
Karachentsev, I. D., Makarov, D. I., & Kaisina, E. I. 2013, AJ, 145, 101
Kormendy, J., & Ho, L. C. 2013, ARA&A, 51, 511
Latif, M. A., & Ferrara, A. 2016, PASA, 33, e051
Lemons, S. M., Reines, A. E., Plotkin, R. M., Gallo, E., & Greene, J. E. 2015, ApJ, 805, 12
Martin-Navarro, I., & Mezcua, M. 2018, ArXiv e-prints, arXiv:1802.07277
Mezcua, M. 2017, International Journal of Modern Physics D, 26, 1730021
Mezcua, M., Civano, F., Fabbiano, G., Miyaji, T., & Marchesi, S. 2016, ApJ, 817, 20
Mezcua, M., Civano, F., Marchesi, S., et al. 2018, MNRAS, arXiv:1802.01567
Moran, E. C., Shahinyan, K., Sugarman, H. R., Vélez, D. O., & Eracleous, M. 2014, AJ, 148, 136
Mortlock, D. J., Warren, S. J., Venemans, B. P., et al. 2011, Nat, 474, 616
Nguyen, D. D., Seth, A. C., den Brok, M., et al. 2017, ApJ, 836, 237
Nucita, A. A., Manni, L., De Paolis, F., Giordano, M., & Ingrosso, G. 2017, ApJ, 837, 66
Nyland, K., Marvil, J., Wrobel, J. M., Young, L. M., & Zauderer, B. A. 2012, ApJ, 753, 103
Nyland, K., Young, L. M., Wrobel, J. M., et al. 2016, MNRAS, 458, 2221
Nyland, K., Davis, T. A., Nguyen, D. D., et al. 2017, ApJ, 845, 50
Nyland, K., Harwood, J. J., Mukherjee, D., et al. 2018, ApJ, 859, 23
Padovani, P. 2016, A&A Rev., 24, 13
Penny, S. J., Masters, K. L., Smethurst, R., et al. 2018, MNRAS, 476, 979
Plotkin, R. M. & Reines, A. E. 2018, Science with a Next-Generation Very Large Array, ASP Monograph 7, ed. E. J. Murphy (San Francisco, CA: ASP), 719
Prieto, J., Escala, A., Volonteri, M., & Dubois, Y. 2017, ApJ, 836, 216
Reines, A. E., & Comastri, A. 2016, PASA, 33, e054
Reines, A. E., Greene, J. E., & Geha, M. 2013, ApJ, 775, 116
Reines, A. E., Sivakoff, G. R., Johnson, K. E., & Brogan, C. L. 2011, Nat, 470, 66

Reines, A. E., & Volonteri, M. 2015, ApJ, 813, 82
Sartori, L. F., Schawinski, K., Treister, E., et al. 2015, MNRAS, 454, 3722
Shirakata, H., Kawaguchi, T., Okamoto, T., et al. 2016, MNRAS, 461, 4389
Silk, J. 2017, ApJ, 839, L13
Trebitsch, M., Volonteri, M., Dubois, Y., & Madau, P. 2018, MNRAS, arXiv:1712.05804
Volonteri, M. 2010, A&A Rev., 18, 279
Volonteri, M., Reines, A. E., Atek, H., Stark, D. P., & Trebitsch, M. 2017, ApJ, 849, 155
Wrobel, J. M., & Ho, L. C. 2006, ApJ, 646, L95

Science with a Next-Generation Very Large Array
ASP Conference Series, Monograph 7
Eric J. Murphy, ed.

Radio Jet-ISM Feedback on Sub-Galactic Scales

Kristina Nyland,[1] Dipanjan Mukherjee,[2] Mark Lacy,[1] Isabella Prandoni,[4] Jeremy J. Harwood,[3] Katherine Alatalo,[5] Geoffrey Bicknell,[6] and Bjorn Emonts[1]

[1]*National Radio Astronomy Observatory, Charlottesville, VA, 22903;* knyland@nrao.edu, mlacy@nrao.edu

[2]*Dipartimento di Fisica Generale, Universita degli Studi di Torino, Via Pietro Giuria 1, 10125 Torino, Italy;* dipanjan.mukherjee@unito.it

[3]*Centre for Astrophysics Research, School of Physics, Astronomy and Mathematics, University of Hertfordshire, College Lane, Hatfield AL10 9AB, UK;* jeremy.harwood@physics.org

[4]*INAF-Istituto di Radioastronomia, via P. Gobetti 101, 40129, Bologna, Italy;* prandoni@ira.inaf.it

[5]*Space Telescope Science Institute, 3700 San Martin Dr., Baltimore, MD 21218, USA;* kalatalo@stsci.edu

[6]*Australian National University, Research School of Astronomy & Astrophysics, Cotter Rd, Weston, ACT 2611, Australia;* geoff.bicknell@anu.edu.au

Abstract. Energetic feedback by active galactic nuclei (AGNs) plays an important evolutionary role in the regulation of star formation (SF) on galactic scales. However, the effects of this feedback as a function of redshift and galaxy properties such as mass, environment, and cold gas content remain poorly understood. Given its unique combination of frequency range, angular resolution, and sensitivity, the ngVLA will serve as a transformational new tool in our understanding of how radio jets affect their surroundings. By combining broadband continuum data with measurements of the cold gas content and kinematics, the ngVLA will quantify the energetic impact of radio jets hosted by gas-rich galaxies as the jets interact with the star-forming gas reservoirs of their hosts.

1. Introduction

AGN feedback may be driven by radiative winds launched by the accretion disks of powerful quasars or spurred by radio jets/lobes as they heat, expel, or shock their surroundings. Observational evidence for both modes of feedback have been reported (e.g., Fabian 2012, and references therein). Villar Martín et al. (2014) show that, on average, radio jets appear to be capable of producing more extreme gas outflows than accretion disk winds. However, the relative importance of each mode, and the depen-

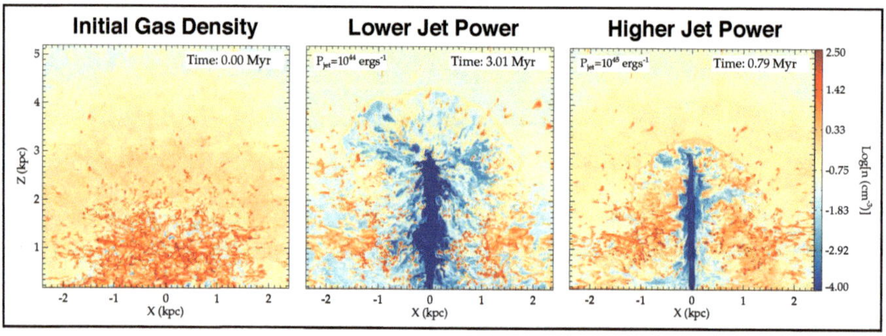

Figure 1. Snapshots from the relativistic hydrodynamic radio jet simulations (Mukherjee et al. 2016, 2017) showing the effect on an identical initial ISM (left) made by a radio jet with $P_{\rm jet} = 10^{44}$ erg s^{-1} (center) and $P_{\rm jet} = 10^{45}$ erg s^{-1} (right). The more powerful radio jet is able to more quickly "drill" through the ISM of its host galaxy, while the lower-power radio jet is trapped by the ISM and able to disrupt the surrounding gas for a longer time period and over a larger volume.

dence on redshift and other factors, such as galaxy mass and environment, remain open areas of research.

We also lack a fundamental understanding of exactly how radio jets transfer energy to their surroundings, how much energy is transferred to the different gas phases, and under what conditions significant positive/negative feedback is produced. We know that radio jets may deposit energy into their surroundings through a variety of mechanisms including heating, shocks, and/or turbulence (Fabian 2012; Alatalo et al. 2015; Soker 2016), and may also directly couple to gas in their surroundings and physically expel it (e.g., Morganti et al. 2013). However, the details of these processes – and under which conditions and environments different mechanisms dominate – are poorly understood.

Identifying jet-ISM feedback and constraining its impact on galaxy evolution requires observations with high sensitivity, angular resolution, and broad frequency coverage, and thus poses challenges to existing radio telescopes. With its unprecedented design consisting of ~214 × 18 m antennas operating from 1.2 to 116 GHz with baselines out to several hundred km in the main array (see Selina et al. 2018, this volume p. 15), the ngVLA will dramatically improve our understanding of how radio jets influence their surroundings.

2. Observing Jet-ISM Feedback with the ngVLA

The ngVLA will complement source morphologies and energetics constraints from deep, high-resolution continuum observations with spectral line data that encode information on the ISM content and conditions. The combination of broadband continuum and spectral line imaging will allow the ngVLA to uniquely probe the energetic impact of radio jets on the ambient cold gas. Spectral line measurements of molecular and atomic gas on comparable angular scales can be used to identify AGN-driven outflows (as well as gas inflow associated with fueling), perform detailed kinematic studies to gauge the amount of energy injected into the gas via feedback, and address the future evolutionary impact on local/global scales caused by the AGN feedback. Molecular gas and continuum estimates of the energetics of the outflow and jet can be directly compared with state-of-the art simulations, such as those shown in Figure 1, to deeply probe the underlying feedback physics.

Continuum + cold gas ngVLA studies would – for the first time – provide constraints on the prevalence and energetic importance of jet-ISM feedback in the dominant population of low-luminosity ($L_{1.4\,\mathrm{GHz}} < 10^{24}$ W Hz^{-1}) AGNs residing in "normal" galaxies. This is a particularly exciting prospect given recent observational evidence that lower-power radio AGNs may be able to significantly affect the interstellar medium (ISM) conditions of their hosts through feedback from sub-galactic-scale radio jets (e.g., Alatalo et al. 2011, 2015; Davis et al. 2012; Nyland et al. 2013; Godfrey & Shabala 2016; Querejeta et al. 2016; Zschaechner et al. 2016). From a theoretical standpoint, recent relativistic hydrodynamic simulations of radio jets propagating through a dense ISM (Mukherjee et al. 2016, 2017; Figure 1) provide further support for this possibility, demonstrating that while powerful radio jets rapidly "drill" through the ISM, lower-power jets become entrained in the ISM and are ultimately able to transfer energy over a much larger volume and for a longer period of time.

2.1. Sub-Galactic-Scale Radio Jets

In Figure 2, we illustrate the redshift dependence of the observed angular jet extent for a wide range of radio jet size scales ranging from sub-parsec jets to giant radio galaxies with Mpc-scale lobes. The maximum angular resolution (defined as $\theta_{\mathrm{max}} = \lambda/B_{\mathrm{max}}$) for each of the proposed ngVLA observing bands (assuming $B_{\mathrm{max}} \approx 500$ km) is also shown in this figure. Future ngVLA studies of radio jets with intrinsic extents of a few pc to a few kpc will be able to fully utilize the unique combination of angular resolution, collecting area, and frequency coverage of the ngVLA over a wide range of redshifts. This range of angular scales, combined with the frequency range of the ngVLA, highlights the suitability of the ngVLA for studies of jet-ISM feedback associated with lower-power radio AGNs with sub-galactic jets.

Observations of more extended radio jets that may be engaged in feedback on intergalactic or intracluster medium scales will be possible with suitable combinations of weighting and (u, v)-tapering. The inclusion of the proposed short baseline array of ~19 6 m dishes in addition to the main array of 214 18 m dishes would provide additional short-spacing information that would be beneficial for imaging studies of AGNs with large-scale, low-surface-brightness radio emission.

2.2. ISM Content and Conditions

2.2.1. Ionized Gas and the Magneto-Ionic Medium

Faraday rotation of polarized radio emission provides a unique probe of the line of sight combination of magnetic fields and ionized gas towards radio-emitting plasma. Jet-ISM interactions, particularly at high redshift, can result in material with very high Faraday rotations (e.g., Carilli et al. 1994). Such high rotation measures need high frequency, high angular resolution observations to resolve the Faraday components, which can be mixed in (entrained) with the radio plasma, be external in a region of shocked ISM surrounding the radio source, or correspond to dense ISM within the host galaxy. These components require Faraday Synthesis (Brentjens & de Bruyn 2005) to disentangle them. The ngVLA, with its wide frequency coverage and high angular resolution, will be perfect for this work.

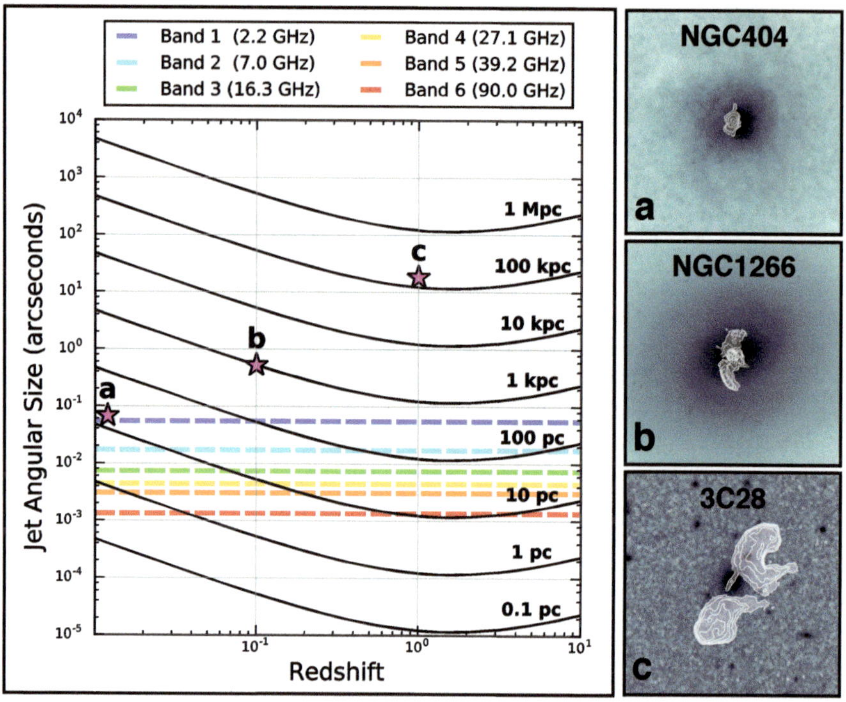

Figure 2. Jet angular size as a function of redshift. The black solid lines trace the redshift dependence of the angular extent of a jetted AGN for intrinsic jet sizes (measured from end to end along the major axis of the jet) from 0.1 pc to 1 Mpc. The maximum angular resolution of the ngVLA at the center of each of the ngVLA bands as defined in Selina & Murphy (2017) is denoted by the dashed colored lines. The magenta stars and thumbnails to the right of the main figure indicate three jetted radio AGNs representing a wide range of jet size scales: **a)** the dwarf galaxy NGC 404 with a jet extent of 10 pc, **b)** the jet-driven feedback host NGC 1266 with a jet extent of 1 kpc, and **c)** the radio galaxy 3C28 with a jet extent of 150 kpc. The redshifts of the representative sources correspond to simulated ngVLA maps from Nyland et al. (2018) at $z \approx 0$ ($d = 10$ Mpc), $z = 0.1$, and $z = 1.0$, respectively. Figure adapted from Nyland et al. (2018).

2.2.2. Atomic Gas

The absorption of atomic hydrogen at 21 cm against background continuum emission associated with a radio AGN provides a powerful means of directly identifying jet-driven outflows and quantifying their effect on the cold ISM (e.g., Morganti & Oosterloo 2018). HI absorption offers a key advantage over studies of the HI line in emission in terms of detectability, since the detection of HI absorption is independent of redshift and depends solely on the underlying strength of the background continuum source. In addition, the relatively low spin temperature of HI of ~100–150 K (Condon & Ransom 2016) makes the detection of emission at high angular resolution difficult or impossible due to brightness temperature sensitivity limitations (see §2.2.4). HI absorption observations, on the other hand, depend only on the brightness temperature of

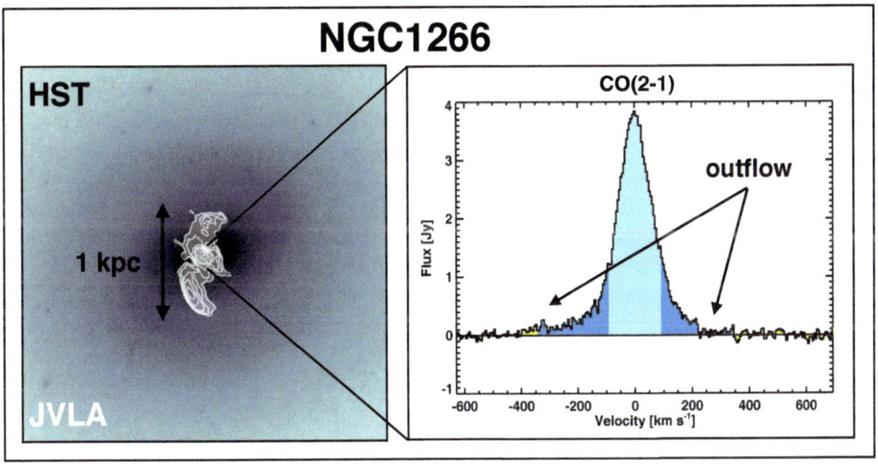

Figure 3. The radio jet and outflow of the nearby galaxy NGC1266. (*Left:*) The background colorscale image illustrates the *HST J*-band data (WFC3, F140W; Nyland et al. 2013) and the filled white contours trace VLA 5 GHz continuum data from Nyland et al. (2016). (*Right:*) The CO(2–1) data from CARMA highlights the molecular outflow that was originally identified in Alatalo et al. (2011) based on the presence of excess emission in the wings of the spectrum (dark blue).

the background continuum source, and may therefore be performed on much smaller (e.g., milliarcsecond) scales. In the context of jet-driven feedback, the detection of a blue-shifted spectral component in HI absorption is a signature of an outflow, which can be unambiguously distinguished from other possibilities, such as inflow or rotation. Additionally, kinematic constraints from HI absorption observations probe the gas conditions by providing direct measurements of the kinetic energy of any outflow components (e.g., Nyland et al. 2013; Morganti et al. 2013) as well as characterizing the degree of turbulence (Lacy et al. 2017).

The ability of the ngVLA to observe the HI line will ultimately depend on the lower frequency cutoff of its observing range. Assuming the ngVLA will observe down to 1.2 GHz, HI studies would be limited to nearby galaxies ($0 < z < 0.1$). HI absorption surveys of bright (1.4 GHz fluxes $\gtrsim$ a few tens of mJy), nearby radio AGNs with existing radio telescopes have reported detection rates of $\sim$30% (Geréb et al. 2014; Maccagni et al. 2017), suggesting that blind surveys of HI absorption in even lower-luminosity systems may be possible. The possibility of extending the ngVLA's frequency range below 1 GHz (e.g., Taylor et al. 2017) would greatly expand the redshift range over which HI would be observable with the ngVLA.

2.2.3. Molecular Gas

Identifying jet-driven molecular outflows is crucial for understanding the multiphase nature of AGN feedback (e.g., Rupke & Veilleux 2013; Emonts et al. 2014; Sakamoto et al. 2014; Alatalo et al. 2015; Morganti et al. 2015a). Molecular outflows may be identified on the basis of their spectral line shapes, such as the presence of broad wings or a shifted component (e.g., NGC 1266; Figure 3) or a P Cygni profile (Sakamoto et al. 2009). A survey of the cold gas properties of a large statistical sample of AGNs spanning a wide range of environments, host galaxy morphologies, SMBH masses, and nuclear activity classifications (e.g., high/low Eddington ratios, jetted/radio quiet, Compton thick/unobscured, etc.) would ultimately help establish an observationally-motivated model for the cosmic importance of outflows launched by active nuclei. We emphasize that improving our understanding of AGN feedback through ngVLA molecular gas observations will not be limited to objects with detectable outflows. Observations of more subtle feedback effects, such as a significant increase in the turbulence of the gas, or a substantial change in star formation efficiency/depletion time in the vicinity of the AGN, will also be possible (e.g., Alatalo et al. 2015; Oosterloo et al. 2017).

The lowest energy transitions of the CO molecule, CO(1–0), CO(2–1), and CO(3–2) at rest frequencies of 115, 230, and 345 GHz, respectively, trace the total molecular gas reservoir at relatively low densities ($n_{\rm H_2} \sim 10^3$ cm^{-3}). The CO(1–0) line will be accessible to the ngVLA over the redshift ranges $0 < z \lesssim 0.5$ and $z \gtrsim 1.5$. The gap from $z \approx 0.5$ to 1.5 is due to the high telluric opacity of molecular oxygen that precludes ground-based observations from 52 to 68 GHz. Observations of the CO(2–1) line will be possible from $1 \lesssim z \lesssim 2$ and $z \gtrsim 3.5$, and the CO(3–2) line will be accessible over the range $2 \lesssim z \lesssim 4$ and also at $z \gtrsim 6$ (though see §2.2.4 regarding important caveats). We note that none of the low-J CO lines will be observable from $z = 0.5 - 1.0$, though transitions of other species probing denser gas, such as SiO and CS, will be accessible. For a graphical description of the redshift dependence of a wide variety of molecular gas species and transitions observable with the ngVLA, we refer readers to Figures 2 and 9 in Casey et al. (2015).

In addition, imaging of the cold molecular gas in the vicinity of the AGN can provide important clues on the fueling of the central BH. Depending on the geometry and kinematics of this gas (disks, patches or filaments), the fueling mechanism and efficiency can be determined, in comparison with models (e.g., Gaspari et al. 2015).

2.2.4. Important Caveats

The minimum excitation temperature of the CO(1–0) line is 5.53 K, and radio telescopes must therefore have sufficient brightness temperature sensitivity to detect this line. With maximum ngVLA baselines of ~500 km in the north-south direction, the naturally-weighted angular resolution at 93 GHz would be ~10 mas, corresponding to a brightness temperature sensitivity[1] of $\sigma_{T_{\rm B}} \sim 350$ K for an integration time of 1 hr, a channel width of 10 km s^{-1}, and robust weighting (Selina & Murphy 2017). Thus, significant tapering of the data, as well as the application of new weighting schemes, will be necessary for studying jet-ISM feedback through ngVLA observations of the low-J CO transitions.

At high redshifts, the increasing influence of the cosmic microwave background (CMB) may also hinder the detectability of CO. The increasing CMB temperature at

[1]The brightness temperature sensitivity of a point source is $\sigma_{T_{\rm B}} = \left(\frac{S}{\Omega_{\rm A}}\right)\frac{\lambda^2}{2k}$, where S is the flux density in units of W m^{-2} Hz^{-1}, k is the Boltzmann constant (1.38×10^{-23} Jy K^{-1}), and λ is the observing wavelength in meters. The quantity $\Omega_{\rm A}$ is the beam solid angle, $\Omega_{\rm A} = \frac{\pi\theta^2_{\rm FWHM}}{4\ln(2)}$, where $\theta_{\rm FWHM}$ is the angular resolution in radians.

high redshifts[2] both reduces the contrast of the CO emission against the background (particularly in cold molecular clouds with $T_{\mathrm{kinetic}} \sim 20$ K) and changes the shape of the CO spectral line energy distribution by exciting a greater proportion of higher-J rotational levels (e.g., da Cunha et al. 2013; Zhang et al. 2016). This may be problematic for ngVLA observations of cold molecular clouds in non-starbursting galaxies at $z \sim 4$, where $T_{\mathrm{CMB}} = 13.65$ K. However, CMB heating is expected to be less problematic in molecular clouds in the vicinity of AGNs and jets, since AGN heating may increase the gas kinetic temperature to hundreds of degrees (Matsushita et al. 1998; Krips et al. 2008; Viti et al. 2014; Glenn et al. 2015; Richings & Faucher-Giguère 2018).

3. Multiwavelength Synergy

The unique capabilities of the ngVLA will facilitate exciting advancements in our understanding of AGN feedback and its broader connection to galaxy evolution, particularly when combined with multiwavelength data from other state-of-the-art instruments. In terms of synergy with current radio telescopes, observations with the Atacama Large Millimeter and Sub-millimeter Array (ALMA) at frequencies above the ngVLA's limit of 116 GHz will provide key insights into the energetic and chemical impact of jet-driven feedback on the dense gas phase of the ISM. In the low radio frequency regime, the Square Kilometre Array (SKA) and its pathfinders will probe the 21cm line out to higher redshifts (though at lower spatial resolution) than the ngVLA (Morganti et al. 2015b). The combination of constraints on both the atomic and molecular gas conditions from ngVLA and SKA observations will be important for studying the full impact of energetic jet-driven feedback on the ISM.

References

Alatalo, K., Blitz, L., Young, L. M., et al. 2011, ApJ, 735, 88
Alatalo, K., Lacy, M., Lanz, L., et al. 2015, ApJ, 798, 31
Brentjens, M. A., & de Bruyn, A. G. 2005, A&A, 441, 1217
Carilli, C. L., Owen, F. N., & Harris, D. E. 1994, AJ, 107, 480
Casey, C. M., Hodge, J. A., Lacy, M., et al. 2015, ArXiv e-prints, arXiv:1510.06411
Condon, J. J., & Ransom, S. M. 2016, Essential Radio Astronomy
da Cunha, E., Groves, B., Walter, F., et al. 2013, ApJ, 766, 13
Davis, T. A., Krajnović, D., McDermid, R. M., et al. 2012, MNRAS, 426, 1574
Emonts, B. H. C., Piqueras-López, J., Colina, L., et al. 2014, A&A, 572, A40
Fabian, A. C. 2012, ARA&A, 50, 455
Gaspari, M., Brighenti, F., & Temi, P. 2015, A&A, 579, A62
Geréb, K., Morganti, R., & Oosterloo, T. A. 2014, A&A, 569, A35
Glenn, J., Rangwala, N., Maloney, P. R., & Kamenetzky, J. R. 2015, ApJ, 800, 105
Godfrey, L. E. H., & Shabala, S. S. 2016, MNRAS, 456, 1172
Krips, M., Neri, R., García-Burillo, S., et al. 2008, ApJ, 677, 262
Lacy, M., Croft, S., Fragile, C., Wood, S., & Nyland, K. 2017, ApJ, 838, 146
Maccagni, F. M., Morganti, R., Oosterloo, T. A., Geréb, K., & Maddox, N. 2017, A&A, 604, A43
Matsushita, S., Kohno, K., Vila-Vilaro, B., Tosaki, T., & Kawabe, R. 1998, ApJ, 495, 267

[2]The redshift dependence of the CMB temperature follows the relation $T_{\mathrm{CMB}} = T^{z=0}_{\mathrm{CMB}} \times (1 + z)$, where $T^{z=0}_{\mathrm{CMB}} = 2.73$ K is the CMB temperature at $z = 0$.

Morganti, R., Frieswijk, W., Oonk, R. J. B., Oosterloo, T., & Tadhunter, C. 2013, A&A, 552, L4
Morganti, R., & Oosterloo, T. 2018, ArXiv e-prints, arXiv:1807.01475
Morganti, R., Oosterloo, T., Oonk, J. B. R., Frieswijk, W., & Tadhunter, C. 2015a, A&A, 580, A1
Morganti, R., Sadler, E. M., & Curran, S. 2015b, Advancing Astrophysics with the Square Kilometre Array (AASKA14), 134
Mukherjee, D., Bicknell, G. V., Sutherland, R., & Wagner, A. 2016, MNRAS, 461, 967
—. 2017, MNRAS, 471, 2790
Nyland, K., Alatalo, K., Wrobel, J. M., et al. 2013, ApJ, 779, 173
Nyland, K., Young, L. M., Wrobel, J. M., et al. 2016, MNRAS, 458, 2221
Nyland, K., Harwood, J. J., Mukherjee, D., et al. 2018, ApJ, 859, 23
Oosterloo, T., Raymond Oonk, J. B., Morganti, R., et al. 2017, A&A, 608, A38
Querejeta, M., Schinnerer, E., García-Burillo, S., et al. 2016, A&A, 593, A118
Richings, A. J., & Faucher-Giguère, C.-A. 2018, MNRAS, 474, 3673
Rupke, D. S. N., & Veilleux, S. 2013, ApJ, 775, L15
Sakamoto, K., Aalto, S., Combes, F., Evans, A., & Peck, A. 2014, ApJ, 797, 90
Sakamoto, K., Aalto, S., Wilner, D. J., et al. 2009, ApJ, 700, L104
Selina, R., & Murphy, E. J. 2017, Next Generation Very Large Array Memo Series, 17, http://library.nrao.edu/ngvla.shtml
Selina, R. J., Murphy, E. J., McKinnon, M. et al. 2018, Science with a Next-Generation Very Large Array, ASP Monograph 7, ed. E. J. Murphy (San Francisco, CA: ASP), 15
Soker, N. 2016, NewAR, 75, 1
Taylor, G., Dowell, J., Malins, J., et al. 2017, ArXiv e-prints, arXiv:1708.00090
Villar Martín, M., Emonts, B., Humphrey, A., Cabrera Lavers, A., & Binette, L. 2014, MNRAS, 440, 3202
Viti, S., García-Burillo, S., Fuente, A., et al. 2014, A&A, 570, A28
Zhang, Z.-Y., Papadopoulos, P. P., Ivison, R. J., et al. 2016, Royal Society Open Science, 3, 160025
Zschaechner, L. K., Walter, F., Bolatto, A., et al. 2016, ApJ, 832, 142

Science with a Next-Generation Very Large Array
ASP Conference Series, Monograph 7
Eric J. Murphy, ed.

Unwinding Star Formation a with Next-Generation Galactic Plane Survey

Erik Rosolowsky,[1] Adam Ginsburg,[2] and Adam Leroy[3]

[1]*University of Albera, Edmonton, AB, Canada;* `rosolowsky@ualberta.ca`

[2]*National Radio Astronomy Observatory, Socorro, NM, USA;* `adam.g.ginsburg@gmail.com`

[3]*The Ohio State University, Columbus, OH, USA;* `leroy.42@osu.edu`

Abstract. The ngVLA will enable for efficient surveys of a large fraction of the Galactic plane in a suite of molecular gas tracers from the 3mm band, including both low (CO) and high (HCN, HCO^+, N_2H^+) critical density transitions. The key motivation for a next generation Galactic Plane survey is spanning a wide dynamic range of scales at high sensitivity. Such a survey will determine the connections between the kiloparsec scale galactic processes like feedback and instabilities down through the turbulent cascade to the thermal scales where individual stellar cores form (0.1 pc). While current facilities must stitch together these effects over different regions, only the ngVLA will have the mapping efficiency required to map square-degree areas with the $1''$ resolution to deduce how galactic physics regulates all aspects of star formation.

1. Introduction

Star formation is the dominant process in the secular evolution of galaxies, and understanding how galaxies form stars through cosmic time remains a vital open question in understanding the evolution of the Universe. The next-generation Very Large Array (ngVLA) is uniquely able to provide the definitive census of the dominant physical and chemical processes that shape the molecular ISM and govern how stars form. Historically, studies of star formation in the context of galaxy evolution have focused on the rate of star formation as a function of gas properties and local galactic environment. However, the properties of the molecular gas must set the other outputs of star formation: the initial mass function, cluster masses and properties, and distribution of binary masses and orbital properties. All of these outputs establish the resulting stellar populations, the patterns of enrichment, and the nature of feedback into a galaxy.

One of the major data products from the ngVLA should be a survey of a large fraction of the Milky Way disk at spatial scales resolving the thermal dissipation length in molecular clouds (0.1 pc). This survey will (1) conclusively identify the critical scales governing star formation and (2) definitively link variations in local Galactic properties to changes in star formation for all the outputs of the process (IMF, binary properties, clusters). While we routinely observe on scales from < 0.1 pc in nearby clouds to 10s of pc scales across the disk, we are lacking a unified survey that covers the full range of these scales over a substantial fraction of the Galactic volume. Using the full suite of molecular line diagnostics available throughout the ngVLA observational bands, we will directly measure the flow of energy, mass and momentum through the interstellar medium into newly forming stars and their feedback on the surrounding gas.

2. Scientific Motivation

The molecular ISM is shaped by gravitation, turbulence, magnetism, radiation, and chemistry. These effects play out on a wide range of scales from the scale length of the disk (kpc) down to the thermal dissipation scale (< 0.1 pc). Since the molecular medium hosts all star formation in the modern universe, understanding how the conditions in the ISM establish star formation has been a subject of study for decades. While vast progress has been made, the subject remains unsolved and the primary difficulty in making progress arises because the process is both *multi-physics* and *multi-scale*. With the ngVLA, it will be possible to establish the definitive observational data set that will lay the groundwork for overcoming these barriers. This possibility comes because only the ngVLA has the capability of making a multi-tracer study of the Galactic plane (to address multi-physics) over a vast spatial dynamic range (to address multi-scale). Unlike other proposed facilities the ngVLA can efficiently observe the smallest scales over a large part of the disk. By observing a suite of molecular gas tracers, we will be able to establish the definitive inventory of physics at play in disk galaxies.

Wide Spatial Dynamic Range: The ngVLA makes it possible to measure *all* the relevant scales in the molecular medium and calculate the interplay between the governing physical effects. This will reveal whether the outer scales of the molecular medium (the clouds) represent a significant feature in the hierarchy of the ISM. The proposed observations will trace this hierarchy down to the turbulent dissipation scales across the disk, identifying how the balance of physical forces evolves with scale throughout the galaxy, making a direct connection to the regulation of the star formation process. This wide range of scales is required since the best probes of turbulence require characterization over a wide range. Critically, these data will test conjectures that the properties of the turbulent flows account for variations in modes of star formation through the disk. Given the sensitivity and range of scales proposed by the ngVLA, this survey would enable the definitive and ultimate tests of the physics of galaxy-scale star formation given our current understanding.

To date, the molecular ISM has already been surveyed at these different scales in different objects (Figure 1). Studies of nearby molecular clouds probe down to the thermal scales of the emission (e.g., Ridge et al. 2006; Kong et al. 2018). Surveys of the Milky Way disk span larger scales that can reach the scale height and length of the disk (e.g., Dame et al. 2001; Jackson et al. 2006; Dempsey et al. 2013) but are limited in their dynamic range at the small scales. While these disjoint data have been at the foundation of our understanding of star formation, we are hampered by having to synthesize across different galactic environments. Specifically, changes in the efficiency of the star formation process arise in different galactic environments, but our understanding of how star formation plays out at small scales is extrapolated from our observations of local clouds.

Efficient Observation of Multiple Tracers: The centimeter to millimeter bands provide tracers of all the relevant physical processes in star formation. With wide instantaneous bandwidth, it becomes possible for the ngVLA to sample these all simultaneously. While molecular hydrogen is difficult to observe directly, our understanding of ISM properties comes through different chemical species. We can trace the location of the material through relatively uniform bulk gas tracers like the isotopologues of CO, which trace $n \sim 10^2\ \mathrm{cm}^{-3}$ gas (Shirley 2015). With excellent spectral resolution, we use the velocity information to understand ISM dynamics and therefore turbulence and gravitation. High-dipole-moment species such as HCN, CS, N_2H^+, and CN provide similar observations, but only in the highest density gas associated with star formation ($n \sim 10^{4.5}$-$10^5\ \mathrm{cm}^{-3}$; Shirley 2015). Molecular line excitation also directly traces the influence of the radiation field (e.g., ^{12}CO excitation in Figure 1). Recombination lines and centimetre-wave continuum data directly identify regions of high mass star formation (e.g., see Balser et al. 2018, this volume p. 431; Murphy et al. 2018, this volume p. 421). Polarized continuum data can unwind the structure of the magnetic field. While all these are accessible to the ngVLA, here we focus on the utility of the millimetre-wave molecular line toolkit.

3. ngVLA Observations and Complementary Facilities

This chapter proposes the next generation Galactic Plane Survey (ngGPS) in several tracers across the 3 mm band. This science case is built around the idea that the ngVLA will be able to survey several spectral lines separated by up to ~ 10 GHz. With this assumption, the proposed survey focuses on two spectral windows: the CO isotopologues near 110 GHz (^{12}CO, ^{13}CO, C^{18}O as well as CN) and a suite of dense gas tracers (HCN, HCO^+, CS, N_2H^+) near 90 GHz. The molecular disk observable by the ngVLA spans ~ 300 deg^2 which should be imaged from the largest possible angular scales down to 1″. The maximum resolution corresponds to two resolution elements across the 0.1 pc thermal scale at a distance of 10 kpc. We target 500 mK sensitivity in an 0.2 km/s channel, representative of detecting faint tracers of dense gas at the thermal line width for H_2. Using the ngVLA reference design, reaching this depth and resolution requires 70 seconds of integration time. With a primary beam size of 0.4 arcmin^2, covering 1 deg^2 would take 175 hours per tuning. For reference, using ALMA to map 1 deg^2 would require 2300 hours of 12 m array time. Thus, individual regions can be mapped in the scope of PI projects on the ngVLA and an extra-large program can provide the ultimate Galactic plane survey.

The high surface brightness sensitivity of the ngVLA over a wide range of spatial scales is what enables the ngGPS. While ALMA can observe many of the same properties, it will be restricted to small scale studies in the Milky Way. A broad census of the entire Galaxy remains prohibitive to other facilities, especially spanning the wide range of scales. However, it is precisely the wide area survey that will provide the progress toward understanding star formation.

In overlapping portions of the sky, ALMA can readily probe the higher rotational transitions of the target species over small regions. However, the ngVLA is uniquely able to probe regions in first and second Galactic quadrants, including the nearest signs of interaction with the spiral structure found in the Cygnus X region. The complementary access to higher-energy transitions probes the thermodynamics of GMCs using CO spectral-line energy distributions. ALMA and potentially the Origins Space Telescope (OST) will necessarily focus on small regions whereas OST will have relatively poor resolution. The ngVLA data set provides the best available channel to stitch together these complementary views of the molecular ISM. No other facility would be able to provide the wide-dynamic-range imaging of the Galaxy in these spectral lines.

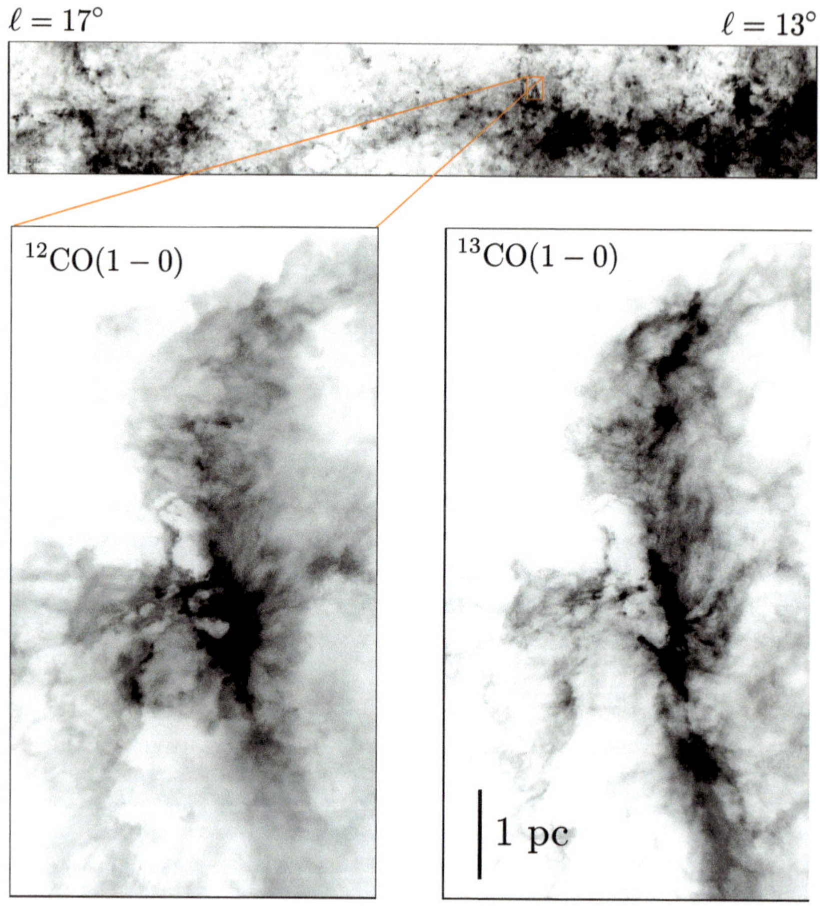

Figure 1. Example data quality that would be provided through ngGPS observations of the Galactic plane based on example data sets. The top panel shows observations taken from the JCMT CO High Resolution survey in ^{12}CO(3-2) which is the highest resolution molecular line image of the Galactic Plane available (14″; Dempsey et al. 2013). The bottom panels show integrated intensity maps for ^{12}CO(1-0) and ^{13}CO(1-0) of the Integral Shaped Filament in Orion obtained through the CARMA-NRO Orion Survey (Kong et al. 2018). The ngGPS would provide a high-resolution, multi-tracer survey of the Galactic plane with linear resolution comparable to the Orion data.

4. Scientific Strategy

The ngGPS aims to characterize the star forming interstellar medium. The broadest impact of such a survey would be from the release of survey data products for general use, immediately becoming the definitive ground truth survey of the molecular medium in the Milky Way.

Used as a survey, the data can be used to infer the physical properties of the molecular medium throughout the Milky Way. The CO isotopologues reveal the opacity and excitation conditions within molecular clouds, which is challenging to do from single line studies alone. With a wide dynamic range in spatial resolutions, observing low opacity lines is essential for a complete census of high column density regions (e.g., $C^{18}O$). Tracers that are only excited in high density regions (e.g., HCN, HCO^+, N_2H^+) are key for tracing mass and momentum flow into the regions that immediately host star formation. By tracing down to 0.1 pc scales, the ngGPS will identify the *core* populations throughout the visible Galactic plane, which will provide a link to IMF variations and show how clusters form and evolve.

While the 3 mm band provides a robust physical toolkit, the key advantage of a Galactic plane survey is to use this full toolkit to trace how star forming gas evolves in different Milky Way environments. With a progressively refined understanding of the longitude-velocity space founded on trigonometric parallax (Reid et al. 2014), it will be possible to map molecular emission to three dimensional positions in the Galaxy. The three key contrasts are dynamical environment (arm vs. interarm vs. center), radial (tracing different stellar potentials, metallicities) and vertical (tracing different pressure levels). The survey will trace the mass distribution, and the flow of momentum (through Doppler shift analyses) and energy flux (through Doppler shift and molecular line cooling studies). While it is difficult to predict the exact questions that will directly drive a Galactic plane survey after ngVLA construction, such a data set would immediately answer a myriad of questions about Galactic evolution that currently face the community. By directly tracing gas motion in the molecular medium throughout the galaxy, we could directly measure how turbulence regulates star formation (McKee & Ostriker 2007) by relating its power spectrum, anisotropy, and solenoidal fraction to local star formation rates. With a broad census of molecular clouds, we could trace the life cycle of molecular clouds including their thermodynamics, and chemistry. We would have the data required to measure variations of the core mass function in time and evolution, establishing the understanding the stellar initial mass function. Furthermore, we would also know the angular momentum distribution within molecular clouds and determining how this regulates binary star formation.

Depending on the flexibility of the ngVLA instrument, commensal science goals could easily be achieved through a flexible deployment of correlator and receiver resources. For example, continuum observations will reveal compact Hɪɪ regions, bright dust emission features, and non-thermal events. Stacking radio recombination line emission will provide opacity-free measurements of the dynamics of ionized regions as well as tracers of the ionizing radiation field. Finally, a multi-epoch survey strategy may be ideal for revealing transient sources for further study.

Acknowledgments. ER acknowledges the support of the Natural Sciences and Engineering Research Council of Canada (NSERC), funding reference number RGPIN-2017-03987.

References

Balser, D. S., Anderson, L. D., Bania, T. M., et al. 2018, Science with a Next-Generation Very Large Array, ASP Monograph 7, ed. E. J. Murphy (San Francisco, CA: ASP), 431

Dame, T. M., Hartmann, D., & Thaddeus, P. 2001, ApJ, 547, 792

Dempsey, J. T., Thomas, H. S., & Currie, M. J. 2013, ApJS, 209, 8

Jackson, J. M., Rathborne, J. M., Shah, R. Y., et al. 2006, ApJS, 163, 145

Kong, S., Arce, H. G., Feddersen, J. R., et al. 2018, arXiv:1803.11522
McKee, C. F., & Ostriker, E. C. 2007, ARA&A, 45, 565
Murphy, E. J., Condon, J. J., Alberdi, A., et al. 2018, Science with a Next-Generation Very Large Array, ASP Monograph 7, ed. E. J. Murphy (San Francisco, CA: ASP), 421
Reid, M. J., Menten, K. M., Brunthaler, A., et al. 2014, ApJ, 783, 130
Ridge, N. A., Di Francesco, J., Kirk, H., et al. 2006, AJ, 131, 2921
Shirley, Y. L. 2015, PASP, 127, 299

Science with a Next-Generation Very Large Array
ASP Conference Series, Monograph 7
Eric J. Murphy, ed.

Neutral Atomic Hydrogen in the Local Universe

D.J. Pisano,[1,2] Fabian Walter,[3] and Snežana Stanimirović[4]

[1]*Dept. of Physics & Astronomy and Center for Gravitational Waves & Cosmology, West Virginia University, Morgantown, WV, USA*

[2]*Adjunct Astronomer at Green Bank Observatory, Green Bank, WV, USA;* djpisano@mail.wvu.edu

[3]*Max Planck Institute für Astronomie, Heidelberg, Germany;* walter@mpia.de

[4]*Dept. of Astronomy, U. Wisconsin-Madison, Madison, WI, USA;* sstanimi@astro.wisc.edu

Abstract. One of the outstanding questions in astronomy today is how gas flows from the circumgalactic medium (CGM) onto the disks of galaxies and then transitions from the diffuse atomic medium into molecular star-forming cores. For studies of the CGM, the next generation Very Large Array (ngVLA) will have the sensitivity and resolution to measure the sizes of the neutral atomic hydrogen (HI) disks of galaxies and complete a census of the HI content around galaxies. Within galaxies, the ngVLA will be able to resolve HI clouds in large numbers of galaxies beyond the Local Group providing measurements of the physical conditions of gas across a wide range of galaxy types. Finally, within our own Milky Way, the ngVLA will provide a dense grid of HI absorption spectra in the cold and warm neutral medium constraining the temperature and density of atomic gas as it transitions into molecular gas. Combined with radio continuum and molecular line data from the ngVLA plus multi-wavelength data from other planned facilities, ngVLA will have a key role in understanding star-formation in the local universe while complementing future studies with the Square Kilometre Array.

1. Introduction

The next generation Very Large Array (ngVLA) will play a crucial role in the studies of 21-cm atomic neutral hydrogen (HI) emission in and around galaxies in the local Universe. The unprecedented sensitivity and resolution of the ngVLA will provide key information on the atomic gas reservoir in the circumgalactic medium (CGM) and its flow onto the gas disks of galaxies. The ngVLA will connect this large scale view of HI to individual star-forming cores to study the relatively unexplored transition from atomic to molecular gas in nearby galaxies and ultimately how gas is converted into stars.

In order to understand how gas flows onto galaxies and is then converted into molecular gas and, eventually, stars, there are three regimes to study: HI around galaxies in the circumgalactic medium; HI within nearby galaxies; and HI in the Milky Way. We will take these up in turn.

2. HI in the Circumgalactic Medium

The extent, morphology and dynamics of the gas in the very far outskirts of galaxies are still essentially unexplored observationally. These outer parts of galaxies are the interface between the inner star-forming disk and the cosmic web, and likely these are also the regions where galaxies accrete gas, possibly through the "cold accretion" process (e.g. Kereš et al. 2005). Measuring the very extended gas distributions in galaxies, and its connection to the cosmic web, will be a main science driver for the ngVLA.

To date, most of the exploration of the CGM has come through UV absorption line studies. The COS-HALOS project (Tumlinson et al. 2013; Werk et al. 2014) has used background quasars to study the Lymanα absorption in the halos of low redshift galaxies. The project has found that HI absorption at $N_{HI} \gtrsim 10^{14}\mathrm{cm}^{-2}$ is ubiquitous out to 150 kpc for star-forming galaxies and present in 75% of passive galaxies as well (Tumlinson et al. 2013). This cool CGM gas represents 25%-45% of the total baryon mass within the virial radius of the galaxy (Werk et al. 2014)[1]. Unfortunately, above $N_{HI} \sim 10^{16}\mathrm{cm}^{-2}$ saturation of absorption lines makes it difficult to get an accurate measure of N_{HI}; these are the Lyman Limit Systems. While below $N_{HI} \sim 10^{16-17}\mathrm{cm}^{-2}$, HI absorption is common, particularly in the intergalactic medium, it is impossible to image in 21-cm HI emission. Obtaining deep HI emission observations with the ngVLA is the only way forward: while Lyman-α absorption observations can reach very low column densities, their pencil-beam nature make it extremely difficult to reconstruct the full gas distribution and its kinematics.

In galaxy disks the HI surface density is declining as a function of radius, but observations with current L-band facilities are typically not sensitive enough (reaching column density sensitivities just below $10^{20}\mathrm{cm}^{-2}$) to map the HI distribution at larger radii and correspondingly lower column densities. Does the azimuthally averaged HI surface density profile continue to decline at the same rate as in the inner disk, or do other effects start to play a role in determining the shape of the very outer radial HI profile? For example, when HI column densities decrease, one expectation is that the density of the neutral gas is no longer sufficient for self shielding and this could cause the majority of the gas to become ionized by the intergalactic radiation field. Models that include this ionization from the intergalactic radiation field predict that the steep radial decline of neutral gas surface density will transition, below column densities of $\sim 10^{18}\mathrm{cm}^{-2}$, to a more extended, low column density outer disk with a flatter radial profile (e.g. Maloney 1993; Popping et al. 2009; Braun 2012), but many of the details are model-dependent. Gas accretion is not included in these models, and it is currently not known to what degree this process will affect the outskirts of the gaseous disks of galaxies. Simulations and some tentative observations indicate the presence of low-column density, kpc-sized gas features near galaxy disks (e.g. Braun & Thilker 2004; Popping et al. 2009; Wolfe et al. 2013, 2016). It is certainly conceivable that these could also have a big impact on the HI distribution at very low column densities. In order to detect such features, it is important to have both sensitive observations and resolution that is well matched to the size of the emitter. This can be seen in Figure 1. The ngVLA will have both the sensitivity and resolution to map this diffuse HI at low N_{HI} for a large number of galaxies.

---

[1]Our current understanding of the state of the CGM is well-summarized in the review by Tumlinson et al. (2017)

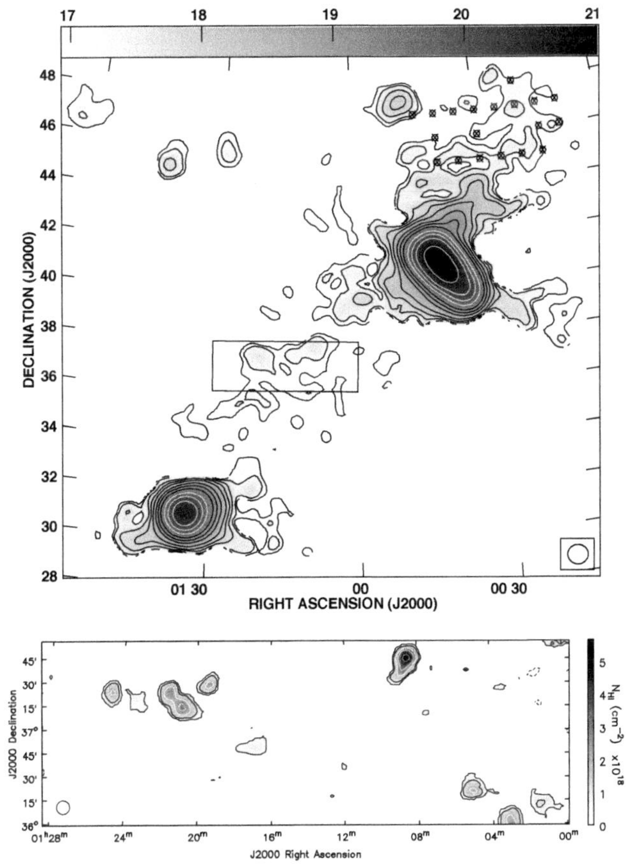

Figure 1. (*Top:*) A map of the total Hɪ emission associated with M 33 (lower left) and M 31 (upper right) from Braun & Thilker (2004). The contours are at log N_{HI} – 17.0,17.3, 17.7, 18.0, 18.3, 18.7, 19.0, 19.3, 19.7, 20.0, 20.3, and 20.7 [cm^{-2}]. The beamsize is shown in the lower right of the panel. The box shows the region mapped by Wolfe et al. (2016) with the GBT. (*Bottom:*) The GBT Hɪ map from Wolfe et al. (2016). The contours are at -1, 1, 2, 4, 6, and 10 times 5×10^{17}cm^{-2}. The beam size is the circle in the lower left of the image. Note that the Hɪ structures detected by Braun & Thilker (2004) are revealed to be much smaller, higher-N_{HI} features by Wolfe et al. (2016), illustrating the critical importance of resolution and sensitivity.

Besides characterizing the sizes of the gas disks in nearby galaxies, observing the extended HI distribution will at the same time constrain the numbers of individual HI clouds around the target galaxies. Based on the relative kinematics and morphologies/locations, one can distinguish between tidal dwarfs, compact high velocity clouds (HVCs), and tidal material that could be present around the galaxies under consideration. This will provide a link to the missing galaxy problem, i.e., satellites with clumps of cool HI, but no star-formation. Cloud sizes and masses will then provide key insights into the physical state of gas in the CGM, which can be compared to cosmological simulations of the structure of the CGM of galaxies (e.g. Shen et al. 2013). Such simulations indicate that the most compelling signature of the cold accretion phenomenon onto galaxies is a high covering factor ~20-30% of Lyman limit system absorption in the CGM on scales out to ~100 kpc (scales that can be easily covered with the ngVLA with small mosaics).

In addition to this deep ngVLA HI imaging, newly developed techniques are now available to obtain azimuthally averaged measurements at previously inaccessible radii based on the shifting and stacking technique described in Ianjamasimanana et al. (2012). This method uses an extrapolation of the (typically flat) rotation curve to create a model velocity field at large radii well beyond that of the directly detected HI disk. This is then used to shift all spectra in the data cube in velocity by the amount indicated by the extrapolated velocity field at that position. This ensures that all (potential) HI signals are lined up at the same reference velocity. Using the same tilted-ring model that was used for the shifting step, the spectra are then selected in annuli of increasing galaxy radius and are stacked. The stacked spectrum will have a lower noise value, and if any underlying HI signal has been lined up through the shift process, the signal-to-noise ratio will have increased and signatures of otherwise undetected very low column density HI can be detected. Testing this methodology using The HI Nearby Galaxy Survey (THINGS; Walter et al. 2008) and HALOGAS data has demonstrated that the HI column density limit can be pushed down by an order of magnitude (Ianjamasimanana et al. 2018). The combination of ngVLA's sensitivity, together with the above stacking technique thus will enable measurements at unprecedented HI columns ($\ll 10^{17}$cm^{-2}), that have only been accessible via Lyman α absorption measurements. In summary, through the combination of ngVLA observations and new techniques to enhance HI column density sensitivities through stacking will reach depths in HI column densities that were previously inconceivable. This measurement will provide the currently completely unexplored connection to the cosmic HI web.

3. HI in Nearby Galaxies

Understanding the processes that drive star formation in galaxies is one of the most challenging astrophysical topics, both observationally and theoretically. Of particular interest are studies of how stars form out of the interstellar medium (ISM) on scales comparable to the Jeans mass and Jeans length of neutral atomic clouds, and how, in turn, these stars shape the structure and physical properties of their ambient ISM. Of key importance n any such study are high resolution observations of the different tracers of the ISM (in particular HI and CO observations) and the stellar population (e.g., broad band and Hα imaging). For such studies, resolutions of 5 – 150 pc are needed to resolve the giant molecular cloud (GMC) complexes and the related individual sites of star formation. It is also on these scales where the feedback of individual supernovae and/or massive stars is expected to be most profound.

Although arcsecond resolutions are routinely achieved using ground–based optical telescopes, it is observations of similar resolutions at radio wavelengths, in particular the HI, that are lacking. In fact, to date, this is the limiting factor of all coherent ISM studies on these small, critical scales. In the era of ALMA, observations of the molecular gas phase (using CO as a tracer) are now routinely pushed to arcsecond resolutions. These CO observations are typically limited to the central few arcminutes of the target galaxies (given by the primary beam size of millimeter interferometers) and do not recover the atomic gas phase – they thus cannot be used to understand global star formation processes in galaxies. Recently, observations of OH emission in the Milky Way have been shown to trace molecular gas that is unseen in CO emission (Allen et al. 2012, 2015). The ngVLA may be able to detect such molecular gas clouds in other galaxies.

On the contrary, most nearby galaxies are small enough to fit within the primary beam of existing radio interferometers ($FWHM_{VLA} \sim 30'$) for HI observations and the angular resolution is the limiting factor. Typically resolutions of order 15″ are achieved; only the VLA has been pushing these resolutions down to ~6″ for large samples of galaxies. THINGS demonstrated that 6″ resolution can be achieved for a sample of 35 galaxies of any size (i.e., dwarfs to spirals) with the VLA. Only a couple of galaxies have been imaged in HI emission at higher resolution, ~1.6″, with the VLA (van Zee et al. 2006), but this requires an extremely large time commitment with the VLA. The ngVLA will fundamentally change this situation. The possibility of mapping nearby galaxies at linear resolutions of 5 – 150 pc opens the prospects of exploring processes in the ISM on scales which have hitherto been inaccessible. Although some notable exceptions exist, such as: the LMC: (Kim et al. 2003); the SMC: (Staveley-Smith et al. 1997); IC 10: (Wilcots & Miller 1998); and other Local Group dwarfs: (Begum et al. 2006), but these galaxies are not representative of the galaxy main sequence at low redshift. Figure 2 shows a THINGS image of a typical spiral galaxy at ~90 pc resolution; ngVLA will achieve better resolution and sensitivity for larger samples of galaxies in the same amount of time that the VLA took for THINGS.

(Sub-)arcsec resolution is commonly reached through massive optical/NIR imaging efforts [ground and space based, currently with *HST*, and in the near future, *JWST* (dust/PAH emission, embedded SFR tracers)]. Resolution-matched HI imaging will enable measurements on the ISM conditions required for H_2 formation (Krumholz et al. 2009; Krumholz 2013), and the distribution of, e.g., the HI/H_2 ratio will constrain the evolution of molecular clouds. For example, recent sub-pc scale studies of GMCs in the Milky Way have shown that the HI surface density and the H_2 fraction agree well with the equilibrium model predictions (e.g. Lee et al. 2012). ngVLA HI observations of Local Group galaxies will provide crucial tests for the HI surface density and the H_2 fraction on pc-scales in galaxies of varying metallicity and interstellar radiation field required to test both equilibrium and non-equilibrium models for H_2 formation and molecular cloud evolution. Observations at GMC-scale resolution are also crucial to study turbulence in the ISM. It is at these scales that energy gets injected into the ISM which then cascades down to smaller scales. We can test if the HI power spectra indeed follow the predicted power-law scales for turbulence like that found in the SMC (Stanimirovic et al. 1999; Elmegreen et al. 2001) in a large range of extragalactic environments.

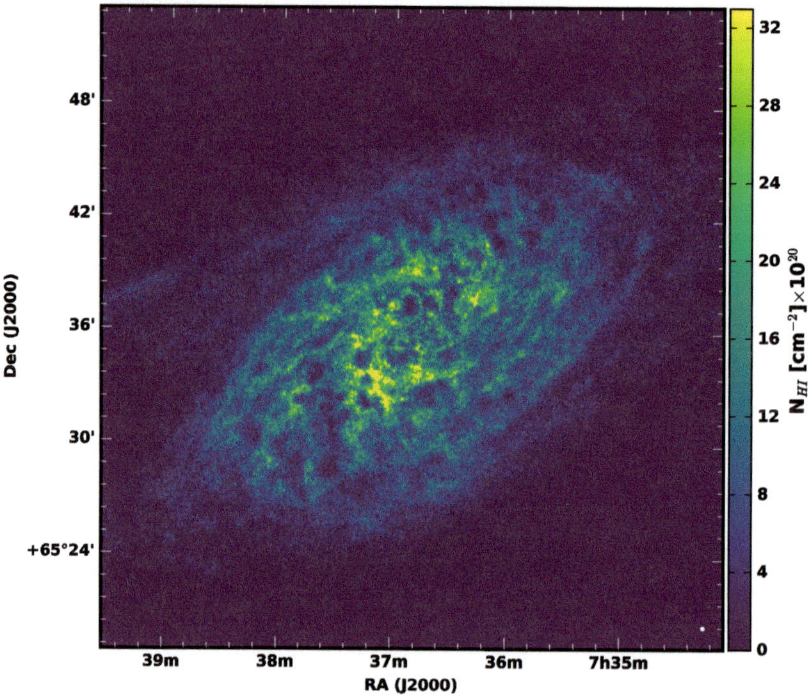

Figure 2. A total HI intensity (moment 0) image of NGC 2403 from THINGS (Walter et al. 2008). These data were imaged with robust weighting resulting in a resolution of ~6″. At the distance of NGC 2403, 3.2 Mpc, this corresponds to a physical resolution of 93 pc. The ngVLA will be able to achieve better resolution and sensitivity for larger samples of galaxies in the same amount of time as the VLA.

ngVLA GMC-resolution imaging of the HI emission will allow us to study the fine scale structure of the ISM in galaxies of different types. To date, the only galaxies for which similar linear resolutions have been achieved have been Local Group dwarfs, notably the LMC/SMC (e.g. Kim et al. 2003). These studies have shown how the emission breaks up in filaments and shells, even at the smallest currently accessible scales. The presence of fractal-like structure in the HI emission is also evident from recent HI imaging of our Galaxy (Beuther et al. 2016, in particular the large VLA THOR project). The ngVLA will reach sensitivities and column densities that this work can be replicated in galaxies across the Hubble sequence, including main sequence galaxies in the local universe. GMC-scale resolution represents a critical and fundamental scale as it samples the typical sizes of individual giant molecular and atomic clouds that are the precursors of star formation. Observations at this resolution will also enable us to directly study the feedback/energy input from individual supernovae due to massive stars.

A unique feature of the high-resolution Hɪ imaging is that they will, for the first time, enable us to study non-circular motions in the Hɪ disks of galaxies at scales that are directly relevant for the ΛCDM cusp/core controversy. Hɪ observations have the immediate advantage that the kinematics in the very centers of galaxies can be immediately linked to their outskirts. The large range of Hubble types that can be studied (including the important dwarf galaxies whose kinematics/DM content cannot be studied through ALMA imaging using CO) will also enable precise quantification of the effects of the non-circular motions on the dark matter profile as a function of e.g., disk mass, depth of the potential, etc. Measuring the strength of shocks and other non-rotation velocity components at these scales (e.g., across spiral arms/bars) will allow us to derive independent estimates of the mass in the disk (see e.g., descriptions in Weiner et al. 2001b,a), another crucial input in the derivation of dark matter profiles in galaxies. This information will also provide important insights in assessing the local stability of gas clouds against gravitational collapse.

GMC-scale resolution data will resolve the giant atomic cloud complexes for the first time in galaxies beyond the local group and will allow investigations of if and how the peak Hɪ temperature of the warm Hɪ changes of a function of galaxy type. Of particular interest is the question if peak Hɪ temperature is different in low-metallicity dwarfs as compared to more metal-rich spirals. Such studies will directly constrain the physical conditions of the interstellar medium on sub-100 pc scales. In addition, background continuum sources will be bright enough to perform Hɪ absorption studies. In this context, ngVLA's 1″ resolution is essential to isolate the absorption from the surrounding emission. This then opens the possibility to determine directly the Hɪ spin temperature, something which has largely only been possible in Local Group galaxies (e.g. Dickey & Brinks 1993; Braun 1997).

4. Cloud Scale Imaging of Hɪ in the Milky Way

The increased sensitivity of the ngVLA will be crucial for addressing major outstanding questions in ISM physics. The diffuse neutral ISM exists in two flavors: the cold neutral medium (CNM) and the warm neutral medium (WNM). The CNM is easily studied with the 21-cm line absorption and its properties have been measured extensively, while only several measurements of the WNM temperature exist so far. The main reason for this observational paucity is the low optical depth of the WNM, $\tau \sim 10^{-3}$. This makes the WNM one of the least understood phases of the ISM. To constrain theoretical and numerical models of the ISM, temperature distributions over the full temperature range from ~20 to ~10^4 K are essential. Using the upgraded Very Large Array (project 21-SPONGE), Murray et al. (2014, 2015, 2017) demonstrated that already the VLA can detect Hɪ absorption lines with a peak optical depth ~10^{-3}. However, the existing sample size of the current measurements is of order ~50 directions – way too low to start building proper statistical samples for a comparison with numerical simulations and understanding of how the CNM-WNM phase mix vary with interstellar environments.

In addition, the Hɪ excitation processes are still not fully understood. Using a stacking analysis, Murray et al. (2014) detected statistically the presence of a widespread WNM population with T_s=7200 K. This demonstrated that the non-collisional excitation of Hɪ is significant even at high Galactic latitudes. As Lyα scattering is the most likely candidate for additional excitation of HI, the Murray et al. results show that the fraction of Lyα photons, and/or the photon propagation throughout the ISM, are likely more complicated than what is commonly assumed. For example, both a theoretical study by Liszt (2001) and state-of-the-art numerical simulations (Kim et al. 2014) assume a uniform flux of Lyα photons throughout the ISM and result is the expected $T_s < 4000$ K. In that regard, the high sensitivity enabled by the ngVLA will allow direct measurements of WNM spin temperature and its spatial variations in the Milky Way by observing hundreds to thousands of directions with $\sigma_T < 0.05$. By stacking selectively, different interstellar environments will be probed for the first time. Finer scale variations may also be possible to measure by imaging resolved continuum sources to map out variations on sub-parsec scales (e.g. Faison et al. 1998; Lazio et al. 2009; Roy et al. 2012).

Simultaneously, the ngVLA will be able to map out the raw fuel for H_2 formation across major GMCs in the Milky Way. While it is commonly assumed that the CNM is the key ingredient for making H_2 because of its high density, the observational evidence whether GMC precursors are in the form of dense HI clouds, diffuse HI, or smaller molecular clouds, is still lacking (Dobbs & Pringle 2013). In addition, numerical simulations suggest that GMCs grow via accretion of the CNM from the envelopes. Recent Green Bank Telescope results (Allen et al. 2012, 2015) have shown that OH is tracing diffuse molecular gas that is not seen via CO measurements. This may represent an intermediate stage in forming GMCs. Constraining where the fuel for H_2 formation is in and around GMCs, what are its physical properties, and what is the accretion rate of this material onto the GMCs, are essential next steps. Due to sensitivity limitations, almost all HI absorption measurements (e.g., 21- SPONGE) have focused on random directions, utilizing strong background radio sources. The ngVLA can provide, for the first time, dense grids of HI absorption spectra in the direction of selected GMCs. The ngVLA has thus potential to revolutionize our understanding of the ISM physics, in particular the critical phase transition from the WNM to the CNM to H_2 formation.

5. Limitations of Current Astronomical Instrumentation

Current radio interferometers lack the combination of resolution and surface brightness sensitivity. In addition, the larger dishes of the VLA and WSRT restrict the field of view limiting the region of study. These all combine to make it near impossible to map the low-N_{HI} gas within the virial radius of nearby galaxies, while being able to resolve any features. While single-dish telescopes, like the GBT, have the surface brightness sensitivity, they lack the resolution to map filamentary structures beyond the Local Group. For high resolution studies of HI in nearby galaxies or in the Milky Way, the sparse nature of the VLA or WSRT mean that a lot of flux is missing in these data.

In terms of planned interferometers, MeerKAT is nearing completion and will be well-suited to mapping the diffuse, low-N_{HI} gas around nearby galaxies; the MHONGOOSE survey (de Blok et al. 2017) will do exactly this for a sample of 30 galaxies. The IMAGINE survey, currently being undertaken with the Australia Telescope Compact Array (ATCA) and the Parkes radio telescope is also studying the diffuse HI around galaxies, but both of these surveys will only reach the needed sensitivity over a beam size of $\sim 30''$. Neither survey will have the high resolution, $\sim 1''$ that will be needed to study the internal structure of galaxies at sufficient detail.

6. Unique Capabilities of the ngVLA for H I Science

The planned technical capabilities of the ngVLA at 1.4 GHz should be matched by the SKA. The SKA, however, will be located in the southern hemisphere leaving many nearby galaxies (e.g., M 31) and the outer portions of the Milky Way inaccessible. Furthermore, current plans are for the SKA to operate in a survey mode. The ngVLA has the opportunity to be PI-driven allowing for a wider variety of science to be done. In addition, for sources that extend over a large area, total power H I measurements will be required. While the ngVLA has a dense core and will be able to recover flux on much larger angular scales than the VLA, short spacing data will still be needed for some observations. This can be provided with existing instrumentation using the Green Bank Telescope (Frayer 2017) or the planned ngVLA Short Baseline Array (see Selina et al. 2018, this volume p. 15). Finally, since ngVLA will be able to operate at higher frequencies than the SKA, much of the science that requires complementary multi-wavelength data (see below) will require the ngVLA, even if the H I data comes from SKA observations.

7. Experimental Layout

The reference configuration of the ngVLA, Spiral214, can be divided into three parts: the "core" array, the "plains" array, and the full array. For the science described here, only the core array (with B_{max} ~1 km) and the plains array (with B_{max} ~30 km), will be useful (ngVLA memo 17). Some H I absorption studies, such as using background radio galaxies to map small-scale, high-N_{HI} gas, may use the full array. Using the reference design sensitivity, the core array will be able to reach $\sigma_{N_{HI}} \sim 10^{18}\mathrm{cm}^{-2}$ in 1 hour over a 10 $\mathrm{km\,s^{-1}}$ channel with a beamsize of $\sim 40''$, while the plains array will reach $\sigma_{N_{HI}} \sim 6\times10^{20}\mathrm{cm}^{-2}$ for the same time and channel width over a beam of $\sim 1''$.

For studies of the CGM of nearby galaxies, astronomers will want to use the ngVLA to study H I at $N_{HI} \sim 10^{17-18}\mathrm{cm}^{-2}$ for a $\sim 20\ \mathrm{km\,s^{-1}}$ linewidth. To get a 3σ detection of H I at $N_{HI} = 10^{18}\mathrm{cm}^{-2}$ will require ~30 hours. To reach $N_{HI} = 10^{17}\mathrm{cm}^{-2}$ would require ~3000 hours at a resolution of $43''$. If we degrade the resolution to $\sim 1'$, then we could do this in about 600 hours for a single target. These time requirements roughly match those of the SKA1-MID telescope (see de Blok et al. 2017, for examples). Given the planned, semi-random distribution of antennas in the core array, we can anticipate that the sensitivity will improve with additional tapering. As diffuse H I emission could be extended over large angular scales, single-dish, total power H I data may be needed. This can easily be provided by the GBT using an array receiver, such as FLAG (Roshi et al. 2018). Since the H I filaments expected from models of cold flow accretion are on the scale of 1–10 kpc across (e.g. Popping et al. 2009), we can probe to $N_{HI} \lesssim 10^{17}\mathrm{cm}^{-2}$ for a small sample of nearby galaxies ($d < 10$ Mpc) with relatively modest time requirements (<1000 hours) and we can study a much larger sample of galaxies out to larger distances at $N_{HI} = 10^{18}\mathrm{cm}^{-2}$ in a similar amount of time.

For studies of H I within nearby galaxies, we will want the high angular resolution of the plains array ($\theta \sim 1''$). Samples of galaxies at distances of 1 Mpc< d <30 Mpc are ideal targets–and a wide range of different types of galaxies can be studied in this volume. At these distances, $1''$ resolution corresponds to spatial resolutions of 5 – 150 pc. In 1 hour, for a 10 $\mathrm{km\,s^{-1}}$ channel width, the ngVLA will yield $\sigma_{N_{HI}} \sim 6\times10^{20}\mathrm{cm}^{-2}$. Even this sensitivity will allow astronomers to study the high-N_{HI} associated with the gas transitioning between the atomic and molecular phases. In a couple of days of observing we could then redo a survey like THINGS at much higher physical resolution and sensitivity.

In addition to direct observations of HI in nearby galaxies and the Milky Way, we would want to get commensal observations of OH, radio recombination lines, and radio continuum with full polarization. These data will help map out the full transition of gas from the atomic to the molecular to the ionized phase.

8. Complementarity

First and foremost, the ngVLA study of HI within nearby galaxies, including the Milky Way, will be complemented by the ngVLA's studies at other wavelengths of other molecules in these systems, such as CO, HCN, NH_3, etc., as well as the radio continuum associated with star formation and heated dust. This will yield one of the most detailed pictures of how gas is accreted onto galaxies, transitions into molecular clouds, and then turns into stars. Complementing the data from ngVLA will be a broad range of facilities. In the infrared, from the near IR through the far IR, will be *JWST*, *Euclid*, SPICA, WFIRST, and ALMA. These telescopes will be able to study warm dust, important cooling lines in the ISM (such as [CII]), protostars and obscured star formation in general.

In the study of the CGM around galaxies, new X-ray facilities such as eROSITA and Athena should allow for the study of the hot halos of galaxies, either directly or via stacking. Unfortunately, at present there are no planned UV telescopes that could probe the CGM and IGM in absorption, tracing the WHIM with better sensitivity and towards a denser grid of background sources around many galaxies. Such a mission would allow astronomers to both map the hot gas in the X-rays, the warm-hot ionized gas in the CGM using UV absorption lines, and then diffuse, warm neutral hydrogen with the ngVLA. This would give us a complete census of the gas available to accrete onto galaxies and fuel future star formation in the local universe.

Acknowledgments. The authors thank Nickolas Pingel for his assistance in making figures.

References

Allen, R. J., Hogg, D. E., & Engelke, P. D. 2015, AJ, 149, 123
Allen, R. J., Ivette Rodríguez, M., Black, J. H., & Booth, R. S. 2012, AJ, 143, 97
Begum, A., Chengalur, J. N., Karachentsev, I. D., Kaisin, S. S., & Sharina, M. E. 2006, MNRAS, 365, 1220
Beuther, H., Bihr, S., Rugel, M., et al. 2016, A&A, 595, A32
Braun, R. 1997, ApJ, 484, 637
—. 2012, ApJ, 749, 87
Braun, R., & Thilker, D. A. 2004, A&A, 417, 421
de Blok, W. J. G., Adams, E. A. K., Amram, P., et al. 2017, ArXiv e-prints, arXiv:1709.08458
Dickey, J. M., & Brinks, E. 1993, ApJ, 405, 153
Dobbs, C. L., & Pringle, J. E. 2013, MNRAS, 432, 653
Elmegreen, B. G., Kim, S., & Staveley-Smith, L. 2001, ApJ, 548, 749
Faison, M. D., Goss, W. M., Diamond, P. J., & Taylor, G. B. 1998, AJ, 116, 2916
Frayer, D. T. 2017, ArXiv e-prints, arXiv:1706.02726
Ianjamasimanana, R., de Blok, W. J. G., Walter, F., & Heald, G. H. 2012, AJ, 144, 96

Ianjamasimanana, R., Walter, F., de Blok, W. J. G., Heald, G. H., & Brinks, E. 2018, ArXiv e-prints, arXiv:1803.10291
Kereš, D., Katz, N., Weinberg, D. H., & Davé, R. 2005, MNRAS, 363, 2
Kim, C.-G., Ostriker, E. C., & Kim, W.-T. 2014, ApJ, 786, 64
Kim, S., Staveley-Smith, L., Dopita, M. A., et al. 2003, ApJS, 148, 473
Krumholz, M. R. 2013, MNRAS, 436, 2747
Krumholz, M. R., McKee, C. F., & Tumlinson, J. 2009, ApJ, 693, 216
Lazio, T. J. W., Brogan, C. L., Goss, W. M., & Stanimirović, S. 2009, AJ, 137, 4526
Lee, M.-Y., Stanimirović, S., Douglas, K. A., et al. 2012, ApJ, 748, 75
Liszt, H. 2001, A&A, 371, 698
Maloney, P. 1993, ApJ, 414, 41
Murray, C. E., Stanimirović, S., Kim, C.-G., et al. 2017, ApJ, 837, 55
Murray, C. E., Lindner, R. R., Stanimirović, S., et al. 2014, ApJ, 781, L41
Murray, C. E., Stanimirović, S., Goss, W. M., et al. 2015, ApJ, 804, 89
Popping, A., Davé, R., Braun, R., & Oppenheimer, B. D. 2009, A&A, 504, 15
Roshi, D. A., Shillue, W., Simon, B., et al. 2018, AJ, 155, 202
Roy, N., Minter, A. H., Goss, W. M., Brogan, C. L., & Lazio, T. J. W. 2012, ApJ, 749, 144
Selina, R. J., Murphy, E. J., McKinnon, M., et al. 2018, Science with a Next-Generation Very Large Array, ASP Monograph 7, ed. E. J. Murphy (San Francisco, CA: ASP), 15
Shen, S., Madau, P., Guedes, J., et al. 2013, ApJ, 765, 89
Stanimirovic, S., Staveley-Smith, L., Dickey, J. M., Sault, R. J., & Snowden, S. L. 1999, MNRAS, 302, 417
Staveley-Smith, L., Sault, R. J., Hatzidimitriou, D., Kesteven, M. J., & McConnell, D. 1997, MNRAS, 289, 225
Tumlinson, J., Peeples, M. S., & Werk, J. K. 2017, ARA&A, 55, 389
Tumlinson, J., Thom, C., Werk, J. K., et al. 2013, ApJ, 777, 59
van Zee, L., Cannon, J. M., & Skillman, E. D. 2006, in Bulletin of the American Astronomical Society, Vol. 38, American Astronomical Society Meeting Abstracts, 1136
Walter, F., Brinks, E., de Blok, W. J. G., et al. 2008, AJ, 136, 2563
Weiner, B. J., Sellwood, J. A., & Williams, T. B. 2001a, ApJ, 546, 931
Weiner, B. J., Williams, T. B., van Gorkom, J. H., & Sellwood, J. A. 2001b, ApJ, 546, 916
Werk, J. K., Prochaska, J. X., Tumlinson, J., et al. 2014, ApJ, 792, 8
Wilcots, E. M., & Miller, B. W. 1998, AJ, 116, 2363
Wolfe, S. A., Lockman, F. J., & Pisano, D. J. 2016, ApJ, 816, 81
Wolfe, S. A., Pisano, D. J., Lockman, F. J., McGaugh, S. S., & Shaya, E. J. 2013, Nat, 497, 224

Science with a Next-Generation Very Large Array
ASP Conference Series, Monograph 7
Eric J. Murphy, ed.

Gas Density Across the Universe

Adam K. Leroy,[1] Alberto Bolatto,[2] Amanda Kepley,[3] David Meier,[4] and Tony Wong[5]

[1]*Ohio State University, Department of Astronomy, Columbus, OH 43210;* leroy.42@osu.edu

[2]*Department of Astronomy, University of Maryland, College Park, MD 20742*

[3]*National Radio Astronomy Observatory, 520 Edgemont Road, Charlottesville, VA 22903-2475*

[4]*New Mexico Institute of Mining and Technology, 801 Leroy Place, Socorro, NM 87801*

[5]*Department of Astronomy, University of Illinois, Urbana, IL 61801*

Abstract. Gas density is widely believed to play a governing role in star formation. However, the exact role of density in setting the star formation rate remains debated. We also lack a general theory that explains how the gas density distribution in galaxies is set. The primary factor preventing the resolution of these issues is the limited number of observations of the gas density distribution across diverse environments. Centimeter- and millimeter-wave spectroscopy offer the most promising way forward in this field, but the key density-sensitive transitions are faint compared to the capabilities of current telescopes. In this chapter, we describe how a next-generation Very Large Array (ngVLA) represents the natural next step forward in this sensitivity-limited field. Such a facility would provide a crucial link between the "Milky Way" and "Extragalactic" views of star formation and dramatically advance our understanding of the drive and role of gas density in galaxies, building on current results from ALMA, NOEMA, the Green Bank Telescope, and other current facilities working in this area.

1. Is Density Destiny in Star Formation?

Gas density and star formation are closely linked. Observationally, several lines of evidence show a strong relationship between gas density and star formation. Within the Milky Way, the amount of dense gas is the best predictor of the current rate of star formation in nearby clouds (e.g., Evans et al. 2014). Specifically, star formation appears associated with dense, filamentary substructure within these clouds (see review by André et al. 2014). On larger scales, integrated measurements of entire galaxies demonstrate a linear correlation between the rate of recent star formation and the amount of dense molecular gas traced by high critical density emission lines, especially HCN and HCO^+ (e.g., Gao & Solomon 2004).

This close association makes sense, given that gas density plays a key role in nearly every theory of star formation. But so far our observations of dense gas have been severely limited by the sensitivity of our telescopes. The key lines to observe dense gas are more than an order of magnitude fainter than the workhorse CO lines. In the absence of the kind of sensitive, resolved observations of dense gas that will be made by the ngVLA, the exact role played by density in star formation remains relatively unconstrained.

Recent theories of star formation disagree about whether the mean density, only the fraction of dense gas, or the distribution of densities is the key factor setting the rate of star formation. In the first category, the governing timescale to convert gas into stars is the gravitational free-fall time at some characteristic scale. Then this timescale, and so the normalized rate of star formation, is set by the mean density of the ISM at that scale (e.g., Krumholz & Dekel 2012; Elmegreen 2018). In a theory like this, for example, the mean density of molecular cloud will govern its ability to form stars. In a second class of models, gas above some threshold density forms stars in a semi-universal way (e.g., Gao & Solomon 2004; Lada et al. 2012). Such models focus on the formation of dense gas, however it happens, as the key limiting step to form stars. A third category of theories considered the distribution of densities within a cloud (often treated as turbulent, isothermal gas following, e.g., Padoan & Nordlund 2002; Krumholz & McKee 2005) to play the key role in determining its ability to form stars (e.g., Hennebelle & Chabrier 2011; Federrath & Klessen 2012).

We also know that the density of gas changes in response to the environment within a galaxy. For example, major mergers (such as the local ultra-luminous infrared galaxy population) appear to have a much higher mean gas density than normal spiral galaxies (Gao & Solomon 2004; García-Burillo et al. 2012). The central parts of disk galaxies — including the Milky Way's own Central Molecular Zone — show higher gas densities than the more quiescent outer parts of disks (e.g., Longmore et al. 2013; Usero et al. 2015; Chen et al. 2015; Bigiel et al. 2016; Gallagher et al. 2018). Indeed, gas density also appears to change systematically across spiral galaxies, with denser gas in high surface density, gas rich environments. Though there are suggestions of a link to interstellar gas pressure (e.g., Helfer & Blitz 1997), we lack a predictive theory that explains how the structure and properties of a galaxy set its gas density distribution.

The central role of density in star formation theory, the close association of molecular gas density and star formation, and the observation that gas density changes in response to environment prompts a series of questions:

1) What is the distribution of gas volume densities in the molecular interstellar medium?

In theoretical work, this is often discussed as the density probability distribution function (PDF). Knowledge of the density PDF is central to understand star formation (see above), feedback (e.g., Krumholz. et al. 2016), and emission from molecular lines (e.g., Leroy et al. 2017). Based on theoretical work, a lognormal, power law, or some hybrid distribution is often assumed (see Padoan & Nordlund 2002; Krumholz & McKee 2005). However, observational constraints on this quantity remain weak. The best observational measurements currently consist of column density measurements in local clouds (e.g., Abreu-Vicente et al. 2015). However, as discussed above, observations clearly indicate that the mean gas density changes across the universe, and resolved observations of local clouds offer a window into only a very narrow subset of the physical conditions found across the universe.

2) How does gas density relate to star formation in different environments?

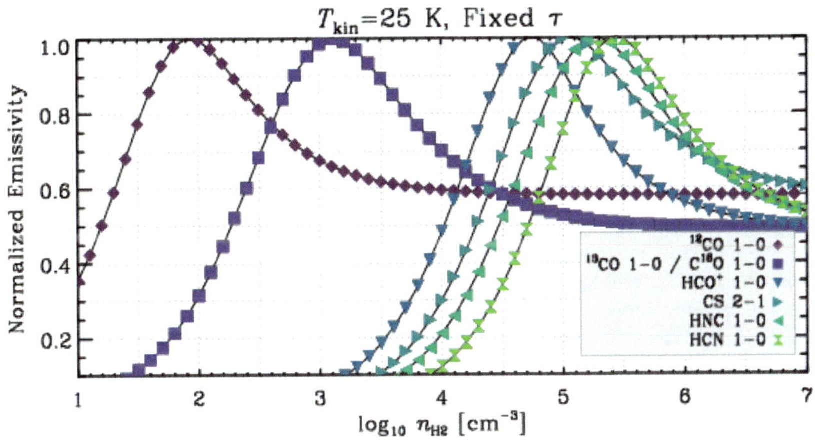

Figure 1. **Emissivity vs. Density for Some cm- and mm-Wave Lines.** Emissivity (intensity per column density) as a function of collider volume density for a subset of lines in the ngVLA high frequency window (following Leroy et al. 2017). The ngVLA covers low-*J* lines (which are relatively less sensitive to excitation effects) that emit effectively under a wide range of gas densities. Combing these lines offers the prospect to constrain the density distribution point by point across galaxies. The sensitivity, bandwidth (allowing heavy multiplexing), and resolution of the ngVLA are unmatched for this kind of work, giving it the prospect to act as an unparalleled density mapping machine.

As described above, observations suggest a strong link between gas density and star formation. However, the quantitative relationship between the gas density distribution and the formation of stars and clusters remains weakly constrained by observations. The strongest possible test of many current theories of star formation in molecular clouds would be detailed observations of the density distribution, dynamical state, and star formation rate of molecular clouds in a wide range of environments and evolutionary states. These observations would clearly resolve whether there some threshold density above which star formation proceeds in a universal way and whether the gravitational free-fall time is indeed the governing timescale for star formation over many scales.

3) How does gas density relate to galactic structure and dynamics?

The gas density distribution clearly changes across the universe, with denser gas found in environments with deep potential wells and strong compressive effects (e.g., galaxy mergers, spiral arms, gas flows along bars). Our quantitative knowledge of how gas density responds to environment, and vice versa, is in its infancy, however.

2. Sensitive cm-and mm-Wave Spectroscopy as the Path Forward

Thee preceding questions remain major open issues because observing molecular gas density is incredibly challenging. The densest regions of molecular clouds are typically, ~ 0.1–1 pc in scale (see Lada & Lada 2003; André et al. 2014). In Milky Way molecular clouds, one can achieve the resolution to pick out individual dense substructures, and so reconstruct the density distribution from imaging. Observations in the Milky Way, and the Solar Neighborhood, however, probe only a limited subset of the environments found across the universe.

Even with the ngVLA, is simply infeasible to image gas at 0.1–1 pc resolution beyond the handful of nearest galaxies. As a result, cm- and mm-wave spectroscopy is our best tool to study the density distribution in more distant systems. The cm- to mm-wave part of the spectrum contains rotational transitions from a host of interstellar molecules. These transitions, in turn, have a wide range of critical densities. The emissivity of gas in any particular molecular transition falls off quickly with decreasing density below the effective critical density. By measuring the relative intensities of lines with a wide range of effective critical densities, one can thus probe a wide range of different gas densities. We can use these observations to constrain the distribution of mass as a function of gas volume density within a beam (i.e., to constrain the density PDF in the beam). This technique works without any need to resolve individual sub-parsec scale dense substructures because the spectroscopic tracers already preferentially trace gas near the effective critical density of the line.

This approach holds huge promise: the ability to measure the distribution of densities in each cloud, or even each moderate size region, across a galaxy will allow us to concretely link this key driver for star formation on cloud scales (density) to galactic environment. This push towards measuring physical conditions in the cold gas across galaxies is a major thrust of star formation studies in ~ 2020 and is sure to remain a key area moving in to the era of the ngVLA.

3. The Need for the ngVLA

The main obstacle to deploying this approach widely is the faintness of the high critical density transitions available in the cm- and mm-wave window. The brightest transitions in this window include rotational transitions of HCN and HCO^+. These molecules are often $\sim 10^8$ times less abundant than H_2 and $\sim 10^4$ times less abundant than CO (e.g., Martín et al. 2006), and the dense gas often makes up only ~ 1–10% of a cold cloud by mass (e.g., Lada et al. 2010). As a result, even the brightest high critical density lines ("dense gas tracers") are extremely faint compared to the standard molecular gas-tracing CO lines. Line ratios between HCN (the brightest dense gas tracer) and CO seldom exceed 0.1 across a large part of a galaxy and typical values in galaxy disks are ~ 0.02 to < 0.01 (Usero et al. 2015; Bigiel et al. 2016; Gallagher et al. 2018).

To harness the power of mm-wave spectroscopy to trace density across galaxies, we need an instrument that simultaneously achieves high spatial resolution, very high surface brightness sensitivity, and large spectral coverage (to capture many transitions at once). The ngVLA is exactly this instrument. The reference design review version of the ngVLA achieves rms sensitivity 0.01 K per 10 km s^{-1} channel at $1''$ resolution in 1 hour. The beam size of $1''$ matches the sizes of individual molecular clouds for galaxies out to the Virgo Cluster (e.g., see Sun et al. 2018). At this resolution and for a characteristic cloud line width of 10 km s^{-1}, ~ 0.01 K is a typical brightness for HCN in a star-forming galaxy. Finally, the ngVLA bandwidth simultaneously captures almost all of the density-sensitive diagnostics in the $\lambda = 2.6$–4 mm window (depending on the exact distribution of the 20 GHz of bandwidth across the band), allowing a stunning gain in efficiency to carry out this fundamentally multi-line science. Previous telescopes require many tunings to cover the suite of lines seen in Figure 1.

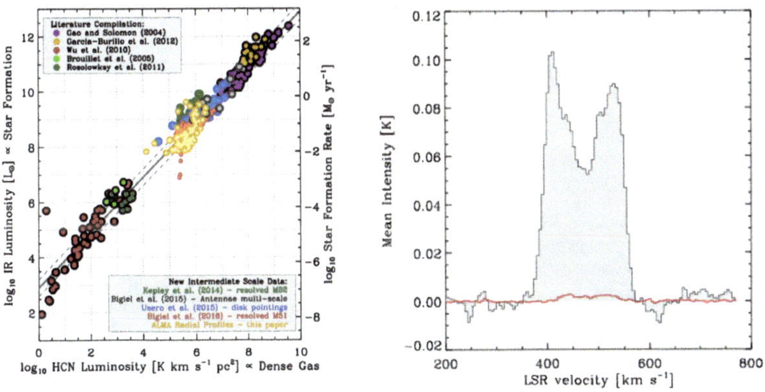

Figure 2. **The Promise and Challenge of Mapping Dense Gas.** (*Left:*) From Gallagher et al. (2018), the correlation between recent star formation and dense gas seen over 15 years of HCN studies (following Gao & Solomon 2004; Wu et al. 2005, 2010). Across decades in scale, tracers of high density gas correlate well with star formation, though the detailed physics have yet to be well-constrained by observations. (*Right:*) The integrated HCN spectrum of M51 (from Bigiel et al. 2016, and see Chen et al. 2015) plotted in red over the integrated CO spectrum of the galaxy in gray. Even the brightest "dense gas tracers" are ~ 10–100 times fainter than CO emission. As a result, the sort of density-probing spectroscopy discussed in this chapter strains the capabilities of current facilities. The sensitivity of the ngVLA will greatly surpass the sensitivity of these facilities, enabling us to map dense gas tracers at high resolution across the local universe (~ Virgo).

Put more simply, Figure 3 shows a ~ 1″ resolution CO (2-1) map of a nearby star-forming spiral galaxy, M 99 (NGC 4254), from ALMA's PHANGS survey (see Sun et al. 2018, data from A. K. Leroy, E. Schinnerer et al. 2019, in preparation). In the right panel, we scale the CO map by a typical HCN-to-CO ratio and saturate the color scale at the rms line sensitivity achieved in one hour of integration (this level is also shown by a white line). This exercise shows that the fiducial sensitivity of the ngVLA in an hour is of the right order to detect faint, high critical density tracers. Integrations of a few hours or ten hours towards a part of a local galaxy should recover a large suite of faint lines in each beam, allowing the reconstruction of cloud-by-cloud density distribution across galaxies.

To be more concrete, the ngVLA at 90 GHz could cover the inner, gas rich part of M99 (the illustrated galaxy in Figure 3) in about 10 pointings. With an integration time of ~ 10 hours per pointing and 10 pointings, a survey of the active part of this galaxy would cost ~ 100 hours. Such a survey would have a sensitivity of ~ 0.003 K per 10 km s^{-1} channel at physical resolution ~ 75 pc, covering at least half a dozen lines with different effective critical densities (CN, HCN, HCO^+, HNC, CS, their isotopologues, etc.). This would allow one to detect the dense gas in an individual molecular cloud ($M \sim 5 \times 10^5\ M_\odot$) at reasonable signal to noise down to dense gas fractions of a few percent in multiple tracers. This cloud-by-cloud view of the density distribution across a whole galaxy is exactly the kind of observation needed to address the key questions outlined above. It would be the kind of leap forward for dense gas that the

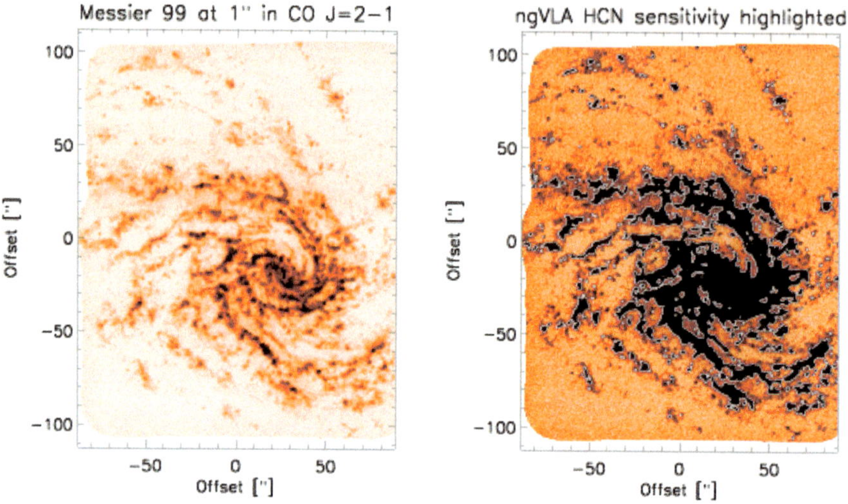

Figure 3. **The ngVLA Will Spectroscopically Image Faint Lines at High Resolution With Huge Multiplexing Capability.** CO (2-1) peak temperature map (in a 10 km s^{-1} channel) for a local (Virgo cluster) star forming spiral at 1″ resolution, sufficient to distinguish individual clouds, spiral arms, and distinct environments. On the right we show a version of the map scaled by a typical HCN-to-CO ratio (Usero et al. 2015) and saturated at the 1 hour, 1″, 10 km s^{-1} line sensitivity of the ngVLA reference design, which is also shown by a white contour. The exercise shows that *the ngVLA will be able to detect high critical density lines on the scale of individual clouds in a few hours of integration. While doing so, it simultaneously observes a huge part of the cm- and mm-wave spectrum and covers a large area.* These capabilities make the ngVLA a density-mapping machine unmatched by any current or planned facility. In the text, we sketch how a ~ 100 hour program could be expected to measure the density distribution in individual molecular clouds across the whole active part of the galaxy.

pioneering PAWS survey (Schinnerer et al. 2013) made for molecular gas: the first simultaneously sensitive, wide area, and high resolution view of gas (here dense gas, for PAWS CO emission) across a whole large part of galaxy.

4. Complementary Science — Chemistry and Physical Conditions Across Galaxies

Density is not the only factor at play in the cm- and mm- part of the spectrum. Chemistry depends on environment and varies the abundance of each species as a function of environment (e.g., see van Dishoeck 2014). Both gas excitation and radiative transfer through clouds can play key roles in what we observe. A strength of the ngVLA's large proposed bandwidth is the ability to observe many transitions in one observation. Contrasting results from lines with similar critical densities will allow us to isolate the effects of density and chemistry. In addition, each density-probing observation will also be a line survey with potential sensitivity to the presence of shocks, the relative contributions of PDRs, XDRs, and CRDRs to that part of the galaxy, and excitation in the gas. Though we highlight density as a goal, this direction is inextricably linked to astrochemistry and the broader topic of physical conditions and excitation in the interstellar medium.

5. Interface with Existing and Planned Facilities

Many telescopes are making critical contributions to this topic right now including ALMA, NOEMA, the GBT, IRAM 30m, Nobeyama 45m, and the LMT. While no existing or planned facility approaches the power of the proposed ngVLA at the key frequencies $\nu < 116$ GHz, these other facilities are likely to do key work over the next decade that will inform future science directions with the ngVLA and have strong synergies with the ngVLA when it comes online.

Currently, ALMA is the best instrument available to the density-mapping experiments described above (soon to be paired with NOEMA). We expect that the state of the field when ngVLA comes online will likely be set by these two facilities. However, ALMA necessarily works at coarser resolution with less sensitivity and less bandwidth to map these lines. For matched resolution ($1''$), ALMA achieves ~ 9 times worse sensitivity in an hour with only a slightly bigger field of view. This calculation means that to match the sensitivity achieved by the ngVLA in 1 hour, ALMA would have to integrate almost 100 hours, rendering large surveys impossible. ALMA also covers only 8 GHz of bandwidth (though this is likely to be improved by upgrades). The ngVLA's proposed 20 GHz instantaneous bandwidth would allow vastly improved line multiplexing, covering a large part of the spectrum in a single observation. The sensitivity and bandwidth of the ngVLA will allow us to survey many lines in many different types of systems with relatively high resolution. The ngVLA promises a similar upgrade over the full NOEMA, which offers a wide (16 GHz) instantaneous bandwidth that allows for great multiplexing, but cannot match the raw sensitivity of the ngVLA due to the vast mismatch in collecting area.

Even so, ALMA and NOEMA will certainly blaze the trail towards the kind of science described above. Even with the ngVLA, ALMA will remain the key tool to study excitation, including the interface of excitation and density (likely by contrasting with ngVLA studies). This ability to combine high-J (with ALMA) and low-J (with ngVLA) observations will open an extremely powerful window into the physical state of gas in galaxies.

Similarly, the Green Bank Telescope, the IRAM 30-m telescope, the Nobeyama 45-m telescope, and the Large Millimeter Telescope will all make key contributions in this area over the next years. These single dish telescopes, however, necessarily emphasize wide field mapping, leveraging their surface brightness sensitivity, but do not resolve galaxies into individual clouds or physically distinct regions. These telescopes, especially if upgraded with new instrumentation, are likely to dramatically improve our integrated inventory of dense gas across the universe. These results will put the ngVLA in the position to resolve the physics related to density in discrete parts of galaxies: clouds, arms, and galaxy nuclei.

In thinking about the interplay of these facilities, a key point to keep in mind is that the stunning combination of bandwidth and collecting area for the ngVLA is guaranteed to open new windows into the physical state of gas across galaxies. Even in the most optimistic scenario, in which HCN and HCO^+ are mapped across many galaxies over the next ten years by the GBT, NOEMA, and ALMA, a suite of more sophisticated diagnostics — optically thin isotopologues, chemicals that appear when CO freezes out, etc. — lie just out of the most optimistic reach of current telescopes. The ngVLA will be an incredible tool to access these lines.

Acknowledgments. We thank the referee of this article, Fabian Walter, for helpful comments that improved the content. AKL gratefully acknowledges partial support by the National Science Foundation under Grants No. 1615105, 1615109, and 1653300.

References

Abreu-Vicente, J., Kainulainen, J., Stutz, A., Henning, T., & Beuther, H. 2015, A&A, 581, A74
André, P., Di Francesco, J., Ward-Thompson, D., et al. 2014, Protostars and Planets VI, 27
Bigiel, F., Leroy, A. K., Jiménez-Donaire, M. J., et al. 2016, ApJ, 822, L26
Chen, H., Gao, Y., Braine, J., & Gu, Q. 2015, ApJ, 810, 140
Elmegreen, B. G. 2018, ApJ, 854, 16
Evans, II, N. J., Heiderman, A., & Vutisalchavakul, N. 2014, ApJ, 782, 114
Federrath, C., & Klessen, R. S. 2012, ApJ, 761, 156
Gallagher, M. J., Leroy, A. K., Bigiel, F., et al. 2018, ArXiv e-prints, arXiv:1803.10785
Gao, Y., & Solomon, P. M. 2004, ApJ, 606, 271
García-Burillo, S., Usero, A., Alonso-Herrero, A., et al. 2012, A&A, 539, A8
Helfer, T. T., & Blitz, L. 1997, ApJ, 478, 162
Hennebelle, P., & Chabrier, G. 2011, ApJ, 743, L29
Krumholz, M. R., & Dekel, A. 2012, ApJ, 753, 16
Krumholz., M. R., Kruijssen, J. M. D., & Crocker, R. M. 2016, ArXiv e-prints, arXiv:1605.02850
Krumholz, M. R., & McKee, C. F. 2005, ApJ, 630, 250
Lada, C. J., Forbrich, J., Lombardi, M., & Alves, J. F. 2012, ApJ, 745, 190
Lada, C. J., & Lada, E. A. 2003, ARA&A, 41, 57
Lada, C. J., Lombardi, M., & Alves, J. F. 2010, ApJ, 724, 687
Leroy, A. K., Usero, A., Schruba, A., et al. 2017, ApJ, 835, 217
Longmore, S. N., Bally, J., Testi, L., et al. 2013, MNRAS, 429, 987
Martín, S., Mauersberger, R., Martín-Pintado, J., Henkel, C., & García-Burillo, S. 2006, ApJS, 164, 450
Padoan, P., & Nordlund, Å. 2002, ApJ, 576, 870
Schinnerer, E., Meidt, S. E., Pety, J., et al. 2013, ApJ, 779, 42
Sun, J., Leroy, A. K., Schruba, A., et al. 2018, ApJ, 860, 172
Usero, A., Leroy, A. K., Walter, F., et al. 2015, AJ, 150, 115
van Dishoeck, E. F. 2014, Faraday Discussions, 168, 9
Wu, J., Evans, II, N. J., Gao, Y., et al. 2005, ApJ, 635, L173
Wu, J., Evans, II, N. J., Shirley, Y. L., & Knez, C. 2010, ApJS, 188, 313

Science with a Next-Generation Very Large Array
ASP Conference Series, Monograph 7
Eric J. Murphy, ed.

Extreme Scattering Events and Symmetric Achromatic Variations

H. K. Vedantham

Cahill Center for Astronomy and Astrophysics, California Institute of Technology, 1200 E. California Blvd., Pasadena, CA 91125, USA

Abstract. Radio variability in quasars has been seen on timescales ranging from days to years due to both intrinsic and propagation induced effects (Kellermann & Pauliny-Toth 1968; Rickett 1990). Although separating the two is not always straightforward, observations of singular 'events' in radio light curves have led to two compelling, and thus far unresolved mysteries in propagation induced variability— extreme scattering events (ESE) that are a result of plasma lensing by AU-scale ionized structures in the interstellar medium, and symmetric achromatic variability (SAV) that is likely caused by gravitational lensing by $\gtrsim 10^3\,M_\odot$ objects. Nearly all theoretical explanations describing these putative lenses have remarkable astrophysical implications. In this chapter we introduce these phenomena, state the unanswered questions and discuss avenues to answer them with a ~weekly-cadence flux-monitoring survey of $10^3 - 10^4$ flat-spectrum radio quasars with the ngVLA.

1. Extreme Scattering Events

1.1. Introduction and Phenomenology

Extreme scattering events (ESE) were first identified by Fiedler et al. (1987) as rare ($\approx 7 \times 10^{-3}\,\mathrm{src}^{-1}\,\mathrm{yr}^{-1}$) ~ month-long 'U'-shaped dips in radio light curves of some sources monitored by the Green Bank survey (Figure 1a). The dip is bracketed by two high amplification spikes reminiscent of caustic crossing from gravitational lensing parlance. ESEs are highly chromatic due to changes in source size as well as the scattering properties of the intervening medium with spectral frequency (Figures 1a, 1c). The time-symmetry and singular nature of these events belied explanations based on intrinsic variability. Instead, transverse passage of a small, dense 'plasma lens' was identified as the cause (Fiedler et al. 1987; Romani et al. 1987). The refractive index of plasma varies with wavelength as λ^2 explaining the frequency structure, and ray tracing through a simple Gaussian lens profile can satisfactorily explain the light curve phenomenology. The anticipated variations in dispersion measures during an ESE (Coles et al. 2015; Kerr et al. 2018), as well as multiple lensed images of a quasar during an ESE have been detected (Figure 1b) firmly establishing the plasma lensing hypothesis. A closely related phenomenon is extreme scintillation (or intra-day variability), the most poignant example of which was seen in J1819+3845 (Figure 1d). Though not a singular event like an ESE, the length-scale and electron densities in plasma structures

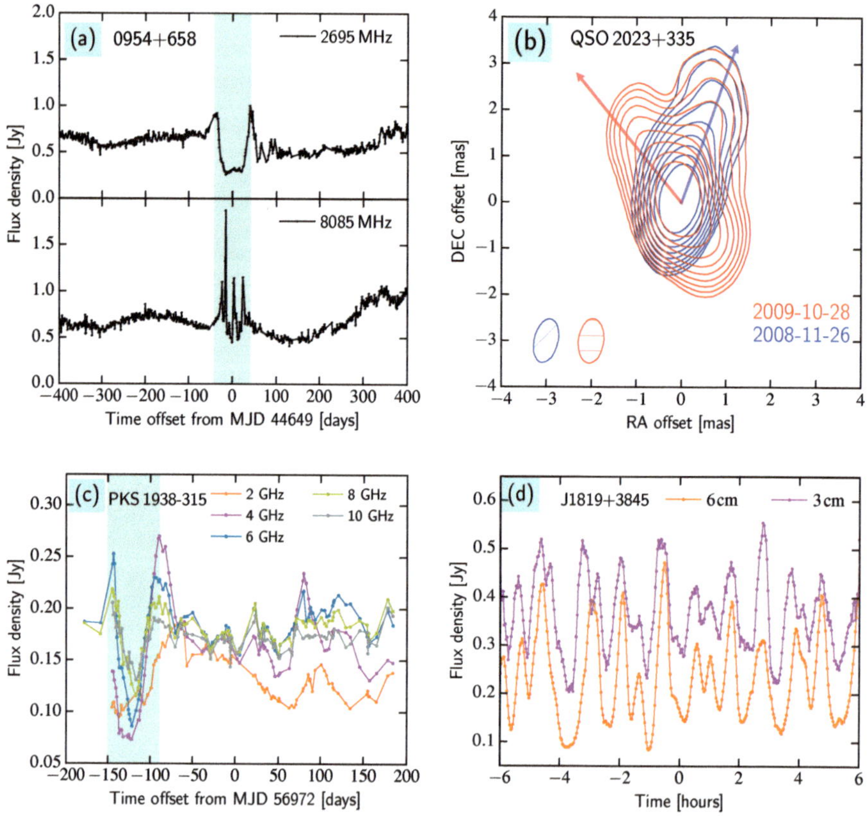

Figure 1. Observational manifestations of extreme interstellar scattering. (a) discovery of ESEs in QSO 0954+658 showing highly achromatic and time-symmetric U-shaped features. (b) VLBI images of QSO 2023+335 during (red) and prior (blue) to an ESE observed in its 15 GHz light curve (not shown for brevity), clearly showing the presence of multiple lensed images along an axis (red arrow) not aligned with the jet axis (blue arrow; Lister et al. 2009). (c) Real-time wide-band detection of an ESE in PKS 1938-315 that ruled out converging lens models (Bannister et al. 2016). (d) Extreme scintillation in J1819+3845 ($\sim 30\,\%$ flux-density modulation on timescales of tens of minutes caused by plasma clouds with density $n_e \gg 10^2\,\mathrm{cm}^{-3}$ within $1-3$ pc from the Sun (de Bruyn & Macquart 2014).

that cause such scintillation is akin to that required for ESEs.[1] We now show that the inferred ESE plasma properties are quite dramatic.

1.2. A New Component of the ISM?

A plasma cloud of radius r and electron column N_e advances the phase of a electromagnetic wave of wavelength λ by $\phi = \lambda r_e N_e$ where r_e is the classical electron radius. This causes to refraction of light through an angle $\theta_s \approx \lambda^2 N_e r_e/(2\pi r)$. The ratio of the refraction angle to that subtended by the lens, $\alpha = D\theta_s/r$ (D is the lens distance) is a measure of the lensing strength. For the cloud to have an observable impact, θ_s must be of the same order as the cloud-size ($\alpha \gtrsim 1$). To appreciate the lensing strength of ESE clouds, consider a typical ESE that last about $\sim$ 1 month which for typical Galactic proper motion of $v \sim 100\,\mathrm{km\,s^{-1}}$ sets the characteristic cloud size of order $r \sim 1$ AU. The electron density in the lens must therefore be

$$n_e = \frac{N_e}{r} \sim 10^3\,\mathrm{cm^{-3}}\,\frac{r}{\mathrm{AU}}\left(\frac{\lambda}{10\,\mathrm{cm}}\right)^{-2}\left(\frac{D}{1\,\mathrm{kpc}}\right)^{-1}, \tag{1}$$

with a column density of

$$N_e \sim 10^{17}\,\mathrm{cm^{-2}}\left(\frac{r}{\mathrm{AU}}\right)^2\left(\frac{\lambda}{10\,\mathrm{cm}}\right)^{-2}\left(\frac{D}{1\,\mathrm{kpc}}\right)^{-1}. \tag{2}$$

The gas pressure in the lens is over four orders of magnitude larger than the ambient ISM pressure ($\sim 0.1\,\mathrm{cm^{-3}}$ in the warm ionized medium). This pressure-support problem can be circumvented in two ways: (a) an edge on sheet-like geometry can provide the required column, but extreme sheet axial ratios of order 10^4 are required. (b) the lenses could be ionized sheaths of self-gravitating $\sim 10^2$ AU-scale clouds of (presumably) molecular gas (similar to cometary knots in the Helix nebula (Vorontsov-Velyaminov 1968; Walker et al. 2017). Reconciling the observed ESE rate requires these knots to have a total mass that rivals the Galactic stellar mass – an exotic, if not unpalatable conclusion!

1.3. ESEs with the ngVLA

The molecular knots hypothesis can be tested via absorption spectroscopic signatures from associated dust and/or molecular ro-vibrational lines during an ESE. On the other hand, the non-axisymmetric lensing potential of edge-on sheets can potentially be discriminated from axi-symmetric models by multi-epoch position and morphology of lensed images with VLBI. Implementing these in practice requires a sample of ESEs with real-time detection for multi-wavelength and $\lesssim 1$ mas resolution VLBI follow-up (see, e.g., Bannister et al. 2016). VLBI detection is crucial since the lensing strength and source size (relative to the lens) are highly degenerate.[2] A weaker lens occulting a smaller source can produce similar light curves to a stronger lens (even by factor of

[1]Ideally one should use the phrase ESE for extreme scintillation (J1819+3845) and call the U-shaped events as extreme lensing events (ELE), but we have retained the historic naming convention here.

[2]Current non-parametric lens modeling techniques typically assume a point-like source and absence of caustic formation (see, e.g., Tuntsov et al. 2016).

$\sim 10^2$) occulting a relatively larger source (Vedantham et al. 2017a). The main discriminant then becomes the ray deflection angles that scale linearly with lensing strength and can be readily distinguished by VLBI observations of multiple images expected on ~ 1 mas scales.

Adopting the ESE rate of 7×10^{-3} src^{-1} yr^{-1} (Fiedler et al. 1987) and observational parameters from the successful ESE survey of (Bannister et al. 2016), we foresee monitoring observations of ~ 3000 compact flat spectrum radio quasars from the CRATES catalog (Healey et al. 2007) with measured spectral index > -0.4, declination north of $-30°$, and 3.6 cm flux-density in excess of 100 mJy. With 70 subarrays of three elements each, the flux-density of such sources can be measured in each 64 MHz channels (for ascertaining spectral modulations indicative of ESEs) between ≈ 4 and 20 GHz (Bands 2 and 3), to better than 30σ in just 30 s integrations. Assuming an overhead of 30%, all sources can be observed within two hours. Hence an ngVLA ESE survey will yield a $\gtrsim 1$ high quality ESEs for VLBI and multi-wavelength follow-up each month with just 0.6% of time commitment. As the target sources are bright ($\gtrsim 0.1$ Jy), follow-up observations are feasible with existing VLBI infrastructure (VLBA and/or EVN).

2. Symmetric Achromatic Variations

2.1. Introduction and Phenomenology

A recent addition to the mystery of U-shaped light curves is the phenomenon of symmetric achromatic variations (SAV) reported by Vedantham et al. (2017b). SAV was identified in the OVRO 40 m blazar monitoring survey, as a time-symmetric year-long U-shaped feature in the light curve of PKS1413+135 (Figure 2a). The light-curve profile is reminiscent of ESEs, but the phenomenon is observed to be achromatic from 15 GHz to 234 GHz— a spectral span that constitutes a factor of ~ 400 in plasma lensing strength. This makes the ESE hypothesis highly problematic (Vedantham et al. 2017a) and Vedantham et al. (2017b) considered gravitational lensing (which is achromatic) as a cause. The light curve can be fit with a physical lens model that consists of two point-like masses with mass ratio ~ 0.01 in the presence of an external mass sheet presumably from the galaxy that hosts the point-like masses (Figures 2b, 2c). Radio sources such as PKS1413+135 are known to have relativistically moving compact ($\sim \mu$as-scale) components. The lensing magnification modulates the flux-density of these components as they cross the lens yielding the observed light curve. The phenomena is similar to optical microlensing of Galactic stars (Udalski et al. 1993; Alcock et al. 1993), but probes a significantly larger mass-scale at cosmic distances. For instance, in case of PKS1413+135, the lens mass is constrained to lie between 10^3 and $10^6\,M_\odot$ depending on the precise lensing geometry. Finally, for strong lensing to be viable, the lenses must be particularly compact ($\gtrsim 10^3\,M_\odot\,\mathrm{pc}^{-2}$). Examples of objects with such compactness include globular clusters, dense molecular cores, and certain dark matter candidates including primordial black holes (see, e.g., Carr et al. 2017).

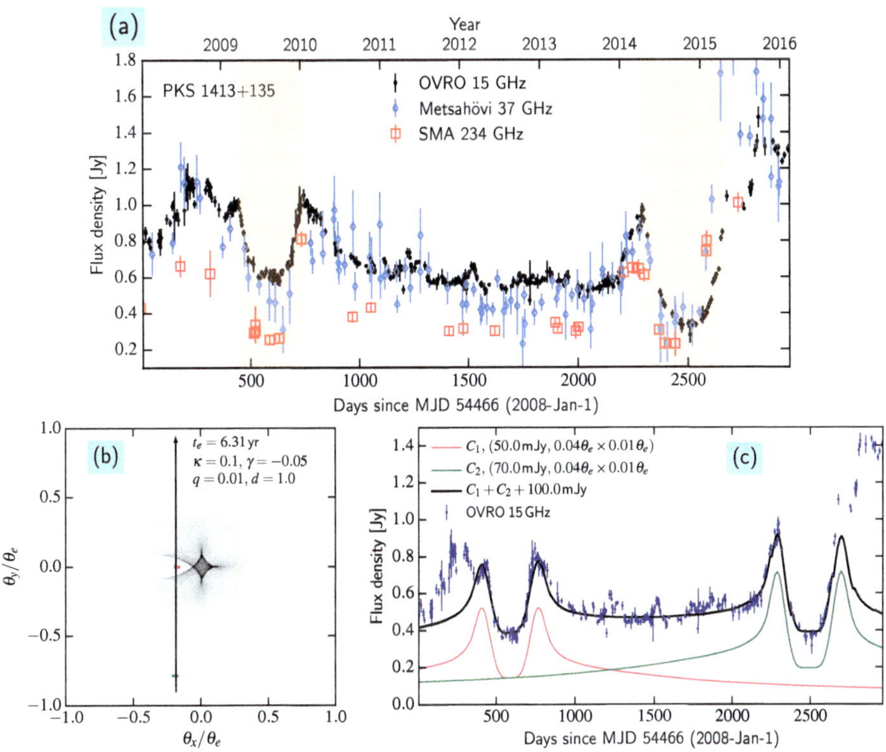

Figure 2. Symmetric achromatic variations in PKS1413+135 (Vedantham et al. 2017b, Copyright AAS; reproduced with permission). (a) Light curve showing time symmetric U-shaped features that are achromatic from 15 GHz to 234 GHz. (b) Magnification map of a possible lens model with two point like masses (q, d: mass ratio and separation) with and external smooth mass sheet from the host galaxy mass (κ, γ: convergence and shear parameter), and path of two relativistically moving source components that can reproduce the data. (c) Light curves of the two components (red and blue) and the total light curve fit to the 15 GHz data (blue points; linear trend removed). The data also admits oblique source trajectories and other lens configurations (see Figure 4 of Vedantham et al. 2017b).

2.2. A Population of Intermediate Mass Lenses?

Gravitational lensing leads to multiple imaging on angular scales of $\Delta\theta \approx \theta_e$ where the Einstein radius, θ_e depends on the lens mass M_l and the lensing geometry:

$$\theta_e \approx 90\,\mu\text{as}\left(\frac{M_l}{10^3\,M_\odot}\right)^{1/2}\left(\frac{D_l D_s/D_{ls}}{1\text{Gpc}}\right)^{-1/2}.$$

where D_l, D_s and D_{ls} are the observer–lens, observer–source and lens–source distance respectively. To obtain a distinct SAV signature, the source must be no larger than a few percent of the Einstein radius. The angular size of compact steep-spectrum sources is constrained by inverse Compton losses: their brightness temperatures are bound to be $\lesssim 10^{12}$ K, while relativistic effects can lead to apparent brightness temperature of $\sim 10^{13}$ K (Johnson et al. 2016). Observations of SAV therefore places a lower bound on the Einstein radius and hence places the mass-scale of the putative lens in the intermediate mass regime:

$$M_l \gtrsim 10^3\,M_\odot\left(\frac{S}{0.1\text{ Jy}}\right)\left(\frac{\lambda}{1\text{ cm}}\right)^2\left(\frac{D_l D_s/D_{ls}}{1\text{Gpc}}\right)$$

The compact μas-scale components (unresolved by VLBI) typically have relativistic velocities and will cross such a lens in a timescale of

$$t \approx 1.4\,\text{yr}\left(\frac{M_l}{10^3\,M_\odot}\right)^{1/2}\left(\frac{D_l D_s/D_{ls}}{1\text{Gpc}}\right)^{1/2}.$$

This underscores the specific utility of relativistic radio sources. While typical virial speeds of $\sim 300\,\text{km s}^{-1}$ can only cross a $10^3\,M_\odot$ lens in time-scales of centuries, relativistic sources can cross on ~ year-long timescales which allows for detection of lensing candidates with flux-density monitoring campaigns alone. We note that, while we are not aware of any other mechanism that can explain SAVs, the lensing hypothesis must await confirmation from detection of lensed images in VLBI follow-up of future SAVs. Though challenging, observations with existing global mm-wave VLBI techniques have the required resolution of several tens of μas.

2.3. SAVs and the ngVLA

The main focus to solve the SAV puzzle is in real-time identification of an SAV for mm-wave VLBI follow-up. SAVs are rare: the events in PKS1413+135 were found from a sample of 10^3 sources $\times 8$ yr of data (cadence of twice per week). Moreover, as in the case of PKD1413+135, light curves at widely spaced frequencies are required to discriminate SAVs from ESEs. Bearing these in mind, we foresee an ngVLA monitoring survey of about 8000 flat-spectrum sources ($\alpha > -0.5$) north of $-30°$ declination in the the CRATES catalog once every fortnight[3], in bands 3 and 5, to find ~ 1 year-long SAV candidate per year for multi-epoch mm-wave VLBI follow-up. We end by noting that confirmation of gravitational lensing as a cause of SAVs would open up a new sub-field of study into cosmic matter on small mass-scales, which can potentially have important implication for evidencing the nature of dark matter.

[3]With a lack of chromatic signatures, time-symmetry is key to differentiating intrinsic variations and, if possible a cadence of one per week is ideal.

References

Alcock, C., Akerlof, C. W., Allsman, R. A., et al. 1993, Nature, 365, 621
Bannister, K. W., Stevens, J., Tuntsov, A. V., et al. 2016, Science, 351, 354
Carr, B., Raidal, M., Tenkanen, T., Vaskonen, V., & Veermäe, H. 2017, Phys. Rev. D, 96, 023514
Coles, W. A., Kerr, M., Shannon, R. M., et al. 2015, ApJ, 808, 113
de Bruyn, A. G., & Macquart, J.-P. 2014, VizieR Online Data Catalog, 357,
Fiedler, R. L., Dennison, B., Johnston, K. J., & Hewish, A. 1987, Nat, 326, 675
Healey, S. E., Romani, R. W., Taylor, G. B., et al. 2007, ApJS, 171, 61
Johnson, M. D., Kovalev, Y. Y., Gwinn, C. R., et al. 2016, ApJ, 820, L10
Kellermann, K. I., & Pauliny-Toth, I. I. K. 1968, ARA&A, 6, 417
Kerr, M., Coles, W. A., Ward, C. A., et al. 2018, MNRAS, 474, 4637
Lister, M. L., Aller, H. D., Aller, M. F., et al. 2009, AJ, 137, 3718
Rickett, B. J. 1990, ARA&A, 28, 561
Romani, R. W., Blandford, R. D., & Cordes, J. M. 1987, Nat, 328, 324
Tuntsov, A. V., Walker, M. A., Koopmans, L. V. E., et al. 2016, ApJ, 817, 176
Udalski, A., Szymanski, M., Kaluzny, J., et al. 1993, Acta Astronomica, 43, 289
Vedantham, H. K., Readhead, A. C. S., Hovatta, T., et al. 2017a, ApJ, 845, 90
—. 2017b, ApJ, 845, 89
Vorontsov-Velyaminov, B. A. 1968, in IAU Symposium, Vol. 34, Planetary Nebulae, ed. D. E. Osterbrock & C. R. O'dell, 256
Walker, M. A., Tuntsov, A. V., Bignall, H., et al. 2017, ApJ, 843, 15

Science with a Next-Generation Very Large Array
ASP Conference Series, Monograph 7
Eric J. Murphy, ed.

Molecular Clouds in Galaxies

Adam K. Leroy,[1] Alberto Bolatto,[2] Erik Rosolowsky,[3] and Eva Schinnerer[4]

[1]*Ohio State University, Department of Astronomy, Columbus, OH 43210;* `leroy.42@osu.edu`

[2] *Department of Astronomy, University of Maryland, College Park, MD 20742*

[3] *University of Alberta, Department of Physics, 4-183 CCIS, Edmonton AB T6G 2E1, Canada*

[4] *Max Planck Institute für Astronomie, Königstuhl 17, 69117, Heidelberg, Germany*

Abstract. Stars form in cold clouds of predominantly molecular (H_2) gas. We are just beginning to understand how the formation, properties, and destruction of these clouds varies across the universe. In this chapter, we describe how the thermal line imaging capabilities of the proposed next generation Very Large Array (ngVLA) could make major contributions to this field. Looking at CO emission, the proposed ngVLA would be able to quickly survey the bulk properties of molecular clouds across the whole nearby galaxy population. This includes many unique very nearby northern targets (e.g., Andromeda) inaccessible to ALMA. Such surveys offer a main observational constraint on the formation, destruction, lifetime, and star formation properties of clouds. Targeting specific regions, the ngVLA will also be able to heavily resolve clouds in the nearest galaxies. This will allow detailed studies of the substructure and kinematics — and so the internal physics — of clouds across different chemical and dynamical environments.

1. The Fundamental Unit of the Star-Forming Interstellar Medium

Stars form in massive, cold clouds of predominantly molecular (H_2) gas (e.g., see reviews by Fukui & Kawamura 2010; Kennicutt & Evans 2012; Heyer & Dame 2015). Understanding the formation, evolution, and destruction of these giant molecular clouds (GMCs) is a major focus of millimeter wave astronomy, and we are still learning how their properties are imprinted on the resulting stellar population.

In theory, these clouds form through a mixture of large scale instabilities, conversion of atomic to molecular gas, and agglomeration of smaller structures (e.g., see Dobbs et al. 2014). Their internal motions are dominated by supersonic turbulence, and they have a large amount of sub-structure. For example, local clouds have mean H_2 column densities of $n_{H2} \sim 10^2$ cm^{-2} but stars form in dense filaments and cores with $n_{H2} > 10^5$ cm^{-2} (e.g., see Lada et al. 2010; André et al. 2014). In our current understanding, over one or more crossing or collapse times, a mixture of turbulence and external triggers collect low density gas into high density sub-structures. Then stars form inside these dense sub-structures. Stellar feedback from these newly formed stars then deposits energy and momentum back into the parent cloud, perhaps even destroying it.

But these broad statements cover a large amount of ignorance. Many of our basic questions about GMCs still lack satisfactory answers. These include:

1. What is the dominant formation mechanism for GMCs in different environments?
2. How much molecular gas is actually in bound structures, and how much exists in a "diffuse" non star-forming phase?
3. How do the dynamical state, character of turbulence (e.g., strength, driving scale, and type of driving), internal structure, and chemical state of GMCs depend on environment? How do they evolve with time?
4. Are the star-forming sub-structures within GMCs created mainly by turbulence or by interactions with the large-scale environment (e.g., triggered by cloud-cloud collisions or spiral arm passage)?
5. What is the main destruction mechanism for GMCs? Stellar feedback, galactic gravitational forces, or something else? How long do they live once they have been formed?
6. Does the fraction of gas that a GMC turns into stars over its lifetime depend on its properties? How does the rate at which stars form depend on the clouds' properties? What about the fraction of stars formed in bound star clusters?

The key to progress in these areas are observations that measure the demographics and internal structure of GMCs over the full diversity of dynamical, energetic, and chemical environments found in the universe. We aim to understand how GMCs work in environments spanning from low metallicity, shallow potential well dwarf galaxies to violent, merger-induced starbursts or the highly stable gas disks in early-type and post-starburst galaxies.

In each of these environments, complete demographics of the cloud population will allow astronomers to constraint the life cycle, destruction, and formation mechanisms of clouds. Meanwhile, detailed chemical, structural, and kinematic observations of individual clouds offer the prospect to to illuminate the physics at play inside these crucial objects.

2. Millimeter-Wave Spectroscopic Imaging as the Key to Understand GMCs

High physical resolution millimeter spectroscopic imaging is the key to understand GMCs. Specifically, the low-J rotational lines of CO, the second most abundant molecule (after the effectively invisible H_2), trace the mass, structure, and kinematics of GMCs. CO imaging has provided most of the basis for what we currently know about GMCs in other galaxies, and this situation seems likely to hold heading in to the next decades.

The problem is that isolating GMCs requires resolution of $\approx$ 50–100 pc, while resolving clouds and measuring their internal structure requires resolution $\approx$ 1–10 pc. CO line emission from a GMC has brightness temperature of order $\sim$ 1 K and line width $\approx$ 1–20 km s^{-1} (e.g., see Sun et al. 2018). At the $\approx$ 5–20 Mpc distance of the nearest

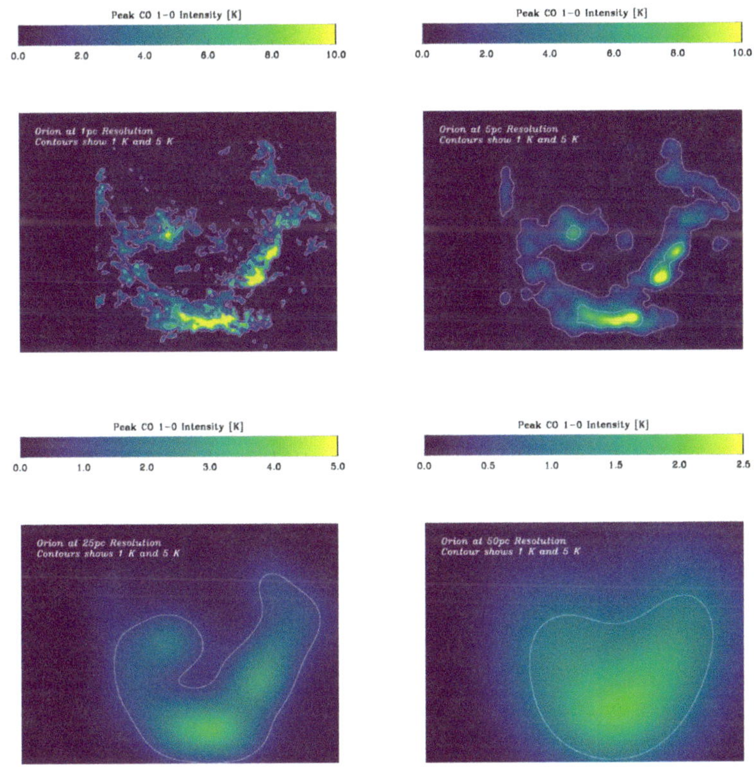

Figure 1. **The ngVLA View of Clouds in Distant Galaxies — Orion at 1, 5, 25, and 50 pc Resolution:** The Orion Molecular clouds, as seen in CO $J = 1 \rightarrow 0$ by Dame et al. (2001). (*Top Left:*) At 1 pc resolution, as the ngVLA could see molecular clouds in Local Group galaxies. The contours show 1 and 5 Kelvin brightness temperature. The ngVLA will achieve $\sim$ 1 K rms noise per 10 km s^{-1} channel at $0\farcs1$ resolution in 1 hour on source, so that this detailed view of clouds (in fact $\sim$ 2 to 3 times sharper than this) will be possible for Local Group targets, including the spiral galaxies M33 and M31. (*Top Right:*) A similar view at 5 pc resolution, resolution readily achievable in the nearest groups of galaxies, e.g., for the M81+M82 group. These high resolutions will allow detailed study of how cloud subtructure and internal kinematics change as a function of environment. (*Bottom Left* and *Bottom Right:*) Progressively lower resolution views. The bottom two panels, at resolution 25 pc and 50 pc show the view the ngVLA can achieve in $\sim$ 1 *minute integration time* targeting galaxies at $\sim$ 5–20 Mpc. This allows the fantastic possibility of cloud demographics across the whole local galaxy population. This figure the peak brightness temperature structure, but spectroscopic imaging also returns information on the cloud kinematics, giving access to bulk energetics (at low resolution) and diagnostics of turbulence and internal flows (at high resolution).

massive, star-forming galaxies, this translates to a required resolution of $\sim 0\farcs5$–$2''$

to isolate GMCs and study their population. And higher resolution of $\sim 0\farcs1$–$0\farcs4$ is required to resolve GMC substructure even in these nearby spiral galaxies. For rarer systems that lie at greater distances (e.g., major galaxy mergers and rare phases of galaxy evolution) the angular resolution requirements will be even stricter.

Interferometers suffer from a well-known tradeoff between surface brightness sensitivity and resolution, and for the pre-ALMA generation of interferometers, imaging a ~ 1 K line at these resolutions was daunting. Heroic efforts by the pre-NOEMA PdBI obtained the first cloud-scale ($\theta = 1'' = 40$ pc) map of CO of M51 (e.g., Schinnerer et al. 2013) and CARMA imaged the inner regions of many nearby galaxies (Donovan Meyer et al. 2013). But these efforts required large time investments (dozens to hundreds of hours of the highest quality time available to the telescopes) and often remained sensitive only to the most massive GMCs.

These pre-ALMA studies already conclusively demonstrate that the classic picture of a universal population of GMCs does not hold across all environments. Rather, GMCs appear to vary in their surface density, density, and line width (e.g., Heyer et al. 2009; Hughes et al. 2013; Donovan Meyer et al. 2013). And the data strongly suggest coupling between the local disk structure, ISM conditions, and the properties of clouds (e.g., Field et al. 2011; Hughes et al. 2013).

In the ALMA era, this picture has become even clearer. Observations show real, significant variations of GMC-scale properties across the local universe (e.g., see Leroy et al. 2015; Davis et al. 2017; Sun et al. 2018; Egusa et al. 2018, among many others). The primary sense of this variation appears to be huge variations in the internal pressure of molecular gas (Sun et al. 2018), but more modest – though still crucial – changes in the dynamical state of gas are also evident (Davis et al. 2017; Egusa et al. 2018; Sun et al. 2018). In addition, the mass function of clouds has long been known to vary within and among galaxies (e.g., Rosolowsky & Blitz 2005; Colombo et al. 2014; Freeman et al. 2017)

These observations drive the need for large-scale comparative demographic studies of clouds. They also underline the need for detailed studies of clouds in different environments – if the populations of clouds changes from place to place, do the governing physics also change? How do the turbulence and sub-structure compare across these different cloud populations? And how do the stars and clusters formed vary as the density, turbulence, and dynamical state of the ISM also vary?

3. The Next Generation Very Large Array — Molecular Cloud Populations Well Beyond Virgo

Answering these questions requires observing CO, or similar, line emission with sub-Kelvin sensitivity at angular resolutions $\theta = 0\farcs1$–$1''$. This requires a massive, sensitive, extended-configuration interferometer like the proposed next-generation Very Large Array (ngVLA; Carilli et al. 2015). Because CO emission has significant power at spatial scales reaching up to many hundreds of pc (e.g., Pety et al. 2013), these observations require good sensitivity at all spatial scales, including short spacing information.

The thermal line imaging capabilities of the ngVLA are exactly suited to this kind of science. In a single field in an hour, the ngVLA reaches rms noise of 0.01 K at

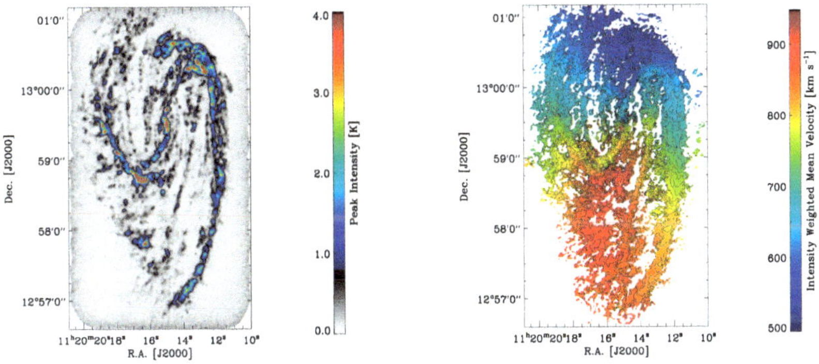

Figure 2. **The ngVLA View of Clouds in Distant Galaxies — Whole Galaxies at ~ 50 pc Resolution.** Peak CO (2-1) intensity (*left*) and intensity-weighted velocity (*right*) emission from the nearby barred spiral NGC 3627 (Messier 66) at 50 pc (≈ 1.3″) resolution. At this resolution, the gas in the galaxy breaks up into individual cloud-sized patches of emission and the line widths, masses, energetics of individual clouds can be estimated. Sun et al. (2018) used these data to show a strong dependence of line width, surface density, and internal pressure on dynamical environment across the galaxy and variations among galaxies. Targeting the fundamental CO (1-0) transition, the ngVLA can map a field to comparable sensitivity in less than a minute. This allows surveys of clouds across a whole big galaxy in only a few hours out to beyond the Virgo Cluster, and farther using the ngVLA in high resolution modes. This ability is currently new with ALMA, and is ushering molecular cloud studies into the era of survey science. *The ngVLA will build on ALMA's progress in this field, observing the fundamental transition, targeting unique nearby northern targets, and mapping with potentially a few times faster speed than ALMA.*

$\theta = 1''$ over a 10 km s^{-1} line width, typical for a GMC[1]. This allows the prospect to reach ~ 0.1 K sensitivity in under a minute. This will allow the ngVLA to easily mosaic many hundreds of fields in a few hours, measuring the CO surface brightness, luminosity, and line width of all clouds across a galaxy.

Capable of a noise of ~ 1 K in one hour at $\theta \approx 0\farcs1$, the ngVLA can resolve the molecular cloud populations in galaxies at distances of 20 – 100 Mpc. This will allow us to study the GMCs in environments vastly different from the ones present in our own Galaxy, such as Luminous and Ultra-Luminous Infrared Galaxies and galaxies in the nearest galaxy clusters, including those undergoing quenching and gas loss processes such as ram pressure stripping.

This allows the prospect to statistically characterize how GMC populations depend on environment. Does the mass function of clouds shift in response to the local Jean's mass or the mean pressure of the ISM? Does the dynamical state (ratio of kinetic to potential energy) change between galaxy centers, regions of strong gas flow like bars

[1]Taking the 93 GHz values for the reference design to apply to CO $J = 1 \rightarrow 0$ at 115 GHz.

or arms (e.g., Meidt et al. 2013; Egusa et al. 2018), and more quiescent disk galaxies? How do the mean density and mean internal pressure of clouds vary in response to changing disk conditions?

All of these questions can be effectively addressed with high completeness (and thus high sensitivity) surveys that estimate the mass, kinetic energy, and size of a patch of the molecular interstellar medium on the scale of an individual GMC. To be effective, these surveys need to cover large, well-characterized parts (or all area) of many galaxies, so that any individual field must be covered quickly. For this application, reaching resolutions approximately matched to GMCs is more important than heavily resolving them, and here the ngVLA "core" will be incredibly powerful. There are a few hundred moderately massive northern star forming galaxies within $\sim$ 20 Mpc. At a few hours apiece, the ngVLA could survey CO emission at cloud scale from all of them. Expanding to cover all of the local (but more distant) starbursts (e.g., the GOALS targets; Armus et al. 2009) and a sample of nearer, low metallicity dwarf galaxies (e.g., many of the LVL targets; Lee et al. 2009) would add another few hundred hours. With order $\sim 1,000$ hours, the ngVLA would build definitive atlas of GMCs across all local environments.

Beyond just empirically understanding the influence of environment on clouds, these population studies are the crucial element to constrain the physics governing the formation, destruction, and evolution of clouds. In the last few years there has been large interest in resolving the ISM down to the scales of individual star-forming regions and using tracers that probe different phases of the star formation and feedback process (e.g., Kawamura et al. 2009; Schruba et al. 2010; Kruijssen & Longmore 2014; Corbelli et al. 2017). With the ngVLA it will, for example, be possible to image a whole galaxy at cloud-scales using CO to trace molecular gas, HCN and HCO^+ to trace dense gas, 21-cm HI line emission to trace atomic gas, and free-free continuum emission to trace young stars and ionized gas. The joint distributions of these tracers at high resolution are the key observation that encodes the life cycle of GMCs: their birth, life (including the formation of stars), and ultimate destruction.

Because of the wide instantaneous bandwidth of the ngVLA, such a survey would also be able to trace the chemical and physical conditions inside each cloud. Although the exact receiver set-up is not decided upon yet, simultaneous coverage of CO isotopologue lines (^{13}CO, $C^{18}O$, or even $C^{17}O$) could offer constraints on bulk gas conditions, optical depth, and maybe even nucleo-synthesis (e.g., Sliwa et al. 2017). Covering high critical density lines (see gas density-focused white papers in this book) would provide probes of how gas density depends on cloud properties (or on time, as above). And comparing, both statistically and directly, the properties of clouds to local tracers of recent star formation will allow a direct link between gas properties, star formation efficiencies, star formation rates, and star formation timescales.

4. The Next Generation Very Large Array — The Local Group in Stunning Detail

When focusing on a single field, the ngVLA reaches rms noise ~ 1 K at $\theta = 0\farcs1$ resolution in 1 hour. This provides the prospect to dissect individual clouds in stunning detail out to many Mpc – complementing the survey science described above with detailed studies of the physics playing out inside individual clouds. This also allows the Local Group spirals M31 and M33 to be observed with detail currently reserved for Milky

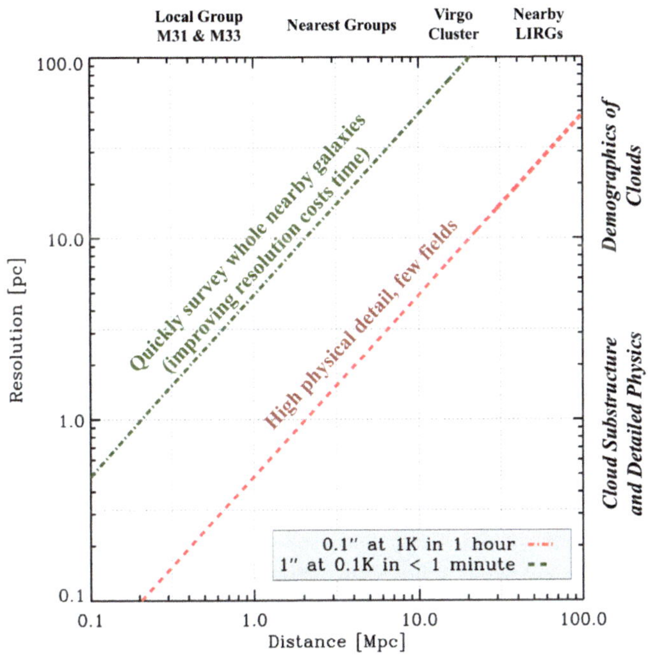

Figure 3. **Schematic View of ngVLA Capabilities Targeting Molecular Clouds.**. Physical resolution (y-axis) achieved at $1''$ (green) and $0\rlap{.}''1$ (red) resolution according to the ngVLA reference design. The green line shows a mode in which the ngVLA can survey whole large areas of nearby galaxies in a few areas. This will allow detailed studies in the Local Group and cloud demographics out to beyond the Virgo Cluster. The red line shows higher resolution mapping, still leveraging the ngVLA core, that can be used to study a few fields at high detail in a few hours. This will be crucial to study the internal physics of clouds in exquisite detail in the nearest galaxies. It is also the key mode to survey bright, but distant targets like LIRGs, which are rare systems that probe crucial extreme physics.

Way molecular clouds (similar to what is currently done by ALMA targeting the Magellanic Clouds, e.g., see Wong et al. 2017). Conversely, large surveys of Local Group galaxies at $\theta \approx 1''$ would yield ≈ 4 pc linear resolution over huge areas.

Heavily resolving the internal structure of molecular clouds in other galaxies is *terra incognita* and offers a huge opportunity to gain insight into how the physics of star formation change (and stay the same) across the universe. Resolution $\theta = 0\rlap{.}''1$ translates to $\sim$ 0.5 pc at 1 Mpc. The ngVLA would be able to measure the internal (turbulent) size-linewidth relation (e.g., Heyer et al. 2004) as well as more sophisticated turbulence diagnostics (e.g., Burkhart et al. 2015) in clouds across M31, M33, the Local Group dwarfs, and (with somewhat more effort) galaxies in the nearest groups (including M81 and M82, $d \sim 3.5$ Mpc). The same observations would help constrain the column density distribution function, which has been the topic of intense interest in the Milky

Way in recent years (e.g., Kainulainen et al. 2009; Lombardi et al. 2015) because of its close link to star formation.

These hyper-resolved observations would offer the prospect to search other galaxies for dense filamentary structures that have captivated Milky Way astronomers both within clouds (e.g., André et al. 2014) and on larger scales (Jackson et al. 2010). These filaments have been argued to be fundamental to star formation (André et al. 2014) and perhaps to serve as the "bones" of the Milky Way's interstellar medium (Zucker et al. 2017). Other studies have emphasized biases in earlier analysis (Panopoulou et al. 2017), questioning whether these structures are indeed universal. An external, hyper-resolution (< 1 pc) view of GMCs in other galaxies offers huge potential to tell us how the cold, dense ISM is structured.

5. Relationship to Other Facilities — Building on ALMA's Legacy and a Close Complement to *JWST*

Studying higher J CO lines, CO $J = 2 \rightarrow 1$ and $J = 3 \rightarrow 2$, ALMA is currently revolutionizing this field. Its capabilities in these higher J transitions resemble those of the ngVLA in CO $J = 1 \rightarrow 0$. As discussed above, ALMA's findings so far suggest that there is significant depth to this topic, and in the nearby universe the north holds a set of key, very nearby targets. So the ngVLA is highly likely to pick up an active field and push it farther, exploring the effects of excitation and digging in to a new set of key targets. The Local Group spirals M33 and M31 stand out. These are the nearest star-forming spiral galaxies to the Milky Way. Because of this proximity, they will always be the place where one can explore ISM physics at high resolution and in greater detail than in more distant targets. Along similar lines, the M81 group, also too far north to be seen by ALMA, represents the nearest major violent interaction. It hosts the prototype merger-induced starburst M82. And M51, M101, and NGC 6946 have all been key targets to study gas and star formation due to their proximity and activity. But sit at declination $\delta > 30^\circ$ (M33 may be surveyed by ALMA, the others are all at $\delta > 40^\circ$).

In the north, the main other relevant facility will be NOEMA, which can and will pursue many of the same science goals. Establishing the basic structure and evolution of the molecular ISM is likely to be such a deep vein that all three telescopes (ALMA, NOEMA, and the ngVLA) can be expected to make major contributions.

In the early 2020s, NASA's *James Webb Space Telescope* (*JWST*) will launch. By observing near-IR recombination lines and the mid-IR dust continuum, *JWST* will be an unparalleled instrument to survey the earliest, most heavily embedded phases of star formation at high angular resolution. In this sense *JWST* and the ngVLA are extraordinarily complementary: ngVLA will observe the cold, star-forming gas and *JWST* will observe how it forms stars.

6. Likely Impact

With these capabilities, the ngVLA would immediately become the premier instrument to study molecular cloud populations and structure in other galaxies. It would represent a powerful northern complement to ALMA. Its observations would naturally complement observations of recent, embedded star formation by the *JWST* (as well as ngVLA continuum observations of recent star formation).

Acknowledgments. AKL is partially supported by the National Science Foundation under Grants No. 1615105, 1615109, and 1653300.

References

André, P., Di Francesco, J., Ward-Thompson, D., et al. 2014, Protostars and Planets VI, 27
Armus, L., Mazzarella, J. M., Evans, A. S., et al. 2009, PASP, 121, 559
Burkhart, B., Collins, D. C., & Lazarian, A. 2015, ApJ, 808, 48
Carilli, C. L., McKinnon, M., Ott, J., et al. 2015, ArXiv e-prints, arXiv:1510.06438
Colombo, D., Hughes, A., Schinnerer, E., et al. 2014, ArXiv e-prints, arXiv:1401.1505
Corbelli, E., Braine, J., Bandiera, R., et al. 2017, A&A, 601, A146
Dame, T. M., Hartmann, D., & Thaddeus, P. 2001, ApJ, 547, 792
Davis, T. A., Bureau, M., Onishi, K., et al. 2017, MNRAS, 468, 4675
Donovan Meyer, J., Koda, J., Momose, R., et al. 2013, ApJ, 772, 107
Egusa, F., Hirota, A., Baba, J., & Muraoka, K. 2018, ApJ, 854, 90
Field, G. B., Blackman, E. G., & Keto, E. R. 2011, MNRAS, 416, 710
Freeman, P., Rosolowsky, E., Kruijssen, J. M. D., Bastian, N., & Adamo, A. 2017, MNRAS, 468, 1769
Fukui, Y., & Kawamura, A. 2010, ARA&A, 48, 547
Heyer, M., & Dame, T. M. 2015, ARA&A, 53, 583
Heyer, M., Krawczyk, C., Duval, J., & Jackson, J. M. 2009, ApJ, 699, 1092
Heyer, M. H., Corbelli, E., Schneider, S. E., & Young, J. S. 2004, ApJ, 602, 723
Hughes, A., Meidt, S. E., Schinnerer, E., et al. 2013, ApJ, 779, 44
Jackson, J. M., Finn, S. C., Chambers, E. T., Rathborne, J. M., & Simon, R. 2010, ApJ, 719, L185
Kainulainen, J., Beuther, H., Henning, T., & Plume, R. 2009, A&A, 508, L35
Kawamura, A., Mizuno, Y., Minamidani, T., et al. 2009, ApJS, 184, 1
Kennicutt, R. C., & Evans, N. J. 2012, ARA&A, 50, 531
Kruijssen, J. M. D., & Longmore, S. N. 2014, MNRAS, 439, 3239
Lada, C. J., Lombardi, M., & Alves, J. F. 2010, ApJ, 724, 687
Lee, J. C., Gil de Paz, A., Tremonti, C., et al. 2009, ApJ, 706, 599
Leroy, A. K., Bolatto, A. D., Ostriker, E. C., et al. 2015, ApJ, 801, 25
Lombardi, M., Alves, J., & Lada, C. J. 2015, A&A, 576, L1
Meidt, S. E., Schinnerer, E., García-Burillo, S., et al. 2013, ApJ, 779, 45
Panopoulou, G. V., Psaradaki, I., Skalidis, R., Tassis, K., & Andrews, J. J. 2017, MNRAS, 466, 2529
Pety, J., Schinnerer, E., Leroy, A. K., et al. 2013, ApJ, 779, 43
Rosolowsky, E., & Blitz, L. 2005, ApJ, 623, 826
Schinnerer, E., Meidt, S. E., Pety, J., et al. 2013, ApJ, 779, 42
Schruba, A., Leroy, A. K., Walter, F., Sandstrom, K., & Rosolowsky, E. 2010, ApJ, 722, 1699
Sliwa, K., Wilson, C. D., Aalto, S., & Privon, G. C. 2017, ApJ, 840, L11
Sun, J., Leroy, A. K., Schruba, A., et al. 2018, ApJ, 860, 172
Wong, T., Hughes, A., Tokuda, K., et al. 2017, ApJ, 850, 139
Zucker, C., Battersby, C., & Goodman, A. 2017, ArXiv e-prints, arXiv:1712.09655

Science with a Next-Generation Very Large Array
ASP Conference Series, Monograph 7
Eric J. Murphy, ed.

Precision Gas-dynamical Mass Measurement of Supermassive Black Holes with the ngVLA

Benjamin D. Boizelle,[1] Kristina Nyland,[2] and Timothy A. Davis[3]

[1]*George P. and Cynthia Woods Mitchell Institute for Fundamental Physics and Astronomy, Department of Physics and Astronomy, Texas A&M University, College Station, TX 77843;* `bboizelle@tamu.edu`

[2]*National Radio Astronomy Observatory, Charlottesville, VA 22903;* `knyland@nrao.edu`

[3]*School of Physics & Astronomy, Cardiff University, Queens Buildings, The Parade, Cardiff, Wales, UK* `DavisT@cardiff.ac.uk`

Abstract. Emission line observations of circumnuclear gas disks in the ALMA era have begun to resolve molecular gas tracer kinematics near supermassive black holes (BHs), enabling highly precise mass determination in the best cases. The ngVLA is capable of extremely high spatial resolution imaging of the CO(1−0) transition at 115 GHz for nearby galaxies. Furthermore, its high (anticipated) emission line sensitivity suggests this array can produce benchmark BH mass measurements. We discuss lessons learned from gas-dynamical modeling of recent ALMA data sets and also compare ALMA and ngVLA CO simulations of a dynamically cold disk. While only a fraction of all local galaxies likely possess sufficiently bright, regularly-rotating nuclear molecular gas, in such cases the ngVLA is expected to more efficiently resolve such emission arising at a projected 50−100 mas from the central BH.

1. Introduction

The observed properties of massive, bulge-dominated galaxies support the existence of a strong co-evolution between the formation and growth of their stellar spheroids and the supermassive black holes (BHs) residing in their nuclei (e.g., Kormendy & Ho 2013; Heckman & Best 2014). Accurate BH mass measurements are necessary to investigate BH growth over cosmic time as well as to improve constraints on theoretical models of their formation at high redshift. Despite the importance of these measurements to our understanding of galaxy evolution, the full distribution of BH masses at fixed galaxy properties such as morphology, redshift, and stellar mass – particularly for lower-mass central BHs – remains poorly constrained (e.g., Greene et al. 2016). ALMA is now beginning to have its long-anticipated (e.g., Maiolino 2008) revolutionary impact on circumnuclear kinematic studies, already facilitating precision BH mass measurements (e.g., Barth et al. 2016a; Davis et al. 2017) using more common molecular gas tracers. In this science case, we first describe the benefits and challenges associated with cold molecular gas emission as a kinematic tracer, and afterwards detail the promise of

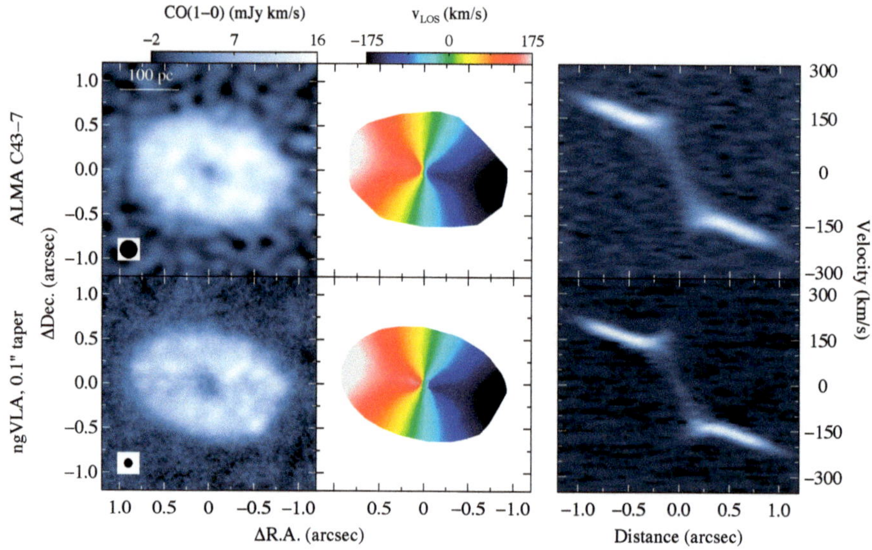

Figure 1. Line moment maps and position-velocity diagrams taken from full-cube ALMA 0.″2 (*top row*) and ngVLA 0.″1 [*bottom row*; after applying an outer (u, v)-taper)] simulations of CO(1−0) disk rotation. The randomly generated clumpy circumnuclear disk at a distance of ~ 30 Mpc has a central hole (with radius ~ 0.2″ ≈ 30 pc) in CO-bright gas and a plausible host galaxy gravitational potential (similar to the case found in Davis et al. 2018), with a BH mass of $2 \times 10^8\,M_\odot$ (and $r_g \sim 0.''3$). The same 2 hr on-source integration time reaches 0.46 and 0.14 mJy/beam rms sensitivities in 10 km s^{-1} channels for the ALMA and ngVLA data sets, respectively. Simulations show that the ngVLA enables better identification and isolation of the faint central CO emission. This primary kinematic signature, buried in higher line moments of the ALMA model cube, becomes more distinct in the simulated ngVLA line-of-sight centroid velocity field. Even in cases where the molecular gas does not originate very deep within r_g, ngVLA observations will promote more confident M_{BH} determination (e.g., see Barth et al. 2016a,b). When the CO emission tracers Keplerian rotation, high angular resolution imaging with this proposed array will enable benchmark BH mass measurements.

the ngVLA to enable even more sensitive, higher angular resolution exploration of the central gravitational potentials of nearby galaxies.

2. Measuring BH Masses with Molecular Gas Tracers

Gas-dynamical modeling of a thin coplanar disk within a galaxy nucleus is an appealing technique to determine its central BH mass (M_{BH}). In addition to conceptual and computational simplicity, idealized rotating disks are also free from most of the potentially serious systematics that affect stellar-dynamical M_{BH} measurements (e.g., van den Bosch & de Zeeuw 2010, McConnell et al. 2013) and can be applied to systems with larger-scale stellar non-axisymmetries (such as bars and mergers). Many early studies used observations of ionized atomic gas disks to map the gravitational potentials of galaxy nuclei, but modeling complications (especially gas turbulence; see van der Marel & van den Bosch 1998, Walsh et al. 2010) limit the confidence of $M_{\rm BH}$ determinations from such kinematically warm tracers (Kormendy & Ho 2013). Dynamically cold molecular emission is a promising avenue as such relaxed systems retain the benefits of disk-like rotation while avoiding the primary systematics afflicting the treatment of ionized gas disks. In most cases, these dynamically cold disks are found to be nearly ideal central probes of nearby galaxies. The centers of many late and early-type galaxies (LTGs/ETGs), possessing both active and quiescent nuclei, contain significant molecular gas reservoirs ($\gtrsim 10^7\ M_\odot$; Bolatto et al. 2017, Alatalo et al. 2013), indicating that BH masses in nearly all types of galaxies may be measured in the same manner by observing and modeling cold molecular gas rotation.

Unambiguously detecting a BH's kinematic impact hinges on (at least approximately) resolving its sphere of influence $r_{\rm g} \approx GM_{\rm BH}/\sigma_\star^2$ wherein $M_{\rm BH}$ dominates over the extended mass contributions and gives rise to an elevated stellar velocity dispersion $\sigma_\star$. Even moderate ($0\farcs3 - 0\farcs5$) resolution, sensitive molecular line observations with ALMA have the potential to provide dynamical BH mass measurements (or strong constraints) for perhaps several hundred nearby ($\lesssim 100$ Mpc) galaxies (Davis 2014), thereby greatly increasing the number of $M_{\rm BH}$ determinations made through dynamical means. Precision BH mass measurements (at the few percent level or better) require resolving kinematic tracers well within $r_{\rm g}$, where the observed rotation is maximally sensitive to the BH influence. In the best (but rare) examples, extremely well-resolved extragalactic H_2O megamaser disks trace out clear Keplerian rotation and yield "gold standard" $M_{\rm BH}$ values (generally $\sim 10^7\ M_\odot$) for these active galaxies (e.g., Miyoshi et al. 1995; Kuo et al. 2011).

Disentangling the central rapid gas velocities of more common molecular tracers such as CO has been largely unachievable by the previous generation of mm/sub-mm arrays due to beam smearing: for example, a $10^8\ M_\odot$ BH will have a projected $r_{\rm g} \sim 0\farcs2$ when viewed at a distance of 20 Mpc. For one such nearby galaxy, Davis et al. (2013) used the CARMA array to just barely resolve CO(2−1) emission originating within the sphere of influence and permit the first BH mass constraint using CO kinematics. Now, ALMA facilitates much deeper, higher angular resolution imaging of this and other tracers in a fraction of the time. Data sets that just marginally resolve $r_{\rm g}$ have already produced gas-dynamical BH mass measurements with total uncertainties (on the order of 20 – 30%, including systematics; Barth et al. 2016b; Onishi et al. 2017; Davis et al. 2017, 2018) that are commensurate with those derived from stellar-dynamical modeling. Observations that more fully resolve $r_{\rm g}$ can produce benchmark $M_{\rm BH}$, enabling tests on the assumptions used in other techniques when applied to measure the same BH mass (e.g., Barth et al. 2016a; Rusli et al. 2011).

3. Lessons From ALMA Observations

Over an order of magnitude higher maximum angular resolution imaging at ~ 115 GHz than ALMA initially suggests that the ngVLA will make possible correspondingly more precise BH mass measurements. This array is theoretically capable of resolving the spheres of influence for a Milky Way-analog BH observed at ~ 100 Mpc, as well as for $\sim 40\%$ of the galaxies in SDSS. However, not all nearby galaxies possess bright circumnuclear molecular gas emission originating from within $r_{\rm g}$. Among those that do, some fraction will exhibit relaxed central gas kinematics. Given the relatively smaller angular $r_{\rm g}$ sizes of LTGs, kinematic studies that resolve $r_{\rm g}$ for these targets are still nascent. We here focus on ETGs as they provide a clearer illustration. Roughly half of all luminous E/S0 hosts harbor detectable CO emission, although perhaps in only ~ 10% does the gas exhibit dynamically cold rotation (Alatalo et al. 2013). Recent ALMA observations reveal that commonly utilized molecular species tend to be dissociated (or are at least underluminous) in the central 10-100 pc (Onishi et al. 2017; Barth et al. 2016a; Boizelle et al. 2017; Davis et al. 2018).

Boizelle et al. (2017) estimate that perhaps a few percent of all ETGs will be prime candidates for precision BH mass measurement using ALMA. A handful of these (and some unknown fraction of LTGs) may still show strong, unresolved Keplerian-like rotation at the edge of ALMA capabilities, warranting even higher angular resolution line imaging with the ngVLA. Obtaining such well-resolved observations, with several independent resolution elements across r_g, alleviates nearly all of the remaining degeneracies (e.g., between M_{BH} and the gas turbulent velocity dispersion; Barth et al. 2016a,b) and systematics (e.g., uncertainty in the circular velocity profile due to central dust obscuration of the luminous matter distribution; Barth et al. 2016b; Boizelle et al. 2018, in prep.). We believe this proposed cm/mm array will be optimally utilized when following up on previous interferometric observations (with $\theta_{1/2} \lesssim 0\farcs25$) that clearly identify gas emission within r_g. Its anticipated sensitivity will also aid in detecting faint central emission that is otherwise beneath the limit of previous ALMA observations.

4. ngVLA Unique Capabilities

High-resolution observations of the CO(1−0) transition (with line excitation temperature $T_{ex} \sim 5 - 10$ K) using ALMA are limited by extended configuration overheads and brightness temperature sensitivities. Typical CO surface brightness measurements suggest that ALMA is capable of resolving CO(1−0) emission interior to (with about four independent resolution elements across) r_g for nearby galaxies with large projected spheres of influence (~ $0\farcs2 - 0\farcs3$; e.g., Boizelle et al. 2017; Davis et al. 2017, 2018). An integrated line signal-to-noise ratio (S/N) $\gtrsim 10$ ensures confident detection of kinematic features in these disks with low intrinsic line widths ($\sigma \sim 5-20$ km s^{-1}; see Barth et al. 2016a). Achieving this S/N level requires rms sensitivities ~ 0.2 – 0.4 mJy/beam in 10 km s^{-1} channels, with the resulting ALMA 115 GHz on-source (total) integrations times ranging between 3 – 14 (9 – 32) hours. As a result of these requirements, kinematic studies of local galaxies with ALMA often focus on the CO(2−1) transition to better balance limiting sensitivities, weather constraints, and achievable angular resolution. Based on current performance metrics, ngVLA CO(1−0) observations with the same channel spacing are expected to reach of order ~ 100 μJy/beam line rms (producing S/N~20) in a single hour (on-source) on 100 mas scales. These expected results also improve on ALMA Band 6 integration time forecasts, more than halving the time allocation to accurately measure CO line properties at ~ $0\farcs1$ resolution.

In Figure 2, we show our simulated ALMA and ngVLA full-cube CO(1−0) observations of a model molecular disk to compare array capabilities. The combination of higher angular resolution and lower limiting sensitivities enable detection of additional faint, high-velocity central emission that promotes a more confident M_{BH} determination. Slightly higher angular resolution ngVLA imaging of such circumnuclear disks should be possible on 50 mas scales for reasonable (a few hours) integration times. On these angular scales, moderately accurate (~ 20%) M_{BH} determinations of ~ 10^9 $M_\odot$ should be possible for galaxies at distances of up to 300 Mpc, with precision ($\lesssim$ 10%)

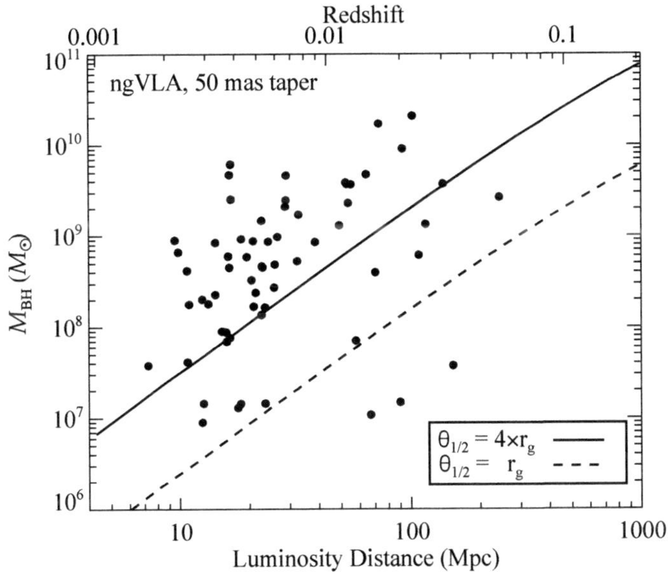

Figure 2. Minimum BH mass detectable using the ngVLA at ~ 115 GHz [with a (u,v) taper to achieve 50 mas angular resolution] as a function of luminosity distance assuming a standard cosmology (Planck Collaboration et al. 2016). Data points represent the elliptical and classical bulge galaxies on which the $M_{\rm BH} - \sigma_*$ relation is currently based (Kormendy & Ho 2013). Curves show the mass limit reachable by the ngVLA using CO(1−0) observations that just marginally resolve the BH sphere of influence (*dashed*; as is typical for most stellar/gaseous $M_{\rm BH}$ determinations) and that more highly resolve r_g (*solid*); assuming detectable CO emission in regular rotation within r_g, expected total mass uncertainties are about 20 – 30% and $\lesssim$ 10%, respectively.

measurements to nearly 75 Mpc (see Figure 2). Almost two orders of magnitude higher resolution CO(1−0) observations are *technically* feasible with this cm/mm array. However, the typical CO emission deficit at radii deep within r_g and prohibitively long integration times suggest that ngVLA extreme angular resolution observations are not likely to be very useful for BH studies in most cases.

The above scenario focused on galaxies with large angular r_g sizes, tending to favor nearby ETGs. The anticipated array performance suggests that ngVLA CO imaging is likely able to identify lower mass (~ 10^7 $M_\odot$) BH kinematic signatures in late-type targets out to at least 20 Mpc. High-resolution cold molecular gas observations of such (often dusty) nuclei will result in more confident mass determinations than typically obtained using ionized atomic tracers (Beifiori et al. 2009) as well as better constraints on the low-mass end of the BH/host galaxy scaling relationships. This proposed array is also likely to be impactful in extragalactic H_2O megamaser BH mass measurements in active galaxies that provide direct geometrical measurement of H_0 (see Braatz et al. 2018, this volume p. 821). While its longest baselines are not able to map out the 22 GHz emission at sub-mas resolution, the ngVLA will provide an important large-collecting area element in new and revised VLBI campaigns (i.e., with the ngVLA Long Baseline Array) at greater sensitivity.

5. Multiwavelength Synergy

The next generation of giant segmented mirror telescopes (such as the E-ELT and the TMT) will offer angular resolutions down to mas scales, comparable to those technically achievable using the ngVLA in its high-frequency bands. These optical/NIR observatories will dramatically increase the number of unobscured nuclei for which it is possible to directly probe stellar and gaseous kinematics within the BH sphere of influence. Space-based NIR observations with the *James Webb Space Telescope* are expected to expand the sample of active galaxies with quality $M_{\rm BH}$ values using techniques such as reverberation mapping. Ongoing ALMA work is now providing tight BH mass constraints for galaxies with obscured nuclei, and the ngVLA contributions are expected to increase both the precision and number of these dynamical measurements. These highly complimentary avenues promise to more fully fill in the BH census with high-quality measurements spanning BH mass and host galaxy properties. Increasing the number of candidates for detailed dynamical modeling will similarly increase the number of cases where multiple techniques are applicable to the same target. Such precision $M_{\rm BH}$ values will enable crucial studies of the underlying assumptions and systematics inherent to each method.

6. Summary

The ngVLA will expand on ALMA's revolutionary capability to probe cold molecular gas rotation within $r_{\rm g}$. Anticipated factors of at least $\sim 2-4$ improvements in line sensitivity with the ngVLA should enable detection of additional faint circumnuclear CO(1−0) emission, and an estimated maximum angular resolution of about 50 mas will better resolve Keplerian rotation for galaxies out to a greater distance. The ngVLA is therefore expected to drive the field of BH mass measurement towards more highly resolving cold gas kinematics that in turn facilitate more precise $M_{\rm BH}$ determinations.

References

Alatalo, K., Davis, T. A., Bureau, M., et al. 2013, MNRAS, 432, 1796
Barth, A. J., Boizelle, B. D., Darling, J., et al. 2016a, ApJ, 822, L28
Barth, A. J., Darling, J., Baker, A. J., et al. 2016b, ApJ, 823, 51
Beifiori, A., Sarzi, M., Corsini, E. M., et al. 2009, ApJ, 692, 856
Boizelle, B. D., Barth, A. J., Darling, J., et al. 2017, ApJ, 845, 170
Bolatto, A. D., Wong, T., Utomo, D., et al. 2017, ApJ, 846, 159
Braatz, J. A., Wilson, A. S., & Henkel, C. 1996, ApJS, 106, 51
Braatz, J., Pesce, D., Condon, J., & Reid, M. 2018, Science with a Next-Generation Very Large Array, ASP Monograph 7, ed. E. J. Murphy (San Francisco, CA: ASP), 821
Davis, T. A. 2014, MNRAS, 443, 911
Davis, T. A., Bureau, M., Cappellari, M., Sarzi, M., & Blitz, L. 2013, Nat, 494, 328
Davis, T. A., Bureau, M., Onishi, K., et al. 2017, MNRAS, 468, 4675
—. 2018, MNRAS, 473, 3818
Greene, J. E., Seth, A., Kim, M., et al. 2016, ApJ, 826, L32
Heckman, T. M., & Best, P. N. 2014, ARA&A, 52, 589

Kormendy, J., & Ho, L. C. 2013, ARA&A, 51, 511
Kuo, C. Y., Braatz, J. A., Condon, J. J., et al. 2011, ApJ, 727, 20
Maiolino, R. 2008, New Astronomy Reviews, 52, 339
McConnell, N. J., Chen, S.-F. S., Ma, C.-P., et al. 2013, ApJ, 768, L21
Miyoshi, M., Moran, J., Herrnstein, J., et al. 1995, Nat, 373, 127
Onishi, K., Iguchi, S., Davis, T. A., et al. 2017, MNRAS, 468, 4663
Planck Collaboration, Ade, P. A. R., Aghanim, N., et al. 2016, A&A, 594, A13
Rusli, S. P., Thomas, J., Erwin, P., et al. 2011, MNRAS, 410, 1223
van den Bosch, R. C. E., & de Zeeuw, P. T. 2010, MNRAS, 401, 1770
van der Marel, R. P., & van den Bosch, F. C. 1998, AJ, 116, 2220
Walsh, J. L., Barth, A. J., & Sarzi, M. 2010, ApJ, 721, 762

Science with a Next-Generation Very Large Array
ASP Conference Series, Monograph 7
Eric J. Murphy, ed.

Accurate Probes of Density and Temperature in Star-Forming Gas

Tony Wong,[1] David S. Meier,[2] and Erik Rosolowsky[3]

[1]*U. of Illinois, Urbana, IL 61801, USA;* `wongt@illinois.edu`

[2]*New Mexico Institute of Mining and Technology, Socorro, NM 87801, USA;* `david.meier@nmt.edu`

[3]*U. of Alberta, Edmonton, AB T6G 2E1, Canada;* `rosolowsky@ualberta.ca`

Abstract. We discuss how observations of ammonia and formaldehyde in nearby galaxies using the next generation Very Large Array (ngVLA) would advance our understanding of star formation physics across different environments by providing robust estimates of gas density and temperature. Currently such observations are mainly performed using filled-aperture antennas that, at best, only crudely distinguish nuclear from disk regions. The ngVLA will map these lines in emission or absorption across the nearest galaxies, and prove especially powerful in combination with ALMA.

1. Introduction and Motivation

In the Milky Way, observations at sub-parsec scales probe the physical processes in molecular clouds that lead to star formation, while in external galaxies, the focus of star formation studies shifts to the large-scale rate of star formation and how it responds to the supply and conditions of the molecular gas (see review by Kennicutt & Evans 2012). The nearest galaxies, at distances of $\lesssim$10 Mpc, occupy an important transitional regime where parsec-level resolution can be achieved with modern instruments, while still allowing a wide variety of environmental conditions to be probed. Since density and temperature are two of the factors that determine a cloud's susceptibility to collapse, measurements of density and temperature in star-forming regions of nearby galaxies can play a crucial role in linking physical processes to the star formation rate.

Elsewhere in this volume, Leroy et al. (2018, p. 483) discuss the use of density-sensitive molecular lines at millimeter wavelengths (such as HCN and HCO^+) to constrain the distribution of mass as a function of volume density, i.e., the density PDF. As discussed in that contribution, mapping of multiple lines can reveal much more about the state of the interstellar medium (ISM) than CO observations alone. However, because molecular abundances can vary significantly due to chemical and photodissociation effects, and because the brightest lines also tend to be optically thick, millimeter line fluxes serve as imperfect density probes. Even for optically thin lines, such as isotopologues of CO, the effects of density and temperature are often highly degenerate. A complementary approach, that can also be pursued with the ngVLA, is to robustly measure density and temperature using sensitive observations of spectral lines of ammonia (NH_3) and formaldehyde (H_2CO). Such measurements can be used to test theories for the rate and efficiency of star formation (e.g., Krumholz & McKee 2005), many of

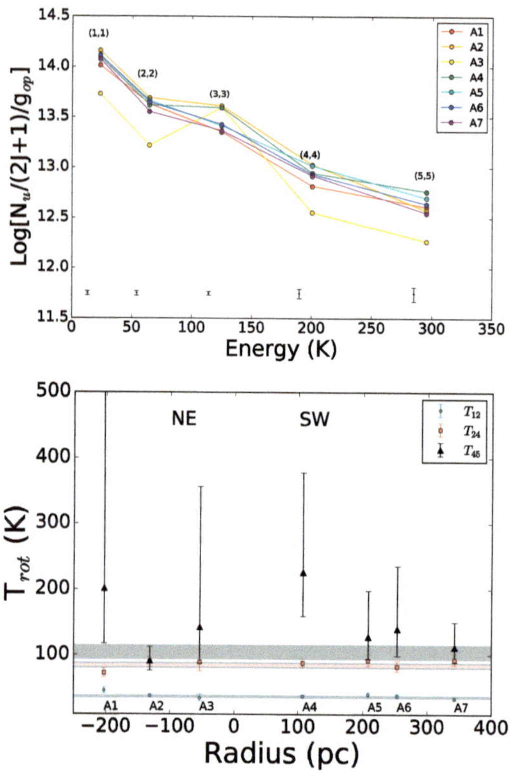

Figure 1. (*Top:*) Boltzmann diagram including 5 metastable transitions of NH_3 observed with the VLA towards seven positions (corresponding to colored lines) in NGC 253. Error bars are plotted along the bottom and correspond to unweighted averages. (*Bottom:*) Rotation temperatures derived from metastable line pairs, plotted against distance from the galaxy center. The horizontal bars represent single temperature fits for each pair, showing that T_{12} traces gas at ~40 K while T_{24} and T_{45} trace gas at ~100 K. Adapted from Gorski et al. (2017).

which invoke the Jeans scale and the free-fall time in the dense gas (two quantities which themselves depend on density and temperature). Density and temperature measurements can also be used to constrain the impact of star formation on its surrounding interstellar environment through mechanical and radiative feedback, which are thought to be important drivers of galaxy evolution.

The greatest leverage will be provided by studies which combine ngVLA and ALMA data to study both the moderate density gas (traced by ALMA CO imaging) and the dense star-forming gas (traced by the ngVLA in HCN, HCO^+, NH_3 and H_2CO). ALMA observations of the temperature-sensitive (sub-)mm transitions of H_2CO will also be highly complementary to the ngVLA.

2. NH_3 and H_2CO as Probes

Metastable inversion states of NH_3 trace gas at densities > 10^3 cm^{-3}, have level populations controlled largely by collisional processes, and lead to a large number of lines, sampling a wide range of temperatures, within the relatively small frequency range of 23–27 GHz—easily covered by a single observing configuration (see also the article by Di Francesco et al. 2018, this volume, p. 299). NH_3 is thus a useful probe of gas kinetic temperature from ~20 K to ~300 K (Ho & Townes 1983). For each pair of metastable states, a rotational temperature can be derived from the Boltzmann diagram (e.g., Figure 1, from Gorski et al. 2017), and can be converted to kinetic temperature once corrections for transitions involving non-metastable states are applied (e.g., Walmsley & Ungerechts 1983). Excluding the (3,3) transition which is susceptible to masing, one can thus obtain multiple $T_{\rm rot}$ measurements per sight line. Pairs with larger energy separations probe higher temperatures, so multiple values of $T_{\rm rot}$ at a given location (as seen in NGC 253, Figure 1 bottom) may indicate the presence of multiple temperature components along the line of sight.

Formaldehyde is present throughout the ISM at fairly high abundance and shares some similarities to ammonia, being a slightly asymmetric rotor. Like NH_3, for which radiative transitions between K ladders are forbidden, H_2CO offers multiple transitions (with $\Delta J = 1$, at mm or sub-mm wavelengths) which are close in frequency but belong to different K ladders, making these an excellent probe of temperature. In addition, however, the K-doublet ($\Delta J = 0$) transitions of H_2CO at 4.8 and 14.4 GHz possess unique density sensitivity, transitioning from absorption (of the CMB or any continuum source) to emission when crossing a density near $10^{5.5}$ cm^{-3} (e.g., Mangum et al. 2008). Densities near the transition occur only in the densest cloud cores in the Milky Way, but may be much more common in starburst galaxies. Moreover, measurement of both of these lower frequency H_2CO lines, with independent knowledge of kinetic temperature from NH_3, provides unique solutions for the (ensemble average) gas density and H_2CO column density (Mangum et al. 2008, 2013a), from which the thermal gas pressure can also be derived. Figure 2, from Mangum et al. (2013a), demonstrates the importance of independent $T_{\rm kin}$ measurements for this work: adopting a $T_{\rm kin}$ based on infrared dust emission leads to much higher inferred densities for starbursts than direct (NH_3-based) $T_{\rm kin}$ estimates.

3. Why the ngVLA is Needed

Given relatively low opacity and abundance ratios, and the low beam filling factor of the emission in normal galaxies, both high surface brightness sensitivity and high resolution are required for detection of these molecules. For ammonia, the ngVLA could achieve comparable surface brightness sensitivity to Effelsberg but with a 5″ rather than 40″ beam, allowing resolved mapping of galaxies where extended emission is present. The ngVLA could also detect very bright (T_b ~ 60 K, Beuther et al. 2005) pc-scale (0.″1) clumps similar to NGC 6334 I(N) at a distance of 2 Mpc. Recent observations by the VLA (Gorski et al. 2017, 2018) show the current state-of-the-art, which is adequate for only the brightest or most compact sources within starburst environments.

For the 4.8 and 14 GHz formaldehyde lines, the ability to map the lines in absorption will be key (e.g., Baan et al. 1997). Beam-matched observations of both lines

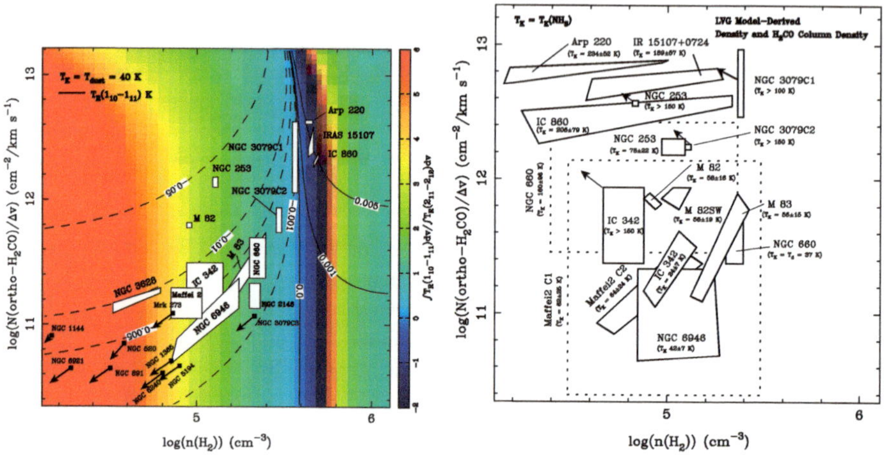

Figure 2. (*Left:*) From Mangum et al. (2013a), LVG model predictions for gas density $n(H_2)$ and ortho-H_2CO column density per velocity gradient for galaxies observed with the GBT, assuming $T_{kin} = T_{dust} \approx 40$ K. Colors are model line ratios, contours are modeled intensities (dashed being negative). The transition from absorption to emission at $\log n \approx 5.5$ is clearly seen. (*Right:*) Same plot using NH_3-measured values of T_{kin}. The higher temperatures in starbursts imply more moderate densities than inferred in the left panel.

are most valuable for constraining density, and would require the use of different array configurations.

No current or planned facility will have comparable collecting area or resolution at 1 cm for ammonia observations. For the lowest frequency (4.8 GHz) formaldehyde transition, it would be possible to make sensitive observations with a large single dish (e.g. future upgrade to FAST). Such observations would complement ngVLA given the importance of flux recovery when both emission and absorption are present. Both the 4.8 and 14 GHz transitions of H_2CO may lie within reach of the mid-frequency Square Kilometre Array (SKA), but the SKA is likely to be less suited for making deep pointed observations of nearby galaxies.

4. Experimental Setup

Based on the ngVLA Reference Design, we assume that the ngVLA will achieve, with a 1″ beam, a brightness sensitivity of ~100 mK at 20 GHz using 10 km s^{-1} channels after 4 hours of on-source integration.

- **NH_3**: The NH_3 (1,1) to (7,7) lines have frequencies ranging from 23.694 GHz to 25.715 GHz (e.g., Mangum et al. 2013b). The (non-metastable) NH_3 (2,1) line at 23.099 GHz provides additional constraints on radiative excitation. For extragalactic work, spectral resolutions of ~0.1 MHz (~1 km/s) are generally adequate; the very narrow (~10 kHz) lines seen in Galactic dark clouds (Rydbeck et al. 1977) will be blended over many clouds. The sizes of the emitting regions are of order ~1 pc for Milky Way-like clouds (0.′′1 at 2 Mpc) but may be extended up to ∼500 pc in starburst galaxies. For actively star-forming galaxies, typical integrated intensities per line are 0.5 to 5 K km/s for beam sizes of 40″ (NGC 253 and IC 342, Mauersberger et al. 2003; M51, Takano et al. 2013). For a line width of 100 km/s, this corresponds to a beam-averaged surface brightness up to ∼50 mK and a flux density up to ∼40 mJy. The ngVLA will be sensitive to this average surface brightness even at a resolution of 1″, and thus achieve very high S/N ratios toward compact emission peaks.

- **H_2CO**: Formaldehyde has well-studied *K*-doublet transitions at 4.8 GHz (1_{10}–1_{11}), 14.4 GHz (2_{11}–2_{12}), 29 GHz (3_{12}–3_{13}), and 48 GHz (4_{13}–4_{14}). The lower-frequency transitions have more modest excitation requirements and thus provide cleaner probes of density (i.e., are less sensitive to assumed excitation temperature). A channel width of 16 kHz (∼0.4 km/s) and a bandwidth per window of 100 MHz (∼2000 km/s) should be adequate for extragalactic work. The low-frequency lines (as observed with the GBT; Mangum et al. 2008, 2013a) are most often detected in absorption, with depths of ∼1–10 mK and ∼200 km/s line widths on a 50″ scale. If the absorbing material is clumpy, of course, much deeper absorption lines should be seen by the ngVLA, given sufficient background emission (e.g., ∼1 K absorption depths have been detected towards NGC 253 with a 15″ VLA beam by Baan et al. 1997).

Acknowledgments. This document grew out of discussions within the Galactic Ecosystems Working Group. We owe a particular debt to Adam Leroy, who led ngVLA Memo No. 7, as well as Cornelia Lang and Alberto Bolatto for organizing discussions. We thank Jeff Mangum for useful comments on an earlier draft and Mark Gorski for providing Figure 1.

References

Baan, W. A., Bragg, A. E., Henkel, C., & Wilson, T. L. 1997, ApJ, 491, 134
Beuther, H., Thorwirth, S., Zhang, Q., et al. 2005, ApJ, 627, 834
Di Francecso, J., Keown, J., Friesen, R., Bourke, T., & Caselli, P. 2018, Science with a Next-Generation Very Large Array, ASP Monograph 7, ed. E. J. Murphy (San Francisco, CA: ASP), 299
Gorski, M., Ott, J., Rand, R., et al. 2017, ApJ, 842, 124
Gorski, M., Ott, J., Rand, R., et al. 2018, ApJ, 856, 134
Ho, P. T. P., & Townes, C. H. 1983, ARA&A, 21, 239
Kennicutt, R. C., & Evans, N. J. 2012, ARA&A, 50, 531
Krumholz, M. R., & McKee, C. F. 2005, ApJ, 630, 250
Leroy, A. K., Bolatto, A., Kepley, A., Meier, D., & Wong, T. 2018, Science with a Next-Generation Very Large Array, ASP Monograph 7, ed. E. J. Murphy (San Francisco, CA: ASP), 483
Mangum, J. G., Darling, J., Menten, K. M., & Henkel, C. 2008, ApJ, 673, 832
Mangum, J. G., Darling, J., Henkel, C., & Menten, K. M. 2013a, ApJ, 766, 108
Mangum, J. G., Darling, J., Henkel, C., et al. 2013b, ApJ, 779, 33
Mauersberger, R., Henkel, C., Weiß, A., Peck, A. B., & Hagiwara, Y. 2003, A&A, 403, 561
Rydbeck, O. E. H., Sume, A., Hjalmarson, A., et al. 1977, ApJ, 215, L35
Takano, S., Takano, T., Nakai, N., Kawaguchi, K., & Schilke, P. 2013, A&A, 552, A34
Walmsley, C. M., & Ungerechts, H. 1983, A&A, 122, 164

Science with a Next-Generation Very Large Array
ASP Conference Series, Monograph 7
Eric J. Murphy, ed.

Astrometry and Long Baseline Science

Mark Reid,[1] Laurent Loinard,[2] and Thomas Maccarone[3]

[1]*Harvard-Smithsonian CfA, Cambridge, MA, USA;* `reid@cfa.harvard.edu`

[2]*Centro de Radioastronomía – UNAM, Morelia, Michoacán, Mexico;* `l.loinard@crya.unam.mx`

[3]*Texas Tech University, Lubbock, TX, USA;* `Thomas.Maccarone@ttu.edu`

Abstract. Recent advances in VLBI have led to astrometric accuracy exceeding that of the *Gaia* mission goals. This chapter describes some important astrophysical problems that can be addressed with sub-milliarcsecond imaging and μas astrometry using the ngVLA with long baselines.

1. Introduction

Astrometry at centimeter to millimeter wavelengths has advanced dramatically over the last decade. VLBI techniques have been perfected and relative positional accuracy of $\approx 10\,\mu$as is routinely achieved for compact sources relative to background quasars (Reid & Honma 2014). Indeed, this demonstrated astrometric accuracy exceeds that of the *Gaia* mission goal and, since radio waves can penetrate even hundreds of magnitudes of visual extinction, provides unique opportunities to fully explore the Galactic plane and deeply embedded sources in star forming regions. Long baseline radio interferometry has produced some remarkable results, including 1) detailed mapping of accretion sub-parsec scale disks around supermassive black holes in galaxies well into the Hubble flow, yielding "gold standard" masses for the black holes and direct estimates of the Hubble constant independent of all other methods (Gao et al. 2016), and 2) measurements of trigonometric parallax to masers in star forming regions across large portions of the Milky Way, revealing its true spiral structure, size, and rotation curve (Reid et al. 2014).

Many potential applications of radio astrometry are currently limited by the sensitivity of long baseline interferometry with, for example, ten antennas of 25-m diameter. The ngVLA Long Baseline Array (LBA), with its baselines comparable to the size of North America, could provide a sensitivity increase of more than an order of magnitude, which translates to the ability to detect > 30-times more sources (for $N \propto S^{-3/2}$), both for astrometric targets and background quasars. The greatly increased background source counts project to a decrease in target-quasar separation by a factor of $\sqrt{30}$, which directly translates to improved astrometric accuracy by the same factor. Thus, the ngVLA could potentially achieve astrometric accuracies of $\approx 1\,\mu$as! In this chapter we explore some of the applications of the astounding capabilities of the ngVLA with long baselines based on our current knowledge of the Universe. Of course, by

2030 and later, new and unknown applications are almost certain to be important for the advancement of astrophysics.

2. Gravitational Wave Sources

The ngVLA has the potential to directly observe the last stages of the inspiral of compact binaries, as well as post merger activity, across the Universe. The merger of two neutron stars, such as recently observed by *LIGO*, or a neutron star and a black hole, can produce emission across the electromagnetic spectrum. Even at distances of hundreds of Mpc, the ngVLA could make movies of expanding ejecta or jets, since at cosmological source distances, scales of $\sim$ 1 pc correspond to $\sim$ 1 mas (Alexander et al. 2017). One can also expect to witness the inspiral of supermassive black holes (SMBHs), complementing a future space-based gravitational wave detector like *LISA*, by providing the only way to achieve *resolved* images of electromagnetic counterparts.

The International Pulsar Timing Array goal is to detect gravitational waves by timing an array of about 30 pulsars. While distances to these pulsars can be solved for from the timing data, having to solve for these terms introduces a noise term and parameter correlations. Obtaining independent pulsar distances via trigonometric parallaxes will be possible with long baselines on the ngVLA, and this could substantially increase the sensitivity of gravitational wave detections. (See also §4 for the importance of accurate distance for the interpretation of the effects of gravitation radiation.)

3. Black Holes

Fermi has detected an amazing variety of energetic phenomena, including beams of particles of unexpectedly high energy. Some of the most energetic sources in the universe are found in AGN. One of the most important and long-standing problems in astrophysics is to answer how do cosmic accelerators work and what are they accelerating? VLBA imaging of AGNs has revealed that gamma-ray emission comes from jet components (Hodgson et al. 2017), and not always from the immediate vicinity of SMBHs. Still, important questions remain: Are jets proton-electron or positron-electron plasma? What powers and collimates AGN jets and how do these jets evolve and interact with their host galaxy? Long baseline radio interferometry provides the highest angular resolution possible to make movies of these amazing phenomena on scales down to the ergospheres of nearby SMBHs. Note that recent advances in theory and computation have opened the possibility of fully relativistic, magneto-hydrodynamic simulations in 3-dimensions. High sensitivity and resolution observations with long baselines on the ngVLA will be critical to guide simulations and finally arrive at a deep understanding of these fundamental astrophysical phenomena.

H_2O masers in AGN accretion disks have been used to measure accurate SMBH masses and the Hubble constant. But they also have the potential to determine physical properties of these sub-parsec scale disks. Of particular interest is mapping the magnetic fields in these disks via spatially resolved polarization measurements (Zeeman effect). Currently, observations have provided upper limits to magnetic field strength (Modjaz et al. 2005), but, with the greatly increased sensitivity of the ngVLA, detection should be possible. Indeed, it will be the only telescope that can make such resolved measurements in the foreseeable future.

Stellar mass black holes pose equally fascinating questions. Do these black holes form from supernova (SN) or by direct collapse without explosions? What is the origin of black hole spin? The ngVLA will be able to provide unique clues that address these questions by measuring parallax and proper motions. For example, the long-standing controversy over the distance to Cyg X-1 (the source of the famous wager between Thorne and Hawking) was recently resolved by long baseline radio astrometry; its accurate parallax distance of 1.86 ± 0.12 kpc (Reid et al. 2011) allowed a precise determination of the mass of the compact object of $15{\pm}1\ M_{\odot}$ (Orosz et al. 2011), clearly indicating a black hole, and was key to determining that the black hole is spinning maximally with $a* > 0.92$ (Gao et al. 2011). This binary is too young for accretion to have appreciably spun up the black hole, indicating it was born with great spin. Regarding its birth, its measured distance and proper motion (peculiar motion of only 20 km/s) matches that of the Cyg OB3 star-forming cluster, establishing that it was born in this young cluster. Since a SN explosion would have disrupted the region, the black hole probably formed with a quiet, prompt stellar collapse (Mirabel & Rodrigues 2003). But this is only one source and similar measurements of many more, weaker sources with the ngVLA are needed to understand their complete demographics and determine how they formed. It is important to note that most known black hole X-ray binaries are both too far and too extincted for precise *Gaia* parallaxes.

At the present time, measurements of astrometric wobble of the black hole (induced by the secondary star in the binary) are marginally possible in one X-ray binary, Cygnus X-1 (Reid et al. 2011). With the ngVLA, if it has enough collecting area on long baselines, proper tracing of orbits should become possible for a few X-ray binaries, both due to the increased sensitivity and the ability to work at higher frequencies, which will allow both better angular resolution and more easily manageable systematics due to less extended emission from jets. This is vital, because the biggest uncertainty in black hole mass measurements usually comes from the precision of the estimation of the inclination angle of the orbit, since the inferred mass scales as $\sin^3 i$. Additional uncertainties come into play from having incorrect inclination angles using either the disk continuum method for spin estimation (e.g., Steiner et al. 2017) or the reflection method (e.g., Garcia et al., 2013).

Understanding the distribution of masses of black holes and neutron stars is one of the few means we have to probe the actual process of supernova explosions. At the present time, there appears to be a gap between the lowest mass black holes and the highest mass neutron stars, which would imply that whatever instability causes supernovae to actually blow up must be relatively rapid (Belczynski et al. 2012), but it remains quite possible that the gap is an artifact of the biases in inclination angle measurements from ellipsoidal modulations (Kreidberg et al. 2012). Having even a few systems where direct inclination angle measurements from astrometric wobble can be used to calibrate the ellipsoidal modulation of the secondary's light would be vital for building samples of well-understood black hole masses and spins.

4. Fundamental Physics

Pulsar parallaxes and proper motions precisely locate these stellar remnants in the Galaxy and provide full phase-space information. Coupled with rotation and dispersion measurements, this can be used to model the magnetic field and electron density of the Milky Way (Cordes et al. 1991). Peculiar motions (after removing Galactic rotation) give direct information on "kicks" received at birth by asymmetrical SN explosions. Additionally, knowing distances to pulsars allows accurate mass measurements. Pulsar mass is the key parameter to discriminating among competing models for the equation-of-state of material at the extreme density of neutron stars.

One interesting problem critically dependent on the Galaxy's fundamental parameters (e.g., R_0 & Θ_0; see the Milky Way section below) involves the interpretation of the orbital decay of the Hulse-Taylor binary pulsar system, owing to gravitational radiation as predicted by General Relativity. In 1993, the Nobel Prize in Physics was awarded for this measurement. In order to properly interpret the observed decay rate, one needs to account for the accelerations of the Sun (Θ_0^2/R_0) and the pulsar [$\Theta(R)^2/R$] from their Galactic orbits. In 1993, uncertainties in the values of R_0 and Θ_0 limited the Relativity test to ±0.23%. With improved values based on maser parallaxes, this uncertainty was reduced by a factor of 3. Interestingly, with these values, there is a 3σ discrepancy from General Relativity, using the pulsar distance of 9.9 kpc assumed in 1993. This discrepancy would vanish if the pulsar distance is 7.2 kpc (Reid et al. 2014). An accurate pulsar parallax is critical to test this prediction, but may require the sensitivity of the ngVLA as the pulsar is weak. Of course there are other binaries that can be used to test General Relativity, and distances to these coupled with accurate models of the Galaxy are needed and could be supplied by the ngVLA with long baselines.

Are physical "constants", such as the fine-structure constant and the proton-to-electron mass ratio, different in the early Universe? Does the cosmic microwave background temperature evolve as predicted with redshift? These and other questions (see section on Galaxy Evolution) can be addressed by imaging molecular clouds at high redshift in absorption against bright AGNs. However, high sensitivity and angular resolution (~ 1 mas) are crucial to detecting and isolating (resolving) individual pc-scale clouds, since observations at lower resolutions blend together clouds that have a wide range of physical conditions (Sato et al. 2013).

5. Galaxy Evolution

The Local Group offers a critical opportunity to study the formation and evolution of galaxies with extremely high resolution (i.e., local cosmology). Long baseline observations have been able to measure the proper motions of two of Andromeda's satellite galaxies, M 33 (Brunthaler et al. 2005) and IC10 (Brunthaler et al. 2007). Figure 1 schematically shows the locations of the Milky Way, M 31 (Andromeda) and M 33, along with the motion of M 33. These constrain values of the distance and mass of Andromeda. The key parameter for determining the the distribution of dark matter and the history and fate of the Local Group is the proper motion of Andromeda itself, which would give its 3-D velocity relative to the Milky Way.

Current estimates of Andromeda's proper motion vary considerably, from values near zero to over 100 km/s. This range allows scenarios in which Andromeda directly hits the Milky Way in a few billion years or instead they could orbit each other for a very long time. The most direct and accurate measurement to solve this problem would be astrometric measurements of the AGN in Andromeda, M 31*, with μas accuracy over several years. This may only be possible with the ngVLA LBA, since M 31* is extremely weak, ~ 10 μJy.

How early do SMBHs form and how do they relate to their protogalaxies? When studying protogalaxies, it is crucial to understand what portion of the emission is at-

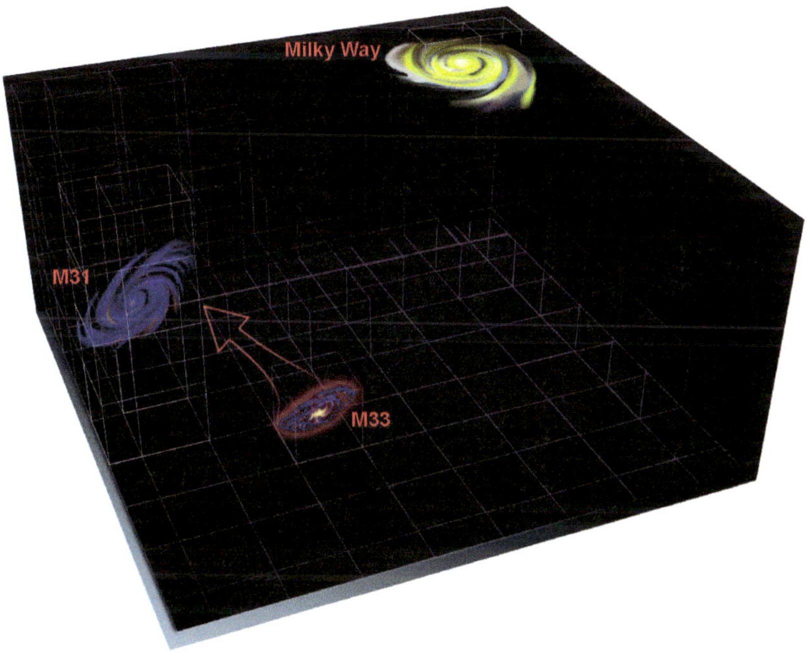

Figure 1. Local Group proper motions. Schematic view of the Local Group of Galaxies with the Milky Way in the upper right and the Andromeda galaxy (M 31) in the left. The measured 3-D motion of M 33 is indicated with the red arrow. Only the radial component, approaching the Milky Way at $\approx$ 110 km s^{-1}, of the motion of Andromeda is known. If Andromeda's proper motion is small it will collide with the Milky Way in a few billion years. However, if it has a substantial proper motion ($\sim$ 100 km s^{-1}), Andromeda and the Milky Way will orbit each other for a very long time.

tributable to a SMBH and what comes from a more extended star-burst. At $z \approx 6$, a 100 pc sized star-burst nucleus subtends 18 mas (and smaller at intermediate redshifts). For a radio-bright SMBH, the best telescope to resolve and separate AGN from star-bursts would be the ngVLA with long baselines.

Are molecular clouds seen toward high redshift galaxies different than those seen locally? Molecular absorption line observations at the high angular resolution afforded by long baselines on the ngVLA may be the only foreseeable way to probe the density, temperature, and structure (filamentary?) of molecular clouds in the early Universe.

6. Milky Way Structure

VLBI parallax and proper motions of masers have been measured for $\approx$ 200 high mass star formation regions (HMSFRs), as shown in Figure 2. These clearly trace spiral structure across large portions of the Milky Way (Reid et al. 2014), and provide the most accurate estimates of the distance to the Galactic center, R_0 (currently with about 2% accuracy), the rotation speed of the Galaxy at the Sun, Θ_0, and the Galaxy's rotation curve. However, with current sensitives, very few sources are detected that are beyond the Galactic center. So, roughly half of the Milky Way remains "terra incognito." The ngVLA LBA, with baselines of thousands of km, will have the sensitivity and angular resolution to complete a map of the Milky Way. The greater sensitivity of the ngVLA, compared to the current VLBA, allows the use of weaker QSOs that are then closer in angle to the target sources. This yields improved parallax accuracy which coupled with an increased sample size should allow estimation of R_0 to 1% accuracy.

Pulsar parallaxes can provide an accurate map of their locations in the Galaxy. Coupled with rotation and dispersion measurements, they can be used to model the magnetic field and electron density of the Milky Way. Since radio astrometry improves dramatically at frequencies above a few GHz (to minimize the turbulent and difficult to model ionosphere), the greatly improved sensitivity of the ngVLA, compared to current long baseline interferometers, is needed to do astrometry on these very steep-spectrum sources.

7. Young Stars and Stellar Systems

The evolution of a star is almost entirely determined by its mass (and, to a much smaller extent, its chemical composition). Thus, direct stellar mass measurements are of great importance to constrain theoretical stellar evolution models. Such measurements can be obtained in multiple stellar systems if the orbital motions are monitored with sufficient accuracy and over a period covering a significant fraction of an orbital period. For main and pre-main sequence stars, this is best achieved using optical and near-infrared imaging techniques (adaptive optics, aperture masking interferometry, etc.). However, for protostellar objects one must turn to longer wavelengths on account of the high extinction that affects such deeply embedded objects. Millimeter and centimeter interferometric observations are critical, since they can reach both very high angular resolution and high astrometric accuracy.

At the moment, monitoring of orbital motions has only been possible for a small number of very young (class 0) stellar systems: e.g., IRAS 16293–2422 (Loinard et al. 2002), L1551 IRS5 (Rodriguez et al. 2003), and YLW15 (Curiel et al. 2002), via

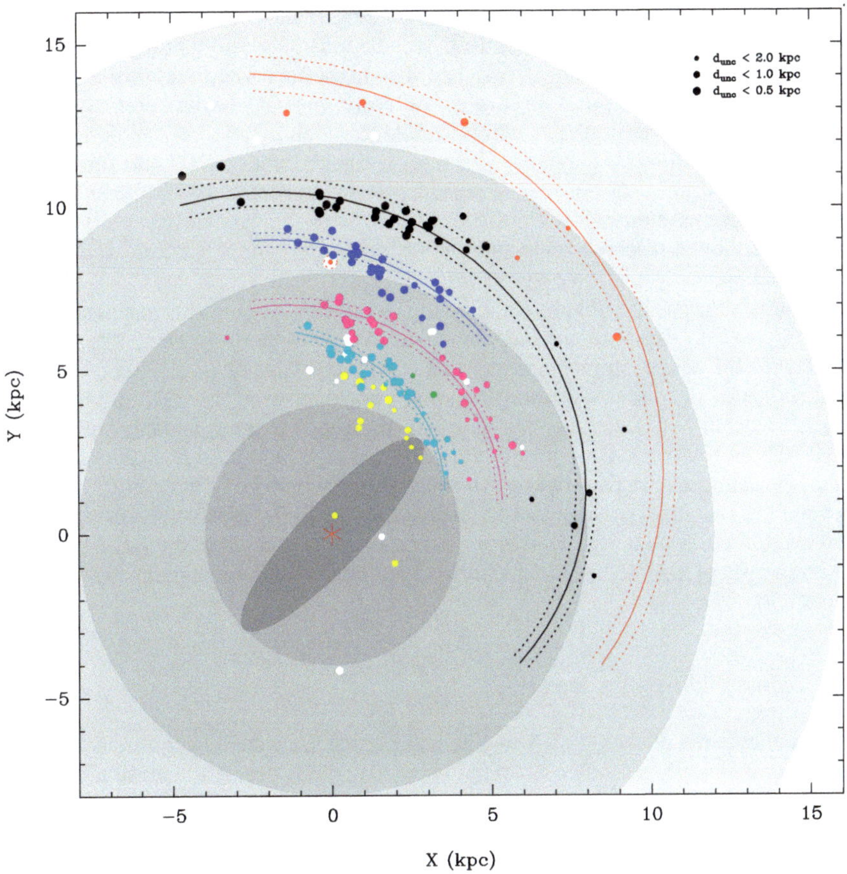

Figure 2. Plan view of the Milky Way showing the locations of high-mass star forming regions (HMSFRs) with trigonometric parallaxes measured with VLBI (Reid et al, in preparation). The Galactic center (*red asterisk*) is at (0,0) and the Sun (*red Sun symbol*) is at (0,8.34). HMSFRs were assigned to spiral arms based primarily on association with structure seen in $\ell - V$ plots of CO and hydrogen emission (and not based on the measured parallaxes): Inner Galaxy sources, *yellow dots*; Scutum arm, *cyan octagons*; Sagittarius arm, *magenta hexagons*; Local arm, *blue pentagons*; Perseus arm, *black squares*; Outer arm, *red triangles*. Distance errors are indicated in the legend. The background grey disks provide scale, with radii corresponding in round numbers to the Galactic bar region ($\approx$ 4 kpc), the solar circle ($\approx$ 8 kpc), co-rotation of the spiral pattern and Galactic orbits ($\approx$ 12 kpc), and the end of major star formation ($\approx$ 16 kpc). The "long" bar is indicated with a shaded ellipse. The *solid* curved lines trace the centers (and *dotted* lines the 1σ widths) of the spiral arms. Note that there are few measurements for sources beyond the Galactic center (negative Y values); long baselines with the ngVLA will provide the sensitivity and angular resolution to help finish this map.

multi-epoch VLA observations. However, none of these observations cover a sufficient fraction of an orbit to enable a reliable mass estimate. The ngVLA, having baselines longer than 300 km and a sensitivity 10-times higher than the VLA, will permit significant progress in two ways. First, relative positions between the members of binary systems could be measured to at least 10-times higher accuracy, and this would result in a much higher precision in orbit (and, therefore, mass) determination. Second, the higher resolution of the ngVLA would enable the identification of much tighter systems (with separations as small as $\sim$ 10 mas, compared with $\sim$ 100 mas for the VLA) with commensurably shorter orbital periods. Reliable mass measurements would be attained for such systems. It is worth emphasizing that the very early evolution of stars is still a very open research topic and that accurate mass measurements at the earliest protostellar stages would have a great impact.

Protostellar jets are ubiquitous in young stellar objects, and two mechanisms have been proposed to explain their launching. In magnetospheric models, they are associated with the accretion mechanisms that occur on scales of $\sim$ 0.1 AU ($\sim$ 1 mas at the distance of nearby star-forming regions). For disk-wind models, in contrast, the launching occurs on scales comparable with those of the inner accretion disk region ($\sim$ 10 AU; or 100 mas). Baselines of at least 300 km would provide an angular resolution in the centimeter regime of about 10 mas, which would be sufficient to discriminate between the two mechanisms, and to resolve the jets in the transversal direction if the disk-wind hypothesis holds. However, baselines of at least a few thousand kilometers (i.e.,the ngVLA LBA) are required to resolve jets launched through magneto-centripetal acceleration.

8. Commercial Applications

While the benefits of the ngVLA to classical astrophysics have been discussed above, there may be significant other benefits to society. VLBI observations supply accurate telescope locations on a global scale. This is used to calibrate GPS positioning, which is an integral part of our economic activity. Also VLBI has demonstrated that it can track spacecraft with exceptional accuracy. Had VLBI been used to track the Martian lander that crashed into its surface, the metric-vs-English units error that lead to the multi-billion dollar disaster could have been caught and rectified. As the USA aims to send robotic crafts and humans to Mars and beyond, it seems foolish (and dangerous) not to use the best independent method of spacecraft tracking, which the ngVLA LBA could provide.

References

Alexander, K. D. et al. 2017, ApJ, 848, 21
Belczynski, K., Wiktorowicz, G., Fryer, C. L. et al. 2012, ApJ, 757, 91
Brunthaler, A. et al. 2005, Sci, 307, 1440
Brunthaler, A. et al. 2007, A&A, 462, 101
Cordes, J. M. et al. 1991, Nat, 354, 121
Curiel, S., Girart, J. M., Rodriguez, L. F. & Canto, J. 2002, ApJ, 638, 878
Gao, L. et al. 2011, ApJ, 742, 85
Gao, F. et al. 2016, ApJ, 817, 128
Garcia, J., Dauser, T. Reynolds, C. S. et al. 2013, ApJ, 768, 146
Hodgson, J. A. et al. 2017, A&A, 597, 80

Kreidberg, L., Bailyn, C. D., Farr, W. M. & Kalogera, V. 2012, ApJ, 757, 36
Loinard, L. 2002, RMxAA, 38, 61
Mirabel, I. F. & Rodrigues, I. 2003, Sci, 300, 1119
Modjaz, M., Moran, J. M., Kondratko, P. T. & Greenhill, L. J. 2005, ApJ, 626, 104
Orosz, J. A. et al. 2011, ApJ, 742, 84
Reid, M. J. et al. 2011, ApJ, 742, 83
Reid, M. J. & Honma, M. 2014, ARA&A, 52, 339
Reid, M. J. et al. 2014, ApJ, 700, 137
Rodriguez, L. F. et al. 2003, ApJ, 583, 330
Sato, M., Reid, M. J., Menten, K. M. & Carilli, C. L. 2013, ApJ, 764, 132
Steiner, J. F., Garcia, J. A., Eikmann, W. et al. 2017, ApJ, 836, 119

Part VII

Galaxy Formation and Evolution

Science with a Next-Generation Very Large Array
ASP Conference Series, Monograph 7
Eric J. Murphy, ed.

Imaging Molecular Gas at High Redshift

C. L. Carilli and Y. Shao

National Radio Astronomy Observatory, Socorro, NM 87801, U.S.A.;
`ccarilli@nrao.edu, yshao@nrao.edu`

Abstract. We perform simulations of the capabilities of the next generation Very Large Array in the context of imaging low order CO emission from typical high redshift star forming galaxies at ~ 1 kpc resolution. We adopt as a spatial and dynamical template the CO 1-0 emission from M 51, scaled accordingly for redshift, transition, and total gas mass. The molecular gas masses investigated are factors of 1.4, 3.5, and 12.5 larger that of M 51, at z = 0.5, 2, and 4.2, respectively. The z = 2 galaxy gas mass is comparable to the lowest mass galaxies currently being discovered in the deepest ALMA and NOEMA cosmological CO line surveys, corresponding to galaxies with star formation rates ~ 10 to 100 $M_{\odot}$ yr^{-1}. The ngVLA will perform quality imaging at 1kpc resolution of the gas distribution and dynamics over this disk. We recover the overall rotation curve, galaxy orientation properties, and molecular ISM internal velocity dispersion. The model at z = 4.2 corresponds to a massive star forming main sequence disk (SFR ~ 130 $M_{\odot}$ yr^{-1}). The ngVLA can obtain 1kpc resolution images of such a system in a reasonable integration time, and recover the basic galaxy orientation parameters, and, asymptotically, the maximum rotation velocity. We compare the ngVLA results with capabilities of ALMA and the Jansky VLA. ALMA and the VLA can detect the integrated low order CO emission from these galaxies, but lack the sensitivity to perform the high resolution imaging to recover the dynamics at 1kpc scales. To do so would require of order 1000 hrs per galaxy with these current facilities. We investigate a 'minimal' ngVLA configuration, removing the longest baselines and much of the very compact core, and find good imaging can still be performed at 1 kpc resolution.

1. Introduction

The next generation Very Large Array (ngVLA), is being considered as a future large radio facility operating in the 1.2 – 116 GHz range. The current design involves ten times the effective collecting area of the VLA and ALMA, with more than ten times longer baselines, providing mas-resolution, plus a dense core on km-scales for high surface brightness imaging. The ngVLA opens unique new parameter space in the imaging of thermal emission from cosmic objects ranging from protoplanetary disks to distant galaxies, as well as unprecedented broad band continuum polarimetric imaging of non-thermal processes (McKinnon et al. 2016; Carilli et al. 2015; Selina & Murphy 2017; Bolatto et al. 2017).

One of the primary science drivers for the ngVLA involves study of the molecular gas content of galaxies throughout cosmic time (Carilli et al. 2015). Cool molecular gas provides the fuel for star formation in galaxies, and hence represents a key constituent in the study of the Baryon cycle during galaxy evolution (Carilli & Walter 2013). A

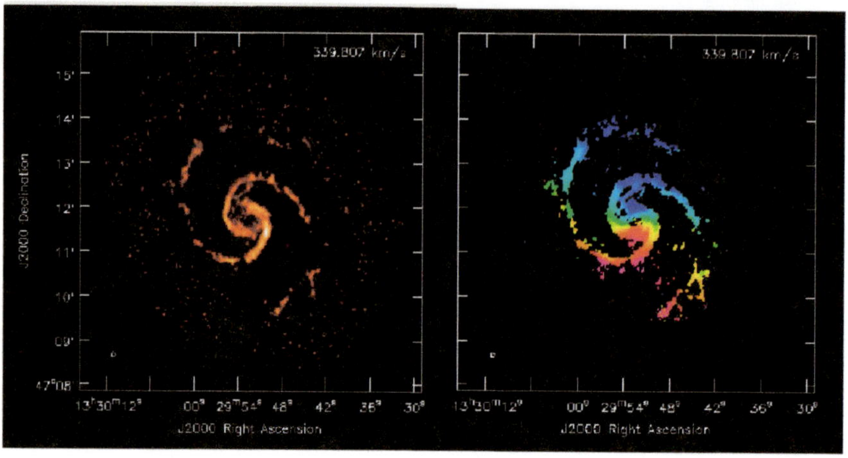

Figure 1. BIMA SONG CO 1-0 observations of M 51 at 200 pc resolution ($5\farcs5$; (Helfer et al. 2003)

standard method for deriving the molecular gas mass in galaxies involves an empirical calibration of the relationship between the CO 1-0 tracer and total gas mass (Bolatto et al. 2013). The high sensitivity and wide fractional bandwidth of the ngVLA provides a powerful tool to study the evolution of the molecular content of galaxies, through observations of the low order CO transitions.

An area that is currently poorly explored is the detailed distribution and kinematics of the molecular gas in early galaxies. Progress has been made using ALMA, the VLA, and NOEMA, but these studies are time consuming (tens of hrs per galaxy), and remain limited in spatial resolution (few kpc). The ngVLA will revolutionize this field, through sub-kpc resolution observations of the molecular gas down to very low brightnesses in galaxies out to the highest redshifts (Carilli & Walter 2013). In particular, the full frequency range from 30 – 100 GHz, covers both low and high order CO transitions at high redshift, with $10\times$ the sensitivity of ALMA and the VLA.

We investigate the ability of the ngVLA to image the molecular ISM in high redshift galaxies at ≤ 1 kpc resolution. This scale corresponds to the giant molecular star forming clumps being discovered in active star forming galaxies during the peak epoch of cosmic star formation, around $z \sim 2$ (Genzel et al. 2011). We compare the capabilities of the ngVLA to observe typical 'main sequence' star forming galaxies, with those of ALMA and the VLA. We also consider a minimal array design with which such measurements could be made with reasonable accuracy.

2. Template: CO 1-0 Emission from M 51

We adopt as a spatial template the CO 1-0 emission from the nearby star forming disk galaxy, M 51. M 51 is one of the best studied galaxies in molecular gas (Schinnerer et al. 2013). M 51 is at a distance of 7.6 Mpc (recessional velocity = 463 km/s), with an integrated CO 1-0 luminosity of $L'_{CO} = 1.6\times10^9$ K km/s pc^2, implying a molecular gas mass of 5.5×10^9 $(\alpha/3.4)\,M_\odot$, where $\alpha = 3.4$, corresponding to the Galactic conversion

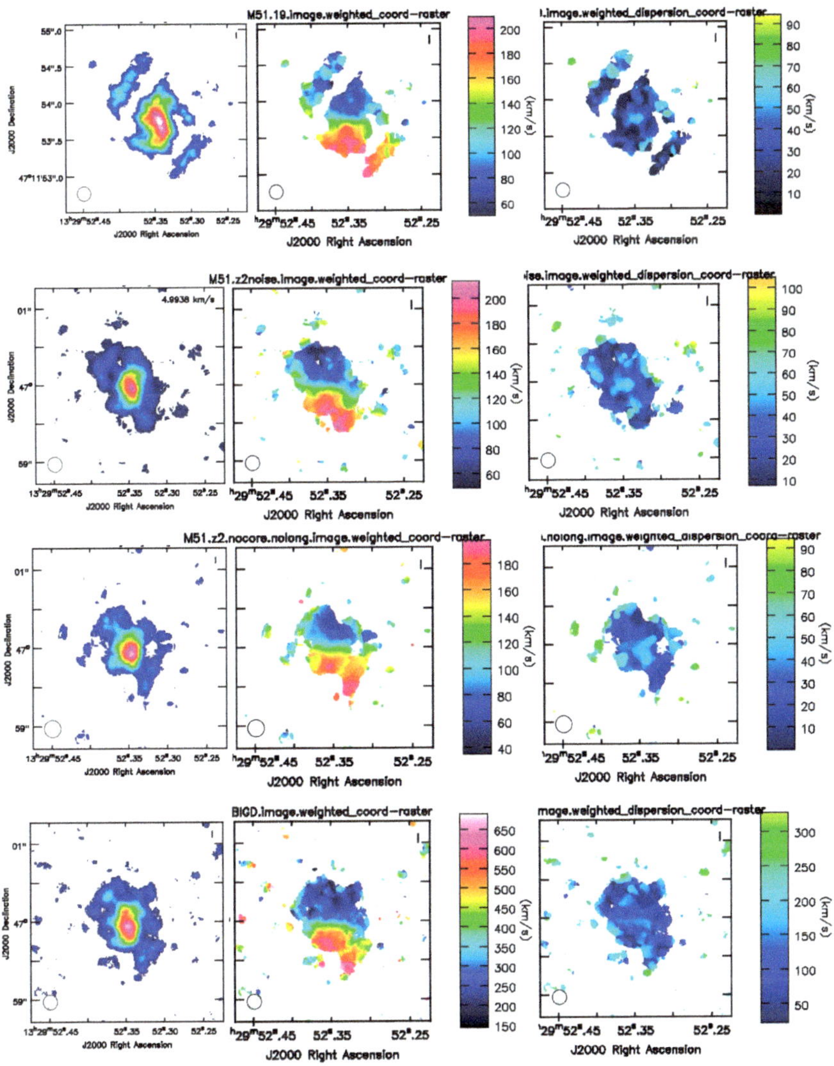

Figure 2. (*Top Row:*) Simulated ngVLA observations using at 77 GHz of the CO 1-0 emission from an M 51 analog at z = 0.5 for a 30 hr observation ($SFR \sim 7\,M_\odot$ yr^{-1}). The left frame shows the velocity integrated emission. The middle shows the intensity weighted mean velocity, and the right shows the velocity dispersion. (Frames are the same in all four examples). The spatial resolution is 0.″19. (*Second row:*) Simulated ngVLA observation at 77 GHz of the CO 2-1 emission from the z = 2 model galaxy (SFR ~ 25 $M_\odot$ yr^{-1}). The spatial resolution is 0.″19. (*Third row:*) Simulated observation using a 'minimal configuration' of the ngVLA (only 1/3 of the core, and no antennas at radii outside the original VLA Y), at 77 GHz of the CO 2-1 emission from the z = 2 model. The spatial resolution is 0.″19. (*Bottom row:*) Simulated ngVLA observation at 44 GHz of the CO 2-1 emission from a massive star forming disk galaxy at z = 4.2 (SFR ~ 130 $M_\odot$ yr^{-1}). The spatial resolution is 0.″22.

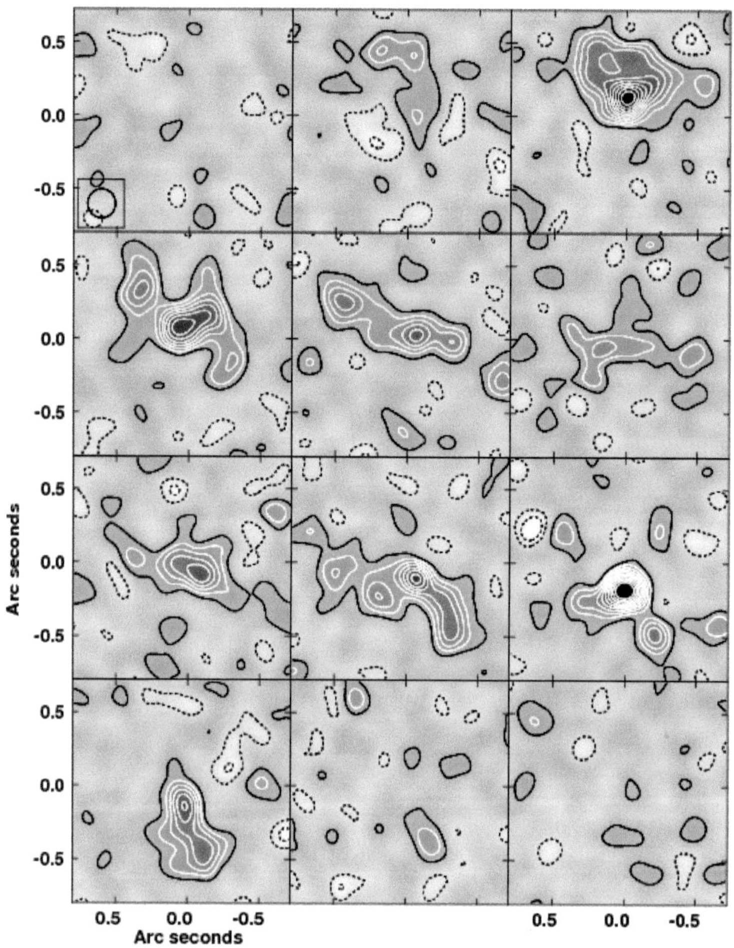

Figure 3. Spectral channel images of the CO 2-1 emission from star forming disk galaxy at $z = 2$ (SFR $\sim 25\ M_\odot\ \mathrm{yr}^{-1}$), from the 30 hr ngVLA observations tapered to $0\farcs19$ resolution.The first contour level is 26 μJy beam^{-1}, and the contours are linear. Negative contours are dashed.

factor for CO 1-0 luminosity to total molecular gas mass. We give the galaxy properties for M 51 and the higher redshift models in Table 1. The rotation curve of M 51 is well studied, and the numbers are given in our comparitive dynamical analysis below. We employ the public BIMA SONG CO 1-0 data cubes (Helfer et al. 2003), as the starting point of the models. The moment 0 and 1 images (velocity-integrated CO emission and intensity weighted mean velocity) for these data are shown in Figure 1.

We then redshift the source to $z = 0.5$ and $z = 2$, and generate mock observations of the CO emission with the ngVLA and ALMA. For the $z = 0.5$ calculation, we adopt the 1-0 line, redshifted to 77 GHz. For the $z = 2$ calculation, we adopt the CO

Table 1. Galaxy Model parameters

Model	CO flux Jy km s^{-1}	CO line	z	d_L Mpc	L'_{CO} 10^9 K km/s pc^2	M_{H2} $(\alpha/3.4)10^{10}\,M_\odot$	L_{IR} $10^{10}\,L_\odot$	SFR $M_\odot$ yr^{-1}
M 51	10150	1-0	0.0015	7.6	1.6	0.55	4.7	4.7
z=0.5	0.15	1-0	0.5	2860	2.3	0.8	7.4	7.4
z=2	0.1	2-1	2.0	15800	5.8	2.0	25	25
z=4.2	0.1	2-1	4.2	38800	20	6.9	130	130

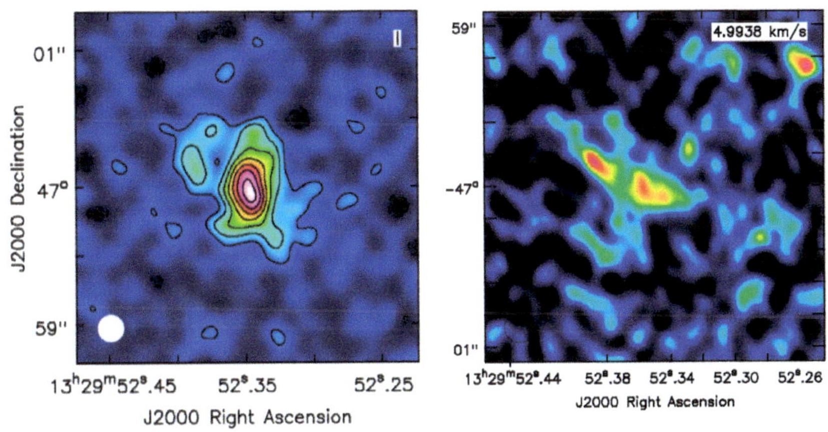

Figure 4. (*Left:*) The ngVLA velocity integrated CO emission as per Figure 2, row 2, for comparison. (*Right:*) Simulated observations using ALMA at 77 GHz of the velocity-integrated total CO 2-1 emission from the $z = 2$ model.

2-1 line, redshifted to 77 GHz. We assume thermalized line excitation at least to the 2-1 line, and hence increase the line flux densities (in Jy), by a factor four over the 1-0 flux densities. The CO 1-0 luminosity has been one of the primary diagnostics for deriving total molecular gas mass in galaxies (Bolatto et al. 2013). One of the issues in the study of high redshift galaxies historically is that many observations of CO emission have employed higher order transitions, thereby requiring an assumption about the excitation ladder of CO to derive a total molecular gas mass. Observing CO 1-0 directly avoids this extrapolation. Moreover, it has been found that the 2-1 line is almost always thermally excited (i.e., a luminosity four times that of 1-0, in Jy km s^{-1}), in high z galaxies, hence a factor 4 extrapolation is justified. However, as shown by Casey et al. (2015), going to higher order leads to substantial uncertainty in extrapolation to the 1-0 luminosity due to varying sub-thermal excitation in different galaxies. Hence, the importance of observing the low order transitions (Daddi et al. 2015; Carilli & Walter 2013).

An important recent discovery with regards to the molecular gas in galaxies is the observation of a rapid increase in the gas mass to stellar mass ratio with redshift. Studies of main sequence star forming disks galaxies show an order of magnitude increase in this ratio from $z = 0$ to 2. The peak epoch of cosmic star formation corresponds to an epoch when the baryon content of typical star forming disk galaxies is dominated by gas, not stars (Tacconi et al. 2010, 2018; Genzel et al. 2015; Schinnerer et al. 2016; Daddi et al. 2010; Geach et al. 2011; Scoville et al. 2017; Decarli et al. 2016). To make some allowance for this increase in gas mass fraction with redshift, we increase the CO luminosity by a factor 1.4 for the $z = 0.5$ model, and 3.5 for the $z = 2$ model, relative to M 51.

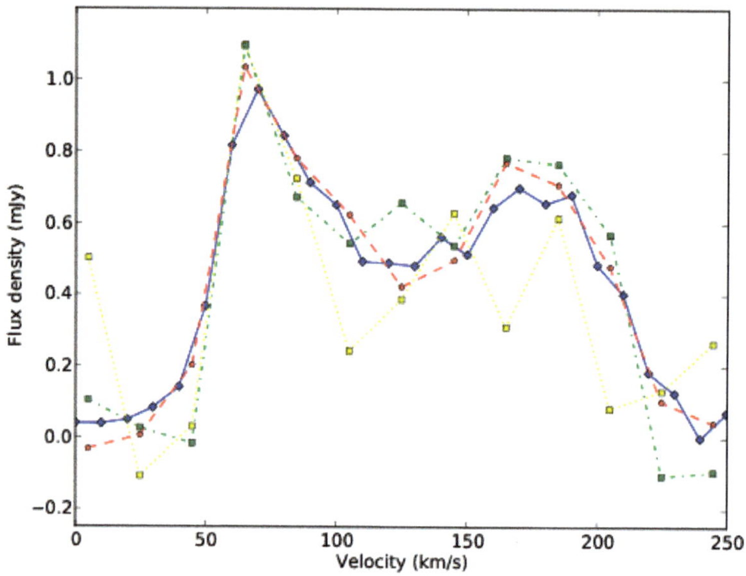

Figure 5. The spatially integrated CO 2-1 spectrum of the $z = 2$ model galaxy. Blue is the input model with no noise added, and 10 km s^{-1} channel^{-1}. The red is for the tapered ngVLA 30hr observation shown in Figure 2, row 2, at 20 km s^{-1} channel^{-1}. The yellow is for the ALMA observation shown in Figure 5, at 20 km s^{-1} channel^{-1}. The green is for the minimal ngVLA array (30% core, Y-only), 30hr observation shown in Figure 2, row 3, at 20 km s^{-1} channel^{-1}.

The CO 2-1 luminosity for the $z = 2$ model is then: $L'_{CO} = 5.8 \times 10^9$ K km/s pc^2, implying a molecular gas mass of 2.0×10^{10} $(\alpha/3.4)$ $M_\odot$. This is at the extreme low end of the galaxies currently being discovered in the deepest ALMA, VLA, and NOEMA cosmological surveys for molecular line emission from $z \sim 2$ galaxies (Walter et al. 2016, 2014; Decarli et al. 2016). Using the standard relationship between IR luminosity and CO luminosity for main sequence galaxies leads to: $L_{IR} \sim 2.5 \times 10^{11}$ $L_\odot$, and a star formation rate ~ 25 $M_\odot$ yr^{-1} (Daddi et al. 2010).

We also adopt the CO 2-1 line at $z = 2$ because it will be less affected by brightness contrast relative to the CMB (Zhang et al. 2016). Since all observations are made relative to the mean background of the CMB, the increasing temperature of the CMB with redshift can depress the observed surface brightness of the CO and dust emission of high redshift galaxies, with the effect becoming more noticeable for low order transi-

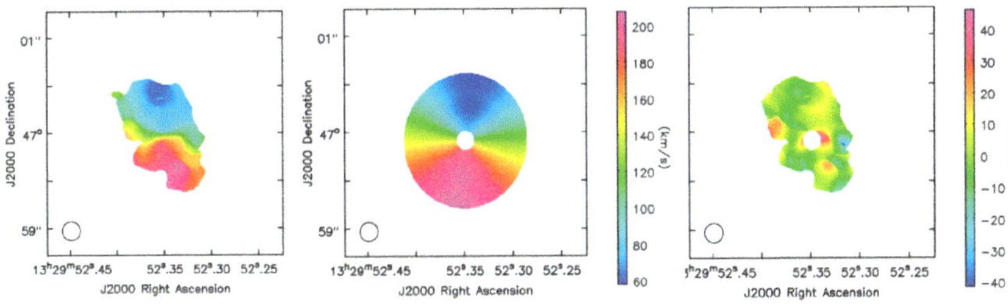

Figure 6. (*Left:*) the mean velocity of the CO emission from the $z = 2$ model derived from the ngVLA using the XGAUS technique described in §5.3. (*Center:*) the model for the rotation curve derived using ROTCUR. (*Right:*) the residuals from differencing the model and observation.

tions. Their calculation shows that at $z \sim 2$, the effect on the CO 2-1 brightness should be minimal.

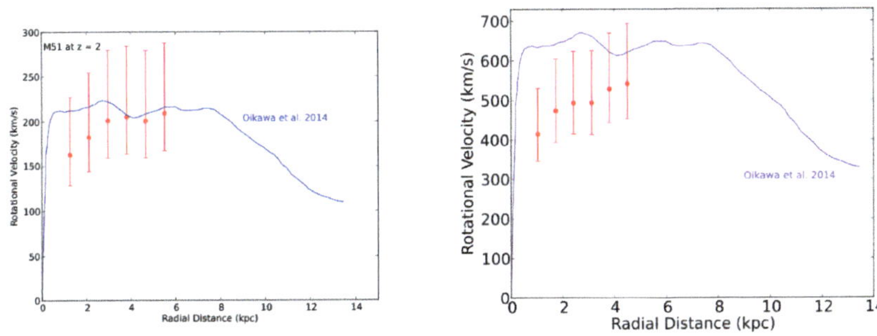

Figure 7. (*Left:*) The rotation curve derived for the $z = 2$ model from the ngVLA data using ROTCUR (red points plus errors). The blue line shows the true rotation curve (Oikawa & Sofue 2014). (*Right:*) The rotation curve derived for the $z = 4.2$ massive star forming disk galaxy, derived from the ngVLA data using ROTCUR (red points plus errors). The blue curve shows the input model scaled by a factor 3 (Oikawa & Sofue 2014).

As a final comparison, we consider a galaxy at $z = 4.2$ with 3.5× higher CO luminosity than the $z = 2$ model (12.5× M51), and a total velocity width a factor three larger. We adopt the CO 2-1 line, redshifted to 44 GHz. This system would correspond to the more massive star forming disk galaxies seen at high redshift (SFR $\sim 130\,M_\odot$ yr^{-1}; Casey et al. 2014).

3. Configuration

We employ the Southwest configuration, distributed across New Mexico and Chihuahua[1]. The array includes 45% of the antennas in a core of diameter ~ 1 km, cen-

[1]These simulations were performed with a slightly older, 300 antenna configuration, vs. the latest 214 antenna configuration. However, the (u, v)-weighting employed down weights both the longer spacings

tered on the VLA site. Then some 35% of the antennas out to VLA A array baselines of 30 km, and the rest to baselines as long as 500 km, into Northern Mexico.

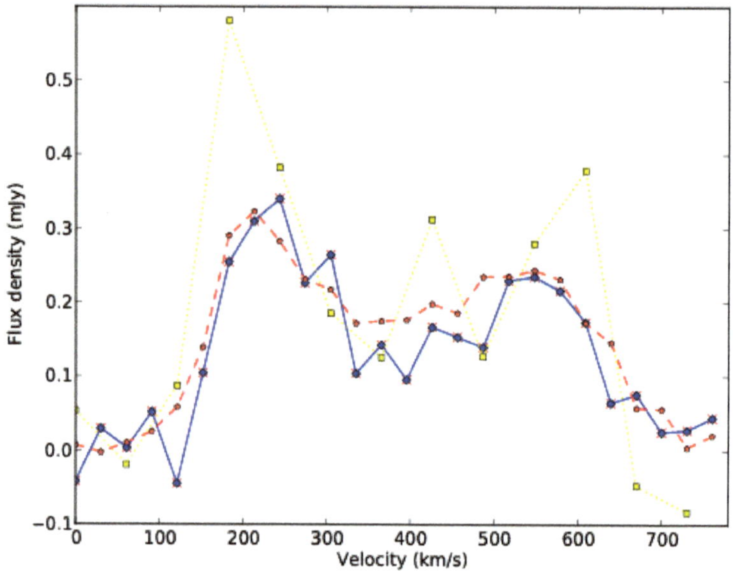

Figure 8. The integrated spectrum of the CO 2-1 emission from the $z = 4.2$ massive star forming disk galaxy. Blue is the input model with no noise and 30 km s^{-1} channel^{-1}. The red is for the tapered ngVLA 30hr observation shown in Figure 2, row 4, at 30 km s^{-1} channel^{-1}. The yellow is for a 30hr VLA observation, at 60 km s^{-1} channel^{-1}.

For the ngVLA noise calculation at 77 GHz, we assume the radiometer equation, using an 18m diameter antenna, with 70% efficiency, 70K system temperature at 77 GHz, and a 30 hr observation. For the ngVLA noise calculation at 44 GHz, we use an 18m diameter antenna, with 75% efficiency, 55K system temperature, and a 30 hr observation.

For the ALMA simulations, we use the ALMA-out18 and ALMA-out22 configurations. These have 50 antennas extending to baselines that give a naturally weighted beam at 77 GHz of 0.″45 and 0.″19, respectively. For the ALMA noise calculation, we assume antennas of 12m diameter, with an efficiency of 75%, and 70K system temperature, and a 30hr integration. Note that these observations require ALMA band 2, currently under design.

(beyond 30 km), and the core (within 1 km), while the intermediate spacing array is much the same as the latest configuration. Hence, the results will not change appreciably using the current 214 antenna main array.

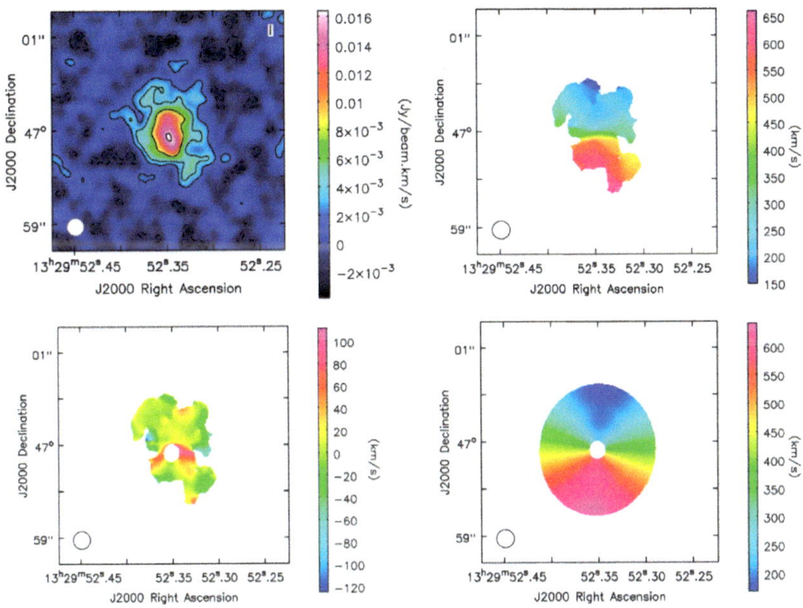

Figure 9. (*Upper Left:*) the velocity integrated CO 2-1 emission from the z = 4.2 galaxy, (*Upper Right:*) the mean velocity, derived from the ngVLA using the XGAUS technique described in §5.3. (*Lower Right:*) the model for the rotation curve derived using ROTCUR. (*Lower Left:*) the residuals from the model and observation difference.

For the VLA simulation, we employ the B configuration, with 27 antennas of 25m diameter, with an efficiency of 35%, and 65K system temperature, and a 30 hr observing time.

The noise in the final images is also dictated by the channel width and visibility weighting, with natural weighting (robust = 2) giving the optimal noise performance. The different ALMA configurations are designed to give a reasonable synthesized beam for natural weighting.

The ngVLA has a very non-uniform antenna distribution. The naturally weighted beam for this centrally condensed distribution leads to a PSF with a high resolution core of a few mas width at 77 GHz, plus a broad, prominent pedestal or plateau in the synthesized beam with a response of ~ 50% over ~ 1″ scale. This prominent pedestal leads to severe problems when trying to image complex structure. The imaging process entails a balance between sensitivity and synthesized beam shape, using visibility tapering and Briggs robust weighting (Carilli et al. 2018; Cotton & Condon 2017; Carilli 2017, 2016; Carilli et al. 2016; Carilli 2018).

4. Mechanics of the Simulation

The starting model is the BIMA SONG observations of CO 1-0 at about 5″.5 resolution (Helfer et al. 2003). This corresponds to 0.20 kpc at the distance of M 51. The channel width is 10 km s^{-1} = 3.8 MHz at 115GHz. Details of the process can be found in Carilli & Shao (2017). In brief, we adjust the frequency, pixel size, and flux density of the model to match the line luminosity and galaxy size as a function of redshift.

We employed the SIMOBSERVE task within CASA to generate uv data sets. Instructions on how this is done are found on the ngVLA web page. We simulated a 30 hr observation, made up of a series of 3 hr scheduling blocks around transit.

Table 2. Imaging parameters

Array	Freq.	Channel Width	Robust	(u, v)-Taper	Resolution	rms noise
	GHz	MHz		arcsec	arcsec	μJy bm^{-1} ch^{-1}
ngVLA CO 1-0	77	5.1	0.5	0.15	0.22	16
ngVLA CO 2-1	77	5.1	0.25	0.15	0.19	16
ALMA-out22	77	5.1	2		0.19	75
ALMA-out18	77	5.1	2		0.45	75
ngVLA-min	77	5.1	1	0.15	0.21	24
ngVLA	44	4.5	0.5	0.1	0.19	10
VLA	44	4.5	2		0.22	80

In all cases we employed the CLEAN algorithm with Briggs weighting, and adjusted the robust parameter and (u, v)-taper to give a reasonable synthesized beam and noise performance. Our target resolution was $\sim$ 0.″2, which corresponds to $\sim$ 1 kpc physical for $z \geq 0.5$. Synthesized beams for different weighting schemes are given in Table 2. We use a cell size of 0.″01 throughout.

5. Results

5.1. $z = 0.5$

We focus our more detailed analysis on the $z = 2$ model. However, for reference, we have performed a mock observation of M 51 at $z = 0.5$. Figure 2, row 1 shows the CO 1-0 moment images for a 30hr observation for the $z = 0.5$ model with the ngVLA. The observing frequency is 77 GHz. Our target resolution is $\sim$ 0.″2 $\sim$ 1 kpc at $z = 0.5$. To achieve a well behaved synthesized beam (i.e., no pedestal), but retain reasonable sensitivity, we employ the imaging parameters listed in Table 1. The rms per 20 km s^{-1} channel, is about 16 μJy beam^{-1}.

We generate moment images of the CO 1-0 emission blanking at surface brightnesses below about 1.5σ. The results in Figure 2 show that the ngVLA will make quality images of the velocity integrated intensity, as well as the overall velocity field. The emission is dominated by the central regions, but emission from the spiral arms remains clear, as well as the large-scale galaxy velocity structure in the extended regions. The minimum measured velocity dispersions are 20km/s, which is at the limit of the measurements.

5.2. $z = 2.0$

We next generate a mock observation of the CO 2-1 emission from the model galaxy at $z = 2$. Again, 30hrs integration is assumed, and the observing frequency is 77 GHz. The imaging parameters are given in Table 2. Figure 3 shows the resulting CO 2-1 channel images for the $z = 2$ model. The rms per 20 km s^{-1} channel (5.1 MHz), is 16 μJy beam^{-1}. The corresponding 3σ gas mass limit per channel is $\sim 2 \times 10^8$ $(\alpha/3.4)\,M_\odot$.

Emission is clearly detected over the full velocity range of the galaxy, starting with the northern arm at low velocity, through the main gas distribution at the galaxy center, to the southern arm at high velocity.

Figure 2, row 2 shows the moment images for this system derived from the ngVLA observations. While the spiral arms are less prominent, the emission is still seen over most of the disk, and the velocity field is reasonable clear.

Figure 4 shows the integrated spectrum of the CO emission from the galaxy. The red curve is derive from the mock ngVLA observation. The blue curve shows the intrinsic emission derived from a model with no noise added. The integrated emission profile is very well recovered, with a mean flux density of 0.65 mJy beam^{-1}, and a total velocity width of 150 km s^{-1}.

The CO 2-1 luminosity is $L'_{CO} = 5.8 \times 10^9$ K km/s pc^2, implying a molecular gas mass of 2.0×10^{10} $(\alpha/3.4) M_\odot$. This corresponds to a CO luminosity a factor 3.5× larger than M 51 itself.

We next consider a similar 30hr observation using ALMA. To achieve a similar spatial resolution of ~ 1 kpc, we employ the 'out22' configuration and natural weighting of the visibilities. This leads to an rms noise of 75 μJy beam^{-1} per 20 km s^{-1} channel. Figure 5 shows the velocity integrated emission from the ALMA observation, compared to that from the ngVLA observation. ALMA barely detects the velocity integrated emission at this resolution. To recover the dynamics at the level of detail seen in Figure 2, row 2 would required about 700 hrs of integration time on ALMA.

We then use a smaller ALMA array (out18 to achieve 0.″45 resolution), to avoid over-resolving the emission, and again integrating for 30hrs. The spatially integrated spectrum is shown as the yellow curve in Figure 4. The spatially integrated emission is detected in the spectrum, although still at relatively low signal-to-noise.

5.3. Rotation Curve Analysis

We have analyzed our mock CO 2-1 ngVLA observations of the $z = 2$ disk using the standard tools for galaxy dynamical analysis available in AIPS and GIPSY. The details of the process can be found in Jones et al. (2017a).

In brief, we generate an optimal moment 1 image (intensity weighted mean velocity), using the XGAUS routine in AIPS. This routine fits a Gaussian function in frequency (velocity) at each pixel in the cube to obtain the mean velocity.

We then run the resulting moment 1 image through the ROTCUR program in GIPSY. This program assumes the gas has a pure circular rotation in a gas disk, and fits the velocity field with a series of tilted rings as a function of radius with a given thickness (Rogstad et al. 1974). Each ring has a number of parameters that are used in the fitting, including rotation velocity, inclination angle of the normal to the line of sight, and position angle of the projected major axis of the ring on the sky.

Figure 6 shows the results, including the moment 1 image from XGAUS, the resulting fit dynamical disk, and the residuals (fit - observed). The data are reasonably well fit by the tilted ring model. The fit position angle of the major axis is 172°, as compared to the 'true' value for M 51 of 162° (Oikawa & Sofue 2014). The fit inclination angle is 21°, as compared to 24° for the nominal 'true' value of M 51 (Shetty et al.

2007). Note that M 51 has a warped disk, but the warp starts at radius ~ 7 kpc, which is outside our sensitivity area (Oikawa & Sofue 2014).

The fit rotation velocities (corrected for inclination) for the rings versus radius are shown in Figure 7a. The blue curve is the rotation curve of M 51 from (Oikawa & Sofue 2014). While the error bars are substantial, the magnitude in the outer rings matches well the 'true' rotation velocity of ~ 200 km s^{-1}. This plot also shows the limitations of such a process: recovering the inner steep rise in the rotation curve within ~ 1 to 2 kpc radius remains problematic with data that has only 1kpc resolution.

5.4. $z = 4.2$

As a final example, we present CO 2-1 emission from a massive disk galaxy at $z = 4.2$. Note that at this redshift, the combination of CMB excitation, and the fact that observations always measure the surface brightness relative to the CMB, may become issues. At $z \sim 4$, for cold molecular clouds (20K), the 'CMB effect' could lower the observed 2-1 surface brightnesses by up to 60%, while for warmer gas (50 K), the CMB effect is $\leq$ 20% (Zhang et al. 2016). Given the galaxy in question is a massive star forming galaxy, we assume the latter.

We again assume 30hr integration. The observing frequency is 44 GHz. The imaging parameters are given in Table 1, resulting in a PSF with a FWHM = 0.19" and no pedestal. The resulting rms noise per 30 km s^{-1} channel, is 10 μJy $beam^{-1}$.

Figure 2, row 4 shows the moment images for this system derived from the ngVLA observations. While the spiral arms are less prominent, the emission is still seen over most of the disk, including the north and south arms, and the overall velocity field is clear.

Figure 8 shows the integrated spectrum of the CO emission from this galaxy. The red curve is derived from the mock ngVLA observation. The blue curve shows the intrinsic emission derived from a model with no noise added. The integrated emission profile is very well recovered, with a mean flux density of 0.22 mJy $beam^{-1}$, and a total velocity width of 450 km s^{-1}.

The implied CO 2-1 luminosity is: $L'_{CO} = 2.0 \times 10^{10}$ K km/s pc^2, implying a molecular gas mass of 6.9×10^{10} $(\alpha/3.4)$ $M_{\odot}$. The implied star formation rate for this galaxy based on the standard star formation law for disk galaxies would be ~ 130 $M_{\odot}$ yr^{-1} (Daddi et al. 2010). Such a galaxy corresponds to a more massive star forming disk galaxy ('main sequence'), observed at high redshift (Carilli & Walter 2013; Casey et al. 2014).

We next consider a 30hr observation using the current VLA. To achieve a similar spatial resolution of ~ 1 kpc, we employ the B-configuration and natural weighting of the visibilities, but include a taper parameter of 0.″15. This leads to an rms noise of 80 μJy $beam^{-1}$ per 30 km s^{-1} channel, and a spatial resolution of 0.″22). The signal is not detectable above the noise in the channel images. We created a spatially integrated spectrum, as shown as the yellow curve in Figure 8, smoothed to 60 km s^{-1} $channel^{-1}$. The emission can be seen in this integrated spectrum, but at marginal signal-to-noise. To reach the required signal-to-noise to perform the imaging in Figure 2, row 4 would require 2000 hrs with the current VLA.

We again perform a rotation curve analysis using XGAUS and ROTCUR on the ngVLA most observations. Figure 9 shows the results, including the moment images and the resulting fit dynamical disk, and the residuals (observed - fit). The fit position angle of the major axis is 173°, as compared to the 'true' value for M 51 of 162°

(Oikawa & Sofue 2014). The fit inclination angle is 26°, as compared to 24° for the nominal 'true' value of M 51 (Shetty et al. 2007).

The fit rotation velocities (corrected for inclination) for the rings versus radius are shown in Figure 7b. The blue curve is the rotation curve of M 51 from (Oikawa & Sofue 2014), scaled up by a factor three. In this case, the measured rotation curve appears to asymptote to the maximum rotation velocity at 6kpc radius. This under-estimation over much of the radius may result from the marginal measurement of the velocity field on the southern (fainter) spirals arms of the galaxy. These results indicate the limitations of this kind of analysis, for a given signal to noise and spatial resolution.

5.5. Minimal Array

We investigate a 'miminal array configuration' that would be adequate to perform the CO imaging at $z = 2$ considered in $5.2.

We revised the Southwest configuration as follows. First, we remove all antennas outside the area covered by the current VLA Y configuration (all antennas beyond 15km radius of the array core). These antennas are heavily down-weighted due to the desire to obtain ~ 0.″2 resolution, which corresponds to baselines of about 4km at 77 GHz and 8km at 44 GHz. This removes about 1/3 of the antennas. Then we remove 2/3 of the core antennas, or 82 antennas. The core antennas were also down-weighted by the use of the robust parameter in the imaging analysis above. This leaves us with 114 antennas (out of the original 300), in an array that has good baseline coverage on scales out to 30km.

We then perform the same mock observation as above for the $z = 2$ model (30hrs). In this case, we could employ robust = 1, along with the same cell and taper as before, to achieve similar resolution (0.″21). The rms noise per 20km s^{-1} channel is now 24 μJy beam^{-1}.

The resulting moment images are shown in Figure 2, Row 3. The resolution is slightly lower, but the signal to noise is still adequate to recover all the main structures of the galaxy as seen with the original SW array. The green curve on Figure 4 shows the resulting integrated spectrum. This matches well the true spectrum shown in the blue.

6. Conclusions

We have performed a detailed analysis of the capabilities of the ngVLA to image at 1kpc resolution the low order CO emission from high redshift main sequence star forming galaxies. The template for the galaxy size and dynamics is M 51. The model galaxies include: (i) a galaxy with CO luminosity 40% higher than M 51 at $z = 0.5$ (star formation rate ~ 7 $M_\odot$ yr^{-1}), (ii) a modest star forming main sequence disk galaxy at $z = 2$ (SFR ~ 25 $M_\odot$ yr^{-1}), and (iii) a massive star forming disk at $z = 4.2$ (SFR ~ 130 $M_\odot$ yr^{-1}). The main results are as follows:

- The overall gas distribution and dynamics are recovered in all cases by the ngVLA.
- Detailed analysis of the mock rotation curves recover the main galaxy orientation parameters, such as inclination angle and position angle of the major axis, to within 10% in all cases.

- The rotation velocity versus radius is determined reasonably out to the maximum radius of the CO disk (6 kpc) for the $z = 0.5$ and $z = 2$, with the exception of the rapid rise in the rotation curve in the inner 2 kpc, which cannot be recovered at 1kpc resolution.

- The rotation curve analysis for the $z = 4.2$ galaxy asymptotes to the maximum rotation velocity of the system. This is likely due to the more limited signal-to-noise on the southern spiral arm emission.

- ALMA and the VLA can detect the integrated emission from these galaxies, but neither can recover the dynamics at high spatial resolution. To do so would require of order 1000 hrs per galaxy with these current facilities.

- We explore a 'minimal array' to perform this measurement involving 1/3 of the dense 1km core, and no antennas beyond the current VLA foot-print, giving an array of 114 antennas to radii of about 15 km. We can still recover the source dynamics at $0\farcs22$ resolution, with a noise figure higher by about 50% relative to the full array when suitably tapered and weighted.

Observing detailed galaxy dynamics is one of the critical, and yet under explored, elements in studies of high redshift galaxies. Multiple techniques are now being employed, with the Hα and IFU specroscopy in the near-IR employed on large telescopes for galaxies at $z \sim 2$ (Forster-Schreiber et al. 2018), and [CII] 158μm spectroscopy with ALMA at $z \geq 3$ (Jones et al. 2017b). The CO observations play a complementary role, tracing the molecular gas, as opposed to the ionized or neutral atomic gas.

Acknowledgments. The National Radio Astronomy Observatory is a facility of the National Science Foundation operated under cooperative agreement by Associated Universities, Inc.

References

Bolatto, A. et al. 2013, ARAA, 51, 207
Bolatto, A. et al. 2017, ngVLA Memo No. 19
Carilli, C. 2018, ngVLA Memo No. 47
Carilli, C., et al. 2018, ngVLA Memo No. 35
Carilli, C. 2018, ngVLA Memo No. 47
Carilli, C. 2017, ngVLA Memo No. 16
Carilli, C., et al. 2016, ngVLA Memo No. 11
Carilli, C. 2016, ngVLA Memo No. 12
Carilli, C. & Shao, Y. 2017, ngVLA Memo No. 13
Carilli, C., et al. 2015, ngVLA Memo No. 5 (arXiv:1510.06438)
Carilli, C.L. & Walter, F. 2013, ARAA 51, 105
Casey, C. et al. 2014, PhR, 541, 45
Cotton, W. & Condon, J. 2017, ngVLA Memo No. 30
Daddi, E. et al. 2015, A& A, 577, 46
Daddi, E et al. 2010, ApJ, 713, 686
Decarli, R. et al. 2016, ApJ, 833, 70
Forster-Schreiber et al. 2018, arXiv:1802.07276
Geach, J. et al. 2011, ApJ, 730, L19
Genzel, R. et al. 2015, ApJ, 800, 20
Genzel, R. et al. 2011, ApJ, 733, 101
Helfer, T. et al. 2003, ApJS, 145, S259

Jones, G.C. et al. 2017, ApJ, 850, 180
Jones, G.C. et al. 2017, ApJ, 845, 175
McKinnon, M. et al. 2016, SPIE, 9906, 27
Oikawa & Sofue 2014, PASJ, 66, 77
Rogstad et al. 1974, ApJ, 193, 309
Schinnerer, E. et al. 2016, ApJ, 833, 112
Schinnerer, E. et al. 2013, ApJ, 779, 42
Shetty et al. 2007, ApJ, 665, 1138
Selina, R. & Murphy, E. 2017, ngVLA Memo. No. 17
Tacconi, L. et al. 2010, Nature, 463, 781
Tacconi, L. et al. 2018, ApJ, 853, 179
Scoville, N. et al. 2017, ApJ, 837, 150
Walter, F. et al. 2016, ApJ, 833, 67
Walter, F. et al. 2014, ApJ, 782, 79
Zhang et al. 2016, Royal Society Open Science, Volume 3, Issue 6, id.160025

Science with a Next-Generation Very Large Array
ASP Conference Series, Monograph 7
Eric J. Murphy, ed.

The Star Formation History of the Universe

Amy J. Barger,[1] Kotaro Kohno,[2] Eric J. Murphy,[3] Mark T. Sargent,[4] and James J. Condon[3]

[1] *University of Wisconsin-Madison, Madison, WI, USA;* barger@astro.wisc.edu

[2] *University of Tokyo, Tokyo, Japan;* kkohno@ioa.s.u-tokyo.ac.jp

[3] *National Radio Astronomy Observatory, Charlottesville, VA, USA;* emurphy@nrao.edu, jcondon@nrao.edu

[4] *University of Sussex, Brighton, UK;* Mark.Sargent@sussex.ac.uk

Abstract. Deep continuum surveys at radio wavelengths are able to cover large areas, yield high angular resolution, and do not suffer from dust extinction, thus providing a robust way to measure the star formation history of the universe. However, with the current sensitivities of extant radio telescopes, it remains challenging to detect galaxies that dominate the cosmic star formation history even with extremely long integrations. With the ngVLA, a new portion of parameter space will be opened up for radio continuum surveys: deep ($5\sigma \approx 1\,\mu$Jy bm^{-1}; SFR $\approx 30\,M_{\odot}\,$yr^{-1} at $z \approx 2$), large-area ($\sim$1 deg^2), sub-arcsecond surveys at high frequencies ($\sim$8 GHz), where the observed radio emission from high-z galaxies should be dominated by free-free emission, providing a robust measurement for the star formation history of the universe. By being able to image the star formation activity, unbiased by dust, for a large, homogeneous sample of galaxies with a wide range of luminosities into the epoch of reionization, such surveys with the ngVLA will be highly complementary to those conducted by *JWST*, which will only be sensitive to un-obscured star formation.

1. Introduction

The rest-frame UV light density has been successfully mapped to beyond $z \sim 10$ using the *Hubble Space Telescope*. However, very luminous galaxies emit much of their light at far-infrared (FIR) to millimeter (mm) wavelengths and may be extremely faint in the UV/optical due to large amounts of obscuring dust. Thus, long-wavelength observations are needed to determine how much star formation we are missing from UV/optical surveys in these populations of heavily enshrouded systems. Indeed, the universe produces comparable amounts of energy in the FIR and optical (Puget et al. 1996; Fixsen et al. 1998), requiring a detailed understanding of the FIR population at all redshifts to understand how galaxies grow and evolve over cosmic time.

Uniformly selected samples are key for determining the dusty star formation history, and there are various ways we can try and construct such samples. One route is to observe at submm wavelengths, which are unique in that they are equally sensitive to high and low redshifts due to the negative *K*-correction (i.e., with increasing redshift, the fixed submm bandpass samples higher on the Rayleigh-Jeans tail of the galaxy's dust emission, which compensates for the redshift dimming). Imaging surveys with single-dish submm and mm telescopes and cameras have the advantage of being able to cover large fields and produce uniformly selected samples in a single bandpass to a well-established flux limit. Their main disadvantage is low resolution. This is particularly problematic for blank field observations, since the confusion limit[1] can be reached at fairly high fluxes. In contrast, interferometric submm and mm imaging surveys enjoy high spatial resolution and sensitivity, but their fields-of-view are small.

An alternative route is to conduct surveys at radio wavelengths. The advantages of such surveys are large areas, high spatial resolution, and no extinction. However, at current sensitivities, the positive K-correction at these wavelengths results in redshift limitations ($z \lesssim 3$) for even the most luminous galaxies. Differentiating star-forming galaxies from active galactic nuclei (AGNs) at radio wavelengths is also challenging.

Fortunately, the ngVLA is poised to revolutionize radio continuum surveys and make it possible to construct a complete star formation history. In this chapter, we describe an ngVLA survey that could complement the *James Webb Space Telescope (JWST)* deep fields by providing high-resolution imaging of star formation, unbiased by dust, for a large, homogeneous sample of galaxies with a wide range of luminosities and star formation rates (SFRs) that probe into the epoch of reionization (i.e., $z \gtrsim 6$).

2. Scientific Importance and Survey Considerations

Single-dish submm imaging surveys have revealed a significant population of extremely luminous, dusty star-forming galaxies at the highest redshifts (e.g., see Casey 2014). These galaxies have dramatically changed our understanding of the star formation history of the universe and have challenged cosmological simulations in producing them. At some point, the submm selected population will begin to overlap in SFR with the optically selected population. To obtain large samples of highly obscured star-forming galaxies at the depths necessary to observe the overlap with the optical population, we can use next generation radio continuum surveys that cover large areas, have high spatial resolution, and are insensitive to extinction. In order to determine the optimal frequency continuum survey to be made with the ngVLA, we need to consider the dominant physical mechanisms that contribute to the radio emission (e.g., see Murphy et al. 2018, this volume, p. 421) at high redshifts.

From the well-established FIR-radio correlation (e.g., Helou et al. 1985; Condon 1992), we know that non-thermal synchrotron emission is a valuable tracer of the total amount of star formation in galaxies, unbiased by dust. Synchrotron emission is dominant in lower frequency radio observations ($\sim$ 1 GHz). The large fields-of-view at these frequencies are an advantage for measuring substantial numbers of very high-redshift galaxies; however, AGNs are a contaminant in the star-forming galaxy samples. Moreover, there are uncertainties in the conversion from radio flux density to SFR. The sensitivity of current facilities at these frequencies is very limiting. For example, one of the deepest existing VLA images ($5\sigma \approx 11\,\mu$Jy of the GOODS-N field at 1.4 GHz; Owen 2018) is only able to probe luminosities of ultraluminous infrared galaxies or

[1]Confusion refers to the blending of sources and/or the noise being dominated by unresolved contributions from fainter sources.

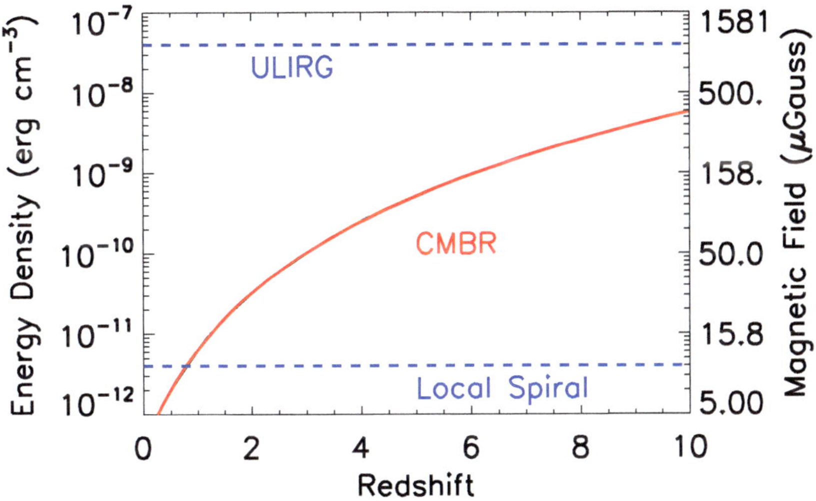

Figure 1. Energy density of the CMBR plotted against redshift. The corresponding magnetic field strength is shown on the right-hand axis. Also shown is the typical magnetic field energy density for a local spiral and ULIRG (Carilli 2001; McBride et al. 2014).

ULIRGs ($L_{IR} \gtrsim 10^{12}\, L_\odot$; SFR $\gtrsim 125\, M_\odot\, \mathrm{yr}^{-1}$) to $z \sim 3$ and luminous infrared galaxies or LIRGs ($L_{IR} \gtrsim 10^{11}\, L_\odot$; SFR $\gtrsim 12\, M_\odot\, \mathrm{yr}^{-1}$) to $z \sim 1$ (see Figure 14 in Barger 2014).

Inverse Compton (IC) scattering of relativistic electrons off of cosmic microwave background radiation (CMBR) photons may also have an impact on synchrotron emission. When the CMBR energy density exceeds the galaxy magnetic energy density, the IC scattering dominates (e.g., Condon 1992; Murphy 2009). There is a redshift dependence in the amount of IC scattering that occurs, since the energy density of the CMBR goes as $\sim (1+z)^4$. There is also a luminosity dependence, since the synchrotron emission will be highly quenched in moderate SFR galaxies at high redshifts, but possibly never in ULIRGs where the magnetic energy density is higher. This effect is illustrated in Figure 1, where we plot energy density versus redshift for a local spiral and a ULIRG (from McBride et al. 2014 following Carilli 2001) and for the redshift-dependent CMBR. We show the magnetic field strength on the right-hand vertical axis.

Consequently, this means that the emission detected from high-redshift galaxies with moderate SFRs should be dominated by free-free emission, modulo the possible contribution from anomalous dust emission (e.g., Murphy et al. 2010, 2018; Planck Collaboration et al. 2015), which may arise from spinning dust grains (e.g., Draine 1998). This in turns results in a rather ideal situation for attempting to cleanly measure star formation rates form high-redshifts systems. Free-free emission, which is dominant at rest-frame tens of GHz (Condon 1992), is directly proportional to the production rate

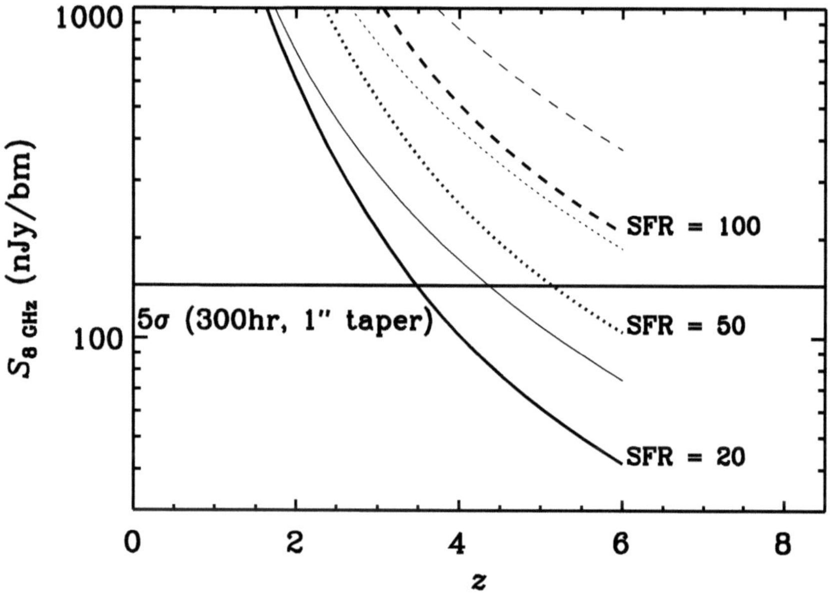

Figure 2. Telescope sensitivity in nJy bm^{-1} at 8 GHz plotted against redshift indicating the expected brightness of a 4 kpc disk galaxy forming stars at a rate of 20, 50, and 100 $M_\odot$ yr^{-1}. The heavy-weighted lines include estimates for synchrotron dimming due to IC scattering of cosmic ray electrons in galaxies due to the increasing CMBR temperature with redshift, while the corresponding lighter-weighted lines indicate the expected brightnesses in the absence of any synchrotron dimming due to CMB effects. Given the current sensitivity specifications, by tapering to a 1″ synthesized beam, the ngVLA will have enough brightness temperature sensitivity to detect galaxies forming stars at a rate of > 50 $M_\odot$ yr^{-1} into the Epoch of Reionization. Using the existing VLA, the detection of such a galaxy would take $\gtrsim$10,000 hr!

of ionizing photons by young, massive stars, making it extremely clean for measuring accurate SFRs (see, e.g., Murphy et al. 2015).

3. Anticipated Results and Astronomical Impact

The ngVLA will open up completely new parameter space for radio continuum surveys, allowing for the extremely deep, large-area mappings to be made at much higher frequencies than traditional surveys at ~1 GHz. For instance, by covering the anticipated maximum size of potential *JWST* deep fields ($\lesssim$1 deg^2), which are the most heavily studied regions on the sky, to a sensitivity of 0.2 μJy per beam at ≈8 GHz (1″ resolution), highly robust, dust-unbiased SFRs of the galaxies in these fields can be obtained. In particular, such a survey will allow the detection of many thousands of L^* (i.e., ~ $2 \times 10^{12}\,L_\odot$; SFR $\gtrsim$ 25 $M_\odot$ yr^{-1}) galaxies at $z \sim 2$ to 4 as well as galaxies with $L_{IR} \gtrsim 5 \times 10^{12}\,L_\odot$; SFR $\gtrsim$ 625 $M_\odot$ yr^{-1} out to $z \sim 8$, where we currently have little constraints on the potential numbers of such luminous galaxies. Furthermore, using the same data, but applying a different imaging weighing scheme to create finer resolution maps (i.e., 0$\farcs$1, or even higher for brighter systems), one can obtain information on morphologies of these systems, revealing highly obscured regions that could be missed in rest-frame optical/UV light form *JWST* imaging.

For a single-field, ultra-deep integration (i.e., ≈300 hr; rms ≈ 30 nJy bm^{-1} at 8 GHz) the ngVLA should be able to detect a galaxy at the 5σ level with a SFR of $\sim 20\,M_\odot\,\mathrm{yr}^{-1}$ out to $z \sim 3$, and a galaxy with a SFR of $\sim 100\,M_\odot\,\mathrm{yr}^{-1}$ to $z \sim 7$. This is illustrated in Figure 2, which shows the observed 8 GHz brightness of 4 kpc diameter disk galaxies with a range of SFRs. The 5σ 8 GHz rms of the ngVLA tapered to a 1$''$ synthesized beam (i.e., ≈ 5 kpc at $z \approx 7$) is based on the updated sensitivities given in (Selina & Murphy 2017)[2]. Consequently, given the current sensitivity specifications, by tapering to a 1$''$ synthesized beam the ngVLA will have enough brightness temperature sensitivity to detect galaxies forming stars at a rate of $> 50\,M_\odot\,\mathrm{yr}^{-1}$ into the epoch of reionization. Using the existing VLA, the detection of such a galaxy would take ~10,000 hr.

By combining the data for a large sample of 8 GHz selected galaxies with observations at other frequencies from 1 to 50 GHz that could also be made with the ngVLA, one could map the radio spectral energy distributions (SEDs) of the galaxies, which would make it possible to robustly account for the different thermal and non-thermal energetics by separating the free-free and potential synchrotron contributions. Another interesting outcome of mapping the radio SEDs for a number of galaxies of a given luminosity is to determine at which redshift the spectrum becomes purely free-free emission, resulting in a rough measurement for the magnetic field strength of galaxies at that luminosity by equating it to the CMBR energy density at that redshift (see Figure 1). And, while not the focus of this chapter, it is worth pointing out that with such a large frequency range, we could make measurements of low-J CO lines to obtain redshifts for many of the galaxies—particularly the more dusty galaxies where it is very difficult to measure redshifts from optical and near-infrared spectroscopy—and to provide information on the gas masses of the galaxies (e.g., see Decarli et al. 2018, this volume, p. 581).

4. Summary

Single-dish submm images have low resolution, and blank field observations hit the confusion limit at relatively bright flux densities (~ 2 mJy). Interferometric submm/mm images have high spatial resolution and sensitivity but suffer from small fields-of-view. The deepest existing continuum surveys with the VLA are limited to finding complete samples of ULIRGs to $z \sim 3$ and LIRGs to $z \sim 1$ (e.g., see Figure 14 in Barger 2014), while Milky Way luminosities can only be detected in the local region ($z \ll 0.5$). The unprecedented sensitivity of continuum imaging surveys with the ngVLA will push the boundaries of galaxy evolution studies considerably (see Figure 2). Moreover, with the ngVLA's broad high-frequency coverage, we will be able to estimate the fraction of light being produced by free-free versus synchrotron as a function of frequency. The ngVLA observations will add enormous value to the *JWST* deep fields by providing high-resolution, dust-free images of the galaxies to study the morphologies of the star formation. They will also provide complementary observations to SKA1, which will

[2] `http://ngvla.nrao.edu/page/refdesign`

not be able to detect similar populations of galaxies at these higher frequencies, allowing a better understanding of the radio production mechanisms over cosmic time.

References

Barger, A. J., Cowie, L. L., Chen, C.-C., et al. 2014, ApJ, 784, 9
Carilli, C. L. 2001, in Starburst Galaxies: Near and Far, ed. L. Tacconi and D. Lutz, 309
Casey, C. M., Narayanan, D., Cooray, A. 2014, Physics Reports, 541, 45
Condon, J. J. 1992, ARA&A, 30, 575
Decarli, R., Carilli, C., Casey, C., et al. 2018, ASP Monograph 7, Science with a Next-Generation Very Large Array, ed. E. J. Murphy (San Francisco, CA: ASP), 581
Draine, B. T., & Lazarian, A, 1998, ApJ, 494, L19
Fixsen, D. J., Dwek, E., Mather, J. C., Bennett, C. L., & Shafer, R. A. 1998, ApJ, 508, 123
Helou, G., Soifer, B. T., & Rowan-Robinson, M. 1985, ApJ, 298, L7
McBride, J., Quataert, E., Heiles, C., Bauermeister, A. 2014, ApJ, 780, 182
Murphy, E. J. 2009, ApJ, 706, 482
Murphy, E. J., Helou, G., Condon, J. J., et al. 2010a, ApJ, 709, L108
Murphy, E. J., Sargent, M. T., Beswick, R. J., et al. 2015, Advancing Astrophysics with the Square Kilometre Array (AASKA14), 85
Murphy, E. J., Linden, S. T., Dong, D., et al. 2018, ApJ, 862, 20
Murphy, E. J., Condon, J. J., Alberdi, A., et al. 2018, ASP Monograph 7, Science with a Next-Generation Very Large Array, ed. E. J. Murphy (San Francisco, CA: ASP), 421
Owen, F. N. 2018, ApJS, 235, 34
Planck Collaboration, Adam, R., Ade, P. A. R., Aghanim, N., et al. 2015, A&A, 594, A10
Puget, J.-L., Abergel, A., Bernard, J.-P., et al. 1996, A&A, 308, L5
Selina, R. and Murphy, E. J., 2017, ngVLA Memo #17

Science with a Next-Generation Very Large Array
ASP Conference Series, Monograph 7
Eric J. Murphy, ed.

[CII] 158μm Emission from $z \geq 10$ Galaxies

C. L. Carilli,[1,2] E. J. Murphy,[3] A. Ferrara,[4] and P. Dayal[5]

[1]*National Radio Astronomy Observatory, Socorro, NM 87801;*
ccarilli@nrao.edu

[2]*Cavendish Astrophysics, Cambridge, UK*

[3]*National Radio Astronomy Observatory, Charlottesville, VA 22903;*
emurphy@nrao.edu

[4]*Scuola Normale Superiore, Piazza dei Cavalieri 7, I-56126 Pisa, Italy*

[5]*Kapteyn Astronomical Institute, University of Groningen, P.O. Box 800, 9700 AV Groningen, The Netherlands*

Abstract. We consider the capabilities of ALMA and the ngVLA to detect and image the [CII] 158 μm line from galaxies into the cosmic 'dark ages' ($z \sim$ 10 to 20). The [CII] line may prove to be a powerful tool in determining spectroscopic redshifts, and galaxy dynamics, for the first galaxies. In 40 hr, ALMA has the sensitivity to detect the integrated [CII] line emission from a moderate metallicity, active star-forming galaxy [$Z_A = 0.2\,Z_\odot$; star formation rate (SFR) = 5 $M_\odot$ yr^{-1}], at $z = 10$ at a significance of 6σ. The ngVLA will detect the integrated [CII] line emission from a Milky-Way like star formation rate galaxy ($Z_A = 0.2\,Z_\odot$, SFR = 1 $M_\odot$ yr^{-1}), at $z = 15$ at a significance of 6σ. Imaging simulations show that the ngVLA can determine rotation dynamics for active star-forming galaxies at $z \sim 15$, if they exist. The [CII] detection rate in blind surveys will be slow (of order unity per 40 hr pointing).[1]

1. Introduction

The $z \sim 15$ Universe is at the edge of our current understanding. A handful of theoretical studies have speculated on the cosmic star formation rate (SFR) density at these redshifts (Mashian et al. 2016; Duffy et al. 2017; Chary & Pope 2010; Dayal et al. 2014; Yue et al. 2015; Topping & Shull 2015). Observational constraints on extreme redshift galaxies are poor, based on extrapolation of the few galaxies and AGN known at $z \sim 7$ to 8, and the even fewer galaxy candidates at $z \sim 8$ to 11.

Encouraging results come from observations of a relatively mature interstellar medium, and active star formation, in some of the very high redshift sources discovered to date. The last few years have seen an explosion in the number of [CII] 158μm detections at high redshift, including high resolution imaging of the gas dynamics on kpc-scales in both AGN host galaxies and in more normal star-forming galaxies at

[1]This paper is a brief synopsis of a paper presented in the Astrophysical Journal (Carilli et al. 2017). We refer the interested reader to the Journal article for more detail.

$z \sim 5.5$ to 7.5 (see Carilli et al. 2017 for a summary). The most recent results include the detection of the [OIII] 88μm fine structure line and/or the [CII] line, from galaxies at $z = 7.2$, 8.4, and 9.1 (Laporte et al. 2017; Hashimoto et al. 2018a,b) and the detection of strong [CII] and dust continuum emission from a quasar host galaxy at $z = 7.5$ (Venemans et al. 2017). While encouraging, observations remain sparse, and the most basic questions remain on the nature, and even existence, of galaxies at $z \sim 15$.

Given the uncertainty in our knowledge of galaxies at extreme redshifts, in this study we focus on a few simple of questions: if such extreme redshift galaxies exist, what kind of facility is required to detect, and possibly image, the [CII] 158 μm line emission? How do the prospects depend on basic galaxy properties, such as metallicity and star formation rate? And based on what little we know of galaxy demographics at very early epochs, what kind of numbers can we expect in blind cosmological spectral deep fields?

2. The [CII] 158μm line

The [CII] 158 μm line is one of the brightest spectral line from star-forming galaxies at far-infrared wavelengths and longer, carrying between 0.1% to 1% of the total far infrared luminosity of star forming galaxies (Stacey et al. 1991; Carilli & Walter 2013). The [CII] fine structure line traces both neutral and ionized gas in galaxies, and is the dominant coolant of star-forming gas in galaxies (Velusamy et al. 2015). While the line is only visible from space in the nearby Universe, it becomes easier to observe with increasing redshift, moving into the most sensitive bands of large ground based millimeter telescopes, such as NOEMA[2], and the ALMA[3].

As a predictor for the [CII] 158 μm luminosity from early galaxies we use the Vallini et al. (2015) relationship (their Equation 8). This theoretical and observational analysis considers in detail the relationships between star formation rate, galaxy metallicity, and [CII] luminosity. We adopt a few representative galaxy characteristics, including the main parameters of star formation rate, metallicity, redshift, and [CII] luminosity, and compare these to the capabilities of the given facilities.

3. Telescopes

The relevant ALMA bands are 3, 4, and 5, corresponding to frequencies of 84 – 116 GHz, 125 – 163 GHz, and 163 – 211 GHz, respectively. These bands then cover the [CII] line (1900.54 GHz rest frequency), between $z = 10$ and 20, almost continuously, with gaps of a few MHz due to atmospheric O_2 absorption at 118 GHz and 183 GHz. The current bandwidth for ALMA is 8 GHz, with a upgrade to 16 GHz or 32 GHz being considered sometime in the future. We employ the ALMA sensitivity calculator, under good weather conditions (3rd octile), with 50 antennas. For the sake of illustration, we adopt a fiducial line width of 100 km s^{-1} (see below) and an on-source integration time of 40 hr. The rms sensitivity per channel is 21 μJy beam^{-1} channel^{-1}, roughly

---

[2]`http://iram-institute.org/EN/noema-project.php`

[3]`http://www.almaobservatory.org`

independent of frequency due to the increasing channel width in Hz for a fixed velocity resolution, offsetting decreasing system sensitivity with increasing frequency.

Table 1. Facilities

Telescope	Redshifts	Frequencies (GHz)	rms (μJy beam^{-1})	Bandwidth (GHz)
ngVLA	15 – 20	116 – 90	2.0	26
ALMA	10 – 15	173 – 116	21	8 (32)

For the ngVLA we employ the "Southwest" configuration, and we adopt sensitivity parameters consistent with ngVLA memo 17 (Selina & Murphy 2017). The maximum redshift we consider is $z = 20$, so we only consider frequencies between 90 and 116 GHz. The current reference design has a nominal maximum bandwidth of 20 GHz, although broader bandwidths are under investigation. For the sake of number counts, the $z = 18.8$ to 20 range (90 GHz to 96 GHz), contributes very little to the total number of sources detected in blind searches. For the purpose of estimating the sensitivity of the ngVLA for realistic observations, and to explore the imaging capabilities in the event of the discovery of any relatively luminous sources, we have employed the CASA simulation tools (Carilli & Shao 2017), developed for the ngVLA project.

We simulate a 40 hr observation, and we employ the CLEAN algorithm with Briggs weighting. We adjust the ROBUST parameter, the (u, v)-taper, and the cell size, to give a reasonable synthesized beam and noise performance. Our target resolution is $\sim 0\farcs4$ for detection, and $\sim 0\farcs2$ for imaging.

We adopt as a spatial and dynamical template, the observed CO 1-0 emission from the nearby star-forming disk galaxy, M 51 (Helfer et al. 2003). We arbitrarily reduce the physical size of the disk by a factor three, with the idea that very early galaxies are likely smaller than nearby galaxies. Again, this exercise is for illustrative purposes, and the input model is just a representative spatial/dynamical template for a disk galaxy, with the relevant parameters being size, velocity, and luminosity. We then adjust the line luminosity per channel per beam, to achieve a given integrated [CII] 158 μm luminosity at a given redshift.

4. Results

4.1. Spectroscopic Confirmation of $z \gtrsim 10$ Candidates

An obvious application of the [CII] 158 μm line search will be to determine spectroscopic redshifts for near-IR dropout candidate galaxies at $z \sim 10$ to 20. We start with the relationship between the [CII] velocity integrated line flux, in the standard flux units of Jy km s^{-1}, versus redshift. We adopt a metallically of $Z_A = 0.2\, Z_\odot$, and star formation rates of $1\, M_\odot\, \mathrm{yr}^{-1}$ and $5\, M_\odot\, \mathrm{yr}^{-1}$. Figure 1 shows the predicted [CII] line flux versus redshift for the two models, along with the 1σ sensitivity of ALMA and the ngVLA.

This image simulation shows that, in 40 hr, the ngVLA will be able to detect the integrated [CII] line emission from moderate metallicity and star formation rate galaxies ($Z_A = 0.2$, SFR $= 1\, M_\odot\, \mathrm{yr}^{-1}$), at $z = 15$ at a significance of 6σ. This significance reduces to 4σ at $z = 20$.

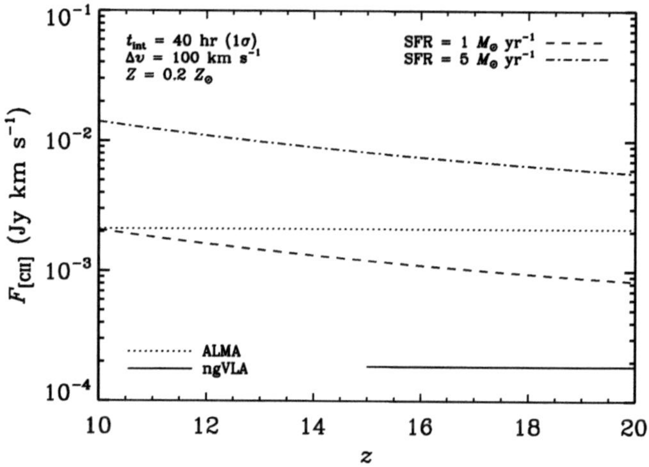

Figure 1. [CII] 158 μm velocity integrated line flux versus redshift for galaxies with star formation rates of 1 $M_{\odot}$ yr^{-1} and 5 $M_{\odot}$ yr^{-1}, and metallicity of 0.2 $Z_{\odot}$, based on the relationship given in Equation 12 of Vallini et al. (2015). The rms sensitivity in a 100 km s^{-1} channel and 40 hr integration is shown for both ALMA and the ngVLA.

In 40 hr, ALMA will be able to detect the integrated [CII] line emission from a higher star formation rate galaxy ($Z_A = 0.2\,Z_{\odot}$, SFR = 5 $M_{\odot}$ yr^{-1}), at $z = 10$ at a significance of 6σ. This significance reduces to 4σ at $z = 15$. ALMA will be hard-pressed to detect a moderate metallicity ($Z_A = 0.2\,Z_{\odot}$), lower star formation rate (1 $M_{\odot}$ yr^{-1}) galaxy, requiring 1000 hr for a 5σ detection of the velocity integrated line flux, even at $z = 10$.

We next consider dependence on metallicity. Figure 3 shows the relationship between [CII] luminosity (in Solar units), to star formation rate, for three different metallicities: $Z_A = 0.04$, 0.2, and 1.0 $Z_{\odot}$, for a galaxy at $z = 15$. Again shown are the ALMA and ngVLA sensitivities in 40 hr, 100 km s^{-1} channels. The Vallini et al. (2015) model has the [CII] luminosity as a strong function of metallicity. If the gas has Solar metallicity, the ALMA detection threshold (4σ) reduces to a galaxy with a star formation rate of 2.5 $M_{\odot}$ yr^{-1} (compared to 5 $M_{\odot}$ yr^{-1} for $Z_A = 0.2$), while that for the ngVLA reduces to 0.4 $M_{\odot}$ yr^{-1} (compared to 1 $M_{\odot}$ yr^{-1} for $Z_A = 0.2$). Conversely, for a low metallicity galaxy of $Z_A = 0.04\,Z_{\odot}$, these values increase to 100 $M_{\odot}$ yr^{-1} and 10 $M_{\odot}$ yr^{-1}, respectively.

4.2. Kinematics of $z \gtrsim 10$ Galaxies

We investigate the potential for obtaining kinematic information from such galaxies using the ngVLA. We find that the best even the ngVLA can do for a $z = 15$, $Z_A = 0.2\,Z_{\odot}$, and SFR = 1 $M_{\odot}$ yr^{-1} galaxy, in 40 hr is a 5.5σ detection of the integrated emission, with little or no dynamical information.

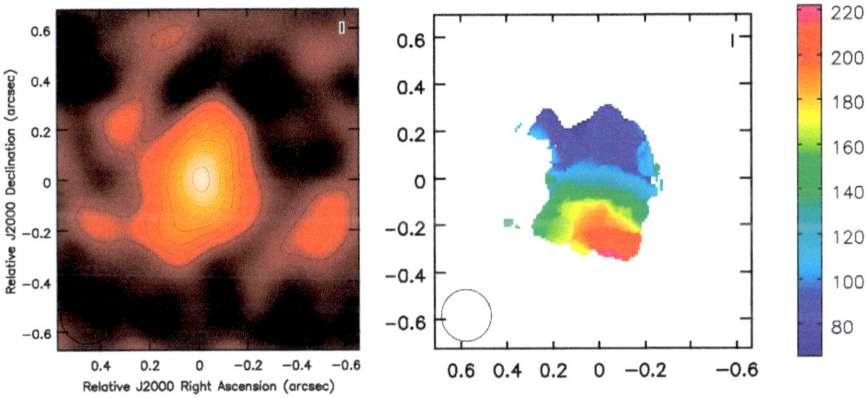

Figure 2. (*Left:*) A simulated image of the velocity integrated [CII] 158 μm emission from a $z = 15$ galaxy with a star formation rate of 5 $M_\odot$ yr^{-1}, and a metallicity of 0.2 $Z_\odot$, assuming for a 40 hr observation with the ngVLA. The contour levels are -6, -3, 3, 6, 9, 12, 15, 18, 21 μJy beam^{-1}. The rms noise on the image is about 1.8 μJy beam^{-1}, and the synthesized beam FWHM is 0.″22. (*Right:*) The intensity weighted mean [CII] velocity (moment 1).

For a higher SFR galaxy ($z = 15$, $Z_A = 0.2\,Z_\odot$, 5 $M_\odot$ yr^{-1} galaxy), in a 40 hr observation, the ngVLA can recover the overall rotational dynamics of the system. The imaging results for the velocity integrated emission (mom 0), and the intensity weighted mean velocity (mom 1), are shown in Figure 2. The beam size in this case is FWHM $\sim$ 0.″2, and the channel images at 20 km s^{-1} channel^{-1} have an rms noise of 4.5 μJy beam^{-1}.

4.3. The Potential for Blind Searches of $z \gtrsim 10$ Galaxies

Another application for the [CII] line will be blind cosmological deep fields. The advent of very wide bandwidth spectrometers has led to a new type of cosmological deep field, namely, spectral volumetric deep fields, in which a three dimensional search for spectral lines can be made, with redshift as the third dimension (Walter et al. 2016).

Table 2. Number of Detections per 40 hr Pointing

Model	ngVLA $15 < z < 16$	ngVLA $15 < z < 20$	ALMA $10 < z < 10.5$ (8 GHz)	ALMA $11 < z < 14$ (32 GHz)
CP10, 1 $M_\odot$ yr^{-1}	0.29	1.3	–	–
CP10, 5 $M_\odot$ yr^{-1}	0.11	0.48	0.29	0.68
Dayal14, 1 $M_\odot$ yr^{-1}	0.36	0.64	–	–
Dayal14, 5 $M_\odot$ yr^{-1}	6.9×10^{-4}	7.3×10^{-4}	2.8	1.4

Given the large uncertainty in the predicted galaxy luminosity function beyond $z \sim 10$, we investigated two theoretical predictions for the [CII] luminosity function with very different methodologies. The first method used star forming galaxy number counts of Chary & Pope (2010, CP10). These galaxy counts are based on backward-evolving

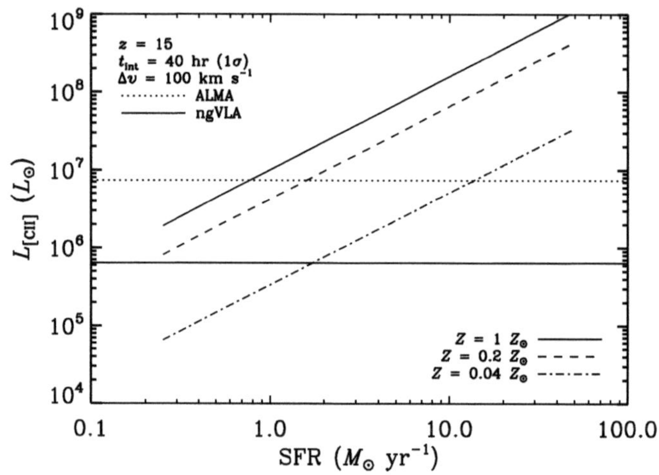

Figure 3. [CII] 158 μm line luminosity versus star formation rate and metallicity, based on the relationship given in Equation 12 of Vallini et al. (2015). Three different metallicities are shown. Also shown is the rms sensitivity of ALMA and the ngVLA for a galaxy at $z = 15$, assuming a 100 km s^{-1} channel and 40 hr integration.

models for the infrared luminosity function of Chary & Elbaz (2001), anchored by a variety of observational data from *Spitzer* and *Herschel*.

The second method employed the calculations of high redshift galaxy formation of Dayal et al. (2014). This model aims at isolating the essential physics driving early galaxy formation via a merger-tree based semi-analytical model including the key physics of star formation, supernova feedback, and the growth of progressively more massive systems (via halo mergers and gas accretion). This model reproduces well both the slope and amplitude of the UV LF from $z = 5$ to $z = 10$.

The two models predict the cummulative co-moving number density of star-forming galaxies above a given star formation rate as a function of redshift. These values can be converted to cummulative number densities of [CII] emitting galaxies as a function of line flux, using the models of Vallini et al. (2015). The [CII] number densities vs. flux can then be turned into the number of observed galaxies in a given integration time, bandwidth, and field of view, using the sensitivities, field sizes, and bandwidths of the ngVLA and ALMA, as discussed in §3.

The ngVLA covers the 90 – 116 GHz range, corresponding to $z = 20$ to 15. We also consider just the number of galaxies between $z = 15$ and 16. ALMA has receivers that will cover from $z = 10$ to 15, or frequencies from 173 GHz to 116 GHz. Currently, the bandwidth is limited to 8 GHz. We consider an 8 GHz blind search in the Band 5 from 165 GHz to 173 GHz ($z = 10.5$ to 10), and one covering most of Band 4 with a hypothetical 32 GHz bandwidth system, from 126 GHz to 158 GHz ($z = 11$ to 14).

In Table 2, we tabulate the number of galaxies detected in [CII] emission per 40 hr integration per frequency tuning, for the ngVLA and ALMA, and for the different models. For the ngVLA, and for SFR $\geq 1\,M_\odot\,\mathrm{yr}^{-1}$, the models predict that one to two independent pointings will be required to detect one galaxy over the full redshift range, on average. For the CP10 model, these sources have a broader redshift distribution, with 22% of the sources at z = 15 to 16. For the Dayal14 model, the majority (64%), of the sources are in this lowest redshift bin.

For ALMA and an SFR $\geq 5\,M_\odot\,\mathrm{yr}^{-1}$, the predicted number of detections differs significantly between models. For the 8 GHz bandwidth search in Band 5 (z = 10 to 10.5), the CP10 model requires about three pointings for a single detection, on average, while the Dayal14 model has more low redshift, brighter galaxies, with three sources per pointing expected. For the hypothetical 32 GHz bandwidth search in Band 4 (z = 11 to 14), the values are roughly two pointings needed for a single detection for the CP10 model, and one pointing needed for the Dayal14 model.

Overall, the detection rates in blind surveys will be slow (of order unity per 40 hr pointing). However, the observations are well suited to commensal searches on all programs employing the very wide bands that may be available in future.

A key issue in blind searches is spurious detections and verifying sources, especially given the large number of voxels searched for emission in the proposed surveys (see Carilli et al. 2017). Recent blind line searches have developed some techniques for making statistical corrections to number counts based on e.g., comparing the number of negative and positive detections at a give level (Decarli et al. 2016; Walter et al. 2016; Aravena et al. 2016). However, the problem still remains as to how to verify that a given detection is associated with a $z > 10$ galaxy. One possible method will be broad band near-IR colors from e,g., *JWST*, or large ground based telescopes. Likewise, follow-up spectroscopy with large ground and space-based telescopes may reveal atomic lines (Barrow et al. 2017). Lastly, ALMA could be used to search for [OIII] 88μm emission, in cases of low metallicity galaxies (Cormier et al. 2015; Hashimoto et al. 2018a,b).

5. Conclusions

We have considered observing [CII] 158μm emission from z = 10 to 20 galaxies. The [CII] line may prove to be a powerful tool to determine spectroscopic redshifts, and galaxy dynamics, for the first galaxies at the end of the dark ages, such as identified as near-IR dropout candidates by *JWST*.

In 40 hr, the ngVLA has the sensitivity to detect the integrated [CII] line emission from moderate metallicity and (Milky-Way like) star formation rate galaxies ($Z_A = 0.2$, SFR = $1\,M_\odot\,\mathrm{yr}^{-1}$), at $z = 15$ at a significance of 6σ. This significance reduces to 4σ at $z = 20$. In 40 hr, ALMA has the sensitivity to detect the integrated [CII] line emission from a higher star formation rate galaxy ($Z_A = 0.2\,Z_\odot$, SFR = $5\,M_\odot\,\mathrm{yr}^{-1}$), at $z = 10$ at a significance of 6σ. This significance reduces to 4σ at $z = 15$. We also consider dependencies on metallically and star formation rate.

We perform imaging simulations using a plausible model for the gas dynamics of disk galaxies, scaled to the sizes and luminosities expected for these early galaxies. The ngVLA will recover rotation dynamics for active star-forming galaxies ($\gtrsim 5\,M_\odot\,\mathrm{yr}^{-1}$ at $z \sim 15$), in reasonable integration times.

We adopt two models for very high redshift galaxy formation, and calculate the expected detection rate for [CII] emission at $z \sim 10$ to 20, in blind, wide bandwidth, spectroscopic deep fields. The detection rates in blind surveys will be slow (of order unity per 40 hr pointing). However, the observations are well suited to commensal searches on all programs employing the very wide bands that may be available in future.

Acknowledgments. PD acknowledges support from the European Research Council's starting grant ERC StG-717001 and from the European Commission's and University of Groningen's CO-FUND Rosalind Franklin program. The National Radio Astronomy Observatory is a facilites of the National Science Foundation operated under cooperative agreement by Associated Universities, Inc.. We thank Ranga-Ram Chary for discussions on the models and the paper, and N. Scoville for bringing up the possibility of very high z [CII] in the ngVLA context.

References

Aravena, M.; Decarli, R.; Walter, F.; Bouwens, R.; Oesch, P. A. et al. 2016, ApJ, 833, 71
Barrow, Kirk S. S.; Wise, John H.; Norman, Michael L.; O'Shea, Brian W.; Xu, Hao 2017, MNRAS, 469, 4863
Carilli, C. L. & Shao, Y. 2017, Next Generation Very Large Array Memo No. 13 (http://library.nrao.edu/ngvla.shtml)
Carilli, C. L., & Walter, F. 2013, ARA&A, 51, 105
Carilli, C. L., Murphy, E., Ferrara, A., Dayal, P. 2017, ApJ, 848, 49
Chary, R.-R., & Pope, A. 2010, arXiv:1003.1731
Chary, R. & Elbaz, D. 2010, ApJ, 556, 562
Cormier, D., Madden, S. C., Lebouteiller, V., et al. 2015, A&A, 578, A53
Dayal, P., Ferrara, A., Dunlop, J. S., & Pacucci, F. 2014, MNRAS, 445, 2545
Decarli, Roberto; Walter, Fabian; Aravena, Manuel; Carilli, Chris; Bouwens, Rychard et al. 2016, ApJ, 833, 69
Duffy, A. R., Mutch, S. J., Poole, G. B., et al. 2017, arXiv:1705.07255
Helfer, T. T., Thornley, M. D., Regan, M. W., et al. 2003, ApJS, 145, 259
Hashimoto, T., Laporte, N., Mawatari, K. et al. et al. 2018, Nature, 557, 392
Hashimoto, T., Inoue, A., Mawatari, K. et al. et al. 2018, PASJ, submitted
Laporte, N., Ellis, R. S., Boone, F., et al. 2017, arXiv:1703.02039
Mashian, N., Oesch, P. A., & Loeb, A. 2016, MNRAS, 455, 2101
Selina, R. & Murphy, E. 2017, Next Generation Very Large Array Memo No. 17 (http://library.nrao.edu/ngvla.shtml)
Stacey, G. J., Geis, N., Genzel, R., et al. 1991, ApJ, 373, 423
Topping, M.W. & Shull, J.M. 2015, ApJ, 800, 97
Vallini, L., Gallerani, S., Ferrara, A., Pallottini, A., & Yue, B. 2015, ApJ, 813, 36
Velusamy, T., Langer, W. D., Goldsmith, P. F., & Pineda, J. L. 2015, A&A, 578, A135
Venemans, B. P., Walter, F., Decarli, R., Banados, E., Carilli, C. et al. 2017, ApJ, 851, 8
Walter, Fabian; Decarli, Roberto; Aravena, Manuel; Carilli, Chris; Bouwens, Rychard et al. 2016, ApJ, 833, 67
Yue, B., Ferrara, A., Pallottini, A., Gallerani, S., & Vallini, L. 2015, MNRAS, 450, 3829

Science with a Next-Generation Very Large Array
ASP Conference Series, Monograph 7
Eric J. Murphy, ed.

Cold Gas in High-z Galaxies: The Molecular Gas Budget

R. Decarli,[1,2] C. Carilli,[3,4] C. Casey,[5] B. Emonts,[6] J.A. Hodge,[7] K. Kohno,[8] D. Narayanan,[9] D. Riechers,[10] M.T. Sargent,[11] and F. Walter[2,3]

[1]*INAF – Osservatorio di Astrofisica e Scienza dello Spazio, via Gobetti 93/3, 40129 Bologna, Italy;* `roberto.decarli@inaf.it`

[2]*Max Planck Institut für Astronomie, Königstuhl 17, 69117 Heidelberg, Germany*

[3]*National Radio Astronomy Observatory, Pete V. Domenici Array Science Center, P.O. Box O, Socorro, NM 87801, USA*

[4]*Cavendish Laboratory, University of Cambridge, 19 J.J. Thomson Avenue, Cambridge CB3 0HE, UK*

[5]*Department of Astronomy, The University of Texas at Austin, 2515 Speedway Blvd Stop C1400, Austin, TX 78712*

[6]*National Radio Astronomy Observatory, 520 Edgemont Road, Charlottesville, VA 22903, USA*

[7]*Leiden Observatory, Niels Bohrweg 2, 2333 CA Leiden, The Netherlands*

[8]*Institute of Astronomy, School of Science, The University of Tokyo, 2-21-1 Osawa, Mitaka, Tokyo 181-0015, Japan*

[9]*Department of Astronomy, University of Florida, 211 Bryant Space Science Center, Gainesville, FL 32611, USA*

[10]*Cornell University, 220 Space Sciences Building, Ithaca, NY 14853, USA*

[11]*Astronomy Centre, Department of Physics and Astronomy, University of Sussex, Brighton, BN1 9QH, UK*

Abstract. The goal of this chapter is to accurately pin down the molecular gas content of high redshift galaxies. By targeting the CO ground transition, we circumvent uncertainties related to CO excitation. The ngVLA can observe the CO(1-0) line at all redshifts, thus exposing the evolution of gaseous reservoirs from the earliest epochs down to the peak of the cosmic history of star formation. The order-of-magnitude improvement in the number of CO detections with respect to state-of-the-art observational campaigns will provide a unique insight on the evolution of galaxies through cosmic time. We show that the ngVLA will perform a precision measurement of the dense gas history of the Universe, the fuel for star formation, back to the epoch of cosmic reionization and the formation of the first galaxies.

1. Scientific Rationale

Our understanding of the evolution of galaxies, and in particular of the cosmic history of star formation, is limited by our ignorance of the amount of cold gas that galaxies could use to fuel star formation in various cosmic epochs. The bulk of the Universe's cold gas reservoir is comprised of hydrogen gas in the form of atomic hydrogen, HI, and molecular hydrogen, H_2, the latter of which is responsible for star formation in cold, condensed molecular clouds. A key goal of the ngVLA will be to assess the molecular gas content in high-redshift galaxies.

Although abundant, H_2 is not directly observable under normal circumstances in the star-forming medium. The bright transitions of carbon monoxide (CO), the second most abundant molecule in the universe, are thus the workhorse for investigating the molecular content of galaxies through cosmic time. Indeed, the vast majority of the few hundred molecular line detections reported so far at $z > 1$ have been of CO (see more in reviews of Solomon & vanden Bout 2005, Carilli & Walter 2013). Due to frequency shifting at sufficiently high-redshift, millimeter-operating facilities like IRAM's PdBI/NOEMA, CARMA, and ALMA can only observe the higher-J transitions of CO. The intrinsic brightness of these lines depends not only on the amount of molecular gas, but primarily on the excitation of the CO molecules, which is affected by the temperature and density of the gas. The so-called CO Spectral Line Energy Distribution (SLED) might vary substantially in different regions of a galaxy by up to factors of five, and even between the relatively low-J transitions from CO(3-2) to CO(1-0) — this implies that high-J transitions cannot be used to gauge the molecular gas mass. On the other hand, the lower-J transitions, including the ground state CO(1-0) transition, are only accessible to longer wavelength radio observatories like the VLA (see Figure 1).

From the CO luminosity one can infer molecular gas masses M_{H2} via the CO–to–H_2 conversion factor, X_{CO} or α_{CO}. This conversion varies by up to a factor of five depending on the conditions of the gas in the ISM, including metallicity. Recent work has focused on using dust continuum measurements to scale to gas masses (Magdis et al. 2011, 2012; Santini et al. 2014; Scoville et al. 2014, 2016). However, this technique typically assumes that the dust-to-gas ratio is fixed across a wide range of galaxies and redshifts, which has been shown to be a poor assumption in some cases (Rémy-Ruyer et al. 2014; Capak et al. 2015). In particular, the dust-to-gas ratio is known to depend strongly on metallicity (e.g., Issa et al. 1990; Lisenfeld & Ferrara 1998; Draine et al. 2007; Bolatto et al. 2013; Berta et al. 2016). Furthermore, the important kinematic signatures that come along with molecular line measurements are missed in dust continuum measurements.

The ideal probe of the cold molecular gas reservoir is therefore the ground state transition of CO, CO(1-0), where the scaling to molecular hydrogen mass has only the uncertainty in X_{CO} with which to contend. Constraining X_{CO} in high-z galaxies is possible via empirical relations linking the conversion factor with the dynamical mass and stellar mass, which further reduce the uncertainty on gas mass. The Jansky VLA, GBT, and ATCA have detected CO(1-0) in a number of high-redshift starburst galaxies, albeit a few at a time with long integrations (Ivison et al. 2011; Hodge et al. 2012; Papadopoulos et al. 2012; Greve et al. 2014; Emonts et al. 2014). These initial detections of CO(1-0) at high-redshift have not only constrained the total molecular gas mass of hydrogen in these high-z systems, but also highlighted the importance of understanding the spatial distribution of gas in distant galaxies. While the submillimeter-luminous galaxy population, i.e., dusty star forming galaxies (DSFGs) have been found to be

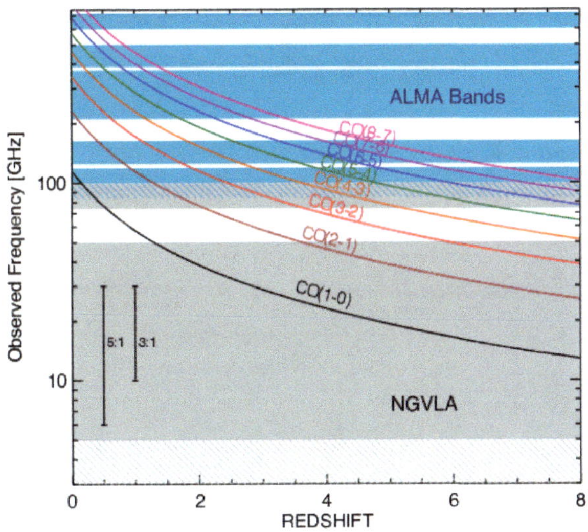

Figure 1. The observed frequencies of various transitions of CO from z=0 to z=8. While the ALMA bands (blue shading) are sensitive to many transitions of CO at low redshift, the low-J CO lines redshift out of the bands beyond $z \sim 0.5 - 2$. These low-J CO lines are critical for characterization of the total molecular gas reservoir in high-z galaxies and are only currently accessible for high-redshift galaxies using the VLA. The gray region indicates the ngVLA coverage of 5–100 GHz, with a gap around the 59–75 GHz water band. The gray hashed region represents the reach capabilities for ngVLA at low frequencies. With the advent of very wide-bandwidth setups for the ngVLA, we will be able to obtain multiple CO transitions for galaxies beyond $z \sim 3.5$ in one correlator setup, which will be revolutionary for studying dust and gas in very high-redshift galaxies. Figure adapted from Casey et al. (2015).

relatively compact in high-J transitions (Tacconi et al. 2008; Bothwell et al. 2010), consistent with the idea that they are scaled-up analogs to local ultra-luminous infrared galaxies (ULIRGs; Sanders & Mirabel 1996), their CO(1-0) maps are dramatically different, showing gas extending as far as 16 kpc from the galaxy centers (Ivison et al. 2011; Spilker et al. 2015; see also Decarli et al. 2016b). Studies of high-redshift protocluster radio galaxies even revealed bright CO(1-0) reservoirs out to distances of ~60 kpc from the central galaxy (Emonts et al. 2014). While these larger sizes are more suggestive of disk-like rotation dynamics and widespread molecular gas reservoirs in the halo environment of massive galaxies, the number of galaxies surveyed at the current VLA sensitivity makes it difficult to infer large-scale population dynamics or draw comparisons between low-z and high-z galaxy populations — a situation that will not significantly change even after ALMA starts to offer Band 1 & 2 observations. The ngVLA is desperately needed to dramatically increase the number of galaxies surveyed in CO(1-0) at high-redshift — by factors in the thousands, in line with what is currently observed from high-redshift galaxies via direct starlight. Figure 2 illustrates the anticipated improvement in ngVLA sensitivity over the current VLA and ALMA depths across 10–120 GHz.

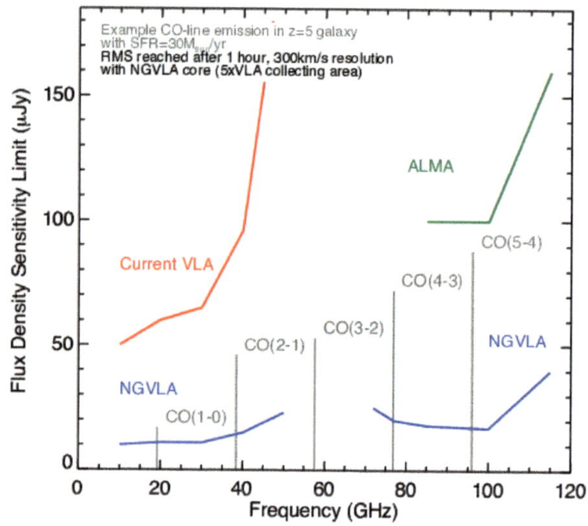

Figure 2. The projected sensitivity limits of the proposed ngVLA core, which will have five times the effective collecting area of the current VLA. This plot also shows the sensitivity limits of ALMA alongside the current VLA and projected ngVLA as a function of frequency (the gap from 50–75 GHz is due to atmosphere). Overplotted are the anticipated CO line strengths, from CO(1-0) to CO(5-4) of a typical $z = 5$ galaxy (with SFR = 30 $M_\odot\,\mathrm{yr}^{-1}$). This type of galaxy is currently well below the detection limit of existing facilities, including ALMA, yet is crucial for understanding the star forming budget of the Universe's first galaxies. Figure adapted from Casey et al. (2015).

To demonstrate the dramatic impact of ngVLA on detection rates of CO(1-0) in high-z samples, we generated a mock observation with a 3:1 bandwidth ratio tuned to 11–33 GHz. This frequency range will pick up the CO(1-0) transition from $2.5 < z < 9.5$ *in a single frequency setting.* The rms reached after a one hour integration with five times the collecting area of the current VLA is $\sim$ 10 μJy beam^{-1} rms in 2 MHz bins, sufficient to resolve individual CO lines into 3–5 spectral bins across the entire bandwidth.

To estimate the number of galaxies accessible to CO(1-0) detection within one ngVLA pointing in a 11–33 GHz frequency scan, we computed the flux distribution of CO emitters expected based on state-of-the-art semi-analytical models of galaxy formation and evolution (Popping et al. 2012, 2014, 2016). These models couple a semi-analytic model of galaxy formation with a radiative transfer code to make predictions for the luminosity function of CO. They treat the interstellar medium as a density distribution for each galaxy, and assume a log-normal density distribution for the gas within clouds. They then consider only the UV contribution to the heating (no X-rays), and model the CO chemistry based on a fit to results from the photodissociation region code of Wolfire et al. (2010). The value of $\alpha_{\rm CO}$ is left free to vary from galaxy to galaxy based on the various galaxy properties (mass, compactness, SFR, etc.). We transform the luminosity function of CO(1-0) computed at various redshifts into flux distributions, focusing on the redshift interval $z > 2.5$. We consider the volume corresponding to the ngVLA primary beam (assuming 25 m antennas) integrated along the line of sight in various contiguous redshift slices. The total volume of the universe sampled with such a scan is ~88,500 comoving Mpc^3. By scaling the CO luminosity / flux distributions to this volume, we obtain the expected number of CO-emitting galaxies at various redshifts in our spectral scan. If we restrict our search to the limit of $\sim$ 10 μJy beam^{-1} rms in 2 MHz bins, we consider a typical line width of 200 km s^{-1}, and we require secure > 5-σ detections, we expect 33 CO(1-0) detections at $z \sim 2$; 43 at $z \sim 4$; and even 2 at $z \sim 6$ (see Figure 3). To first order, these estimates are in general agreement with the relatively coarse observational constraints on the CO luminosity functions at high redshift available so far (Walter et al. 2014; Decarli et al. 2016a) and with predictions based on recent estimates of the integrated infrared luminosity function (Casey et al. 2014; Vallini et al. 2016) which is roughly complete out to $z \sim 2$, translated into a best-guess CO(1-0) luminosity function using a conservative estimate of a ULIRG-type $L'_{\rm CO(1-0)}$-to-$L_{\rm IR}$ scaling (Bothwell et al. 2013) and assuming a line width of 200 km s^{-1}. Further support to these estimates is derived from the empirical predictions for the number of line emitters in a CO deep field by da Cunha et al. (2013a), based on a physically motivated spectral energy distribution model along with empirical relations between the observed CO line and infrared luminosities of star-forming galaxies. These estimates capitalized on the deepest available optical/near-infrared data for the Hubble Ultra Deep Field (UDF). Given all of the assumptions that went into the models, and particularly since the estimates were derived in very different ways, this relative agreement is encouraging. Intriguingly, preliminary results from larger campaigns blindly searching for molecular gas emission in deep fields suggest that the predictions based on semi-analytical models might underestimate the number of CO–bright sources at the high–luminosity end of the luminosity functions (see results from ASPECS LP in Decarli et al. 2018, in preparation; and from COLDz in Riechers et al. 2018, in preparation). If these preliminary results are confirmed, the expected number of CO detections per ngVLA pointings might exceed 100.

The mock observation described in this paper will not only be sensitive to CO(1-0), but also CO(2-1) emission at early epochs. However, the sensitivity and field of view for CO(2-1) will be significantly different for the same redshift regime given the differences in frequency, and will more closely mirror the coverage for the low-redshift CO(1-0) emitters. The survey sensitivity with respect to CO(2-1) depends on source excitation as the ratio of $S_{\rm CO(2-1)}$=$S_{\rm CO(1-0)}$ can vary from 2–4. Perhaps counter-intuitively, the depth of CO(2-1) observations at high-z will be greater than for CO(1-0), although the number of expected sources per single pointing will not differ substantially from the expected CO(1-0) emitters given the limited field of view. Line degeneracy may then be of concern, since it will be difficult to disentangle low-redshift CO(1-0) emitters and high-redshift CO(2-1) emitters. Other multiwavelength information (and perhaps parallel ALMA observations) would be necessary to break this degeneracy.

The estimates discussed so far refer to an individual pointing with 25 m antennas. Clearly, smaller antenna dishes would correspond to an increased surveyed area, and therefore larger number of detections at a given survey depth. In case of a mosaic, we note that weaving a series of single pointings together will have unique advantages with such a wide bandwidth. For a single pointing example above, the low frequency space probes the largest volumes, but at shallower luminosity limits at high-redshifts. Tiling a mosaic together by maximizing overlap for the low-frequency observations

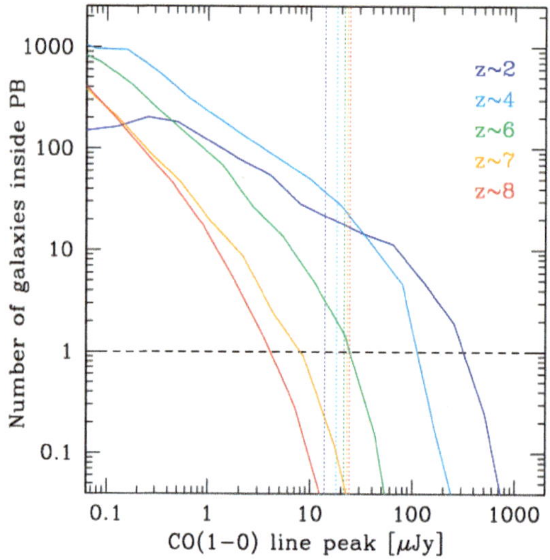

Figure 3. Predicted CO(1-0) flux distributions in various redshift bins, according to the semi-analytical models by Popping et al. (2012, 2014, 2016). The vertical lines mark the sensitivity that we will reach with ngVLA in the example of molecular deep field outlined in the text. Tens of galaxies at $z = 2 - 4$ will be detected within a single pointing. Recent molecular deep field investigations (ASPECS: Decarli et al. 2016a, 2018, in preparation; COLDz: Riechers et al. 2018, in preparation) suggest that semi-analytical models tend to underpredict the bright end of the CO luminosity function, i.e., the expected detection counts per ngVLA pointing might exceed 100. For a comparison, at the present sensitivity only a handful of detections per pointings are expected.

will dramatically reduce the rms at higher-redshifts, while mapping a large area. Figure 4 shows the effect of mosaicing on the CO(1-0) luminosity limit out to high-redshifts, sensitive to objects in the gray parameter space.

While direct molecular gas detections are certainly the main focus of future ngVLA extragalactic deep fields, it is worth noting that these surveys will be done in well-surveyed legacy fields, where there is already substantial ancillary data including spectroscopic redshifts. With large numbers of spectroscopic redshifts, stacking to measure the median molecular gas reservoir in LBGs would become possible as a function of redshift, environment, or other physical factors.

An important caveat is that, at sufficiently high redshift, the temperature of the CMB is non-negligible and can contribute to dust continuum and gas heating within galaxies. While this extra heating of the gas (and dust) will boost the intrinsic fluxes, the CMB also becomes a stronger background against which these fluxes must be measured. The net effect of these two competing effects on CO emission is less straightforward than for the dust continuum emission (da Cunha et al. 2013b), as the regions of cold molecular gas may not be in local thermal equilibrium, or the kinetic temperature of the gas may not be thermally coupled to the dust in an obvious way, particularly in early-Universe galaxies. However, the overall effect of CMB heating will be to cut

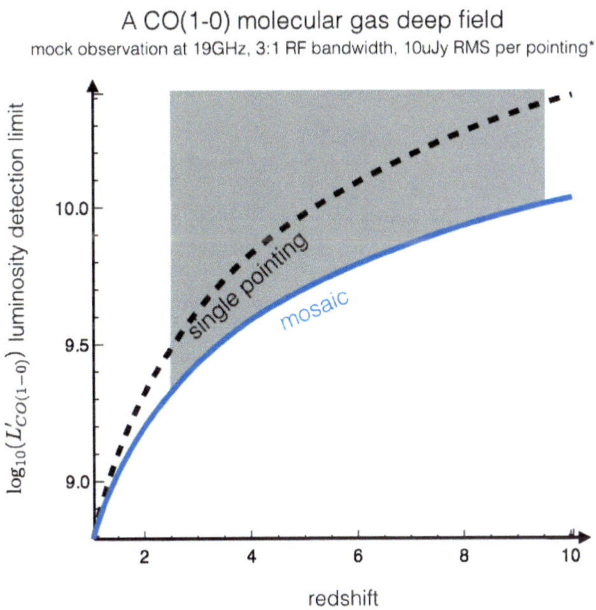

Figure 4. Luminosity limit per pointing reached in a molecular deep field initiative, with a single pointing and with a mosaic. By tiling different pointings with appropriate overlap, we significantly improve on the final depth of the data in any given region. Figure adapted from Casey et al. (2015).

down the number of direct detections in the $z \gtrsim 6$ Universe at low gas temperatures ($T \lesssim 20$ K).

On a technical note, using phased array feeds (PAFs) with multiple pixels instead of single pixel feeds would allow the ngVLA to survey areas of sky corresponding to volumes of ~0.1 Gpc^3, covering the full range of cosmic environments. For example, if we were to use a 3×3 pixel PAF we could survey the entire 2 deg^2 COSMOS field to this depth in a few hundred hours, detecting $\sim 10^5$ CO emitting galaxies. This would allow us, for example, to compute 2-point correlation functions within the CO emitting population, and cross-correlations between CO emitting galaxies and other populations such as galaxies with large stellar masses, galaxy clusters and AGN, all of which can be compared directly with the output of simulations. This is critical for assessing the collapse of large scale structure and environmental impact on star formation in galaxies.

1.1. Dense Gas History of the Universe

The long-term goal is to have a precise measure of the dense gas history of the Universe, to complement the detailed measurements of the cosmic star formation rate and stellar mass density. The ngVLA is the only facility capable of performing such a measurement, due to the order-of-magnitude increase in sensitivity, coupled with extreme bandwidth, needed for deep surveys of molecular gas.

In Figure 5 we show the current best published limits to the dense gas history of the Universe from the ALMA ASPECS and IRAM CO deep field surveys (Decarli et

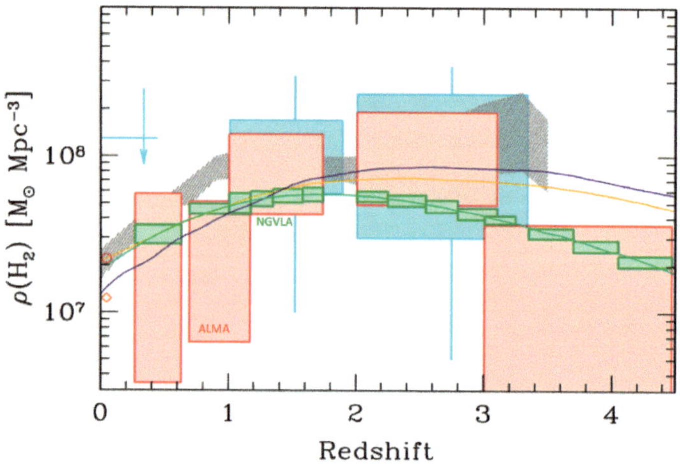

Figure 5. The dense gas history of the Universe as measures by ALMA (red), PdBI (blue), and the projected constraints for the ngVLA (green), with a comparable observing time (10s to 100 hours; Decarli et al. 2016b).

al. 2016b). While there is a general rise and fall with redshift, paralleling that in the star formation history of the Universe, the error bars are such that one cannot even rule-out a flat distribution with redshift.

Figure 5 also shows the constraints that will be obtained with a CO deep field with the ngVLA using a similar observing time (tens to 100 hours). Note that these estimates are tied to a specific model, for lack of better information. We see that the ngVLA will provide the observations of the dense gas that are required to obtain a complete picture of the baryon cycle of gas to stars across cosmic time.

1.2. Measurements Required

The science goals presented here require the detection of faint lines distributed over large volumes. The key features will be the sensitivity to faint lines, and the wide simultaneous coverage in terms of both field of view and redshift (i.e., frequency) space.

1.3. Uniqueness to ngVLA Capabilities:

The key features needed in order to successfully address the proposed science goals are:

1. sensitivity to faint line observations in the 12–50 GHz range, in order to sample CO(1-0) at the highest redshifts;

2. imaging capabilities to 1–3″ scales, in order to unambiguously identify counterparts without out-resolving the faint line emission;

3. large bandwidth: $\Delta\nu/\nu > 1$, in order to maximize the simultaneous CO redshift coverage and therefore the survey speed;

4. large field of view: > 1 arcmin^2 per pointing at any frequency, again to insure rapid coverage of large volumes.

The combination of requirements 1 and 2 forces us to restrict to radio interferometers. The VLA in principle has the power to achieve these goals, but lacks the required sensitivity (see Figure 2). ALMA performs better in terms of sensitivity at ~3 millimeter wavelengths, but it lacks the frequency coverage below 80 GHz, which is crucial to observe CO(1-0) at high redshift.

1.4. Longevity/Durability: with Respect to Existing and Planned (>2025) Facilities

The molecular scan efforts will inevitably focus on well studied fields where exquisitely deep observations at various wavelengths are already in place, and more will come in the coming years. E.g., all the key extragalactic fields (HDF-N, HUDF, GOODS, COSMOS, EGS, etc.) have been heavily hammered with *HST*, *Spitzer*, *Herschel*, ALMA, *Chandra*, *XMM*, VLT, Keck, and will be natural targets for extensive investigations with upcoming facilities and observatories, such as *JWST*, WFIRST, TMT/E-ELT. Optical/NIR images will be used to characterize the stellar populations in the galaxies detected in CO(1-0) in the ngVLA molecular scans. Optical/NIR spectroscopy will provide precise redshifts that can be used to further push the sensitivity of the ngVLA observations via stacks. For the brightest sources, optical/NIR spectra will also provide information on, e.g., the gas metallicity and other properties of the ISM thanks to rest-frame optical diagnostics. X-ray observations will allow to separate AGN from quiescent galaxies, thus allowing us to put the ngVLA measurements into the broader context of galaxy and black hole joint evolution. The multi-wavelength photometric coverage will provide precise stellar masses and star formation rates, which will be crucial to complement the molecular gas information provided by ngVLA. *JWST* data will enormously improve on the existing data, both in terms of wavelength coverage, sensitivity, and angular resolution (especially in the MIR), and in terms of spectroscopic capabilities at NIR and MIR wavelengths. Similarly, the advent of 30 m class telescopes will open up the possibility of spectroscopically characterizing sources that are barely detected in broad-band photometry with the state-of-the-art instrumentation — again fostering an obvious synergy with the ngVLA effort.

References

Berta S., Lutz D., Genzel R., Förster-Schreiber N.M., Tacconi L.J., 2016, A&A, 587, 73

Bolatto A.D., Wolfire M., Leroy A.K. 2013, ARA&A, 51, 207

Bothwell M.S., Chapman S.C., Tacconi L., Smail I., Ivison R.J., Casey C.M., Bertoldi F., Beswick R., et al. 2010, MNRAS, 405, 219

Bothwell M.S., Smail I., Chapman S.C., Genzel R., Ivison R.J., Tacconi L.J., Alaghband-Zadeh S., Bertoldi, F., et al. 2013, MNRAS, 429, 3047

Capak P.L., Carilli C., Jones G., Casey C.M., Riechers D., Sheth K., Carollo C.M., Ilbert O., et al. 2015, Nature, 522, 455

Carilli C.L. & Walter F., 2013, ARA&A, 51, 105

Casey C.M., Narayanan D., Cooray A. 2014, PhR., 541, 45

Casey C.M., Hodge J.A., Lacy M,. et al. 2015, arXiv:1510.06411

da Cunha E., Walter F., Decarli R., Bertoldi F., Carilli C., Daddi E., Elbaz D., Ivison R., et al., 2013a, ApJ, 765, 9

da Cunha E., Groves B., Walter F., Decarli R., Weiß A., Bertoldi F., Carilli C., Daddi E., Elbaz D., Ivison R., et al., 2013b, ApJ, 766, 13

Decarli R., Walter F., Aravena M., Carilli C., Bouwens R., da Cunha E., Daddi E., Ivison R.J., et al. 2016a, ApJ, 833, 69

Decarli R., Walter F., Aravena M., Carilli C., Bouwens R., da Cunha E., Daddi E., Elbaz D., et al. 2016b, ApJ, 833, 70
Draine B.T., Dale D.A., Bendo G., Gordon K.D., Smith J.D.T., Armus L., Engelbracht C.W., et al. 2007, ApJ, 663, 866
Emonts B.H.C., Norris R.P., Feain I., Mao M.Y., Ekers R.D., Miley G., Seymour N., Röttgering H.J.A., Villar-Martín M., Sadler E.M. 2014, MNRAS, 438, 2898
Greve T.R., Leonidaki I., Xilouris E.M., Weiß A., Zhang Z.-Y., van der Werf P., Aalto S., Armus L., et al. 2014, ApJ, 794, 142
Hodge J.A., Carilli C.L., Walter F., de Blok W.J.G., Riechers D., Daddi E., Lentati L. 2012, ApJ, 760, 11
Issa M.R., MacLaren I., Wolfendale A.W. 1990, A&A, 236, 237
Ivison R.J., Papadopoulos P.P., Smail I., Greve T.R., Thomson A.P., Xilouris E.M., Chapman S.C. 2011, MNRAS, 412, 1913
Lisenfeld U. & Ferrara A. 1998, ApJ, 496< 145
Magdis G.E., Daddi E., Elbaz D., Sargent M., Dickinson M., Dannerbauer H., Aussel H., Walter F., et al. 2011, ApJ, 740, L15
Magdis G.E., Daddi E., Sargent M., Elbaz D., Gobat R., Dannerbauer H., Feruglio C., Tan Q., et al. 2012, ApJ, 758, L9
Papadopoulos P.P., van der Werf P.P., Xilouris E.M., Isaak K.G., Gao Y., Mühle S. 2012, MNRAS, 426, 2601
Popping G., Caputi K.I., Somerville R.S., Trager S.C. 2012, MNRAS, 425, 2386
Popping G., Somerville R.S., Trager S.C., 2014a, MNRAS, 442, 2398
Popping G., van Kampen E., Decarli R., Spaans M., Somerville R.S., Trager S.C. 2016, arXiv:1602.02761
Rémy-Ruyer A., Madden S.C., Galliano F., Galametz M., Takeuchi T.T., Asano R.S., Zhukovska S., Lebouteiller V., et al. 2014, A&A, 563, A31
Sanders D.B. & Mirabel I.F. 1996, ARA&A, 34, 749
Santini P., Maiolino R., Magnelli B., Lutz D., Lamastra A., Li Causi G., Eales S., Andreani P., et al. 2014, A&A, 562, A30
Scoville N., Aussel H., Sheth K., Scott K.S., Sanders D., Ivison R., Pope A., Capak P., et al. 2014, ApJ, 783, 84
Scoville N., Sheth K., Aussel H., Vanden Bout P., Capak P., Bongiorno A., Casey C.M., Murchikova L., et al. 2016, ApJ, 820, 83
Solomon P.M. & vanden Bout P.A., 2005, ARA&A, 43, 677
Spilker J.S., Aravena M., Marrone D.P., Béthermin M., Bothwell M.S., Carlstrom J.E., Chapman S.C., Collier J.D., et al. 2015, ApJ, 811, 124
Tacconi L.J., Genzel R., Smail I., Neri R., Chapman S.C., Ivison R.J., Blain A., Cox P., et al. 2008, ApJ, 680, 246
Vallini L., Gruppioni C., Pozzi F., Vignali C., Zamorani G. 2016, MNRAS, 456, L40
Walter F., Decarli R., Sargent M., Carilli C., Dickinson M., Riechers D., Ellis R., Stark D., et al. 2014, ApJ, 782, 79
Wolfire M.G., Hollenbach D., McKee C.F. 2010, ApJ, 716, 1191

Science with a Next-Generation Very Large Array
ASP Conference Series, Monograph 7
Eric J. Murphy, ed.

Cold Gas in High-z Galaxies: The Dense ISM

R. Decarli,[1,2] C. Carilli,[3,4] C. Casey,[5] B. Emonts,[6] J.A. Hodge,[7] K. Kohno,[8] D. Narayanan,[9] D. Riechers,[10] M.T. Sargent,[11] and F. Walter[2,3]

[1]*INAF – Osservatorio di Astrofisica e Scienza dello Spazio, via Gobetti 93/3, 40129 Bologna, Italy;* `roberto.decarli@inaf.it`

[2]*Max Planck Institut für Astronomie, Königstuhl 17, 69117 Heidelberg, Germany*

[3]*National Radio Astronomy Observatory, Pete V. Domenici Array Science Center, P.O. Box O, Socorro, NM 87801, USA*

[4]*Cavendish Laboratory, University of Cambridge, 19 J.J. Thomson Avenue, Cambridge CB3 0HE, UK*

[5]*Department of Astronomy, The University of Texas at Austin, 2515 Speedway Blvd Stop C1400, Austin, TX 78712*

[6]*National Radio Astronomy Observatory, 520 Edgemont Road, Charlottesville, VA 22903, USA*

[7]*Leiden Observatory, Niels Bohrweg 2, 2333 CA Leiden, The Netherlands*

[8]*Institute of Astronomy, School of Science, The University of Tokyo, 2-21-1 Osawa, Mitaka, Tokyo 181-0015, Japan*

[9]*Department of Astronomy, University of Florida, 211 Bryant Space Science Center, Gainesville, FL 32611, USA*

[10]*Cornell University, 220 Space Sciences Building, Ithaca, NY 14853, USA*

[11]*Astronomy Centre, Department of Physics and Astronomy, University of Sussex, Brighton, BN1 9QH, UK*

Abstract. The goal of this chapter is to study physical conditions of the interstellar medium (ISM) in distant galaxies. In particular, its densest component is associated with the inner cores of clouds – this is where star formation takes place. Carbon monoxide is usually used to trace molecular gas emission; however, its transitions are practically opaque, thus preventing astronomers from piercing through the clouds, into the deepest layers that are most intimately connected with the formation of stars. Other dense gas tracers are required, although they are typically too faint to be effectively observed in high redshift main sequence galaxies. The ngVLA will offer for the first time the sensitivity at radio frequencies that is needed to detect $[\mathrm{C\,I}]_{1-0}$ at $z \geq 5$, as well as the ground transitions of dense gas tracers of the ISM such as HCN, HNC, HCO^+ at $z > 1$, beyond the tip of the iceberg of the hyper-luminous sources that can be studied now. These new tools will critically contribute to our understanding of the intimate interplay between gas clouds and star formation in different environments and cosmic epochs.

1. Scientific Rationale

While undoubtedly the carbon monoxide (CO) rotational emission lines are extremely useful to gauge the molecular gas mass in distant galaxies, they have a limited power in exposing the diversity of properties and the complexity of the multi-phased ISM. For instance, both observational and theoretical studies revealed substantial amount of diffuse molecular gas that is not associated with CO emission (e.g., Langer et al. 2014; Glover & Smith 2016). On the other hand, CO transitions reach optical depth values close to unity even at very modest column densities, i.e., CO-emitting regions appear opaque. Furthermore, the critical density required to collisionally de-excite CO is relatively low ($n_{\rm H2} \sim 10^2 - 10^3$ cm^{-3} for the lower-J transitions), meaning that CO is a fairly poor tracer of the dense molecular cores, where star-formation within distant galaxies is ultimately taking place.

The neutral carbon fine structure lines [CI] at 609 and 370 μm arise in the external parts of molecular clouds, where the radiation field is too intense for CO molecules to survive, but at not enough to ionize the carbon atoms (optical extinction $A_V = 1 - 5$ mag). The 370 μm transition has been detected even at $z \approx 7$ (Venemans et al. 2017). The two neutral carbon fine structure lines are ideal tracers of the more diffuse molecular ISM (see, e.g., Glover et al. 2015). In particular, because of the different critical densities, their ratio is an excellent tracer of the gas density. Furthermore, in combination with other lines (such as CO or [CII]) they can be used to infer physical properties of the ISM such as the gas density and the intensity of the incident radiation field (see, e.g., Meijerink et al. 2008; Popping et al. 2018). [CI] is also considered a complementary tracer of the cold gas mass, which can be used to infer either an independent measurement of the CO–to–H_2 conversion factor, $\alpha_{\rm CO}$, or to measure the neutral carbon abundance, $X_{\rm CI}$ (see, e.g., Bothwell et al. 2017; Popping et al. 2017).

To date, only about 40 galaxies at $z > 1$ have been detected in [CI]. This list mostly consists of sub-mm galaxies and quasar host galaxies. E.g., Bothwell et al. (2017) use [CI] 609 μm observations, in combination with other ISM lines, to infer molecular gas mass, density and intensity of the radiation field in 13 dusty, star–forming galaxies at $2 < z < 5$ using ALMA. They find a higher density and stronger UV field for the star–forming medium in these galaxies than estimated by previous studies of similar sources. At higher redshifts, however, the $[\rm CI]_{1-0}$ transition is shifted outside the ALMA bands at $z > 4.8$. This ground transition is a better mass tracer than the $[\rm CI]_{2-1}$ transition (observable with ALMA at these high redshifts), which is also sensitive to the excitation temperature. Furthermore, only the detection of both these lines provides a direct measure of the neutral gas density. The detection of the $[\rm CI]_{1-0}$ line at higher redshift thus requires sensitive observations in the frequency ranges that will be covered by the ngVLA.

In addition to neutral carbon, other molecular gas tracers can provide us with precious complementary information than CO. E.g., high dipole moment molecules like hydrogen cyanide (HCN) are only collisionally de-excited at very high densities, making them much more reliable tracers of the very dense gas directly associated with the formation of individual stars. Some studies of HCN in the nearby universe have even found evidence that the ratio of HCN luminosity to far-infrared (FIR) luminosity remains constant over >8 orders of magnitude in HCN luminosity, suggesting that HCN may be a fundamental direct probe of star forming 'units', and that the only difference between star formation on different scales and in different environments is the number

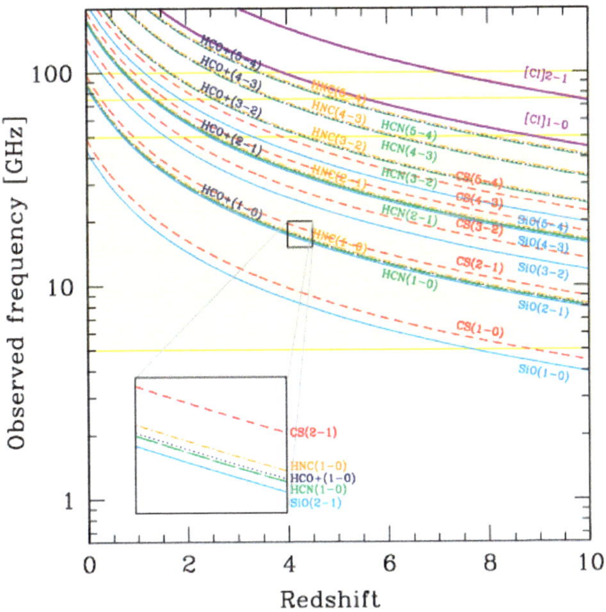

Figure 1. Redshifted frequencies of dense gas tracers, many of which will be accessible in single-tuning setups with the ngVLA at high-redshift. In particular, the 1-0 and 2-1 transitions of HNC, HCO+, HCN and SiO lines are spaced very closely in frequency (see inset zoom-in). While ALMA can detect the mid- and high-*J* transitions of these molecules at high-redshift, significant uncertainties regarding excitation mean that the low-*J* transitions accessible with the ngVLA are more suitable as tracers of the total dense gas mass.

of these fundamental star-forming units (e.g., Gao & Solomon 2004; Wu et al. 2005, 2010; Zhang et al. 2014).

However, not much is known about dense gas tracers in the more distant universe. Owing to the fact that they only trace the densest regions of the ISM and are therefore less abundant, emission from the rotational transitions of molecules like HCN is usually an order of magnitude fainter than CO, complicating efforts to detect and study these tracers at high-z. As a result, only a few high-z galaxies have been detected in dense gas tracers to-date — most of which are either intrinsically extremely luminous, or strongly lensed (e.g., Solomon et al. 2003; Vanden Bout et al. 2004; Carilli et al. 2005; Gao et al. 2007; Riechers et al. 2007, 2011; Danielson et al. 2011). With ALMA now online, the situation will clearly improve dramatically in the near future. However, as with CO, ALMA will only be able to detect the mid- and high-*J* transitions of dense gas tracers like HCN, HNC, and HCO+. These higher-level transitions are less directly tied to the total dense gas mass, requiring assumptions about the (highly uncertain) excitation ratios. In addition, these higher-*J* transitions are more likely to be affected by IR pumping, which local studies find may be common in ultraluminous galaxies (e.g., Aalto et al. 2007). Thus, while necessary to understand the overall excitation

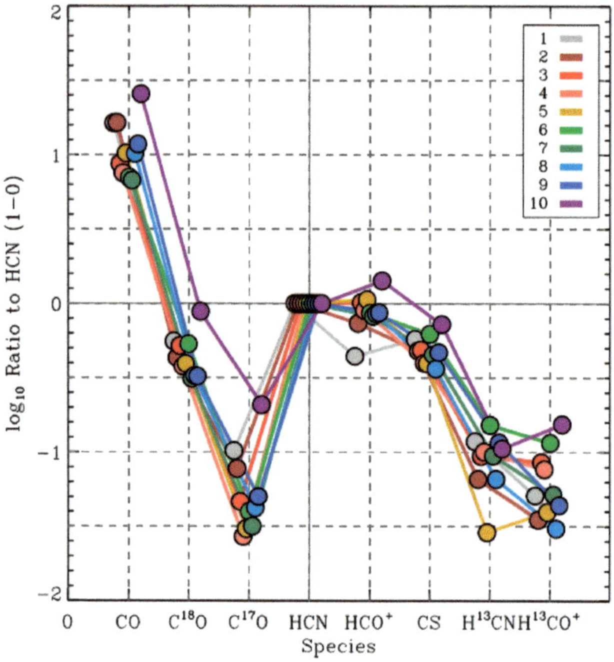

Figure 2. The intensity (in units of temperature) of the ground transitions of CO, C^{18}O, C^{17}O, HCN, HCO$^+$, CS, and H^{13}CNH^{13}CO$^+$ in various clouds (labeled 1–10) in the local starburst galaxy NGC 253, normalized to the observed HCN(1-0) flux. CO is at least 10× brighter than other dense gas tracers – in order to detect these other lines at high redshift, the unprecedented sensitivity of ngVLA at radio frequencies will be instrumental. Figure adapted from Leroy et al. (2015).

properties of high-z sources, the ALMA-detectable transitions may be unsuitable as tracers of the total dense gas mass.

The current Jansky VLA probes the right frequency range to detect the crucial low-J transitions of these high critical density molecules (Figure 1). However, its limited sensitivity means that the current state-of-the-art for high-z detections consists of a smattering of strongly lensed hyper-starbursting quasar hosts (e.g., Vanden Bout et al. 2004; Riechers et al. 2007, 2011). With significantly increased sensitivity, the ngVLA would extend studies of the dense gas mass at high-z beyond this handful of extreme objects for the first time. In addition to tracing the dense gas mass at high-z, such studies are critical for constraining models of star formation based on the gas density PDF, as these models make testable predictions about the FIR-HCN relation in FIR-luminous objects (e.g., Krumholz & Thompson 2007; Narayanan et al. 2008). Finally, while angular resolution is not a priority for these photon-starved studies, the brightest objects could even be spatially resolved by a ngVLA on ~kpc scales (requiring baselines on the order of the current VLA). The ngVLA would thus enable detailed studies of the dense gas at high-redshift such as are currently only possible in the local universe.

1.1. Measurements Required

This science goal requires integrated flux measurements of faint lines in high redshift galaxies. The sensitivity and the coverage in the radio bands are prime technical requirements for these observations. The targeted lines will be $>10\times$ fainter than CO (in temperature units). A galaxy with stellar mass $M_* \approx 10^{11}\,M_\odot$ on the massive main sequence of star–forming galaxies at $z \sim 2$ has SFR $\approx 100\,M_\odot\,\mathrm{yr}^{-1}$ (Santini et al. 2017). This corresponds to an IR luminosity $L_{\mathrm{IR}} \approx 5\times10^{11}\,L_\odot$. Assuming the CO–to–IR luminosity empirical relation from Carilli & Walter (2013), we infer an expected CO(1-0) luminosity of 7.0×10^9 K km s^{-1} pc^2, or 345,000 $L_\odot$. For a CO(1-0)/HCN(1-0)$\approx$10 (see Figure 2), we infer an expected flux density for the HCN(1-0) transition of 7 μJy (assuming a line width of 300 km s^{-1}). Such a line can be detected at 3σ significance with 26 hr of integration, assuming that the ngVLA will have at least $5\times$ the collecting area of the current VLA. No existing instrument has the sensitivity at radio frequencies to perform the proposed observations.

Imagining requirements are limited at $\sim 1-3''$ resolution, in order to avoid confusion with other sources in the field, and at the same time limit the risk of out-resolving the faint emission. A large bandwidth would allow to simultaneously cover multiple ISM tracers (see Figure 1) in one shot, i.e., it would be ideal although it is not a strict requirement.

1.2. Longevity/Durability: with Respect to Existing and Planned (>2025) Facilities

There are obvious synergies with other key observatories to envision in the context of the characterization of the ISM in high-z galaxies. In particular, ALMA can provide coverage of the intermediate and high-J transitions of the dense gas tracers, as well as of the [CI]$_{2-1}$ line, whereas ngVLA will focus on the lower-J lines and on the [CI]$_{1-0}$ transition (see Figure 1). The combinations of CO transitions ranging from low to high J, together with the denser gas tracers targeted with ngVLA + ALMA, will enable an unprecedentedly detailed study of the properties of the densest parts of the ISM, which are not currently accessible. This will enable a direct link between the gas properties and the on-going star formation, given that the dense gas tracers are intimately connected with birthplace of stars. *JWST* and the 30 m class telescope will provide sensitive spectroscopy of rest-frame UV/optical/NIR ISM tracers, mostly associated with the ionized phase of the gas, thus paving the way for a coherent description of the gas cycle in galaxies throughout cosmic history.

References

Aalto S., Spaans M., Wiedner M.C., Hüttemeister S. 2007, A&A, 464, 193

Bothwell M.S., Aguirre J.E., Aravena M., et al. 2017, MNRAS, 466, 2825

Carilli C.L., Solomon P., Vanden Bout P., Walter F., Beelen A., Cox P., Bertoldi F., Menten K.M., Isaak K.G., Chandler C.J., Omont A. 2005, ApJ, 618, 586

Carilli C.L. & Walter F., 2013, ARA&A, 51, 105

Danielson A.L.R., Swinbank A.M., Smail I., Cox P., Edge A.C., Weiß A., Harris A.I., Baker A.J., De Breuck C., Geach J.E., et al. 2011, MNRAS, 410, 1687

Decarli, R., Walter, F., Aravena, M. et al. 2016, ApJ, 833, 69

Gao Y., Solomon P.M. 2004, ApJ, 606, 271

Gao Y., Carilli C.L., Solomon P.M., Vanden Bout P.A. 2007, ApJ, 660, L93

Glover S.C.O., Clark P.C., Micic M., Molina F. 2015, MNRAS, 448, 1607

Glover S.C.O., Smith R.J. 2016, MNRAS, 462, 3011
Krumholz M.R., Thompson T.A. 2007, ApJ, 669, 289
Langer W.D., Velusamy T., Pineda J.L., Willacy K., Goldsmith P.F. 2014, A&A, 561, A122
Leroy A., Bolatto A.D., Ostriker E.C., et al. 2015, ApJ, 801, L25
Meijerink R., Glassgold A.E., Najita J.R. 2008, ApJ, 676, 518
Narayanan D., Cox T.J., Hernquist L. 2008, ApJ, 681, L77
Popping G., Decarli R., Man A.W.S., et al. 2017, A&A, 602, A11
Popping G., Narayanan D., Somerville R.S., et al. 2018, arXiv:1805.11093
Riechers D.A., Walter F., Carilli C.L., Bertoldi F. 2007, ApJ, 671, L13
Riechers D.A., Walter F., Carilli C.L., Cox P., Weiß A., Bertoldi F., Menten K.M. 2011, ApJ, 726, 50
Santini P., Fontana A., Castellano M., et al. 2017, ApJ, 847, 76
Solomon P.M., Vanden Bout P., Carilli C., Guelin M. 2003, Nature, 426, 636
Vanden Bout P.A., Solomon P.M., Maddalena R.J. 2004, ApJ, 614, L97
Venemans B.P., Walter F., Decarli R., Ferkinhoff C., Weiß A., Findlay J.R., McMahon R.G., Sutherland W.J., Meijerink R. 2017, ApJ, 845, 154
Wu J., Evans N.J. II, Gao Y., Solomon P.M., Shirley Y.L., Vanden Bout P.A. 2005, ApJ, 635, L173
Wu J., Evans N.J. II, Shirley Y.L., Knez C. 2010, ApJS, 188, 313
Zhang Z.-Y., Gao Y., Henkel C., Zhao Y., Wang J., Menten K.M., Güsten R. 2014, ApJ, 784, L31

Science with a Next-Generation Very Large Array
ASP Conference Series, Monograph 7
Eric J. Murphy, ed.

Cold Gas in High-z Galaxies: CO as Redshift Beacon

R. Decarli,[1,2] C. Carilli,[3,4] C. Casey,[5] B. Emonts,[6] J.A. Hodge,[7] K. Kohno,[8] D. Narayanan,[9] D. Riechers,[10] M.T. Sargent,[11] and F. Walter[2,3]

[1]*INAF – Osservatorio di Astrofisica e Scienza dello Spazio, via Gobetti 93/3, 40129 Bologna, Italy;* roberto.decarli@inaf.it

[2]*Max Planck Institut für Astronomie, Königstuhl 17, 69117 Heidelberg, Germany*

[3]*National Radio Astronomy Observatory, Pete V. Domenici Array Science Center, P.O. Box O, Socorro, NM 87801, USA*

[4]*Cavendish Laboratory, University of Cambridge, 19 J.J. Thomson Avenue, Cambridge CB3 0HE, UK*

[5]*Department of Astronomy, The University of Texas at Austin, 2515 Speedway Blvd Stop C1400, Austin, TX 78712*

[6]*National Radio Astronomy Observatory, 520 Edgemont Road, Charlottesville, VA 22903, USA*

[7]*Leiden Observatory, Niels Bohrweg 2, 2333 CA Leiden, The Netherlands*

[8]*Institute of Astronomy, School of Science, The University of Tokyo, 2-21-1 Osawa, Mitaka, Tokyo 181-0015, Japan*

[9]*Department of Astronomy, University of Florida, 211 Bryant Space Science Center, Gainesville, FL 32611, USA*

[10]*Cornell University, 220 Space Sciences Building, Ithaca, NY 14853, USA*

[11]*Astronomy Centre, Department of Physics and Astronomy, University of Sussex, Brighton, BN1 9QH, UK*

Abstract. The goal of this chapter is to address the use of the ngVLA as a CO redshift machine for dust-obscured high-redshift galaxies which lack of clear counterparts at other wavelengths. Thanks to its unprecedented large simultaneous bandwidth and sensitivity, the ngVLA will be able to detect multiple low-J CO transitions at all redshifts > 1. In particular, at $z \geq 4.76$ two CO transitions will be covered in a single frequency setting, thus ensuring unambiguous line identification. The ngVLA capabilities fill in a redshift range where other approaches (e.g., photometric redshifts, search for optical/radio counterparts, etc.) typically fail due to the combination of intrinsically faint emission and increasing luminosity distance. This will allow us to explore the formation of massive galaxies in the early cosmic times.

1. Scientific Rationale

Since the identification of the first sub-mm galaxies and dusty star-forming galaxies (DSFGs; e.g., Hughes et al. 1998; Ivison et al. 1998, 2002; Smail et al. 2002; Chapman et al. 2005), it has been clear that obtaining precise redshift measurements for these sources would represent a major challenge, especially at the highest redshifts (e.g., Riechers et al. 2013; Marrone et al. 2018). A combination of intrinsic faintness, high redshift, and dust reddening limits the detectability of rest-frame optical/UV spectral features that could be used to pin down the redshift to ~50% of the DSFGs studied so far (e.g., Danielson et al. 2017). This is particularly severe for those galaxies for which the estimated unobscured star formation is only a tiny fraction of the total star–formation rate. The peak wavelength of the dust spectral energy distribution can be used as a tentative redshift proxy (see, e.g., Riechers et al. 2013; da Cunha et al. 2015), although it is highly degenerate with the dust temperature. The search for radio continuum counterparts biases the selection against high redshifts (as the radio continuum does not benefit of the negative *k*-correction that is in place at sub-mm wavelengths). As a result, the intrinsic redshift distribution of these sources is still under debate (see Casey et al. 2014, for a review). The question of whether the unconfirmed DSFGs sit at similar redshifts or higher redshifts than the others is yet to be clarified, with the incompleteness being a strong function of redshift. The discovery of bright dusty galaxies like HFLS3 (Riechers et al. 2013) and SPT0311–58 (Marrone et al. 2018), and the four [CII]–bright galaxies in the field of four $z > 6$ quasars (Decarli et al. 2017) hint at the existence of such a population at very high redshifts, but so far, confirming such sources (especially at lower luminosities or in the absence of gravitational lensing) in a systematic way has not been efficient.

A promising way forward consists in searching for multiple CO transitions. Since DSFGs are very luminous at IR wavelengths, and the CO luminosity correlates with $L_{\rm IR}$ (see, e.g., Carilli & Walter 2013), CO transitions are expected to be bright in DSFGs. CO transitions have rest frequencies of $\sim 115.2 \times J_{\rm up}$ GHz. The detection of two transitions might leave some degeneracy in the redshift interpretation [e.g., if one line is observed at $2\times$ the frequency of the other, they could be CO(1-0) and CO(2-1), or CO(2-1) and CO(4-3), or CO(3-2) and CO(6-5), etc]. However, by combining the line detection with ancillary information (e.g., broad-band optical/NIR photometry) one could significantly restrict the number of possible identifications. Furthermore, the CO spectral energy distribution is typically fairly regular, so that if one observe, e.g., CO(2-1) and CO(4-3), the flux of the CO(3-2) can be predicted pretty accurately. The presence or lack of intermediate transitions can then further trim the parameter space of allowed interpretations.

The main drawback of CO-based redshifts is that large frequency ranges need to be sampled in order to find one or more lines. At any redshift, CO transitions are spaced by several tens of GHz in the observed frame — thus representing an observational challenge, as multiple frequency setting need to be combined in order to effectively sample a large bandwidth (see Figure 1). Weiß et al. (2013) used existing capability of ALMA to scan a large fraction of the 3mm band, and searched for high-*J* CO emission in 26 DSFGs. Each source was observed in 5 different frequency settings, for a total of ~10 min on source. However, the high detection rate (~90% of sources have been detected in at least 1 CO transition) of such short observations was only possible thanks to the extreme brightness of the strongly-magnified sources in their sample. A similar effort with PdBI targeting the non-lensed galaxy HDF850.1 resulted in the first redshift

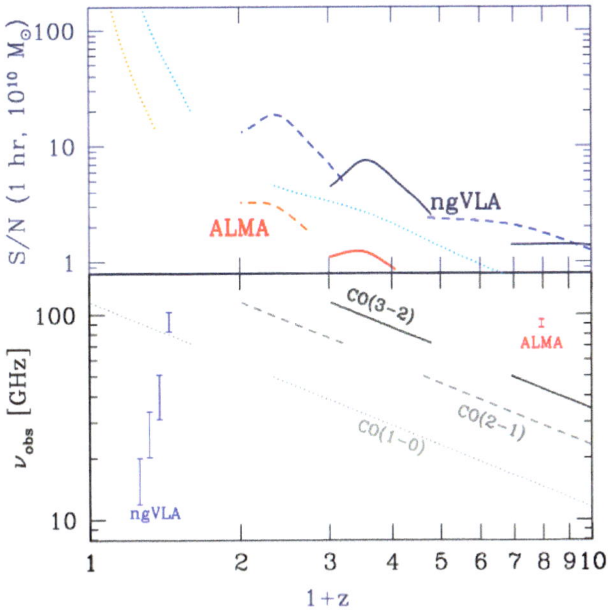

Figure 1. The power of ngVLA as a redshift beacon, compared to ALMA. (*Top:*) The expected S/N of CO(1-0) (dotted), CO(2-1) (dashed), and CO(3-2) (solid lines) emission for a galaxy with molecular mass $M_{H2} = 10^{10}\,M_\odot$, as a function of redshift. We assume 1 hr of integration, a line width of 300 km s^{-1}, a CO–to–H_2 conversion of $\alpha_{CO} = 3.6\,M_\odot\,(\mathrm{K\,km\,s^{-1}\,pc^2})^{-1}$ (Daddi et al. 2010), and a CO excitation typical of main sequence galaxies (Daddi et al. 2015). The ngVLA (blue lines) will provide 5–10 × higher S/N for the same integration time than ALMA (red lines) in the overlapping frequency ranges, and will expand the redshift coverage of low–J CO transitions to much higher redshifts, virtually covering every redshift $z > 1$. (*Bottom:*) The observed frequency of the same CO transitions as a function of redshift, compared with the simultaneous bandwidth of ngVLA and ALMA (shown as vertical bars). The large bandwidth of ngVLA will enable to sample wide CO redshift ranges in a single frequency setting.

determination of this archetypal source ($z = 5.183$); this however required ~100 hr of integration (e.g., Walter et al. 2012). Even with full ALMA, redshift scans are going to be expensive in "normal" DSFGs that are unlensed. Additionally, by focusing on wavelengths equal to or larger than 3mm, we can only sample intermediate to high-J CO transitions, thus the method would be applicable only to highly excited sources.

The ngVLA is going to revolutionize this line of search, as 1) the large simultaneous bandwidth coverage will maximize the probability of detecting the CO(1-0) line at $z > 2$ from any source with only a few frequency settings, even if only very coarse constraints on the redshift are available; 2) by focusing on the lowest-J transitions, it will be insensitive to excitation bias, i.e., the method will be applicable to any CO-emitting gaseous reservoir, not only to the most excited; 3) at $z > 4.76$, it will provide coverage of both the CO(1-0) and CO(2-1) lines. This will provide unambiguous redshift identification for all the DSFGs at the highest redshift, where the optical/NIR spectroscopy and the radio continuum follow-up are extremely incomplete.

This is quantitatively demonstrated in Figure 1, where we show the expected signal-to-noise ratio (S/N) of CO(1-0), CO(2-1), and CO(3-2) in a galaxy with molecular gas mass $M_{\rm H2}$=10^{10} $M_\odot$, line width Δv=300 km s^{-1}, a CO–to–H_2 conversion factor $\alpha_{\rm CO}$=3.6 $M_\odot$ (K km s^{-1} pc^2)$^{-1}$ (Daddi et al. 2010), and a CO excitation typical for main sequence galaxies (Daddi et al. 2015). In only a few hours of integration, the ngVLA will be able to securely detect CO(1-0) in two wide redshift windows ($z < 0.5$ and $2.2 < z < 10$), as well as CO(2-1) and/or CO(3-2) at virtually any redshift $z > 1$. For a comparison, ALMA will only target intermediate to high-J transitions at $z > 3$, and their detection will require several hours per frequency settings, and multiple frequency settings to compensate for the smaller bandwidth. The ngVLA will therefore represent the main facility for CO redshift searches in DSFGs over most of the cosmic history in the future decades. For a comparison, the Square Kilometre Array will only be competitive with ngVLA in this line of search if it will have a dedicated high–frequency (5–50 GHz) band, which is still under debate: The intermediate frequency band (0.3–10 GHz) becomes sensitive to CO only at $z > 10.5$, where we do not expect a significant population of dusty galaxies to be already in place, and other redshift tracers than CO (e.g., fine structure lines at rest–frame far–infrared wavelengths, which are shifted in the ALMA 1–3 mm bands at these redshifts) become more effective redshift tracers.

1.1. Measurements Required

We need to pin down the redshift of dusty galaxies with unknown or poorly constrained redshift via the observation of the CO(1-0) [and possibly CO(2-1)] line.

1.2. Uniqueness to ngVLA Capabilities (e.g., frequency coverage, resolution, etc.)

Thanks to the large simultaneous band coverage, the ngVLA will be an unparalleled facility to blindly search for lines associated with dusty, high-redshift galaxies. For any galaxy at $z > 2$, the coverage of one/two lines is ensured irrespective of the redshift. By targeting the lowest J transitions, this line search circumvents any potential excitation bias (that would inevitably affect a similar endeavor with, e.g., ALMA). Finally, thanks to the improved sensitivity with respect to the VLA, this approach will become applicable even to relatively CO-faint sources, whereas to date it has only been used for the very brightest sources.

1.3. Longevity/Durability: with Respect to Existing and Planned (>2025) Facilities

The advent of *JWST* and the 30 m class telescopes will significantly push the sensitivity of optical/NIR/MIR spectroscopic investigations, thus mitigating the spectroscopic incompleteness that currently prevents us from characterizing DFSGs, especially at high redshift. However, even these facilities will have a hard time in measuring redshifts of the reddest sources at the dawn of galaxy formation. For these galaxies, the ngVLA will play a unique role as a redshift machine.

References

Carilli C.L. & Walter F., 2013, ARA&A, 51, 105

Casey C.M., Narayanan D., Cooray A. 2014, PhR., 541, 45

Chapman S.C., Blain A.W., Smail I., Ivison R.J. 2005, ApJ, 622, 772

da Cunha E., Walter F., Smail I.R., Swinbank A.M., Simpson J.M., Decarli R., Hodge J.A., Weiß A., et al. 2015, ApJ, 806, 110

Daddi E., Bournaud F., Walter F., Dannerbauer H., Carilli C.L., Dickinson M., Elbaz D., Morrison G.E., et al., 2010, ApJ, 713, 686

Daddi E., Dannerbauer H., Liu D., Aravena M., Bournaud F., Walter F., Riechers D., Magdis G., et al. 2015, A&A, 577, 46

Danielson A.L.R., Swinbank A.M., Smail I., Simpson J.M., Casey C.M., Chapman S.C., da Cunha E., Hodge J.A., et al. 2017, ApJ, 840, 78

Decarli R., Walter F., Venemans B.P., Bañados E., Bertoldi F., Carilli C., Fan X., Farina E.P., et al. 2017, Nature, 545, 457

Hughes D.H., Serjeant S., Dunlop J., Rowan-Robinson M., Blain A., Mann R.G., Ivison R., Peacock J., et al. 1998, Nature, 394, 241

Ivison R.J., Smail I., Le Borgne J.-F., Blain A.W., Kneib J.-P., Bezecourt J., Kerr T.H., Davies J.K. 1998, MNRAS, 298, 583

Ivison R.J., Greve T.R., Smail I., Dunlop J.S., Roche N.D., Scott S.E., Page M.J., Stevens J.A., et al. 2002, MNRAS, 337, 1

Marrone D.P., Spilker J.S., Hayward C.C., Vieira J.D., Aravena M., Ashby M.L.N., Bayliss M.B., Béthermin M., et al. 2018, Nature, 553, 51

Riechers D.A., Bradford C.M., Clements D.L., Dowell C.D., Pérez-Fournon I., Ivison R.J., Bridge C., Conley A., et al. 2013 Nature, 496, 329

Smail I., Ivison R.J., Blain A.W., Kneib J.-P. 2002, MNRAS, 331, 495

Walter F., Decarli R., Carilli C., Bertoldi F., Cox P., da Cunha E., Daddi E., Dickinson M., et al., 2012, Nature, 486, 233

Weiß A., De Breuck C., Marrone D.P., Vieira J.D., Aguirre J.E., Aird K.A., Aravena M., Ashby M.L.N., Bayliss M., Benson B.A., et al. 2013, ApJ, 767, 88

Science with a Next-Generation Very Large Array
ASP Conference Series, Monograph 7
Eric J. Murphy, ed.

The Molecular High-z Universe on Large Scales: Low-Surface-Brightness CO and the Strength of the ngVLA Core

Bjorn Emonts,[1] Chris Carilli,[2] Desika Narayanan,[3,4,5] Matthew Lehnert,[6] and Kristina Nyland[1]

[1]*National Radio Astronomy Observatory, 520 Edgemont Road, Charlottesville, VA 22903* (bemonts@nrao.edu)

[2]*National Radio Astronomy Observatory, P.O. Box 0, Socorro, NM 87 801*

[3]*Department of Astronomy, University of Florida, 211 Bryant Space Science Center, Gainesville, FL 32611*

[4]*University of Florida Informatics Institute, 432 Newell Drive, CISE Bldg E251, Gainesville, FL 32611*

[5]*Cosmic Dawn Centre (DAWN), University of Copenhagen, Julian Maries vej 30, DK-2100, Copenhagen, Denmark*

[6]*Sorbonne Université, CNRS, UMR 7095, Institut d'Astrophysique de Paris, 98bis bvd Arago, 75014, Paris, France*

Abstract. The next-generation Very Large Array (ngVLA) will revolutionize our understanding of the Early Universe by tracing the coldest phase of molecular gas – the raw ingredient for star formation – in the most distant galaxies and galaxy-clusters. The km-scale core of the ngVLA will be densely packed with antennas, making it a prime instrument for imaging low-surface-brightness emission from large-scale molecular gas in the high-z circum- and inter-galactic medium (CGM/IGM). Recent studies indicate that large amounts of cold molecular gas are hiding in the 10s–100 kpc environments of distant galaxies, but that technical limitations on existing telescope arrays have prevented us from efficiently detecting these large molecular reservoirs. This may have led to a severly biased view of the molecular Universe. In this chapter we explore of potential for low-surface-brightness CO observations of the Early Universe, and how the core of the ngVLA will reveal the cold molecular Universe to limits and at scales not currently detectable by radio telescopes. As such, the ngVLA core will be a powerful instrument for studying the cold baryon cycle that drives the early evolution of galaxies and clusters.

1. Introduction

The evolution of galaxies is tightly linked to processes in the circum- and inter-galactic medium (CGM/IGM). Unfortunately, most of the baryons in the CGM/IGM are too faint to be easily detected. At high-z, we view glimpses of dark baryonic halos through quasar absorption lines, or cooling-radiation emitted as Lyα. Absorption-line work inferred the presence of large, ~100 kpc halos of warm/cool (T~10^4 K), metal-enriched gas around high-z quasars (e.g., Prochaska et al. 2014; Lau et al. 2016). However, a direct connection to the stellar growth of massive galaxies remains missing, because we have yet to identify the ultimate reservoir of halo gas that has sufficient mass to fuel widespread star-formation, namely the cold molecular gas (T∼10–100 K).

The massive Spiderweb Galaxy at $z = 2.2$ (Miley et al. 2006) revealed the first evidence for the existence of a cold molecular CGM in the distant Universe (Emonts et al. 2016, Figure 1). CO(1-0) observations with very compact array configuration of the Australia Telescope Compact Array (ATCA) traced a 70 kpc reservoir of molecular halo gas across the CGM of this massive forming proto-cluster galaxy (Emonts et al. 2016). The extended CO follows diffuse blue light from star formation that occurs in-situ within the CGM (Hatch et al. 2008, see Figure 1). This is evidence that there is a massive ($M_{H2} \sim 10^{11}$ $M_{\odot}$) reservoir of gas in the CGM that cooled well beyond the temperature of Lyα-emitting gas (T ∼ few 10^3 – 10^4 K), and is actively feeding star formation across the halo. More recent observations of [C I] $^3P_1$-$^3P_0$ and CO(4-3) with the Atacama Large Millimeter Array (ALMA) revealed that the cold CGM has densities of several 100 cm^{-3}, with average carbon abundance and excitation conditions resembling those of starforming galaxies (Emonts et al. 2018). This does not corroborate models of efficient and direct stream-fed accretion of relatively pristine gas (Dekel et al. 2009). Instead, the [CI] and CO properties agree with more complex recycling models, where the gas in the IGM is a melange from various sources – metal-enriched outflows, mass transfer among galaxies, gas accretion, and mergers (see Emonts et al. 2018).

Other studies using very compact interferometers also found CO-emitting gas on scales of tens of kpc around high-z massive galaxies (e.g., Emonts et al. 2014; Cicone et al. 2015; Emonts et al. 2015; Ginolfi et al. 2017; Dannerbauer et al. 2017). These results are starting to reveal a picture that the CGM is truly multi-phase, and that the Early Universe may contain much more molecular gas hiding outside galaxies than has thus far been observed. However, many questions remain about the nature of the cold molecular CGM/IGM, and its role in early galaxy evolution. How common is the presence of a cold CGM, in particular among the more general population of high-z galaxies? Is the cold gas diffuse or associated with star-formation? What is its origin (tidal debris, outflows, mixing, or accretion)? And how can we observe this cold CGM/IGM?

In this article, we will explain that technical limitations on array designs have mostly prevented us from reaching the sensitive low-surface-brightness levels needed to efficiently image widespread molecular gas across the CGM/IGM. This likely produced a severely biased view of the molecular Universe. The core of the next-generation VLA (ngVLA) will be a powerful instrument for tracing the molecular CGM, allowing us to study the cold baryon cycle that governs the early growth of massive galaxies.

2. Low-Surface-Brightness CO

Some of the strongest tracers for molecular gas are the various rotational transitions of carbon-monoxide, ^{12}CO(J,J-1). With the ALMA, it has become routine to observe CO, and other molecular species, in high-z galaxies. However, imaging the cold molecular medium outside galaxies on tens to hundred kpc scales has been limited by two observational challenges:

- **Excitation conditions of the molecular medium.** The critical density of CO(J,J-1) scales roughly with $n_{crit} \propto J^3$. The ground-transition CO(1-0) (ν_{rest} = 115.27 GHz) has an effective critical density of only several 100 cm^{-3}, and the $J = 1$ level of CO is

Figure 1. The cold circum-galactic medium of the Spiderweb Galaxy ($z = 2.2$). A total intensity image of the CO(1-0) emission is shown in light-blue in the main panel, as detected with the ATCA (Emonts et al. 2016). Superposed in brown are various channel maps of atomic carbon, [C sc i] $^3P_1 - {}^3P_0$, as detected with ALMA (Emonts et al. 2018). A bright spot of H_2O emission is shown in yellow (Gullberg et al. 2016). The background image is taken with the *Hubble Space Telescope* Advanced Camera for Surveys (*HST*/ACS) in the F475W and F814W filters (Miley et al. 2006). The radio source is shown in white contours, as observed with the VLA in the C-configuration at 35 GHz (Emonts et al. 2016). The in-set in the top-left corner shows the combined *HST*/ACS F475W+F814W image of the same region as the main panel (Miley et al. 2006), but optimized to visualize the diffuse optical light from in-situ star formation (Hatch et al. 2008).

substantially populated down to T ~10 K (Papadopoulos et al. 2004). However, these values increase by an order of magnitude or more for the high-J transitions, like CO(3-2) and higher. This means that these high-J transitions trace the warmer and denser molecular gas in starburst and AGN regions, and may severely underestimate masses from colder, lower density, and sub-thermally excited gas components (Papadopoulos et al. 2000). To get robust estimates of the *total* molecular gas mass in galaxies at $z > 1.5$, it is vital that CO(1-0) or CO(2-1) are observed in the 20-50 GHz regime. *This is particularly important for widespread CO-emitting reservoirs of molecular gas in the circum- and inter-galactic medium, as the excitation conditions and temperature of the molecular gas in these environments may be low compared to the ISM of galaxies.*

• **Short-spacing problem of radio interferometers.** Radio interferometers sample the sky at discrete intervals, set by the length of antennas pairs ('baselines'), frequency coverage and scan-time. Because large-scale features are filtered out by widely separated antennas, *low-surface-brightness emission can only be imaged with compact*

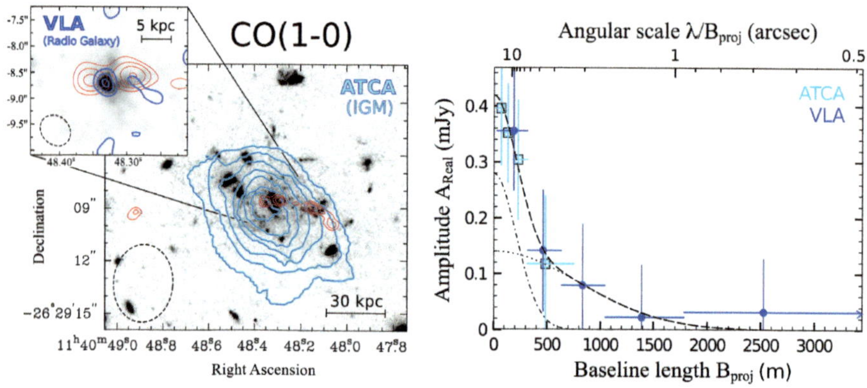

Figure 2. Low-surface-brightness CO(1-0) from Figure 1 across the halo of the Spiderweb Galaxy (z = 2.2), reproduced from Emonts et al. (2016). (*Left:*) ATCA CO(1-0) contours (blue) are overlaid onto the *HST*/ACS F475W+F814W image from Miley et al. (2006). Contour levels: 0.020, 0.038, 0.056, 0.074, 0.092, 0.110, 0.128 $\mathrm{Jy\,beam^{-1} \times km\,s^{-1}}$. The inset shows the CO(1-0) total-intensity contours from the VLA in DnC+CnB-configuration at 3.5, 4.5σ. The full-resolution VLA data reveal only a third of the total CO imaged on large scales with ATCA. (*Right:*) Visibility-amplitudes of the CO(1-0) emission are plotted as function of the projected baseline length in data from the ATCA (light-blue squares) and VLA DnC+CnB (dark-blue dots). As described in Emonts et al. (2016): "The amplitude plotted on the Y-axis is the real part of the complex interferometer visibility for the CO-signal, when vector-averaged across the velocity range -100 to +200 km/s and over all baselines covered by the horizontal error bars. The $\pm 1\sigma$ uncertainty in the amplitude of the averaged signal is indicated with the vertical error bars. The dashed line shows a model where we superimposed a Gaussian distribution with FWHM = 3″ (dash-dotted line) and a Gaussian distribution with FWHM = 0″.8 (dotted line)." This plot illustrates that it is essential to use short-baseline configurations to recover extended CO emission.

array-configurations at relatively low spatial resolution. This problem is particularly severe when studying extended gas reservoirs in the mm regime. For example, to recover CO at z ~2 on scales ≥15 kpc requires antennas to be placed <1 km apart at 35 GHz, and less than a few 100 m for the lower ALMA bands! This problem is visualized in Figure 2.

Unfortunately, millimeter observatories, like ALMA, generally operate at frequencies >85 GHz. This means that the ground-transition CO(1-0) rapidly redshifts out of the receiver bands. Several of the traditional cm-observatories, like the Very Large Array (VLA) and ATCA, can target redshifted CO(1-0) and CO(2-1) in the 20-50 GHz regime, but they crucially lack sensitivity on very short (≪ 1 km) baselines for efficiently imaging the cold molecular CGM/IGM (Figure 2).

3. The Strength of the ngVLA Core

The core the ngVLA (Carilli 2016) will be densely packed with antennas that will vastly supersede the current VLA D-configuration in number of short baselines. The

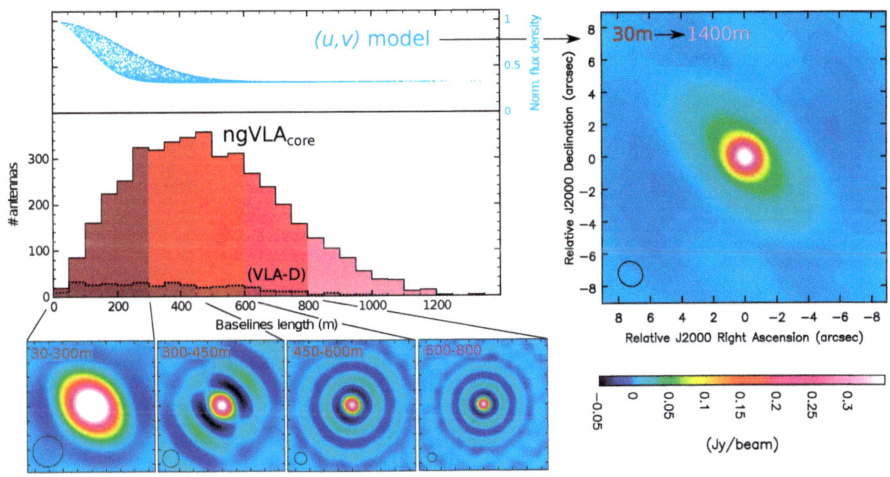

Figure 3. Simulated observations with the ngVLA core. (*Top Left:*) Red-colored histogram of the baseline distribution of the 18 m antennas in the ngVLA core, using the reference design of Carilli (2018). The colors from dark to light red contain 24% (30–300m), 23% (300–450m), 22% (450–600m), 20% (600–800m) and 11% (800–1400m) of the total number of baselines in the core. The overlaying black histogram at the bottom shows the baseline distribution of the VLA in D-configuration. The blue dots at the top of the figure are drawn on the same x-axis and represent a simulated snap-shot observation of a target model at the zenith, using the simulation tools in the Common Astronomy Software Applications (CASA) software (McMullin et al. 2007). This model consists of a Gaussian function that represents extended emission with a superposed delta-function of an unresolved point-source component, both placed at the phase center. The Gaussian has FWHM $b_{maj} = 6''$ and $b_{min} = 3''$ with PA = 45°, which corresponds to about 50 × 25 kpc at $z = 2$. The signal is normalized and placed in a single channel of width 30 MHz, with 2/3rd (1/3rd) of the total flux in the Gaussian (point-source) component. No noise has been added to the simulation and a basic clean of 1000 iterations down to 0.1 × the peak flux was applied. (*Top Right:*) Total intensity image of the simulated emission of the target model using the ngVLA core with an exposure time of 8h. The image includes all the baselines in the range 30–1400m. (*Bottom:*) The small panels show the total intensity image when restricting the (u, v)-coverage to the baselines indicated in the colored parts of the histogram (i.e., 30–300 m, 300–450 m, 450–600 m and 600–800 m). The dimensions and color-scaling for all four panels are the same as for the main image of the target model in the top right panel. This figure illustrates that the shortest (≤ several 100m) baselines in the core of the ngVLA are crucial for recovering and imaging extended low-surface-brightness emission.

current reference design for the core includes 94 antennas with a diameter of 18 m that cover baselines out to about 1 km (Carilli 2018). While the longer ngVLA baselines will not be able to detect extended CO because they resolved out the emission, the ngVLA core will be a revolutionary new instrument for unlocking the cold molecular Universe. This is further visualized in Figure 3, where we show the baseline coverage of the ngVLA core (Carilli 2018), as well as simulated imaging of extended emission. The simulation shown in Figure 3 is simplistic, and does not take into account noise or gas kinematics and clumpiness. Nevertheless, it shows the need for sufficient sensitivity on baselines less than several × 100m in the core of the ngVLA, in order to optimize low surface brightness observations.

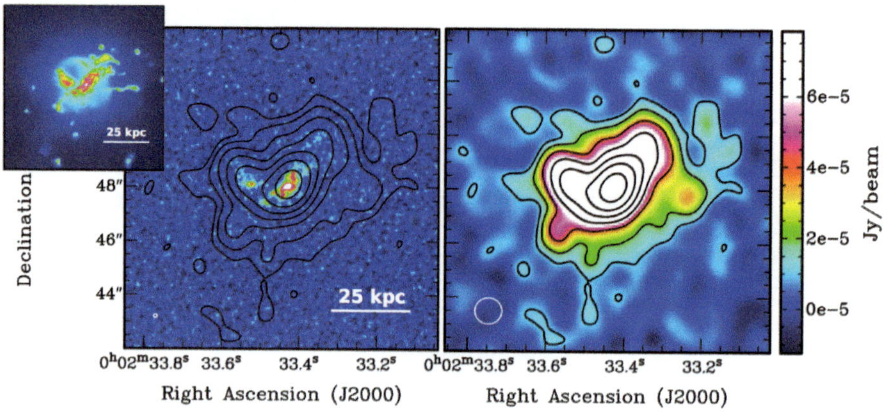

Figure 4. Simulated CO(1-0) observation of a merging system with extended halo from the cosmological simulations of Narayanan et al. (2015) (top-left inset), using the ngVLA configuration described in Carilli (2018). The simulations were performed in CASA, and assume $z = 2$, exposure time $t_{\rm exp} = 48$ h, channel width of 9 MHz ($\sim$ 70 km s^{-1}), and total CO(1-0) luminosity $I_{\rm CO(1-0)} = 0.21$ Jy beam^{-1} km s^{-1} (i.e., similar to that of the Spiderweb Galaxy in Figure 2; Emonts et al. 2016). Noise has been added to the visibilities, based on a theoretical root-mean-square (rms) noise of 2.3 μJy beam^{-1} when taking the inner 168 antennas of the array and using natural weighting. (*Left:*) Simulated CO(1-0) image at a resolution of 0.″126 ($\sim$1 kpc), using a robust weighting of 0 and (u, v)-tapering of 0.″09. This taper provides a resolution relevant only to the antennas that are distributed across the Plains of San Augustin with baselines $\leq$30 km, while down-weighting the core. The synthesized beam is shown in white in the bottom-left corner. The rms noise level is 3.0 μJy beam^{-1}, with a 3σ surface brightness limit of 0.5 K, and 3σ limit for H_2 mass clumps of 5.6 $(\alpha_{\rm CO}/0.8) \times 10^8\ M_\odot$. (*Right:*) Simulated CO(1-0) image at resolution of 1″, using a robust weighting of 0.5 and (u, v)-tapering of 0.″85. This taper provides a resolution relevant to the km-scale core of the array, and down-weights the longer spacings. The synthesized beam is shown in white in the bottom-left corner. The rms noise level is 3.4 μJy beam^{-1}, with a 3σ surface brightness limit of 9 mK. The halo gas is detected at 3σ across ~80 kpc, and includes ~60% more CO(1-0) flux than the 0.″13 plot on the left. The black contours start at 3σ (10 μJy) and increase by factors of 2. For visualization purposes, the same contours are also overlaid in the left plot. More details are given in Carilli (2018).

A more realistic simulation is shown in Figure 4. Here we simulate the expected CO(1-0) emission observed with the ngVLA in a massive ($\sim 10^{12}\ M_\odot$) high-z merger system. The CO luminosities are calculated using the Narayanan et al. (2012) scaling relations between $X_{\rm CO}$ and $\Sigma_{\rm H2}$, and the Narayanan & Krumholz (2014) model relationship between CO excitation ladders and $\Sigma_{\rm SFR}$. For the purposes of modeling the CO flux, the system is placed at $z = 2$ and has its overall flux scaled to the CO(1-0) intensity observed in the Spiderweb Galaxy (Figure 2). The system contains bright merging components across ~30 kpc and a halo stretching twice as far out. We simulate a 48 h observation with the ngVLA (Carilli 2018), and apply different tapers to increase the surface brightness sensitivity. When going from a moderately tapered 0.″13 resolution to a heavily tapered 1″ resolution, the surface brightness sensitivity increases by almost two orders of magnitude, reaching an rms noise level of 3 mK at 1″ resolution. At this brightness sensitivity, we detect the CO(1-0) across the halo environment on a scale of ~80 kpc. In the simulated 1″ image we detect ~60 % more CO(1-0) flux at a level above 3σ than in the 0.″13 image.

Low-surface-brightness observations of molecular gas with the ngVLA core will be complemented by ALMA, provided that ALMA is used in its most compact configurations, because the baseline length *in* $k\lambda$ increases rapidly at the higher frequencies. ALMA can target not only the high-J transition of CO, but also atomic carbon [C I], which is expected to be concomitant with CO(1-0) and can serve as an important alternative mass tracer (Papadopoulos et al. 2004, see also Figure 1). With the advent of Band 1 (Huang et al. 2016), ALMA will be able to observe down to 35 GHz, where it can target CO(1-0) out to $z \leq 2.3$. This is also roughly the redshift out to which the Square Kilometre Array (SKA) in its phase-1 will be sensitive for imaging the 21 cm line of neutral hydrogen in the most massive galaxies. For targeting the low-J CO lines at higher redshifts (15 – 50 GHz regime), where we can unravel the cold gas content of the very Early Universe, the ngVLA is needed.

As a note of caution, CO lines become dimmer when observed at higher redshifts as a result of reduced contrast against the increasing temperature of the Cosmic Microwave Background (da Cunha et al. 2013; Zhang et al. 2016). For extended CO(1-0), this dimming can reach anywhere from the typical 30-40% expected for the coldest (T~20 K) component of the ISM at $z \sim 2$ to up to factors of a few for gas in the CGM at higher redshifts. A detailed investigation into the dimming of the low-J CO lines is beyond the scope of this chapter. However, to account for this dimming by the CMB, the vastly increased sensitivity of the ngVLA core over existing instruments, like the VLA, is critical.

Therefore, the ngVLA will be the only instrument in the foreseeable future than can target the crucial CO(1-0) transition at $z \geq 2.3$ with a core-configuration that has the brightness sensitivity to reveal the full extent of the cold molecular Universe.

4. Concluding remarks

While the next-generation VLA will be a powerful instrument to study the radio Universe with unprecedented resolution, it is crucial to consider its surface brightness performance during the design phase. The core of the ngVLA, i.e., the baselines in the inner km, will act as a world-leading instrument for unlocking the large-scale cold gas content of the Early Universe. To make optimal use of this capability, it is important to consider the following specifications:

• To maximize the surface-brightness sensitivity of the ngVLA, it is essential to optimize the sensitivity on the shortest (≤ several 100m) baselines.

• To utilize the ngVLA core as a unique stand-alone instrument, it is worth considering sub-array functionality between the inner and outer antennas.

Currently, the Short Baseline Array (SBA) is included as part of the ngVLA reference design (see Selina et al. 2018, this volume p. 15). The SBA aims at recovering emission on the largest spatial scales, and would be analogous to the Atacama Compact Array (a.k.a. Morita Array) of the ALMA observatory. The current reference design has the SBA consisting of 19 antennas of 6 m diameter and using 4 18 m diameter dishes in total-power mode, with interferometric baselines ranging from 11–56 m (Mason et al 2018). It will be essential to investigate to what extent the limited sensitivity of this Short Baseline Array will contribute to efficiently mapping faint, low-surface-brightness CO emission at high redshifts, but this is beyond the scope of this chapter.

Concluding, the core of the ngVLA will reveal the cold molecular medium in the Early Universe to limits and at scales not currently detectable with existing millimeter observatories. As such, it will be the most powerful instrument for investigating the cold baryon cycle that drives the early evolution of galaxies and clusters.

Acknowledgments. We thank Brian Mason for his feedback on the Short Baseline Array. The National Radio Astronomy Observatory is a facility of the National Science Foundation operated under cooperative agreement by Associated Universities, Inc.

References

Carilli, C. L. 2016, Next Generation VLA Memo Series, 12
Carilli, C. L. 2018, Next Generation VLA Memo Series, 41
Cicone, C., Maiolino, R., Gallerani, S., et al. 2015, A&A, 574, A14
da Cunha, E., Groves, B., Walter, F., et al. 2013, ApJ, 766, 13
Dannerbauer, H., Lehnert, M. D., Emonts, B., et al. 2017, A&A, 608, A48
Dekel, A., Birnboim, Y., Engel, G., et al. 2009, Nature, 457, 451
Emonts B. H. C., et al., 2014, MNRAS, 438, 2898
Emonts, B. H. C., Mao, M. Y., Stroe, A., et al. 2015, MNRAS, 451, 1025
Emonts B. H. C., et al., 2016, Science, 354, 1128
Emonts, B. H. C., et al. 2018, MNRAS, 477, L60
Ginolfi, M., et al. 2017, MNRAS, 468, 3468
Gullberg, B., et al. 2016, A&A, 591, 73
Hatch N. A., et al., 2008, MNRAS, 383, 931
Huang, Y. D. (., Morata, O., Koch, P. M., et al. 2016, Proceeding of the SPIE, 9911, 99111V
Lau, M. W., Prochaska, J. X., & Hennawi, J. F. 2016, ApJS, 226, 25
Mason, B. 2018, Next Generation VLA Memo Series, 43.
McMullin, J. P., et al. 2007, ASP Conference Series, 376, 127
Miley G. K., et al., 2006, ApJL, 650, L29
Narayanan, D., Krumholz, M. R., Ostriker, E. C., & Hernquist, L. 2012, MNRAS, 421, 3127
Narayanan, D., & Krumholz, M. R. 2014, MNRAS, 442, 1411
Narayanan D., et al., 2015, Nature, 525, 496
Papadopoulos, et al. 2000, ApJ, 528, 626
Papadopoulos P., Thi W., Viti S., 2004, MNRAS, 351, 147
Prochaska J. X., Lau M., Hennawi J., 2014, ApJ, 796, 140
Selina, R. J., Murphy, E. J., McKinnon, M. et al. 2018, Science with a Next-Generation Very Large Array, ASP Monograph 7, ed. E. Murphy (San Francisco, CA: ASP), 15
Zhang, Z.-Y., Papadopoulos, P. P., Ivison, R. J., et al. 2016, Royal Society Open Science, 3, 160025

Science with a Next-Generation Very Large Array
ASP Conference Series, Monograph 7
Eric J. Murphy, ed.

Young Radio AGN in the ngVLA Era

Pallavi Patil,[1,2] Kristina Nyland,[1] Jeremy J. Harwood,[3] Amy Kimball,[4] and Dipanjan Mukherjee[5]

[1]*National Radio Astronomy Observatory, Charlottesville, VA 22903;* `knyland@nrao.edu` ; `ppatil@nrao.edu`

[2]*University of Virginia, Department of Astronomy, Charlottesville, VA 22903, USA;* `pp3uq@virginia.edu`

[3]*Centre for Astrophysics Research, School of Physics, Astronomy and Mathematics, University of Hertfordshire, College Lane, Hatfield AL10 9AB, UK;* `jeremy.harwood@physics.org`

[4]*National Radio Astronomy Observatory, Socorro, NM 87801;* `akimball@nrao.edu`

[5]*Dipartimento di Fisica Generale, Universita degli Studi di Torino, Via Pietro Giuria 1, 10125 Torino, Italy;* `dipanjan.mukherjee@unito.it`

Abstract. Most massive galaxies are now thought to go through an Active Galactic Nucleus (AGN) phase one or more times. Yet, the cause of triggering and the variations in the intrinsic and observed properties of AGN population are still poorly understood. Young, compact radio sources associated with accreting supermassive black holes (SMBHs) represent an important phase in the life cycles of jetted AGN for understanding AGN triggering and duty cycles. The superb sensitivity and resolution of the ngVLA, coupled with its broad frequency coverage, will provide exciting new insights into our understanding of the life cycles of radio AGN and their impact on galaxy evolution. The high spatial resolution of the ngVLA will enable resolved mapping of young radio AGN on sub-kiloparsec scales over a wide range of redshifts. With broad continuum coverage from 1.2 to 116 GHz, the ngVLA will excel at estimating ages of sources as old as 30 – 40 Myr at $z \sim 1$. In combination with lower-frequency ($\nu < 1$ GHz) instruments such as ngLOBO and the Square Kilometre Array, the ngVLA will robustly characterize the spectral energy distributions of young radio AGN.

1. Introduction

Young, compact radio sources associated with accreting massive black holes (MBHs) represent a key phase in the life cycles of jetted active galactic nuclei (AGN). These objects have recently (re)ignited their central engines within the past 10^2 to 10^4 years, resulting in sub-galactic jet extents of ~10 pc to a few kpc. One of the unique characteristics of young radio AGN is their convex radio spectral energy distributions (SEDs) and characteristic spectral turnovers that peak at frequencies in the range of hundreds of MHz to a few GHz. Well-studied classes of young radio AGN (O'Dea 1998; Orienti 2016, and references therein) include Compact Steep-Spectrum (CSS) sources,

Gigahertz-peaked Spectrum (GPS) sources, and High-frequency Peakers (HFP). These young radio AGN may eventually evolve to become large and powerful FR I/FR II galaxies (Fanaroff & Riley 1974; Snellen et al. 2000; Kunert-Bajraszewska et al. 2010).

An open question regarding young radio AGN is the level of their energetic impact on the interstellar medium (ISM) of the host galaxy. Since their radio jets are fully contained within the host galaxy, spatially-resolved observations are necessary. A growing number of studies have reported detections of multiphase outflow signatures associated with young radio AGN engaged in jet-ISM feedback (e.g., Chandola et al. 2011; Holt et al. 2011; Morganti et al. 2013), but the importance of this feedback in the context of galaxy evolution remains unclear. The ngVLA will address this issue through both continuum and spectral line observations. Broad-band continuum measurements of the turnover frequency of the radio SEDs of young radio sources will also directly constrain the density distribution of the ISM for sources at low redshift (§2.3; Bicknell et al. 1997, 2017; Jeyakumar 2016). Follow-up spectral line observations probing the conditions of the ISM in the vicinity of the jets of the young radio AGN will further constrain the kinematics of the gas, thus probing the energetic impact of the jet-driven feedback.

Young radio AGN also hold vital clues about radio triggering and duty cycles, which are still poorly understood (Tadhunter 2016). An improved understanding of MBH duty cycles and MBH accretion mechanisms will help explain the impact of MBH growth on galaxy growth and evolution (Shulevski et al. 2015). Molecular gas line or HI absorption diagnostics can establish a link between fueling and different triggering mechanisms (Maccagni et al. 2014). A significant amount of dense gas is seen in young radio AGN (O'Dea 1998), and several studies have found that HI is detected more frequently in GPS and CSS sources compared to other types of radio AGN (Geréb et al. 2014; O'Dea 1998 and references therein). Therefore, the investigation of these newborn radio AGN can give us a direct view of many transient processes and better inform simulations of galaxy formation and evolution.

2. Anticipated Results

2.1. Resolving Young Radio AGN with the ngVLA

In Figure 1, we show examples of young radio AGN candidates with jet extents of 60 to 85 pc. The observations are taken with the Very Long Baseline Array at C-band (Patil et al., in preparation). These radio AGN were drawn from the sample of extremely infrared luminous sources identified in Lonsdale et al. (2015) that are believed to be in a unique evolutionary stage just after the (re)ignition of the radio AGN but while the host galaxy is going through a starburst phase due to a recent merger. The angular resolution, frequency range, and sensitivity offered by the ngVLA are optimal for efficiently resolving the structures of young radio AGN, such as the sources highlighted in Figure 1, and constraining their ages through broadband continuum observations (§2.2).

Simulations of relativistic jets interacting with ISM (Mukherjee et al. 2016; Bicknell et al. 2017; Mukherjee et al. 2018) have shown that jets drive shocks in the ISM via an energy bubble. Furthermore, Mukherjee et al. (2018) have found that jet powers, inclinations, and densities affect AGN-driven outflows, thus motivating the need for high-resolution radio imaging to obtain spatially-resolved structural, spectral, and polarimetric information on the radio jets. In addition, direct measurements of flux

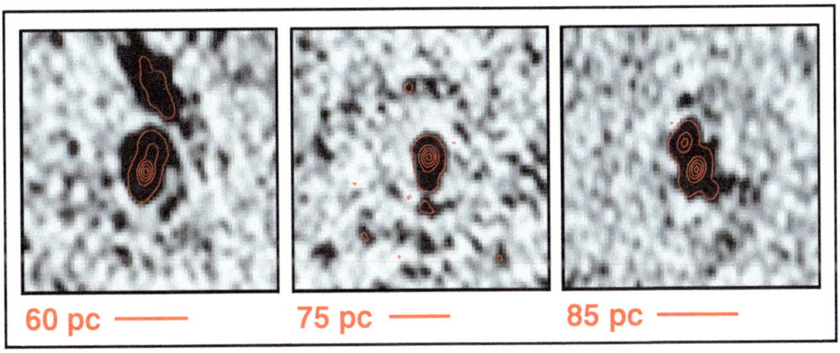

Figure 1. Example Very Long Baseline Array continuum images (Nyland et al. 2018; Patil et al. 2019, in preparation) of young, compact radio AGN with extents of tens of pc. These sources are drawn from the sample presented in Lonsdale et al. (2015), which selects heavily obscured, powerful AGN embedded in luminous starbursting galaxies at intermediate redshifts with compact radio emission based on their properties from the Wide-Field Infrared Survey Explorer (Wright et al. 2010) and the NRAO–VLA Sky Survey (Condon et al. 1998). The high collecting area and angular resolution of the ngVLA will facilitate deep, spatially-resolved surveys of compact radio jets with comparable extents in a wide range of systems, providing new insights into the energetic impact of young radio jets under a variety of host galaxy conditions and properties.

densities and source sizes provide estimates of the energy density and pressure in the synchrotron plasma, which can then constrain ISM densities and the rate of the advancing shock front into the ambient medium. Limits on Faraday rotation measures obtained through polarimetry help constrain models of the nuclear environment.

The planned longest baselines of the ngVLA of up to ~1000 km are required to provide the subarcsecond resolution needed to study the population of young radio AGN. The ngVLA will be able to easily detect radio AGN with sub-kiloparsec structures over a wide redshift range (Nyland et al. 2018). The significant increase in point source sensitivity compared to Jansky Very Large Array (VLA) will enable the efficient detection of radio emission in young radio AGN. An interesting point of comparison is the Legacy eMERLIN Multi-band Imaging of Nearby Galaxies survey (LeMMINGs; Baldi et al. 2018) that aims to probe the radio properties of nearby AGN at sub-arcsecond spatial resolution. In the same on-source integration time used in the LeMMINGs survey (48 minutes), the ngVLA would be able to reach a depth of about 100 times deeper (~0.7 μJy beam^{-1}) in its Band 1 (1.2 – 3.5 GHz).

2.2. Radio Spectral Ages

For a fixed magnetic field, if the initial energy distribution of a population of electrons follows a power law, given by $N(E) = N_0 E^{-\delta}$, the energy losses scale as $\tau = \frac{E}{dE/dt} \propto 1/E \propto 1/\nu^2$. This results in a preferential cooling of higher energy electrons. Assuming no other mode of particle acceleration, the resulting spectrum becomes increasingly curved over time. This behavior can be used to determine the characteristic age of a radio source (e.g., Myers & Spangler 1985; Harwood et al. 2013).

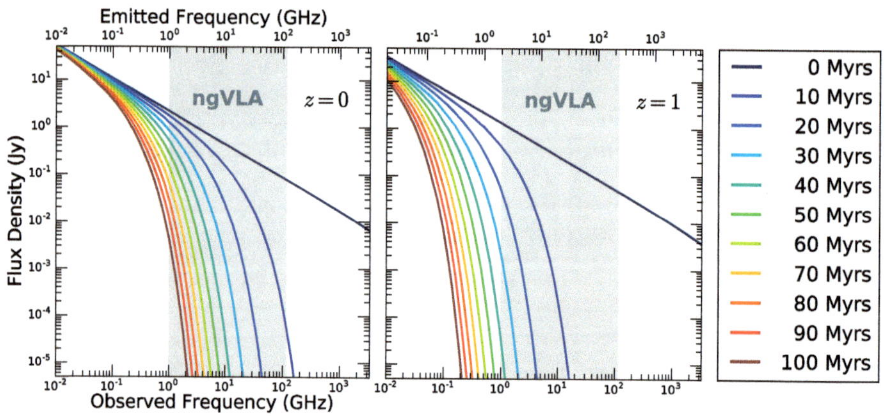

Figure 2. Example of JP model (Jaffe & Perola 1973) spectral ages calculated using the BRATS software (Harwood et al. 2013) demonstrating the need for ngVLA observations spanning a wide range of frequencies. The left and center panels correspond to redshifts of 0 and 1, respectively. The flux density values shown on the y-axis have been arbitrarily scaled. Because of its advantages of the wide frequency range and angular resolution compared to the SKA, the ngVLA will uniquely excel in studies of low-redshift radio AGN that are young, or higher-redshift AGN that are embedded in dense environments. Adapted from Nyland et al. (2018).

In Figure 2, we show an example of spectral ages calculated using the BRATS software[1] (Harwood et al. 2013) for an arbitrary jetted radio AGN at $z = 0$ and $z = 1$. As illustrated by the spectral age curves in the figure, the ngVLA will excel in studies of radio AGN spanning a broad range of ages at low redshift, as well as young or embedded radio AGN at higher redshifts. The results from the Australia Telescope 20 GHz (AT20G) survey (Murphy et al. 2010) indicate that continuum measurements in the tens of GHz range are needed to model the radio spectral energy distributions adequately (Sadler et al. 2006, 2008). These observations are particularly crucial for modeling the ages of young, low-redshift sources less than 10 Myrs old. Lower frequency radio continuum data in the MHz range are important for constraining the ages of high-z sources; however, the inclusion of the lowest-frequency ngVLA bands down to ~1 GHz would provide sufficient frequency coverage for measuring the ages of sources as old as 30 – 40 Myrs at $z \sim 1$.

2.3. Characterizing Spectral Turnover

CSS, GPS, and HFP sources show an inverse correlation between the spectral peak frequency and their linear sizes (O'Dea 1998; Orienti & Dallacasa 2014). The spectra may be peaked due to the absorption by either a relativistic plasma or an external screen of ionized gas. Simulations by Bicknell et al. (2017) have shown that the location of each source on the turnover-linear size relation is dependent on the warm ISM density, suggestive of an evolutionary link between all three classes of young radio AGN. However,

---

[1] http://www.askanastronomer.co.uk/brats/

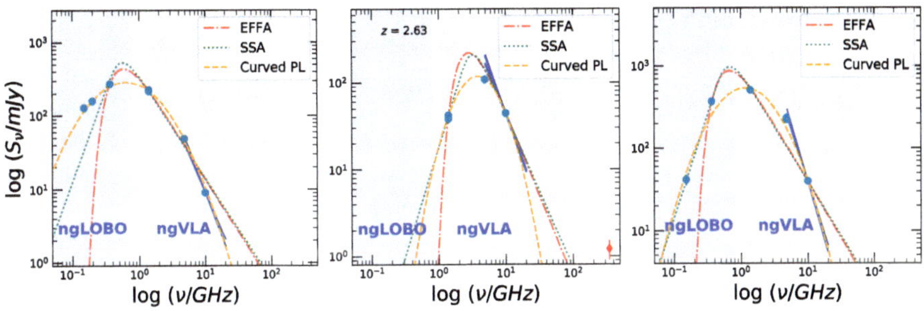

Figure 3. An example of radio SED fitting in GPS/HFP sources. These sources are taken from a parent sample of heavily obscured and hyper-luminous quasars with compact radio emission (Lonsdale et al. 2015). The data points at 10 GHz are from a high-resolution VLA imaging study of young radio AGN (Patil et al. 2019, in preparation). The blue line segments show the in-band spectral indices of the VLA data. The remaining data points are from published radio surveys. The source in the second panel has a known redshift, and the red diamond is the ALMA Band 7 observation taken in cycle 0. The curves are the best-fit solutions of different radio SED models. A red dashed dot line is for the external free-free absorption model (EFFA), a dotted green line corresponds to the synchrotron self-absorption model (SSA), and the yellow dashed line is for the generic curved power law model. The ngVLA and ngLOBO will map the radio SED in these sources and will be able to distinguish between different theoretical models.

the details of the absorption model remain unclear (Callingham et al. 2015). Therefore, high-resolution observations sampling both sides of the peak/turnover frequency are essential for identifying the absorption process.

As shown in Figure 3, the inclusion of low-frequency observations around 1 GHz (or lower) would enhance studies of the turnover in the optically-thick radio SEDs of young radio AGN. This analysis would provide new insights into the dominant physical mechanism responsible for the turnover (e.g., synchrotron self-absorption vs. free-free absorption; Kellermann 1966) in young radio AGN over a larger parameter space. The next generation LOw Band Observatory (ngLOBO; Taylor et al. 2017), a proposed commensal facility that would leverage ngVLA infrastructure, would extend the frequency coverage to below 1 GHz. NgLOBO would enable more robust radio SED and spectral aging model studies with the ngVLA, particularly for slightly older (> 10 Myr) and more distant ($z > 1$) sources.

3. Uniqueness to ngVLA Capabilities

The broad frequency range (1.2–116 GHz) of the ngVLA will facilitate surveys of the radio spectral ages of *young* AGN residing in gas-rich galaxies at low and intermediate redshifts. We emphasize that unlike other next-generation radio facilities that will operate primarily at lower frequencies (e.g., the SKA), the ngVLA will be uniquely well-suited for measuring the ages of young radio AGN potentially engaged in jet-ISM feedback.

4. Multiwavelength Synergy

Young radio AGN are an important class of objects for studying the interplay between galaxies and their central engines. Upcoming or current facilities in the radio regime, as well as at other wavelengths, will be essential for probing the morphologies, evolutionary stages, and energetics of young radio AGN, and will ultimately provide new insights into radio jet physics and its impact on galaxy growth and evolution.

4.1. SKA

The ngVLA and the Square Kilometre Array (SKA) will both allow broadband studies of radio SEDs over their respective wide frequency ranges, thus providing the temporal information from spectral aging models necessary for interpreting the evolutionary stage of the source. The SKA will be able to detect emission associated with fading radio lobes, tracing the regime of radio AGN with older spectral ages ("AGN archaeology"; Morganti 2017). In Figure 2, we show that the 1.2–116 GHz observing range of the ngVLA is advantageous for studies of low-redshift radio AGN, or intermediate redshift sources that are very young. Such young sources are often embedded in dense environments, and are known to drive multi-phase gas outflows (e.g., Holt et al. 2008; Geréb et al. 2015). Broadband radio continuum surveys with both the ngVLA and SKA will therefore be needed to construct a complete picture of the life cycles of radio AGN and their connection to galaxy evolution.

4.2. ALMA

The frequency range of the ngVLA samples non-thermal synchrotron emission associated with radio AGN and recent star formation as well as thermal emission from current star-forming regions. The higher frequency bands of the Atacama Large Millimeter/Submillimeter Array (ALMA) are sensitive to the dust continuum, which can be used to help determine robust star formation rates. In terms of spectral line emission, the ngVLA will be able to detect the bulk of the molecular ISM through the low-J transitions of the CO molecule, while ALMA will be sensitive to higher-J CO transitions as well as a wide variety of additional molecular species and transitions that will characterize the ISM conditions. Thus, the combination of the ngVLA and ALMA will be important for constraining the impact of young radio AGN on the ISM and star formation on galactic scales.

4.3. Optical and Infrared Telescopes:

Next-generation optical and infrared telescopes such as the *James Webb* Space Telescope *(JWST)*, Thirty Meter Telescope (TMT), Giant Magellan Telescope (GMT), and the Wide-Field Infrared Survey Telescope *(WFIRST)* will provide a wealth of complementary information about the physical conditions of young radio AGN and the properties of their hosts. This includes resolving host morphologies, studying the kinematics of the warm ionized gas via spectroscopy, delineating star-formation and AGN via mid-infrared imaging and SED analyses, and tracing AGN-driven outflows and their excitation mechanisms.

References

Baldi, R. D., Williams, D. R. A., McHardy, I. M., et al. 2018, MNRAS, 476, 3478

Bicknell, G. V., Dopita, M. A., & O'Dea, C. P. O. 1997, ApJ, 485, 112
Bicknell, G. V., Mukherjee, D. D., Wagner, A. Y., & Sutherland, R. S. 2017, MNRAS, submitted
Callingham, J. R., Gaensler, B. M., Ekers, R. D., et al. 2015, ApJ, 809, 168
Chandola, Y., Sirothia, S. K., & Saikia, D. J. 2011, MNRAS, 418, 1787
Condon, J. J., Cotton, W. D., Greisen, E. W., et al. 1998, AJ, 115, 1693
Fanaroff, B. L., & Riley, J. M. 1974, MNRAS, 167, 31P
Geréb, K., Maccagni, F. M., Morganti, R., & Oosterloo, T. A. 2015, A&A, 575, A44
Geréb, K., Morganti, R., & Oosterloo, T. A. 2014, A&A, 569, A35
Harwood, J. J., Hardcastle, M. J., Croston, J. H., & Goodger, J. L. 2013, MNRAS, 435, 3353
Holt, J., Tadhunter, C. N., & Morganti, R. 2008, MNRAS, 387, 639
Holt, J., Tadhunter, C. N., Morganti, R., & Emonts, B. H. C. 2011, MNRAS, 410, 1527
Jaffe, W. J., & Perola, G. C. 1973, A&A, 26, 423
Jeyakumar, S. 2016, MNRAS, 458, 3786
Kellermann, K. I. 1966, Australian Journal of Physics, 19, 195
Kunert-Bajraszewska, M., Gawroński, M. P., Labiano, A., & Siemiginowska, A. 2010, MNRAS, 408, 2261
Lonsdale, C. J., Lacy, M., Kimball, A. E., et al. 2015, ApJ, 813, 45
Maccagni, F. M., Morganti, R., Oosterloo, T. A., & Mahony, E. K. 2014, A&A, 571, A67
Morganti, R. 2017, Nature Astronomy, 1, 596
Morganti, R., Frieswijk, W., Oonk, R. J. B., Oosterloo, T., & Tadhunter, C. 2013, A&A, 552, L4
Mukherjee, D., Bicknell, G. V., Sutherland, R., & Wagner, A. 2016, MNRAS, 461, 967
Mukherjee, D., Bicknell, G. V., Wagner, A. Y., Sutherland, R. S., & Silk, J. 2018, ArXiv e-prints, arXiv:1803.08305
Murphy, T., Sadler, E. M., Ekers, R. D., et al. 2010, MNRAS, 402, 2403
Myers, S. T., & Spangler, S. R. 1985, ApJ, 291, 52
Nyland, K., Harwood, J. J., Mukherjee, D., et al. 2018, ApJ, 859, 23
O'Dea, C. P. 1998, PASP, 110, 493
Orienti, M. 2016, Astronomische Nachrichten, 337, 9
Orienti, M., & Dallacasa, D. 2014, MNRAS, 438, 463
Sadler, E. M., Ricci, R., Ekers, R. D., et al. 2008, MNRAS, 385, 1656
—. 2006, MNRAS, 371, 898
Shulevski, A., Morganti, R., Barthel, P. D., et al. 2015, A&A, 583, A89
Snellen, I. A. G., Schilizzi, R. T., Miley, G. K., et al. 2000, MNRAS, 319, 445
Tadhunter, C. 2016, A&A Rev., 24, 10
Taylor, G., Dowell, J., Malins, J., et al. 2017, ArXiv e-prints, arXiv:1708.00090
Wright, E. L., Eisenhardt, P. R. M., Mainzer, A. K., et al. 2010, AJ, 140, 1868

Science with a Next-Generation Very Large Array
ASP Conference Series, Monograph 7
Eric J. Murphy, ed.

Probing Obscured MBH Accretion and Growth since Cosmic Dawn

Wiphu Rujopakarn,[1,2,3] Kristina Nyland,[4] Amy E. Kimball,[5]
and Isabella Prandoni[6]

[1]*Department of Physics, Faculty of Science, Chulalongkorn University, 254 Phayathai Road, Pathumwan, Bangkok 10330, Thailand;* `wiphu.r@chula.ac.th`

[2]*National Astronomical Research Institute of Thailand (Public Organization), Don Kaeo, Mae Rim, Chiang Mai 50180, Thailand*

[3]*Kavli Institute for the Physics and Mathematics of the universe (WPI), The University of Tokyo Institutes for Advanced Study, The University of Tokyo, Kashiwa, Chiba 277-8583, Japan*

[4]*National Radio Astronomy Observatory, Charlottesville, VA 22903, USA;* `knyland@nrao.edu`

[5]*National Radio Astronomy Observatory, Socorro, NM 87801, USA;* `akimball@nrao.edu`

[6]*INAF-Istituto di Radioastronomia, via P. Gobetti 101, 40129, Bologna, Italy;* `prandoni@ira.inaf.it`

Abstract. Most of the stars today reside in galactic spheroids, whose properties are tightly tied to the supermassive black holes (MBHs) at their centers, implying that the accretion activity onto MBHs leaves a lasting imprint on the evolution of their host galaxies. Despite the importance of this so-called MBH-galaxy co-evolution, the physical mechanisms responsible for driving this relationship – such as the dominant mode of energetic feedback from active galactic nuclei (AGN) – remain a poorly understand aspect of galaxy assembly. A key challenge for identifying and characterizing AGN during the peak epoch of galaxy assembly and beyond is the presence of large columns of gas and dust, which fuels the growth of their MBHs but effectively obscures them from view in optical and X-ray studies. The high sensitivity of the ngVLA will capture emission from AGN in an extinction-free manner out to $z \sim 6$ and beyond. At lower-redshifts ($z \sim 2$), the high angular resolution of the ngVLA will enable spatially-resolved studies capable of localizing the sites of actively growing MBHs within their host galaxies during the peak epoch of cosmic assembly.

1. Introduction

The convergence of theory and observation in galaxy evolution is a major milestone in astrophysics. With appropriate sets of parameters, models of galaxy formation can reproduce, for instance, the luminosity and mass distributions of today's galaxies and the cosmic star formation rate (SFR) history since the epoch of reionization. However, many of the most fundamental processes in the model are not well understood, especially down to galactic scales, where the current frontier questions in galaxy evolution lie. Central among these issues is the empirical link between galaxy assembly and supermassive black hole (MBH) accretion and growth. Advancing our understanding of this symbiotic relationship requires a complete census of the first generation of AGN at cosmic dawn as well as a spatially-resolved study of individual galaxies at the peak of their assembly, $1 < z < 3$, to map their star formation (SF) and MBH accretion activities (i.e., active galactic nuclei; AGN). Fundamental breakthroughs in these areas thus require sub-arcsecond resolution and an extinction-independent tracer of *both* AGN and SF.

Centimetric radio continuum observations uniquely meet these observational demands. Deep, high-resolution continuum surveys with current and next-generation facilities will therefore play an essential role in advancing the current galaxy formation and evolution paradigm in the decades to come. In addition to tracing AGN and SF activity in an extinction-free manner, centimetric radio continuum observations of AGN and SF emission are also relatively stable in time. While other AGN indicators such as hard X-ray emission or emission lines can fluctuate rapidly over timescales of days to months (Hickox et al. 2014), giving only an instantaneous indication of the level of activity, radio emission has a synchrotron timescale of $10^6 - 10^7$ years (Beck & Krause 2005) to serve as a beacon localizing the AGN. Obviously, a prerequisite to utilizing these characteristics is to confirm the AGN origin of the radio emission, which, by nature, traces both AGN and SF activities. Thanks to the Atacama Large Millimeter/submillimeter Array (ALMA), this is now possible out to $z \sim 10$ with the help of the negative K-correction. As a result, we are reaching the start of an era when centimetric radio continuum can be utilized to the fullest extent to study MBH growth.

2. Current Observational Prospects

2.1. Identifying Obscured AGN at $z \gtrsim 6$

Luminous AGN have been found out to $z = 7.5$ with $M_{\rm BH} \sim 10^9\ M_\odot$ (Bañados et al. 2018). Such extreme objects are rare (only two with spectroscopic confirmation at $z > 7$ so far), yet provide crucial evidence that MBH can grow rapidly up to a massive size within a Gyr after the Big Bang. If the Bañados et al. (2018) quasar is at the bright-end of the quasar luminosity function, it remains a question whether more typical MBHs are growing rapidly elsewhere during this epoch. Typical, actively growing MBHs at $z > 6$ appear to be beyond the present observational capabilities: none of the spectroscopically confirmed $z = 6-8$ galaxies in the *Hubble* Ultra-Deep Field (HUDF) are detected in the individual or stacked *Chandra* 4 and 7 Ms images (Giallongo et al. 2015; Luo et al. 2017). Treister et al. (2013) postulate that these galaxies may not contain MBH, that MBH are not growing despite their large sSFR, or that MBHs are growing elsewhere in lower-mass or obscured galaxies not identified by the current *Hubble* and *Chandra* observations. This missing typical population of AGN could have important implications on the cosmic accretion history, feedback on the intergalactic medium, reionization, and to our understanding of how MBHs and galactic bulges coevolved.

If the typical sites of MBH growth in the early universe are indeed in low-mass galaxies, the current optical, IR, and X-ray indicators will have difficulties identifying them. They will be too faint in the optical/near-IR, contain too little dust to power mid-IR emission, and emit at too low of a luminosity for direct X-ray detection at high redshift (also not identified in large numbers in other bands to enable X-ray stacking).

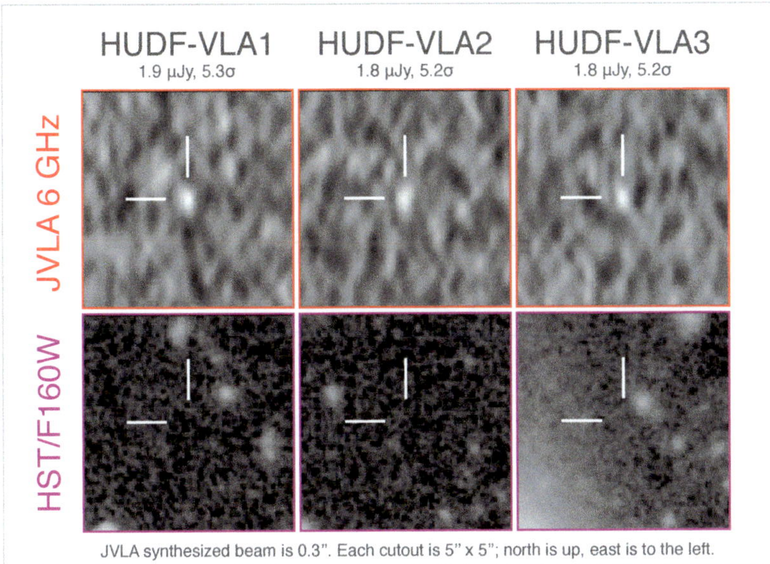

Figure 1. The $\geqslant 5\sigma$ sources in the HUDF JVLA 6 GHz survey with absolutely no counterparts in any other bands (*Hubble* 1.6 μm shown here). An object has to be at $z \gtrsim 5$ in order to evade detection in ultra-deep *Hubble, Spitzer*, and ground-based imaging sensitivity of the HUDF. There are no sources in the negative map at this significance — it is extremely unlikely that these are noise peaks.

On the other hand, early MBHs could be accreting in a different mode that confers on them the observed rapid growth (e.g., Volonteri et al. 2015), but that the radiation is trapped (Coughlin & Begelman 2014) and conceals them from current survey approaches. In either case, centimetric radio observations have the potential to identify them because of the symbiotic nature of the accretion onto MBH and the resulting relativistic jets that are a signpost of AGN in the radio (e.g., Begelman et al. 1984). Radio observations are also extinction-free and have recently become sensitive enough to find typical AGN at $z \gg 4$ with the advent of the Jansky Very Large Array (VLA). This is therefore a potentially groundbreaking means to identify typical AGN at high z.

Recently, Rujopakarn et al. (2016) has conducted a sensitive single-pointing survey of the HUDF using the VLA at 6 GHz (0.3 μJy beam^{-1} rms). They have found that 99% of radio sources detected above 5σ have optical or near-IR counterparts. The remaining 1% – three sources within the central 100$''$ of the primary beam – are of great interest (Figure 1). They are extremely unlikely to be noise peaks because the noise distribution from the source-free area is well-fitted by a Gaussian and there are no sources in the inverted map above 5σ in the same area. They are also persistent over the two year period to acquire 177 hours of VLA observations, effectively ruling out the possibilities of transient or non-astrophysical artifacts. To evade detection in a field with such a deep suite of ground and space-based imaging, reaching 30.5 mag (5σ) at 1 μm, an object must either be at moderate redshifts, $z \sim 1-3$, but extremely dust-obscured, which is ruled out by the lack of detections with *Spitzer* and ALMA (the

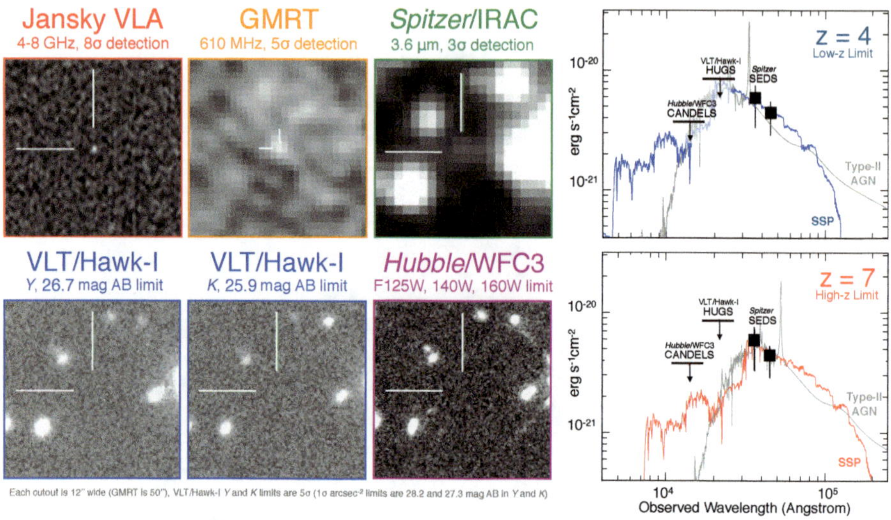

Figure 2. A high-z AGN candidate detected in the pilot survey of Rujopakarn et al. (2016) that is only detected in the VLA and an ultra-deep follow-up imaging with the GMRT and *Spitzer*/IRAC, but not any other bands; the most probable redshift is $z \sim 4 - 7$. Follow-up spectroscopy with the ESO VLT/X-Shooter with 6-hours integration finds no Lyα, typical for $z > 6$ candidates, highlighting the necessity for next-generation facilities (e.g., *JWST*) to study these populations.

1.3 mm deep field; Dunlop et al. 2017); or have low M_* (e.g., $10^7 - 10^8\ M_\odot$, which will be an interesting class of AGN in itself); or they must be at higher z (i.e., $z \gtrsim 5 - 6$).

Characterization of these high-z radio AGN candidates has proven to be extremely challenging for current facilities. An example is a candidate shown in Figure 2 from the pilot survey of Rujopakarn et al. (2016). In this case, a faint *Spitzer*/IRAC counterpart is found, but the non-detection in deep *Hubble* and ground-based near-IR images effectively ruled out hosts with $M_* \gtrsim 10^8\ M_\odot$ at $z \lesssim 4$. This radio source was followed-up with the X-Shooter instrument (simultaneous UV, optical, and near-IR spectroscopy) on the ESO Very Large Telescope for 6 hours to search for a Lyα line to confirm the redshift. No line was detected, suggesting that this could be a case of the low Lyα detection rate at $z \gtrsim 6.5$, where the reionization is not yet completed (Schenker et al. 2014). Their characterization is beyond the capability of current optical/IR observatories and will require next-generation facilities (e.g., *JWST*).

2.2. MBH Localization at $z \sim 2$

Despite the multiple lines of evidence showing a link between galaxy assembly and MBH growth, there is no consensus on the causality of this relationship. Direct links on a global scale between a MBH accretion rate (BHAR) and the star formation rate (SFR) of its host are elusive. Although the two processes appear to be correlated, this result may only reflect that both depend on the stellar mass of the host galaxy (Xu et al. 2015; Stanley et al. 2017). For local galaxies, there is evidence for *localized* enhanced SF close to an AGN, related to the BHAR (e.g., Diamond-Stanic & Rieke

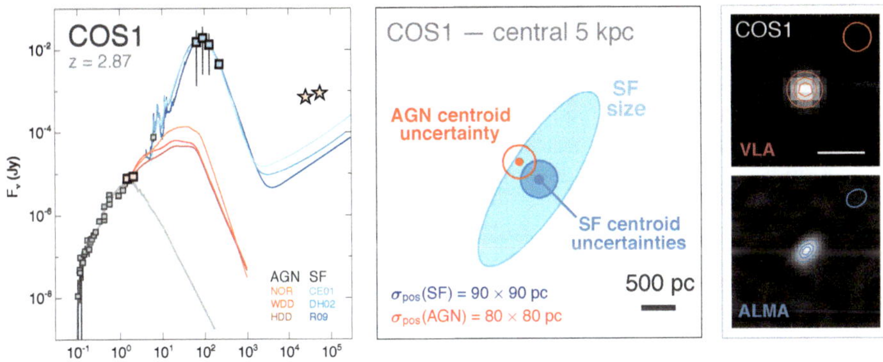

Figure 3. A new technique to accurately pinpoint the sites of intense SF and MBH growth at $z \sim 2$. (*Left:*) strong radio excess over the level associated with SFR is AGN-dominated, whereas the submillimeter emission captured by ALMA is virtually free of the AGN emission and hence trace the distribution of SF. (*Middle:*) Using this technique, Rujopakarn et al. (2018) found AGN to occur within the compact regions of intense SF. (*Right:*) Accurate localization of SF and AGN are enabled by sub-arcsec ALMA and VLA observations. Postage stamps are $4'' \times 4''$; the synthesized beam size is shown in the respective image.

2012). Improving our understanding of this link at higher redshifts requires higher resolution imaging to determine whether the sites of MBHs are also associated with the most vigorous SF. Yet, such images have until very recently been impossible at $z \sim 2$, at the peak of both SFR and BHAR and thus potentially the formative era for the current relation.

The VLA provides sub-arcsecond imaging that can penetrate dust to trace synchrotron emission associated with SF as well as emission from any AGN core and jets. The submillimeter continuum at, e.g., 870 μm traces the cold dust emission associated with SF. This emission is virtually free of AGN contribution because the IR spectra of AGN tori plummet rapidly longward of 40 μm (Lyu & Rieke 2017). Likewise, extrapolating the radio slopes from ~1 GHz to the ALMA band shows that the non-thermal emission from all but the most radio-loud AGN is likely to be 2 – 3 orders of magnitude fainter than the thermal dust emission of their hosts. Any galaxy where the radio emission is enhanced to well above the level implied by the far-IR/radio correlation for SF galaxies presents a robust radio AGN signature (Figure 3). In these galaxies with strong radio enhancement, the VLA traces AGN and ALMA traces nearly pure SF, so sub-arcsecond VLA and ALMA images can localize — at sub-kpc resolution — the sites of SF relative to the MBH in active galaxies.

Central to this technique is that (1) at $z \sim 2$, AGN commonly have strong radio sources, unresolved at sub-arcsec resolution (cf. parsec-scale cores and jets; Zensus 1997); (2) at GHz rest frequencies, the central radio emission within a sub-arcsec beam accurately pinpoints the AGN core (Nagar et al. 2002; Pushkarev & Kovalev 2012); (3) the uncertainty of the centroid position, $\sigma_{\rm pos}$, of a radio or submillimeter source is $\sigma_{\rm pos} \approx \theta_{\rm beam}/(2 \times {\rm S/N})$, where $\theta_{\rm beam}$ is the synthesized beam size and S/N is the peak signal-to-noise ratio (Condon 1997). For example, a 10σ point source observed with a 0.″15 beam can be localized to about 8 milliarcsecond (mas); and, lastly (4) the astrometric references of the VLA and ALMA are tied to their respective phase calibrator positions, which are in turn tied to the International Coordinate Reference Frame (ICRF) to within $\lesssim 10$ mas. That is, the astrometry of VLA and ALMA are tied to the common reference frame and can be directly compared.

Rujopakarn et al. (2018) demonstrated this technique in three objects with $\geqslant 5\times$ radio excess over the level predicted by the far-IR/radio correlation, and with the ALMA observations resolving the sub/millimeter emission. In all three galaxies, the AGN are located within the $\lesssim 1$ kpc regions of gas-rich, heavily obscured, intense nuclear SF (Figure 3). If the star-formation and AGN-driven outflows that may be present (e.g., Mullaney et al. 2013) do not completely disrupt the cold gas supply, and both types of activity proceed for the duration of their host's typical stellar mass doubling time of $\simeq 0.2$ Gyr, then the newly formed stellar mass within the central kpc will be of order $10^{11}\ M_{\odot}$. Likewise, if the MBH is allowed to accrete at a rate predicted by the correlation between SFR and the average BHAR relation (e.g., Chen et al. 2013) for the same duration, the accreted MBH mass will be $10^{6.7} - 10^{7.8}\ M_{\odot}$, similar to those found in local massive galactic bulges. This is consistent with *in situ* galactic bulge and MBH growth, and may represent the dominant process regulating the bulge–MBH relationship through which all massive galaxies may pass.

3. Anticipated Results with the ngVLA

3.1. An All-Sky Survey with the ngVLA

Although the VLA has proven to be an excellent tool for the detection of high-redshift AGN, such observations are time consuming, particularly in the case of all sky surveys. A survey with the ngVLA carried out in a similar manner to the VLA Sky Survey (VLASS) would offer improvements in efficiency due to the wider field of view of the 18 m dishes and higher collecting area compared to the VLA. An all-sky survey with the same total observing time as VLASS ($\approx$6000 hours) could be $\approx$10 times deeper (7 μJy beam^{-1}), thereby capturing high-redshift AGN populations only detectable in the radio. Such a survey would be expected to identify $\sim 10^7$ AGN at sub-arcsecond angular resolution (Nyland et al. 2018).

3.2. Ultra-deep ngVLA Square Degree Survey

The AGN localization study with the VLA described in Rujopakarn et al. (2018) offered an important demonstration of the need for both high resolution and depth for pinpointing the sites of actively growing MBHs. However, the Rujopakarn et al. (2018) study covered only a single VLA pointing at C-band, and required 177 hours of integration time. The ngVLA will allow this technique to be applied to much larger sky areas and offer an even higher degree of positional accuracy. The milliarcsecond resolution of ngVLA will push the localization accuracy of the AGN to the sub-milliarcsecond regime using the technique described in §2.2. Most importantly, it will resolve the morphologies of these sub-kpc jets at $z \sim 2$ to understand their role in regulating the assembly of the bulge (Nyland et al. 2018). Because there are no local counterparts to gas-rich bulge-forming galaxies, the ngVLA will be the only avenue for providing theorists and simulators with constraints on the nature of typical AGN-driven feedback at the peak of galaxy assembly.

An ngVLA 8 GHz survey over 1 deg^2 with a 1σ depth of 300 nJy $beam^{-1}$ and an angular resolution of $0\farcs1$ would provide an ideal combination of area, depth, and resolution to robustly pinpoint the relative locations of galactic bulge and MBH growth. Extrapolating from the current ngVLA reference design, such a survey would require $\approx$ 200 hours. We note that this survey could be conducted as a high-resolution component to that described in Barger et al. (2018, this volume p. 551), and would also provide as an added bonus information on extinction-free morphologies of SF.

References

Bañados, E., Venemans, B. P., Mazzucchelli, C., et al. 2018, Nat, 553, 473

Barger, A. J., Kohno, K., Murphy, E. J., Sargent, M. T., & Condon, J. J. 2018, Science with a Next-Generation Very Large Array, ASP Monograph 7, ed. E. J. Murphy (San Francisco, CA: ASP), 551

Beck, R., & Krause, M. 2005, Astronomische Nachrichten, 326, 414

Begelman, M. C., Blandford, R. D., & Rees, M. J. 1984, Reviews of Modern Physics, 56, 255

Chen, C.-T. J., Hickox, R. C., Alberts, S., et al. 2013, ApJ, 773, 3

Condon, J. J. 1997, PASP, 109, 166

Coughlin, E. R., & Begelman, M. C. 2014, ApJ, 781, 82

Diamond-Stanic, A. M., & Rieke, G. H. 2012, ApJ, 746, 168

Dunlop, J. S., McLure, R. J., Biggs, A. D., et al. 2017, MNRAS, 466, 861

Giallongo, E., Grazian, A., Fiore, F., et al. 2015, A&A, 578, A83

Hickox, R. C., Mullaney, J. R., Alexander, D. M., et al. 2014, ApJ, 782, 9

Luo, B., Brandt, W. N., Xue, Y. Q., et al. 2017, ApJS, 228, 2

Lyu, J., & Rieke, G. H. 2017, ApJ, 841, 76

Mullaney, J. R., Alexander, D. M., Fine, S., et al. 2013, MNRAS, 433, 622

Nagar, N. M., Falcke, H., Wilson, A. S., & Ulvestad, J. S. 2002, A&A, 392, 53

Nyland, K., Harwood, J. J., Mukherjee, D., et al. 2018, ArXiv e-prints, arXiv:1803.02357

Pushkarev, A. B., & Kovalev, Y. Y. 2012, A&A, 544, A34

Rujopakarn, W., Dunlop, J. S., Rieke, G. H., et al. 2016, ApJ, 833, 12

Rujopakarn, W., Nyland, K., Rieke, G. H., et al. 2018, ApJ, 854, L4

Schenker, M. A., Ellis, R. S., Konidaris, N. P., & Stark, D. P. 2014, ApJ, 795, 20

Stanley, F., Alexander, D. M., Harrison, C. M., et al. 2017, MNRAS, 472, 2221

Treister, E., Schawinski, K., Volonteri, M., & Natarajan, P. 2013, ApJ, 778, 130

Volonteri, M., Silk, J., & Dubus, G. 2015, ApJ, 804, 148

Xu, L., Rieke, G. H., Egami, E., et al. 2015, ApJ, 808, 159

Zensus, J. A. 1997, ARA&A, 35, 607

Science with a Next-Generation Very Large Array
ASP Conference Series, Monograph 7
Eric J. Murphy, ed.

From Megaparsecs To Milliparsecs: Galaxy Evolution and Supermassive Black Holes with NANOGrav and the ngVLA

Stephen R. Taylor[1] and Joseph Simon[2]

[1]*Theoretical AstroPhysics Including Relativity (TAPIR) Group, MC 350-17, California Institute of Technology, Pasadena 91125, CA, USA;* `srtaylor@caltech.edu.edu`

[2]*Jet Propulsion Laboratory, California Institute of Technology, 4800 Oak Grove Drive, Pasadena 91109, CA, USA;* `joseph.j.simon@jpl.nasa.gov`

Abstract. The dynamical evolution of supermassive black-hole binary systems is tethered on large scales to the merger rate of massive galaxies, and on small scales to the stellar and gaseous environments of galactic cores. The population of these systems will create an ensemble gravitational-wave signal at nanohertz frequencies that is detectable by Pulsar Timing Arrays (PTA), such as the North American Nanohertz Observatory for Gravitational Waves (NANOGrav). By measuring nanohertz gravitational waves, we will be able to definitely demonstrate that supermassive black-hole binaries reach milliparsec separations, characterize their sub-parsec dynamical evolution, constrain scaling relationships between the bulge and central black-hole mass, and understand how gas inflow at large scales is fed to central galactic regions. The ngVLA will be an essential complement to this search; its goal resolution will allow for dual AGN to be distinguished, allowing for merger-rate estimates to constrain PTA searches. Using its intercontinental VLBI capabilities (i.e., the ngVLA Long Baseline Array), parsec-scale resolution of binary evolution may be possible, indicating whether binaries stall or are driven efficiently to sub-parsec separations. NANOGrav gravitational-wave analysis, in concert with observations by the ngVLA and complementary facilities, will paint a multi-messenger portrait of galaxy evolution and the dynamics of the most massive black-holes in the Universe.

1. Scientific Goals

Within the current paradigm of hierarchical cosmological structure formation, galaxy growth occurs through a continuous process of gas and dark matter accretion, interspersed with major and minor mergers (e.g., Blumenthal et al. 1984). The fact that most massive galaxies are expected to harbor a supermassive black hole (SMBH) (Kormendy & Richstone 1995; Magorrian et al. 1998) has led to the understanding that central SMBHs share a common evolutionary history with their host galaxies. The prevalence of major mergers and the rate of gas accretion, both of which can contribute to the growth of SMBH and galaxy alike, are, as of yet, poorly constrained across cosmic time. One avenue for understanding the relative roles of these mechanisms is the study of post-merger galaxies, which are expected to harbor a "dual" SMBH system composed of the two black holes from each of the merging galaxies. Understanding the dynamical evolution of SMBH binaries in the rich physical environment of post-

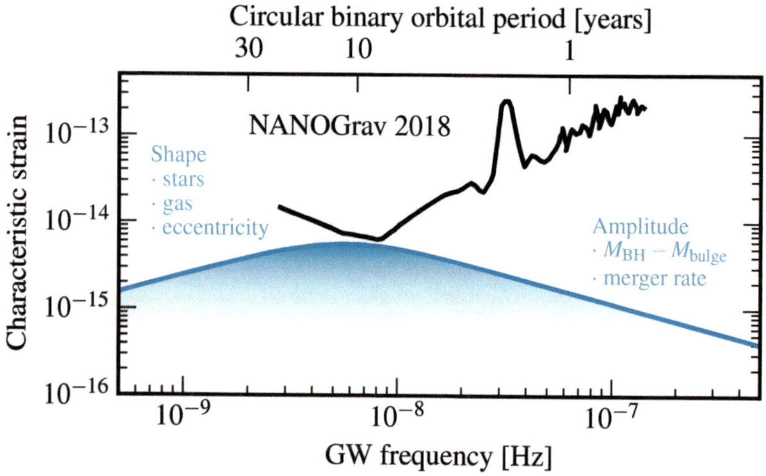

Figure 1. An illustration of the various astrophysical factors affecting the shape and amplitude of the characteristic strain spectrum of GWs emitted by a population of inspiraling supermassive black-hole binaries. This is contrasted against the per-frequency sensitivity of NANOGrav's most recent dataset, consisting of 45 millisecond pulsars with precision timing over an 11-year timescale. The spike at a frequency of 1 yr^{-1} corresponds to a loss in PTA sensitivity due to fitting each pulsar's position.

merger galactic environments is key to solving the puzzles of how SMBHs and galaxies co-evolve, how SMBHs grow over cosmic time, and how (or even if) the final stages of SMBH mergers take place in galaxy centers.

Precision timing of Galactic millisecond pulsars over the last decade has allowed the North American Nanohertz Observatory for Gravitational Waves (NANOGrav, and other collaborations around the globe) to use Arecibo, GBT, and VLA observations to forge a network of kiloparsec-spaced "clocks" that can respond to transiting nanohertz gravitational waves (GWs) through induced Doppler shifts of radio-pulse arrival rates. The dominant source of these nanohertz GWs is likely to be a population of SMBH binaries that are formed naturally in post-merger galaxies. The SMBHs from each galaxy will sink to the center of the common merger remnant through interactions with the galactic gas, stars and dark matter, and eventually become gravitationally bound as binaries (Begelman et al. 1980). Through continued interaction with the environment, the binaries' orbits will tighten and GW emission will become an increasingly dominant factor in their continued evolution. After sufficiently strong environmental interactions the binaries will eventually reach milliparsec orbital separation, at which point their GWs will be emitted in the Pulsar Timing Array (PTA) band at roughly nanohertz frequencies.

NANOGrav seeks to understand the evolution of galaxies and massive black-holes through cosmic time, as well as the dynamics of massive black-hole binaries formed through galaxy mergers. These goals intersect with the ngVLA science case in various ways; the most pertinent being:

- **Detection of a stochastic background of GWs from inspiraling SMBH binary systems**, thus proving that dynamical interactions in galactic centers can mitigate the "final parsec problem" of SMBH evolution, allowing SMBHs to merge in galaxy centers. *The ngVLA 214 antenna main array has a goal resolution of ≈5 mas at 30 GHz, equivalent to* $\sim$ 50 pc *at* $z \sim 1$*, which is sufficient to distinguish dual AGN and convert observed fractions of dual/multiple AGN into SMBH merger rates. These rate estimates can inform future NANOGrav observing strategies.*
- **Spectral characterization of the GW strain signal** which will lead to an inference of the dynamical processes influencing SMBH binaries at sub-parsec scales. The dominant process is expected to be scattering of stars on orbits that intersect the central galactic regions, or interaction with a viscous circumbinary disk. *Intercontinental VLBI capabilities provided by the ngVLA Long Baseline Array (LBA; see Selina et al. 2018, p. 15 in this volume) will achieve ~pc-scale resolution of central galactic regions out to* $z \sim 0.1$. *This will allow a large statistical sample of SMBH binary separations to be collected, thereby informing whether binaries stall at ~pc separations, or are driven efficiently inward.*
- **Constrain the scaling relationships between galactic bulge properties and central SMBHs** (Ferrarese & Merritt 2000; Gebhardt et al. 2000), thereby illuminating the symbiotic evolution between the two, and shedding light on the growth of galaxies/SMBHs through cosmic time. *The ngVLA will be able to resolve the influence radius of* $10^9 M_\odot$ *supermassive black holes out to* $z \sim 0.1$*, drastically improving mass estimates that may lead to revisions of the* $M_{\rm BH} - M_{\rm bulge}$ *relationship.*
- **Understanding how accretion of gas onto SMBHs in the galactic center is related to the inflow of gas at larger scales.** There is some question of whether accretion of material from a circumbinary disk is an important mechanism of SMBH binary hardening, given that the resident massive galaxies should in general be gas poor. But this is highly uncertain. *ngVLA observations of gas reservoirs and inflow at larger scales, as well as gas dynamics in the innermost galaxy regions, will provide crucial constraints on gaseous accretion onto a central SMBH binary.*

2. SMBH Dynamical Evolution in Post-merger Galaxies

The dynamical evolution of SMBH binaries in post-merger galaxies is complex, involving a series of physical processes handing off to the next at ever smaller scales as the SMBHs become gravitationally bound, then tighten to milliparsec separations. We now describe the series of physical processes within post-merger galaxies that drive SMBHs to become gravitationally bound, and then to evolve as a binary into the nanohertz-emitting regime where PTAs are sensitive. At each stage, we comment on how ngVLA goals and observations interface with NANOGrav requirements.

2.1. Dynamical Friction

When massive galaxies merge together, their resident SMBHs sink to the center of the resulting galactic remnant through *dynamical friction* (Merritt & Milosavljević 2005). This is the consequence of many weak and long-range gravitational scattering events within the surrounding stellar, gas, and dark matter distributions, creating a drag that causes the SMBHs to decelerate and transfer energy to the ambient media (Chandrasekhar 1943). For systems with extreme mass ratios ($\lesssim 10^{-2}$) or very low total masses, dynamical friction may not be effective at forming a bound binary from the two SMBH within a Hubble time. In this case the pair might become "stalled" at larger separations, with one of the two SMBH left to wander the galaxy at ~ kpc separations (Dvorkin & Barausse 2017). It is possible that a non-negligible fraction of galaxies may have such wandering SMBH, some of which may be observable as offset-AGN.

- With the ngVLA LBA capabilities, the improvement in angular resolution over existing instruments will allow for direct imaging of systems that may contain offset AGN to extend to much higher redshifts (Casey et al. 2015), thereby potentially gathering a large enough statistical sample to test this stalling hypothesis.

2.2. Stellar Loss-Cone Scattering

At parsec separations, dynamical friction hands off to individual 3-body scattering events between stars in the galactic core and the SMBH binary (Begelman et al. 1980). Stars slingshot off the binary, which can extract orbital energy from the system (Mikkola & Valtonen 1992; Quinlan 1996). However, only stars in centrophilic orbits with very low angular momentum have trajectories which bring them deep enough into the galactic center to interact with the binary. The region of stellar-orbit phase space that is occupied by these types of stars is known as the "loss cone" (LC) (Frank & Rees 1976). Stars which extract energy from the binary in a scattering event tend to be ejected from the core, depleting the LC. As with dynamical friction at larger scales, binaries can also stall here, at parsec scales, due to inefficiency of the LC-refilling —a phenomenon referred to as the "final parsec problem" (Milosavljević & Merritt 2003). Generally, binaries which don't reach sub-parsec separations will be unable to merge via GW emission within a Hubble time (McWilliams et al. 2014).

Various mechanisms have been explored to see whether the LC can be efficiently refilled or populated to ensure continuous hardening of the binary down to milliparsec separations. In general, any form of bulge morphological triaxiality will ensure a continually-refilled LC that can mitigate the final-parsec problem (Vasiliev & Merritt 2013). Isolated galaxies often exhibit triaxiality, and given that the SMBH binaries of interest are the result of galactic mergers, triaxiality and general asymmetries can be expected as a natural post-merger by-product. Also, post-merger galaxies often harbor large, dense molecular clouds that can be channeled into the galactic center, acting as a perturber for the stellar distribution that will refill the LC (Young & Scoville 1991), or even directly hardening the binary (Goicovic et al. 2017).

- Current instruments cannot directly resolve molecular gas clouds within post-merger galaxies. However, the ngVLA will provide the sub-parsec resolution that is necessary to image these gas clouds (Bolatto et al. 2017). This will provide a much-needed causal link explaining how the LC can remain filled in different kinds of post-merger galaxies.

2.3. Viscous Circumbinary Disk Interaction

At centiparsec to milliparsec separations, viscous angular momentum exchange to a gaseous circumbinary disk may play an important role in hardening the binary (Begelman et al. 1980; Kocsis & Sesana 2011). This influence will depend on the details of the dissipative physics of the disk. The simplest model is that of a coplanar prograde α-disk (Shakura & Sunyaev 1973), where the binary torques the disk and carves out a cavity where no gas flows. The individual SMBHs may be surrounded by smaller "mini-disks". More complex disk structures are possible, such as the existence of several physically-distinct regions (Shapiro & Teukolsky 1986). Additionally, high-density disks (equivalently: high-accretion rates) may provide rapid hardening, but are potentially unstable. Furthermore, hardening rates will change as the system passes through different types of migration based on the interaction between the secondary SMBH and the disk — at larger separations the secondary SMBH is driven along with the gas flow, whereas when the mass enclosed by the binary orbit becomes comparable to the secondary SMBH the system is driven by binary dynamics.

- All of these complicated processes are happening on scales that are beyond the reach of current observational instruments. The ngVLA should have the capability to resolve the cold gas reservoirs feeding these disks and the subsequent AGN (Nyland et al. 2018).

2.4. Gravitational-Wave Inspiral

The emission of gravitational radiation will always dominate binary orbital evolution at the smallest separation scales ($\lesssim$ milliparsec). This is when the binary has decoupled from the galactic environment and can be considered as an isolated physical system. In this case, the dissipation of orbital energy will depend only on the constituent SMBH masses, the orbital semi-major axis, and the binary's eccentricity.

- This stage is below the angular resolution scale of the ngVLA, and will require PTA observations of nanohertz GWs to complete the picture of SMBH dynamical evolution.

3. Anticipated Complementary Results

NANOGrav's key goal is the detection and characterization of GWs at nanohertz frequencies, with the target being a population of inspiraling SMBH binaries. The stochastic gravitational wave background (GWB) signal produced by a population of inspiraling SMBH binaries can be described by its characteristic-strain spectrum, $h_c(f)$. The strain budget at different frequencies depends on various cosmological and astrophysical factors (Sesana et al. 2004), which are summarized in Figure 1 and in the following. In the simple case of a population of circular binaries whose orbital evolution is driven entirely by the emission of GWs, $h_c(f) \propto f^{-2/3}$.

Interaction of a binary with its surrounding galactic stellar distribution, as described in §2.2, will lead to low-frequency attenuation of the characteristic strain spectrum of GWs (i.e., a spectral turnover). This can be separated into two distinct effects: (1) the direct coupling leads to a faster reduction in the orbital separation than GW-driven inspiral, such that the amount of time spent by each binary at low frequencies is reduced; (2) extraction of angular momentum by stellar slingshots can excite eccentricity, which leads to faster GW-driven inspiral, and (again) lower residence time at low frequencies. The excitation of binary eccentricity by interaction with stars will further attenuate the strain spectrum at low frequencies, leading to an even sharper turnover (e.g., Taylor et al. 2017). Coupling of a viscous circumbinary disk with a SMBH binary, as discussed in §2.3, will also lead to attenuation of the characteristic strain spectrum of GWs through both direct coupling, and excitation of eccentricity.

3.1. Detection of Stochastic Background

NANOGrav expects to detect the nanohertz GWB within the next 2-3 years (Taylor et al. 2016). This initial detection will likely be of a long-timescale stochastic process, present in all pulsars and correlated between them with a signature unique to GWs. But it is unlikely that the shape of the spectrum will be very well constrained. Nevertheless, detection alone is sufficient to provide the first direct evidence that the final parsec problem of massive black-hole evolution is mitigated, and even with poorly constrained parameters, there will be many models of post-merger galactic environments that will be ruled out.

3.2. Spectral Characterization of Background

Over the next decade, NANOGrav's sensitivity will improve to the stage at which the slope of the GW strain spectrum will be well constrained. This will allow for robust inference of the dominant dynamical process for SMBH binary evolution. However, the entire population will consist of a mixture of galaxy types and dynamical environmental processes. Thus, it will only be through the combination of NANOGrav data with models built from the in-depth surveys of galactic cores, like those the ngVLA will conduct, that a complete understanding of the conditions under which the final parsec problem is mitigated will be gained.

3.3. Constraining the Scaling Relationships Between Galaxies and Central SMBHs

NANOGrav limits on the stochastic GWB can already be expressed directly as limits on various model parameters, and the scaling relationship between galactic bulge properties and central SMBH masses has been shown to be one of the most important model parameters (Simon & Burke-Spolaor 2016; Arzoumanian et al. 2018). In the coming years, NANOGrav will place tight constraints on the galaxy-SMBH scaling relationship, specifically for SMBHs in post-major-merger galaxies. As with constraints on the strain-spectrum slope, these constraints on the scaling relation can only be properly used to make a clear statement about galaxy growth when combined with electromagnetic observations of the associated galactic bulges. To this end, the ngVLA's proposed capabilities will allow resolution down to the influence radius of $10^9 M_\odot$ SMBHs out to $z \sim 0.1$, which is more than sufficient to characterize properties of galactic bulges and to measure the SMBH mass.

3.4. Accretion onto SMBH

NANOGrav's spectral characterization of the GWB signal will illuminate the important dynamical processes that harden SMBH binaries to milliparsec separations. This will allow a measurement of various astrophysical parameters, such as the mass density of loss-cone stars in galactic centers, binary eccentricities, and SMBH accretion rates. Inference of the latter will of course depend on the assumed disk dynamics, but will ultimately describe the relevance of gas dynamics are in this overall picture of SMBH dynamical evolution. It may be the case that the kinds of SMBHs that we are looking for reside in gas-poor massive galaxies, in which case stellar scattering will control the evolution at sub-parsec separations until GW-driven inspiral begins to dominate. We currently lack sufficient observational constraints on circumbinary disk dynamics to inform our intuition of its importance. The ngVLA will measure the accretion and build-up of HI gas in galactic outskirts, track gas inflow to central regions, and (with with the ngVLA LBA) be able to resolve down to the ~pc scale out to $z \sim 0.1$. It will therefore give crucial insight into the importance of gas dynamics in SMBH evolution.

4. Summary & Synergies

Pulsar-timing arrays (such as NANOGrav) will detect the ensemble gravitational-wave signal from many inspiraling supermassive black-hole binaries throughout the Universe within the next 3 – 7 years (Taylor et al. 2016). The statistical properties of this signal will reflect the dynamical history of these supermassive black-holes as they evolve to form a bound system and reach milliparsec orbital separations. NANOGrav will constrain the environments of supermassive black-hole binaries (Taylor et al. 2017) through detailed studies of the gravitational-wave strain spectrum in the nanohertz band. The ngVLA 214 antenna main array has a goal resolution of $\approx$ 5 mas, equivalent to ~ 50 pc at $z \sim 1$. This resolution is sufficient to distinguish dual AGN, allowing estimates of SMBH merger rates to be formed that can act as prior constraints for PTA analysis. Furthermore, using the intercontinental VLBI capabilities of the LBA, the ngVLA will achieve ~pc-scale resolution out to $z \sim 0.1$, allowing a large measured sample of SMBH binary separations to inform whether binaries stall at ~pc separations or are driven efficiently to the sub-pc regime by dynamical interactions. The intercontinental ngVLA LBA will also allow the influence radius of supermassive black-holes to be resolved, which will drastically improve mass estimation and thus the fidelity of derived $M_{\rm BH} - M_{\rm bulge}$ relationships. Finally, ngVLA observations of galactic gas reservoirs, large-scale inflows, and gas dynamics in central galactic regions will provide crucial insight into the physics of SMBH binary accretion processes.

There are several current and future facilities that will complement the ngVLA in the aforementioned areas. These include the *Atacama Large Millimeter/submillimeter Array* (ALMA)[1], whose angular resolution at 230 GHz with baseline length of 15 km is 15 mas, which is comparable to the ngVLA's goal resolution for the 214 antenna main array of $\approx$ 5 mas at 30 GHz with a maximum baseline length of $\approx$ 1000 km (see Selina et al. 2018, this volume p. 15). Likewise, the ngVLA should have similar angular resolution to adaptive-optics corrected images from the *Giant Magellan Telescope* (GMT)[2], *Thirty Meter Telescope* (TMT)[3], and the *Extremely Large Telescope* (ELT)[4]. This range of complementarity in angular resolution will allow for validation and comparison studies of sub-kiloparsec galaxy scales out to $z \sim 1$, which is sufficient to distinguish dual AGN for the purpose of estimating SMBH binary merger rates. On the other hand, the

[1] `http://www.eso.org/public/usa/teles-instr/alma`

[2] `https://www.gmto.org`

[3] `https://www.tmt.org`

[4] `https://www.eso.org/public/usa/teles-instr/elt`

SKA[5] is expected to operate concurrently with the ngVLA, but its point-source sensitivity will be only be half that of the ngVLA in their frequency-overlap band (Casey et al. 2015). Similarly, *JWST*[6] will operate at similar wavelengths to GMT/TMT/ELT, but have lower angular resolution than those instruments and the ngVLA.

NANOGrav gravitational-wave analysis, in concert with observations by the ngVLA and complementary facilities, will paint a multi-messenger portrait of galaxy evolution and the dynamics of the most massive black-holes in the Universe.

Acknowledgments. The NANOGrav project receives support from NSF Physics Frontier Center award number 1430284. SRT acknowledges support from the NANOGrav Physics Frontier Center. JS acknowledge support from the JPL RTD program. Some of this research was undertaken at the Jet Propulsion Laboratory, California Institute of Technology, under a contract with NASA.

References

Arzoumanian, Z., Baker, P. T., Brazier, A., et al. 2018, ApJ, 859, 47
Begelman, M. C., Blandford, R. D., & Rees, M. J. 1980, Nat, 287, 307
Blumenthal, G. R., Faber, S. M., Primack, J. R., & Rees, M. J. 1984, Nat, 311, 517
Bolatto, A. D., Chatterjee, S., Casey, C. M., et al. 2017, ArXiv e-prints, arXiv:1711.09960
Casey, C. M., Hodge, J. A., Lacy, M., et al. 2015, NRAO Next Generation Very Large Array Memos Series, arXiv:1510.06411. `http://library.nrao.edu/ngvla.shtml`
Chandrasekhar, S. 1943, ApJ, 97, 255
Dvorkin, I., & Barausse, E. 2017, MNRAS, 470, 4547
Ferrarese, L., & Merritt, D. 2000, ApJ, 539, L9
Frank, J., & Rees, M. J. 1976, Monthly Notices of the Royal Astronomical Society, 176, 633
Gebhardt, K., Bender, R., Bower, G., et al. 2000, ApJ, 539, L13
Goicovic, F. G., Sesana, A., Cuadra, J., & Stasyszyn, F. 2017, MNRAS, 472, 514
Kocsis, B., & Sesana, A. 2011, MNRAS, 411, 1467
Kormendy, J., & Richstone, D. 1995, ARA&A, 33, 581
Magorrian, J., Tremaine, S., Richstone, D., et al. 1998, AJ, 115, 2285
McWilliams, S. T., Ostriker, J. P., & Pretorius, F. 2014, ApJ, 789, 156
Merritt, D., & Milosavljević, M. 2005, Living Reviews in Relativity, 8, astro-ph/0410364
Mikkola, S., & Valtonen, M. J. 1992, MNRAS, 259, 115
Milosavljević, M., & Merritt, D. 2003, in American Institute of Physics Conference Series, Vol. 686, The Astrophysics of Gravitational Wave Sources, ed. J. M. Centrella, 201–210
Nyland, K., Harwood, J. J., Mukherjee, D., et al. 2018, ArXiv e-prints, arXiv:1803.02357
Quinlan, G. D. 1996, New Astronomy, 1, 35
Selina, R. J., Murphy, E. J., McKinnon, M., et al. 2018, Science with a Next-Generation Very Large Array, ASP Monograph 7, ed. E. J. Murphy (San Francisco, CA: ASP), 15
Sesana, A., Haardt, F., Madau, P., & Volonteri, M. 2004, ApJ, 611, 623
Shakura, N. I., & Sunyaev, R. A. 1973, A&A, 24, 337
Shapiro, S. L., & Teukolsky, S. A. 1986, Black Holes, White Dwarfs and Neutron Stars: The Physics of Compact Objects (John Wiley & Sons), 672
Simon, J., & Burke-Spolaor, S. 2016, ApJ, 826, 11
Taylor, S. R., Simon, J., & Sampson, L. 2017, Physical Review Letters, 118, 181102
Taylor, S. R., Vallisneri, M., Ellis, J. A., et al. 2016, ApJ, 819, L6
Vasiliev, E., & Merritt, D. 2013, ApJ, 774, 87
Young, J. S., & Scoville, N. Z. 1991, ARA&A, 29, 581

---

[5]`https://www.skatelescope.org`

[6]`https://www.jwst.nasa.gov`

Science with a Next-Generation Very Large Array
ASP Conference Series, Monograph 7
Eric J. Murphy, ed.

High-resolution Imaging of Radio Jets Launched by Active Galactic Nuclei: New Insights on Formation, Structure, and Evolution Enabled by the ngVLA

Matthew L. Lister,[1] Kenneth I. Kellermann,[2] and Preeti Kharb[3]

[1]*Dept. of Physics and Astronomy, Purdue University, West Lafayette, IN, USA;* `mlister@purdue.edu`

[2]*NRAO, Charlottesville, VA, USA;* `kkellerm@nrao.edu`

[3]*NCRA-TIFR, Pune, India;* `kharb@ncra.tifr.res.in`

Abstract. Jetted plasma outflows from active galactic nuclei (AGN) represent the most energetic phenomena in the known universe, and play a key role in regulating galaxy formation through feedback processes (Harrison et al. 2018). The Jansky VLA and VLBA have played an indispensable role in understanding the physics of these powerful jets and their environments, via high angular resolution full polarization imaging and astrometric studies. By bridging the current interferometric gap between the VLA and VLBA with intermediate baselines, the ngVLA offers exciting new opportunities to explore the intermediate regions downstream of the high Lorentz factor pc-scale AGN jets imaged by VLBI, where entrainment, deceleration, collimation and particle acceleration all take place. The ability to image exceedingly faint radio emission on scales of 10s to 100s of milliarcseconds will lead to new breakthroughs in resolving some of the most pressing questions regarding the formation, structure, and evolution of AGN jets. These are in turn crucial for a more complete understanding of the formation of supermassive black holes and galaxies in the early universe and their subsequent evolution.

1. Magnetic Field Structure

Since AGN jets are fundamentally magnetically-driven phenomena, polarimetric observations provide numerous physical insights otherwise unattainable via total intensity imaging. As we describe in the specific science topic sections below, the full (linear and circular) polarization capabilities of the ngVLA represent a major advance over existing instruments, as applied to AGN jet studies. Specifically, Faraday depth, rotation, and depolarization measurements offer a direct means of probing the properties of jet plasma and its surrounding medium, as well as the overall magnetic field. VLBA polarimetric imaging has revealed complex pc-scale magnetic phenomena, such as MHD waves (Cohen et al. 2014), stationary features (Jorstad et al. 2017), and structures consistent with planar and conical shocks (Marscher et al. 2008). On kiloparsec scales, polarization studies of radio lobes (Figure 1) have been a crucial element supporting the unification of radio galaxies and quasars (Garrington et al. 1988), and provide one of the only means of studying large scale magnetic fields in galaxy clusters (Bonafede et al. 2010). Although difficult to measure with existing radio telescope arrays, the de-

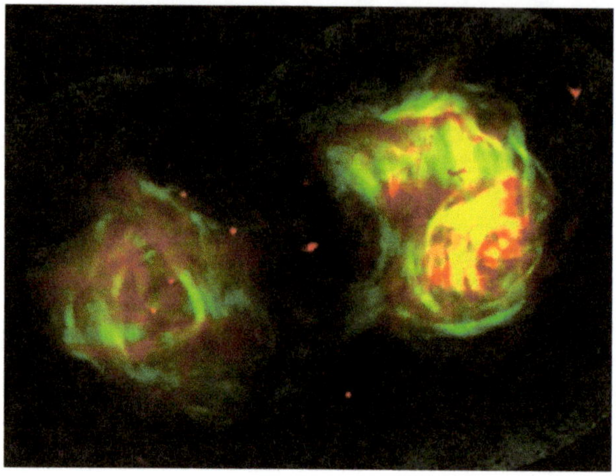

Figure 1. VLA image of the radio galaxy Fornax A. In this hue-intensity image the brightness corresponds to the intensity of the lobes while the highly polarized structure is saturated white and regions of increasing depolarization are shaded red. Image courtesy of NRAO/AUI.

tections of circular polarization (Homan & Lister 2006) and Faraday rotation gradients (Gabuzda et al. 2017) in a few pc-scale AGN jets have revised our views of internal jet physics over the last decade, and have prompted many detailed numerical studies to better understand the jet launching and collimation process (e.g., Broderick & McKinney 2010; Tchekhovskoy & Bromberg 2016). With the new 50 – 250 km baselines and substantially improved sensitivity afforded by the ngVLA, it will be possible to study the magnetic field structure of lobes and terminal jet hotspots in AGN in unprecedented detail. This will lead to a better understanding of particle acceleration mechanisms and how relativistic jet flows are decelerated.

2. Launching and Collimation Mechanisms

Current theory and numerical simulations for the launching of jets favor scenarios that typically involve a supermassive black hole and accretion disk in the nucleus of the host galaxy. Magnetic fields play an important role, due to the large kinetic power of the outflows, and the lack of sufficient thermal and radiation pressure in the accretion disk region. A toroidal field component can provide a means of producing the highly collimated jets imaged by VLBI. Magnetic fields also provide the means of extracting energy and angular momentum from the accretion disk and a rotating black hole.

High resolution radio imaging of nearby AGN jets has revealed important details regarding collimation near the base of the jet. In the case of the nearby AGN M 87, the jet initially has a wide opening angle, which is sharply collimated at ~ 100 Schwarzschild radii (R_s) to a roughly parabolic profile (Hada et al. 2013). Only at $\sim 10^5\ R_s$ does the jet take on a conical shape expected for simple adiabatic expansion (Asada & Nakamura 2012). Some magnetic launching models (e.g., Komissarov et al.

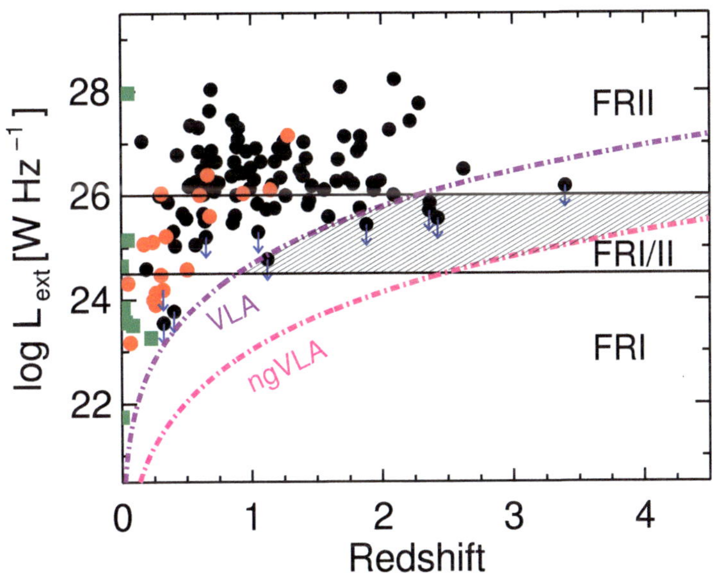

Figure 2. The 1.4 GHz extended luminosity with respect to redshift for the original MOJAVE sample (Kharb et al. 2010b). Black and red circles denote quasars and BL Lac objects, respectively, while green squares denote radio galaxies. Core-only sources are represented as upper limits with downward arrows. The solid lines indicate the FRI–FRII divide (extrapolated from 178 MHz to 1.4 GHz assuming a spectral index, α=0.8), following Ledlow & Owen (1996) and Landt et al. (2006). The purple line denotes the sensitivity limit for the historical VLA which was used to carry out the MOJAVE study, while the magenta line denotes the sensitivity limit for the ngVLA (see http://ngvla.nrao.edu/page/refdesign). ngVLA will be able to detect more than twice as many "hybrid FRI/II" sources, compared to previous studies, revolutionizing the study of radio-loud AGN.

2007) predict strong collimation once the jet flow passes through the fast-magnetosonic point, when the bulk conversion of magnetic to kinetic energy takes place. Multi-epoch VLBI imaging of the inner M 87 jet has also shown an evolution of the flow velocity that can be explained in the framework of MHD jet acceleration and this type of energy conversion (Mertens et al. 2016; Walker et al. 2018a). Since M 87 is unusually close (16 Mpc), the ngVLA will be the only instrument capable of probing faint structure in the jet further downstream, including the intriguing *HST*-1 knot region, which has been seen to exhibit much higher internal proper motions (up to ~ 4 c) than the (sub-luminal) inner jet (Giroletti et al. 2012). This type of analysis can also be extended to many other nearby radio galaxies such as NGC 6251 (Tseng et al. 2016) and PKS 0227-369 (Potter & Cotter 2013). At present, the only instrument capable of probing these scales is e-MERLIN, which, having only 7 antennas, cannot provide the dual-sensitivity needed to probe the diffuse and compact structures that are present in these nearby jets. The

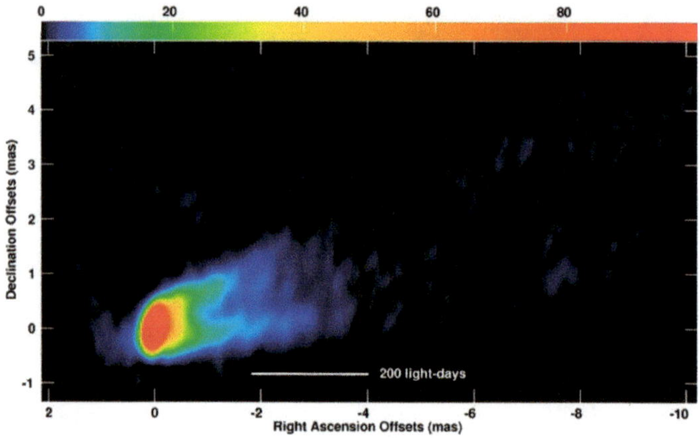

Figure 3. VLBA image of the radio jet in M 87. With a high resolution of 430 by 210 μas (1.3 by 0.64 light-months), it reveals that the jet appears brighter at its edge compared to the center, and collimation of the jet is occurring within 100 Schwarzschild radii. Image courtesy of NRAO/AUI.

ngVLA will provide 10 times the continuum imaging sensitivity of e-MERLIN at 5 GHz in one tenth the integration time.

3. Jet Structure

With its unprecedented sensitivity and baselines out to 1000 km, the ngVLA 214 main array holds the exciting prospect of transversely resolving for the first time the jets of many nearby radio galaxies such as NGC 315, 3C 66B, 3C 264, and NGC 4261 out to large projected distances. Since radio galaxy jets are oriented closer to the plane of the sky, they suffer less from projection and relativistic beaming effects. In a handful of cases with current instruments, the (de-boosted) receding jet can be imaged and compared with the approaching jet, giving an accurate measurement of the viewing angle, which is often impossible to determine by other means. This technique also provides detailed information on the velocity field and magnetic field configuration in the jets, and has been used extensively by Laing & Bridle (2014) to make important discoveries regarding the accelerations of jet plasma and its interaction with the external medium at the jet boundary. Although currently limited to slower and less beamed (FR-I class) jets, with the ngVLA it will be possible to study the receding jets and velocity fields in the most powerful (FR-II class) AGN (Figure 2). In particular, more examples of enhanced radio emission at the jet boundaries will be found, which have been studied so far in only the nearest AGN jets such as M 87 (Figure 3), Mrk 501 (Giroletti et al. 2008), and 3C 84 (Nagai et al. 2014). At present it is unclear whether this phenomenon is a result of interaction with the external medium, or differential beaming associated with a transverse jet velocity profile, a corkscrew-like rotation of the flow (Walker et al. 2018a) or helical magnetic fields (Clausen-Brown et al. 2011).

4. Bending

Many jets have 'S'-shaped or more complex curvatures which have been variously attributed to nozzle precession, hydrodynamic instabilities, jet-cloud collisions, or ram pressure from the external medium. By tracing the jet at high resolution from pc to kpc scales, the ngVLA studies will help to distinguish between these scenarios for individual objects, and determine for the first time the range and form of bending in the AGN jet population (Figure 4). The ngVLA will also probe a much larger variety of 'U'-shaped galaxies, which have been used successfully to identify galaxy clusters and directly study their intergalactic medium (Paterno-Mahler et al. 2017). Opportunities will also exist for studying large scale hydrodynamic instability modes in AGN jets, which manifest themselves as apparent helical structures on $\sim$ 100 pc scales (Hardee et al. 2005). It will be possible to trace out to large projected distances the extreme apparent bends in highly aligned AGN jets (blazars), which on occasion directly cross our line of sight, providing useful data for testing radiative transfer and jet emission models (Homan et al. 2002; Alberdi et al. 1997).

5. Precession

A recent discovery from long-term VLBA studies is that bright superluminal features in pc-scale AGN jets will frequently emerge in a particular direction that changes slowly on decadal timescales (Lister et al. 2013). The net result is that at any given time, only the 'energized' portion of the jet containing recently ejected features is visible with current interferometers. Over time, subsequent ejections 'fill in' the jet, revealing its true conical shape in stacked images (Pushkarev et al. 2017). Evidence of helical-type jet morphology has been seen on kpc-scales with the VLA that is suggestive of precession, and a particularly intriguing class of 'X-shaped' radio galaxies may represent cases where a dramatic 90° or more shift has occurred in the jet direction over time. The ngVLA will be vital for connecting the dynamic morphology seen on pc-scales with emission immediately downstream, and for identifying cases where the jet direction is actively shifting.

6. Proper Motions and Astrometry

The VLBA is a remarkable tool for precision jet astrometry, being able to measure the speeds and accelerations of thousands of individual jet features on decade or more timescales. Large surveys are able to directly witness jet kinematics through time-lapse imaging, and have revealed that powerful AGN jets are still undergoing collimation and acceleration on scales of tens to hundreds of parsecs downstream of the central engine (Homan et al. 2015; Jorstad et al. 2017), and occasionally in abrupt fashion (Homan et al. 2003). Recent work with *HST* and VLA data on two AGN has shown that this type of astrometry can also be performed on kiloparsec scale jet features (Meyer et al. 2017a). The unprecedented sensitivity and > 50 km baselines of the ngVLA, in combination with improved resolution afforded by *JWST*, will extend these measurements to a much larger set of more distant and more powerful AGN. Such studies are crucial for constraining numerical models that seek to understand how and where jets decelerate (e.g., Gourgouliatos & Komissarov 2018).

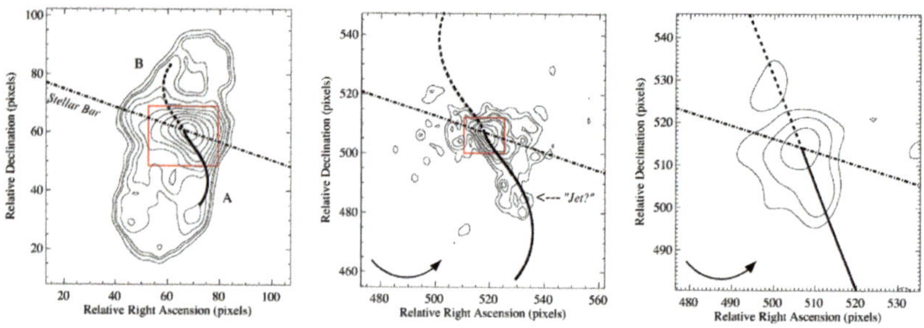

Figure 4. A precessing jet model fit to the radio emission from sub-kiloparsec to parsec scales in the Seyfert galaxy NGC 6764 (Kharb et al. 2010a). The left panel shows a VLA 1.4 GHz image, the middle panel a VLA 5 GHz image, and the right panel a VLBA 1.6 GHz image. The red boxes indicate the region spanned by the next image panel in the sequence.

7. Knots and Hotspots

Considerable uncertainties still exist in our understanding of the emission processes in bright jet knots in kpc-scale jets (Figure 5). Recent multi-wavelength studies with the VLA, ALMA, *HST*, and *Chandra* have shown that simple inverse-Compton scattering models with single-electron populations cannot adequately account for the spectral energy distributions of jet knots (e.g., Meyer et al. 2017b; Jester et al. 2006). In some cases, such as M 87, jet knots have been postulated to be the sites of variable gamma-ray emission, but detailed supporting data are lacking. Multi-epoch, multi-frequency polarimetric imaging with the longest ngVLA baselines will make it possible to perform spectral aging, Faraday depth and magnetic field analyses of jet knots with unprecedented resolution and sensitivity. The terminal hotspots of AGN jets, where the flow undergoes abrupt deceleration while interacting the intergalactic medium, represent important targets for ngVLA studies of feedback mechanisms, and for resolving why some AGN display hybrid jet morphologies (Figure 5). In hybrid AGN, one jet appears to be of high power, with a terminal hotspot, while the other has properties more consistent with a lower power flow that is decelerated gradually along its length by entrainment with the external medium (Gopal-Krishna & Wiita 2000). Hotspots and radio lobes are also useful for probing the past activity history of the jets, especially at longer radio wavelengths which sample the older electron population.

8. Radio-Loud and Radio-Quiet Quasars

Fifty years after the discovery of quasars, the distinction between radio-loud quasars (RLQ) and radio quiet quasars (RQQ) is still being debated. Some authors have reported a clear bi-modal separation in radio/optical luminosity ratio R while others have claimed that the distribution is continuous with no evidence for two populations (see Kellermann et al. 2016 for a review). Using the VLA, Kimball et al. (2011) and Kellermann et al. (2016) have detected radio emission from essentially all 176 quasars in a volume-limited sample from the SDSS contained within a narrow range of redshift. Only about 10 – 15% of the SDSS quasars fall in the radio-loud classification, irre-

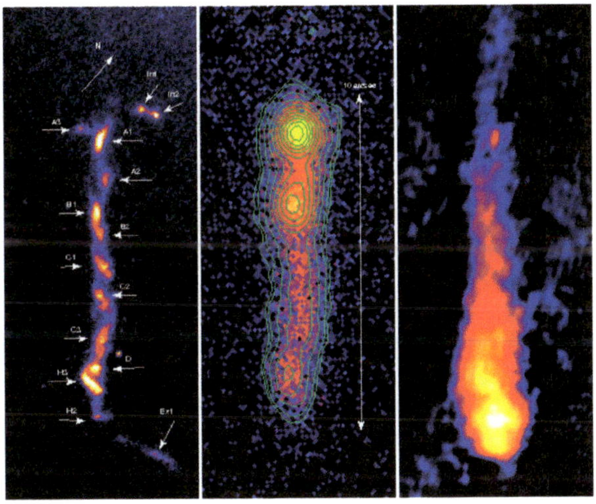

Figure 5. Images of the 30 kpc-long jet in the quasar 3C 273. From left to right, the images are optical (*HST*), X-ray (*Chandra*), and radio (MERLIN). (Credit: Optical: NASA/STScI X-ray: NASA/CXC Radio: MERLIN) http://chandra.harvard.edu/photo/2000/0131/0131_multi.tif

spective of whether the RLQs are defined by *R* or by luminosity. Condon et al. (2013) have shown that the radio emission from the RQQ is due to star formation in the host galaxy and is not directly related to the SMBH which gives rise to the AGN. All quasars are thought to harbor a SMBH needed to support the huge optical-IR luminosity which defines the category of quasars. Why then is only a small fraction, 10-15% of quasars RL? Many explanations have been offered, including intermittent activity, absorption of the radio emission by intervening medium, host galaxy properties, magnetic fields, and black hole spin or accretion rate. A particularly elegant interpretation was suggested by Scheuer & Readhead (1979) that RLQ are just the subset of quasars whose relativistic beams are oriented nearly along the line of sight. But at the time, it appeared that too large a fraction of optically selected quasars appeared radio loud and that the observed detection rate as a function of decreasing flux density was not consistent with simple beaming models. Using more modern radio observations of optically selected quasars and now knowing that relativistic jets often have finite width and significant bends, as well as recognizing that optically selected quasars are not necessarily randomly oriented in the sky, the observed fraction (10-15%) of RLQ and the observed number flux density relations now appear consistent with relativistic beaming models. However, the extended radio emission seen in more than half of RLQs represents a clear challenge to the relativistic beaming interpretation of the RL/RQ dichotomy. Kinematic observations of kpc (arcsec) scale are difficult as the expected motions are typically only a few tenths of a milliarcsec per year. Currently there is evidence of modest superluminal motion in 7 mm observations of M 87 (Walker et al. 2018b), and in the kpc-scale jets of a few AGN (§6). High-resolution-high sensitivity proper motion observations with the ngVLA will give further insight to the importance of relativistic beaming in kpc scale radio jets and its implications for the RL/RQ dichotomy.

9. Low-luminosity/Radio-quiet AGN

More than 80% of AGN, comprising Seyfert and Low Ionization Nuclear Emission Line Region (LINER) galaxies, do not possess collimated plasma outflows on kpc-scales. Weak AGN outflow emission is often indistinguishable from outflows due to galactic stellar activity or nuclear starbursts (e.g., Gallimore et al. 2006). Disentangling AGN and stellar contributions to the radio emission are essential to understanding the nature of the AGN in low luminosity AGN (LLAGN), and in discerning why the vast majority of AGN do not produce large-scale jets. Multi-scale observations from parsec to kpc-scales can trace the connection between an AGN jet and kpc-scale lobes observed in many LLAGN (Figure 4). Polarization-sensitive imaging can further distinguish between the stellar and AGN outflows, due to presumably more organized magnetic field structures and higher degrees of polarization in AGN outflows compared to stellar outflows. The ngVLA will detect fainter radio emission in LLAGN and in doing so will trace AGN outflows as they are launched from the black hole-accretion-disk systems and propagate through the ISM of their host galaxies.

References

Alberdi, A., Krichbaum, T. P., Graham, D. A., et al. 1997, A&A, 327, 513
Asada, K., & Nakamura, M. 2012, ApJ, 745, L28
Bonafede, A., Feretti, L., Murgia, M., et al. 2010, A&A, 513, A30
Broderick, A. E., & McKinney, J. C. 2010, ApJ, 725, 750
Clausen-Brown, E., Lyutikov, M., & Kharb, P. 2011, MNRAS, 415, 2081
Cohen, M. H., Meier, D. L., Arshakian, T. G., et al. 2014, ApJ, 787, 151
Condon, J. J., Kellermann, K. I., Kimball, A. E., Ivezić, Ž., & Perley, R. A. 2013, ApJ, 768, 37
Gabuzda, D. C., Roche, N., Kirwan, A., et al. 2017, MNRAS, 472, 1792
Gallimore, J. F., Axon, D. J., O'Dea, C. P., Baum, S. A., & Pedlar, A. 2006, AJ, 132, 546
Garrington, S. T., Leahy, J. P., Conway, R. G., & Laing, R. A. 1988, Nat, 331, 147
Giroletti, M., Giovannini, G., Cotton, W. D., et al. 2008, A&A, 488, 905
Giroletti, M., Hada, K., Giovannini, G., et al. 2012, A&A, 538, L10
Gopal-Krishna, & Wiita, P. J. 2000, A&A, 363, 507
Gourgouliatos, K. N., & Komissarov, S. S. 2018, Nature Astronomy, 2, 167
Hada, K., Kino, M., Doi, A., et al. 2013, ApJ, 775, 70
Hardee, P. E., Walker, R. C., & Gómez, J. L. 2005, ApJ, 620, 646
Harrison, C. M., Costa, T., Tadhunter, C. N., et al. 2018, Nature Astronomy, 2, 198
Homan, D. C., & Lister, M. L. 2006, AJ, 131, 1262
Homan, D. C., Lister, M. L., Kellermann, K. I., et al. 2003, ApJ, 589, L9
Homan, D. C., Lister, M. L., Kovalev, Y. Y., et al. 2015, ApJ, 798, 134
Homan, D. C., Wardle, J. F. C., Cheung, C. C., Roberts, D. H., & Attridge, J. M. 2002, ApJ, 580, 742
Jester, S., Harris, D. E., Marshall, H. L., & Meisenheimer, K. 2006, ApJ, 648, 900
Jorstad, S. G., Marscher, A. P., Morozova, D. A., et al. 2017, ApJ, 846, 98
Kellermann, K. I., Condon, J. J., Kimball, A. E., Perley, R. A., & Ivezić, Ž. 2016, ApJ, 831, 168
Kharb, P., Hota, A., Croston, J. H., et al. 2010a, ApJ, 723, 580
Kharb, P., Lister, M. L., & Cooper, N. J. 2010b, ApJ, 710, 764
Kimball, A. E., Kellermann, K. I., Condon, J. J., Ivezić, Ž., & Perley, R. A. 2011, ApJ, 739, L29
Komissarov, S. S., Barkov, M. V., Vlahakis, N., & Königl, A. 2007, MNRAS, 380, 51
Laing, R. A., & Bridle, A. H. 2014, MNRAS, 437, 3405

Landt, H., Perlman, E. S., & Padovani, P. 2006, ApJ, 637, 183
Ledlow, M. J., & Owen, F. N. 1996, AJ, 112, 9
Lister, M. L., Aller, M. F., Aller, H. D., et al. 2013, AJ, 146, 120
Marscher, A. P., Jorstad, S. G., D'Arcangelo, F. D., et al. 2008, Nat, 452, 966
Mertens, F., Lobanov, A. P., Walker, R. C., & Hardee, P. E. 2016, A&A, 595, A54
Meyer, E., Sparks, W., Georganopoulos, M., et al. 2017a, Galaxies, 5, 8
Meyer, E. T., Breiding, P., Georganopoulos, M., et al. 2017b, ApJ, 835, L35
Nagai, H., Haga, T., Giovannini, G., et al. 2014, ApJ, 785, 53
Paterno-Mahler, R., Blanton, E. L., Brodwin, M., et al. 2017, ApJ, 844, 78
Potter, W. J., & Cotter, G. 2013, MNRAS, 429, 1189
Pushkarev, A. B., Kovalev, Y. Y., Lister, M. L., & Savolainen, T. 2017, MNRAS, 468, 4992
Scheuer, P. A. G., & Readhead, A. C. S. 1979, Nat, 277, 182
Tchekhovskoy, A., & Bromberg, O. 2016, MNRAS, 461, L46
Tseng, C.-Y., Asada, K., Nakamura, M., et al. 2016, ApJ, 833, 288
Walker, R. C., Hardee, P. E., Davies, F. B., Ly, C., & Junor, W. 2018a, ApJ, 855, 128
—. 2018b, ArXiv e-prints, arXiv:1802.06166

Science with a Next-Generation Very Large Array
ASP Conference Series, Monograph 7
Eric J. Murphy, ed.

Imaging Cold Gas to 1 kpc Scales in High-Redshift Galaxies with the ngVLA

Caitlin M. Casey,[1] Desika Narayanan,[2,3] Chris Carilli,[4] Jaclyn Champagne,[1] Chao-Ling Hung,[1] Romeel Davé,[5] Roberto Decarli,[6] Eric J. Murphy,[7] Gergo Popping,[8] Dominik Riechers,[9] Rachel S. Somerville,[10,11] and Fabian Walter[8]

[1] *Department of Astronomy, The University of Texas at Austin, 2515 Speedway Blvd, Austin, TX 78712, USA;* `cmcasey@utexas.edu`

[2] *Department of Astronomy, University of Florida, 211 Bryant Space Science Center, P.O. Box 112055, Gainesville, FL 32611, USA*

[3] *Cosmic Dawn Center (DAWN), Niels Bohr Institute, University of Copenhagen, DK-2100 Copenhagen ; DTU-Space, Technical University of Denmark, DK-2800 Kgs. Lyngby*

[4] *National Radio Astronomy Observatory, 1003 Lopezville Rd, Socorro, NM 87801, USA*

[5] *Institute for Astronomy, University of Edinburgh, Royal Observatory, Blackford Hill, Edinburgh EH9 3HJ, UK*

[6] *INAF - Osservatorio di Astrofisica e Scienza dello Spazio di Bologna, via Cobetti 93/3, 40129 Bologna, Italy*

[7] *National Radio Astronomy Observatory, 1180 Boxwood Estate Rd, Charlottesville, VA 22903, USA*

[8] *Max Planck Institute for Astronomy, Königstuhl 17, 69117 Heidelberg, Germany*

[9] *Department of Astronomy, Cornell University, 616A Space Science Building, Ithaca, NY 14850, USA*

[10] *Department of Physics & Astronomy, Rutgers University, 136 Frelinghuysen Rd, Piscataway, NJ 08854, USA*

[11] *Center for Computational Astrophysics, 162 5th Ave, New York, NY 10010, USA*

Abstract. The next generation Very Large Array (ngVLA) will revolutionize our understanding of the distant Universe via the detection of cold molecular gas in the first galaxies. Its impact on studies of galaxy characterization via detailed gas dynamics will provide crucial insight on dominant physical drivers for star-formation in high redshift galaxies, including the exchange of gas from scales of the circumgalactic medium down to resolved clouds on mass scales of $\sim 10^5\ M_\odot$. In this study, we employ a series of high-resolution, cosmological, hydrodynamic zoom simulations from the Mufasa simulation suite and a CASA simulator to generate mock ngVLA observations of a $z \sim 4.5$ gas-rich star-forming galaxy. Using the Despotic radiative transfer code that encompasses simultaneous thermal, chemical, and statistical equilibrium in calculating the molecular and atomic level populations, we generate parallel mock observations of low-J to high-J transitions of CO from ALMA for comparison. We find that observations of CO(1-0) are especially important for tracing the systemic redshift of the galaxy and the total mass of the well-shielded molecular gas reservoir, while even CO(2-1) can predominantly trace denser gas regions distinct from CO(1-0). The factor of 100 times improvement in mapping speed for the ngVLA beyond the Jansky VLA and the proposed ALMA Band 1 will make these detailed, high-resolution imaging and kinematic studies of CO(1-0) routine at $z \sim 2 - 5$.

1. Introduction

Cold molecular gas is the fuel for cosmic star-formation (Solomon & Vanden Bout 2005; Carilli & Walter 2013). This gas is most commonly probed through observations of CO, the most abundant molecule after molecular hydrogen, and a molecule which is easily excitable at low temperatures. Resolved CO studies of distant galaxies can not only provide the basic measurement of a galaxy's total molecular gas mass, but one can also extract detailed kinematics of that cold gas reservoir. The kinematics give direct constraints on the internal dynamics of galaxies out to high-z and insight into the physics of the interstellar medium (e.g., Engel et al. 2010; Bothwell et al. 2010; Ivison et al. 2012) whether or not they are dispersion dominated, agitated by a dynamical interaction, or rotating in a disk. Ideally, these kinematic tracers would be available for a large sample of early Universe galaxies to address important questions in galaxy formation and evolution, like assessing the relative role of major mergers in driving the growth of massive galaxies over cosmic time.

It is not immediately clear that all tracers of molecular gas in distant galaxies would lead to the same observational conclusions about galaxy dynamics, as each traces fundamentally different physical states of the gas. For example, the lowest ground state transition of CO, CO(1-0), should trace the most diffuse and massive gas reservoir in galaxies, while higher-J transitions probe sequentially denser environments. Very few galaxies in the early Universe have resolved observations from *multiple* molecular gas tracers, because the time required to spatially map each transition is currently prohibitive for large samples of galaxies, even on the most sensitive interferometers like ALMA. And yet, there is some indication that different tracers do point to fundamentally different gas reservoirs, that could lead to differences of interpretation of their dynamics (Bothwell et al. 2013; Hodge et al. 2012).

In this study, we investigate quantitative differences in observations of different CO transitions in a hydrodynamic simulation of a early Universe massive galaxy. Based on a direct comparison between the inferred results from our mock observations and the cosmological simulations, we investigate the capabilities of ALMA and the proposed next generation Very Large Array (ngVLA; Carilli 2015) to constrain the mode of star formation, dynamical mass, and molecular gas kinematics in individual high-redshift galaxies (see Casey et al. 2015, for a broader discussion of ngVLA's use in high-z extragalactic astrophysics). This work highlights the complexity in inferring these important quantities in early Universe galaxies, and the need for the next generation facilities.

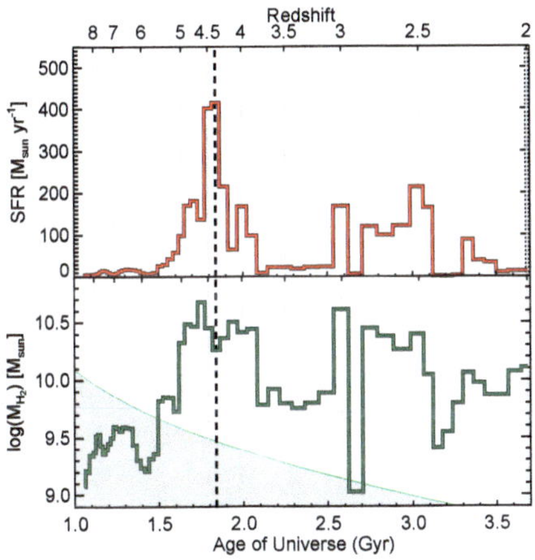

Figure 1. The star-formation history (*top*) and molecular hydrogen mass (*bottom*) of the evolving massive galaxy in formation. The zoom-in simulation is run from high-redshift to $z = 2$, at which point it has a stellar mass of $2\times10^{11}\ M_\odot$. The system undergoes two major star-forming events, and we analyze the snapshot where the SFR is maximized near 400 $M_\odot$/yr at $z = 4.37$ (marked with a vertical dashed line). The thin green line represents the 5σ detection limit for the ngVLA in one hour; to resolve the kinematics of the galaxy across ~10 beams, $10\times$ the observing time is required to reach adequate analysis depth.

2. Hydrodynamic Zoom-In Simulation Case Study

We employ a series of high-resolution, cosmological, hydrodynamic zoom simulations (Narayanan et al. 2018b,a) from the MUFASA simulation suite (Davé et al. 2016, 2017) and a CASA simulator to generate mock ngVLA observations. Using the DESPOTIC radiative transfer code (Krumholz 2014) that encompasses simultaneous thermal, chemical and statistical equilibrium in calculating the molecular and atomic level populations (Narayanan & Krumholz 2017), we generate parallel mock observations of low-J to high-J transitions of CO. For this study we take a single snapshot of one zoom simulation. The galaxy being formed is an extremely massive $M_\star > 4\times10^{11}\ M_\odot$ at $z = 0$; the zoom in itself is only run from high-z down to $z = 2$, at which point the stellar mass is already $2\times10^{11}\ M_\odot$, which is $\approx M_\star$ at $z = 2$. The system undergoes two 'starbursting' episodes during its history at $z \sim 4.5$ and $z \sim 2$; the first is driven by the infall of several small galaxies (i.e., minor mergers) while the later burst is driven by a major merger. We have chosen to focus our study on the earlier $z \sim 4.5$ snapshot, which represents the strongest starburst episode of the galaxy's history. The galaxy's star formation history and evolving molecular gas mass is shown in Figure 1. At the sampled snapshot, the galaxy's total stellar mass is $7\times10^{10}\ M_\odot$, its mass of molecular hydrogen is $2\times10^{10}\ M_\odot$, and its 50 Myr-averaged SFR is 415 $M_\odot\,\mathrm{yr}^{-1}$. The simulated galaxy's gas kinematic structure is inherently complex owing to the minor interactions, though morphologically the galaxy presents as a stable disk.

We note that this source sampled at this particular time is expected to be among the brighter galaxies detectable with the ngVLA. Figure 1 shows the effective 5σ de-

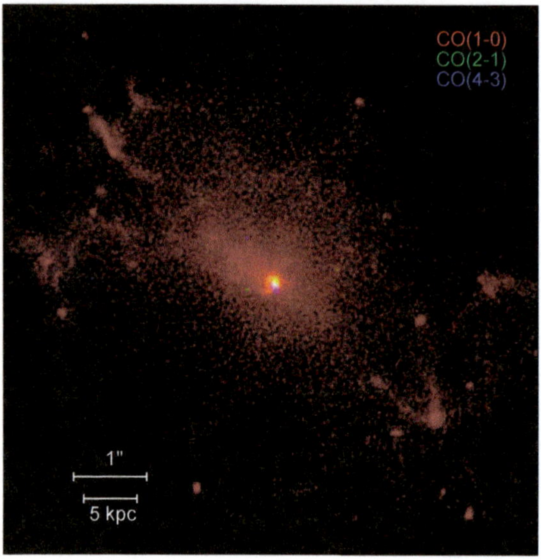

Figure 2. A false-color image of molecular gas emission in the z = 4.37 snapshot of the massive galaxy in formation. CO(1-0) emission is traced in red, CO(2-1) in green and CO(4-3) in blue. The luminosity scale is weighted to match the peak line intensity at the galaxy's core; so while the higher-J transitions are intrinsically brighter lines, this scaling shows the physical scale over which each line is emitting. The disk of molecular gas extending out beyond ~8 kpc from the core is only prominent in CO(1-0) emission.

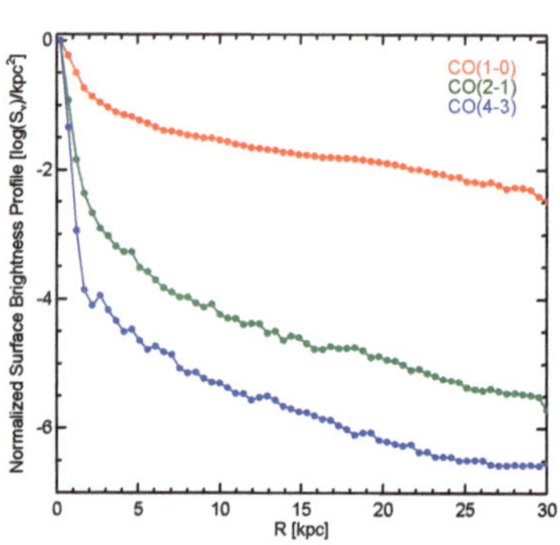

Figure 3. The intrinsic surface brightness profiles of CO(1-0), CO(2-1) and CO(4-3) for the simulated galaxy shown in Figure 2, differing by 3-4 orders of magnitude at large galactocentric radii.

tection limit as a function of redshift for a 1 hour ngVLA observation on CO(1-0). This limit is determined by working backward from a target rms of 10 μJy/beam per 2 MHz channel in 1 hour of observations, an assumed average 200 km s^{-1} line width, and a CO-to-H_2 conversion factor of $\alpha_{CO} = 1\,M_\odot\,(\mathrm{K\,km\,s^{-1}\,pc^2})^{-1}$. In other words, this is the limit at which the total integrated CO(1-0) line would have a 5σ significance if it were 200 km s^{-1} wide. The galaxy snapshot studied here would be detectable with a total signal-to-noise ratio (SNR) of ~32 in 1 hour with the ngVLA. In contrast, GN20, at a similar luminosity and redshift, at $z = 4.05$ required a total of 120 hours with the current VLA. If, as in this study, one wishes to spatially resolve the CO line across ~10 beams, 10 hours of integration are required to reach gas masses at this limit. If this signal needs to be further resolved spectrally into N spectral bins (where N has a minimum value of ~4 to kinematically resolve rotation), the rms per channel goes up as $\sqrt{N}$ and the required observing time scales with N.

For this particular galaxy snapshot, the integrated CO luminosity is $2\times10^{10}\,L_\odot$. Distributed across ~10 beams and ~7 spectral bins, the average SNR per beam is anticipated to be ~10 in a 1 hour ngVLA integration of CO(1-0). The required observing time would be approximately 30 minutes for the ~4× brighter CO(2-1) transition. The same system as observed with ALMA in CO(4-3) at matched SNR per spatial and spectral bin would require ~12 hours of on-source integration.

Figure 2 shows a false-color image tracing the molecular gas luminosity of the simulated galaxy itself, where CO(1-0) emission is shown in red, relative to CO(2-1) in green and CO(4-3) in blue[1]. The entire gas reservoir emits somewhat uniformly in CO(1-0), indicative of diffuse gas in the galaxy's interstellar and circumgalactic mediums, rendering this tri-color image largely red over most of the galaxy's observed surface area. The radial surface brightness profiles of CO emission is shown in Figure 3; the contrast of peak emission at the core to diffuse in the outskirts is only a factor of 10-100 in CO(1-0) while it is $>10^{4-6}$ for higher-J CO lines, including CO(2-1).

The kinematic breakdown of the gas in this galaxy is shown at two different angular resolutions in both CO(1-0) and CO(2-1) in Figure 4. The simulated kinematic breakdown in CO(4-3) is very similar to the CO(2-1) profile. All diagrams are limited to a dynamic range of ~150 from peak intensity, which not only mimics realistic user-limited depths of future datasets, but avoids complex calibration needed for high dynamic range interferometric datasets. In other words, even though there is an extended component of emission in CO(2-1) and higher J transitions, they are unlikely to be detected given the significant surface brightness discrepancy between the core and extended disk, as shown in Figure 3. The first two rows show the gas profile at 40 mas resolution, equivalent to a physical resolution scale of 270 pc at this redshift. Without any simulated noise, these diagrams represent the intrinsic kinematic structure of the simulated galaxy. Certainly the complexity of the gas reservoir is quite clear: this system is not a smoothly rotating disk, but rather a turbulent, gas-rich, dispersion dominated system. The difference between the kinematic profiles in CO(1-0) vs CO(2-1) is striking. The compact CO(2-1), which would only be observable on spatial scales <1″, probes the galaxy's core on ~2-3 kpc scales. Though one would conclude that this galaxy is dominated by gas turbulence from both CO(1-0) and CO(2-1) kinematics, only the CO(1-0) data shed light on the more diffuse gas on large physical scales.

[1]Note that at this redshift, CO(3-2) is not observable as its frequency falls in an opaque atmospheric frequency range.

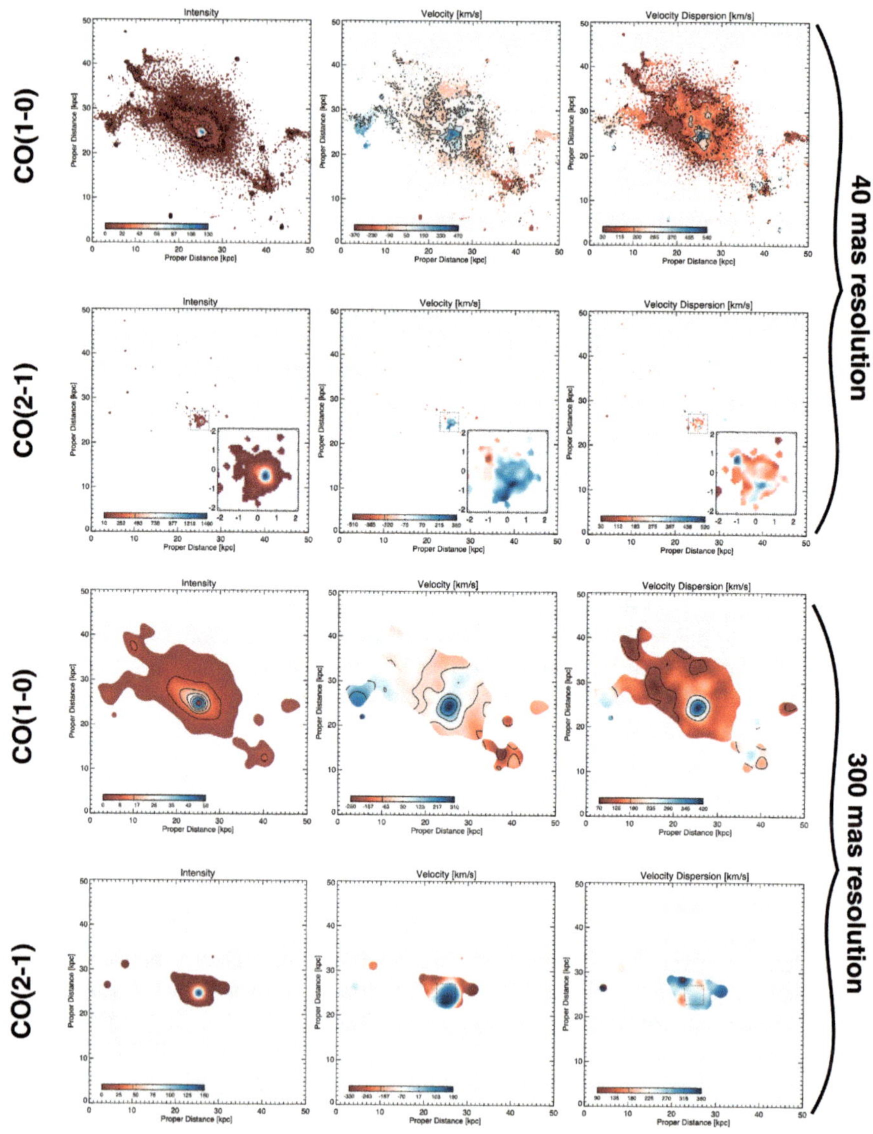

Figure 4. CO intensity, velocity, and velocity dispersion diagrams for the galaxy snapshot under analysis, in both CO(1-0), first and third rows, and CO(2-1), second and fourth rows. The first two rows show the intrinsic kinematic structure of the simulated galaxy with 40 mas resolution, corresponding to physical scales of 270 pc. At bottom, the same diagrams are shown at 300 mas resolution, physically 2 kpc resolution.

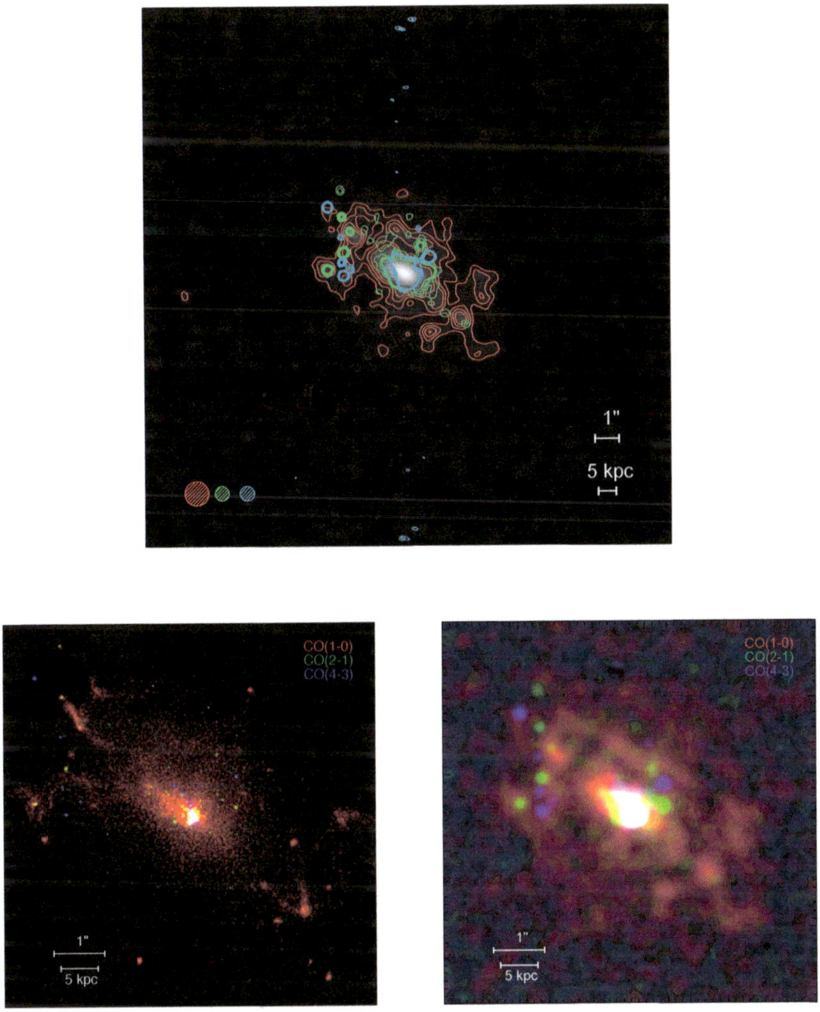

Figure 5. (*Top:*) the ngVLA CO(1-0) moment 0 map overlaid with contours of CO(1-0), CO(2-1) (both from the ngVLA) and CO(4-3) emission (from ALMA). The respective beamsizes, $0\farcs5$ beam in CO(1-0) and $0\farcs3$ beam in both CO(2-1) and CO(4-3), are illustrated in the lower lefthand corner. (*Bottom:*) side-by-side comparison of the raw CO emission in the simulated galaxy against what would be observable with the ngVLA in tricolor. Note that ALMA can observe CO(4-3), and the sole difference is the expected required on-source time to reach a matched depth: <1 hour on source with ngVLA vs ~12 hours on source with current ALMA.

The third and forth rows of Figure 4 show lower resolution maps of the same kinematic structure, with a 0.″3 beam. It is reassuring that the underlying kinematics are intact at this low resolution, and the same physical conclusions would be reached in analyzing these maps as those with much higher spatial resolution: that the system is dispersion dominated. While the two views of this system at different resolutions may provide consistent interpretations, it is important to emphasize that work on artificially redshifting $z \sim 0$ high-SNR, high-resolution galaxy CO data cubes to the same resolution and SNR as typical high-z galaxies has shown that it is easy to draw different conclusions as to the physical drivers of a galaxy's evolution (Hung et al. 2015). Much like the higher resolution images, the CO(2-1) fails to probe the total underlying gas reservoir.

3. Simulating Observables with CASA

We pass the simulated position-position-velocity maps (with very high spatial resolution and velocity resolution) through CASA to simulate the imaging capabilities of the ngVLA. Imaging for these high-redshift galaxies is needed on ~0.″1 scales and so does not require the extremely long ngVLA baselines. However, diffuse gas emission on large scales is of general interest and so the more compact baselines, in concert with the extended baselines on the plateau, will be crucial to the recovery of good image quality.

Figure 5 shows a direct comparison between the observable molecular gas reservoir in the three transitions and tri-color and the underlying intrinsic gas distribution. The mock ngVLA observations mimic the simulated data, also showing the extended reservoir of CO(1-0) beyond the core that is quite luminous in CO(2-1) and CO(4-3). The ngVLA images also show detection of several knots of higher-J CO emission, which trace smaller molecular gas complexes in the galaxy's halo.

4. Kinematically Distinct Gas Reservoirs

In this particular case study, the moment zero line intensity maps shown above already highlight fundamental differences between different CO tracers in a high-z starbursting system. However, these differences are not isolated to the spatial extent of the gas. Figure 6 shows the spectra for each of the three lines and highlight how they are also kinematically distinct. The ground-state transition of CO, extended ~13 kpc along the semi-major axis of the galaxy, is centered on 0 km s^{-1} with a 300 km s^{-1} FWHM. CO(2-1) exhibits a very different kinematic signature, comprised of two components – gas that is kinematically aligned with the CO(1-0) emission and gas that is substantially offset at +350 km s^{-1}, as well as significantly brighter. Higher-J lines, like the ALMA-observed CO(4-3) traces only this offset component, which is perfectly aligned with the galaxy's nucleus.

This has very important implications on deriving dynamical masses, inferring kinematics of galaxies, and measuring gas masses, depletion times, and star-formation efficiencies for galaxies at high-z, because the vast majority of high-z galaxies are and will be characterized by their high-J CO tracers, and not CO(1-0). This demonstrates a key need for the ngVLA: tracing the fundamental cold gas reservoir in high-z galaxies as a core focus of extragalactic astrophysics.

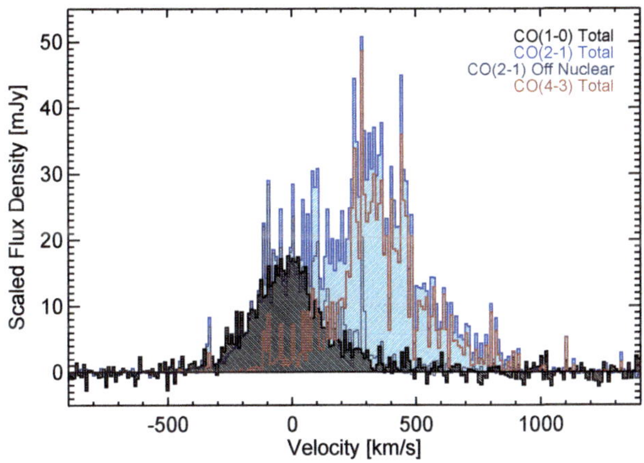

Figure 6. The CO spectra of the simulated galaxies centered on the systemic redshift, which is well-probed by the peak line flux of CO(1-0) (black). All spectra have been renormalized for clear viewing of the kinematic components of the gas reservoir. CO(2-1) is double peaked, with an extended emission component coincident with the CO(1-0) emission and compact nuclear emission, which is both brighter and offset $\sim$400 km s^{-1} from the systemic velocity. CO(4-3) emission primarily traces this nuclear offset peak with both low and high dispersion gas.

Acknowledgments. This work was completed with the generous support of the National Radio Astronomy Observatory next generation VLA community students program grant. CMC, JBC, and CLH thank UT Austin for support through the College of Natural Sciences.

References

Bothwell, M. S., Chapman, S. C., Tacconi, L., et al. 2010, MNRAS, 405, 219
Bothwell, M. S., Aguirre, J. E., Chapman, S. C., et al. 2013, ApJ, 779, 67
Carilli, C. 2015, Advancing Astrophysics with the Square Kilometre Array (AASKA14), 171
Carilli, C. L., & Walter, F. 2013, ARA&A, 51, 105
Casey, C. M., Hodge, J. A., Lacy, M., et al. 2015, ArXiv e-prints, arXiv:1510.06411
Davé, R., Rafieferantsoa, M. H., Thompson, R. J., & Hopkins, P. F. 2017, MNRAS, 467, 115
Davé, R., Thompson, R., & Hopkins, P. F. 2016, MNRAS, 462, 3265
Engel, H., Tacconi, L. J., Davies, R. I., et al. 2010, ApJ, 724, 233
Hodge, J. A., Carilli, C. L., Walter, F., et al. 2012, ApJ, 760, 11
Hung, C.-L., Rich, J. A., Yuan, T., et al. 2015, ApJ, 803, 62
Ivison, R. J., et al. 2012, MNRAS, 425, 1320
Krumholz, M. R. 2014, MNRAS, 437, 1662
Narayanan, D., Conroy, C., Dave, R., Johnson, B., & Popping, G. 2018a, ArXiv e-prints, arXiv:1805.06905
Narayanan, D., Davé, R., Johnson, B. D., et al. 2018b, MNRAS, 474, 1718
Narayanan, D., & Krumholz, M. R. 2017, MNRAS, 467, 50
Solomon, P. M., & Vanden Bout, P. A. 2005, ARA&A, 43, 677

Science with a Next-Generation Very Large Array
ASP Conference Series, Monograph 7
Eric J. Murphy, ed.

An ngVLA Wide Area AGN Survey

Allison Kirkpatrick,[1,2] Kirsten Hall,[3] Kristina Nyland,[4] Mark Lacy,[4] and Isabella Prandoni[5]

[1]*University of Kansas, Lawrence, KS, USA;* akirkpatrick@ku.edu

[2]*Yale University, New Haven, CT, USA*

[3]*Johns Hopkins University, Baltimore, MD, USA*

[4]*National Radio Astronomy Observatory, Charlottesville, VA, USA*

[5]*INAF - IRA, Via P. Gobetti 101, 40129 Bologna, Italy*

Abstract. The next generation Very Large Array (ngVLA) will have unprecedented sensitivities and mapping speeds at 1.2 – 8 GHz. We discuss how the active galactic nuclei (AGN) community can benefit from a wide-area, medium depth ngVLA survey. We propose a 10 deg^2 survey in the Stripe 82 field using the 8 GHz band with an rms depth of 1 μJy beam^{-1}. We will detect ~130,000 galaxies, including radio-quiet AGN out to $z \sim 7$. We can measure the luminosity and space density evolution of radio-quiet and radio-loud AGN. We can also measure AGN evolution through clustering of both populations using cross-correlation functions. A wide area ngVLA survey will benefit from existing multiwavelength AGN populations, particularly in the Stripe 82 field, as well as new information from next-generation optical and infrared survey instruments such as LSST and *WFIRST*.

1. Introduction

Likely every massive galaxy has a supermassive black hole at its center, and this black hole is predicted to have a regulatory effect on the buildup of stellar mass due to the tight relationship between stellar mass and black hole mass (e.g., Magorrian et al. 1998) On a cosmic scale, the star formation rate density (SFRD) peaks at $z \sim 1 - 3$, and this is an epoch where the black hole accretion rate density peaks as well (Madau & Dickinson 2014), although how much of this stellar and black hole mass assembly is occurring within the same galaxies is currently unknown. Probing stellar and black hole mass assembly within similar populations requires wide-area deep surveys, due to the necessity of detecting faint star formation and black hole accretion. Futhermore, studying the large-scale environments and clustering properties of active galactic nuclei (AGN; the phase in which the central black hole is actively accreting) and quasars can elucidate their co-evolution with their host galaxies as well as their greater role in galaxy evolution.

The next generation of radio telescopes (ngVLA and SKA) will enable very deep radio surveys, allowing astronomers to detect *all* AGN, not only the tiny fraction of classical radio-loud AGN, where the radio emission is due to large scale jets. The current picture of radio emission emanating from AGN is that there are two populations: radiatively efficient AGN (quasars/Seyferts) and radiatively inefficient AGN.

Radiatively inefficient AGN (accretion rates <1%) are mainly radio-loud (RL), and in the local universe are typically associated with massive early-type galaxies. Radio-mode feedback (so-called "maintenance mode") is observed in these massive red systems. The enhanced radio emission can be used to calculate the mechanical energy output into the intergalactic medium (Willott et al. 1999), which is observed to heat the centers of clusters and likely serves as the key mechanism that prevents star formation. A common approach to constraining the importance of radio-mode feedback has been to measure the evolution of the luminosity density of RL AGN, assuming that the evolving luminosity functions directly trace the evolving kinetic energy output (Smolčić et al. 2017). Currently unknown is how much maintenance mode feedback could arise from radio-quiet (RQ) AGN through sub-kpc jets. This is particularly relevant at $z > 2$, where the RQ AGN luminosity function dominates over RL AGN (Smolčić et al. 2017).

Radiatively efficient AGN are a more enigmatic population. They are mainly RQ ($L_{1.4\,GHz} < 10^{24}$ W/Hz) quasars and Seyferts (only 10% of quasars/Seyferts are RL), so the radio emission may come entirely from star formation. Typically, they are AGN identified at other wavelengths and then followed up with radio observations. However, 30% of radio-selected RQ AGN have a radio excess (that is a brighter radio flux than would be expected from the standard far-IR/radio correlation), raising the question of whether small-scale radio jets can contribute to the emission of RQ AGN[1](Delvecchio et al. 2017; Smolčić et al. 2017). The radio emission is commonly used to calculate the star formation rate (SFR) in RQ AGN, while AGN emission is seen to contaminate other wavelengths. The radio spectral slope in RQ AGN may differ non-negligibly from star forming galaxies (SFGs), so that a correction is required before converting to an SFR. Relatively little is known about the RQ population, due to the necessity of a deep and wide survey, high angular resolution to separate the AGN from the host, and differing AGN selection techniques. Here, we outline how ngVLA capabilities can accomplish such a survey in 200 hrs, and what the benefit will be.

2. Radio Surveys: Past and Future

Building up statistical samples of AGN requires large-area surveys, since AGN are bright but rare. The VLA FIRST survey covered 10,000 deg^2 at 20 cm at a resolution of 5″ and an rms of 0.15 mJy beam^{-1}. From this survey, strong radio emission was used to find a population of extremely dust-reddened quasars, predicted to be an early evolutionary stage before feedback from the AGN blows out the majority of a galaxy's dust (Glikman et al. 2007; Urrutia et al. 2009). The wide area also allowed the detection of a $z \sim 6$ radio-loud quasar, demonstrating how rare these objects are (Zeimann et al. 2011).

Currently, the VLA Sky Survey (VLASS) is imaging the entire sky visible to the VLA at 2-4 GHz at a resolution of 2.5″ down to an rms of 69 μJy beam^{-1}. This all sky survey is well suited to probing how radio emission might be heating the cores of

[1]Another possibility for the origin of the radio emission in RQ AGN is shocks associated with quasar-driven winds (Zakamska et al. 2014).

clusters through combination with X-ray all sky surveys from ROSAT, which trace hot gas in clusters.

MIGHTEE is a wide-area survey currently being planned with MeerKAT (the precursor to SKA), which will survey 20 deg^2 down to μJy beam^{-1} sensitivities (Jarvis et al. 2017). A survey of this size is designed to detect the environments of low-power radio AGN ($L < 10^{24}$ W Hz^{-1}), which are experiencing radiatively inefficient accretion, in contrast with RQ AGN (Best & Heckman 2012).

The high sensitivity and angular resolution of the ngVLA will capture emission from AGN in an extinction-free manner at the peak epoch of cosmic assembly. The ngVLA will have 6 receivers spanning 1.2 - 116 GHz, maximum baselines of up to 1000 km, and sub-arcsecond resolution. Current and past radio surveys, including MeerKAT, are limited to ~5″ resolution, and SKA Phase-I is expected to have resolutions of 0.″5 – 1″. The unprecedented resolution of the ngVLA will greatly enhance our ability to distinguish high surface brightness AGN emission from jets, and to spatially isolate AGN cores from lower surface brightness emission from star formation, which is critical for identifying low-power radio AGN and for calculating SFR. The bands are wide enough (the 8 GHz band spans 3.5 – 12.3 GHz) that we will be able to separate the flat spectrum of AGN cores from extended steep spectrum emission from star formation or jets. We describe the details of an ngVLA survey below as well as outlining two science use cases – mass assembly and clustering.

2.1. A 10 deg^2, 1 μJy ngVLA Survey

The 8 GHz receiver is optimal for a wide-field survey due to its 7.′3 field of view and sub-arcsecond resolution (~0.″3 with a 36 km baseline). The dust continuum does not contaminate 8 GHz observations until $z > 8$ making synchrotron emission from radio-loud AGN easily distinguishable from star forming galaxies (SFGs). For a 10-square-degree survey at 8 GHz tapered to a resolution of 0.″1 with a 1σ depth of 1 μJy beam^{-1} (given the current ngVLA reference design[2], we estimate an integration time of ≈200 hrs, not including overheads), we will be able to detect L^* SFGs at $z < 3$ (Figure 1a). We calculate L^* using an evolving IR luminosity function (Gruppioni et al. 2013) as this is better constrained than radio L^* evolution (Padovani et al. 2011). We use M 82, a standard SFG, to calculate the predicted radio emission, as the radio to far-IR ratio is consistent for SFGs and RQ AGN (Padovani et al. 2011). RL AGN will dominate the survey at $S_\nu > 500\,\mu$Jy (Bonzini et al. 2013). Figure 1b illustrates the depth of our proposed ngVLA survey in the context of current and future radio surveys. The ngVLA wide survey (10 deg^2) nicely complements the deeper but narrower SKA surveys planned for SKA Band 5 (1 and 0.008 deg^2 respectively), that will have similar angular resolution.

2.2. The Case for Stripe 82

The Sloan Digital Sky Survey (SDSS, York et al. 2000; Abazajian et al. 2009; Eisenstein et al. 2011; Dawson et al. 2013; Alam et al. 2015) Stripe 82 field is a natural choice for a 10 deg^2 legacy survey. 15 deg^2 of Stripe 82 has contiguous *Chandra* or *XMM-Newton* coverage (LaMassa et al. 2016), detecting all the brightest X-ray sources ($L_X > 10^{44}$ erg s^{-1}), which are the sources most likely to be radio-loud. Stripe 82 has

[2]http://ngvla.nrao.edu/page/refdesign

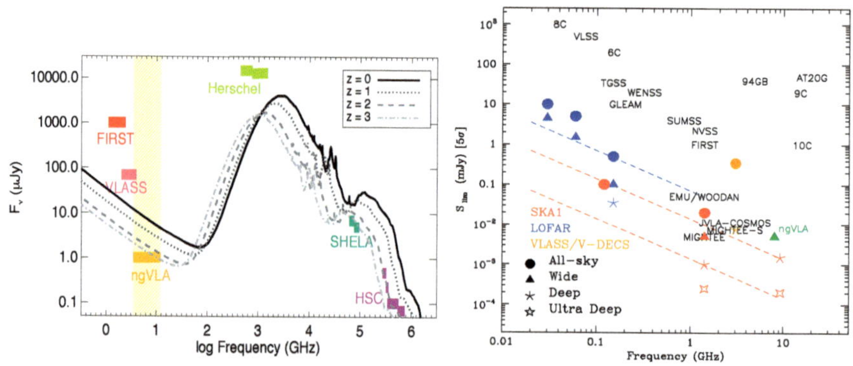

Figure 1. (*Left:*) An L^* SFG (represented by M 82) at $z = 0.5$ (solid line), $z = 1$ (dotted line), $z = 2$ (dashed line), and $z = 3$ (dot-dashed line). The ngVLA 8 GHz band is shaded in orange. The 1 μJy beam^{-1} rms limit (orange bar) corresponds to an L^* galaxy at $z = 3$. We illustrate the flux limits of other Stripe 82 surveys: FIRST (red), VLASS (pink), it Herschel (chartreuse), SHELA IRAC (aquamarine), and HSC (purple). (*Right:*) 5σ flux limit vs. frequency for ngVLA in comparison with existing and planned surveys. LOFAR, VLASS, and SKA1 reference surveys are highlighted in blue, orange and red respectively. Different symbols refer to different survey coverage: all-sky (filled circles); wide tiers (filled triangles); deep tiers (asterisks); ultra deep tiers (starred symbols). The red and blue dashed lines indicate a slope of ν^{-1} for different 1.4 GHz fluxes.

extensive overlapping multiwavelength coverage, in particular the *Spitzer* SPIeS (Timlin et al. 2016) and SHELA (Papovich et al. 2016) surveys, *Herschel* SPIRE (Viero et al. 2014), Subaru HyperSuprimeCam (HSC; Miyazaki et al. 2018), as well as VLA FIRST (Becker et al. 1995) and VLASS (Hales 2013). We illustrate the various flux density limits in Figure 1.

2.3. Expected Number of Sources

We calculate the expected number of sources in a 1 μJy beam^{-1}, 10 deg^2 survey by extrapolating from the JVLA-COSMOS 3 GHz survey. JVLA-COSMOS covered 1.77 deg^2 with a uniform rms noise of 2.3 μJy beam^{-1}. The final catalog consists of 10,830 sources detected with signal-to-noise ratio (S/N) >5. Novak et al. (2017) calculate 1.4 GHz luminosity functions out to $z \sim 5$ for SFGs[3], and Smolčić et al. (2017) does the same for radio excess AGN. Both works assume an $\alpha = -0.7$ to convert from 3 GHz to 1.4 GHz.

The number of sources in a given spherical shell volume is

$$N = \Phi(L, z) \frac{dV_c}{dz\, d\Omega} \Omega\, \Delta z\, \Delta \log L \tag{1}$$

[3]In this subsection, we follow the convention in Novak et al. (2017) and use the term SFG to mean any source where the radio emission arises primarily from star formation. In practice, this can consist of SFGs and RQ AGN.

We extrapolate $\Phi(L, z)$ for AGN and SFGs to the proposed area and flux limit of our survey, setting a detection threshold at S/N>3. We use $\alpha = -0.7$ to convert between 1.4 GHz and 8 GHz luminosities. We anticipate observing 112,000 SFGs and 18,000 AGN, for a total of 130,000 sources.

3. Luminosity Evolution of all AGN out to $z \sim 7$

To date, the best constraints on the evolving radio luminosity function come from deep observations in the COSMOS field (Smolčić et al. 2017; Novak et al. 2017) and CDFS field (Padovani et al. 2011). Both surveys are hampered by small number statistics, particularly at $z > 3$, owing to the relatively small survey areas. Although JVLA-COSMOS covers 2.6 deg^2, only 1.77 deg^2 contains optical/NIR counterparts (Laigle et al. 2016), which are needed to measure redshifts and identify AGN.

Our significantly wider field will greatly improve number statistics at high redshift. We will detect ~300 RL AGN at $z > 3.3$, an order of magnitude more than JVLA-COSMOS. For SFGs, JVLA-COSMOS reached a limiting redshift of $z \sim 5.7$. Our 10 deg^2 survey will detect ~60 SFGs at $z > 5.7$, with 4 of those being at $z > 6.8$, enabling us to constrain the high-z radio luminosity function. The luminosity density describes how the radio population has evolved over cosmic time. For SFGs, this luminosity density translates directly to a star formation rate density (SFRD; e.g., Yun et al. 2001; Murphy et al. 2011). For RL AGN, it corresponds to a mechanical energy output arising from feedback (Willott et al. 1999). We predict the luminosity density evolution observed with an ngVLA survey by assuming SFGs and AGN follow a pure luminosity evolution:

$$\Phi[L, z] = \Phi_0 \left[\frac{L}{(1+z)^{\alpha_L + z\beta_L}} \right] \quad (2)$$

We fit for α_L and β_L using our extrapolated SFG and AGN luminosity functions (see §2.3). We then integrate $\Phi(L, z)$ to obtain the luminosity density (Figure 2).

For RL AGN, a wider field may not allow us to detect significantly higher redshift AGN than what have already been found due to the evolution of their space density. Smolčić et al. (2017) predict that the space density (and corresponding luminosity density) falls off significantly at $z > 3$, hampering their detectability even in wide fields. However, the luminosity and density evolution of RL AGN is still highly uncertain, particularly beyond $z > 3$, so our improved number statistics will enable us to more robustly determine the high-z luminosity function.

For radio SFGs, we will detect galaxies up to $z \sim 7.5$, beyond the capabilities of any current survey. In JVLA-COSMOS, Novak et al. (2017) divide sources into either radio SFGs or radio AGN, based on the ratio of radio emission to the infrared. While the radio emission in SFGs can be attributed to star formation, it does not mean that these galaxies are purely star-forming. They may be selected as RQ AGN, if multiwavelength AGN identification techniques are used, as was done in the CDFS field (Padovani et al. 2011; Bonzini et al. 2013). Our sub-arcsecond resolution will enable us to measure the radio emission in the central kpc and outskirts of every galaxy out to the redshift limit of this survey, cleaning separating the amount of radio emission due to star formation and AGN.

Measuring the SFRD of RQ AGN will determine how much of star formation and black hole growth is truly coeval. Bonzini et al. (2013) calculate the fraction of SFGs and RQ AGN as a function of flux density in the VLA CDFS survey, and Padovani et al. (2011) estimates that RQ AGN follow a similar space density evolution as SFGs. We use the fractions in Bonzini et al. (2013) to estimate how the SFRD curve might break up into the SFR in true SFGs and the SFR in RQ AGN in Figure 2. 30% of RQ AGN have a radio emission component likely due to the AGN itself, rather than star formation, so this component can then be converted to a mechanical energy output, as has been done for RL AGN (Willott et al. 1999). The sub-arcsecond resolution of an ngVLA survey will allow for a complete picture of mass assembly and energy output in radiatively efficient and inefficient AGN and SFGs.

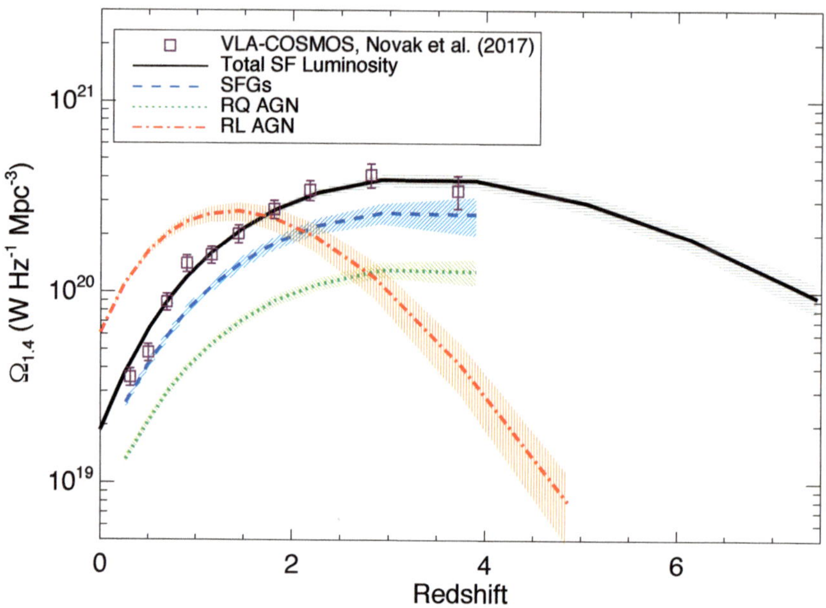

Figure 2. This figure predicts the radio luminosity functions we would measure with the ngVLA. We have converted 8 GHz to 1.4 GHz to allow easy comparison with the literature. The black curve predicts $\Omega_{1.4}$ from star formation, which we have divided into a component from SFGs (blue) and RQ AGN (green). The current best measurement of the SFG $\Omega_{1.4}$ is shown as the purple squares and comes from VLA-COSMO (Novak et al. 2017). The RL AGN $\Omega_{1.4}$ measurements from Smolčić et al. (2017) reach the same redshift as illustrated by the red curve. With a wider survey, we will be able to test whether the RL AGN luminosity density really declines so strongly after $z \sim 3$.

4. Cross Correlation Functions

Astronomical sky surveys allow for the computation of galaxy correlation functions. Of particular use and interest are two point correlation functions: the probability above that of a Poisson distribution of finding two objects within some distance of one another. Quantifying auto- and cross-correlation functions of different populations of galaxies yields information about the large-scale structure of the universe, or the way that galaxies cluster together. Analyzing correlation functions as a function of redshift tells about

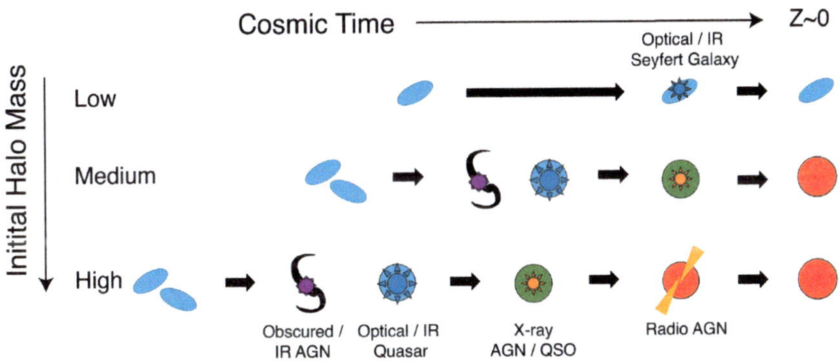

Figure 3. Figure adapted from Hickox et al. (2009) showing how different wavelength selection of AGN corresponds to a simple evolutionary sequence. The AGN phase begins after halos (dark ovals) reach a critical mass. Rapid black hole growth is initially obscured, followed by an unobscured X-ray phase. After quenching of star formation, red ellipticals are observed as radio AGN, primarily in the centers of massive halos. Measuring clustering of different AGN populations at different redshifts can test this scenario since clustering depends on halo mass.

the growth of structure and the evolution of galaxy clusters as a function of cosmic time, since clustering strength of a population should increase as time passes (Groth & Peebles 1977).

The Sloan Digital Sky Survey covers over 35% of the sky and has spectroscopically catalogued hundreds of thousands of quasars yielding accurate redshifts (out to $z \sim 7$) for each. Quasars are excellent tracers of large scale structure because they inhabit massive galaxies in small groups and clusters (Wold et al. 2000; Coldwell & Lambas 2006; Trainor & Steidel 2012), though they have low spatial density. Statistical samples from surveys such as the SDSS and the 2dF QSO Redshift Survey (Croom et al. 2004) enabled the measurement of the quasar auto-correlation function from $z \sim 0.5$ (e.g., Porciani et al. 2004; Croom et al. 2005) out to $z \gtrsim 2.9$ (Shen et al. 2007).

Quantifying the clustering properties of different types of AGN – X-ray, radio-loud, radio-quiet, IR bright – can be informative of any fundamental differences between these populations and how they impact galaxy evolution. Clustering studies inform us of the dark matter halo masses in which objects reside as well as how they occupy their dark matter halos (Peacock & Smith 2000; Scoccimarro et al. 2001; Berlind & Weinberg 2002). For example, Mandelbaum et al. (2009) find that RL AGN live, on average, in halos that are an order of magnitude higher than the RQ population. Powell et al. (2018) cross correlate X-ray selected AGN with 2MASS near-IR galaxies and find that at $z < 0.1$ AGN live in group environments, and that obscured AGN tend to occupy denser environments than the unobscured population.

A Stripe 82 ngVLA survey provides a unique opportunity to test the evolution of AGN, their host galaxies, and their clustered environments. Hickox et al. (2009) measure the clustering and accretion rates of multiwavelength AGN at $0.25 < z < 0.8$ and find that RL AGN are the most strongly clustered, while IR AGN are the least strongly clustered. They propose an evolutionary sequence that we illustrate with Figure 3. With an ngVLA survey, we will be able to measure the clustering of RL AGN, red quasars, and X-ray luminous AGN out to $z \sim 4$ (depending on the RL AGN space density), for the first time measuring how the clustering of the different populations evolves over cosmic time. The proposed survey will even be useful in quantifying physical properties of the RQ population by cross-correlating the survey map with a catalog of known RQ quasars, which by the time of the ngVLA survey will be much deeper than what we have currently as a product of future optical surveys. Our survey will be able to test at what halo mass the quasar phase occurs, and whether immediately after the IR/optical (currently measured with WISE and *Spitzer*/IRAC) AGN phase, the galaxy evolves into a RL AGN, producing maintenance mode feedback (Hickox et al. 2009).

5. Summary and Future Directions

A 10 deg^2, 1 μJy beam^{-1} ngVLA survey will detect 130,000 galaxies in ~200 hrs, allowing us to characterize the luminosity function of SFGs, RQ AGN, and RL AGN to $z \sim 7$. Existing X-ray, optical, and IR counterparts yield photo-z's, useful for calculating the clustering of multiwavelength AGN populations. However, by the time the ngVLA begins operations (2030s), we will have much deeper optical data than current SDSS. We will have robust photo-z measurements from deep surveys such as HSC and LSST that would be better matched to the increased sensitivity of the ngVLA. LSST will help with AGN identification/characterization through optical variability measurements (e.g., Choi et al. 2014). *WFIRST* (if launched) will provide sub-arcsecond resolution IR imaging that would be well-matched to that of the ngVLA and trace the rest-frame optical emission of high-z galaxies, allowing us to separate emission in the central kpc of galaxies, which should be dominated by the AGN, from the outskirts. eROSITA will provide the first all sky X-ray survey up to 10 keV, enabling mapping of the hot gas in clusters heated by AGN feedback. Potentially *Lynx* will provide sub-arcsecond X-ray observations capable of directly imaging supermassive black holes in the early universe. By combining next generation telescopes with the ngVLA, we will uncover the hidden history of black hole growth through cosmic time.

Acknowledgments. The authors thank Jeyhan Kartaltepe for the use of Figure 3, which she adapted from Hickox et al. (2009) and Vernesa Smolčić for helpful conversations.

References

Abazajian, K. N., Adelman-McCarthy, J. K., Agüeros, M. A., et al. 2009, ApJS, 182, 543
Alam, S., Albareti, F. D., Allende Prieto, C., et al. 2015, ApJS, 219, 12
Becker, R. H., White, R. L., & Helfand, D. J. 1995, ApJ, 450, 559
Berlind A. A., Weinberg D. H., 2002, ApJ, 575, 587
Best P. N. & Heckman T. M., 2012, MNRAS, 421, 1569
Bonzini, M., Padovani, P., Mainieri, V., et al. 2013, MNRAS, 436, 3759
Choi, J., Conroy, C., Moustakas, J., et al. 2014, ApJ, 792, 95
Coldwell, G. V. & Lambas, D. G 2006, MNRAS, 371, 786
Croom, S. M., Smith, R. J., Boyle, B. J., et al. 2004, MNRAS, 349, 1397
Croom, S. M., Boyle, B. J., Shanks, T., et al. 2005, MNRAS, 356, 415
Dawson, K. S., Schlegel, D. J., Ahn, C. P., et al. 2013, AJ, 145, 10
Delvecchio, I., Smolčić, V., Zamorani, G., et al. 2017, A&A, 602, 3
Eisenstein, D. J., Weinberg, D. H., Agol, E., et al. 2011, AJ, 142, 72
Glikman, E., Helfand, D. J., White, R. L., et al. 2007, ApJ, 667, 673

Groth, E. J. and Peebles, P. J. E. 1977, ApJ, 217, 385
Gruppioni, C., Pozzi, F., Rodighiero, G., et al. 2013, MNRAS, 432, 23
Hales, C. A. 2013, ArXiv e-prints, arXiv:1312.4602
Hickox, R. C., Jones, C., Forman, W. R., et al. 2009, ApJ, 696, 891
Jarvis M. J., Taylor, R., Agudo, I., et al. 2017, preprint (arXiv:1709.01901)
Laigle, C., McCracken, H. J., Ilbert, O., et al. 2016, ApJ, 224, 24
LaMassa, S. M., Urry, C. M., Cappelluti, N., et al. 2016, ApJ, 817, 172
Madau, P. & Dickinson, M. 2014, ARA&A, 52, 415
Magorrian, J., Tremaine, S., Richstone, D., et al. 1998, AJ, 115, 2285
Mandelbaum, R., Li, C., Kauffmann, G., & White, S. D. M. 2009, MNRAS, 393, 377
Miyazaki, S., Komiyama, Y., Kawanomoto, S., et al. 2018, PASJ, 70, S1
Murphy, E. J., Condon, J. J., Schinnerer, E., et al. 2011, ApJ, 737, 67
Novak, M., Smolčić, V., Delhaize, J., et al. 2017, A&A, 602, A5
Padovani, P., Miller, N., Kellermann, K. I., et al. 2011, ApJ, 740, 20
Papovich, C., Shipley, H. V., Mehrtens, N., et al. 2016, ApJS, 224, 28
Peacock J. A., Smith R. E., 2000, MNRAS, 318, 1144
Porciani, C., Magliocchetti, M., & Norberg, P. 2004, MNRAS, 355, 1010
Powell, M. C., Cappelluti, N., Urry, C. M., et al. and Koss, M. 2018, ApJ, 858, 110
Scoccimarro R., Sheth R. K., Hui L., Jain B., 2001, ApJ, 546, 20
Shen, Y., Strauss, M. A., Oguri, M., et al. 2007, AJ, 133, 2222
Smolčić, V., Novak, M., Delvecchio, I., et al. 2017, A&A, 602, A6
Timlin, J. D., Ross, N. P., Richards, G. T., et al. 2016, ApJS, 225, 1
Trainor, R. F. & Steidel, C. C. 2012, ApJ, 752, 39
Urrutia, T., Becker, R. H., White, R. L., et al. 2009, ApJ, 698, 1095
Viero, M. P., Asboth, V., Roseboom, I. G, et al. 2014, ApJS, 210, 22
Willott, C. J., Rawlings, S., Blundell, K. M., & Lacy, M. 1999, MNRAS, 309, 1017
Wold, M., Lacy, M., Lilje, P. B., & Serjeant, S. 2000, MNRAS, 316, 267
York, D. G., Adelman, J., Anderson, Jr., J. E. et al. 2000, AJ, 120, 1579
Yun, M. S., Reddy, N. A., & Condon, J. J. 2001, ApJ, 554, 803
Zakamska, N. L. & Greene, J. E. 2014, MNRAS, 442, 784
Zeimann, G. R., White, R. L., Becker, R. H., et al. 2011, ApJ, 736, 57

Science with a Next-Generation Very Large Array
ASP Conference Series, Monograph 7
Eric J. Murphy, ed.

The Sunyaev-Zeldovich Effect from Quasar and Starburst Winds

Mark Lacy,[1] Suchetana Chatterjee,[2] Avinanda Chakraborty,[2] Brian Mason,[1] Craig Sarazin,[3] Amy Kimball,[4] Kristina Nyland,[1] Graca Rocha,[5] and Barney Rowe[6]

[1]*NRAO, Charlottesville, VA, USA;* mlacy@nrao.edu, bmason@nrao.edu, knyland@nrao.edu

[2]*Presidency University, Kolkata, India*

[3]*Department of Astronomy, University of Virginia, 530 McCormick Rd., Charlottesville, VA 22904, USA*

[4]*NRAO, Socorro, NM, USA*

[5]*Jet Propulsion Laboratory, Pasadena, CA, USA*

[6]*University College, London, UK*

Abstract. The Sunyaev-Zeldovich Effect (SZE) can be used to detect the hot bubbles in the intergalactic medium blown by energetic winds from AGN and starbursts. By directly constraining the kinetic luminosity and total energy of the outflow, it offers the promise of greatly increasing our understanding of the effects of wind feedback on galaxy evolution. Detecting the SZE in these winds is very challenging, at the edge of what is possible using existing facilities. The scale of the signal (10-100 kpc) is, however, well matched to interferometers operating at mm wavelengths for objects at $z \sim 1$. Thus this could become a major science area for the ngVLA, especially if the design of the core is optimized for sensitivity on angular scales of $> 1''$ in the 90 GHz band.

1. Introduction

Winds from AGN and starbursts are one of the major sources of feedback in galaxy evolution. Their interaction with the interstellar medium (ISM) of the galaxy can inject turbulence, dissociate molecular gas, or even drive the gas out of the galaxy completely (Silk & Rees 1998; Bower et al. 2006; Hopkins et al. 2006; Croton et al. 2006; Richardson et al. 2016). Estimating the energy in a wind or outflow is very difficult, however, as, in most models, the predominant phase in the outflowing gas is hot ($\sim 10^7$K) gas with low density. Emission from such gas can only be detected in the X-ray. The dependence of X-ray emission on the square of the density, and confusion with point sources such as AGN, make any detection extremely difficult (e.g., Powell et al. 2018). There is another way to detect the host gas in these winds other than through their emission though, and that is through the Sunyaev-Zeldovich Effect (SZE). This is the distortion to the spectrum of the cosmic microwave background (CMB) radiation when it travels through hot gas. Two terms exist, one thermal and one kinetic. The thermal compo-

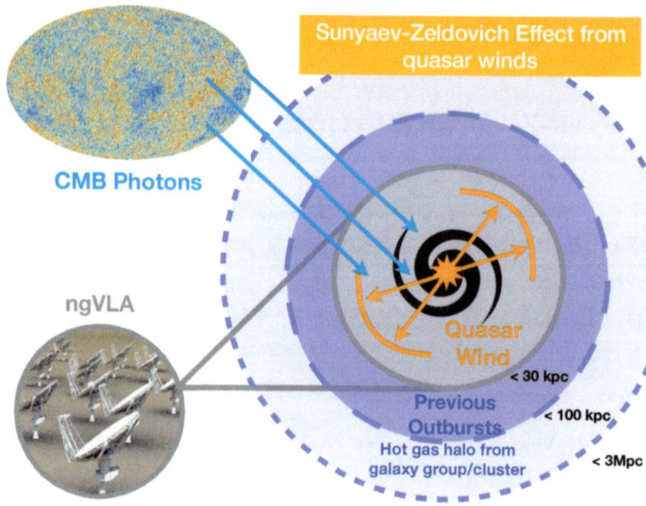

Figure 1. Schematic of the observation of a quasar wind using the ngVLA. Microwave background photons are upscattered in energy through their interaction with fast moving electrons in an expanding quasar wind. Contributions come both from the bulk motion of the electrons towards or away from the observer, and from the thermal motion of the electrons in the hot wind. Relics of previous activity also contribute to the SZE on scales ~ 100 kpc, along with any hot gaseous halo from a group or cluster of galaxies surrounding the quasar on Mpc scales.

nent (the tSZE) is proportional to the line of sight electron pressure through the gas, while the kinetic component (the kSZE) is proportional to the line of sight bulk velocity of the electrons. While the tSZE is now routinely detected in galaxy clusters (e.g., Bleem et al. 2015), the kSZE has only been convincingly detected in the cluster MACS J0717.5+3745 (Sayers et al. 2013; Adam et al. 2017).

A wind from a quasar or starburst produces an expanding bubble of hot gas that is highly overpressured compared to the surrounding medium (interstellar, intragroup or intracluster; see Figure 1). Thus, it should be possible to detect the SZE towards powerful quasars and starbursts (Chatterjee & Kosowsky 2007; Chatterjee et al. 2008; Rowe & Silk 2011; hereafter RS11). Statistical studies using stacked data from single dish telescopes have detected significant signals from quasar hosts (Chatterjee et al. 2010; Ruan et al. 2015; Crichton et al. 2016), but it is unclear whether these results are affected by contamination of the SZE signal within the large single beam size by either the intragroup or intracluster medium around the quasar (below the null in the tSZE at 218 GHz; Figure 2), or from star formation in the quasar host at high frequencies (above 218 GHz) (Cen & Safarzadeh 2015; Soergel et al. 2017). The SZE signal is independent of redshift, making it a powerful tool for studying the intergalactic medium (IGM) at high redshifts.

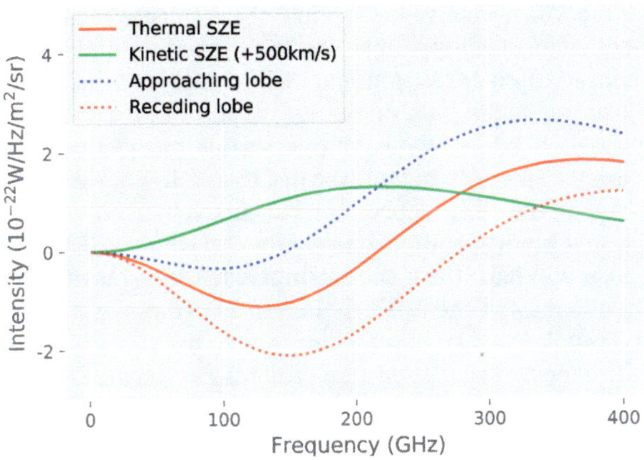

Figure 2. The intensity of the SZE as a function of frequency. The purely thermal model (red) assumes a Compton Parameter $y = 10^{-5}$ (see §3.1), and shows the typical tSZE signal from a quasar wind bubble, with a decrement in the microwave background observed shortward of 218 GHz and an increment at higher frequencies. The kinetic model (green) is the signal from electrons with a bulk velocity of 500 km/s towards the observer with the same line of sight optical depth as the thermal population (assuming a temperature of 10^7K for the thermal elections). The two dashed lines show the situation when the outflow consists of two lobes, one approaching the observer and one receding (Lacy et al. 2018). The thermal decrement signal from the receding lobe is enhanced by the bulk motion of the outflow. For the approaching lobe, the decrement signal is almost canceled by the kinetic signal, but the increment at high frequencies is enhanced.

2. Direct Detection Experiments

Attempts to directly detect the SZE from quasar- or starburst-driven winds can be made with interferometers such as the Atacama Large Millimetre/submillimetre Array (ALMA; e.g., Chatterjee et al. 2008; RS11), though the observations are very challenging, requiring long (tens of hours) integration times to detect or constrain even the most powerful outflows. At $z \gtrsim 1$, the predicted size scale of the SZE signal, $\sim 10 - 100$ kpc, is well matched to the angular resolution of ALMA or the VLA in their most compact configurations, and thus to the core of the proposed ngVLA. The good match of the interferometer beam to the expected scale of the signal reduces the effect of beam dilution compared to single dish observations (though the largest outflows will be well matched to the resolution of the Greenbank Telescope (GBT), $\approx 10''$ at 90 GHz). The need to subtract contaminating emission from the central object (AGN or starburst) driving the flow also means that high angular resolution is needed. This can be achieved either by taking data in a larger configuration, or, in the case of a relatively well-filled aperture like ALMA or the ngVLA core, weighting the visibilities in such a way as to enhance the resolution.

2.1. The Contribution of the Kinetic SZE

In galaxy clusters, the kSZE, due to the motion of the cluster relative to the CMB, is typically much smaller in magnitude than the tSZE from the thermal motions of the electrons, and is hard to disentangle from the tSZE signal. In winds though, the kSZE signal can be very strong as the bulk motion of the electrons can be several thousand $\mathrm{km\,s^{-1}}$. Even in post-shocked gas, the bubble expansion rate can be $\gg 100\,\mathrm{km\,s^{-1}}$ and contribute significantly to the SZE signal (see discussion in Lacy et al. 2018). Furthermore, because the sign of the kSZE depends on the bulk velocity of the electrons relative to the line of sight, in a bipolar outflow one side will have the tSZE and kSZE adding together, and one side will have them subtracting, resulting in asymmetric decrements at the observing frequencies of the ngVLA (Figure 2). To provide further evidence of a kSZE detection, complementary observations above the tSZE null at 218 GHz will be needed to see the expected change in sign of the tSZE signal. Detection of both the kSZE and tSZE signal will allow the accurate characterization of the hot wind, including its kinetic luminosity, age, and mass outflow rate.

2.2. Relativistic Jets

Feedback from relativistic plasma jets (as opposed to thermal winds), in the "radio" mode (e.g., Fabian 2012) has also been suggested as an important influence on galaxy formation. The potentially long lifetime of the low-power radio jets powering radio galaxies with luminosities $\sim 10^{24}\,\mathrm{W\,Hz^{-1}}$ ($\gtrsim 10^8$ yr) makes this form of feedback an attractive option for heating the intragroup or intracluster medium consistently during gaps between bursts of AGN activity at high accretion rates ($\sim 0.1 - 1 \times$ Eddington). Feedback by short-lived ($\sim 10^7$ yr) powerful FRII radio jets (e.g., Nesvadba et al. 2017) is easy to detect (though understanding its effects on the ISM of the host is more complex, e.g., Mukherjee et al. 2016), but the radio mode typically involves radio sources of much lower power ($L_{1.4\mathrm{GHz}} \sim 10^{24}\,\mathrm{W\,Hz^{-1}}$).

As electrons become relativistic, the strength of the SZE as a function of electron pressure drops. So a relic radio source can be detected as a local reduction of the SZE signal in the intracluster medium. This opens up the possibility of finding evidence for ancient ($> 10^8$yr) episodes of radio mode feedback via the SZE (Pfrommer et al. 2005), even if the synchrotron emission has faded. Indeed, as the morphology and scale size of any thermal wind signal is likely to differ from that of a well-collimated relativistic jet, it might be possible to detect both an enhancement in the SZE from a thermal wind and a reduction from bubbles fed by relativistic plasma from radio jets in a single object. This makes the lower frequency capabilities of the ngVLA important to this case too - sensitive observations at cm wavelengths and subarcsecond resolution will be needed to image any jets so we can understand the relative contributions of thermal winds and relativistic jets to the feedback process.

3. Experimental Design

3.1. Signal Strength and Scale Size

The intensity of the SZE spectral distortion observed at frequency ν (wavelength λ), assuming an isothermal plasma and defining $x = h\nu/kT_{\mathrm{CMB}}$, is given by e.g., Sazonov

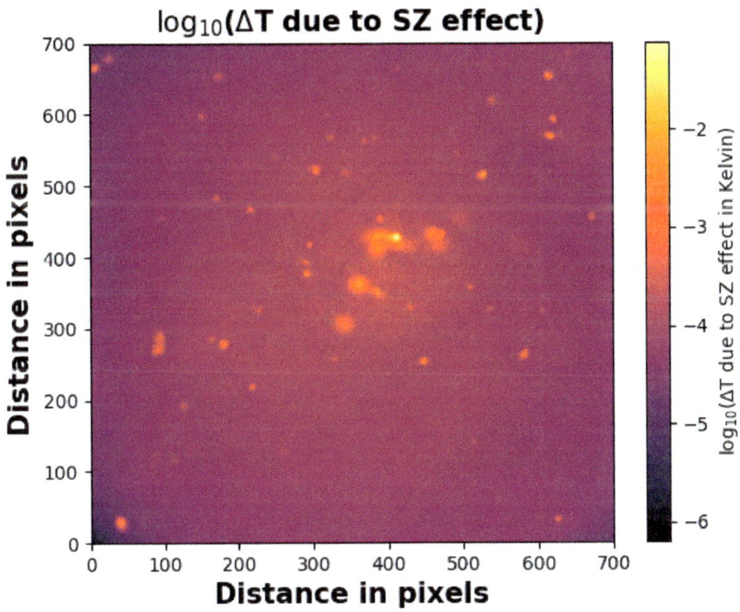

Figure 3. Simulation of the tSZE signal from Chakraborty (2017) using the hydrodynamic simulations of Khandai et al. (2015). The central object is a $2 \times 10^9 M_\odot$ quasar with a bolometric luminosity of $3 \times 10^{11} L_\odot$. Each pixel is 0.14 kpc in size; the image is $\approx$100 kpc on a side.

& Sunyaev (1998):

$$\Delta I(x) = \frac{2kT_{\rm CMB}}{\lambda^2} \frac{x^2 e^x}{(e^x - 1)^2} \tau \left[\frac{kT_e}{m_e c^2} f_1(x) + \frac{v_r}{c} \right]. \quad (1)$$

The left-hand term inside the brackets on the right-hand side of Equation 1 represents the tSZE component, and the right-hand term the kSZE. τ is the Thompson scattering optical depth through the plasma, $\tau = \sigma_T \int dl\, n_e(l)$; and $f_1(x) = x \coth(x/2) - 4$ expresses the frequency dependence of the tSZE. The v_r/c term due to the kSZE is proportional to the line of sight velocity v_r. This expression neglects higher order terms in v_r/c and $kT_e/m_e c^2$. The leading order relativistic correction in T_e is a 4% correction to the tSZE magnitude for a $T_e = 5$ keV plasma at $\nu = 140$ GHz (Itoh & Nozawa 2004). The strength of the tSZE is often expressed in terms of the Compton parameter, $y = \tau k_{\rm B} T_{\rm e}/(m_{\rm e} c^2)$.

Both analytical and numerical models (e.g., Figure 3) provide some guidance on the likely size scale and detectability of the tSZE from winds. RS11 give an analytic prescription for starburst winds using a simple adiabatic model (which works equally well for AGN winds as the model assumes a spherically symmetric outflow from a small central source). They obtain a peak value for the y-parameter in the bubble,

$$y_{\rm peak} = 1.13 \times 10^{-5} (L_{\rm W}/10^{12} L_\odot)^{3/5} (t/10^7 {\rm yr})^{1/5} (1+z)^{6/5}, \quad (2)$$

where L_W is the kinetic power of the wind, and t the age of the outflow, with an estimated radius of the bubble of

$$R = 48\left(\frac{L_W}{10^{12}L_\odot}\right)^{1/5}\left(\frac{t}{10^7\mathrm{yr}}\right)^{3/5}(1+z)^{-3/5}\mathrm{kpc}. \qquad (3)$$

The gas temperature in the post-shock region behind the edge of the bubble, $T_s = 3\mu m_p v_s^2/(16k)$, is typically high ($\sim 10^7$K), and the densities low ($\sim 10^{-5}\mathrm{m}^{-3}$), resulting in long cooling times $\sim 10^9$ yr. The contribution from the kSZE is dependent on geometry and bubble expansion speed, as well as the observing frequency, but can be comparable to that from the tSZE (Lacy et al. 2018).

Hydrodynamic simulations by Chatterjee et al. (2008) emphasize the cumulative effect of outbursts over the lifetime of the host galaxy. These indicate a larger halo of hot gas, $\sim 0.1 - 1$ Mpc in size and formed from the superposition of multiple AGN outbursts, that can be detected by telescopes with high surface brightness sensitivity and less filtering of emission on large angular scales than the ngVLA interferometer. For this science case in particular, a large single dish addition to the ngVLA would be optimal.

The peak signal from the tSZE decrement (which is relevant for the ngVLA frequency range) is at $\approx$ 130 GHz, higher than the proposed maximum operating frequency (though accessible to ALMA). The very large collecting area of the ngVLA, however, will make it more effective than ALMA at this science even operating at 100 GHz, where the tSZE decrement signal remains 87% of the peak.

3.2. Observing Strategy and Possible Targets for the ngVLA

As the strength of the tSZE signal is proportional to the wind luminosity to the 3/5 power, it makes sense to target the most luminous examples of starbursts and quasars known. The size scale of the wind bubble of a few arcsec makes it well suited to interferometers rather than the single dishes used in most SZE experiments, and is only a weak function of the luminosity of the wind, mostly depending on the age of the outflow.

Possible starburst targets are discussed in RS11. Quasar targets include the most luminous quasars in the northern hemisphere, with bolometric luminosities (L_{bol}) close to $10^{15}L_\odot$ (e.g., HS 1700+6421; z = 2.75). Another strategy would be to target luminous quasars with known winds in their absorption line spectra (Broad Absorption Line quasars). Even using these luminous targets, detection of any signal will be challenging. For the hyperluminous southern quasar HE 0515-4414 (which also has $L_{\mathrm{bol}} \approx 10^{15}L_\odot$), we detected a signal at 3.5σ with a $y_{\mathrm{peak}} = 2\times10^{-5}$ using a deep ALMA (12.4 hr time on source) observation. This detection corresponds to a kinetic luminosity in the winds of about 0.01% of L_{bol} (Lacy et al. 2018). Nevertheless, the outflow in HE 0515-4414 still has a total enthalpy $\approx 3\times10^{53}$ W, comparable to the largest bubbles excavated by radio jets in radio mode feedback, and also seems to be long-lived (with a cooling time $\sim$ 0.6 Gyr), and so could be providing feedback in a similar manner to radio sources.

For a given τ, T_e and v_r, the SZE signal from winds is independent of redshift, making these observations particularly valuable for studying winds at $z \gg 1$, which would be almost impossible to detect in the X-ray. The simple RS11 wind model in Equations 2 and 3 suggest that a wind at high redshift would have both a higher decrement and a smaller size (making it more suitable for follow-up with an interferometer).

However, the RS11 model assumes that the IGM density around the quasar is scaling only with redshift, which may not be the case in practice, and also, at higher redshift, the signal is less likely to have been enhanced by earlier outbursts of activity. Further observations are thus needed to investigate how the detectability of the SZE signal from winds depends on redshift.

Yet another approach is to select the most massive quiescent elliptical galaxies that must have supported luminous starburst and/or quasar activity in the past on the basis that – assuming a fixed fraction of the radiative energy goes into the wind – then these object will have the largest remnant bubbles, even if they are not currently active. An example of this approach applied to a stacking analysis is the 3.6σ detection of an SZE signal in stacked luminous red galaxies (Spacek et al. 2016).

The larger collecting area and wide bandwidth of ngVLA make it potentially much more sensitive than ALMA for detecting these winds, however, the current proposed configuration ("ngvla-revB"), where the core is offset to the north of the overall antenna distribution is not optimal for sensitivity on angular scales > 1-arcsec at 100 GHz (Carilli 2018), so in practice the detection sensitivity to a 50 kpc outflow signal at $z \sim 1$ is comparable to ALMA. Where the ngVLA will come into its own is in spatially resolving the most powerful outflows, where $\approx$ 2 arcsec resolution will be possible at 41 GHz without severe sensitivity losses from tapering. This will allow the accurate measurement of wind properties in the most energetic outflows, and a detailed comparison of the feedback from thermal winds versus non-thermal radio jets.

Acknowledgments. SC and AC would like to thank Rudrani Kar Chowdhury and Alankar Dutta for help with the simulation code. The National Radio Astronomy Observatory is a facility of the National Science Foundation operated under cooperative agreement by Associated Universities, Inc.

References

Adam, R., Bartalucci, I., Pratt, G. W., et al. 2017, A&A, 598, A115
Bîrzan, L., McNamara, B. R., Nulsen, P. E. J., Carilli, C. L., & Wise, M. W. 2008, ApJ, 686, 859
Bleem, L. E., Stalder, B., de Haan, T., et al. 2015, ApJS, 216, 27
Bower, R. G., Benson, A. J., Malbon, R., et al. 2006, MNRAS, 370, 645
Carilli, C. 2018, ngVLA Memo 47
Cen, R., & Safarzadeh, M. 2015, ApJ, 809, L32
Chatterjee, S., & Kosowsky, A. 2007, ApJ, 661, L113
Chatterjee, S., Di Matteo, T., Kosowsky, A., & Pelupessy, I. 2008, MNRAS, 390, 535
Chatterjee, S., Ho, S., Newman, J. A., & Kosowsky, A. 2010, ApJ, 720, 299
Chakraborty, A. 2017, Master's Thesis, Presidency University.
Crichton, D., Gralla, M. B., Hall, K., et al. 2016, MNRAS, 458, 1478
Croton, D. J., Springel, V., White, S. D. M., et al. 2006, MNRAS, 365, 11
Fabian, A. C. 2012, ARA&A, 50, 455
Hopkins, P. F., Hernquist, L., Cox, T. J., et al. 2006, ApJS, 163, 1
Itoh, N., & Nozawa, S. 2004, A&A, 417, 827
Khandai, N., Di Matteo, T., Croft, R., et al. 2015, MNRAS, 450, 1349
Lacy, M. et al. MNRAS submitted
Mukherjee, D., Bicknell, G. V., Sutherland, R., & Wagner, A. 2016, MNRAS, 461, 967
Nesvadba, N. P. H., Drouart, G., De Breuck, C., et al. 2017, A&A, 600, A121
Pfrommer, C., Enßlin, T. A., & Sarazin, C. L. 2005, A&A, 430, 799
Powell, M. C., Husemann, B., Tremblay, G. R., et al. 2018, arXiv:1807.00839
Richardson, M. L. A., Scannapieco, E., Devriendt, J., et al. 2016, ApJ, 825, 83

Rowe, B., & Silk, J. 2011, MNRAS, 412, 905 (RS11)
Ruan, J. J., McQuinn, M., & Anderson, S. F. 2015, ApJ, 802, 135
Sayers, J., Mroczkowski, T., Zemcov, M., et al. 2013, ApJ, 778, 52
Sazonov, S. Y., & Sunyaev, R. A. 1998, Astronomy Letters, 24, 553
Silk, J., & Rees, M. J. 1998, A&A, 331, L1
Spacek, A., Scannapieco, E., Cohen, S., Joshi, B., & Mauskopf, P. 2016, ApJ, 819, 128

Science with a Next-Generation Very Large Array
ASP Conference Series, Monograph 7
Eric J. Murphy, ed.

Characterizing Feedback Through Molecular Outflows Across Cosmic Time

Justin Spilker[1] and Kristina Nyland[2]

[1]*University of Texas at Austin, Austin, TX, USA;* `spilkerj@gmail.com`

[2]*National Radio Astronomy Observatory, Charlottesville, VA, USA;* `knyland@nrao.edu`

Abstract. Galactic winds are ubiquitously observed in galaxies both locally and in the high-redshift Universe. While these winds span many orders of magnitude in both temperature and density, observations of nearby galaxies show that the cold molecular phase tends to dominate both the mass and momentum carried. The capabilities of the ngVLA for the study of molecular outflows at low redshift are described elsewhere in this volume; here we focus on the ability of the ngVLA to detect and image such outflows in the high-redshift Universe via deep observations of low-J transitions of the CO molecule. The ngVLA is capable of detecting molecular outflows from typical galaxies on the star-forming sequence with $\log M_{\rm star}/M_{\odot} \gtrsim 10.5$ to $z \sim 3$, and galaxies with higher star formation rates to beyond $z \sim 4$. The ngVLA will enable an understanding of the feedback processes that shape galaxies throughout the epoch of galaxy assembly when the bulk of the stars in the Universe were formed. While the emission associated with outflows is faint in comparison to the emission from the galaxy, deep observations are also required for high-resolution dynamical studies, allowing for the routine simultaneous detection and imaging of the outflows.

1. Galaxy Feedback Through Cosmic Time

One of the most important realizations of the last fifteen years is the vital role that feedback processes must play in the evolution of galaxies. The energetics of feedback processes are critical unknowns in theoretical models of galaxy evolution. Energy and momentum from active galactic nuclei (AGN), Type II supernovae, and/or radiation pressure from massive stars are input in ad hoc manners below the resolution limit of simulations. These feedback processes play critical roles in the evolution of galaxies by regulating the growth of stellar mass in relation to dark matter halos, establishing a connection between stellar velocity dispersion and supermassive black hole mass, and enriching the circumgalactic medium with metals out to large radii.

At the extreme end, high-resolution hydrodynamical simulations currently favor dramatic processes such as starbursts triggered by gas-rich mergers or disk instabilities to deplete, expel, or heat the gas supply and prevent star formation (e.g., Hopkins et al. 2008; Ceverino et al. 2015; Wellons et al. 2015). Feedback also remains critical even for less extreme galaxies – state-of-the-art treatments of feedback yield 'bursty' star formation histories, with star formation rates (SFRs) that vary by an order of magnitude or more on timescales of $\sim 10 - 100\,$Myr (e.g., Hopkins et al. 2014). This burstiness is related to the ability of feedback to quickly and efficiently disrupt star-forming gas until the gas can cool and resume forming stars.

Observational evidence for the role of feedback has grown substantially in recent years thanks to the discovery of ubiquitous gas outflows in local and moderate-redshift galaxies. Powerful outflows have been observed in both ionized (e.g., Förster Schreiber et al. 2014; Genzel et al. 2014) and neutral gas (e.g., Shapley et al. 2003; Rupke et al. 2005). Critically, however, the discovery of ubiquitous massive, energetic *molecular* winds in low-redshift star-forming galaxies, (Ultra)Luminous Infrared Galaxies [(U)LIRGs], and AGN host galaxies has provided evidence that feedback can at least temporarily suppress star formation by acting *directly* on the fuel from which stars form (e.g., Feruglio et al. 2010; Veilleux et al. 2013; Cicone et al. 2014; Leroy et al. 2015).

Because the spectral signatures of outflows are very faint, very few detections of molecular outflows are known at higher redshifts (e.g., Cicone et al. 2015; Spilker et al. 2018). Nevertheless, the high-redshift regime ($1 \lesssim z \lesssim 4$) represents a critical epoch in the history of the Universe: the cosmic SFR density peaked at $z \sim 2-3$, and most of the stars in the Universe were formed prior to $z = 1$ (e.g., Madau & Dickinson 2014). Understanding the role of feedback in regulating the growth of galaxies at this epoch is thus of great importance. The prospects for the ngVLA to characterize galactic feedback processes in low-redshift galaxies are described elsewhere in this volume (i.e., Bolatto et al. 2018, p. 441), and we refer the reader to that chapter for additional discussion of the importance of feedback processes and key outstanding questions to be addressed by observations of molecular outflows. Here, we focus on the capabilities of the ngVLA to detect and image molecular outflows across the peak of cosmic star formation.

2. High-Redshift Discovery Space for the ngVLA

As currently envisioned, the ngVLA will have $\approx 10\times$ the sensitivity of either the existing Karl G. Janksy VLA or ALMA. This large increase in sensitivity, especially in the 3-4 mm atmospheric window, is critical for the detection of molecular outflows at moderate redshift. Molecular outflows have been primarily observed through far-infrared absorption lines of OH and, of relevance for the ngVLA, low-J CO lines. In the case of CO, the signature of winds is evident as additional emission at high relative velocities to the galaxy systemic velocity. If the spatial resolution of the observations is low, outflows can be identified from the integrated CO spectrum alone – in order to reasonably expect to detect the faint outflowing gas, the galaxy itself will be very significantly detected with a very well-characterized line profile. If the spatial resolution is high, the outflow can additionally be distinguished because it is not expected to follow the predominant galactic kinematics (e.g., the outflow is likely to be launched from the minor axis perpendicular to a galaxy's rotation curve).

2.1. Molecular Outflow Detection Limits

To demonstrate the capability of the ngVLA to detect molecular winds at high redshift, we perform a simple simulation. As a template, we use the outflows observed in low-redshift U/LIRGs in CO by Cicone et al. (2014). These galaxies have typical SFRs $\sim 100\, M_\odot$/yr and outflow masses $M_{\rm out} \sim$ few $\times\, 10^8\, M_\odot$. The molecular mass contained in the outflows typically comprises 1–10% of the total galaxy molecular gas mass.

Because CO(1–0) is not accessible for all redshifts, we also simulate CO(2–1) and (3–2) emission, assuming line brightness temperature ratios to CO(1–0) of r_{21} = 0.8 and r_{31} = 0.5. We note that many scaling relations between outflow properties and galaxy properties are poorly-constrained even at low redshift and are completely unknown at high redshifts; this exercise is merely meant to be illustrative.

We use the current best estimates for the array performance, assuming the array sensitivity for 1″ imaging in order to avoid resolving out the outflows as much as possible. This effectively corresponds to considering only the antennas that comprise the 'core' of the array on baselines $\lesssim$1.3 km. We assume a 10 hr integration time and 50 km s^{-1} channelization of the data. The result is shown in Figure 1 for three hypothetical galaxies at redshifts z =0.5, 1.5, and 2.5, with the outflow observed in CO(1–0), CO(2–1), and CO(3–2), respectively. In each case, high-velocity emission corresponding to the molecular wind is easily detected (the emission from the galaxy itself is also detected at a signal-to-noise of many tens per channel).

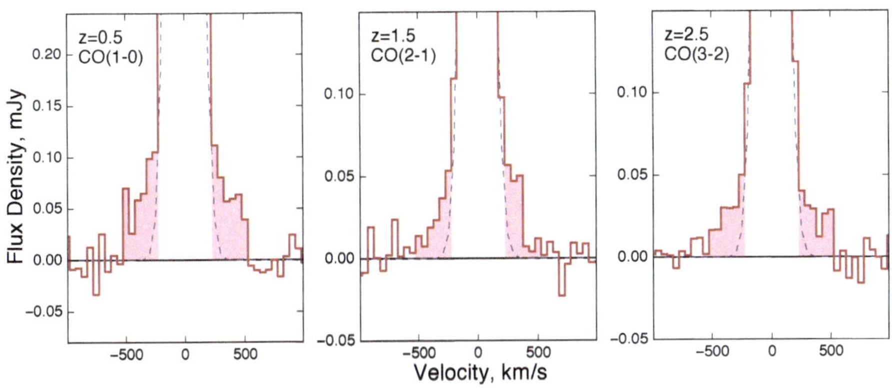

Figure 1. Three example CO spectra of galaxies at high redshift, showing the capability for the ngVLA to detect molecular outflows via high-velocity wings in the spectra that are distinct from the emission from the galaxy (illustrated as Gaussian fits to the spectra core, which extend well above the upper y-axis limits). We use outflows observed in low-redshift U/LIRGs (SFR ≈100 $M_\odot$/yr, M_{out} ~ few ×10^8 $M_\odot$; Cicone et al. 2014) as templates to create these spectra, assume a 10 hr integration, and use the expected ngVLA sensitivity for 1″ imaging. In each case high-velocity emission is well-detected.

Detections of similar outflows using existing facilities would be virtually impossible: the current VLA lacks receivers that can access the 3–4 mm atmospheric window where the CO emission from outflows is brightest, and reaching equivalent depths with ALMA would require of order 300 – 500 hours of observation. Only the order-of-magnitude increase in sensitivity afforded by the ngVLA makes the detection of outflows such as these possible in the distant Universe.

In Figure 2 we show the 5σ detection limits for molecular outflows as a function of redshift and targeted CO transition. We assume a CO-H_2 conversion factor α_{CO} = 0.8 $M_\odot$ (K km s^{-1} pc^2)$^{-1}$ and a 500 km s^{-1} outflow velocity width. For all redshifts z > 1, more sensitive mass limits can be reached using CO(2–1) or CO(3–2), even if the CO excitation is low (little is known about the CO excitation in existing detected winds). Because the outflow mass M_{out} is a rather abstract quantity, we label the right-hand axis of Figure 2 in terms of the approximate corresponding galaxy SFR, assuming a linear correlation between $M_{\rm out}$ and SFR. We assume the outflow mass constitutes 3% of the total galaxy gas mass, and convert to an SFR assuming a gas depletion time $t_{\rm dep} = M_{\rm gas}/{\rm SFR} = 150\,{\rm Myr}$. This 'conversion' is obviously extremely simplistic and uncertain, and the true scaling relations between outflow properties and galaxy properties would be a key outcome of a future ngVLA campaign. We merely aim to place the detectable outflows in terms more familiar to the extragalactic community using known characteristics of low-redshift outflows.

Figure 2 demonstrates that outflows similar to those in the nearby universe are detectable by the ngVLA out to high redshift. In particular, the ngVLA can detect outflows even from 'typical' massive galaxies – we illustrate the SFR of galaxies on the star-forming ("main") sequence for galaxies with $\log M_{\rm star}/M_{\odot} = 10.5$ and 11 as a grey shaded region in Figure 2 (Speagle et al. 2014). The ngVLA will place outflows from normal galaxies within reach, and should comparatively easily detect outflows from more highly star-forming objects. At the very highest redshifts, $z > 4$, only winds from the most highly star-forming objects are likely to be accessible (e.g., the population of high-SFR dusty star-forming galaxies; Spilker et al. 2018), a consequence of the inexorable cosmological dimming of spectral features.

2.2. Prospects for High-Resolution Imaging of Outflows

The previous subsection demonstrated the detectability of molecular winds assuming fairly low-resolution $1''$ imaging. Such observations effectively only include the core of the ngVLA, with $\approx$44% of the total collecting area on baselines <1.3 km. The second 'tier' of longer baselines extends to $\approx$30 km and adds a further $\approx$35% of the total collecting area. High-resolution observations are thus possible, at the expense of spreading the fixed total source flux over multiple resolution elements.

Precious little is known about the structure and geometry of molecular winds even at low redshifts, and even less is known in the distant universe. The ngVLA will have the capability to directly image molecular winds at high redshift, if the CO emission is sufficiently bright and/or concentrated into bright clumps. At $0\farcs1$ resolution, for example, the ngVLA will reach sensitivities ~20% better than at $1''$ resolution, but this is insufficient to compensate for the 100× smaller beam area. Thus, the surface brightness sensitivity of the ngVLA will necessarily limit the imaging of outflows to only the brightest, most massive clumps contained within the winds. Current simulations predict extremely clumpy outflows due to hydrodynamical instabilities (e.g., Richings & Faucher-Giguère 2018; Schneider et al. 2018), so the detection of wind substructure on scales accessible to the ngVLA may yet prove feasible.

We note also that one of the key science drivers of the ngVLA is the ability to image low-excitation CO emission at high resolution and sensitivity (see, for example, Carilli & Shao 2018, in this volume p. 535). Such observations face challenges very similar to those we identify here – namely, faint CO emission requires significant integration times to detect at high resolution. Carilli & Shao present simulations that assume 30 hr observations in order to characterize the dynamical properties of an M51-like galaxy at redshifts $z = 0.5 - 4.2$. Such observations would also be sensitive to subkiloparsec-scale clumps of molecular material in a molecular wind with a mass limit $\approx$ 2× below the thresholds shown in Figure 2. Thus, perhaps the most promising pathway to detailed imaging outflows by the ngVLA will be to simply make use of observations with a primary goal of characterizing the detailed dynamics of galaxies.

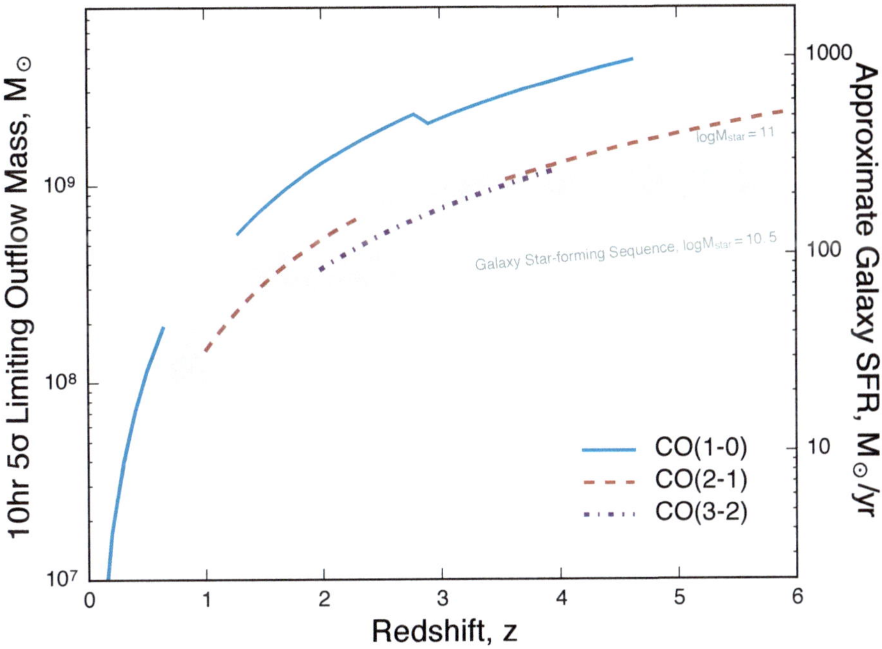

Figure 2. The approximate 5σ outflow mass detection limits as a function of redshift. Gaps in the redshift coverage for a given CO transition are due to the telluric oxygen complex near 60 GHz. Because the outflow mass M_{out} is a rather abstract quantity, the right-hand axis shows approximate corresponding galaxy SFRs (see text for details); this axis is merely intended to place the detectable outflows in more familiar context. We also show the SFR corresponding to the galaxy star-forming sequence for galaxies with stellar masses of $\log M_{\text{star}}/M_{\odot} = 10.5$ and 11 as a grey shaded region (Speagle et al. 2014). The ngVLA will detect molecular outflows from typical massive galaxies on the star-forming sequence out to $z \sim 4$ as well as from starburst galaxies well above the star-forming sequence.

3. Towards a Multi-phase Understanding of Outflows: Complementarity with Future Facilities

Given its fundamental importance in the evolution of galaxies, feedback will continue to be a focus of study for the foreseeable future. The ngVLA will have the ability to detect, and in some cases image at high resolution, the molecular phase of outflows at high redshift. Outflows are however complex, spanning many orders of magnitude in temperature and density, and so different phases of the winds are observable across the electromagnetic spectrum (e.g., Leroy et al. 2015). Again, we refer the reader elsewhere in this volume for a more detailed discussion of the capabilities of envisioned future facilities in the study of outflows (see Bolatto et al. 2018, p. 441). Here we merely highlight a few facilities in the context of high-redshift work.

By the time the ngVLA enters operations, 30 m-class ground-based optical telescopes will have made great advances studying the warm ionized phase of outflows through observations of Hα and [OIII]. The Giant Magellan Telescope, Thirty-Meter

Telescope, and European Extremely Large Telescope all have plans for near-infrared integral field units, allowing studies of ionized outflows at matched or better resolution than will be feasible with the ngVLA to $z \sim 2.5$. Meanwhile, the *James Webb Space Telescope* (*JWST*) will have observed outflows in a similar fashion to even higher redshift at somewhat worse spatial and spectral resolution. The winds detected by *JWST* are likely to become well-studied 'template' outflows, and the galaxies hosting such winds will be natural early targets for ngVLA observations of the molecular phase. These same facilities will also likely constrain the neutral atomic component of outflows through the detection of low-ionization metal absorption lines (e.g., Na I, Mg II), although these lines are far fainter than the bright optical strong lines.

Finally, the proposed NASA flagship *Origins Space Telescope* (OST) would allow for comprehensive views of the large majority of outflowing gas to very high redshift ($z \sim 5$). OST would provide a thousandfold increase in spectral sensitivity and mapping speed compared to previous far-infrared facilities, allowing the detection of outflows in higher-redshift, lower-mass, lower-SFR galaxies. While the spatial resolution of OST would necessarily be much lower than that of the ngVLA, OST will observe molecular, neutral, and warm ionized phases of outflows in thousands of star-forming galaxies through the detection of P Cygni profiles in far-infrared lines of OH and high-velocity wings of fine structure lines. Molecular winds detected by OST would be prime targets for ngVLA followup. The ngVLA will allow for both high-resolution imaging of the outflowing gas, and provide better estimates of the outflow masses and mass loss rates because CO observations are not limited to only the line-of-sight component of the wind as in absorption studies.

Acknowledgments. J.S. thanks the University of Texas at Austin for support through a Harlan J. Smith Fellowship.

References

Bolatto, A., Armus, L., Leroy, A., et al. 2018, Science with a Next-Generation Very Large Array, ASP Monograph 7, ed. E. J. Murphy (San Francisco, CA: ASP), 441
Carilli, C., & Shao, Y. 2018, Science with a Next-Generation Very Large Array, ASP Monograph 7, ed. E. J. Murphy (San Francisco, CA: ASP), 535
Ceverino, D., Dekel, A., Tweed, D., & Primack, J. 2015, MNRAS, 447, 3291
Cicone, C., Maiolino, R., Sturm, E., et al. 2014, A&A, 562, A21
Cicone, C., Maiolino, R., Gallerani, S., et al. 2015, A&A, 574, A14
Feruglio, C., Maiolino, R., Piconcelli, E., et al. 2010, A&A, 518, L155
Förster Schreiber, N. M., Genzel, R., Newman, S. F., et al. 2014, ApJ, 787, 38
Genzel, R., Förster Schreiber, N. M., Rosario, D., et al. 2014, ApJ, 796, 7
Hopkins, P. F., Hernquist, L., Cox, T. J., & Kereš, D. 2008, ApJS, 175, 356
Hopkins, P. F., Kereš, D., Oñorbe, J., et al. 2014, MNRAS, 445, 581
Leroy, A. K., Walter, F., Martini, P., et al. 2015, ApJ, 814, 83
Madau, P., & Dickinson, M. 2014, ARA&A, 52, 415
Richings, A. J., & Faucher-Giguère, C.-A. 2018, MNRAS, 474, 3673
Rupke, D. S., Veilleux, S., & Sanders, D. B. 2005, ApJS, 160, 115
Schneider, E. E., Robertson, B. E., & Thompson, T. A. 2018, ApJ, 862, 56
Shapley, A. E., Steidel, C. C., Pettini, M., & Adelberger, K. L. 2003, ApJ, 588, 65
Speagle, J. S., Steinhardt, C. L., Capak, P. L., & Silverman, J. D. 2014, ApJS, 214, 15
Spilker, J. S., Aravena, M., Béthermin, M., et al. 2018, Science, 361, 1016
Veilleux, S., Meléndez, M., Sturm, E., et al. 2013, ApJ, 776, 27
Wellons, S., Torrey, P., Ma, C.-P., et al. 2015, MNRAS, 449, 361

Science with a Next-Generation Very Large Array
ASP Conference Series, Monograph 7
Eric J. Murphy, ed.

Probing Dark Matter and Galaxy Evolution with Proper Motions of Galaxies

Thomas J. Maccarone[1] and Anthony H. Gonzalez[2]

[1] *Texas Tech University, Lubbock, TX, USA;* thomas.maccarone@ttu.edu

[2] *University of Florida, Gainesville, FL, USA*

Abstract. We discuss several problems that can be solved by measuring the proper motions of galaxies in clusters of galaxies, something that should be feasible in the Next Generation Very Large Array era. First, in the Virgo Cluster, the Laser Interferometric Space-based Antenna should provide standard siren distances for a substantial number of galaxies which contain black hole-black hole binaries. Measuring the proper motions of these galaxies will give 6-dimensional phase space information about the orbits of these galaxies, providing a unique set of tracers of cluster dynamics. Additionally, in other clusters proper motion measurements should allow the determination of whether spiral galaxies are on their first infall into the cluster, and measurements of the proper motions of galaxies in merging clusters will allow direct measures of the speed of the merger, which can then be compared with estimates from shocks.

1. Description of the Problem

Understanding the nature of the dark matter and how it impacts structure formation is a key problem in astrophysics. On the scales of individual galaxies, rotation curves can provide a good understandings of the relationship between distance from the center of mass of the galaxies, and the mass enclosed, albeit with some uncertainty about deviations from sphericity of the Galactic halo. Over the next $\sim$ 15 years we can expect to see the launch of the LISA mission, which will have the sensitivity to detect stellar-mass double black holes in the Virgo Cluster of galaxies, and from which we expect to see a non-negligible number of sources (Kremer et al. 2018; Jin 2018). These sources should have positional accuracies on the order 1 square degree and, crucially, standard siren distances accurate to about 1%, allowing them to be associated uniquely with their host galaxies. In other contexts, the proper motions of galaxies alone, even without the precise distances expected from gravitational wave sources, should allow fundamental tests of the dark matter hypothesis in nearby groups. Even without standard siren distance measurements, proper motions should yield useful estimates of the merger speed in merging clusters of galaxies, and determination of whether galaxies in clusters are on their first infalls, and how this might correlate with galaxy morphology and gas content.

2. Scientific Importance and Astronomical Impact

We are now at the start of an era in which *Gaia* is revolutionizing our understanding of the structure of the Milky Way galaxy by obtaining 6-D phase space measurements for a large number of objects. At the present time, no such information is available for any galaxies in clusters of galaxies, or even in nearby groups of galaxies. The next generation Very Large Array (ngVLA) has the potential to measure proper motions for large numbers of active galactic nuclei in nearby clusters and groups of galaxies. Given sufficient angular resolution, radio data can also provide precise, direct geometric distances to a small number of galaxies with either expanding supernova shells or with bright masers in disks. Excitingly, around the time we anticipate the ngVLA to be commissioned, we also anticipate that LISA will launch, opening up the era of standard siren distances to nearby galaxies which can be used to obtain precise distances to many galaxies, and also to provide precise calibration of the distances to the others.

Understanding the spatial distribution of dark matter in clusters of galaxies is one of the keys to understanding structure formation. Probes of dark matter halos typically rely on observables that are integrated along the line of sight (e.g., gravitational lensing shear, X-ray and Sunyaev-Zel'dovich surface brightness profiles), which must be transformed modulo physical assumptions to yield a mass distribution.

Collecting "test masses" within clusters of galaxies with well-specified positions and three-dimensional velocities can give precision information about the gravitational potential at specific locations within the potential. This information can then be combined with information collected through other techniques to eliminate degeneracies associated with line-of-sight projection and provide a far more precise determination of the mass distribution than is currently possible. The above technique is precisely the approach taken in measuring the mass distribution of the Milky Way. The observations described below would probe the mass distribution of Virgo at nearly the fidelity that has been possible for the Milky Way prior to Gaia, taking a first key step towards a detailed mass model for the Virgo cluster.

Moreover, measuring proper motions of galaxies is of great value for understanding a variety of other issues related to understanding galaxy evolution and cluster evolution. Precision proper motions of galaxies can be measured via masers in ordered disks, for the galaxies where they are found (Reid et al. 2018, this volume p. 523; see also Reid et al. 2009). Here we focus on alternative tracers of the galaxies' positions that can be used at large distances, even for galaxies without masers.

3. Anticipated Results

Galaxies within clusters already have their positions on the sky and their radial velocities well measured. This information is sufficient to obtain estimates of the mass enclosed within a given radius based on some assumptions about how the clusters deproject themselves. To improve upon these results, precise distances and proper motions are necessary. We can expect in the next 20 years to have a good set of standard siren distances for many galaxies in the Virgo Cluster, and perhaps a few galaxies in the Fornax Cluster. Obtaining the proper motions requires the combination of sensitivity and angular resolution of the ngVLA, and benefits strongly from having significant sensitivity on baselines well in excess of 300 km.

The Laser Interferometric Gravitational-wave Antenna (LISA) will detect double black hole binaries within its period range as continuously emitting sources of gravitational waves. From the gravitational waves themselves, it should be possible to measure standard siren distances to a few percent, and to identify the host galaxy from within an error volume of about 1 Mpc^3 or less. Once the host galaxy has been correctly identified, its location can used to refine the distance estimate, bringing the error in distance down to about 1%. This will rely on having small gravitational accelerations of the merging black holes; however, this is expected, because even in models where they form in globular clusters, they are expected to be ejected (Kremer et al. 2018). Furthermore, for the shortest period binaries in the LISA band, the cluster acceleration will affect the measurement of the period derivative by only a few percent.

Obtaining proper motions of the host galaxies of the gravitational wave sources then fills out the six-dimensional phase space measurements. The velocity dispersion of galaxies in the Virgo Cluster is about 700 km/s, while the velocity dispersions within the more massive galaxies can be $\sim$ 300 – 400 km/s. As a result, the use of optically bright tracers (e.g., red giants, planetary nebulae or globular clusters) to measure the proper motions of the massive galaxies that are most likely to contain the binary black holes providing standard siren distances is problematic – even with precise measurements of the velocities of individual objects, several thousand of them would be required to centroid the broad velocity dispersions to 10 km/s precision. While some of the largest galaxies in the Virgo Cluster have thousands of globular clusters, some fraction of these (especially in M 87) are likely to be intracluster, intergalactic globular clusters projected against the host galaxy, and many of the clusters, whether members or not, will be too faint for precise proper motion measurements, even with *JWST*. Furthermore, the uncertainties on these measurements will not improve substantially with a long time baseline, because the primary source of uncertainty will be how well the velocity distribution has been sampled, rather than the uncertainties on the individual measurements. Given that the standard siren distances will be accurate to a few percent, it is of great value to obtain similar precision in the proper motions in order not to have them dominate the total uncertainty budget for the phase space measurements. Making these measurements thus requires the use of the galaxies' nuclei, rather than their stellar populations.

A few of the nearby groups of galaxies are also excellent candidates for studies of dark matter halos on different scales. Oehm et al. (2017) have shown that proper motion dispersions of the centers of mass of the galaxies of tens of μas per year, corresponding to hundreds of km/s are expected among the galaxies in the M 81 group, and that the relative magnitudes of these space velocities can provide tests of whether dynamical friction is happening as predicted by dark matter theory (Kroupa 2015). Here, the smaller galaxies, M82 and NGC 3077 may not contain bright enough active galactic nuclei for estimating their proper motions, but as dwarf galaxies, with small internal velocity dispersions, other radio sources may be effective tracers of their motions. These tracers could include planetary nebulae and the brightest HII regions, and, in some cases, individual luminous blue variable stars, through their wind emission.

Finally, at a lower precision level, merging clusters and merging groups of galaxies can be studied to determine the relative merger speeds, and hence to determine the ratio between the masses of the two clusters. The Bullet Cluster (Clowe et al. 2006), for example, has easily detectable AGN in both subclusters (Malu et al. 2016), and with a relative velocity of 4500 km/s at $z = 0.296$, would be expected to show a space velocity of about 1 μas/year, as would Abell 2034, which has a merger speed which is somewhat slower, but which is located more nearby. The Bullet Cluster itself is too far south for ngVLA. In the near future, with *eROSITA*, *WFIRST* and *Euclid*, we can expect new merging clusters and merger compact groups of galaxies to be discovered, allowing for precise tests of the extremes of the dynamic intermediate redshift Universe.

4. Ancillary Astrophysics: Understanding the Morphology-Density Correlation in Clusters

In addition to enabling modeling of the dark matter distribution, proper motions for cluster galaxies provide an incredibly powerful tool for disentangling the physical factors impacting the evolution of these galaxies. In particular, it has long been known that spiral galaxies are quite rare in dense clusters of galaxies, especially in their cores (Oemler 1974). The subset of spirals that do exist in cluster cores therefore provide an important sample for understanding in detail the processes by which these galaxies are quenched. These galaxies are most likely on their first approach into the cluster, but full 6-D phase space information is necessary to determine their orbits, locations relative to the cluster center, and full orbital velocities. All of these quantities are critical for understanding the impact of cluster-specific quenching processes. For instance, quantifying the impact of present (future) ram pressure requires knowledge of the orbital velocity (and orbit). Such information currently does not exist – proper motions are required.

Even for galaxies that lack standard siren distance measurements, full 3-D velocities and angular positions combined with a model for the mass distribution are sufficient to identify substructures and constrain orbital motions at a fidelity that enables one to disentangle whether a galaxy is infalling for the first time, is associated with an accreted substructure, or had resided in the cluster core for Gyrs. Consequently, it becomes possible to infer the distribution of environments that galaxy has encountered within the cluster.

5. Limitations of Current Astronomical Instrumentation

The instrument best suited to address this problem at the current time would be the VLBA. It has 0.185 mJy noise in one hour at 40 GHz and 0.2 mas resolution. This leads to two key disadvantages relative to the ngVLA. First, the exposure times start to become prohibitive (although some long observations to provide a first epoch at the present time, so that the ngVLA has a long time baseline for proper motions would be valuable). Second, the systematics with the VLBA are likely to be significantly worse than with ngVLA Long Baseline Array (LBA); systematics increase with the separation between the source and the phase calibrator, and with approximately ten times better sensitivity the angular separation of the nearest good calibrator will typically be much smaller. Optical telescopes can obtain the combination of signal-to-noise and resolution nominally needed to address these questions, but then rely strongly on extremely precise flat-fielding and modeling of the point spread function that goes well beyond what has currently been established to be possible. With continental baselines, relative astrometry of AGN to μas accuracy per epoch in the radio should be straightforward to achieve, while optical astrometry appears to be limited by systematics at a level roughly 10-100 times worse.

6. Connection to Unique ngVLA Capabilities

As discussed above, this work can be done only with radio-based proper motion measurements of the active galactic nuclei of Virgo Cluster galaxies. The excellent sensitivity of the ngVLA is needed to make these measurements at high frequency where the systematics due to core shifts are least important, and with the ability to use moderately bright phase calibrators located very close to the source. Some aspects of this work can be done with the Southwest configuration for the ngVLA, but many of these projects would require, or at least strongly benefit from, the ngVLA LBA.

7. Experimental Layout

7.1. Virgo Cluster

The radio flux densities of the large galaxies in the Virgo Cluster are mostly ~ 1 mJy or more (Merloni et al. 2003). We take this as a baseline value. For an AGN at this brightness, we can expect a noise level of about 0.3 μJy/beam in one hour with the ngVLA, working at 40 GHz with a beam size of 2 mas. For a 1 mJy source, this then gives signal to noise of about 3000, meaning that the positional accuracy of the AGN will be good to about 0.7 μas, which is likely to be approximately the systematic limit. Obtaining a better handle on the systematics, so that the statistical limits can be reached, provides a strong motivation here for the ngVLA LBA; still, taking more observations and over a longer time baseline can also be done to help ensure that the results are robust. Positions measured to this accuracy will yield proper motions accurate at the level of ≈ 50 km/s on a one-year time baseline, so that within 5 years, proper motions could be measured to an accuracy of 10 km/s. Because the observations needed are quite short this also allows for extra checks to be done to ensure that systematics such as core shifts are not problematic, by allowing extra observations at different frequencies, and observations over a range of source fluxes. Additionally, use of the ngVLA LBA would improve both the systematics and the sensitivity.

While only the galaxies with LISA binaries are likely to be useful for the six-dimensional phase space measurements, even having good five-dimensional phase space measurements for all the large galaxies will provide very valuable information. We thus believe it would make most sense to observe the 50 or so most luminous galaxies. This also has the advantage of not requiring that the LISA detections have already been made at the time the ngVLA starts collecting data. For astrometric projects, given the importance of having a long time baseline, this is vital. We thus expect that conducting this project would require about 50 hours per epoch to observe the 50 radio-brightest galaxies. It may occur that LISA uncovers some stellar-mass binary black holes in fainter galaxies, which may lead to a bit of extra time being required, but as a first order estimate, we can expect this program to require about 250 hours of ngVLA time, spread over five years.

7.2. M 81 Group

For the M 81 group, the test of whether dark matter is producing dynamical friction proposed by Oehm et al. (2017) relies on making precise proper motions of M 81, M 82 and NGC 3077, and precision of 2 μas/year is needed. To obtain such precision over a 5 year baseline, one needs precision of 7 μas per measurement (slightly worse precision can be tolerated using more than 2 measurements).

7.3. Merging Clusters

As a template object, we consider Abell 2034. This is a merging cluster of galaxies with an estimated merger speed of about 2000 km/s (Owers et al. 2014), mostly across the plane of the sky (Monteiro-Oliveira et al. 2018). The cluster is at a redshift of $z = 0.115$, or a distance of 433 Mpc. The proper motion should thus be 0.8μas per year. Thus, with systematics that should be achievable in a straightforward manner with the ngVLA LBA, it should be possible to measure the proper motion at 10σ in a little over a decade. In fact, it is not yet certain if this cluster is in the early or late phases of its merger at the present time (Monteiro-Oliveira et al. 2018). At the present time, only relatively low angular resolution radio images have been obtained for this galaxy. It is likely, though that there should be at least a few AGN with flux densities above 50 μJy in each sub-cluster, allowing for μas positional accuracy in reasonable exposure times. This system is clearly the best case right now for a merger cluster which is near enough, with fast enough tangential velocities, and far enough North for proper motions to be useful. It is likely, though, that in the near future, new surveys will uncover some additional candidates.

Acknowledgments. We are extraordinarily grateful to Matt Benacquista for valuable discussions and providing some of the foundational ideas that led to this work and to Joan Wrobel, Pau Amaro-Seaone and Robyn Sanderson for additional useful discussions.

References

Clowe D., Bradač, Gonzalez A.H., Markevitch M., Randall S.W., Jones C., Zaritsky D., 2006, ApJ, 648L, 109

Jin, D., 2018, PhD Thesis, University of Texas-Rio Grande Valley

Kremer K., Chatterjee S., Breivik K., Rodriguez C.L., Larson S.L., Rasio F.A., 2018, PhRvL, 120, 1103

Kroupa P., 2015, in *Lessons from the Local Group: A Conference in honour of David Block and Bruce Elmegreen*, Springer International Publishing, Switzerland, p. 337

Malu S., Datta A., Sandhu P, 2016, Ap&SS, 361, 255

Merloni A., Heinz S., Di Matteo T., 2003, MNRAS, 345, 1057

Monteiro-Oliveira R., Cypriano E.S., Vitorelli A.Z., Ribeiro A.L.B., Sodré L., Dupke R., Mendes de Oliveira C., 2018, MNRAS, in press

Oehm W., Thies I., Kroupa P., 2017, MNRAS, 467, 273

Oemler A., 1974, ApJ, 194, 1

Owers M.S., et al., 2014, ApJ, 780, 163

Reid M., Brunthaler A., Menten K.M., Loinard L., Wrobel J., 2009, "Motions of Galaxies in the Local Group and Beyond: an Astro2010 Science White Paper"

Reid, M., Loinard, L., & Maccarone T. 2018, Science with a Next-Generation Very Large Array, ASP Monograph 7, ed. E. J. Murphy (San Francisco, CA: ASP), 523

Science with a Next-Generation Very Large Array
ASP Conference Series, Monograph 7
Eric J. Murphy, ed.

A Formaldehyde Deep Field

Jeremy Darling

Center for Astrophysics and Space Astronomy, Department of Astrophysical and Planetary Sciences, University of Colorado, 389 UCB, Boulder, CO 80309-0389, USA

Abstract. Formaldehyde (H_2CO) is often observed at centimeter wavelengths as an absorption line against the cosmic microwave background (CMB). This is possible when energy level populations are anti-inverted to the point where line excitation temperatures fall below the local CMB temperature. Collisions with molecular hydrogen "pump" this anti-maser excitation, and the cm line ratios of H_2CO provide a measurement of the local H_2 density. H_2CO absorption of CMB light provides all of the benefits of absorption lines (no distance dimming) but none of the drawbacks: the CMB provides uniform illumination of all molecular gas in galaxies (no pencil beam sampling), and all galaxies lie in front of the CMB (no fortuitous alignments with background light sources are needed). A formaldehyde deep field (FDF) would therefore provide a blind, mass-limited survey of molecular gas across the history of star formation and galaxy evolution. Moreover, the combination of column density and number density measurements may provide geometric distances in large galaxy samples and at higher redshifts than can be done using the Sunyaev Zel'dovich effect in galaxy clusters. We present a possible ngVLA FDF that would span redshifts $z = 0$–7 and provide H_2CO line ratios over $z = 1.36$–3.14.

1. Background

The formaldehyde molecule (H_2CO) has peculiar non-thermal excitation properties in the physical conditions typical of star-forming regions. Similar to molecules that can be induced to form masers via population inversion through a pumping mechanism, H_2CO is often "pumped" into an anti-inverted state by collisions with molecular hydrogen (H_2). Anti-inversion is an over-population of a lower-energy state compared to thermal, forming a kind of ultra-cold anti-maser that can absorb cosmic microwave background (CMB) photons if the line excitation temperature drops below the local CMB temperature. Townes (1997) glibly called this effect the "dasar" — darkness amplification by stimulated absorption of radiation[1]. Unlike maser excitation, which requires special local conditions, this anti-inversion of H_2CO is nearly ubiquitous in the Galaxy and in external galaxies (e.g., Ginsburg et al. 2011; Mangum et al. 2013) and seems to be the natural state of the molecule for a wide range of physical conditions (e.g., Darling & Zeiger 2012). Moreover, H_2CO is observed wherever CO is present and is not strictly a high density molecular gas tracer. The anti-inverted K-doublet lines

[1]Note that darkness is not in fact amplified, nor is absorption stimulated.

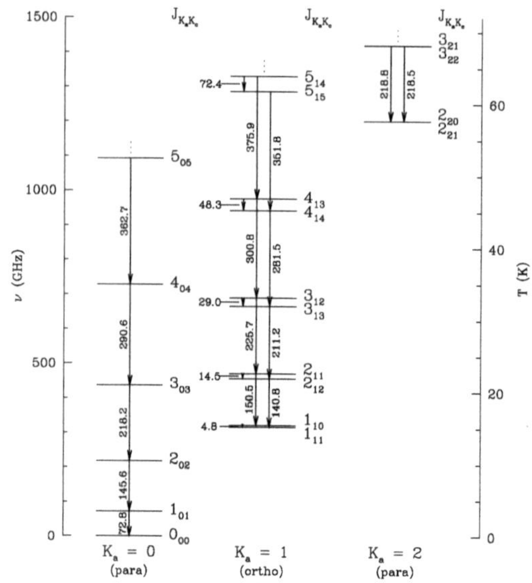

Figure 1. Formaldehyde energy levels for the first three rotation ladders and $J \leq 5$ (after Darling & Zeiger 2012). Allowed transition frequencies are listed in GHz. The anti-inverted cm lines are the K-doublet transitions in the $K_a = 1$ ortho-H_2CO rotation ladder at 4.8, 14.5, 29.0, 48.3, and 72.4 GHz (6, 2, 1, 0.6, and 0.4 cm, respectively).

are not optically thick and their ratio is a measure of the H_2 number density (via collisional pumping). The dasar effect can thus be used to make a census of molecular gas mass and gas density, independent of redshift (as described below).

The goal of this work is to examine the feasibility of and science enabled by an ngVLA formaldehyde deep field (FDF). We assume a flat cosmology with H_0 = 70 km s^{-1} Mpc^{-1}, $\Omega_m = 0.3$, and $\Omega_\Lambda = 0.7$.

1.1. Formaldehyde Anti-Inversion

H_2CO is an asymmetric top molecule, which creates a splitting of the primary rotation states (denoted by quantum number J; see Figure 1). H_2 collisions can overpopulate the lower energy states of these K-doublet rotation states, creating excitation temperatures below the local cosmic microwave background (CMB) temperature by roughly 1–2 K. The observational signature of this anti-inversion is absorption against the CMB in the centimeter wavelength $K_a = 1$ ortho-H_2CO lines primarily at 1, 2, and 6 cm.

This effect is strongest for the lowest energy cm lines and is insensitive to the local gas kinetic temperature. The observed line excitation temperature is also insensitive to the local CMB temperature [which scales from its value of $T_0 = 2.73$ K today as $T_{\rm CMB}(z) = (1 + z)\,T_0$]. The anti-inversion favors a fairly wide range of H_2 density, roughly 10^2–10^5 cm^{-3}, and the cm line ratios indicate the local gas number density. See Darling & Zeiger (2012) for detailed modeling versus redshift, Mangum et al. (2008) and Mangum et al. (2013) for physical conditions derived from observations of H_2CO in nearby star-forming galaxies, and Zeiger & Darling (2010) for a study of H_2CO anti-inversion in the gravitational lens B0218+357 at $z = 0.68$.

1.2. Absorption of CMB Light

The absorption of CMB photons by H_2CO implies that the line strength in beam-matched observations is independent of distance. And for $z \gtrsim 1$, beam-matching is no longer a strong function of distance because angular sizes become flat for $z \simeq 1$–3 and grow thereafter (Figure 2). The unusual circumstances presented by H_2CO anti-inversion have some compelling consequences:

(1) Absorption lines do not require fortuitous alignment of the object of interest with an illuminating light source. The CMB lies behind every galaxy and therefore every galaxy with molecular gas may be studied in H_2CO absorption.

(2) Unlike traditional absorption line studies, the illuminating "beam" is not a pencil beam that samples a subset of the intervening galaxy or gas cloud. The CMB provides an illuminating screen that is uniform to parts in 10^5 in the CMB rest frame. All gas is sampled in a manner similar to emission line observations (but absorption does not diminish with distance).

(3) The consequence of the above two points and the distance-independent nature of absorption lines is that it is possible to survey all H_2CO gas in the universe in a mass-limited fashion, provided one can beam-match to molecular gas regions in galaxies while achieving sub-Kelvin surface brightness sensitivity.

(4) H_2CO absorption lines provide the column density of gas, and H_2CO line ratios provide the local gas number density. This implies that Sunyaev-Zel'dovich-like distance measurements are possible using the molecular gas in galaxies (Darling & Zeiger 2012). In contrast to S-Z measurements of the X-ray gas in clusters that extend to $z \sim 1$, the H_2CO geometric distance measurement may be possible in galaxies up to $z \sim 6$.

Given these consequences of the peculiarities of the H_2CO molecule, one can therefore consider a formaldehyde "deep field" for the ngVLA: a blind, gas mass-limited survey of star-forming galaxies across the history of cosmic star formation.

2. Angular Resolution and Redshift Coverage

In order to avoid beam dilution against the CMB light screen, observations should be beam-matched to the size of star-forming regions in galaxies. Figure 2 shows angular size tracks versus redshift for 10, 5, 2, and 1 kpc scales. It also shows the angular resolution of the ngVLA versus redshift for the 6, 2, and 1 cm H_2CO lines and the redshift range covered by each ngVLA band for each line. When the beam size for a given band and redshift is below a given angular size track then that physical scale is resolved. Remarkably, the ngVLA can resolve or roughly beam-match 1 kpc at all redshifts in the 1 and 2 cm lines. Scales of 2 kpc or greater can be resolved in the 6 cm line up to $z = 3$. The ngVLA can therefore beam-match the physical scales relevant to star-forming galaxies including compact starbursts or major mergers (but see §3 for a discussion of brightness temperature sensitivity). Here we examine possible band choices for a FDF:

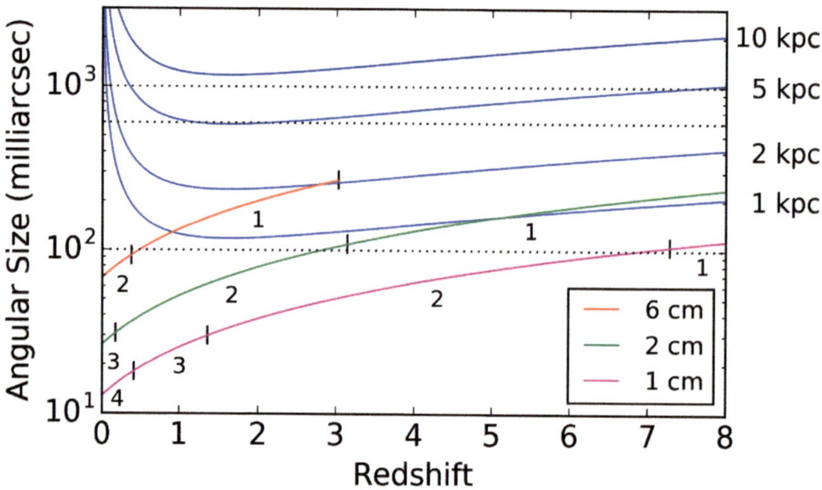

Figure 2. Angular size of various linear size scales (blue lines) and ngVLA beams versus redshift (red, green, and magenta for the 6, 2, and 1 cm lines, respectively). Three H_2CO lines are included, and the ngVLA bands are labeled by number. When a ngVLA beam lies below a galaxy size scale (of a disk or starburst nucleus, for example), then it is possible to resolve or beam-match that scale. If most cases, the ngVLA will not suffer beam dilution when observing star-forming galaxies or starburst nuclei. Black dotted lines indicate the fiducial 1 arcsec, 600 mas, and 100 mas resolutions used for surface brightness calculations (§3).

2.1. Band 1

The ngVLA Band 1 (1.2–3.5 GHz) spans $z = 0.38$–3.02 in the 6 cm line and $z = 3.14$–11.07 in the 2 cm line (Figure 2). In one observation covering the full 2.3 GHz bandwidth, it would be possible to nearly span the history of star formation, from 9.3 Gyr to only 0.4 Gyr after the Big Bang (although molecular gas is unlikely to be detected earlier than $z \sim 7$, or ~0.7 Gyr). Band 1 would be able to beam-match ≥ 2 kpc scales in the 6 cm line and ≥ 1 kpc scales in the 2 cm line spanning the above redshift ranges. Band 1 alone would not provide any H_2CO line ratios, so a FDF using only Band 1 would provide a census of star formation and molecular gas mass (modulo abundance variations between galaxies), but it would not provide H_2 densities.

2.2. Band 2

Band 2 (3.5–12.3 GHz) spans $z \leq 0.38$ in the 6 cm line, $z = 0.18$–3.14 in the 2 cm line, and $z = 1.36$–7.28 in the 1 cm line. The redshift overlap of the 2 cm and 1 cm lines, spanning $z = 1.36$–3.14, would enable the multi-line diagnostics described above (§1.1) near the peak of cosmic star formation history. There is also a small low-redshift shell, $z = 0.18$–0.38, where the 6:2 cm line ratio will be available. Band 2 observations would resolve sub-kpc physical scales at all redshifts (Figure 2).

2.3. Bands 1 and 2

The strongest anti-inverted H_2CO line is likely to be the 6 cm line, but Band 2 only captures this line at low redshift. If Bands 1 and 2 were both observed as part of a FDF, the 6:2 cm line ratio would be available for $z = 0.18$–3.02, the 2:1 cm line ratio for $z > 1.36$ (to redshifts where no molecular gas is expected to exist), and the 6:1 cm line ratio for $z = 1.36$–3.02. It would therefore be possible to measure $n(H_2)$ for nearly all redshifts, $z > 0.18$, spanning the history of cosmic star formation and galaxy evolution. The combined Band 1 and Band 2 FDF would be able to beam-match or resolve ≥ 2 kpc scales in all cases. The Band 2 field of view would limit the overlap between the two bands (24.4 arcmin versus 7.3 arcmin FWHM at the frequency centers of Bands 1 and 2, respectively).

3. Sensitivity

The observed line temperature depends on the difference between the line excitation temperature and the background continuum brightness temperature, redshifted to the observer's reference frame, and on the optical depth. For the H_2CO K-doublet lines, the continuum is the CMB (but may also include the host galaxy continuum) at the host galaxy's redshift:

$$\Delta T_{\rm Obs} = \frac{T_{\rm x}(z) - T_{\rm CMB}(z)}{1+z}(1 - e^{-\tau}). \quad (1)$$

Darling & Zeiger (2012) showed that the observed temperature decrement $\Delta T_{\rm Obs}$ is insensitive to redshift or the local gas temperature and spans a large range in local gas density, 10^2 cm^{-3} $\lesssim n(H_2) \lesssim 10^5$ cm^{-3}. At low density, the temperature decrement approaches zero (the line excitation temperature equilibrates with the CMB), and at high density, the line excitation temperature thermalizes to the local gas temperature. Typical temperature decrements $[T_{\rm x}(z) - T_{\rm CMB}(z)]/(1+z)$ are ~2 K for the 6 cm line and ~1 K for the 1 cm line. The detection of anti-inverted cm lines will therefore rely critically on the filling factor of molecular gas and the line optical depths, both of which will combine to manifest as an effective optical depth. For $\tau_{\rm eff} = 0.1$, $\Delta T_{\rm Obs} \simeq 0.1$–$0.2$ K.

The 100 mas-resolution reference design brightness temperature sensitivity for spectral lines, scaled to 100 km s^{-1} and 100 hours, is 71 K rms at 2.4 GHz (Band 1), 4.4 K at 8 GHz (Band 2), and 0.65 K at 16 GHz (Band 3). However, this resolution would correspond to less than 1 kpc for all redshifts and therefore over-resolve the molecular gas in galaxies (Figure 2). In order to beam-match 5 kpc scales at all redshifts, the angular resolution would need to be roughly 600 mas. At this resolution, the rms brightness temperatures are reduced to roughly 2.0 K (Band 1), 122 mK (Band 2), and 18 mK (Band 3). If one reduces the resolution to 1 arcsecond, which will resolve 10 kpc scales at all redshifts, the rms line brightness temperatures for 100 km s^{-1} channels and a 100-hour integration become 0.9 K (Band 1), 55 mK (Band 2), and 8 mK (Band 3).

4. A ngVLA Formaldehyde Deep Field

Given the angular resolution, redshift coverage for single lines and line ratios, and sensitivity considerations above, the best compromise ngVLA FDF could be:

- A single 100-hour pointing,
- 100 km s^{-1} channels to adequately sample the velocity span of molecular gas in galaxies (~300 km s^{-1}),
- Full synthesis of Band 2, which will include the 6, 2, and 1 cm H_2CO lines spanning the molecular history of the universe, z = 0–7, and
- Angular resolution of $0\farcs6$, enabling beam-matching to 5 kpc scales at all redshifts.

Line ratios will be available near the peak of the cosmic star formation history, from z = 1.36 to z = 3.14. Single-line detections will have to be disambiguated using ancillary data such as photometric redshifts, but this will be straightforward with a carefully-selected FDF location. While the addition of Band 1 would be highly desirable, particularly for density measurements at nearly all redshifts, the brightness temperature sensitivity requirements would indicate a prohibitive integration time.

5. Conclusions

The ngVLA is uniquely capable of observing a formaldehyde deep field (FDF), which would provide a distance-independent mass-limited census of molecular gas across the history of star formation and galaxy evolution. H_2CO line ratios in the FDF will provide a measurement of the local H_2 gas density. It may also be possible to make geometric distance measurements over a large redshift range based on the H_2CO K-doublet line depths and line ratios (Darling & Zeiger 2012). A FDF would complement flux-limited "blind" molecular emission line surveys (e.g., Pavesi et al. 2018) and could break the usual degeneracy between molecular gas temperature and density encountered in line excitation studies.

References

Darling, J. & Zeiger, B. 2012, ApJ, 749, L33
Ginsburg, A., Darling, J., Battersby, C., Zeiger, B., & Bally, J. 2011, ApJ, 736, 149
Mangum, J. G., Darling, J., Menten, K. M., & Henkel, C. 2008, ApJ, 673, 832
Mangum, J. G., Darling, J., Henkel, C., & Menten, K. M. 2013, ApJ, 766, 108
Pavesi, R. Sharon, C. E., Riechers, D. A., et al. 2018, ApJ, 864, 49
Townes, C. H. 1997, Quantum Electronics, 27(12), 1031
Zeiger, B., & Darling, J. 2010, ApJ, 709, 386

Part VIII

The Formation and Evolution of Black Holes in the Era of Multi-Messenger Astronomy

Science with a Next-Generation Very Large Array
ASP Conference Series, Monograph 7
Eric J. Murphy, ed.

Supermassive Black Hole Pairs and Binaries

Sarah Burke-Spolaor,[1,2] Laura Blecha,[3] Tamara Bogdanović,[4] Julia M. Comerford,[5] Joseph Lazio,[6] Xin Liu,[7] Thomas J. Maccarone,[8] Dominic Pesce,[9,10] Yue Shen,[7] and Greg Taylor[11]

[1]*Department of Physics and Astronomy, West Virginia University, Morgantown WV 26505, USA;* `sarah.spolaor@mail.wvu.edu`

[2]*Center for Gravitational Waves and Cosmology, West Virginia University, Morgantown WV 26501, USA*

[3]*Department of Physics, University of Florida, Gainesville, FL 32611, USA;* `lblecha@ufl.edu`

[4]*Center for Relativistic Astrophysics, Georgia Institute of Technology, Atlanta, GA 30332, USA;* `tamarab@gatech.edu`

[5]*Department of Astrophysical and Planetary Sciences, University of Colorado, Boulder, CO 80309, USA;* `Julie.Comerford@colorado.edu`

[6]*Jet Propulsion Laboratory, California Institute of Technology, Pasadena CA 91109, USA*

[7]*Department of Astronomy, University of Illinois at Urbana-Champaign, Urbana, IL 61801, USA;* `xinliuxl@illinois.edu`, `shenyue@illinois.edu`

[8]*Department of Physics & Astronomy, Texas Tech University, Lubbock, TX 79409-1051, USA*

[9]*Department of Astronomy, University of Virginia, 530 McCormick Road, Charlottesville, VA 22904, USA*

[10]*National Radio Astronomy Observatory, 520 Edgemont Road, Charlottesville, VA 22903, USA;* `dpesce@virginia.edu`

[11]*Department of Physics and Astronomy, University of New Mexico, Albuquerque, NM 87131, USA;* `gbtaylor@unm.edu`

Abstract. The next-generation Very Large Array (ngVLA) has the potential to be a workhorse for the discovery and study of paired supermassive black holes either at large separations (dual) or in tightly bound systems (binary). In this chapter, we outline the science case for the study of these supermassive pairs, and summarize discovery methods that can be used at radio wavelengths to discover them: including morphological, spectral, and time-domain identifications. One critical aspect of this work is that multi-messenger binary black hole studies may be possible with the ngVLA when combined with gravitational-wave searches using pulsar timing array techniques. However, long-baseline interferometry (i.e., $\gg$ 1000 km with the ngVLA Long Baseline Array) will make this possibility more likely by expanding the redshift range at which radio emission arising from two separate black holes may be resolved and studied.

1. The Importance and Impact of Binary Supermassive Black Hole Observations

The ngVLA will be a powerful probe of dual and binary supermassive black holes (SMBHs), performing detailed studies of their population and evolution, and enabling powerful multi-messenger science when combined with gravitational-wave detection. The ngVLA will enable such studies by an unprecedented combination of sensitivity, frequency coverage, and particularly if equipped with long baselines, essentially unparalleled angular resolution.

Dual ($\lesssim 10\,$kpc separation) and binary ($\lesssim 10\,$pc separation) SMBH systems can form during major galaxy mergers. Figure 1 illustrates the formation and evolution of double SMBH systems, from the initial stages of a merger of galaxies to the emission of gravitational waves to an SMBH merger (Begelman et al. 1980). When one or both SMBHs power an active galactic nucleus, multi-wavelength emission can directly mark the presence and locations of the SMBHs. Jets produced by these active nuclei will generate radio emission. Because jet cores closely trace the location of a SMBH, and because of the extended lifetime of synchrotron jet observability, radio observations both at high resolution and of larger-scale jet structures provide excellent tools to trace the current and past dynamics of SMBHs in a merging system.

SMBH pairs confidently mark ongoing galaxy mergers and imminent SMBH coalescences, and are therefore a strong probe of redshift-dependent merger rates, occupation fractions of dual SMBHs in galaxies, and post-merger dynamical evolution. If we determine the rate and environments of systems containing two or more widely separated active galactic nuclei (AGN), we can explore merger-induced activity of nuclei and susbequent SMBH growth. These dual AGN probe the critical regime during which the black holes and their associated galactic-scale stellar cores are virialized (e.g., within $\sim$10 kpc black hole separation). During this phase, both star formation and AGN activity peak during galaxy mergers, according to cosmological simulations (Blecha et al. 2013). Having a sample of dual AGN is key to testing predictions from simulations, such as whether the most luminous AGN are preferentially triggered in mergers and whether there is a time-lag between star formation triggering and AGN triggering in mergers (Hopkins 2012; Hopkins et al. 2014).

A systematic survey of dual and binary SMBHs also allows a direct prediction of the projected gravitational wave signals in the low and very low frequency regime of gravitational radiation. Both pulsar timing arrays (PTAs) and future space-based laser interferometers have binary SMBHs as a key target. They may detect both "discrete" individual targets, and a stochastic background of small-orbit SMBH binaries.

The ngVLA will also be coming online at a time when PTAs like the North American Nanohertz Observatory for Gravitational Waves (NANOGrav) expects to have already detected a number of discrete binary SMBHs (e.g., Mingarelli et al. 2017; Kelley et al. 2017b). LISA is expected to directly constrain SMBH binary coalescence rates soon after its launch (Klein et al. 2016; Amaro-Seoane et al. 2017). Gravitational-wave and multi-messenger astronomy with SMBHs will thus be a leading pursuit of the ngVLA era. Particularly with long (VLBI) baselines of the Long Baseline Array (LBA),

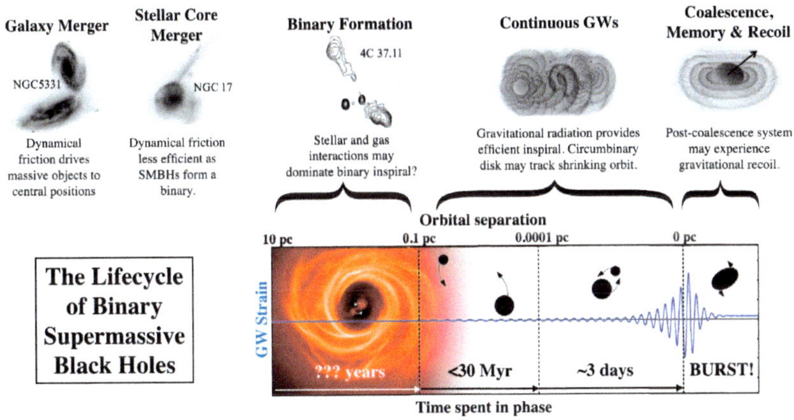

Figure 1. Figure from Burke-Spolaor et al. (2018), summarizing the binary SMBH life-cycle. A major unknown in binary evolution theory is the efficiency of inspiral from ~10 pc down to ~0.1 pc separations, after which the binary can coalesce efficiently due to gravitational waves. The ngVLA will discover widely separated binaries, and depending on its long-baseline sensitivity, could discover targets detectable by pulsar timing arrays (PTAs). PTAs such as NANOGrav can detect supermassive ($> 10^8\ M_\odot$) binaries within ~0.1 pc separation (second panel in the lower figure). On rare occasion, PTAs may detect the permanent space-time deformation (GW memory) caused by a binary's coalescence (Favata 2009). LISA will detect lower-mass binaries, up to $\sim 10^7\ M_\odot$, in the weeks or months leading up to coalescence. Image credits: Galaxies, Hubble/STSci; 4C37.11, Rodriguez et al. (2006); Simulation visuals, C. Henze/NASA; Circumbinary accretion disk, C. Cuadra.

the ngVLA will have the potential to identify the hosts of emitting gravitational-wave systems, enabling broad-scoped science that includes:

- Precise orbital tracking of binary SMBHs via parsec-scale jet morphology and core tracking (Bansal et al. 2017).
- Use of SMBHs as a standard ruler (D'Orazio & Loeb 2017).
- Measurement of merger-induced accretion rates and imaging of small-scale AGN feedback (Müller-Sánchez et al. 2015).
- Detailed multi-wavelength and multi-messenger studies of circumbinary disk evolution and binary inspiral timescales (e.g., Kelley et al. 2017a).
- Calibration of SMBH-host relations up to moderate/high redshifts via direct SMBHB mass measurements (host identification breaks the distance/mass degeneracy encountered in PTA discrete source detection).

In the remainder of this chapter, we describe the various approaches that can be used to identify SMBH pairs with the ngVLA (§ 2), compare those with past searches in § 3, and discuss synergies with multi-wavelength observations in § 4. Finally, § 5 lays out several example ngVLA studies and their general requirements for success.

2. Identifying Paired SMBHs with Radio Emission

The ngVLA could be used to conduct surveys for SMBH pairs, follow up candidates identified in other surveys, or both. Either way, in order to reveal dual or binary SMBH

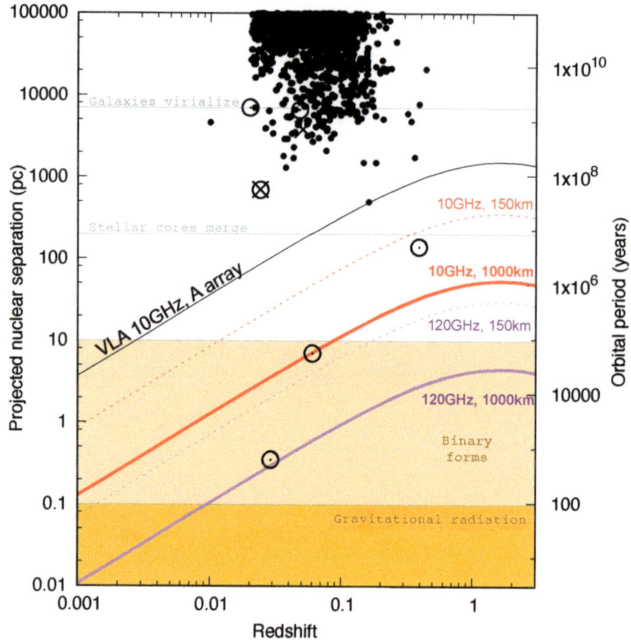

Figure 2. The observed population of candidate AGN pairs to date compared with the resolution of the ngVLA. Each point represents a dual AGN discovered via radio (X), X-ray (circles), or optical/near-IR (dots; from Liu et al. 2011). Approximate lines for critical stages in binary formation and evolution are marked as horizontal lines. The red, green, and purple curves indicate the resolution limit of 10 GHz, 50 GHz, and 120 GHz center observing bands, respectively; for each observing set-up, one cannot resolve a binary orbit below that line. The dashed curves show the resolution limit of 150 km baselines, while the solid curves show a nominal 1000 km extended-baseline array. The ngVLA's LBA can resolve, and hence directly image and identify, double supermassive black holes at sub-10 pc separations. Longer baselines and higher frequencies have the potential to resolve multi-messenger SMBH binaries in the nearby Universe (indicated by the curves that cross the dark yellow region).

candidates with the ngVLA, a few identification methods are available. The ngVLA can identify pairs through studies of morphology, spectra, and/or time-dependence of the bulk relativistic flow of particles from the SMBHs.

2.1. Direct Core Imaging

Radio imaging of multiple self-absorbed synchrotron cores in the ~1–50 GHz frequency range is one of the most direct routes to identify dual SMBHs. It is exceedingly rare to find bright multiple flat-spectrum objects within close proximity (e.g., Rodriguez et al. 2006; Burke-Spolaor 2011). The radio "core" that is seen in SMBHs represents the shock front near the base of the radio jet, which is compact and in a region of considerable optical depth. These can show a flat spectrum (with spectral index $\alpha \gtrsim -0.5$, where $S \propto f^{\alpha}$) up to moderate GHz frequencies (e.g., Blandford & Königl 1979). This technique relies on detecting more than one flat-spectrum, compact radio source within a common merger host. Confident confirmation of the nature of such detections requires modelling of broad-band continuum spectra, high-resolution morphological modelling, and ideally long-term studies of proper motion. For instance, in the object 4C37.11 (Figure 1), there are two cores at a projected separation of 7 pc that remain compact down to sub-mas resolutions, with larger-scale diffuse emission that support the interpretation of the two knots as distinct, but gravitationally interacting, SMBHs. Long-term astrometric observations of this object have led to the first tentative "orbital tracking" for a resolved binary, although its orbital period is fairly long at $\sim 10^4$ yr (Bansal et al. 2017).

Figure 2 shows a census of known and candidate resolved (imaged) dual SMBHs. Uniquely, radio identifications have probed separations within a few hundred parsecs: the radius of a typical galactic stellar bulge (Merritt & Milosavljević 2005). Double black hole systems with separations smaller than a few tens of parsecs are *direct precursors* to the gravitational-wave emitters detectable by PTAs and space-based laser interferometers, so are of utmost interest to the target science outlined in § 1. There is only one confident binary in this range (Rodriguez et al. 2006), and only a scant list of candidates (Deane et al. 2014; Kharb et al. 2017).

Binary SMBHs are relatively rare, and performing a complete blind search for dual radio cores requires relatively high dynamic range and thorough (u, v)-coverage at high resolution. Present facilities have aspects of these but not all. While the VLA has excellent sensitivity and (u, v)-coverage, it lacks sufficient resolution to discover close pairs. While current long-baseline facilities have excellent resolution, they lack the snapshot sensitivity and (u, v)-coverage to thoroughly probe cores (e.g., Deller & Middelberg 2014). The ngVLA has a unique potential to reach a balance between these factors (Bansal et al. 2018).

2.2. Jet Morphology

Identification of dual AGN via large-scale jet morphology is somewhat fraught because of the difficulty in discerning external vs. internal influences on the jet(s). Dual AGN emitting on long time scales can produce quadruple jets (Figure 3), and binary interactions can result in precessing jets that produce large-scale S-shaped radio jet morphologies. Parabolic SMBH-SMBH encounters and binary coalescences can rapidly reorient a jet, potentially producing X-shaped morphologies (Merritt & Ekers 2002). Previous large-scale radio surveys (e.g., NVSS, FIRST) have had a strong output of candidate binary or post-merger SMBHs identified through their large-scale jet morphologies. Recent surveys of such radio jet morphologies include 87 X-shaped radio jets found with the VLA (Roberts et al. 2018). However, past studies have revealed that many such morphologies can be described by backwards-flows of expelled jet material caused by fluid dynamics, rather than originating from a dual system (Saripalli & Subrahmanyan 2009).

Still, some (as shown in Figure 3) have been demonstrated either conclusively or extensively as candidate genuine dual AGN. Broad searches for such features require low-surface-brightness sensitivity coupled with resolution.

Large-scale periodic knots along radio jets have also been hypothesized to be periodic accretion episodes fueled by a binary; however, these are difficult to distinguish from flow instabilities (Godfrey et al. 2012).

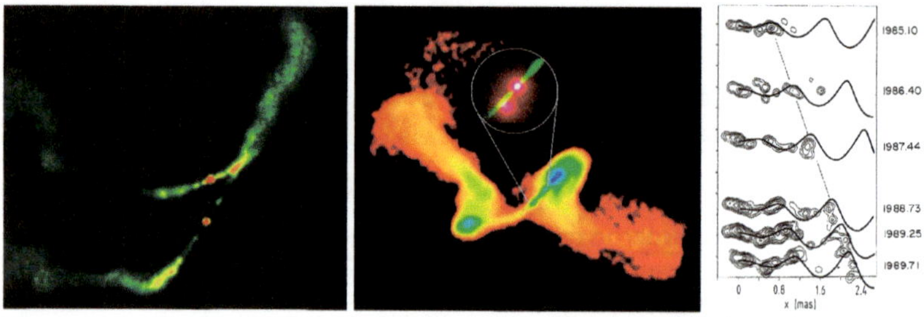

Figure 3. Three examples of candidate dual AGN identified through their static or time-dependent radio jet morphology (§§ 2.2, 2.3). (*Left:*) 3C 75, the dual cores of which have a 7 kpc projected separation and exhibits large-scale jet structures that have misaligned but correlated morphology. (Credit: NRAO/AUI; F. N. Owen, C. P. O'Dea, M. Inoue, & J. Eilek). (*Center:*) NGC 326, which has an X-shaped structure that inspired its dual nucleus interpretation. The cores were confirmed at other wavelengths, as shown. (Credit: NRAO/AUI; STScI [inset]). (*Right:*) The helical jet of 1928+738 shows a 2.9-year cycle (Roos et al. 1993).

2.3. Variability in Morphology or Flux

Some radio quasars show light curves that appear to be periodic or sinusoidal. At high spatial resolutions, some jets have morphological variability: e.g.,1928+738 has jets that trace out helical patterns as a function of time (Figure 3; Roos et al. 1993). A common interpretation of both of these effects is that an accreting, jet-producing black hole has a jet axis whose angle is being modified by a secondary, orbiting SMBH. Confident identification of a periodicity in a light curve requires long-term variability monitoring (over more than a few cycles) to ensure that the observed variability is genuinely cyclical, and not simply the detection of a red noise process (Vaughan et al. 2016; Charisi et al. 2018). Thus, light curve variability programs require a long-term time investment for candidate binary identification.

As with the tracking of 4C37.11 (Bansal et al. 2017), in an ideal case we will be able to track the orbital movements of one or two distinct cores. However, making the simple but suitable assumption that binaries are circular and obey Kepler's third law, their approximate largest physical separation scales with binary period as $a \propto P^{2/3}$. This means that the majority of resolvable binaries will have orbital periods well beyond human lifetimes. Tracking the orbital motion will only be possible for long baselines, and for the massive compact systems at low redshifts. With long-baseline (~10,000 km) arrays, such observations will be possible for $10^9\ M_\odot$ radio-emitting binaries with few parsec separations out to $z \sim 0.1$ on timescales of a few years. This clearly underscores the importance of the ngVLA LBA. If ngVLA baselines were to be limited to 300 km, then the timescales for resolving orbital properties will increase by a factor of 200, essentially making the ngVLA unsuitable for such studies despite its terrific snapshot sensitivity.

2.4. Maser Dynamics

Long-baseline observations of masers have been instrumental in finding SMBH masses and analyzing accretion disk dynamics (e.g., Miyoshi et al. 1995, Braatz et al. 2010, Kuo et al. 2013, Gao et al. 2016). Deviations from Keplerian motion in the maser rotation curves can indicate that the potential is not that of a point source (e.g., Kuo et al. 2011), and they can therefore be sensitive to close SMBH binaries (i.e., orbital separations of roughly ~0.01–0.1 pc).

For wider SMBH binaries (~1–100 pc separation), the maser rotation curve is not expected to be substantially different from Keplerian. In this case, one can compare the barycenter velocity derived from the maser rotation curve (which unambiguously traces the SMBH velocity) to that of the galaxy on large scales (~10's of kpc, traced using stellar or HI dynamics). Significant differences between these two velocities are then indicative of binary or recoiling SMBHs, although a measurement of a velocity offset cannot by itself distinguish between the two possibilities (Pesce et al. 2018).

3. Current Radio Searches and Limitations

In the past, most radio-identified candidate binary SMBHs have mainly been found serendipitously. A few concerted efforts have attempted to survey directly for binary SMBHs. Here we discuss these, their successes, and limitations.

A number of blind or semi-blind searches have been made for dual radio cores, which as described in § 2.1 is most easily realized through spectral imaging to search for multiple compact, flat-spectrum radio components. Burke-Spolaor (2011) searched all available historical data from the Very Long Baseline Array with dual-frequency imaging; despite the large sample size (~3200 AGN), this search was limited primarily by sensitivity. The VIPS survey of radio cores (Helmboldt et al. 2007) originally discovered 4C37.11 as an extended flat-spectrum object; this source was later evidenced to be a binary SMBH (Rodriguez et al. 2006). Tremblay et al. (2016) performed multi-frequency imaging of the flux-limited ~1100 AGN included in VIPS, all of which were pre-selected to be bright and flat-spectrum. This search did not identify any new binary systems from the VIPS sample, and concluded that 15 candidates remained unconfirmed. Finally, Condon et al. (2011) performed deep observations of 834 radio-bright 2MASS galaxies, under the hypothesis that the most massive galaxies must have undergone a major merger in their history, and therefore might contain a binary or recoiling SMBH. So far from these surveys, only one resolved dual core, 4C37.11, was re-detected. On simple arguments of merger rates, inspiral timescales, and radio AGN luminosity functions, at least tens of SMBH pairs must exist within these samples; thus, it is clear that these past experiments have been limited by a combination of (u, v)-coverage, sensitivity, dynamic range, and perhaps sample size.

4. Current Multi-Wavelength Synergies and Limitations

Dual and binary SMBH signatures have been proposed in a number of wavebands (see, e.g., the review in Burke-Spolaor 2013), and searches pursuing those signatures have revealed a number of binary candidates. We note that in all cases SMBH binaries have been challenging to identify because of their small separation on the sky, the uncertainties related to the uniqueness of their observational signatures, and the limited baseline of monitoring campaigns. Direct radio imaging, however, can provide an excellent follow-up to explore the nature of objects identified with these signatures.

The most prolific binary search method thus far has been the use of photometric and spectroscopic optical signatures to reveal candidate SMBH binaries. We describe that work here given its relevance to upcoming synoptic sky surveys that will be contemporary to ngVLA.

Multi-wavelength (radio, optical, and X-ray) searches for SMBH systems with large separations, corresponding to early stages of galactic mergers, have so far successfully identified a number of dual and offset AGN, as shown in Figure 3 (Komossa et al. 2003; Fu et al. 2011; Koss et al. 2011, 2016; Liu et al. 2013; Comerford et al. 2015; Barrows et al. 2016; Müller-Sánchez et al. 2015, and others).

SMBHs with even smaller (parsec and sub-parsec) separations are representative of the later stages of galactic mergers in which the two SMBHs are sufficiently close to form a gravitationally bound pair. Direct imaging using VLBI can confidently detect such a system, however requires deep observations to do so successfully. In this regard, optical surveys can help to identify a sample of SMBH binaries for VLBI followup (see Bogdanović 2015, for a review). Current studies have been exploring periodic quasar variability as a potential indicator for binary SMBH emissions; these and future surveys of this kind using synoptic instruments like PAN-STARRS and LSST will reveal many candidate binary systems ideal for follow-up with the ngVLA.

Current photometric surveys have already uncovered about 150 SMBH binary candidates (e.g., Valtonen et al. 2008; Graham et al. 2015; Charisi et al. 2016; Liu et al. 2016). In this approach, the quasi-periodic variability in the lightcurves of monitored quasars is interpreted as a manifestation of binary orbital motion. Because of the finite temporal extent of the surveys, most of these candidates have orbital periods of order a few years, hence would be in the regime of gravitational-radiation-dominated inspiral (Figure 1). However, studies of this type suffer the same red-noise issues noted in § 2.3. A recent, initially quite promising binary SMBH candidate from a quasar light curve, PG1302, has recently been losing traction. Additional data from the All-Sky Automated Survey for Supernovae (ASAS-SN), extending the light curve to present-day, shows that the periodicity does not persist and disfavors the binary SMBH interpretation (Liu et al. 2018). Furthermore, the upper limit placed by PTAs on the gravitational wave background largely rules out the amplitude implied by the $\sim$ 150 photometric binary candidates, implying that some significant fraction of them are unlikely to be SMBH binaries (Sesana et al. 2017). While this will lead to a downward revision in the number of photometrically identified SMBH binary candidates, it provides a nice example of the effectiveness of multi-messenger techniques, when they can be combined.

Spectroscopic searches for SMBH binaries have so far identified several dozen candidates. They rely on the detection of a velocity shift in the emission-line spectrum of an SMBH binary that arises as a consequence of the binary orbital motion (Gaskell 1983, 1996; Bogdanović et al. 2009). The main complication of this approach is that the velocity-shift signature is not unique to SMBH binaries; it may also arise from outflowing winds near the SMBH, or even from a gravitational-wave recoil kick (e.g., Popović 2012; Eracleous et al. 2012; Barth et al. 2015). To address this ambiguity, spectroscopic searches have been designed to monitor the offset of broad emission-line profiles over multiple epochs and target sources in which modulations in the offset are consistent with binary orbital motion (Bon et al. 2012, 2016; Eracleous et al. 2012; Decarli et al. 2013; Ju et al. 2013; Liu et al. 2014; Shen et al. 2013; Runnoe et al. 2015, 2017; Li et al. 2016; Wang et al. 2017). Two drawbacks of this approach however remain: (a) temporal baselines of spectroscopic campaigns cover only a fraction of the SMBH binary orbital cycle and (b) intrinsic variability of the line profiles may mimic or hide radial velocity variations, making these measurements more challenging and less deterministic (Runnoe et al. 2017).

5. Example ngVLA Studies

Dual and binary SMBH studies can be realized by a number of specific studies enabled by the ngVLA:

Blind surveys. The ngVLA can survey many sources (thousands) with high sensitivity, high dynamic range, and high two-dimensional image fidelity (i.e.,relatively complete (u, v)-coverage) in the frequency range 1–50 GHz with moderate resolution. These abilities would enable a blind survey for resolved dual cores in the $\gtrsim$few parsec separation regime. Such a survey could also reveal single, offset cores that may arise from a gravitational-wave recoil kick following a SMBH merger. Candidates can be cross-matched with future optical/near-IR or X-ray (e.g., Athena) surveys to measure spectra, redshifts, and disturbed host galaxy structures. While the multi-wavelength observations might not have the resolution of the ngVLA, they would provide a more complete understanding of a candidate and whether it is a genuine dual AGN, a gravitational lens, a chance projection, or even a recoiling SMBH.

Targeted surveys to test candidate binaries. Many binary SMBH candidates are likely to be identified in the coming decades by numerous instruments (including PTAs, LSST and the extremely large optical telescopes currently being planned). The ngVLA will be the premier instrument to follow up these candidates in radio, with the goal of understanding their long-term dynamics through large-scale jet morphologies (at low GHz frequencies, given sufficient sensitivity to low surface brightness emission). The ngVLA LBA will enable observation and tracking of resolved dual cores, which will truly confirm a binary and allow in-depth study of SMBH binary evolution.

Core variability. The frequency range of ngVLA makes it well-suited to observe AGN that are dominated by core emission. Past work has shown that in blind surveys at ~10 GHz, radio cores are more dominant than diffuse lobes (Massardi et al. 2011). Thus, the high sensitivity of the ngVLA at tens of GHz frequencies can permit easier time-resolved radio quasar light curve monitoring, and thus have the potential to identify more examples that might exhibit periodic activity due to variable accretion or orbital precession.

Multi-messenger searches. GW experiments can only poorly localize their detections (Ellis 2013). In concert with multi-wavelength facilities, the ngVLA could perform efficient follow-up surveys for all galaxies within a given error radius, identifying a host by looking for the signatures described above. With sufficiently high sensitivity on long baselines, the ngVLA could observe and track the orbital motions of radio cores for the most nearby GW candidates, thus enabling specific multi-messenger studies. In particular, PTAs will preferentially detect the most massive ($M_{\rm BH} \gtrsim 10^9\, M_\odot$), nearby ($z \lesssim 0.1$) objects that are accessible to Earth-based VLBI (e.g., Kelley et al. 2017a; Mingarelli et al. 2017).

Long-baseline benefits. The ngVLA LBA has clear benefits for this field of research. As envisioned, with each LBA station consisting of a clusters of $\sim 2-4$ antennas, the ngVLA will yield several major benefits. First, much of the imaging of double black holes is resolution-limited, meaning that going to longer baselines would immediately allow the probing of closer pairs of black holes. Second, long baselines would allow the maser experiments described in § 2.4. Third, the clusters themselves can be used for pulsar timing when the ngVLA core is not being used as part of a VLBI project, giving access to the pulsars too far north for SKA timing.

The above studies enabled by the ngVLA will provide unique probes of SMBH binary evolution in regimes inaccessible with other methods. This includes crucial data on binaries in the truly gravitationally bound phase, such as information about their separations, inspiral timescales, luminosities and variability even when not in a rapidly-accreting "quasar" state. Moreover, studies of radio jet morphology and energetics will provide unprecedented constraints on the mechanisms by which SMBH feedback drives host evolution and contributes to the origin of the observed SMBH-galaxy correlations. In concert with complementary SMBH binary studies by PTAs and LISA,[1] the ngVLA will be a key cornerstone in the coming era of multimessenger astronomy. The synergy between these electromagnetic and gravitational-wave probes of the binary regime has the potential to unravel the full puzzle of binary SMBH evolution and its close ties to the evolution of galaxies.

Acknowledgments. SBS is supported by NSF award #1458952. SBS and TJWL are members of the NANOGrav Physics Frontiers Center, which is supported by NSF award #1430284. T.B. acknowledges support by the National Aeronautics and Space Administration under Grant No. NNX15AK84G issued through the Astrophysics Theory Program and by the Research Corporation for Science Advancement through a Cottrell Scholar Award. L.B. is supported by NSF award #1715413. D. P. acknowledges support from the NSF through the Grote Reber Fellowship Program administered by Associated Universities, Inc./National Radio Astronomy Observatory. Y.S. acknowledges support from an Alfred P. Sloan Research Fellowship and NSF grant AST-1715579. Part of this research was carried out at the Jet Propulsion Laboratory, California Institute of Technology, under a contract with the National Aeronautics and Space Administration.

References

Amaro-Seoane, P., Audley, H., Babak, S., & et al. 2017, ArXiv e-prints, arXiv:1702.00786
Bansal, K., Taylor, G. B., Peck, A. B., Zavala, R. T., & Romani, R. W. 2017, ApJ, 843, 14
Barrows, R. S., Comerford, J. M., Greene, J. E., & Pooley, D. 2016, ApJ, 829, 37
Barth, A. J., Bennert, V. N., Canalizo, G., & et al. 2015, ApJS, 217, 26
Begelman, M. C., Blandford, R. D., & Rees, M. J. 1980, Nat, 287, 307
Blandford, R. D., & Königl, A. 1979, ApJ, 232, 34
Blecha, L., Loeb, A., & Narayan, R. 2013, MNRAS, 429, 2594
Bogdanović, T. 2015, in Astrophysics and Space Science Proceedings, Vol. 40, Gravitational Wave Astrophysics, ed. C. F. Sopuerta, 103
Bogdanović, T., Eracleous, M., & Sigurdsson, S. 2009, New Astronomy reviews, 53, 113

[1]See also the chapter on ngVLA/LISA synergies in Ravi et al. (2018, this volume p. 781)

Bon, E., Jovanović, P., Marziani, P., & et al. 2012, ApJ, 759, 118
Bon, E., Zucker, S., Netzer, H., & et al. 2016, ApJS, 225, 29
Braatz, J. A., Reid, M. J., Humphreys, E. M. L., et al. 2010, ApJ, 718, 657
Burke-Spolaor, S. 2011, MNRAS, 410, 2113
—. 2013, Classical and Quantum Gravity, 30, 224013
Burke-Spolaor, S., Hazboun, J., Kelley, L., et al. 2018, in prep.
Charisi, M., Bartos, I., Haiman, Z., & et al. 2016, MNRAS, 463, 2145
Charisi, M., Haiman, Z., Schiminovich, D., & D'Orazio, D. J. 2018, MNRAS, arXiv:1801.06189
Comerford, J. M., Pooley, D., Barrows, R. S., & et al. 2015, ApJ, 806, 219
Condon, J., Darling, J., Kovalev, Y. Y., & Petrov, L. 2011, ArXiv e-prints, arXiv:1110.6252
Deane, R. P., Paragi, Z., Jarvis, M. J., & et al. 2014, Nat, 511, 57
Decarli, R., Dotti, M., Fumagalli, M., & et al. 2013, MNRAS, 433, 1492
Deller, A. T., & Middelberg, E. 2014, AJ, 147, 14
D'Orazio, D. J., & Loeb, A. 2017, ArXiv e-prints, arXiv:1712.02362
Ellis, J. A. 2013, Classical and Quantum Gravity, 30, 224004
Eracleous, M., Boroson, T. A., Halpern, J. P., & Liu, J. 2012, ApJS, 201, 23
Favata, M. 2009, ApJ, 696, L159
Fu, H., Zhang, Z.-Y., Assef, R. J., & et al. 2011, ApJ, 740, L44+
Gao, F., Braatz, J. A., Reid, M. J., et al. 2016, ApJ, 817, 128
Gaskell, C. M. 1983, in Liege International Astrophysical Colloquia, ed. J.-P. Swings, Vol. 24, 473–477
Gaskell, C. M. 1996, ApJ, 464, L107
Godfrey, L. E. H., Lovell, J. E. J., Burke-Spolaor, S., & et al. 2012, ApJ, 758, L27
Graham, M. J., Djorgovski, S. G., Stern, D., & et al. 2015, MNRAS, 453, 1562
Helmboldt, J. F., Taylor, G. B., Tremblay, S., & et al. 2007, ApJ, 658, 203
Hopkins, P. F. 2012, MNRAS, 420, L8
Hopkins, P. F., Kocevski, D. D., & Bundy, K. 2014, MNRAS, 445, 823
Ju, W., Greene, J. E., Rafikov, R. R., & et al. 2013, ApJ, 777, 44
Kelley, L. Z., Blecha, L., & Hernquist, L. 2017a, MNRAS, 464, 3131
Kelley, L. Z., Blecha, L., Hernquist, L., Sesana, A., & Taylor, S. 2017b, submitted, arXiv:1711.00075
Kharb, P., Lal, D. V., & Merritt, D. 2017, Nature Astronomy, 1, 727
Klein, A., Barausse, E., Sesana, A., et al. 2016, Phys.Rev.D, 93, 024003
Komossa, S., Burwitz, V., Hasinger, G., & et al. 2003, ApJ, 582, L15
Koss, M., Mushotzky, R., Treister, E., & et al. 2011, ApJ, 735, L42
Koss, M. J., Glidden, A., Baloković, M., & et al. 2016, ApJ, 824, L4
Kuo, C. Y., Braatz, J. A., Reid, M. J., et al. 2013, ApJ, 767, 155
Kuo, C. Y., Braatz, J. A., Condon, J. J., et al. 2011, ApJ, 727, 20
Li, Y.-R., Wang, J.-M., Ho, L. C., & et al. 2016, ApJ, 822, 4
Liu, T., Gezari, S., Burgett, W., & et al. 2016, ApJ, 833, 6
Liu, T., Gezari, S., & Miller, M. C. 2018, ArXiv e-prints, arXiv:1803.05448
Liu, X., Civano, F., Shen, Y., et al. 2013, ApJ, 762, 110
Liu, X., Shen, Y., Bian, F., Loeb, A., & Tremaine, S. 2014, ApJ, 789, 140
Liu, X., Shen, Y., Strauss, M. A., & Hao, L. 2011, ApJ, 737, 101
Massardi, M., Ekers, R. D., Murphy, T., & et al. 2011, MNRAS, 412, 318
Merritt, D., & Ekers, R. D. 2002, Science, 297, 1310
Merritt, D., & Milosavljević, M. 2005, Living Reviews in Relativity, 8, astro-ph/0410364
Mingarelli, C. M. F., Lazio, T. J. W., Sesana, A., & et al. 2017, Nature Astronomy, 1, 886
Miyoshi, M., Moran, J., Herrnstein, J., et al. 1995, Nat, 373, 127
Müller-Sánchez, F., Comerford, J. M., Nevin, R., & et al. 2015, ApJ, 813, 103
Pesce, D. W., Braatz, J. A., Condon, J. J., & Greene, J. E. 2018, ApJ, 863, 149
Popović, L. Č. 2012, New Astronomy Reviews, 56, 74
Ravi, V. 2018, Science with a Next-Generation Very Large Array, ASP Monograph 7, ed. E. J. Murphy (San Francisco, CA: ASP), 781

Roberts, D. H., Saripalli, L., Wang, K. X., & et al. 2018, ApJ, 852, 47
Rodriguez, C., Taylor, G., Zavala, R., et al. 2006, ApJ, 646, 49
Roos, N., Kaastra, J. S., & Hummel, C. A. 1993, ApJ, 409, 130
Runnoe, J. C., Eracleous, M., Mathes, G., & et al. 2015, ApJS, 221, 7
Runnoe, J. C., Eracleous, M., Pennell, A., & et al. 2017, MNRAS, 468, 1683
Saripalli, L., & Subrahmanyan, R. 2009, ApJ, 695, 156
Sesana, A., Haiman, Z., Kocsis, B., & Kelley, L. Z. 2017, ArXiv e-prints, arXiv:1703.10611
Shen, Y., Liu, X., Loeb, A., & Tremaine, S. 2013, ApJ, 775, 49
Tremblay, S. E., Taylor, G. B., Ortiz, A. A., & et al. 2016, MNRAS, 459, 820
Valtonen, M. J., Lehto, H. J., Nilsson, K., & et al. 2008, Nat, 452, 851
Vaughan, S., Uttley, P., Markowitz, A. G., & et al. 2016, MNRAS, 461, 3145
Wang, L., Greene, J. E., Ju, W., et al. 2017, ApJ, 834, 129

Science with a Next-Generation Very Large Array
ASP Conference Series, Monograph 7
Eric J. Murphy, ed.

Compact Binary Mergers as Traced by Gravitational Waves

A. Corsi,[1] D. A. Frail,[2] D. Lazzati,[3] D. Carbone,[1] E. J. Murphy,[4] B. J. Owen,[1] D. J. Sand,[5] and R. O'Shaughnessy[6]

[1]*Department of Physics and Astronomy, Texas Tech University, Box 1051, Lubbock, TX 79409, USA;* `e-mail:alessandra.corsi@ttu.edu`

[2]*National Radio Astronomy Observatory, P.O. Box O, Socorro, NM 87801, USA*

[3]*Department of Physics, Oregon State University, 301 Weniger Hall, Corvallis, OR, 97331, USA*

[4]*National Radio Astronomy Observatory, 520 Edgemont Rd, Charlottesville, VA 22903, USA*

[5]*Department of Astronomy and Steward Observatory, University of Arizona, 933 N Cherry Ave, Tucson, AZ 85719, USA*

[6]*Center for Computational Relativity and Gravitation, Rochester Institute of Technology, Rochester, NY 14623, USA*

Abstract. In light of the recent dazzling discovery of GW170817, we discuss several new scientific opportunities that would emerge in multi-messenger time-domain astrophysics if a facility like the next generation Very Large Array (ngVLA) were to work in tandem with ground-based gravitational wave (GW) detectors. These include probing wide-angle ejecta and off-axis afterglows of neutron star (NS)-NS mergers; enabling direct size measurements of radio ejecta from NS-NS mergers; and unraveling the physics behind the progenitors of compact binary mergers via host galaxy studies at radio wavelengths. Our results show that, thanks to its unprecedented sensitivity and resolution, the ngVLA will enable transformational results in the multi-messenger exploration of the transient radio sky.

1. Introduction

Thanks to its unprecedented sensitivity and resolution, the next generation Very Large Array (ngVLA) has the potential to enable transformational results in the exploration of the dynamic radio sky (Bower et al. 2015). In light of the recent dazzling discovery of GW170817, a binary neutron star merger whose gravitational wave (GW) chirp (Abbott et al. 2017a) was accompanied by light at all wavelengths (see, e.g., Abbott et al. 2017b, and references therein), here we discuss several new scientific opportunities that would emerge in multi-messenger time-domain astrophysics if a facility like the ngVLA[1] were to work in tandem with ground-based GW detectors:

1. Probing wide-angle ejecta and off-axis afterglows of neutron star (NS)-NS, and black hole (BH)-NS mergers in low density ISM via Stokes I continuum and polarization measurements;

2. Enabling direct size measurements and dynamical constraints of radio ejecta from NS-NS and/or BH-NS mergers with Very Long baseline interferometry (VLBI) using the ngVLA Long Baseline Array (LBA; see Selina et al. 2018, this volume p. 15);

3. Unraveling the physics behind the progenitors of BH-BH, BH-NS, and NS-NS mergers via host galaxy studies at radio wavelengths.

In what follows, we briefly describe the scientific landscape expected to be realized when the ngVLA may become operational (§ 2). Then, we discuss radio studies of NS-NS/BH-NS ejecta in Stokes I continuum, linear polarization fraction, and resolved imaging (§§ 3 and 4). We show how the ngVLA could perform resolved host galaxy studies of binary NS-NS and BH-NS mergers (§ 5). Finally, in § 6, we summarize and conclude.

2. Multi-Messenger Astronomy in the Post-2027 Scientific Landscape

During their first two observing runs (O1/O2), the advanced Laser Interferometer Gravitational wave Observatory (LIGO) and Virgo made the big leagues by detecting several BH-BH binaries (Abbott et al. 2016a,b, 2017c,d) as well as GW170817, a NS-NS merger with an electromagnetic (EM) counterpart (Abbott et al. 2017e,a,b). Thanks to these discoveries we expect that about 10 years from now, in the ngVLA era, the field of time-domain astronomy will have fully transitioned to time-domain GW astrophysics. Based on current projections, the network of ground-based GW detectors in the ngVLA era could include Virgo (operating at nominal advanced sensitivity), the two advanced LIGO detectors likely in their so-called plus configuration (which foresees a factor of ~ 5 increase in event rates with respect to advanced LIGO at full sensitivity), the Kamioka Gravitational Wave Detector (KAGRA), and LIGO India. This world-wide network of detectors will be identifying potentially tens to hundreds of GW in-spirals and mergers in the local universe, with localization areas of order $\lesssim 10\,\mathrm{deg}^2$ (a factor of $\gtrsim 10$ better than today; Abbott et al. 2016c).

As demonstrated by the massive broad-band observational effort that unveiled the EM counterpart of GW170817 (see Abbott et al. 2017b, and references therein), spectroscopic and multi-wavelength observations of GW localization areas are going to be key to finding EM counterparts to GW triggers, removing potential false positives, identifying host galaxies, and constraining several key aspects of the physics of compact binary mergers. Radio and optical observations, for example, crucially complement

---

[1]Hereafter we assume that the ngVLA will have $\approx 10\times$ the sensitivity of the Jansky VLA, operate from $\sim$1.2 GHz (35 cm) to 116 GHz (2.6 mm) with up to 20 GHz of bandwidth with a compact core for high surface-brightness sensitivity, and extended baselines of at least hundreds of kilometers and ultimately across the continent for high-resolution imaging (Bolatto et al. 2017).

each other: while optical emission traces the slower thermally-emitting material, radio probes the non-thermal fastest-moving ejecta. Indeed, in the case of GW170817 (Abbott et al. 2017e,a), optical/IR observations revealed a kilonova and the r-process nucleosynthesis, while the late-time radio and X-ray emissions probed a completely different component, namely, a fast ejecta observed at large angles (e.g Abbott et al. 2017e,b; Coulter et al. 2017; Evans et al. 2017; Hallinan et al. 2017; Kasliwal et al. 2017; Lazzati et al. 2017, 2018; Mooley et al. 2018; Troja et al. 2017; Valenti et al. 2017).

Hereafter, we make the reasonable assumption that by the time the ngVLA becomes operational, the community will have collected statistically significant samples of various types of compact binary mergers, and that for at least some of these host galaxies will have been identified (either directly through the detection of EM counterparts, or indirectly through potential anisotropy in their spatial distribution; see, e.g., Raccanelli et al. 2016, and references therein). It is in this context that the value of a PI-driven radio telescope such as the ngVLA, with a sensitivity and resolution matched to (or encompassing that of) other post-2027 facilities, is best understood. Here we focus on topics that make use of the superior ngVLA sensitivity and resolution at frequencies in between those for which SKA1-MID ($\nu \lesssim 1$ GHz) and ALMA ($\nu \gtrsim 100$ GHz) will be the premier radio facilities.

3. Probing Wide-Angle Ejecta and Off-Axis Jets of NS-NS/BH-NS Mergers

Broadly speaking, three major classes of radio counterparts of BH-NS/NS-NS mergers are thought to exist (e.g., Nakar & Piran 2011; van Eerten et al. 2012; Hotokezaka et al. 2016; Lazzati et al. 2017, 2018; Nakar & Piran 2018): (i) Counterparts associated with sub-relativistic merger ejecta producing radio remnants on timescales of a few years; (ii) ultra-relativistic jets that produce short gamma-ray bursts (GRBs) and (on-axis) radio afterglows in the direction of the jet (evolving on timescales of a few days); (iii) mildly-relativistic, wide-angle "cocoons", also referred to as jet "wings", whose emission evolves on timescales of weeks, and which may or may not be accompanied by the contribution of a successful but (initially) off-axis ultra-relativistic jet. A more detailed description of these various components can be found in the contribution to this science book by Nicole Lloyd-Ronning et al. (2018, this volume p. 701).

While the presence of fast jets that successfully break out from the merger ejecta [scenario (ii) above] is predicted by models positing NS-NS mergers as central engines of short GRBs, the temporal evolution of GW170817 radio afterglow has given evidence for the presence of emission coming from a wide-angle outflow [scenario (iii) above; Kasliwal et al. 2017; Hallinan et al. 2017; Lazzati et al. 2017; Mooley et al. 2018; Lazzati et al. 2018] The key question of whether a relativistic, short GRB-like jet also emerged successfully, but was not observed as a bright, fast-decaying afterglow in GW170817 because misaligned with our line of sight (off-axis), remains open. Answering this question is crucial to ultimately understand whether short GRBs track the NS-NS merger rate, or if instead a larger variety of ejecta outcomes is possible in these mergers, including so-called "choked" jets.

Building a large sample of NS-NS/BH-NS mergers with radio counterparts is the first step into answering the above question. To this end, we certainly need a radio telescope like the ngVLA: as evident from the top panel of Figure 1, only with the ngVLA can we detect GW170817-like radio counterparts (in Stokes I continuum) up to

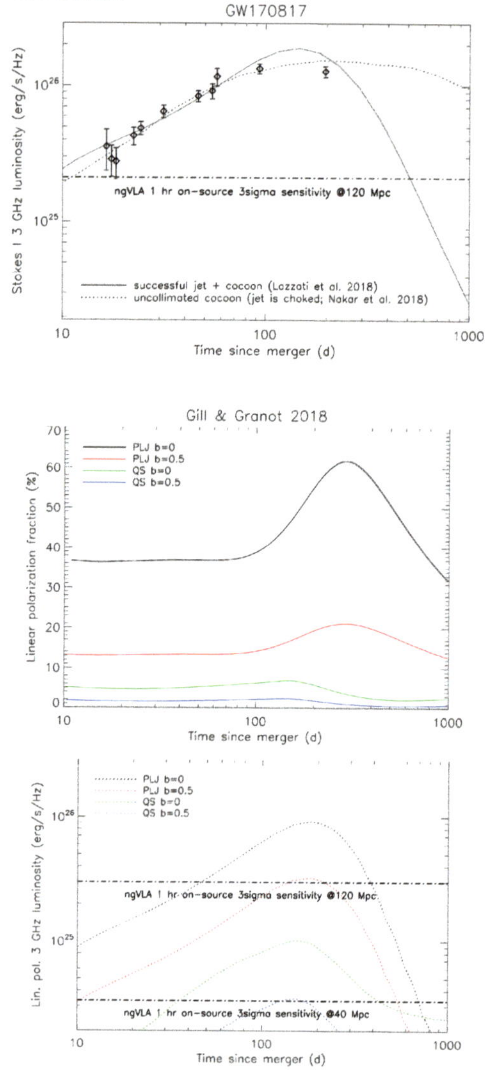

Figure 1. (*Top:*) GW170817 3 GHz light curve (diamonds; Dobie et al. 2018) compared with the best fit models in a successful jet plus cocoon scenario (solid line; Lazzati et al. 2018), and in a cocoon with choked jet scenario (dotted line; Nakar et al. 2018). Only with the ngVLA (rms sensitivity $\approx 0.4\,\mu$Jy/beam in 1 hr at 2 − 3 GHz) we can probe GW170817-like radio counterparts up to the distance horizon for highly significant triple coincidence detections by LIGO and Virgo at their full sensitivty ($d_L \approx 120$ Mpc, 3× farther than GW170817, for an event rate about 30× higher; see Abbott et al. 2016c). (*Center:*) Predictions by Gill & Granot (2018) for the linear polarization fraction $\sqrt{Q^2 + U^2}/I$ of the GHz radio flux for different types of ejecta and magnetic field structures. The quasi-spherical ejecta (QS) best represents a cocoon with a choked jet; the power-law jet (PLJ) best represents the case of a successful jet with a cocoon; the parameter b describes the structure of the magnetic field: for $b = 0$ the field is completely in the plane of the shock, and for $b = 0.5$ the field component in the direction of the shock normal also contributes. See Gill & Granot (2018) for more details. (*Bottom:*) Best fit Stokes I continuum at 3 GHz from the top panel (solid and dotted lines), multiplied by the linear polarization fraction shown in the center panel for different ejecta and magnetic fields. The ngVLA will probe a large variety of ejecta and magnetic field structures, which are inaccessible to current radio facilities.

distances comparable to the horizon for highly significant triple (LIGO+Virgo) coincidence triggers of current ground-based detectors at their full sensitivity ($d_L \approx 120$ Mpc or 3× farther than GW170817, or an event rate about 30× higher; Abbott et al. 2016c). This conclusion has been confirmed more generally in the recent study by Carbone & Corsi (2018), which shows that even considering a larger variety of possible radio afterglow outcomes from NS-NS mergers [see (i)-(iii) above], the ngVLA will be key to enabling discoveries within a volume large enough for statistically meaningful studies ($\gtrsim 10$ events per year).

While radio continuum monitoring of NS-NS merger ejecta cannot fully remove model degeneracies (see, e.g., the top panel of Figure 1 where the dotted and continuum line largely overlap until a few hundred days since merger; Nakar & Piran 2018; Lazzati et al. 2018), radio polarimetry offers the potential of providing a much stronger model discriminant (see Figure 1 central panel, and e.g., Rossi et al. 2004; Gill & Granot 2018; Nakar et al. 2018, for further discussion). Indeed, the sensitivity of an array like the ngVLA will offer us the unprecedented opportunity of probing directly the angular and radial structure of NS-NS merger ejecta, and of their magnetic fields, via radio polarization studies (Corsi et al. 2018). As shown in the central panel of Figure 1, a large degree of linear polarization can be considered a smoking gun for the presence of a successful misaligned jet accompanying a large angle outflow in a NS-NS merger. On the other hand, a quasi-spherical outflow that could result from a choked jet would produce linearly polarized emission at a much lower level.

As evident from the bottom panel of Figure 1, we need the sensitivity of the ngVLA to track the temporal evolution of the linear polarization fraction of GW170817-like radio counterparts up to $d_L \sim 120$ Mpc. The degree of linear polarization can help constrain outflow structure models and, when paired with direct imaging (which can constrain the ejecta structure regardless of the details of magnetic field structure), also enable us to gain critical insight into the possible magnetic field configurations in NS-NS merger ejecta (b parameter in the central panel of Figure 1).

4. Direct Size Measurement of Radio Ejecta from NS-NS and BH-NS Mergers

Resolved imaging of the radio ejecta associated with BH-NS or NS-NS mergers, which requires VLBI techniques, represents the only direct way to map the speed distribution of merger ejecta, and distinguish e.g., collimated relativistic fireballs observed off-axis from quasi-spherical relativistic ejecta. Imaging, coupled with linear polarization studies, could break the degeneracy between outflow and magnetic field structure (see previous section and Figure 1 central panel).

The image of a NS-NS merger ejecta will generally depend on the details of the interaction of the fastest ejecta component with with the slower, neutron-rich material, and several different outcomes are possible (see, e.g., Nakar et al. 2018). However, following Lazzati et al. (2018), here we consider two extreme cases that are likely to roughly "bracket" the variety of possible outcomes: an uncollimated (spherical) relativistic ejecta whose image would appear to be ring-like - brighter near the edge and dimmer near the center (see the top-left panel of Figure 3; see also Granot et al. 1999); and a top-hat (i.e., uniform) collimated relativistic jet whose axis is misaligned with respect to the observer (see the middle- and bottom-left panels of Figure 2).

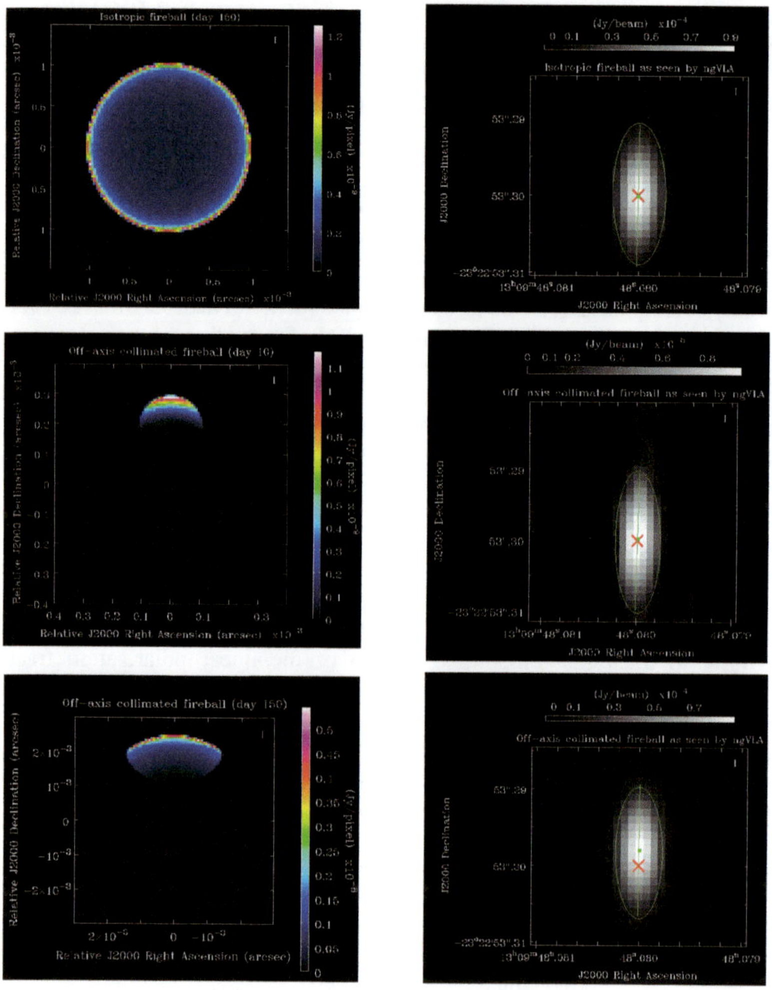

Figure 2. (*Top:*) Predicted image (left) and simulated observation (right) at 2.4 GHz of an isotropic NS-NS merger ejecta at day $\approx$ 150 since merger with isotropic energy $E_{\rm iso} = 2.5\times10^{49}$ erg, fraction of energy in electrons of $\epsilon_e = 0.03$, fraction of energy in magnetic fields of $\epsilon_B = 0.0002$, expanding in an ISM of density $n_{\rm ISM} = 6.8 \times 10^{-4}\,{\rm cm}^{-3}$ (see Lazzati et al. 2018, for more details). (*Center:*) Predicted image (left) and simulated observation (right) at 2.4 GHz of a top-hat collimated (θ_j = 23 deg) fireball from a NS-NS merger observed off-axis ($\theta_{\rm view}$ = 39 deg), at day $\approx$ 10 since merger, with $E_{\rm iso} = 5.7 \times 10^{50}$ erg, $\epsilon_e = 0.01$, $\epsilon_B = 0.003$, $n_{\rm ISM} = 3.2 \times 10^{-4}\,{\rm cm}^{-3}$. (*Bottom:*) Same as center but for an observing epoch of 150 d since merger. In all right panels we have assumed the April 2018 ngVLA-VLBA configuration (see text), 4 hr integration time at a central frequency of 2.4 GHz, 2.3 GHz bandwidth, and noise rms of 0.2 μJy/beam (natural weighting). As evident, the off-axis collimated ejecta would appear to have an emission centroid (green dot) offset from the location of the emission centroid that one would measure at early times and/or e.g., at optical wavelengths (red cross). Moreover, for a collimated off-axis fireball the emission centroid is observed to move over time (compare the location of the green dot in the center- and bottom-right panels relative to the red cross).

Using the April 2018 updated VLBA configuration of the ngVLA[2], we have simulated what the ngVLA would see given the emission models shown in the left panels of Figure 2. We have assumed a 4 hr-long observation at 2.4 GHz, and an rms sensitivity of about 0.2 μJy/beam (for natural weighting). As shown in the right panels of Figure 2, an off-axis collimated ejecta would appear to have a time-dependent emission centroid increasingly offset from the location of the counterpart that one would measure at very early times and/or e.g., at optical wavelengths (compare the time-dependent location of the green dot in the middle- and bottom-right panels relative to the fixed red cross). On the other hand, the emission centroid of an isotropic ejecta would be time-invariant and coincide with the optical location at all times (top-right panel in Figure 2). Moreover, as shown in Figure 3, the VLBA configuration of the ngVLA will give us enough sensitivity and resolution for mapping directly radio ejecta of NS-NS (and BH-NS) mergers as bright as GW170817 ($\approx$ 100 μJy) around $\approx$ 150 d since merger. Specifically, an analysis of the visibility as a function of baseline will distinguish off-axis collimated outflows from isotropic ones. We note that the situation will only be improved using the current LBA configuration, which includes additional antenna stations and increased collecting area than the April 2018 VLBA configuration.

Table 1. Host galaxy properties of short GRBs. From left to right the columns are: GRB name, redshift, host galaxy (optical) SFR, effective host galaxy radius (in optical/IR), expected galaxy radio FWHM at $d_L \approx 120$ Mpc, 2.4 GHz luminosity as derived from the optical SFR (see text for more details), 2.4 GHz radio flux density at $d_L = 120$ Mpc, and derived radio brightness temperature at 2.4 GHz and 120 Mpc (see text for more details).

GRB	z	SFR ($M_\odot$/yr)	$r_{e,z}$ ($''$)	r_e (kpc)	FWHM 120 Mpc ($''$)	$L_{2.4\,\mathrm{GHz}}$ (erg s^{-1} Hz^{-1})	$F_{2.4\,\mathrm{GHz}}$ 120 Mpc (μ Jy)	$T_{2.4\,\mathrm{GHz}}$ (K)
061201	0.111	0.14	1.09	2.2	4.0	1.5×10^{27}	8.8×10	1.2
070429B	0.902	1.1	0.65	5.1	9.2	1.2×10^{28}	6.9×10^{2}	1.7
070714B	0.922	0.44	0.34	2.7	4.9	4.6×10^{27}	2.8×10^{2}	2.5
070724A	0.457	2.5	0.63	3.7	6.7	2.6×10^{28}	1.6×10^{3}	7.5
071227	0.381	0.6	0.91	4.8	8.7	6.3×10^{27}	3.8×10^{2}	1.1
090510	0.903	0.3	0.93	7.3	13	3.1×10^{27}	1.9×10^{2}	0.23
090515	0.403	0.1	1.19	6.5	12	1.0×10^{27}	6.3×10	0.097
130603B	0.356	1.7	0.62	3.1	5.6	1.8×10^{28}	1.1×10^{3}	7.3

5. Resolving the Host Galaxies of Compact Binary Mergers

GW observations of compact object binaries (NS-NS, BH-NS, or BH-BH) can constrain the properties of the compact object themselves, such as masses, spins, and merger rates. However, understanding their progenitors and formation channels (i.e., how do compact binary systems actually form and evolve) requires the identification of host galaxies, and more generally a detailed study of the merger environment.

[2]This configuration, which predates that of the current ngVLA LBA (see Selina et al. 2018, this volume p. 15), includes the full 214 antennas of the ngVLA main array, plus 11 continental-scale antennas: five 18 m antennas randomly placed at the Green Bank site and individual 18 m antennas located at the existing Saint Croix, Hancock, North Liberty, Brewster, Owens Valley, Mauna Kea VLBA sites.

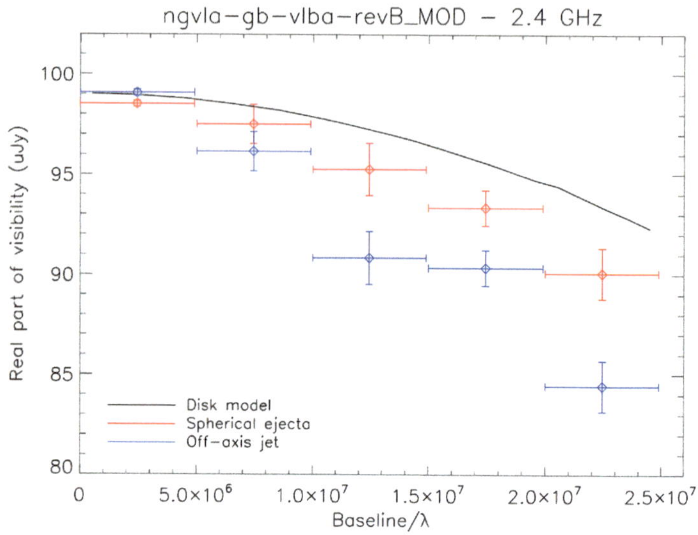

Figure 3. Real part of the ngVLA visibility as a function of baseline for the two models shown in the top and bottom panels of Figure 2 (isotropic and off-axis collimated fireballs at at 150 d since explosion when the 2.4 GHz flux density is of order $\approx 100\,\mu$Jy) compared to that of a uniform disk of 2 mas diameter and total flux density of $100\,\mu$Jy (comparable to the 3 GHz peak flux of GW170817). We assume the April 2018 ngVLA-VLBA configuration (see text), 4 hr integration time at a central frequency of 2.4 GHz, a nominal 2.3 GHz bandwidth, and noise rms of $0.2\,\mu$Jy/beam (natural weighting). As evident, the ngVLA can resolve and distinguish different ejecta structures.

A first constraint on the progenitors of compact binary mergers, and their age distribution, would be provided by the demographics of host galaxies. In fact, the distribution of merger timescales impacts the mix of early- and late-type hosts: generally speaking, the smaller the merger delay, the stronger the connection with (recent) star formation (SF) rather than with stellar mass alone, and the larger the late-type fraction (e.g., Zheng & Ramirez-Ruiz 2007).

Radio continuum emission from galaxies traces the SF rate (SFR) during the last $\sim 50 - 100$ Myr (e.g., Murphy et al. 2011), and could thus help track the connection between mergers and recent SFR rate. In fact, due to the age-dust-metallicity degeneracy, the optical/UV emission alone is an unreliable measure of SFR in dusty galaxies. Of course, radio measurements are subject to AGN contamination. That said, radio measurements have been demonstrated to be an important complementary tool to constrain the SFR in the past, for example for short GRB hosts. Short GRB mergers are believed to be cosmological compact binary mergers containing at least one NS, so are immediately pertinent comparisons. Radio observations of the host galaxies of GRB 071227 and GRB 120804A provided clear evidence for SFRs at least an order of magnitude larger than those derived using optical data alone (Nicuesa Guelbenzu et al. 2014), suggesting either a starburst origin or an AGN contribution. The host galaxy

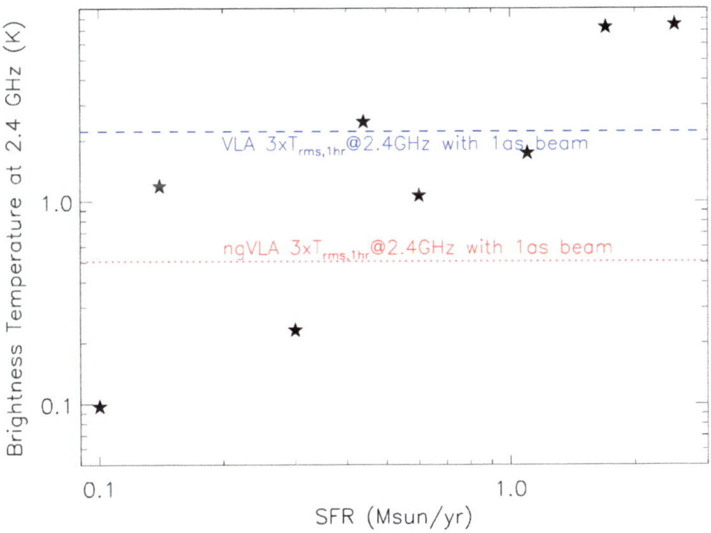

Figure 4. Predictions for the radio brightness temperature of NS-NS and BH-NS binary merger host galaxies located at 120 Mpc. These predictions are based on our current knowledge of short GRB hosts (see Table 1). The ngVLA will reach a surface brightness sensitivity of $\approx$ 0.1688 K in 1 hr with $\approx 1''$ resolution at 2.4 GHz, thus resolving all of these hosts at $d_L \lesssim 120$ Mpc (red dotted line). More than half of the short GRB-like host galaxies at $d_L \lesssim 120$ Mpc are inaccessible to the current Jansky VLA in 1 hr (blue dashed line).

NGC4993 of the recently detected binary NS merger GW170817, associated with the short GRB 170817A, shows a radio SFR of $\approx 0.1\,M_\odot$/yr, approximately 10× higher than the one estimated using the galaxy broad-band photometry, indicative of an AGN dominating the radio emission (Blanchard et al. 2017). We note that while the FIR is also an extinction free SFR tracer, ground-based extinction-free radio observations can be advantageous compared to satellite-based FIR observations.

The study of sub-galactic environments (achievable via *resolved* multi-wavelength studies of the host galaxies) can be used to gain other clues to the progenitors of compact binary mergers, such as the presence or absence of a spatial association with star formation or stellar mass, and the distribution of offsets of EM counterparts with respect to their host galaxy light (which can be used to map natal kicks, e.g., Belczynski et al. 2006; Behroozi et al. 2014, and references therein).

While currently most of the cosmological short GRB host galaxies remain undetected in the radio, a sensitive array like the ngVLA will be able to resolve short GRB-like hosts within the LIGO horizon distance. To show this, in Table 1 we collected the (optical/UV) SFR and optical sizes of short GRB hosts available in the literature (Berger 2014). We use the measured SFR to calculate the expected host galaxy luminosity at 1.4 GHz via the Murphy et al. (2011) SFR-to-non-thermal radio emission calibration relation: $\left(\frac{\mathrm{SFR}_{1.4\,\mathrm{GHz}}}{M_\odot \mathrm{yr}^{-1}}\right) = 6.35 \times 10^{-29} \left(\frac{L_{1.4\,\mathrm{GHz}}}{\mathrm{erg\,s^{-1}Hz^{-1}}}\right)$. To estimate the corresponding luminosity at $\approx$ 2.4 GHz (the central frequency of the ngVLA lowest frequency band), we extrapolate from the 1.4 GHz luminosity using a spectral index of ≈ -0.75 (as appropriate for non-thermal galaxy emission). Since radio emission from local galaxies is observed to be more concentrated than in the optical (Condon et al. 2002; Murphy et al. 2017), we then use the effective galaxy radius r_e from optical/IR observations as FWHM size of the galaxy at radio wavelengths (Fong & Berger 2013), and calculate the brightness temperature as: $T = 1.36\frac{\lambda^2}{\theta^2}S$, where λ is the observed wavelength in cm, θ is the FWHM radio size in arcsec, and S is the flux density in mJy at the considered wavelength ($\lambda \approx 12.5$ cm for observations at 2.4 GHz). Our results are plotted in Figure 4. The ngVLA will reach a surface brightness sensitivity of ≈ 0.1688 K in 1 hr with $\approx 1''$ resolution at 2.4 GHz, thus resolving most of these hosts at $d_L \lesssim 120$ Mpc. The current Jansky VLA in its A configuration would offer a comparable resolution of $\approx 1''$ at 2.4 GHz, but with a worse surface brightness sensitivity in 1 hr of ≈ 0.742 K. More than half of the host galaxies in Figure 4 have a brightness temperature $\lesssim 3 \times 0.742\,\mathrm{K} \approx 2.2$ K, thus the ngVLA is likely to greatly enlarge the sample of resolvable short GRB-like host galaxies.

6. Summary and Conclusion

We have investigated three new scientific opportunities that would emerge in time-domain astrophysics if a facility like the ngVLA were to work in tandem with ground-based GW detectors, and shown that:

1. The ngVLA, thanks to its superior sensitivity, will substantially extend the reach of the current Jansky VLA for discovering and characterizing radio afterglows of NS-NS/BH-NS mergers, and their linear polarization fractions, to distances 3× as large (for which event rates are a factor of $\sim 30\times$ larger);

2. The ngVLA LBA, thanks to its superior sensitivity at the longest baselines, can enable direct size measurements of radio afterglows from NS-NS and/or BH-NS mergers, probe directly the dynamics of merger ejecta and, when VLBI is paired with linear polarization studies, constrain the magnetic field structure;

3. The ngVLA, thanks to its improved surface brightness sensitivity at lower frequencies, can also enable resolved studies of NS-NS and/or BH-NS mergers host galaxies within distances of ≈ 120 Mpc.

Acknowledgments. A.C. thanks Chris Carilli and Remy Indebetouw for helping with ngVLA simulations in CASA. A.C. and D.C. acknowledge support form NSF CAREER award #1455090. The NRAO is a facility of the National Science Foundation operated under cooperative agreement by Associated Universities, Inc.

References

Abbott, B. P., Abbott, R., Abbott, T. D., et al. 2016a, Phys. Rev. Lett., 116, 241103
—. 2016b, Phys. Rev. Lett., 116, 061102
—. 2016c, Living Reviews in Relativity, 19, 1
—. 2017a, Phys. Rev. Lett., 119, 161101
—. 2017b, ApJL, 848, L12
—. 2017c, Phys. Rev. Lett., 118, 221101

—. 2017d, Phys. Rev. Lett., 119, 141101
—. 2017e, ApJL, 848, L13
Behroozi, P. S., Ramirez-Ruiz, E., & Fryer, C. L. 2014, ApJ, 792, 123
Belczynski, K., Perna, R., Bulik, T., et al. 2006, ApJ, 648, 1110
Berger, E. 2014, ARA&A, 52, 43
Blanchard, P. K., Berger, E., Fong, W., et al. 2017, ApJ, 848, L22
Bolatto, A., et al. 2017, ngVLA Memo Series, 19
Bower, G. C., et al. 2015, ngVLA Memo Series, 9
Carbone, D., & Corsi, A. 2018, ApJ, submitted
Condon, J. J., Cotton, W. D., & Broderick, J. J. 2002, AJ, 124, 675
Corsi, A., Hallinan, G. W., Lazzati, D., et al. 2018, ArXiv e-prints, arXiv:1806.03136
Coulter, D. A., Foley, R. J., Kilpatrick, C. D., et al. 2017, ArXiv e-prints, arXiv:1710.05452
Dobie, D., Kaplan, D. L., Murphy, T., et al. 2018, ArXiv e-prints, arXiv:1803.06853
Evans, P. A., Cenko, S. B., Kennea, J. A., et al. 2017, ArXiv e-prints, arXiv:1710.05437
Fong, W., & Berger, E. 2013, ApJ, 776, 18
Gill, R., & Granot, J. 2018, ArXiv e-prints, arXiv:1803.05892
Granot, J., Piran, T., & Sari, R. 1999, ApJ, 513, 679
Hallinan, G., Corsi, A., Mooley, K. P., et al. 2017, Science, 358, 1579
Hotokezaka, K., Nissanke, S., Hallinan, G., et al. 2016, ApJ, 831, 190
Kasliwal, M. M., Nakar, E., Singer, L. P., et al. 2017, Science, 358, 1559
Lazzati, D., Deich, A., Morsony, B. J., & Workman, J. C. 2017, MNRAS, 471, 1652
Lazzati, D., Perna, R., Morsony, B. J., et al. 2018, ArXiv e-prints, arXiv:1712.03237
Lloyd-Ronning, N. 2018, Science with a Next-Generation Very Large Array, ASP Monograph 7, ed. E. J. Murphy (San Francisco, CA: ASP), 701
Mooley, K. P., Nakar, E., Hotokezaka, K., et al. 2018, Nature, 554, 207
Murphy, E. J., Momjian, E., Condon, J. J., et al. 2017, ApJ, 839, 35
Murphy, E. J., Condon, J. J., Schinnerer, E., et al. 2011, ApJ, 737, 67
Nakar, E., Gottlieb, O., Piran, T., Kasliwal, M. M., & Hallinan, G. 2018, ArXiv e-prints, arXiv:1803.07595
Nakar, E., & Piran, T. 2011, Nat, 478, 82
—. 2018, ArXiv e-prints, arXiv:1801.09712
Nicuesa Guelbenzu, A., Klose, S., Michaowski, M., et al. 2014, ApJ, 789, 45
Raccanelli, A., Kovetz, E. D., Bird, S., Cholis, I., & Muñoz, J. B. 2016, Phys.Rev.D, 94, 023516
Rossi, E. M., Lazzati, D., Salmonson, J. D., & Ghisellini, G. 2004, MNRAS, 354, 86
Selina, R. J., Murphy, E. J., McKinnon, M., et al. 2018, Science with a Next-Generation Very Large Array, ASP Monograph 7, ed. E. J. Murphy (San Francisco, CA: ASP), 15
Troja, E., Piro, L., van Eerten, H., et al. 2017, Nat, 551, 71
Valenti, S., David, Sand, J., et al. 2017, ApJ, 848, L24
van Eerten, H., van der Horst, A., & MacFadyen, A. 2012, ApJ, 749, 44
Zheng, Z., & Ramirez-Ruiz, E. 2007, ApJ, 665, 1220

Science with a Next-Generation Very Large Array
ASP Conference Series, Monograph 7
Eric J. Murphy, ed.

Radio Emission from Short Gamma-ray Bursts in the Multi-messenger Era

Nicole Lloyd-Ronning

Los Alamos National Lab, Los Alamos, NM, USA 87545; `lloyd-ronning@lanl.gov`

University of New Mexico, Los Alamos, NM, USA 87544

Abstract. We examine the expected radio emission from short GRBs detectable by the ngVLA, guided by the observed distributions of their afterglow parameters. We give estimates for the peak emission from both the forward and reverse shock, as well as the late-time off-axis jet emission. In the context of neutron star merger model for short GRBs, we also estimate the radio emission from the dynamical ejecta of these double neutron star systems. The latter two quasi-spherical radio emission components are particularly important given the advent of gravitational wave (GW) detections of binary compact object mergers. We discuss how the ngVLA has potential to shed light on the physics behind short GRBs, and estimate the fraction of GRBs detectable in the radio that may be coincident with a GW signal from a neutron star (NS) merger.

1. Introduction

The most robust model for short gamma-ray bursts (sGRBs) is the merger of two compact objects, such as two neutron stars (NS-NS) or a neutron star and a black hole (NS-BH). The timescales and energetics involved in the merger have always made these systems plausible progenitors for sGRBs (Eichler et al. 1989; Narayan et al. 1992), but other clues including the location of these bursts in their host galaxies, the lack of associated supernovae, and the observed sGRB rates have provided convincing evidence that these bursts are associated with the older stellar populations expected of compact objects (Rosswog & Ramirez-Ruiz 2003; Fox et al. 2005; Soderberg et al. 2006a; Lee & Ramirez-Ruiz 2007; Berger et al. 2009; Kocevski et al. 2010; Leibler & Berger 2010; Fong et al. 2010; Berger 2010; Fong et al. 2013, 2014). The recent detection of gravitational waves from a neutron star merger (Abbott et al. 2017) and the associated short gamma-ray burst (Abbott et al. 2017b) have provided the smoking-gun evidence that at least some sGRB progenitors result from binary neutron star mergers.

There has been a concerted effort to follow up short GRBs with the goal of detecting the afterglow and potentially learning more about this class of gamma-ray bursts (for a review, see Berger 2014). To date, about 93%, 84%,, and 58% of sGRBs have been followed up in the X-ray, optical, and radio respectively (Fong et al. 2015). Of these follow-up efforts, 74% have an X-ray afterglow, 34% have been seen in the optical, and only 7% in the radio.

Radio emission in particular has been suggested as a useful tool to help elucidate sGRBs (Nakar & Piran 2011; Metzger & Berger 2012; Berger 2014; Hotokezaka et al. 2016; Resmi 2017). In the classic fireball model (Meszaros & Rees 1997; Sari, Piran, & Narayan 1998), the afterglow emission comes from synchrotron radiation from the forward (and sometimes reverse) shock of the blast wave; the radio component tends to peak at later times, farther from the central engine, more clearly probing the circumstellar environment of the progenitor (compared to the optical and X-ray occurring at earlier times, closer to the cataclysmic event). In addition, the radio afterglow can in principle be detected when the jet has decelerated (so that the radiation is not strongly beamed), and be detected even when the observer is not aligned with the axis of the GRB jet (Levinson et al. 2002; Totani & Panaitescu 2002), or potentially (if there is enough baryon contamination) from a hot cocoon surrounding the merger site (Ramirez-Ruiz et al. 2002; Murguia-Berthier et al. 2014; Lazzati et al. 2016). Finally, models of compact object mergers predict a significant amount of mass tidally ejected during the merger process (Rosswog 2005; Oechslin et al. 2007; Korobkin et al. 2012) which can radiate in several ways, including via shocks with the external medium (Nakar & Piran 2011). These latter two emission scenarios are particularly exciting as they are potential electromagnetic counterparts (and hence carry additional important information) to a gravitational wave signal from a double neutron star (DNS) merger event.

The ngVLA – with its broadband frequency coverage and superb sensitivity – is the ideal instrument to study the radio emission components of sGRBs. In particular, the ngVLA frequency range of 1.2 to 116 GHz is a prime bandpass for catching the early radio reverse shock emission, early-time self-absorbed forward shock emission, later-time optically thin forward shock and off-axis jet emission. As we discuss below, discerning these emission components is the key to constraining the physical parameters that determine the emission of a compact object merger.

2. Radio Emission Components from Compact Object Mergers

In the standard picture of a relativistic external blast wave, the onset of the afterglow occurs around the deceleration time – i.e., when the blast wave has swept up enough external material to begin to decelerate $t_{dec} \propto (E/n)^{1/3}\Gamma^{-8/3}$ (Blandford & McKee 1976). Assuming the emission mechanism is primarily synchrotron radiation one can calculate the characteristic synchrotron break frequencies at this time, depending on the global and microphysical parameters of the burst. These characteristic frequencies are ν_a, the synchrotron self-absorption frequency, ν_m, the frequency corresponding to the "minimum" (characteristic) electron energy, and ν_c, the frequency corresponding to the energy at which an electron loses most of its energy to radiation These expressions are given in Table 2 of Granot & Sari (2002) for both a constant density and wind medium.

In general, $\nu_a < \nu_m < \nu_c$ for the forward shock component. We can, in the context of this model, calculate how the frequencies evolve with time and when they enter the radio band. For optically thin emission, ν_m is usually the most relevant – i.e., the flux density is brightest at this frequency and it should dominate the spectrum in the radio band (here, we focus on optically thin emission keeping in mind this provides upper limits to what we should detect in the radio band, although see the brief discussion of the effects of self-absorption below). The cooling frequency ν_c generally stays well above the radio band for an extended time (this may in any case be an oversimplification of the acceleration and cooling processes – see, e.g., Lloyd & Petrosian 2000, for a discussion on how realistic continual acceleration of particles eliminates this characteristic frequency). For the reverse shock, the minimum electron frequency is roughly $\nu_{m,RS} \approx \nu_m/\Gamma^2$ (assuming the fraction of energy in the magnetic field is roughly the same for the forward and reverse shock, as explained below), and thus peaks in the radio at earlier times than the forward shock. We discuss the radio emission from the different components of a compact object merger below.

2.1. Jet Forward Shock

For a constant density medium, we consider the peak flux density at ν_m when it enters the radio band. This frequency is given by:

$$\nu_m = 10^{15}\ \text{Hz} \cdot \left\{3.73(p - 0.67)(1 + z)^{1/2}\epsilon_e^2\epsilon_B^{1/2}E_{52}^{1/2}t_{days}^{-3/2}\right\} \tag{1}$$

where p is the electron energy power-law index, z is the redshift, ϵ_e is the fraction of energy in the electrons, ϵ_B is the fraction of energy in the magnetic field, E_{52} is the isotropic equivalent energy normalized to 10^{52} ergs and t_{days} is the time normalized to a day.

Notice this expression is particularly sensitive to the fraction of energy in the electrons. The peak flux at this frequency is given by:

$$f_p(\nu_m) = 9930\ \mu\text{Jy} \cdot \left\{(p + 0.14)(1 + z)\epsilon_B^{1/2}n^{1/2}E_{52}d_{L28}^{-2}\right\}. \tag{2}$$

2.2. Jet Reverse Shock

There have been many studies of the reverse shock from a relativistic blast wave (e.g., Meszaros & Rees 1997; Sari & Piran 1999; Kobayashi 2000; Zhang et al. 2003; Kobayashi & Zhang 2003; Zou et al. 2005 and references therein), and the early-time radio flare observation of GRB 990123 has been attributed to the reverse shock (Kulkarni, Frail & Sari 1999; Nakar & Piran 2005). In addition, Soderberg & Ramirez-Ruiz (2003) examined the expected strength of the reverse shock in six long GRBs, and were able to constrain the hydrodynamic evolution and bulk Lorentz factors of these bursts from this component.

As pointed out by these references and others, the evolution of the flux density and break frequencies in the reverse shock depends on whether the blast wave is Newtonian or relativistic (among other factors), which in turn is related to the shell thickness Δ estimated from the observed prompt gamma-ray burst duration T by $\Delta \sim cT/(1 + z)$. For a thick shell, $\Delta > l/2\Gamma^{8/3}$, where l is the Sedov length in an ISM $\equiv (3E/4\pi nm_pc^2)^{1/3}$, the reverse shock has time to become relativistic and the standard Blandford-McKee solution applies. For a thin shell, the reverse shock remains Newtonian and the Lorentz factor of this shock evolves as $\Gamma_{RS} \sim r^{-g}$, with $g \sim 2$ (Kobayashi 2000). Short bursts with $T < 1$ s are likely in the thin shell – and therefore Newtonian – regime. However, we note that for a range of g values, the time evolution of the flux density and characteristic frequencies are fairly similar between the relativistic and Newtonian regimes.

Generally speaking, because of the higher mass density in the shell, the peak flux density in the reverse shock $f_{p,RS}$ will be higher by a factor of Γ relative to the forward shock,

$$f_{p,RS} \approx \Gamma f_{p,FS} \tag{3}$$

but the minimum electron frequency in the reverse shock $\nu_{m,RS}$ will be lower by a factor of Γ^2,

$$\nu_{m,RS} \approx \nu_{m,FS}/\Gamma^2 \tag{4}$$

assuming the forward and reverse shock have the same fraction of energy in the magnetic field (not necessarily a well-justified assumption). At early times, when this emission is brightest, this frequency falls right in the range of the ngVLA bandpass for typical sGRB parameters; the ngVLA gives us the opportunity for the first time to catch the peak of this reverse shock emission, breaking the degeneracies among the various parameters that determine sGRB emission.

2.2.1. Self-Absorbed Reverse Shock

At radio wavelengths, synchrotron self-absorption should be considered – under certain conditions lower energy photons are self-absorbed, and the flux density is suppressed. Self-absorption may be particularly relevant in the region of reverse shock, where the density is higher relative to the forward shock region. Resmi & Zhang (2016) calculated the relevance of the self-absorption frequency and flux density in the reverse shock, before and after shock crossing. For later time radio emission, we consider the frequencies and flux densities after the shock crosses the thin shell (but see their Appendix A.1 for expressions in all ranges of parameter space).
Roughly, at the high radio frequencies we are considering here, the flux density at the time of the peak can be obtained from equation 30 of Resmi & Zhang (2016):

$$f_{p,RS} = f_{p,RS,\nu_m}(\nu_{a,RS}/\nu_{m,RS})^{-\beta} \tag{5}$$

where $\beta = (p-1)/2$.
The reverse shock flux density is suppressed at a minimum by factors ranging from about 0.3 to 0.01. We emphasize again, therefore, that our estimates are upper limits to the emission from the reverse shock.

2.3. Jet Off-axis Emission

Because GRB outflows are relativistic, off-axis emission is highly suppressed due to the relativistic beaming of the radiation. However, once the Lorentz factor decreases to a value on the order of the inverse of the viewing angle of the observer, $\Gamma \sim 1/\theta_v$, the flux density is similar to an on-axis observer (Rhoads 1997; Granot et al. 2002; see also Figures 7 and 8 of Granot & van der Horst 2014).

As a simple estimate of the peak of the off-axis emission, we calculate the flux density at a time when the blast wave is no longer relativistic, so that the radiation is no longer relativistically beamed and is detectable to an observer well off-axis. This time - which occurs after the jet break time - is also an upper limit to the time of detection of the unbeamed flux density. The blast wave becomes non-relativistic around ~ 1 yr $(E_{52}/n)^{1/3}$ (Piran 2004); at this time, ν_m is usually well below the typical observed radio band. Hence, to estimate the radio emission we need the flux density above ν_m at a time when the blast wave has decelerated enough to become non-relativistic. Using Frail, Waxman, Kulkarni (2000), the flux density at 8.46GHz at the non-relativistic transition is:

$$\begin{aligned} F_{\nu>\nu_m} = 1635\,\mu\text{Jy}\,\{&(1+z)\epsilon_e\epsilon_B^{3/4}n^{3/4}E_{52}(t/t_{NR})^{3(1-p)/2+3/5} \\ &d_{L28}^{-2}(\nu/8.46\text{ GHz})^{(1-p)/2}(\theta_j/0.1)^2\} \end{aligned} \tag{6}$$

where θ_j is the physical opening angle of the jet. Note there are many ways to model off-axis emission (see, e.g., Nakar, Piran, & Granot 2002; Soderberg et al. 2006b; Waxman

2004; Oren et al. 2004), depending on the underlying assumptions of the behavior of the jet. This emission - the magnitude and time of the of the peak flux density in particular - depends of course on the observer viewing angle (for a recent estimation of radio flux density dependence on viewing angle, see Carbone & Corsi 2018). We emphasize again that we have taken the very conservative estimate of the flux density at a time when the blast wave is non-relativistic, and the emission is isotropic and observable to viewers at all angles.

2.4. Tidal or Dynamical Ejecta

In addition to emission from the jet component, there is also emission from mass either dynamically ejected or from winds during the merger process (Rosswog et al. 1999; Ruffert & Janka 2001; Yamamoto, Shibata, & Taniguchi 2008; Rezzolla et al. 2010). This mass (typically in the range from $\sim 0.01 - 0.1\,M_{\odot}$) is ejected with a velocity $\beta_i = v/c \sim 0.1$ (Nakar & Piran 2011), and can shock with the external medium, emitting in the radio band. We note that some simulations have found higher values for the tidal ejecta velocity, $\sim 0.2c - 0.3c$, (e.g., Radice et al. 2016), which will increase the values of the peak flux density; however, this value depends strongly on the equation of state of the neutron star, and a definitive value for the ejecta velocity from neutron star mergers is far from settled (for a brief discussion of this issue, see Metzger 2017). We note importantly that the bulk of the ejecta mass (which likely came from the wind component of the merger) from GW170817 was fit with a velocity of $0.08c$ (Troja et al. 2017).

If the observed frequency is above the minimum electron frequency ν_m and the self-absorption frequency ν_a – both expected to be less than around 1 GHz for canonical values of energies and densities for mergers – then the flux density will peak at a time (Nakar & Piran 2011; Nakar, Piran, & Rosswog 2013):

$$t_{dec} \approx 300 \text{ days } E_{51}^{1/3} n^{-1/3} \beta_i^{-5/3} \tag{7}$$

where $t_{dec} = R_{dec}/c\beta_i$ is the deceleration time [when the ejecta has swept up mass comparable to its own at a radius R_{dec}; after t_{dec}, the flow decelerates in Sedov-Taylor flow $\beta \approx \beta_i (R/R_{dec})^{-3/2}$].

The peak of the specific flux density occurs at t_{dec} and is given by (Nakar & Piran 2011):

$$F_{\nu>\nu_m} \approx 4012\,\mu\text{Jy } E_{52} n^{\frac{p+1}{4}} \epsilon_B^{\frac{p+1}{4}} \epsilon_e^{p-1} \beta_i^{\frac{5p-7}{2}} d_{28}^{-2} \left(\frac{\nu_{obs}}{8.46GHz}\right)^{-\frac{p-1}{2}} \tag{8}$$

Note our change of units compared to Nakar & Piran (2011) and Berger (2014), that ϵ_B and ϵ_e are the absolute, non-normalized values, and ν_{obs} is the observed frequency normalized to 8.46 GHz.

3. Predicted Radio Emission from a Compact Merger

Figure 1 shows the predicted radio flux density at 8.46 GHz from the forward shock (blue circles), reverse shock (red stars), off-axis emission (green squares), and tidal ejecta (light blue diamonds) for different models, described in the figure caption. For the top panels of Figure 1, we drew each of the physical GRB parameters from a Gaussian distribution with a normalized mean of 1, except for the electron energy power-law

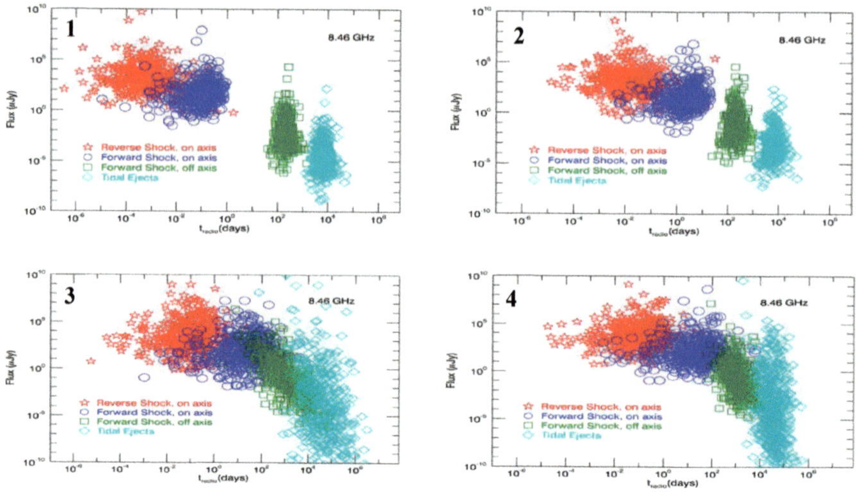

Figure 1. (*Top Left:*) Model 1 – Flux as a function of time using densities from a distribution centered around 0.1 cm^{-3}, a mean energy of $E = 10^{50}$ erg, a mean Lorentz factor of 50, a mean redshift of 0.3, a fixed value of $p = 2.3$, and ϵ_e and ϵ_B drawn from a Gaussian with a mean of 0.01. For the tidal ejecta component, we assumed an ejecta velocity $\beta_i = 0.1$. (*Top Right:*) Model 2 – Same as Model 1 in left panel, but with ϵ_e and ϵ_B drawn from a Gaussian with a mean of 0.1 and 0.01 respectively. (*Bottom Left:*) Model 3 – Flux as a function of time using distributions of parameters mimicking those from the multi-wavelength fits of Fong et al. (2015). Densities are from a distribution centered around 0.005 cm^{-3} with a large spread (~ 2.5 orders of magnitude), a mean energy of $E = 10^{51}$ erg, a mean redshift of 0.5, a mean value of $p = 2.43$. The parameters ϵ_e and ϵ_B are both drawn from a Gaussian with a mean of 0.1. The Lorentz factor employed is $\Gamma = 50$. (*Bottom Right:*) Model 4 – Same as Model 3, but with less spread in the density distribution (a standard deviation of 1.25 orders of magnitude as opposed to 2.5 orders of magnitude).

index which was fixed at $p = 2.3$, and, in the case of the tidal ejecta component, an ejecta velocity of $\beta_i = 0.1$. We used a mean density of 0.1 cm^{-3}, a mean energy of $E = 10^{50}$ erg, a mean Lorentz factor of 50, and a mean redshift of 0.3. The averages of the parameters ϵ_e and ϵ_B varied from 0.1 to 0.01. These plots show that the variation in GRB parameters can give a wide range of flux density values.

For the bottom panels in Figure 1, we plot the predicted radio flux densities, where spectral parameters were taken from distributions that mimic the observed or fitted distributions from Fong et al. (2015). The difference between these two bottom panels is the choice of standard deviation of the density distribution in log space (2.5 and 1.5 for the left and right panels respectively). Again, there is a large distribution in flux density from the various emission components for our choice of spectral parameters. In principle, we may be able to distinguish these components from the time at which they peak in the radio. This is relatively well delineated, with the reverse shock peaking earliest, then the forward shock, off-axis emission and finally, the dynamical ejecta component. As is always the case with gamma-ray bursts, the more multi-band tem-

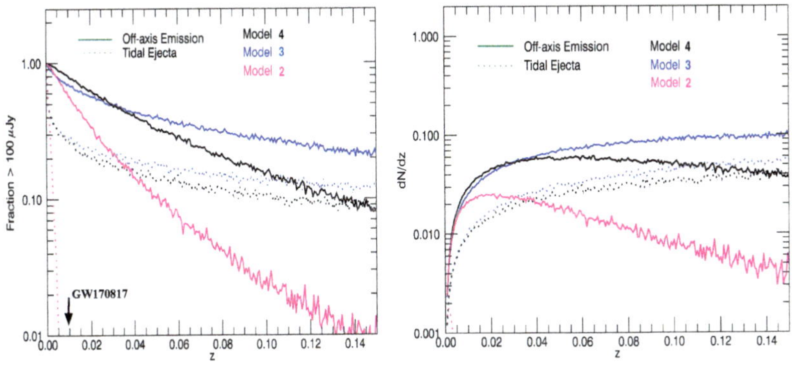

Figure 2. (*Left:*) Fraction of GRBs with off-axis emission (solid lines) and tidal ejecta emission (dotted lines) above 100 μJy at low redshifts (relevant to the sensitivity of aLIGO in detecting gravitational waves from neutron star mergers), for models 2 - 4 described in Figure 1. (*Right:*) The fraction from the left panel multiplied by the sGRB differential distribution as a function of z. Details on the distribution used are described in the text. Note we have marked the distance of the neutron star merger/gamma-ray burst GW170817 for reference.

poral and spectral information we can obtain for each burst, the better chance we have of putting constraints on their emission and therefore underlying spectral parameters. In particular, multi-wavelength observations will help us constrain the characteristic frequencies (and their corresponding flux densities) of the spectrum, helping to break some of the degeneracy between various physical GRB parameters. In addition to this, however, there is a real need from a theoretical standpoint to better understand the microphysics of relativistic shocks (through, for example, long timescale particle-in-cell simulations) and place tighter constraints on the electron energy index p, as well as the parameters ϵ_e and ϵ_B, both of which can in principle take on a wide range of values (spanning orders of magnitude) and contribute to large uncertainty in the expected flux density.

3.1. Prospects for "Orphan" Components, and EM Detections Coincident with aLIGO

Because the off-axis and tidal ejecta components are quasi-isotropic, we can ask what fraction of these components we might detect in the radio, if the GRB jet is not directed toward us (such that the prompt gamma-ray signal is not detected). With the successful detection of gravitational waves from astrophysical events (Abbott et al. 2016), and in particular the detection of gravitational waves and electromagnetic emission from GW170817 (Abbott et al. 2017), we have definitively entered a new era of multi-messenger astronomy.

The left panel of Figure 2 shows the fraction of GRBs with tidal ejecta (dotted lines) and off-axis components (solid lines) that fall above a conservative limit of 100 μJy out to a redshift $z \sim 0.15$ (relevant to aLIGO's design sensitivity in detecting a GW signal from a DNS). We show curves for models 2, 3 and 4 in Figure 1, under the assumption that we have full time coverage (i.e., that we are pointing our tele-

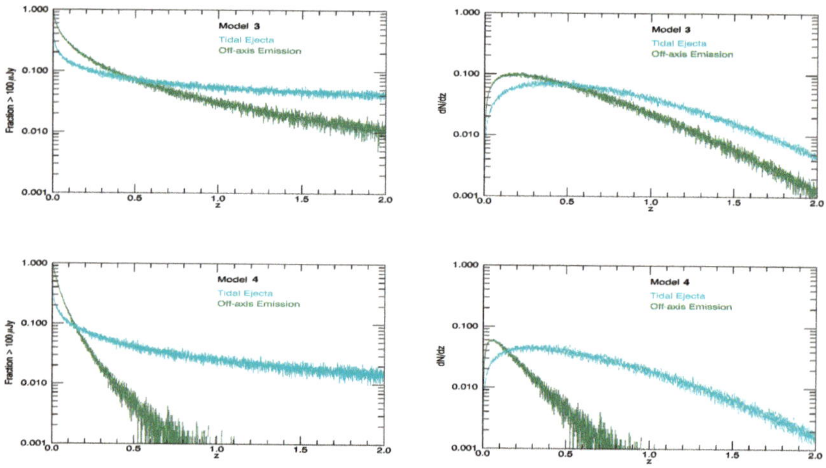

Figure 3. (*Left:*) Fraction of GRBs with off-axis emission (green) and tidal ejecta emission (cyan) above 100 μJy as a function of redshift, for the Models 3 (top) and 4 (bottom) described in Figure 1. (*Right:*) The fraction from the left panels multiplied by the sGRB differential distribution as a function of z. Details on the distribution used are described in the text.

scopes at the GRB at the time of the peak). We find that for the models that mimic the observed/fitted sGRB parameters from Fong et al. (2015), we expect to detect the quasi-isotropic radio emission component in $\sim 10-20\%$ of events that occur at distances to which aLIGO will be (at design specification) sensitive to NS mergers ($\sim$ 100 Mpc). Note that at the distance of GW170817 (40 Mpc or $z = 0.009783$), we expect to detect the off-axis radio component from the sGRB jet and it appears that this component was indeed detected (Alexander et al. 2017b). The right panel of Figure 2 shows this fraction multiplied by a fiducial sGRB differential redshift distribution, modeled after the solid line in Figure 4 of Guetta & Piran (2006) – a distribution of redshift that peaks at about $z = 0.5$ and falls off gradually to zero by $z \sim 2$. The left and right panels of Figure 3 are similar to Figure 2, but extended to a redshift of 2. See the chapter by Corsi et al. (2018, this volume p. 689) for further discussion of coincident gravitational and electromagnetic waves from compact binary mergers.

4. The Future of sGRBs with ngVLA

The ngVLA - with its broad spectral range and sensitivity - will allow us for the first time to better discriminate between different components of sGRB emission and elucidate the physics behind these fascinating events.

We can get additional important information on short gamma-ray bursts if there is rapid follow-up (< 1 day) in the radio - this will give the best chance of detecting the reverse shock emission component, and the ngVLA may afford us this opportunity. The circumburst density must also be low enough to allow for a slow-cooling reverse shock (as mentioned in Laskar et al. 2013, 2016), but such densities are expected for compact object binary progenitors of sGRBs.

The detectable radio components to DNS or NS-BH or mergers are not necessarily easy to distinguish given the observed range of sGRB physical parameters (in other words, the distributions of sGRB parameters based on afterglow fits give a wide spread in their radio flux density values). However, detecting the time of the peak of each emission component is perhaps the biggest distinguishing factor (see Figure 1), and spectral information constraining the peak flux density of each component is crucial. The ngVLA's wider spectral coverage will be crucial in separating these components.

The fraction of sGRBs with an off-axis "orphan" component detectable in the radio (at late times when the blast wave is non-relativistic and the emission is no longer beamed) and potentially coincident with a gravitational wave signal from a DNS merger is ~ 0.1 at distances to which aLIGO would be sensitive to such a signal. This statement is of course model dependent. However, for a reasonable set of models based on the Fong et al. (2015) fits to sGRB afterglow data, *we expect to detect the quasi-isotropic radio emission for about* 10% *(for the tidal ejecta emission) and up to* 20% *(for the off-axis emission) of sGRBs at a redshift of* $z \sim 0.15$. For shorter distances (e.g., the distance of GW170817 of 40 Mpc or $z = 0.009783$), there is a very high probability to detect the radio emission from the off-axis jet component for most fiducial models. At larger distances, the detectable fraction of these quasi-isotropic components falls off significantly.

The radio emission is a key piece of the puzzle in understanding GRB emission and combining ngVLA observations with additional multi-wavelength follow-up will allow us to constrain the underlying physics of the outflow producing short gamma-ray bursts. Efforts in this vein are particularly timely in light of potential additional detections by aLIGO of gravitational wave emission from a double neutron star merger. A better understanding of the various components of electromagnetic emission from these objects will provide a more complete picture of these systems and ultimately help us understand their role in the context of stellar evolution in the universe.

Acknowledgments. We thank Dieter Hartmann, Chris Fryer and Brandon Wiggins for discussions related to this work. Work at LANL was done under the auspices of the National Nuclear Security Administration of the US Department of Energy at Los Alamos National Laboratory LA-UR-17-24900

References

Abbott, B.P., et al. 2016, Phys. Rev. Lett, 116, 6
Abbott, B.P., et al. 2017, ApJ, 848, L43
Abbott, B.P., et al. 2017, ApJ, 848, L13
Alexander, K.D., et al. 2017, ApJ, 848, L21
Berger, E. , Cenko, S.B, Fox, D.B., & Cucchiara, A. 2009, ApJ, 704, 877
Berger, E. 2010, ApJ, 722, 1946
Berger, E. 2014, ARA&A, 52, 43
Blandford, R.D. & McKee, C.F. 1976, Phys.Fluids, 19, 1130
Carbone, D. & Corsi, A. 2018, arXiv 1801.02361
Corsi, A., Frail, D. A., Lazzati, D., et al. 2018, Science with a Next-Generation Very Large Array, ASP Monograph 7, ed. E. J. Murphy (San Francisco, CA: ASP), 689
Eichler, D., Livio, M., Piran, T., & Schramm, D. N. 1989, Nature, 240, 126
Fong, W., Berger, E., & Fox, D.B., 2010, ApJ, 708, 9
Fong, W., et al. 2013, ApJ, 769, 56
Fong, W., et al. 2014, ApJ, 780, 118
Fong, W., Berger, E., Margutti, R., & Zauderer B.A., 2015, ApJ, 815, 102

Fox, D.B., et al. 2005, Nature, 437, 845
Frail, D.A., Waxman E., Kulkarni, S.R., 2000, ApJ, 537, 191
Granot, J. & Sari, R. 2002, ApJ, 568, 820
Granot, J., Panaitescu, A., Kumar, P. & Woosley, S. 2002,
Granot, J. & van der Horst, A.J. 2014, PASA, 31, 8
Guetta, D. & Piran, T. 2006, A&A, 453, 823
Hotokezaka, K. et al. 2016, ApJ, 831, 190
Kobayashi, S.2000, ApJ, 545, 807
Kobayashi, S. & Zhang, B. 2003, ApJ, 597, 455
Kocevski, D., et al. 2010, MNRAS, 404, 963
Kopac, D. 2015, ApJ, 806, 179
Korobkin, O., Rosswog, S., Arcones, A., Winteler, C. 2012, MNRAS, 426, 1940
Kulkarni, S., Frail, D.A. & Sari, R. 1999, ApJ, 522, L97
Laskar, T. et al. 2013, ApJ, 776, 119
Laskar, T. et al. 2016, ApJ, submitted, arXiv1606.08873
Lazzati, D. et al. 2016, MNRAS, submitted, arXiv1610.01157
Lee, W.H. & Ramirez-Ruiz, E. 2007 NJPh, 9, 17
Leibler, C.N. & Berger, E. 2010, ApJ, 725, 1202
Levinson, A., Ofek, E. O., Waxman, E., Gal-Yam, A. 2002, ApJ, 576, 923
Lloyd, N.M. & Petrosian, V. 2000, ApJ, 543, 722
Meszaros, P. & Rees, M.J. 1997, ApJ, 476, 231
Metzger, B.D. & Berger, E. 2012, ApJ, 746, 48
Metzger, B.D. 2017, arXiv 1710.05931
Murguia-Berthier, A., et al. 2014, ApJ, 788, L8
Nakar, E. & Piran, T. 2005, ApJ, 619, L147
Nakar, E. & Piran, T. 2011, Nature, 478, 82
Nakar, E., Piran, T. & Granot, J. 2002, ApJ, 579, 699
Nakar, E., Piran, T. & Rosswog, S. 2013, MNRAS, 430, 2121
Narayan, R., Paczynski, B., & Piran, T. 1992, ApJ, 395, L83
Oechslin, R., Janka, H.-T., & Marek, M. 2007, A&A, 467, 395
Oren, Y., Nakar, E., & Piran, T. 2004, MNRAS, 353, L35
Piran, T. 2004, RvMP, 76, 1143
Radice, D., et al. 2016, MNRAS, 460, 3255
Ramirez-Ruiz, E., Celotti, A., Rees, M.J. 2002, MNRAS, 337, 1349
Resmi, L. & Zhang, B. 2016, ApJ, 825, 48
Resmi, L. 2017 JApA, 38, 56
Rezzolla, L., Baiotii, L., Giacomazzo, B., Link, D., Font, J.A. 2010, Classical Quantum Gravity, 27, 114105
Rhoads, J.E. 1997, ApJ, 478, L1
Rosswog, S., et al. 1999, A&A, 341, 499
Rosswog, S. & Ramirez-Ruiz, E. 2003, MNRAS, 343, L36
Rosswog, S. 2005, ApJ, 634, 1202
Ruffert, M. & Janka, H.-T. 2001, A& A 380, 544
Sari,R. & Piran, T. 1999, ApJ, 520, 641
Sari,R., Piran, T., & Narayan, R. 1998, ApJ, 497, L17
Soderberg, A. et al. 2006, ApJ, 638, 930
Soderberg, A. & Ramirez-Ruiz, E. 2003, MNRAS, 345, 854
Soderberg, A. et al. 2006, ApJ, 650, 261
Totani, T. & Panaitescu, A. 2002, ApJ, 576, 120
Troja, E. et al. 2107, Nature, in press
Waxman, E. 2004, ApJ, 602, 886
Yamamoto, T., Shibata, M. & Taniguchi, K. 2008, Phys. Rev. D. 78, 064054
Zhang, B., Kobayashi, S. & Meszaros, P. 2003, ApJ, 595, 950
Zou, Y.C, Wu, X.F. & Dai, Z.G. 2005, MNRAS, 363, 93

Science with a Next-Generation Very Large Array
ASP Conference Series, Monograph 7
Eric J. Murphy, ed.

Revealing the Galactic Population of BHs

Thomas J. Maccarone,[1] Laura Chomiuk,[2] Jay Strader,[2] James Miller-Jones,[3] and Greg Sivakoff[4]

[1]*Texas Tech University, Lubbock, TX, USA;* Thomas.Maccarone@ttu.edu

[2]*Michigan State University, East Lansing, MI, USA;* chomiuk@pa.msu.edu, strader@pa.msu.edu

[3]*Curtin University, Perth, WA, Australia* james.miller-jones@curtin.edu.au

[4]*University of Alberta, Edmonton, Alberta, Canada* sivakoff@ualberta.ca

Abstract. We discuss the case for using the Next Generation Very Large Array both to discover new black hole X-ray binaries astrometrically, and to characterize them. We anticipate that the ngVLA will be able to find $\sim$100 new black hole X-ray binaries, as well as a host of other interesting radio stars, in a few hundred hour survey. Parallax and astrometric wobble measurements will be achievable in feasible follow-up surveys especially using long baseline capabilities. The ngVLA's high angular resolution, high survey speed, and high frequency sensitivity give it a unique range of parameter space over which it is sensitive.

1. Description of the Problem

The ngVLA will illuminate the formation processes of black holes (BHs) in close binaries through high angular resolution, high sensitivity radio observations. A multi-epoch Galactic survey with the ngVLA will answer key questions like: How many binaries host BHs? What kinds of cosmic cataclysms produce BHs? How do BHs end up in close systems with other compact objects, in order to merge and produce gravitational waves? Such a survey is timely, as we currently stand at the dawn of gravitational wave astrophysics.

In recent years, we have seen the discoveries of stellar-mass BH-BH binaries merging, as well as one double neutron star merger (Abbott et al. 2016, 2017). While LIGO demonstrates the existence of inspiraling BH binaries (and can measures key parameters like mass and spin for these objects), understanding the origin of these systems requires information that can only come from electromagnetic studies of BH binary progenitors. Key questions remain about whether these objects form via standard binary stellar evolution (e.g., Belczynski et al. 2007), triple star evolution (e.g., Rodriguez & Antonini 2018), chemically homogeneous binary evolution (de Mink et al. 2009), or dynamically in globular clusters (e.g., Miller & Hamilton 2002). The rates of these mergers are dependent on a variety of poorly constrained parameters, including, but not limited to, common envelope efficiency, BH natal kicks distributions, and the range of masses of BHs in binaries. Understanding the populations of BHs in binaries in our own Galaxy will yield crucial constraints on all of these.

Furthermore, the same data that will yield the measurements needed to understand double compact object mergers will give vital insights into how the supernovae themselves explode. We currently know of only ≈20 dynamically confirmed stellar-mass BHs in our Milky Way. The known objects have mostly been discovered via bright X-ray outbursts, and then followed up spectroscopically in quiescence. This selection function creates a strong bias toward systems with long orbital periods (e.g., Arur & Maccarone 2018) Building a sample of accreting BHs in quiescence through a sensitive radio survey will yield a sample with much better understood selection effects. The ngVLA can build such a sample, with suffiicient sensitivity to detect quiescent BH X-ray binaries over a large volume of our Galaxy, and sufficient resolution to measure proper motions and thereby filter out background extragalactic contaminants.

2. Scientific Importance and Astronomical Impact

The total number of accreting stellar-mass BHs in our Galaxy is wildly unconstrained, potentially ranging from $100 - 10^8$ (Tetarenko et al. 2016), with clear dynamical signatures of about 20 (Casares & Jonker 2014) and about 60 strong candidates (Corral-Santana et al. 2016). Arguments based on the metal enrichment of the Galaxy suggest that there should be $\sim 10^8 - 10^9$ stellar-mass BHs in the Galaxy (e.g., Samland 1998), and this number is consistent with the upper limits from microlensing searches (e.g., Wyrzykowski et al. 2016). Typically 1–2 new stellar-mass black holes in outburst are discovered per year, and almost all of the known stellar-mass BHs are on the near side of the Milky Way Galaxy. These facts, plus modelling of selection effects suggest that the total number of BH low-mass X-ray binaries is likely to be a few thousand or more (Arur & Maccarone 2018).

The number with high mass donor stars is even more uncertain, in part because wind-fed systems may evade discovery because they do not host X-ray outbursts (see e.g Tetarenko et al. 2016). There may, in fact, be a population of systems very similar to the canonical high mass X-ray binary BH system Cygnus X-1, except with wider separations so that they have lower accretion rates and X-ray luminosities. These systems are poor bets to be discovered in X-ray and optical/IR surveys because of the broad similarities between their expected appearance and the expected appearance of generic massive stars. However, they should be readily distinguished at radio wavelengths by their spectral index; the flat radio spectra produced by an accreting BH's jets will contrast with the optically-thick thermal spectrum expected for a massive star:

An improved understanding of the population size of accreting BHs in the Milky Way will place strong constraints on some of the largest open questions in stellar astrophysics today. It will be a crucial constraint on the efficiency of common envelope evolution in close binary systems (e.g., Ivanova & Chaichenets 2011). It will also constrain the strength of kicks granted to BHs in their natal supernovae.

The amplitude of BH kicks can illuminate the primary channel of stellar-mass BH formation. It has been speculated that BHs may form from a prompt collapse of the entire massive star at the end if its lifetime or, alternatively, from fallback accretion onto a neutron star. The latter case should apply a much stronger natal kick to the BH at the time of formation, both because of the temporary presence of the neutron star (e.g., Kalogera 1996) and the symmetric mass loss from a moving object in a binary

(Blaauw 1961). The measurement of BH kicks is also important for determining if LIGO BH binaries are produced through normal binary evolution or dynamical interaction (i.e., in dense star clusters). The gravitational waveform reveals the misalignment of spin – of the black holes relative to one another, and relative to the orbital plane. Dynamically formed binaries should have an approximately random distribution of spin orientations, while binaries formed through binary evolution should have some preference for aligned spins – unless very large kicks take place. It is thus vital to determine whether typical BH kicks are large enough to misalign BH spins substantially from their orbital planes in order to understand how robustly misaligned spins indicate a globular cluster formation scenario. The scale heights of black holes and neutron stars in current samples seem to be quite similar (Repetto & Nelemans 2015)... However, there does seem to be at least one case, Cygnus X-1, where the applied kick appears quite small (Wong et al. 2012). An ngVLA survey for accreting BHs will constrain BH kicks in two ways: by measuring the number of BHs in close binaries in our Galaxy, and by directly measuring their proper motions.

We can also place constraints on the formation of LIGO BH mergers by comparing the BH populations in the Galactic field with those in dense stellar environments (globular clusters, Galactic center), and estimating the relative importance of binary and dynamical channels for forming close BH binaries. At the present time, there are some strong radio-selected BH candidates in globular clusters (e.g., Strader et al. 2012; Chomiuk et al. 2013). Both new detections and confirmations of existing candidates' membership can be made via astrometry.

A larger, more representative sample of black holes will enable a high-quality measurement of the BH mass distribution, and thereby constrain some of the most poorly understood aspects of supernova explosions. In particular for core-collapse supernovae, there is still no consensus about why the explosions actually take place, with the leading models being a standing accretion shock instability and the Rayleigh-Taylor instability (e.g., Belczynski et al. 2012, and references within). Because the process takes place while a thick envelope covers the stellar core, and the supernova light curves and spectra are largely sensitive only to the total energy input and the mass of the envelope, other approaches to understanding the events are vital. In principle, gravitational waves and neutrinos provide information at the time of explosion, but these are only detectable at the time of explosion and within small horizon volumes. The compact remnants of supernovae are therefore the best observational clues as to the processes driving supernovae—and specifically, the distribution of the masses of the compact stellar remnants is very information-rich. Belczynski et al. (2012) argue that models in which the explosion proceeds on a timescale of less than about 0.2 seconds after the core collapse will produce a substantial gap in masses between the heaviest neutron stars and the lightest BHs, while explosions which happen on timescales of order 0.5 seconds or more will lead to a continuous distribution of compact object masses. At the present time, there appears to be such a gap when the mass distribution is modelled as the sum of Gaussians (Özel et al. 2010; Farr et al. 2011), but other functional forms for the mass distribution allow a continuous range of masses (Farr et al. 2011).

The present size of the BH mass sample is simply not sufficient to establish firmly whether there is a mass gap. Furthermore, some biases in our estimates of the inclination angles of the binaries may be partially or fully responsible for the apparent presence of this mass gap (Kreidberg et al. 2012). Astrometric wobble measurements of even a few binaries would provide a gold-standard calibration for ellipsoidal modulation measurements, and would be possible with high sensitivity, high frequency VLBI measurements (see also the article by Reid et al. 2018, this volume p. 523). With the identification of a substantial population of accreting BHs, we will have in hand an ideal sample for follow-up observations enabled by both the ngVLA itself along with other multi-wavelength facilities available in the next decades. This larger sample will also be unbiased, including the full population of accreting stellar-mass black holes to enable a measurement of the BH mass function and the amplitude of BH natal kicks.

3. Anticipated Results

The ngVLA can detect accreting stellar-mass BHs over a large volume of our Galaxy, and distinguish them from interlopers. An astrometric survey would quickly yield the ability to separate background AGN from foreground X-ray binaries while simultaneously giving good measurements of their proper motions. Numerous other classes of Galactic sources will also be present in a deep, wide astrometric survey, and the combination of the radio properties with other multi-wavelength data sets will not only allow identification of which objects are stellar-mass BHs in binaries, but also will allow characterization of these other populations. We expect to detect bright flare stars, cataclysmic variables, planetary nebulae, and pulsars (most of which should identifiable based on radio spectra, radio variability, angular extent of radio emission or obvious associations with bright foreground stars), and also neutron star X-ray binaries, transitional millisecond pulsars and colliding wind binaries (which may require more careful follow-up work). A particularly interesting class of other objects that could be found in such a survey is isolated black holes accreting from the interstellar medium (Maccarone 2005; Fender, Maccarone & Heywood 2013).

The radio survey could be conducted in a reasonable exposure time, since objects like the radio-faintest known X-ray binaries (e.g., A0620-00 & XTE J1118+480; Gallo et al. 2006, 2014) could be detected in about 6 minutes at 6σ at a distance of $2-4$ kpc with the proposed ngVLA sensitivity. BHs accreting at higher rates, like V404 Cyg in quiescence, are a factor of ~ 15 more luminous (Miller-Jones et al. 2009) and can be observed well past the distance of the Galactic Center. We thus expect to be able to detect $\sim 10\%$ of the short period BH X-ray binaries in a 10 deg^2 survey region near the Galactic Center. This region contains about 10% of the stellar mass of the Galaxy, meaning that we would expect to detect at least 1% of the total number of short period BH X-ray binaries in the Milky Way, and the majority of the long period, V404 Cyg-like BH X-ray binaries. We thus expect that ~ 100 new BH X-ray binaries should be discovered in this proposed survey, based on population estimates from X-ray studies (e.g., Arur & Maccarone 2018), and perhaps much larger numbers based on the surprising discovery of a single strong BH candidate with a parallax placing it in front of M 15 (Tetarenko et al. 2016). We emphasize further that this proposed program would require a modest amount of time, ado a broad range of additional science, and be easily extendable over time both by extending the field of view and making deeper observations.

At a distance of 8 kpc, with 10 mas resolution (requiring significant collecting area on baselines of at least 1000 km), an uncertainty on proper motion of 20 km/s would result from a nominal two-year time baseline for 6σ detections. Thus, the uncertainty will be much less than the typical stellar velocity dispersion in the Solar Neighborhood, so only a small fraction of objects should be expected to have small enough proper motions to be confused with background AGN, and this fraction should decrease further if the majority of systems form with strong natal kicks.

Having long baselines (of approximately 1000 km or more) is thus crucial in order to allow sufficiently precise proper motion measurements to make the proper motions diagnostics sufficient both for ruling out background AGN sources, and allowing the measurements to yield useful information about natal kicks. While the proper motion uncertainty increases for sources behind the Galactic Center, if such sources follow the Galactic rotation curve, they will have space velocities of about 400 km/sec because of the rotation relative to the Earth's motion, so even sources at such distances, if bright enough to be detectable, should show measureable proper motions, even in the absence of natal kicks.

4. Limitations of Current Astronomical Instrumentation

Current long baseline arrays are simply not sensitive enough to meet our science goals. It would require 23 hours per pointing to reach our per-epoch sensitivity goal with the LBO – a factor of 4000 longer than with the ngVLA, and to cover $10\,\mathrm{deg}^2$ at 6 GHz would require about 640 pointings, so that it would take about 7 years at 6 hours per day to conduct this survey. The European long baseline facilities are too far north to do this work effectively (e.g., e-Merlin sees the Galactic Center region as a maximum elevation of 7°). The VLBA also runs into some problems with source scattering in the most scattered parts of the sky so that only a small fraction of its baselines become useful in such regions.

While the *Gaia* mission holds some promise for identifying BHs in very wide binaries from their astrometric wobble (Barstow et al. 2014; Mashian & Loeb 2017), it will struggle to obtain results on the optically faint quiescent binaries in the crowded, dust-extinguished region of the Galactic Bulge. This dense and populous region is essential for understanding the Galactic BH population and how its characteristics depend on density. Furthermore, the orbital period range probed by *Gaia* through astrometric wobble is much longer than the orbital period range probed by detecting accreting systems – if *Gaia* does discover many stellar mass BHs as wobblers, this information is complementary to what radio observations could discover. Only about 10% of the strong candidate BH X-ray binaries known are bright enough optically for *Gaia* to detect them, despite the strong biases in existing samples toward nearby, unreddened objects.

5. Connection to Unique ngVLA Capabilities

The long baselines (~1000 km) and high resolution of the ngVLA 214 antenna main array are critical to this astrometric survey. In addition, relatively high frequencies (> 4 GHz) are necessary to obtain good image quality in the Galactic Plane and the Galactic Bulge, as low frequencies will be affected by scattering. A broad range of baselines is needed for these studies in order that even in the most confused regions, like those in Cygnus and those very close to the Galactic Center, good angular resolution is possible without over-resolving heavily scatter-broadened sources.

5.1. Benefits to This Science From a More Extended Configuration

In addition, continental and global baselines would be of tremendous interest for follow-up of many of the sources discovered through an astrometric survey. Most of the known X-ray binaries are too faint optically for *Gaia* to provide parallax distances. Absolute astrometry with the *James Webb Space Telescope* is expected to be limited to about 0.5 mas, even at arbitrarily large signal to noise, meaning that geometric parallax distances will be limited to about 1 kpc. As a result, the best geometric parallax distances for these sources will come from radio data, rather than from optical data. Follow-up could be done with the ngVLA Long Baseline Array (LBA). While the follow-up would require more time per target, it could be done at higher frequencies to improve positional accuracy in a given integration time and mitigate the effects of scattering, and it could be done only for the sources of interest. It would make an excellent usage of the portion of the array that would not be needed when most of the core antennae are being used for low surface brightness extended source projects. To do this follow-up though, a sample of X-ray binaries with known radio brightness must first be collected.

6. Experimental Layout

We plan to observe in the Band 2 covering 3.5–12.3 GHz, as a compromise between field of view, resolution, and sensitivity (we expect the spectra of accreting BHs to be flat, $S_\nu \propto \nu^0$). We estimate that the required sensitivity of each epoch of the survey is 0.8 μJy beam^{-1} (achievable in 6 minutes on source). We require at least three visits covering the survey area, in order to obtain high-quality proper motion measurements. We anticipate achieving an image resolution of ~10 mas by working at medium frequency with 1000 km baselines.

By covering 10 deg^2 of the Galactic bulge, we will survey the BH population over a substantial fraction of the Galaxy and span a range of environments, including the dense central regions and star-forming regions and the Galactic field. We estimate that each epoch will require 100 hr on source, implying that the entire survey can be carried out in 300 hr.

After doing preliminary classification of the radio sources based solely on their radio properties, the AGN, pulsars, planetary nebulae, cataclysmic variables and isolated massive stars should have been filtered out of the data sample.

Many of these other classes of Galactic stellar radio sources will be interesting in their own rights. Remaining after filtering will be various classes of BH and neutron star X-ray binaries, active binaries and colliding wind binaries. For these, multi-wavelength follow-ups will be necessary. The active binaries will be the most challenging class of contaminant of the X-ray binary population, but the bulk of these should be identifiable from circular polarization for relatively high signal-to-noise sources. For fainter sources, radio/X-ray ratios, which will be high enough to be X-ray binaries only in flaring states, will be useful, since the source are highly unlikely to be flaring in all epochs.

7. Complementarity with Searches at Other Wavelengths

Non-simultaneous coordination with the WFIRST microlensing planet search fields will provide high cadence infrared data and thereby ellipsoidal modulation measurements to estimate orbital periods and inclination angles for many of the objects.[1] BH mass estimates can be obtained using optical/IR spectroscopy to measure the width of emission lines, in conjunction with the Casares (2016) relation. A healthy diverse portfolio of optical/IR telescopes spanning diameters, 4 – 30 m, would be ideal for the spectroscopic follow-up, but it is likely that once the sources of interest are identified that most should be bright enough that the Casares relation or some infrared equivalent will be suitable for estimating the radial velocity amplitude from emission lines.

Proposed future X-ray missions such as eROSITA and Athena (which are scheduled to proceed), and Lynx and STAR-X can be expected to deliver substantial populations of faint candidate X-ray binaries, interspersed with large numbers of members of other X-ray source classes. Because large fractions of these will be radio sources as well, particularly at the sensitivity levels we discuss here, the relatively uncrowded radio data, which will carry with them proper motion information, will help sort out the identification of these sources.

8. Additional Astrometric Stellar-Mass BH Science

Apart from the value of a large, dedicated Galactic survey, the ngVLA with the Long Baseline Array, would be vital for a variety of other stellar-mass BH science. In particular, with μJy sensitivity on >5000 km baselines, at 40 GHz would allow for geometric parallax distance measurements for most of these same stellar-mass BHs the survey would find, and for many accreting neutron stars, even in heavily scattered regions. With 0.2 mas resolution, and 6 σ detections, and assuming good phase calibration, one could expect parallax measurements accurate to about 15% for sources at the Galactic Center distance, and the accuracy would improve linearly with both signal to noise, and decreasing distance. Crucially, these objects will be predominantly too faint for *Gaia* even to detect, especially given their large extinctions. This thus represents an area where radio astrometry will provide crucial capabilities.

9. A Next Stage, More Ambitious Plan for Detecting More Sources

The plan outlined above shows that with 300 hours of ngVLA time, one would be able to do some important first order work on understanding the populations of X-ray binaries. It is important to consider, though, that this survey would discover $\sim$ 100 BH binaries, and would only reach Galactic Center distances for the brightest ones. To build a sample of many hundreds of BH binaries, one could easily expand this survey through a combination of deeper observations on the same part of the sky, and coverage of a wider swath of the Galactic Plane. Since observations within 1° of the Galactic Plane will reach a height of 500 pc only at the furthest reaches of the Galaxy, we can expect that increasing the exposure time will increase the number of sources roughly linearly,

[1] Because of the very large number of W UMa stars, using the WFIRST data alone is unlikely to produce good catalogs of BH and neutron star candidates.

since the volume included will scale as $t^{3/4}$, and until the most numerous source classes can be reached at the distance of the Galactic Center, the space density of sources should be rising with increased distance. Thus to reach a number of $\sim$ 1000 sources, we could conduct a survey which detects objects like A0620-00 at the distance of the Galactic Center, which would require quadrupling the exposure times for one epoch.

References

Abbott, B. P., et al. 2016, PRL, 6, 041015
Abbott, B. P., et al. 2017, PRL, 119, 161101
Arur K., Maccarone T.J., 2018, MNRAS, 474, 69
Barstow M., et al., 2014, "White paper: Gaia and the end states of stellar evolution", arXiv/1407.6163
Belczynski K., Taam R.E., Kalogera V., Rasio F.A., Bulik T., 2007, ApJ, 662, 504
Belczynski K., Wiktorowicz G., Fryer C.L., Holz D.E., Kalogera V., 2012, ApJ, 757, 91
Blaauw A., 1961, BAN, 15, 265
Casares J., 2016, ApJ, 822, 99
Casares J. & Jonker P.G., 2014, SSRv, 182, 223
Chomiuk L., Strader J., Maccarone T.J., Miller-Jones J.C.A., Heinke C. Noyola E., Seth A.C., Ransom S., 2013, ApJ, 777, 69
Corral-Santana J.M., Casares J., Muñoz-Darias T., Bauer F.E., Martínez-Pais I.G., Russell D.M., 2016, A&A, 587, 61
de Mink S., Cantiello M., Langer N., Pols O.R., Brott I., Yoon S.-C., 2009, A&A, 497, 243
Farr W., Sravan N., Cantrell A., Kreidberg L., Bailyn C.D., Mandel I., Kalogera V., 2011, ApJ, 741, 103
Fender R.P., Maccarone T.J., Heywood I., 2013, MNRAS, 430, 1538
Gallo E., Fender R.P., Miller-Jones J.C.A., Merloni A., Jonker P.G., Heinz S., Maccarone T.J., van der Klis M., 2006, MNRAS, 370, 1351
Gallo E., et al., 2014, MNRAS, 445, 290
Ivanova N., Chaichenets S., 2011, ApJ, 731L, 36
Kalogera V., 1996, ApJ, 471, 352
Kreidberg L., Bailyn C.D., Farr W.M., Kalogera V., 2012, ApJ, 757, 36
Maccarone T.J., 2005, MNRAS, 360L, 30
Mashian N., Loeb A., 2017, MNRAS, 470, 2611
Miller M.C., Hamilton D.P., 2002, MNRAS, 330, 232
Miller-Jones J.C.A., Jonker P.G., Dhawan V., Brisken W., Rupen M.P., Nelemans G., Gallo E., 2009, ApJ, 706L, 230
Özel F., Psaltis D., Narayan R., McClintock J.E., 2010, ApJ, 725, 1918
Reid M.J., McClintock J.E., Narayan R., Gou L., Remillard R.A., Orosz J.A., 2011, ApJ, 742, 83
Reid, M., Loinard, L., & Maccarone, T. 2018, Science with a Next-Generation Very Large Array, ASP Monograph 7, ed. E. J. Murphy (San Francisco, CA: ASP), 523
Repetto S., Nelemans G., 2015, MNRAS, 453, 3341
Reynolds C.S., 2014, SSRv, 183, 277
Rodriguez, C. L. & Antonini, F. 2018, arXiv 1805.08212
Samland M., 1998, ApJ, 496, 155
Strader J., Chomiuk L., Maccarone T.J., Miller-Jones J.C.A., Seth A.C., 2012, Nature, 490, 71
Tetarenko B.E., et al., 2016, ApJ, 825, 10
Wong T.-W., Valsecchi F., Fragos T., Kalogera V., 2012, ApJ, 747, 111
Wyrzykowski, L., et al., 2016, MNRAS, 458, 3012
Zhang S., Cui W., Chen W., 1997, ApJ, 482, L155

Science with a Next-Generation Very Large Array
ASP Conference Series, Monograph 7
Eric J. Murphy, ed.

Local Constraints on Supermassive Black Hole Seeds from a Next Generation Very Large Array

Richard M. Plotkin[1] and Amy E. Reines[2]

[1]*International Centre for Radio Astronomy Research - Curtin University, GPO Box U1987, Perth, WA 6845, Australia;* `richard.plotkin@curtin.edu.au`

[2]*Department of Physics, Montana State University, Bozeman, MT 59717, USA;* `amy.reines@montana.edu`

Abstract. Determining the mechanisms that formed and grew the first supermassive black holes is one of the top priorities in extragalactic astrophysics. Observational clues can be inferred from the demographics of massive black holes (in the ten thousand through million Solar mass range) in nearby low-mass galaxies. This chapter of the next generation Very Large Array (ngVLA) Science Book describes how an ngVLA can play a prominent role in developing large samples of weakly accreting active galactic nuclei in low-mass galaxies (out to nearly 1 Gpc), which will help constrain the types of objects that originally seeded the growth of supermassive black holes.

1. Introduction

Determining the formation mechanism(s) of the first black holes is one of the highest priority research areas in extragalactic astrophysics today. In this chapter we describe how a next generation Very Large Array (ngVLA) will provide key new constraints on the types of objects that originally "seeded" the growth of supermassive black holes. In particular, when combined with other multiwavelength probes, an ngVLA will be a crucial player in performing population studies of black holes in the $10^4 \lesssim M_{\rm BH} \lesssim 10^6\, M_\odot$ range, objects that we will refer to as massive black holes (mBHs).

Radio observations provide a window into one of the most energetically important components of accreting black holes, their relativistic jets. Jets can carry large amounts of mechanical power away from a black hole accretion disk, making them one of the primary channels through which black holes can influence their large-scale (kpc) environments (e.g., Fabian 2012). As strong synchrotron emitters, jets release large amounts of radiative power in the radio waveband. Thus, jetted radio emission can serve as a signpost that signals the presence of an accreting black hole. With its order of magnitude improvement in sensitivity and exquisite spatial resolution, an ngVLA will be a powerful machine for identifying and studying accreting mBHs, including studies on how mBHs influence their environments [for details on the latter, see the chapter by Nyland & Alatalo (2018, this volume p. 451).Combining ngVLA radio signatures with radiation at other wavebands (e..g., X-ray spectral and timing studies, high spatial-resolution infrared/optical spectroscopy and imaging, etc.) will provide a significantly improved understanding on the local mBH population, on mBH accretion and feedback, and on how black hole growth was seeded in the early Universe.

2. The First Black Holes

The existence of high-redshift quasars powered by $\sim 10^9\,M_\odot$ black holes at $z \gtrsim 7$ suggests that the first black hole seeds were 'heavy', with masses $M_{\rm BH} \sim 10^5\,M_\odot$ (which could have formed from the direct collapse of large clouds of gas, e.g., Loeb & Rasio 1994; Begelman et al. 2006). Otherwise, $10^9\,M_\odot$ black holes could not have grown so massive only 700 Myr after the Big Bang (assuming growth through Eddington limited accretion, e.g., Mortlock et al. 2011; Bañados et al. 2018). However, whether or not $10^5\,M_\odot$ seeds are the exception or the rule remains uncertain, and it is possible that some fraction of black holes could have been seeded with lighter objects in the $\sim 10^2\,M_\odot$ range (i.e., remnants from Population III stars, e.g., Haiman & Loeb 2001; Madau & Rees 2001; Madau et al. 2014).

While observations of high-redshift quasars provide important constraints and boundary conditions, it is not feasible with current facilities to directly observe black holes as small as $10^5\,M_\odot$ in the very high-redshift Universe (Volonteri & Reines 2016). To study low-mass black holes, we instead rely on clues that are embedded within black hole populations found in nearby galaxies. In particular, the fraction of galaxies hosting a nuclear black hole (as a function of galaxy stellar mass) and black hole/host galaxy scaling relations at low masses provide local diagnostics on the seed black hole population (see, e.g., Volonteri 2010; Greene 2012; Reines & Comastri 2016; Mezcua 2017 for reviews).

2.1. Low-Mass Black Holes in Nearby Dwarf Galaxies

Because of their relatively quiet evolutionary histories, and the ability for supernova feedback to stunt black hole growth, we do not expect black holes in nearby dwarf galaxies[1] to have grown much within a Hubble time (e.g., Dubois et al. 2015; Habouzit et al. 2017). In turn, characterizing mBHs in dwarf galaxies provides a powerful lever arm for constraining the seed black hole population (e.g., Volonteri & Natarajan 2009; van Wassenhove et al. 2010; Bellovary et al. 2011). In the past decade we have witnessed an explosion in sample sizes of mBHs in dwarf galaxies, growing from a handful of isolated cases 10-15 years ago (e.g., Filippenko & Ho 2003; Barth et al. 2004), to now homogeneously selected samples reaching $\sim 10^2$ objects in dwarfs (e.g., Reines et al. 2013; Mezcua et al. 2018) and other low-mass galaxies (e.g., Greene & Ho 2004, 2007; Dong et al. 2012). mBHs in dwarf galaxies typically have masses in the $10^4 - 10^6\,M_\odot$ range, with the smallest mBH discovered so far residing in the dwarf galaxy RGG 118, weighing in at only $5 \times 10^4\,M_\odot$ (Baldassare et al. 2015).

Discovering mBHs in dwarf galaxies relies on exploiting radiative signatures from accreting mBHs (see Reines & Comastri 2016 for a recent review). That is, we are sensitive only to mBHs that happen to be shining as active galactic nuclei (AGN). Arguably the most efficient discovery method so far has been based on optical signatures using emission line diagnostics (utlizing data from large-scale spectroscopic surveys; Reines

[1]Following Reines et al. (2013), we define dwarf galaxies as having stellar masses $M_\star < 3 \times 10^9\,M_\odot$, which is comparable to the Large Magellanic Cloud.

et al. 2013).[2] Optical selection, however, is biased toward AGN with low extinction that are accreting relatively rapidly ($L_{\rm bol}/L_{\rm Edd} > 0.1$).

Low-mass AGN can also be found through X-ray surveys (e.g., Kamizasa et al. 2012; Schramm et al. 2013; Lemons et al. 2015; Mezcua et al. 2016, 2018), although X-ray selection is biased against highly obscured systems. Also, a practical limitation with (current) X-ray surveys is that economically surveying large areas of sky generally requires single-epoch X-ray 'snapshots' with our most sensitive facilities. These types of X-ray observations rarely provide enough photons to pursue high signal-to-noise spectroscopic or timing studies. Without spectral and timing information, current X-ray facilities are not usually capable (on their own) of differentiating between accreting mBHs and luminous X-ray binary systems in cases with X-ray luminosities $L_X \lesssim 10^{40} \mathrm{erg\,s^{-1}}$, except for in a handful of bright systems with long X-ray exposure times, or in some cases with a sufficient number of epochs to monitor variability.

3. Revealing Low-Mass AGN with an ngVLA

Some of the current practical limitations to pursuing large population studies can be overcome by exploiting jetted radio emission from accreting mBHs. Radio observations have already discovered a few mBHs in dwarfs (Reines et al. 2011, 2014), and a new survey of dwarf galaxies with the Jansky Very Large Array (VLA) is revealing many new candidates (Reines et al. 2019, in preparation). As described below, once radio facilities start reaching ngVLA sensitivities, mBH AGN selection can become efficient through combined multiwavelength efforts.

At low Eddington ratios ($<10^{-2} L_{\rm Edd}$), all AGN are believed to launch compact, partially self-absorbed synchrotron jets that will appear point-like even at ngVLA resolutions (Blandford & Königl 1979; Ho 2008). In this weak accretion regime, the ratio of radio:X-ray luminosity scales in a predictable way with the mass of the accreting black hole according to the fundamental plane of black hole activity, $\log L_{\rm R} = 0.6 \log L_{\rm X} + 0.78 \log M_{\rm BH} + 7.33$ (e.g., Merloni et al. 2003; Falcke et al. 2004; Plotkin et al. 2012), where $L_{\rm R}$ is at 5 GHz, $L_{\rm X}$ is from 2-10 keV, and $M_{\rm BH}$ is in Solar units.

Since radio jets are expected to be ubiquitous at low accretion rates, a radio survey can be effective at recovering populations of mBH candidates if undertaken with a sensitive enough telescope. Advantages of a radio survey include:

- targeting mBHs in a weak accretion regime will recover objects missed by optical surveys (where the latter rely on diagnostics that are expected to be present mostly at higher accretion rates). Note that for any sensible luminosity function, the majority of AGN accrete at low Eddington ratios;

- jetted radio emission will be detectable from highly absorbed AGN, thereby allowing radio selection to uncover some objects that would be missed by complementary X-ray and/or optical approaches;

- at Mpc distances, contamination in the radio waveband from X-ray binary systems should be negligible. Except, at ngVLA sensitivities ($5\sigma_{\rm rms} \approx 1\,\mu$Jy/beam)

---

[2] Infrared searches have also yielded large sample sizes, although see Hainline et al. (2016) for a discussion on the purity of infrared selected samples.

it is plausible to (rarely) detect emission from transient radio flares produced by accreting stellar mass black holes out to $\approx$30 Mpc.[3]

We note that, in this low-luminosity regime, radio surveys will have to contend with radio emission from star formation (i.e., free-free emission in H II regions) and from supernova remnants. However, jetted radio emission from weakly accreting mBHs will be compact and unresolved at Mpc distances. Thus, with long ngVLA baselines ($\sim 10^3$ km), point-like radio emission would be a strong indicator of a viable AGN candidate (e.g., Reines & Deller 2012).

Figure 1 displays expected radio flux densities at 8 GHz for $10^{-5} < L_X/L_{Edd} < 10^{-3}$ AGNs powered by 10^4 (red swath), 10^5 (blue swath) and $10^6\,M_\odot$ (grey swath) mBHs as a function of distance.[4] We also show the current VLA $5\sigma_{rms}$ detection limit (dotted horizontal line) and an ngVLA $5\sigma_{rms}$ detection limit (dashed horizontal line) assuming 1 hour integrations. The shaded regions require 2-10 keV X-ray fluxes $F_X > 10^{-15}$ erg s^{-1} cm^{-2} to illustrate the types of AGN that can be accessed with an ngVLA combined with an X-ray facility like *Chandra*. An ngVLA would be able to detect relatively massive mBHs ($10^6\,M_\odot$) out to nearly 1 Gpc (almost a factor of four farther than accessible by the current VLA within reasonable exposure times). Excitingly, an ngVLA could detect jetted radio emission from $10^4\,M_\odot$ mBHs accreting at $L_X \gtrsim 10^{-4}\,L_{Edd}$, or $10^5\,M_\odot$ mBHs at $\gtrsim 10^{-6}\,L_{Edd}$, at the distance of the Virgo cluster (16.4 Mpc; corresponding to 2-10 keV X-ray fluxes $F_X \gtrsim 4\times10^{-15}$ and $\gtrsim 4\times10^{-16}$ erg s^{-1} cm^{-2}, respectively), which we expand on in §5.

3.1. Very Long Baseline Capabilities

The inclusion of very long baselines (i.e., $\sim 10^4$ km that are provided by the ngVLA Long Baseline Array) will deliver an exciting opportunity to measure proper motions from nearby AGN. We generally expect compact radio emission in the weak accretion regime to remain unresolved at GHz frequencies, even with very long baselines, such that we can expect $\approx 10\,\mu$as positional accuracy for bright sources. With such accuracy, we could measure a radio-emitting mBH in Virgo move by 5σ (i.e., $\approx 50\,\mu$as) in a time baseline of only four years (if we assume a projected velocity of $\approx$1000 km s^{-1} at 16.4 Mpc, which is a typical peculiar velocity in Virgo). Over 10-25 years it would be feasible to measure proper motions out to 50 Mpc, for velocities from 500-1000 km s^{-1}. Such precision astrometry would allow us to employ AGN proper motion measurements (in conjunction with multiwavelength radiative signatures) for distinguishing nearby mBHs from more distant AGN. In some lucky and nearby cases, there is even the potential to measure mBH motion from recoil kicks imparted by gravitational wave emission after a merger.

---

[3]We estimate this limit by scaling an extreme 20 Jy radio flare from the Galactic black hole X-ray binary Cyg X-3 (Corbel et al. 2012).

[4]We note that we expect all AGN accreting at $\lesssim 10^{-2}\,L_{Edd}$ to emit compact radio emission. We adopt the range $10^{-5} - 10^{-3}\,L_{Edd}$ in Figure 1 simply for illustrative purposes, to keep the figure from appearing too cluttered.

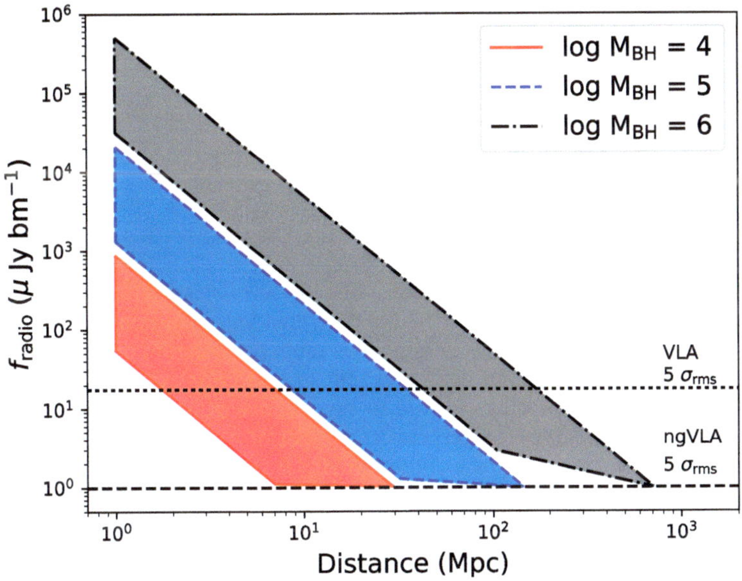

Figure 1. Distances to which an ngVLA could detect compact radio emission from an accreting mBH with 1 hour integrations, if the mBH falls on the fundamental plane of black hole activity (Merloni et al. 2003), and if the mBH has a 2-10 keV X-ray flux $F_X > 10^{-15}$ erg s^{-1} cm^{-2}. The red shaded region bounded by red solid lines illustrates the expected radio flux density (at 8 GHz) for a 10^4 $M_\odot$ mBH accreting between $10^{-5} < L_X < 10^{-3}$ $L_{\rm Edd}$; the blue shaded region (bounded by dashed lines) and the grey shaded region (bounded by dashed-dotted lines) illustrate the same for a 10^5 and 10^6 $M_\odot$ mBH, respectively. $5\sigma_{\rm rms}$ radio detection limits for an ngVLA and for the VLA are shown as dashed and dotted lines, respectively (both assuming 1 hour on source). An ngVLA could detect relatively massive mBHs out to nearly 1 Gpc, and it could detect mBHs as small as 10^4 $M_\odot$ at the distance of the Virgo cluster of galaxies (~16.4 Mpc). Note that the Merloni et al. (2003) fundamental plane has an intrinsic scatter of 0.88 dex in radio luminosity, which is not included in this figure.

4. Jet Physics and Mass Scaling

We describe in the next section one way in which revealing new samples of mBHs through an ngVLA would observationally constrain the masses of black hole seeds. In this section, we stress that more sensitive radio observations would also allow studies on the physics of jets in the low-accretion rate and low-mass regime. For example, the sensitivity of an ngVLA would open up economic time-domain studies for bright sources, and its wideband frequency capabilities would allow radio spectral studies (up to 10^2 GHz) for an unprecedented large sample of mBHs (which might even allow detailed decomposition of mBH radio jets from radio emission produced by star formation for the brightest objects). An ngVLA would help constrain the role of AGN feedback in dwarf galaxies and at low accretion rates in the so-called 'kinetic' feedback mode (e.g., Xie et al. 2017; Dashyan et al. 2018). By populating observations of extragalactic jets launched by black holes in the $10^4 - 10^6 M_{\rm BH}$ range, an ngVLA would also provide empirical constraints to drive studies on the scalability of black hole accretion and jet launching, through comparisons of mBHs to more massive black holes ($\gtrsim 10^6\, M_\odot$) at the centers of large galaxies, and to lower-mass stellar black holes ($\sim 10\, M_\odot$) in X-ray binary systems.

5. Multiwavelength Studies of Black Hole Seeds

5.1. Multiwavelength Coordination

Constraining black hole seeds through AGN in dwarf galaxies is a multiwavelength endeavor, and it is not a problem that can be tackled looking within only a single band of the electromagnetic spectrum. However, in a multiwavelength context, we expect a radio facility like the ngVLA to play a prominent role. For example, as discussed earlier, optical selection so far has proved to be effective at higher Eddington ratios ($\gtrsim 0.1\, L_{\rm Edd}$), but optical surveys largely miss the types of weakly accreting objects for which combined radio/X-ray searches are optimal. It is not obvious whether (in the near future) large high-spatial resolution optical telescopes will outperform radio surveys in the weak accretion regime: weakly accreting AGN do not always have obvious optical counterparts in dwarf galaxies (e.g., Reines et al. 2011, 2014), and optical AGN diagnostics, like broad emission lines and narrow high-excitation lines, are expected to become substantially weaker at lower Eddington ratios (e.g., Padovani et al. 2017, and references therein). Thus, efficient optical selection of weakly accreting mBHs would require routine detections of low-excitation narrow AGN emission lines, a method that is possible in principle but has not yet revealed large numbers of AGN candidates in dwarf galaxies. Combining radio and optical samples will therefore reveal AGN over a wider range of Eddington ratios than can be recovered individually.

It is challenging for current (and near-future) X-ray facilities to economically survey large numbers of dwarf galaxies, especially with sufficient sensitivity to unambiguously control for contamination from X-ray binaries. Advantages of an ngVLA are that it can *efficiently* reach useful sensitivities in relatively short (<1 hour) exposures, and contamination from X-ray binaries will be rare for nearby galaxies, and negligible at $\gtrsim$30 Mpc. Furthermore, with its long baselines, contamination from star formation can be minimised by selecting only point-like radio emission.

Radio-selected AGN candidates would then await confirmation by other telescopes, including high spatial-resolution X-ray facilities like the *Chandra X-ray Observatory* or the *Athena X-ray Observatory* (Figure 1 demonstrates the discovery space that is currently feasible via *Chandra* followup to ngVLA AGN candidates). Some of the X-ray brighter systems (e.g., $F_{\rm X} \gtrsim 10^{-13}$ erg s^{-1} cm^{-2}) will also appear in future all sky surveys (e.g., eROSITA on the *Spectrum-Roentgen-Gamma* satellite). High spatial-resolution infrared spectroscopy, e.g., with the *James Webb Space Telescope (JWST)*, may also provide line diagnostics to confirm weakly accreting and/or optically obscured AGN (e.g., Satyapal et al. 2009).

5.2. Local Constraints on Black Hole Growth

Identifying AGN in dwarf galaxies is still a young field observationally. The progression of connecting observations of mBHs to theoretical models of seed black hole formation will require continuing to systematically assemble even larger samples over the next decades. As an example, consider the AGN Multiwavelength Survey of Early-Type Galaxies (AMUSE; Gallo et al. 2008; Miller et al. 2012), which performed a *Chandra* X-ray survey of 100 early-type galaxies in the Virgo cluster and 100 non-cluster galaxies at comparable distances to Virgo. By considering the fraction of nuclear X-ray detections within the 200 object AMUSE sample, Miller et al. (2015) provide a framework demonstrating how black hole occupation (and seed formation) can be constrained through local observations of weakly accreting AGN. They place a limit that >20% of local low-mass galaxies ($M_\star < 10^{10} M_\odot$) are occupied by an mBH, and they cannot rule out full occupation. Amongst the results from Miller et al. (2015) we note:

- their 200 object sample is not large enough to place stringent constraints on occupation (although see their Figure 4 for how uncertainties improve with sample size);
- with the types of X-ray sensitivities achievable if aiming to pursue $N \sim 10^2$ sample sizes, contamination from X-ray binaries is an important systematic to be controlled for.

Despite the above, we stress that quantitative methods are already fully developed by Miller et al. (2015), who demonstrate that local calculations of the occupation fraction are in reach. An ngVLA would help improve sample sizes, and it would help provide confidence that contamination from X-ray binaries is negligible.

Since one expects full occupation at the high galaxy (stellar) mass end regardless of the seed black hole mass distribution, diagnosing the seed formation channels will require focusing on lower-mass galaxies (Miller et al. 2015). A typical *Chandra* X-ray exposure time utilized by the AMUSE survey at the distance of Virgo was ~1.4 h, which allowed identification of a $10^5\ M_\odot$ mBH accreting at nearly $10^{-5}\ L_{\rm Edd}$ (a regime that is just barely accessible with the current VLA; see Figure 1). With a comparable or shorter time investment on an ngVLA (depending on overheads), radio observations could identify a $10^5\ M_\odot$ mBH accreting an order of magnitude more weakly ($10^{-6}\ L_{\rm Edd}$), which would clearly improve the sample size (the corresponding X-ray flux would be $\approx 4 \times 10^{-16}$ erg s^{-1} cm^{-2}, which is feasible given that only a subset of objects would be so faint). Thus, identification of AGN in dwarf galaxies (at ngVLA sensitivites) can be expected to place significantly improved constraints on the local black hole occupation fraction and mBH mass distribution, thereby helping to place observational constraints on the population of objects that seeded the growth of black holes in the early Universe.

References

Bañados, E., Venemans, B. P., Mazzucchelli, C., et al. 2018, Nat, 553, 473
Baldassare, V. F., Reines, A. E., Gallo, E., & Greene, J. E. 2015, ApJ, 809, L14
Barth, A. J., Ho, L. C., Rutledge, R. E., & Sargent, W. L. W. 2004, ApJ, 607, 90
Begelman, M. C., Volonteri, M., & Rees, M. J. 2006, MNRAS, 370, 289
Bellovary, J., Volonteri, M., Governato, F., et al. 2011, ApJ, 742, 13
Blandford, R. D., & Königl, A. 1979, ApJ, 232, 34
Cann, J. M., Satyapal, S., Abel, N. P., et al. 2018, ArXiv e-prints, arXiv:1805.09351
Corbel, S., Dubus, G., Tomsick, J. A., et al. 2012, MNRAS, 421, 2947
Dashyan, G., Silk, J., Mamon, G. A., Dubois, Y., & Hartwig, T. 2018, MNRAS, 473, 5698
Dong, X.-B., Ho, L. C., Yuan, W., et al. 2012, ApJ, 755, 167

Dubois, Y., Volonteri, M., Silk, J., et al. 2015, MNRAS, 452, 1502
Fabian, A. C. 2012, ARA&A, 50, 455
Falcke, H., Körding, E., & Markoff, S. 2004, A&A, 414, 895
Filippenko, A. V., & Ho, L. C. 2003, ApJ, 588, L13
Gallo, E., Treu, T., Jacob, J., et al. 2008, ApJ, 680, 154
Greene, J. E. 2012, Nature Communications, 3, 1304
Greene, J. E., & Ho, L. C. 2004, ApJ, 610, 722
—. 2007, ApJ, 670, 92
Habouzit, M., Volonteri, M., & Dubois, Y. 2017, MNRAS, 468, 3935
Haiman, Z., & Loeb, A. 2001, ApJ, 552, 459
Hainline, K. N., Reines, A. E., Greene, J. E., & Stern, D. 2016, ApJ, 832, 119
Ho, L. C. 2008, ARA&A, 46, 475
Inami, H., Armus, L., Charmandaris, V., et al. 2013, ApJ, 777, 156
Kamizasa, N., Terashima, Y., & Awaki, H. 2012, ApJ, 751, 39
Lemons, S. M., Reines, A. E., Plotkin, R. M., Gallo, E., & Greene, J. E. 2015, ApJ, 805, 12
Loeb, A., & Rasio, F. A. 1994, ApJ, 432, 52
Madau, P., Haardt, F., & Dotti, M. 2014, ApJ, 784, L38
Madau, P., & Rees, M. J. 2001, ApJ, 551, L27
Merloni, A., Heinz, S., & di Matteo, T. 2003, MNRAS, 345, 1057
Mezcua, M. 2017, International Journal of Modern Physics D, 26, 1730021
Mezcua, M., Civano, F., Fabbiano, G., Miyaji, T., & Marchesi, S. 2016, ApJ, 817, 20
Mezcua, M., Civano, F., Marchesi, S., et al. 2018, MNRAS, arXiv:1802.01567
Miller, B., Gallo, E., Treu, T., & Woo, J.-H. 2012, ApJ, 747, 57
Miller, B. P., Gallo, E., Greene, J. E., et al. 2015, ApJ, 799, 98
Mortlock, D. J., Warren, S. J., Venemans, B. P., et al. 2011, Nat, 474, 616
Nyland, K. & Alatalo, K. 2018, Science with a Next-Generation Very Large Array, ASP Monograph 7, ed. E. J. Murphy (San Francisco, CA: ASP), 451
Padovani, P., Alexander, D. M., Assef, R. J., et al. 2017, A&A Rev., 25, 2
Plotkin, R. M., Markoff, S., Kelly, B. C., Körding, E., & Anderson, S. F. 2012, MNRAS, 419, 267
Reines, A. E., & Comastri, A. 2016, PASA, 33, e054
Reines, A. E., & Deller, A. T. 2012, ApJ, 750, L24
Reines, A. E., Greene, J. E., & Geha, M. 2013, ApJ, 775, 116
Reines, A. E., Plotkin, R. M., Russell, T. D., et al. 2014, ApJ, 787, L30
Reines, A. E., Sivakoff, G. R., Johnson, K. E., & Brogan, C. L. 2011, Nat, 470, 66
Satyapal, S., Böker, T., Mcalpine, W., et al. 2009, ApJ, 704, 439
Schramm, M., Silverman, J. D., Greene, J. E., et al. 2013, ApJ, 773, 150
van Wassenhove, S., Volonteri, M., Walker, M. G., & Gair, J. R. 2010, MNRAS, 408, 1139
Volonteri, M. 2010, A&A Rev., 18, 279
Volonteri, M., & Natarajan, P. 2009, MNRAS, 400, 1911
Volonteri, M., & Reines, A. E. 2016, ApJ, 820, L6
Xie, F.-G., Yuan, F., & Ho, L. C. 2017, ApJ, 844, 42

Science with a Next-Generation Very Large Array
ASP Conference Series, Monograph 7
Eric J. Murphy, ed.

Accretion and Jets in Local Compact Objects

Deanne L. Coppejans,[1] James C.A. Miller-Jones,[2] Elmar G. Körding,[3] Gregory R. Sivakoff,[4] and Michael P. Rupen[5]

[1]*Center for Interdisciplinary Exploration and Research in Astrophysics (CIERA) and Department of Physics and Astronomy, Northwestern University, Evanston, IL 60208;* deanne.coppejans@northwestern.edu

[2]*International Centre for Radio Astronomy Research – Curtin University, GPO Box U1987, Perth, WA 6845, Australia;* james.miller-jones@curtin.edu.au

[3]*Department of Astrophysics/IMAPP, Radboud University, PO Box 9010, NL-6500 GL Nijmegen, the Netherlands;* elmar@koerding.eu

[4]*Department of Physics, University of Alberta, CCIS 4-183, Edmonton, AB T6G 2E1, Canada;* sivakoff@ualberta.ca

[5]*Herzberg Astronomy and Astrophysics Research Centre, 717 White Lake Road, Penticton, BC, V2A 6J9, Canada;* michael.rupen@nrc-cnrc.gc.ca

Abstract. Despite the prevalence of jets in accreting systems and their impact on the surrounding medium, the fundamental physics of how they are launched and collimated is not fully understood. Radio observations of local compact objects, including accreting stellar mass black holes, neutron stars and white dwarfs, probe their jet emission. Coupled with multi-wavelength observations, this allows us to test the underlying accretion-outflow connection and to establish the relationship between the accretor properties and the jet power, which is necessary to accurately model jets. Compact accretors are nearby, numerous and come in a range of accretor properties, and hence are ideal probes for the underlying jet physics. Despite this, there are a number of key outstanding questions regarding accretion-driven outflows in these objects that cannot be answered with current radio observations. The vastly improved sensitivity, polarization capabilities, spatial resolution and high-frequency coverage of the ngVLA will be crucial to answering these, and subsequently determining the fundamental physics behind accretion and jets at all physical scales.

1. Introduction

Jets are ubiquitous, occurring in accreting systems as diverse as proto-stars and Active Galactic Nuclei (AGN). This intimate relation between accretion and directed outflow suggests that jets play an important role in re-directing angular momentum, energy, and magnetic fields from the infalling gas, while the consequent injection of mass and energy into the accretion reservoir is likely important for phenomena ranging from the initial mass function of stars to the evolution of galaxies. Despite this, we still do not fully understand the mechanism whereby jets are launched and collimated or how this is affected by simple physical properties of the accretor. One of the best ways to study the fundamental link between jets and accretion is through observations of accreting compact objects in the local universe. These accreting black holes, neutron stars, and white dwarfs evolve through accretion states on observable time-scales (days to months) and are common enough that there are nearby objects for detailed study, as well as large numbers available for statistical comparisons. Furthermore, standard techniques in the field of stellar binaries can directly determine fundamental parameters such as mass and binary separation, and even constrain the evolutionary history of these systems. Through comparative studies of the different classes of compact accretors, we can determine how the properties of the accretor (mass, spin, accretion rate, absence or presence of a solid surface, strength of magnetic field, importance of relativistic effects) are related to the jet power, and can probe the underlying connection between the accretion flow and the jets. These are very complex systems, with a rich variety of emission from the accretor, its companion, the accretion flow, and the jet. In many systems, radio observations are unique in providing a view of the jet (via synchrotron emission) unconfused by all the other emission sources, and this combined with the high angular resolution of radio interferometers has made radio data indispensable to accretion/outflow studies of compact objects. In this article we will highlight some of these questions for black hole (§2), neutron star (§3) and white dwarf (§4) compact binaries.

Even in an era where the SKA will likely be operating, the ngVLA will play a critical role answering these questions. In addition to its greater sensitivity and northern sky coverage, ngVLA observations at greater than 10 GHz will detect jet emission closer to where jet particles are first accelerated. At these frequencies, high angular resolution imaging will expose (transient) jet ejecta and better resolve the compact, steady jet associated with sources emitting at a low fraction of their maximum (Eddington-luminosity limited) mass accretion rate. High frequency observations will also mitigate issues of confusion and scattering that might occur at lower frequencies. The large spectral lever arm allowed by ngVLA will also provide much better constraints on the radio spectra, and hence disentangle different emission mechanisms.

2. Black Hole X-ray Binaries

Black hole X-ray binaries (BHXRBs) are ideal testbeds for studying how the mass accretion rate and accretion flow geometry (as inferred from strong X-ray emission) affect the launching of outflows (jets and winds) from compact objects. A direct empirical link between accretion properties (and their changes on timescales of days to months) and radio emission from the jets has been observed in numerous outbursts of some tens of BHXRBs, and holds great promise for answering fundamental questions regarding how and why accretion geometry changes, and how this affects jet properties. However, current instrumentation has limited our progress in answering these fundamental questions; more sensitive observations with broader spectral coverage are necessary.

2.1. The Accretion-Outflow Connection and the Fundamental Plane of Black Hole Activity

Low-mass BHXRBs show repeated outbursts in which the accretion rate and hence the source luminosity increases by several orders of magnitude from quiescence ($L < 10^{-5.5} L_{\rm Edd}$, where $L_{\rm Edd}$ is the Eddington luminosity), while the other parameters of the system (e.g., mass, inclination, spin) stay effectively constant. At luminosities up to

$L \sim 10^{-1}L_{\rm Edd}$, the accretion scenario is best described by a thin disk that is likely truncated at a few tens to hundreds of gravitational radii (e.g., Esin et al. 1997). Within this radius is a hot, optically-thin, vertically-extended flow (such as an advection-dominated accretion flow), whose hot electrons comptonize the disk and accretion flow photons, leading to the observed power-law X-ray emission. During this phase of an outburst, the radio emission is strongly correlated with the X-ray luminosity (the radio/X-ray correlation, Hannikainen et al. 1998, see Figure 1), interpreted as a relation between accretion rate and outflow in a compact, steady jet. This correlation is a major foundation upon which was built the (broadly used) Fundamental Plane of Black Hole Activity (FPBHA), which connects radio luminosity, X-ray luminosity, and black hole mass across all mass scales from BXHRBs to AGN (Merloni et al. 2003; Falcke et al. 2004).

Around $L \sim 10^{-1}L_{\rm Edd}$, the X-ray spectrum softens rapidly as the accretion geometry transforms to a thin disk that extends all the way to the innermost stable circular orbit (ISCO; see Shakura & Sunyaev 1973). During this transition, the compact core radio jet is observed to be quenched (Tananbaum et al. 1972) and instead one often observes bright, transient ejecta moving away from the central binary at relativistic speeds (e.g., Mirabel & Rodríguez 1994; Tingay et al. 1995). Being well-separated from the core, the observed radio emission then no longer falls on the FPBHA. When the source moves back towards quiescence, the core radio emission reappears and the source returns to the radio/X-ray correlation (and the FPBHA).

With better coverage, many sources have deviated from the standard FPBHA, seeming instead to follow a 'radio-quiet' branch of the radio/X-ray correlation at $10^{-3} \lesssim L/L_{\rm Edd} \lesssim 10^{-1}$ (see, e.g., Corbel et al. 2004; Soleri & Fender 2011). This directly challenges the simple picture presented of the FPBHA with its notion of a uniform disc-jet connection in accreting black holes across all masses. The mean slope of the radio spectrum is currently our only known difference between the radio-quiet and radio-loud branches in BHXRBs (Espinasse & Fender 2018). Meier (2001) showed that systems with a thick disk and a rapidly rotating geometry (i.e., a Kerr BH as opposed to a Schwarzschild BH) produce stronger radio-emitting jets for a given accretion power, and that this supports the paradigm where radio-quiet AGN are powered by non-spinning supermassive BHs. While this might suggest 'radio-quiet' BHXRBs are powered by Schwarzschild stellar-mass BHs, several radio-quiet BHXRBs have been observed to transition back to the the main radio/X-ray correlation at $L \sim 10^{-3}L_{\rm Edd}$ (e.g., H1743−322; Coriat et al. 2011). Light curves of a sample of BHXRBs over a large frequency range when $L \sim 10^{-3}L_{\rm Edd}$ (and radio luminosity is relatively low) are needed to determine if the radio-quiet track continues, or if all radio-quiet BHXRBs transition back to the the main radio/X-ray correlation. These observations will determine the nature of the radio-quiet branch and inform us on when the FPBHA is valid and what additional parameters (such as black hole spin or residual inner disk emission) might be needed.

The ngVLA can also help determine how jets switch off. The sensitivity of current radio telescopes is insufficient to determine whether the compact steady radio jet is fully quenched or only less luminous when the thin accretion disk reaches the ISCO (e.g., Tananbaum et al. 1972; Drappeau et al. 2017). Current observations indicate that the emission is quenched by more than 2.5 orders of magnitude relative to when the inner accretion flow is geometrically thick (e.g., Russell et al. 2011). Given the gain in sensitivity, ngVLA observations of a sample of BHXRBs in the soft state will probe this quenching factor to more than an order of magnitude deeper than current limits, and significantly deeper than that observed between radio-loud and radio-quiet AGN. An important caveat is that the coronal emission from the secondary star, or possibly winds, could be detectable at these limits, and the source of any detected radio emission would need to be confirmed using the improved ngVLA spectral and polarization constraints. Careful selection of targets with less active secondary stars would also help to mitigate this. If the quenching is any stronger, new suppression mechanisms (e.g., low radiative efficiency, Poynting flux dominated jet, etc.) would be required beyond the geometrical factor suggested by Meier (2001).

Currently, it is not clear how BHXRBs behave in quiescence (the state in which BHXRBs and super-massive black holes spend most of their time). In particular, does the radio/X-ray correlation and FPBHA universally extend towards the lowest luminosities? Answering this will help to break the degeneracies from X-ray spectral fitting in determining the nature of the accretion flow, and will enable us to interpret the radio-bright sources seen in and towards globular clusters (Strader et al. 2012; Tetarenko et al. 2016). The ngVLA will extend the radio/X-ray correlation more than 2 orders of magnitude towards lower luminosities.

2.2. Jet Properties

Fundamental properties of the jets in BHXRBs are currently unknown. For instance, when and how does the jet decelerate and expand as it moves away from the central black hole? What is the shape (or shapes) of jets? What is the radiative (and kinetic) power of jets? What are the magnetic field configurations of jets? And how do all of these vary as accretion flow properties (e.g., the accretion geometry) change? To date, studies of a handful of individual sources have allowed us to piece some of this information together, although questions still remain, and a lack of spectral coverage, spatial resolution and insufficient sensitivity to monitor the jet over sufficiently short time-scales is preventing progress. The ngVLA will be transformative in enabling simultaneous studies of all of these questions in individual sources.

For instance, the value of high-cadence multi-wavelength observations in determining jet properties was recently demonstrated through observations of the BHXRB V404 Cygni in outburst. Via the use of subarrays, high time resolution light curves at multiple frequencies enabled Tetarenko et al. (2017) to use physical modeling to determine some of the key jet parameters. However, V404 Cygni is a nearby source and one of the few that reached the Eddington luminosity, so higher-sensitivity radio observations are necessary to extend such studies to a sample of BHXRBs. In particular, ngVLA observations with wide spectral coverage (potentially using subarrays) are needed to monitor variability propagating down through the frequency bands as they move down the jet, determining the jet velocity from the time delays between bands, and the opening angle from the variability timescales at different frequencies. The high spatial resolution, and great sensitivity at high frequencies will allow direct imaging of jets to much lower surface brightnesses and thus farther away from the central source, given that the transient ejecta fade due to adiabatic expansion as they move away from the core. This will track the expansion and deceleration of the jet.

By resolving the magnetic field direction and fractional linear polarization, polarization studies can reveal the geometry of the compact steady jet and discrete jet ejects (e.g., Curran et al. 2015), even when those geometries are not directly imaged. Such studies have been strongly sensitivity limited to date, as the polarization fractions are typically of the order of a few percent in compact jets to several tens of percent in transient ejecta. The ngVLA will extend these studies to a wider (radio-fainter) range of BH and NS XRB systems, with the potential to trace jet changes to accretion changes. With long baselines, ngVLA will not only verify the use of unresolved polarimetry for tracing jet geometry, but also test whether spine-sheath jet geometries invoked in blazars (e.g., Attridge et al. 1999; Pushkarev et al. 2012) apply to XRBs.

By observing a sample of sources over the wide frequency coverage of the ngVLA, we can additionally observe how the jet break frequency depends on the properties of the accretion flow. Observations of the BHXRBs MAXI J1836-194 and MAXI J1659-152 suggest that the jet break evolves to lower radio frequencies as X-ray spectrum softens and the accretion flow geometry changes during an outburst (Russell et al. 2013, 2014; van der Horst et al. 2013). The jet-break frequency is thus expected to move through the ngVLA bands during the hard-to-soft state transition.

With its high-frequency coverage, significantly enhanced sensitivity, improved spatial resolution, and good (u, v)-coverage, ngVLA images of BHXRBs will be dramatically improved. We expect to see the sort of images that we have of extragalactic sources (e.g., Cyg A), revealing details of the complex interactions between the jet and the surrounding medium. This is akin to the qualitative difference that came in studies of extragalactic jets with the advent of the VLA. With the ngVLA we can make these images and study objects that evolve on human (and even undergraduate) timescales.

3. Neutron Star X-ray Binaries

Neutron star X-ray binaries (NSXRBs) are compact binaries with a neutron star accretor. In contrast to the BHXRBs, the accretor has a surface and a stellar magnetic field. It is currently unknown how these properties directly affect jet formation and collimation, so comparative studies between these two classes of object are a powerful tool to determine this. At present these studies are severely hampered by a lack of sensitivity, as NSXRBs are fainter than their black hole counterparts at radio wavelengths. Increased sensitivity with higher frequency coverage and better spatial resolution are necessary to determine which classes of NSXRBs launch jets (and hence which system parameters enable/prevent jet launching) and how the accretion-outflow connection differs from the BHXRBs.

In the transitional millisecond pulsars (tMSPs; a class of rapidly-rotating neutron stars that switch between accretion and rotation powered phases during the last stages of recycling), it is currently not known whether the observed radio emission is produced by a jet or another mechanism. In at least one tMSP, anti-correlated radio and X-ray emission has been detected that cannot readily be explained by traditional accretion-outflow models (Bogdanov et al. 2017). The key to determining the source of the radio emission is simultaneous X-ray and ngVLA radio observations of a sample of tMSPs with sufficient sensitivity to track the spectral evolution over the radio variability timescales (~ 10 s). At 8 GHz, the ngVLA can reach the sensitivities required to track the in-band spectral evolution of the observed $\sim 200\,\mu$Jy flares in less than a second[1] and effectively sample the flaring emission.

Similarly, it is not known whether strong magnetic fields ($> 10^{12}$ G) inhibit the formation of jets. The lack of observed jets in strongly magnetic NSXRBs (e.g., Fender &

[1]assuming a 0.21 μJy/beam rms for a 1 hour observation using the full bandwidth (ngVLA memo #17)

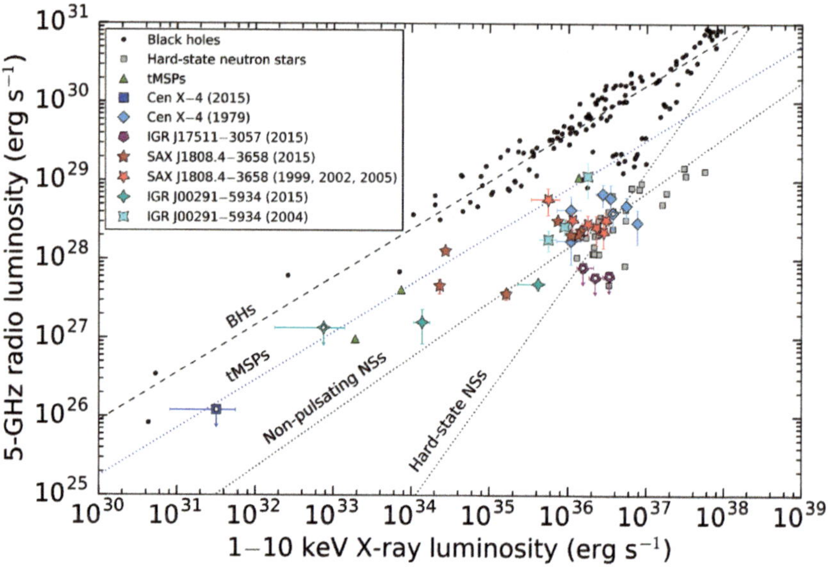

Figure 1. Black hole accretors (from BHXRBs to AGN) show a relation between the radio luminosity (L_r) and the X-ray luminosity (L_X) of the form $L_r \propto L_X^{\beta}$, which relates the jet power to the accretion rate. This is referred to as the radio/X-ray correlation. The question of whether neutron star systems show a similar relation remains open, although NSXRBs are radio quiet in comparison to their BHXRB counterparts (e.g., Fender & Kuulkers 2001; Migliari & Fender 2006). Unlike current instruments, the ngVLA will have the sensitivity to study NSXRBs down to the crucial phase space $L_X \leq 10^{36}$ erg s^{-1}. Credit: Figure 6 from Tudor et al. (2017).

Hendry 2000) has resulted in a number of jet formation models that suppress jet formation at high magnetic field strengths (e.g., Massi & Kaufman Bernadó 2008). Recently however, radio detections of a few such systems have been found that are consistent with jet synchrotron emission (van den Eijnden et al. 2018a,b). These detections of nearby systems are too faint (10–100 μJy) to distinguish jet emission from other mechanisms. Other possibilities include emission from shocks in the interaction between the accretion flow and magnetosphere or a propeller-driven outflow, but for Her X-1, van den Eijnden et al. state that these two scenarios are unlikely. Stronger constraints on the linear and circular polarization fractions from the improved sensitivity of the ngVLA will be important in determining the relevant emission mechanism, as will the improved spectral resolution and access to higher frequencies to search for characteristic spectral features such as jet breaks.

The key to comparing the accretion-outflow connection in BHXRBs and NSXRBs is through the radio/X-ray plane (see Figure 1) via measurements of the radio (L_r) and X-ray luminosity (L_X). As described in §2, this gives the relationship between the jet power and accretion rate. At present it is unclear whether the different classes or states of neutron star show a single behavior. Radio measurements of systems with $L_X \leq 10^{36}$ erg s^{-1} are necessary to determine the relation/s, but there are very few detections in this range. The ngVLA sensitivity is necessary to effectively sample this phase-space, as the NSXRBs are comparatively radio-quiet to their BH counterparts. Gallo et al. (2018) found that the BH systems are a factor $\sim$ 22 more radio-loud than the NSXRBs for a given x-ray luminosity. Clarifying this NSXRB behavior on the radio/x-ray plane will make it possible to determine what role the radiative efficiency of the accretion plays in NSXRBs and BHXRBs.

4. Cataclysmic Variables

Cataclysmic Variables (CVs) are binary stars in which a white dwarf accretes material from a low-mass main-sequence star via Roche-lobe overflow (see Warner 1995). In the 'non-magnetic' CVs (the subject of this section and hereafter referred to simply as 'CVs'), the transfered material accretes via an accretion disk, as the magnetic field strength of the white dwarf (at $\lesssim 10^6$ G) is not sufficient to channel the accretion flow from the disk. Until recently, it was believed that CVs were one of the few classes of compact accretors to not launch jets and they have consequently been used to constrain jet-launching models in compact accretors (e.g., Soker & Lasota 2004). Other classes of accreting white dwarfs such as the novae (e.g., Sokoloski et al. 2008) and symbiotic stars (e.g., Brocksopp et al. 2004) are known to launch jets, but CVs were thought to be the exception. In 2008 however, Körding et al. showed that the same accretion-outflow cycle observed in XRBs could be mapped to CVs. This and subsequent studies (Miller-Jones et al. 2011; Russell et al. 2016; Mooley et al. 2017) indicated a likely jet in the prototypical CV SS Cyg. If CVs are launching jets this will have significant implications for jet-launching models in compact accretors because CVs were previously used to constrain these models. They are numerous, nearby, comparatively better understood than the black hole and neutron star accretors and are non-relativistic. Additionally, the *Gaia* survey (Gaia Collaboration et al. 2016) has published the distances to many nearby CVs, making it is possible to convert observables into physical parameters and do comparative studies between systems. For all these reasons, if CVs are launching jets, they will be important probes of accretion and jet physics in compact accretors. The enhanced sensitivity, higher frequency coverage, and higher resolution of the ngVLA are necessary to establish whether this is the case.

CVs are faint at radio frequencies (on the order of 10s to 100s of μJy at 300 pc), and only with the recent VLA upgrade have we had the sensitivity required to detect even the handful of very nearby systems (Körding et al. 2011; Coppejans et al. 2015, 2016). With the notable exception of the extraordinarily bright CV SS Cyg, the sensitivity of current instruments is insufficient to determine the spectral indices, put strong constraints on the polarization fraction or determine the spatial extent of the emitting region (through VLBI observations) of CVs.

There are five key parts to the question of whether CVs launch jets that require ngVLA observations to make progress. *First*, we need to determine the broad spectral behavior to ascertain the emission mechanism. *Second*, the radio spectrum needs to be observed out to 116 GHz to look for a possible jet break (analogously to the XRBs). *Third*, observations mapping the spectral evolution, timing and polarization properties of the ~minute time-scale flares equivalent to those of SS Cyg (Mooley et al. 2017) are needed for a sample of CVs. This is necessary to be able to distinguish jet emission from alternative mechanisms such as magnetic reconnections in the disk (see Meintjes 2016). Importantly, current observations show different polarization behavior during the flares of the 'magnetic' and 'non-magnetic' CVs, indicating different radio emission mechanisms in these classes. In the 'magnetic CVs'[2], highly circularly polarized flaring emission characteristic of electron-cyclotron maser emission has been detected in some systems, while in others the low polarization limits are more indicative of gyrosynchrotron emission (Abada-Simon et al. 1993; Dulk et al. 1983; Chanmugam 1987; Wright et al. 1988; Pavelin et al. 1994; Meintjes & Venter 2005; Mason & Gray 2007; Barrett et al. 2017). In contrast only one 'non-magnetic CV' has shown a highly circularly polarized flare (Coppejans et al. 2015). Deep ngVLA limits on the polarization fractions of the flares will help to reveal the impact of the magnetic field strength of the white dwarf on the emission and consequently determine the radio emission mechanisms at play in these objects. *Fourth*, to conclusively determine if CVs launch jets, observations that spatially resolve the jet emission are necessary. The resolution at 8 GHz (assuming natural weighting) is 49 mas (ngVLA memo #17), which will probe physical scales on the order of 10^{14} cm (1000 times the orbital separation) at 300 pc. VLBA observations of SS Cyg have shown indications of extended emission at a 4σ level (Russell et al. 2016), but the majority of the CV population are too faint for comparable observations at present. *Fifth*, a larger sample of detected CVs are necessary to sample the full range of CV properties (e.g., accretor mass, accretion rate and magnetic field strength) to determine how these properties affect jet-launching. Assuming a 0.21 μJy/beam rms for a 1 hour observation at 8 GHz (ngVLA memo #17), we will detect emission from even the faintest observed CVs at a 5σ level out to $\sim$ 1700 pc. Using a lower-limit on the space density of CVs of $2 \times 10^{-6}\,\mathrm{pc}^{-3}$ (Pretorius & Knigge 2012), this puts a lower-limit on the number of CVs that can be detected at $\gtrsim$7000. This is a minimum of a factor 700 enlargement of our current sample, which will enable conclusive statistical studies. Given the detection volume of CVs, and the large ngVLA field-of-view particularly at lower frequencies, CVs will also be detected in ngVLA survey data and transient searches using multiple beams. Unlike with current instruments, these systems will be detected serendipitously and will not exclusively require targeted detection experiments.

We end this section on a related note, by briefly highlighting two white dwarf binaries (AE Aqr and AR Sco) for which radio observations are driving significant advances in our understanding of the physics of matter transfer onto rapidly rotating, strongly magnetic compact stars; in both cases the ngVLA will significantly further progress. AE Aqr is a CV with a white dwarf that is highly magnetic and has a short spin period such that the spinning magnetosphere causes the majority of the transfered material to be flung out of the system rather than accreted (a propeller phase; Wynn et al. 1997; Meintjes & de Jager 2000; Meintjes & Venter 2005). AR Sco consists of a white dwarf and M-dwarf binary in a short orbital period (3.55 h), but unlike CVs there is no evidence of accretion. It has been called the first (and to date, only) white dwarf pulsar as it is powered by the spin-down of the white dwarf rather than via accretion (Marsh et al. 2016; Buckley et al. 2017; Stiller et al. 2018, see however Potter & Buckley 2018). Studies of AE Aqr and AR Sco have had important implications for our understanding of the evolutionary stages of mass accretion in white dwarfs, and the modelling of these mass transfer stages is likely relevant to the higher mass compact binaries such as the accreting millisecond pulsars (Campana et al. 2018). For both of these objects, the improved sensitivity of the ngVLA will enable higher time- and spectral resolution

[2]CVs where the magnetic field strength of the white dwarf exceeds $\approx 10^6$ G and truncates the disk at the Alv́en radius. The strongest field systems prohibit the formation of a disk.

studies with finer sampling of the polarized flares, for more detailed modelling of the radio emission. Another promising avenue is the serendipitous discoveries of additional objects like these for more context on these processes.

5. Conclusion

ngVLA multi-band observations will track the spectral, spatial, polarization and variability properties of jets and outflows in compact accretors on scales and luminosities that are currently unreachable. To probe the accretion-ejection connection, complementary optical, UV and X-ray observations are additionally necessary to monitor the accretion flow. In the XRBs, this will be accomplished through simultaneous X-ray observations. In the CVs, emission from the accretion flow peaks at optical, UV and X-ray wavelengths (in order of decreasing distance to the white dwarf). Current, planned and possible missions that can provide this complementary coverage of the accretion flow include Evryscope, LSST, WFIRST, LUVOIR, Athena, STROBE-X, the successor to TESS (possibly FINESSE), and the Lynx X-ray Surveyor. The vast improvement of radio observations offered by the ngVLA, as well as this complementary coverage from multi-wavelength future missions, are key to answering the outstanding questions related to accretion and jet physics in compact accretors, and by extension other classes of accretors.

Acknowledgments. We thank the referee for comments and suggestions that have significantly improved this paper. GRS acknowledges support by an NSERC Discovery Grant. JCAM-J is the recipient of an Australian Research Council Future Fellowship (FT140101082).

References

Abada-Simon, M., Lecacheux, A., Bastian, T. S., Bookbinder, J. A., & Dulk, G. A. 1993, ApJ, 406, 692
Attridge, J. M., Roberts, D. H., & Wardle, J. F. C. 1999, ApJ, 518, L87
Barrett, P. E., Dieck, C., Beasley, A. J., Singh, K. P., & Mason, P. A. 2017, AJ, 154, 252
Bogdanov, S., et al. 2017, ArXiv e-prints
Brocksopp, C., Sokoloski, J. L., Kaiser, C., Richards, A. M., Muxlow, T. W. B., & Seymour, N. 2004, MNRAS, 347, 430
Buckley, D. A. H., Meintjes, P. J., Potter, S. B., Marsh, T. R., & Gänsicke, B. T. 2017, Nature Astronomy, 1, 0029
Campana, S., Stella, L., Mereghetti, S., & de Martino, D. 2018, A&A, 610, A46
Chanmugam, G. 1987, Ap&SS, 130, 53
Coppejans, D. L., Körding, E. G., Miller-Jones, J. C. A., Rupen, M. P., Knigge, C., Sivakoff, G. R., & Groot, P. J. 2015, MNRAS, 451, 3801
Coppejans, D. L., et al. 2016, MNRAS, 463, 2229
Corbel, S., Fender, R. P., Tomsick, J. A., Tzioumis, A. K., & Tingay, S. 2004, ApJ, 617, 1272
Coriat, M., et al. 2011, MNRAS, 414, 677
Curran, P. A., et al. 2015, MNRAS, 451, 3975
Drappeau, S., et al. 2017, MNRAS, 466, 4272
Dulk, G. A., Bastian, T. S., & Chanmugam, G. 1983, ApJ, 273, 249
Esin, A. A., McClintock, J. E., & Narayan, R. 1997, ApJ, 489, 865
Espinasse, M., & Fender, R. 2018, MNRAS, 473, 4122
Falcke, H., Körding, E., & Markoff, S. 2004, A&A, 414, 895
Fender, R. P., & Hendry, M. A. 2000, MNRAS, 317, 1
Fender, R. P., & Kuulkers, E. 2001, MNRAS, 324, 923

Gaia Collaboration, et al. 2016, A&A, 595, A1
Gallo, E., Degenaar, N., & van den Eijnden, J. 2018, MNRAS, 478, L132
Hannikainen, D. C., Hunstead, R.W., Campbell-Wilson, D., & Sood, R.K. 1998, A&A, 337, 460
Körding, E., Rupen, M., Knigge, C., Fender, R., Dhawan, V., Templeton, M., & Muxlow, T. 2008, Science, 320, 1318
Körding, E. G., Knigge, C., Tzioumis, T., & Fender, R. 2011, MNRAS, 418, L129
Marsh, T. R., et al. 2016, Nat, 537, 374
Mason, P. A., & Gray, C. L. 2007, ApJ, 660, 662
Massi, M., & Kaufman Bernadó, M. 2008, A&A, 477, 1
Meier, D. L. 2001, ApJ, 548, L9
Meintjes, P. 2016, in Proceedings of Accretion Processes in Cosmic Sources. 5-10 September 2016 Saint Petersburg, Russia (APCS2016), Online at `http://pos.sissa.it/cgi-bin/reader/conf.cgi?confid=288`, id.39, 39
Meintjes, P. J., & de Jager, O. C. 2000, MNRAS, 311, 611
Meintjes, P. J., & Venter, L. A. 2005, MNRAS, 360, 573
Merloni, A., Heinz, S., & di Matteo, T. 2003, MNRAS, 345, 1057
Migliari, S., & Fender, R. P. 2006, MNRAS, 366, 79
Miller-Jones, J. C. A., et al. 2011, in IAU Symposium, Vol. 275, IAU Symposium, ed. G. E. Romero, R. A. Sunyaev, & T. Belloni, 224
Mirabel, I. F., & Rodríguez, L. F. 1994, Nat, 371, 46
Mooley, K. P., et al. 2017, MNRAS, 467, L31
Pavelin, P. E., Spencer, R. E., & Davis, R. J. 1994, MNRAS, 269, 779
Potter, S. B., & Buckley, D. A. H. 2018, MNRAS, 478, L78
Pretorius, M. L., & Knigge, C. 2012, MNRAS, 419, 1442
Pushkarev, A. B., Hovatta, T., Kovalev, Y. Y., Lister, M. L., Lobanov, A. P., Savolainen, T., & Zensus, J. A. 2012, A&A, 545, A113
Russell, D. M., Miller-Jones, J. C. A., Maccarone, T. J., Yang, Y. J., Fender, R. P., & Lewis, F. 2011, ApJ, 739, L19
Russell, D. M., et al. 2013, ApJ, 768, L35
Russell, T. D., et al. 2016, MNRAS, 460, 3720
Russell, T. D., Soria, R., Miller-Jones, J. C. A., Curran, P. A., Markoff, S., Russell, D. M., & Sivakoff, G. R. 2014, MNRAS, 439, 1390
Shakura, N. I., & Sunyaev, R. A. 1973, A&A, 24, 337
Soker, N., & Lasota, J.-P. 2004, A&A, 422, 1039
Sokoloski, J. L., Rupen, M. P., & Mioduszewski, A. J. 2008, ApJ, 685, L137
Soleri, P., & Fender, R. 2011, MNRAS, 413, 2269
Stiller, R. A., Littlefield, C., Garnavich, P., Wood, C., Hambsch, F.-J., & Myers, G. 2018, arXiv:1802.04323
Strader, J., Chomiuk, L., Maccarone, T. J., Miller-Jones, J. C. A., & Seth, A. C. 2012, Nat, 490, 71
Tananbaum, H., Gursky, H., Kellogg, E., Giacconi, R., & Jones, C. 1972, ApJ, 177, L5
Tetarenko, A. J., et al. 2017, MNRAS, 469, 3141
Tetarenko, B. E., et al. 2016, ApJ, 825, 10
Tingay, S. J., et al. 1995, Nat, 374, 141
Tudor, V., et al. 2017, MNRAS, 470, 324
van den Eijnden, J., Degenaar, N., Russell, T. D., Miller-Jones, J. C. A., Wijnands, R., Miller, J. M., King, A. L., & Rupen, M. P. 2018a, MNRAS, 474, L91
van den Eijnden, J., Degenaar, N., Russell, T. D., Miller-Jones, J. C. A., Wijnands, R., Miller, J. M., King, A. L., & Rupen, M. P. 2018b, MNRAS, 473, L141
van der Horst, A. J., et al. 2013, MNRAS, 436, 2625
Warner, B. 1995, Cambridge Astrophysics Series, 28
Wright, A. E., Stewart, R. T., Nelson, G. J., Slee, O. B., & Cropper, M. 1988, MNRAS, 231, 319
Wynn, G. A., King, A. R., & Horne, K. 1997, MNRAS, 286, 436

Science with a Next-Generation Very Large Array
ASP Conference Series, Monograph 7
Eric J. Murphy, ed.

Tidal Disruption Events

Sjoert van Velzen,[1] Geoffrey C. Bower,[2] and Brian D. Metzger[3]

[1]*New York University, New York, NY 10003, USA;* sjoert@nyu.edu

[2]*Academia Sinica Institute of Astronomy and Astrophysics, Hilo, HI 96720, USA*

[3]*Columbia Astrophysics Laboratory, New York, NY 10027, USA*

Abstract. The tidal disruption and subsequent accretion of a star by a supermassive black hole can be used as a laboratory to study the physics of relativistic jets. The ngVLA is the only planned instrument that can both discover and characterize a large number of these short-lived radio sources. In particular the high-frequency capabilities of the ngVLA enable this important leap forward. Multi-frequency radio follow-up observations (3 – 100 GHz) of tidal disruption events found in optical or X-ray surveys will provide a measurement of the jet efficiency as a function of black hole spin, thus enabling a direct test of the prediction that relativistic jets require high spin. Hundreds of tidal disruption jets will be discovered in a blind ngVLA survey for radio transients. By including VLBI observations with the ngVLA Long Baseline Array, we can resolve some of these sources, obtaining a robust measurement of the jet launch date and the magnetic field strength. From the thermal emission of the tidal disruption flare we can measure the accretion rate at this launch date, thus providing another unique opportunity to identify the conditions that lead to jet production.

1. Introduction: Thermal and Non-Thermal Emission

A rare glimpse into the properties of normally quiescent massive black holes is afforded when a star passes sufficiently close that it is torn apart by the black hole's tidal gravitational field (Hills 1975). The process of disruption leaves a significant fraction of the shredded star gravitationally bound to the black hole (Rees 1988; Evans & Kochanek 1989) and this stellar debris has been predicted to power a thermal flare at optical, UV, and X-ray wavelengths. Over a dozen such thermal tidal disruption flare (TDF) candidates have now been identified (Komossa 2015; van Velzen 2018 and references therein). Based on the current detection rate in optical surveys, we expect that the Large Synoptic Survey Telescope (LSST) will yield thousands of thermal TDFs per year (Gezari et al. 2009; van Velzen et al. 2011).

Radio follow-up observations of thermal TDFs have mostly yielded non-detections (Bower et al. 2013; van Velzen et al. 2013; Arcavi et al. 2014). Only two low-redshift thermal TDFs ($z < 0.02$) have been detected in radio follow-up observations (van Velzen et al. 2016; Alexander et al. 2016, 2017). The origin of this radio emission is long debated, both a jet (van Velzen et al. 2011), a disk-wind (Alexander et al. 2016), or unbound stellar debris (Krolik et al. 2016) have been proposed. For one event (ASASSN-14li; Holoien et al. 2016), the detection of a cross-correlation between

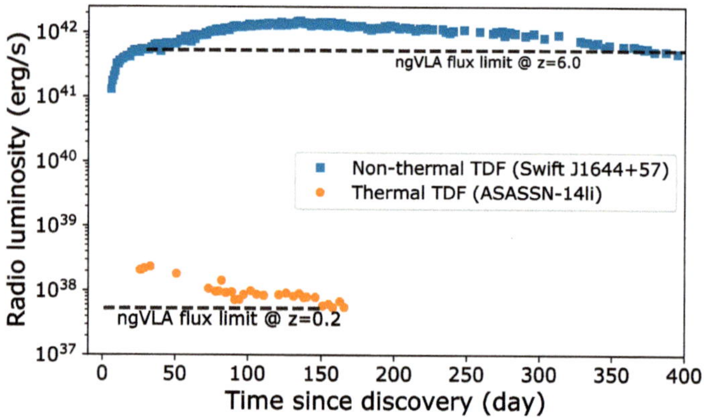

Figure 1. Radio light curve (15 GHz) of two archetype radio-emitting TDFs. The dashed line indicates a 5σ detection threshold of a potential ngVLA transient survey using 10 minute snapshots at 8 GHz (Band 2) to a reach an image rms of 1 μJy/beam. If most thermal TDFs are accompanied by radio emission similar to ASASSN-14li, a blind ngVLA survey covering 300 deg^2 should detect ~ 10 of these events per year, the expected detection rate of events similar to Swift J1644+57 is an order of magnitude higher.

the X-ray and radio light curves provides evidence that the radio emission is produced internal to a moderately relativistic jet (Pasham & van Velzen 2018).

While most known TDFs are dominated by thermal emission, a second class of events is discovered via non-thermal γ-ray emission. The most famous example in this class is Swift J1644+57 (Bloom et al. 2011; Levan et al. 2011; Burrows et al. 2011; Zauderer et al. 2011). This source was characterized by powerful ($\nu L_\nu \sim 10^{47}$ erg s^{-1}) hard X-ray emission. The long duration of Swift J1644+57 and a position coincident with the nucleus of a previously quiescent galaxy led to the conclusion that it was powered by rapid accretion onto the central black hole following a stellar disruption. The rapid X-ray variability suggested an origin internal to a jet that is relativistically beaming its radiation along our line of sight, similar to the blazar geometry of normal active galactic nuclei (Bloom et al. 2011).

Swift J1644+57 was also characterized by luminous synchrotron radio emission, that brightened gradually over the course of several months (Berger et al. 2012; Zauderer et al. 2013), as shown in Figure 1. This radio emission resulted from the shock interaction between the stellar tidal disruption jet and the external gas surrounding the black hole (Giannios & Metzger 2011), similar to a gamma-ray burst afterglow. Two more jetted tidal disruption events with similar X-ray and radio properties to Swift J1644+57 have been found (Cenko et al. 2012; Pasham et al. 2015; Brown et al. 2015).

2. Using Tidal Disruption Events to Study the Physics of Jet Launching

When a central object gathers mass through an accretion disk, jets are ubiquitous. Jets transport energy away from the disk, thus greatly enhancing the scale at which the accreting object (neutron star, black hole, or protostar) can influence its environment. As discussed in detail in earlier chapters of this book (e.g., Nyland et al. 2018, this volume p. 457), jets from supermassive black holes can provide strong negative feedback onto their host galaxy. However, adding jet mode feedback to cosmological galaxy simulations (e.g., Vogelsberger et al. 2014) requires a key ingredient that is currently missing: the jet duty cycle and lifetime. If powerful, long-lived jets can only be produced for a small fraction of active galactic nuclei (AGN), jet feedback will be limited to the subclass of galaxies that host these black holes. On the other hand, AGN may experience multiple cycles of jet production, implying a higher efficiency of inter-galaxy jet feedback. These two scenarios, long-lived versus intermittent jets, broadly correspond to two opposing theories of jet production: the spin paradigm and the accretion paradigm.

In the spin paradigm, jets are powered by the rotational energy of the black hole, which can be tapped using the magnetic field lines that thread the black hole horizon (Blandford & Znajek 1977). Only rapidly spinning black holes are observed to be radio loud. Alternatively, the low number of radio-loud quasars (e.g., Kellermann et al. 1989) can be explained by intermittent jet production due to a state change of the accretion disk (Falcke et al. 2004; Körding et al. 2006). This accretion-paradigm is motivated by observations of stellar mass black holes in X-ray binaries, where jet production triggered by a state change of the accretion disk can be observed directly (Fender, Belloni, & Gallo 2004).

The current body of observations of AGN is not sufficient to solve the problem of jet launching because state changes of these systems are nearly[1] impossible to observe directly due to the long timescales associated with the feeding of AGN. A tidal disruption event provides an opportunity to observe jets from supermassive black holes in real time. Since the fallback rate onto the black hole changes from super-Eddington to sub-Eddington in a few months to years, we may expect jet production due to state changes of the accretion disk (Giannios & Metzger 2011; van Velzen et al. 2011). However our current radio observations of tidal disruption are not yet sufficient to make unambiguous inference on jet production. First of all, the number of radio-detected sources is very small. Second, the nature of radio emission from thermal TDFs is unclear. And third, the event rate of the non-thermal TDFs with powerful jets is unknown.

The ngVLA will overcome all these limitations at once. A key element that makes this possible is the increased sensitivity at high-frequency ($\nu > 3$ GHz). Due to synchrotron self-absorption, TDF jets – like most synchrotron-powered transients – are very faint at low radio frequencies. Only when the source grows large enough, it becomes bright at $\nu \sim 1$ GHz. But at this large size, the typical timescale of the radio emission is long (~ 1 yr); since this timescale is similar to the cadence of most surveys, the amplitude of most synchrotron transients observed at low frequencies is small. Finally, with the high angular resolution of the ngVLA Long Baseline Array we can resolve radio emission from TDFs and thus watch these newborn jets grow with time.

[1]The so-called changing look AGN (e.g., LaMassa et al. 2015) could present an exception to normal AGN variability (Graham et al. 2017) and can be used to study the jet-disk connection over ~ 10 yr timescales (e.g., Koay et al. 2016).

In the next two subsections we discuss how the ngVLA can revolutionize TDF research using two different modes of observation: follow-up and discovery.

2.1. Follow-up of TDFs with the ngVLA

Radio follow-up observations with current facilities have only yielded detections for low-redshift TDFs ($z < 0.05$). Current optical surveys for thermal TDFs are nearly complete for these nearby events, finding about one per year. While the LSST and other optical or X-ray surveys will increase the detection rate of thermal TDFs to thousands per year, the majority of these will be far too distant to be detectable with Jansky VLA follow-up observations. This implies the detection rate of radio-emitting thermal TDFs will fall behind the overall detection rate of TDFs.

A tenfold increase to the sensitivity of the Jansky VLA will push the maximum redshift for detection (within 10 min of integration time per source) of radio emission from thermal TDFs to $z = 0.2$. The order of magnitude increase to the detection rate will enable measurements of the incidence of radio emission as a function of host galaxy properties. Most importantly, we can measure the jet production efficiency as a function of black hole spin. Disruptions by black holes more massive than $10^8\,M_\odot$ are visible only when the black hole is spinning (Stone et al. 2018). Such events are rare, but the first TDF candidate from a black hole above this critical mass was recently detected (Dong et al. 2016; Leloudas et al. 2016; Margutti et al. 2017; Krühler et al. 2018). To conclude, ngVLA follow-up observations of thermal TDFs will yield a measurement of the radio luminosity as a function of black hole spin, thus providing a crucial test of the spin paradigm of jet production.

2.2. Discovery of TDFs in Blind ngVLA surveys

Thanks to their high luminosity, radio emission from non-thermal TDF jets (such as Swift J1644+57) can be detected by current radio surveys well into the dawn of galaxy formation (Metzger, Williams, & Berger 2015). As demonstrated in van Velzen et al. (2013), the detection rate of these events in surveys for radio transients can be estimated using the observed light curve of Swift J1644+57. This estimate requires the volumetric event rate, which can in turn be estimated from the observed rate of on-axis tidal disruption jets ($\sim 10^{-2}\,\mathrm{Gpc}^{-3}\mathrm{yr}^{-1}$, based on the *Swift* data; Brown et al. 2015; Levan et al. 2016) and applying a beaming correction (δ) to obtain the isotropic rate. The beaming correction can both be estimated from the requirement that the intrinsic X-ray luminosity should not exceed the Eddington limit, or by modeling the radio light curves. Both of these approaches yield $\delta \sim 10^2$, hence the volumetric rate of events similar to Swift J1644+57 can be estimated to be $\sim 1\,\mathrm{Gpc}^{-3}\mathrm{yr}^{-1}$.

Using a model to predict the off-axis light curve of events similar to Swift J1644+57, the resulting detection rate of off-axis TDF jets in VLASS is about 10 per year. However the cadence of the VLASS is low ($\approx$ 16 months), hence most sources will be detected after maximum light. The cadence of surveys at $\approx$ 1 GHz by SKA precursors, such as VAST (Murphy et al. 2013) and ThunderKAT (Fender et al. 2017), is higher. The detection rate for off-axis TDF jets in these surveys is also about 10 per year (Metzger et al. 2015). However, due to the very shallow amplitude of the 1 GHz light curves, the identification of these events using only SKA data will be difficult; coordinated observations at higher frequencies (X-ray or optical) will be required (Donnarumma & Rossi 2015; Donnarumma et al. 2015). Follow-up observations of SKA-discovered TDFs with the VLA or ALMA are difficult because the emission at these higher radio frequencies fades within months to weeks, while the emission at 1 GHz is delayed by about a year. Therefore most TDFs candidates found in SKA surveys will have monochromatic radio light curves.

The high frequency capabilities for the ngVLA will enable both the discovery *and* characterization of TDFs in blind radio surveys. Using observations from 3 to 100 GHz, we can measure the total energy of the jet and estimate the expansion velocity. By comparing the event rate from this blind radio survey to the observed rate of TDFs found by *Swift* we obtaind the beaming factor (δ). The detection of a low beaming factor will imply a super-Eddington luminosity for the hard X-ray, as suggested in some models of Swift J1644+57 (Kara et al. 2016). Scaling from estimates by Metzger et al. (2015) for the number of TDFs in VLASS (using 10^4 deg^2 at an rms of 0.1 mJy/beam), we can expect $\sim 10^2$ events per year for an ngVLA transient survey in Band 1 (3 GHz).

For some low-redshift TDFs with powerful jets we can resolve the new radio source. With ngVLA observations in Band 4 ($\approx$ 30 GHz, yielding a maximum resolutoin of a few mas) we can likely resolve jets similar to Swift J1644+57 at $z < 0.1$ within one year after launch. The number of detections at this redshift is low ($<$ 1 per year); including VLBI to the ngVLA would signficantly increase the detection rate. Resolving a newly created jet presents a unique oppurtunity, allowing a measurement of the magnetic field of the radio-emitting region without using the equipartition assumption (e.g., Falcke et al. 1999). Furthermore, by extrapolating the source size measurements backwards in time, we obtain the launch date of the jet. This launch date can be compared to the light curve of the TDF disk emission, providing a direct test of the prediction that jet launching requires super-Eddington accretion rates (e.g., Tchekhovskoy et al. 2014).

Comparing the rate of tidal disruption jets detected in ngVLA surveys to the rate of TDFs detected by their thermal disk emission (in X-ray or optical surveys) we obtain the jet efficiency (i.e., the fraction of disruptions that lead to jets). Using multi-frequency radio light curves we can measure jet efficiency as a function of the total jet energy. Within the accretion paradigm we would expect that the efficiency is constant because the jet energy reflects the amount of stellar debris that was accreted. If, on the other hand, jets are powered by spin, we would expect a few TDFs with a total jet energy that exceeds the rest-mass energy of the accreted debris.

In this chapter we focused on using radio observations of TDFs to constrain models of jet formation. Yet the radio light curves of these sources can also be used as tools to measure the gas density in nuclei of their host galaxies (Zauderer et al. 2011; Berger et al. 2012; Generozov et al. 2017; Eftekhari et al. 2018). Since we can detect sources similar to Swift J1644+57 to very high redshifts (Figure 1), we can use ngVLA observations to probe the environments of black holes in the earliest stages of their growth.

References

Alexander, K. D., Berger, E., Guillochon, J., Zauderer, B. A., & Williams, P. K. G. 2016, ApJ, 819, L25

Alexander, K. D., Wieringa, M. H., Berger, E., Saxton, R. D., & Komossa, S. 2017, ApJ, 837, 153

Arcavi, I., Gal-Yam, A., Sullivan, M., et al. 2014, ApJ, 793, 38

Berger, E., Zauderer, A., Pooley, G. G., et al. 2012, ApJ, 748, 36

Blandford, R. D., & Znajek, R. L. 1977, MNRAS, 179, 433

Bloom, J. S., Giannios, D., Metzger, B. D., et al. 2011, Science, 333, 203

Bower, G. C., Metzger, B. D., Cenko, S. B., Silverman, J. M., & Bloom, J. S. 2013, ApJ, 763, 84
Brown, G. C., Levan, A. J., Stanway, E. R., et al. 2015, MNRAS, 452, 4297
Burrows, D. N., Kennea, J. A., Ghisellini, G., et al. 2011, Nat, 476, 421
Cenko, S. B., Bloom, J. S., Kulkarni, S. R., et al. 2012, MNRAS, 420, 2684
Dong, S., Shappee, B. J., Prieto, J. L., et al. 2016, Science, 351, 257
Donnarumma, I., & Rossi, E. M. 2015, ApJ, 803, 36
Donnarumma, I., Rossi, E. M., Fender, R., et al. 2015, Advancing Astrophysics with the Square Kilometre Array (AASKA14), 54
Eftekhari, T., Berger, E., Zauderer, B. A., Margutti, R., & Alexander, K. D. 2018, ApJ, 854, 86
Evans, C. R., & Kochanek, C. S. 1989, ApJ, 346, L13
Falcke, H., Körding, E., & Markoff, S. 2004, A&A, 414, 895
Falcke, H., Bower, G. C., Lobanov, A. P., et al. 1999, ApJ, 514, L17
Fender, R., Woudt, P., Armstrong, R., et al. 2017, 1711.04132
Fender, R. P., Belloni, T. M., & Gallo, E. 2004, MNRAS, 355, 1105
Generozov, A., Mimica, P., Metzger, B. D., et al. 2017, MNRAS, 464, 2481
Gezari, S., Heckman, T., Cenko, S. B., et al. 2009, ApJ, 698, 1367
Giannios, D., & Metzger, B. D. 2011, MNRAS, 416, 2102
Graham, M. J., Djorgovski, S. G., Drake, A. J., et al. 2017, MNRAS, 470, 4112
Hills, J. G. 1975, Nat, 254, 295
Holoien, T. W.-S., Kochanek, C. S., Prieto, J. L., et al. 2016, MNRAS, 455, 2918
Kara, E., Miller, J. M., Reynolds, C., & Dai, L. 2016, Nat, 535, 388
Kellermann, K. I., Sramek, R., Schmidt, M., Shaffer, D. B., & Green, R. 1989, AJ, 98, 1195
Koay, J. Y., Vestergaard, M., Bignall, H. E., Reynolds, C., & Peterson, B. M. 2016, MNRAS, 460, 304
Komossa, S. 2015, Journal of High Energy Astrophysics, 7, 148
Körding, E. G., Jester, S., & Fender, R. 2006, MNRAS, 372, 1366
Krolik, J., Piran, T., Svirski, G., & Cheng, R. M. 2016, ApJ, 827, 127
Krühler, T., Fraser, M., Leloudas, G., et al. 2018, A&A, 610, A14
LaMassa, S. M., Cales, S., Moran, E. C., et al. 2015, ApJ, 800, 144
Leloudas, G., Fraser, M., Stone, N. C., et al. 2016, Nature Astronomy, 1, 0002
Levan, A. J., Tanvir, N. R., Cenko, S. B., et al. 2011, Science, 333, 199
Levan, A. J., Tanvir, N. R., Brown, G. C., et al. 2016, ApJ, 819, 51
Margutti, R., Metzger, B. D., Chornock, R., et al. 2017, ApJ, 836, 25
Metzger, B. D., Williams, P. K. G., & Berger, E. 2015, ApJ, 806, 224
Murphy, T., Chatterjee, S., Kaplan, D. L., et al. 2013, 30, 6
Nyland, K., Harwood, J. J., Mukherjee, D. et al. 2018, Science with a Next-Generation Very Large Array, ASP Monograph 7, ed. E. J. Murphy (San Francisco, CA: ASP), 457
Pasham, D. R., & van Velzen, S. 2018, ApJ, 856, 14
Pasham, D. R., Cenko, S. B., Levan, A. J., et al. 2015, ApJ, 805, 68
Rees, M. J. 1988, Nat, 333, 523
Stone, N. C., Kesden, M., Cheng, R. M., & Velzen, S. v. 2018, arXiv:1801.10180v1
Tchekhovskoy, A., Metzger, B. D., Giannios, D., & Kelley, L. Z. 2014, MNRAS, 437, 2744
van Velzen, S. 2018, ApJ, 852, 72
van Velzen, S., Frail, D. A., Körding, E., & Falcke, H. 2013, A&A, 552, A5
van Velzen, S., Körding, E., & Falcke, H. 2011, MNRAS, 417, L51
van Velzen, S., Farrar, G. R., Gezari, S., et al. 2011, ApJ, 741, 73
van Velzen, S., Anderson, G. E., Stone, N. C., et al. 2016, Science, 351, 62
Vogelsberger, M., Genel, S., Springel, V., et al. 2014, Nat, 509, 177
Zauderer, B. A., Berger, E., Margutti, R., et al. 2013, ApJ, 767, 152
Zauderer, B. A., Berger, E., Soderberg, A. M., et al. 2011, Nat, 476, 425

Science with a Next-Generation Very Large Array
ASP Conference Series, Monograph 7
Eric J. Murphy, ed.

Intermediate-Mass Black Holes in Globular Cluster Systems

J. M. Wrobel,[1] J. C. A. Miller-Jones,[2] K. E. Nyland,[3] and T. J. Maccarone[4]

[1]*National Radio Astronomy Observatory, P.O. Box O, Socorro, NM 87801, USA;* jwrobel@nrao.edu

[2]*International Centre for Radio Astronomy Research, Curtin University, GPO Box U1987, Perth, WA 6845, Australia;* james.miller-jones@curtin.edu.au

[3]*National Radio Astronomy Observatory, Charlottesville, VA 22903, USA;* knyland@nrao.edu

[4]*Department of Physics and Astronomy, Texas Tech University, Box 41051, Lubbock, TX 79409-1051, USA;* thomas.maccarone@ttu.edu

Abstract. Using the next generation Very Large Array (ngVLA), we will make a comprehensive inventory of intermediate-mass black holes (IMBHs) in hundreds of globular cluster systems out to a distance of 25 Mpc. IMBHs have masses $M_{IMBH} \sim 100 - 100,000\ M_{\odot}$. Finding them in globular clusters would validate a formation channel for seed black holes in the early universe and inform event predictions for gravitational wave facilities. Reaching a large number of globular clusters is key, as Fragione et al. (2018) predict that only a few percent will have retained their gravitational-wave fostering IMBHs.

1. Scientific Motivation

Theory suggests that globular clusters (GCs) of stars can host intermediate-mass black holes (IMBHs) with masses $M_{IMBH} \sim 100 - 100,000\ M_{\odot}$ (Miller & Hamilton 2002; Gurkan et al. 2004; Portegies Zwart et al. 2004; Giersz et al. 2015; Mezcua 2017). Finding IMBHs in GCs would validate a formation channel for seed black holes (BHs) in the early universe (e.g., Volonteri 2010; Sakurai et al. 2017); populate the mass gap between the well-understood stellar-mass BHs and the well-studied supermassive BHs (e.g., Kormendy & Ho 2013; Tetarenko et al. 2016a); and test if scaling relations between stellar systems and their central BHs extend into poorly-explored mass regimes (e.g., Graham & Scott 2015). Also, the GC system of a typical galaxy contains several hundred GCs (Harris 2016). Thus studying such a system could constrain the mass spectrum of IMBHs in its GCs and, related, the ability of its GCs to retain their IMBHs and foster gravitational wave (GW) events (e.g., Holley-Bockelmann et al. 2008; Fragione et al. 2018). These properties could vary from galaxy to galaxy, so many GC systems should be studied. Importantly, a GC system with a low fraction of IMBHs at present could be linked to a high rate of GW events in the past.

To search for IMBHs in GCs, one looks for evidence that the IMBHs are affecting the properties of their GC hosts. For GCs in the Local Group, a common approach is to use optical or infrared data to look for the dynamical signatures of IMBHs on the orbits of stars in the GCs. Such sphere-of-influence studies have a contentious history (e.g., Baumgardt 2017; Mezcua 2017, and references therein), even leading to differing IMBH masses when using the orbits of stars or of radio pulsars in the same GC (e.g., Gieles et al. 2017; Perera et al. 2017). Having 3-dimensional velocity fields should improve the dynamical searches (Drukier & Bailyn 2003). Still, they are fundamentally limited by shot noise due the small number of orbits influenced (e.g., van der Marel 2013). The dynamical searches are also susceptible to measuring high concentrations of stellar remnants rather than an IMBH (e.g., den Brok et al. 2014).

An independent approach that bypasses these issues is to look for the accretion signatures of IMBHs in GCs (Maccarone 2016, and references therein). This approach leverages on decades of studies of the signatures of accretion onto both stellar-mass and supermassive BHs (Fender & Munoz-Darias 2016, and references therein). Here, we apply a synchrotron radio model to search for the signatures of low rates of accretion onto IMBHs in entire GC systems. The model was developed for GCs in the Local Group (Maccarone 2004; Maccarone & Servillat 2008, 2010; Strader et al. 2012) and is summarized in §2. We have used the U.S. National Science Foundation's Karl G. Jansky Very Large Array (VLA; Perley et al. 2011) to demonstrate the feasibility of a radio search for IMBHs in one nearby GC system (Wrobel et al. 2016) and summarize that effort in §3. In §4 we demonstrate the breakthrough role that the Next Generation VLA (ngVLA; Selina & Murphy 2017) will have in searching for IMBHs in hundreds of GC systems hosted in nearby galaxies. We close, in §5, by linking these searches to related studies using facilities contemporary with the ngVLA.

2. Synchrotron Radio Model

We invoke a semi-empirical model to predict the mass of an IMBH that, if accreting slowly from the tenuous gas supplied by evolving stars, is consistent with the synchrotron radio luminosity of a GC (Strader et al. 2012, and references therein). Following Strader et al. (2012), we assume gas-capture at 3% of the Bondi rate (Pellegrini 2005; Merloni & Heinz 2007) for gas at a density of 0.2 particles cm^{-3} as measured by Freire et al. (2001), and at a constant temperature of 10,000 K as justified by Scott & Rose (1975). We also assume that accretion proceeds at less than 2% of the Eddington rate, thus involving an advection-dominated accretion flow with a predictable, persistent X-ray luminosity.[1] We then use the empirical fundamental-plane of BH activity (Merloni et al. 2003; Falcke et al. 2004; Plotkin et al. 2012) to predict the synchrotron radio luminosity. The radio emission is expected to be persistent, flat-spectrum, jet-like but spatially unresolved, and located near the dynamical center of the GC.

From parameter uncertainties, Strader et al. (2012) estimate that the IMBH mass associated with a given radio luminosity could be in error by 0.39 dex, thus a factor of 2.5. To improve the robustness of such masses, one could fold in more recent data, especially regarding the highly uncertain gas-capture parameter. It could also be profitable to examine the model's underlying framework. For example, guided by Scott & Rose (1975), the model assumes that the classic Bondi flow toward a point mass, the IMBH, is isothermal. Yet the classic Bondi accretion rate would be lower if the flow

[1] An IMBH accreting at higher than 2% of the Eddington rate would enter an X-ray-luminous state (Maccarone 2003) and be easily detectable in existing surveys. But no such X-ray sources exist in Galactic GCs.

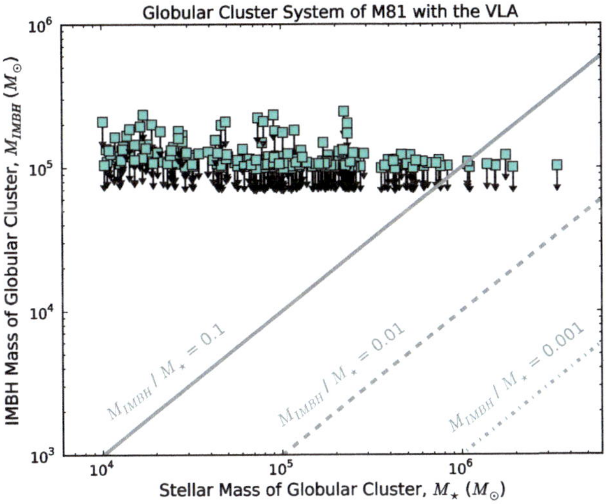

Figure 1. 3σ upper limits to the mass of the IMBH, M_{IMBH}, according to the Strader et al. (2012) model, as a function of the stellar mass, $M_\star$, for probable GCs in M 81 (Wrobel et al. 2016). The grey diagonal lines of constant $M_{IMBH}/M_\star$ convey fiducial ratios of IMBH mass to stellar mass.

could be argued to be adiabatic (Pellegrini 2005; Perera et al. 2017). On the other hand, models for an isothermal flow toward an IMBH embedded in a realistic GC potential appear to achieve higher accretion rates than classic Bondi flows that are isothermal (Pepe & Pellizza 2013).

3. Globular Cluster System of M 81 with the VLA

We searched for the radiative signatures of IMBH accretion from 206 probable GCs in M 81 (Wrobel et al. 2016), a spiral galaxy at a distance of 3.63 Mpc (Freedman et al. 1994). Forty percent of the probable GCs are spectroscopically confirmed, with the balance deemed to be good GC candidates (Nantais et al. 2011). Our search used a four-pointing VLA mosaic at a wavelength of 5.5 cm and a spatial resolution of 1.5 arcsec (26.4 pc). It achieved 3σ upper limits of $3\times(4.3-51)$ μJy for point sources in individual GCs, depending on their location in the mosaic. Weighted-mean image stacks yielded 3σ upper limits of 3×0.43 μJy for all GCs and 3×0.74 μJy for the 49 GCs with stellar masses $M_\star \gtrsim 200,000\,M_\odot$ based on the Nantais et al. (2011) photometry. We applied the Strader et al. (2012) synchrotron model to predict the IMBH mass, M_{IMBH}, consistent with a given luminosity at 5.5 cm. Figure 1 shows upper limits on M_{IMBH} and $M_\star$ for the individual GCs.

For 7 GCs in M 81 the ratios of their IMBH masses to stellar masses, $M_{IMBH}/M_\star$, appear to be less than 0.03-0.09. These upper limits are competitive with the value of 0.02 for the five-times-closer G1, the only extragalactic GC with an IMBH inferred by spatially resolving its sphere of influence on cluster stars (Gebhardt et al. 2005). [Note, though, that G1's X-ray and radio properties suggest a ratio of less than 0.01

(Miller-Jones et al. 2012).] Also, the M 81 stacks correspond to mean IMBH masses of less than 42,000 $M_\odot$ for all GCs and less than 51,000 $M_\odot$ for the 49 GCs with high stellar masses. The VLA is thus making inroads into the difficult-to-observe regime of IMBHs in extragalactic GCs. Further significant progress demands deeper radio observations of individual GCs in M 81 and in other nearby GC systems. Such efforts will admittedly have poorer mass sensitivities than possible for individual Galactic GCs (e.g., Strader et al. 2012). But an extragalactic observation can capture many GCs and its mass sensitivity can be improved by stacking those GCs.

4. Globular Cluster Systems with the ngVLA

We consider using Band 3 of the ngVLA (Selina & Murphy 2017) to examine globular cluster systems in the local universe. Band 3 has a central frequency of 17 GHz and a bandwidth of 8.4 GHz. We approximate its central wavelength as 2 cm. The Harris et al. (2013) compilation of GC systems involves 422 galaxies. The distribution of the galaxies' distances shows two peaks. A minor peak contains tens of galaxies with distances out to 10 Mpc. A major peak contains hundreds of galaxies with distances between 10 and 25 Mpc. A typical galaxy's GC system holds several hundred GCs spread over an effective diameter of a few tens of kpcs (Harris 2016; Kartha et al. 2016). Notably, the major peak in distance contains tens of thousands of GCs in total. Figure 2 shows a potentially rich GC system at 23 Mpc (Brodie et al. 2014).

Figure 2. GC system of the early-type galaxy NGC 4365 at a distance of 23 Mpc. The small circles mark the GC candidates in the inner 18′×17′ of a *gri* Suprime-Cam image. 1′= 6.7 kpc. From Brodie et al. (2014).

We applied the Strader et al. (2012) synchrotron model to predict the luminosity at 2 cm as a function of the mass, M_{IMBH}, of a putative IMBH in a GC. The associated point-source flux densities, S_{2cm}, were then derived for GCs at distances of 10 and 25 Mpc. In Figure 3, the sloping lines show how to convert from S_{2cm} to M_{IMBH} for the two distances, while the vertical lines show 3σ detections with the ngVLA, assuming tapered and robust weighting, with integrations of 1, 10 and 100 hours (Selina & Murphy 2017). At higher signal-to-noise ratios, the wide frequency coverage could test the flat-spectrum prediction, as well as raise flags about steep-spectrum contaminants.

At 10 Mpc the synchrotron model predicts a flux density of $S_{2cm} = 0.35\ \mu$Jy from an IMBH of 83,000 $M_{\odot}$. The ngVLA can make a 3σ detection with a 10-hour integration and a tapered, robustly-weighted resolution of 100 mas (5 pc). This resolution matches the half-starlight diameter of a GC (Brodie & Strader 2006). From Selina & Murphy (2017), the field of view (FOV) of the ngVLA is a circle of diameter 3.′4 (10 kpc) at full width half maximum. Most of a GC system can therefore be encompassed in a few FOVs. Each FOV can simultaneously capture many GCs. Undetected GCs can also be stacked. A stacking performance as for M 81 (§3) can improve the IMBH mass sensitivity by a factor of two. At 25 Mpc the synchrotron model predicts a flux density of $S_{2cm} = 0.35\ \mu$Jy from an IMBH of 163,000 $M_{\odot}$. The ngVLA can make a 3σ detection with a 10-hour integration, localize the source to the GC, and encompass most of the GC system in a few FOVs. Each FOV can simultaneously capture many GCs. Stacking can also be done, and is expected to reach the mass sensitivity mentioned above for an individual GC at 10 Mpc.

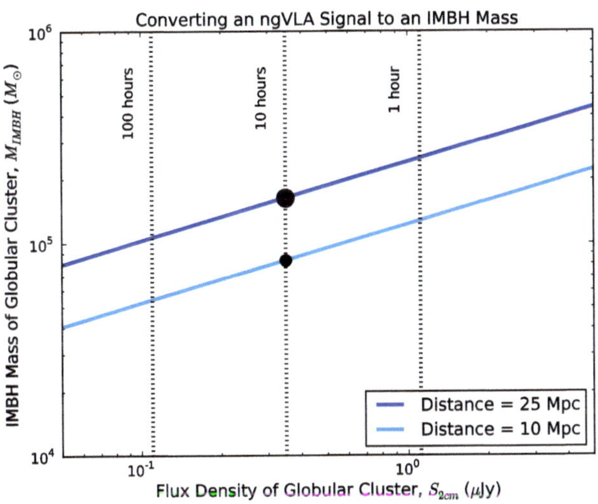

Figure 3. ngVLA signals, S_{2cm}, from IMBH masses, M_{IMBH}, in GCs at distances of 10 and 25 Mpc, according to the Strader et al. (2012) model. The small and big black dots highlight 3σ mass sensitivities at 10 and 25 Mpc, respectively, after 10 hours on target. Hundreds of GC systems have distances between 10 and 25 Mpc, and hold tens of thousands of GCs in total. Reaching large numbers of GCs is key, as Fragione et al. (2018) predict that only a few percent will have retained their IMBHs.

Regarding possible radio contaminants in extragalactic GCs, guidance comes from radio studies of X-ray binaries in the Galaxy and M 31. Dozens of persistent radio emitters are known in the Galaxy, and their range of radio luminosities suggests negligible contamination beyond the Local Group (e.g., Tetarenko et al. 2016b). One radio-flaring X-ray binary has been identified in M 31 (Middleton et al. 2013). A radio analog of it could be detected out to 25 Mpc, but it would fade after a few months so not be mistaken for a persistent emitter. Three Galactic X-ray binaries have similarly strong radio emission. In two cases this is likely related to massive donar stars, absent from GCs. The third case, GRS 1915+105, is an unusually long-lived transient that is both radio and X-ray bright (Fender & Belloni 2004). An analog of GRS 1915+105 would be detectable at both wavelengths out to 25 Mpc, but the fundamental plane of BH activity would unmask it as a stellar-mass BH. Also, when searching tens of thousands of GCs, we should beware of possible radio contamination from background source populations. Simulated source counts near 2 cm suggest that star forming galaxies will dominate at μJy levels (Wilman et al. 2008). But such galaxies have steep specta and finite sizes, so will not be mistaken for IMBHs that have flat spectra and are point-like.

In summary, with its sensitivity, bandwidth, spatial resolution, and FOV, the ngVLA at 2 cm will efficiently probe IMBH masses in hundreds of GC systems out to a distance of 25 Mpc.

5. The ngVLA and Its Contemporary Facilities in the 2030s

Gravitational Wave Facilities. Fragione et al. (2018) explored the fate of primordial GCs, each born with a central IMBH. They modelled the evolution of the GCs in their host galaxy, and of the IMBHs undergoing successive, GW-producing mergers with stellar-mass BHs in the GCs. For primordial GCs that survived to the present day, they found that a few percent retained their IMBHs and the balance lost their IMBH when a GW recoil ejected it from the GC host. Once ejected, the IMBHs are no longer able to foster GW events. They used these results to predict GW event rates for the *Laser Interferometer Space Antenna* (*LISA*; Amaro-Seoane et al. 2017) and Europe's Einstein Telescope (Hild et al. 2011). The rates for the latter facility also apply to the U.S.'s similarly-scoped Cosmic Explorer (Abbott et al. 2016). IMBHs with masses between 1000 and 10,000 $M_\odot$ yielded mergers at rates that could be detected by all three GW facilities. IMBHs with masses $\gtrsim 10,000\ M_\odot$ yielded mergers at rates that could be detected only by *LISA*. If the ngVLA searches do not find the expected mix of IMBHs in GC systems, it could challenge the framework underlying the GW predictions. ngVLA searches for point-like emission from IMBHs would be easy to conduct during Early Science, notionally starting in 2028. Guided by such early ngVLA results, Fragione et al. (2018) could revisit their GW predictions.

Electromagnetic Wave Facilities. A key science driver for extremely large telescopes (ELTs) in the 30 m class is to measure, at a distance of 10 Mpc, a BH mass as low as 100,000 $M_\odot$ by spatially resolving its sphere of influence in its GC host (Do et al. 2014). For example, if the Infrared Imaging Spectrometer (IRIS)[2] on the Thirty Meter Telescope (TMT) can achieve the diffraction limit of 18 mas at 2 μm, then this approach will yield a sample of IMBHs in GCs out to a distance of 10 Mpc. A sphere-of-influence study with IRIS must be done one GC at a time, a shortcoming that makes it expensive to inventory the range of IMBH masses in a galaxy's GC system. The

[2]`https://www.tmt.org/page/science-instruments`

TMT's Infrared Multi-object Spectrometer (IRMS)[3] in spectroscopy mode will have a FOV of 2′.0 × 0′.6. This being a tenth of the ngVLA FOV, a sphere-of-influence study with IRMS would require ten pointings to cover one ngVLA pointing. Regardless of the situation at 10 Mpc, the ELT approach cannot reach the hundreds of GC systems with distances between 10 and 25 Mpc.

The *Chandra* X-ray mission and its proposed successors, *Lynx*[4] and the *Advanced X-ray Imaging Satellite*,[5] feature spatial resolutions of 300 to 500 mas. These will suffice to roughly localize X-ray sources to GCs out to a distance of 25 Mpc. But an X-ray–only search for the accretion signatures of IMBHs in GCs will be hindered by confusion from X-ray binaries in GCs (e.g., Joseph et al. 2017). Specifically, X-ray–only detections of GCs cannot discriminate between X-ray binaries and IMBHs. Fortunately, the empirical fundamental-plane of BH activity (Merloni et al. 2003; Falcke et al. 2004; Plotkin et al. 2012) implies that the persistent radio emission from IMBHs is expected to be several hundred times greater than that from X-ray binaries. Thus ngVLA imaging can be used to separate X-ray detections into bins for X-ray binaries and for IMBHs. X-ray binaries are known to be time-variable in both the radio and X-ray bands, so this radio–X-ray synergy would be strengthened by simultaneous observations with the ngVLA and the X-ray mission.

The deployment baseline of SKA1-Mid (Dewdney et al. 2015; Borjesson 2017) will offer a spatial resolution of 57 mas at 3.3 cm with uniform weighting. This resolution suffices to search for the accretion signatures of IMBHs in GCs with declinations south of 10°. But as only 67 SKA1-Mid antennas will be available at 3.3 cm, the effective collecting area of that telescope will only be about that of the current VLA, which is insufficient for our purposes.

Many of the GC systems in Harris et al. (2013) will need further optical imaging before finding charts are available for all of their GCs. Such images can mostly be acquired with current ground-based telescopes in the 4-m class (e.g., Hargis & Rhode 2014) or 8-m class (e.g., Brodie et al. 2014). But for some particularly confused regions, it may be necessary to obtain images from space-based telescopes. For example, the Near Infrared Camera on the *James Webb Space Telescope* will offer spatial resolutions of 80 and 160 mas in its short- and long-wavelength channels, respectively.[6]

Acknowledgments. NRAO is a facility of the National Science Foundation, operated under cooperative agreement by Associated Universities, Inc.

References

Abbott, B. P., et al. 2016, arXiv:1607.08697
Amaro-Seoane, P., et al. 2017, Laser Interferometer Space Antenna, arXiv:1702.00786
Baumgardt, H. 2017, MNRAS, 464, 2174
Borjesson, L. 2017, Notes from the Chair of the SKA Board from the meeting of 718-192017
Brodie, J. P., & Strader, J. 2006, ARA&A, 44, 193
Brodie, J. P., Romanowsky, A. J., Strader, J., et al. 2014, ApJ, 796, 52

---

[3]https://www.tmt.org/page/science-instruments

[4]http://cxc.harvard.edu/cdo/cxo2lynx2017/index.html

[5]http://axis.astro.umd.edu

[6]https://jwst.stsci.edu/files/live/sites/jwst/files/home/instrumentation/technical\documents/jwst-pocket-guide.pdf

Dewdney, P., Turner, W., Braun, R. 2015, SKA-TEL-SKO-0000308
den Brok, M., van de Ven, G., van den Bosch, R., & Watkins, L. 2014, MNRAS, 438, 487
Do, T., Wright, S., Barth, A. J., et al. 2014, AJ, 147, 93
Drukier, G. A., & Bailyn, C. D. 2003, ApJ, 597, L125
Falcke, H., Kording, E., & Markoff, S. 2004, A&A, 414, 895
Fender, R., & Belloni, T. 2004, ARA&A, 42, 317
Fender, R., & Munoz-Darias, T. 2016, in Astrophysical Black Holes, Lecture Notes in Physics, Volume 905, eds. F. Haardt et al. (Springer International Publishing: Switzerland), 65
Fragione, G., Ginsburg, I., & Kocsis, B. 2018, ApJ, 856, 92
Freedman, W. L., Hughes, S. M., Madore, B. F., et al. 1994, ApJ, 427, 628
Freire, P. C., Kramer, M., Lyne, A. G., et al. 2001, ApJ, 557, L105
Gebhardt, K., Rich, R. M., & Ho, L. C. 2005, ApJ, 634, 1093
Gieles, M., Balbinot, E., Yaaqib, R. I. S. M., et al. 2017, MNRAS, 473, 4832
Giersz, M., Leigh, N., Hypki, A. Lutzgendorf, N., & Askar, A. 2015, MNRAS, 454, 315
Graham, A. W., & Scott, N. 2015, ApJ, 798, 54
Gurkan, M. A., Freitag, M., & Rasio, F. A. 2004, ApJ, 604, 632
Hargis, J. R., & Rhode, K. L. 2014, ApJ, 796, 62
Harris, W. E., Harris, G. L., & Alessi, M. 2013, ApJ, 772, 82
Harris, W. E. 2016, AJ, 151, 102
Hild, S., et al. 2011, Class. Quantum Grav., 2011, 28, 094813
Holley-Bockelmann, K., Gültekin, K., Shoemaker, D., & Yunes, N. 2008, ApJ, 686, 829
Joseph, T. D., Maccarone, T. J., Kraft, R.P., & Sivakoff, G. R. 2017, MNRAS, 470, 4133
Kartha, S. S., Forbes, D. A., Alabi, A. B., et al. 2016, MNRAS, 458, 105
Kormendy, J., & Ho, L. C. 2013 ARA&A, 51, 511
Maccarone, T. J. 2003, å, 409, 697
Maccarone, T. J. 2004, MNRAS, 351, 1049
Maccarone, T. J., & Servillat, M. 2008, MNRAS, 389, 379
Maccarone, T. J., & Servillat, M. 2010, MNRAS, 408, 2511
Maccarone, T. J. 2016, Mem. S. A. It., 87, 559
Merloni, A., Heinz, S., & DiMatteo, T. 2003, MNRAS, 345, 1057
Merloni, A., & Heinz, S. 2007, MNRAS, 381, 589
Mezcua, M. 2017, Int. J. Mod. Phys. D, 26, 1730021
Middleton, M. J., Miller-Jones, J. C. A., Markoff, S., et al. 2013, Nat, 493, 187
Miller, M. C., & Hamilton, D. P. 2002, MNRAS, 330, 232
Miller-Jones, J. C. A., Wrobel, J. M., Sivakoff, G. R., et al. 2012, ApJ, 755, L1
Nantais, J. B., Huchra, J. P., Zezas, A., Gazeas, K., & Strader, J. 2011, AJ, 142, 183
Pellegrini, S. 2005, ApJ, 624, 155
Pepe, C., & Pellizza, L. J. 2013, MNRAS, 430, 2789
Perera, B. B. P., Stappers, B. W., Lyne, A. G., et al. 2017, MNRAS, 468, 2114
Perley, R. A., Chandler, C. C., Butler, B. J., & Wrobel, J. M. 2011, ApJ, 739, L1
Plotkin, R. M., Markoff, S., Kelly, B.C., Koerding, E., Anderson, S.F. 2012, MNRAS, 419, 267
Portegies Zwart, S. F., Baumgardt, H., Hut, P., Makino, J., McMillian, S. L. W. 2004, Nat, 428, 724
Sakurai, Y., Yoshida, N., Fujii, M. S., & Hirano, S. 2017, MNRAS, 472, 1677
Scott, E. H., & Rose, W. K. 1975, ApJ, 197, 147
Selina, R. & Murphy, E. 2017, Next Generation Very Large Array Memo No. 17, ngVLA Reference Design Development & Performance Estimates
Strader, J., Chomiuk, L., Maccarone, T. J., et al. 2012, ApJ, 750, L27
Tetarenko, B. E., Sivakoff, G. R., Heinke C. O., & Gladstone, J. C. 2016a, ApJS, 222, 15
Tetarenko, B. E., Bahramian, A., Arnason, R. M., et al. 2016b, ApJ, 825, 10
van der Marel, R. 2013, SnowPAC 2013 - Black Hole Fingerprints: Dynamics, Disruptions and Demographics
Volonteri, M. 2010, Astron. Astrophys. Rev., 18, 279
Wilman, R. J., Miller, L., Jarvis, M. J., et al. 2008, MNRAS, 388, 1335
Wrobel, J. M., Miller-Jones, J. C. A., & Middleton, M. J. 2016, AJ, 152, 22

Science with a Next-Generation Very Large Array
ASP Conference Series, Monograph 7
Eric J. Murphy, ed.

Science with Pulsar Timing Arrays and the ngVLA

The North American Nanohertz Observatory for Gravitational Waves Collaboration
info@nanograv.org

Abstract. Pulsar timing arrays (PTAs) can be used to detect and study gravitational waves in the nanohertz band (i.e., wavelengths of order light-years). This requires high-precision, decades-long data sets from sensitive, instrumentally stable telescopes. NANOGrav and its collaborators in the International Pulsar Timing Array consortium are on the verge of the first detection of the stochastic background produced by supermassive binary black holes, which form via the mergers of massive galaxies. By providing Northern hemisphere sky coverage with exquisite sensitivity and higher frequency coverage compared to the SKA, the ngVLA will be a fundamental component in the next phase of nanohertz GW astrophysics, enabling detailed characterization of the stochastic background and the detection of individual sources contributing to the background, as well as detections of (or stringent constraints on) cosmic strings and other exotica. Here we summarize the scientific goals of PTAs and the technical requirements for the ngVLA to play a significant role in the characterization of the nanohertz gravitational wave universe.

1. Introduction: Science with Pulsar Timing Arrays

The recent detections of binary black hole (Abbott et al. 2016) and binary neutron star (Abbott et al. 2017) mergers by the LIGO-Virgo collaboration provide spectacular confirmation of the existence of gravitational waves (GWs) at frequencies of a few hundred hertz. These discoveries have measured the speed of gravity to phenomenal precision, determined the origin of half of the elements of the periodic table, and probed relativistic explosions, as well as changing our understanding of stellar evolution and binary interactions.

GWs at nanohertz frequencies – with wavelengths on the order of light-years – can be detected by a pulsar timing array (PTA), where a collection of stable millisecond pulsars (MSPs) is timed over a period of years to decades. These low frequency GWs produce pulse arrival-time perturbations that are *spatially correlated* with a quadrupolar pattern on the sky (e.g., Hellings & Downs 1983), enabling the identification of a GW signal against a background of other effects including pulsar timing noise, stochastic pulse propagation in the interstellar and interplanetary media, and uncertainties in the position of the solar system barycenter. The most promising sources of GWs at nanohertz frequencies are supermassive black hole binaries (SMBHBs). Our current understanding of the formation of galaxies and the history of mass assembly in the universe requires the mergers of galaxies and, very probably, mergers of the supermassive black holes they host. Figure 1 shows the SMBHB life-cycle. A detection of the stochastic background of GW emission produced by the mergers of SMBHBs is im-

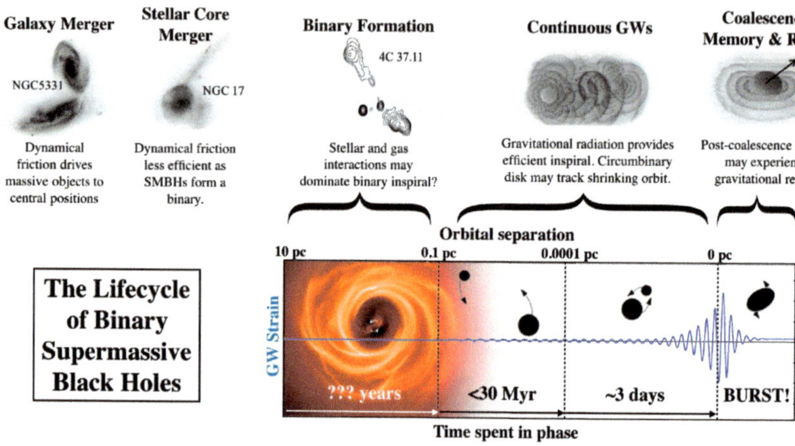

Figure 1. Plot of the binary SMBH life-cycle, adapted from NANOGrav Astrophysics Working Group et al. (2018). A significant unknown in binary evolution theory is the efficiency of inspiral from ~10 pc down to ~0.1 pc separations–the "final parsec" problem–after which the binary can coalesce efficiently due to gravitational wave emission. Through sky imaging techniques, the ngVLA itself may discover widely separated binaries, and depending on its long-baseline sensitivity, could discover targets that are detectable by PTAs. Separately, the ngVLA could contribute to the sensitivity of PTAs, allowing the detection of GW emission from the same systems. PTAs such as NANOGrav can detect supermassive ($> 10^8\,M_\odot$) black hole binaries within ~0.1 pc separation (second panel in the lower figure). On rare occasion, PTAs may detect the permanent space-time deformation (GW memory) caused by a binary's coalescence (Favata 2009). Image credits: Galaxies, *Hubble*/STScI; 4C37.11, Rodriguez et al. (2006); Simulation visuals, C. Henze/NASA; Circumbinary accretion disk, C. Cuadra.

minent. The next step is studying individual binary systems through multimessenger probes. By allowing the flexible allocation of collecting area to time pulsars of interest on a sustained and regular basis, the ngVLA will contribute significantly to the sensitivity of PTAs. Since all-sky coverage is essential, the ngVLA complements the Southern hemisphere coverage of the planned SKA. Further, as we show below, the optimal frequencies for timing distant, high dispersion measure pulsars are significantly higher than 1 GHz, making the ngVLA an ideal instrument for the expanding PTAs of the future.

In addition, through sky imaging techniques, the ngVLA may uncover widely separated SMBHBs, and depending on its long-baseline sensitivity, could discover targets that are simultaneously emitting gravitational waves in the PTA band. This would make the ngVLA a premier multi-messenger astronomy facility *on its own*. Thus, it is possible that ngVLA could be a leading facility in the detection and study of binary SMBHs. It may contribute unique information to the full characterization of SMBH evolution that will be performed by current and future observatories such as LISA, LSST, ELT-class facilities, and the North American Nanohertz Observatory for Gravitational Waves (NANOGrav).

2. Current State of the Art and Scientific Impact

NANOGrav is a collaboration of astronomers using the world's most sensitive pulsar timing facilities (the Arecibo Observatory and the Green Bank Telescope) to monitor several tens of millisecond pulsars: we are expanding and improving a Galactic-scale detector for low-frequency GWs. NANOGrav's recent 11-year data release (Arzoumanian et al. 2018a) has allowed stringent limits to be placed on the stochastic background of GWs from SMBHB mergers (Arzoumanian et al. 2018b), as well as meaningful constraints on more speculative source classes such as primordial GWs and cosmic strings that can form during phase transitions in the early Universe.

At present NANOGrav both competes and collaborates with its international partners, the European PTA and the Parkes PTA. As new telescopes (such as FAST and CHIME) come on line, we expect that collaboration under the umbrella of the International Pulsar Timing Array (IPTA) will play a more significant role. The leadership role of NANOGrav depends on continued access to telescope facilities for pulsar timing, along with new instrumentation that enables improvement in the sensitivity to gravitational waves.

In the next decade, we expect that the stochastic GW background will be successfully detected, either by NANOGrav or as an IPTA effort. The scientific focus will then shift to the precise spectral characterization of the stochastic GW background, measurement of its anisotropy, detection of continuous waves and bursts with memory from individual GW sources, joint observations of their electromagnetic counterparts, as well as constraints on the predictions from more exotic physics. This will help address pressing questions in the joint evolution of supermassive black holes and their host galaxies (see NANOGrav Astrophysics Working Group et al. 2018 for more details).

For instance, galaxy merger rates, SMBH-host co-evolution, dynamical relaxation timescales including the potential that SMBHBs may stall at wide separations, can all affect the amplitude scaling of the GW background. The GW background spectrum might be detected with a shallow slope at frequencies $\lesssim$10 nHz if SMBHBs have strong interactions with their environments (stars and gas) during the late stages of orbital evolution. PTA constraints on or measurement of the background's amplitude and spectral shape can inform all of the above astrophysical uncertainties about the ensemble SMBHB population. A detection of the SMBHB background would provide the first comprehensive proof of the consensus view that SMBHs reside in most or all massive galaxies. Moving past global characterizations, constraints on the background anisotropy may highlight actively interacting galaxy clusters or large-scale cosmic features.

Beyond the detection of the stochastic background, PTAs will reveal individually-resolvable continuous-wave sources as massive and nearby systems can be disentangled from the background. These detections will provide the most direct probe of the early-inspiral stage of a SMBHB, and can provide measurements of the binary's position, phase, and an entangled estimate of chirp mass and luminosity distance ($\mathcal{M}/D_{\rm L}$). If the host of a continuous-wave source can be identified with electromagnetic data, the mass/distance degeneracy can be broken by a redshift measurement; furthermore, this will enable a "calibration" of black hole mass/host galaxy relations at intermediate redshifts by the precise measurement of the central mass in these galaxies. If detected, the pulsar term can permit temporal aperture synthesis, also allowing $\mathcal{M}$ and $D_{\rm L}$ to be disentangled. Evolution of the waveform over PTA experimental durations is unlikely for SMBHBs; however, this would also disentangle the $\mathcal{M}/D_{\rm L}$ term.

In our roadmap for the next 15 to 20 years, the ngVLA will provide Northern hemisphere coverage for PTAs with superb sensitivity and be a fundamental component of the multi-telescope program needed for nanohertz GW science that will be conducted by NANOGrav and the IPTA.

Figure 2 shows our GW sensitivity goal with the ngVLA in terms of the gravitational wave strain versus frequency. The top curve shows the sky-averaged 95% upper limits for the NANOGrav 11-yr data set (our most current data set). The bottom two curves (sky-averaged and best sky-location 95% upper limit curves) are computed from observation simulations that assume a continuation of our current observing program followed by 10 years of ngVLA observations. The black points show the brightest 10 sources in 500 simulations of the SMBHB population in the universe. Depending on the details of the simulation parameters, an average of 5-10 SMBHB lie above the upper limit curve for our best sky location using the ngVLA. This estimate is also in broad agreement with Mingarelli et al. (2017), who base their predictions on local massive galaxies.

Figure 3 shows the improvement in our detection range with the ngVLA in terms of the luminosity distance as a function of the supermassive black hole binary total mass for two of our most sensitive frequencies. In the range of supermassive black hole binary masses relevant for pulsar timing arrays, the luminosity distance (our astrophysical reach) improves by over an order of magnitude. This increases the volume of the universe that is being probed for individual systems of nanohertz GW emission by a factor of over 1000.

Below we detail the improvements necessary to increase the sensitivity of our GW PTA detectors to achieve the scientific goals described here, and outline a future observational program. These improvements are then translated into a summary of the technical and usage requirements for the ngVLA to play a leading role in PTA science.

3. Improving Nanohertz GW Detectors

The key features of the ngVLA that are relevant to PTAs are its high sensitivity, Northern hemisphere sky-coverage, broad spectral coverage, and the ability to multiplex its sensitivity into subarrays. The last is a game-changing advantage over large single-dish telescopes. We discuss two of the more important advances possible with the ngVLA below.

3.1. Subarrays and the Fight Against Jitter

The subarray capability of the ngVLA is particularly important because, unlike traditional imaging observations, pulsar timing observations do not allow sensitivity to be traded against integration time without constraints. Pulsar timing involves time-tagging of pulse profiles obtained by averaging large numbers ($N \sim 10^5$ to 10^6) of individual pulses. While such averages are highly stable, the amplitude and pulse-phase jitter of individual pulses causes small deviations of the average shape, which translate into arrival time errors $\propto N^{-1/2}$, as illustrated in Figure 4.

Pulse jitter therefore sets a floor on the minimum number of pulses that must be averaged to achieve a specified arrival-time precision for the millisecond pulsars whose arrival time errors are dominated by jitter (rather than signal-to-noise ratio considerations). For those objects, the required integration time does not decrease with increased

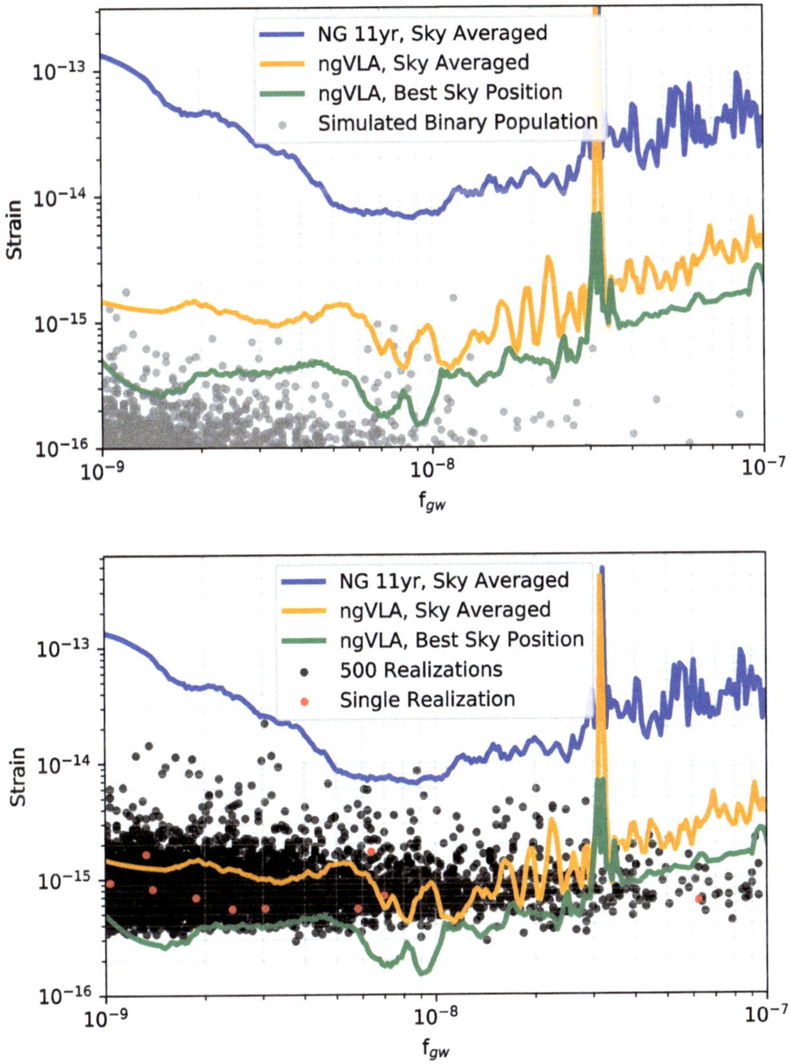

Figure 2. Plot of current sensitivity and sensitivity improvements for simulated observations using the ngVLA. In each panel, the top (blue) curve shows the current sky-averaged NANOGrav 11-yr 95% upper limits. The bottom curves show the sky-averaged (orange) and best sky location (green) 95% upper limits for simulated observations that assume a continuation of our current observing program followed by 10 years of ngVLA observations. (*Upper Panel:*) The gray points show a single simulation of the supermassive binary black hole population, the majority of which lie below our sensitivity limits. (*Lower Panel:*) The black points show the brightest 10 sources (only) in each of 500 simulations of the supermassive binary black hole population. Red points show the brightest 10 sources in one of these simulations.

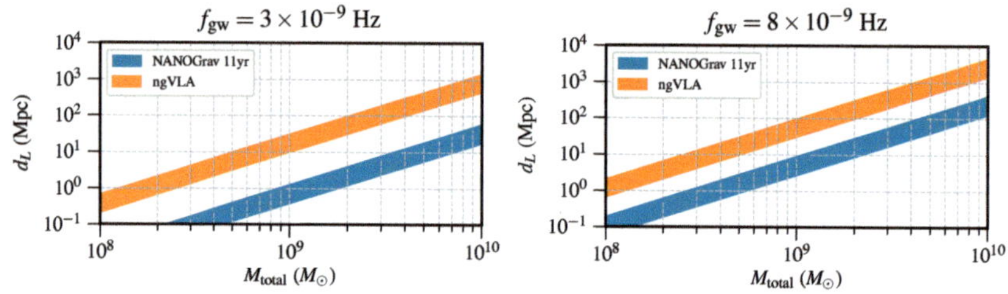

Figure 3. Plot of the luminosity distance versus the supermassive binary black hole total mass for our most current observations (the NANOGrav 11-yr data set, blue) and simulated ngVLA observations (orange). The luminosity distance (our astrophysical reach) increases by over an order of magnitude over all mass ranges.

telescope sensitivity. For faint pulsars where radiometer noise currently dominates timing errors, increased sensitivity can improve their arrival time estimation precision and in some cases they will then be jitter dominated. Therefore the full sensitivity of the ngVLA can be used to improve the timing errors of weaker MSPs, while brighter MSPs lend themselves to observations with subarrays large enough so that the arrival times are jitter dominated, thus optimizing pulsar timing throughput with the ngVLA.

3.2. Frequency Coverage and the Fight Against the ISM

The ionized interstellar medium causes chromatic perturbations of arrival times that must be removed using multifrequency observations at frequencies that are optimized for each MSP. Currently, there is not much flexibility in the choice of observing bands used for timing measurements, implying that there are significant opportunities for improving arrival times and the sensitivity to GWs with future instrumentation and at the ngVLA.

Pulsars are typically brighter at lower frequencies ($S_\nu \propto \nu^\alpha$; $\alpha \sim -1.6$) but considerations of receiver and sky background noise along with interstellar effects drive the range of observation frequencies (0.4 to 2 GHz) currently used by NANOGrav at Arecibo and the GBT. Typically, a given MSP is observed at both a low frequency (typically < 1 GHz) and a high frequency (> 1 GHz) in order to fit for the dispersion measure (DM), which corresponds to the integrated line-of-sight electron column density ($\mathrm{DM} \equiv \int_0^S n_e\, ds$). For the very high precision timing required for PTAs, the time variability of the DM requires contemporaneous measurements over a wide frequency range at each observation epoch.

The optimization of pulsar timing observations with respect to radio observing bands has been investigated in detail by Lam et al. (2017), leading to the conclusion that MSP timing can be improved significantly by using higher frequencies and larger bandwidths. As shown in Figure 5, higher frequencies and larger bandwidths can produce higher precision time-of-arrival (TOA) measurements than currently achieved. The effect is especially pronounced for pulsars at higher DMs, where propagation effects are increasingly important (Figure 5, right panel). As the NANOGrav PTA expands, newer pulsar discoveries are likely to be fainter and more distant, so higher sensitivities and higher frequencies will become critical over time. However, access to frequencies down to 1–1.5 GHz remains essential; we note that with wide enough available bandwidths,

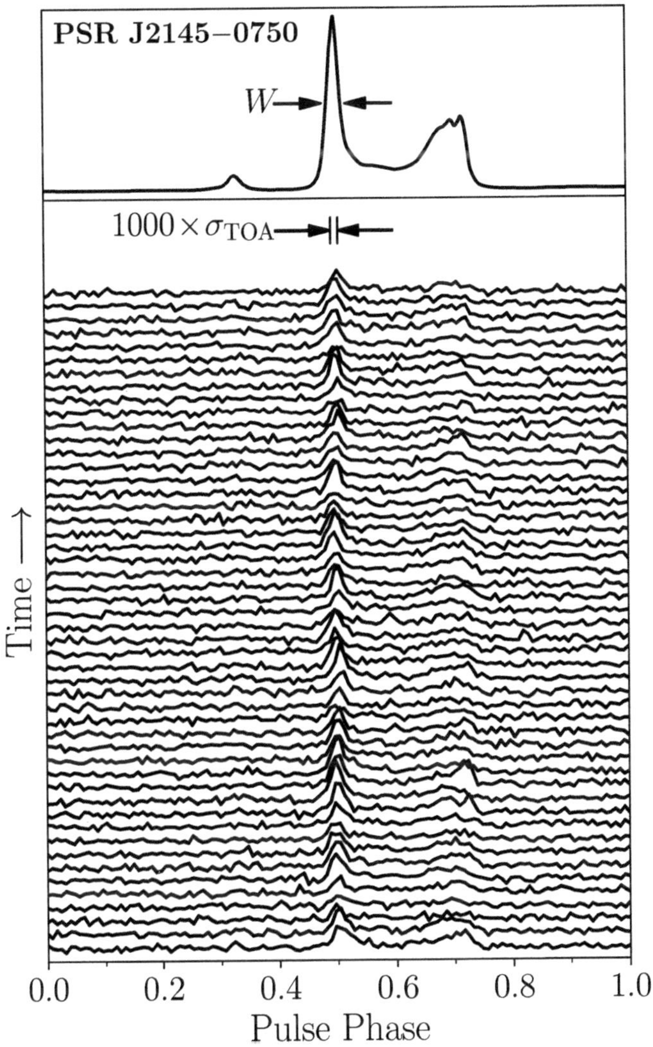

Figure 4. Pulse profiles for PSR J2145−0750, a 16.05 ms pulsar, at 820 MHz. The long-term average profile is shown at the top. The waterfall plot consists of a sequence of 10-pulse average profiles, illustrating the effect of pulse jitter, which reduces with the (number of pulses)$^{-1/2}$. The attained TOA precision $\sigma_{\rm TOA} \sim 0.1\ \mu$s, much smaller than the pulse width $W \sim 395\ \mu$s. High-sensitivity, broad-bandwidth observations with the ngVLA can reduce the arrival-time uncertainties drastically for many of our pulsars, allowing for improved GW characterization in the future.

dual-frequency observations may no longer be required for DM estimation, leading to large gains in efficiency.

These requirements can be addressed using the ngVLA's continuous frequency coverage in the overall ~ 1 to 10 GHz range of interest for PTAs.

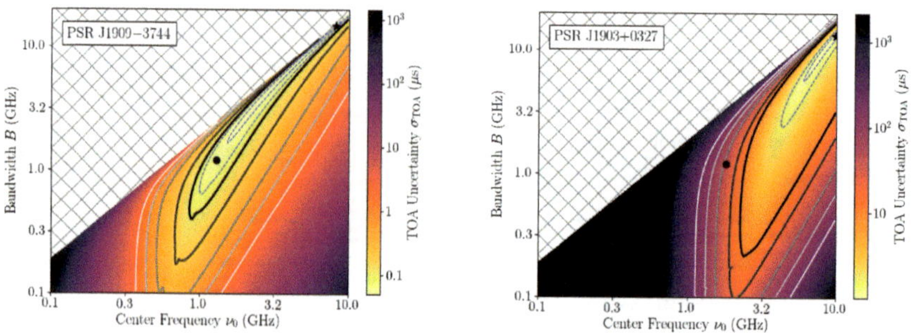

Figure 5. The uncertainty in pulse time of arrival measurements as a function of observing center frequency and bandwidth for two millisecond pulsars currently being observed by NANOGrav, adapted from our work in Lam et al. (2017). (*Left:*) PSR J1909–3744, which is one of the best-timed pulsars. Solid contours indicate TOA uncertainties of 2, 1, 0.5, 0.2, and 0.1 μs, in order of increasing darkness and thickness. The minimum TOA uncertainty (black star) is $\sigma_{\rm TOA}(\nu_0 = 8.1 \text{ GHz}, B = 14.8 \text{ GHz}) = 50$ ns and the estimate given the current frequency coverage (black circle) is $\sigma_{\rm TOA}(\nu_0 = 1.3 \text{ GHz}, B = 1.2 \text{ GHz}) = 59$ ns. The two dashed blue contours represent a 10% and 50% increase above the minimum $\sigma_{\rm TOA}$. (*Right:*) PSR J1903+0327, which is the pulsar with the highest dispersion measure (297.52 pc cm^{-3}) currently observed by NANOGrav. Contours indicate TOA uncertainties of 200, 100, 50, 20, and 10 μs, in order of increasing darkness and thickness. The minimum TOA uncertainty (black star) is $\sigma_{\rm TOA}(\nu_0 = 9.8 \text{ GHz}, B = 13.2 \text{ GHz}) = 1.0\,\mu$s and the estimate given the current frequency coverage (black circle) is $\sigma_{\rm TOA}(\nu_0 = 1.8 \text{ GHz}, B = 1.2 \text{ GHz}) = 44.0\,\mu$s. Higher observing frequencies with larger bandwidths potentially allow significant improvements in timing precision.

4. Observational Program

The current NANOGrav timing program (Arzoumanian et al. 2018a) uses integration times of 0.25–0.5 hr per pulsar at Arecibo and Green Bank with cadences of weekly (for the best pulsars) to once every three weeks, and allocating similar integration times with a selected fraction of the ngVLA collecting area offers more flexibility and higher efficiency compared to steering the entire collecting area of a large single dish to a sequence of pulsar timing targets.

Thus we envisage two usage modes of the ngVLA for PTA observations: (1) A sole-user mode, where the array is used in full on a single object or with up to ~ 5 subarrays, each observing a pulsar of interest; or (2) A shared-user mode, where a single phased subarray $\gtrsim$20% of the ngVLA collecting area (i.e., comparable to the current GBT or more), depending on the pulsar flux density, is used to observe a single pulsar. As an aside, we note that sub-arraying reduces the net computational load, since baselines are not correlated across sub-arrays. Thus, as long as the correlator resources can be flexibly re-deployed, no extra computation resources will be required for shared-user or multi-target phased sub-array operation.

Frequency and Bandwidth: As described above, frequency agility in the overall 1 to 10 GHz range is needed.

Observing Cadence: NANOGrav currently observes ~70 pulsars, with each pulsar in the array being observed (approximately) every three weeks. Beyond the detection of the stochastic GW background, the future PTA scientific program requires both an increasing number of pulsars in the array and an improved observing cadence.

A straw-man future observing program might involve ~200 pulsars distributed over the sky. If each pulsar can be timed with a 20% sub-array of the ngVLA for ~0.5 hr every week, the total NANOGrav observing program may require ~20 hr/week of the full array. However, such estimates require two important caveats. (1) The PTA observing program needs to be sustained for years—see, e.g., our recent data release spanning 11 years (Arzoumanian et al. 2018a). (2) As the PTA expands, future pulsar additions to the array may be disproportionately fainter, requiring larger integration times. Instead, it appears more likely that NANOGrav will rely on the ngVLA to time the most critical and faintest pulsars, with the remainder being timed at partner facilities.

Pulsar Searching: The PTA requirements for the ngVLA described here do *not* include the capability of blind, full field-of-view pulsar searches, which impose enormous computational loads. However, sensitivity to GWs increases with the number of pulsars, and requires a distribution of sources all over the sky for a good GW detector response function. It is likely that the discovery of new pulsars will be dominated by single dish telescopes and by hybrid methods, where multi-wavelength imaging is used to identify candidate compact sources that are then searched for pulsations. (A pulsar that is suitable for timing as part of a PTA will be easily identified as a compact source in radio sky survey images, or potentially via a high-energy counterpart.)

Complementarity with the SKA: In contrast to the ngVLA, pulsar searching will be part of the core scientific mission of the planned SKA, and it will detect new pulsars to be timed as part of the future PTA. However, the varied scientific demands on the SKA imply that it is unlikely to accommodate regular pulsar timing observations at the required cadence for an all-sky PTA with a large number of pulsars. The ngVLA program described here will provide unique Northern hemisphere coverage, as well as the high frequencies needed to optimally observe distant, high-DM pulsars, as shown in Figure 5. We expect to continue our current practice of cross-observing of the best few pulsars (e.g., PSR J1909−3744 is currently observed by Australian, European, and North American telescopes) in order to address systematics and build a robust, fully-integrated international timing program.

Multi-messenger Coverage: Elsewhere in this volume, chapters by Taylor and Simon (2018, "From Megaparsecs To Milliparsecs: Galaxy Evolution and Supermassive Black Holes with NANOGrav and the ngVLA," p. 611) and Burke-Spolaor et al. (2018, "Supermassive Black Hole Pairs and Binaries," p. 677) discuss the complementary capabilities of the ngVLA to not only contribute to the detection of GWs, but to observe the radio signatures of the selfsame targets. This capability potentially makes the ngVLA a comprehensive facility for multi-messenger studies of binary SMBHs by itself. In addition, the next generation of Extremely Large Telescopes and space-based optical/infrared and high energy surveys will reveal electromagnetic counterparts to the SMBHB population probed by PTA observations.

5. Technical Requirements for PTA Observations with the ngVLA

Given the more general programmatic issues described above, here we summarize the technical requirements for the ngVLA in order to enable PTA observations. We note that at present, none of these requirements represents a "tall pole" or driver for the ngVLA specifications.

Phased Sub-Arrays: Multiple independent sub-arrays are required, up to ~10, with each sub-array providing an independent phased array beam. It is preferable if each sub-array can maintain phase coherence over the duration of a pulsar timing observation, possibly using real-time self-calibration strategies with in-field calibrator sources.

Frequency coverage and Bandwidth: As described above and investigated further in Lam et al. (2017), wide-band frequency coverage is required and coverage down to 1–1.5 GHz is essential. We note that some pulsars with large and time-variable DMs may be suited to simultaneous dual-band observations using two independent sub-arrays, a capability of the ngVLA that is simply not available at any single-dish facility.

Correlator and Computation needs: Correlator specifications for pulsar observations are described in an ngVLA memo by Demorest et al., but in broad outline, phased array beams that can be sampled at 50 μs with 0.5 MHz channels are sufficient for PTA requirements. The output of the phased array beam will be coherently de-dispersed (i.e., a digital filter will be applied to remove the known average pulse dispersion as a function of frequency) in real time, before sampling. Given that cross-correlation for imaging will not be routinely required (and in any case, dishes observing different fields will never be correlated against each other), PTA observations using sub-arrays will be far less computationally challenging than full field-of-view, full-resolution imaging observations.

Polarization Calibration: Emission from millisecond pulsars is typically polarized, with linear polarizations at the few–50% level (e.g., Yan et al. 2011). The phased array beams will thus require polarization calibration (and more importantly, polarization stability) at the 5–10% level, preferably better.

Clock Stability: The long-term clock stability requirement for pulsar timing observations is currently met by tying observatory masers to GPS time. On the short term, clock stability requirements for pulsar timing are exceeded by the requirements for high-frequency VLBI.

Data Management and Curation: PTA observations are a long-term enterprise, and the scientific value of the data set increases with time baseline as wider ranges of GW frequencies are probed to higher sensitivities. NRAO has an exemplary track record of incorporating infrastructure for data management over decade-long timescales at the VLA, but pulsar observations (e.g., at the GBT) have typically not been included in these plans. For the ngVLA, the volume of PTA observations is expected to be dwarfed by full-field visibility storage requirements, and a long term archival and curation plan is both essential and straightforward.

6. Conclusions

PTAs have the potential to open a new window on the gravitational wave spectrum, probing nanohertz emission from the supermassive binary black hole mergers that accompany mass assembly in the universe, as well as other, more exotic, sources. Due to its exquisite sensitivity and northern hemisphere sky-coverage, the ngVLA will play a key role in PTA observations as long as certain key requirements are met: most importantly, independently phased sub-arrays and frequency coverage down to 1–1.5 GHz. None of the requirements or constraints pose a significant obstacle given the existing specifications of the ngVLA.

Acknowledgments. The NANOGrav Physics Frontiers Center is supported by the National Science Foundation award number 1430284.

References

Abbott, B. P., Abbott, R., Abbott, T. D., et al. 2016, Physical Review Letters, 116, 061102
—. 2017, Physical Review Letters, 119, 161101
Arzoumanian, Z., Brazier, A., Burke-Spolaor, S., et al. 2018a, ApJS, 235, 37
Arzoumanian, Z., Baker, P. T., Brazier, A., et al. 2018b, ApJ, 859, 47
Burke-Spolaor, S., Blecha, L., Bogdanović, T., et al. 2018, Science with a Next-Generation Very Large Array, ASP Monograph 7, ed. E. J. Murphy (San Francisco, CA: ASP), 677
Favata, M. 2009, ApJ, 696, L159
Hellings, R. W., & Downs, G. S. 1983, ApJ, 265, L39
Lam, M. T., McLaughlin, M. A., Cordes, J. M., Chatterjee, S., & Lazio, T. J. W. 2017, ApJ, submitted, arXiv:1710.02272
Mingarelli, C. M. F., Lazio, T. J. W., Sesana, A., et al. 2017, Nature Astronomy, 1, 886
NANOGrav Astrophysics Working Group, et al. 2018, in prep.
Rodriguez, C., Taylor, G. B., Zavala, R. T., et al. 2006, ApJ, 646, 49
Taylor, S. R. & Simon, J. 2018, Science with a Next-Generation Very Large Array, ASP Monograph 7, ed. E. J. Murphy (San Francisco, CA: ASP), 611
Yan, W. M., Manchester, R. N., van Straten, W., et al. 2011, MNRAS, 414, 2087

Science with a Next-Generation Very Large Array
ASP Conference Series, Monograph 7
Eric J. Murphy, ed.

Offset Active Galactic Nuclei

Laura Blecha,[1], Walter Brisken,[2] Sarah Burke-Spolaor,[3,4] Francesca Civano,[5] Julia Comerford,[6] Jeremy Darling,[6] T. Joseph W. Lazio,[7] and Thomas J. Maccarone[8]

[1]*University of Florida, Gainesville, FL, USA;* lblecha@ufl.edu

[2]*Long Baseline Observatory, Socorro, NM, USA;*wbrisken@lbo.us

[3]*Department of Physics and Astronomy, West Virginia University, P.O. Box 6315, Morgantown, WV 26506, USA*

[4]*Center for Gravitational Waves and Cosmology, West Virginia University, Chestnut Ridge Research Building, Morgantown, WV 26505, USA;* sarah.spolaor@mail.wvu.edu

[5] *Harvard Smithsonian Center for Astrophysics, Cambridge, MA, USA;* fcivano@cfa.harvard.edu

[6]*University of Colorado, Boulder, CO, USA;* julie.comerford@colorado.edu, jeremy.darling@colorado.edu

[7]*Jet Propulsion Laboratory, California Institute of Technology, Pasadena, CA USA;* Joseph.Lazio@jpl.nasa.gov

[8]*Texas Tech University, Lubbock, TX, USA;*thomas.maccarone@ttu.edu

Abstract. Gravitational-wave (GW) and gravitational slingshot recoil kicks, which are natural products of SMBH evolution in merging galaxies, can produce active galactic "nuclei" that are offset from the centers of their host galaxies. Detections of offset AGN would provide key constraints on SMBH binary mass and spin evolution and on GW event rates. Although numerous offset AGN candidates have been identified, none have been definitively confirmed. The ngVLA offers unparalleled capabilities to identify and confirm candidate offset AGN from sub-parsec to kiloparsec scales, opening a new avenue for multi-messenger studies in the dawn of low-frequency GW astronomy.

1. Offset AGN as Signposts for Supermassive Black Hole Mergers

High-angular-resolution, high-sensitivity radio observations with the next generation Very Large Array (ngVLA) can uniquely probe a poorly understood aspect of supermassive black hole (SMBH) evolution: how often are SMBHs offset from the nuclei of their host galaxies? Our nomenclature for accreting SMBHs – namely, active galactic nuclei, or AGN – reflects their nominal location, but SMBHs are not always centrally located. In ongoing galaxy mergers, for example, offset AGN are produced when one of the two SMBHs is active; many such objects have been found (e.g., Barrows et al. 2016). Here we focus instead on other sources of offset SMBHs: gravitational-wave

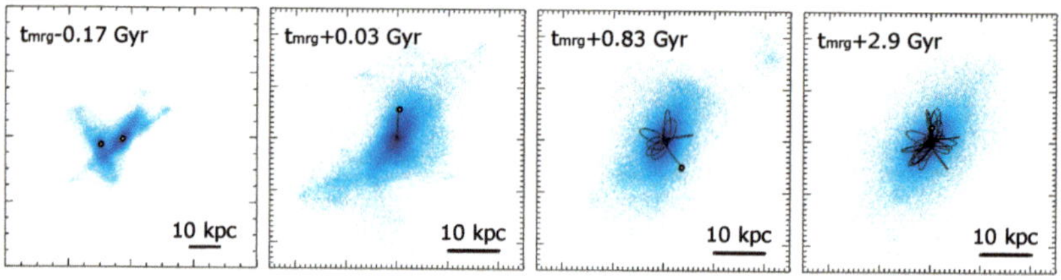

Figure 1. Illustrative example of merging galaxies and subsequent recoiling SMBH trajectory at four time snapshots, from a GADGET hydrodynamics simulation of an isolated galaxy merger from Blecha et al. (2011). The blue color scale shows the stellar density; black dots denote the SMBH position. The left panel shows the two galaxy nuclei shortly before coalescence; subsequent panels show the (empty) merged nucleus and the trajectory of the offset recoiling SMBH after being kicked at 80% of the central escape speed.

(GW) recoil and gravitational slingshot kicks, which can displace SMBHs from nuclei or even eject them from galaxies entirely. Such systems are more than a mere curiosity. Confirmed offset SMBHs can constrain the evolution of binary and merging SMBHs, including mass and spin evolution in the final stages of the merger. Ejected and offset SMBHs also affect BH-galaxy co-evolution and introduce scatter into BH-bulge correlations (Volonteri 2007; Blecha et al. 2011; Sijacki et al. 2011).

1.1. Recoiling Black Holes

Asymmetric gravitational wave (GW) emission during a BH merger can impart a kick of up to 5000 km s^{-1} to the BH merger remnant (Campanelli et al. 2007; Lousto et al. 2010), which can eject SMBHs from even galaxy clusters. Such extreme recoil kicks should be exceedingly rare, but kicks of even a few hundred km s^{-1} can cause SMBHs to wander for more than a Gyr and can produce detectable offsets (e.g., Gualandris & Merritt 2008, Blecha & Loeb 2008; Figure 1). GW recoil velocities depend sensitively on the BH spins just before merger—if the BHs are non-spinning or have perfectly aligned spins, the maximum possible kick is only 175 km s^{-1}. Few constraints on SMBH spins currently exist, but detections of offset, rapidly-recoiling SMBHs would provide strong evidence for misaligned, spinning progenitor SMBHs.

Alternately, if binary SMBH inspiral timescales are longer than the typical time between galaxy mergers, a subsequent galaxy merger may introduce a third BH into an existing binary system. If a close triple system forms, the three-body encounter will often eject the lightest SMBH, producing a gravitational slingshot kick and possibly driving the remnant binary SMBH to rapid merger (e.g., Hoffman & Loeb 2007; Bonetti et al. 2016, 2018). A key distinction between GW and slingshot recoil is that the latter leaves more than one BH remaining in the system after the kick.

1.2. Observable Signatures of Offset AGN

In either case, if the recoiling SMBH is actively accreting at the time of the kick, it will carry along its accretion disk, broad emission line (BL) region, and radio-emitting core (for most recoils, everything within $\sim 10^4$–10^5 gravitational radii will typically remain bound to the SMBH). The recoiling SMBH could then be observed as an "offset AGN" for up to tens of Myr (e.g., Madau & Quataert 2004; Loeb 2007; Blecha et al. 2011).

In cases where the SMBH is not ejected from the galaxy entirely, larger offsets are still favored because it spends most time near the apocenter of its orbit. Depending on the recoil trajectory and the type of observation, spatial and/or velocity offsets could be observed.

Unambiguous confirmation of recoiling AGN has proven tricky with existing observations due to a number of factors, including the large samples needed for a discovery, possible confusion with an inspiraling pre-merger AGN, and the limited resolution and sensitivity of infrared (IR), optical, and ultraviolet (UV) telescopes. The ngVLA will be able to resolve spatial offsets at sub-pc scales, such that the sensitivity to GW recoiling AGN will be limited primarily by the accuracy of relative astrometry with optical or IR imaging of the host galaxy centroid. At least when host nuclei are comprised mainly of older stellar populations, astrometric centroiding in the near-IR can be quite accurate. Condon et al. (2017) conducted a search for offset AGN with the VLBA, using 2MASS data to centroid the host bulge to within $\sim$ 100 mas. They found an offset AGN traveling away from a brightest cluster galaxy (BCG) core at $\gtrsim$ 2000 km s^{-1}, but they were able to identify a secondary compact galaxy associated with the offset AGN; this and surrounding tidal debris strongly indicate tidal stripping of a satellite galaxy by the BCG rather than a recoiling AGN.

In contrast, a slingshot recoil, in which a close binary SMBH (or single merged SMBH) remains behind as a dual or single radio source, could be resolved down to the limit of relative ngVLA astrometry ($\lesssim$ 1% of the beam FWHM for moderately bright sources). Moreover, the ngVLA offers the entirely new possibility of measuring the proper motions of rapidly-recoiling AGN in nearby systems. An ngVLA detection of a displaced AGN with a proper motion measurement would constitute the gold standard for a confirmed recoiling AGN.

The BL region remains gravitationally bound to the recoiling SMBH, while the narrow emission line region (on $\lesssim$ kpc scales) is left behind in the galaxy. Consequently, one observable signature of a recoiling SMBH would be a velocity offset between the broad and narrow lines. The relative rarity of large BL offsets in SDSS quasars (Eracleous et al. 2012) is in tension with the predicted number of observable recoiling AGN, *if* most progenitor SMBHs have randomly oriented spins (Blecha et al. 2016). This suggests that spin alignment occurs in some merging SMBH binaries. Assuming that gas-rich galaxy mergers form massive circumbinary disks that torque BH spins into alignment, but that gas-poor mergers have misaligned spins, the cosmological models of Blecha et al. (2016) predict that hundreds of spatially offset AGN may be detected at the sensitivity and resolution of large-area surveys with, e.g., LSST, WFIRST, or Euclid (see also Volonteri & Madau 2008). Figure 2 suggests that an ngVLA survey could detect a much higher space density of recoiling AGN with smaller spatial offsets. Moreover, these predictions for optical/near-IR surveys do not account for possible nuclear obscuration, which is prevalent in merger-triggered AGN hosts (e.g., Veilleux et al. 2009; Kocevski et al. 2015; Ricci et al. 2017; Blecha et al. 2018). While nuclear obscuration in the remnant galaxy will not affect recoiling AGN with large spatial offsets, the ngVLA will be uniquely sensitive to offset AGN on < pc-to-kpc scales that are still embedded in obscuring nuclear gas and dust.

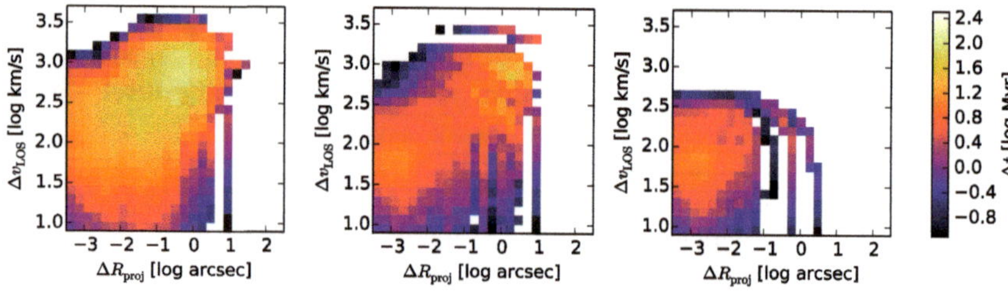

Figure 2. Time-weighted distributions of recoiling AGN offsets across cosmic time, based the Illustris cosmological simulations and the recoil model of Blecha et al. (2016). Only GW recoil is considered (not slingshot recoil), and only AGN that would be optically detected at the depth of *HST*-COSMOS are included. (*Left:*) Pre-merger spins are randomly oriented. (*Middle:*) A "hybrid" spin model is assumed in which BH spins are aligned only in gas-rich mergers. (*Right:*) Pre-merger BH spins are aligned to within 5°. The ngVLA will be sensitive to recoiling AGN with both small and large spatial offsets, and for highly aligned BH spins, GW recoils may be detectable only with long-baseline radio imaging. ngVLA detections of spatial offsets coupled with line-of-sight velocity offsets from optical spectra would provide strong evidence for recoiling BHs. (From Blecha et al. 2016.)

2. Current Searches for Offset AGN and Limitations

2.1. Spectroscopic and kpc-scale Spatial Offsets

A growing number of candidate recoiling AGN have been found over the last decade. Some have been identified spectroscopically, via offset BLs separated by $\gtrsim 1000\ \mathrm{km\,s^{-1}}$ from the narrow emission lines (e.g., Komossa et al. 2008; Robinson et al. 2010). Eracleous et al. (2012) have identified a sample of 88 SDSS quasars with large BL offsets; some of these could be interpreted as recoiling AGN candidates, though offset BLs can also arise from binary SMBHs, unusual double-peaked emitters, or gas outflows (e.g., Decarli et al. 2014; Runnoe et al. 2017).

Other recoil candidates have been identified via spatial offsets on kpc scales (Jonker et al. 2010; Koss et al. 2014; Markakis et al. 2015; Kalfountzou et al. 2017). If an offset AGN is well-separated from a galaxy and is not embedded in a secondary galaxy of its own, it cannot be explained by dual BH scenarios. However, if the host galaxy of an inspiraling, offset AGN has been significantly tidally stripped by a larger galaxy (as with some ultra-compact dwarfs; e.g., Seth et al. 2014, or in the cores of galaxy clusters; e.g., Condon et al. 2017), the offset host galaxy and tidal streams may be difficult to detect beyond the local Universe. For nearby offset AGN, high-resolution optical or near-IR imaging can place stringent constraints on the presence of an associated stellar nucleus component that is separate from the primary (central) galaxy (e.g., Koss et al. 2014). Note that the presence or absence of disturbed morphology in a galaxy does not itself distinguish between a recoiling SMBH and an inspiraling dual SMBH, as tidal features are typically present for hundreds of Myr before and after the SMBH merger.

It is also possible for unusual long-lived Type IIn supernovae to mimic offset AGN signatures, however; SDSS 1133 is a striking example of such a candidate (Koss et al. 2014). Further monitoring and high-resolution follow-up should be able to distinguish between a recoiling AGN and a Type IIn SN in the near future. More generally, time domain surveys such as Pan-STARRS and LSST can be used to identify offset point sources and distinguish those that are stochastically variable from transient SN flares (Kumar et al. 2015).

Arguably the most promising candidates are those with both spatial and velocity offset signatures (Civano et al. 2010, 2012; Chiaberge et al. 2017); CID-42 is one such example (Figure 3). Simulations indicate that this system can be well-modeled as a rapidly recoiling BH with a plausible offset AGN lifetime (Blecha et al. 2013). These combined lines of evidence present a compelling case for a rapidly recoiling BH in CID-42, but the data could also be explained as an inspiraling, kpc-scale BH pair in which one BH is quiescent or intrinsically faint.

In fact, none of the recoil candidates discovered to date has been confirmed. If the recoil candidate is superimposed on the host galaxy, the dual BH scenario is difficult to exclude, but high-quality resolved stellar or gas kinematics (with, e.g., ALMA, *JWST* or optical integral field spectroscopy) could distinguish between a pre- and post-merger nucleus. High-sensitivity ngVLA observations will provide strong constraints on AGN offsets and on the possible presence of a secondary, faint AGN in the host nucleus.

2.2. Parsec-scale Spatial Offsets

If a recoiling BH is not ejected entirely from its host galaxy, it will eventually return to the nucleus, possibly several Gyr after the recoil event, where it may undergo long-lived, small-amplitude oscillations about the galaxy center (e.g., Gualandris & Merritt 2008; Blecha & Loeb 2008). Batcheldor et al. (2010) and Lena et al. (2014) found significant optical photometric offsets between the AGN and galaxy nucleus on scales < 10 pc in six core elliptical galaxies (out of 14 observed, including M87), suggesting that such small-scale offsets are quite common in core ellipticals. However, in four of these galaxies, the offset is approximately aligned with the axis of coincident radio emission, which cannot be readily explained by a physical mechanism. And in fact, the measured offset in M87 has recently been attributed to a transient jet outburst; more recent data show no evidence of a spatial offset (López-Navas & Prieto 2018). In order to determine the true prevalence of small-scale AGN offsets, high-sensitivity observations ngVLA are needed to probe low-luminosity and radio-quiet AGN in galaxies with diverse morphologies.

3. Unique ngVLA Capabilities

The high resolution, high sensitivity, and broad frequency coverage of the ngVLA will enable a new class of observations to identify and follow up candidate recoiling AGN. In particular, the increase in resolution relative to the VLA along with increased sensitivity will allow more rapid detection of small-scale spatial offsets and small-scale extended features in a large number of objects, which will be critical for both systematic searches and for follow up of offset AGN candidates. The increase in sensitivity over the VLA and VLBA will allow the first large systematic radio search for offset AGN, and for known candidates, the ngVLA will place stringent limits on low-surface-brightness extended features and secondary sources. The broadband spectral coverage of the ngVLA combined with these capabilities will provide crucial data for confirming or eliminating candidate recoiling AGN in cases where their true nature cannot be discerned via other means. We describe several promising possible

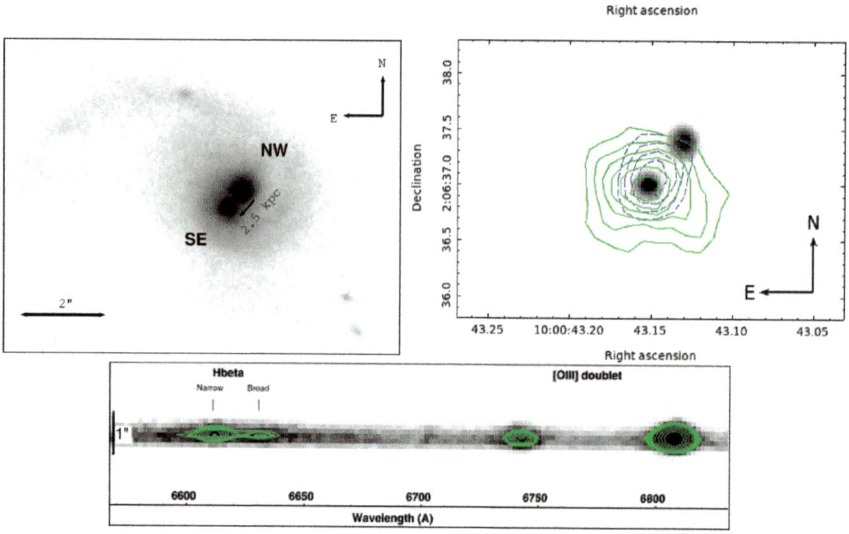

Figure 3. Observations of recoiling AGN candidate CID-42. (*Top Left:*) The F814W *HST* image of this source reveals two optical nuclei separated by 2.5 kpc and a prominent tidal tail. The NW (upper right) nucleus is marginally resolved; the SE (lower left) nucleus is a point source. (From Civano et al. 2010.) (*Top right:*) Combined image of the nucleus of CID-42 including *HST* optical (gray), Chandra X-ray (green), and VLA 3 GHz radio (blue). The X-ray and radio emission are consistent with a point source originating entirely from the SE source, which is the putative recoiling AGN. (From Civano et al. 2012 and Novak et al. 2015.) (*Bottom:*) Magellan/IMACS optical slit spectrum of CID-42, showing an offset of 1300 km s^{-1} between the broad and narrow lines. (From Civano et al. 2010.)

studies here:

Survey for spatially-offset AGN

The ngVLA will have the capability to carry out a systematic survey of thousands of galaxies with high sensitivity, which will enable a search for compact radio cores offset from their host nuclei on sub-pc to kpc scales. Such a survey would be well-aligned with the goals of a complementary search for ~ pc-scale and sub-pc-scale binary SMBHs (see the chapter on SMBH binaries by Burke-Spolaor et. al. 2018, p. 677 in this volume). High dynamic range observations with the ngVLA would allow either detections of dual compact radio cores with unequal luminosities or strong limits on the absence of a second radio core; the latter would indicate a possible GW recoiling SMBH. Furthermore, unlike optical instruments, the ngVLA will be able to detect AGN with small-scale (< pc - kpc) offsets that are embedded in heavily obscured nuclei, which are common in merger remnants. Because the accretion rate will decrease monotonically with time for an isolated, recoiling AGN, high sensitivity observations with the ngVLA can also detect offset AGN that have traveled to separations $\gtrsim$ 1–10 kpc from the host nucleus.

Based on the models of Blecha et al. (2016, Figure 2), we estimate that spatially offset, GW-recoiling AGN detectable with the ngVLA (with a SNR of 10 and spatial resolution of 10 mas) have a density of > 0.1 deg^{-2} out to $z \sim 0.5$. This of course assumes that the host centroid can be measured with sufficient accuracy, which would be difficult for small offsets at these redshifts. In the nearby universe ($z \lesssim 0.05$), we estimate that up to ~ 60-70 spatially offset GW-recoiling AGN could be detected with the ngVLA under the same assumptions. This also assumes that SMBH binary spins are not always highly aligned prior to merger.

If spins *are* always efficiently aligned to within a few degrees, GW recoil velocities are much lower ($\lesssim$ 300 km s^{-1}), and spatial offsets $\gtrsim$ 0.05" are very rare (Figure 2). Crucially, this means that long-baseline ngVLA observations are about the *only* means by which spatially offset, GW-recoiling AGN could be detected. If spins are always aligned to within 5°, then using the same assumptions as above we expect 1–2 detectable spatially offset AGN within $z < 0.05$, and > 0.01 deg^{-2} out to $z \sim 0.5$. The spatial offsets of any GW-recoiling AGN identified with the ngVLA would therefore constrain the pre-merger spins of SMBH binaries.

Time variability and spectral information can be used to determine whether an offset radio source is a true offset radio AGN core or simply a knot in a radio jet. Specifically, spectral index measurements can distinguish between a flat-spectrum core (indicating ongoing AGN activity) and a steep-spectrum jet (with an older electron population). The wide frequency coverage and high resolution of the ngVLA is ideal for this kind of measurement, and its broadband capabilities allow spectral indices to be measured within a single band.

Galaxies with a single apparent stellar nucleus that also have disturbed morphology will be promising targets to search for offset or dual radio cores. Core elliptical galaxies with shallow central density profiles are also promising targets, as recoil oscillations about the nucleus should be longer lived in these cases; however, multi-wavelength astrometry of the host nucleus is more difficult in such systems. Low-mass galaxies are of great interest as well, because of their potential to constrain the population of isolated and merging low-mass SMBHs that are most relevant for LISA. Although the galaxy merger rate is lower for low-mass galaxies, their lower escape speeds greatly increase the probability that a recoil kick will produce a measurable spatial offset.

Follow-up of known candidates

The ngVLA will provide an excellent means of following up known recoil candidates. In many cases, high dynamic range, high sensitivity radio observations will be crucial for distinguishing between a kpc-scale, inspiraling dual AGN (in which one AGN is intrinsically faint) and a single, offset recoiling AGN (cf. CID-42, Figure 3). Even when the offset source is well-separated from the galaxy, detection of a compact radio core coincident with an optical, IR, or X-ray point source would provide strong support for the recoil scenario, and would be crucial for confirming offsets that are marginally resolved at other wavelengths.

LSST, WFIRST, and Euclid have the potential to identify a large population of spatially offset AGN (Blecha et al. 2016). LSST will also be able to distinguish stochastically-variable offset AGN from transient offset sources (i.e., SNe; cf. Kumar et al. 2015). ngVLA variability studies will also be important for confirmation of recoiling AGN candidates identified in single- or multi-epoch data, and they can be used to rule out apparent small-scale spatial offsets produced by transient jet phenomena (cf. López-Navas & Prieto 2018).

In addition to the sample of known quasars with large BL offsets (Eracleous et al. 2012), more such candidates will be identifiable in newer spectroscopic surveys (e.g., SDSS eBOSS and DESI). ngVLA observations of these sources will be uniquely capable of resolving pc-scale binary SMBHs or detecting offset SMBHs on pc to kpc scales.

Proper motion measurements

Long baselines (i.e., $\gtrsim 1000$ km) afforded by the ngVLA Long Baseline Array will make the telescope uniquely capable of providing the most direct type of confirmation of a recoiling AGN: proper motion studies of recoil. Relative astrometric precision of $< 1\%$ of the beam FWHM will enable proper motions of $\sim 1\mu$as yr^{-1} to be detected out to $\sim$ 200 Mpc over a 5–10 yr time baseline for transverse recoil velocities $\gtrsim 1000$ km s^{-1}. We predict that up to $\sim$ 10 detectable recoiling AGN with such a high transverse velocity could be found within this volume (if spins are not always efficiently aligned prior to merger—no high-velocity recoils will occur in that case).

In practice, achieving this precision for single objects (i.e., GW recoiling AGN) requires accounting for other sources of proper motion that can be of similar magnitude, such as secular aberration drift, secular extragalactic parallax, transverse peculiar motion, and jet plasma motion. However, peculiar motion of the host galaxy will generally be much less than 1000 km s^{-1} except in cluster environments, and other sources of secular proper motion can be accounted for in cases where a nearby source is present. Repeated multi-band observations of promising candidates can also be used to correct for intrinsic motion of a jet. Moreover, in nearby galaxies with high star formation rates (tens of $M_\odot$ yr^{-1}, as may be expected in the final stages of a gas-rich major merger hosting a recoiling SMBH), we expect tens of HII regions to be luminous enough for $\lesssim 10\ \mu$as astrometric precision, which would greatly constrain the bulk proper motion of the galaxy and reduce uncertainty on the relative proper motion of the AGN.

Slingshot recoils present easier targets for proper motion studies. The single or binary SMBH remaining in the galaxy nucleus would allow precise relative astrometry with the full capabilities of the long-baseline ngVLA, such that high-velocity proper motions out to $\sim$ 200 Mpc could be more easily identified. The prevalence of slingshot recoils resulting from three-body SMBH interactions is not known, but theoretical studies suggest that they are not uncommon (Kelley et al. 2017; Kulkarni & Loeb 2012). Promising monitoring targets would be galaxies that exhibit morphological and kinematic signatures of recent mergers (such as tidal shells and streams) in which two radio AGN are at least marginally resolved. Because slingshot recoil requires that a second galaxy merger occurs before the SMBH binary from a previous merger has had time to coalesce, likely hosts should also include gas-poor, bulge-dominated galaxies in which SMBH binary inspiral is likely to stall, as well as galaxies in rich group environments.

Searches for LISA electromagnetic counterparts

A large fraction of merging SMBHs should be in gas-rich environments where they can be expected to accrete near the Eddington luminosity (e.g., Bode et al. 2010; Farris et al. 2014; Bowen et al. 2017). Because LISA will rarely give adequate time to look for massive merging SMBHs beforehand, an "after-the-fact" technique for identifying the EM counterparts is essential for finding the true host galaxies and thereby using these events for standard siren measurements of the expansion of the

Universe. Eddington-limited $10^7\,M_\odot$ SMBHs at $z \approx 1$ should have core radio flux densities of $\approx 10\,\mu$Jy, meaning that ~ 100 of them could be well-measured in 10 hours with the ngVLA. Because the SMBH merger product will be less massive that the sum of the two progenitor SMBHs, the accretion disk will temporarily fail to reach the innermost stable circular orbit (ISCO) of the new system and will therefore suddenly cease to produce a strong jet; accretion will restart on the much longer (many-year) viscous timescale. The amplitude of this variability should be far larger than typically seen from normal AGN activity. Monitoring could then be used to identify the object whose core turns off completely, starting at high frequencies and gradually progressing to lower frequencies. This is expected to happen on timescales of a few weeks at 100 GHz and a few years at 1 GHz. The other high-resolution radio-based technique for identifying the hosts of recent SMBH mergers is that, if the pre-merger BH spins were misaligned with one another, there should be two hot spots rather than one in the AGN lobe.

Other multi-wavelength and multi-messenger synergy

ngVLA observations will be a crucial component of multi-wavelength campaigns to identify and confirm candidate recoiling AGN. High-resolution optical or IR imaging of the host stellar light is needed to measure the position of the host centroid and to identify signatures of a recent merger (tidal tails, shells, etc.); this includes ground-based adaptive optics imaging with current instruments and future thirty-meter class telescopes, as well as with *HST* and *JWST*. Resolved optical, IR, or sub-mm spectroscopy can reveal the nuclear stellar and gas dynamics and better constrain the dynamical center and merger stage of the galaxy. In addition to radio spectral data, multi-wavelength imaging can also be used rule out the possibility of a knot in a radio jet mimicking an offset radio core. Optical spectra also provide BL velocities. X-ray imaging and spectra, particularly in hard X-ray bands, will complement radio observations of offset AGN and constrain the nuclear column density (and hence the possible presence of a second, obscured AGN).

Any ngVLA detections of recoiling (or binary) SMBHs in advance of LISA science observations can be used to constrain the LISA event rate. Moreover, even non-detections will constrain pre-merger spin evolution and thus the likelihood that LISA GW waveforms will be precessing. ngVLA and other multi-wavelength detections of offset AGN, combined with GW detections of SMBHs binaries and mergers with PTAs and LISA, have the potential to transform our understanding of SMBH evolution in merging galaxies.

The ngVLA studies outlined here offer unique capabilities for identifying and confirming offset AGN that cannot be achieved with other observations alone. These studies will constrain the mass and spin evolution of SMBHs in mergers as well as the GW source population. ngVLA searches for recoiling AGN have strong synergy with imaging and spectroscopic observations from sub-mm to X-ray wavelengths, and they are a natural complement to searches for binary SMBHs with the ngVLA and other instruments. The dynamical evolution of SMBHs in merging galaxies is a pressing open question in the dawn of low-frequency GW astronomy, and ngVLA studies of offset AGN will be a key component of the multi-messenger campaign to unravel this mystery.

Acknowledgments. LB acknowledges support by NSF award #1715413. JD acknowledges support from the NSF grant AST-1411605. Part of this research was carried out at the Jet Propulsion Laboratory, California Institute of Technology, under a contract with the National Aeronautics and Space Administration.

References

Barrows, R. S., Comerford, J. M., Greene, J. E., & Pooley, D. 2016, ApJ, 829, 37
Batcheldor, D., et. al. 2010, ApJ, 717, L6
Blecha, L., Civano, F., Elvis, M., & Loeb, A. 2013, MNRAS, 428, 1341
Blecha, L., Cox, T. J., Loeb, A., & Hernquist, L. 2011, MNRAS, 412, 2154
Blecha, L., & Loeb, A. 2008, MNRAS, 390, 1311
Blecha, L., et. al. 2016, MNRAS, 456, 961
Blecha, L., Snyder, G. F., Satyapal, S., & Ellison, S. L. 2018, MNRAS, 478, 3056
Bode, T., et. al. 2010, ApJ, 715, 1117
Bonetti, M., Haardt, F., Sesana, A., & Barausse, E. 2016, MNRAS, 461, 4419
— 2018, MNRAS, 477, 3910
Bowen, D. B., et. al. 2017, ApJ, 838, 42
Burke-Spolaor, S., Blecha, L., Bogdanović, T., et al. 2018, Science with a Next-Generation Very Large Array, ASP Monograph 7, ed. E. J. Murphy (San Francisco, CA: ASP), 677
Campanelli, M., et. al. 2007, Physical Review Letters, 98, 231102
Chiaberge, M., et. al. 2017, A&A, 600, A57
Civano, F., et. al. 2012, ApJ, 752, 49
Civano, F., et. al. 2010, ApJ, 717, 209
Condon, J. J., Darling, J., Kovalev, Y. Y., & Petrov, L. 2017, ApJ, 834, 184
Decarli, R., et. al. 2014, MNRAS, 445, 1558
Eracleous, M., Boroson, T. A., Halpern, J. P., & Liu, J. 2012, ApJS, 201, 23
Farris, B. D., Duffell, P., MacFadyen, A. I., & Haiman, Z. 2014, ApJ, 783, 134
Gualandris, A., & Merritt, D. 2008, ApJ, 678, 780
Hoffman, L., & Loeb, A. 2007, MNRAS, 377, 957
Jonker, P. G., et. al. 2010, MNRAS, 407, 645
Kalfountzou, E., Santos Lleo, M., & Trichas, M. 2017, ApJ, 851, L15
Kelley, L. Z., Blecha, L., & Hernquist, L. 2017, MNRAS, 464, 3131
Kocevski, D. D., et. al. 2015, ApJ, 814, 104
Komossa, S., Zhou, H., & Lu, H. 2008, ApJ, 678, L81
Koss, M., et. al. 2014, MNRAS, 445, 515
Kulkarni, G., & Loeb, A. 2012, MNRAS, 422, 1306
Kumar, S., et. al. 2015, ApJ, 802, 27
Lena, D., et. al. 2014, ApJ, 795, 146
Loeb, A. 2007, Physical Review Letters, 99, 041103
López-Navas, E., & Prieto, M. A. 2018, MNRAS, 480, 4099
Lousto, C. O., et. al. 2010, Classical and Quantum Gravity, 27, 114006
Madau, P., & Quataert, E. 2004, ApJ, 606, L17
Markakis, K., et. al. 2015, A&A, 580, A11
Novak, M., et. al. 2015, MNRAS, 447, 1282
Ricci, C., et. al. 2017, MNRAS, 468, 1273
Robinson, A., et. al. 2010, ApJ, 717, L122
Runnoe, J. C., et. al. 2017, MNRAS, 468, 1683
Seth, A. C., et. al. 2014, Nat, 513, 398
Sijacki, D., Springel, V., & Haehnelt, M. G. 2011, MNRAS, 414, 3656
Veilleux, S., et. al. 2009, ApJS, 182, 628
Volonteri, M. 2007, ApJ, 663, L5
Volonteri, M., & Madau, P. 2008, ApJ, 687, L57

Science with a Next-Generation Very Large Array
ASP Conference Series, Monograph 7
Eric J. Murphy, ed.

Serendipitous Fast Transient Science with the ngVLA

C. J. Law,[1] G. C. Bower,[2] S. Burke-Spolaor,[4,5] B. J. Butler,[3] P. Demorest,[3] T. J. W. Lazio,[6] and J. D. Linford[4,5]

[1] *UC Berkeley,* claw@astro.berkeley.edu

[2] *Academia Sinica Institute of Astronomy and Astrophysics, 645 N. A'ohoku Place, Hilo, HI 96720, USA*

[3] *National Radio Astronomy Observatory, Socorro, NM 87801, USA*

[4] *Department of Physics and Astronomy, West Virginia University, Morgantown, WV 26506, USA*

[5] *Center for Gravitational Waves and Cosmology, West Virginia University, Chestnut Ridge Research Building, Morgantown, WV 26505*

[6] *Jet Propulsion Laboratory, California Institute of Technology, Pasadena, CA 91109, USA*

Abstract. Fast radio transients are rich with physical information about the coherent emission mechanisms that generate them and the media they propagate through. We describe the science potential of fast transients at the ngVLA with a focus on the new class of extragalactic transient known as Fast Radio Bursts (FRBs). The rarity of the bursts makes them challenging targets for dedicated observing, but their short duration and high brightness opens the possibility of detection via real-time, commensal searches. We advocate for a correlator architecture that provides access to high-time resolution data products and allows for the potential addition of dedicated computing for real-time analysis. The novel frequency coverage of the ngVLA will test origin models and potentially find FRBs in novel environments.

1. Introduction

Time-domain astronomy was identified as a discovery frontier in the "New Worlds, New Horizons" Decadal Review of Astronomy & Astrophysics. Just as new detector technology opened new spectral domains in prior decades, new computational power is opening the time domain. A new generation of high-energy, optical, radio, and gravitational wave surveys are being conducted with a focus on sensitivity to time-domain phenomena like supernovae, gamma-ray bursts, and relativistic mergers (Gehrels et al. 2004; LSST Science Collaboration et al. 2009). The Very Large Array Sky Survey (VLASS) will survey the sky in three epochs in order to support transient science (Lacy et al. 2019, in preparation; Mooley et al. 2016).

The study of fast radio transients is important for a number of fundamental problems in astrophysics, including the discovery of new classes of object like the Fast Radio Burst (FRB, Lorimer et al. 2007), detecting the hidden baryonic matter in the intergalactic medium (McQuinn 2014), and the search for habitable exoplanets (Lazio et al. 2004). These systems are very rare, so large surveys are conducted for a chance to find a single source. However, even a single discovery can produce a great scientific payoff by offering a unique view of the physics at play.

Real-time, commensal data analysis creates discovery potential by searching an otherwise inaccessible data stream. At the VLA, we have developed an architecture for dedicated fast transients searches[1]. Commensal data analysis is creating multiplicative benefit to large projects like the VLASS and could do so for the ngVLA. Here we describe the science benefit from a set of technical capabilities at the ngVLA:

1. Fast-sampled visibilities — Fast transients are accessible via wide-field searches with millisecond-scale visibilities. In the case of the VLA, this was made possible by a flexible correlator design and networking support for recording data rates up to 1 TB hour^{-1}.

2. *In situ* analysis — An extensible correlator design would allow new computing resources to be dedicated to novel data analysis, such as image-domain transient searches. At the VLA, real-time computing reduces the recorded data rate by a factor of 10^3 and allows rapid response to new transient detections.

3. Commensal observing — *In situ* analysis makes it possible to distribute a high-speed copy of all VLA observations to the transient search system (Pokorny & Law 2017). This dramatically increases the amount of observing time that can be searched.

2. Fast Transients

Radio transients are commonly considered "fast" if they last less than about 1 second. This time scale crudely separates slowly evolving transients (due to synchrotron emission or extrinsic variability) from fast, coherent emission processes. The best known class of fast radio transient is the pulsar, a rotating neutron star that is detected through brief intense pulses of light (Hewish et al. 1968; Hulse & Taylor 1975). On fast timescales, radio light propagation effects, such as dispersion, can be measured. Propagation effects have been used to infer the distances and sizes of transients (Bower et al. 2014) and model Galactic structure (Cordes & Lazio 2002).

Many other astronomical objects also are detectable as fast radio transients. Active stars (Bastian et al. 1998; Osten & Bastian 2006), substellar objects (Hallinan et al. 2008; Kao et al. 2018), and (potentially) exoplanets (Bastian et al. 2018) have strong magnetic fields and coherent emission that can manifest on fast timescales[2]. A variety of coherent emission mechanisms are responsible and the radio signals are rich with physical information (e.g., plasma density, magnetic field strength).

FRBs are a new class of fast radio transient that originate at cosmological distances (Tendulkar et al. 2017). The energies required to power the bursts are extreme (10^{40} erg in radio alone; Katz 2017; Law et al. 2017), so they likely trace exotic systems that are rare in the local Universe (e.g., millisecond magnetars; Kashiyama & Murase 2017;

[1] The *realfast* project, `http://realfast.io`.

[2] See also `http://sites.nationalacademies.org/SSB/CurrentProjects/SSB_180659`.

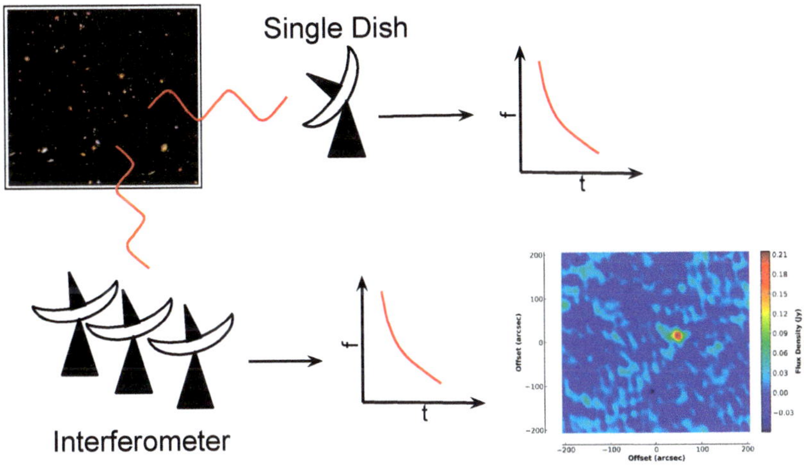

Figure 1. Single dish telescopes and interferometers can both detect transients in fields of order 10′. However, only interferometers can localize sources to arcsecond precision and find multiwavelength counterparts.

Metzger et al. 2017). Single-dish telescopes have discovered dozens of FRBs[3], but their localization precision (~arcminutes) is too poor to find unique multiwavelength associations such as host galaxies (Figure 1).

Interferometers are emerging as the best platform to pursue FRB science through localization (Johnston et al. 2008; Bandura et al. 2014; Bailes et al. 2017; Law et al. 2018). Arcsecond-scale localizations are needed to associate FRBs to their host galaxies and measure distances. In the future, a key goal is to detect, localize, and characterize hundreds of FRBs to use them to constrain progenitor models (Loeb et al. 2014; Pen & Connor 2015), host environments (Piro 2016; Cordes et al. 2017), properties of the intergalactic medium (McQuinn 2014; Akahori et al. 2016), and potentially even cosmological parameters (Deng & Zhang 2014; Yang & Zhang 2016).

Here, we focus on motivation for FRB discovery with interferometers, which is computationally demanding and can be easily extended to fast transient discovery in general.

3. Connection to Unique ngVLA Capabilities

Progress in understanding FRBs and their utility as probes is limited by the number of detections with well-characterized host galaxies. At present, only one FRB has a confident association with a host galaxy (Tendulkar et al. 2017) and it was found due to its repetition. The estimated volumetric rate of FRBs is comparable to that of core-

---

[3]See `http://frbcat.org` for the latest FRB list (Petroff et al. 2016).

collapse supernovae ($\sim 5 \times 10^{-5}$ Mpc^{-3} yr^{-1}; Nicholl et al. 2017; Law et al. 2017), but this estimate is likely highly biased by uncertainties in beaming angle and repetition of bursts. The observed (projected) rate is better defined, and shows that an area-time product on the order of hundreds of hours-deg^2 is required to discover a single source blindly ($\sim$ 600 sky^{-1} day^{-1} above 1 Jy; Lawrence et al. 2017). However, a commensal, real-time transient search capability can find and localize FRBs instantaneously without dedicated observing campaigns.

The ngVLA will be highly effective at detecting and localizing FRBs if the correlator architecture is designed in the manner that we describe. The most distant FRBs ($z > 1$) will be the most useful as probes of galaxy formation and cosmology, but they will require more precise localization (sub-arcsecond) to unambiguously identify host galaxy counterparts (Eftekhari & Berger 2017). Localization to *milliarcsecond* precision has even been used to associate an FRB to an individual star-forming region *within* a host galaxy to study its immediate environment (Bassa et al. 2017). The cumulative FRB event rate scales with flux limit as a powerlaw ($R_{>S} \propto S^{\alpha}$) with index –1 to –1.5 (Champion et al. 2016; Lawrence et al. 2017), so the high sensitivity of the ngVLA will increase the rate of detection by an order of magnitude compared to the VLA.

The novel frequency coverage of the ngVLA will test origin models and potentially find FRBs in different environments. Some models predict that the youngest and most active FRBs only will be detectable at frequencies above a few GHz (Margalit et al. 2018), but no blind surveys have been conducted there. The best studied FRB (121102) has a peak brightness that is largely independent of radio frequency from 1 to 8 GHz (Gajjar et al. 2018), which makes it similar to active magnetars (Gelfand et al. 2017).

4. Experimental Layout

High time resolution visibilities are required to be sensitive to dispersed radio pulses. Cold plasma in the interstellar and intergalactic media induce a frequency-dependent arrival time of $\tau = 4.15\text{ms DM } (\nu_1^{-2} - \nu_2^{-2})$, with DM in pc cm^{-3} and $\nu_{1,2}$ in GHz. A typical cm-wavelength detection of an FRB must correct for dispersive delays of order 100 ms in order to maintain sensitivity to the ms-duration impulse. At the VLA, the *realfast* system dedisperses visibilities and thus can form images with ideal sensitivity for ms-duration transients (Law et al. 2018).

After generating fast-sampled visibilities the second step is to create wide-field images and search for transients. This search can be performed off-line as part of a dedicated observation. The search algorithm parallelizes well by splitting data in time, which allows *realfast* to distribute computing over a small cluster of CPUs or GPUs. We demonstrated this in a VLA search for the repeating FRB (Spitler et al. 2016) in which we recorded fast-sampled visibilities at a rate of 1 TB hour^{-1}, searched them, and localized the FRB at subarcsecond precision (Chatterjee et al. 2017; Marcote et al. 2017).

However, the high-rate visibility data need not be saved. As demonstrated by *realfast* at the VLA (see Figure 2), visibility data can be streamed from the correlator to a dedicated computing system (Pokorny & Law 2017). Real-time analysis with dedicated computing allows us to trigger the recording of data only when candidate transients are found. This has been found to reduce the data rate by two to three orders of magnitude under typical observing conditions.

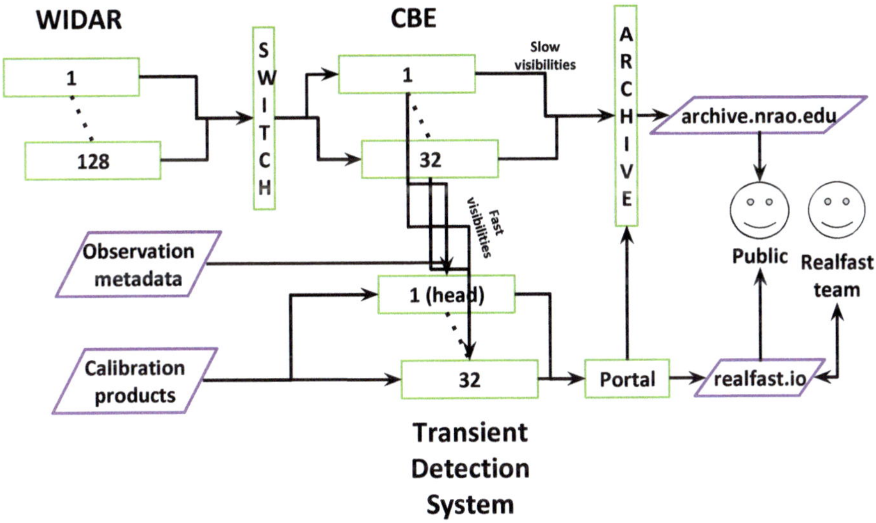

Figure 2. *Realfast* as an example of a commensal, fast transient search system. At the VLA, WIDAR and the CBE produce the raw data for both primary and commensal observers. The Transient Detection system runs the *realfast* search pipeline and produces products such as visibilities for candidate transients.

The *realfast* architecture can be thought of as a prototype for commensal fast transient searches at the ngVLA. In this sense, there is no "experimental layout" beyond support for fast-sampled visibilities, commensality, and dedicated computation. These features allow new science to be done with the same telescope. At the VLA, the *realfast* system is being treated as an enhancement to primary observing, with new, public data products made available by default.

5. Complementarity

The host galaxy for FRB 121102 was discovered to be a faint (R = 25 mag) dwarf galaxy at z = 0.2 (Tendulkar et al. 2017). To find host galaxies of the most distant FRBs, we will require much more sensitive optical/infrared observations that will only be available with next-generation telescopes (TMT/E-ELT/*JWST*).

FRB dispersion measures may have significant contributions from their host environments that compromises their utility as probes (Shao & Zhang 2017; Walters et al. 2018). Measuring properties of the intergalactic medium or constraining cosmological parameters will require detailed characterization of their host galaxy environments. The requirements on optical and infrared spectroscopy will push toward high spatial resolution observations on galaxies fainter than the host of FRB 121102 (Bassa et al. 2017).

A blind survey of fast transients and FRBs at the ngVLA will provide a complementary sample to surveys planned with SKA and its precursors. Observations at frequencies above $\sim$ 2 GHz will be less affected by free-free absorption and may also find younger, more active FRBs.

Acknowledgments. This work has been supported in countless ways by the the staff of the VLA. The National Radio Astronomy Observatory is a facility of the National Science Foundation operated under cooperative agreement by Associated Universities, Inc.. C.J.L. acknowledges support the National Science Foundation under grant 1611606. SBS and JDL gratefully acknowledge support by the National Science Foundation through grant 1714897. Part of this research was carried out at the Jet Propulsion Laboratory, California Institute of Technology, under a contract with the National Aeronautics and Space Administration.

References

Akahori, T., Ryu, D., & Gaensler, B. M. 2016, ApJ, 824, 105
Bailes, M., Jameson, A., Flynn, C., et al. 2017, Publications of the Astronomical Society of Australia, 34, e045
Bandura, K., Addison, G. E., Amiri, M., et al. 2014, in Ground-based and Airborne Telescopes V, Vol. 9145, 914522
Bassa, C. G., Tendulkar, S. P., Adams, E. A. K., et al. 2017, ApJ, 843, L8
Bastian, T. S., Benz, A. O., & Gary, D. E. 1998, Annual Review of Astronomy and Astrophysics, 36, 131
Bastian, T. S., Villadsen, J., Maps, A., Hallinan, G., & Beasley, A. J. 2018, ApJ, 857, 133
Bower, G. C., Deller, A., Demorest, P., et al. 2014, ApJ, 780, L2
Champion, D. J., Petroff, E., Kramer, M., et al. 2016, MNRAS, 460, L30
Chatterjee, S., Law, C. J., Wharton, R. S., et al. 2017, Nat, 541, 58
Cordes, J. M., & Lazio, T. J. W. 2002, ArXiv Astrophysics e-prints, arXiv:astro-ph/0207156
Cordes, J. M., Wasserman, I., Hessels, J. W. T., et al. 2017, ApJ, 842, 35
Deng, W., & Zhang, B. 2014, ApJ, 783, L35
Eftekhari, T., & Berger, E. 2017, ApJ, 849, 162
Gajjar, V., Siemion, A. P. V., Price, D. C., et al. 2018, ArXiv e-prints, arXiv:1804.04101
Gehrels, N., Chincarini, G., Giommi, P., et al. 2004, ApJ, 611, 1005
Gelfand, J. D., Ransom, S., Kouveliotou, C., et al. 2017, ApJ, 850, 53
Hallinan, G., Antonova, A., Doyle, J. G., et al. 2008, ApJ, 684, 644
Hewish, A., Bell, S. J., Pilkington, J. D. H., Scott, P. F., & Collins, R. A. 1968, Nat, 217, 709
Hulse, R. A., & Taylor, J. H. 1975, ApJ, 195, L51
Johnston, S., Taylor, R., Bailes, M., et al. 2008, Experimental Astronomy, 22, 151
Kao, M. M., Hallinan, G., Pineda, J. S., Stevenson, D., & Burgasser, A. 2018, The Astrophysical Journal Supplement Series, 237, 25
Kashiyama, K., & Murase, K. 2017, ArXiv e-prints, 1701.04815, arXiv:1701.04815
Katz, J. I. 2017, MNRAS, 467, L96
Lacy, M., Law, C., & Myers, S. in prep, ArXiv e-prints
Law, C. J., Abruzzo, M. W., Bassa, C. G., et al. 2017, ApJ, 850, 76
Law, C. J., Bower, G. C., Burke-Spolaor, S., et al. 2018, The Astrophysical Journal Supplement Series, 236, 8
Lawrence, E., Vander Wiel, S., Law, C., Burke Spolaor, S., & Bower, G. C. 2017, AJ, 154, 117
Lazio, W., T. J., Farrell, W. M., Dietrick, J., et al. 2004, ApJ, 612, 511
Loeb, A., Shvartzvald, Y., & Maoz, D. 2014, MNRAS, 439, L46

Lorimer, D. R., Bailes, M., McLaughlin, M. A., Narkevic, D. J., & Crawford, F. 2007, Science, 318, 777
LSST Science Collaboration, Abell, P. A., Allison, J., et al. 2009, ArXiv e-prints, arXiv:0912.0201
Marcote, B., Paragi, Z., Hessels, J. W. T., et al. 2017, ApJ, 834, L8
Margalit, B., Metzger, B. D., Berger, E., et al. 2018, ArXiv e-prints, arXiv:1806.05690
McQuinn, M. 2014, ApJ, 780, L33
Metzger, B. D., Berger, E., & Margalit, B. 2017, ApJ, 841, 14
Mooley, K. P., Hallinan, G., Bourke, S., et al. 2016, ApJ, 818, 105
Nicholl, M., Williams, P. K. G., Berger, E., et al. 2017, ApJ, 843, 84
Osten, R. A., & Bastian, T. S. 2006, ApJ, 637, 1016
Pen, U.-L., & Connor, L. 2015, ArXiv e-prints, arXiv:1501.01341
Petroff, E., Barr, E. D., Jameson, A., et al. 2016, PASA, 33, e045
Piro, A. L. 2016, ApJ, 824, L32
Pokorny, M., & Law, C. J. 2017, vysmaw: Fast visibility stream muncher, Astrophysics Source Code Library, ascl:1710.001
Shao, L., & Zhang, B. 2017, Phys.Rev.D, 95, 123010
Spitler, L. G., Scholz, P., Hessels, J. W. T., et al. 2016, Nat, 531, 202
Tendulkar, S. P., Bassa, C. G., Cordes, J. M., et al. 2017, ApJ, 834, L7
Walters, A., Weltman, A., Gaensler, B. M., Ma, Y.-Z., & Witzemann, A. 2018, ApJ, 856, 65
Yang, Y.-P., & Zhang, B. 2016, ApJ, 830, L31

Science with a Next-Generation Very Large Array
ASP Conference Series, Monograph 7
Eric J. Murphy, ed.

Flares from Coalescing Black Holes in the Centimeter-Wavelength Transient Sky

Vikram Ravi

Cahill Center for Astronomy and Astrophysics MC 249-17, California Institute of Technology, Pasadena CA 91125, USA; vikram@caltech.edu

Abstract. Radio observations have resulted in some of the most fundamental and exciting discoveries in time-domain astronomy. Here I demonstrate the potential for the next generation Very Large Array (ngVLA) to localize prompt centimeter-wavelength counterparts to the coalescing binary supermassive black holes (SMBHs) to be detected by the LISA mission. The resulting redshift measurements of the hosts of LISA-detected coalescences will populate a Hubble diagram up to the epoch of the earliest galaxies. Analysis of the host environments and the electromagnetic coalescence signatures will provide crucial insight into the interactions between binary SMBHs and their environments, and the mechanisms of SMBH formation and growth

1. Introduction

Centimeter-wavelength observations of extragalactic cataclysms and outbursts have resulted in some of the most exciting discoveries in time-domain astronomy. Most recently, the VLA-discovered radio afterglow of the first binary neutron-star merger event to be detected in gravitational waves (GWs; Abbott et al. 2017) has proven decisive in determining the geometry, energetics and composition of the pre- and post-event ejecta (Alexander et al. 2017; Hallinan et al. 2017; Mooley et al. 2018). The VLA also provided the first interferometric localization of a Fast Radio Burst (FRB), conclusively establishing the distant-extragalactic nature of the phenomenon and identifying an FRB host galaxy (Chatterjee et al. 2017). Observations of evolving radio emission associated with the tidal-disruption and accretion of stars by supermassive black holes (tidal disruption events; TDEs) have revealed a diversity of outcomes, from sub-relativistic outflows (Alexander et al. 2016) to relativistic jets (e.g., Zauderer et al. 2011; Mattila et al. 2018). Radio monitoring of the afterglows of γ-ray bursts (GRBs; Frail et al. 2001) have enabled accurate calorimetry of the explosions, and the characterization of beaming in GRBs. In turn, the presence or absence of rapidly evolving radio emission has proven to be a crucial discriminant between stellar explosions that result in transient relativistic jets, and those that do not (e.g., Soderberg et al. 2006).

This paper outlines the means by which the ngVLA can discover the electromagnetic (EM) counterparts to binary supermassive black holes (SMBHs) caught in the act of coalescence, using their GW emission. The Laser Interferometer Space Antenna (LISA; Amaro-Seoane et al. 2017), and its potential companion missions (Luo et al. 2016; Hu & Wu 2017), will detect GWs from a few to a few hundred coalescing binary SMBHs each year. Following the success of the LISA Pathfinder mission (Armano et al. 2016), the European Space Agency selected a three-satellite LISA mission for its

L3 launch in 2034. From its beginning, LISA will be a bona fide observatory for GW astrophysics. For example, several known ultracompact Galactic white-dwarf binaries will appear like radio-frequency birdies in the $10^{-4} - 0.1$ Hz LISA band (Stroeer & Vecchio 2006), and the unresolved background of such binaries will form the dominant broadband LISA "noise" source (Farmer & Phinney 2003). However, LISA detections of inspiralling and coalescing binary SMBHs will probe the wholly unknown formation and growth mechanisms of SMBH seeds, and help unravel the rich astrophysics that governs the fates of binary SMBHs in merging galaxies (for reviews, see Haiman 2013; Colpi 2014).

Much of the uncertainty in population synthesis models for the binary SMBHs to be detected by LISA reflects our lack of knowledge of the formation and evolution of these systems (Klein et al. 2016). For example, the two leading models for SMBH seeding [remnants of $O(10^2 M_\odot)$ Pop. III stars, or $O(10^5 M_\odot)$ black holes formed through the gravitational collapse of primordial-gas disks] result in detection rates that vary by an order of magnitude for some LISA configurations. The characteristics of the $10^{-4} - 0.1$ Hz LISA sensitivity curve, combined with the nature of GW emission from binary systems, makes LISA most sensitive to the coalescences of binaries with masses $M_B \sim 10^5 - 10^8 M_\odot$ at redshifts up to $z \sim 10$. GWs from binary SMBHs will sweep through the LISA frequency band for days to years, and localization error regions of $\sim 1 - 10\,\mathrm{deg}^2$ are expected in the days to weeks prior to coalescence (Lang & Hughes 2008).

EM identifications of LISA-detected binaries are required to realize their scientific promise. GW-only detections of SMBH-SMBH coalescence events will supply component masses and redshifts with $O(10\%)$ accuracies (Hughes 2002). However, redshifts based on host-galaxy identifications will fully specify the parameters of the coalescing systems, and enable their use as alternative probes of cosmological expansion (Tamanini et al. 2016). The evolution of the gravitational waveforms of binary SMBHs exactly encode the (redshifted) mass of the system, which in turn provides GW luminosity estimates with no scatter. Thus, redshift measurements of LISA-detected binaries will lead to the assembly of a Hubble diagram with unprecedented accuracy up to $z \sim 10$. The characterization of a sample of coalescing-SMBH host galaxies will also provide crucial insight into the environments and mechanisms conducive to the formation and orbital decay of binary SMBHs (Colpi 2014). SMBH formation scenarios will be refined (Klein et al. 2016), in a complementary manner to other instruments such as *JWST* (Natarajan et al. 2017). The nature of the EM signature itself will further test models for interactions between the SMBHs and their environments, such as the formation and sustenance of accretion disks and relativistic jets (Schnittman 2011).

In this article, I focus on prompt EM signatures of coalescing binary SMBHs, which will enable contemporaneous multi-messenger studies of LISA detections. I do not consider the possibility of detecting EM signatures of binary-SMBH coalescences *independently* of GW observations, because the low event rate and faintness of the potential signatures would necessitate an impracticably large survey.[1] In §2, I describe predictions for relativistic-jet launching upon SMBH-SMBH coalescence, and the estimated radio counterparts. In §3, I assess the rate of background/interloper events, and conclude in §4 with expectations for ngVLA observations.

---

[1]Burke-Spolaor et al. (2018) describe the EM signatures of binary SMBHs prior to coalescence elsewhere in this volume (p. 677).

2. Relativistic Jets Launched Upon SMBH-SMBH Coalescence

Binary SMBHs will likely form in environments rich in dynamically cold gas (Kelley et al. 2017), and thus be embedded in accretion disks. Prior to coalescence, a binary SMBH will have caused its accretion disk to retreat to a radius where viscous torques in the disk balance the gravitational torques. Periodic accretion episodes will nonetheless result from gas crossing into the hollow center of the disk (Roedig et al. 2011), and the final violent accretion of this residual gas may cause a flare at $\sim$ 1 day before the merger (Chang et al. 2010). The near-field perturbations of space caused by the coalescing SMBH will partially dissipate into any gas present, resulting in a prompt (tens of minutes) thermal EM flare that can be comparable to the Eddington luminosity of the system (Krolik 2010). Rapidly time-variable emission may also be caused by the gravitational recoil of the post-coalescence SMBH perturbing the accretion disk (Anderson et al. 2010).

Several groups have simulated the evolution of fossil accretion disks around coalescing SMBHs. Palenzuela et al. (2010) adopted the force-free approximation for a magnetically dominated tenuous plasma binary-SMBH environment, and found that collimated Poynting-flux outflows were launched from each SMBH during the inspiral phase, with a luminosity of $10^{43}[M/(10^8 M_\odot)]^2(v/c)^2$ erg s^{-1} that peaked at coalescence. An updated assessment of the Poynting flux emanent from such a system by Moesta et al. (2012) showed that the dual jets were sub-dominant by a factor of $\sim$ 100 to a quadrupolar Poynting outflow. However, substantively different results were obtained using full GRMHD simulations of initially matter-dominated accretion flows (plasma $\beta = p_{gas}/p_{mag} = 40$) onto coalescing SMBHs (Giacomazzo et al. 2012; Kelly et al. 2017), which likely better represent reality (e.g., Noble et al. 2012). The capability of the latter simulations to trace magnetic flux-freezing, and resolve MHD turbulence, enabled the compression and amplification of the magnetic field in plasma accreting onto the SMBHs to be identified. Unlike simulations in the force-free approximation, a single collimated Poynting outflow along the angular-momentum vector of the binary was observed, with a few-hour peak in luminosity at the time of coalescence, prior to settling into a higher-luminosity single-SMBH outflow at $\tau_{final} \sim 10^6[M/(10^8 M_\odot)]$ s after coalescence (M is the initial SMBH mass, assuming an equal mass-ratio system). The magnetic field amplification resulted in a stronger prompt luminosity peak of $L_{prompt} = 5 \times 10^{47}[M/(10^8 M_\odot)]^2$ erg s^{-1}, and a final outflow luminosity of $L_{final} = 10^{48}[M/(10^8 M_\odot)]^2$ erg s^{-1}. These quantities are robust to variations in the initial magnetic-field energy density within a reasonable range, but scale with the initial matter density, ρ_0, as $\rho_0/(10^{-11}$ g cm$^{-3})$.

Collimated Poynting-flux outflows are the leading model for the launching of relativistic astrophysical jets (e.g., Spruit 2011). I interpret the results of Giacomazzo et al. (2012) and Kelly et al. (2017) as indicative of the genesis of relativistic jets upon SMBH-SMBH coalescences in plasma-rich environments. The main uncertainty in the total jet power is the typical matter density in the immediate surrounds of binaries; I assume the presence of a radiatively inefficient accretion flow onto the binary during its final inspiral, with the fiducial value of $\rho_0 = 10^{-11}$ g cm^{-3}. The mechanisms by which initially Poynting-dominated jets attain their mass loading, and hence their finite Lorentz factors $\Gamma_0 = (1 - \beta^2)^{-1/2}$, are unknown, as are the characteristic length scales on which this occurs (Sikora et al. 2005). However, based on VLBI observations of superluminal AGN jets, I adopt $\Gamma_0 = 10$ (Hovatta et al. 2009), and assume that particle entrainment occurs on sub-parsec scales (Pushkarev & Kovalev 2015). Assuming full efficiency in the conversion of Poynting flux to jetted matter/energy, I model the Poynting-luminosity increase coincident with SMBH-SMBH coalescence as a $\tau_{\rm prompt} = 10^3[M/(10^8 M_\odot)]$ s transient jet with a total energy output of $E_{\rm prompt} = 5 \times 10^{50}[M/(10^8 M_\odot)]^3$ erg. The properties of the final jet ($\tau_{\rm final}$ and $L_{\rm final}$) are as above. The model is schematically illustrated in Figure 1.

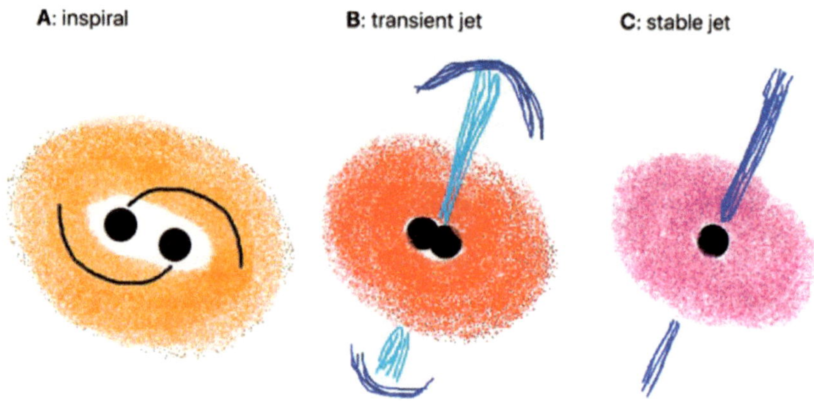

Figure 1. Cartoon illustration of the proposed EM signature of binary-SMBH coalescence events, to be targeted by the ngVLA. (*Panel A:*) LISA will detect and localize inspiralling binary SMBHs in the days to weeks prior to coalescence. (*Panel B:*) Upon coalescence, a prompt transient jet is launched, which drives a shock in the circum-nuclear medium. Radio synchrotron emission from this shock is detectable with the ngVLA. (*Panel C:*) The system eventually settles into a stable state, with a persistent jet that is also possibly radio-detectable.

Relativistic jets have several observational manifestations. In this paper, I predict their centimeter-wavelength radio signatures. The persistent twin jets observed in the binary-SMBH simulations of Palenzuela et al. (2010) led Kaplan et al. (2011) and O'Shaughnessy et al. (2011) to model the associated radio emission, assumed to be due to relativistic electrons advected with each jet, simply as a fixed fraction of the total jet luminosity. However, the transient nature of the jet model that I adopt here necessitates a different approach.

2.1. Predictions for Centimeter-Wavelength Transients

Radio emission due to relativistic jets and outflows is generated by electrons emitting synchrotron radiation. For on-axis AGN (e.g., blazars, BL Lacs), a correlation exists between the bolometric luminosity $L_{\rm jet}$ (for which the γ-ray luminosity is a rough proxy) and the radio-synchrotron luminosity $L_{\rm rad} \sim \nu L_\nu$, where ν is the radio frequency and L_ν is the radio spectral luminosity at its approximate peak (Ghirlanda et al. 2011). The bolometric luminosity can in turn be estimated using the total jet kinetic power $P_{\rm jet}$ (Nemmen et al. 2012). I use these empirical results to estimate the peak radio flux density of the persistent final jet associated with a binary-SMBH coalescence event as

$$F_{\rm final} \approx 15a^3[M/(10^8 M_\odot)]^2[d_L/(45\,{\rm Gpc})]^{-2}\,{\rm mJy}, \quad (1)$$

where $a = (1 - \beta)/(1 - \beta\cos\theta)$ [$\beta = (1 - \Gamma^{-2})^{-1/2}$] is a factor that accounts for an off-axis observer orientation at an angle θ, d_L is the luminosity distance, and I assume $\nu_p = 10\,\mathrm{GHz}$ for the rest-frame spectral peak.[2] Equation 1, and the assumption of a constant-Γ jet, implies that the final jet is only important for small viewing angles θ. For example, $\theta = 60°$ implies $a = 0.01$, reducing the fiducial F_{final} to 15 nJy. Note that the GW strain amplitudes emitted by face-on binaries are a factor of four larger than those emitted by edge-on binaries, and that small values of θ are therefore more likely to be observed (Wahlquist 1987).

The $L_{\mathrm{jet}} - L_{\mathrm{rad}}$ correlation, which is stronger when only variable emission is considered (albeit with a time-lag), is interpreted as all the EM emission originating from particles accelerated in internal shocks within jets. In the case of newly launched transient jets such as in γ-ray bursts, particle acceleration also occurs in the external shock at the interface between the jet and the circum-nuclear medium (CNM; Piran 2004). The initial transient jet will drive a relativistic shock through the CNM, which becomes Newtonian once the kinetic energy of the swept-up CNM is equivalent to that of the jet. The final jet will in turn power a shock within the initial jet, with the radio luminosity estimated above, that will ultimately further accelerate the forward CNM shock.

I model the radio emission associated with the external CNM shock using the semi-analytic calculations of Leventis et al. (2012) implemented in their *Spherefit* code. Although these results were derived for a spherical outflow, they are applicable to an on-axis observer of a relativistic jet. To evaluate the off-axis emission, I apply straightforward relativistic corrections relevant to a point-mass in linear motion. The shock Lorentz factor scales approximately as $\Gamma_0(t/\tau_{\mathrm{prompt}})^{-(3-k)/2}$ in the initial relativistic phase, where k is the index of a power-law describing the radial variation in the CNM density: $\rho(r) \propto r^{-k}$. For an off-axis observer, the time is given by t/a, the observed frequency is given by νa, and the observed flux-density is given by $a^3 F_\nu$. I neglect any contributions to the observed radio emission from a reverse shock propagating backwards through the transient jet. Observations of stellar tidal disruption events (e.g., Alexander et al. 2016) motivate a fiducial CNM number-density profile of $\rho(r) = 10[r/(10^{17}\,\mathrm{cm})]^{-1.5}\,\mathrm{cm}^{-3}$. Finally, I make the standard assumption for the accelerated electron power-law energy spectral index of $p = 3$. Predicted lightcurves for the external-shock emission are shown in Figure 2.

Emission in EM bands besides the radio is unlikely to be of significant importance for the jet model considered here, unless the jet is viewed on-axis. Synchrotron emission from the CNM shock will be most readily detected in the radio band (Sari et al. 1998). Relativistic jets powered by SMBHs are also most easily detected in the radio band at high redshifts (e.g., Miley & De Breuck 2008). The wider landscape of (non-jetted) EM signatures of SMBH-SMBH coalescence is a topic of intense investigation (Schnittman 2011). Nonetheless, in analogy with TDEs such as that observed in the galaxy merger Arp 229 (Mattila et al. 2018), dust obscuration may significantly affect signatures in important wavelength ranges (optical, soft X-ray) besides the radio.

[2]Compact radio sources with spectral peaks in the tens of GHz range are thought to be the youngest instances of active galaxies (O'Dea 1998). This motivates our fiducial choice of $\nu_p = 10\,\mathrm{GHz}$ for the spectral peak of the persistent final jet.

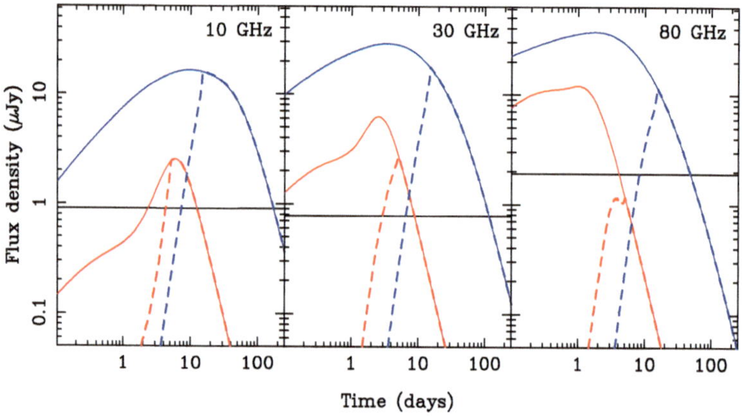

Figure 2. Predicted lightcurves at 10 GHz, 30 GHz and 80 GHz of emission corresponding to the external CSM shock due to the prompt transient jet predicted to occur during SMBH-SMBH coalescence events. The blue (higher-flux) curves correspond to the merger of two $10^8 M_\odot$ SMBHs at $z = 5$, and the red (lower-flux) curves correspond to the merger of two $10^7 M_\odot$ SMBHs at $z = 1$. The solid curves were calculated for on-axis observers, while the dashed curves were calculated for a viewing angle of 60° to the jet direction. The horizontal lines are the 6σ detection thresholds for 9 h ngVLA observations in each band (ngVLA Memo #5).

3. The Rate of Background Events

The ngVLA will need to survey GW localization regions of several deg^2 to detect radio counterparts to binary-SMBH coalescences. Radio emission associated with prompt transient jets will vary on timescales of days to weeks, while emission associated with persistent final jets will turn on over weeks to months. It is important therefore to understand the background rate of transient and variable radio sources with flux densities and variability timescales comparable to the objects of interest. Variability in the radio sky is not empirically characterized at these faint flux densities. Below, I consider in turn the contributions from cataclysmic events, scintillating compact AGN, and intrinsic variability in AGN.

Cataclysmic Events: The afterglows of core-collapse supernovae are the most common cataclysmic events in the radio sky (for a compilation, see Mooley et al. 2016). Assuming flux-density statistics consistent with a non-evolving population in Euclidean space, 6 deg^{-2} events are expected peaking over 10 μJy. However, besides the relativistic explosions that form $\sim$ 1% of the supernova population, the variability timescales of radio supernovae will be one to two orders of magnitude larger than the prompt binary-SMBH coalescence counterparts. Furthermore, the low radio luminosities of radio-supernova afterglows imply a nearby population even for the ngVLA, at redshifts $z \lesssim 0.3$. Approximately 0.01 – 0.1 deg^{-2} on-axis GRBs and jetted tidal disruption events are expected, which will more closely mimic the predicted prompt counterparts. Thus, at most $\sim$ 1 relativistic explosion is expected to form an interloper in ngVLA observations with 10 μJy sensitivity of any given 10 deg^2 binary-SMBH coalescence localization region. The host galaxies and redshifts of each interloper radio event will need to be identified, and compared with coarse redshift information from

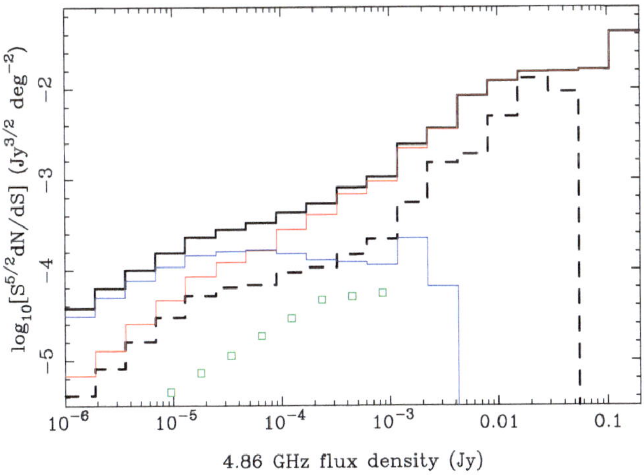

Figure 3. Simulated radio-source differential counts at 4.86 GHz from the SKA Simulated Skies project (Wilman et al. 2008, see http://s-cubed.physics.ox.ac.uk). The thick black trace corresponds to all sources, and the thick dashed black trace indicates sources larger than 50 mas. The ngVLA will thus be able to refine transient/variable candidate lists by factors of $\gtrsim 4$ in a single epoch by excluding extended sources. The green squares correspond to a simulation of the maximum possible counts of sources with angular extents $< 5\,\mu$as, which are likely to refractively scintillate at moderate to high Galactic latitudes (Ravi et al. 2019, in preparation). Note that the most common cataclysmic-event interloper, core-collapse supernovae, will have $S^{5/2}dN/dS \approx 2 \times 10^{-7}$ Jy$^{3/2}$ deg^{-2} (Mooley et al. 2016).

the GW detection, to isolate the true binary-SMBH counterpart. Accurate radio localization to $\sim$ 10 mas (to match *Gaia*-based optical astrometric accuracy) may also enable off-nuclear sources to be discarded.

Scintillating AGN: Spatio-temporal density variations in the Milky Way ionized-ISM cause refractive scintillations of compact extragalactic radio sources. Maximum modulations (modulation indices of order unity) are observed for sources smaller than a few μas, at frequencies between $1-10$ GHz off the Galactic plane (Walker 1998), on timescales of several hours. Although compact sources (e.g., observable with VLBI) are a subdominant population below flux densities of 1 mJy (Middelberg et al. 2013), any optically thick synchrotron source fainter than 1 mJy at a few GHz will compact enough to scintillate. Simulations (Ravi et al. 2018, in preparation) suggest that up to 500 deg^{-2} scintillating sources $> 10\,\mu$Jy will be observable at 10 GHz. These sources can be distinguished from the prompt counterparts of binary-SMBH coalescences by (a) higher-frequency observations, where modulation indices will typically be lower, (b) wide-band monitoring to identify unusual instantaneous spectral shapes caused by scintillation (e.g., Bannister et al. 2016; Alexander et al. 2017), and (c) longer-term monitoring to identify re-brightening episodes.

Variable AGN: Intrinsic variability in compact AGN may be a more insidious interloper class. Mooley et al. (2016) suggest that $\sim$ 1 deg^{-2} AGN is expected to vary on timescales of days to years above 0.3 mJy, possibly due to the propagation of shocks internal to jets (Marscher & Gear 1985). This implies the existence of a large interloper population at the tens of μJy level, given the still-increasing AGN source counts in this flux-density range (e.g., Condon et al. 2012, and Figure 3 here). Having discarded all interlopers that are not temporally coincident with the GW events, and those that are outside the GW-derived redshift bounds, the only way to distinguish between an internal shock within a pre-existing jet and a newly formed jet shocking an external CNM is by careful modeling of the evolution of the radio source. Such modeling could, for example, distinguish between a shock propagating through a dense, radially varying CNM, and a sparse jet.

4. Localizing LISA Events with the ngVLA

Radio observations will continue to be of great value in classifying and characterizing enigmatic classes of extragalactic transients in the ngVLA era. Dedicated ngVLA observations of the few to a few hundred coalescing binary SMBHs to be detected annually by LISA are required for the discovery of their EM counterparts. The localization of the EM counterparts will result in their host galaxies and redshifts being identified, and will unlock their rich astrophysical and cosmological potential. LISA will provide localization regions of $1-10\,\mathrm{deg}^2$ in the days to weeks prior to coalescence, and redshift measurements accurate to $O(10\%)$. GRMHD simulations of inspiralling SMBHs embedded in realistic accretion flows predict a prompt jet with energy $E_{\mathrm{prompt}} = 5\times10^{50}[M/(10^8 M_\odot)]^3$ erg lasting a few hours upon coalescence. The systems will then, on timescales $\tau_{\mathrm{final}} \sim 10^6[M/(10^8 M_\odot)]$ s, launch stable jets with lumonisities $L_{\mathrm{final}} = 10^{48}[M/(10^8 M_\odot)]^2\,\mathrm{erg\,s^{-1}}$. Examples of the radio lightcurves expected due to the prompt transient jets shocking the CNM are shown in Figure 2, and the final jet radio flux density is estimated in Equation 1. For example, for a fiducial off-axis observer orientation, coalescences of $10^8 M_\odot$ SMBHs at $z = 5$ will produce prompt jets with radio flux densities peaking above $10\,\mu$Jy at 10, 30, and 80 GHz, within $\sim$ 10 days of coalescence. The final jets will generally be observable only in more favorable orientations, depending on their Lorentz factors.

From an observational point of view, I have provided specific motivation for the use of the ngVLA to search for individual faint transient/variable sources in up to few-deg^2 regions. Some general technical considerations for this task are discussed below.

Point-Source Sensitivity: The mooted performance of the ngVLA at centimeter wavelengths, in particular above the 14 GHz limit of the baseline design for SKA1-mid, is highly desirable. Although lower frequencies (1–10 GHz) are typically used for time-domain work, synchrotron emission from expanding CNM shocks will have self-absorption spectral peaks that are brighter at higher frequencies at earlier times. Centimeter-wavelength observations therefore enable the proposed LISA-event counterparts to be detected sooner, and allow for the evolution of the self-absorbed spectrum to be monitored, thus characterizing the expansion velocity, total energy, and the CNM density profile.

Survey Speed: The ngVLA as defined in Memo #5 will require several pointings to survey few-deg^2 regions for transient/variable sources. For the LISA case, observations to depths of a few μJy will be required. The approximate, optimistic ngVLA survey speeds corresponding to 5 μJy rms continuum noise in the 2, 10, 30, and 80 GHz bands are 5.3, 0.9, 0.14, and 0.002 deg^2/hr. Dwell times of 20 – 130 s per pointing are required, which may motivate an on-the-fly mosaicking approach in some cases. However, it is evident that deep cadenced surveys of few-deg^2 regions will only be feasible with the ngVLA in the 2 GHz and 10 GHz bands, and possibly the 30 GHz band in exceptional cases. In cases where rms noise levels of $\gtrsim 10\,\mu$Jy are sufficient, the dwell times per pointing become small enough that sub-arraying to cover a larger frequency range becomes a possiblity. Otherwise, higher-frequency follow-up of individual sources of interest detected at lower frequencies will be a more practicable strategy.

Angular Resolution and Astrometric Accuracy: Wide-field imaging with the highest angular resolution and corresponding astrometric accuracy achievable with the ngVLA is desirable for this science case. The majority of sources near the detection threshold of a naturally weighted image with sub-100 mas angular resolution and few-μJy sensitivity will be star-forming galaxies and extended (e.g., Condon et al. 2012), and thus separable from the point-like sources of interest. This is demonstrated in Figure 3, where I compare simulated source counts of < 50 mas objects with the total radio-source population. Astrometric accuracy of $\lesssim 10$ mas will further allow for accurate radio-optical image registration in the post-*Gaia* era, enabling the rejection of some interloper radio-transient events.

Spectral Coverage: The characterization of any detected LISA-event counterpart over the full ngVLA band would be an ideal outcome. In particular, it is important to identify and monitor any continuum spectral peak. Wide-band data are also useful in rejecting interloper events. For example, scintillation of compact sources is mitigated at higher frequencies (typically $\gtrsim 10$ GHz), and can sometimes be identified by unusual spectral shapes. In-band spectral indices will also be useful in source identification.

Triggered, cadenced observing. Finally, the ngVLA science case presented here will require support for triggered, cadenced observing. Procedures for such observing modes have been honed on the VLA. However, triggered observations have been more difficult to implement with ALMA (Alexander et al. 2017). It is important that policies enabling fast-turnaround proposals and rapidly scheduled observations are included in ngVLA operations planning. Further, tools that enable rapid data reduction by the user community will also be necessary.

Acknowledgments. I thank S. Phinney, S. Kulkarni, G. Hallinan, J. Lazio, and S. Burke-Spolaor for useful discussions.

References

Abbott, B. P., Abbott, R., Abbott, T. D., et al. 2017, PRL, 119, 161101
Alexander, K. D., Berger, E., Guillochon, J., Zauderer, B. A., & Williams, P. K. G. 2016, ApJL, 819, L25
Alexander, K. D., Berger, E., Fong, W., et al. 2017, ApJL, 848, L21
Alexander, K. D., Laskar, T., Berger, E., et al. 2017, ApJ, 848, 69
Alexander, K. D., Berger, E., Bower, G., et al. 2017, arXiv:1703.04692
Amaro-Seoane, P., Audley, H., Babak, S., et al. 2017, arXiv:1702.00786

Anderson, M., Lehner, L., Megevand, M., & Neilsen, D. 2010, Phys. Rev. D, 81, 044004
Armano, M., Audley, H., Auger, G., et al. 2016, Physical Review Letters, 116, 231101
Bannister, K. W., Stevens, J., Tuntsov, A. V., et al. 2016, Science, 351, 354
Burke-Spolaor, S., Blecha, L., Bogdanović et al. 2018, Science with a Next-Generation Very Large Array, ASP Monograph 7, ed. E. J. Murphy (San Francisco, CA: ASP), 677
Chang, P., Strubbe, L. E., Menou, K., & Quataert, E. 2010, MNRAS, 407, 2007
Chatterjee, S., Law, C. J., Wharton, R. S., et al. 2017, Nature, 541, 58
Colpi, M. 2014, SSR, 183, 189
Condon, J. J., Cotton, W. D., Fomalont, E. B., et al. 2012, ApJ, 758, 23
Farmer, A. J., & Phinney, E. S. 2003, MNRAS, 346, 1197
Frail, D. A., Kulkarni, S. R., Sari, R., et al. 2001, ApJL, 562, L55
Ghirlanda, G., Ghisellini, G., Tavecchio, F., Foschini, L., Bonnoli, G. 2011, MNRAS, 413, 852
Giacomazzo, B., Baker, J. G., Miller, M. C., Reynolds, C. S., & van Meter, J. R. 2012, ApJL, 752, L15
Haiman, Z. 2013, The First Galaxies, 396, 293
Hallinan, G., Corsi, A., Mooley, K. P., et al. 2017, Science, 358, 1579
Hovatta, T., Valtaoja, E., Tornikoski, M., & Lähteenmäki, A. 2009, A&A, 494, 527
Hu, Wen-Rui & Wu, Yue-Liang 2017, National Science Review, 4, 685
Hughes, S. A. 2002, MNRAS, 331, 805
Kaplan, D. L., O'Shaughnessy, R., Sesana, A., & Volonteri, M. 2011, ApJL, 734, L37
Kelley, L. Z., Blecha, L., & Hernquist, L. 2017, MNRAS, 464, 3131
Kelly, B. J., Baker, J. G., Etienne, Z. B., Giacomazzo, B., & Schnittman, J. 2017, arXiv:1710.02132
Klein, A., Barausse, E., Sesana, A., et al. 2016, Phys. Rev. D, 93, 024003
Krolik, J. H. 2010, ApJ, 709, 774
Lang, R. N., & Hughes, S. A. 2008, ApJ, 677, 1184-1200
Leventis, K., van Eerten, H. J., Meliani, Z., & Wijers, R. A. M. J. 2012, MNRAS, 427, 1329
Luo, J., Chen, L.-S., Duan, H.-Z., et al. 2016, Classical and Quantum Gravity, 33, 035010
Marscher, A. P., & Gear, W. K. 1985, ApJ, 298, 114
Mattila, S., Pérez-Torres, M., Efstathiou, A., et al. 2018, arXiv:1806.05717
Middelberg, E., Deller, A. T., Norris, R. P., et al. 2013, A&A, 551, A97
Miley, G., & De Breuck, C. 2008, A&ARv, 15, 67
Moesta, P., Alic, D., Rezzolla, L., Zanotti, O., & Palenzuela, C. 2012, ApJL, 749, L32
Mooley, K. P., Hallinan, G., Bourke, S., et al. 2016, ApJ, 818, 105
Mooley, K. P., Nakar, E., Hotokezaka, K., et al. 2018, Nature, 554, 207
Natarajan, P., Pacucci, F., Ferrara, A., et al. 2017, ApJ, 838, 117
Nemmen, R. S., Georganopoulos, M., Guiriec, S., et al. 2012, Science, 338, 1445
Noble, S. C., Mundim, B. C., Nakano, H., et al. 2012, ApJ, 755, 51
O'Dea, C. P. 1998, PASP, 110, 493
O'Shaughnessy, R., Kaplan, D. L., Sesana, A., & Kamble, A. 2011, ApJ, 743, 136
Palenzuela, C., Lehner, L., & Liebling, S. L. 2010, Science, 329, 927
Piran, T. 2004, Reviews of Modern Physics, 76, 114
Pushkarev, A. B., & Kovalev, Y. Y. 2015, MNRAS, 452, 4274
Roedig, C., Dotti, M., Sesana, A., Cuadra, J., & Colpi, M. 2011, MNRAS, 415, 3033
Sari, R., Piran, T., & Narayan, R. 1998, ApJL, 497, L17
Schnittman, J. D. 2011, Classical and Quantum Gravity, 28, 094021
Sikora, M., Begelman, M. C., Madejski, G. M., & Lasota, J.-P. 2005, ApJ, 625, 72
Soderberg, A. M., Kulkarni, S. R., Nakar, E., et al. 2006, Nature, 442, 1014
Spruit, H. C. 2011, American Institute of Physics Conference Series, 1381, 227
Stroeer, A., & Vecchio, A. 2006, Classical and Quantum Gravity, 23, S809
Tamanini, N., Caprini, C., Barausse, E., et al. 2016, JCAP, 4, 002
Wahlquist, H. 1987, General Relativity and Gravitation, 19, 1101
Walker, M. A. 1998, MNRAS, 294, 307
Wilman, R. J., Miller, L., Jarvis, M. J., et al. 2008, MNRAS, 388, 1335
Zauderer, B. A., Berger, E., Soderberg, A. M., et al. 2011, Nature, 476, 425

Part IX

Cosmology and Fundamental Physics

Science with a Next-Generation Very Large Array
ASP Conference Series, Monograph 7
Eric J. Murphy, ed.

Galactic Center Pulsars with the ngVLA

Geoffrey C. Bower,[1] Shami Chatterjee,[2] Jim Cordes,[2] Paul Demorest,[3] Julia S. Deneva,[4] Jason Dexter,[5] Michael Kramer,[6] Joseph Lazio,[7] Scott Ransom,[8] Lijing Shao,[6] Norbert Wex,[6] and Robert Wharton[6]

[1]*Academia Sinica Institute of Astronomy and Astrophysics, Hilo, HI 96720, USA;* gbower@asiaa.sinica.edu.tw

[2]*Department of Astronomy, Cornell University, Ithaca, NY 14853, USA;* shami@astro.cornell.edu, cordes@astro.cornell.edu

[3]*NRAO, Socorro, NM 87801, USA;* pdemores@nrao.edu

[4]*George Mason University, resident at the Naval Research Laboratory, Washington, DC 20575, USA;* iulia.deneva.ctr@nrl.navy.mil

[5]*Max Planck Institute for Extraterrestrial Physics, Garching, Germany;* jdexter@mpe.mpg.de

[6]*Max-Planck-Institut für Radioastronomie, Auf dem Hügel 69, D-53121 Bonn, Germany;* mkramer@mpifr-bonn.mpg.de, lshao@mpifr-bonn.mpg.de, wex@mpifr-bonn.mpg.de, wharton@mpifr-bonn.mpg.de

[7]*Jet Propulsion Laboratory, California Institute of Technology, CA 91109, USA* Joseph.Lazio@jpl.nasa.gov

[8]*NRAO, Charlottesville, VA 22903, USA;* sransom@nrao.edu

Abstract. Pulsars in the Galactic Center (GC) are important probes of General Relativity, star formation, stellar dynamics, stellar evolution, and the interstellar medium. Despite years of searching, only a handful of pulsars in the central $0\overset{\circ}{.}5$ are known. The high-frequency sensitivity of ngVLA will open a new window for discovery and characterization of pulsars in the GC. A pulsar in orbit around the GC black hole, Sgr A*, will provide an unprecedented probe of black hole physics and General Relativity.

1. Scientific Goals

Currently, only six pulsars in the central $30'$ of the Galaxy have been detected. The most spectacular of these, J1745−2900, is a transient magnetar located only 0.1 pc in projection from Sgr A*. This small number stands in sharp contrast to predictions for the number of pulsars based on the rapid star formation rate and high density of young massive stars that can serve as pulsar progenitors. Numerous searches at a wide range of wavelengths have been carried out.

Until recently, the standard explanation for the absence of pulsars was the presence of hyperstrong interstellar scattering that smeared pulses over a timescale of $> 10^2$ s at low frequencies. Higher frequency searches reduced scattering effects but suffered from lower sensitivity due to the typical steep spectrum of pulsars. The discovery of J1745−2900 in 2013, however, suggested that scattering effects from the hyperstrong screen cannot fully account for the absence of detected pulsars, leading to the "missing pulsar" problem. Possible solutions to the absence of observed pulsars include more complex scattering models, stellar population synthesis arguments, and mechanisms for the suppression of the pulsar emission mechanism.

The population of millisecond pulsars (MSPs), which are the most prized targets for dynamical studies, has not been probed at all by existing observations. Even with the much reduced scattering strength inferred by J1745−2900, nearly all MSPs would remain undetected by existing surveys through a combination of time smearing and low flux density. ngVLA Galactic Center (GC) pulsar surveys will probe both the slow pulsar and MSP populations.

The ngVLA will make dramatic changes to our understanding of the GC pulsar population and potentially lead to the discovery of a pulsar in a short-period bound orbit to Sgr A*. This will come primarily through the factor of 10 improvement in sensitivity at high frequencies that the ngVLA promises. Specific science goals are:

- **Searching for and Timing a Pulsar Bound to Sgr A*:** It has been demonstrated that pulsars orbiting Sgr A* will be superb probes for studying the properties of the central supermassive black hole. It is sufficient to find and time a normal, slowly rotating pulsar in a reasonable orbit, in order to measure the mass of Sgr A* with a precision of 1 $M_{\odot}$(!), to test the cosmic censorship conjecture to a precision of about 0.1% and to test the no-hair theorem to a precision of 1%. These tests are possible even with a rather modest timing precision of 100 μs due to the large mass of Sgr A* and the measurement of relativistic and classical spin-orbit coupling, including the detection of frame-dragging.

- **Searching for and Characterizing the GC Pulsar Population:** Finding GC pulsars will not only lead to unique studies of the General Relativistic description of black holes, but we also gain invaluable information about the GC region itself: the characteristic age distribution of the discovered pulsars will give insight into the star formation history; MSPs can be used as accelerometers to probe the local gravitational potential; the measured dispersion and scattering measures (and their variability) would allow us to probe the distribution, clumpiness and other properties of the central interstellar medium; this includes measurements of the central magnetic field using Faraday rotation. Proper motions of young pulsars can be used to point back to regions of recent star formation and/or supernova remnants. Broadly, we define the GC pulsar population to be that which falls within the Central Molecular Zone (diameter ~ 250 pc).

- **Characterizing the Scattering Environment:** Interstellar scattering prevents detection of pulsars at low frequencies. This scattering appears to be strongly variable as a function of position towards the GC. Observations of low frequency masers, extragalactic background sources, as well as known pulsars can characterize the scattering screen and provide insights into the window function for pulsar detection.

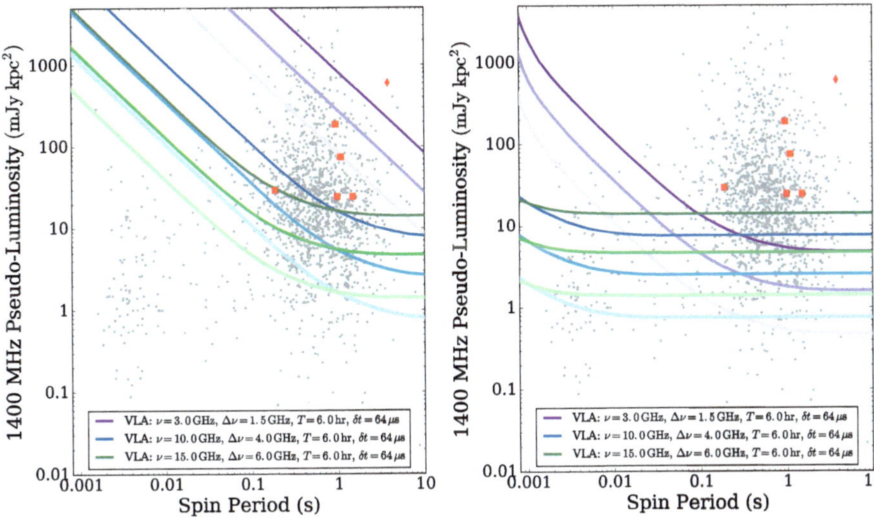

Figure 1. Field and GC pulsars (grey dots and red symbols, respectively) in period-pseudo-luminosity space for hyperstrong scattering (left) and magnetar-like scattering (right). The red diamond indicates the GC magnetar J1745−2900. Curves indicate sensitivity of the 1, 3, and 10× the VLA at three frequencies (darker to lighter curves). Sensitivity curves include nominal VLA and ngVLA system performance and scaling for average pulsar spectral index.

The proximity of the Galactic Center and the power of the ngVLA combine to provide an unprecedented and unique opportunity to study the environment of a galactic nucleus through pulsars.

2. Pulsar Searching in the GC

Numerous searches for GC pulsars have been carried out over a wide range of frequencies. Prior to 2013, these searches had led to the detection of only five pulsars in the central 0°.5 (e.g., Kramer et al. 2000; Johnston et al. 2006; Deneva et al. 2009; Macquart et al. 2010; Siemion et al. 2013; Eatough et al. 2013a). Theoretical expectations based on the high star formation rate and the density of high mass stars in the central molecular zone suggest that there should be hundreds or thousands of detectable pulsars within this region (Pfahl & Loeb 2004). In particular, the immediate environment of the Sgr A* contains a $10^5\ M_\odot$ cluster of massive stars with ages $\sim$ 4 Myr–8 Myr bound to the black hole, which are possible progenitors to neutron stars and pulsars (Blum et al. 2003; Pfuhl et al. 2011).

The strong interstellar scattering towards the GC may account for the absence of detected pulsars, especially at low radio frequencies. Turbulent electrons along the line of sight will produce angular broadening of sources in the image domain and temporal broadening of sources in the time domain. The temporal scattering scale for a thin scattering screen is related to the angular broadening by the following equation:

$$\tau_s = 6.3 \text{ s} \times \left(\frac{D}{8.5\text{kpc}}\right)\left(\frac{\theta_s}{1.3 \text{ arcsec}}\right)^2 \left(\frac{D}{\Delta} - 1\right)\nu^{-4}, \tag{1}$$

where $D = 8.3 \pm 0.3$ kpc (Genzel et al. 2010) is the distance to the GC, Δ is the distance from the GC to the scattering medium, and ν is the observing frequency in GHz (Cordes & Lazio 1997). The observed angular diameter θ_s is extrapolated to a frequency of 1 GHz using a scaling of ν^{-2}. For Kolmogorov turbulence with an inner scale $r_{in} > b_{max}$, we expect $\theta_s \propto \nu^{-2}$ and $\tau_s \propto \nu^{-4}$. Several pieces of evidence demonstrate that scattering may be important in the GC. Sgr A* itself is observed to show substantial angular broadening, as are OH masers and some extragalactic background sources. These pieces of evidence have been used to argue for a hyperstrong scattering screen that is located within the central few hundred parsecs of the GC (Lazio & Cordes 1998a). If the source to screen distance is small, an observed angular broadening implies a large temporal broadening. In the case of the hyperstrong model, $\tau_s \gtrsim 10^3$ s at 1 GHz, sufficient to make even slow pulsars undetectable up to ~ 10 GHz.

The discovery and characterization of the GC magnetar J1745−2900 has indicated that the hyperstrong scattering model cannot hold for all of the GC (Eatough et al. 2013b; Spitler et al. 2014; Bower et al. 2014). The GC magnetar, with $P = 3.76$ s is detected to frequencies as low as ~ 1 GHz, with a characteristic $\tau_s = 1.3$ s at 1 GHz. The GC magnetar also shows angular broadening identical to that of Sgr A*. Together, these results imply a source to screen distance of order 6 kpc, well outside of the GC, and the ability to readily detect pulsars in the GC at frequencies of a few GHz. It remains unclear whether the magnetar-like scattering screen describes all of the GC region or may be patchy. Characterization of the other GC pulsars show distances to the scattering screen $\lesssim 2$ kpc (Dexter et al. 2017). Other environmental effects may also alter the neutron star population in the GC (Dexter & O'Leary 2014).

In the case of both hyperstrong and magnetar-like scattering, there is still a substantial benefit in reduction of scattering that is achieved by going to higher frequencies. This benefit has to be offset against the steep spectral index of both slow pulsars and MSPs and decreasing sensitivity of many search telescopes with increasing frequency. The average slow pulsar has a spectral index of $\alpha \approx -1.4$ (Bates et al. 2013) although some objects including magnetars have flat spectra to very high frequencies (Torne et al. 2017).

Figure 1 shows the distribution of known field pulsars along with sensitivity curves for the VLA and the ngVLA at a range of frequencies and for hyperstrong and magnetar-like scattering. These curves reveal that for magnetar-like scattering the ngVLA will have sensitivity to discover the majority of field pulsars at the GC and a significant fraction of MSPs. In the more challenging case of hyperstrong scattering, the high frequency ngVLA is essential for obtaining sensitivity to slow pulsars.

3. Constraints on Black Hole Parameters

In order to estimate the impact of ngVLA on timing of a pulsar around Sgr A* and testing the no-hair theorem, we have extended the work by Wex & Kopeikin (1999) and Liu et al. (2012). We have set up a numerical integration scheme for orbits around Sgr A*, using MCMC parameter estimation code; we extract the parameter uncertainties and

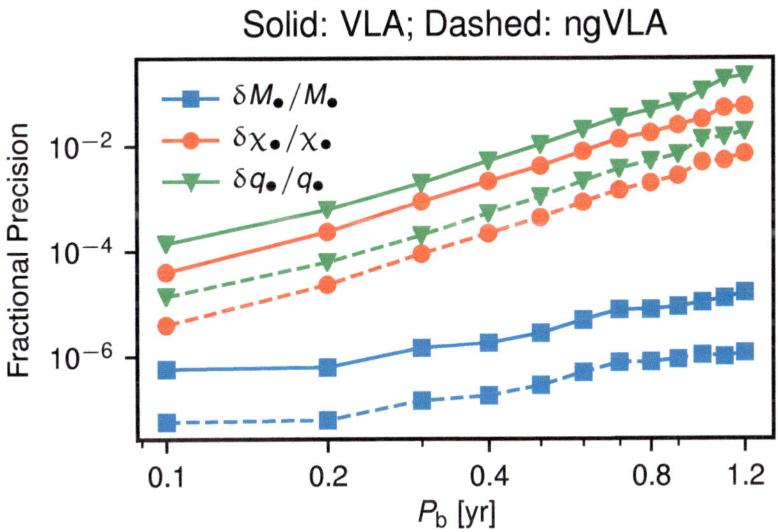

Figure 2. Constraints on black hole parameters from timing observations of a pulsar in a bound orbit for the VLA (solid lines) and for the ngVLA (dashed lines). Constraints are given for black hole mass ($M_\bullet$), spin ($\chi_\bullet$), and quadrupole ($q_\bullet$). These constraints were calculated for a $P = 0.5$ s pulsar with orbital eccentricity $e = 0.8$ and timing residuals of $\sigma_{\rm TOA} = 1$ ms and 0.1 ms for the VLA and ngVLA, respectively. Simulations include weekly observations over a 5 yr interval.

covariances. Figure 2 shows estimates of the uncertainty in black hole mass, spin, and quadrupole moment for a set of pulsar orbits timed with the VLA and the ngVLA. Results for a more complete set of orbits will be presented in a future publication (Shao et al. 2019, in preparation). Effectively, the precision on these parameters in this regime scales directly with increased collecting area.

The parameter constraints are governed primarily by the rms time of arrival (TOA) residuals, $\sigma_{\rm TOA}$. For given intrinsic pulsar emission properties and scattering properties, $\sigma_{\rm TOA}$ scales inversely with collecting area. Figure 3 shows estimates of the timing error for two scattering models (hyperstrong and magnetar) as a function of frequency for the VLA, ngVLA with 107 antennas, and ngVLA with 214 antennas.

General Relativistic effects are best measured in systems where the observational duration is greater than the orbital period, but constraints on General Relativity and BH properties can still be obtained for longer period systems. We consider the case of a pulsar with a 15 yr period observed over a 5 yr interval during pericenter. The BH mass can be constrained with ngVLA to an accuracy $\sim 10^{-5}$. Lense-Thirring precession parameters will be found with an signal-to-noise ratio between 10 and 1000, depending on the details of orbit and the observations. At large radii, perturbations to the pulsar orbit through stellar interactions will limit the accuracy with which General Relativity constraints can be made. Next generation adaptive optics on large optical telescopes imaging the stellar cluster may enable modeling of these perturbations.

Another limitation of current GC pulsar searches is the insensitivity to a large fraction of binary pulsars. The minimum detectable pulsar pseudo-luminosity (L =

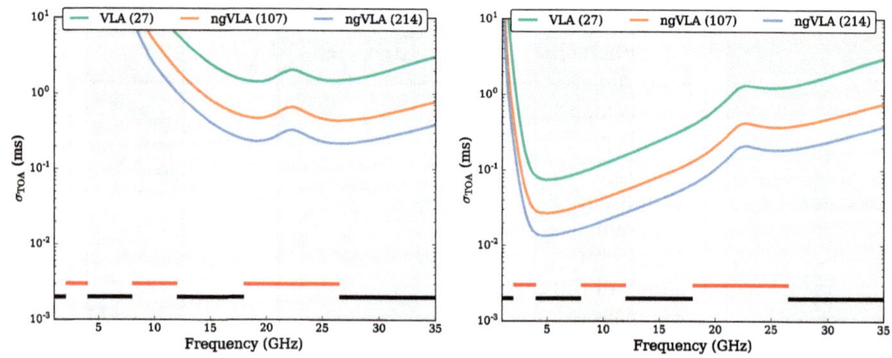

Figure 3. TOA residuals for a P = 0.5 s pulsar with a 10 ms pulsar width, using an integration time of 4 hr and a bandwidth of 1 GHz. The residuals are shown for hyperstrong scattering with τ = 2300 s at 1 GHz (*left*) and for a level of scattering comparable to the magnetar J1745−2900, with τ = 1.3 s at 1 GHz (*right*).

$S\,d^2$) for a pulsar survey scales roughly as

$$L_{\rm min} \propto d^{-2}G^{-1}T_{\rm obs}^{-1/2} \tag{2}$$

where d is distance, G is the telescope gain, and $T_{\rm obs}$ is the observing time. Because the GC is so far away, GC searches have typically observed for as long as possible ($\approx$ 6 hr at the VLA). While this increases the sensitivity to faint isolated pulsars, it can also reduce the effectiveness of typical binary pulsar search methods. For example, the acceleration search technique is only effective for orbital periods with $P_{\rm orb} \gtrsim 10T_{\rm obs}$ (Johnston & Kulkarni 1991; Ransom et al. 2003). A VLA GC pulsar search that observes for 6 hr would be most sensitive to binary pulsars with orbits longer than about 60 hr, which includes only about 50% of known pulsar binaries (and only 30% of binaries in globular clusters).

The increased sensitivity of the ngVLA means that flux density limits achieved by the VLA can be reached in much shorter observing times, thus allowing for the possibility of detecting much shorter period binary pulsars. For $(G_{\rm ng}/G_{\rm VLA}) = 5$, the ngVLA could reach the same flux density limit as a 6 hr VLA observation in only $T_{\rm ng} = T_{\rm VLA}/\sqrt{5} \approx 15$ min. In this shorter observing span, an acceleration search could be used to find pulsars with orbits as short as 2.5 hr, which would include about 95% of all known pulsar binaries.

Long baseline astrometry of the pulsar relative to Sgr A* with the ngVLA can also contribute to parameter constraints for long-period systems. For the case of 100 μas astrometric error (Bower et al. 2015), we can measure the orientation of the pulsar orbit in the sky, i.e., the longitude of the ascending node with sufficient precision to improve General Relativistic constraints.

4. Other Probes of Black Hole Physics and the Uniqueness of ngVLA

There are a number of experiments currently seeking to characterize Sgr A*, the properties of the SMBH, and General Relativity. These include the Event Horizon Telescope

(EHT) and its millimeter wavelength imaging of Sgr A* and large optical telescope campaigns to measure stellar orbits near Sgr A*. ngVLA timing of a pulsar in a bound orbit compares favorably against these techniques and provides important complementary information (Psaltis et al. 2016).

EHT constraints will primarily arise from modeling the static and time-domain images of the immediate region surrounding Sgr A* out to a radius of $\sim 10R_S$ (Broderick et al. 2014). General Relativity and BH properties introduce image distortions that can be translated into parameter constraints. These measurements, however, must be interpreted in light of the complex astrophysics of accretion, jet launching, and particle acceleration in the vicinity of the BH. Simulations have shown that EHT results will be able to measure the BH mass, spin, and quadrupole moment.

Stellar astrometry with existing and future large optical telescopes has the potential to measure orbital precession. Current constraints on the black hole mass from stellar orbits have an accuracy of order 10% (e.g., Genzel et al. 2010) and future measurements with existing facilities are projected to improve on these results substantially (Waisberg et al. 2018). The closest star, S2, has a period of 15 yr and is currently going through periastron passage. Improved instruments may discover fainter stars in shorter orbits leading to greater precision. Sub-milliarcsecond astrometry provides a length-scale precision that is orders of magnitude less precise than millisecond-accuracy pulsar timing, leading to complementary but correspondingly less accurate BH and General Relativity parameter constraints.

5. The Accretion Flow, ISM, and Stars

Pulsed emission provides a powerful probe of the plasma and magnetic field along the line of sight through a variety of measurements. The dispersion measure (DM) gives the total electron column density. The rotation measure (RM) gives the integrated line of sight magnetic field and electron density. The scattering measure (SM) characterizes turbulent plasma, which is typically localized to individual regions along the line of sight.

A short period pulsar will orbit inside of the Bondi radius ($\sim 10^5 R_S$; $P \sim 10^3$ yr) of Sgr A*. Polarimetry of Sgr A* and the magnetar J1745−2900 has shown that a substantial fraction of the observed Sgr A* RM originates from the accretion flow on scales between 10 R_S and 10^5 R_S, providing strong constraints on the accretion rate and the nature of the radiatively inefficient accretion flow (Bower et al. 2003; Marrone et al. 2007). A pulsar in an eccentric orbit would provide a measure of the radial profile of the accretion flow, inaccessible to any other technique.

On larger angular scales, a collection of GC pulsars would provide multiple probes of the interstellar medium (ISM) in the GC region (Schnitzeler et al. 2016). With a sufficient density of pulsars, the ngVLA can map out magnetized and ionized structures throughout the central molecular zone. Astrometry of these pulsars can provide origins of individual pulsars in known supernova remnants or young stellar clusters.

ngVLA pulsar studies will complement ngVLA and SKA studies of the GC ISM through other means. The scattering medium has been studied in the past through imaging of stellar masers and extragalactic background sources (van Langevelde & Diamond 1991; van Langevelde et al. 1992; Frail et al. 1994; Lazio & Cordes 1998b). Sensitivity and angular resolution constraints have limited observations, however, to a small number of sight lines. The ngVLA will provide the sensitivity and resolution to measure the angular sizes of a wide range of maser species and a large number of background sources and map out the spatial variations of the scattering towards the inner degree of the GC. This will provide important insight into both the source of scattering and characterization of the scattering that can lead to optimization of pulsar searches.

6. Implications for Array Design

A number of elements of the array design are constrained by our scientific goals.

- *Correlator and beamformer requirements:* Searches for pulsars will make use of beamforming capabilities as well as high-time resolution imaging with the correlator. Different regimes of the search space will be optimized through tradeoffs in field of view and bandwidth. Timing observations will rely primarily on wide-band receivers and beamformer backends.

- *Effect of the various receiver options on GC searches:* Pulsar searching requires high instantaneous sensitivity that cannot always be traded off against greater integration time. High frequency ($\sim$ 30 GHz) sensitivity is likely to be required to avoid the effects of scattering for detection of MSPs.

- *Effect of array configuration on GC searches:* Searching and imaging primarily require a compact configuration. Timing observations can be obtained with beamforming of extended configurations but these will introduce challenges in calibration and availability due to weather. Characterization of the scattering environment and pulsar astrometry will require the ngVLA Long Baseline Array (i.e., $>$ 1000 km baselines) to resolve compact sources and achieve sub-milliarcsecond positional accuracy.

7. Summary

The order of magnitude increase in high frequency sensitivity that the ngVLA provides will be essential for the discovery and characterization of pulsars in the Galactic Center, the nearest and best studied galactic nucleus. Characterization of the population within the Central Molecular Zone will give new insights into the ISM and stellar populations of one of the most dynamic regions in the Galaxy. Discovery of a pulsar in orbit around Sgr A* will provide an unprecedented opportunity for fundamental physics in the environment of a black hole.

Acknowledgments. SC, JMC, PD, JL, and SR are members of the NANOGrav Physics Frontier Center, which is supported by the National Science Foundation award 1430284. MK, LS, NW, and RW acknowledge financial support by the European Research Council (ERC) for the ERC Synergy Grant BlackHoleCam under contract no. 610058. JSD acknowledges support by NASA under grant DPR S–15633–Y. Part of this research was carried out at the Jet Propulsion Laboratory, California Institute of Technology, under a contract with the National Aeronautics and Space Administration.

References

Bates, S. D., Lorimer, D. R., & Verbiest, J. P. W. 2013, MNRAS, 431, 1352
Blum, R. D., Ramírez, S. V., Sellgren, K., & Olsen, K. 2003, ApJ, 597, 323
Bower, G. C., Wright, M. C. H., Falcke, H., & Backer, D. C. 2003, ApJ, 588, 331
Bower, G. C., Deller, A., Demorest, P., et al. 2014, ApJ, 780, L2
—. 2015, ApJ, 798, 120
Broderick, A. E., Johannsen, T., Loeb, A., & Psaltis, D. 2014, ApJ, 784, 7
Cordes, J. M., & Lazio, T. J. W. 1997, ApJ, 475, 557
Deneva, J. S., Cordes, J. M., & Lazio, T. J. W. 2009, ApJ, 702, L177
Dexter, J., & O'Leary, R. M. 2014, ApJ, 783, L7
Dexter, J., Deller, A., Bower, G. C., et al. 2017, MNRAS, 471, 3563
Eatough, R. P., Kramer, M., Klein, B., et al. 2013a, in IAU Symposium, Vol. 291, IAU Symposium, 382–384
Eatough, R. P., Falcke, H., Karuppusamy, R., et al. 2013b, Nat, 501, 391
Frail, D. A., Diamond, P. J., Cordes, J. M., & van Langevelde, H. J. 1994, ApJ, 427, L43
Genzel, R., Eisenhauer, F., & Gillessen, S. 2010, Reviews of Modern Physics, 82, 3121
Johnston, H. M., & Kulkarni, S. R. 1991, ApJ, 368, 504
Johnston, S., Kramer, M., Lorimer, D. R., et al. 2006, MNRAS, 373, L6
Kramer, M., Klein, B., Lorimer, D., et al. 2000, in Astronomical Society of the Pacific Conference Series, Vol. 202, IAU Colloq. 177: Pulsar Astronomy - 2000 and Beyond, ed. M. Kramer, N. Wex, & R. Wielebinski, 37
Lazio, T. J. W., & Cordes, J. M. 1998a, ApJ, 505, 715. http://adsabs.harvard.edu/cgi-bin/nph-bib_query?bibcode=1998ApJ...505..715L&db_key=AST
—. 1998b, ApJS, 118, 201. http://adsabs.harvard.edu/cgi-bin/nph-bib_query?bibcode=1998ApJS..118..201L&db_key=AST
Liu, K., Wex, N., Kramer, M., Cordes, J. M., & Lazio, T. J. W. 2012, ApJ, 747, 1
Macquart, J.-P., Kanekar, N., Frail, D. A., & Ransom, S. M. 2010, ApJ, 715, 939
Marrone, D. P., Moran, J. M., Zhao, J.-H., & Rao, R. 2007, ApJ, 654, L57
Pfahl, E., & Loeb, A. 2004, ApJ, 615, 253
Pfuhl, O., Fritz, T. K., Zilka, M., et al. 2011, ApJ, 741, 108
Psaltis, D., Wex, N., & Kramer, M. 2016, ApJ, 818, 121
Ransom, S. M., Cordes, J. M., & Eikenberry, S. S. 2003, ApJ, 589, 911
Schnitzeler, D. H. F. M., Eatough, R. P., Ferrière, K., et al. 2016, MNRAS, 459, 3005
Siemion, A., Bailes, M., Bower, G., et al. 2013, in IAU Symposium, Vol. 291, IAU Symposium, 57–57
Spitler, L. G., Lee, K. J., Eatough, R. P., et al. 2014, ApJ, 780, L3
Torne, P., Desvignes, G., Eatough, R. P., et al. 2017, MNRAS, 465, 242
van Langevelde, H. J., & Diamond, P. J. 1991, MNRAS, 249, 7P
van Langevelde, H. J., Frail, D. A., Cordes, J. M., & Diamond, P. J. 1992, ApJ, 396, 686
Waisberg, I., Dexter, J., Gillessen, S., et al. 2018, MNRAS, arXiv:1802.08198
Wex, N., & Kopeikin, S. M. 1999, ApJ, 514, 388

Science with a Next-Generation Very Large Array
ASP Conference Series, Monograph 7
Eric J. Murphy, ed.

Weak Gravitational Lensing with CO Galaxies

Philip Bull,[1] Ian Harrison,[2] and Eric Huff[3]

[1]*Radio Astronomy Laboratory, University of California Berkeley, Berkeley, CA 94720, USA;* philbull@berkeley.edu

[2]*Jodrell Bank Centre for Astrophysics, School of Physics & Astronomy, The University of Manchester, Manchester M13 9PL, UK;* ian.harrison-2@manchester.ac.uk

[3]*Jet Propulsion Laboratory, California Institute of Technology, 4800 Oak Grove Drive, Pasadena, CA, USA;* Eric.M.Huff@jpl.nasa.gov

Abstract. Optical weak lensing surveys have become a powerful tool for precision cosmology, but remain subject to systematic effects that can severely bias cosmological parameter estimates if not carefully removed. We discuss the possibility of performing complementary weak lensing surveys at radio/microwave frequencies, using detections of CO-emitting galaxies with resolved continuum images from the ngVLA. This method has completely different systematic uncertainties to optical weak lensing shear measurements (e.g., in terms of blending, PSF, and redshift uncertainties), and can provide additional information to help disentangle intrinsic alignments from the cosmological shear signal. A combined analysis of optical and CO galaxy lensing surveys would therefore provide an extremely stringent validation of highly-sensitive future surveys with *Euclid*, LSST, and WFIRST, definitively rejecting biases due to residual systematic effects. A lensing survey on the ngVLA would also provide valuable spectral (kinematic) and polarimetric information, which can be used to develop novel cosmological analyses that are not currently possible in the optical.

1. Introduction and Scientific Motivation

As light travels from distant galaxies, it is distorted by the gravitational pull of matter that it encounters along the way. While some light rays pass close enough to dense concentrations of matter to be strongly lensed, producing arcs and multiple images, the majority of rays are only slightly affected. This weak gravitational lensing effect can be measured in a number of ways, but the most common is to look for lensing shear, which causes coherent correlations in the ellipticity of galaxies over angular scales of a few arcminutes or larger (Bartelmann & Schneider 2001; Munshi et al. 2008; Hoekstra & Jain 2008; Kilbinger 2015; Mandelbaum 2017). This has motivated the development of a slew of large galaxy surveys, predominantly in the optical, that seek to measure the shapes of many tens or even hundreds of millions of galaxies across cosmic time in order to characterize the statistical properties of the weak lensing 'shear' signal.

Weak lensing has rapidly developed into one of the pillars of observational cosmology. This is due in large part to its extremely high information content – the lensing signal not only encodes information about the total matter distribution (including cold dark matter and baryons) along the line of sight, but also the growth rate of large-scale structure and the geometry of space (Van Waerbeke et al. 1999; Jain & Taylor 2003). These quantities are instrumental in attempts to understand some of the biggest problems in cosmology and fundamental physics, such as the nature of cosmic acceleration and the Cosmological Constant problem, and the nature of gravity on extremely large distance scales. Lensing can also be used to understand the connection between galaxies and the dark matter structures that they are embedded in (Hoekstra et al. 2004; Sheldon et al. 2004).

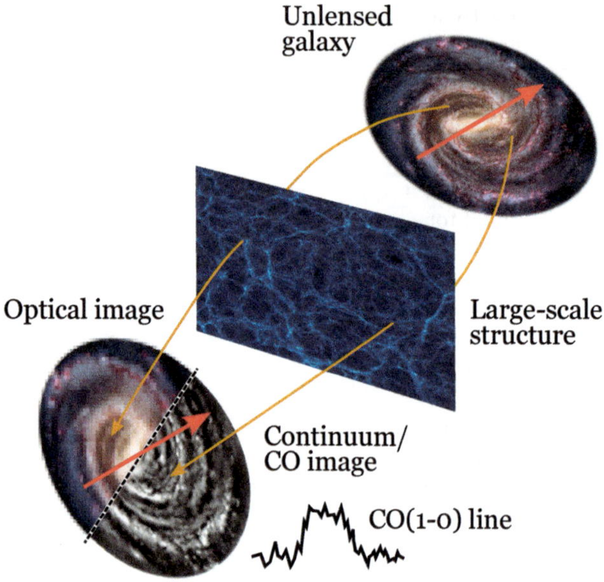

Figure 1. Illustration showing the weak gravitational lensing of CO galaxies. Light rays (orange lines) are bent and sheared as they pass through the large-scale structure, resulting in a distortion of the observed shape of the galaxy. The galaxy can be imaged in multiple bands, including optical and radio continuum/CO line emission. The advantage of the CO/radio continuum strategy advocated here is that it preserves redshift information through detection of the CO(1–0) line, and has different PSF characteristics from the optical. The fact that lensing leaves the polarization angle unaffected (red arrows) can also be used to reduce shape noise and solve for the intrinsic alignment effect. (Composite image. Credit: P. Bull/NASA/S. Gottlober/MultiDark.)

This information does not come easily however. Weak lensing observations remain some of the most challenging in all of astronomy, with precise control over myriad instrumental and astrophysical systematics required to even detect the effect, let alone measure its statistical properties with the sub-percent precision required to beat down errors on cosmological parameters. Despite a series of enhancements in analysis techniques (e.g. Kaiser et al. 1995; Bernstein & Jarvis 2002; Miller et al. 2007; Huff & Mandelbaum 2017), low-level systematics can still remain in the data, subtly biasing the resulting cosmological constraints (Heavens et al. 2000; Catelan et al. 2001; Hirata & Seljak 2003; Huterer et al. 2006; Massey et al. 2013; Mandelbaum 2017).

The development of a rigorous and independent cross-check of the optical weak lensing signal is therefore of vital importance – without it, the unprecedented precision of flagship datasets such as the LSST lensing sample could be undermined. Radio weak lensing (Figure 1) represents a highly promising solution to this problem (Chang et al. 2004; Brown 2012; Patel 2016; Camera et al. 2017) – the galaxy shape measurement process is fundamentally different in the radio, owing to the almost-deterministic nature of synthesized beams, the lack of atmospheric seeing effects, and differences in galaxy morphologies between the radio and optical (Patel et al. 2010; Tunbridge et al. 2016). As such, many weak lensing systematics are expected to be strongly suppressed by radio-optical cross-correlations (Demetroullas & Brown 2016). A consistent picture between the radio and optical cross- and auto-power spectra would also be strong evidence that systematics are under control, and could therefore thoroughly validate the weak lensing methodology.

In addition to providing a cross-check on optical surveys, radio surveys also open up several exciting possibilities to measure extra information about the lensing galaxy sample that would be difficult, if not impossible, to obtain in the optical. Spectroscopic redshifts can be measured for large fractions of the sample for example, an extremely costly exercise in the optical. As well as removing photometric redshift calibration as a key source of uncertainty, this would allow more information to be recovered from the lensing signal in the radial direction, potentially significantly improving constraints on some cosmological parameters. Polarization information can also be readily obtained in the radio, allowing the intrinsic orientations of galaxies to be inferred – again removing an important source of systematic error (intrinsic alignments).

In this chapter, we examine the possibility of measuring the weak lensing shear signal from a large sample of CO-emitting galaxies in several spectroscopic redshift bins across a substantial sky area and redshift range. This requires a survey of a large number of moderately-resolved CO-emitting galaxies. The galaxy shapes can be measured using continuum emission integrated over a few wide channels in each band, while the galaxy redshifts will be determined from spectroscopic detections of the CO(1-0) line. The resulting shear catalog can be cross-correlated with optical catalogs to greatly reduce the impact of systematics. We also discuss some of the novel analysis techniques that are enabled by radio observations, including polarization-based measurements of intrinsic alignments, and kinematic lensing using resolved spectral lines.

2. Systematics Mitigation with Radio-Optical Cross-Correlations

Cosmological constraints from weak lensing surveys require the creation of a shear map on the sky. The spatially varying shear γ must be estimated from the shapes of individual galaxies which sample the shear field, and is typically done using a simple (often weighted) average over galaxies in a sky patch small enough such that the shear is constant within the patch. Because weak lensing causes such a small change in the ellipticity of an individual galaxy, even small systematic biases in the estimation of its shape can quickly overwhelm the cosmological signal and cause significant errors in e.g., measuring the dark energy equation of state. For a linear model of bias on shear $\gamma_{\rm meas} = (1 + m)\gamma_{\rm true} + c$ it is necessary for $m < O(10^{-2})$ and $c < O(10^{-3})$ for Stage III surveys such as DES and SKA1 and $m < O(10^{-3})$ and $c < O(10^{-4})$ for Stage IV surveys such as LSST and SKA2 (Amara & Refregier 2008).

Weak lensing with interferometers at radio wavelengths provides a unique opportunity to mitigate many of these systematics. Cosmological constraints are typically derived by taking two-point statistics of shear maps; if these power spectra are made by cross-correlating radio and optical shear maps, wavelength-dependent systematics either disappear or can be calibrated out (see Camera et al. 2017). Some of the most important systematics for optical surveys are as follows:

PSF systematics. Estimating γ from galaxy shapes requires deconvolution of the shape of the Point Spread Function (PSF) from each individual source. For ground-based optical surveys, atmospheric seeing produces a PSF which spatially varies in a non-deterministic way. Significant efforts are put into PSF reconstruction and strategies for precise and accurate deconvolution. Even for space-based surveys, challenges remain due to instabilities in the optical systems, non-trivial color dependence of the PSF, and effects within the CCD (Mandelbaum 2015).

In contrast, interferometer dirty beams may be seen as a stable and highly deterministic forward model, allowing for either precise determination of galaxy shapes in images or directly in the (u, v)-plane. Even if shape measurement systematics remain, they will be uncorrelated with those from optical surveys, meaning systematic uncertainties from PSF removal should be highly suppressed in radio-optical cross-correlations.

Blended (confused) sources. A substantial fraction of the galaxies detected in deep optical surveys like LSST are expected to be overlapping on the sky. Such 'blended' sources can be hard to identify, especially when the resolution limits imposed by atmospheric seeing are taken into account. Unidentified blends mostly add random noise to the measured shear signal, as the blended galaxies should have no preferred orientation. This source of noise can be removed simply by removing the blended sources (if they can be identified), although this can substantially reduce the effective number density of galaxies in the sample. More problematic is the effect of blends on photometric redshift estimation; trying to fit a photometric redshift to two sources at very different redshifts can result in catastrophic failure of the photo-z algorithm, which can ultimately cause biases in cosmological parameter estimates (Samuroff et al. 2018). The high angular resolution afforded by the long baselines of the ngVLA may be useful in at least identifying blended sources for some of the galaxies in overlapping optical surveys from continuum images, while the degree of blending/confusion in the radio survey will only depend on instrumental resolution (since the atmospheric seeing is a much smaller effect). The availability of CO line redshifts will also help mitigate catastrophic photometric redshift errors caused by blends. The extent to which the ngVLA data is useful for identifying blends in optical data will depend on the final array layout and depth of the survey, but if suitable optimizations are made, the impact could be considerable – especially if photo-z estimation algorithms are unable to reduce their sensitivity to blends.

Intrinsic alignments. The simple average estimator for γ relies on the assumption that galaxy shapes were uncorrelated before lensing (i.e., in the absence of lensing, the average ellipticity should be zero). However, the realities of galaxy formation processes mean that this is not true in the real Universe. Failing to account for these intrinsic galaxy alignments may bias cosmological parameter estimation by many standard deviations (Troxel & Ishak 2014). By measuring both the lensed total intensity galaxy position angle and a quantity which is not affected by lensing, such as polarization position angle (Brown & Battye 2011) or spatially-resolved kinematics (Morales 2006), intrinsic alignments can be mapped and accounted for in the lensing analysis. Being able to account for this directly has significant advantages over current strategies applied in optical weak lensing surveys, which typically involve either removing data (seriously degrading cosmological constraints) or the application of poorly-motivated models for the intrinsic alignment signal. Such a map of intrinsic alignments would not only be useful for an ngVLA weak lensing survey itself, but could also be used to model the intrinsic alignments within optical surveys (if the same galaxy populations are being probed).

3. Limitations of Current Astronomical Instrumentation

While pathfinder weak lensing observations have been made with existing radio facilities (e.g. Chang et al. 2004; Patel et al. 2010), they are far from achieving parity with optical surveys – by any metric. The current state of the art is SuperCLASS[1], a $\sim 1.75\ \mathrm{deg}^2$ survey on e-MERLIN and VLA that targets several Abell clusters with overlapping Subaru imaging. While these instruments have comparatively slow survey speeds, somewhat sparse (u, v)-coverage (even when combined), and operate in a band ($\sim 1.4 - 1.5$ GHz) that lacks emission lines suitable for providing redshift information, SuperCLASS is nevertheless useful as a way of proving the radio weak lensing method and providing a testing ground for future techniques.

The Square Kilometre Array (SKA) and its precursors MeerKAT and ASKAP have better sensitivity and much denser (u, v)-coverage, making imaging and shape measurement much easier. The precursors lack sufficiently long baselines to resolve higher-redshift continuum galaxies however, and so the first large-area radio weak lensing survey will probably have to wait until SKA1 comes online. A $\sim 5,000\ \mathrm{deg}^2$ continuum survey on SKA1-MID Band 2 (950 – 1750 MHz) is expected to be competitive with the Dark Energy Survey (Brown et al. 2015; Harrison et al. 2016), and a small fraction of the detected galaxies will have 21cm line detections and therefore redshifts. Unless they can be cross-matched with optical galaxies (with spectroscopic or photometric redshifts), the large fraction of sources without a 21cm detection will have effectively no redshift information, resulting in what would effectively be a 2D lensing survey (containing significantly less information about cosmological evolution).

As discussed in the previous section, optical surveys also have some inherent limitations when it comes to measuring the weak lensing signal, mostly centered around uncertainties in the shape, stability, and frequency-dependence of the point spread function, and a lack of knowledge about the intrinsic shapes of the galaxies. A number of these can be mitigated by some features of the ngVLA and radio weak lensing observations more generally, as we discuss in the next section.

---

[1] `http://www.e-merlin.ac.uk/legacy/projects/superclass.html`

4. Connection to Unique ngVLA Capabilities

The ngVLA will have a unique combination of spectral/spatial resolution and wide bandwidth, allowing it to detect large numbers of semi-resolved galaxies in both continuum and CO line emission at low and intermediate redshifts. Other radio weak lensing surveys have been proposed, for example using the SKA at lower frequencies (~ 1.5 GHz), but these mostly lack redshift information except for a small fraction of sources that also have a detectable 21cm line. A CO galaxy sample would also be better matched to the LSST and DES samples than a 21cm sample, as it will preferentially contain normal star-forming galaxies that will make up the bulk of the optical samples. The higher angular resolution of the ngVLA will also reduce the occurrence of blended/confused sources, which are a significant source of noise in lensing measurements (Samuroff et al. 2018).

The ngVLA will also be able to operate in fundamentally different survey modes from optical experiments. A major advantage over optical telescopes is the ease with which polarized observations can be made. Weak lensing leaves the polarization angle of a galaxy unchanged. Since polarization traces the disk, this can be used to measure the intrinsic orientation of the galaxy, i.e., its orientation before it was sheared. As discussed in §2, this helps to mitigate the effects of intrinsic alignments, which are correlations in the ellipticities of galaxies caused by non-lensing effects like tidal shears that are a significant nuisance effect for weak lensing cosmology (e.g. Joachimi et al. 2015).

The high spectral resolution of the ngVLA can be used to similar effect. If the CO lines of some subset of galaxies can be spectrally resolved, one can obtain highly complementary information about their circular velocity. As in the polarized case, this can be used to infer the intrinsic ellipticity and thus greatly suppress shape noise (Morales 2006). If a significant decrease in shape noise can be achieved, the number density of galaxies required to obtain a high signal-to-noise ratio(SNR) detection of the lensing signal can also be greatly reduced, making the ngVLA a much more efficient weak lensing survey instrument without the need to substantially alter its design.

5. Measurement Technique and Experimental Layout

In this section, we give a high-level overview of how a CO weak lensing survey could be carried out in practice. Two measurements are required per galaxy:

(a). The galaxy's shape (ellipticity), which can be obtained with high signal-to-noise through imaging of the continuum emission;

(b). A spectroscopic determination of its redshift through a detection of the CO (1-0; 115.2712 GHz) line.

The first measurement requires that the target galaxies be sufficiently resolved that their ellipticities can be measured. This suggests a target resolution of $\sim 1''$ for galaxies at $z \sim 1$ (see Figure 2). A wide instantaneous bandwidth is preferred to maximize continuum SNR, ideally breaking up the full available bandwidth into a handful of wide channels to allow for the correction of frequency-dependent beam effects. Note that the galaxies need only be semi-resolved in order to measure their ellipticities accurately enough. Beam smearing effects will be important when measuring the shape of the

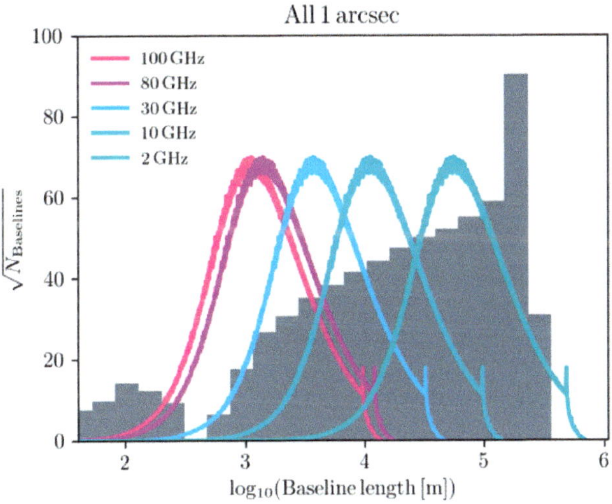

Figure 2. Comparison of the angular resolution of the ngVLA to the angular scales affected by the cosmic shear signal, as a function of frequency band (redshift). The colored curves show the (normalized) shear signal for different observed CO(1–0) line frequencies. They were calculated by taking a representative field of view populated with galaxies, smoothed to 1″ resolution, applying an appropriate level of weak lensing shear, and then calculating the power spectrum of the difference between the sheared and unsheared images (note the Nyquist artefact in the tails). The gray histogram is the orientation-averaged ngVLA baseline distribution, assuming the Conway array layout.

galaxy, hence the need to break up the band into smaller chunks; the PSF must be known very accurately, including its frequency dependence, and time-variation of the PSF should also be strongly controlled. An accurate flux density calibration is not strictly required, except to apply cuts to select certain galaxy sub-populations. Imaging may not be strictly necessary if visibility-space shape measurement methods can be used (Chang & Refregier 2002; Patel et al. 2014; Rivi et al. 2016), although this will require gridded visibilities to be stored (Harrison & Brown 2015).

The second measurement requires just enough spectral resolution to identify and measure the redshifted CO line integrated over the solid angle of the galaxy. In the simplest case, there is no requirement to accurately measure the line flux, line width, or other spectroscopic properties, so a low-SNR line detection is sufficient. Depending on typical linewidths, the spectral resolution required will likely be a few hundred km/s.

The SNR of the weak lensing measurement strongly depends on the number density of sources for which shapes can be measured, as well as the spatial volume that can be surveyed. A minimal proof of concept survey would need to target a few hundred square degrees with source number densities of order a few per arcmin^2. An ideal survey for studying lensing systematics would be matched to the WFIRST footprint (~2,200 deg^2), with a high source number density ($\sim 60 - 70$ per arcmin^2). An ideal cosmological survey would cover a large fraction of the *Euclid* or LSST footprints (~15,000 deg^2) with 20-30 sources per arcmin^2. Lower number densities could be sustained with the addition of spectroscopic redshift information, which compensates by providing access to radial Fourier modes that LSST and other photometric surveys effectively discard. A continuum-only weak lensing survey would also be possible, negating the need to carry out measurement (b), but is unlikely to be competitive with optical surveys with photometric redshifts.

An alternative to increasing the source number density is to measure (c) the polarization angle of the galaxies; or (d) accurately measure their linewidths. These would allow the methods described in §4 to be used to recover the intrinsic galaxy shapes/alignment, thus significantly cancelling important noise sources such as shape noise and intrinsic alignments. Carrying out these much more detailed measurements would likely necessitate a much smaller survey area, with greater depth per pointing. The trade-offs between these different survey modes will need to be explored within the context of a realistic model of the (polarized) luminosity, size, and circular velocity distribution of CO galaxies.

Finally, a significant hurdle is the small field of view of the ngVLA, which substantially reduces its survey speed. The $\sim 1\,\mathrm{deg}^2$ primary beam of SKA gives it a significant advantage in this respect. Array designs that increase the field of view of the ngVLA and optimize its baseline distribution for sensitivity at $\sim 1''$ scales (c.f. Figure 2) are therefore highly desirable for this science case. Examples of how to achieve this might include reducing dish sizes and adopting a more compact layout with a denser core, perhaps in the context of a 'hybrid' design with a dense core of small dishes at the center and larger dishes spread out to longer baseline lengths. An advantage of such a design is that the core and outer sets of dishes could be operated independently if desired, allowing large surveys and high-resolution pointed observations to be performed concurrently.

6. Complementarity

While a radio weak lensing survey performed with the ngVLA on its own would be a interesting and potentially valuable cosmological test, the power of performing such a survey lies firmly in the ability to jointly analyze it with an overlapping optical survey. The cross-correlation of a radio weak lensing sample with contemporary optical datasets (e.g., from LSST, WFIRST, and *Euclid*) would provide an extremely robust validation of the weak lensing methodology that so much of the observational cosmology program will rely upon in the next decade. This represents an immensely valuable synergy with practically all of the large cosmological survey instruments planned for the beyond-2025 timeframe, and would represent the "last word" in the interpretation of weak lensing data.

This science case also complements and extends the radio weak lensing science case of the SKA to higher frequencies, higher spatial resolutions, and arguably a more appropriate sample for optical cross-correlations, as well as providing high-value spectroscopic redshift information, and the ability to use novel techniques such as kinematic lensing and intrinsic alignment removal with polarization. Additionally, the large sample of galaxies that would be obtained in performing a radio weak lensing survey would clearly be valuable for a wide array of astrophysical applications too, such as studies of galaxy formation and evolution, and galaxy structure and dynamics (if spectrally resolved/polarized observations are prioritized), providing a further connection to future NASA flagship missions such as the Origins Space Telescope and LUVOIR.

Acknowledgments. We are grateful to Stefano Camera and Joe Lazio for extensive discussions.

References

Amara, A. & Refregier, A. 2008, MNRAS 391, 228
Bartelmann, M. & Schneider, P. 2001, Phys. Rept. 340, 291
Bernstein, G. M. & Jarvis, M. 2002, Astron. J. 123, 583
Brown, M. L., & Battye, R. A. 2011, MNRAS, 410, 2057
Brown, M. L. 2001, arXiv:1206.4437
Brown, M. L. et al. (JVLA Collaboration) 2013, arXiv:1312.5618
Brown, M. L. et al. 2015, PoS AASKA 14, 023
Camera, S., Harrison, I., Bonaldi, A. & Brown, M. L. 2017, MNRAS 464, 4, 4747
Catelan, P., Kamionkowski, M. & Blandford, R. D. 2001, MNRAS 320, L7
Chang, T.-C., & Refregier, A. 2002, ApJ, 570, 447
Chang, T. C., Refregier, A. & Helfand, D. J. 2004, Astrophys. J. 617, 794
Demetroullas, C. & Brown, M. L. 2016, MNRAS 456, 3100
Harrison, I., & Brown, M. L. 2015, arXiv:1507.06639
Harrison, I., Camera, S., Zuntz, J., & Brown, M. L. 2016, MNRAS 463, 3574
Heavens, A., Refregier, A. & Heymans, C. 2000, MNRAS 319, 649
Hirata, C. M. & Seljak, U. 2003, MNRAS 343, 459
Hoekstra, H., Yee, H. K. C. & Gladders, M. D. 2004, ApJ 606, 67
Hoekstra, H. & Jain, B. 2008, Ann. Rev. Nucl. Part. Sci. 58, 99
Huff, E. M., Krause, E., Eifler, T., George, M. R. & Schlegel, D. 2013, arXiv:1311.1489
Huff, E. & Mandelbaum, R. 2017, arXiv:1702.02600.
Huterer, D., Takada, M., Bernstein, G. & Jain, B. 2006, MNRAS 366, 101
Jain, B. & Taylor, A. 2003, Phys. Rev. Lett. 91, 141302
Joachimi, B. et al. 2015, Space Sci. Rev. 193, 1, 1
Kaiser, N., Squires, G. & Broadhurst, T. J. 1995, Astrophys. J. 449, 460
Kilbinger, M. 2015, Rept. Prog. Phys. 78, 086901
Mandelbaum, R. 2015, JINST 10, 05, C05017
Mandelbaum, R. 2017, arXiv:1710.03235
Massey, R., Hoekstra, H., Kitching, T., et al. 2013, MNRAS 429, 661
Miller, L., Kitching, T. D., Heymans, C., et al. 2007, MNRAS 382, 315
Morales, M. F. 2006, Astrophys. J. 650, L21
Munshi, D., Valageas, P., Van Waerbeke, L. & Heavens, A. 2008, Phys. Rept. 462, 67
Patel, P., Bacon, D. J., Beswick, R. J. et al. 2010, MNRAS, 401, 2572
Patel, P., Abdalla, F. B., Bacon, D. J., et al. 2014, MNRAS, 444, 2893
Patel, P. 2016, arXiv:1602.07482
Rivi, M., Miller, L., Makhathini, S., & Abdalla, F. B. 2016, MNRAS, 463, 1881
Samuroff, S., Bridle, S. L., Zuntz, J., et al. 2018, MNRAS, 475, 4524
Sheldon, E. S. et al. (SDSS Collaboration) 2004, AJ127, 2544
Troxel, M. A. & Ishak, M. 2014, Phys. Rept. 558, 1
Tunbridge, B., Harrison, I. & Brown, M. L. 2016, MNRAS 463, 3339
Van Waerbeke, L., Bernardeau, F. & Mellier, Y. 1999, Astron. Astrophys. 342, 15

Science with a Next-Generation Very Large Array
ASP Conference Series, Monograph 7
Eric J. Murphy, ed.

Extragalactic Proper Motions: Gravitational Waves and Cosmology

Jeremy Darling, Alexandra Truebenbach, and Jennie Paine

Center for Astrophysics and Space Astronomy, Department of Astrophysical and Planetary Sciences, University of Colorado, 389 UCB, Boulder, CO 80309-0389, USA; `jeremy.darling@colorado.edu`

Abstract. Extragalactic proper motions can reveal a variety of cosmological and local phenomena over a range of angular scales. These include observer-induced proper motions, such as the secular aberration drift caused by the solar acceleration about the Galactic Center and a secular extragalactic parallax resulting from our motion with respect to the cosmic microwave background rest frame. Cosmological effects include anisotropic expansion, transverse peculiar velocities induced by large scale structure, and the real-time evolution of the baryon acoustic oscillation. Long-period gravitational waves can deflect light rays, producing an apparent quadrupolar proper motion signal. We review these effects, their imprints on global correlated extragalactic proper motions, their expected amplitudes, the current best measurements (if any), and predictions for *Gaia*. Finally, we describe a possible ngVLA Long Baseline Array program to measure or constrain these proper motion signals. In most cases, the ngVLA can surpass or complement the expected end-of-mission performance of the *Gaia* mission.

1. Background

The universe is dynamic, as we know from cosmological redshifts induced by the Hubble expansion, but astronomers tend to treat extragalactic objects as fixed in the sky with fixed apparent velocities. If measured with enough precision, however, nothing is constant: all objects will change their redshifts and positions at a rate of order $H_0 \simeq 7 \times 10^{-11}$ yr^{-1} $\simeq 15$ μas yr^{-1}. The secular redshift drift caused by a non-constant expansion (Sandage 1962) is of order 0.3 cm s^{-1} yr^{-1} at $z \simeq 1$ and may be measured by optical telescopes using the Lyα forest or by radio telescopes using H I 21 cm or molecular absorption lines (Loeb 1998; Darling 2012). Proper motions of extragalactic objects may be caused by peculiar motions induced by large scale structure (Darling 2013; Darling & Truebenbach 2018), by primordial gravitational waves (e.g., Pyne et al. 1996; Gwinn et al. 1997; Book & Flanagan 2011; Darling et al. 2018), by the recession of fixed objects such as the baryon acoustic oscillation, or by anisotropic expansion (Fontanini et al. 2009; Quercellini et al. 2009; Titov 2009; Darling 2014).

Observer-induced proper motions are also possible: these include the secular aberration drift caused by acceleration of the solar system barycenter about the Galactic Center (e.g., Bastian 1995; Eubanks et al. 1995), observations from a rotating reference frame, and secular extragalactic parallax caused by motion with respect to the cosmic microwave background (CMB; Ding & Croft 2009).

Proper motions depict a discretely sampled vector field on the celestial sphere. In order to detect and characterize correlated motions, it is natural to describe vector fields using vector spherical harmonics (VSH), which are the vector equivalent of the scalar spherical harmonics used to describe signals such as the CMB temperature pattern, the geoid, or equipotentials (Mignard & Klioner 2012). VSH are characterized by their degree ℓ and order m, and resemble electromagnetic fields. They can therefore be separated into curl-free (E-mode) and divergenceless (B-mode) vector fields that are typically connected to distinct physical phenomena. The general method for characterizing a correlated proper motion field is to fit VSH to the observed proper motions and to calculate the power in and significance of each degree ℓ for each mode.

2. Expected (and Possible) Signals

Table 1 summarizes the expected and possible global extragalactic proper motion signals. This summary is likely to be incomplete. Here we provide a brief description of each physical or observer-induced effect, and §3 discusses the impact the ngVLA might have on detecting or constraining these phenomena.

2.1. Secular Aberration Drift

Aberration of light is caused by the finite speed of light and the motion of the observer with respect to a light source. The resulting deflection of light scales as $\vec{v}/c$. If the observer accelerates, then the aberration exhibits a secular drift, and objects appear to stream in the direction of the acceleration vector. The solar system barycenter accelerates at roughly 0.7 cm s^{-1} yr^{-1} as the Sun orbits the Galactic Center, resulting in an apparent ~5 μas yr^{-1} E-mode dipole converging on the Galactic Center (Figure 1). This has been detected in the proper motions of radio sources, first by Titov et al. (2011), using *a priori* knowledge of the expected effect for data trimming, and recently without priors by Truebenbach & Darling (2017). Titov & Krásná (2018) have further developed the VLBI-specific methodology for extracting this signal.

2.2. Secular Parallax

The CMB shows a temperature dipole of 3.4 mK, which is caused by the motion of the solar system barycenter with respect to the CMB rest frame (e.g., Hinshaw et al. 2009). This amounts to a relative motion of 369 ± 0.9 km s^{-1} or 78 AU yr^{-1}. This motion will induce a maximum secular parallax of 78 μas yr^{-1} Mpc^{-1}: nearby galaxies in directions perpendicular to the CMB poles will show a reflex motion opposite our motion when compared to distant galaxies. Precise proper motion measurements of galaxies within ~20 Mpc will allow a statistical detection of the distance-dependent E-mode dipole caused by secular parallax. When referenced to this average dipole, it will be possible to detect the peculiar motions of individual galaxies as well as their geometric distances. Secular parallax may provide a distance ladder-free method for measuring distances in the local universe.

2.3. Rotation

As observers in a rotating reference frame, we have traditionally used quasars to define a fixed, non-rotating reference frame in order to measure and monitor the rotation of the Earth (e.g., McCarthy & Petit 2004). We thus have no means to detect non-terrestrial rotation that has an axis close to polar, but it is possible to detect or constrain rotation about axes that are not aligned with the polar axis. While the question of a rotating universe can be nettlesome to contemplate and violates our assumption of cosmological isotropy, one can nonetheless make precise, sub-μas yr^{-1} measurements of the effect because it would manifest as a B-mode dipole in the proper motion vector field.

2.4. Anisotropic Expansion

In an isotropically expanding universe, objects move radially away from every observer (modulo small peculiar motions due to local density perturbations; see §2.6), so there will be no global correlated proper motions caused by isotropic Hubble expansion. However, anisotropic expansion would cause objects to stream across the sky toward the directions of fastest expansion and away from directions of slowest expansion. Assuming a simple triaxial anisotropy, one can show that the resulting celestial proper motion pattern can be completely described by an E-mode VSH quadrupole (Darling 2014). Fitting a curl-free quadrupole to a proper motion field can therefore measure or constrain the (an)isotropy of the Hubble expansion without *a priori* knowledge of H_0. We can express the Hubble constant in terms of an angular rate, $H_0 \simeq 15$ μas yr^{-1}, which means that a 10% anisotropy would produce a quadrupole with amplitude 1.5 μas yr^{-1}.

2.5. Gravitational Waves

Stochastic gravitational waves deflect light rays in a quadrupolar (and higher ℓ) pattern with equal power in the E- and B-modes (Figure 1; Pyne et al. 1996; Gwinn et al. 1997; Book & Flanagan 2011). The gravitational waves that will produce extragalactic proper motions lie in the frequency range 10^{-18} Hz $< f < 10^{-8}$ Hz (H_0 to 0.3 yr^{-1}), which overlaps the pulsar timing and CMB polarization regimes, but uniquely covers about seven orders of magnitude of frequency space between the two methods (Darling et al. 2018). The cosmic energy density of gravitational waves Ω_{GW} can be related to the proper motion variance as

$$\Omega_{\rm GW} \sim \langle \mu^2 \rangle / H_0^2 \tag{1}$$

and to the quadrupolar power P_2 as

$$\Omega_{\rm GW} = \frac{6}{5}\frac{1}{4\pi}\frac{P_2}{H_\circ^2} = 0.00042\,\frac{P_2}{(1\ \mu{\rm as\ yr^{-1}})^2}\,h_{70}^{-2} \tag{2}$$

(Gwinn et al. 1997; Book & Flanagan 2011; Darling et al. 2018). Measuring or constraining the proper motion quadrupole power can therefore detect or place limits on primordial gravitational waves in a unique portion of the gravitational wave spectrum.

2.6. Large Scale Structure

The mass density distribution of large-scale structure reflects the mass power spectrum, the shape and evolution of which relies on cosmological parameters. The transverse peculiar motions of extragalactic objects can be used to measure the density perturbations from large-scale structure without a reliance on precise distance measurements. While line-of-sight velocity studies use distances to differentiate Hubble expansion from peculiar velocity, peculiar motions across the line-of-sight are separable from Hubble expansion because no proper motion will occur in a homogeneous expansion (Nusser et al. 2012; Darling 2013). Thus, one can employ pairs of galaxies as "cosmic rulers" to measure the real-time change in the apparent size of the rulers caused by the cosmic expansion and to detect structures that have decoupled from the Hubble flow (Darling 2013).

Given the definition of angular diameter distance, $\theta = \ell/D_A$, where a "ruler" of proper length ℓ subtends small angle θ at angular diameter distance D_A, cosmic expansion and a changing ℓ can produce an observed fractional rate of change in θ:

$$\frac{\Delta\theta/\Delta t_\circ}{\sin\theta} \equiv \frac{\dot{\theta}}{\sin\theta} = \frac{-\dot{D}_A}{D_A} + \frac{\dot{\ell}}{\ell} = \frac{-H(z)}{1+z} + \frac{\dot{\ell}}{\ell} \tag{3}$$

where $H(z) = H_\circ\sqrt{\Omega_{M,\circ}(1+z)^3 + \Omega_\Lambda}$ in a flat universe, $\Delta t_\circ$ is the observer's time increment, $\dot{\theta}$ is the *relative* proper motion, and $\dot{\ell}$ is the observed change in proper length, $\Delta\ell/\Delta t_\circ$, related to the physical (rest-frame) transverse velocity as $v_\perp = \dot{\ell}\,(1+z)$.

If ℓ is not a gravitationally influenced structure and grows with the expansion, then $\dot{\ell}/\ell = H(z)/1+z$, exactly canceling the first term in Equation (3). In this case, $\dot{\theta} = 0$, and there is no proper motion for objects co-moving with an isotropically expanding universe, as expected. If ℓ is decoupled from the expansion, however, then for most reasonable gravitational motions, $\dot{\ell}/\ell$ is a minor modification to the expansion contribution to $\dot{\theta}/\theta$ because the expansion, except for small redshifts or small structures, dominates (Darling 2013). The deviation of $\dot{\theta}/\theta$ from the null signal of pure Hubble expansion can be used to probe the mass distribution of large-scale structure and, thus, to test the shape of the mass power spectrum without a dependence on precise distance measurements or a "distance ladder."

2.7. Baryon Acoustic Oscillation Evolution

The baryon acoustic oscillation (BAO) is a standard ruler arising from pre-CMB density fluctuations, which can be observed as an overdensity of galaxies on the scale of ~150 comoving Mpc (Eisenstein et al. 2005). At redshift $z = 0.5$, the BAO scale subtends $\theta_{BAO} = 4^\circ\!.5$, which is equivalent to VSH degree $\ell \sim 40$. Taking the time derivative, we obtain an expression for proper motion on these scales:

$$\mu_{BAO} = \frac{\Delta\theta_{BAO}}{\Delta t_\circ} \simeq -\theta_{BAO}\, H_0 \simeq -1.2\,\mu\text{as yr}^{-1}. \tag{4}$$

The BAO evolution will therefore manifest as a convergent E-mode signal around $\ell \sim 40$ (Figure 1). To first order, the BAO scale depends on the expansion rate $H(z)$ and the angular diameter distance $D_A(z)$ at the observed redshift, but the rate of change of this standard ruler is dominated by its recession ("receding objects appear to shrink") and depends to first order on $H_0/D_A(z)$. Detection of this effect relies critically on the sky density of sources, which must adequately sample angular scales smaller than $4^\circ\!.5$.

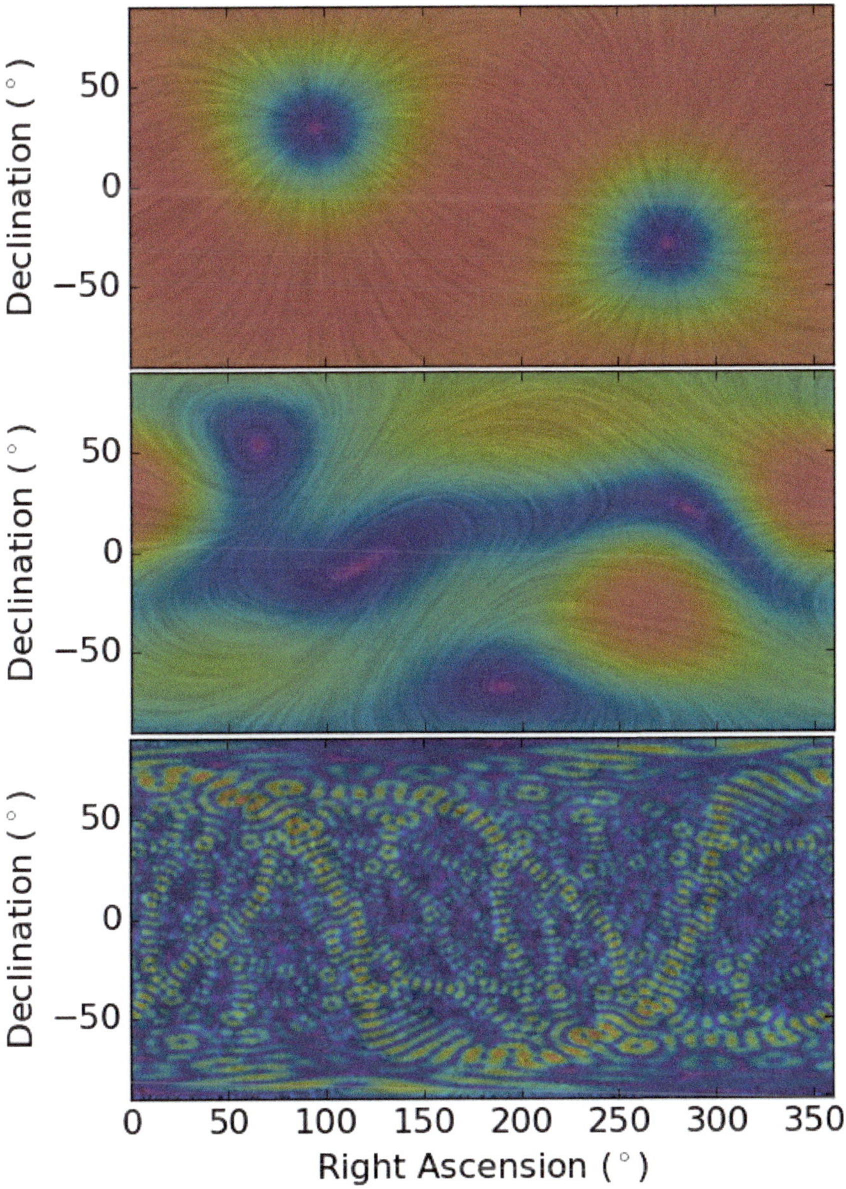

Figure 1. All-sky stream plots. Streamlines indicate the vector field direction, and the colors indicate the vector amplitude, from violet (zero) to red (maximum). (*Top:*) Secular aberration drift dipole detected by (Truebenbach & Darling 2017). (*Middle:*) Randomly generated gravitational wave stream plot, after Darling et al. (2018). (*Bottom:*) Randomly generated BAO streamlines.

Table 1. Global Correlated Proper Motion Signals

Effect	ℓ	Mode	Amplitude (μas yr^{-1})	Recent Measurement (μas yr^{-1})	Ref	*Gaia* (Predicted)	ngVLA (Predicted)	Prev. Work
Secular Aberration Drift	1	E	~5	5.2 ± 0.2	1	10σ	50σ	2,3,4,5
Secular Parallax	1	E	78 Mpc^{-1}	...	...	$\sim 10\sigma$	$\sim 10\sigma$ (1)	6
Rotation	1	B	Unknown	0.45 ± 0.27	7	$< 0.5\ \mu$as yr^{-1}	$< 0.1\ \mu$as yr^{-1} (2)	2,5,7
Anisotropic Expansion	2	E	Unknown	< 7%	8	< 3%	< 0.7%	8,9,10,11
Gravitational Waves	≥ 2	E+B	Unknown	$\Omega_{GW} < 6.4 \times 10^{-3}$	7	$\Omega_{GW} < 4 \times 10^{-4}$	$\Omega_{GW} < 10^{-5}$	1,7,12,13,14,15,16
Large Scale Structure	$\gtrsim 5$	E	-15 to $+5$	8.3 ± 14.9	17	10σ	$\sim 20\sigma$ (3)	17,18
BAO Evolution	~40	E	-1.2 at $z = 0.5$	...	...	4σ	$\sim 10\sigma$ (4)	...

Notes: 1 – Detection of the secular parallax is limited by the number of compact radio sources that can (and would) be monitored in the local volume. 2 – ngVLA observations will only be sensitive to rotation axes that are not aligned with the Earth's rotation axis. 3 – Detection of peculiar velocities associated with large scale structure will depend on the number of close pairs of radio sources. 4 – Detection of the BAO evolution signal will depend strongly on the sky density of proper motion measurements.

References: 1 – Titov & Krásná (2018) 2 - Titov et al. (2011); 3 – Xu et al. (2012); 4 – Titov & Lambert (2013); 5 – Truebenbach & Darling (2017); 6 – Ding & Croft (2009); 7 – Darling et al. (2018); 8 – Darling (2014); 9 – Chang & Lin (2015); 10 – Bengaly (2016); 11 – Paine et al. (2018); 12 – Braginsky et al. (1990); 13 – Pyne et al. (1996); 14 – Kaiser & Jaffe (1997); 15 – Gwinn et al. (1997); 16 – Book & Flanagan (2011); 17 – Darling (2013); 18 – Truebenbach, A., 2018, ApJ, in prep.

3. ngVLA Science

The ngVLA can detect or constrain all of these phenomena and improve upon many of the expected *Gaia* measurements. Table 1 lists the various proper motion signals, their VSH modes, the expected amplitude of the signal (if known), a recent measurement (if any), and predictions for *Gaia* and the ngVLA. We make the following assumptions about a ngVLA astrometry program:

- A sample of 10,000 objects
- Astrometric observations spanning 10 years
- (VLBA-level) astrometric precision: $\pm 10\,\mu$as yr^{-1} per object

Implicit in these assumptions are VLBA-sized baselines (consistent with the ngVLA Long Baseline Array) with roughly ten times the current VLBA collecting area. We further assume that radio jets will pose the same challenges as are found in the current VLBA data. In practice, we simply scale current VLBA-based observations — which include the added intrinsic proper motion "noise" contribution from relativistic jets — by N or $\sqrt{N}$, as appropriate, to the expected ngVLA sample. The proposed sample is roughly ten times larger than current VLBA geodetic monitoring samples, but an increased collecting area would enable the ngVLA (or a long-baseline subarray) to monitor the expanded sample without increasing the observing time commitment. The ngVLA sample would leverage the current VLBA sample and the radio-loud *Gaia* AGN, providing super-decade time baselines for a significant subset of objects.

The proposed observations will enable global detection of correlated signals of $\sim$0.1 μas yr^{-1}, which is $\sim$0.7% of H_0. For most of the phenomena described in §2, the ngVLA would significantly improve on previous work, including the expected *Gaia* performance. While the ngVLA sample size will be a factor of $\sim$50 smaller than the *Gaia* sample, the per-source astrometry will be a factor of $\sim$20–50 times better. We therefore predict that ngVLA observations will substantially improve on *Gaia* global correlated proper motion measurements in most cases. The proposed ngVLA program may not perform as well as our expectations for *Gaia* for measurements requiring fine angular sampling (BAO evolution) or dense volumetric sampling (secular parallax). This is due to the overall physical paucity of compact radio sources compared to optical sources across the sky and in the very nearby universe.

4. Conclusions

The outlook for extragalactic proper motions using the ngVLA is promising, and in most cases it can surpass the expected performance of *Gaia*, despite the added challenge of the intrinsic proper motion caused by radio jets. In particular, the proposed ngVLA astrometry program would provide exquisite precision on the Solar motion in the Galaxy and place the strongest constraints to date on the isotropy of the Hubble expansion in the epoch of dark energy and on the primordial gravitational wave background over roughly 10 decades in frequency.

Acknowledgments. The authors acknowledge support from the NSF grant AST-1411605 and the NASA grant 14-ATP14-0086.

References

Bastian, U. 1995, in ESA Special Publication, Vol. 379, Future Possibilities for Astrometry in Space, ed. M. A. C. Perryman & F. van Leeuwen, 99
Bengaly, C. A. P., Jr. 2016, JCAP, 4, 036
Book, L. G. & Flanagan, É. É. 2011, Phys. Rev. D, 83, 024024
Bower, G. C., Demorest, P., Braatz, J., et al. 2015, ArXiv e-prints, arXiv:1510.06432
Braginsky, V. B., Kardashev, N. S., Polnarev, A. G., & Novikov, I. D. 1990, Nuovo Cimento B, 105, 1141
Chang, Z., & Lin, H.-N. 2015, MNRAS, 446, 2952
Darling, J. 2012, ApJ, 761, L26
Darling, J. 2013, ApJ, 777, L21
Darling, J. 2014, MNRAS, 442, L66
Darling, J., & Truebenbach, A. E. 2018, ApJ, 864, 37
Darling, J., Truebenbach, A., & Paine, J. 2018, ApJ, submitted
Ding, F., & Croft, R. A. C. 2009, MNRAS, 397, 1739
Eisenstein, D. J., Zehavi, I., David W. Hogg, D. W., et al. 2005, ApJ, 633, 560
Eubanks, T. M., Matsakis, D. N., Josties, F. J., et al. 1995, in IAU Symposium, Vol. 166, Astronomical and Astrophysical Objectives of Sub-Milliarcsecond Optical Astrometry, ed. E. Hog & P. K. Seidelmann, 283
Fontanini, M., West, E. J., & Trodden, M. 2009, Phys. Rev. D, 80, 123515
Gwinn, C. R., Eubanks, T. M., Pyne, T., Birkinshaw, M., & Matsakis, D. N. 1997, ApJ, 485, 87
Hinshaw, G., Weiland, J.L., Hill, R.S., et al. 2009, ApJS, 180, 225
Kaiser, N. & Jaffe, A. 1997, ApJ, 484, 545
Loeb, A. 1998, ApJ, 499, L111
McCarthy, D. D., & Petit, G. 2004, IERS Technical Note, 32
Mignard, F. & Klioner, S. 2012, A&A, 547, A59
Nusser, A., Branchini, E., & Davis, M. 2012, ApJ, 755, 58
Paine, J., Darling, J., & Truebenbach, A. 2018, ApJS, submitted
Pyne, T., Gwinn, C. R., Birkinshaw, M., Eubanks, T. M., & Matsakis, D. N. 1996, ApJ, 465, 566
Quercellini, C., Quartin, M., & Amendola, L. 2009, Phys.Rev.Lett, 102, 151302
Sandage, A. 1962, ApJ, 136, 319
Titov, O. 2009, in Proc. 19th European VLBI for Geodesy and Astrometry (EVGA) Working Meeting, ed. G. Bourda, et al., 14
Titov, O., Lambert, S. B., & Gontier, A.-M. 2011, A&A, 529, A91
Titov, O., & Lambert, S. 2013, A&A, 559, A95
Titov, O., & Krásná, H. 2018, A&A, in press
Truebenbach, A. E., & Darling, J. 2017, ApJS, 233, 3
Xu, M. H., Wang, G. L., & Zhao, M. 2012, A&A, 544, A135

Science with a Next-Generation Very Large Array
ASP Conference Series, Monograph 7
Eric J. Murphy, ed.

H_2O Megamaser Cosmology with the ngVLA

Jim Braatz,[1] Dom Pesce,[1,2] Jim Condon,[1] and Mark Reid[3]

[1]*National Radio Astronomy Observatory, Charlottesville, VA 22903*

[2]*University of Virginia, Charlottesville, VA 22903*

[3]*Harvard-Smithsonian Center for Astrophysics, Cambridge, MA 02138*

Abstract. In combination with observations of the Cosmic Microwave Background, a measurement of the Hubble Constant provides a direct test of the standard ΛCDM cosmological model and a powerful constraint on the equation of state of dark energy. Observations of circumnuclear water vapor megamasers at 22 GHz in nearby active galaxies can be used to measure their distances, geometrically, and thereby provide a direct, one step measurement of the Hubble Constant. The measurement is independent of distance ladders and standard candles. With present-day instrumentation, the Megamaser Cosmology Project is expected to reach a ∼4% measurement using the megamaser technique, based on distances to fewer than 10 megamaser systems. A goal of the observational cosmology community is to reach a percent-level measurement of H_0 from several independent astrophysical measurements to minimize the systematics. The ngVLA will provide the sensitivity at 22 GHz required to discover and measure the additional megamaser disks that will enable an H_0 measurement at the ∼1% level.

1. Description of the problem

The expansion rate of the universe at $z \simeq 0$, known as the Hubble Constant (H_0), is the most fundamental parameter in observational cosmology. Historically, measurements of H_0 have been sought to reveal basic properties of the universe, including its age and geometry. Measuring H_0 remains a primary focus of observational cosmology today. The goal of new studies with the ngVLA will be to determine H_0 to ∼1% by measuring geometric distances to H_2O megamasers directly in the Hubble Flow. A measurement at this precision is required to improve our understanding of dark energy and to test the validity of the standard ΛCDM model of cosmology.

2. Scientific Importance

The exquisite observations of the Cosmic Microwave Background (CMB) by the *WMAP* and *Planck* satellites set a powerful framework for precision cosmology. These observations determine the angular-size distance to the surface of last scattering at $z \simeq 1100$ and constrain the geometry of the universe to be very nearly flat. The CMB, however, does not uniquely determine all fundamental cosmological parameters on its own. Observations at $z \simeq 0$, when dark energy is dominant, provide complementary data that constrain critical parameters, including the dark energy equation of state, neutrino mass, and the number of families of relativistic particles. CMB observations can *predict* basic cosmological parameters, including H_0, but only in the context of a specific cosmological model. Comparing CMB predictions to astrophysical measurements of H_0 therefore provides a powerful test of cosmological models.

In the context of the standard model of cosmology, i.e. a geometrically flat ΛCDM universe, *Planck* measurements predict $H_0 = 67.8 \pm 0.9$ km s^{-1} Mpc^{-1} (Planck Collaboration et al. 2016). Measurements from Baryon Acoustic Oscillations combined with SN Ia determine $H_0 = 67.3 \pm 1.0$ km s^{-1} Mpc^{-1} (Alam et al. 2017), in line with the *Planck* prediction. Measurements anchored at low redshift and based on standard candle techniques, however, are in tension with the BAO measurements and the *Planck* predictions: $H_0 \cong 73.24 \pm 1.74$ km s^{-1} Mpc^{-1} (Riess et al. 2016) and $H_0 = 74.3 \pm 2.6$ km s^{-1} Mpc^{-1} (Freedman et al. 2012). Determinations based on observations of gravitationally lensed quasars, meanwhile, give $H_0 \cong 71.9^{+2.4}_{-3.0}$ km s^{-1} Mpc^{-1} (Bonvin et al. 2017).

The measurements of H_0 based on low-z observations and the predictions based on ΛCDM disagree at high significance, and understanding this discrepancy is of fundamental importance. Either the underlying observations suffer from unrecognized uncertainties, against all scrutiny, or adjustments are needed to the ΛCDM standard model of cosmology. A problem of such fundamental importance warrants new, independent, and percent-level measurements of H_0. The long-term goal for the observational cosmology community is to reach percent-level precision in H_0 with agreement across multiple, independent observational methods.

Measuring H_0 with H_2O megamasers is a powerful complement to standard candle methods because it is a one-step measurement, independent of distance ladders and calibrations, and it is a fundamentally geometric method that measures H_0 directly at $z \simeq 0$.

3. Astronomical Impact

Understanding the cosmological model and the nature of dark energy is perhaps the most important problem in fundamental physics. It is unclear whether dark energy is an intrinsic property of the universe (the cosmological constant), or related to a time-variant field with a yet-undiscovered particle. Modifications to General Relativity, the introduction of new relativistic particles, and identification of new modes of interaction between matter and radiation are all also viable options to resolve the outstanding tension between measurements and the standard model.

Refinements to the cosmological model would impact our understanding of all aspects of the universe on its largest scales, including remnant background radiation, the formation of structure, and galaxy and cluster evolution.

4. Anticipated Results

An ngVLA measurement of H_0 can build on methods and results from the Megamaser Cosmology Project (MCP), a multi-year, international project to measure H_0 using the megamaser technique (Reid et al. 2013). The project measures distances to suitable megamaser systems by fitting a warped disk model to observations of 22 GHz water vapor megamasers in the nuclear accretion disk of the host AGNs. Figure 1 shows an

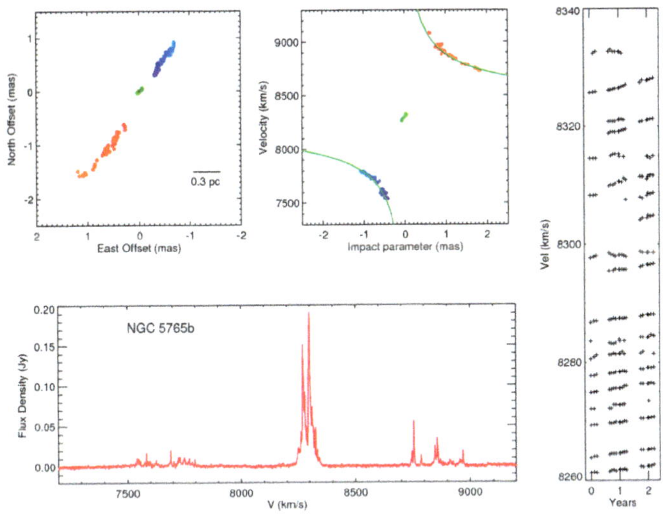

Figure 1. The H_2O Megamaser in NGC 5765b (also see Gao et al. 2016). The maser observations in this galaxy determine its distance at 126.3 ± 11.6 Mpc. The *top left* panel shows the VLBI maser map, with colors of the maser spots representing the line-of-sight velocities. The *top center* panel shows a position-velocity (P-V) diagram. The impact parameter represented on the x-axis is the angular distance measured along the length of the edge-on disk. The solid green lines on the P-V diagram represent a Keplerian fit to the rotation curve. The *bottom panel* shows a representative GBT spectrum. Masers in edge-on AGN accretion disks have a characteristic profile with three groups of maser features, evident in this spectrum. The features centered near 8300 km s^{-1} originate from the front side of the disk while those centered near 7650 km s^{-1} and 8850 km s^{-1} originate from the approaching and receding sides of the edge-on disk, respectively. The *right panel* shows results of GBT spectral monitoring. Each symbol on the plot marks the velocity of a maser peak in the systemic part of the spectrum. The maser velocities increase with time and represent the centripetal acceleration as maser clouds orbit the central supermassive black hole.

example of observations of an edge-on accretion disk megamaser system in the nucleus of the Seyfert 2 galaxy NGC 5765b. Systematic errors in the disk modeling are not likely to be correlated among different galaxies, so the final measurement of H_0 is determined as a weighted mean of measurements to individual galaxies. With present-day instrumentation, the MCP has so far measured H_0 = 69.3 ± 4.2 km s^{-1} Mpc^{-1} (Braatz et al. 2018). When complete, the project should achieve a ~4% total uncertainty.

With the ngVLA, the megamaser-based measurement of H_0 will aim to reach ~1% total uncertainty. If the ngVLA measurement of H_0 aligns with the standard-candle measurements, which are themselves improving, the evidence would become convincingly against the standard ΛCDM model. If it aligns with the *Planck* prediction, additional scrutiny would be warranted to search for systematics among the different astrophysical measurement methods.

5. Limitations of Current Astronomical Instrumentation

Megamaser systems used to measure H_0 are faint, and the megamaser method is limited by the 22 GHz sensitivity of the telescopes available for their study. The MCP uses the most sensitive suite of telescopes working today at 22 GHz, including the GBT for surveys and spectral monitoring observations, and the High Sensitivity Array (the VLBA, GBT, VLA, and 100-m Effelsberg telescope) to map maser disk systems. The final measurement by the MCP will be based upon distances to nine maser disk systems bright enough to measure with these existing facilities. Targeted surveys of over 3000 galaxies were necessary to identify those nine.

6. Connection to Unique ngVLA Capabilities

Only the ngVLA, with its sensitivity approaching an order of magnitude increase over existing facilities, will make a 1% measurement of H_0 plausible using the megamaser method. The measurement, however, will require that the ngVLA meet certain design features to enable precision astrometry of faint masers. Importantly, the ngVLA must have a compact core of antennas containing a substantial fraction of the total collecting area. This core should be concentrated within ~5 km to enable efficient phasing at 22 GHz.

The method also requires the ngVLA Long Baseline Array, which consists of a number of VLBI stations on intercontinental baselines out to over ~5000 km, with substantial collecting area of their own. Long baselines extend both in the E-W and N-S directions. Contiguous frequency coverage from 18-22 GHz within a single receiver band is also important to make surveys efficient for detection of high velocity maser emission. Furthermore, flexible subarray capabilities will be beneficial so that, for example, outer antennas can be utilized for other science while the phased core is operating in a VLBI mode.

7. Experimental Layout

Measuring H_0 with the megamaser technique requires three types of observations. First is a survey to identify the rare, edge-on disk megamasers suitable for distance measurements. Second is sensitive spectral monitoring of those disk megamasers to measure secular drifts in maser lines, indicative of the centripetal accelerations of maser clouds as they orbit the central black hole. And third is sensitive VLBI observations to map the maser features and determine the rotation structure and angular size of the disk. For each megamaser disk being measured, the spectral monitoring would span 1-2 years and include observations on a roughly monthly cadence. Since each VLBI map includes the necessary spectral information needed to track line-of-sight accelerations, a strategy to map and monitor the maser disks simultaneously is feasible with the ngVLA. The positions, velocities, and accelerations of the maser components are then used to constrain a model of a warped disk and determine the distance to the host galaxy.

The precision with which the distance to a megamaser disk can be measured depends on a number of factors, including the richness of its maser spectrum, the layout of observable masers within the disk, and the signal-to-noise of the measurements. For megamasers in the Hubble flow, higher signal-to-noise observations would equate to more precise distances in all cases. So, the first stage of the experimental layout would be to measure distances to all *known* megamasers suitable for such measurements. The MCP surveys and others have discovered ~20 megamaser disk systems whose spectral profile indicates an edge-on disk suitable for a distance measurement. Although the MCP will already have measured distances to nine of those megamasers, those nine would have to be reobserved with the ngVLA to improve their individual measurement uncertainties.

The overall measurement of H_0 based on these ~20 megamasers is not likely to reach the 1% project goal, so new megamaser disks must be discovered. The ngVLA project could aim for 10% distances to ~100 megamasers, or 7% distances to ~50. In practice, the precision for each system is not well known until the disk is observed and modeled. The prototypical megamaser in NGC 4258, at a distance of $7.54 \pm 0.17 \pm 0.10$ Mpc, demonstrates that systematics in an individual system can be < 3% with sufficient sensitivity and angular resolution (Riess et al. 2016).

Ultimately, to reach 1% the second stage of the experiment requires a survey. With a nearly order of magnitude advantage in sensitivity, the ngVLA would be able to discover ~30 times more megamasers than the GBT. It would also extend the practical distance to which megamasers can be measured to a few hundred Mpc, alleviating potential concerns about local variations in the Hubble parameter.

8. Complementarity

Dark energy is the most important unsolved problem in modern physics, and an accurate local measurement of H_0 is the most effective complement to CMB data for constraining the equation of state of dark energy. A number of future facilities, for example WFIRST, will continue to focus on investigations of dark energy and cosmology.

A precise and independent measurement of the Hubble constant also elevates the effectiveness of future CMB observations, such as the "Stage-4" ground-based experiment, CMB-S4. Besides dark energy, a percent-level prior on H_0 also provides the best complement to CMB-S4 experiments for constraining the neutrino mass (Manzotti, Dodelson & Park 2015).

The importance of the Hubble constant demands agreement and verification using several measurement techniques. The long-term goal for observational cosmology is to achieve percent-level measurements that are consistent across independent methods, to minimize the impact of systematic uncertainties.

References

Alam, S., Ata, M., Bailey, S. et al. 2017, *MNRAS*, 470, 2617
Bonvin, V., Coubin, F., Suyu, S. et al. 2017, *MNRAS*, 465, 4914
Braatz, J. Condon, J. Henkel, C. et al. 2018, Proc. IAUS 336, ed. Tarchi, Reid, & Castangia
Freedman, W. L., Madore, B. F., Gobson, B. K., et al. 2012, *ApJ*, 758, 24
Gao, F., Braatz, J. A., Reid, M. J., et al. 2016, *ApJ*, 817, 128
Manzotti, A., Dodelson, S. & Park, Y. 2016, *PhyRevD*, 93, 063009
Planck Collaboration, Ade, P. A. R., Aghanim, N., et al. 2016, *A&A*, 594, 13
Reid, M. J., Braatz, J. A., Condon, J. J., et al. 2013, *ApJ*, 767, 154
Riess, A. G., Macri, L. M., Hoffmann, S. L., et al. 2016, *ApJ*, 826, 56

Author Index

Ábrahám, P., 177
Alatalo, K., 451, 457
Alberdi, A., 421
Anderson, L. D., 431
Andrews, S. M., 137, 155, 193, 209
Armus, L., 441
Asmar, S., 81
Audard, M., 177
Aufdenberg, J., 171

Balick, B., 395
Balser, D. S., 431
Bania, T. M., 431
Barcos-Muñozarcos, L., 421
Barger, A. J., 551
Bartkiewicz, A., 321
Bastian, T. S., 87, 99, 125
Beasley, A., 15
Beltrán, M. T., 249, 293, 309, 381
Bergin, E. A., 209, 217
Beswick, R. J., 421
Bicknell, G., 457
Birnstiel, T., 199
Blake, G. A., 217
Blecha, L., 677, 763
Bogdanović, T., 677
Boizelle, B. D., 509
Bolatto, A. D., 3, 441, 483, 499
Boley, A. C., 171
Bourke, T., 299, 357
Bower, G. C., 737, 773, 793
Braatz, J., 821
Brinks, E., 421
Brisken, W., 763
Brogan, C. L., 321
Brozović, M., 113
Bull, P., 803
Burke-Spolaor, S., 677, 763, 773
Burkhardt, A. M., 217
Butler, B. J., 15, 49, 113, 185, 369, 773

Caballero-Nieves, S., 389
Carbone, D., 689
Carilli, C. L., 15, 369, 535, 557, 565, 575, 581, 587, 629
Carilli, M. T., 369
Carrasco-González, C., 137, 309, 357
Carroll, P. B., 245
Caselli, P., 299, 381
Casey, C. M., 3, 565, 575, 581, 629
Chakraborty, A., 649
Champagne, J., 629
Chandler, C., 155
Chatterjee, S., 3, 649, 751, 793
Chen, B., 99
Chibueze, J. O., 321
Chomiuk, L., 3, 271, 711
Cieza, J., 177
Civano, F., 763
Clark, B., 15
Claussen, M. J., 281
Cleeves, L. I., 217, 223
Comerford, J. M., 677, 763
Condon, J. J., 421, 551, 821
Coppejans, D. L., 727
Cordes, J. M., 257, 793
Cordiner, M. A., 73
Corsi, A., 689
Croft, S., 257
Crosley, M. K., 229
Cruz-Sáenz de Miera, F., 177
Cyganowski, C. J., 321

Dale, D., 3
Darling, J., 669, 763, 813
Davé, R., 629
Davis, T. A., 509
Dayal, P., 557
de Kleer, K., 49
de Pater, I., 3, 49
Decarli, R., 565, 575, 581

Demorest, P., 773, 793
Deneva, J. S., 793
Dexter, J., 793
Di Francesco, J., 3, 299
Dickey, J. M., 431
Dickinson, M., 3
Dong, D., 421
Dong, R., 199
Dunham, M. M., 177
Durand, S., 15

Emonts, B., 457, 565, 575, 581, 587
Erickson, A., 15
Evans, A. S., 421

Ferrara, A., 557
Fleishman, G. D., 99, 125
Frail, D. A., 689
Friesen, R. K., 299, 381

Gündel, M., 177
Galván-Madrid, R., 309, 357
Garrod, R. T., 245, 381
Gary, D. E., 99, 125
Ginsburg, A., 309, 465
Girart, J. M., 357
Glesener, L., 99
Golap, K., 369
Gonzalez, A. H., 663
Grammer, W., 15
Greaves, J. S., 161
Green, J. D., 177
Grosso, N., 177
Gurton, S., 37
Gurwell, M., 49

Hales, A., 177
Hall, K., 639
Hallinan, G., 3, 229
Harper, G. M., 265
Harrison, I., 803
Hartmann, L., 177
Harwood, J. J., 457, 595
Hauschildt, P., 171
Hiriart, R., 15
Hirota, T., 321
Hodge, J. A., 565, 575, 581
Huff, E., 803
Hughes, A. M., 171
Hull, C. L. H, 357
Hung, C.-L., 629
Hunter, T. R., 321

Isella, A., 3, 137, 147, 155, 193, 199

Jackson, J., 15
Johnson, K. E., 421
Johnstone, D., 333, 345

Körding, E. G., 727
Kóspál, Á., 177
Kadam, K., 177
Kastner, J. H., 177, 395
Kellermann, K. I., 619
Kennedy, G., 161
Kennicutt, R. C., 421
Kent, B., 15
Keown, J., 299
Kepley, A., 483
Kharb, P., 619
Kimball, A. E., 595, 603, 649
Kirkpatrick, A., 639
Kohno, K., 3, 551, 565, 575, 581
Kramer, M., 793
Kulkarni, S. R., 3
Kurtz, S., 309

Lützgendorf, N., 389
Lacy, M., 457, 639, 649
Lang, C., 3
Law, C. J., 773
Lazio, T. J. W., 3, 81, 113, 677, 763, 773, 793
Lazzati, D., 689
Lehnert, M., 587
Leroy, A. K., 3, 441, 465, 483, 499
Li, H., 147
Linden, S. T., 421
Linford, J. D., 271, 773
Lister, M. L., 619
Liu, H. B., 309
Liu, S.-F., 147
Liu, X., 677
Lloyd-Ronning, N., 701
Loinard, L., 3, 411, 523
Loomis R. A., 217, 223

Macías, E., 137
Maccarone, T. J., 3, 389, 523, 663, 677, 711, 743, 763

MacLeod, G. C., 321
Margot, J.-L., 113
Mason, B., 15, 649
Matrà, L., 161
Matthews, B. C., 3, 161, 171, 185, 199
Matthews, L. D., 281
McGuire, B. A., 217, 245
McKinnon, M., 15
Meier, D. S., 483, 517
Metzger, B. D., 737
Milam, S., 49
Miller-Jones, J. C. A., 711, 727, 743
Moór, A., 171
Moeckel, C., 49
Morgan, M., 15
Morrison, I. S., 257
Moullet, A., 49
Mukherjee, D., 457, 595
Murphy, E. J., xiii, 3, 15, 421, 551, 557, 629, 689
Mushotzky, R., 441
Muxlow, T. W. B., 421

Naidu, S. P., 113
NANOgrav Collaboration, 751
Narayanan, D., 565, 575, 581, 587, 629
Nita, G. M., 125
Nyland, K., 451, 457, 509, 587, 595, 603, 639, 649, 657, 743

O'Shaughnessy, R., 689
Öberg, K. I., 223
Ojeda, O., 15
Osten, R. A., 3, 229
Owen, B. J., 689

Pérez-Torres, M., 421
Paine, J., 813
Paragi, Z., 257
Pascucci, I., 155
Patil, P., 595
Perez, S., 177
Pesce, D., 677, 821
Pinilla, P., 209
Pisano, D. J., 471
Plotkin, R. M., 719
Popping, G., 629
Postel, A., 177
Prandoni, I., 457, 603, 639

Qi, C., 73

Rab, C., 177
Ransom, S., 793
Ravi, V., 781
Reid, M. J., 3, 411, 523, 821
Reines, A. E., 719
Remijan, A. J., 217
Ricci, L., 147
Riechers, D., 3, 565, 575, 581, 629
Rivilla, V. M., 249, 293
Robishaw, T., 357
Rocha, G., 649
Rodríguez, L. F., 309
Rosero, V., 15
Roshi, D. A., 431
Rosolowsky, E., 465, 499, 517
Rowe, B., 649
Ruiz-Rodriguez, D., 177
Rujopakarn, W., 603
Rupen, M. P., 271, 727

Sahai, R., 395, 403
Sakai, N., 3
Sand, D. J., 689
Sanna, A., 321
Sarazin, C., 649
Sargent, M. T., 421, 551, 565, 575, 581
Sault, R. J., 49
Schinnerer, E., 421, 499
Selina, R. J., 15
Shao, L., 793
Shao, Y., 535
Sharma, R., 345
Sheehan, P., 333, 345
Shen, Y., 677
Shillue, W., 15
Shingledecker, C. N., 217
Siemion, A. P. V., 257
Simon, J., 611
Sivakoff, G. R., 711, 727
Smith, N., 389
Somerville, R. S., 629
Spilker, J., 657
Stanimirović, S., 471
Strader, J., 711
Strugis, S., 15

Tabatabaei, F. S., 421

Tarter, J., 257
Taylor, G., 677
Taylor, S. R., 611
Tobin, J. J., 333, 345
Tollefson, J., 49
Torrelles, J.-M., 321
Truebenbach, A., 813
Turner, J. L., 421

Urbain, D., 15

van der Marel, N., 199
van Velzen, S., 737
Vedantham, H. K., 491
Veilleux, S., 441
von Schill, L., 37
Vorobyov, E. I., 177

Walter, F., 3, 441, 471, 565, 575, 581, 629
Wenger, T. V., 431
Wex, N., 793
Wharton, R., 793
White, J. A., 171, 177
White, S. M., 125, 369
Williams, J. P., 209
Williams, P. K. G., 357
Willis, E. R., 217
Wilner, D. J., 3, 137, 161, 171
Wilson, T. L., 431
Wong, T., 483, 517
Wrobel, J. M., 743
Wyatt, M. C., 161

Zhang, K., 209
Zhang, Q., 293
Zhu, Z., 177, 193
Zijlstra, A., 395

ASTRONOMICAL SOCIETY OF THE PACIFIC

THE ASTRONOMICAL SOCIETY OF THE PACIFIC is an international, nonprofit, scientific, and educational organization. Some 125 years ago, on a chilly February evening in San Francisco, astronomers from Lick Observatory and members of the Pacific Coast Amateur Photographic Association—fresh from viewing the New Year's Day total solar eclipse of 1889 a little to the north of the city—met to share pictures and experiences. Edward Holden, Lick's first director, complimented the amateurs on their service to science and proposed to continue the good fellowship through the founding of a Society "to advance the Science of Astronomy, and to diffuse information concerning it." The Astronomical Society of the Pacific (ASP) was born.

The ASP's purpose is to increase the understanding and appreciation of astronomy by engaging scientists, educators, enthusiasts, and the public to advance science and science literacy. The ASP has become the largest general astronomy society in the world, with members from over 70 nations.

The ASP's professional astronomer members are a key component of the Society. Their desire to share with the public the rich rewards of their work permits the ASP to act as a bridge, explaining the mysteries of the universe. For these members, the ASP publishes the Publications of the Astronomical Society of the Pacific (PASP), a well-respected monthly scientific journal. In 1988, Dr. Harold McNamara, the PASP editor at the time, founded the ASP Conference Series at Brigham Young University. The ASP Conference Series shares recent developments in astronomy and astrophysics with the professional astronomy community.

To learn how to join the ASP or to make a donation, please visit `http://www.astrosociety.org`.

ASTRONOMICAL SOCIETY OF THE PACIFIC

MONOGRAPH SERIES

Published by the Astronomical Society of the Pacific

The ASP Monograph series was established in 1995 to publish select reference titles. For electronic versions of ASP Monographs, please see `http://www.aspmonographs.org`.

INFRARED ATLAS OF THE ARCTURUS SPECTRUM, 0.9-5.3μm
eds. Kenneth Hinkle, Lloyd Wallace, and William Livingston (1995)

VISIBLE AND NEAR INFRARED ATLAS OF THE ARCTURUS SPECTRUM 3727-9300Å
eds. Kenneth Hinkle, Lloyd Wallace, Jeff Valenti, and Dianne Harmer (2000)

ULTRAVIOLET ATLAS OF THE ARCTURUS SPECTRUM 1150-3800Å
eds. Kenneth Hinkle, Lloyd Wallace, Jeff Valenti, and Thomas Ayres (2005)

HANDBOOK OF STAR FORMING REGIONS: VOLUME I THE NORTHERN SKY
ed. Bo Reipurth (2008)

HANDBOOK OF STAR FORMING REGIONS: VOLUME II THE SOUTHERN SKY
ed. Bo Reipurth (2008)

TWENTY YEARS OF ADASS
ed. Ian Evans (2013)

A complete list and electronic versions of ASPCS volumes may be found at `http://www.aspbooks.org`.

All book orders or inquiries concerning the ASP Conference Series, ASP Monographs, or International Astronomical Union Volumes published by the ASP should be directed to:

Astronomical Society of the Pacific
390 Ashton Avenue
San Francisco, CA 94112-1722 USA
Phone: 800-335-2624 (within the USA)
Phone: 415-337-2126
Fax: 415-337-5205
Email: `service@astrosociety.org`

For a complete list of ASP publications, please visit `http://www.astrosociety.org`.